NCCER

Instrumentation Level Two

Upper Saddle River, New Jersey
Columbus, Ohio

NCCER

President: Don Whyte
Director of Curriculum Revision and Development: Daniele Dixon
Instrumentation Project Manager: Deborah Padgett
Production Manager: Debie Ness
Editor: Lori Watson
Desktop Publisher: Rachel Ivines

The NCCER would like to acknowledge the contract service provider for this curriculum:
Topaz Publications, Liverpool, New York.

This information is general in nature and intended for training purposes only. Actual performance of activities described in this manual requires compliance with all applicable operating, service, maintenance, and safety procedures under the direction of qualified personnel. References in this manual to patented or proprietary devices do not constitute a recommendation of their use.

18
ISBN 0-13-047232-8

Preface

This volume was developed by NCCER in response to the training needs of the construction, maintenance, and pipeline industries. It is one of many in NCCER's *Standardized Curricula*. The program, covering training for close to 40 construction and maintenance areas, and including skills assessments, safety training, and management education, was developed over a period of years by industry and education specialists.

NCCER also maintains a National Registry that provides transcripts, certificates, and wallet cards to individuals who have successfully completed modules of NCCER's *Standardized Curricula*, when the training program is delivered by an NCCER Accredited training Sponsor.

The NCCER is a not-for-profit 501(c)(3) education foundation established in 1995 by the world's largest and most progressive construction companies and national construction associations. It was founded to address the severe workforce shortage facing the industry and to develop a standardized training process and curricula. Today, NCCER is supported by hundreds of leading construction and maintenance companies, manufacturers, and national associations, including the following partnering organizations:

PARTNERING ASSOCIATIONS

- American Fire Sprinkler Association
- American Petroleum Institute
- American Society for Training & Development
- American Welding Society
- Associated Builders & Contractors, Inc.
- Association for Career and Technical Education
- Associated General Contractors of America
- Carolinas AGC, Inc.
- Carolinas Electrical Contractors Association
- Citizens Democracy Corps
- Construction Industry Institute
- Construction Users Roundtable
- Design-Build Institute of America
- Merit Contractors Association of Canada
- Metal Building Manufacturers Association
- National Association of Minority Contractors
- National Association of State Supervisors for Trade and Industrial Education
- National Association of Women in Construction
- National Insulation Association
- National Ready Mixed Concrete Association
- National Utility Contractors Association
- National Vocational Technical Honor Society
- North American Crane Bureau
- Painting & Decorating Contractors of America
- Portland Cement Association
- SkillsUSA
- Steel Erectors Association of America
- Texas Gulf Coast Chapter ABC
- U.S. Army Corps of Engineers
- University of Florida
- Women Construction Owners & Executives, USA

Some features of NCCER's *Standardized Curricula* are:

- An industry-proven record of success
- Curricula developed by the industry for the industry
- National standardization providing portability of learned job skills and educational credits
- Credentials for individuals through NCCER's National Registry
- Compliance with Apprenticeship, Training, Employer, and Labor Services (ATELS) requirements for related classroom training (CFR 29:29)
- Well-illustrated, up-to-date, and practical information

Acknowledgments

This curriculum was revised as a result of the farsightedness and leadership of the following sponsors:

Austin Industrial
Cianbro Corporation
Fluor/F&PS
Lee College
Lake Charles ABC

This curriculum would not exist were it not for the dedication and unselfish energy of those volunteers who served on the Authoring Team. A sincere thanks is extended to:

Tommy Arnold
Gordon Hobbs
Russ Michael
Glen Pratt
Jonathan Sacks
Doug Smith
Richard Tunstall

Contents

12201-03 Craft-Related Mathematics 1.i

12202-03 Instrumentation Drawings and Documents, Part Two 2.i

12203-03 Principles of Welding for Instrumentation 3.i

12204-03 Process Control Theory 4.i

12205-03 Detectors, Secondary Elements, Transducers, and Transmitters 5.i

12206-03 Controllers, Recorders, and Indicators 6.i

12207-03 Control Valves, Actuators, and Positioners 7.i

12208-03 Relays and Timers 8.i

12209-03 Switches and Photoelectric Devices 9.i

12210-03 Filters, Regulators, and Dryers 10.i

12211-03 Analyzers and Monitors 11.i

12212-03 Panel-Mounted Instruments 12.i

12213-03 Installing Field-Mounted Instruments 13.i

12214-03 Raceways for Instrumentation 14.i

Instrumentation Level Two . . . . Index

Module 12201-03

Craft-Related Mathematics

COURSE MAP

This course map shows all of the modules in the second level of the Instrumentation curriculum. The suggested training order begins at the bottom and proceeds up. Skill levels increase as you advance on the course map. The local Training Program Sponsor may adjust the training order.

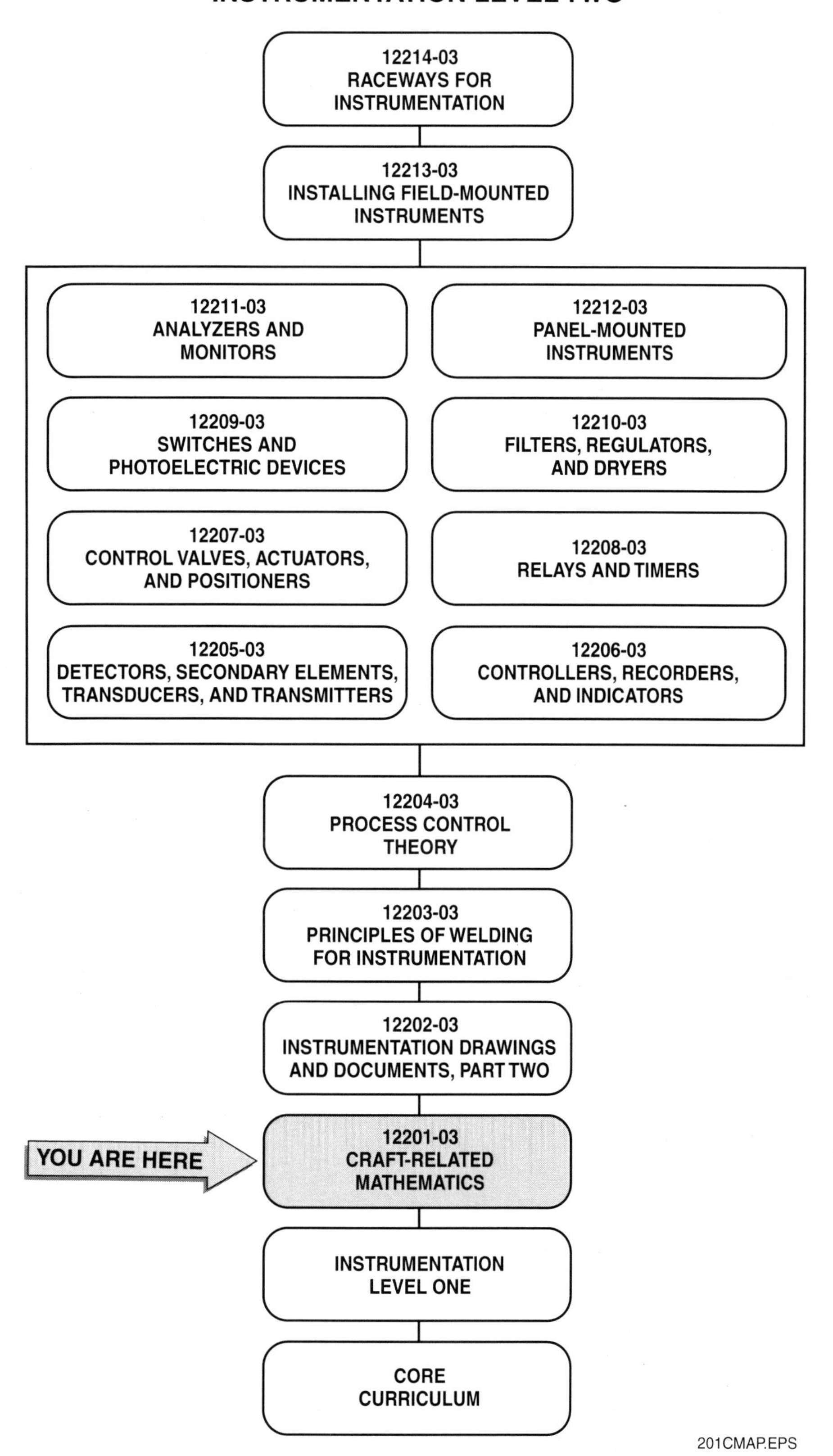

MODULE 12201-03 CONTENTS

1.0.0 INTRODUCTION ... 1.1

2.0.0 METRIC MEASUREMENTS ... 1.2

2.1.0 Converting Lengths ... 1.2

2.2.0 Converting Areas ... 1.2

2.3.0 Converting Volumes ... 1.3

2.4.0 Wet Volume Measurements ... 1.4

2.5.0 Mass Versus Weight ... 1.5

2.5.1 Mass ... 1.5

2.5.2 Weight (Force) ... 1.5

2.6.0 Pressure ... 1.6

2.6.1 The Pascal ... 1.7

2.6.2 Hydrostatic Pressures ... 1.8

2.6.3 Units of Hydrostatic Pressure ... 1.10

2.7.0 Temperature ... 1.13

2.7.1 Temperature Scales ... 1.13

2.7.2 Temperature Conversions ... 1.14

2.8.0 Flow ... 1.14

2.8.1 Units of Flow ... 1.14

2.8.2 Flow Characteristics ... 1.15

3.0.0 HANDHELD CALCULATORS AND INSTRUMENTATION APPLICATIONS ... 1.18

3.1.0 Squares and Square Roots ... 1.18

3.2.0 Auxiliary Functions ... 1.20

3.2.1 Polarity Conversion ... 1.20

3.2.2 Memory ... 1.20

4.0.0 TECHNICAL APPLICATIONS ... 1.20

4.1.0 Converting to a Metric Rule Dipstick ... 1.24

4.2.0 Sight Glass Level Measurement ... 1.24

4.3.0 Conductance Probe Settings ... 1.25

4.4.0 Open Tank Measurement Conversions ... 1.25

4.5.0 Pressurized Tank Measurement Conversions ... 1.26

4.6.0 Temperature Measurement Conversions ... 1.28

SUMMARY ... 1.29

REVIEW QUESTIONS ... 1.30

GLOSSARY ... 1.33

REFERENCES ... 1.34

Figures

Figure 1 Conversion example one 1.2
Figure 2 Conversion example two 1.3
Figure 3 Conversion of units for a square 1.3
Figure 4 Conversion of units for a rectangle 1.3
Figure 5 Conversion of units for a circle 1.4
Figure 6 The measurement of the volume of a box 1.4
Figure 7 Conversion of units for a box 1.4
Figure 8 Conversion of units for a cylindrical tank 1.4
Figure 9 Equal forces applied to different surface areas 1.6
Figure 10 Known forces applied to known surface areas 1.6
Figure 11 Equal downward forces applied to different areas 1.8
Figure 12 Hydrostatic paradox 1.9
Figure 13 Direction of force applied by a liquid 1.9
Figure 14 Actual pressure applied to base surface 1.9
Figure 15 One cubic foot and one cubic inch of water 1.9
Figure 16 Atmospheric pressure 1.10
Figure 17 Open container at sea level 1.11
Figure 18 Sealed containers at sea level 1.11
Figure 19 Torricelli tube 1.12
Figure 20 Equal length tubes at various angles 1.12
Figure 21 Comparison of temperature scales 1.14
Figure 22 Sample temperature unit conversions 1.14
Figure 23 Pressure drop in a reservoir system 1.16
Figure 24 Pressure drop in a pump system 1.16
Figure 25 Velocity profile of laminar flow 1.16
Figure 26 Velocity profile of turbulent flow 1.16
Figure 27 Continuity of flow through a tapered pipe 1.17
Figure 28 Pressure exerted by fluid flowing through a pipe 1.17
Figure 29 Variable head flowmeter restriction 1.17
Figure 30 The basic calculator 1.18
Figure 31 Squaring using a calculator 1.19
Figure 32 Taking a square root using a calculator 1.20
Figure 33 Relative error calculation 1.21
Figure 34 Significant digits 1.22
Figure 35 Significant digits in decimal fractions 1.22
Figure 36 Significant digit examples 1.22
Figure 37 Rounding-off examples 1.23
Figure 38 Precision determinations 1.23
Figure 39 Dip stick level measurement 1.24

Figure 40 Sight glass level measurement 1.24
Figure 41 Conductance probe level measurement 1.25
Figure 42 Open tank application 1.26
Figure 43 Closed or pressurized tank application 1.27

Tables

Table 1 Metric System Prefixes 1.2
Table 2 Volume Conversion Relationships 1.5
Table 3 Mass and Weight Equivalences 1.5
Table 4 Force Conversion Factors 1.7
Table 5 Pressure Conversion Factors 1.7
Table 6 Comparison of Pressure Units 1.13
Table 7 Fluid Pressure Conversion Factors 1.13
Table 8 Units of Flow Rate 1.15
Table 9 Units of Total Flow 1.15
Table 10 Common Squares and Square Roots 1.19

MODULE 12201-03

Craft-Related Mathematics

Objectives

When you have completed this module, you will be able to do the following:

1. Identify similar units of measurement in both the English and metric systems and identify which units are larger.
2. Convert measured values in the English system, using common conversion factor tables, to equivalent metric values.
3. Use a handheld calculator to perform the basic mathematical operations necessary in instrumentation.
4. Use a handheld calculator to square numbers and find the square root of numbers.
5. Perform the mathematical conversions necessary for instrumentation measurements.

Prerequisites

Before you begin this module, it is recommended that you successfully complete the following: Core Curriculum; Instrumentation Level One.

Required Trainee Materials

1. Pencil and paper
2. Calculator
3. Metric rule (20 cm or larger)
4. English ruler (12")

1.0.0 ◆ INTRODUCTION

The **metric system** is a system of weights and measures based on the **meter**. Technicians who have not used the metric system on the job most certainly will use it in the future. One of the basic functions of instrumentation is to measure system parameters of flow, **pressure**, level, and **temperature**.

Most work in science and engineering is based on the exact **measurement** of physical quantities. A measurement is a comparison of one quantity to a defined standard measure of dimension called a **unit**. Whenever a physical quantity is described, the units of the standard to which the quantity was compared must be specified. A number alone is insufficient to describe a physical quantity.

Specifying the unit of measurement for a number used to describe a physical quantity is important because the same physical quantity may be measured using a variety of different units.

All physical quantities can be expressed ultimately in terms of three fundamental parameters—length, **mass**, and time:

- *Length* – The distance from one point to another
- *Mass* – The quantity of matter
- *Time* – The period during which an event occurs

The three most widely used systems of measurement units are:

- The meter-kilogram-second (MKS) system
- The centimeter-gram-second (CGS) system
- The English system

The MKS and CGS units are both parts of the metric system of measure. The English system is probably most familiar to you now. Notice that time is measured in the same units (seconds) in all systems.

In the United States the metric system is used for physics calculations. The English system, however, has been used widely in engineering and construction calculations. Therefore, when working in the process instrumentation industry, it is necessary to have an understanding of both systems of units.

The metric system prefixes are listed in *Table 1*. The use of the metric system is logically arranged. The name of the unit will also represent an order of magnitude (through the prefix) that foot and pound cannot. Transferring American engineering practices and equipment to the metric system over time can be an expensive transition for some industries. The more familiar people become with the metric system, the smoother the transition will be. Most automotive manufacturers have already completed the majority of the transition. Many instrumentation manufacturers publish their technical manuals and instrument data sheets displaying all values in English system units, with the equivalent metric system units in parentheses. At some point in the future, it is possible that only the metric values will be listed. Now is a good time to become familiar with both systems and to learn how to convert from one to the other.

Table 1 Metric System Prefixes

PREFIX	UNIT		
	Fraction	Decimal	Scientific Notation
micro- (μ)	1/1,000,000	.000001	10^{-6}
milli- (m)	1/1,000	.001	10^{-3}
centi- (c)	1/100	.01	10^{-2}
deci- (d)	1/10	.1	10^{-1}
deka- (da)	10	10.	10^{1}
hecto- (h)	100	100.	10^{2}
kilo- (k)	1,000	1,000.	10^{3}
mega- (M)	1,000,000	1,000,000.	10^{6}
giga- (G)	1,000,000,000	1,000,000,000.	10^{9}

201T01.EPS

2.0.0 METRIC MEASUREMENTS

The most common measurements associated with instrumentation can be classified into five categories:

- Dimensional measurements, such as lengths, levels, **areas**, and **volumes**
- Measurements of mass and weight
- Pressure measurements
- Flow measurements
- Temperature measurements

Each will be described in the following sections.

2.1.0 Converting Lengths

Instrumentation technicians may be called on to convert a measurement in one system into the other system's units. For examples, refer to *Figures 1* and *2*.

2.2.0 Converting Areas

Calculating simple areas will be useful later in the module in investigating pressure measurements. In converting from one measurement system to the other, we must remember that every dimension must be converted. So, to convert the dimensions of squares and rectangles, both the length and the width must be converted (refer to *Figures 3* and *4*).

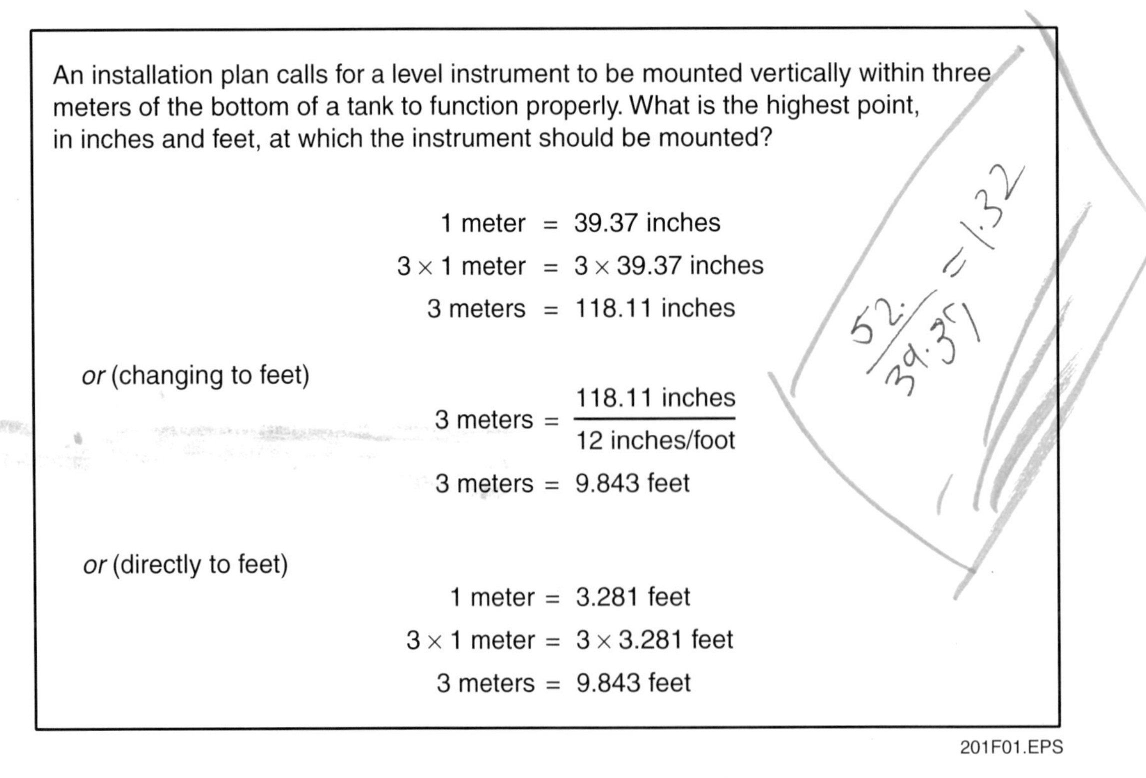
An installation plan calls for a level instrument to be mounted vertically within three meters of the bottom of a tank to function properly. What is the highest point, in inches and feet, at which the instrument should be mounted?

$$1 \text{ meter} = 39.37 \text{ inches}$$
$$3 \times 1 \text{ meter} = 3 \times 39.37 \text{ inches}$$
$$3 \text{ meters} = 118.11 \text{ inches}$$

or (changing to feet)

$$3 \text{ meters} = \frac{118.11 \text{ inches}}{12 \text{ inches/foot}}$$
$$3 \text{ meters} = 9.843 \text{ feet}$$

or (directly to feet)

$$1 \text{ meter} = 3.281 \text{ feet}$$
$$3 \times 1 \text{ meter} = 3 \times 3.281 \text{ feet}$$
$$3 \text{ meters} = 9.843 \text{ feet}$$

201F01.EPS

Figure 1 ◆ Conversion example one.

An installation plan calls for a laser distance-measuring instrument having a range of 150 inches. The manufacturer, however, lists the instrument range on the faceplate and in the instrument manual in meters. Which of the following three instruments would be most appropriate?

a. two meter range

b. three meter range

c. four meter range

$$1 \text{ meter} = 39.37 \text{ inches}$$

$$= \frac{150 \text{ inches}}{39.37 \text{ inches/meter}}$$

$$= 3.81 \text{ meters}$$

Only the four meter range instrument would be appropriate.

201F02.EPS

Figure 2 ◆ Conversion example two.

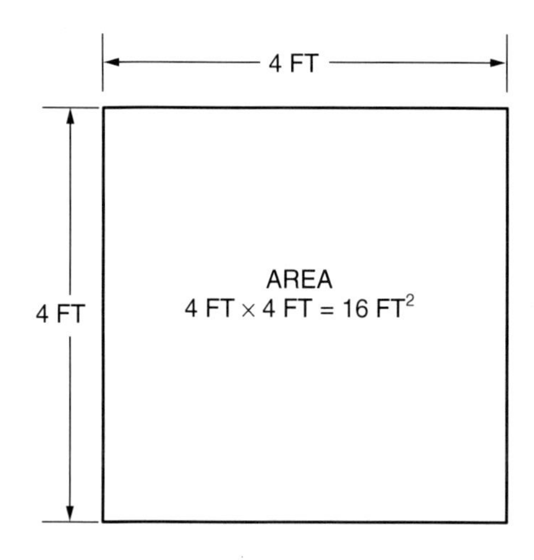

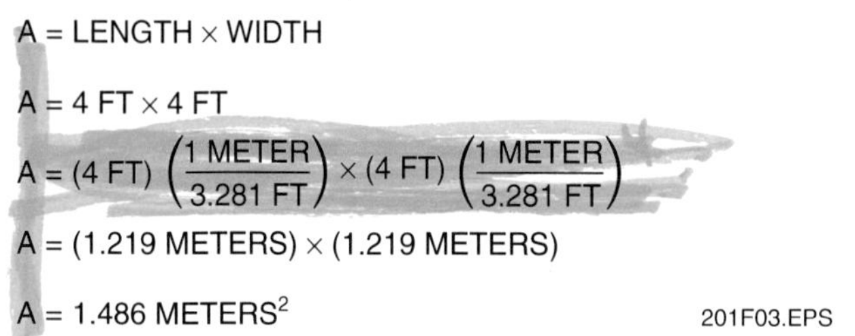

Figure 3 ◆ Conversion of units for a square.

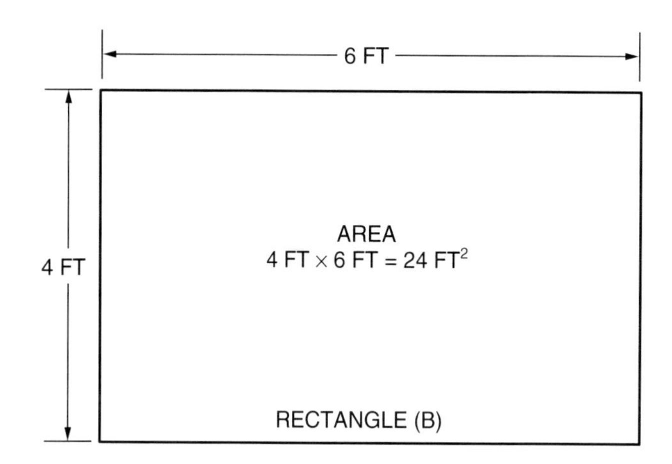

$$A = \text{LENGTH} \times \text{WIDTH}$$

$$A = 4 \text{ FT} \times 6 \text{ FT}$$

$$A = (4 \text{ FT}) \left(\frac{1 \text{ METER}}{3.281 \text{ FT}}\right) \times (6 \text{ FT}) \left(\frac{1 \text{ METER}}{3.281 \text{ FT}}\right)$$

$$A = (1.219 \text{ METERS}) \times (1.829 \text{ METERS})$$

$$A = 2.229 \text{ METERS}^2$$

201F04.EPS

Figure 4 ◆ Conversion of units for a rectangle.

In converting the dimensions of a circle, only the radius has a measured dimension that must be converted. In *Figure 5*, centimeters are converted to inches.

2.3.0 Converting Volumes

Calculating the volumes of box-shaped and cylindrical containers such as tanks and pipes will be useful later in determining how much material there is in a tank and passing through a pipe.

In converting these two volumes from one measurement system unit to the other, remember that every dimension must be independently converted. So, to convert the dimensions of the box in *Figure 6*, the length, width, and depth must be converted as shown in *Figure 7*.

In the example of a cylindrical tank, the same conversion rules apply, as shown in *Figure 8*.

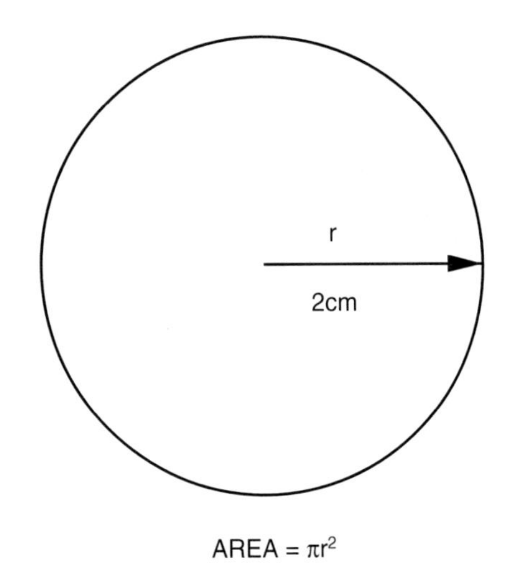

$\text{AREA} = (3.14)\ (2\text{ cm})\ (2\text{ cm})$

$\text{AREA} = (3.14)\ (2\text{ cm})\ (0.3937\text{ in/cm})\ (2\text{ cm})\ (0.3937\text{ in/cm})$

$\text{AREA} = (3.14)\ (0.7874\text{ in})\ (0.7874\text{ in})$

$\text{AREA} = (3.14)\ (0.6200\text{ in}^2)$

$\text{AREA} = 1.95\text{ in}^2$

201F05.EPS

Figure 5 ◆ Conversion of units for a circle.

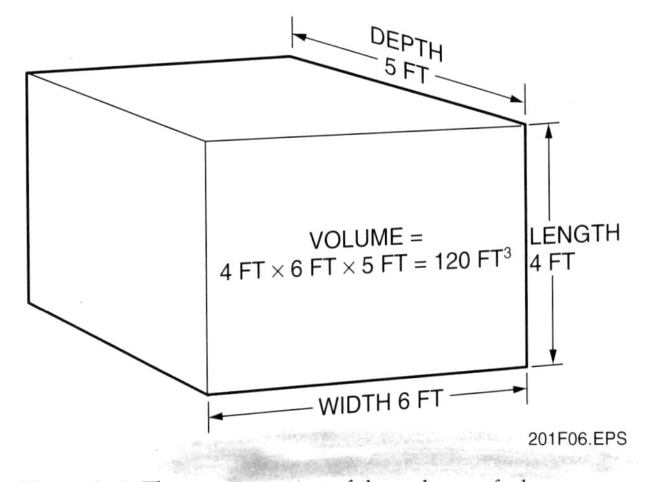

Figure 6 ◆ The measurement of the volume of a box.

2.4.0 Wet Volume Measurements

Previously, the shape of the container, rather than its contents, was used to calculate volume. This is referred to as a dry measure. In practice, trainees are much more used to dealing with wet measures. Wet measures can be defined as the amount of a fluid that would fill the volume, or the shape of, the container. Wet measures in the English system are the pint, quart, and gallon.

The metric system also uses wet measuring units. The **liter** is the most common. It is 5% greater than a US quart.

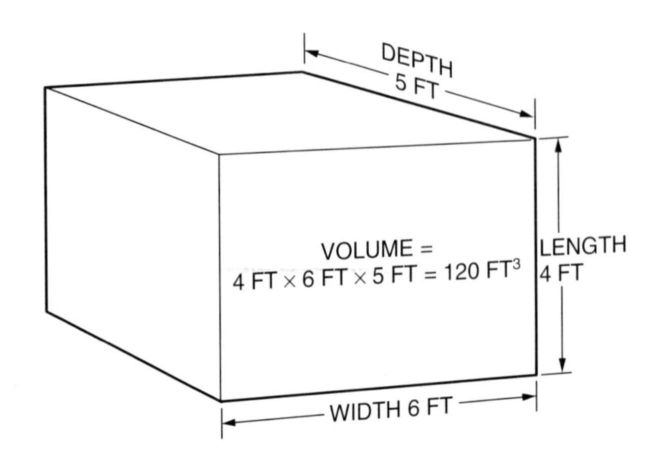

$V = \text{LENGTH} \times \text{WIDTH} \times \text{DEPTH}$

$V = (4\text{ FT}) \left(\frac{1\text{ METER}}{3.281\text{ FT}}\right) \times (6\text{ FT}) \left(\frac{1\text{ METER}}{3.281\text{ FT}}\right) \times (5\text{ FT}) \left(\frac{1\text{ METER}}{3.281\text{ FT}}\right)$

$V = \frac{4\text{ METERS}}{3.281\text{ FT}} \times \frac{6\text{ METERS}}{3.281\text{ FT}} \times \frac{5\text{ METERS}}{3.281\text{ FT}}$

$V = (1.219\text{ METERS}) \times (1.829\text{ METERS}) \times (1.524\text{ METERS})$

$V = 3.398\text{ METERS}^3$

201F07.EPS

Figure 7 ◆ Conversion of units for a box.

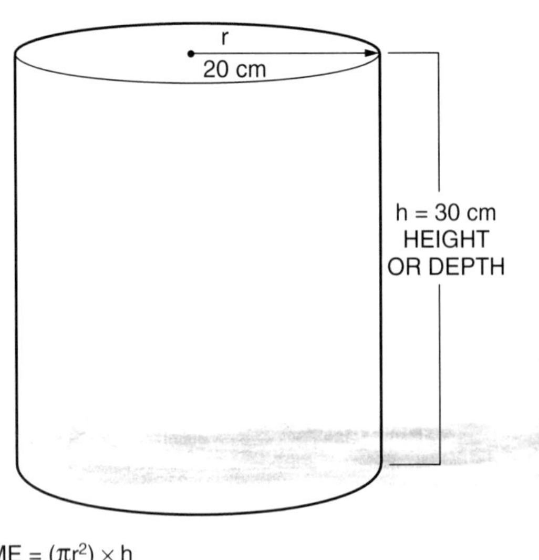

$\text{VOLUME} = (\pi r^2) \times h$

$V = (3.14)\ (20\text{ cm})\ (0.3937\text{ in/cm})\ (20\text{ cm})\ (0.3937\text{ in/cm})\ (30\text{ cm})\ (0.3937\text{ in/cm})$

$V = (3.14)\ (7.874\text{ in})\ (7.874\text{ in})\ (11.811\text{ in})$

$V = 2{,}299\text{ in}^3$

201F08.EPS

Figure 8 ◆ Conversion of units for a cylindrical tank.

By definition, 1 liter is 1 cubic decimeter. In other words, a cube with each side equal to 1 decimeter (or 10 centimeters) will hold 1 liter of a fluid. Knowing the wet measures for a substance allows easy handling and measuring of fluids, because fluids conform to the shape of their containers. Recalculating the amount each time a fluid is moved quickly demonstrates the advantage of the wet measure.

Table 2 shows the volume relationships between the liter and the dry volume of the cubic meter in the metric system and the pint and gallon in the English system.

Table 2 Volume Conversion Relationships

UNIT	m^3	gal	L	pt
CUBIC METER (m^3)	1	0.003785	0.001	0.0004732
U.S. GALLON (gal)	264.2	1	0.2642	0.125
LITER (L)	1000	3.785	1	0.4732
U.S. PINT (pt)	2113	8	2.113	1

201T02.EPS

2.5.0 Mass Versus Weight

Mass is defined as the amount of matter in an object. The term *weight* is often used interchangeably with mass. This is technically incorrect. Weight is actually the **force** on an object that is due to the pull of the Earth's gravity. As a body gets farther away from the Earth, the effect of the Earth's gravity field on that body becomes less and less, making the body weigh less. When climbing a mountain, the climber's actual weight becomes less. The actual mass of the climber's body, however, does not change. Weight and mass are often used interchangeably because they are proportionally the same anywhere on Earth.

2.5.1 Mass

The basic unit of mass is the gram (g). A gram is a relatively small amount of matter. It is equivalent to about 1 milliliter of water.

The same prefixes used with the meter and the liter in the metric system are used with the gram. The most common units are the milligram (mg), gram (g), and kilogram (kg).

2.5.2 Weight (Force)

Weight is actually a force. It is the push exerted on the surface of the Earth due to mass and the pull of the Earth's gravity. A force is a quantity that has two qualities: magnitude and direction.

Weight is a force with a direction always assumed to be downward. In the English system, the most common units for weight are the pound (lb) and the ounce (oz). The English system also uses these units to represent mass. If a distinction is needed, it is so stated, such as pounds-mass or pounds-force.

Table 3 shows the relationship between mass and weight units in both systems. In most instances, they are interchangeable for both mass and weight.

Table 3 Mass and Weight Equivalences

UNIT	kg	lb	oz	g
1 KILOGRAM (kg)	1	2.205	35.27	1000
1 POUND (lb)	0.4536	1	16	453.6
1 OUNCE (oz)	0.02835	0.0625	1	28.35
1 GRAM (g)	0.001	0.002205	0.03527	1

201T03.EPS

In reality, the metric system uses the **newton (N)** as the basic unit of force. The newton is named after Sir Isaac Newton, who is often called the father of classical physics. He discovered the true importance of gravity, created the field of mathematics known as calculus, and defined the properties of matter and bodies in motion.

Newton's First Law of Motion states that every body persists in its state of rest, or of uniform motion in a straight line, unless it is compelled to change that state by forces acting on it. In other words, a body at rest stays at rest and a body in motion stays in motion at a constant **velocity**, unless something (force) acts to influence it.

Force is a quantity having both magnitude and direction. With respect to the body in Newton's First Law, a force can be applied in a direction that aids the moving body to increase its velocity. Conversely, a force can be applied in a direction that opposes the moving body to decrease its velocity. **Acceleration** is the term used to represent an increase in velocity (speed) with respect to time.

Figure 9 shows equal forces applied to two different bodies, which are viewed as two different surface areas.

The surface area that Body A presents to Force A is four times larger that the surface area that Body B presents to Force B. Another way of describing this is to say that Force A is spread over a larger area than Force B. As a result, the force per unit area applied to Body A is ¼ of the force per unit area applied to Body B.

Figure 10 shows the same two bodies with surface areas and forces specified.

In the metric system, pounds per square inch (psi), the most common unit of pressure in the English system, is replaced by the newton per square meter, or N/m^2. The newton per square meter is called a **pascal (Pa)**, after Blaise Pascal, a seventeenth-century French philosopher who contributed greatly to mathematics and engineering.

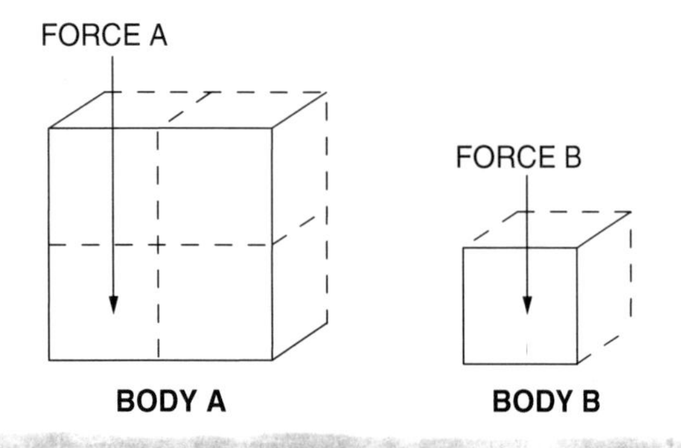

Figure 9 ◆ Equal forces applied to different surface areas.

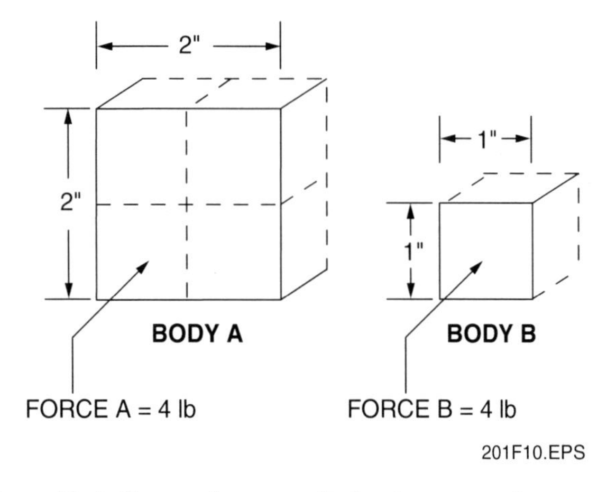

Figure 10 ◆ Known forces applied to known surface areas.

One pascal is not really very much pressure. It is not nearly as much as one psi. If half a cup (about 100 grams) of sugar is spread evenly over the top of a table that is one meter on each side, the pressure or force exerted at any one point on the table is approximately one pascal.

It would take 200,000 Pa (N/m^2) to inflate the ordinary automobile tire to about 28 psi.

A second common metric unit of pressure is the bar (b), which is equal to 100,000 Pa. The bar is used by weather forecasters. Anyone familiar with barometric pressure readings knows that the millibar, which is .001 bar (b), is also used in weather forecasts.

Occasionally in the metric system, the unit **dyne** is used to measure small forces. One newton is equal to 100,000 dynes.

The following sections relate these metric units to the more familiar pressure units of the English system and review the concept of pressure measurement.

2.6.0 Pressure

Pressure is a quantity closely related to force. At a service station, one may put 28 pounds of air in the rear tires of a car. What actually happens is that an amount of air put is put into the tires that can exert 28 more pounds of force per square inch on the inside of the tire than the pressure of the atmospheric air on the outside of the tires.

To cause a specified amount of acceleration, a force must be great enough and applied in such a direction that it overcomes any opposing forces that are present. The mass of a body has a direct effect on the amount of force required.

Newton's Second Law of Motion states: If a body is acted on by a force, that body will be accelerated at a rate directly proportional to the force. The rate is inversely proportional to its mass. In other words, a body with a large mass requires more force to obtain a specified amount of acceleration than does a body with less mass. Mass can be described as the amount of matter contained in a body. The relationship of force, mass, and acceleration is expressed as:

$$\text{Acceleration} = \frac{\text{Force}}{\text{Mass}}$$

or

$$\text{Force} = \text{Mass} \times \text{Acceleration}$$

Two bodies, each made of a different material, can have the same masses. Two bodies, both made of the same material, can have different masses if one body has been made larger by adding more of the material to it.

The basic unit of force in the metric system, the newton, can now be defined as the amount of force required to accelerate one kilogram at the rate of one meter per second per second.

$$\text{Force (one newton)} = \text{Mass (one kilogram)} \times \text{Acceleration (one meter/s/s)}$$

or

$$\text{One newton} = \frac{\text{one kg m}}{s^2}$$

In *Figure 13*, the force per unit area that is applied to Body A is:

$$\frac{\text{Force A}}{\text{Unit Area A}} = \frac{\text{4 pounds}}{\text{4 square inches}} = \text{1 lb per in}^2$$

The force per unit area that is applied to Body B is:

$$\frac{\text{Force B}}{\text{Unit Area B}} = \frac{\text{4 pounds}}{\text{1 square inch}} = \text{4 lb per in}^2$$

The definition of pressure is force per unit area, or:

$$\text{Pressure} = \frac{\text{Force}}{\text{Area}}$$

Pressure represents force applied perpendicular to a surface. The pressure applied to Body A is 1 pound per square inch, or 1 psi. The pressure applied to Body B is 4 pounds per square inch, or 4 psi.

As shown, applying equal forces to bodies of different surface areas results in applying different pressures to the bodies.

2.6.1 The Pascal

When converting to metric system equivalent measures, each dimension must be converted. When dealing with more complex measurements such as pressure, it becomes much easier to use conversion factors directly for the more common units of measure. *Table 4* shows the more common conversion factors used to convert to the newton or dyne directly for general forces. *Table 5* shows the conversion factor used to convert to the pascal or newton/meter^2 or dyne/cm^2 for very small pressures. Because this is such a small number, and it takes many pascals to define a small pressure, the pascal is normally shown as kilo-pascal (kPa).

Table 4 Force Conversion Factors

UNIT	KILOGRAM-FORCE	POUND-FORCE	GRAM-FORCE
NEWTON (N)	9.807	4.448	0.009807
DYNE	980,700	444,800	980.7

201T04.EPS

Therefore, to convert the 1 psi for *Figure 9*:

$$1 \text{ psi} \times \frac{6{,}895 \text{ Pa}}{\text{psi}} = 6{,}895 \text{ Pa}$$

To convert the 4 psi for *Figure 9*:

$$4 \text{ psi} \times \frac{6{,}895 \text{ Pa}}{\text{psi}} = 27{,}580 \text{ Pa or } 27.580\text{kPa}$$

It is more than evident that the pascal is small in comparison to the pound per square inch.

A man who weighs 160 lbs is wearing ice skates. The total surface area of the two skate blades in contact with the ice is:

$$\text{Total Area} = \text{Length} \times \text{Width} \times 2 \text{ blades}$$

or:

$$\text{Total Area} = 12 \text{ inches} \times \tfrac{1}{8} \text{ inch} \times 2$$

or:

$$\text{Total Area} = 3 \text{ square inches}$$

Therefore, the man exerts a pressure on the surface of the ice equal to:

$$\text{Pressure} = \frac{\text{Force}}{\text{Unit Area}}$$

or:

$$\text{Pressure} = \frac{160 \text{ lb}}{3 \text{ in}^2}$$

or:

$$\text{Pressure} = 53.3 \text{ psi or } 367{,}504 \text{ Pa}$$

For another instance, assume that the man who weighs 160 lbs is wearing snow skis. The total surface area of the two skis in contact with the ice is:

$$\text{Total Area} = \text{length} \times \text{width} \times 2 \text{ skis} = 72 \text{ in} \times 3 \text{ in} \times 2$$

or:

$$\text{Total Area} = 432 \text{ in}^2$$

Therefore, the man exerts a pressure on the surface of the ice equal to:

$$\text{Pressure} = \frac{160 \text{ lb}}{432 \text{ in}^2}$$

or:

$$\text{Pressure} = 0.37 \text{ psi or } 2{,}551 \text{ Pa}$$

Table 5 Pressure Conversion Factors

UNIT	GRAM FORCE PER cm^2	POUND-FORCE PER in^2 (psi)	POUND-FORCE PER ft^2	KILOGRAM FORCE PER cm^2
KILO-PASCAL (kPa)	98.070	6.895	.04788	.009807
PASCAL (Pa) NEWTON/METER^2 (N/m^2)	98,070	6,895	47.88	9.807
DYNE/cm^2	980,700	68,950	478.8	98.07

201T05.EPS

A person weighing 160 lbs sees a friend, walking ahead on a frozen lake, break through the ice and fall into the lake. The standing person instinctively knows to lay down flat on the ice immediately. This distributes weight over as large a surface area as possible.

A prone person approximately 6 ft (72 inches) in height and averaging 18 inches wide has a surface area of:

Total Area = 72 in × 18 in

Total Area = 1296 in^2

This means the pressure exerted on the surface of the ice is:

$$\text{Pressure} = \frac{160 \text{ lb}}{1{,}296 \text{ in}^2}$$

Pressure = 0.123 psi or only 851 Pa

That is a reduction in pressure on the ice of over 2,000 times the pressure exerted with ice skates while standing upright.

2.6.2 Hydrostatic Pressures

The previous discussion addressed the mechanical pressure of a solid object coming into contact with another solid object. There are other ways to exert pressure.

Instrumentation is most often used to measure the pressure exerted at the base of the tank by fluids, such as gases or liquids. The most common measure is the pressure exerted by a liquid due to its height or level in the column. This is referred to as the **hydrostatic pressure** of the liquid. Hydrostatic pressure is also referred to as *head pressure*.

The molecules of liquid have mass that is acted on by the Earth's gravity. As explained previously, the mass of the liquid is related to its weight (force) using Newton's Second Law. Therefore, two gallons of water weigh twice as much as one gallon because they have twice the mass of water molecules. One gallon of water weighs more than one gallon of gasoline because the water has more mass per unit volume. Mass per unit volume is known as **density**, and is commonly expressed in *pounds per cubic foot (lb/ft^3)*.

Figure 11(A) shows a container holding one gallon of a liquid. This liquid applies a 5 pound downward force to the 5" by 5" container base. Theoretically, the pressure is:

$$\text{Pressure} = \frac{\text{Force}}{\text{Area}}$$

or:

$$\text{Pressure} = \frac{5 \text{ lb}}{25 \text{ in}^2}$$

or:

Pressure = 0.2 psi

Figure 11(B) shows the same container tipped on its side. With the volume of the liquid still equal to one gallon, the total force applied downward is still 5 pounds. However, this force is applied over a larger area. Theoretically, the pressure is now:

$$\text{Pressure} = \frac{5 \text{ lb}}{50 \text{ in}^2}$$

or:

Pressure = 0.1 psi

The example given in *Figure 11* might indicate that the pressure applied at the base of a liquid container is determined by the area of its base. This is not correct. In fact, the shape of a container has no bearing on the pressure developed by the weight of a liquid. It is the height of the liquid above the base that really matters.

Figure 12 shows two vessels of different shape and capacity. An interconnecting path is provided at the bottom of each vessel to allow flow from one vessel to the other.

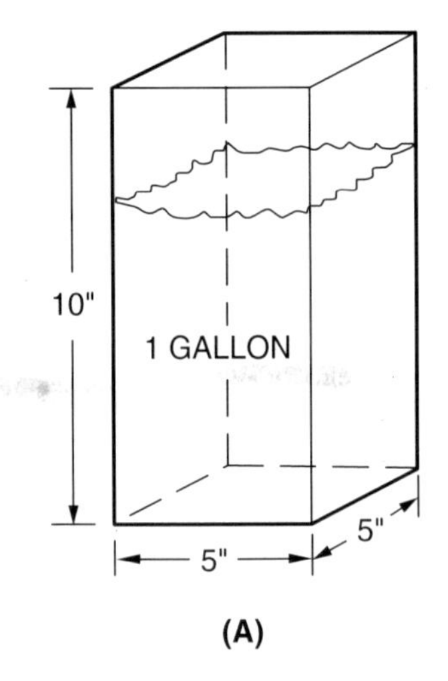

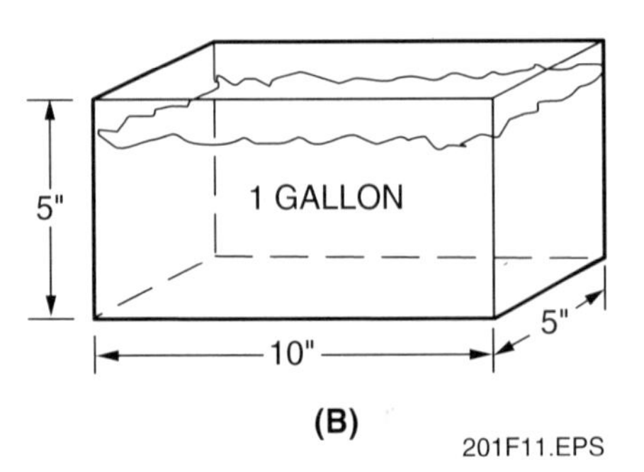

Figure 11 ◆ Equal downward forces applied to different areas.

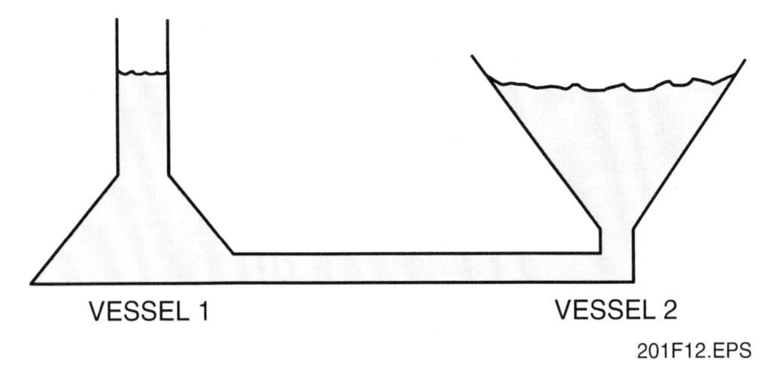

Figure 12 ◆ Hydrostatic paradox.

Vessel 2 contains more liquid mass than Vessel 1. The base area of Vessel 2 is smaller than Vessel 1. These two facts would seem to imply that the pressure applied to the base of Vessel 2 is greater than the pressure applied to the base of Vessel 1. If this was in fact true, then the level in Vessel 1 would be higher than the level in Vessel 2.

However, *Figure 12* shows the levels to be equal in the two vessels, which means equal pressures exist in the vessels. This result baffled scientists for some time. Because of the seemingly unreasonable result, this effect became known as the hydrostatic paradox.

The mystery of the hydrostatic paradox can be explained by analyzing the directions of force that the liquid produces, as shown in *Figure 13*.

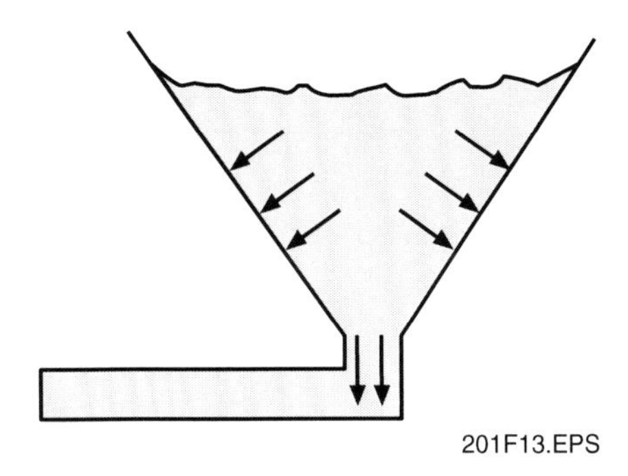

Figure 13 ◆ Direction of force applied by a liquid.

Notice that the directions of force applied to the vessel walls are perpendicular to the walls' surface areas. The pressure applied to the base of the vessel is the result of only the perpendicular forces to its surface area, as shown in *Figure 14*. A liquid cannot sustain a sliding force, as a solid can. A force at any angle other than 90° to the surface merely tends to slide the liquid molecules along the surface without producing a sustained force.

The magnitude of the pressure applied to the base of the vessel depends on the density and height of the liquid column directly over the base, or:

Hydrostatic Pressure = Density × Height

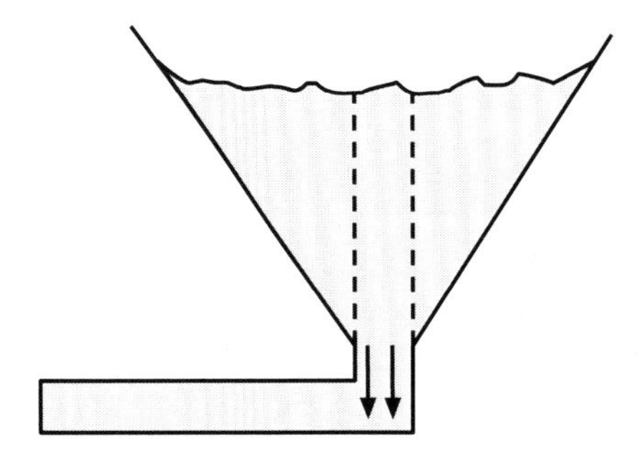

Figure 14 ◆ Actual pressure applied to base surface.

Referring to *Figure 12*, because the liquids in Vessels 1 and 2 have the same density and are at the same height, the hydrostatic pressures at the bases of both vessels are equal. As such, no liquid is displaced from Vessel 2 to Vessel 1.

If in *Figure 12* more liquid is added to one of the vessels, the increase in level of that vessel would cause an increase in hydrostatic pressure. Liquid would be displaced until the levels are again equal, and the hydrostatic pressures would therefore again be equal. It is through this process of equalizing hydrostatic pressures that a liquid seeks its own level.

Figure 15 shows a cubic foot container filled with water.

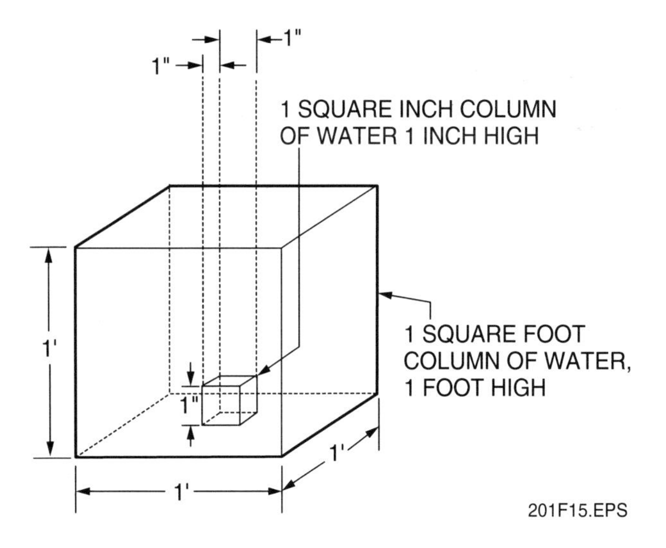

Figure 15 ◆ One cubic foot and one cubic inch of water.

The density of water is 62.4 lb/ft³ at 68°F. The hydrostatic pressure in *Figure 15* is therefore:

Hydrostatic Pressure = Density × Height

$$\text{Hydrostatic Pressure} = \frac{62.4 \text{ lb} \times 1 \text{ ft}}{\text{ft}^3}$$

$$\text{Hydrostatic Pressure} = 62.4 \text{ lb per square foot}$$

or in terms of psi:

$$\text{Hydrostatic Pressure} = \frac{62.4 \text{ lb}}{\text{ft}^2} \times \frac{1 \text{ ft}^2}{144 \text{ in}^2}$$

$$\text{Hydrostatic Pressure} = 0.433 \text{ psi}$$

Therefore a 1-foot head of water is equivalent to a pressure of 0.433 psi.

A 1-inch head of water is equivalent to a pressure of:

$$\text{Hydrostatic Pressure} = \frac{62.4 \text{ lb}}{\text{ft}^3} \times \frac{1 \text{ ft}^3}{1728 \text{ in}^3} \times 1 \text{ in}$$

$$\text{Hydrostatic Pressure} = 0.0361 \text{ psi}$$

A 1-inch head of water is also referred to as a *1-inch water column (1-in w.c.)* because it relates to measuring, or gauging, the amount of pressure required to raise a column of water by 1".

1in w.c. = 0.0361 psi

1 psi = 27.7 in w.c.

A 1-in w.c. column of water produces a head pressure of 0.0361 psi. Conversely, it requires 0.0361 psi to raise the water in the column by one inch.

A 27.7-in w.c. column of water produces a head pressure of 1 psi. Conversely, it requires 1 psi to raise the water in a column by 27.7-in w.c.

A column of water can be used to create pressure. The magnitude of mechanical pressure can also be used to create pressure. A mechanical pressure can be determined by applying it to a water column and measuring the displacement distance.

Other liquids (most commonly mercury) can be used to measure hydrostatic pressure. When liquids are used, the density of the liquid determines the pressure-to-displacement relationship. Mercury is very dense compared to water. Therefore, the same height or column of mercury can be used to measure higher pressures. Both water and mercury (Hg) have become standards for measurement.

2.6.3 Units of Hydrostatic Pressure

Because pressure is derived from the unit of force, pressure must be tied to a point of reference if the magnitude of pressure is to be specified.

There are several reference scales, such as those using water, mercury, and even oil (for very low pressure measurements). The most appropriate reference scale is specified by the manufacturer of the instrumentation. The pressure that is almost always present is the pressure of the atmosphere or air. Understanding **atmospheric pressure** can eliminate significant errors in working with instrumentation systems.

The air surrounding the Earth makes up the atmosphere. The density of the air is based on the number of molecules in a specified volume of the air and is highest at sea level. As the distance increases away from the Earth, the density of the air decreases because the force of gravity decreases with distance. The top of the Earth's atmosphere is that point at which the force of gravity is no longer strong enough to hold any air molecules.

If a column of air 1 inch square and extending from sea level to the top of the Earth's atmosphere could be weighed, it would weigh approximately 14.7 pounds at sea level when the temperature is 32°F.

$$\text{Atmospheric Pressure} = \frac{\text{Force from the Weight of Air}}{\text{Area}}$$

$$\text{Atmospheric Pressure} = \frac{14.7 \text{ lb}}{1 \text{ in}^2}$$

An atmospheric pressure of 14.7 psi is also referred to as 1 atmosphere (atm). *Figure 16* shows a gauge being used to measure the atmospheric pressure at different altitudes.

Notice that the pressure does not change linearly for a given change in distance from Earth. This is because the force of gravity is mathematically nonlinear. Fifty percent of the Earth's atmosphere, in terms of mass, exists between sea level and 3.5 miles up, which is approximately the height of Mt. McKinley in Alaska.

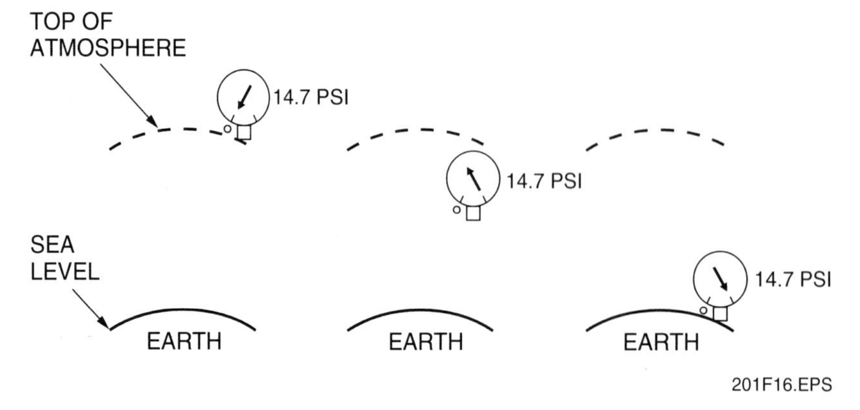

Figure 16 ◆ Atmospheric pressure.

Every surface exposed to air at sea level has an atmospheric pressure of 14.7 psi pressing on it. The door of a house that measures 3' by 7', therefore, has a total force of 22.2 tons applied to one side.

$$36 \text{ inches} \times 84 \text{ inches} = 3{,}024 \text{ inches}$$

$$3{,}024 \text{ inches} \times 14.7 \text{ psi} = 44{,}452.8$$

$$\frac{44{,}452.8}{2{,}000 \text{ lbs}} = 22.2 \text{ tons}$$

The door opens and shuts easily because an equal atmospheric pressure is applied to the other side to cancel the 22.2 tons force.

Figure 17 shows an open container at sea level.

Notice that the container walls have 14.7 psi applied to both the inner and outer sides. The container walls, therefore, have a zero difference in pressure.

Figure 18 shows two sealed containers at sea level.

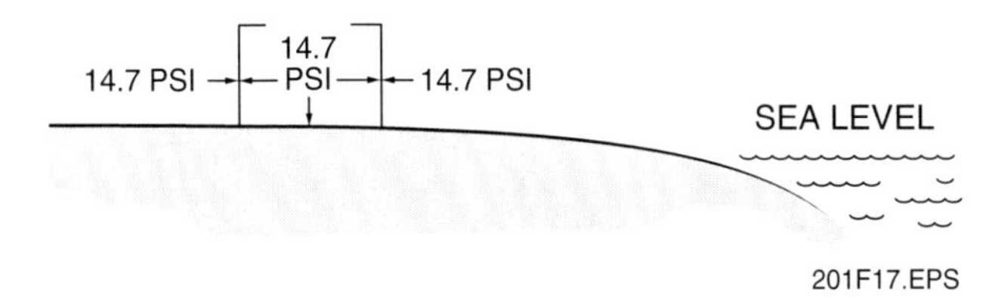

Figure 17 ◆ Open container at sea level.

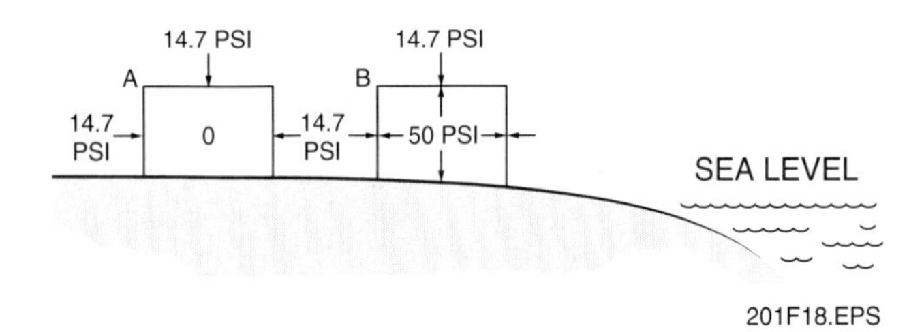

Figure 18 ◆ Sealed containers at sea level.

Container A has had all the air molecules removed. This is referred to as a perfect **vacuum**. The container walls have a 14.7 psi difference in pressure. The directional force of this net pressure is inward.

Air molecules were added to Container B when more air was forced into the container. The container walls have a 35.3 psi difference in pressure, and the directional force of this net pressure is outward.

$$50 \text{ psi} - 14.7 \text{ psi} = 35.3 \text{ psi}$$

If the perfect vacuum of Container A is considered a reference, the pressure in Container B is stated as 50 psi above absolute zero, or no pressure.

If the atmospheric pressure around the containers is considered a reference, the pressure in Container B is stated as 35.3 psi measured, or gauged, above atmospheric.

Several different units are used to express the magnitude of a measured pressure. The unit selected depends on the reference to which the pressure is compared and, in some instances, the magnitude of the pressure.

Absolute pressure is the pressure of a liquid or gas measured in relation to a complete vacuum (zero pressure). The unit of absolute pressure is pounds per square inch absolute, or psia.

Atmospheric pressure is the pressure due to the weight of the atmosphere above. This pressure is approximately 14.7 lb/in^2 at sea level and 32°F. Atmospheric pressure, therefore, varies with elevation from sea level, and with air temperature. Atmospheric pressure is referenced to the 0 psi pressure at the top of the Earth's atmosphere. Therefore, it is a form of absolute pressure and can be expressed as 14.7 psia at sea level.

Gauge pressure is the relative reading of pressure inside a container compared to the atmospheric pressure outside the container when the internal pressure is greater than atmospheric. The term *gauge* refers to the act of measuring the relative pressure above atmospheric. The unit of gauge pressure is pounds per square inch gauge, or psig.

The relationship of absolute, atmospheric, and gauge pressure is:

$$\text{Absolute Pressure} = \text{Gauge Pressure} + \text{Atmospheric Pressure}$$

Referring to *Figure 18*, Container B pressure is:

$$50 \text{ psia} = 35.3 \text{ psig} + 14.7 \text{ psi atmospheric}$$

Inside Container B, the atmospheric pressure could be assumed to equal 14.7 psia, and any further discussions could continue in units of gauge pressure, such as 35.3 psig. However, there is a risk in assuming the atmospheric pressure value. If, for example, the atmospheric pressure decreased to 14.6 psia, the gauge pressure would be:

$$\text{Gauge Pressure} = \text{Absolute Pressure} - \text{Atmospheric Pressure}$$

$$\text{Gauge Pressure} = 50 \text{ psia} - 14.6 \text{ psi atmospheric}$$

$$\text{Gauge Pressure} = 35.4 \text{ psig}$$

This would give an appearance of a 0.1 pound per square inch increase in the pressure inside the container, even though no additional air molecules were added to the container.

Vacuum is the relative reading of pressure inside a container compared to the atmospheric pressure outside the container when the internal pressure is less than atmospheric. A perfect vacuum exists when all air molecules have been removed from a container. Any degree of vacuum other than perfect is known as a partial vacuum. The units of vacuum can be psia, because vacuum is actually the pressure range of 0 psia to 14.7 psia. Usually, however, vacuum is expressed in units of inches of mercury (Hg). This unit is based on absolute pressure and the resultant displacement of mercury in a vertical column.

Torricelli was an Italian scientist who found that a vacuum could be formed by filling a glass tube with mercury, closing it at one end, and then inverting the tube over a pan of mercury (*Figure 19*).

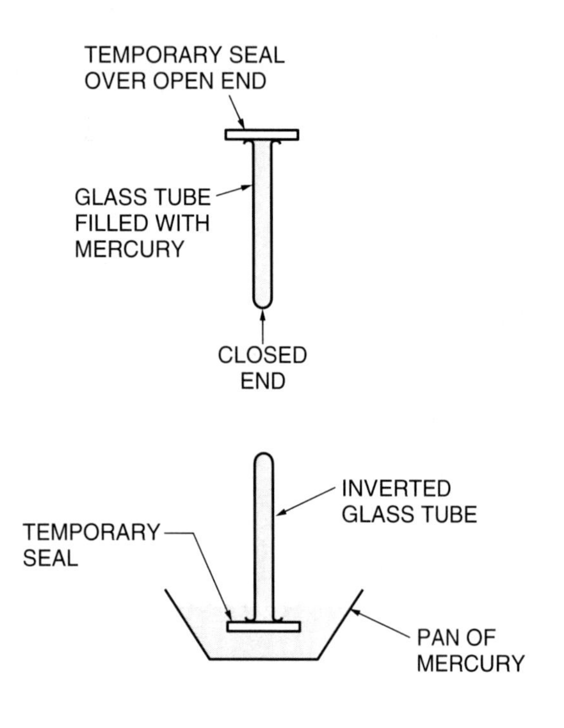

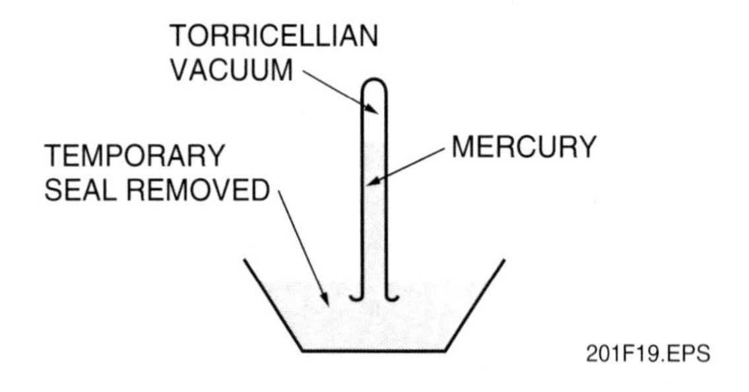

Figure 19 ◆ Torricelli tube.

Torricelli found that some, but not all, of the mercury flows out of the tube and a void or vacuum is formed in the upper portion of the tube. A perfect vacuum, however, is not formed. A small pressure exists due to the vapor pressure of mercury at room temperature. The magnitude of this pressure is about 0.0001 psia at 70°F and can be considered negligible. Other fluids are not suitable for this application because their vapor pressures are higher.

The downward force caused by the gravitational pull on the mass of the mercury causes mercury to flow out of the tube. Atmospheric pressure exerts a downward force on the surface of the mercury in the pan. As mercury flows out of the tube, the downward force it exerts decreases because the mass is decreased. When the downward force due to the weight of the mercury in the tube equals the downward force due to atmospheric pressure, mercury flow out of the tube stops.

Torricelli found that regardless of the length of the tube, the mercury does not exceed a maximum vertical height of 76 cm, or 29.92 inches. He also found that even when the tube is tilted away from the vertical position, the mercury maintains a vertical height of 29.92 inches, as shown in *Figure 20*.

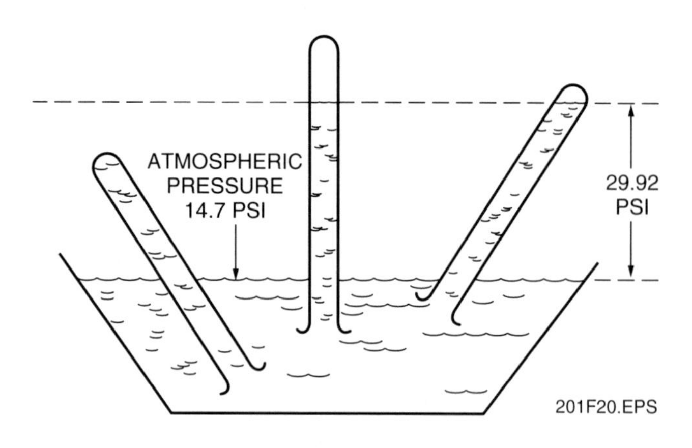

Figure 20 ◆ Equal length tubes at various angles.

Atmospheric pressure of 14.7 psia is, therefore, equal to 29.92 inches of Hg at 32°F.

1 in Hg = 0.4912 psi

1 psi = 2.036 in Hg

The Torricelli tube represents the simplest form of barometer. A barometer is a device used to measure atmospheric pressure. As atmospheric pressure changes, the height of the mercury in the barometer changes. If the mercury pan or reservoir is connected to a vacuum instead of being referenced to atmospheric pressure, the displacement in inches becomes a direct indication of the magnitude of the vacuum.

Table 6 shows the relationship of the different units of pressure and vacuum. *Table 7* shows the common conversion factors to change measurements from the English system to the metric system and back.

Table 6 Comparison of Pressure Units

UNIT	GAUGE PRESSURE	ABSOLUTE PRESSURE	MERCURY COLUMN VACUUM	MERCURY COLUMN PRESSURE	WATER COLUMN PRESSURE	
SYM.	PSIG	PSIA	"Hg	"Hg	"H_2O	
PRESSURE	100.0	114.7	NOT USED	204.0	2774.0	ATMOSPHERIC PRESSURE
	50.0	64.7		102.0	1387.0	
	10.0	24.7		20.4	277.4	
	5.0	19.7		10.2	138.7	
	1.0	15.7		2.04	27.74	
	0.5	15.2		1.02	13.87	
	0.4	15.1		0.82	11.1	
	0.3	15.0		0.61	8.32	
	0.2	14.9		0.41	5.55	
	0.15	14.85		0.31	4.16	
	0.1	14.8		0.2	2.77	
	0.05	14.75		0.1	1.39	
	0.0	14.7	0.0	0.0	0.0	1 atm
VACUUM	-0.05	14.65	0.1	NOT USED	NOT USED	ABSOLUTE ZERO - PERFECT VACUUM
	-0.1	14.6	0.2			
	-0.15	14.55	0.31			
	-0.2	14.5	0.41			
	-0.3	14.4	0.61			
	-0.4	14.3	0.82			
	-0.5	14.2	1.02			
	-1.0	13.7	2.04			
	-5.0	9.7	10.2			
	-10.0	4.7	20.4			
	-14.7	0.0	29.92			

201T06.EPS

2.7.0 Temperature

The text of this module carefully specifies the temperature at which standards of pressure are measured. This is because another way to create pressure is by heating a fluid in a closed container. When fluids are heated, they expand. Fluids in a closed container cannot expand. Therefore, the pressure that the fluid exerts on the walls of the container increases as the fluid tries to expand.

Temperature is normally thought of as how hot or cold an object is. If the temperature of an object is greater than body temperature, it is considered to be hot; if its temperature is lower than body temperature, it is considered to be cold. The hotness or coldness of an object is a result of molecular activity; the greater the activity, the greater the temperature. The activity is an indication of the kinetic energy of the atoms, which make up the molecules. Temperature, therefore, is a measure of the atomic kinetic energy of an object.

The temperature of a substance is one of its most important properties. Both the physical and chemical states of most substances change when the substance is heated or cooled. Therefore, it is essential that the temperature of a process be measured and controlled. This helps ensure the quality of the product and the safety of the process.

2.7.1 Temperature Scales

Temperature is measured in degrees of a temperature scale. In order to establish the scale, a substance such as water is needed so that it can be placed in reproducible conditions from which standard measurements can be determined. The point at which the water freezes at atmospheric pressure is one reproducible condition. The point at which water boils at atmospheric pressure is another.

There are four temperature scales commonly used today. They are the Fahrenheit, Celsius, Rankine, and Kelvin scales.

On the Fahrenheit scale the freezing temperature of water is 32°F and the boiling temperature is 212°F. On the Celsius scale the freezing temperature of water is 0°C and the boiling temperature is 100°C. The Rankine scale and the Kelvin scale are based on the theory that at some condition no molecular activity occurs. The temperature at which this condition occurs is called the *absolute zero* temperature, the lowest temperature possible. Both the Rankine and Kelvin scales have their zero degree points at absolute zero. On the Rankine scale the freezing point of water is 491.7°R and the boiling point is 671.7°R. The increments on the Rankine scale correspond in size to the increments on the Fahrenheit scale; for this reason, the Rankine scale is sometimes referred to as the *absolute Fahrenheit* scale. Both the Rankine and Fahrenheit scales are part of the English system of measurement.

On the Kelvin scale, the freezing point of water is 273K and the boiling point is 373K. The increments on the Kelvin scale correspond to the

Table 7 Fluid Pressure Conversion Factors

UNIT	ATMOSPHERE	kg FORCE PER cm^2	POUND-FORCE PER in^2 (psi)	INCHES H_2O	INCHES Hg	MILLIMETERS Hg
PASCAL (Pa) OR NEWTON/METER^2 (N/m^2)	101,300	98,040	6.895	6.895	6.895	133.3

201T07.EPS

increments on the Celsius scale; for this reason, the Kelvin scale is sometimes referred to as the *absolute Celsius scale*. Both the Kelvin and Celsius scales are parts of the metric system of measurement. Because there are 100 degrees on a standard portion of the Celsius scale, it has also been called *degrees centigrade*.

In the process industry, the scales of primary importance are the Fahrenheit scale and the Celsius scale. The Rankine scale and the Kelvin scale are used primarily in science. A comparison of the scales is illustrated in *Figure 21*.

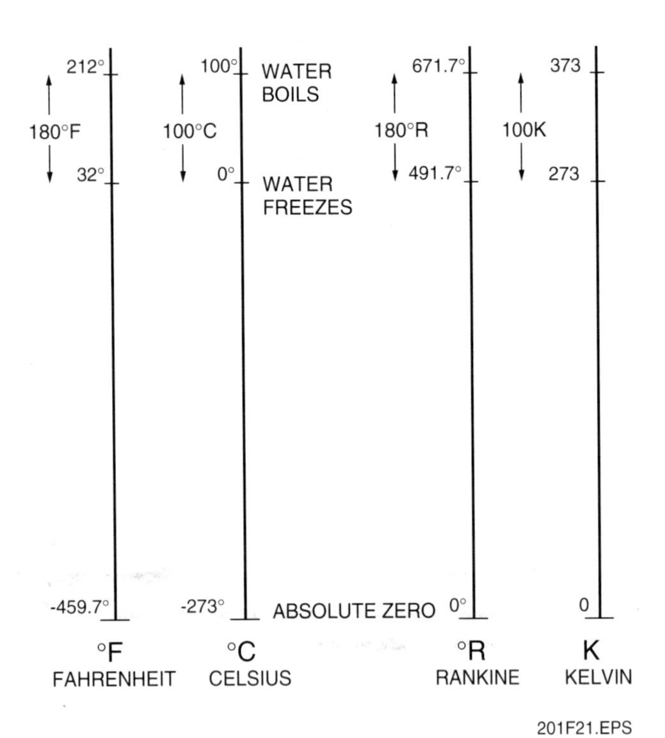

Figure 21 ◆ Comparison of temperature scales.

2.7.2 Temperature Conversions

Because both the Fahrenheit and Celsius scales are used in industry, it sometimes becomes necessary to convert between the two. Be familiar with these conversions.

On the Fahrenheit scale, there are 180 degrees between the freezing temperature and boiling temperature of water. On the Celsius scale, there are 100 degrees between the freezing temperature and the boiling temperature of water. The relationship between the two scales can be expressed as follows:

$$\frac{\text{Fahrenheit Range (freezing to boiling)}}{\text{Celsius Range (freezing to boiling)}} = \frac{180°}{100°} = \frac{9}{5}$$

Therefore, one degree Fahrenheit is $\frac{5}{9}$ of one degree Celsius and conversely, one degree Celsius is $\frac{9}{5}$ of one degree Fahrenheit. To convert a Fahrenheit temperature to a Celsius temperature, it is necessary to subtract 32°, because 32°F corresponds to 0°C, and multiply by $\frac{5}{9}$. To convert a Celsius temperature to a Fahrenheit temperature, multiply by $\frac{9}{5}$, then add 32°C.

$$°C = \frac{5}{9}(°F - 32°)$$
$$°F = \frac{9}{5}(°C) + 32°$$

Practice these calculations to become more comfortable with using them (refer to *Figure 22*).

$°C = \frac{5}{9}(°F - 32°)$

$°F = \frac{9}{5}(°C) + 32°$

Example: 77°F to °C

$$°C = \frac{5}{9}(77° - 32°)$$
$$°C = \frac{5}{9}(45°)$$
$$°C = 25°C$$

Example: 90°C to °F

$$°F = \frac{9}{5}(90°) + 32°$$
$$°F = 162° + 32°$$
$$°F = 194°F$$

201F22.EPS

Figure 22 ◆ Sample temperature unit conversions.

2.8.0 Flow

Flow is the passage of a fixed quantity of a fluid past a point per unit time. Flow measurements are divided into two major categories: **flow rate** measurement and **total flow** or total quantity measurement.

Flow rate is the amount of fluid that moves past a given point at any given instant of time.

Total flow is the amount of fluid that moves past a given point during a specified period of time.

2.8.1 Units of Flow

Flow rate can be expressed in terms of mass, weight, or volume. Mass flow rate is the quantity of material in units of mass flowing past a point at a given instant of time. Flow rate in terms of mass is expressed in English units as pounds-mass per hour, lb_m/hr, or pounds-mass per second, lb_m/s. In metric units, mass flow rate is expressed as kilograms per hour, kg/hr, or kilograms per second, kg/s. Steam flow is a quantity normally measured in units of mass flow rate.

Weight flow rate is the quantity of a material in units of weight such as pounds-force, lb_f, flowing past at a given instant of time. Weight flow rate is expressed as pounds-force per hour, lb_f/hr, or pounds-force per second, lb_f/s. Flow rate in units of weight is most commonly used in chemical plants, paper mills, and in the measurement of liquid fuel flow.

Volumetric flow rate is the quantity of material in cubic units flowing past a point at a given instant of time. For liquid flow measurements the volumetric flow rate is normally expressed in gallons per minute or liters per minute. For gas flow measurements the volumetric flow rate is normally expressed in cubic feet per minute (cfm), or cubic feet per second (cfs).

The volume, pressure, and temperature of a gas are all related. Therefore, a cubic foot cannot be used to define a quantity of gas unless the pressure and temperature of the gas that fills a cubic foot of space are specified. A common unit used for a quantity of gas is the standard cubic foot (scf). A standard cubic foot is the quantity of gas at 68°F and 14.7 psia that is required to fill a volume of one cubic foot. *Table 8* lists the commonly used units of flow rate.

Total flow is also expressed in terms of mass, weight, and volume. Total mass flow is the number of pounds-mass of a material that has passed a point in a given time period. Total weight flow and total volume flow are similar to total mass flow. Total flow is normally employed when it is important to know how much material has flowed, rather than how fast it is flowing. Units used for total flow are given in *Table 9*.

2.8.2 Flow Characteristics

Unbalanced pressure can make a fluid flow from a point of high to low pressure. In watering lawns, the pressure of the water supply affects the flow of water. During summer dry spells, for example, when the water-main pressure drops, the water barely spurts from the nozzle.

Table 8 Units of Flow Rate

MASS FLOW RATE:	lb_m/hr	pound-mass per hour
	kg/hr	kilograms per hour
	kg/s	kilograms per second
WEIGHT FLOW RATE:	lb_f/hr	pounds-force per hour
	lb_f/s	pounds-force per second
VOLUMETRIC FLOW:	scfh	standard cubic feet per hour
	scfm	standard cubic feet per minute
	scfs	standard cubic feet per second
	gpm	gallons per minute
	mgd	million gallons per day

201T08.EPS

Table 9 Units of Total Flow

MASS:	lb_m	pounds-mass
	kg	kilograms
	g	grams
WEIGHT:	lb_f	pounds-force
VOLUME:	pm	gallons per minute
	mgd	million gallons per day
	scf	standard cubic feet

201T09.EPS

In order to cause a fluid to flow, friction or resistance to flow must be overcome. A garden hose offers resistance to the flow of water due to the roughness of the interior of the hose. A ¾" diameter hose delivers more water than a ½" diameter hose of the same length and at the same water-main pressure. Resistance to the flow of water is also due to the internal friction within the liquid, the *viscosity*. Viscosity is the resistance to sliding between two adjacent layers of fluid due to the attraction between the molecules of adjacent layers.

In a process plant, fluids such as water, steam, or air must be moved from one point to another. As in the case of the garden hose, friction or resistance to flow must be overcome.

Several factors determine how much resistance there is to the flow of fluids. Resistance is caused by the fluid rubbing against the interior surfaces of piping. This is called *friction*, as mentioned earlier. Pipes with rough interior surfaces increase the opposition to flow. Changes in the direction of flow will also increase the resistance to flow. Obstructions such as valves or in-line components, which may have smaller diameters than the pipe, increase the opposition to fluid flow. Finally, fluids with high viscosities increase the opposition to flow.

Because pressure is needed to overcome friction or resistance, there is a drop in the pressure of a fluid in a pipe as the fluid flows along the length of the pipe. This pressure drop occurs only when there is a fluid flow, not when the fluid is standing in the pipe. This is illustrated in *Figure 23* by a reservoir and a series of stand pipes. When the outlet valve is shut tight, the resistance of the main pipe to the flow of water has no effect on the water pressure in the main pipes, as indicated by the levels in the stand pipe. This condition is represented by line A. The levels of the water in the columns are exactly equal to the level in the reservoir. If the valve is opened slightly, the resistance of the main pipe to the flow of water causes the water pressure to drop. Line B represents the different pressures that will be established in the stand pipes. The height of the water in the right-hand stand pipe is a measure of the pressure head

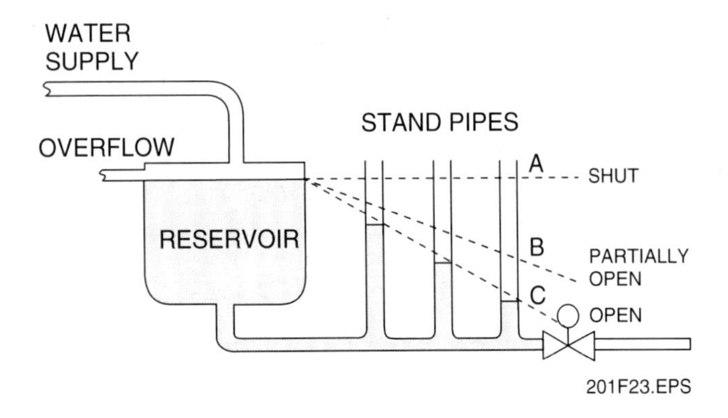

Figure 23 ◆ Pressure drop in a reservoir system.

available at the valve. When the valve is fully open, line C represents the different pressures that will be established in the stand pipes. The water level in the reservoir remains constant because of the water supply and overflow arrangement. This assures a constant pressure of water at the reservoir. As the flow becomes more rapid when the valve is opened, there is a greater loss of pressure due to friction. This loss in pressure is commonly called pressure drop or head loss.

In the above example, the supply pressure was held constant. If a pump is used in place of the tank to increase the supply pressure as shown in *Figure 24*, another fact is evident. With the valve wide open, line D represents a greater pressure drop between the pump discharge and the valve outlet than line E. The flow through the valve will also be greater.

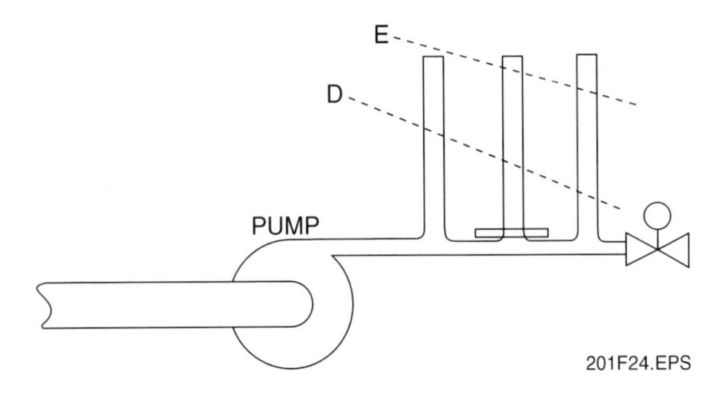

Figure 24 ◆ Pressure drop in a pump system.

Thus, the greater the pressure drop, the greater the flow. The increase in flow also results in a greater total loss of pressure due to friction.

Experiments have shown that this increased loss in pressure is due to the changing characteristics of the flow. The characteristics of flow are velocity-dependent. When the fluid moves through a pipe at relatively low velocities, particles of the fluid move along lines paralleling the pipe walls. Fluid layers slide smoothly over each other. This type of flow is called *laminar* flow. Laminar flow is sometimes referred to as *viscous* or *streamlined* flow. As the velocity of the fluid increases, a point is reached at which the particles begin to move from one layer to another. The approximate velocity at which particle motion becomes more random is called the *critical* velocity. When the velocity increases beyond this relatively narrow range of flow rates, the flow becomes completely random. Eddies and cross-currents move the fluid particles along irregular paths. This type of flow is referred to as *turbulent* flow.

The term *velocity*, when applied to the flow of fluids in pipes, refers to the average flow. The actual velocity of the fluid varies across the section of the pipe. For laminar flow, the velocity of the fluid layer adjacent to the pipe wall is theoretically zero because of friction between the fluid and pipe. A short distance from the wall, velocity increases rapidly and reaches its maximum value at the center of the pipe, as indicated in *Figure 25*.

In turbulent flow, the velocity of the fluid layer along the wall is also increased, but the velocity increases much faster as it moves from the wall than for laminar flow (*Figure 26*). The mixing action of the particles breaks up the fluid layers that result in the parabolic velocity profile of laminar flow. The importance of these velocity profiles becomes evident in discussing measurement of flow.

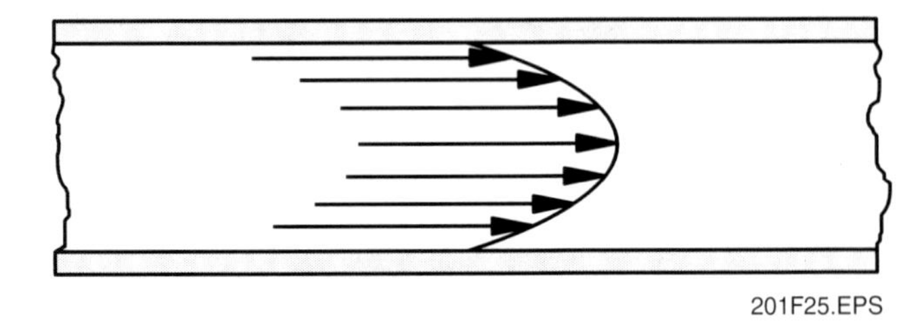

Figure 25 ◆ Velocity profile of laminar flow.

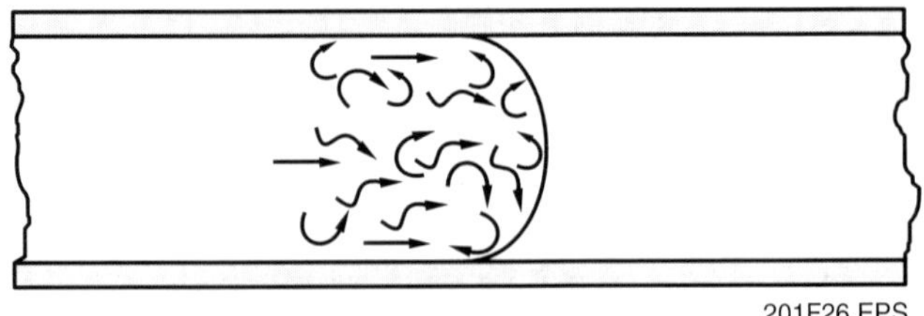

Figure 26 ◆ Velocity profile of turbulent flow.

As fluids flow through a pipe, the mass of the fluid entering the pipe must equal the mass of the fluid leaving the pipe. Consider a situation in which a non-compressible fluid such as water is flowing steadily through a tapered pipe as shown in *Figure 27*. The mass flowing into the pipe, m1, must equal the mass flowing out of the pipe, m2, even though the area through which the fluid flows is decreasing. To maintain steady mass flow, the velocity of the fluid must increase. Mass flow rate through a pipe, therefore, depends on the cross-sectional area of the pipe.

Because the fluid discussed is noncompressible, the density of the fluid does not change as the liquid flows through the pipe. The velocity of the flow varies inversely with the cross-sectional area of the pipe. In the tapered pipe that is shown in *Figure 27*, the velocity of the flow is greater in the narrower section of the pipe than in the wider section.

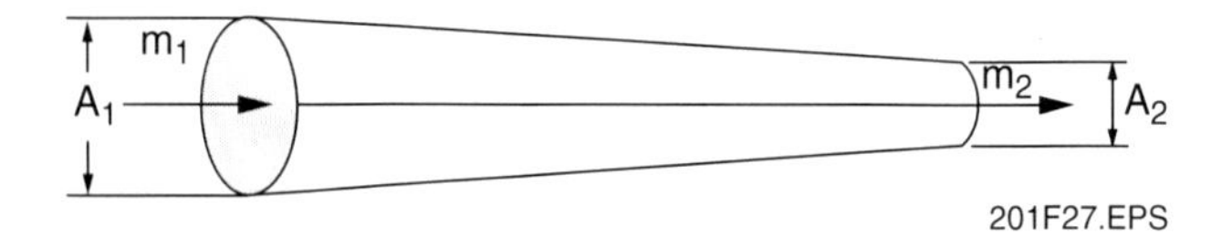

Figure 27 ◆ Continuity of flow through a tapered pipe.

In 1738, in his book *Hydrodynamics*, Daniel Bernoulli first presented an effect that is used universally today to measure flow rates. He stated that if the velocity of a fluid system increases, the pressure of the fluid must decrease. Bernoulli went on to prove this mathematically. A fluid flowing in a pipe exerts pressure in two different directions, as shown in *Figure 28*. Pressure is exerted in the direction of flow and against the walls of the pipe. These pressure measurements can be used to determine flow rate.

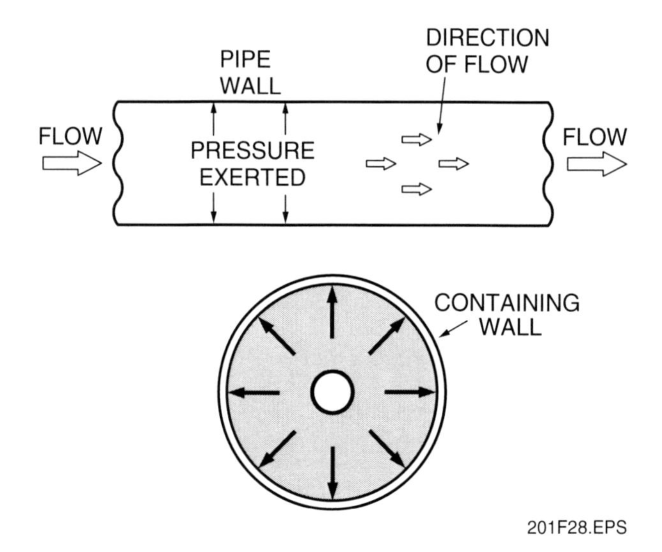

Figure 28 ◆ Pressure exerted by fluid flowing through a pipe.

If a restriction is placed in the pipe, narrowing the flow path as shown in *Figure 29*, the velocity of the fluid increases at and immediately after the restriction. As the velocity of flow increases, the pressure that the fluid exerts on the walls of the pipe decreases. The difference in the pressures before and after the restriction (differential-pressure) is used to determine the flow rate through the pipe.

As the fluid goes through the restriction at point No. 2, the velocity of the fluid must increase in order to move the same amount of mass (water molecules) per second past point No. 2. When the velocity increases, the pressure decreases.

The relationship between differential-pressure and flow is shown by the formula:

$$\text{Flow} = \frac{\sqrt{\text{measured D/P}}}{\sqrt{\text{maximum D/P}}} \times \text{maximum flow}$$

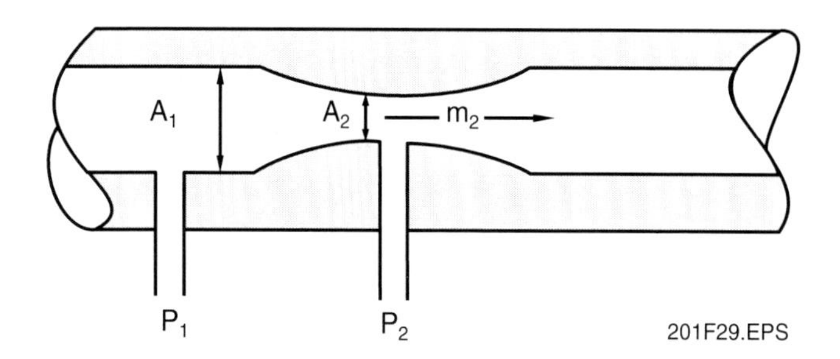

Figure 29 ◆ Variable head flowmeter restriction.

This equation implies that if the area of the restriction and the fluid density are held constant, the mass flow rate is proportional to the square root of the differential-pressure developed across the restriction. Therefore, by measuring the differential-pressure, a signal proportional to the flow squared can be obtained.

Differential-pressure is therefore proportional to the flow squared. The relationship between flow and differential-pressure is nonlinear. This nonlinear relationship is a disadvantage associated with head-type flowmeters. Normally, the square root extraction function required to linearize the flow signal is performed by signal conditioning equipment. Understand that the differential-pressure created by a restriction is related to the flow through the restriction in a nonlinear manner.

The head flowmeter actually measures volume flow rate rather than mass flow rate. Mass flow rates can be calculated or computed easily by determining or sensing temperature and/or pressure. Temperature and pressure affect density, which is one of the terms in determining the mass flow rate. If the volumetric flow rate signal is compensated for changes in temperature and/or pressure, a true mass flow rate signal can be obtained.

The measurement of gas flow presents an additional problem. The density of a compressible fluid, such as a gas, does not remain constant. When a gas passes through a restriction, the velocity increases and the pressure decreases, just as when a liquid flows through a restriction. However, when the pressure of the gas decreases, a significant decrease in density also occurs. This decrease in density means less mass per unit volume. Therefore, because the mass flow rate must remain constant, the velocity of the fluid must increase. This is because of both the reduced area and the decreased density.

Because the density of a compressible fluid changes significantly between the pressure taps, the densities at both taps must be known. The density of the fluid can be determined if the temperature and the pressure of the fluid are known. Gas flow measurements usually take static pressure measurements at one tap and the differential-pressure between the two taps. These measurements assume that no temperature change, or an insignificant temperature change, occurs between the taps. Making this assumption allows the assumption that the density changes occurring between the taps are due to pressure changes only.

In future modules, an in-depth discussion of the principles involved with actual measurements of temperatures, pressures, levels, and flows will be provided. The descriptions in this module are covered to illustrate that both metric and English system units can be applied.

3.0.0 ◆ HANDHELD CALCULATORS AND INSTRUMENTATION APPLICATIONS

In previous modules the basic function of a handheld calculator (*Figure 30*) was discussed. This section explains two of the more advanced functions of a calculator: squares and square roots. Auxiliary functions are also discussed.

3.1.0 Squares and Square Roots

There is a compact notation used to indicate when a number is multiplied by itself. This operation is indicated by placing a small number called an *exponent* above and to the right of another number, called the *base*. The exponent shows how many times the base is multiplied by itself. For example:

6^2 means that 6 is multiplied by itself twice (6×6).

5^3 means that 5 is multiplied by itself 3 times ($5 \times 5 \times 5$).

2^4 means that 2 is multiplied by itself 4 times ($2 \times 2 \times 2 \times 2$).

A special name is given to multiplying a number by itself 2 times. This is called *squaring*. For example:

6^2 is called the *square of 6* or *6 squared*.

$6^2 = 6 \times 6 = 36$.

The square of 6 is 36.

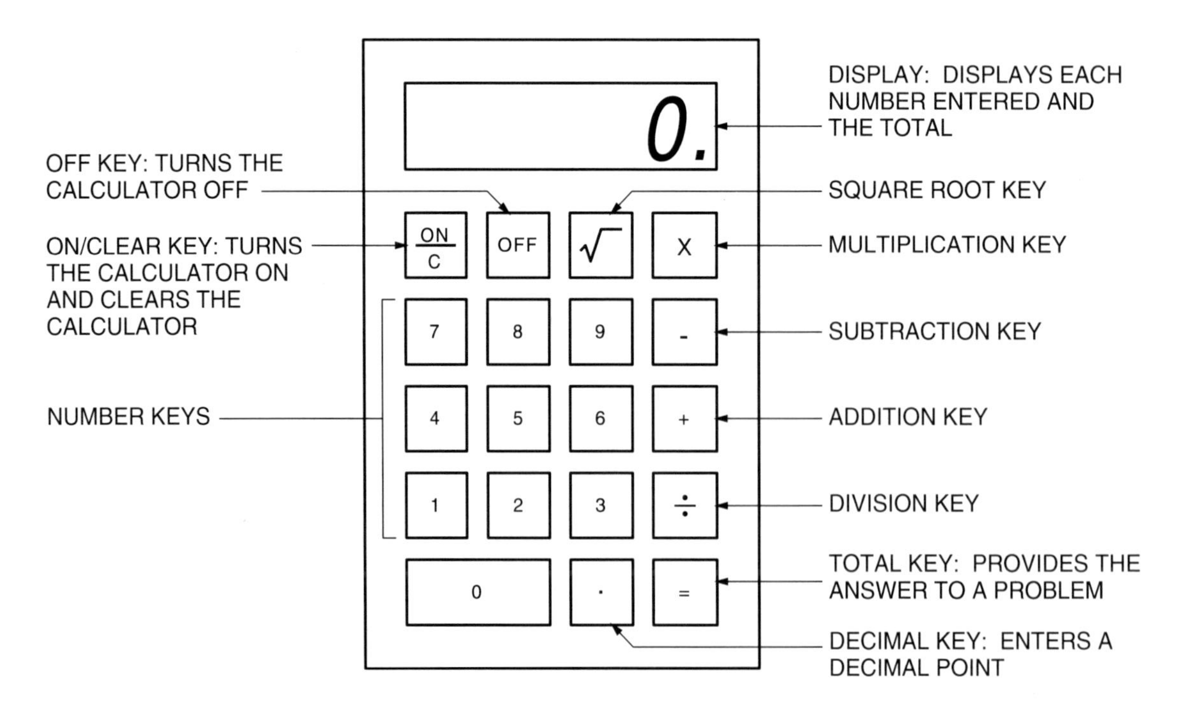

Figure 30 ◆ The basic calculator.

The square root of a number A is the number B which, when multiplied by itself (squared), gives number A as a product. For example:

The square root of 36 is 6 because $6 \times 6 = 36$.

The symbol √ indicates the square root of whatever number is placed under the symbol. For example:

$\sqrt{36} = 6$

Common squares and square roots are shown in *Table 10*.

Handheld calculators make figuring square roots easy. Refer to *Figures 31* and *32*.

Calculating squares by hand can be accomplished relatively easily because a square is only the product of two identical numbers. However, hand-calculating square roots of fractional numbers or large numbers is very cumbersome and impractical.

Practice Calculations

Compute the following using a calculator:

1. $16^2 =$
2. $21.2^2 =$
3. $0.1193^2 =$
4. $8.1^2 =$
5. $6.031^2 =$
6. $\sqrt{49} =$
7. $\sqrt{17.3} =$
8. $\sqrt{81} =$
9. $\sqrt{0.01} =$
10. $\sqrt{5} =$

Answers are located at the end of the module.

Table 10 Common Squares and Square Roots

Number	Square	Number	Square Root
1	1	1	1
2	4	4	2
3	9	9	3
4	16	16	4
5	25	25	5
6	36	36	6
7	49	49	7
8	64	64	8
9	81	81	9
10	100	100	10

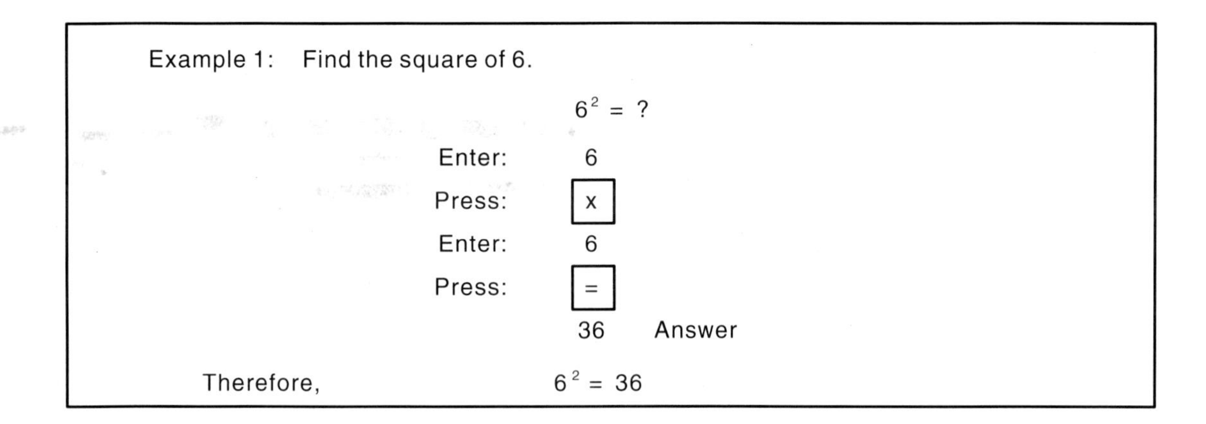

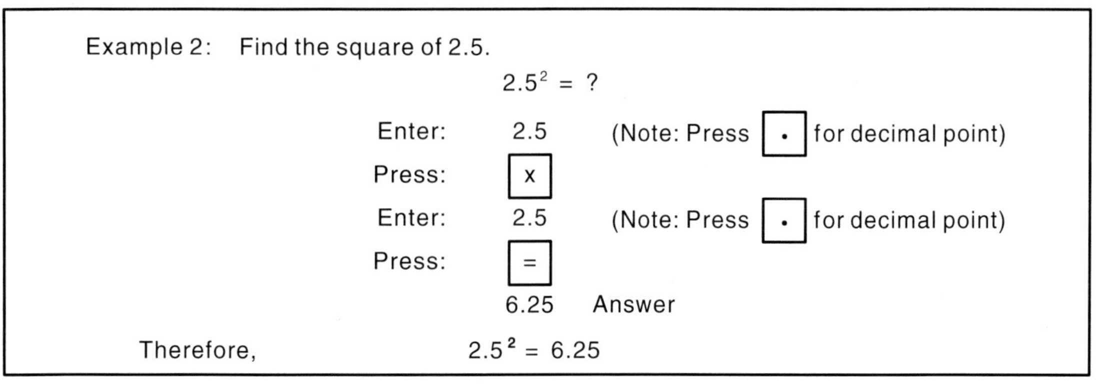

Figure 31 ◆ Squaring using a calculator.

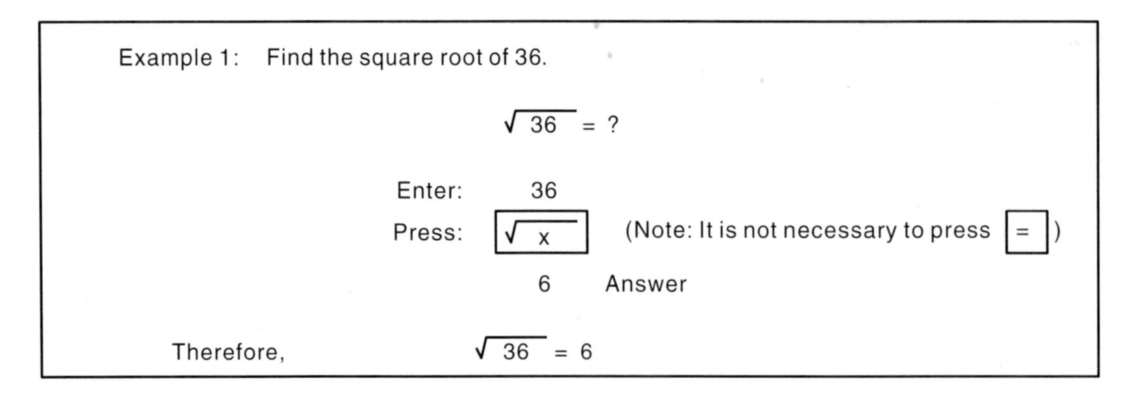

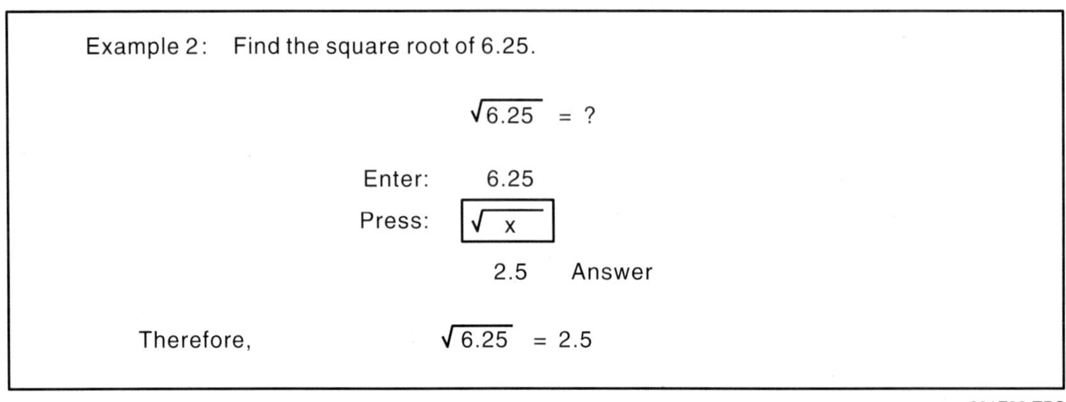

Figure 32 ◆ Taking a square root using a calculator.

3.2.0 Auxiliary Functions

Two auxiliary functions of most handheld calculators are worthy of note. They are polarity conversion and memory capability.

3.2.1 Polarity Conversion

If at any time the value on a calculator must be changed from positive (+) to negative (–) or negative (–) to positive (+), press the polarity reversal key. It is the key indicated as +/– . When subtracting a large number from a small number, the result is correct but negative. Converting back to + will allow the number to be used in further calculations.

3.2.2 Memory

Most calculators have the capability of temporarily storing information (data) in a separate file inside their internal processors. Such a storage file is termed the *memory*. It is like having a second display inside the machine in which values can be entered and saved for later retrieval.

The M+ key is used to add the displayed value to the memory. Used initially to enter or transfer the displayed figure into memory, it can also be used to continue to add displayed values to that memory file. This allows the operator to perform several operations and then sum the results in the memory file.

The M - key is used to subtract the displayed figure from the memory file.

The MRC key is used to recall the amount in the memory file and display it. This figure can be manipulated normally.

There is normally only one memory file, although many newer, more complex calculators have more. Some can be programmed to carry out multiple operations.

Finally, to clear the display, press the clear key C once. To clear the memory file, press the clear key C a second time. Pressing this key twice will erase the memory file.

4.0.0 ◆ TECHNICAL APPLICATIONS

This section presents examples of converting English system values of measurement .

It is unrealistic to expect anyone to remember or memorize all the possible conversion factors and relationships given in the tables in this module. Refer back to any of these tables and use outside references to perform actual unit conversions. It is important to know how to perform the conversion using accurate reference sources normally available in the shop or field.

Unit conversion involves changing from one unit of measurement to another.

The following example illustrates the use of simple conversion factors. In practice, the conversion factor is written directly as the ratio of equivalent physical quantities.

$$\frac{144 \text{ sq in}}{1 \text{ sq ft}} \quad \frac{1 \text{ ft}}{12 \text{ in}} \quad \frac{5{,}280 \text{ ft}}{1 \text{ mile}} \quad \frac{4 \text{ quarts}}{1 \text{ gallon}} \quad \frac{1{,}000 \text{ mm}}{1 \text{ m}}$$

The following technique may be used to make simple conversions.

Step 1 Determine the desired units and the given units.

Step 2 Write a conversion factor relating those two units. The given units should appear in the denominator. The desired units should appear in the numerator. For example, to convert yards to inches, the conversion factor is:

$$\frac{36 \text{ in}}{1 \text{ yd}}$$

Step 3 Multiply the conversion factor by the physical quantity in the given units to find the same physical quantity in the desired units. For example, how many inches are in 2.5 yards?

$$2.5 \text{ yd} \times \frac{36 \text{ in}}{\text{yd}} = 90 \text{ in}$$

Notice that in the last example, the unit *yd* in the numerator and denominator cancel each other, leaving only the unit *in*.

Most problems in science do not involve simple straightforward conversions. If no handbook value is available, it is necessary to make successive conversions. Each conversion is chosen to bring the units closer to the desired units.

The following technique may be used:

Step 1 Write a conversion factor relating two units. The desired units should be in the numerator and the units to be changed should be in the denominator.

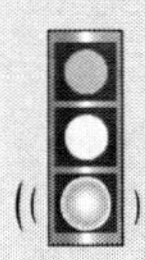

NOTE

In the first pair of parentheses, days cancel days. In the second pair, hours cancel hours. In the third pair, minutes cancel minutes. The units that remain are seconds per year.

Step 2 Successively repeat Step 1 until a conversion factor can be written that results in the final desired units.

Example:

Assume it is necessary to know the number of seconds in a given number of years.

or

(365 days/yr) (24 hrs/day) (60 min/hr) (60 s/min)

or

31,360,000 s/yr

In the review of mathematics and calculator operations, the mechanics of equation and problem solving have been a factor. The numbers used were pure numbers (counting numbers). In the real world, however, measurement is a factor. There are two terms that are applied to the science of measurement: *precision* and *accuracy*. These terms are often used interchangeably, but there is an important distinction. The concept of significant figures is related to the question of precision versus accuracy.

The term accuracy describes the relationship between a measured number and a true value. Refer, for example, to the accuracy of a certain instrument, such as a micrometer. If part of the instrument has been distorted, the measurements taken with this instrument will not agree with the measurements taken with a true, or calibrated instrument. The difference between the true value and the measured value is called error. This difference is defined by a number called the relative error (*Figure 33*).

The relative error is always a positive number, as the absolute-value notation signifies. The absolute

Example : A piece of tubing is known to have an outside diameter of 0.25 inches. A poorly calibrated micrometer measures the diameter as 0.26 inches. What is the relative error of the measurement?

$$\text{Relative Error} = \frac{|0.26 - 0.25|}{0.25} \times 100\% = \frac{0.01}{0.25} \times 100\% = 4\%$$

201F33.EPS

Figure 33 ◆ Relative error calculation.

value of a number is its numerical value when the sign is dropped. The absolute value of either +5 or –5 is 5. Accuracy, then, is determined by the magnitude of the relative error.

The word precision means sharp or clear definition. When performing a repeated set of measurements on the same object, the precision of the measurement depends on the closeness of the individual results. In the example of the distorted micrometer, the results of measurements taken with the instrument can show great precision because repeated measurements will agree. However, the measurements will not be accurate, because the relative error will be large.

In order to help in describing the limits of the precision of a measurement, trainees must understand what is meant by significant figures. Remember that this is about measurement, not counting. In arithmetic, the whole number 2 could just as well be written as 2.00000. However, a measured value of 2 is very different from a measured value of 2.00000. The number of digits following the decimal for a measured quantity implies something about the precision of measurement. The result, 2.001, is not as precise as 2.0001. This section gives the rules for determining the number of significant figures in a measured number, then show why they are of concern. These rules are important because calculators do not take them into account.

Every number has a most significant digit (MSD) and a least significant digit (LSD) (*Figure 34*). The total number of significant digits is the number of digits counted from the MSD to the LSD. For a number that does not contain a decimal point, the left-most non-zero digit is the MSD and the right-most non-zero digit is the LSD.

Notice that zeroes to the right of the number are not significant digits.

For decimal numbers, the left-most non-zero digit is still the MSD (*Figure 35*). However, the right-most digit, even if zero, is the LSD.

These are the rules for determining the MSD, LSD, and the number of significant digits. They are important in arithmetic operations involving measured numbers because measured numbers are frequently combined in some manner. They are added, subtracted, multiplied or divided. For example, one method of determining the speed of an object is to measure the time required to travel a measured distance. The speed, or velocity, is then:

$$v = \frac{s}{t}$$

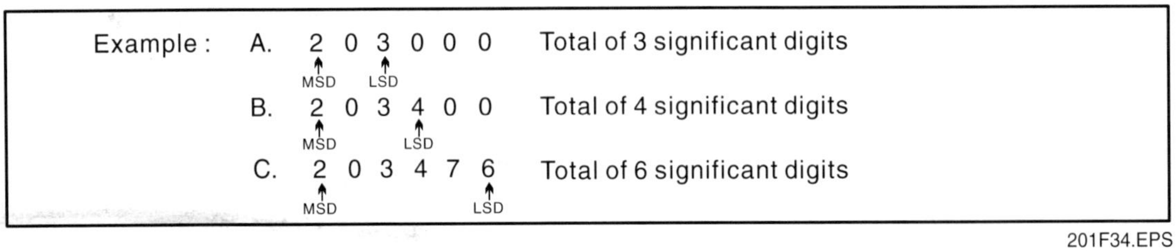

Example: A. 2 0 3 0 0 0 (MSD: 2, LSD: 3) Total of 3 significant digits

B. 2 0 3 4 0 0 (MSD: 2, LSD: 4) Total of 4 significant digits

C. 2 0 3 4 7 6 (MSD: 2, LSD: 6) Total of 6 significant digits

201F34.EPS

Figure 34 ◆ Significant digits.

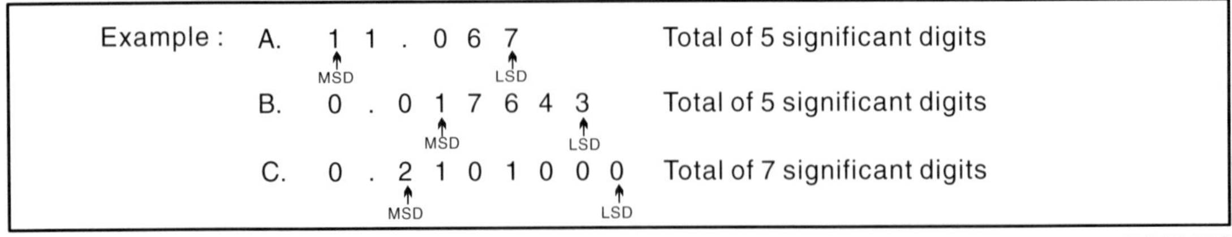

Figure 35 ◆ Significant digits in decimal fractions.

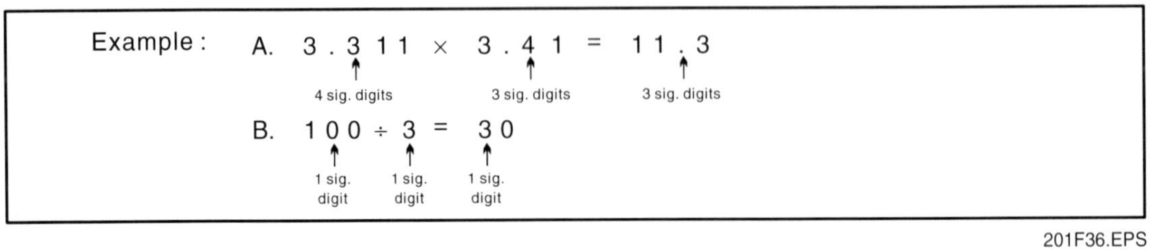

Example: A. 3 . 3 1 1 × 3 . 4 1 = 1 1 . 3 (4 sig. digits; 3 sig. digits; 3 sig. digits)

B. 1 0 0 ÷ 3 = 3 0 (1 sig. digit; 1 sig. digit; 1 sig. digit)

201F36.EPS

Figure 36 ◆ Significant digit examples.

Where:

s = distance

t = time

v = velocity

Thus, the velocity is obtained by dividing one measured number by another measured number: distance by time. Suppose the object travels 100 feet in 3 seconds. Is the speed 33.333333 ft/s? Clearly the answer is no, for it is not possible to obtain a high-precision number simply by dividing two low precision numbers. The same situation occurs in multiplication, because the multiplication process itself increases the apparent precision of the product. The rule for multiplication and division is that the product or quotient is reported with the same number of significant digits as the number with the least number of significant digits of the numbers multiplied or divided (*Figure 36*).

In each case, the answer has the same number of significant figures as the number with the least number of significant figures in the product or quotient.

Now consider the concept of *rounding off*. Rounding off simply means adjusting the final answer so that it has the proper number of significant figures. This is done by eliminating all the digits that exceed the permissible number of significant figures, subject to the following rules:

- If the first digit dropped is less than 5, the last digit is kept as is.
- If the first digit dropped is larger than 5, the last digit kept is increased by 1.
- If the first digit dropped is a 5, special rules apply. When dropping only one digit, a 5, the last digit kept is normally made an even number. When dropping more than one digit and the first digit dropped is a 5, the last digit kept is increased by 1 unless all the remaining digits dropped are zeros. In such cases, the last digit kept is normally made an even number (*Figure 37*).

These rules for rounding off apply to addition, subtraction, multiplication, and division.

When combining measured numbers by addition or subtraction, the rule for determining the precision of the sum or difference is somewhat different than for multiplication or division.

Suppose the distance from one side of a room to the other must be measured. It can be done in two steps. First, measure part of the way with a tape measure accurate to a hundredth of an inch, and find a length of 78.63 inches. Then, pace off the rest of the way and estimate the distance to be 48 inches. Is the length of the room 126.63 inches? The answer must be no, because a high-precision number cannot be obtained by adding a high-precision and low-precision number together. In addition and subtraction, the sum or difference is rounded off so that it has the same precision as the least precise of the numbers added or subtracted. The precision of a number is determined by the place value of its least significant digit. For example:

The LSD of 2.001 is located in the thousandths place. The LSD of 2.0001 is located in the ten-thousandths place. 2.0001 is more precise than 2.001.

Example :

1. Round off the number 32.549 to three significant digits. Because the first digit dropped is 4, the last digit kept, 5, is left as is.

   $32.549 \xrightarrow{\text{rounded off}} 32.5$

2. Round off the number 0.027810 to two significant digits. Because the first digit dropped is 8, the last digit kept, 7, is increased by 1 to 8.

   $0.027810 \xrightarrow{\text{rounded off}} 0.028$

3. Round off the number 32.75 to three significant digits. Because the first digit dropped is 5, the last digit kept, 7, is made an even number, 8.

   $32.75 \xrightarrow{\text{rounded off}} 32.8$

4. Round off the number 0.28521 to two significant digits. Because the first digit dropped is 5 and all the remaining digits dropped are not zeros, the last digit kept, 8, is increased by 1 to 9.

   $0.28521 \xrightarrow{\text{rounded off}} 0.29$

201F37.EPS

Figure 37 ◆ Rounding-off examples .

Therefore, in addition or subtraction, the rounding is done to make the final answer as precise as the least precise of the numbers added or subtracted. In other words, the place value of the LSD in the sum or difference is the same as the place value of the least precise number (*Figure 38*).

Note that the total number of significant figures is not of concern, as in multiplication or division. The location of the least significant digit is what is important.

4.1.0 Converting to a Metric Rule Dipstick

The large tank shown in *Figure 39* is 16 ft high. The English system dipstick indicates the level is 8.5 ft, or 8 ft 6 in. How large a metric rule would be needed to read to the top of the tank? How many meters high would the water level read using the metric rule?

A metric rule at least 16 ft long, or approximately 5 meters long, is required as calculated here:

$$\frac{16 \text{ ft}}{3.281 \text{ ft/meter}} = 4.876 \text{ meters}$$

The level in the tank read by the meter stick would be:

$$\frac{8.5 \text{ ft}}{3.281 \text{ ft/meter}} = 2.59 \text{ meters}$$

4.2.0 Sight Glass Level Measurement

Replace the English system calibrated scale used on the sight glass in *Figure 40* with a metric scale. The bottom of the scale reads 2 ft in the tank. The top of the sight glass reads 6 ft 4 in.

If only a meter stick that is 2 meters long is available, can it be used to replace the sight glass scale?

The length of the calibrated scale needed is from the bottom of the sight glass to the top:

$$6 \text{ ft } 4 \text{ in} - 2 \text{ ft} = 4 \text{ ft } 4 \text{ in}$$

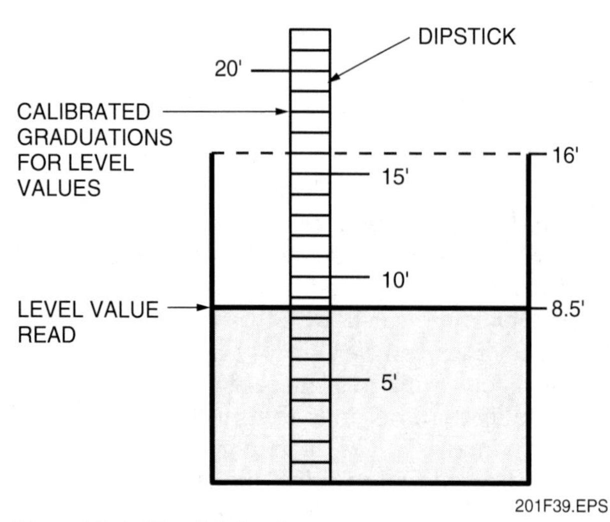

Figure 39 ◆ Dipstick level measurement.

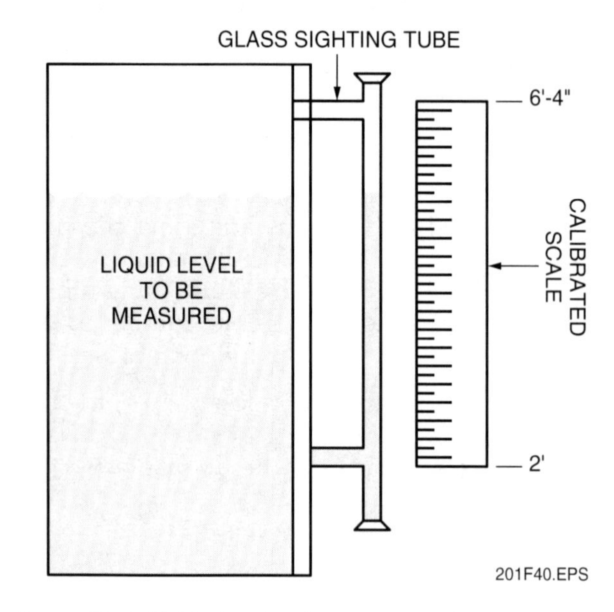

Figure 40 ◆ Sight glass level measurement.

Convert the length of the scale (4 ft 4 in) to inches:

$$\left(4 \text{ ft} \times 12 \frac{\text{in}}{\text{ft}}\right) + 4 \text{ in}$$

$$= 48 \text{ in} + 4 \text{ in}$$

$$= 52 \text{ in}$$

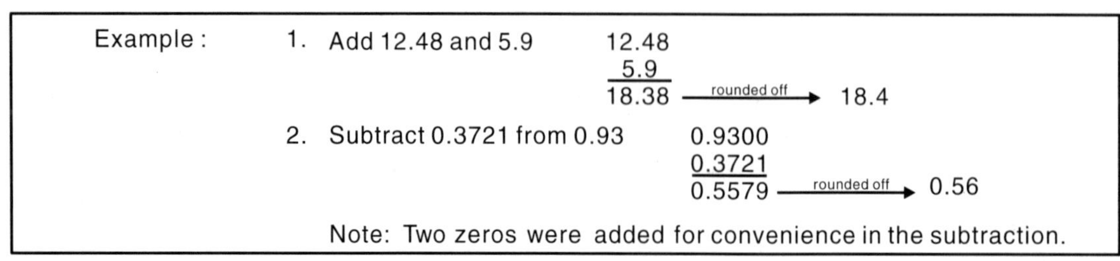

Figure 38 ◆ Precision determinations.

Convert the 52 inches into meters:

$$\frac{52 \text{ in}}{39.37 \text{ in/meter}} = 1.32 \text{ meters}$$

Therefore, a 2-meter rule is large enough. However, it must be relabeled or mounted such that the bottom of the stick and tank are even. Is the 2-meter rule large enough to go from the top of the sight glass to the very bottom of the tank?

If the 2-meter rule is converted to feet and inches, it would be:

$$2 \text{ meters} = 2 \times 3.281 \text{ ft}$$

$$2 \text{ meters} = 6.562 \text{ ft}$$

Converting 0.562 ft into inches:

$$0.562 \text{ ft} \times \frac{12 \text{ in}}{1 \text{ ft}} = 6.7 \text{ in}$$

The 2-meter rule is 6 ft 6.7 in, which is long enough to use in this fashion.

If the rule must be cut to match the sight glass, where would it be cut to accurately indicate the tank level in meters?

Cut at the metric equivalents of 2 ft and 6 ft 4 inches. Mount the metric rule behind the sight glass.

$$2 \text{ ft} = 24 \text{ in}$$

$$\frac{24 \text{ in}}{39.37 \text{ in/meter}} = 0.6096 \text{ meters or } 60.9 \text{ centimeters}$$

$$6 \text{ ft } 4 \text{ in} = (12 \text{ in/ft} \times 6 \text{ ft}) + 4 \text{ in} = 72 \text{ in} + 4 \text{ in} = 76 \text{ in}$$

$$\frac{76 \text{ in}}{39.37 \text{ in/meter}} = 1.930 \text{ meters or } 193.0 \text{ centimeters}$$

4.3.0 Conductance Probe Settings

The conductance probe level measurement system shown in *Figure 41* is basically either an on/off control circuit or an alarm system. As the fluid in the tank passes the tip of the probes, the probe will either conduct electricity or not, based on the level in the tank. Usually, the air is a better insulator than the liquid.

The high-level probe is used as an alarm to warn that the tank is becoming full or to control a valve that allows the liquid to drain out of the tank. The low-level probe warns when the tank is approaching empty or controls valves that either stop flow out of the tank or allow flow into the tank.

A plant is converting to the metric system. In general, all tank alarms (high and low levels) will be set at 10% from the full and empty extremes of the tank. In metric lengths, what size probes are needed for the tank previously discussed?

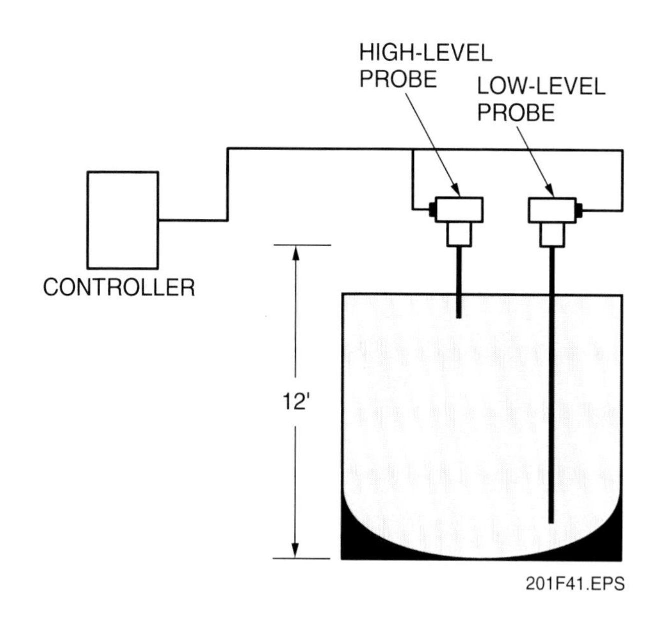

Figure 41 ◆ Conductance probe level measurement.

Assuming the probes are mounted even with the top of the tank, the high-level probe must extend only 10% of the depth of the tank. Therefore,

$$10\% = \frac{10}{100} = 0.1$$

$$0.1 \times 12 \text{ ft} = 1.2 \text{ ft}$$

Converting this to meters:

$$1.2 \text{ ft} \times 0.3048 \text{ meters/ft} = 0.366 \text{ meters or } 36.6 \text{ centimeters}$$

The low-level probe must extend to within 10% of the bottom of the tank, or 90% of the depth of the tank. Therefore,

$$90\% = \frac{90}{100} = 0.9$$

$$0.9 \times 12 \text{ ft} = 10.8 \text{ ft}$$

Converting this to meters:

$$10.8 \text{ ft} \times 0.3048 \text{ meters/ft} = 3.292 \text{ meters or } 329.2 \text{ centimeters}$$

4.4.0 Open Tank Measurement Conversions

The differential-pressure cell shown in *Figure 42* is being used to measure the hydrostatic head (the pressure exerted by the water in the open tank) as a means of determining the water level in the tank.

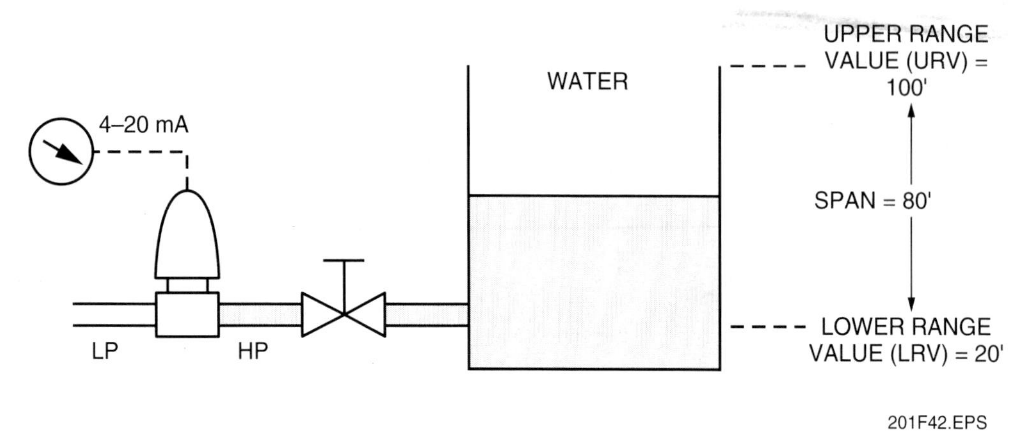

Figure 42 ◆ Open tank application.

The low pressure (LP) side of the cell is vented to atmospheric pressure, which is essentially the same as the pressure exerted by the atmosphere on the top of the water in the tank. The high pressure (HP) side of the cell is connected to a pipe that taps into the tank 20 ft above the bottom of the tank.

If the water level drops to 20 ft or less, the HP side of the cell can't sense any pressure due to the lack of water in the tank. Because atmospheric pressure is felt on both sides of the cell (LP and HP), this is the lowest actual tank level the cell will ever sense. This is called the *lower range value*. It is equal to 20 ft. The pressure cell provides a minimal output signal to a meter of 4mA at this point.

The top of the tank is 100 ft high. The pressure cell will give an output signal of 20mA to the meter at this point. Because the water will overflow the top of the tank above 100 ft, this point is called the *upper range value.*

The difference between the lower range value and the upper range value is 80 ft. This range is called the *span of measurement*. The pressure cell must be able to measure a continuous range of pressure along the span of measurement. The pressure range of this instrument must be from 0 inches of water to 80 ft × 12 in/ft, or 960 inches of water.

Using the tables already provided in this module, select a pressure cell from the warehouse for this application. Half the pressure cells are rated in the old English system, in pounds per square inch, or psi. The other half are rated in pascals (the equivalent metric system pressure units). What range instruments are necessary for each side of the warehouse?

All that is known is that the pressure range has to be large enough to withstand the pressure of a column of water 80 feet (or 960 inches) high. Work is done at atmospheric pressure, given the open nature of the system.

From the discussion of hydrostatic water pressure, 1 psig = 27.7 inches of water column.

Therefore,

$$\frac{960 \text{ in}}{27.7 \text{ in/psig}} = 34.66 \text{ psi}$$

So, in the English system, a pressure cell with a range of 0-35 psi or more is acceptable.

Earlier in this module, the conversion factors for figuring out the range in pascals were discussed. Therefore,

$$34.66 \text{ psi} \times 6{,}895 \text{ Pa/psi} = 238{,}981 \text{ Pa}$$

So, in the metric system, a pressure cell with a range of 240,000 Pa or more is acceptable.

4.5.0 Pressurized Tank Measurement Conversions

Given a similar pressure cell used in the closed system or pressurized tank application shown in *Figure 43*, again determine the range in both the English system and the metric system. The tank has a 50 psi nitrogen pressure blanket applied to it.

First, remember that water is essentially an incompressible fluid. The nitrogen pressure will have no effect on the volume or level of the water.

Second, the valve arrangement allows the nitrogen pressure to be felt on both sides (LP and HP) of the cell. Therefore, it cancels itself out just as atmospheric pressure did in the open tank application.

The next noteworthy item is the way the differential-pressure cell is physically installed in the system. The high pressure connection side is toward the tank. However, the low pressure connection side is not open to the atmosphere. It is connected to a water column (commonly referred to as a reference column). The reference column remains full at all times. The top of this column taps into the tank at the 12-foot mark. If the water level in the tank rises to that 12-foot mark, the pressures sensed on both sides of the cell are the same. The differential-pressure sensed is zero.

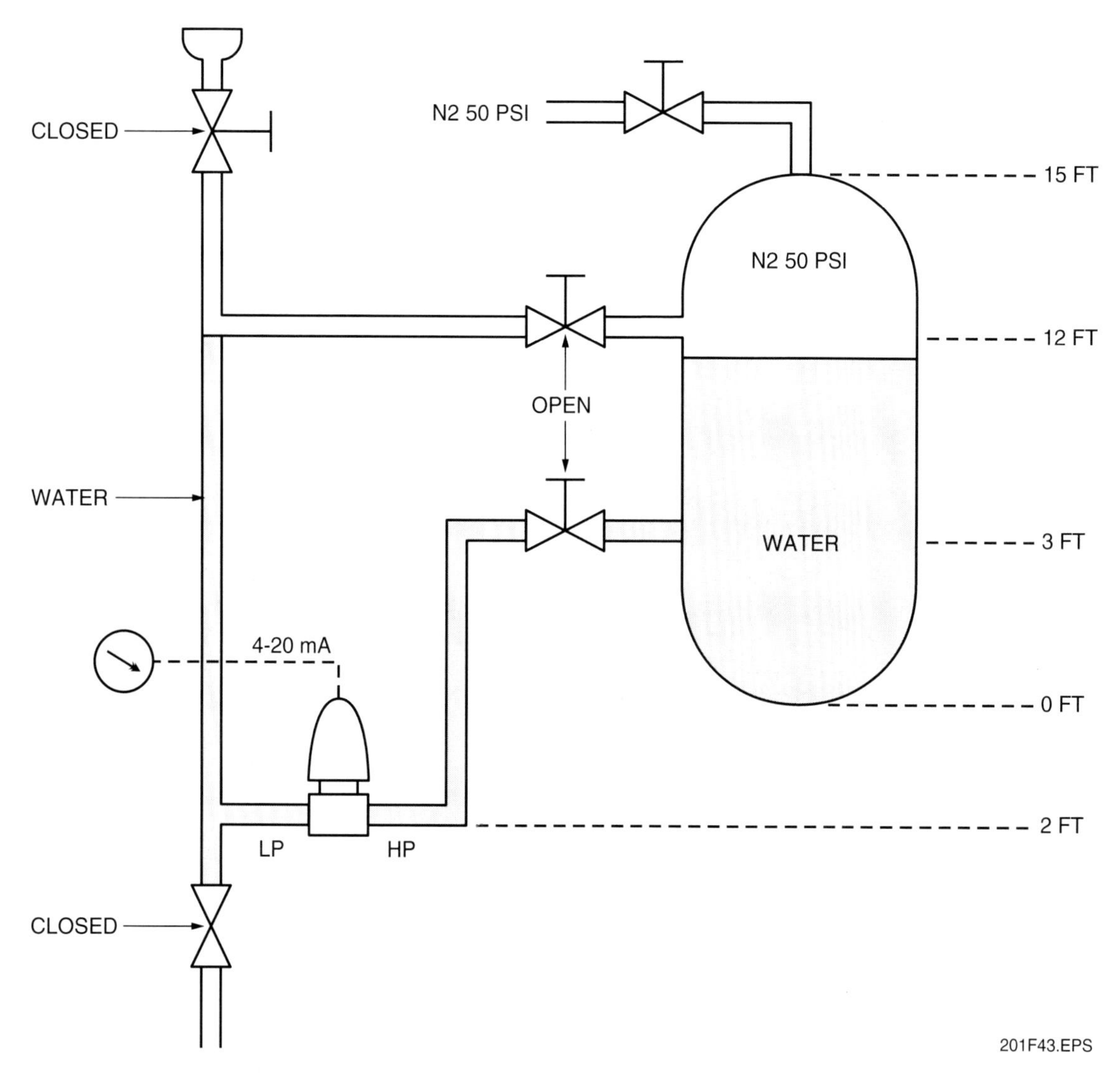

Figure 43 ◆ Closed or pressurized tank application.

As the level of water in the tank decreases from the upper instrument tap to the lower instrument tap, the pressure sensed by the high-pressure side of the cell will also decrease. At the 3-foot mark it will be the lowest pressure. As was noted in the open tank application, any further level decrease cannot be sensed.

What appears unusual about this arrangement is that, as the water level in the tank is increasing, the actual pressure sensed by the high-pressure side of the cell increases, but the overall differential-pressure sensed is decreasing. At zero differential-pressure, the level in the tank is 12 ft (or higher). The differential-pressure is highest when the level in the tank is actually at 3 ft or lower. If the level-indicating meter is the same type used in the open tank application, the voltage increase from the differential-pressure cell would indicate an increase in level. However, in this arrangement, an increasing level would normally produce a lower differential-pressure and a lower voltage output. For the cell to provide the proper output voltage, it must be reverse-acting. For example, as the differential-pressure decreases, the voltage output increases.

Be ready to identify these facts in order to calculate instrument ranges in either the metric or English systems. This will be covered in more detail in future modules, but it is presented here because this application is relatively common. It is used when a low water level is a critical situation requiring immediate or automatic actions. With this arrangement, a mechanical failure, such as a leak in the cell from the high-pressure side to the low-pressure side, will result in a false high-level indication. This prevents automatic action.

The differential-pressure cell in *Figure 43* must be able to measure a continuous range of pressure from the 3 ft to the 12 ft taps, or 9 ft of water. Converting this to inches of water column is easy:

$$9 \text{ ft} \times 12 \text{ in/ft} = 108 \text{ in w.c.}$$

Therefore, the pressure range of the differential-pressure cell needed is 0 inches of water to 108 inches of water, reverse-acting.

Using the tables already provided in this module, select a differential-pressure cell from the warehouse. What instrument range is necessary in the English system side of the warehouse? What instrument range is necessary in the metric system side?

Again, remember:

1 inch water gauge = 0.0361 psi

or

1 psig = 27.7 in w.c.

The instrument's differential-pressure range is represented by a water column of 108 inches. Therefore:

$$108 \text{ in} \times 0.0361 \text{ psi/in} = 3.899 \text{ psig}$$

So, in the English system, a differential-pressure cell with a range of 0–4 psid (pounds per square inch differential) or more is acceptable.

Earlier in this module, the conversion factors for converting the range needed from psi to pascals were discussed. Therefore:

$$3.899 \text{ psi} \times 6{,}895 \text{ Pa/psi} = 26{,}884 \text{ Pa}$$

So, in the metric system, a reverse-acting differential-pressure cell with a range of 27,000 Pa or greater is acceptable.

Finally, remember that the cell casing must be able to withstand the internal static pressure of both the range and the nitrogen blanket.

Therefore, in the English system, the selected cell must be rated to function under static pressures of at least:

$$50 \text{ psig } N_2 + 4 \text{ psid} = 54 \text{ psig}$$

In the metric system, the appropriate cell should be rated to function under static pressures of at least:

$$50 \text{ psig } N_2 \times 6{,}895 \text{ Pa/psig} + 26{,}884 \text{ Pa} = 371{,}634 \text{ Pa}$$

4.6.0 Temperature Measurement Conversions

The application of the metric system to temperature measurements or specifications is very simple. The formula for conversion of degrees Fahrenheit to degrees Celsius is:

$$°C = \tfrac{5}{9} (°F - 32°)$$

The Celsius to Fahrenheit conversion is:

$$°F = \tfrac{9}{5} (°C) + 32$$

The two most common applications of temperature in instrumentation systems deal with the measurement of process temperatures and the environmental temperature conditions.

Two common temperature-measuring devices are resistance temperature detectors (RTDs) and thermocouples (TCs). This section briefly examines the useful ranges of the various types of these measuring devices.

RTDs are commonly made of platinum, nickel, copper, or combinations of these metals with other alloys.

A given device can be used up to a temperature of approximately 300°F. What Celsius temperature would this equal? Use the standard conversion formula:

$$°C = \frac{5}{9} (°F - 32°)$$

$$°C = \frac{5}{9} (300°F - 32°)$$

$$°C = \frac{5}{9} (268°F)$$

$$°C = 0.555 (268°F)$$

$$°C = 148.7°C$$

The nickel RTDs come in two types. Type I can be used up to approximately 400°F, while Type II can be used up to about 600°F. Converting these temperatures to Celsius can provide valuable information.

Type I Nickel RTD:

$$°C = \frac{5}{9} (°F - 32°)$$

$$°C = \frac{5}{9} (400°F - 32°)$$

$$°C = \frac{5}{9} (368°F)$$

$$°C = 0.555 (368°F)$$

$$°C = 204.2°C$$

Type II Nickel RTD:

$$°C = \frac{5}{9} (°F - 32°)$$

$$°C = \frac{5}{9} (600°F - 32°)$$

$$°C = \frac{5}{9} (568°F)$$

$$°C = 0.555 (568°F)$$

$$°C = 315.2°C$$

By far the most expensive RTDs are made of platinum. Platinum RTDs can be used for much higher temperature measurements, such as 1,100°F. In Celsius degrees this would be:

$$°C = \frac{5}{9}(°F - 32°)$$

$$°C = \frac{5}{9}(1{,}100°F - 32°)$$

$$°C = \frac{5}{9}(1{,}068°F)$$

$$°C = 0.555\ (1{,}068°F)$$

$$°C = 592.7°C$$

Convert the temperature ranges of these types of thermocouples to their metric equivalents.

- Type J (Iron-Constantan®): 32° to 1,400°F*
- Type K (Chromel®-Alumel®): 32° to 2,300°F*
- Type T (Copper-Constantan®): –300° to 700°F*
  * Depending on gauge of wire

Type J (Iron-Constantan®): 32°F to 1,400°F:

$$°C = \frac{5}{9}(°F - 32°)$$

$°C = \frac{5}{9}(32°F - 32°)$	$°C = \frac{5}{9}(1{,}400°F - 32°)$
$°C = \frac{5}{9}(0°F)$	$°C = \frac{5}{9}(1{,}368°F)$
$°C = 0°C$	$°C = 0.555\ (1{,}368°F)$
	$°C = 759°C$

Therefore, Type J is 0–800°C.

Type K (Chromel®-Alumel®): 32°F to 2,300°F:

$$°C = \frac{5}{9}(°F - 32°)$$

$°C = \frac{5}{9}(32°F - 32°)$	$°C = \frac{5}{9}(2{,}300°F - 32°)$
$°C = \frac{5}{9}(0°F)$	$°C = \frac{5}{9}(2{,}268°F)$
$°C = 0°C$	$°C = 0.555\ (2{,}268°F)$
	$°C = 1{,}259°C$

Therefore, Type K is 0–1,260°C.

Type T (Copper-Constantan®): –300°F to 700°F

$$°C = \frac{5}{9}(°F - 32°)$$

$°C = \frac{5}{9}(32°F - 32°)$	$°C = \frac{5}{9}(700°F - 32°)$
$°C = \frac{5}{9}(0°F)$	$°C = \frac{5}{9}(668°F)$
$°C = 0°C$	$°C = 0.555\ (668°F)$
	$°C = 370.7°C$

Therefore, Type T is –184°C to 370°C.

One way to double-check the conversion is to use the other formula and convert back.

Summary

Over 95% of the world uses the metric system. The major instrument companies and the U.S. government have adopted the metric system. You must be able to work using it.

A measurement is simply a comparison of a quantity to some definite standard or measure. The metric system is based on the meter, which is equal to 39.37 inches.

All physical quantities can be expressed in terms of length, mass, and time. The metric system measures for these fundamental units are the meter, the gram, and the second, respectively. The metric system is actually easier to use and visualize, because it uses prefixes to denote unit sizes that are multiples of 10.

The most common types of measurements associated with instrumentation can be classified into five categories:

- Dimensional measurements (lengths)
- Mass and/or weight measurements
- Pressure measurements
- Flow measurements
- Temperature measurements

Instrumentation is most often used to measure the pressure of fluids. The most common is the pressure exerted by a liquid due to its height (or depth), which is referred to as the *hydrostatic pressure* or head of the fluid.

Review Questions

1. The three fundamental units of all physical quantities are length, mass, and _____.
   a. pressure
   b. temperature
   c. matter
   d. time

2. The meter-kilogram-second (MKS) system and the centimeter-gram-second (CGS) system are part of the English system of measure.
   a. True
   b. False

3. How many inches are in 1 meter?
   a. 1.52
   b. 12
   c. 39.37
   d. 40

4. When converting the dimensions of a _____, only the radius has a measured dimension that must be converted.
   a. circle
   b. rectangle
   c. cylinder
   d. box

5. How many meters are in 9.843 feet?
   a. 1
   b. 2
   c. 3
   d. 4

6. _____ is the most common wet measuring unit in the metric system.
   a. Quart
   b. Liter
   c. Pint
   d. Gallon

7. The amount of matter in an object is its _____.
   a. weight
   b. force
   c. mass
   d. pressure

8. The newton can be defined as the amount of force required to accelerate one kilogram at the rate of one meter per second per second.
   a. True
   b. False

9. Hydrostatic pressure is also referred to as _____ pressure.
   a. base
   b. interior
   c. vessel
   d. head

10. The Kelvin scale is sometimes referred to as the _____ scale.
    a. Absolute Celsius
    b. Rankine
    c. Fahrenheit
    d. Degrees Centigrade

11. 110°C is equal to _____?
    a. 110°F
    b. 210°F
    c. 230°F
    d. 250°F

12. Match the correct type of flow rate to the corresponding description.

    Flow Rate Type

    A. Mass B. Weight C. Volumetric

    Description

    _C_ The quantity of material in cubic units flowing past a point at a given instant of time.
    _A_ The quantity of material in units of mass flowing past a point at a given instant of time.
    _B_ The quantity of a material in units of weight such as pounds-force, lb_f, flowing past a given instant of time.

13. A fluid flowing in a pipe exerts pressure in two different directions, against both the walls of the pipe and the direction of flow.
    a. True
    b. False

14. Mass flow rates can be calculated or computed easily by determining or sensing temperature and/or _____.
    a. velocity
    b. pressure
    c. volume
    d. density

15. What is the square root of 144?

a. 12
b. 13
c. 14
d. 15

16. Match the correct memory function key to the corresponding description.

Memory Function

A. M+ B. M- C. MRC

Description

 _____ This key is used to subtract the displayed figure from the memory file.
_____ This key is used to recall the amount in memory file and display it.
_____ This key is used to add the displayed value to the memory.

17. _____ conversion involves changing from one unit of measurement to another, always with regards to the same physical quantity.

a. Polarity
b. Temperature
c. Unit
d. Volume

18. For a number that does not contain a decimal point, the left-most non-zero digit is the MSD and the right-most non-zero digit is the LSD.

a. True
b. False

19. The length of a calibrated scale needs to be converted from inches to meters. The scale is 70 inches. What is the length of the scale in meters?

a. .178
b. 1.78
c. 17.8
d. 178

20. Resistance temperature detectors (RTDs) are commonly made of platinum, nickel, or _____.

a. steel
b. aluminum
c. silver
d. copper

GLOSSARY

Trade Terms Introduced in This Module

***Absolute pressure*:** Gauge pressure plus atmospheric pressure. At sea level, it would be gauge pressure plus 14.7. It is referred to as pounds per square inch absolute (psia).

***Acceleration*:** The rate of change of velocity; the process by which a body at rest becomes a body in motion.

***Area*:** The amount of surface in a given plane or two-dimensional shape.

***Atmospheric pressure*:** The pressure caused by the weight of the atmosphere. It is at maximum at sea level and reduces with increasing altitude.

***Density*:** The measure of the amount of mass per unit of volume of a substance. Objects that are more dense weigh more. They therefore can exert a higher pressure per unit of volume.

***Dyne*:** A unit of force equal to $\frac{1}{1,000,000}$ or 0.000001 newton.

***Flow rate*:** The amount of fluid that moves past a given point at any given instant of time.

***Force*:** A push or pull on a surface. Force is commonly considered to be the weight of an object or a fluid.

***Gauge pressure*:** Pressure with reference to atmospheric pressure. When the gauge reads zero, the pressure inside the container is equal to atmospheric pressure. Gauge pressure is expressed as psig.

***Hydrostatic pressure*:** The pressure at the bottom of a column, caused by the weight of the material in the column.

***Liter*:** A standard unit of volume in the metric system equal to 1 cubic decimeter.

***Mass*:** A quantity of matter.

***Measurement*:** A quantity expressed in terms of a definite standard or reference unit.

***Meter*:** A standard unit of length in the metric system equal to 39.37 inches.

***Metric system*:** A system of weights and measures based on the meter.

***Newton (N)*:** The amount of force required to accelerate 1 kilogram at a rate of 1 meter per second squared.

***Pascal (Pa)*:** A unit of pressure equal to the force of one newton exerted on one square meter.

***Pressure*:** Force applied perpendicular to a unit area of surface.

***Temperature*:** A specific degree of heat as measured on a standard scale.

***Total flow*:** The quantity of material that moves past a given point during a specific period of time.

***Unit*:** A definite standard measure of a dimension.

***Vacuum*:** An enclosed space from which air or other gas has been removed.

***Velocity*:** A vector quantity equivalent to speed in a certain direction; the distance moved in a given direction in a given amount of time.

***Volume*:** The amount of space contained in a given three-dimensional shape.

REFERENCES

Additional Resources

This module is intended to present thorough resources for task training. The following reference works are suggested for further study. These are optional materials for continued education rather than for task training.

Thirty Days to Metric Mastery: For People Who Hate Math, D.C. Steinke, House of Charles, 1981.

Thinking Metric, Second Edition, T.F. Gilbert and M.B. Gilbert, John Wiley & Sons, 1978.

NCCER CURRICULA — USER UPDATE

NCCER makes every effort to keep its textbooks up-to-date and free of technical errors. We appreciate your help in this process. If you find an error, a typographical mistake, or an inaccuracy in NCCER's curricula, please fill out this form (or a photocopy), or complete the online form at **www.nccer.org/olf**. Be sure to include the exact module ID number, page number, a detailed description, and your recommended correction. Your input will be brought to the attention of the Authoring Team. Thank you for your assistance.

Instructors – If you have an idea for improving this textbook, or have found that additional materials were necessary to teach this module effectively, please let us know so that we may present your suggestions to the Authoring Team.

NCCER Product Development and Revision
13614 Progress Blvd., Alachua, FL 32615
Email: curriculum@nccer.org
Online: www.nccer.org/olf

❑ Trainee Guide ❑ AIG ❑ Exam ❑ PowerPoints Other ______________________________

Craft / Level: Copyright Date:

Module ID Number / Title:

Section Number(s):

Description:

Recommended Correction:

Your Name:

Address:

Email: Phone:

Module 12202-03

Instrumentation Drawings and Documents, Part Two

COURSE MAP

This course map shows all of the modules in the second level of the Instrumentation curriculum. The suggested training order begins at the bottom and proceeds up. Skill levels increase as you advance on the course map. The local Training Program Sponsor may adjust the training order.

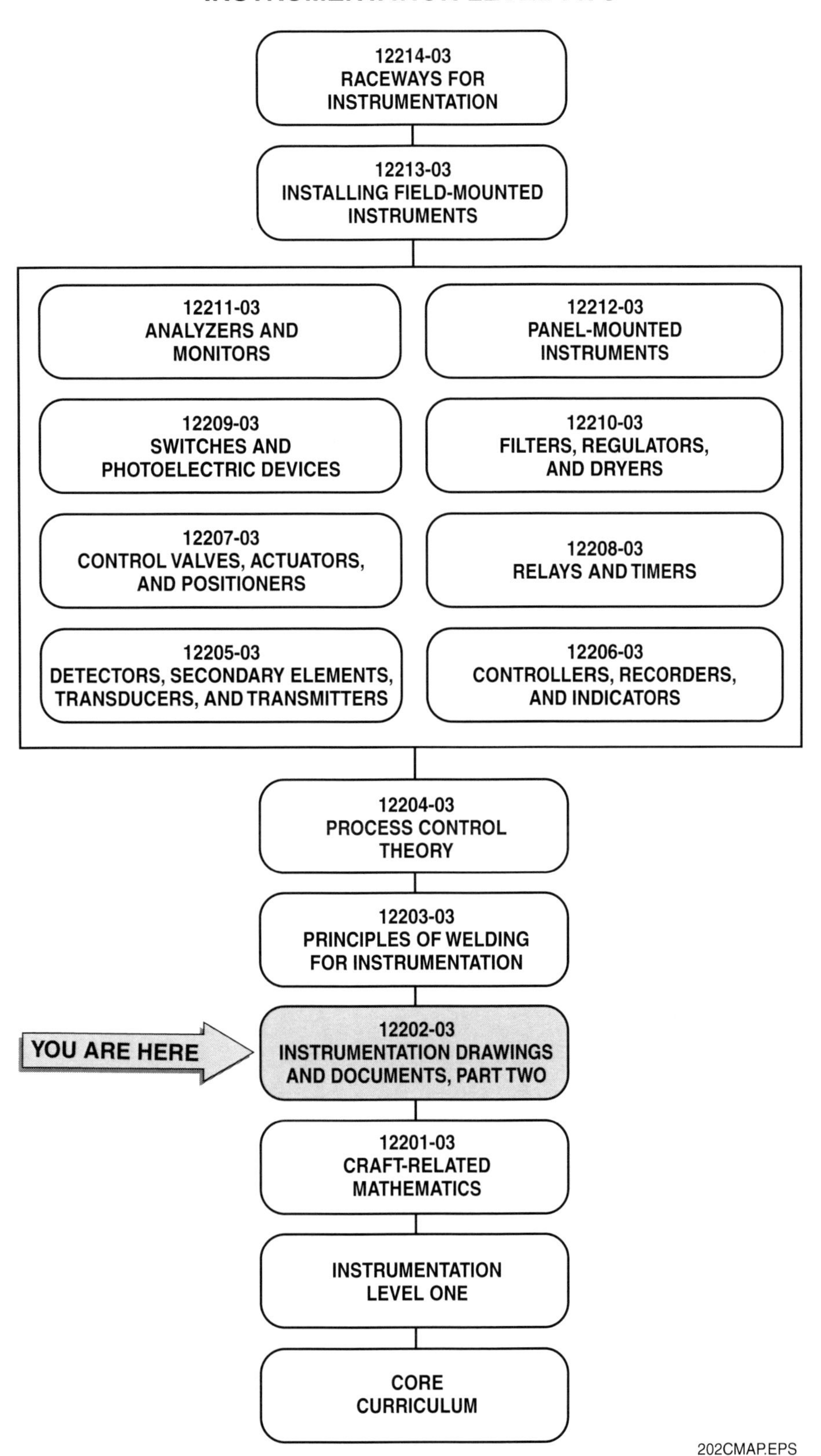

202CMAP.EPS

MODULE 12202-03 CONTENTS

1.0.0 INTRODUCTION 2.1
2.0.0 ELECTRICAL DRAWINGS 2.2
2.1.0 Block Diagrams 2.2
2.2.0 Single- and Three-Line Diagrams 2.2
2.3.0 Wiring Diagrams 2.4
2.3.1 Point-to-Point Method 2.5
2.3.2 Cable Method 2.5
2.3.3 Baseline Method 2.5
2.3.4 Lineless (Wireless) Method 2.5
2.4.0 Raceway Drawings 2.7
2.5.0 Schematic Diagrams 2.7
3.0.0 INSTRUMENTATION DRAWINGS 2.7
3.1.0 P&IDs 2.8
3.2.0 Loop Sheets 2.10
3.2.1 Instrument Tag Numbers 2.10
3.2.2 Location and Connection Section 2.12
3.2.3 Calibration and Specification Section 2.13
3.2.4 Field Checklist 2.13
3.3.0 Logic/Ladder Diagrams 2.13
3.4.0 Equipment Location Drawings 2.14
3.5.0 Installation Detail Drawings 2.14
3.6.0 Flow Drawings 2.17
3.7.0 Instrument Data Sheets 2.17
4.0.0 APPLYING THE DIAGRAMS 2.18
5.0.0 STANDARDIZED DESIGN METHODS 2.20
5.1.0 Electrical Symbols 2.20
5.2.0 Instrumentation Symbols 2.27
5.3.0 Number Systems 2.28
5.4.0 Notes 2.30
5.5.0 Graphic Styles 2.30
SUMMARY 2.33
REVIEW QUESTIONS 2.33
GLOSSARY 2.35
APPENDIX, Standardized Symbols 2.37
REFERENCES & ACKNOWLEDGMENTS 2.57

Figures

Figure 1 Block diagram of a motor circuit 2.2
Figure 2 Block diagram with symbols 2.2
Figure 3 Typical one-line diagram for power equipment (ANSI) 2.3
Figure 4 Three-line diagram of a three-phase motor power circuit 2.4
Figure 5 Wiring diagram of a motor controller 2.4
Figure 6 Point-to-point connection diagram 2.5
Figure 7 Cable method connection diagram 2.6
Figure 8 Baseline connection diagram 2.6
Figure 9 Lineless connection diagram 2.7
Figure 10 Raceway drawing 2.8
Figure 11 Raceway drawing 2.9
Figure 12 P&ID 2.10
Figure 13 Loop sheet 2.11
Figure 14 Loop identification tag number 2.12
Figure 15 Loop sheet location and connection section 2.12
Figure 16 Loop sheet calibration and specification section 2.13
Figure 17 Field checklist 2.13
Figure 18 Ladder diagram 2.14
Figure 19 Instrument location drawing 2.15
Figure 20 Instrument installation drawing 2.16
Figure 21 Bill of materials 2.17
Figure 22 Flow diagram 2.18
Figure 23 Instrument data sheet 2.19
Figure 24 Contacts A and B 2.21
Figure 25 Switch and contacts 2.21
Figure 26 Supplementary contact symbols 2.21
Figure 27 ANSI graphical symbols for electrical diagrams 2.25
Figure 28 Standard wiring diagram symbols 2.26
Figure 29 Standard electronic symbols 2.27
Figure 30 Standard electronic logic symbols 2.28
Figure 31 Electrical drawing symbols 2.29
Figure 32 Drawing number in a title block 2.30
Figure 33 Drawing number 04-I201 . . . . .2.30
Figure 34 Notes from sample P&ID 2.31
Figure 35 Glass box 2.31
Figure 36 Isometric view of a cube 2.31
Figure 37 Making an isometric drawing 2.32
Figure 38 Piping isometric drawing 2.32

Tables

Table 1 Meter Abbreviations 2.20
Table 2 Lamp Color Abbreviations 2.20
Table 3 Relay Abbreviations 2.20
Table 4 Contact and Switch Abbreviations 2.21
Table 5 ANSI Abbreviations for Use on Drawings and in Text 2.22–2.24
Table 6 Department Identifications 2.30
Table 7 Drawing Classifications 2.30

MODULE 12202-03

Instrumentation Drawings and Documents, Part Two

Objectives

When you have completed this module, you will be able to do the following:

1. Identify common types of electrical and instrumentation diagrams and drawings.
2. Read and interpret electrical diagrams used in instrumentation work:
   - Wiring diagrams
   - Ladder diagrams
   - One-line diagrams
   - Motor controller diagrams
3. Read and interpret instrumentation diagrams:
   - P&ID diagrams
   - Loop diagrams
   - Raceway diagrams
4. Draw a loop diagram for a given instrumentation loop.

Prerequisites

Before you begin this module, it is recommended that you successfully complete the following: Core Curriculum; Instrumentation Level One; Instrumentation Level Two, Module 12201-03.

Required Trainee Materials

1. Pencil and paper
2. Straightedge
3. Instrumentation drawing template (if required by instructor)

1.0.0 ◆ INTRODUCTION

Drawings and diagrams provide a universally accepted language for communicating information about simple to complex mechanisms, systems, and processes.

Copies of construction drawings came to be known as blueprints because the ammonia process that was used to reproduce them caused the paper to turn blue. Today, most copies of drawings are black on white, like the copies from a copying machine. It is common now to refer to these as prints.

This module deals with the following topics:

- The types of prints available
- The purpose of each type of print
- The basics of print reading
- Details covering each type of print

The use of prints is the most efficient way to convey information about systems and equipment designs that could not be conveyed by words alone. Through the use of symbols and notes, a large amount of information that might require many pages of written description can be presented in a condensed form on one diagram.

There are many types of drawings and documents associated with the trades. Imagine trying to use a diagram with all the information available for that equipment on the same page. It would be so cluttered that the diagram would be useless. For this reason, many different types of prints have evolved. This module briefly discusses the purpose and application of each of the more common diagrams.

2.0.0 ◆ ELECTRICAL DRAWINGS

Electrical drawings are essential tools in the installation and troubleshooting of electrical systems. The use of standard symbols and terminology allows electrical and instrumentation personnel to follow a designed path in installing or troubleshooting components in a system. They can accomplish these tasks in a universally accepted manner. This section more closely examines the types of electrical diagrams, the symbols and terminology used on these diagrams, and other features associated with the design and content of these diagrams.

2.1.0 Block Diagrams

A block diagram (*Figure 1*) shows the major components and electrical or mechanical interrelations in block form. The lines between the blocks represent the system connections between the components. In electrical block diagrams, a single line may represent one wire or several wires. The block diagram provides an overview of a system. It is a convenient way to show how the pieces of the system fit together, the general operation, and the arrangements of the major components.

Figure 2 is a more detailed type of block diagram. It contains more descriptive information about the equipment, such as size and capacity. It also shows some of the components as symbols.

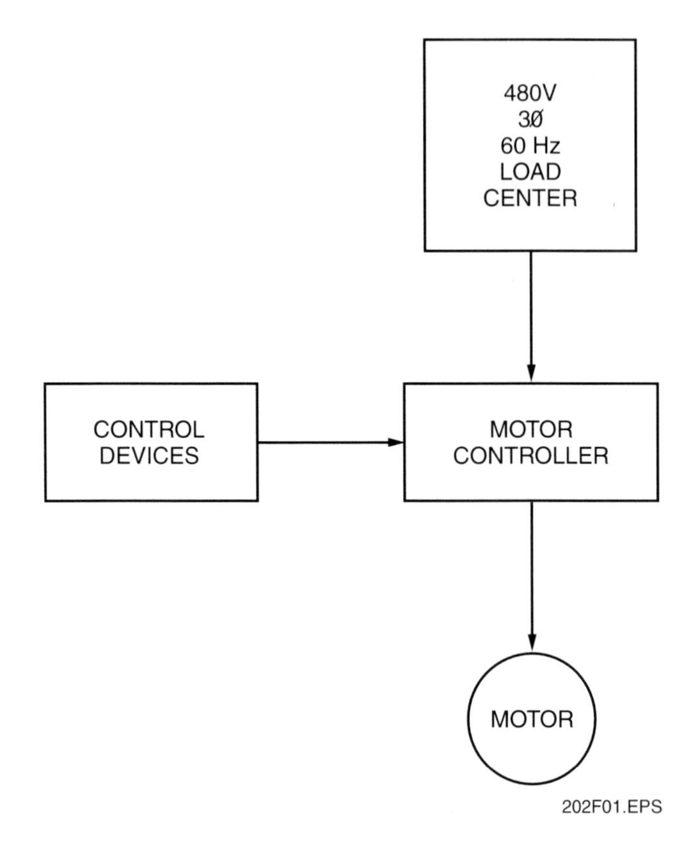

Figure 1 ◆ Block diagram of a motor circuit.

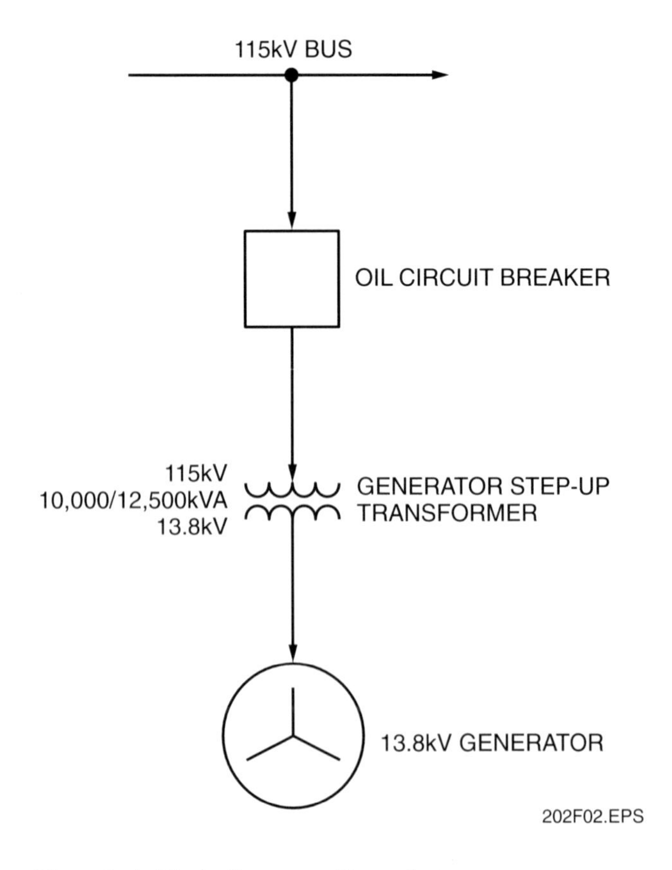

Figure 2 ◆ Block diagram with symbols.

2.2.0 Single- and Three-Line Diagrams

Line diagrams are used to show the interconnection of equipment and components in a system. There are several methods used to make line diagrams. The one-line and three-line methods are more often applied than others. The one-line diagram (*Figure 3*), which is sometimes referred to as a single-line diagram, is used to show the interconnection of the components in a system, but not necessarily the actual connections of the conductors or terminals. It is often drawn using a combination of blocks and symbols to represent the various components in a system. It is more applicable when an overall picture is needed to determine the sequence of component operations.

The line diagram may contain notes, descriptions, **load capacities**, and **National Electrical Manufacturers Association (NEMA)** sizes for the equipment listed. It represents one of the most common tools used in troubleshooting a system. It provides the technician with information that can be used to determine the causes of the symptoms in a system that is experiencing problems. It can also be used to isolate a problem to a particular piece of equipment or component in the system.

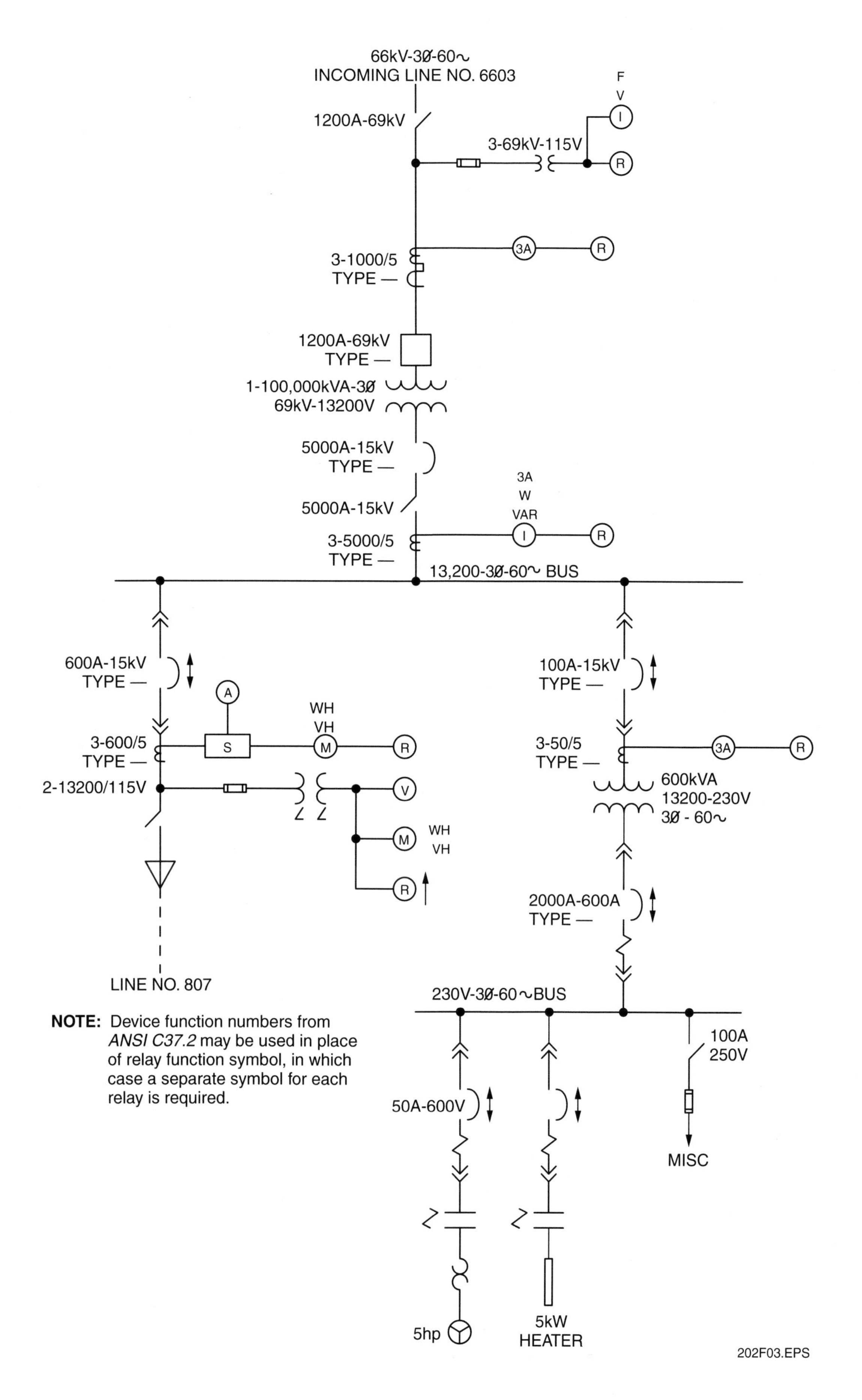

NOTE: Device function numbers from *ANSI C37.2* may be used in place of relay function symbol, in which case a separate symbol for each relay is required.

Figure 3 ◆ Typical one-line diagram for power equipment (ANSI).

Single-line diagrams show the interconnection or circuit path between components in an electrical system or circuit, while three-line diagrams are used to show all of the connections of the individual conductors in the system or circuit.

One of the most common applications of the three-line diagram is on **three-phase systems**. Three-line diagrams also can be drawn to show the conductor connections in three-wire, single-phase applications where a grounded neutral is used as a current-carrying conductor.

The one-line diagram and the three-line diagram complement each other. The single-line diagram is an effective tool in determining the path of the circuit, while the three-line diagram can be applied in troubleshooting the system after the path is determined. *Figure 4* shows a three-line diagram example of a three-phase motor power circuit.

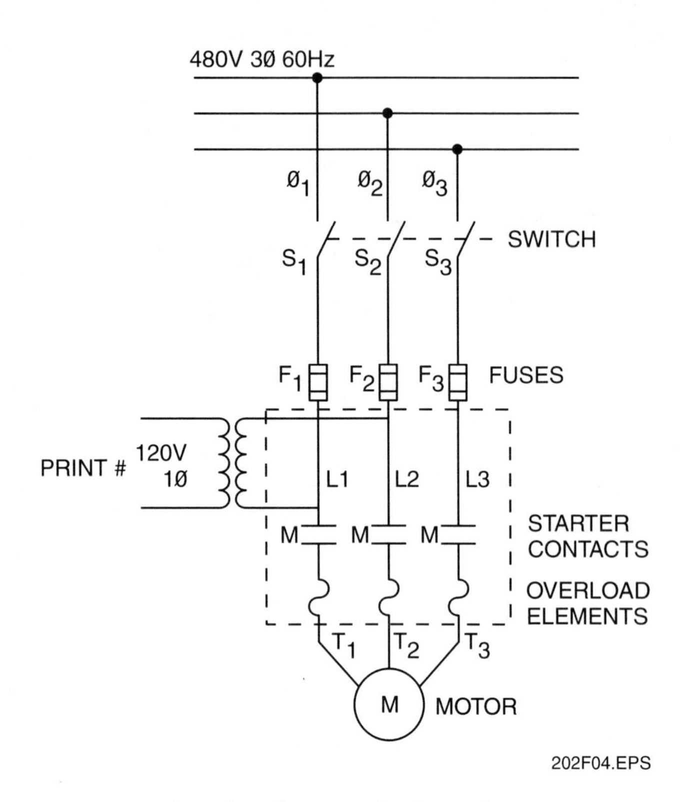

Figure 4 ◆ Three-line diagram of a three-phase motor power circuit.

2.3.0 Wiring Diagrams

The **wiring diagram** is used for installation and troubleshooting. Wiring diagrams show the relative position of various components of the equipment and how each conductor is connected in the circuit. These diagrams are classified in two ways: as internal diagrams, which show the wiring inside a device, and as external diagrams, which show the wiring between the devices in a system. *Figure 5* shows a simple wiring diagram of a motor controller. The wiring diagram is a map of the connections, terminals, conductors, devices, components, and equipment in the circuit or system. It is the final tool used in the installation process or the roadmap for troubleshooting the circuit or system. There are several different methods used by designers to illustrate the wiring connection layout

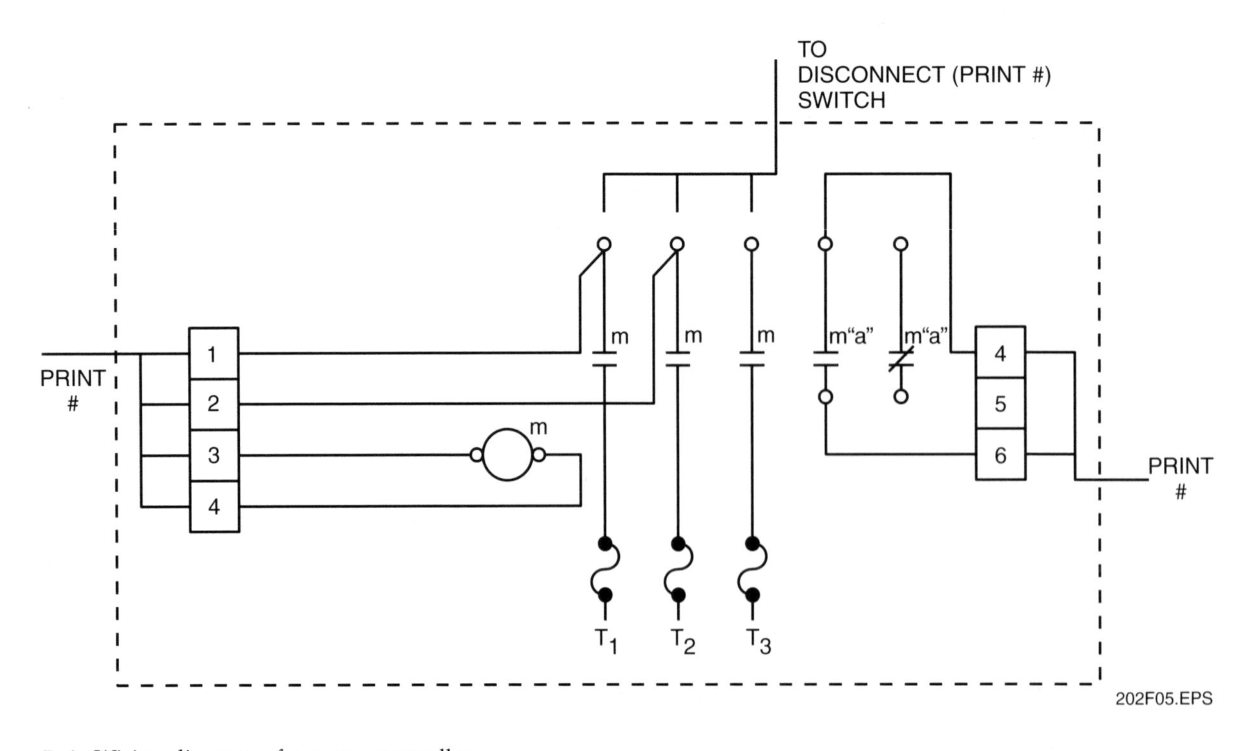

Figure 5 ◆ Wiring diagram of a motor controller.

of a system, including the point-to-point method, cable method, baseline method, and lineless method. The following subsections look more closely into these methods.

2.3.1 Point-to-Point Method

The point-to-point wiring diagram method (*Figure 6*) saves space and helps alleviate clutter on a diagram. The person following a point-to-point wiring diagram generally does not have a description of the equipment, but rather only traces a conductor from one point to another, making sure it follows the correct path. This diagram can be used in troubleshooting, but it is more typically used to terminate wiring during installation. Often technicians responsible for using a point-to-point diagram have no knowledge of the circuitry or even the process involved. Their job is simply to follow the point-to-point diagram and make sure each end of each conductor is terminated at the correct point.

2.3.2 Cable Method

Unlike the point-to-point wiring diagram, the cable wiring diagram method does not show each conductor from end-to-end. It only shows the cable or **cable bundle** that the conductor or conductors are in (*Figure 7*). This method requires much less space than the point-to-point method and allows much more **white space** on the drawing, making it appear less cluttered. One disadvantage of using this method is that the technician or installer does not know the exact path that a conductor takes from one point to the other. He or she only knows that it enters a cable or cable bundle and exits at one or more locations. This diagram requires much more attention to detail and often requires a higher level of conductor termination skills. It is a very common method used in wiring diagram layouts. Its primary function is to serve as an installation tool, not as a troubleshooting tool.

2.3.3 Baseline Method

The baseline method is a variation of the cable method. It is used on complex diagrams. In the baseline diagram, one line is used to represent many wires. The use of the baselines decreases the number of lines in the diagram. *Figure 8* is a baseline connection diagram.

2.3.4 Lineless (Wireless) Method

Lineless diagrams are used by designers on systems that generally have terminal boards that are clearly numbered. A table that lists the wire number, as well as the board and terminal number to which it is connected, must accompany this type of diagram. This is a fairly simple diagram to follow and can be used by less experienced installers

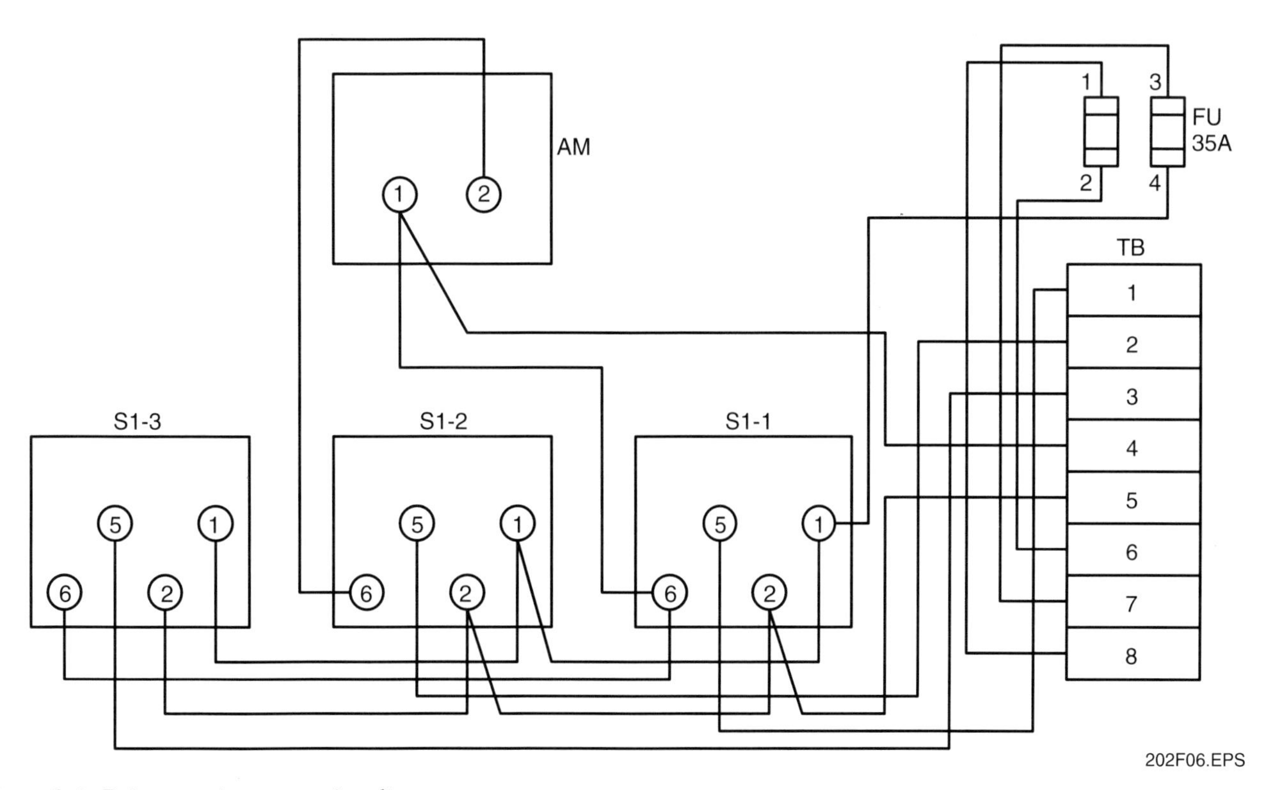

Figure 6 ◆ Point-to-point connection diagram.

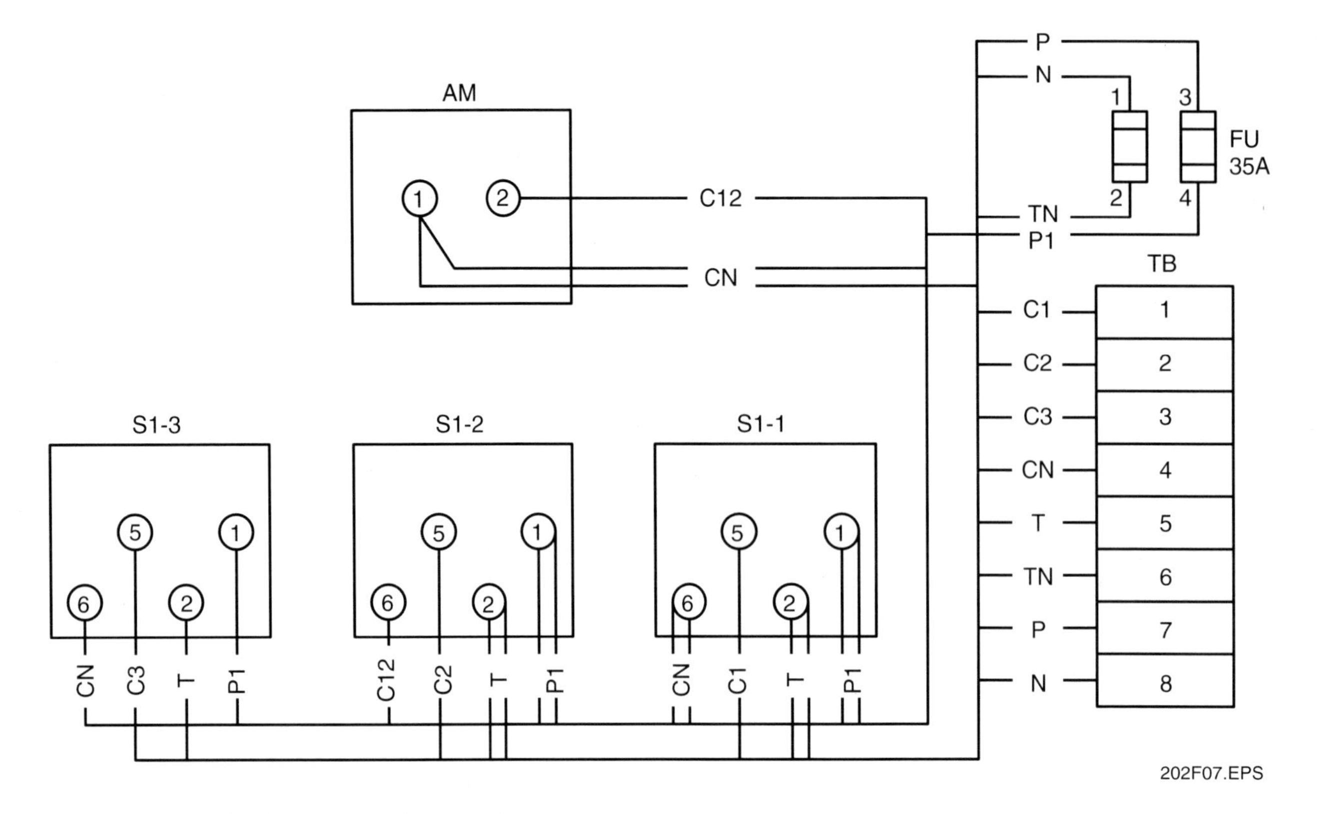

Figure 7 ◆ Cable method connection diagram.

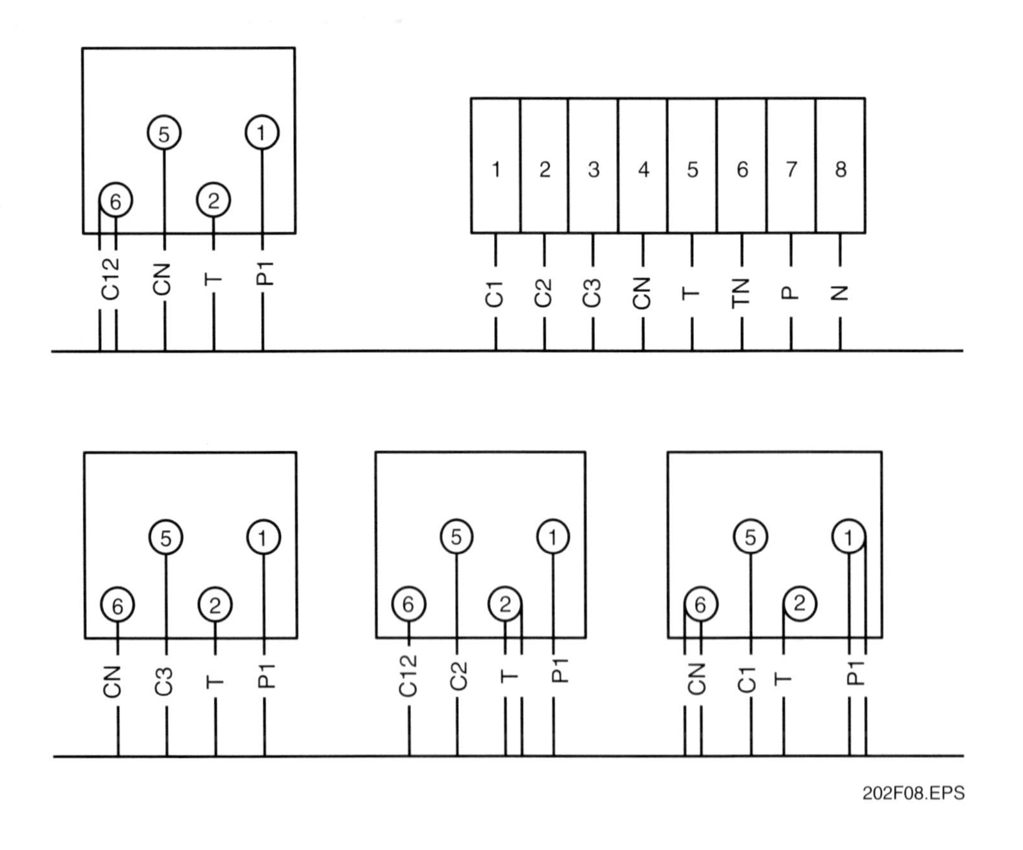

Figure 8 ◆ Baseline connection diagram.

as long as they follow the exact cross-reference of the table. Completed terminations should be double-checked to make sure the correct wire is connected to the correct board and terminal. Using the example shown in *Figure 9*, try to find the wire connection numbered 4 for the end of the 35 amp fuse, using the drawing and the table. Look in the table in the *From* column and find FU4, then look beside FU4 in the *To* column. The table indicates that FU4 is connected to three different points: S1-1,1; S1-2,1; and S1-3,1.

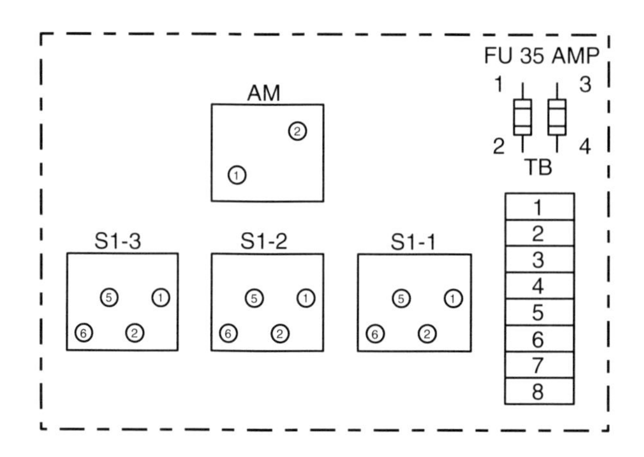

WIRE CONNECTION OF COMPONENTS

WIRE CODE	FROM	TO	WIRE SIZE
C1	TB1	S1-1, 5	
C2	TB2	S1-2, 5	
C3	TB3	S1-3, 5	
CN	TB4	AM1, S1-1, 6; S3-1, 6	AS REQUIRED
T	TB5	S1-1, 2; S1-2, 2; S1-3, 2	
TN	TB6	FU2	
P	TB7	FU3	
N	TB8	FU1	
C12	AM2	S1-2, 6	
P1	FU4	S1-1, 1; S1-2, 1; S1-3, 1	

202F09.EPS

Figure 9 ◆ Lineless connection diagram.

2.4.0 Raceway Drawings

The **raceway drawing** shows the physical layout of raceways. Examples of raceway drawings are shown in *Figures 10* and *11*. A large amount of information is available on these drawings.

For example, on *Figure 10*, the type, identification number, size, elevation, and physical layout of the 24" raceway KT-175 is shown. The size, identification number, and location of conduit drops are also shown in relation to the physical location of the equipment that is hooked up. There are two 30 amp, 600V, 3ph, non-fused disconnect switches with identified conduit runs that are shown with their physical locations.

Figure 11 shows different raceways with the identification numbers KT-174,12", KT-165,36", IT-361,12", and IT-360,12". The KT indicates a power tray, and the IT shows an instrument tray. The last two numbers indicate the width of the tray.

Also shown in this drawing are the conduit drops with identification numbers and the different elevations. There is a note to see drawing 99-D-3373A for continuation. The physical location and direction of the stairs on this level are also shown.

It is important to be able to identify and use all of the information available on raceway drawings, both for installation of systems and for tracing cables during maintenance.

2.5.0 Schematic Diagrams

An electrical schematic diagram is another form of wiring diagram. The electrical schematic shows how the electrical components are connected, but generally does not show physical connections or detailed wiring information. The electrical schematic is primarily a troubleshooting tool. It allows the reader to see how the electrical components interact in a circuit. The three-line diagram in *Figure 4* is also an example of a simple schematic diagram. Note that it shows how the components are electrically connected, but does not show point-to-point wiring information.

Schematic diagrams are usually designed to be read from left to right and from top to bottom. They typically use standard electrical diagram symbols, which are discussed later in this module. The positions of the relay contacts and switches are usually shown as they would be in the de-energized state.

3.0.0 ◆ INSTRUMENTATION DRAWINGS

In addition to the electrical diagrams previously covered, there are several types of diagrams specifically associated with the instrumentation field. These include **piping and instrumentation drawings (P&IDs)**, **loop sheets**, logic/ladder diagrams, **location drawings**, **installation drawings**, **flow diagrams**, and **instrument data sheets**.

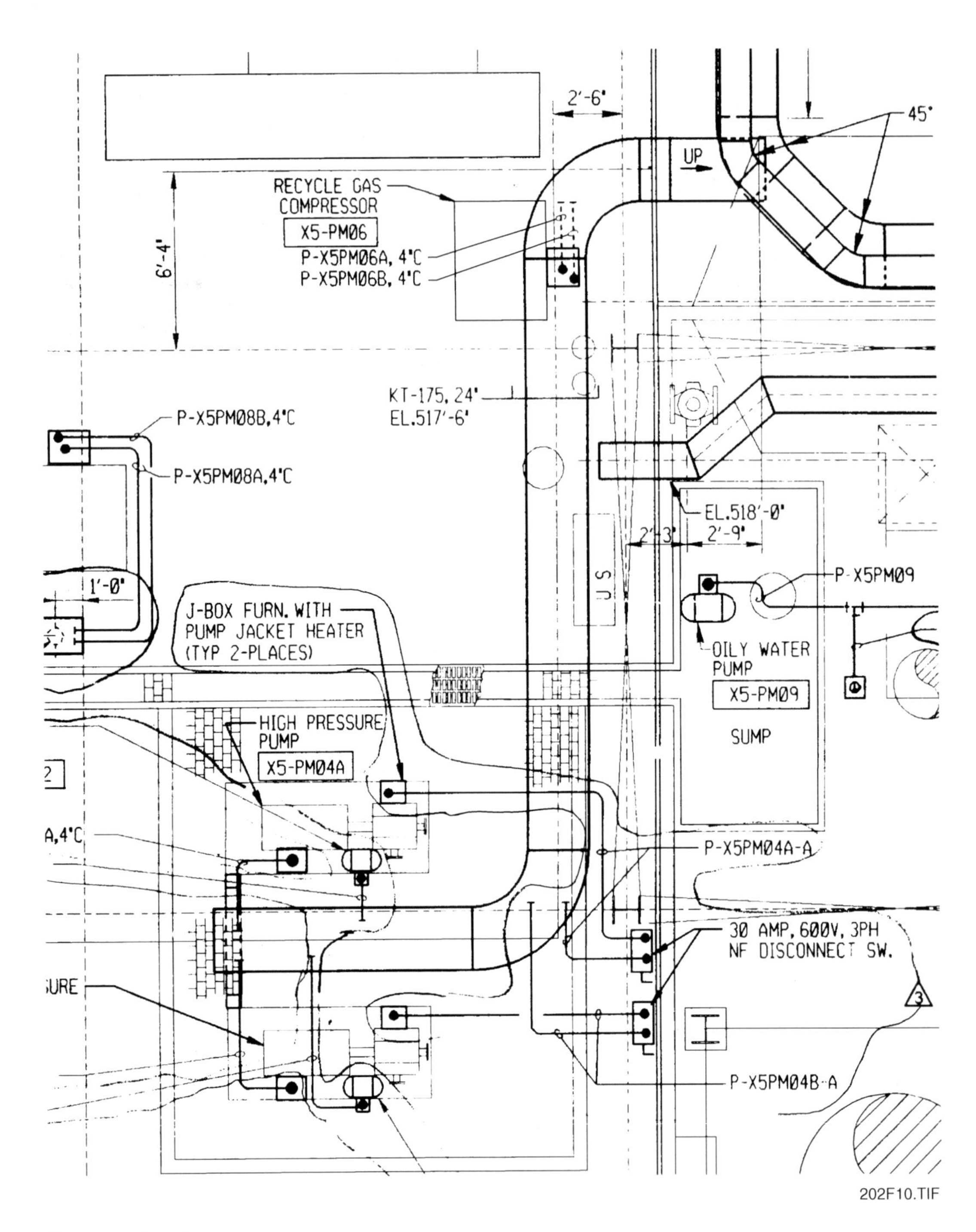

Figure 10 ◆ Raceway drawing.

3.1.0 P&IDs

Piping and instrumentation drawings (P&IDs) illustrate the operational layout of a system, including its piping, valves, and instrumentation. The P&ID shows the relationship between the components, but does not show the physical location. This drawing is often called a bubble diagram by instrumentation personnel because of its use of standard bubble-type symbols in compliance with **Instrumentation, Systems, and Automation Society (ISA)** standards. A P&ID generally shows all components of a particular system. Separate P&IDs are typically used to show different sections or areas of a facility, such as a petrochemical facility, power plant, paper mill, or other industrial site. Information that may be found on a P&ID include pipe sizes, flange sizes, valve sizes, flow direction, and references to other drawings or diagrams. Rather than literally drawing all the com-

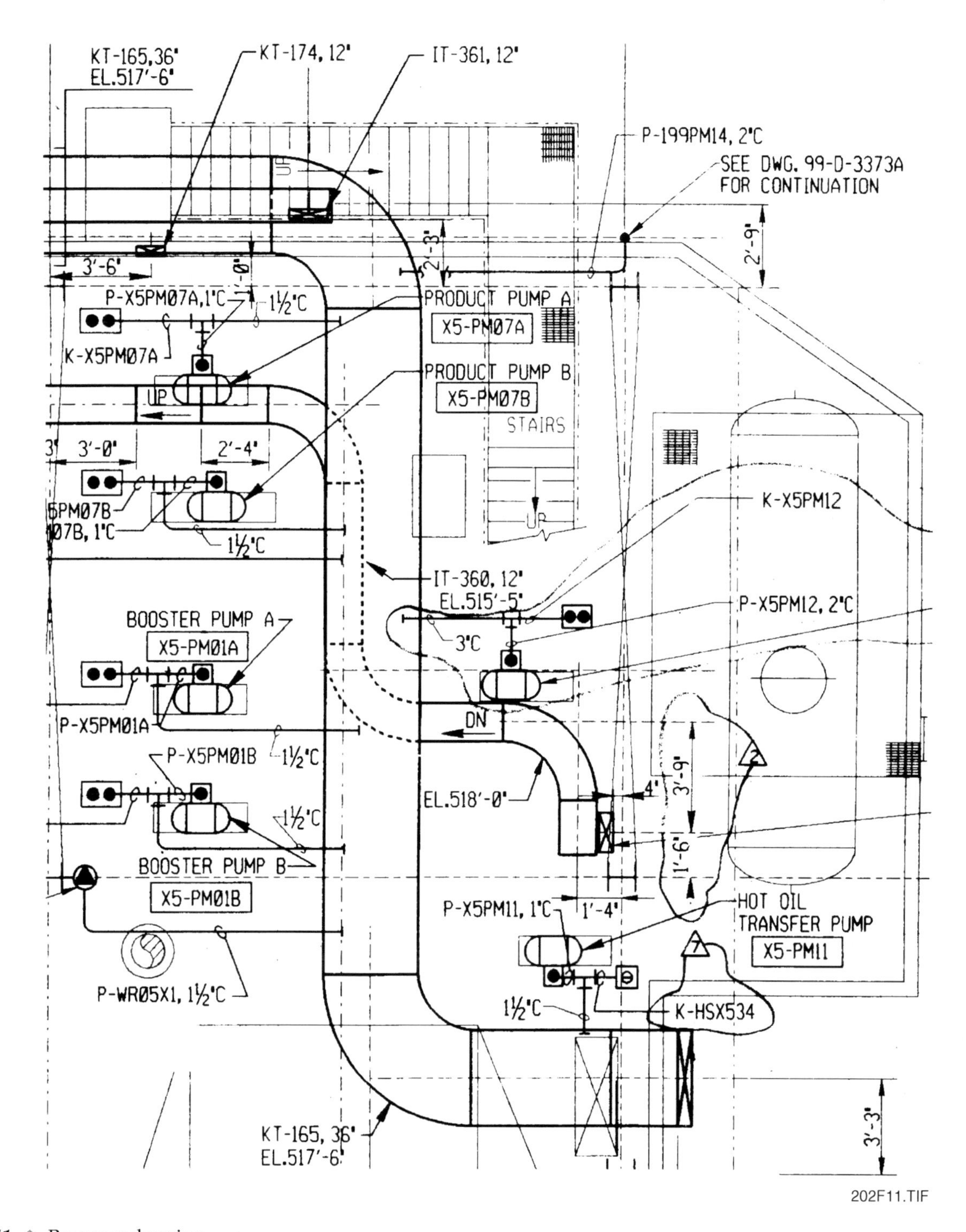

Figure 11 ◆ Raceway drawing.

ponents on the P&ID, standard symbols, including **American National Standards Institute (ANSI)**, ISA, and **American Society of Mechanical Engineer (ASME)** symbols, are generally used by the designer. Even though these standards are used extensively by designers of P&IDs, it is not uncommon to find a **legend** on the P&ID that better identifies the components on the drawing. An example of a P&ID is shown in *Figure 12*. The letters within the bubble represent the type of process (for example, SV for solenoid valve, TI for temperature indicator, PT for pressure transmitter). Under that lettering is the numeric part of the ID, used to differentiate it from similar devices. If a section of a P&ID circuit has several devices grouped together, they all share the same numeric part of the ID. For example, TI203 and TT203 represent a temperature indicator and transmitter/transducer at the same location. P&ID drawings will be discussed in more detail later in the module.

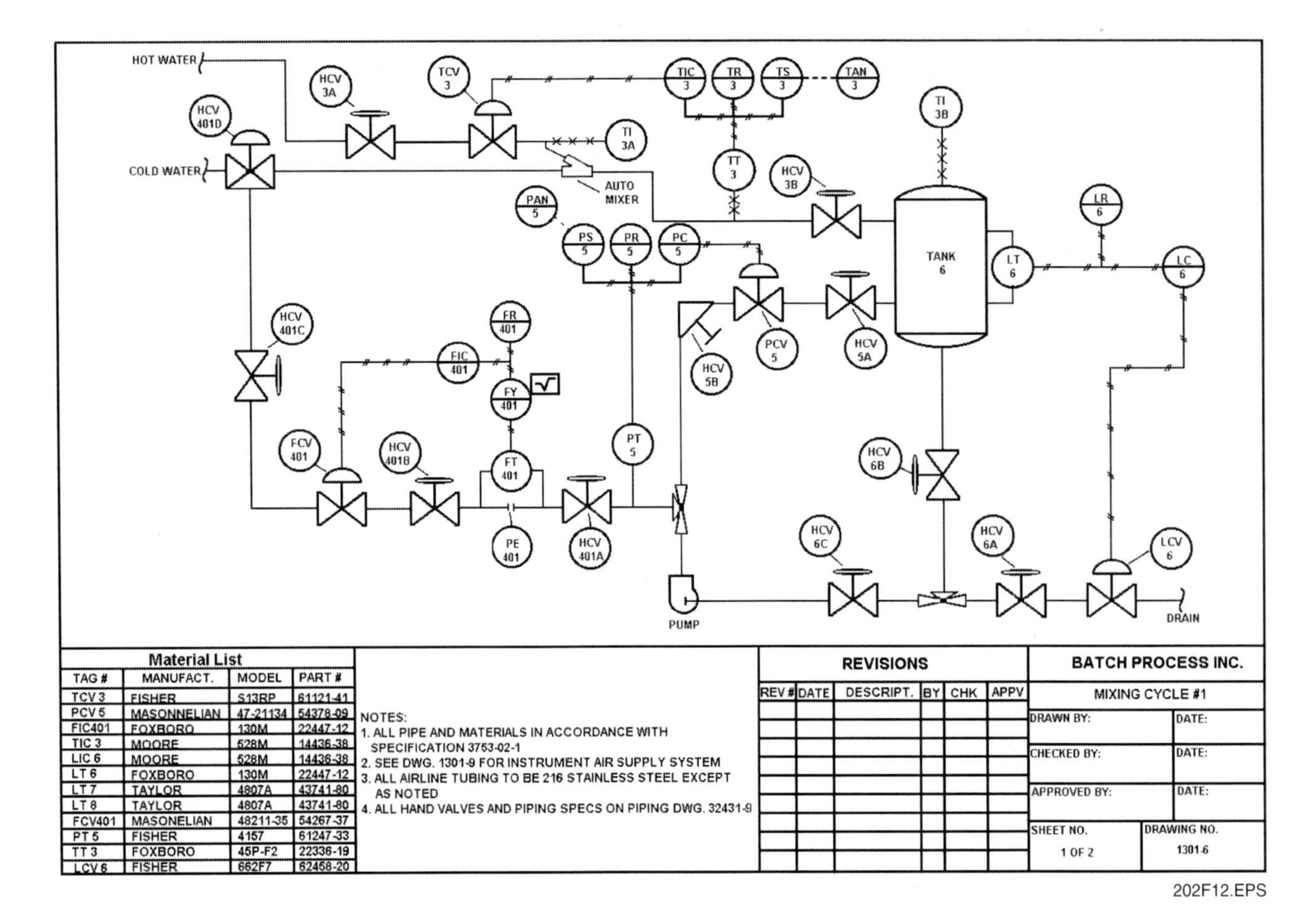

Figure 12 ◆ P&ID.

3.2.0 Loop Sheets

Loop sheets provide information about an instrument loop. By definition, a loop is a combination of two or more instruments or control functions arranged so that signals pass from one to another for the purpose of measurement or control of a process variable. A separate loop sheet is available for each instrument loop. Each loop sheet contains the following:

- Instrument **tag numbers**
- Location and connection section
- Calibration and specification section
- Field checklist

Figure 13 shows a loop sheet.

3.2.1 Instrument Tag Numbers

In most systems, there are many different instruments in a single loop, so tag numbers are a critical source of supplementary information. Loop tag numbers are assigned to each instrument in the loop regardless of its function or location. In addition to specific loop information, the tag number often indicates a location in the plant or a building designation. In some cases, a specific series of numbers may be used to designate a special function. For example, the series 900 to 999 could be used for a safety-related function such as fire protection. When it is necessary to repair or troubleshoot the system, the numbering system can save considerable time and effort. The tag numbers on the loop diagram make it possible to determine which other instruments in the loop might be affected by the problem. With this information, the malfunctioning device can be bypassed and the system can be restored to operation until the device can be repaired or replaced.

Figure 14 shows a loop identification tag number. An instrument tag number on a loop sheet consists of five parts contained in an instrument bubble.

1. The area designation defines the area of the building where the instrument is installed. This identifier is optional and can be omitted.
2. The loop number identifies a specific loop in the plant and is associated with a plant area number.
3. The measured variable identifies the specific instrument used with standard instrument abbreviations.

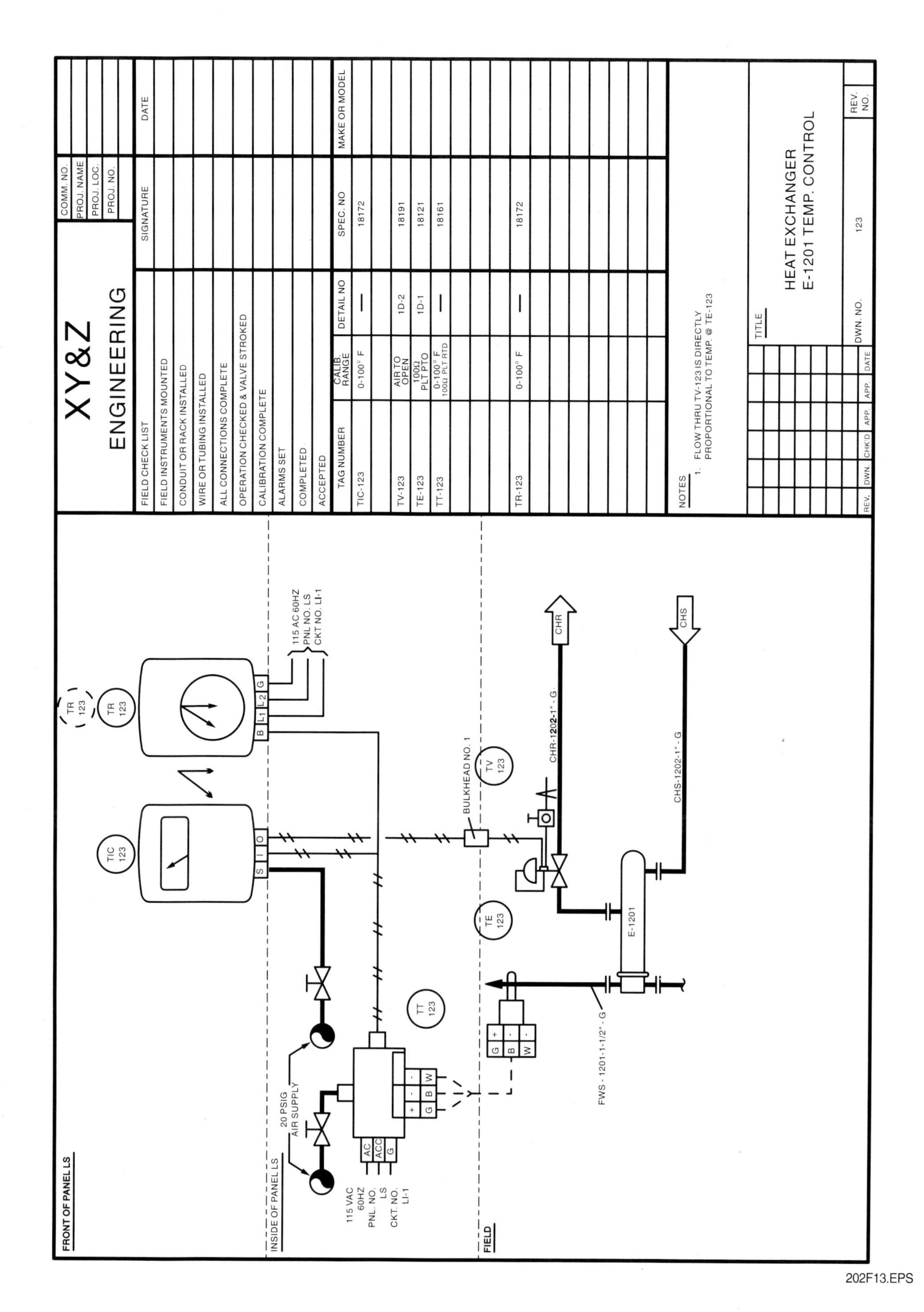

202F13.EPS

Figure 13 ◆ Loop sheet.

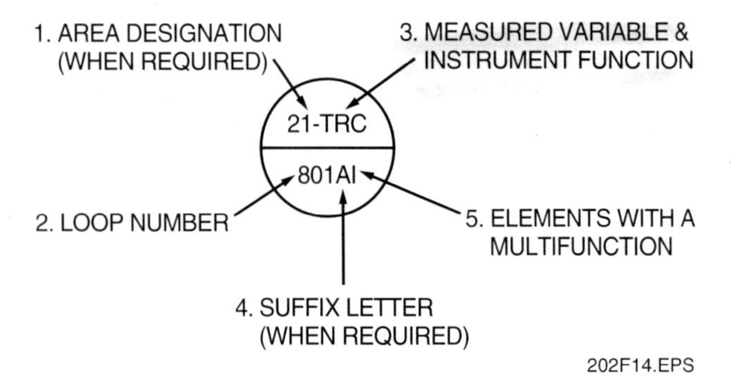

Figure 14 ◆ Loop identification tag number.

4. The suffix number is used to prevent duplication of loop numbers.
5. An element with multifunction is used if the element has a **multi-point function**.

3.2.2 Location and Connection Section

The location and connection section on a loop sheet usually shows three views of the complete instrument loop. This section shows all components of an individual loop, including those components located on the front panel of the control panel, those connected inside the panel, and those that are located in the field.

Symbols on the location or connection section indicate how or where instruments are mounted. Lines, or the absence of lines, on the symbols indicate this information. Line variations include single solid lines, double solid lines, and broken lines. A solid line indicates that the instrument is panel mounted, usually with a group of instruments. In most instances, panel-mounted instruments are easily accessible. Double lines indicate that the instrument is at an auxiliary location, possibly behind a panel, but is normally accessible by the operator. A broken line means the instrument is mounted on the back of the panel, but it is not accessible to the operator. The absence of a horizontal line within a bubble or balloon indicates that the instrument is field-mounted. Field-mounted instruments are usually located near the point of measurement or near a final control element. *Figure 15* shows the location and connection section of a loop sheet.

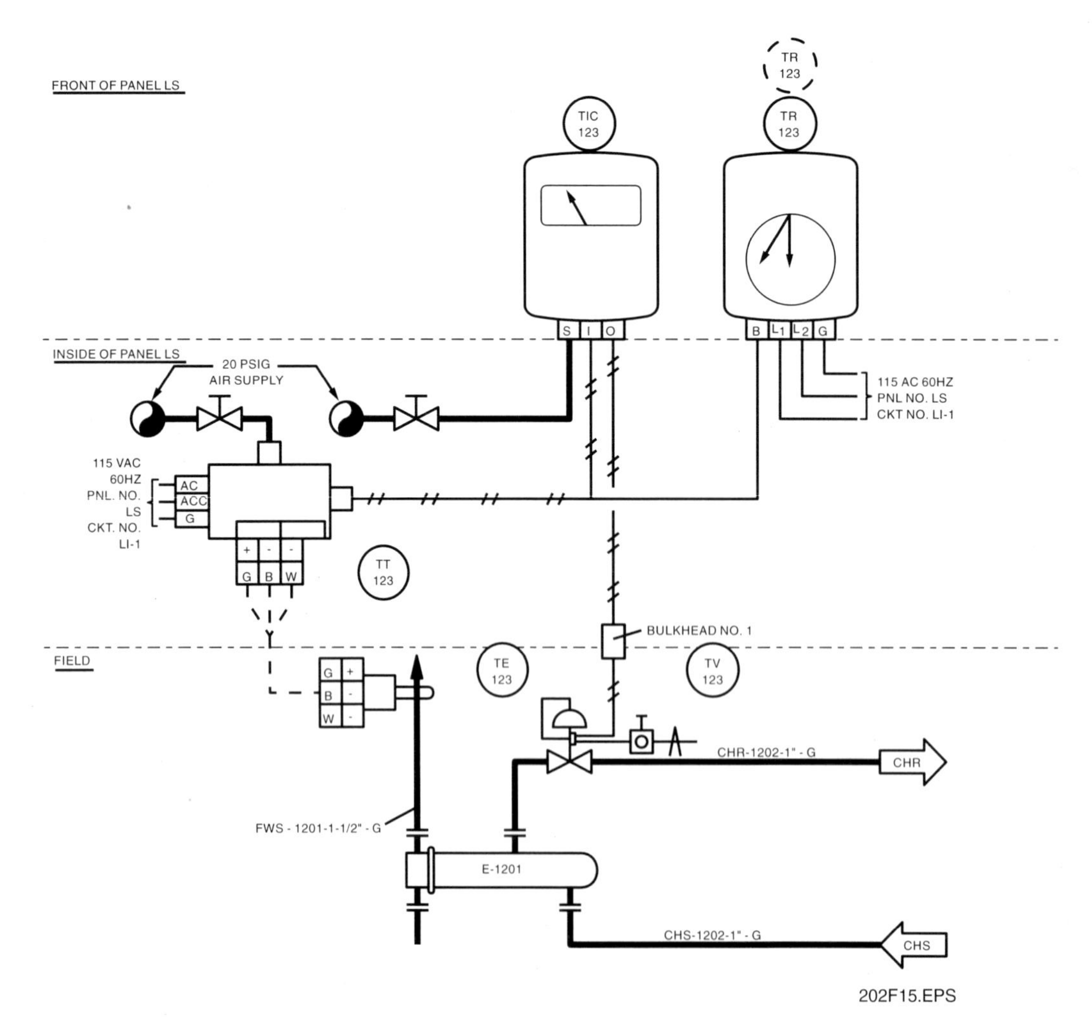

Figure 15 ◆ Loop sheet location and connection section.

3.2.3 Calibration and Specification Section

The calibration and specification section of a loop sheet lists the individual identification number of each instrument in the loop. This section also lists the calibration data, specification number, make, and manufacturer of each instrument. *Figure 16* shows the calibration and specification section of a loop sheet.

TAG NUMBER	CALIB. RANGE	DETAIL NO.	SPEC. NO.	MAKE OR MODEL
TIC-123	0–100°F	—	18172	
TV-123	AIR TO OPEN	ID-2	18191	
TE-123	100Ω PLT PTO	ID-1	18121	
TT-123	0–100°F 100Ω PLT RTD	—	18161	
TR-123	0–100°F	—	18172	

202F16.EPS

Figure 16 ◆ Loop sheet calibration and specification section.

3.2.4 Field Checklist

A field checklist on a loop sheet tracks the original installation and startup checks made on the loop during its construction. This information is very important after the initial installation of the equipment because it gives the technician responsible for maintaining the loop the contact name of the person who performed each item on the checklist. It also provides a tool for the installation contractor to use in determining the percentage of completion of the total project based on the completeness of each field checklist. *Figure 17* shows a typical blank field checklist.

FIELD CHECKLIST	SIGNATURE	DATE
FIELD INSTRUMENTS MOUNTED		
CONDUIT OR RACK INSTALLED		
WIRE OR TUBING INSTALLED		
ALL CONNECTIONS COMPLETE		
OPERATION CHECKED & VALVE STROKED		
CALIBRATION COMPLETE		
ALARMS SET		
COMPLETE		
ACCEPTED		

202F17.EPS

Figure 17 ◆ Field checklist.

3.3.0 Logic/Ladder Diagrams

Ladder diagrams are very useful drawings for troubleshooting. They contain only the logical or sequential circuit paths for the control circuit wiring, including the contacts that control magnetic motor starters, relay coils, pilot lights, and other components. In instrumentation applications that incorporate a PLC, it is not uncommon to see many ladder diagrams used to represent the sequence of operation of the input signals and the output signals. When the ladder diagram is used for such applications, it is often referred to as a logic diagram rather than a ladder diagram. Ladder diagrams are valuable troubleshooting tools. They are easier to use than standard schematic wiring diagrams because they treat each load device as a separate circuit. In a ladder diagram, power is represented as the uprights of the ladder, with L1 the hot side of the single-phase AC power source and L2 the neutral side. Each load device (motor, solenoid, relay coil, light bulb) is shown on a separate rung of the ladder with the switching devices that control the load. If a particular load is not operating when it should, it is easy to identify the devices that could cause the problem. Assume that solenoid 1 (SOL1) controls a valve. If the valve fails to open or close, check to see if PL2 is lighted. If PL2 is lighted, assume that the solenoid is defective. Assuming PL2 is not lighted, the problem is in one of the devices that control the solenoid. For example, flow switch FS4 might have failed. Because there are open contacts of two control relays in series with the solenoid, the load lines for the coils of those relays should be checked if FS4 checks okay. *Figure 18* is an example of a ladder diagram.

Control systems that incorporate programmable logic controllers (PLCs) may have sections of the field wiring drawn as ladder diagrams, as shown in the top image in *Figure 18*. After that field control wiring is interfaced with the PLC, all switches and relay contacts are shown as input devices to the computer program and are always drawn as a set of normally open contacts. This is shown at the bottom of *Figure 18*. This type of diagram is referred to as ladder logic.

In this example there are two control relays, CR1 and CR2. They are being controlled by a series of field control devices, as well as auxiliary

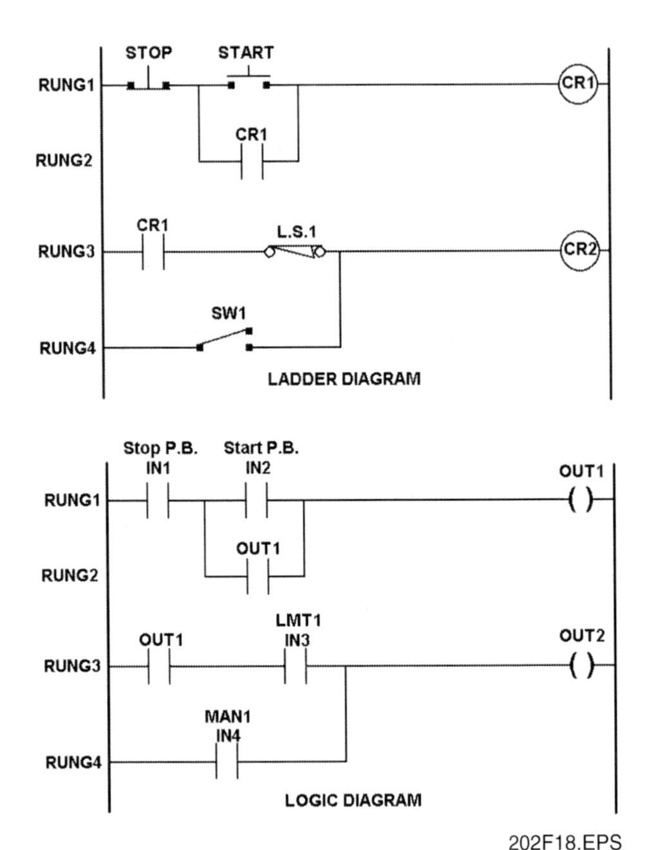

Figure 18 ◆ Ladder diagram.

contacts located on the control relays themselves. The control devices, such as stop/start pushbutton stations, relay contacts, and limit switches shown in the top ladder diagram, are drawn as device symbols and in their normal states. In the bottom drawing, all of these devices are shown as sets of normally open contacts and are referred to as inputs. Likewise, both control relays CR1 and CR2 are illustrated in the top drawing as relay symbols (circles) while in the bottom logic diagram they are drawn as outputs from the computer program (( )). Both types of diagrams are drawn in the ladder format, with two rails and as many rungs as necessary to illustrate the many different loads and paths of the control signals or input signals associated with the system.

Ladder diagrams do not normally address the power circuit, assuming that if the control circuitry is functioning properly, the equipment will operate properly. Ladder diagrams allow for the isolation of the control circuitry by individual rungs and permit a technician to trace and test the control circuitry in a sequential manner until a problem is located.

3.4.0 Equipment Location Drawings

Equipment location drawings show a layout of major equipment on a project. The diagram usually includes either the tag number of the equipment with a description by the tag number or cross-references another drawing that lists the equipment by its technical description. The equipment location diagram and the **specification sheet** for the equipment are often cross-referenced. Layout drawings may also include the interconnections of raceway systems and cable trays.

Instrument location drawings show the positions of process instruments relative to equipment arrangement background flow drawings. A flow drawing represents an actual sequence of events in a process and its related process instrumentation.

Abbreviations and symbols on location drawings identify the physical location of process instruments. These drawings also show the elevation of an instrument by referencing it to ground level, or the floor level of the structure. A location drawing has a title block with all necessary numbers and information. In the location drawing in *Figure 19*, the structural support columns are numbered and lettered to aid field location of equipment.

3.5.0 Installation Detail Drawings

Installation detail drawings are frequently used by systems installers. These drawings generally provide a magnified look at a particular section of an installation to aid in proper installation. These drawings may include notes, a listing of materials needed, and other pertinent information needed necessary to complete the installation. They are generally drawn to scale so that precise measurements can be given. It is not uncommon for an installation project to have many of these drawings. A common method applied to installation detail drawings is to provide a drawing for each unique method of installation, such as wall-mounted, stand mounted, **process mounted**, and panel mounted. Installation detail drawings provide uniform guidelines for proper connection of process instrumentation. They also provide bills of materials, which are necessary for proper installation. *Figure 20* shows an instrument installation drawing.

The bill of materials on the installation drawing lists all the items needed to properly install the particular instrument and any identical to it. The

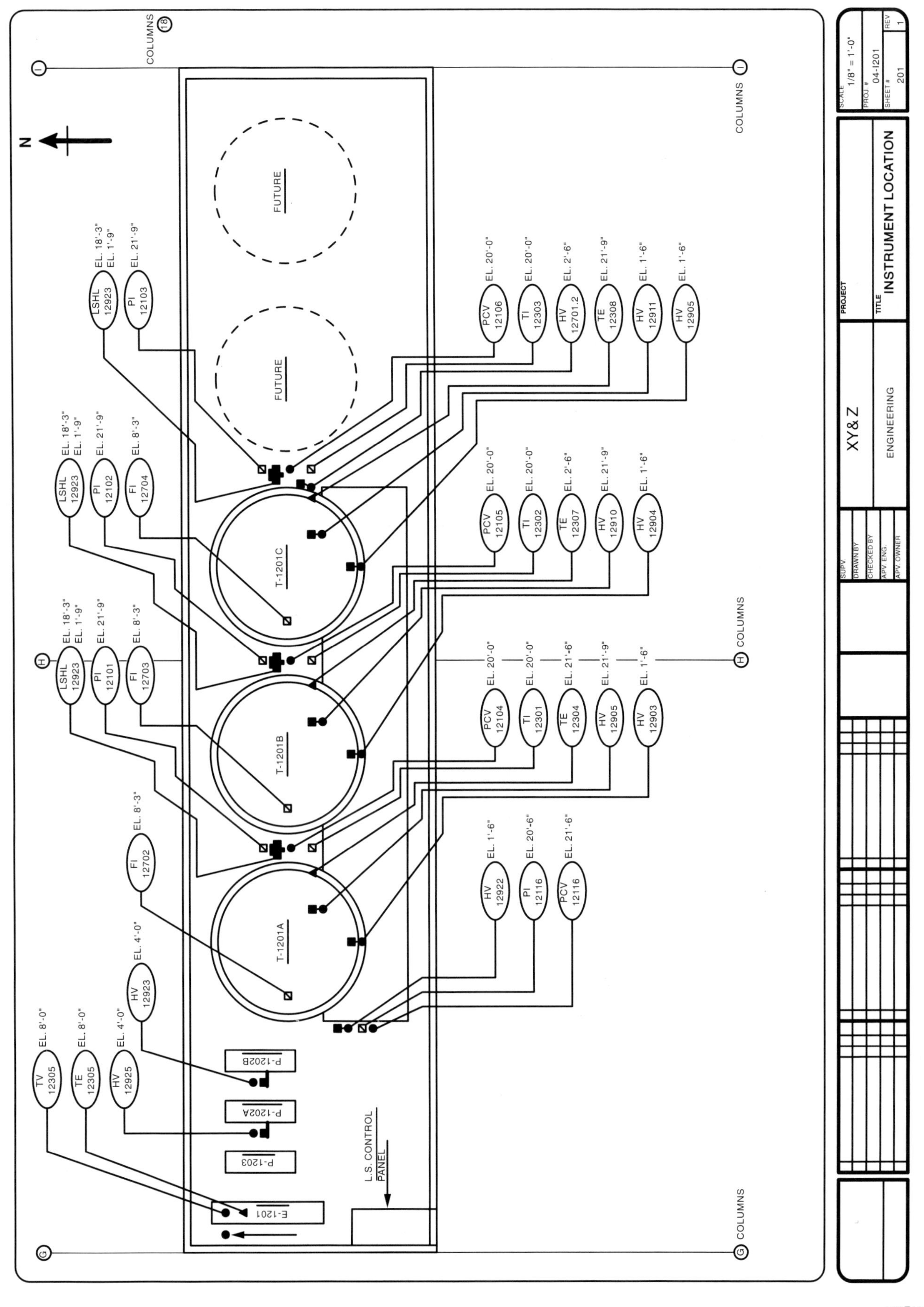

202F19.EPS

Figure 19 ◆ Instrument location drawing.

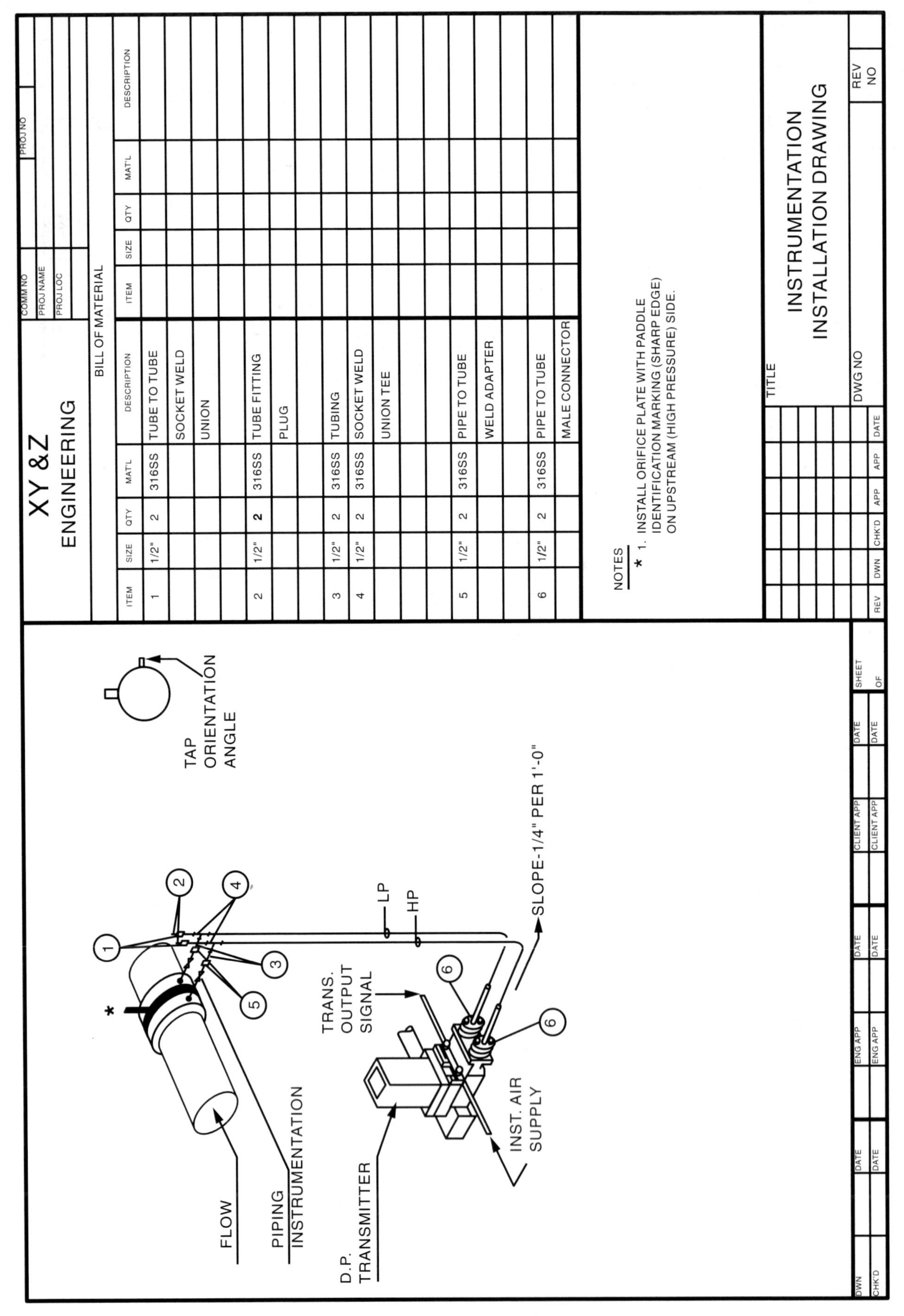

Figure 20 ◆ Instrument installation drawing.

callouts (numbered tags) on the drawing refer to the item section of the bill of materials. The bill of materials also gives required fitting and connection descriptions, sizes, quantities, and the materials of which they are made. *Figure 21* shows a bill of materials.

3.6.0 Flow Drawings

Process flow drawings (PFDs) are often used in conjunction with P&IDs in order to position the instruments in their intended locations in the process. Even though P&IDs often contain the direction of flow, flow diagrams or drawings are used to aid in understanding the overall process. The flow diagrams may be drawn as block diagrams, or they may be drawn using symbols for various vessels, piping, and other equipment located in the process. They may also be contained in the P&ID so that only one drawing is required for both references. *Figure 22* shows a simplified process flow diagram for a basic industrial process system that contains a reactor, vacuum system, freshwater system, and product storage capabilities. By using this drawing in conjunction with a P&ID, it may be easier to locate the instruments and devices in their proper sequence of process flow.

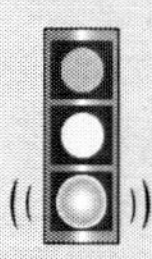

NOTE
Process flow information is usually need-to-know information. Access to the diagrams may be restricted for security issues related to the product or by contract terms. Flow drawings from the actual site should be used for reference.

3.7.0 Instrument Data Sheets

Figure 23 is an example of an instrument data sheet. An instrument data sheets lists all of the technical information that pertains to a particular instrument. This is very important information that applies to different situations.

During the initial installation of a new instrument, the information on the data sheet will show what size fittings are used, what options are attached, and what the diagram on the instrument looks like.

The data sheet lists information needed for maintenance, such as the types of fittings used, the different temperatures and pressures at which the instrument is operated, the *National Electrical Code®* (NEC) hazardous location classification that applies, and other pertinent technical information.

BILL OF MATERIALS

ITEM	SIZE	QTY	MAT'L	DESCRIPTION	ITEM	SIZE	QTY	MAT'L	DESCRIPTION
1	½"	2	316SS	TUBE TO TUBE					
				SOCKET WELD					
				UNION					
2	½"	2	316SS	TUBE FITTING					
				PLUG					
3	½"	2	316SS	TUBING					
4	½"	2	316SS	SOCKET WELD					
				UNION TEE					
5	½"	2	316SS	PIPE TO TUBE					
				WELD ADAPTER					
6	½"	2	316SS	PIPE TO TUBE					
				MALE CONNECTOR					

202F21.EPS

Figure 21 ◆ Bill of materials.

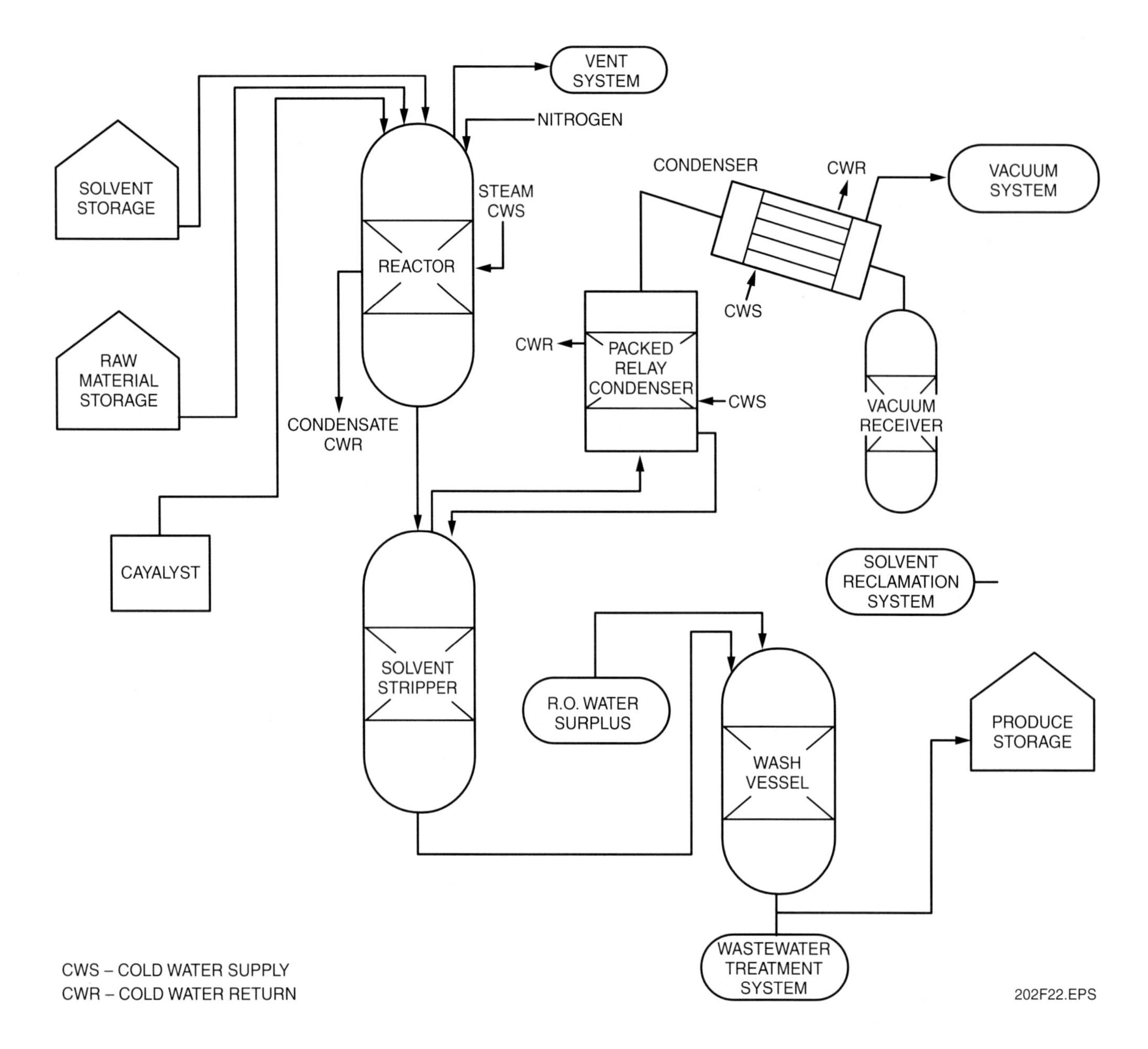

Figure 22 ◆ Flow diagram.

The sample instrument data sheet is a generic sheet that shows the specifications for a level gauge. Instead of listing the actual specifications, the blocks show examples, such as location 1, option 2, pressure 1, and material 4. These items, like location 1, would be taken from a database that lists the actual location that applies to 1. When the actual instrument data sheets are printed, all gauges that are located at location 1 would have the actual location printed in this block.

Instrument data sheets can be used for tasks such as purchasing, installation, maintenance, and calibration.

4.0.0 ◆ APPLYING THE DIAGRAMS

The diagrams, drawings, and sheets covered in this module are often referenced by both electrical and instrumentation workers. The level of diagrams needed depends on the level of information required by the technician or installer. It is not uncommon to require the information found on all the diagrams at some point in either the installation or troubleshooting of a system. If an overall picture of the system is needed in order to isolate the problem area in a process, the block diagram may be required first. After that determination is made, the

XY&Z	CONTRACT NO. 21-3963	INSTRUMENT DATA SHEET	SPEC. NO.		
		FOR AREA	REV.	REV. DATE	DATE
FOR		SERVICE	SPEC. BY		
			CHECKED BY		

TAG NO.	REQUISITION NO.	PURCHASE ORDER NO.	CHANGE ORDER NO.	VENDOR
CATALOG NO.		QUANTITY	ACCOUNT NO.	

	LEVEL GAGE	P&ID 507-0PI_DWG
GENERAL	VESSEL NO.	LINE NO.
	VESSEL NO./SCHEDULE	LINE SIZE
	LINE/VESSEL MATERIAL	LINE MATERIAL
	ITEM 1	OPTION 1
DEVICE DESCRIPTION	GAGE MODEL OR TYPE	MODEL 1
	NUMBER OF SECTIONS	NO. SECTION
	GAGE SIZE	SIZE 1
	VISIBLE RANGE	RANGE 1
	CONNECTION LOCATION	LOCATION 1
	COVER MATERIAL	MATERIAL 1
	CHAMBER MATERIAL	MATERIAL 2
	FASTENER MATERIAL	MATERIAL 3
	GLASS TYPE	MATRL_ELEM
	EXTENSION SIZE	SIZE 2
	EXTENSION MATERIAL	MATRL 4
	VALVE TYPE	TYPE VALVE
	VALVE MATERIAL	MATRL 5
	GAGE CONNECTION	PROCONN 1
	DRAIN CONNECTION	PROCONN 2
	VESSEL CONNECTION	PROCONN 3
	MAX. TEMP./PRESS. RATING	TEMP. MAX 1 / PRESS. MAX 1
	ITEM 2	OPTION 2
	ITEM 3	OPTION 3
	ITEM 4	OPTION 4
	ITEM 5	OPTION 5
SERVICE CONDITIONS	FLUID	FLUID 1
	PRESSURE	PRESS 1
	TEMPERATURE	TEMP 1
	SPECIFIC GRAVITY	SPGRAV 1
	ITEM 6	OPTION 6
	ITEM 7	OPTION 7
OPTIONS	ITEM 8	OPTION 8
	ITEM 9	OPTION 9

NOTES

1. NEC CLASS **NECCLASS** GROUP **NECGRP** DIV. **NECDIV**
2. VENDOR TO PROVIDE PERMANENTLY ATTACHED S.S. TAG WITH TAG NO. HEREIN.

NOTE1 NOTE6
NOTE2 NOTE7
NOTE3 NOTE8
NOTE4 NOTE9
NOTE5 NOTE10

DIAGRAM

APPROX. DIM.: DIM A

APPROVED BY OWNER	DATE	APPROVED BY ENGINEER	DATE	
APPROVED AND PRICED BY VENDOR	DATE	PRICE	MULTIPLIER	NET THIS PAGE
VENDOR PROMISED DELIVERY	RELEASED FOR PURCHASE	DATE	DATE PROMISED	PAGE

202F23.EPS

Figure 23 ◆ Instrument data sheet.

wiring diagrams and ladder diagrams can provide the interconnections or relationships between the equipment in order to further isolate the problem to a particular piece of equipment. Raceway drawings are needed should the tests and measurements indicate a problem in the distribution of the conductors. Loop sheets, installation drawings, location drawings, and bills of material are needed by instrumentation installers.

5.0.0 ◆ STANDARDIZED DESIGN METHODS

Designers must use methods that are universally understood in order to produce drawings and diagrams that can be used by craft workers worldwide. Some of the standardized tools that are applied include **symbology**, numbering systems, notes, and legends. Although some variations may exist from one designer's methods to another's, standards are generally followed. This section examines some design tools including electrical diagram symbols, instrumentation symbols, drawing number systems, notes, and legends.

5.1.0 Electrical Symbols

Every line, symbol, figure, and letter in a diagram must have a specific purpose, and the information being presented must be concise. For example, when the rating of a current transformer is given, an abbreviation such as CT is not needed because the information is implied by the symbol itself. Writing the unit of measure (amp) in this case is also unnecessary, because a current transformer is always rated in amperes. Thus, the numerical rating and the transformer symbol are sufficient.

The key to reading electrical diagrams is understanding and using the electrical legend correctly. The legend explains the abbreviations and symbols used in the diagrams, and also contains general notes and other important information.

There may be some variations between legends developed by different companies. Only the legend specifically designed for a given set of drawings may be used for those drawings.

The legend removes the need to memorize all the symbols and abbreviations presented on a diagram and can be used as a reference for unique symbols. The legend is typically found in the bottom right corner of a print or on a separate drawing.

Abbreviations are an important part of the designer's shorthand. For example, a circle may be used to symbolize a meter, relay, motor, or indicating light. The designer uses a set of standard abbreviations to make the distinction clear. *Table 1* shows the abbreviations used to distinguish what type of meter a given circle represents.

As mentioned earlier, indicating lamps may also be represented by circles. *Table 2* shows the abbreviations used to distinguish the colors of indicating lamps.

Another component commonly represented by circles is relays. *Table 3* shows the abbreviations used for relays.

Still another component commonly represented by a circle is the motor. Motors usually have the horsepower rating in or near the circle representing them. The abbreviation for horsepower is HP.

Any other piece of equipment represented by a circle will be identified in the legend or notes, or will be spelled out on the diagram.

Contacts and switches are also identified by standard abbreviations. *Table 4* lists these abbreviations.

Table 1 Meter Abbreviations

Abbreviation	Meter
A	Ammeter
AH	Ampere-hour meter
CRO	Oscilloscope
DM	Demand meter
F	Frequency meter
GD	Ground detector
OHM	Ohmmeter
OSC	Oscillograph
PF	Power factor meter
PH	Phase meter
SYN	Synchroscope
TD	Transducer
V	Voltmeter
VA	Volt-ammeter
VAR	VAR meter
VARH	VARhour meter
W	Wattmeter
WH	Watthour meter

Table 2 Lamp Color Abbreviations

Abbreviation	Lamp Color
A	Amber
B	Blue
C	Clear
G	Green
R	Red
W	White

Table 3 Relay Abbreviations

Abbreviation	Relay
CC	Closing coil
TD	Time delay relay
CR	Closing/control relay
TDE	Time delay energize
TC	Trip coil
TDD	Time delay de-energize
TR	Trip relay
X	Auxiliary relay

Table 4 Contact and Switch Abbreviations

Abbreviation	Contact/Switch
a	Breaker A contact
b	Breaker B contact
BAS	Bell alarm switch
BLPB	Back-lighted pushbutton
CS	Control switch
FS	Flow switch
LS	Limit switch
PB	Pushbutton
PS	Pressure switch
PSD	Differential-pressure switch
TDO	Time delay open
TDC	Time delay closing
TS	Temperature switch
XSH	Auxiliary switch high
XSL	Auxiliary switch low

The following figures illustrate examples of the previously discussed abbreviations and symbols.

Figure 24 shows contacts A and B in their normally de-energized state. If relay CR is de-energized, contact A is open and contact B is shut. When relay CR is energized, contact A is shut and contact B is open.

Figure 25 shows a control switch and its associated contacts. Contacts 1 through 4 open and close as a result of the operation of control switch #1 (CS1).

In the stop position, contact 2 is shut and the red indicating lamp is lighted. In the start position, contacts 3 and 4 are shut, energizing the M coil and the amber indicating lamp, respectively. When the switch handle is released, the spring returns to the run position. Contact 4 opens, de-energizing the amber lamp. It closes contact 1 to energize the green lamp.

Figure 26 defines abbreviations commonly used to designate relays and switches on prints. The symbols further illustrate descriptions of the abbreviations.

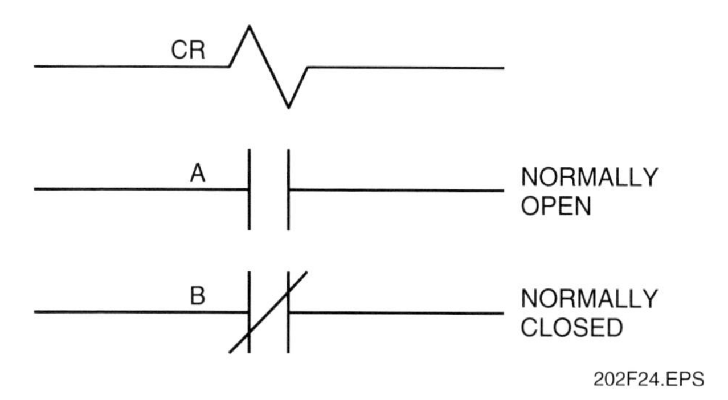

Figure 24 ◆ Contacts A and B.

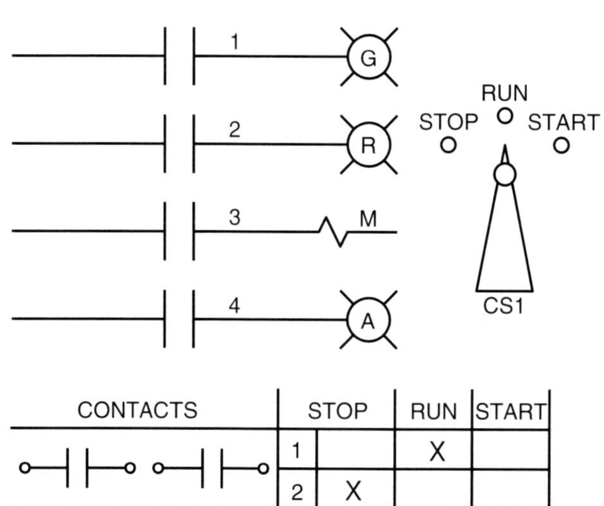

CONTACTS		STOP	RUN	START
	1		X	
	2	X		
	3		X	X
	4			X
SPRING RETURN			←	

CS1

X = CLOSED

202F25.EPS

Figure 25 ◆ Switch and contacts.

Table 5 contains a list of selected standard abbreviations commonly used on prints. This list is not inclusive. Not all engineering companies use this standard, which is based on the American National Standards Institute (ANSI) standard for abbreviations. The list is just a sampling of common abbreviations.

SPST NO		SPST NC		SPDT	
SINGLE-BREAK	DOUBLE-BREAK	SINGLE-BREAK	DOUBLE-BREAK	SINGLE-BREAK	DOUBLE-BREAK
DPST, 2 NO		**DPST, 2 NC**		**DPDT**	
SINGLE-BREAK	DOUBLE-BREAK	SINGLE-BREAK	DOUBLE-BREAK	SINGLE-BREAK	DOUBLE-BREAK

TERMS

SPST - SINGLE-POLE SINGLE-THROW
SPDT - SINGLE-POLE DOUBLE-THROW
DPST - DOUBLE-POLE SINGLE-THROW
DPDT - DOUBLE-POLE DOUBLE-THROW
NO - NORMALLY OPEN
NC - NORMALLY CLOSED

202F26.EPS

Figure 26 ◆ Supplementary contact symbols.

Table 5 ANSI Abbreviations for Use on Drawings and in Text (1 of 3)

Term	Abbreviation
A	
Absolute	ABX
Accessory	ACCESS
Accumulate	ACCUM
Actual	ACT
Adapter	ADPT
Adjust	ADJ
Advance	ADV
Aggregate	AGGR
Air Condition	AIR COND
Allowance	ALLOW
Alloy	ALY
Alternation	ALT
Alternate	ALT
Alternating Current	AC
Aluminum	AL
American National Standard	AMER NATL STD
American Wire Gage	AWG
Ampere	AMP
Amplifier	AMPL
Anneal	ANL
Appendix	APPX
Approved	APPD
Approximate	APPROX
Arc Weld	ARC/W
Area	A
Armature	ARM
Asbestos	ASB
Assembly	ASSY
Authorized	AUTH
Automatic	AUTO
Auto-Transformer	AUTO TR
Auxiliary	AUX
Average	AVG
B	
Babbitt	BAB
Back Feed	BF
Back Pressure	BP
Back to Back	B TO B
Balance	BAL
Ball Bearing	BB
Base Line	BL
Base Plate	BP
Bearing	BRG
Bench Mark	BM
Bent	BT
Between	BET
Between Centers	BC
Between Perpendiculars	BP
Bevel	BEV
Bill Material	B/M
Blank	BLK
Block	BLK
Blueprint	BP
Board	BD
Boiler	BLR
Boiler Feed	BF
Boiler Horsepower	BHP
Bolt Circle	BC
Both Faces	BF
Both Sides	BS
Both Ways	BW
Bottom	BOT
Bracket	BRKT
B	
Brake	BK
Brake Horsepower	BHP
Brass	BRS
Brazing	BRZG
Break	BRK
British Standard	BR STD
British Thermal Units	BTU
Bronze	BRZ
Building	BLDG
Bulkhead	BHD
Burnish	BRNH
Busing	BUSH
Button	BTU
C	
Cabinet	CAB
Calibrate	CAL
Cap Screw	CAP SCR
Capacity	CAP
Carburetor	CARB
Case Harden	CH
Cast Iron	CI
Cast Steel	CS
Casting	CSTG
Cement	CEM
Center	CTR
Center Line	CL
Center of Gravity	CG
Centering	CTR
Chamfer	CHAM
Change	CHG
Channel	CHAN
Check	CHK
Check Valve	CV
Circle	CIR
Circular	CIR
Circumference	CIRC
Clear	CLR
Clearance	CL
Clockwise	CW
Cold Rolled Steel	CRS
Combination	COMB
Combustion	COMB
Company	CO
Concrete	CONC
Connect	CONN
Constant	CONST
Construction	CONST
Contact	CONT
Continue	CONT
Copper	COP
Corporation	CORP
Correct	CORR
Counter	CTR
Counter Clockwise	CCW
Counterbore	CBORE
Counterdrill	CDRILL
Countersink	CSK
Coupling	CPLG
Cover	COV
Cross Section	XSECT
Cubic	CU
Cubic Foor	CU FT
Cubic Inch	CU IN.
Current	CURR

202T05A.EPS

Table 5 ANSI Abbreviations for Use on Drawings and in Text (2 of 3)

Term	Abbreviation
D	
Decimal	DEC
Degree	DEG
Department	DEPT
Design	DSGN
Detail	DET
Diagonal	DIAG
Diagram	DIAG
Diameter	DIA
Dimension	DIM
Discharge	DSCH
Division	DIV
Double	DBL
Dovetail	DVTL
Dowel	DWL
Down	DN
Drafting	DFTG
Draftsman	DFTSMN
Drawing	DWG
Drill or Drill Rod	DR
Duplicate	DUP
E	
Each	EA
East	E
Effective	EFF
Elbow	ELL
Electric	ELEC
Elementary	ELEM
Elevate	ELEV
Elevation	EL
Engine	ENG
Engineer	ENGR
Engineering	ENGRG
Equal	EQ
Equipment	EQUIP
Equivalent	EQUIV
Estimate	EST
Exhaust	EXH
Existing	EXIST
Exterior	EXT
F	
Fabricate	FAB
Fahrenheit	F
Feed	FD
Feet	(')FT
Figure	FIG
Finish	FIN
Flange	FLG
Flat	F
Flat Head	FH
Floor	FL
Fluid	FL
Foot	(')FT
Forward	FWD
Frequency	FREQ
G	
Gage or Gauge	GA
Gallon	GAL
Galvanize	GALV
Gasket	GSKT
General	GEN
Glass	GL
Governor	GOV
Ground	GRD
H	
Half-Round	1/2RD
Hardware	HDW
Head	HD
Heat	HT
Heat Treat	HT TR
Heavy	HVY
High-Pressure	HP
High-Speed	HS
Horizontal	HOR
Horsepower	HP
Hour	HR
Housing	HSG
Hydraulic	HYD
I	
Illustrate	ILLUS
Inch	(")IN.
Inches per Second	IPS
Inside Diameter	ID
Instrument	INST
Internal	INT
Irregular	IRREG
J	
Junction	JCT
K	
Key	K
Keyseat	KST
Keyway	KWY
L	
Lateral	LAT
Left	L
Left Hand	LH
Length	LG
Locate	LOC
Long	LG
Lubricate	LUB
M	
Machine	MACH
Machine Steel	MS
Maintenance	MAINT
Manual	MAN
Manufacturer	MFR
Manufactured	MFD
Manufacturing	MFG
Material	MATL
Maximum	MAX
Mechanical	MECH
Mechanism	MECH
Median	MED
Metal	MET
Meter	M
Miles	MI
Miles per Hour	MPH
Millimeter	MM
Minimum	MIN
Minute	(')MIN
Miscellaneous	MISC
Motor	MOT
Mounted	MTD
Mounting	MTG
Multiple	MULTI

202T05B.EPS

Table 5 ANSI Abbreviations for Use on Drawings and in Text (3 of 3)

Term	Abbreviation
N	
National	NATL
Negative	NEG
Neutral	NEUT
Nominal	NOM
Normal	NOR
North	N
Not to Scale	NTS
Number	NO
O	
Obsolete	OBS
On Center	OC
Opposite	OPP
Original	ORIG
Outlet	OUT
Outside Diameter	OD
P	
Packing	PKG
Part	PT
Patent	PAT
Pattern	PATT
Permanent	PERM
Perpendicular	PERP
Piece	PC
Plate	PL
Plumbing	PLMB
Point	PT
Position	POS
Potential	POT
Pound	LB
Pounds per Square Inch	PSI
Power	PWR
Prefabricated	PREFAB
Preferred	PFD
Pressure	PRESS
Process	PROC
Pushbutton	PB
Q	
Quadrant	QUAD
Quality	QUAL
Quarter	QTR
R	
Radius	R
Received	RECD
Record	REC
Rectangle	RECT
Reduce	RED
Reinforce	REINF
Release	REL
Relief	REL
Remove	REM
Require	REQ
Required	REQD
Return	RET
Reverse	REV
Revolution	REV
Revolutions per Minute	RPM
Right	R
Right Hand	RH
Roller Bearing	RB
Rough	RGH
Round	RD
S	
Schematic	SCHEM
Screw	SCR
Second	SEC
S	
Section	SECT
Separate	SEP
Set Screw	SS
Shaft	SFT
Sheet	SH
Side	S
Single	S
Sleeve	SLV
Slide	SL
Slotted	SLOT
Socket	SOC
Space	SP
Special	SPL
Specific	SP
Spring	SPG
Square	SQ
Standard	STD
Steel	STL
Stock	STK
Straight	STR
Structural	STR
Substitute	SUB
Support	SUP
Surface	SUR
Symbol	SYM
System	SYS
T	
Taper	TPR
Technical	TECH
Template	TEMP
Tension	TENS
Terminal	TERM
Thick	THK
Thousand	M
Thread	THD
Threads per inch	TPI
Through	THRU
Tolerance	TOL
Transfer	TRANS
Typical	TYP
U	
Ultimate	ULT
Unit	U
Universal	UNIV
V	
Vacuum	VAC
Valve	V
Variable	VAR
Versus	VS
Vertical	VERT
Volt	V
Volume	VOL
W	
Washer	WASH
Watt	W
Weight	WT
West	W
Width	W
Working Pressure	WP
X, Y, Z	
Yard	YD

202T05C.EPS

The following figures show symbols commonly used on electrical diagrams. The symbols used throughout the industry are typically based on ANSI standard symbols. *Figure 27* illustrates some of the ANSI symbols for electrical diagrams. *Figure 28* shows some standard wiring diagram symbols. *Figure 29* gives some standard electronic symbols. *Figure 30* shows examples of standard electronic logic symbols. *Figure 31* shows electrical drawing symbols with corresponding physical depictions.

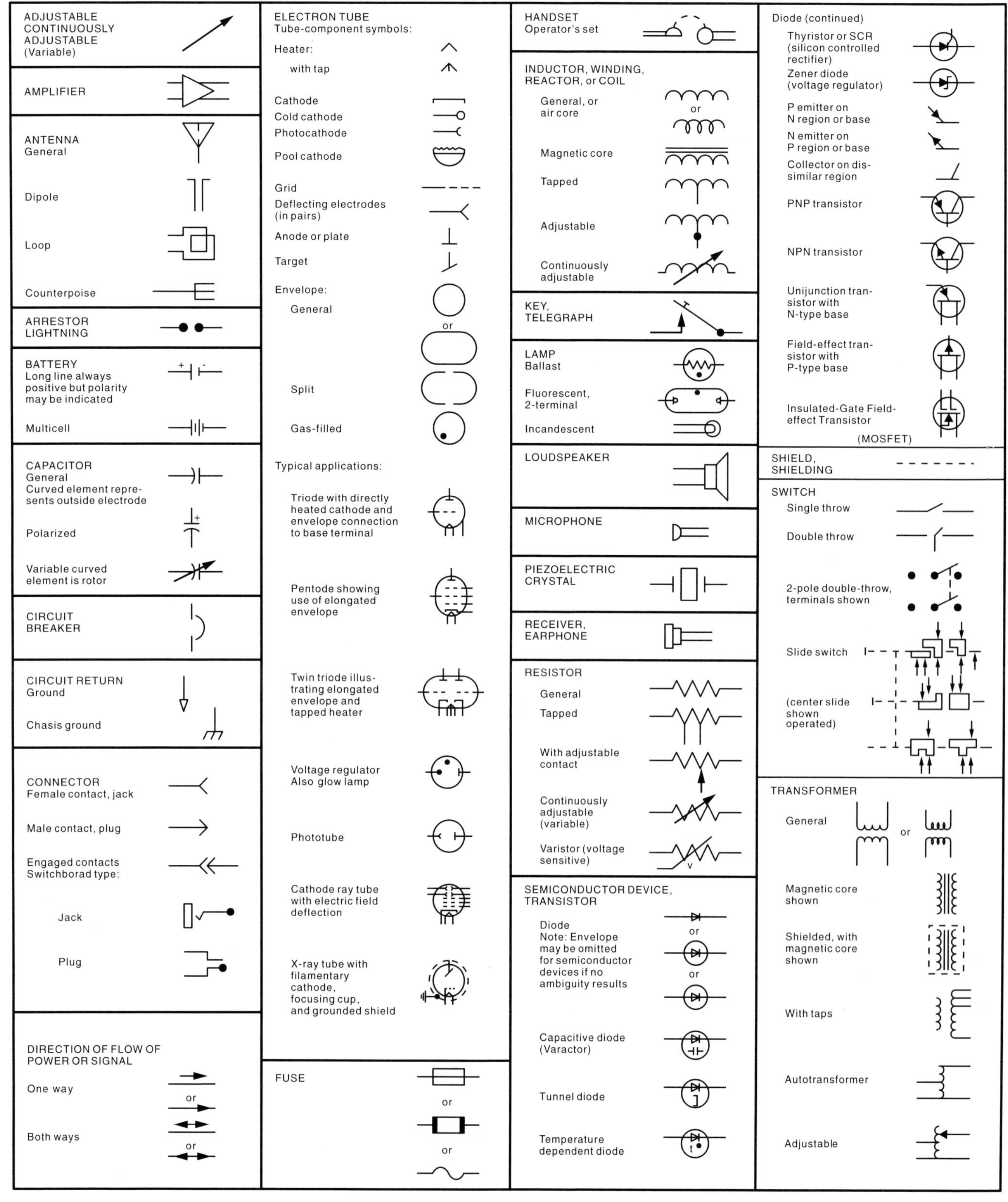

Figure 27 ◆ ANSI graphical symbols for electrical diagrams.

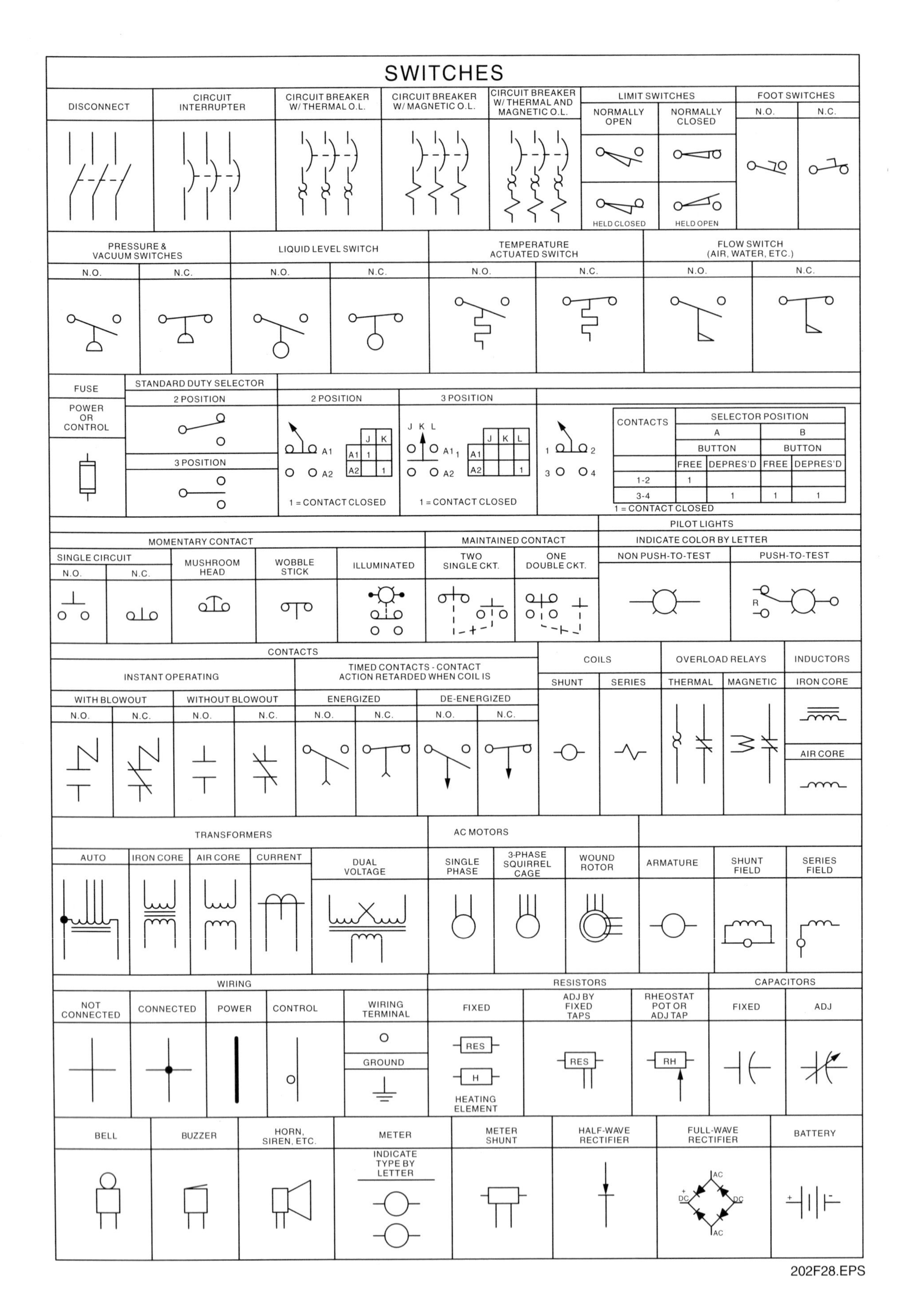

Figure 28 ◆ Standard wiring diagram symbols.

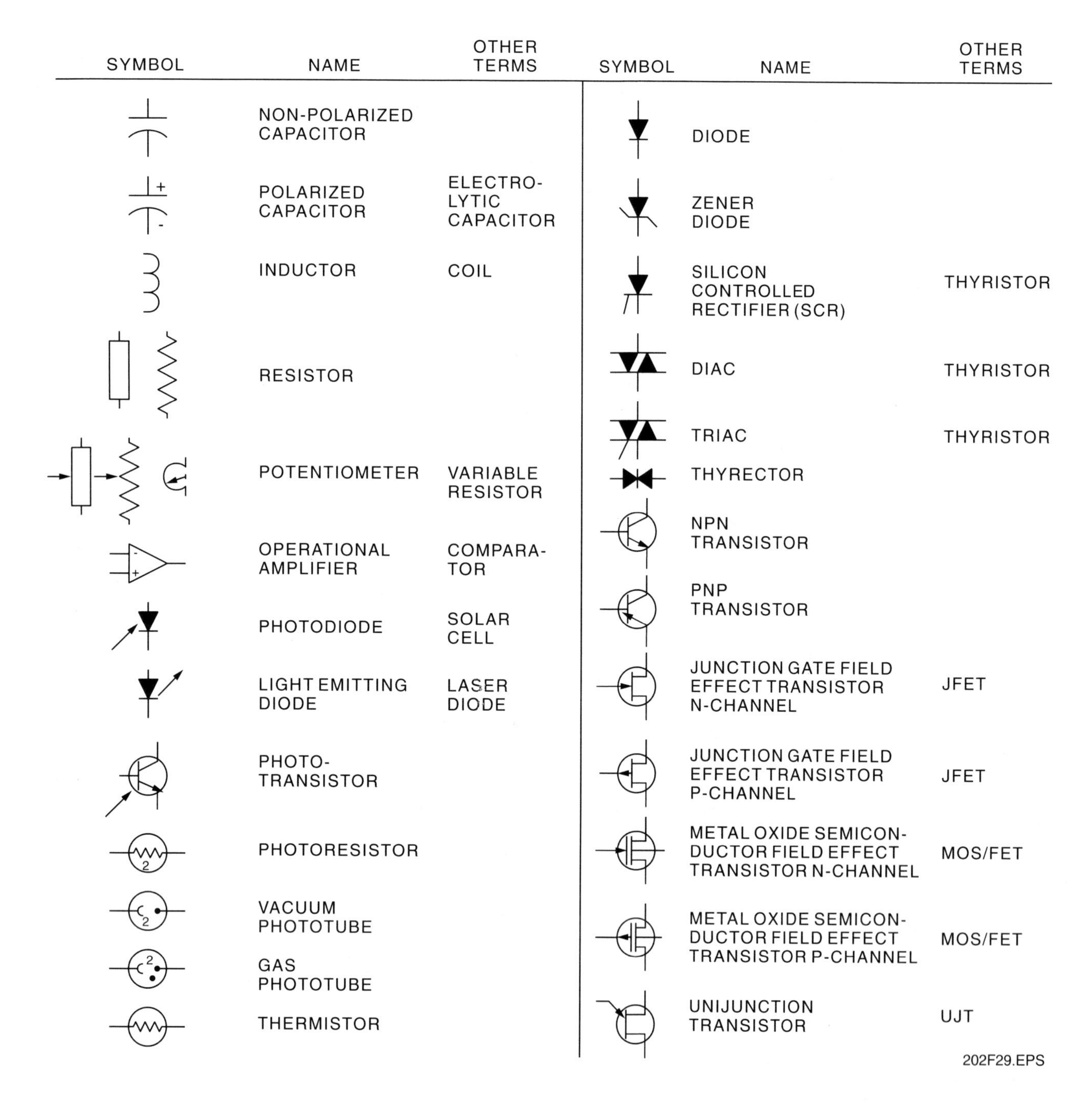

Figure 29 ◆ Standard electronic symbols.

5.2.0 Instrumentation Symbols

Instrumentation diagrams can be used to show a variety of processes, including petrochemical, food, utilities, and mill processes. Regardless of the process shown by the diagram, most instrumentation diagrams use a standard format, such as the one developed by the ISA. The development of such standards is based on the idea that symbols are the language of instrumentation. Understanding and interpreting symbols is easier when everyone speaks the same language. Being able to correctly interpret the symbols and general organization for one drawing makes it possible to interpret most drawings without significant difficulty. Because instrumentation drawings provide an overall view of the process and its instrumentation, they should be considered an important tool in instrumentation. When used properly, instrumentation diagrams make it easier to monitor processes and do routine work. Diagrams can also help save time when it is necessary to develop troubleshooting strategies, because the diagrams illustrate how one component can affect

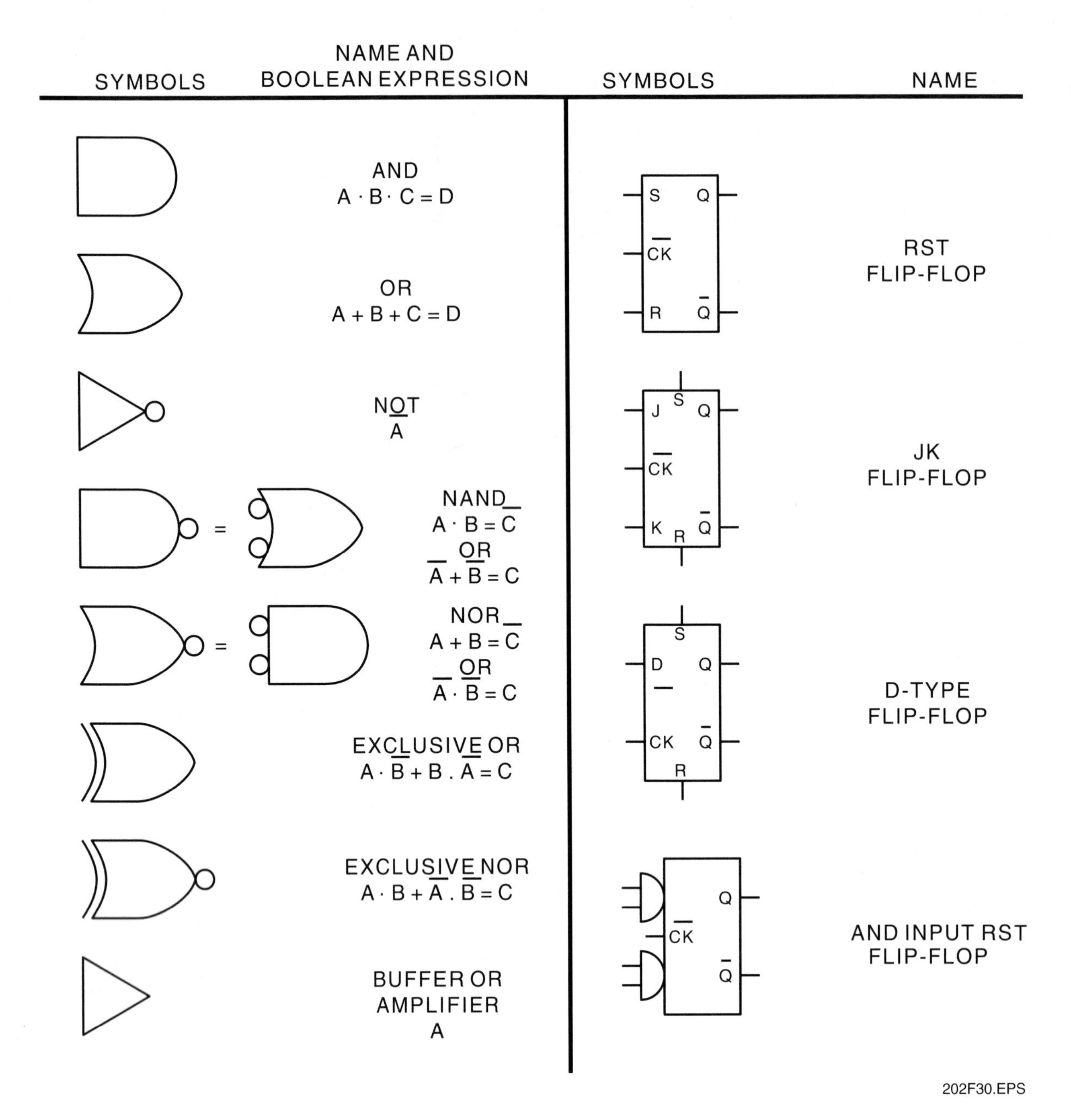

Figure 30 ◆ Standard electronic logic symbols.

others in the system. Instrumentation diagrams also provide a means to determine the safety precautions that must be observed before beginning work on a system.

Instrument symbols, such as circles, letters, numbers, and lines, are used to provide information about the process. Symbols may represent the devices in the system, identify the function of instruments, or indicate how devices are connected to each other or to the process. The appendix in this module contains standardized symbols taken from the ISA's *Instrumentation Symbols and Identification*. Refer to this book to develop a greater understanding of all the symbols associated with instrumentation work.

5.3.0 Number Systems

A drawing numbering system can be used on large or long-term projects that are divided into different areas for special work assignments and drawing identification. In the system used here, all drawings are assigned a six-figure drawing number/letter combination. *Figure 32* shows a drawing number in a title block.

A drawing number provides the following information:

- *Building/area identification* – This number is assigned by the project engineer or design engineering manager. It can also be assigned by the client. The number contains two digits from 01 to 99.

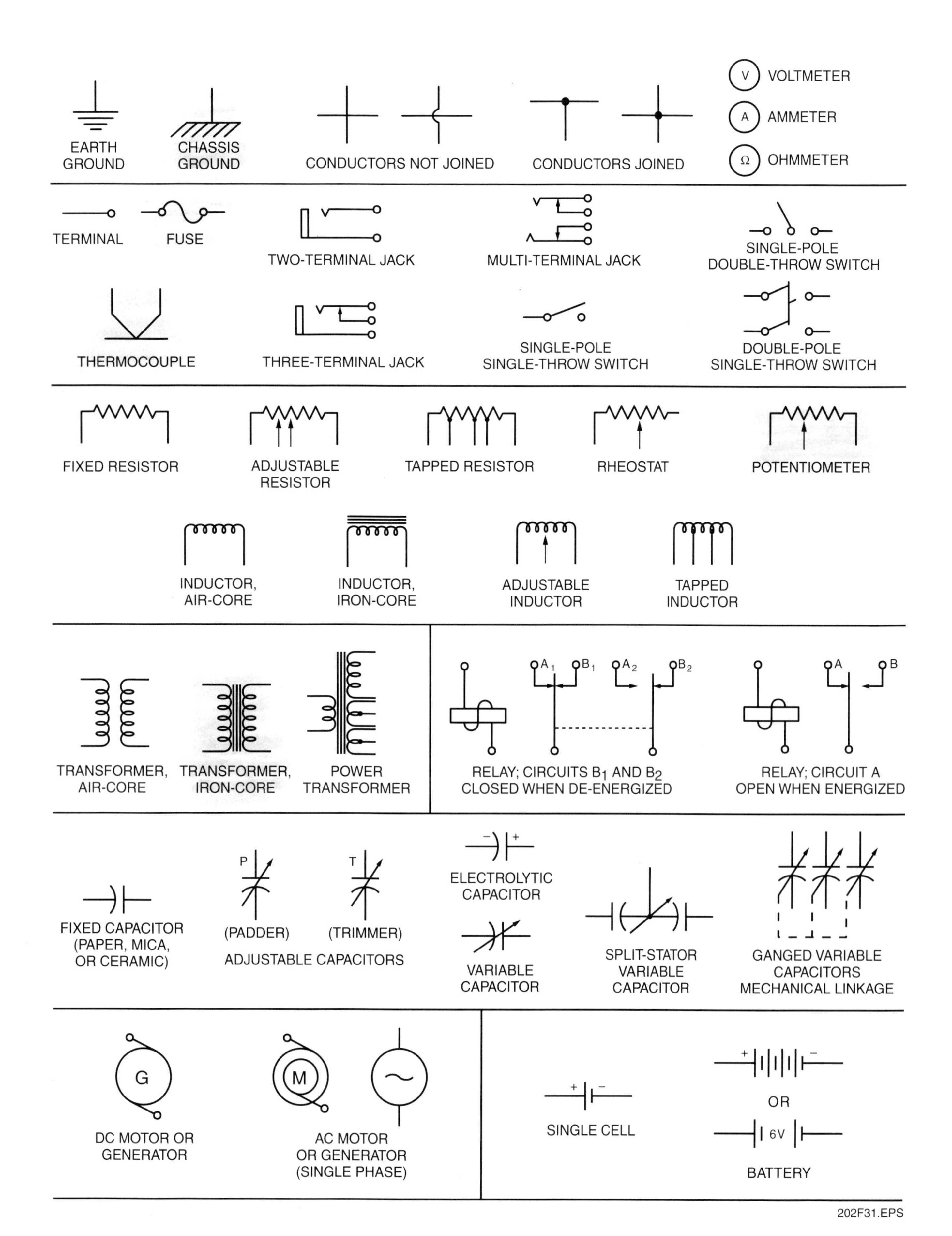

Figure 31 ◆ Electrical drawing symbols.

XY & Z | PROJECT | SCALE 1/8" = 1' - 0"
ENGINEERING | TITLE | DWG # 04-I201
SCALE 201 | 1

202F32.EPS

Figure 32 ◆ Drawing number in a title block.

- *Department identification* – Each department in an engineering group is assigned a department letter. These letters are shown in *Table 6*.
- *Drawing classification* – In each engineering department, drawings are given classification numbers. Each number, from 1 to 9, refers to a different type of drawing. The drawing classifications of a typical instrument department are shown in *Table 7*.
- *Drawing serial number* – This double-digit number is assigned in numerical order, from 01-99, to each drawing in a set of drawings.

Figure 33 shows how drawing number 04-I201 is interpreted. The numbers in drawing number 04-I201 reflect the following information:

- *04* – This is the building/area identification number. It refers to a specific building or an area within a building.
- *I* – This is the letter noting the department responsible for the drawing. *I* identifies the instrumentation department.
- *2* – This is the drawing classification number assigned by the instrument department for an instrumentation P&ID.
- *01* – This is the drawing serial number. The number *01* shows that it is the first drawing in the set of drawings.

5.4.0 Notes

The notes on P&IDs contain information about the drawing. *Figure 34* contains an example of notes from a P&ID. These notes cover symbol explanation, planned future additions to the system, and information on an interlock. The notes contain specific information on the installation that is not shown by the symbols. It is very important to refer to the notes on P&IDs during installation or maintenance of a system.

5.5.0 Graphic Styles

Prints use various graphic styles. The most common graphic styles are diagrams and drawings.

A diagram uses symbols instead of pictures to represent components in a system. In a diagram, there is usually no scale and therefore no correlation to where an item is located in comparison to others on the print.

Table 6 Department Identifications

Abbreviation	Department
C	Civil
A	Architectural
S	Structural
M	Mechanical
F	Fire Protection
P	Piping
E	Electrical
D	Process Design
Q	Equipment
I	Instrumentation
W	Environmental
H	HVAC

Table 7 Drawing Classifications

Number	Type of Drawing
1	Instrumentation loop sheets
2	Instrumentation P&IDs
3	Instrumentation panel arrangement
4	Instrumentation panel tubing diagrams
5	Instrumentation panel wiring diagrams
6	Instrumentation graphic display
7	Installation details
8	Logic diagrams
9	Instrumentation location and routing

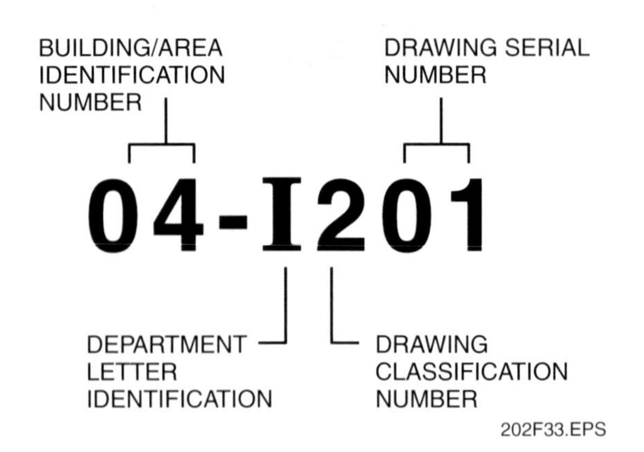

Figure 33 ◆ Drawing number 04-I201.

A drawing is an illustration showing how something actually looks. The two most common types of drawings are the **orthographic drawing**, otherwise known as the multiview projection, and the **isometric drawing**.

The first of these two, the multiview projection, is a means of representing the exact shape of an object in two or more views on planes that are generally at right angles to each other.

NOTES:

1. FOR GENERAL NOTES & SYMBOLS SEE DWG. 01-I201.
2. NOZZLES FOR FUTURE "CLEAN IN PLACE."
3. NOZZLES FOR pH ADJUSTMENT - FLANGE WITH PLUG.
4. STRAINER TO BE DUPLEX STRAINER.
5. VALVE NEEDS TO BE INTERLOCKED WITH LATEX TRANSFER PUMP SO THAT ONLY ONE STORAGE TANK IS FEEDING THE PUMP AT ANY TIME.
6. THIS SPACE AND CONNECTIONS MADE AVAILABLE FOR ADDING TWO (2) MORE LIQUID LATEX STORAGE TANKS AT A FUTURE DATE.
7. 6" SITE GLASS TO BE MOUNTED IN TOP OF TANK.
8. PROVIDE LOW POINT DRAINS.

202F34.EPS

Figure 34 ◆ Notes from sample P&ID.

The easiest way of explaining this concept is using the glass box principle. If planes of projection are placed parallel to the principal faces of the object, they form a glass box, as shown in *Figure 35(A)*. Notice that the observer is always on the outside looking in, so that the object is seen through the planes of projection. Because the glass box has six sides, six views of the object are obtained. Note that the object has three principal dimensions: width, height, and depth.

Because the views of a solid or three-dimensional object must be shown on a flat sheet of paper, the plans must be unfolded so that they will all lie in the same plane, as shown in *Figure 35(B)*. All planes except the rear plane are hinged on the front plane. The rear plane in this case is hinged to the left-side plane. Each plane revolves outward from the original box position until it lies in the front plane, which remains stationary.

The positions of these six planes, after they have been revolved, are shown in *Figure 35(C)*. Carefully identify each of these planes and the corresponding views with its original position in the glass box. Repeat this mental procedure, if necessary, until the revolutions are thoroughly understood. Notice that the front, top, and bottom views all line up vertically and are the same width.

An isometric, or pictorial, view is a three-dimensional drawing of an object drawn at 30-degree angles. *Figure 36* shows an isometric view of a cube. Note that no hidden lines are shown.

The steps in making an isometric drawing of an object composed only of normal surfaces are illustrated in *Figure 37*. Notice that all measurements are made parallel to the main edges of the enclosing box. They are parallel to the isometric axes. No measurement along a diagonal (non-isometric) line on any surface or through the object can be set off directly with the scale. The object may be drawn in the same position by beginning at corner Y, or any other corner, instead of at corner X. An example of a piping isometric drawing is shown in *Figure 38*.

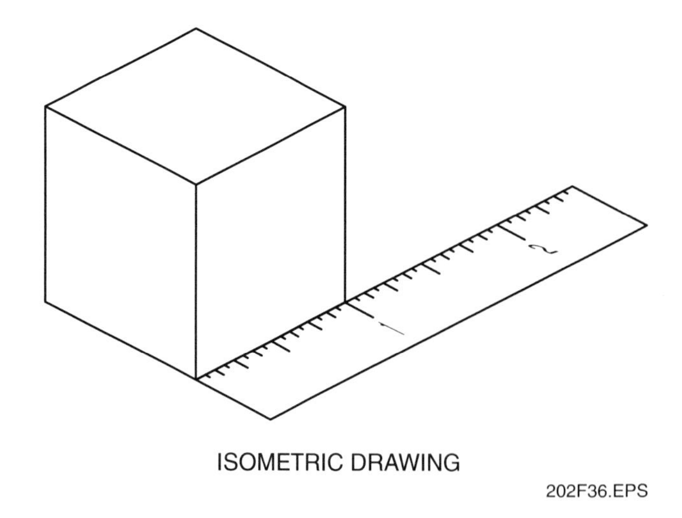

Figure 36 ◆ Isometric view of a cube.

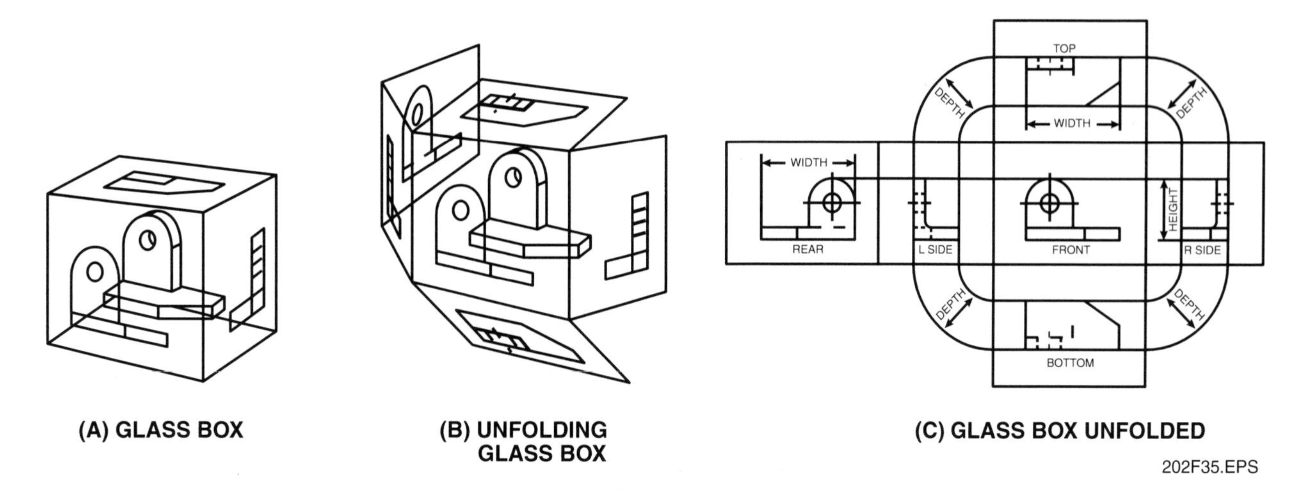

Figure 35 ◆ Glass box.

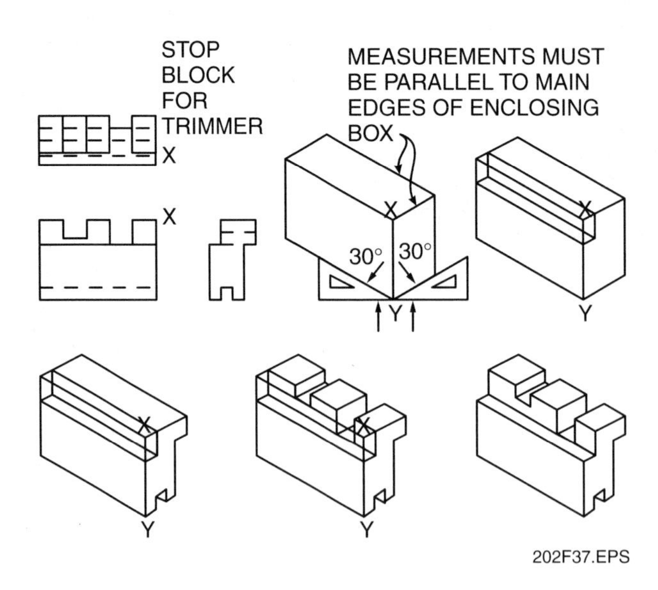

Figure 37 ◆ Making an isometric drawing.

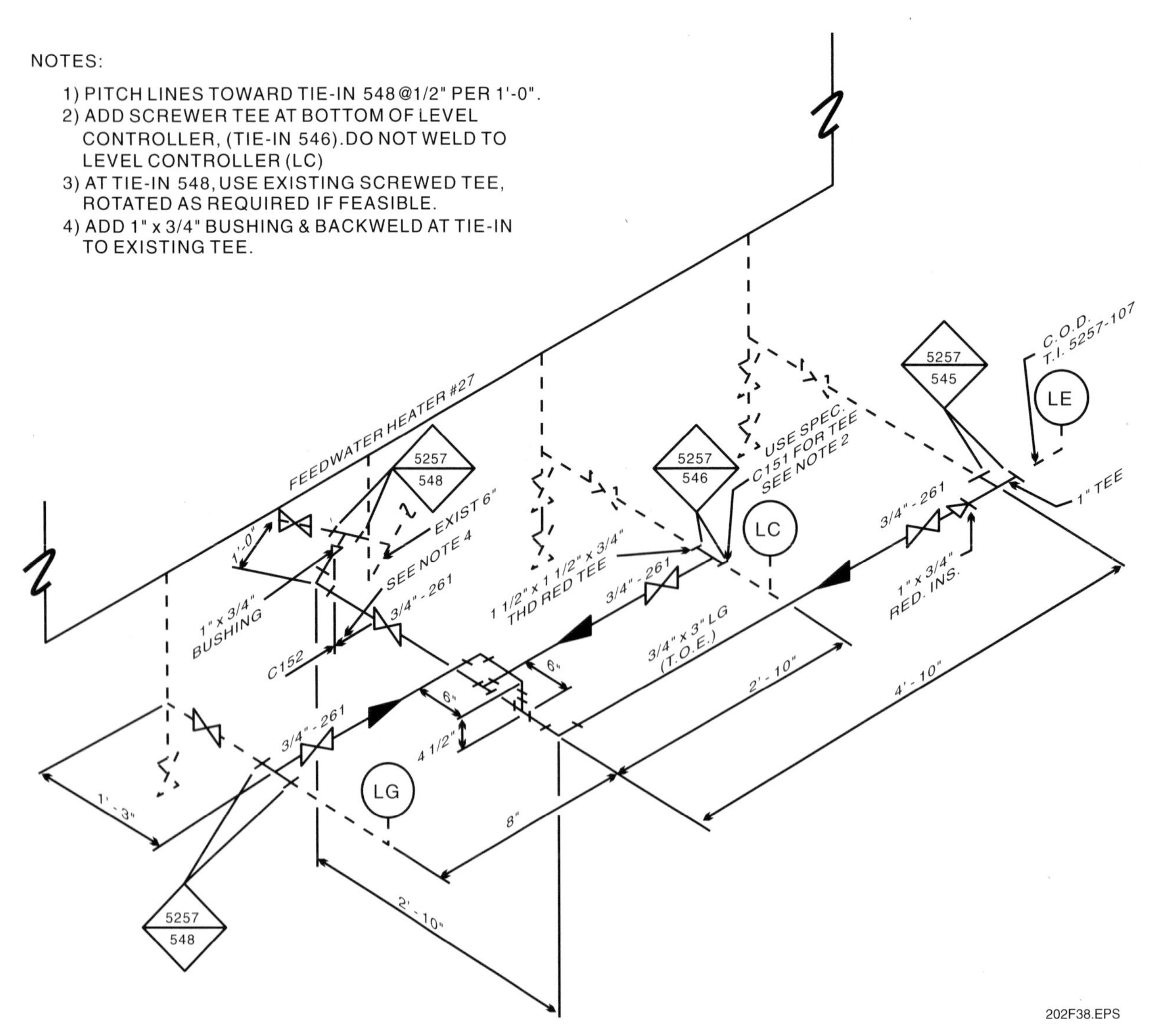

Figure 38 ◆ Piping isometric drawing.

Summary

This module discussed how drawings and diagrams provide a universally accepted language for communicating information about mechanisms, systems, and processes. Through the use of symbols and notes, a large amount of information that might require many pages of written description can be presented in a condensed form on one diagram. Also introduced were the purpose of each type of instrumentation print, the types of prints available, and the basics of print reading.

The key to reading and interpreting prints and diagrams is to understand and use the legends and symbols. The legend shows the symbols used in the diagrams, and also contains general notes and other important information. The text listed abbreviations and symbols used on drawings and in text that are approved by the American National Standards Institute (ANSI).

Review Questions

1. The lines between the blocks on a block diagram represent the _____ between the components.
   a. difference
   b. connections
   c. logic
   d. distance

2. The _____ diagram usually shows all the connections of individual conductors in the system.
   a. three-line
   b. one-line
   c. block
   d. location

3. One-line and three-line electrical wiring diagrams may contain notes, descriptions, _____, and NEMA sizes for the equipment listed.
   a. processes
   b. installation methods
   c. operating instructions
   d. load capacities

4. True or False? In a baseline-type wiring diagram, every conductor is shown as a separate line.
   a. True
   b. False

5. In a raceway diagram, the last two digits of the identification number indicate the _____.
   a. length of the raceway
   b. width of the raceway
   c. location of conduit drops
   d. height of the raceway

6. The letters within the bubble on a bubble diagram (P&ID) represent the type of _____.
   a. instrument supply
   b. process
   c. housing
   d. primary element

7. Information that may be found on a _____ includes pipe sizes, flange sizes, valve sizes, flow direction, and references to other drawings or diagrams.
   a. P&ID
   b. data sheet
   c. loop sheet
   d. location drawing

8. In addition to specific information about each instrument contained in the loop, the _____ often indicates a location in the plant or a building designation.
   a. instrument data sheet
   b. loop identification tag number
   c. legend
   d. calibration sheet

9. On a bubble diagram, a double horizontal line within a bubble indicates that the instrument is _____.
   a. field mounted
   b. panel mounted
   c. at an auxiliary location
   d. to be installed at a later date

10. Instruments that are usually located near the point of measurement or near a final control element are referred to as _____.
    a. panel mounted
    b. auxiliary mounted
    c. field mounted
    d. board mounted

11. Which of the following instrumentation documents can be used by the installation contractor to help determine the percentage of completion of the total project?
    a. Loop sheet
    b. P&ID
    c. Bill of materials
    d. Field checklist

12. In instrumentation applications that incorporate a programmable logic controller, the sequence of operation is often shown on a series of _____.
    a. location drawings
    b. ladder diagrams
    c. loop diagrams
    d. P&IDs

13. True or False? Layout drawings may include the interconnections of raceway systems and cable trays.
    a. True
    b. False

14. In a set of installation drawings, the _____ gives a listing of required fitting and connection descriptions, sizes, quantities, and the material they are made of.
    a. bill of materials
    b. loop sheet
    c. location drawing
    d. ladder diagram

15. Information pertaining to the National Electrical Code's (NEC) hazardous location classification for an instrument to be installed on a project may be found on the _____ sheet.
    a. loop
    b. field check
    c. instrument data
    d. flow

16. True or False? It is uncommon to require the information found on all the diagrams for a single installation.
    a. True
    b. False

17. True or False? Instrumentation and electrical diagrams are easy to read because all designers are required to use the same set of standard symbols.
    a. True
    b. False

Refer to *Figure 1* to answer Question 18.

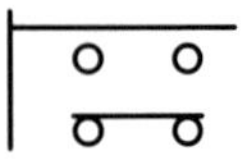

FIGURE 1

18. The symbol shown in *Figure 1* represents a _____ switch.
    a. SPST single break
    b. SPDT double-break
    c. DPDT double-break
    d. DPDT single-break

Refer to *Figure 2* to answer Question 19.

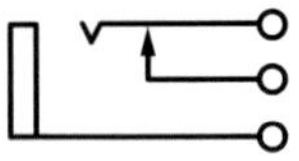

FIGURE 2

19. True or False? The symbol shown in *Figure 2* represents a two-terminal jack.
    a. True
    b. False

20. A multiview projection is also known as a(n) _____ drawing.
    a. orthographic
    b. location
    c. block
    d. non-dimensional

GLOSSARY

Trade Terms Introduced in This Module

***American National Standards Institute (ANSI)*:** An organization of American industry groups that works with the standards committees of other nations to develop standards to facilitate international trade and telecommunications.

***Cable bundle*:** An installation method in which two or more electrical cables are bound together using various forms of binding in order to provide a more orderly installation. The completed bundle is often tagged with a unique number to help identify it and the cables it holds.

***Flow diagram*:** Contains specific information on the manufacturing process, process operating characteristics, and process operating requirements.

***Installation drawing*:** Shows the proper method for mounting an instrument and connecting the instrument to the process.

***Instrument data sheet*:** Lists all of the specific technical information that applies to an instrument.

***Instrumentation, Systems, and Automation Society (ISA)*:** Globally recognized as a standards organization, the ISA develops consensus standards for factory automation, power plants, computer technology, telemetry, and communications. Established in 1945 and accredited by the American National Standards Institute (ANSI), ISA has published more than 125 standards, recommended practices, and technical reports.

***Isometric drawing*:** A drawing that shows three sides of an object and is drawn with the object at 30-degree angles.

***Legend*:** Shows the symbols used in a diagram and contains general notes and other important information.

***Load capacities*:** As used in electrical terms, this defines the maximum currents the equipment is designed to safely tolerate without potential damage to the equipment.

***Location drawing*:** A drawing that shows the position of process instruments relative to the physical location of equipment.

***Loop sheet*:** Provides specific information about an instrument loop and contains the loop identification number, a location and connection section, a calibration and specification section, and a field checklist.

***Multi-point function*:** An element in an instrumentation loop that has more than one function, or that serves or affects more than one component in a system.

***National Electrical Manufacturers Association (NEMA)*:** An organization that sets standards in the United States for electrical equipment, including circuit breaker boxes, wiring, and electrical connectors.

***Orthographic drawing*:** A drawing that represents the exact shape of an object in two or more views on planes that are generally at right angles to each other.

***Piping and instrumentation drawing (P&ID)*:** Shows the functional layout of a fluid system, including the equipment, valves, piping, and instrumentation.

Process mounted: Describes the method of installation in which a device is installed directly onto or into the process being controlled or monitored.

***Raceway drawing*:** Shows the type, identification number, size, elevation, and physical layout of raceways in an installation.

***Specification sheet*:** A listing of equipment, components, materials, and other elements on a project. It specifies properties that they must have in order to satisfy the requirements of the project.

***Symbology*:** The practice of using standardized symbols to represent equipment, elements, components, lines, and other items on a drawing or diagram.

***Tag number*:** A unique number or combination of numbers and letters assigned to a specific piece of equipment or component in a system.

***Three-phase systems*:** Electrical systems that are supplied by three separate power conductors or wires called phase conductors. A majority of industrial motors are powered by three-phase systems.

***White space*:** The areas on a drawing or diagram that contain only paper space without any information. White space often provides room for self-drawn notes.

***Wiring diagrams*:** Shows the relative position of various components of the equipment and how each conductor is connected in the circuit.

APPENDIX

Standardized Symbols

GENERAL INSTRUMENT OR FUNCTION SYMBOLS

	PRIMARY LOCATION ***NORMALLY ACCESSIBLE TO OPERATOR	FIELD MOUNTED	AUXILIARY LOCATION ***NORMALLY ACCESSIBLE TO OPERATOR
DISCRETE INSTRUMENTS	1 * IP1** 	2 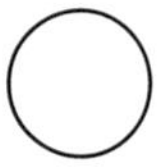	3 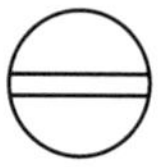
SHARED DISPLAY, SHARED CONTROL	4 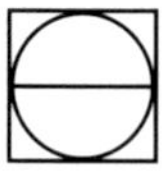	5 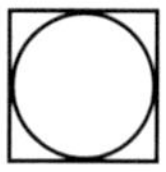	6 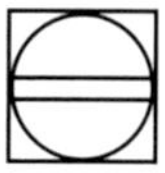
COMPUTER FUNCTION	7 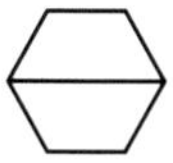	8 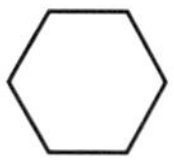	9 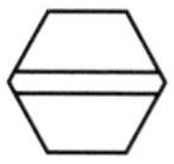
PROGRAMMABLE LOGIC CONTROL	10 	11 	12 

* Symbol size may vary according to the user's needs and the type of document. A suggested square and circle size for large diagrams is shown above. Consistency is recommended.

** Abbreviations of the user's choice, such as IP1 (instrument panel #1), IC2 (instrument console #2), CC3 (computer console #3), may be used when it is necessary to specify instrument or function location.

*** Normally inaccessible or behind-the-panel devices or functions may be depicted by using the same symbol but with dashed horizontal bars, such as:

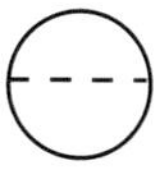

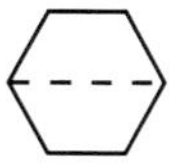

202A01.EPS

GENERAL INSTRUMENT OR FUNCTION SYMBOLS (CONT.)

13	14	15
	6TE 2584-23 INSTRUMENT WITH LONG TAG NUMBER	INSTRUMENTS SHARING COMMON HOUSING*
16 PILOT LIGHT	17 C 12 PANEL-MOUNTED PATCHBOARD POINT 12	18 ** P PURGE OR FLUSHING DEVICE
19 ** R REST FOR LATCH-TYPE ACTUATOR	20 DIAPHRAGM SEAL	21 ** *** I UNDEFINED INTERLOCK LOGIC

* It is not mandatory to show a common housing.

** These diamonds are approximately half the size of the larger ones.

*** For specific logic symbols, see *ANSI/ISA Standard S5.2.*

202A02.EPS

ACTUATOR SYMBOLS

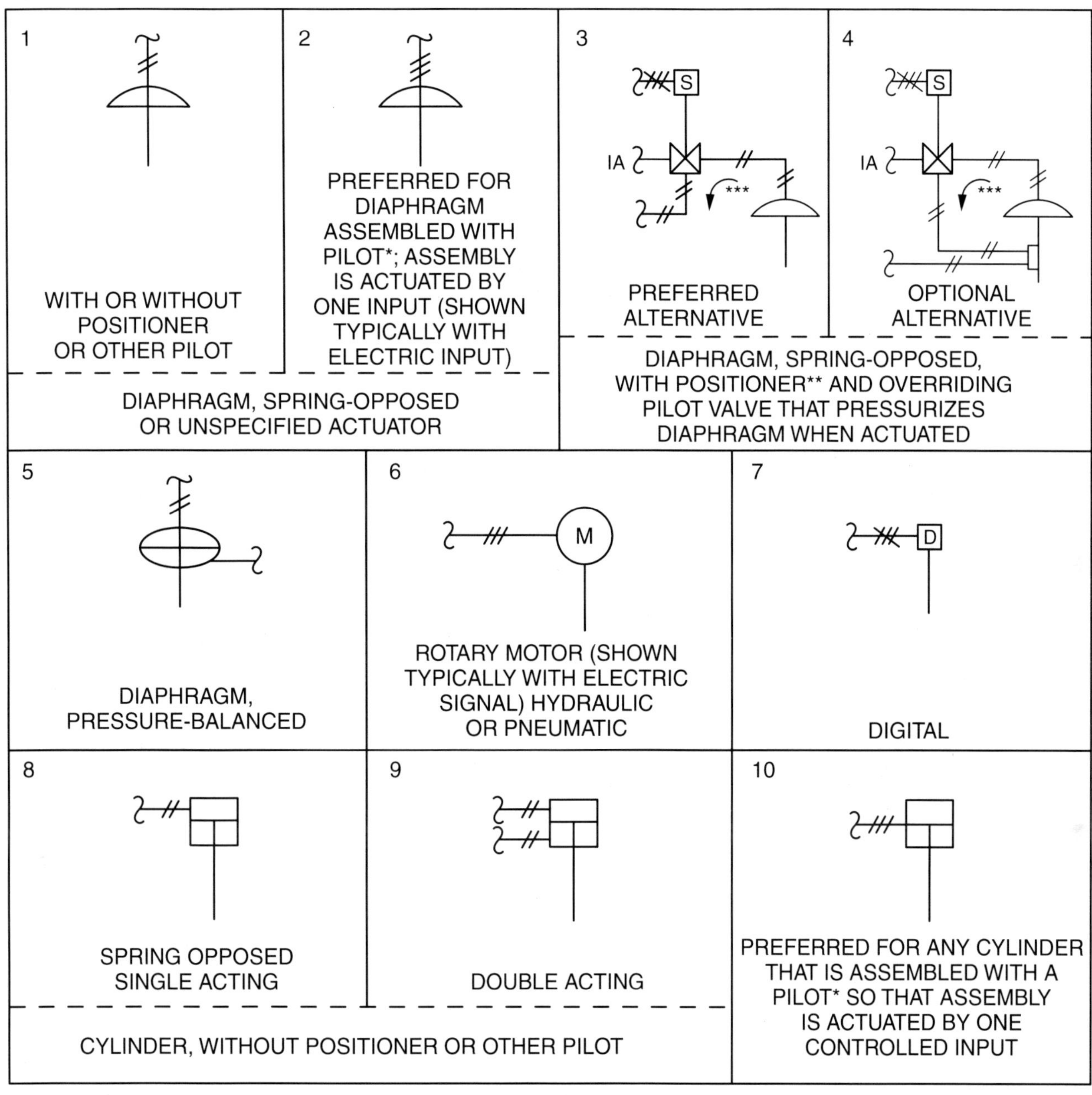

* Pilot may be positioner, solenoid valve, signal converter, or other device.

** The positioner need not be shown unless an intermediate device is on its output. The positioner tagging, ZC, need not be used even if the positioner is shown. The positioner symbol, a box drawn on the actuator shaft, is the same for all types of actuators. When the symbol is used, the type of instrument signal, i.e. pneumatic, electric, etc., is drawn as appropriate. If the positioner symbol is used and there is no intermediate device on its output, then the positioner output signal need not be shown.

*** The arrow represents the path from a common to a fail-open port. It does not correspond necessarily to the direction of fluid flow.

202A03.EPS

ACTUATOR SYMBOLS (CONT.)

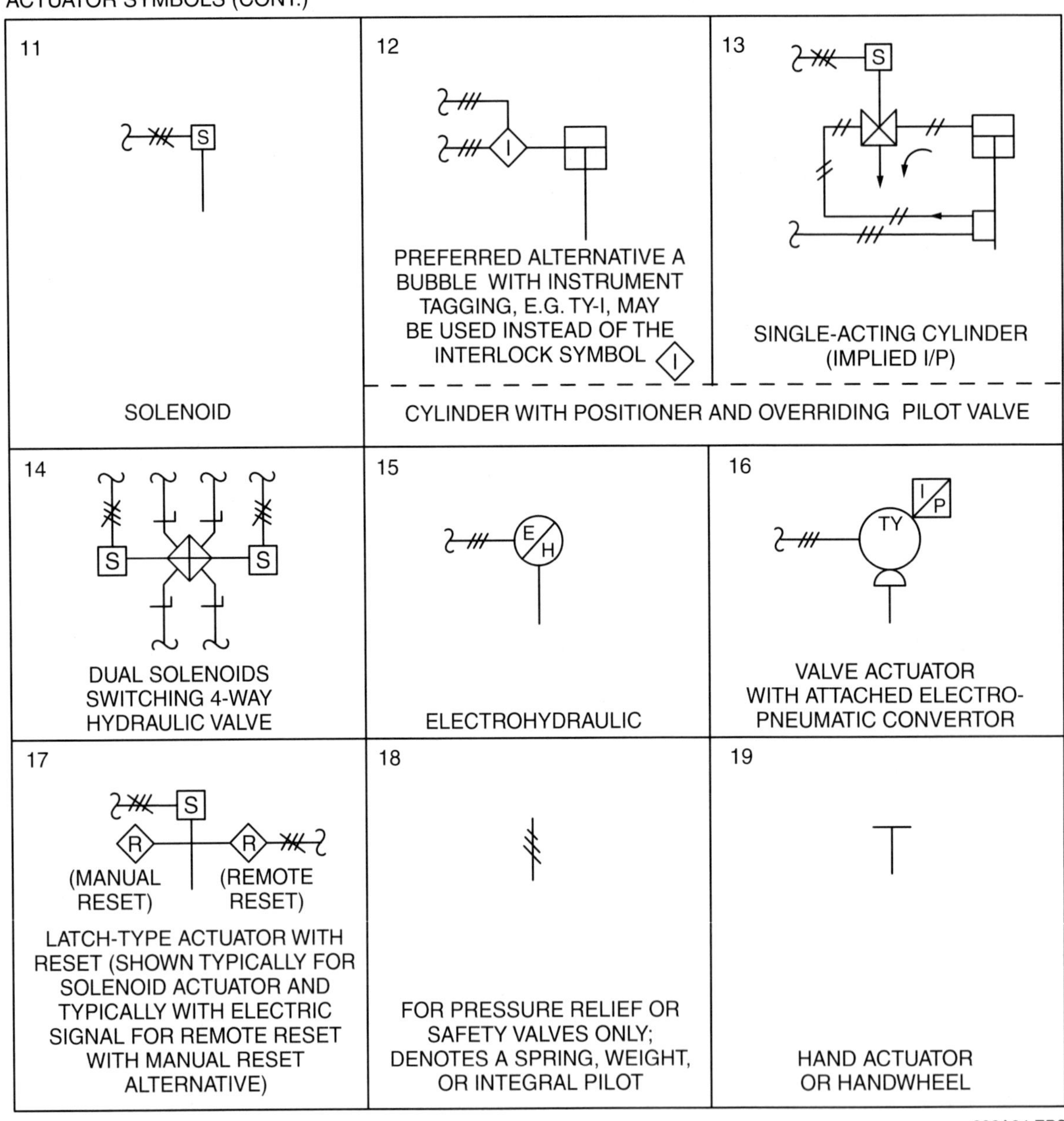

202A04.EPS

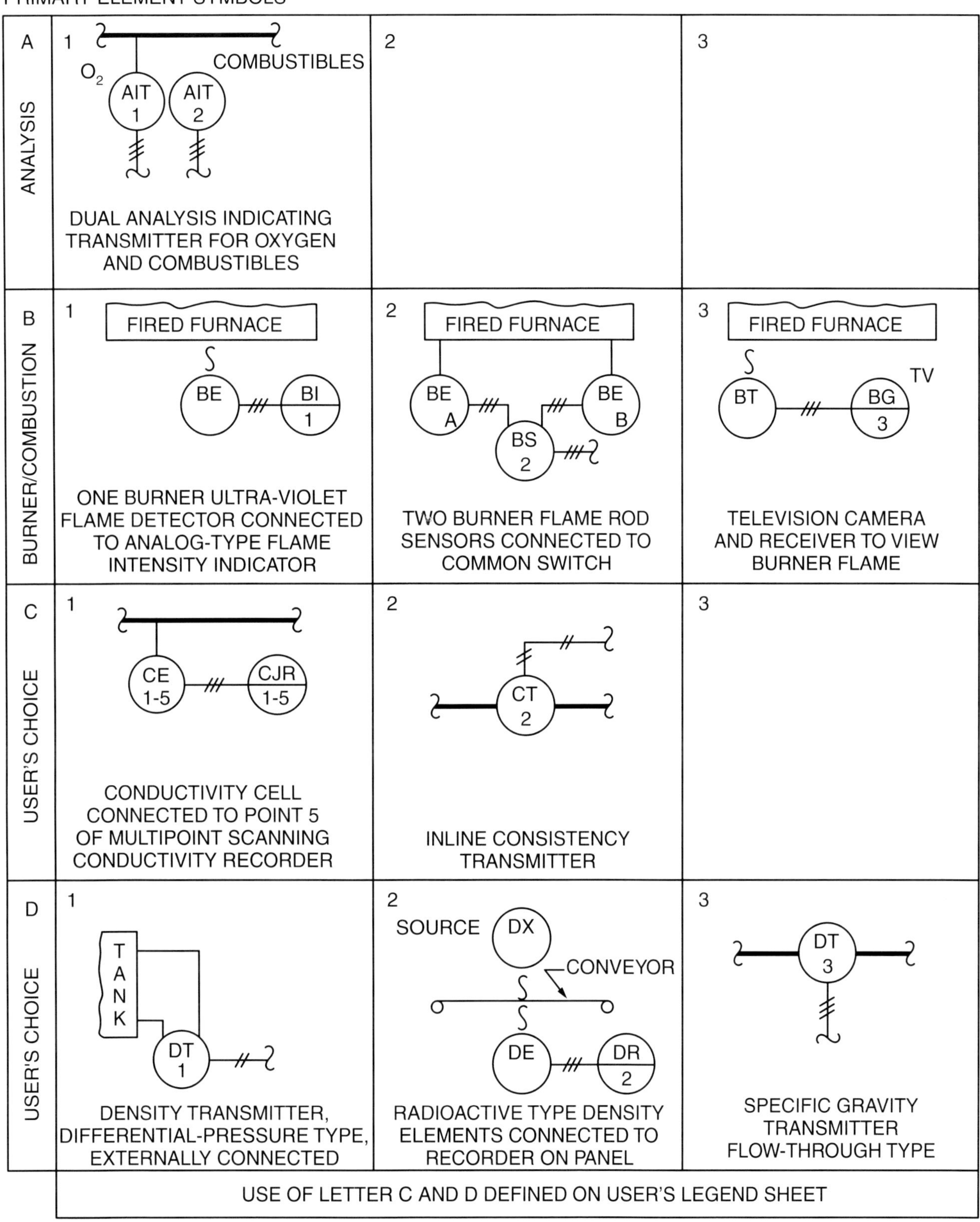
PRIMARY ELEMENT SYMBOLS
A
ANALYSIS
1
COMBUSTIBLES
O2
AIT 1
AIT 2
DUAL ANALYSIS INDICATING TRANSMITTER FOR OXYGEN AND COMBUSTIBLES
2
3
B
BURNER/COMBUSTION
1
FIRED FURNACE
BE
BI 1
ONE BURNER ULTRA-VIOLET FLAME DETECTOR CONNECTED TO ANALOG-TYPE FLAME INTENSITY INDICATOR
2
FIRED FURNACE
BE A
BE B
BS 2
TWO BURNER FLAME ROD SENSORS CONNECTED TO COMMON SWITCH
3
FIRED FURNACE
TV
BT
BG 3
TELEVISION CAMERA AND RECEIVER TO VIEW BURNER FLAME
C
USER'S CHOICE
1
CE 1-5
CJR 1-5
CONDUCTIVITY CELL CONNECTED TO POINT 5 OF MULTIPOINT SCANNING CONDUCTIVITY RECORDER
2
CT 2
INLINE CONSISTENCY TRANSMITTER
3
D
USER'S CHOICE
1
TANK
DT 1
DENSITY TRANSMITTER, DIFFERENTIAL-PRESSURE TYPE, EXTERNALLY CONNECTED
2
SOURCE
DX
CONVEYOR
DE
DR 2
RADIOACTIVE TYPE DENSITY ELEMENTS CONNECTED TO RECORDER ON PANEL
3
DT 3
SPECIFIC GRAVITY TRANSMITTER FLOW-THROUGH TYPE
USE OF LETTER C AND D DEFINED ON USER'S LEGEND SHEET
202A05.EPS

PRIMARY ELEMENT SYMBOLS (CONT.)

F FLOW RATE (CONT.)	22 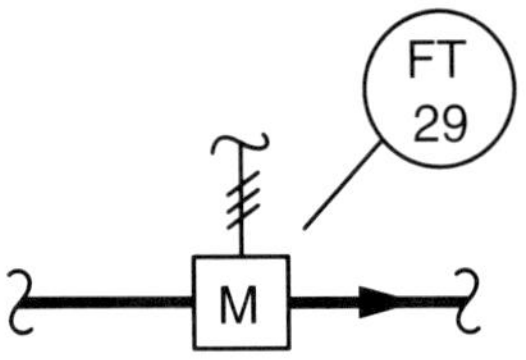 FT 29 M MAGNETIC FLOWMETER WITH INTEGRAL TRANSMITTER	23 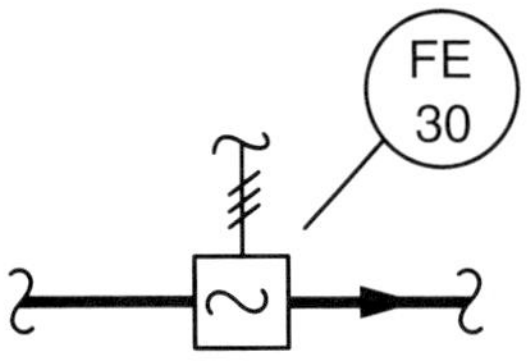 FE 30 SONIC FLOWMETER DOPPLER OR TRANSIT TIME MAY BE ADDED	24
I CURRENT	1 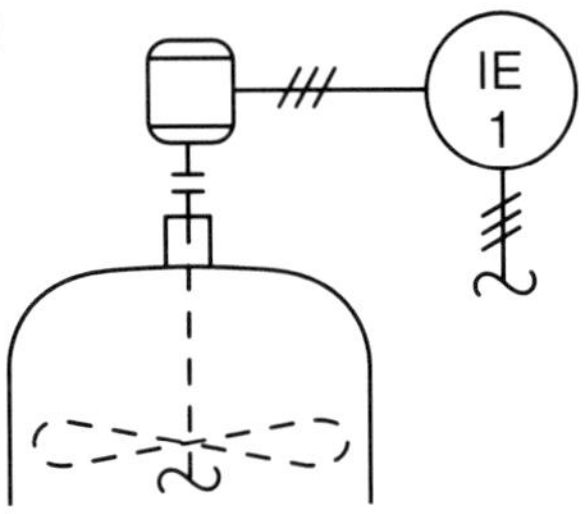 IE 1 CURRENT TRANSFORMER MEASURING CURRENT OF ELECTRIC MOTOR	2	3
J POWER	1 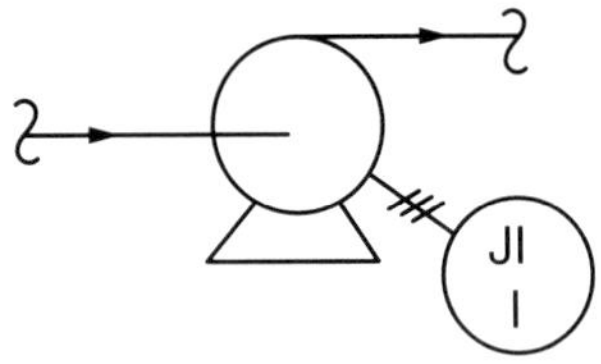 JI I INDICATING WATTMETER CONNECTED TO PUMP MOTOR	2	3
K TIME OR TIME SCHEDULE	1 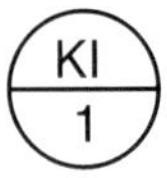 KI 1 CLOCK	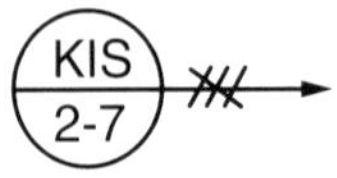 2 KIS 2-7 MULTIPOINT ON-OFF TIME SEQUENCING PROGRAMMER POINT 7	3 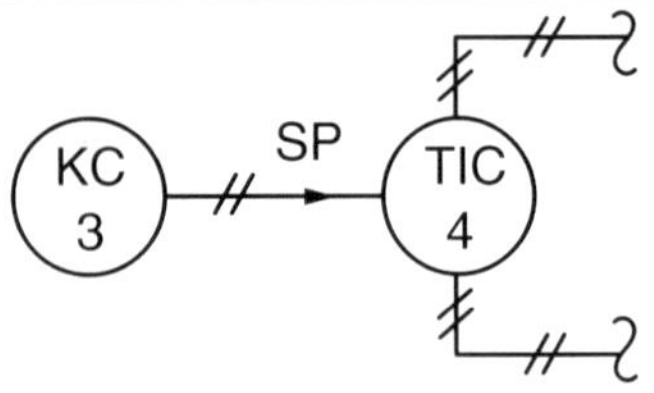 KC 3 SP TIC 4 TIME-SCHEDULE CONTROLLER, ANALOG TYPE, OR SELF-CONTAINED FUNCTION

202A08.EPS

PRIMARY ELEMENT SYMBOLS (CONT.)

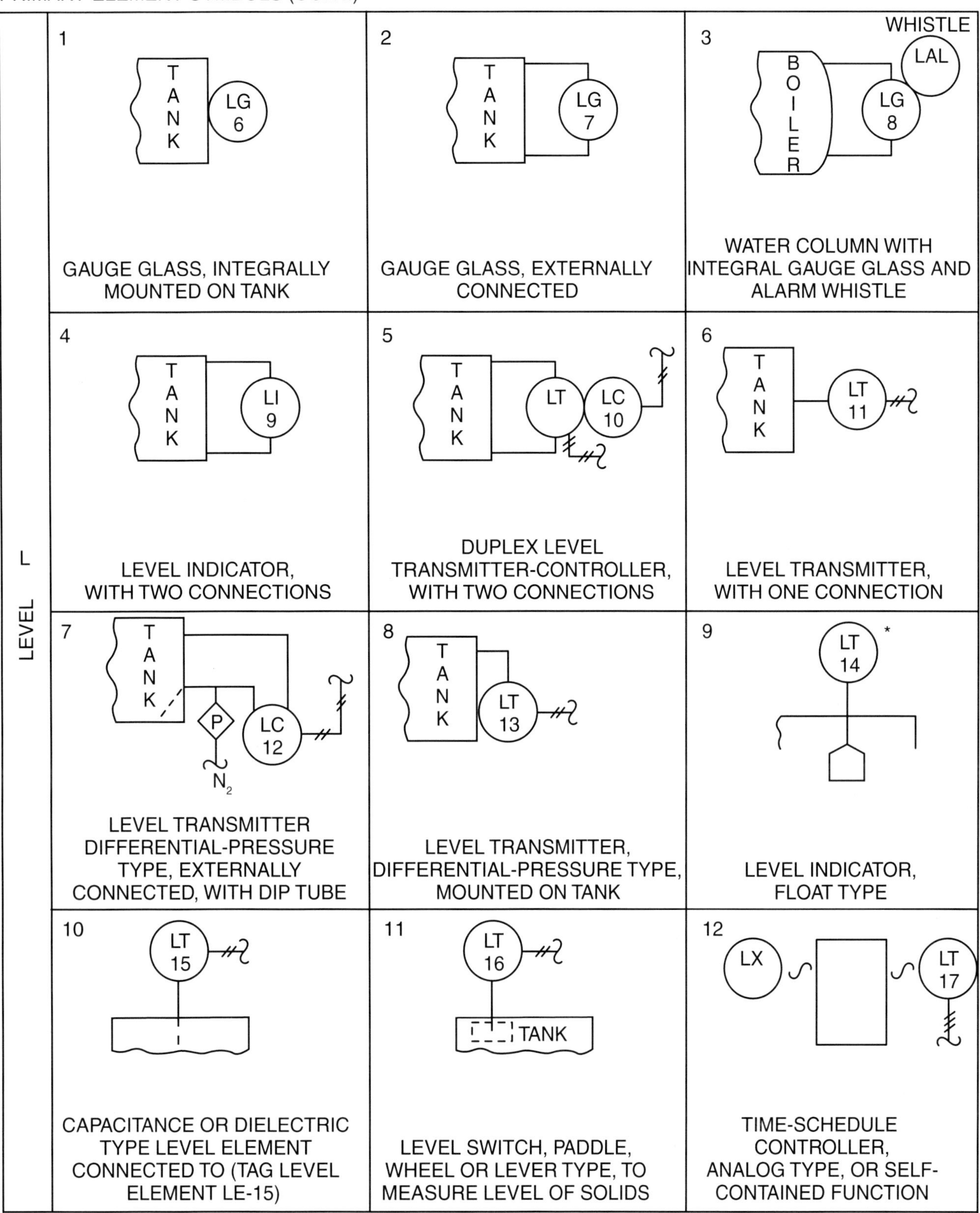

* Notations, such as “mounted at grade,” may be added.

202A09.EPS

PRIMARY ELEMENT SYMBOLS (CONT.)

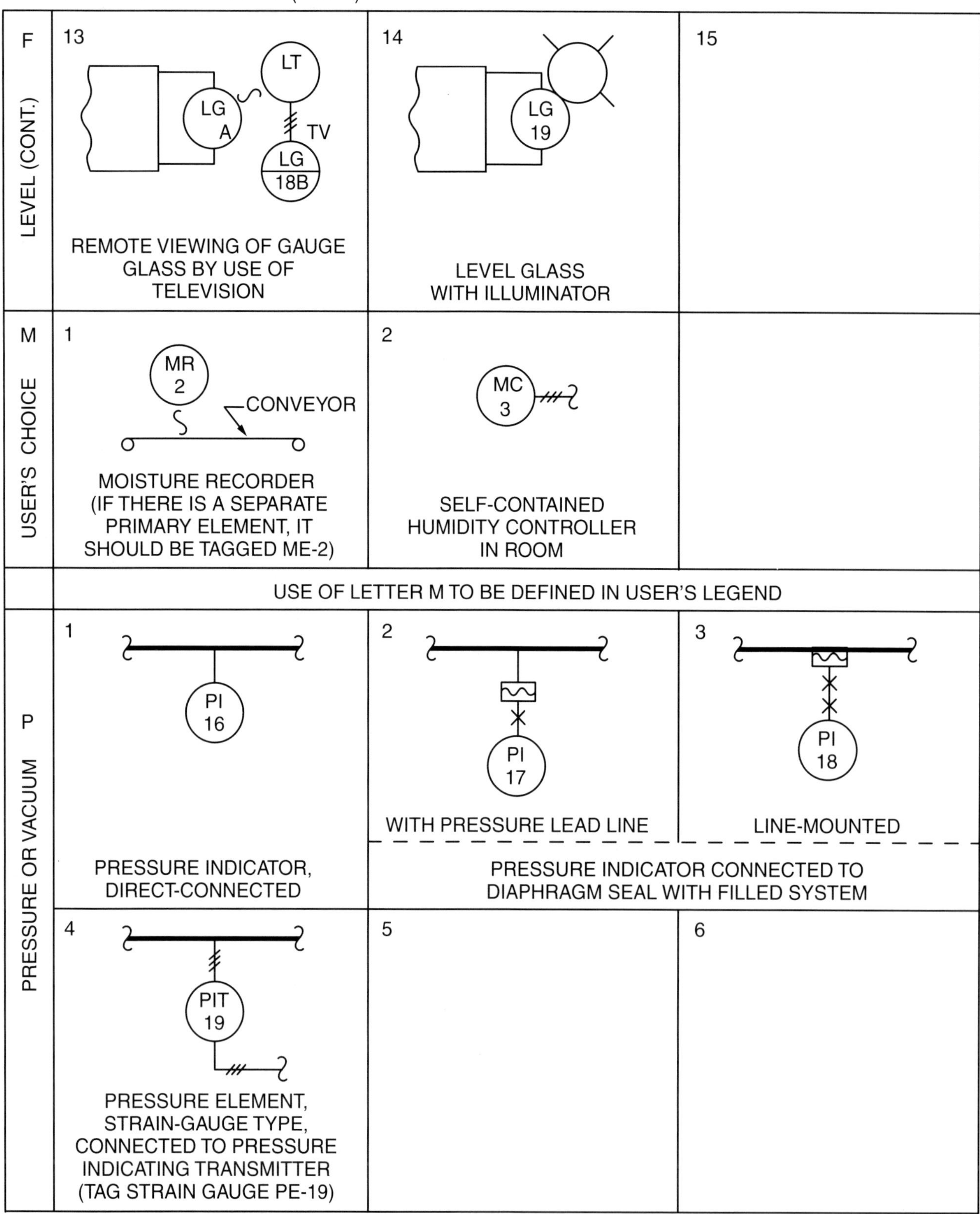

202A10.EPS

PRIMARY ELEMENT SYMBOLS (CONT.)

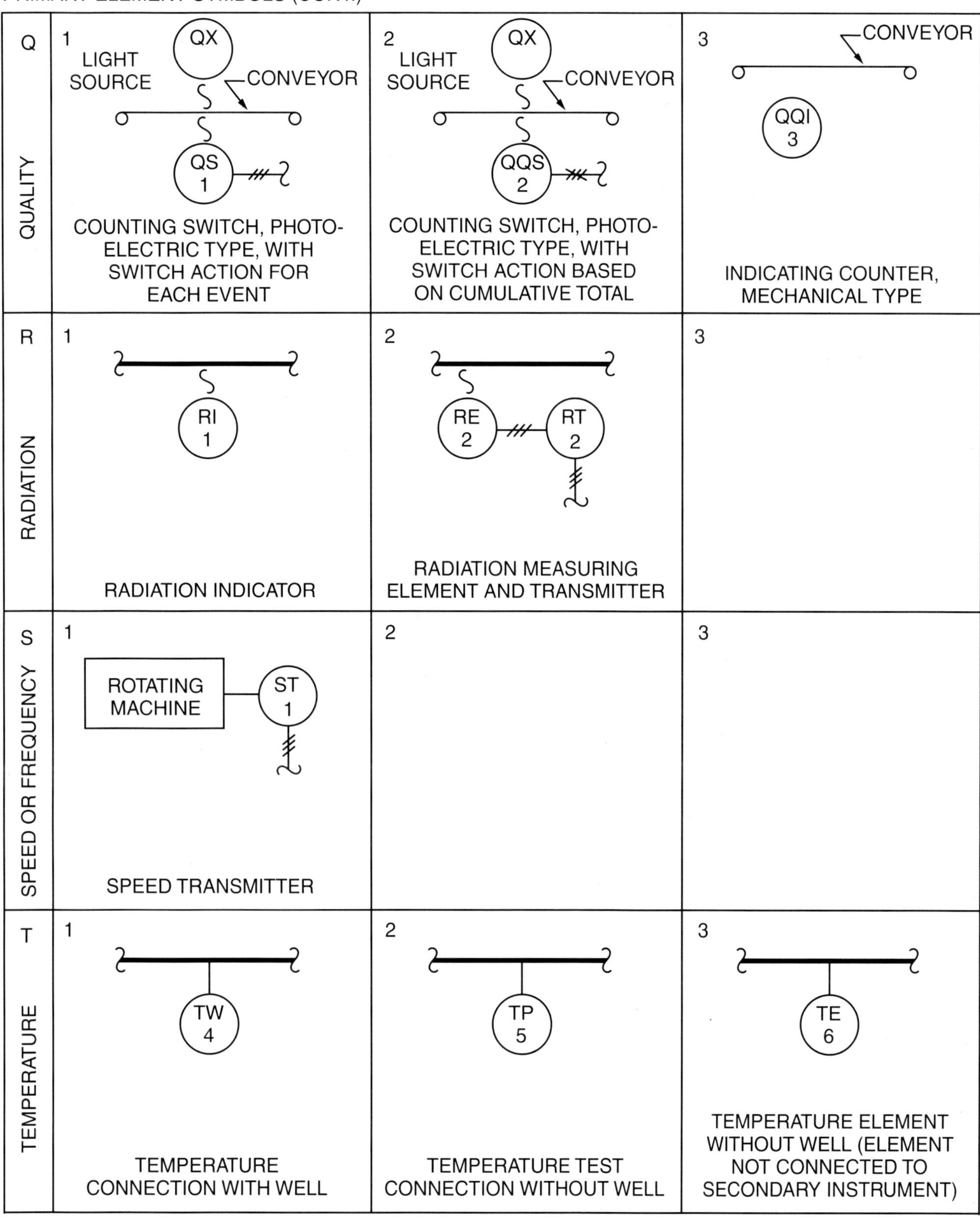

202A11.EPS

PRIMARY ELEMENT SYMBOLS (CONT.)

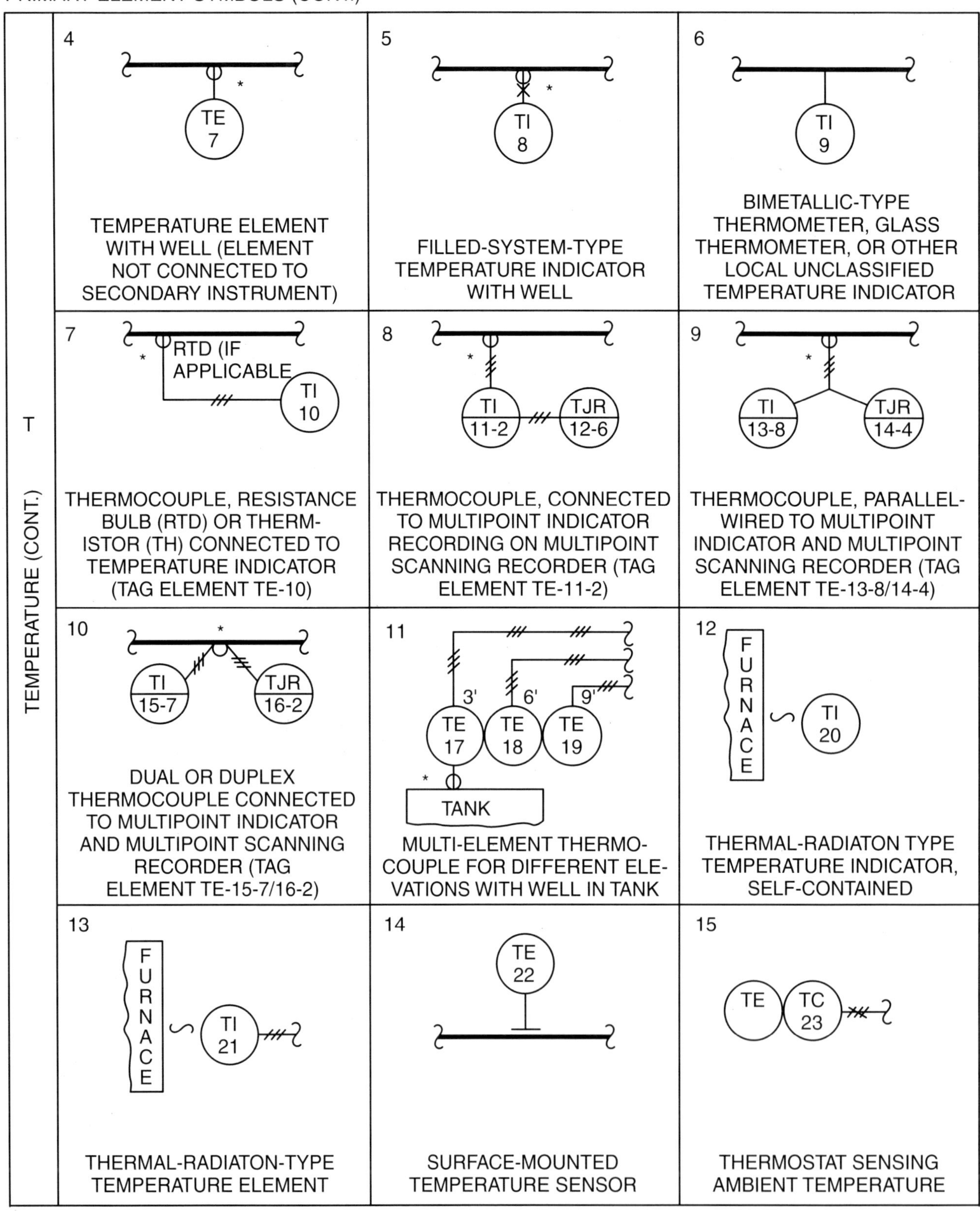

*Use of the thermowell symbol is optional. However, use or omission of the symbol should be consistent throughout a project.

202A12.EPS

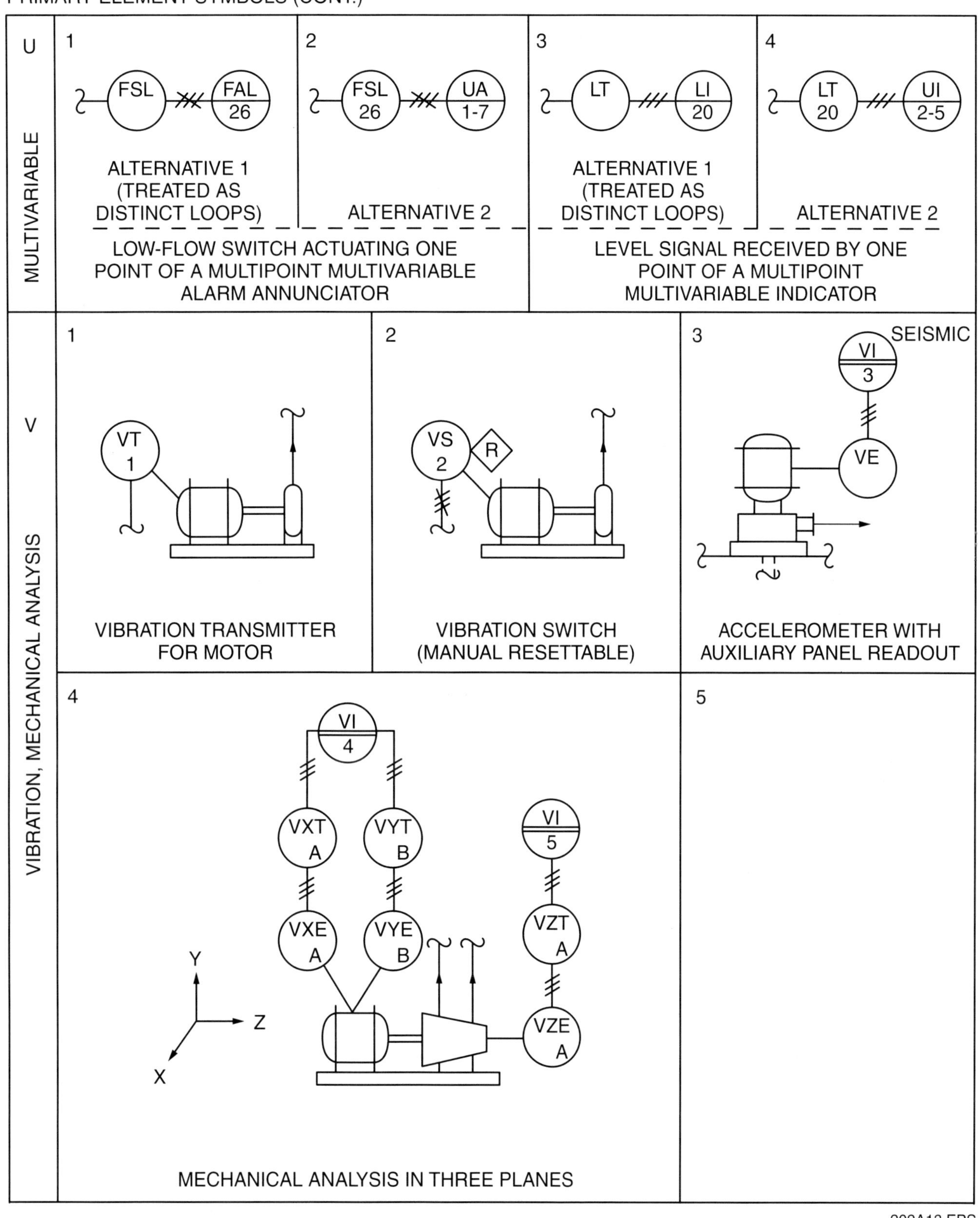
PRIMARY ELEMENT SYMBOLS (CONT.)
U
MULTIVARIABLE
1
FSL
FAL
26
ALTERNATIVE 1
(TREATED AS
DISTINCT LOOPS)
2
FSL
26
UA
1-7
ALTERNATIVE 2
LOW-FLOW SWITCH ACTUATING ONE
POINT OF A MULTIPOINT MULTIVARIABLE
ALARM ANNUNCIATOR
3
LT
LI
20
ALTERNATIVE 1
(TREATED AS
DISTINCT LOOPS)
4
LT
20
UI
2-5
ALTERNATIVE 2
LEVEL SIGNAL RECEIVED BY ONE
POINT OF A MULTIPOINT
MULTIVARIABLE INDICATOR
V
VIBRATION, MECHANICAL ANALYSIS
1
VT
1
VIBRATION TRANSMITTER
FOR MOTOR
2
VS
2
R
VIBRATION SWITCH
(MANUAL RESETTABLE)
3
SEISMIC
VI
3
VE
ACCELEROMETER WITH
AUXILIARY PANEL READOUT
4
VI
4
VXT
A
VYT
B
VXE
A
VYE
B
VI
5
VZT
A
VZE
A
Y
Z
X
MECHANICAL ANALYSIS IN THREE PLANES
5

202A13.EPS

PRIMARY ELEMENT SYMBOLS (CONT.)

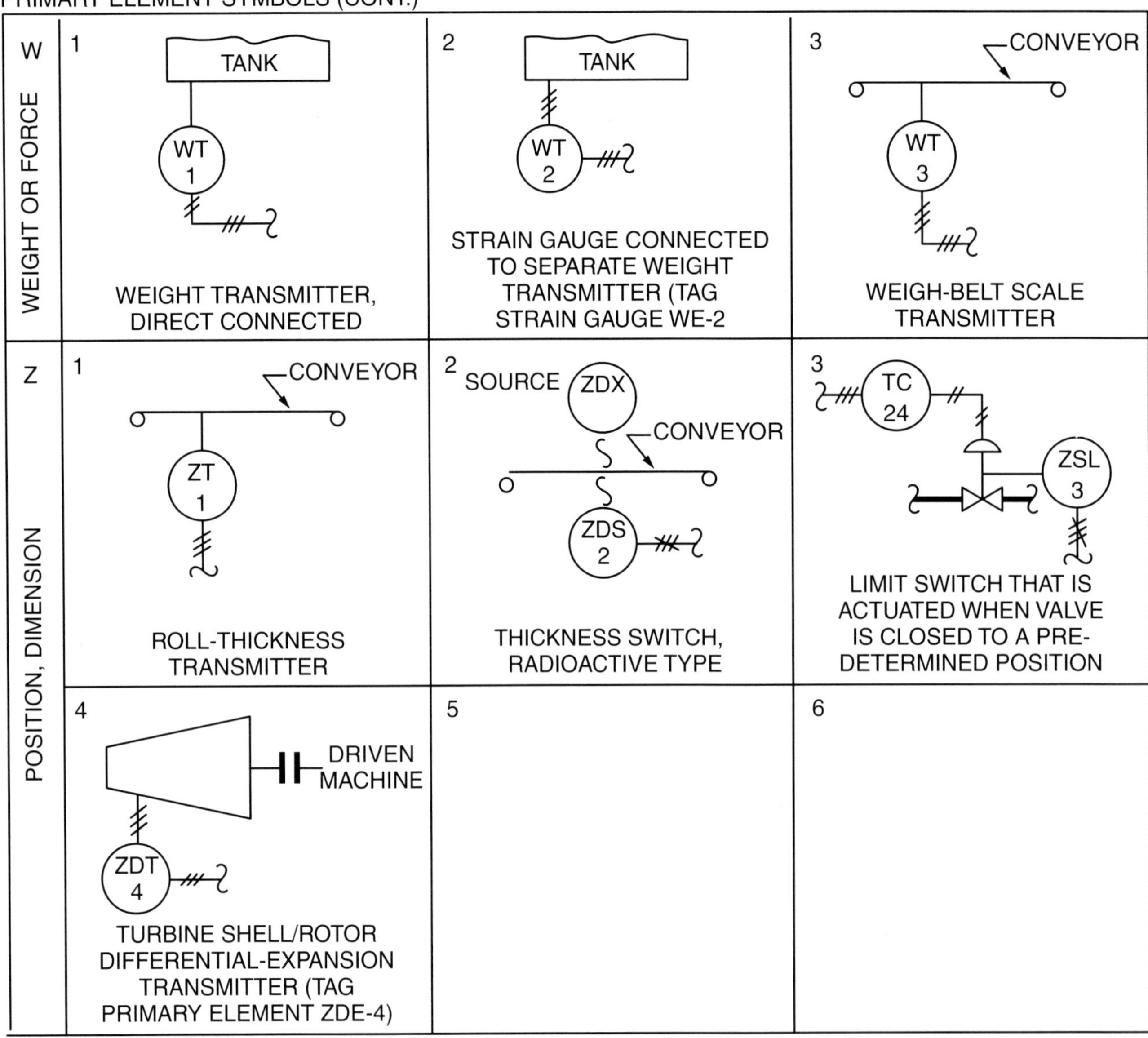

202A14.EPS

SYMBOLS FOR SELF-ACTUATED REGULATORS, VALVES, AND OTHER DEVICES

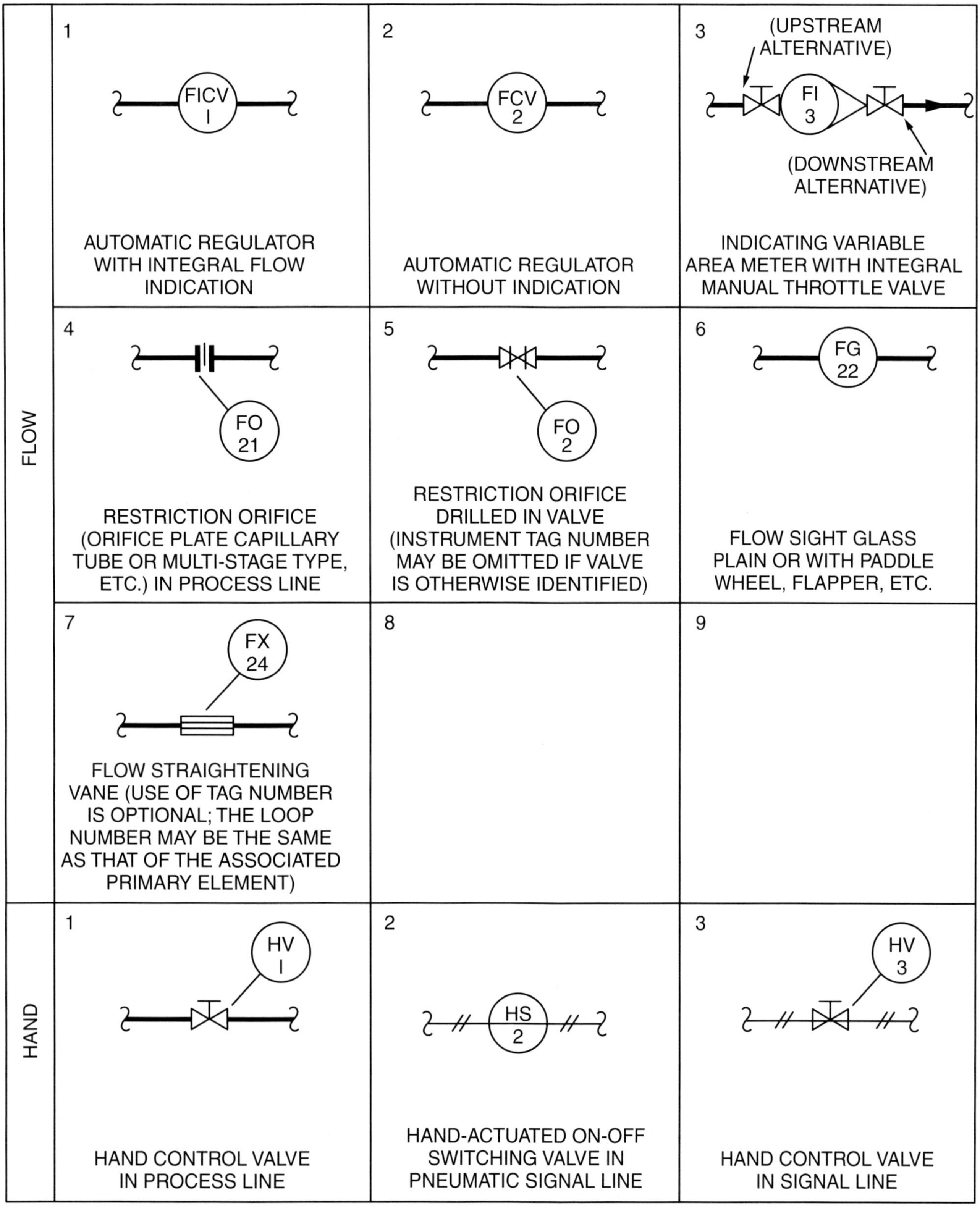

FLOW	1 FICV I AUTOMATIC REGULATOR WITH INTEGRAL FLOW INDICATION	2 FCV 2 AUTOMATIC REGULATOR WITHOUT INDICATION	3 (UPSTREAM ALTERNATIVE) FI 3 (DOWNSTREAM ALTERNATIVE) INDICATING VARIABLE AREA METER WITH INTEGRAL MANUAL THROTTLE VALVE
	4 FO 21 RESTRICTION ORIFICE (ORIFICE PLATE CAPILLARY TUBE OR MULTI-STAGE TYPE, ETC.) IN PROCESS LINE	5 FO 2 RESTRICTION ORIFICE DRILLED IN VALVE (INSTRUMENT TAG NUMBER MAY BE OMITTED IF VALVE IS OTHERWISE IDENTIFIED)	6 FG 22 FLOW SIGHT GLASS PLAIN OR WITH PADDLE WHEEL, FLAPPER, ETC.
	7 FX 24 FLOW STRAIGHTENING VANE (USE OF TAG NUMBER IS OPTIONAL; THE LOOP NUMBER MAY BE THE SAME AS THAT OF THE ASSOCIATED PRIMARY ELEMENT)	8	9
HAND	1 HV I HAND CONTROL VALVE IN PROCESS LINE	2 HS 2 HAND-ACTUATED ON-OFF SWITCHING VALVE IN PNEUMATIC SIGNAL LINE	3 HV 3 HAND CONTROL VALVE IN SIGNAL LINE

202A15.EPS

SYMBOLS FOR SELF-ACTUATED REGULATORS, VALVES, AND OTHER DEVICES (CONT.)

LEVEL	1 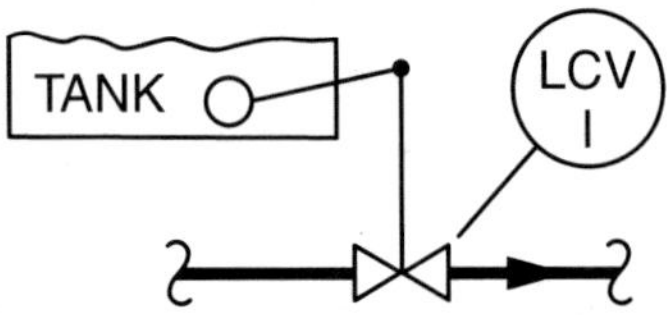 TANK LCV 1 LEVEL REGULATOR WITH MECHANICAL LINKAGE	2	3
PRESSURE	1 PCV 1 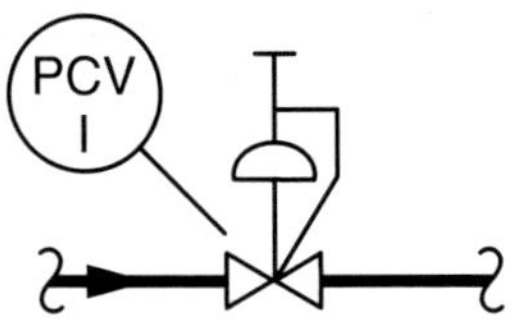 PRESSURE-REDUCING REGULATOR, SELF CONTAINED, WITH HANDWHEEL ADJUSTABLE SET POINT	2 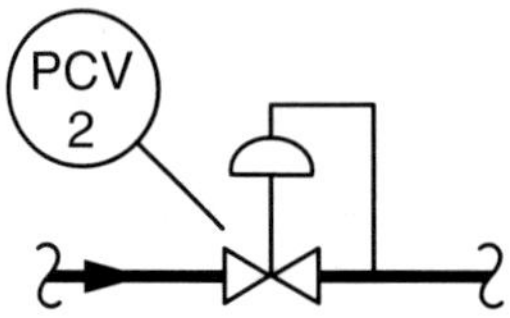 PCV 2 PRESSURE-REDUCING REGULATOR WITH EXTERNAL PRESSURE TAP	3 PDCV 3 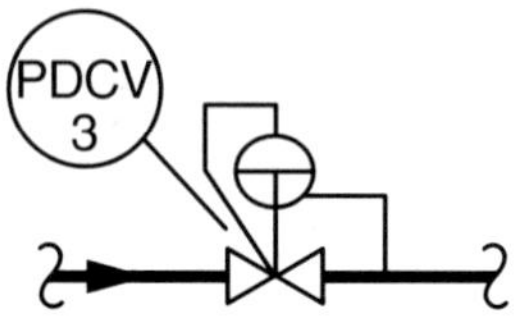 DIFFERENTIAL-PRESSURE-REDUCING REGULATOR WITH INTERNAL AND EXTERNAL PRESSURE TAPS
	4 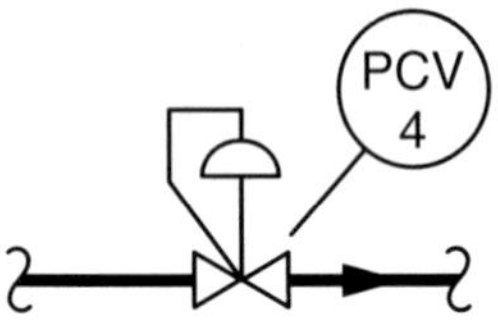 PCV 4 BACKPRESSURE REGULATOR, SELF CONTAINED	5 PCV 5 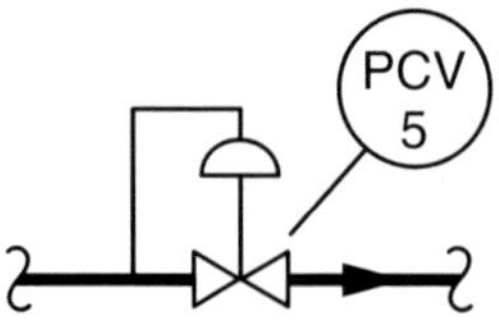 BACKPRESSURE REGULATOR WITH EXTERNAL PRESSURE TAP	6 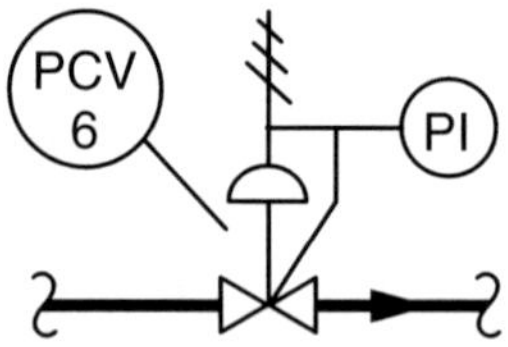 PCV 6 PI PRESSURE-REDUCING REGULATOR WITH INTEGRAL OUTLET PRESSURE RELIEF VALVE AND OPTIONAL PRESSURE INDICATOR (TYPICAL AIR SET)
	7 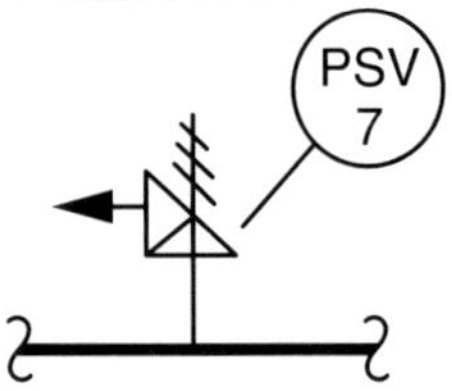 PSV 7 PRESSURE RELIEF OR SAFETY VALVE, GENERAL SYMBOL	8 PSV 8 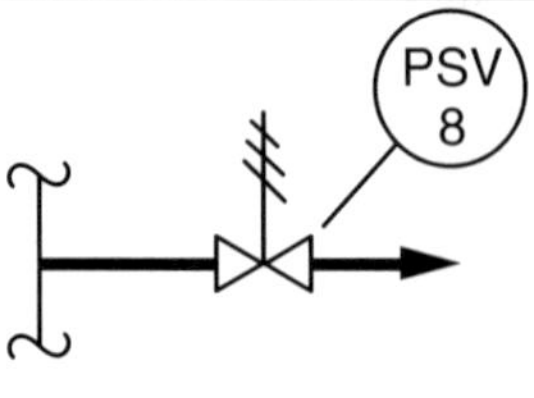 PRESSURE RELIEF OR SAFETY VALVE, STRAIGHT-THROUGH PATTERN, SPRING OR WEIGHT LOADED, OR WITH INTEGRAL PILOT	9 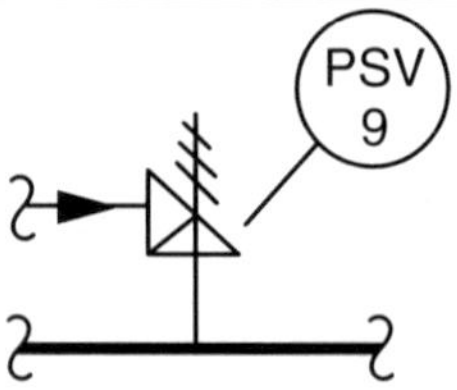 PSV 9 VACUUM RELIEF VALVE, GENERAL SYMBOL

202A16.EPS

SYMBOLS FOR SELF-ACTUATED REGULATORS, VALVES, AND OTHER DEVICES (CONT.)

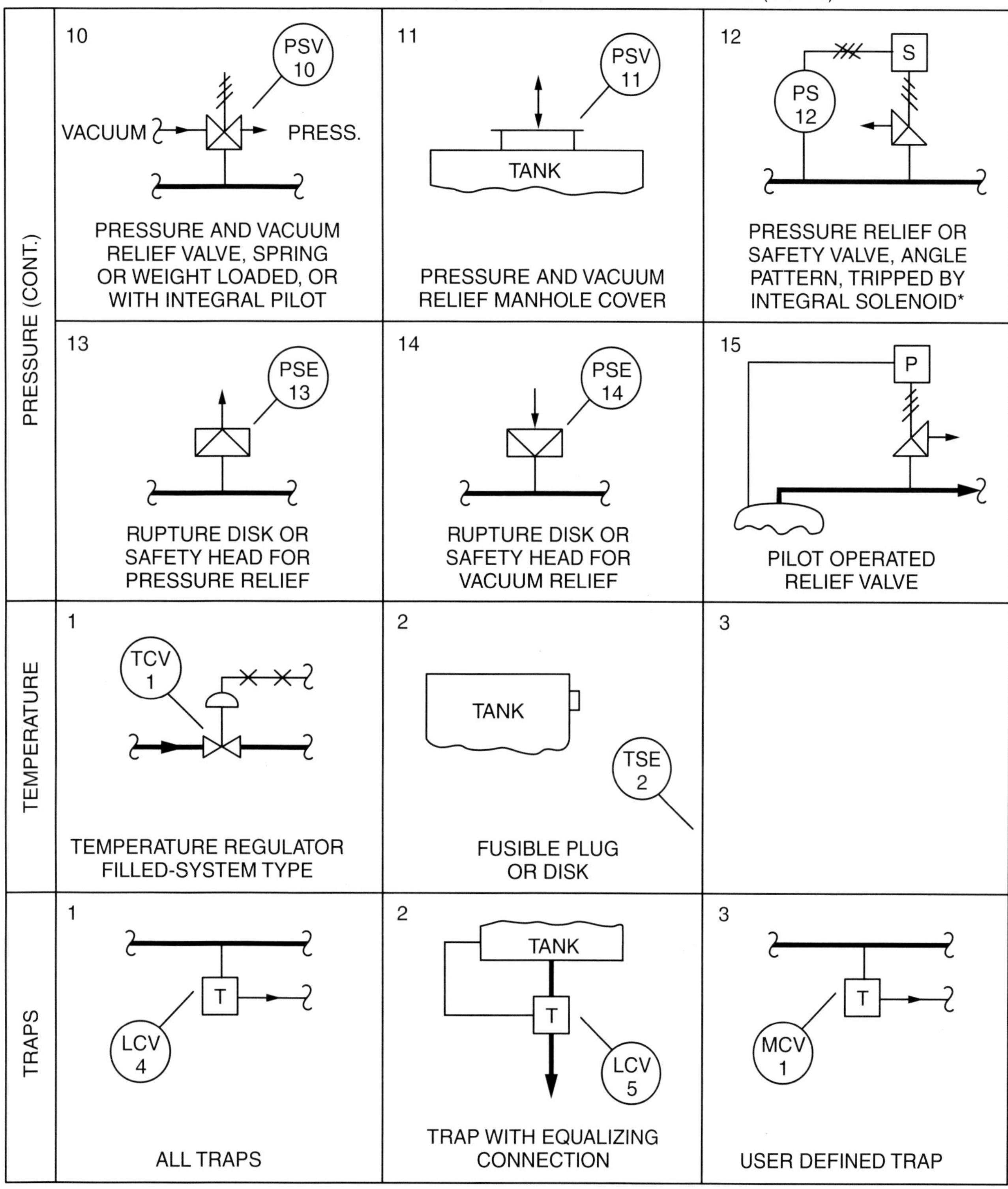

*The solenoid-tripped pressure relief valve is one of the class of power-actuated relief valves and is grouped with the other types of relief valves even though it is not entirely a self-actuated device.

202A17.EPS

CONTROL VALVE BODY SYMBOLS, DAMPER SYMBOLS

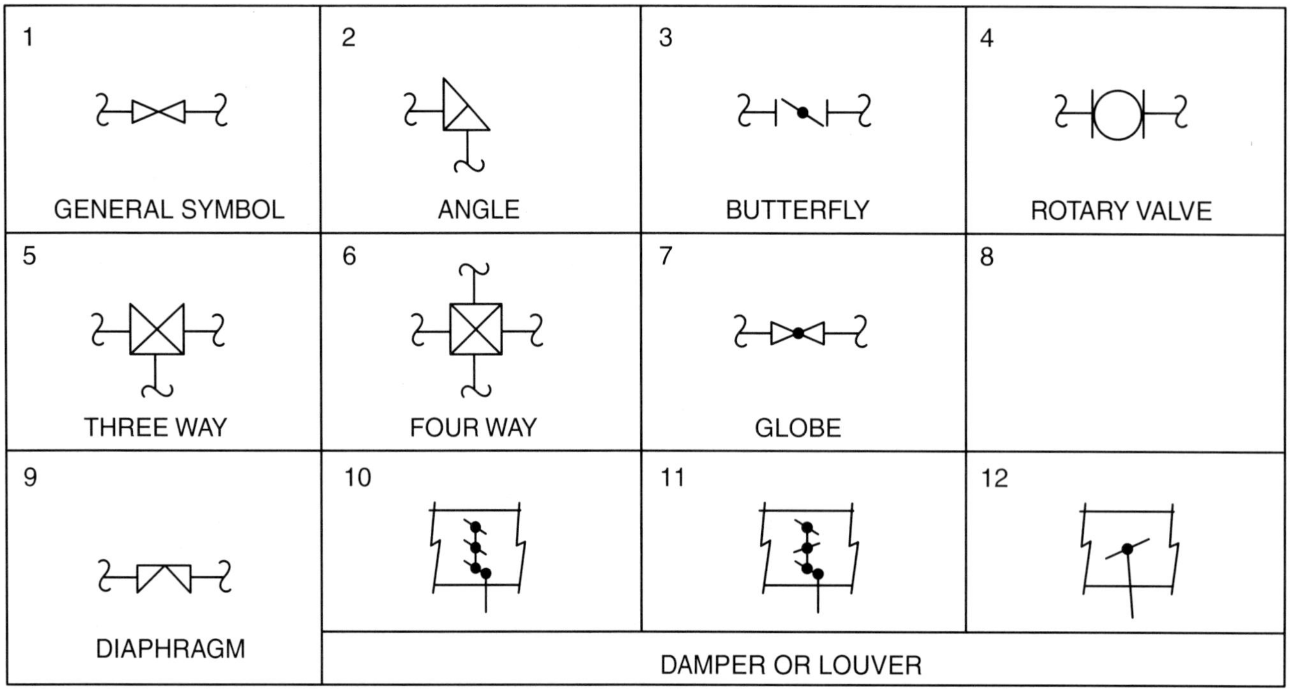

1 GENERAL SYMBOL	2 ANGLE	3 BUTTERFLY	4 ROTARY VALVE
5 THREE WAY	6 FOUR WAY	7 GLOBE	8
9 DIAPHRAGM	10	11	12
	DAMPER OR LOUVER		

Further information may be added adjacent to the body symbol either by note or code number.

202A18.EPS

SYMBOLS FOR ACTUATOR ACTION IN EVENT OF ACTUATOR POWER FAILURE (SHOWN TYPICALLY FOR DIAPHRAGM-ACTUATED CONTROL VALVE)

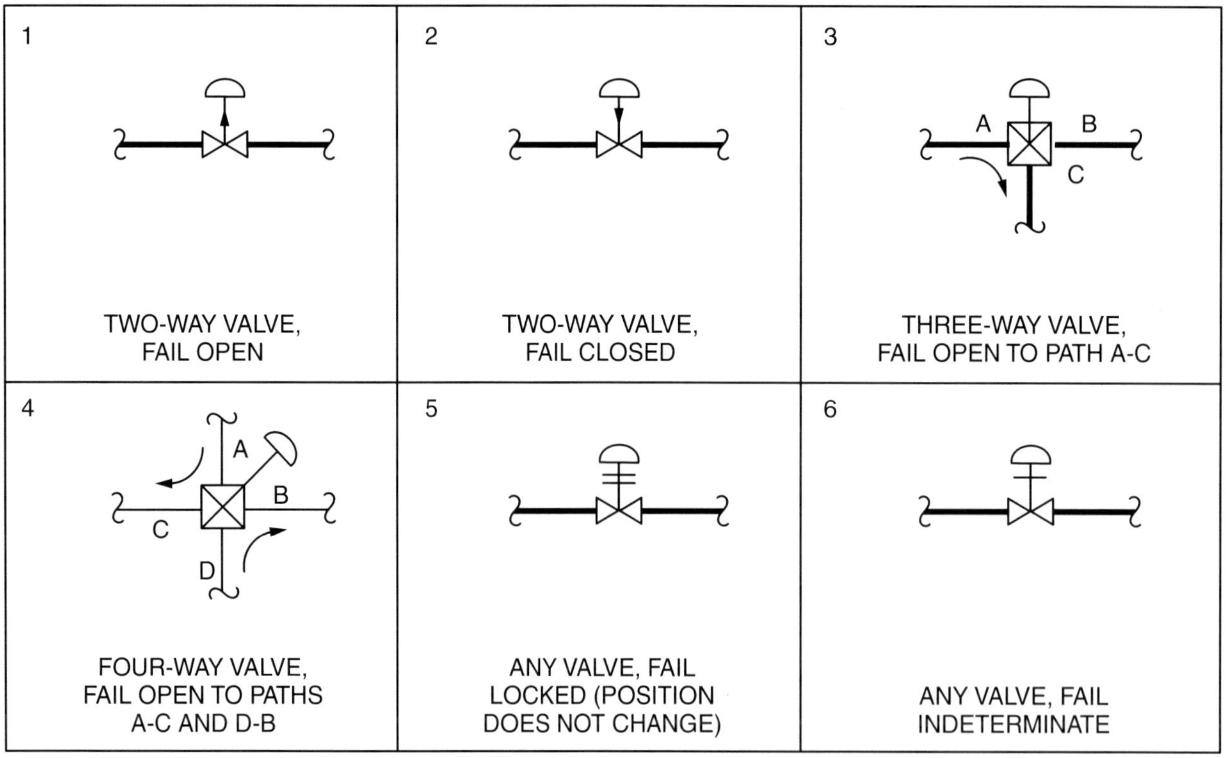

1 TWO-WAY VALVE, FAIL OPEN	2 TWO-WAY VALVE, FAIL CLOSED	3 THREE-WAY VALVE, FAIL OPEN TO PATH A-C
4 FOUR-WAY VALVE, FAIL OPEN TO PATHS A-C AND D-B	5 ANY VALVE, FAIL LOCKED (POSITION DOES NOT CHANGE)	6 ANY VALVE, FAIL INDETERMINATE

The failure modes indicated are those commonly defined by the term *shelf position.* As an alternative to the arrows and bars, the following abbreviations may be employed:

FO - Fail Open
FC - Fail Closed
FL - Fail Locked (last position)
FI - Fail Indeterminate

202A19.EPS

SYMBOLS FOR ACTUATOR ACTION IN EVENT OF ACTUATOR POWER FAILURE (SHOWN TYPICALLY FOR DIAPHRAGM-ACTUATED CONTROL VALVE)

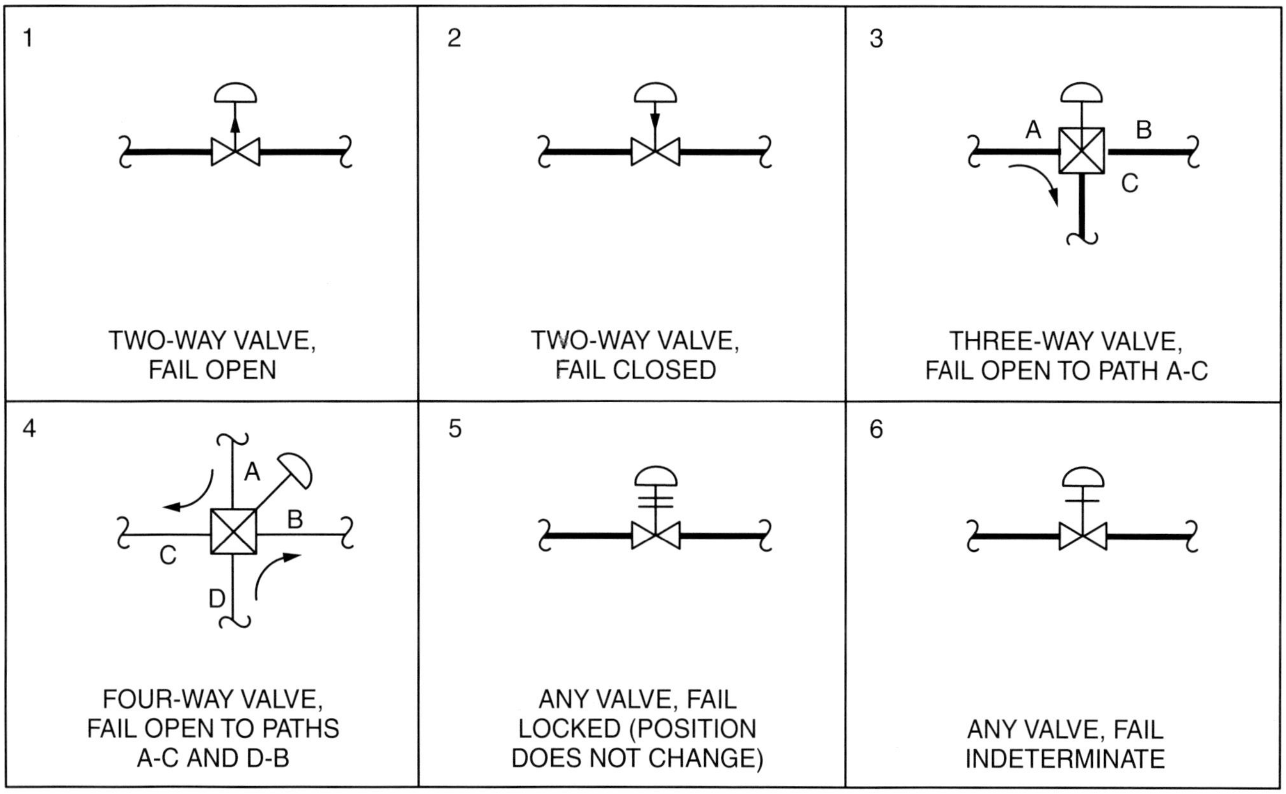

1	2	3
TWO-WAY VALVE, FAIL OPEN	TWO-WAY VALVE, FAIL CLOSED	THREE-WAY VALVE, FAIL OPEN TO PATH A-C
4	5	6
FOUR-WAY VALVE, FAIL OPEN TO PATHS A-C AND D-B	ANY VALVE, FAIL LOCKED (POSITION DOES NOT CHANGE)	ANY VALVE, FAIL INDETERMINATE

The failure modes indicated are those commonly defined by the term *shelf position*. As an alternative to the arrows and bars, the following abbreviations may be employed:

FO - Fail Open
FC - Fail Closed
FL - Fail Locked (last position)
FI - Fail Indeterminate

202A20.EPS

Additional Resources

This module is intended to present thorough resources for task training. The following reference works are suggested for further study. These are optional materials for continued education rather than for task training.

Instrumentation, 1975. F.W. Kirk and N.R. Rimboi, American Technical Society.

ANI/ISA-5.1, *Instrumentation Symbols and Identification*, July 1992.

Acknowledgments

The material for the appendix is excerpted from *ISA-5.1-1984* (R 1992), *Instrumentation Symbols and Identification*, copyright 1984. Used with permission. For more information, visit www.isa.org.

NCCER CURRICULA — USER UPDATE

NCCER makes every effort to keep its textbooks up-to-date and free of technical errors. We appreciate your help in this process. If you find an error, a typographical mistake, or an inaccuracy in NCCER's curricula, please fill out this form (or a photocopy), or complete the online form at **www.nccer.org/olf**. Be sure to include the exact module ID number, page number, a detailed description, and your recommended correction. Your input will be brought to the attention of the Authoring Team. Thank you for your assistance.

Instructors – If you have an idea for improving this textbook, or have found that additional materials were necessary to teach this module effectively, please let us know so that we may present your suggestions to the Authoring Team.

NCCER Product Development and Revision

13614 Progress Blvd., Alachua, FL 32615

Email: curriculum@nccer.org
Online: www.nccer.org/olf

❑ Trainee Guide ❑ AIG ❑ Exam ❑ PowerPoints Other ______________________________

Craft / Level: Copyright Date:

Module ID Number / Title:

Section Number(s):

Description:

Recommended Correction:

Your Name:

Address:

Email: Phone:

Module 12203-03

Principles of Welding for Instrumentation

COURSE MAP

This course map shows all of the modules in the second level of the Instrumentation curriculum. The suggested training order begins at the bottom and proceeds up. Skill levels increase as you advance on the course map. The local Training Program Sponsor may adjust the training order.

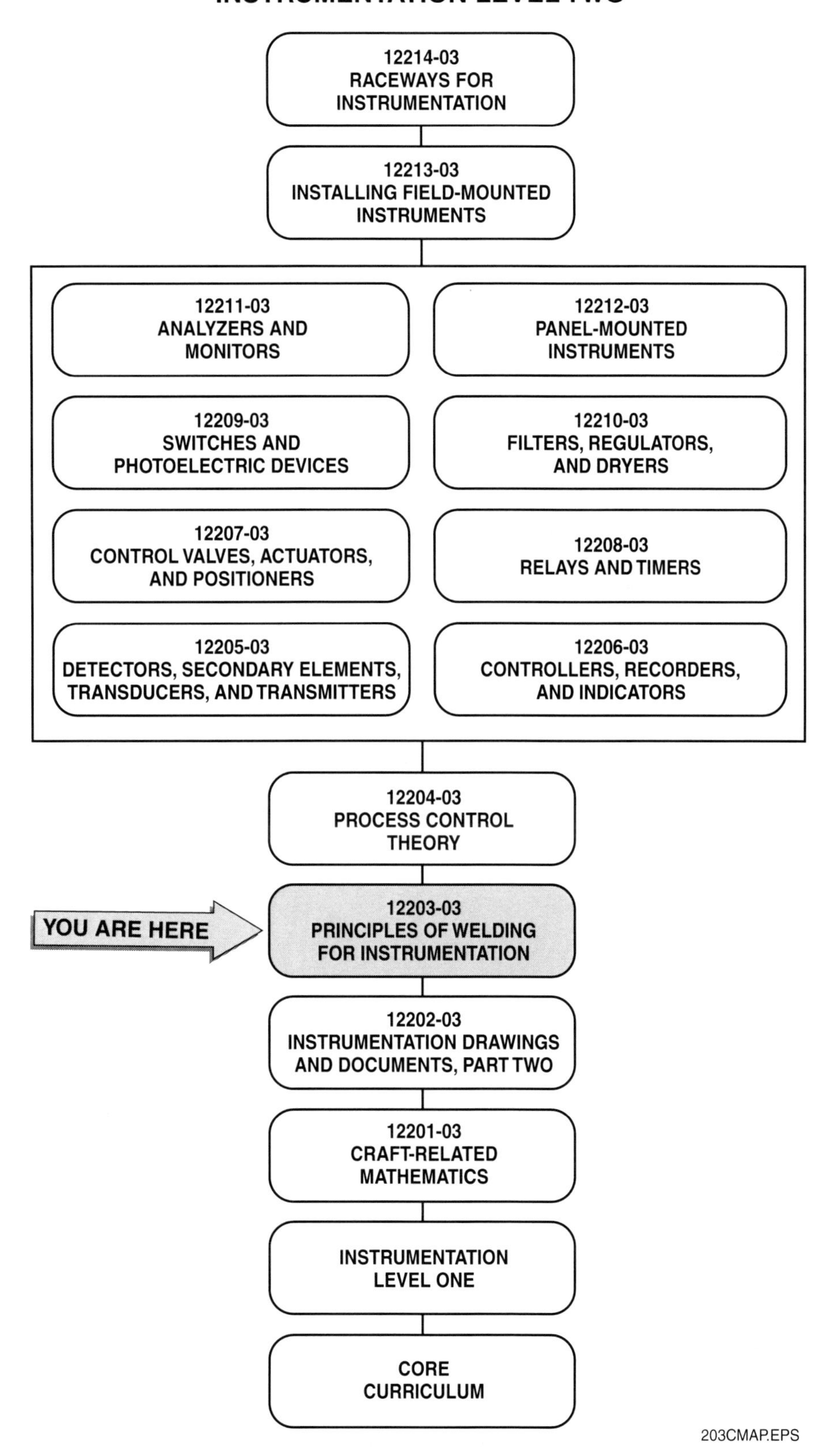

203CMAP.EPS

MODULE 12203-03 CONTENTS

1.0.0 INTRODUCTION . . . 3.1

2.0.0 WELDING SAFETY . . . 3.2

2.1.0 Clothes for Welding . . . 3.2

2.1.1 Protective Leather Clothing . . . 3.3

2.1.2 Welding Gloves . . . 3.3

2.2.0 Eye Protection . . . 3.5

2.3.0 Ear Protection. . . . 3.6

3.0.0 PRINCIPLES AND PROCEDURES OF SHIELDED METAL ARC WELDING (SMAW) . . . 3.6

3.1.0 Equipment, Power Source, and Current. . . . 3.7

3.1.1 Electrodes . . . 3.9

4.0.0 PRINCIPLES OF ORBITAL WELDING . . . 3.9

4.1.0 Orbital Welding Equipment and Accessories . . . 3.10

4.1.1 Orbital Welding Power Supplies . . . 3.10

4.1.2 Orbital Welding Weld Heads. . . . 3.10

4.1.3 Electrodes for Orbital Welding . . . 3.10

4.1.4 Shield Gases and Purge Gases . . . 3.11

4.2.0 Advantages of Orbital Welding . . . 3.11

4.3.0 Applications of Orbital Welding. . . . 3.11

4.4.0 Orbital Welding Basics and Setup . . . 3.12

4.4.1 The Process . . . 3.12

4.4.2 Fit-Up and Alignment . . . 3.12

4.4.3 Parameters . . . 3.12

5.0.0 PRINCIPLES AND PROCEDURES OF GAS TUNGSTEN ARC WELDING (GTAW) . . . 3.13

5.1.0 Methods of Application, Equipment, and Power Source . . . 3.13

5.2.0 Shielding Gases, Electrodes, and Filler Metal . . . 3.14

6.0.0 PRINCIPLES AND PROCEDURES OF GAS METAL ARC WELDING (GMAW) . . . 3.15

6.1.0 Methods of Application, Equipment, and Power Source . . . 3.15

6.1.1 Shielding Gases, Electrodes, and Metal Transfer. . . . 3.16

7.0.0 PRINCIPLES AND PROCEDURES FOR BRAZING AND OXYACETYLENE WELDING . . . 3.19

7.1.0 Method of Operation and Equipment . . . 3.21

7.1.1 Torch Tips . . . 3.23

7.1.2 Regulators . . . 3.23

7.1.3 Gases and Gas Cylinders . . . 3.23

8.0.0 TYPES OF WELDED FITTINGS . . . 3.25

8.1.0 Socket Weld Fittings . . . 3.26

8.2.0 Butt Weld Fittings . . . 3.26

8.3.0 Copper Solder Joint Fittings. . . . 3.26

9.0.0 WELDING APPLICATIONS FOR INSTRUMENTATION 3.28
SUMMARY 3.33
REVIEW QUESTIONS 3.33
GLOSSARY 3.35
REFERENCES & ACKNOWLEDGMENTS 3.36

Figures

Figure 1 Old craft of forging and shaping metal 3.1
Figure 2 Proper work clothing 3.3
Figure 3 Leather protective clothing 3.4
Figure 4 Leather welding gloves 3.5
Figure 5 Gas welding goggles and face shields 3.6
Figure 6 Arc welding helmets 3.6
Figure 7 Earmuffs and earplugs 3.7
Figure 8 Shielded metal arc welding 3.7
Figure 9 Equipment for shielded metal arc welding 3.8
Figure 10 Electrode holders and work lead 3.9
Figure 11 SMAW electrodes 3.9
Figure 12 Orbital welding element 3.10
Figure 13 Field tubing alignment clamps 3.12
Figure 14 Gas tungsten arc welding 3.13
Figure 15 Equipment for gas tungsten arc welding 3.14
Figure 16 Gas metal arc welding 3.15
Figure 17 Equipment for gas metal arc welding 3.16
Figure 18 Wire feed speed versus welding current for several nonferrous metal wire electrodes 3.17
Figure 19 Wire feed speed versus welding current for steel electrode wire 3.17
Figure 20 GMAW transfer methods 3.19
Figure 21 Oxyacetylene welding 3.20
Figure 22 Basic flame types 3.21
Figure 23 Oxyacetylene and oxyfuel outfits 3.22
Figure 24 Striker, cylinder/regulator wrench, and soapstone holders 3.22
Figure 25 General purpose combination oxyacetylene torch and tips 3.23
Figure 26 Oxygen and acetylene regulators 3.24

Figure 27 Typical oxygen cylinder, valve, and safety cap 3.24
Figure 28 Typical acetylene cylinder, valve, safety cap, and safety plugs 3.25
Figure 29 Typical socket and butt weld fittings 3.26
Figure 30 Dimensions for forged steel socket weld fittings 3.27
Figure 31 Butt weld fitting identification 3.28
Figure 32 Butt weld fittings 3.29
Figure 33 Dimensions for cast copper alloy solder fittings 3.29

Tables

Table 1 How Electrode Taper Affects the Arc 3.11
Table 2 Classification of Tungsten Arc Welding Electrodes (*AWS A5.12*) 3.15
Table 3 Shielding Gases and Their Applications for Gas Metal Arc Welding 3.18
Table 4 Electrode Composition and Mechanical Properties Chart (*AWS A15.18* and *A5.28*) 3.20
Table 5 Socket Weld Fitting Chart (Imperial) 3.27
Table 6 Socket Weld Fitting Chart (Metric) 3.28
Table 7 Butt Weld Fitting Chart (Imperial) 3.30
Table 8 Butt Weld Fitting Chart (Metric) 3.31
Table 9 Pressure and Temperature Chart for Solder and Brazed Joints 3.32
Table 10 Cast Copper Alloy Solder Fitting Chart 3.32

Principles of Welding for Instrumentation

Objectives

When you have completed this module, you will be able to do the following:

1. Define stick and orbital tube welding principles, procedures and applications.
2. Identify TIG and MIG welding principles, procedures, and applications.
3. Identify brazing principles, procedures, and applications.
4. Identify types of fittings that are welded.
5. Identify and demonstrate proper use of personal protective equipment.

Prerequisites

Before you begin this module, it is recommended that you successfully complete the following: Core Curriculum; Instrumentation Level One; Instrumentation Level Two, Modules 12201-03 and 12202-03.

Required Trainee Materials

1. Pencil and paper
2. Appropriate personal protective equipment

1.0.0 ◆ INTRODUCTION

Early blacksmiths learned that two pieces of metal could be joined by first heating them, then forcing them together by hammering. For centuries, craftsmen used this practice of forge welding to join metals (*Figure 1*). Unfortunately, the heat and pressures applied in this process were not great enough for complete fusion of the two pieces being joined. In fusion, the metal would be melted sufficiently for a thorough blending of the material in both pieces, hardening as they cooled into a single, continuous, solid mass. In forge welding, however, the effect was more of a mechanical locking-in of the two surfaces in contact. Any fusion was limited to the thin outer skin of the two surfaces. With these conditions, a relatively low force could break the joint and separate the pieces.

203F01.TIF

Figure 1 ◆ Old craft of forging and shaping metal.

The discovery and invention of modern welding processes provided the ideal means of joining metal plates or shapes, castings to castings, forgings to forgings, or forgings to castings. This is how metals should be joined—by fusion—resulting in a complete, permanent union of the two pieces, often with the welded area stronger than either of the pieces joined. With the proper welding materials and techniques, almost any two pieces of metal can be joined into a single unit. **Overlapping** of pieces to be joined is not necessary, and thickness at the weld area can be held to the thickness of the member to either side.

Arc welding today is widely accepted as the best, most economical, most natural, and most practical way to join metals.

Nearly all metals are weldable, provided the proper process and techniques are used. Occasionally, an attempt to weld metal ends in failure because either the proper process or the proper technique has been overlooked.

2.0.0 ◆ WELDING SAFETY

To weld safely, you must always take the following precautions:

- Do not ground welding equipment to instrument stands that have already been installed. Electric shock and equipment damage may result.
- Do not wear clothing made from synthetic fabrics. These will catch fire easily and melt to human skin.
- Make sure gauges are facing away from you when installing and tightening fittings to gas cylinders.
- Use gloves that are specifically designed for TIG welding when performing TIG welding.
- Welding shields must be installed around the welding work area when the welding arc may be visible to personnel in adjacent areas. Welding blankets must be installed when the sparks or hot metal pieces from the welding may come in contact with combustible materials.
- Picking up hot metal with leather welding gloves will burn the leather, causing it to shrivel and become hard. Never handle hot metal with leather welding gloves.
- Shaded safety glasses will not provide sufficient protection from the arc. Looking directly at the arc with shaded safety glasses will result in severe burns to the eyes and face.
- Oil or any type of oil-based lubricants must not be used on or near any oxygen fittings, hoses, lines, regulators, valves, cylinders, or any other component or element associated with the use of oxygen. Exposure of oxygen to oil or oil-based materials can cause severe explosion and personal injury.
- Cylinders must be properly secured on a portable cart, using approved cylinder securing devices, to prevent cylinders from falling and possibly breaking off the cylinder stems or regulators. High-pressure cylinders become dangerous missiles when gas escapes from them. They can cause extreme property damage, severe injury, and even death.
- Do not use oil-based lubricant on any adjustable parts.
- When using a striker to light the torch, aim or position the torch tip away from any part of your body to avoid severe burns. Also aim the torch tip away from any combustible material. Butane or any other type of tobacco lighters should never be used to light torches. They should not be on the person who is performing any type of welding, including oxyacetylene welding, as they may explode or ignite.
- If acetylene is used as the fuel gas, the working pressure must never be allowed to reach 15 psi. Acetylene becomes very dangerous at 15 psi and self-explosive at 29.4 psi.
- Acetone is an extremely volatile liquid chemical and is very explosive. Acetone can be drawn from an acetylene cylinder when it is not upright or if acetylene is withdrawn from an upright tank at flow rates exceeding 1⁄10 of the tank volume. Acetylene cylinders must not be stored on their sides. In unforeseen situations where acetylene cylinders have been turned or stored on their sides, the cylinders must stand upright for a minimum of 30 minutes before use. This allows the acetone to settle to the bottom of the cylinder. Remember that without acetone in the cylinder, acetylene becomes extremely volatile at pressures over 15 psi. Proper storage and handling of cylinders is of the utmost importance.

2.1.0 Clothes for Welding

To avoid burns from sparks and ultraviolet rays, welders must wear appropriate clothing. Never wear polyester or other synthetic fibers. Sparks or intense heat will melt these materials, causing severe burns. Wool or cotton is more resistant to sparks. Dark clothing is also preferred because dark clothes minimize the reflection of arc rays, which could be deflected under the welding helmet.

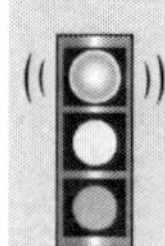

WARNING!
Do not wear clothes made from synthetic fabrics. These will catch fire easily and melt to human skin.

To prevent sparks from lodging in clothing and causing a fire or burns, collars should be kept buttoned and pockets should have flaps that can be buttoned to keep out the sparks. A cap worn with the visor reversed will protect the top of the head and keep sparks from going down the back of your collar. Pants should be cuffless and hang straight down the leg. Pant cuffs will catch sparks and catch on fire. Never wear frayed or fuzzy materials. These materials will trap sparks that will ignite the clothing. Low-top shoes should never be worn while welding. Sparks will fall into the shoes, causing severe burns. Leather work boots at least 8" high should be used to protect against sparks and arc flash. *Figure 2* shows proper work clothing, including leather protection worn over clothing.

2.1.1 Protective Leather Clothing

For additional protection, welders often wear leather apparel over their work clothing. Leather aprons, split-leg aprons, sleeves, and jackets are flexible enough to allow ease of movement while providing additional protection from sparks and heat. Leathers should always be worn when welding out of position or when welding in tight quarters. *Figure 3* shows the different types of leather protective clothing.

2.1.2 Welding Gloves

Leather gauntlet-type gloves are designed specifically for welding. They must be worn when performing any type of arc welding to protect against spattering hot metal and the ultraviolet and infrared arc rays. Leather welding gloves are not designed to handle hot metal. Pliers, tongs, or some other means should be used to handle hot metal. *Figure 4* shows typical welding gloves.

> **WARNING!**
> Picking up hot metal with leather welding gloves will burn the leather, causing it to shrivel and become hard. Never handle hot metal with leather welding gloves..

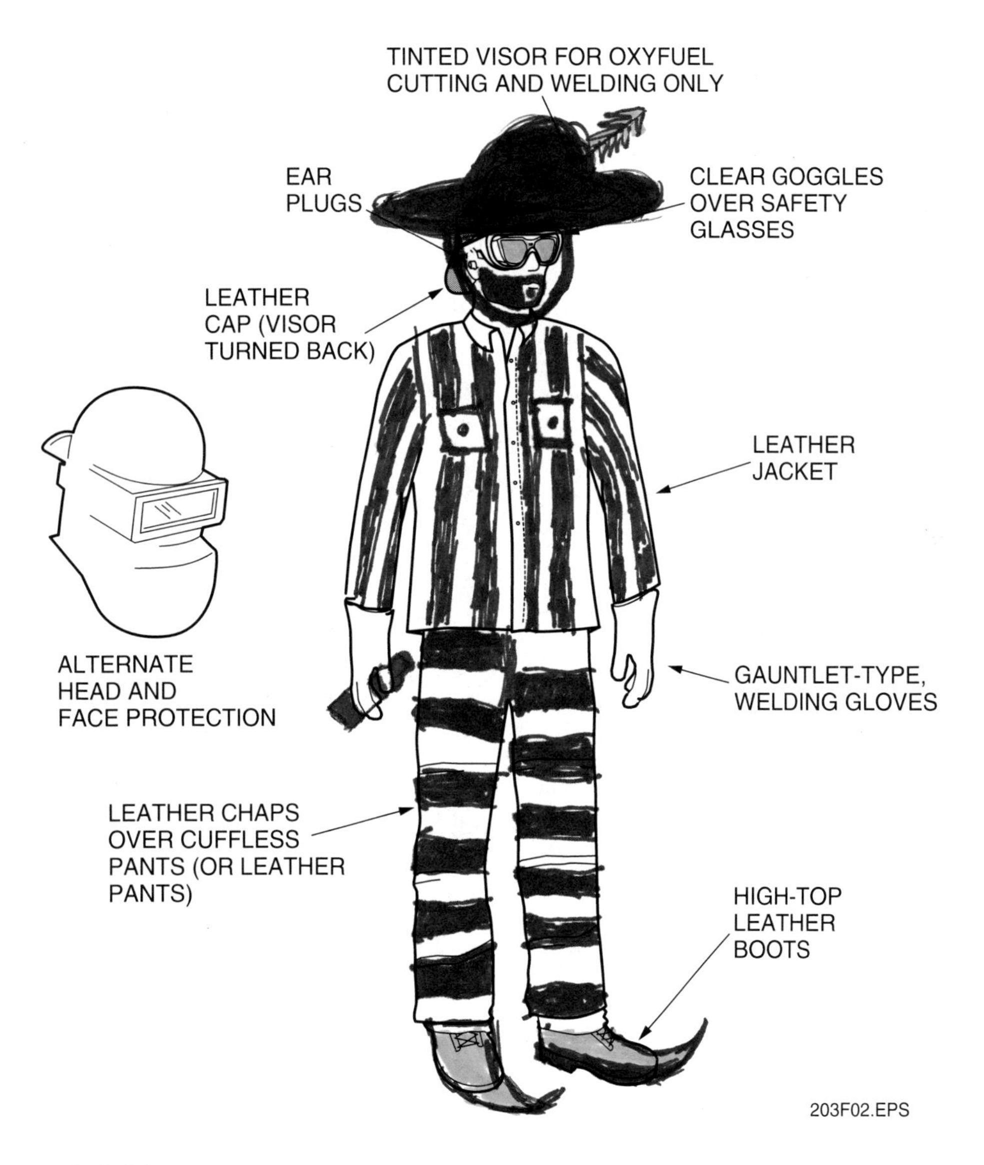

Figure 2 ◆ Proper work clothing.

Figure 3 ◆ Leather protective clothing.

HEAVY-DUTY INSULATED FIRE RESISTANT KEVLAR GLOVES

INSULATING COTTON LINERS FOR GLOVES

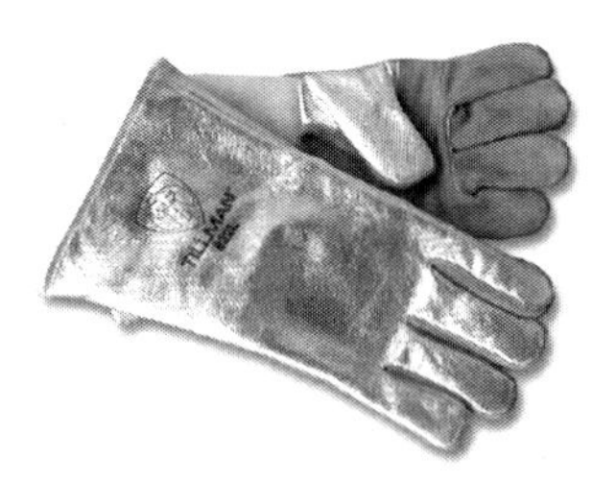

HEAVY-DUTY HEAT REFLECTIVE GLOVES

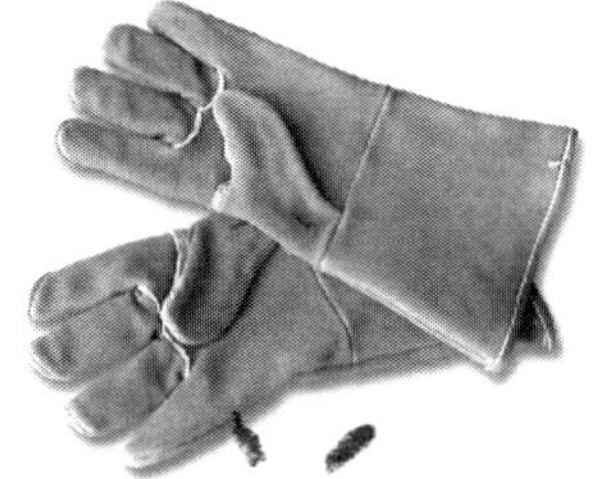

STANDARD LEATHER GLOVES

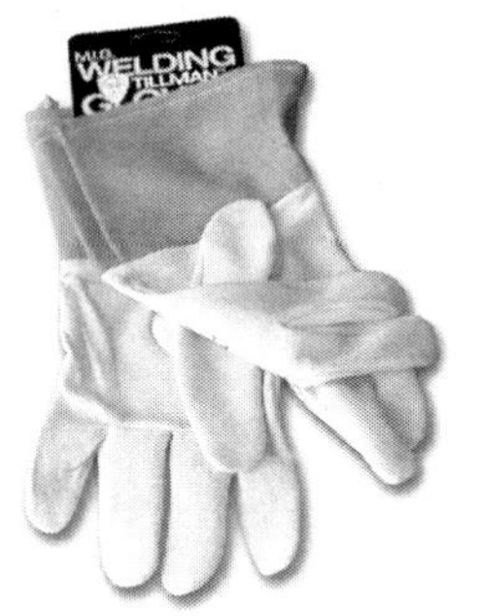

FLEXIBLE, LIGHTWEIGHT TIG WELDING GLOVES

203F04.EPS

Figure 4 ◆ Leather welding gloves.

2.2.0 Eye Protection

Eyes are irreplaceable and can be damaged easily. The welding arc emits ultraviolet and infrared rays that will burn unprotected eyes. The burns caused by the arc are called arc flash. Arc flash causes blistering of the outer eye. This blistering feels like sand or grit in the eyes. It can be caused by looking directly at the arc or by receiving reflected glare from the arc.

Another eye hazard encountered during shielded metal arc welding is flying debris. Grinding and surface cleaning operations create many small particles that are propelled in all directions. Slag, the crusty substance that forms on the deposited weld, is removed with a chipping hammer. During chipping, tiny particles of slag fly in all directions. In order to avoid eye injury from arc flash or flying debris, eye protection must be worn.

Safety glasses and goggles protect the eyes from flying debris. Safety glasses have impact-resistant lenses. Side shields should be fitted to safety glasses to prevent flying debris from entering from the side. Some safety glasses are equipped with shaded lenses to protect against glare.

Safety goggles are contoured to fit the wearer's face. Goggles are a very efficient means of protecting the eyes from flying debris. They are often worn over safety glasses to provide extra protection when grinding or performing surface cleaning. Face shields are worn to protect the entire face from sparks or flying debris. They are typically worn over clear goggles. Goggles or face shields are typically tinted to a shade 4 or 5 for oxyfuel welding and cutting purposes. *Figure 5* shows clear and tinted goggles and face shields.

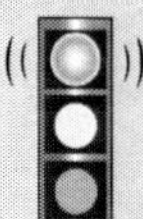

WARNING!

Shaded safety glasses, goggles, or face shields for oxyfuel welding will not provide sufficient protection from an electric arc. Looking directly at an electric arc with shaded lenses for oxyfuel welding will result in severe burns to the eyes and face.

Welding shields (*Figure 6*), also called welding helmets, provide eye and face protection for welders. Some shields are equipped with handles, but most are designed to be worn on the head. They either connect to helmet-like headgear or attach to a hard hat. Shields designed to be worn on the head have pivot points where they attach to the headgear. They can be raised when not needed. Welding shields are made of dark, nonflammable material. The welder observes the arc through a window that is either $2\frac{1}{2}$" by $4\frac{1}{4}$" or $4\frac{1}{2}$" by $5\frac{1}{4}$". The window contains a glass filter plate and a clear glass or plastic safety lens. The safety lens is on the outside to protect the more costly filter plate from damage by spatter and debris. For additional filter plate protection, a clear safety lens is sometimes also placed on the inside of the filter plate. On most welding shields, the window is fixed in the shield. However, some welding shields have a hinge on the $2\frac{1}{2}$" by $4\frac{1}{4}$" window. The hinged window containing the filter plate can be raised, leaving a separate clear safety lens. This protects the welder's face from hot slag during chipping. Many helmets are available with auto-darkening lenses. These lenses change from a light tint to a dark tint within $\frac{1}{25,000}$ of a second when an arc is struck. Some have variable auto-darkening lenses that adjust to the intensity of the arc. The use of this type of helmet eliminates the need to raise the helmet or flip up a lens when welding is started or completed.

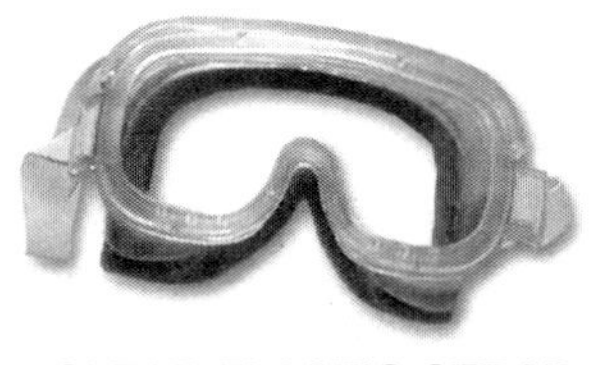

Figure 5 ◆ Gas welding goggles and face shields.

Filter plates come in varying shades. The shade depends on the maximum amount of amperage to be used. The higher the amperage, the darker the filter plate must be. Filter plates are graded by numbers. The larger the number, the darker the filter plate. The following list gives filter plate recommendations for various welding amperage ranges:

- *Shade 10* – 75 to 200 amperes
- *Shade 12* – 200 to 400 amperes
- *Shade 14* – 400 amperes and above

During normal welding operations, the window in the welding shield will become dirty from smoke and weld spatter. To see properly, the window must be cleaned periodically. The safety lens and filter plate can be easily removed and cleaned

Figure 6 ◆ Arc welding helmets.

in the same manner as safety glasses. With use, the outer safety lens will become impregnated with weld spatter. When this occurs, replace it.

2.3.0 Ear Protection

Ear protection is necessary to prevent hearing loss from noise. One source of noise for the welder is pneumatic chipping and scaling hammers. Welders must also protect their ears from flying sparks. This is especially important for out-of-position welding where falling sparks may enter the ear canal, causing painful burns. It is also common for welders who do not protect their ears to suffer from perforated ear drums caused by sparks. Always wear either earmuffs or earplugs for protection. *Figure 7* shows earmuffs and earplugs.

3.0.0 ◆ PRINCIPLES AND PROCEDURES OF SHIELDED METAL ARC WELDING (SMAW)

Shielded metal arc welding (SMAW) is an electric arc welding process in which the heat for welding is generated by an electric arc between a covered metal electrode and the work. The **filler metal** is

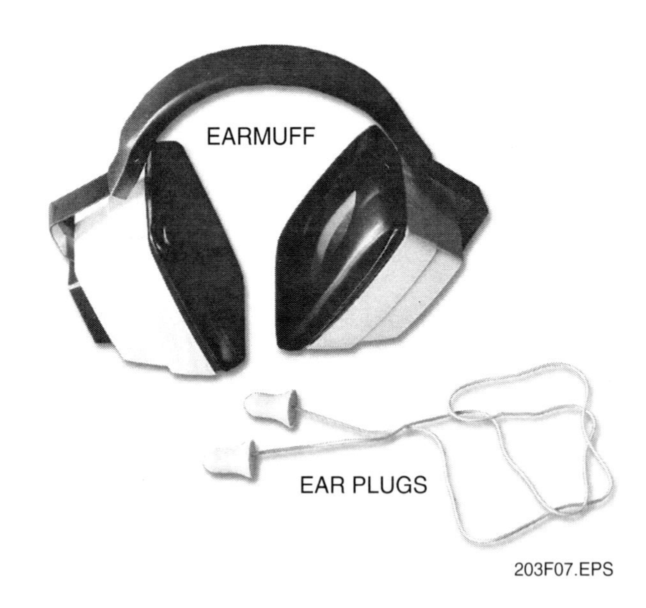

Figure 7 ◆ Earmuffs and earplugs.

deposited from the electrode, and the electrode covering provides the shielding. Some slang names for this process are stick welding and stick electrode welding. A diagram of this process is shown in *Figure 8*.

The SMAW process is the most popular, simplest, and most versatile arc welding process. This process can be used to weld both ferrous and nonferrous metals, and it can weld thicknesses above roughly 18 gauge (1.2 mm) in all positions. The arc is under the control of the welder and is visible. The welding process leaves a slag on the surface of the weld bead which must be removed. The most popular use for this process is for the welding of mild- and low-alloy steels. SMAW equipment is extremely rugged and simple, and the process is flexible in that the welder only needs to take the electrode holder and work lead to the point of welding.

The shielded metal arc welding process is basically a manually operated process. The electrode is clamped in an electrode holder, and the welder manipulates the tip of the electrode in relation to the metal being welded. The arc is struck, maintained, and stopped manually by the welder.

3.1.0 Equipment, Power Source, and Current

Equipment for shielded metal arc welding consists of a power source, welding cable, electrode holder, and work clamp or attachment. This equipment is shown in *Figure 9*.

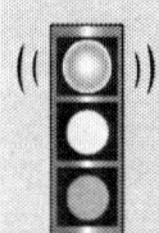

WARNING!
Do not ground welding equipment to instrument stands. Electric shock and equipment damage may result.

The purpose of the power source or welding machine is to provide electric power of the proper current and voltage to maintain a welding arc. Many different sizes and types of power sources are designed for shielded metal arc welding. Most power sources operate on 120V, 240V, or 480V input electric power.

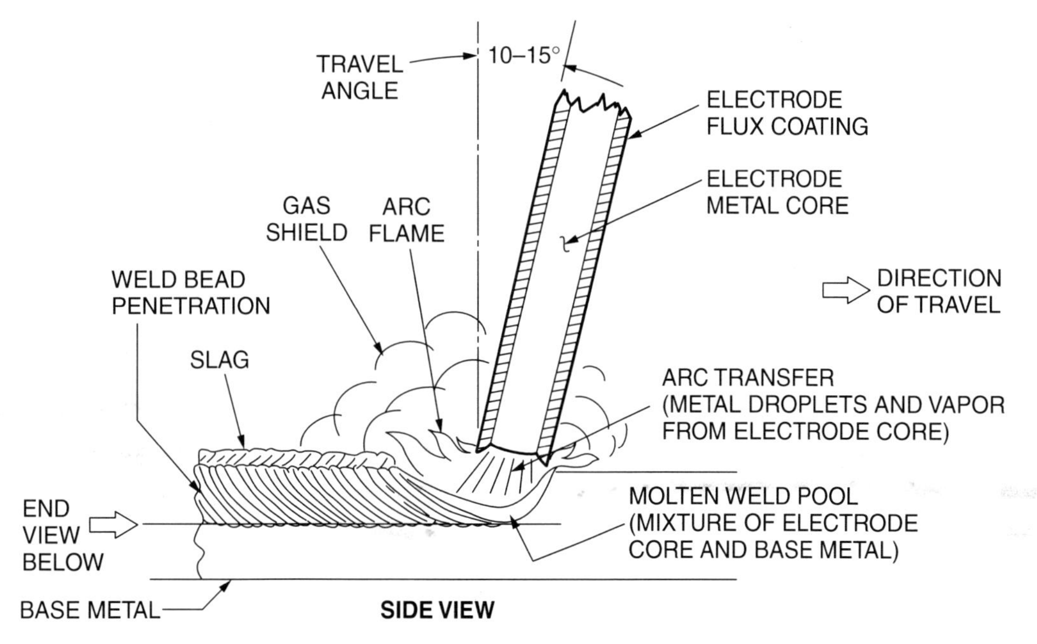

Figure 8 ◆ Shielded metal arc welding.

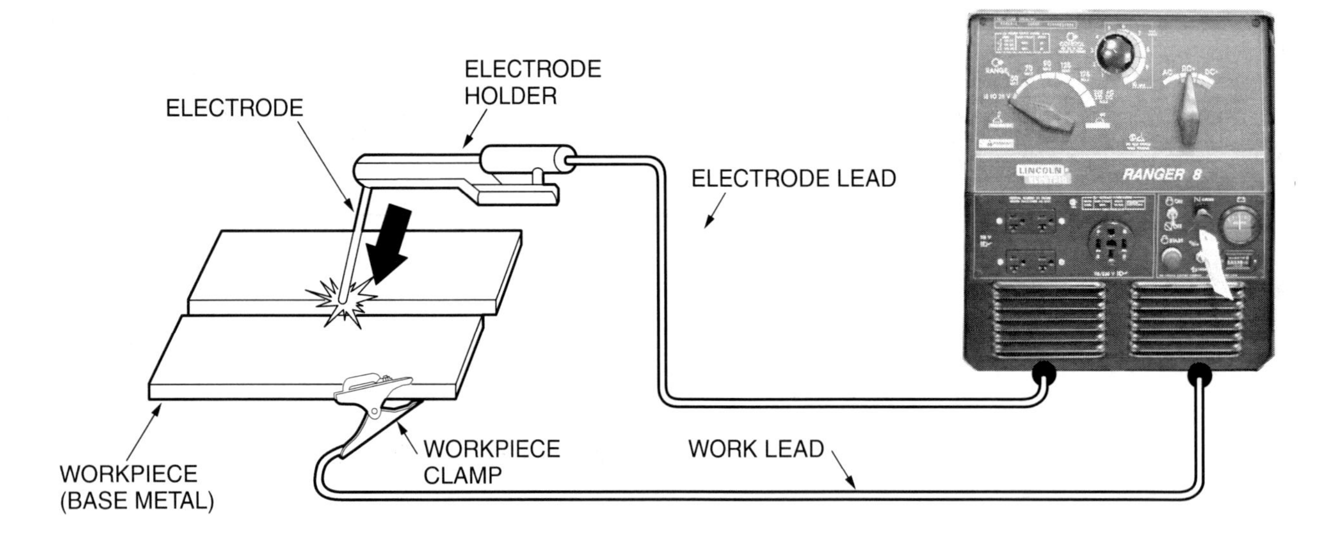

Figure 9 ◆ Equipment for shielded metal arc welding.

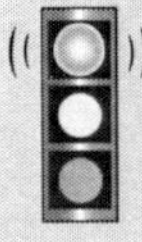

WARNING!

Welding shields must be installed around the welding work area when the welding arc may be visible to personnel in adjacent areas. Welding blankets must be installed when the sparks or hot metal pieces from the welding may come in contact with combustible materials.

Shielded metal arc welding can be accomplished using either direct current (DC) or alternating current (AC). Electrode negative (straight polarity) or electrode positive (reverse polarity) can be used with direct current. Each type of current has distinct advantages, but the selection depends on the availability of equipment and the type of electrode selected.

Direct current flows in one direction continuously through the welding circuit. It has several advantages over alternating current:

- Direct current is better at low currents and with small-diameter electrodes.
- All classes of covered electrodes can be used with satisfactory results.
- Arc starting is generally easier with direct current.
- Maintaining a short arc is easier.
- DC generally produces less weld spatter than alternating current.

Polarity or direction of current flow is important when direct current is used. Electrode negative (straight polarity) is often used when shallower penetration is required. Electrode positive (reverse polarity) is generally used where deep penetration is needed. Normally, electrode negative provides higher deposition rates than electrode positive.

Alternating current is a combination of both polarities that alternates in regular cycles. In each cycle the current starts at zero, builds up to a maximum value in one direction, and returns to zero. Then it builds up to maximum in the opposite direction and returns to zero again. The direction of current changes 120 times per second with the 60 Hz (cycle) current that is used in the United States. Depth of penetration and deposition rates for alternating current are generally between those for DC electrode positive and DC electrode negative. Some of the advantages of alternating current are as follows:

- **Arc blow** is rarely a problem with alternating current.
- Alternating current is well suited for the welding of thick sections using large diameter electrodes.

The electrode holder serves as a clamping device for holding the welding electrode and for transferring the welding current to the electrode. The insulated handle separates the welder's hand from the welding circuit. Two types of electrode holders are available: the clamp type and the collet type (*Figure 10*). Electrode holders come in various sizes and are designated by their current-carrying capacity.

The welding cables and connectors connect the power source to the electrode holder and to the work. These cables are normally made of copper or aluminum. The cable consists of hundreds of fine wires enclosed in an insulated casing of natural or synthetic rubber. The cable that connects the

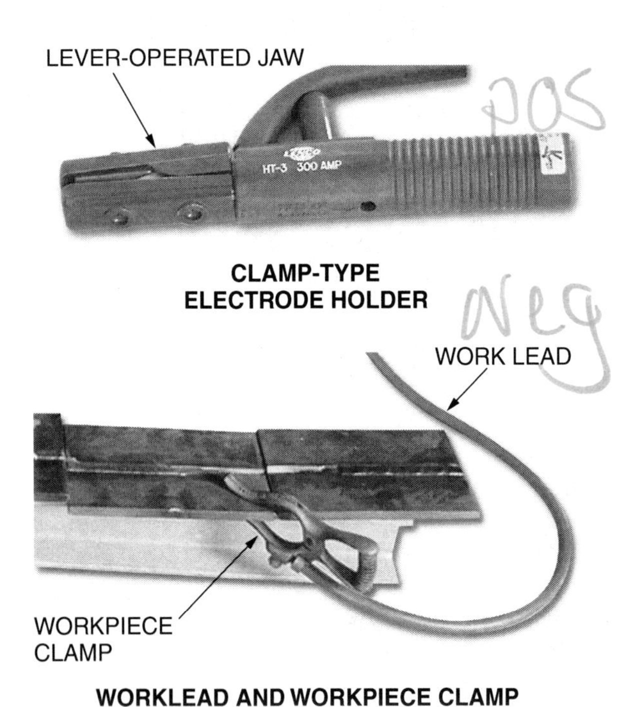

Figure 10 ◆ Electrode holders and work lead.

work to the power source is called the work lead. The work leads are usually connected to the work by a workpiece clamp, as shown in *Figure 10*. The cable that connects the electrode holder to the power source is called the electrode lead.

3.1.1 Electrodes

Electrodes for shielded metal arc welding consist of a core wire with a **flux** coating. See *Figure 11*.

The covering on the electrode dictates the usability of the electrode and provides the following:

- Gas from the decomposition of certain ingredients in the coating to shield the arc from the atmosphere
- Deoxidizers for purifying the deposited weld metal

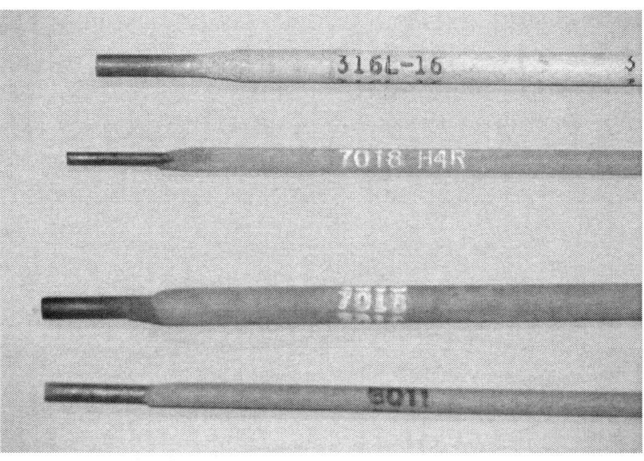

Figure 11 ◆ SMAW electrodes.

- Slag formers to protect the deposited weld metal from oxidation
- Ionizing elements to make the electrode operate more smoothly
- Alloying elements to provide higher strength deposited metal
- Iron powder to improve the productivity of the electrode

The size of the electrode is designated by the diameter of the core wire and the length of the electrode. Covered electrodes are commonly available in sizes from 1⁄16" (1.6 mm) to 5⁄16" (7.9 mm) diameter. Common lengths are 9" (230 mm) to 18" (457 mm), but for special applications electrode lengths up to 36" (914 mm) can be obtained.

The characteristics of different types of electrodes are standardized and defined by the American Welding Society (AWS). An AWS classification number is printed on each electrode for identification purposes.

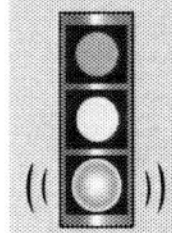

NOTE

Proper storage of welding rods is important. Follow manufacturers' instructions.

4.0.0 ◆ PRINCIPLES OF ORBITAL WELDING

Orbital welding is a process in which a welding machine is clamped to a pipe joint and automatically moves around the joint, making the weld as it goes. Orbital welding was introduced into the aerospace industry in the 1960s. It became more practical for the construction industry in the 1980s when smaller power supplies energized by 120VAC were developed to permit portability of the welding unit. In the construction industry, process piping and instrumentation tubing are clamped in place. An orbital weld head, described later in this section, rotates around the weld joint to make the required weld. New technology on many of these orbital welding systems allows the welder to simply enter the tubing dimensions and material to be welded on the control panel and begin welding. Most orbital welding processes use the gas tungsten arc welding process (GTAW) as the source of the electric arc that melts the base material and forms the weld. In the GTAW process, also referred to as the tungsten **inert gas** (TIG) process, an electric arc is established between a tungsten electrode and the part to be welded. The inert gas, or **shield gas**, is distributed around the weld area through the weld head to shield the weld area from ambient air to prevent oxidation.

In some specialty applications that require small, precise orbital welding, a welding process referred to as plasma welding may be used. Plasma welding uses an electrode to ignite an extremely hot gas within a confined torch tip, which is then forced through a small pinhole orifice, creating a jet of superheated gas called plasma. The plasma is used to liquefy and fuse the metal to be welded. This plasma process can be contained within the weld head of an orbital welder and used for precise welds on small tubing and piping.

4.1.0 Orbital Welding Equipment and Accessories

Orbital welding requires a power supply, orbital weld head, electrode, and shield gas. The power supply, weld head, and electrode are typically matched to one another to assure that the maximum output current rating of the power supply does not exceed the maximum working current ratings of the weld head and electrode. Because the equipment is designed to be used in field operations, branch electrical circuits that supply the current to the power supply must also be rated for the current rating of the power supply.

4.1.1 Orbital Welding Power Supplies

The power supply provides the arc welding current. It also controls welding parameters such as current settings and switching timing, provides the power to drive the motor in the weld head, and switches the shield gas(es) on or off as necessary. Many orbital system power units, including the one shown in *Figure 12*, come equipped with computer-controlled welding parameters that allow the operator to select and set these parameters based on the type and size of material to be welded. The computer maintains these parameters to provide a consistent weld. Power supplies for orbital welding systems are available in many different arrangements and current ratings.

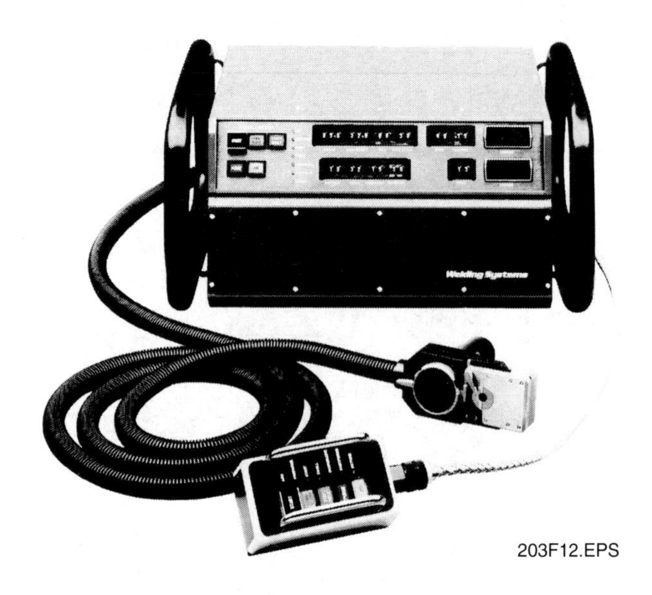

203F12.EPS

Figure 12 ◆ Orbital welding element.

4.1.2 Orbital Welding Weld Heads

Orbital weld heads are normally enclosed within a housing that provides an inert gas chamber around the weld joint. In the instrumentation industry, these enclosed weld heads are practical for welding tube sizes from 1⁄16" to 4". Smaller diameter tubing and pipe may be welded using an orbital by attaching a smaller welding head. The Swagelok® orbital welding system shown in *Figure 12* uses a Series 5 welding head as an attachment in tubing-to-tubing, tubing-to-fitting, fitting-to-fitting, valve-to-fitting, and valve-to-valve applications. The Series 5 welding head provides consistent, quality welds for outside diameters ranging from ⅛" to ⅝". Separate fixture blocks absorb the heat generated during welding, reducing heat input to the weld head. This design also simplifies joint makeup and reduces physical labor to increase productivity.

4.1.3 Electrodes for Orbital Welding

Orbital welding systems all use tungsten electrodes, except for those specialized systems that utilize plasma technology. Manufacturers of tungsten electrodes for orbital welding systems typically add an oxide to pure tungsten in order to improve the arc starting characteristics and the life of the electrodes. The shape and quality of the tungsten electrode tip in an orbital welding process is a vital variable. After a weld procedure has been established, it is important that the tip and surface condition of the electrode be maintained in order to produce a quality weld. The taper of the pointed end of the electrode, which is usually an angle between 14 and 60 degrees, affects the arc shapes. *Table 1* shows how the degree of taper affects the arc. Even though a taper angle with lesser degrees causes the arc to be more stable and easier to start, there is a tradeoff because the life of the electrode is shortened by tapers with lesser angles. In order to produce a quality orbital weld, it is recommended that pre-ground electrodes be used instead of attempting to grind the electrodes to the proper taper on the job site.

Table 1 How Electrode Taper Affects the Arc

Lesser Degrees in Taper of Electrode	Greater Degrees in Taper of Electrode
Easier to start the arc	Harder to start the arc
Arc is more stable	Arc has a tendency to wander
Weaker weld penetration	Stronger weld penetration

4.1.4 Shield Gases and Purge Gases

Shield gas is an inert gas that is distributed around the outside of the weld area, through the weld head. It prevents oxidation caused by oxygen in the ambient atmosphere. Inert gas may also be distributed within the tubing during the welding process and is referred to as **purge gas**. Argon gas is commonly used as a shield or purge gas. CO_2 is used as a purge gas. Helium is sometimes used as shield gas when welding copper material using an orbital welding system.

4.2.0 Advantages of Orbital Welding

Application of the orbital welding system in the instrumentation trade is becoming more popular for many reasons. These include the ability of the operator with minimal welding experience to make consistent welds repeatedly and rapidly. Other advantages in using the orbital welding system include:

- Most welds made using the orbital welding system equal or exceed the quality of manual welds, especially when using a system in which the welding parameters are computer-controlled.
- Orbital welds are consistent, with little or no variance between welds using the same parameter settings.
- Welding certifications are usually not required when using the orbital welding system.
- Orbital welding allows for the welding of piping or tubing in place, without rotating the piping or tubing.
- The weld head is relatively small and permits welding in locations with limited access.

4.3.0 Applications of Orbital Welding

The orbital welding system has many applications in the instrumentation trade, especially in industries that demand both very high quality and higher production in welded piping and tubing. This section identifies some of these industries and discusses how orbital welding can address their welding demands:

- *Food, dairy, and beverage industry* – Welds made on the piping and tubing used in the processes of these industries must be smooth, without any pitting, crevices or cracks, to prevent the trapping or buildup of food particles that can produce bacteria. The consistency of the orbital welds aids in this prevention by providing a constant, high-quality weld as long as the correct parameter settings are selected.
- *Boilers* – Boilers are built with many rows of parallel tubes that are interconnected through the end plates of the boiler, somewhat like a manifold system. These tubes often require welding in place. The orbital welding system provides the ideal configuration for this type of welding.
- *Pharmaceutical industry* – Like the food, dairy, and beverage industries, the pharmaceutical industry is monitored by the Food and Drug Administration. In many of the processes in the pharmaceutical industry, a very high-quality water supply is required. The welds on these piping and tubing systems are often made using the orbital welding system because of its uniformity and consistency in welds. This helps prevent bacterial buildup and contamination, which can be caused by jagged or inconsistent welds.
- *Petrochemical industry* – Much of the piping and tubing in the petrochemical industry is fabricated with some type of stainless steel material. Often, great lengths of process tubing must be installed by the instrumentation fitter or pipe fitter in order to connect the process with the instrument. Because these processes often involve hazardous or flammable chemicals, the job specifications and safety standards may required piping sections to be joined by welding, rather than some other method. Orbital welding produces the consistent, high quality

weld that is necessary to prevent leakage of products that may be considered hazardous to the environment or to humans.

- *Microcircuit manufacturing* – Even the tiniest impurity can affect the operation of solid-state microminiature circuits used in computers and other electronic equipment. Therefore, the manufacture of these devices must be done in clean rooms that are totally free of impurities in the air. Tubing and piping joints and terminations in clean rooms are often done with orbital welding to ensure that nothing escapes through them.

4.4.0 Orbital Welding Basics and Setup

The technician must be familiar with the process, fit-up and alignment, and parameters of orbital welding.

4.4.1 The Process

In order to start an arc in the orbital welding process, the power supply creates a high-voltage signal in a range of 3,500V to 7,000V. This signal is used to break down the insulating properties of the shield gas that protects the weld area from the ambient atmosphere. Whenever the shield gas is exposed to this high-voltage signal, the gas becomes conductive and passes a small amount of current. A capacitor within the power supply is added to this circuit. This reduces the **arc voltage** to a level at which the power supply can provide the necessary current without continually supplying such a high level of voltage. The metal being welded is melted and fused by the heat of the arc.

4.4.2 Fit-Up and Alignment

Parameters are set according to tubing size. Therefore, tubing that is to be welded using an orbital welding system must have uniform OD and ID dimensions in order for the welds to be consistent. The facing of tubing and piping to be welded should be square and flat, with no burrs or chamfers. The gap between two tubes that are butted together to be welded should be less than 5% of the tubing wall thickness. Gaps up to 10% of the wall thickness can be welded using the orbital welding system, but the quality of the weld will suffer and consistency in the welds will be jeopardized. Tubing wall thickness at the area to be welded should be within 5% of the nominal size of the tubing. Mismatches in alignment can be avoided by using appropriate alignment clamps, available through welding accessory suppliers. *Figure 13* shows two sizes of field alignment clamps that may be used to align tubing in the field prior to welding.

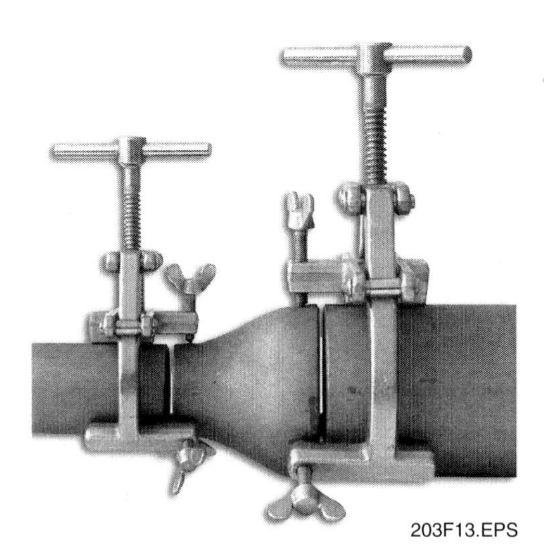

Figure 13 ◆ Field tubing alignment clamps.

4.4.3 Parameters

Many manufacturers of orbital welding equipment offer power supplies with programmed parameters that address a combination of tubing size, wall thickness, and tubing material to be welded. The parameters contained in these programs should be the parameters used to weld the corresponding tubing. However, not every combination of tubing size, thickness, and material can be addressed in these programs. Sometimes the welder/operator must determine the proper settings for the specific weld or welds to be made. This task usually requires a more thorough knowledge of the orbital welding process than is required when using the built-in parameter settings. Elements that must be considered when preparing to weld using an orbital welding system include **arc length**, weld speed, and welding current.

The arc gap depends on the weld current setting. The higher the weld current setting, the greater the arc gap may be. Generally speaking, however, an arc gap base setting of 0.010" plus half the penetration required (which is usually half of the wall thickness expressed in thousandths of an inch) is a good starting point. For example, if the tubing wall thickness to be welded is .020", a starting arc gap might be 0.010" + (½ × 0.020), or 0.020". Automatic arc gap settings are incorporated in the computer-controlled parameter settings.

Welding speed depends on the melting rate of the material to be welded, as well as the wall thickness of the tubing or piping. A good starting speed range, which can be adjusted up or down as needed, is 4" to 10" per minute. The faster speed is used when welding thinner-wall materials. Heavy wall thickness may require a slower welding speed. Again, in systems that use computer-controlled parameters, the speed of the weld head revolution will be controlled by the power supply based on the data entered by the operator.

As with the arc gap and welding speed in systems that use a computer program to control the parameters, the welding current is also controlled by this program. The current level is based on the material to be welded, wall thickness, weld speed, and the shield gas selected. A good starting current level is 1 amp per each 0.001" of wall thickness (based on stainless steel). In the case of stainless steel tubing that has a wall thickness of 0.20", a starting point for current would be 20 amps.

5.0.0 ◆ PRINCIPLES AND PROCEDURES OF GAS TUNGSTEN ARC WELDING (GTAW)

Gas tungsten arc welding (GTAW) fuses the parts to be welded by heating them with an arc between a **nonconsumable** tungsten electrode and the work (*Figure 14*). Filler metal may or may not be used with the process. Shielding is obtained from an inert gas or inert gas mixture. Slang names for the process are TIG welding, heliarc welding, heliwelding, argon-arc welding, and tungsten arc welding.

The GTAW process can be used to weld metals, including steel, stainless steel, aluminum, magnesium, copper, nickel, and titanium. The process can be used on a wide range of metal thicknesses. However, due to the relatively low deposition rates associated with the process, it is most often used to weld thinner materials. GTAW is also popular for depositing the root and hot **passes** on pipe and tubing.

The GTAW process can be used in all welding positions to produce quality welds on almost all metals used in industry. Because a shielding gas is used, the weld is clearly visible to the welder. No spatter is produced, post-weld cleaning is reduced, and slag is not trapped in the weld.

5.1.0 Methods of Application, Equipment, and Power Source

The gas tungsten arc welding process is normally applied using the manual method. The welder controls the torch with one hand and feeds filler metal with the other. The semi-automatic method is also used in some applications. The filler metal

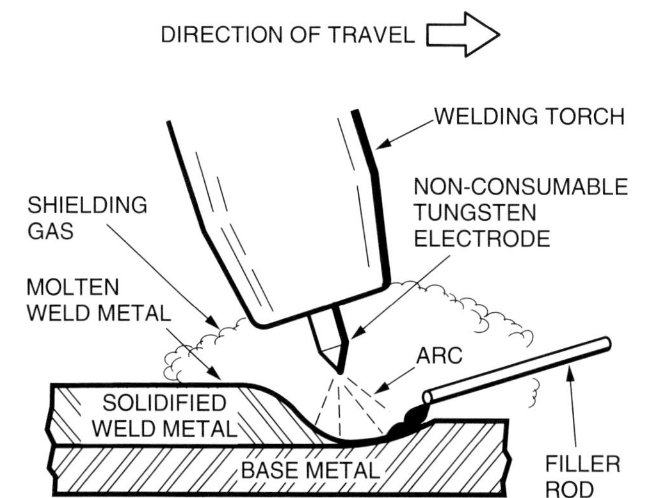

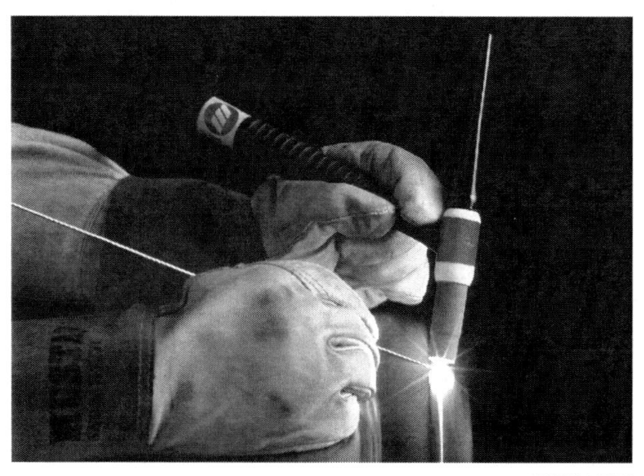

203F14.EPS

Figure 14 ◆ Gas tungsten arc welding.

is fed into the weld puddle by a wire feeder similar to that used in gas metal arc welding. The machine and automatic methods are increasing in popularity. With these systems, the welding operator monitors the welding operation and little welding skill is required. In the manual method, a relatively high degree of welding skill is required.

The major components required for gas tungsten arc welding are shown in *Figure 15*:

- The welding machine or power source
- The gas tungsten arc welding torch, including the tungsten electrode
- The shielding gas and controls
- The filler rod, when required

There are several optional accessories available, including a remote-controlled foot rheostat that permits the welder to control current while welding. Arc timers and controllers, high-frequency units, water circulating systems, and other specialized devices are also available.

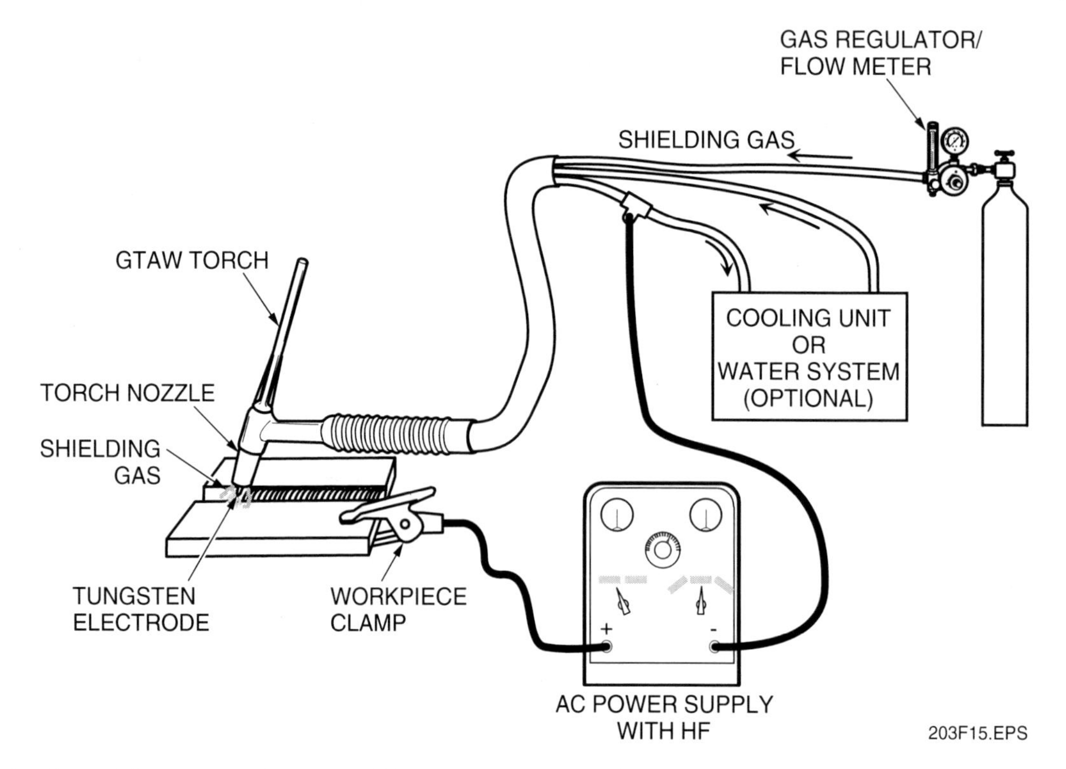

Figure 15 ◆ Equipment for gas tungsten arc welding.

A specially designed welding machine or power source is used for gas tungsten arc welding. Both AC and DC are used. Direct current is recommended for welding stainless steel, carbon steels, copper and its alloys, and nickel and its alloys.

Most machines designed for GTAW are equipped with solenoid valves for controlling the shielding gas and, when used, cooling water. A high-frequency spark gap control is also included in the welding machine, as are special connectors for attaching the welding torch and cable assembly to the machine.

It is possible to use AC or DC power sources designed for SMAW to do GTAW. However, a high-frequency attachment is usually required and the machine must be down rated when welding with alternating current. Best results are obtained with a welding machine specifically designed for GTAW.

The GTAW torch holds the tungsten electrode and directs the shielding gas and welding power to the arc. Torches come in different sizes. The larger sizes are usually water cooled. The torches normally come equipped with a cable assembly that directs the gas, welding power current, and cooling water (when used) from the machine to the torch.

5.2.0 Shielding Gases, Electrodes, and Filler Metal

A shielding gas protects the weld puddle and tungsten electrode from oxidation during welding. Argon and helium are the most common shielding gases used with the GTAW process. Argon produces a lower arc voltage than helium. This makes it especially useful for thin-gauge materials where lower arc heat is necessary to prevent excessive penetration. Helium is normally used on heavy weldments where high heat input is required to produce deeper penetration.

Argon is heavier than air, causing it to remain in the arc area longer when the puddle is located below the gas nozzle. Helium is lighter than air, making it necessary to double the gas flow rates in these positions. Because helium is lighter than air, it is sometimes used in positions that locate the puddle above the gas nozzle, such as the overhead position. In some cases, helium and argon are mixed to produce a shielding gas with the desirable features of both gases.

GTAW electrodes are made of tungsten or tungsten alloys. Tungsten is considered nonconsumable because it has the highest melting point of any metal, 6,170°F (3,410°C). When properly used, the electrode does not touch the molten weld puddle. If the tungsten electrode accidentally touches the weld puddle, it becomes contaminated and must be cleaned immediately. If it is not cleaned, an erratic arc will result. Electrodes are available in three alloys and of pure tungsten. Pure tungsten is the least expensive. Tungsten electrodes are color coded for ease of recognition, as shown in *Table 2*. Tungsten electrodes come in diameters ranging from 0.020" (0.5 mm) up through ¼" (6.4 mm).

Table 2 Classification of Tungsten Arc Welding Electrodes (*AWS A5.12*)

AWS Classification	% Tungsten (min) (by difference)	% Thoria	% Zirconia	Total % of Other Elements (max)	Color Code
EWP	99.5	—	—	0.5	Green
EWTh-1	98.5	0.8 to 1.2	—	0.5	Yellow
EWTh-2	97.5	1.7 to 2.2	—	0.5	Red
EWTh-3(a)	98.95	0.35 to 0.55	—	0.5	Blue
EWZr	99.2	—	0.15 to 0.40	0.5	Brown

The electrodes are available with either a ground finish or a cleaned finish. Tungsten electrodes are normally 3" (75 mm) to 6" (150 mm) long.

GTAW filler metal is solid and bare. The filler metal size is determined by the diameter. Filler metals are available in a wide range of sizes in an approximate range from 0.020" (0.5 mm) to ¼" (6.4 mm). Filler metals are manufactured in straight cut lengths for manual welding and continuous spools for semiautomatic and automatic welding.

Filler metals for GTAW are classified using the same system as GMAW electrodes. The only difference is that gas metal arc wires carry electric current and are considered electrodes (E), while gas tungsten welding wires normally do not carry current and are considered filler rods (R).

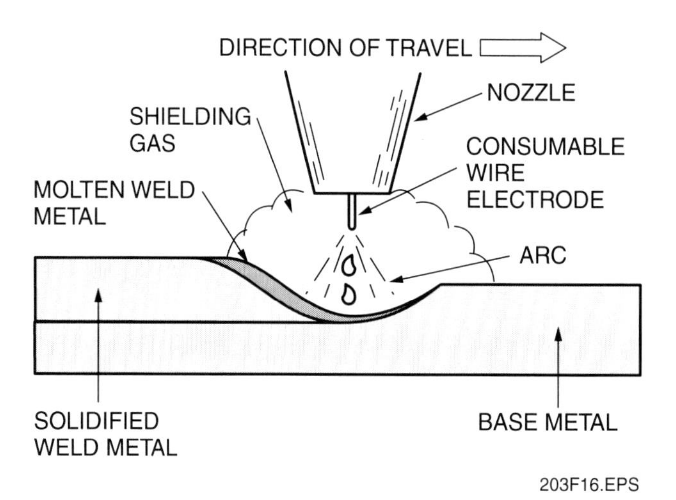

Figure 16 ◆ Gas metal arc welding.

6.0.0 ◆ PRINCIPLES AND PROCEDURES OF GAS METAL ARC WELDING (GMAW)

Gas metal arc welding (GMAW) is an electric arc welding process in which the parts to be welded are fused by heating them with an arc between a solid metal electrode and the work. Filler metal is obtained by melting a wire electrode, which is fed continuously into the arc by the equipment. Shielding is obtained from an externally supplied gas or gas mixture. Slang names for the process include MIG welding, CO_2 welding, micro-wire welding, short arc welding, dip transfer welding, and wire welding. *Figure 16* shows a diagram of this process.

The GMAW process is capable of welding most ferrous and nonferrous metals from thin to thick sections. It can be used in all positions to produce weld deposits with little or no spatter. Higher deposition rates, travel speeds, and welder efficiencies result in less welding time compared to shielded metal arc welding.

6.1.0 Methods of Application, Equipment, and Power Source

GMAW is widely used in the semiautomatic, machine, and automatic modes. Manual welding cannot be done through this process. The most popular method of applying GMAW is semiautomatically, in which the welder guides the gun along the joint and adjusts the welding parameters. The wire feeder continuously feeds the filler wire electrode. The power source maintains the arc length. A relatively low degree of welder skill is required to master the semiautomatic method.

The second most popular method is the automatic method, in which the machinery controls the welding parameters, arc length, joint guidance, and wire feed. The process is merely observed by the operator. GMAW equipment consists of a power source, controls, wire feeder, welding gun, welding cables, and gas shielding system. There are several items that are added to the basic equipment for automatic applications, such as seam followers and devices for providing movement. *Figure 17* shows a diagram of the equipment used for semiautomatic gas metal arc welding.

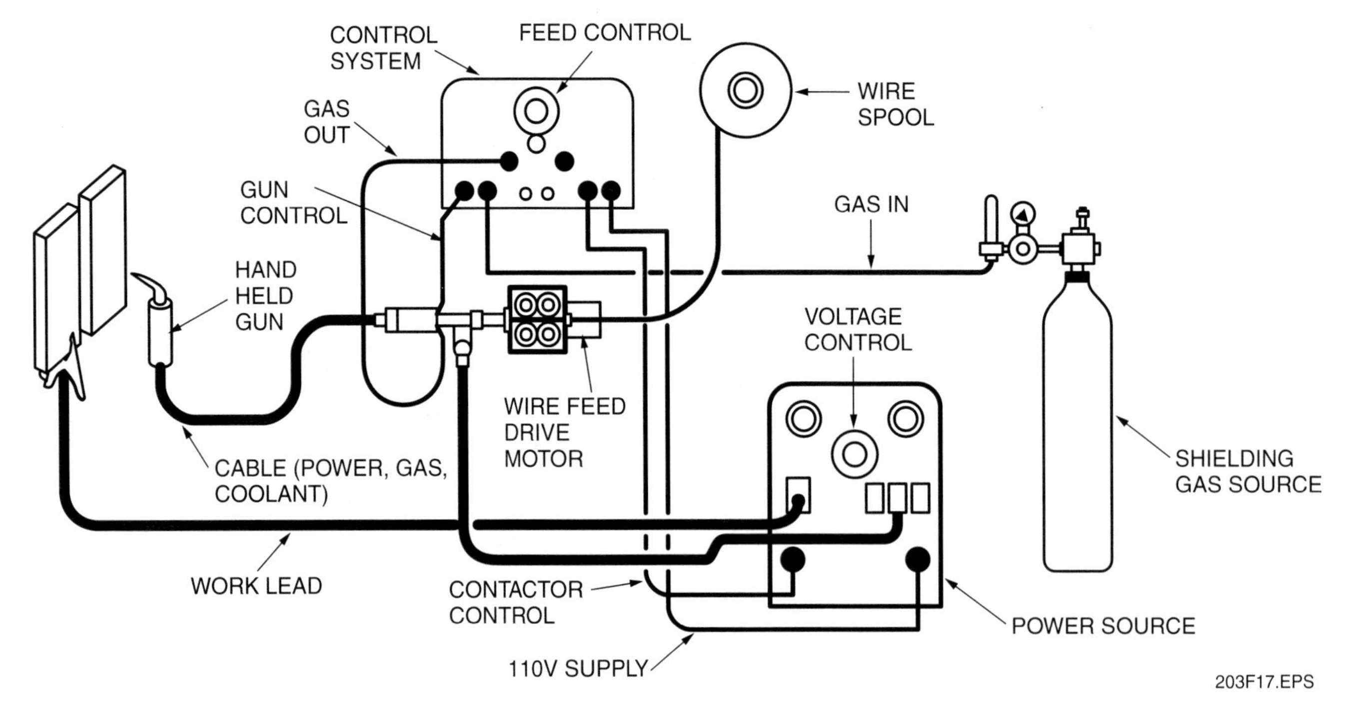

Figure 17 ◆ Equipment for semiautomatic gas metal arc welding.

The wire feeder drives the electrode wire from the wire spool through the cable and gun to the welding arc. The wire feed system must be matched to the power source.

The constant-speed wire feeder is used with the constant voltage power source. In this system, the relationship between electrode melt-off and current must be maintained to produce a stable arc. The current is set by the wire feed control. During welding, the power source provides the proper amount of current to maintain the melt-off rate of the electrode. Because the wire speed is constant, the current fluctuations maintain uniform melt-off to produce a consistent arc length.

A variable speed wire feeder is normally used with a constant current power source. In this system, the wire feed varies to maintain a uniform arc length. A voltage sensing device is incorporated into the system to detect changes in voltage (arc length). Based on the voltage change, the wire feeder speeds up or slows down to correct the change in voltage (arc length).

Figures 18 and *19* depict the relationship between current and wire feed speed for various types and sizes of electrode wire.

6.1.1 *Shielding Gases, Electrodes, and Metal Transfer*

The shielding gas system displaces the air around the arc to prevent contamination of the molten pool of metal by oxygen and nitrogen in the atmosphere. The types of shielding gases and their appropriate application are shown in *Table 3*.

The shielding gas system consists of a gas supply, regulator, flowmeter, and connecting hose.

Shielding gases can be stored in a high-pressure **cylinder** or a bulk storage system for high-volume applications. In some cases, cylinders are manifolded to provide shielding gas to several welding stations. The regulator is normally used on a high-pressure cylinder to reduce the cylinder pressure to a safe working pressure. The flowmeter is used to set the actual flow rate during welding. The hoses are used to carry the shielding gas from the flowmeter to the gun.

This type of metal transfer refers to the method through which molten metal from the electrode crosses the arc to form the weld deposit. Various types of metal transfer are possible with GMAW (*Figure 20*).

One type of metal transfer is commonly called short-circuiting. During short-circuiting transfer, a metal droplet grows at the tip of the electrode wire. As the droplet grows, the wire moves closer to the puddle until it actually comes in contact with the puddle. At this point, a short-circuit is produced that causes the wire to resist heat and pinch off, producing a new arc. This cycle continues many times a second, depending on the current/voltage relationship.

The short-circuit method of metal transfer is commonly used to weld carbon steels, low alloy steels, and some stainless steels in thin to medium thicknesses of plate and pipe. CO_2 and argon/CO_2 mixtures are commonly used as shielding gases.

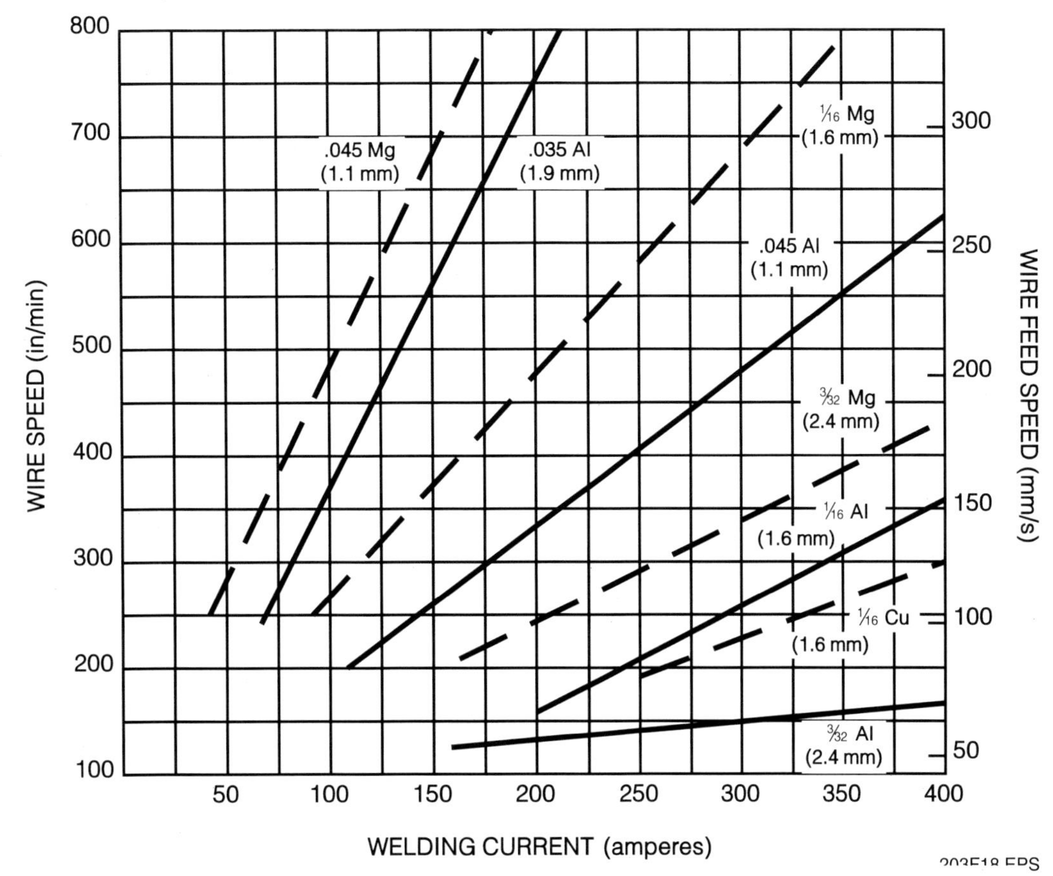

Figure 18 ◆ Wire feed speed versus welding current for several nonferrous metal wire electrodes.

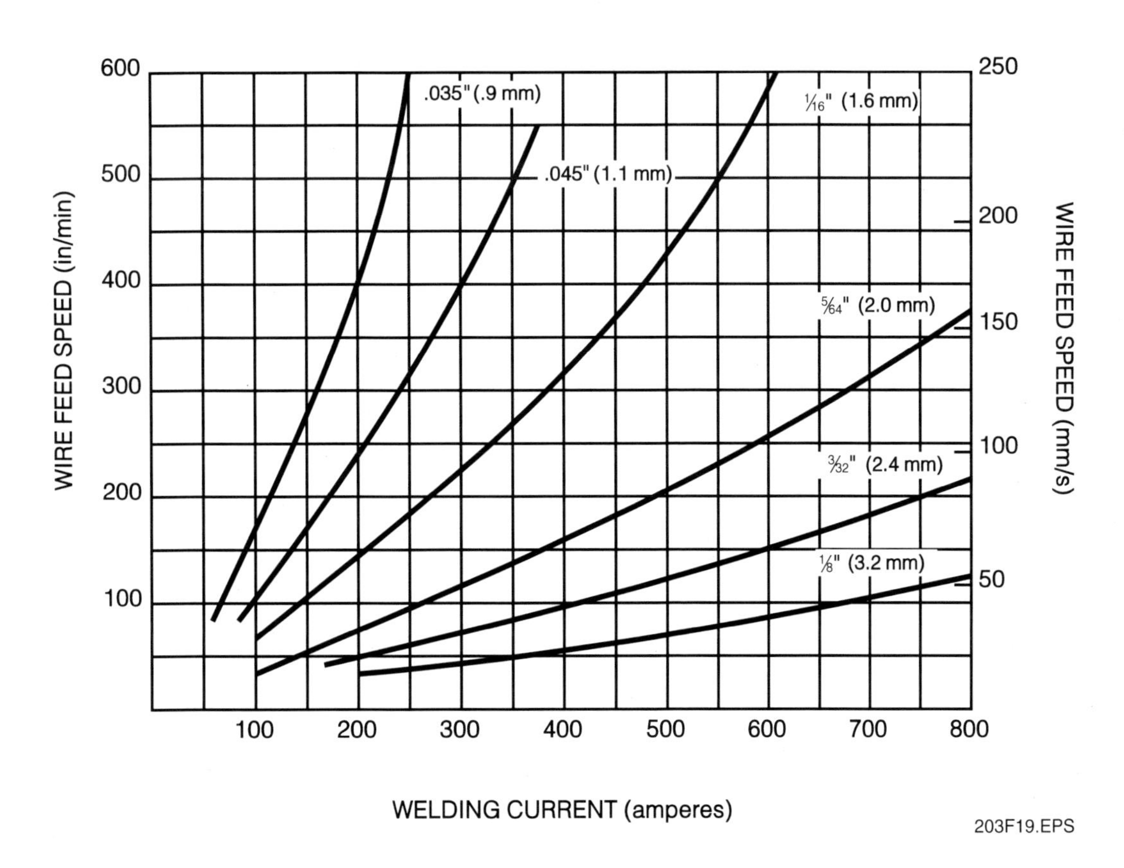

Figure 19 ◆ Wire feed speed versus welding current for steel electrode wire.

Table 3 Shielding Gases and Their Applications for Gas Metal Arc Welding

ARGON	AR	INERT	NONFERROUS METALS	• LEAST EXPENSIVE • INERT GAS PROVIDES SPRAY TRANSFER
ARGON + HELIUM	50% AR 50% HE	INERT	AL, MG, COPPER & THEIR ALLOYS	• HIGHER HEAT IN ARC • USE ON HEAVER THICKNESS • LESS POROSITY • PROVIDES SPRAY TRANSFER
ARGON + OXYGEN	ARGON + 1–2% O	OXIDIZING	STAINLESS STEEL	• OXYGEN PROVIDES ARC STABILITY
ARGON + OXYGEN	ARGON + 3–5% O	OXIDIZING	MILD & LOW ALLOY	• PROVIDES SPRAY TRANSFER
ARGON + CARBON DIOXIDE	75% AR 25% CO_2	SLIGHTLY OXIDIZING	MILD & LOW ALLOY STEELS (ALSO SOME STAINLESS WITH GMAW)	• SMOOTH WELD SURFACE REDUCES PENETRATION • SHORT CIRCUITING
HELIUM + ARGON + CARBON DIOXIDE	90% HE 7.5 AR + 2.5% CO_2	ESSENTIALLY INERT	STAINLESS STEEL AND SOME ALLOY STEELS	• PROVIDES ARC STABILITY • HELPFUL IN OUT OF POSITION WELDING • SHORT CIRCUITING
HELIUM + ARGON	75% HE 25% AR	INERT	AL, MG, COPPER & THEIR ALLOYS	• HIGHER HEAT INPUT THAN AR • MINIMUM POROSITY
CARBON DIOXIDE	CO_2	OXIDIZING	MILD & LOW ALLOY STEELS (ALSO ON SOME STAINLESS STEELS)	• LEAST EXPENSIVE GAS • DEEP PENETRATION • SHORT CIRCUITING OR GLOBULAR
NITROGEN	N_2	ESSENTIALLY INERT	COPPER & COPPER ALLOYS & PURGING STAINLESS STEEL PIPE & TUBING	• HAS HIGH HEAT INPUT • NOT POPULAR IN NORTH AMERICA • GLOBULAR

203T03.EPS

Globular transfer is similar to short-circuiting in that a droplet is formed at the end of the electrode wire. However, during the globular transfer, the molten ball continues to grow until it is larger than the diameter of the electrode wire. Then the droplet detaches and crosses the arc to form the weld deposit. Because of this, the arc is less stable, and more spatter is produced. Globular transfer is commonly used to weld the same metals as those welded in short-circuiting transfer, except in greater thicknesses.

Globular transfer is capable of producing higher deposit rates than short-circuiting, but is limited to the flat and horizontal positions. CO_2 is normally used to shield the arc.

Spray transfer is characterized by small droplets that rapidly transfer the arc. During this metal transfer, the electrode tapers down to a point. The droplets are formed at the tip and pinched off due to electromagnetic forces. Spray transfer is normally used to weld nonferrous metals in all positions. Argon, or a mixture of argon and helium, is used to shield the arc. Spray transfer can be used to weld carbon steels, low alloy steels, and some stainless steels, using a mixture of argon and oxygen (1% to 5% oxygen). Spray transfer on steels is normally used to weld medium to heavy thicknesses of steel in the flat and horizontal positions.

The GMAW electrode wire is solid and bare. Steel wires normally have a thin copper coating to improve electrical pickup and protect the wire from oxidation. The electrode wire size is determined by its diameter. Various diameters are available based on the metal transfer, welding position, and application. The wire is available in

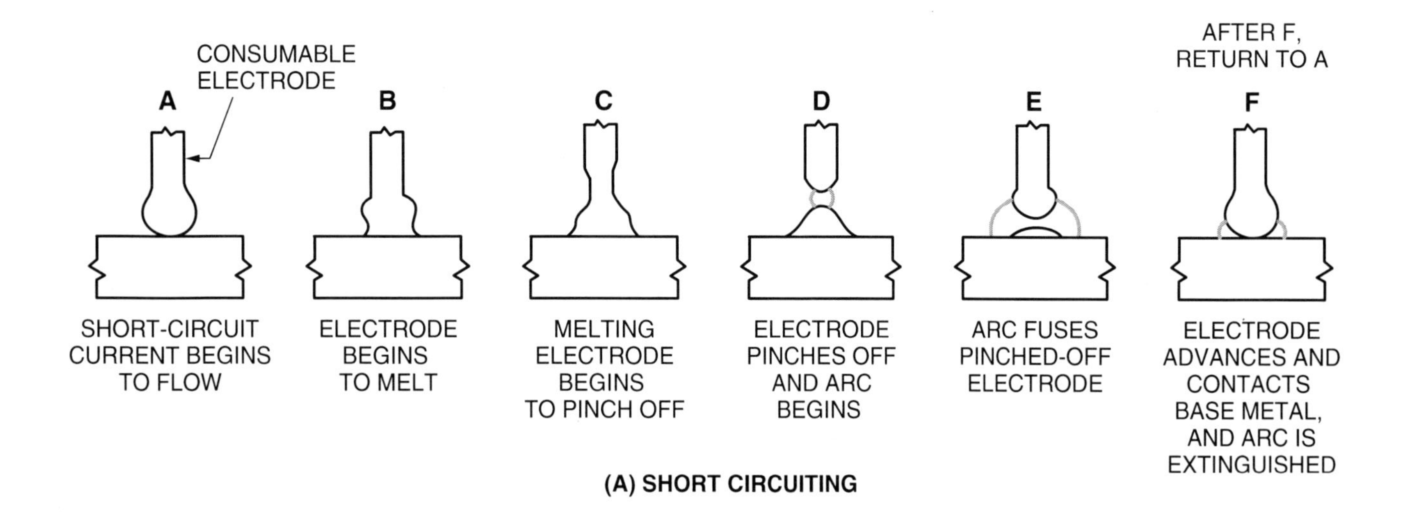

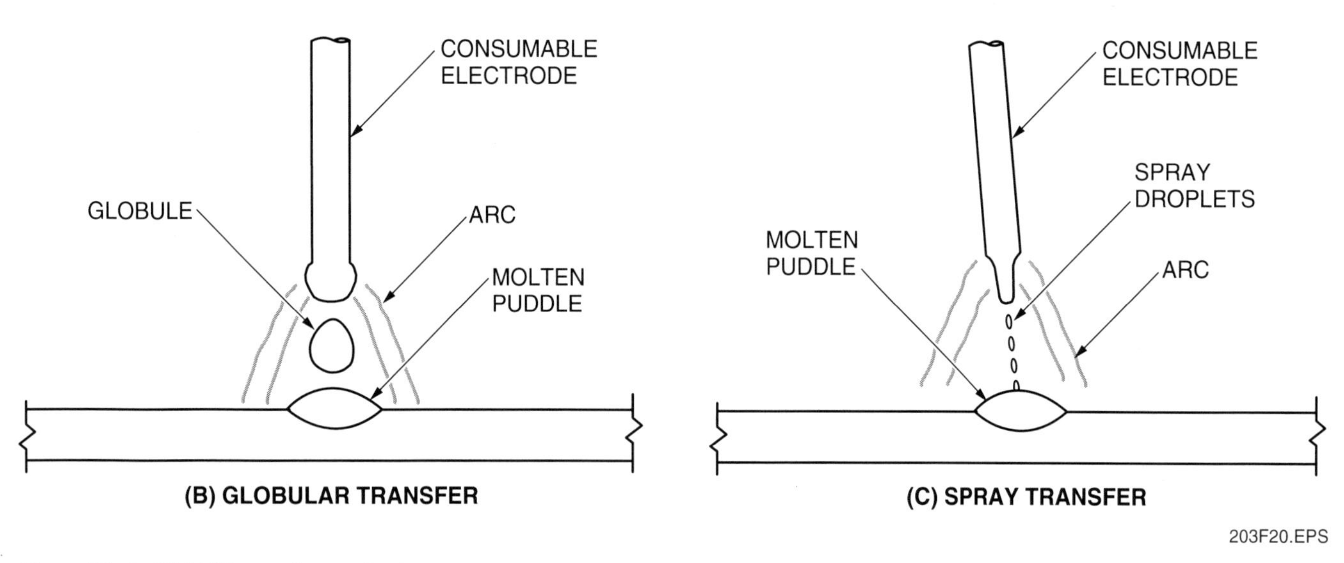

Figure 20 ◆ GMAW transfer methods.

spools, coils, and reels and is packed in special containers for protection against deterioration and contamination.

The American Welding Society classifies electrode wires using a series of numbers and letters, similar to SMAW electrodes. For carbon and low alloy steels, the classifications are based on the mechanical properties of the weld deposits, as well as their chemical compositions. For most other metals, the electrode wire is placed on the spool and/or the filler metal packaging.

A typical steel classification is ER70A-6, where:

The E indicates the filler wire is an electrode that may be used for gas metal arc welding. The R indicates it may also be used as a filler rod for gas tungsten arc or plasma arc welding.

The next two (or three) digits indicate the nominal tensile strength of the filler wire.

The letter to the right of the digits indicates the type of filler metal. An S stands for a solid wire, and a C stands for a metal core wire, which consists of a metal powder core in a metal sheath.

The digit or letters and digit in the suffix indicate the special chemical composition of the filler.

See *Table 4* for electrode compositions and mechanical properties. For stainless steels and the nonferrous metals, the classification is based on the chemical compositions of the electrodes.

7.0.0 ◆ PRINCIPLES AND PROCEDURES FOR BRAZING AND OXYACETYLENE WELDING

Oxyfuel welding (OFW) is a gas welding process used to fuse the parts to be welded by heating them with a gas flame or flames obtained from the combustion of fuel gas and oxygen. The fuel gas may be acetylene, natural gas, propane, or methylacetylene propadiene. Acetylene is normally

Table 4 Electrode Composition and Mechanical Properties Chart (*AWS A15.18* and *A5.28*)

Electrode	C	Mn	Si	P	S	Ni	Cr	Mo	Cu	Other
					CARBON STEELS					
ER70S-2	.07	.90-1.40	.40-.70	.025	.035	–	–	–	.50	Ti, Zr, Al
ER70S-3	.06-.15	.90-1.40	.45-.70	.025	.035	–	–	–	.50	–
ER70S-4	.07-.15	1.00-1.50	.65-.85	.025	.035	–	–	–	.50	–
ER70S-5	.07-.19	.90-1.40	.30-.60	.025	.035	–	–	–	.50	Al
ER70S-6	.07-.15	1.40-1.85	.80-1.15	.025	.035	–	–	–	.50	–
ER70S-7	.07-.15	1.50-2.00	.50-.80	.025	.035	–	–	–	.50	–
ER70S-G		NO CHEMICAL REQUIREMENTS								
					CHROMIUM-MOLYBDENUM STEELS					
ER80S-B2	.07-.12	.40-.70	.40-.70	.025	.025	.20	1.2-1.5	.40-.65	.35	–
ER80S-B2L	.05	.40-.70	.40-.70	.025	.025	.20	1.2-1.5	.40-.65	.35	–
ER90S-B3	.07-.12	.40-.70	.40-.70	.025	.025	.20	2.3-2.7	.90-1.20	.35	–
ER90S-B3L	.05	.40-.70	.40-.70	.025	.025	.20	2.3-2.7	.90-1.20	.35	–
E80C-B2L	.05	.40-1.00	.25-.60	.025	.030	.20	1.00-1.5	.40-.65	.35	–
E80C-B2	.07-.12	.40-1.00	.25-.60	.025	.030	.20	1.0-1.50	.40.-65	.35	–
E90C-B3L	.05	.40-1.00	.25-.60	.025	.030	.20	2.0-2.5	.90-1.20	.35	–
E90C-B3	.07-.12	.40-1.00	.25-.60	.025	.030	.20	2.0-2.5	.90-1.20	.35	–
					NICKEL STEELS					
ER80S-Ni1	.12	1.25	.40-.80	.025	.025	.80-1.10	.15	.15	.35	–
ER80S-Ni2	.12	1.25	.40-.80	.025	.025	2.00-2.75	–	–	.35	–
ER80S-Ni3	.12	1.25	.40-.80	.025	.025	3.00-3.75	–	–	.35	–
E80C-Ni1	.12	1.25	.60	.025	.030	.80-1.10	–	.65	.35	–
E80C-Ni2	.12	1.25	.60	.025	.030	2.00-2.75	–	–	.35	–
E80C-Ni3	.12	1.25	.60	.025	.030	3.00-3.75	–	–	.35	–
					CHROMIUM-MOLYBDENUM STEELS					
ER80S-D2	.07-.12	1.60-2.10	.50-.80	.025	.025	.15	–	.40-.60	.50	–
					OTHER LOW ALLOY STEEL ELECTRODES					
ER100S-1	.08	1.25-1.80	.20-.50	.010	.010	1.40-2.10	.30	.25-.55	.25	V, Ti, Zr, Al
ER100S-2	.12	1.25-1.80	.20-.60	.010	.010	.80-1.25	.30	.20-.55	.35-.65	V, Ti, Zr, Al
ER110S-1	.09	1.40-1.80	.20-.55	.010	1.90-2.60		.50	.25-.55	.25	V, Ti, Zr, Al
ER120S-1	.10	1.40-1.80	.25-.60	.010	2.00-2.80		.60	.30-.65	.25	V, Ti, Zr, Al
ERXXS-G		NO CHEMICAL REQUIREMENTS								
EXXC-G		NO CHEMICAL REQUIREMENTS								

203T04.EPS

used for welding applications. The flame produced by the combustion of acetylene and oxygen is one of the hottest, at approximately 6,300°F (3,482°C). The flame melts the edges of the joint and the filler metal (if used) to produce the weld. Refer to *Figure 21*.

The proportions of oxygen and acetylene determine the type of flame. The three basic flame types are carburizing, neutral, and oxidizing (*Figure 22*). The neutral flame is generally preferred for welding. It has a clean, well-defined white cone indicating the mixture of gases and the fact that no gas is wasted. The carburizing flame, also referred to as a reducing flame, has excess acetylene and adds carbon to the weld. Its white cone has a feathery edge. The oxidizing flame, with an excess of oxygen, has a shorter envelope and a small, pointed, white cone. This flame oxidizes the weld metal and is used only for welding specific metals. Flame cutting is accomplished through the addition of an extra oxygen jet to oxidize the metal being cut.

The oxyacetylene welding process is commonly used to join relatively thin carbon steel, copper, and copper alloys. The process is portable and versatile. The equipment is relatively inexpensive. The weld puddle is clearly visible to the welder, which allows accurate puddle control. The industrial applications of this process are in maintenance repair, small diameter piping, auto body repair, and light manufacturing.

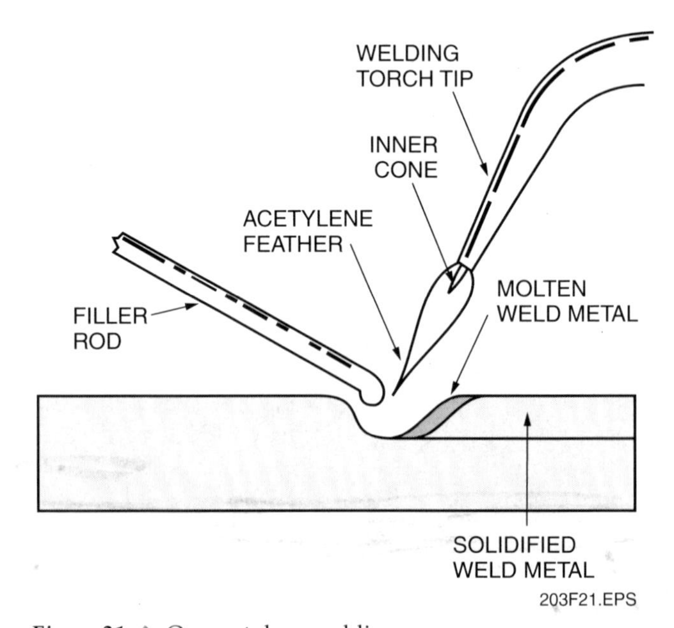

Figure 21 ◆ Oxyacetylene welding.

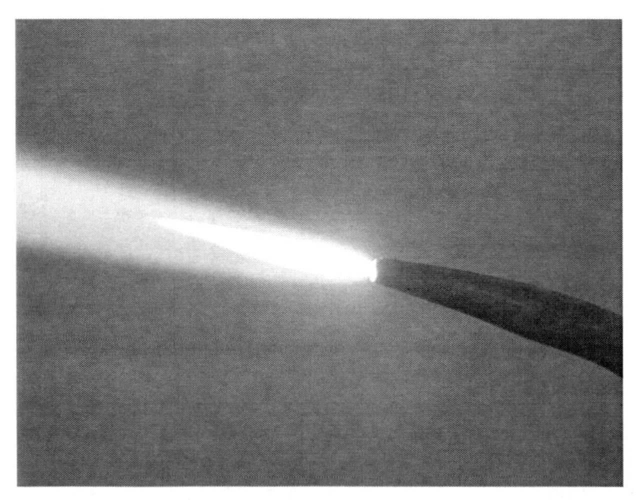
CARBURIZING FLAME

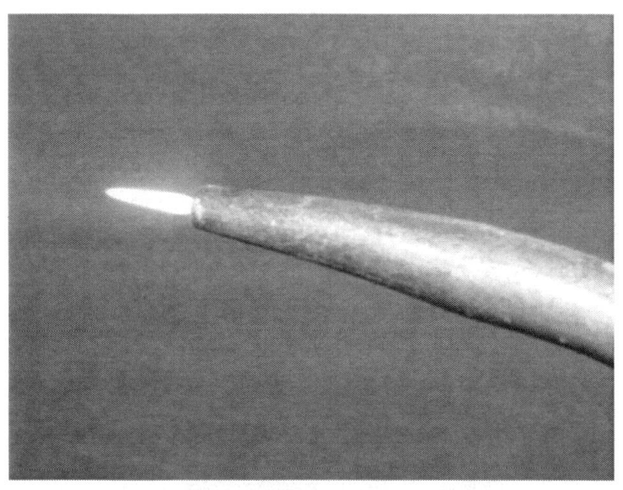
NEUTRAL FLAME

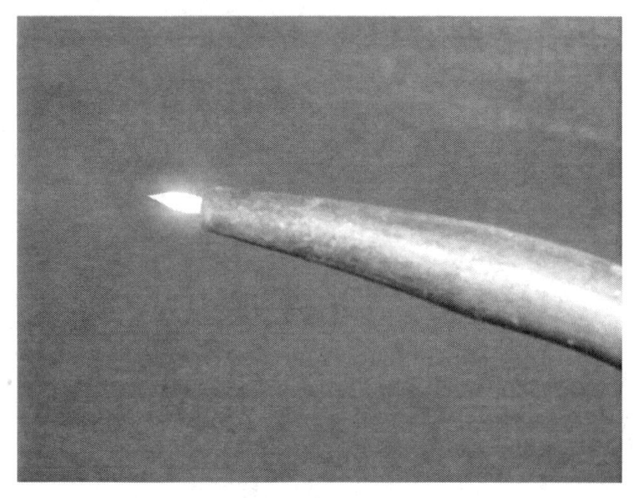
OXIDIZING FLAME

203F22.EPS

Figure 22 ◆ Basic flame types.

7.1.0 Method of Operation and Equipment

Oxyacetylene welding is a manual process that requires a relatively high level of welding experience and practice. It requires a gas-fueled torch to raise the temperature of two similar pieces of metal to their fusion point, allowing them to flow together. A filler rod may be used to deposit additional metal. The gas and oxygen are mixed to correct proportions in the torch. The welder can adjust the torch to produce various types of flames. A quality gas weld is consistent in appearance and shows uniform deposit of weld metal. Some of the factors that must be considered when making a gas weld are edge preparation, spacing and alignment of the parts, temperature control (before, during, and after the welding process), size of the torch tip, size and type of the filler rod, flame adjustment, and rod and torch manipulation.

> **WARNING!**
> Oil or any type of oil-based lubricants must not be used on or near any oxygen fittings, hoses, lines, regulators, valves, cylinders, or any other component or element associated with the use of oxygen. Exposure of oxygen to oil or oil-based materials can cause the oxygen to explode.

Components of an oxyacetylene welding system include the following:

- An oxygen cylinder
- An acetylene cylinder
- An oxygen regulator
- An acetylene regulator
- A dual hose assembly
- A welding/cutting torch with tips

Accessories that may or may not be used with an oxyacetylene welding system follow:

- A **striker** to light the torch
- A cylinder and regulator wrench
- **Soapstone** to mark the pieces to be welded

An oxyacetylene welding unit is often installed on a cart with wheels to make it portable. A shop oxyacetylene welding unit and a hand-portable oxyfuel welding unit are shown in *Figure 23. Figure 24* illustrates a cylinder/regulator wrench combination (also called a torch wrench), a typical striker, and soapstone holders.

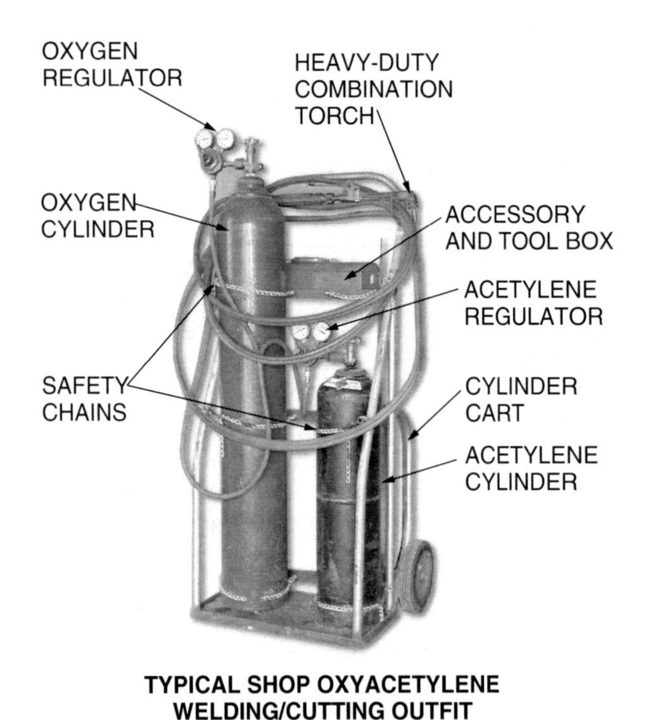

TYPICAL SHOP OXYACETYLENE
WELDING/CUTTING OUTFIT

TYPICAL HAND-PORTABLE OXYFUEL
WELDING/CUTTING OUTFIT

203F23.EPS

Figure 23 ◆ Oxyacetylene and oxyfuel outfits.

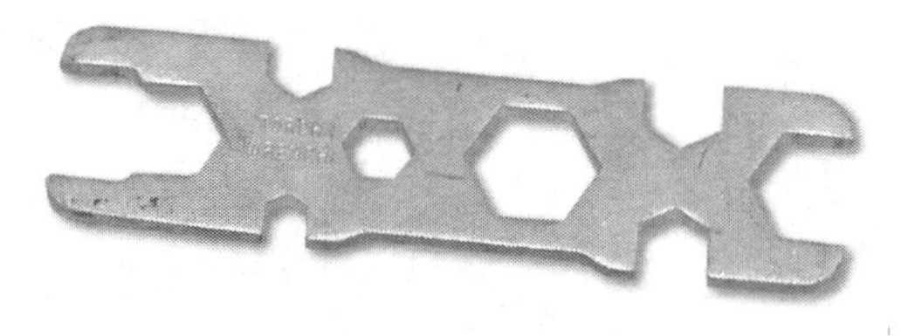

CYLINDER/REGULATOR
(TORCH) WRENCH

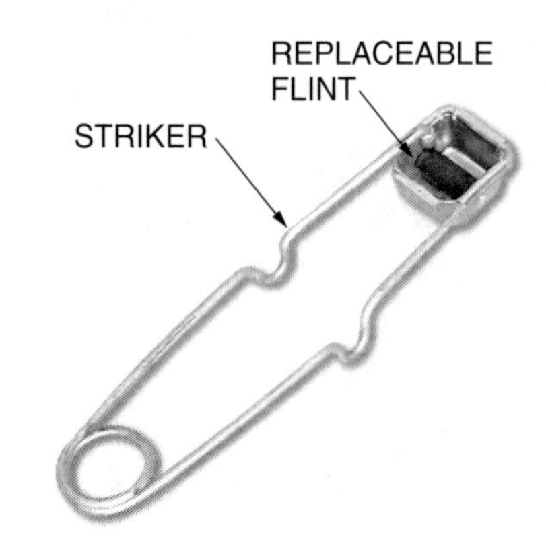

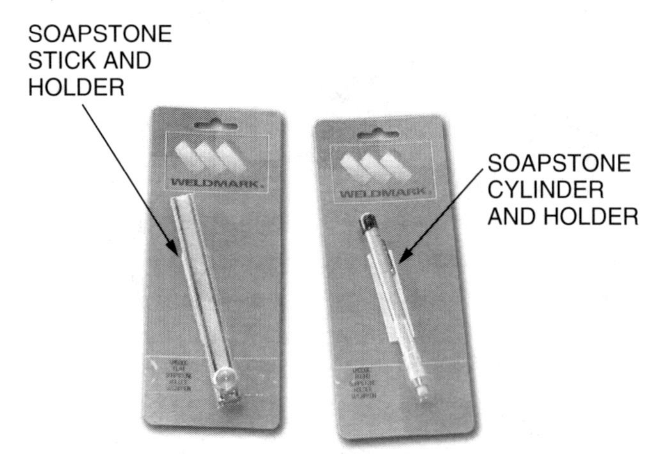

203F24.EPS

Figure 24 ◆ Striker, cylinder/regulator wrench, and soapstone holders.

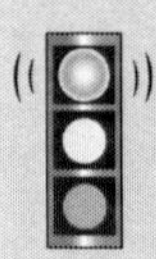

WARNING!

Cylinders must be properly secured to prevent them from falling If the cylinder stems or regulators break off, the cylinders will become dangerous missiles. They can cause extreme property damage, severe injury, and even death.

When using a striker to light the torch, aim or position the torch tip away from any part of your body to avoid severe burns. Also, aim the torch tip away from any combustible material. Butane or any other type of tobacco lighters should never be used to light torches. They should not be on the person who is performing any type of welding, including oxyacetylene welding, as these lighters may explode or ignite.

An oxyacetylene welding torch is designed to mix acetylene and oxygen in the proper proportions. These torches have two needle valves (one for adjusting the oxygen flow and the other for adjusting the fuel flow). Other basic parts include a handle or body, two internal tubes (one for oxygen and another for acetylene), a mixing chamber, and a tip. *Figure 25* shows a combination torch that is used for oxyacetylene welding and cutting. The torch has built-in flashback arrestors and interchangeable tips. The flashback arrestors prevent flames from traveling back up the hoses to the regulators and tanks. All welding torches should be equipped with add-on or internal flashback arrestors.

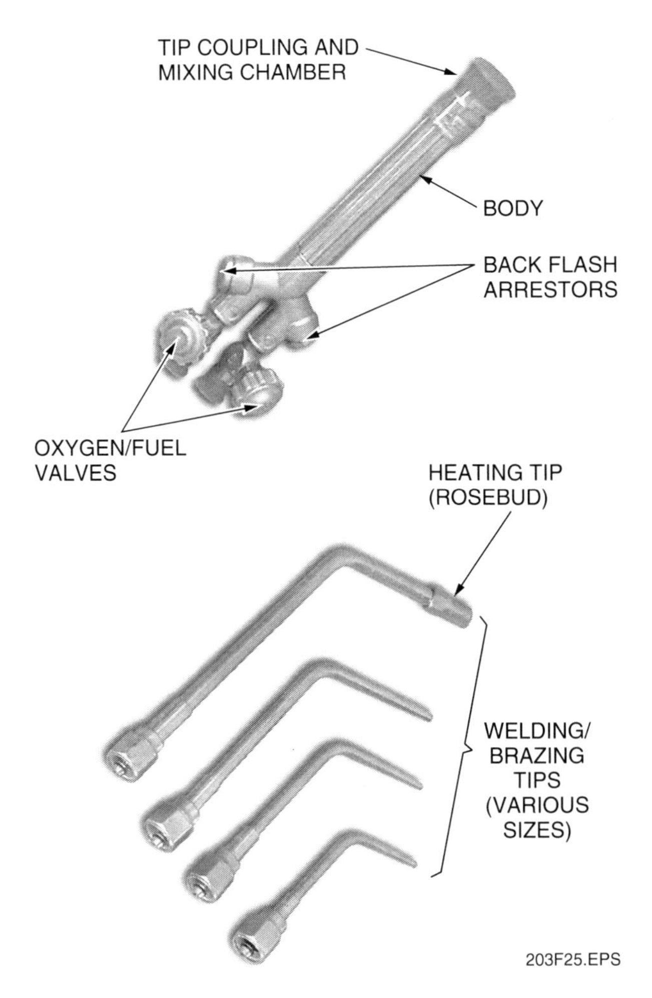

Figure 25 ◆ General purpose combination oxyacetylene torch and tips.

7.1.1 Torch Tips

Welding tips are made from a special copper alloy and are available in different sizes to handle a wide range of uses and plate thicknesses. Most torch tip manufacturers have developed their own unique tip-numbering system, which reflects the size of the orifice in the tip. This number is usually stamped directly on the torch tip. The orifice size determines the amount of fuel gas and oxygen fed to the flame. Therefore, it determines the amount of heat produced by the torch. The larger the orifice, the greater the amount of heat generated. If the torch tip orifice is too small, too little heat will be available to bring the metal to its fusion temperature. If the torch tip is too large, poor welds result, because the weld is made too fast or because the quality of the weld cannot be maintained.

7.1.2 Regulators

The function of the regulators is to reduce the cylinder pressure of the oxygen and acetylene to a working pressure that allows for a controlled welding process. In addition to controlling the pressure, the regulators control the flow of these two gases. The regulators for oxygen and acetylene are different and cannot be interchanged. They each have different thread configurations on the connections to their respective cylinders. Oxygen regulators have **right-hand thread** connections, while acetylene regulators are equipped with **left-hand thread** connections. Regulators are usually designed with two gauges: one that indicates the pressure in the cylinder and one that indicates the working pressure. The working pressures may be regulated to the safe working pressure required for the welding task, but the cylinder pressure indication cannot be regulated. It provides a convenient method of monitoring the degree of fullness of the cylinder. The pressure that is indicated on regulator gauges is called gauge pressure. Cylinders cannot be totally emptied of gas unless they are pumped out using a vacuum pump. *Figure 26* shows an example of an oxygen regulator and an acetylene regulator.

7.1.3 Gases and Gas Cylinders

The two gases used in oxyacetylene welding are oxygen and acetylene. Oxygen is an odorless and colorless gas and is slightly heavier than air. It is nonflammable but is susceptible to combustion when combined with other elements, including oil-based lubricants. The atmosphere is made up of about 21 parts oxygen, 78 parts nitrogen, and fractional parts of rare gases. Oxygen is processed commercially by compressing the air we breathe and cooling it to a point where the gases become liquid, which is approximately –375°F. The temperature is then raised to above –321°F, at which point the nitrogen in the air becomes gas again and is removed. When the temperature of the remaining liquid is raised to –297°F, the oxygen forms gas and is drawn off. The oxygen is further purified and compressed into cylinders for use. In some cases, the amount of oxygen in a compressed gas cylinder can be determined roughly by reading the volume scale (if present) on the high-pressure gauge attached to the regulator.

Acetylene is a flammable fuel gas composed of carbon and hydrogen. When burned with oxygen, acetylene produces a hot flame with a temperature between 5,700°F and 6,300°F. Acetylene is a colorless gas that is lighter than air. It has a disagreeable odor that is readily detected, even when the gas is highly diluted with air.

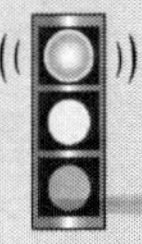

WARNING!

If acetylene is used as the fuel gas, the working pressure must never be allowed to reach 15 psi. Acetylene becomes very dangerous at 15 psi and self-explosive at 29.4 psi.

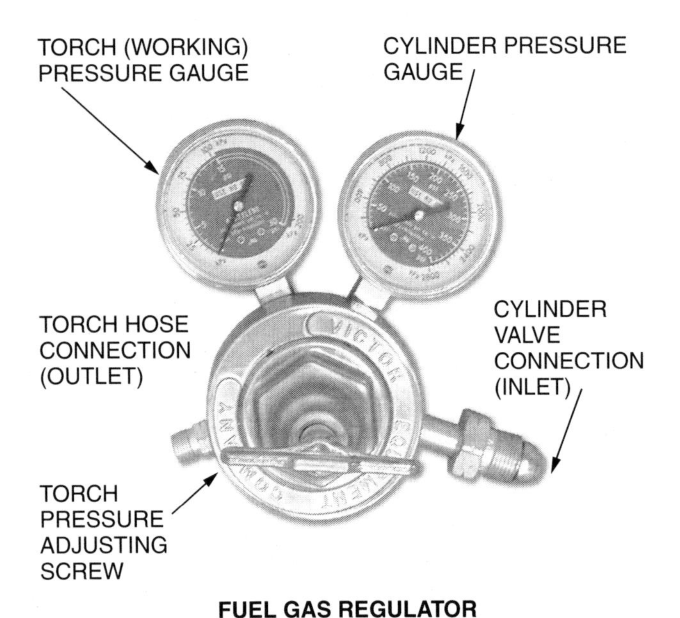

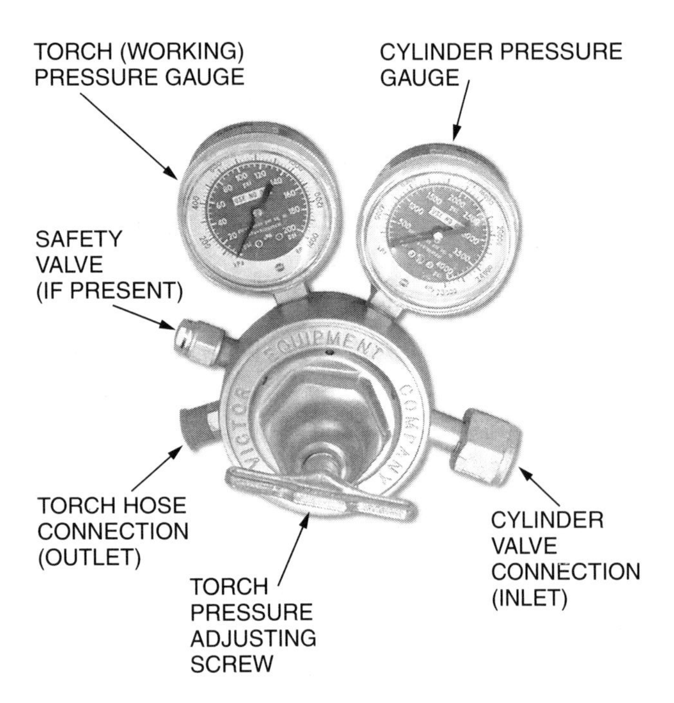

Figure 26 ◆ Oxygen and acetylene regulators .

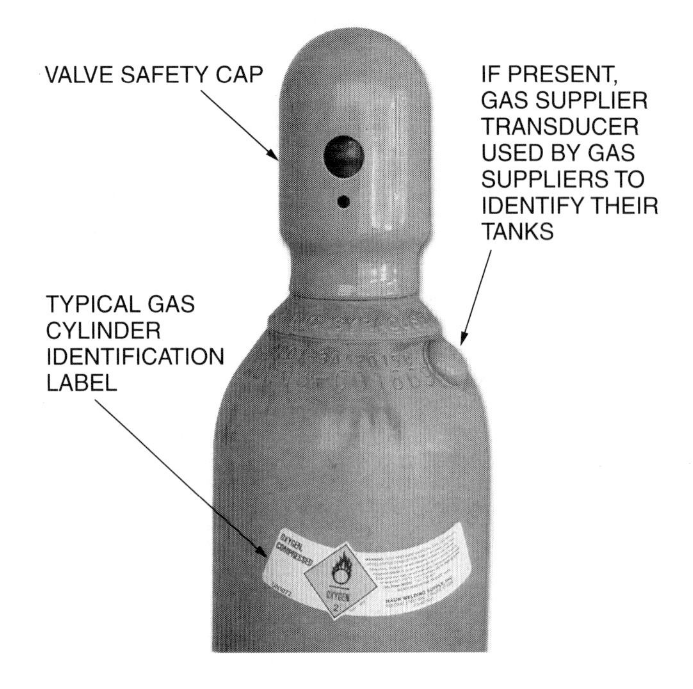

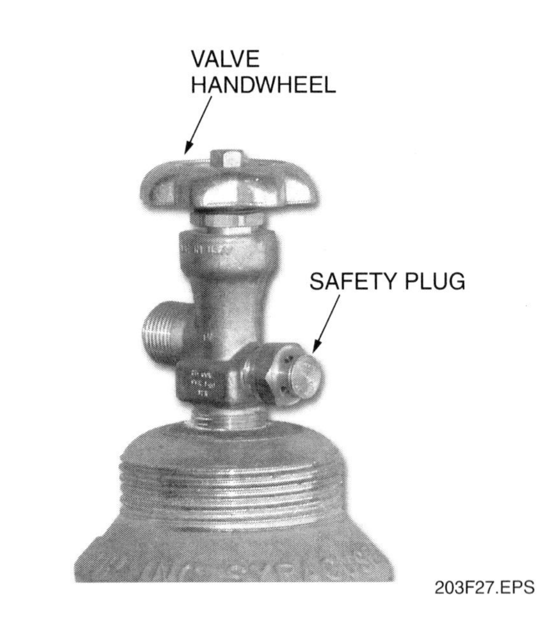

Figure 27 ◆ Typical oxygen cylinder, valve, and safety cap.

Oxygen cylinders used to supply oxygen for oxyacetylene work are seamless steel cylinders. A typical oxygen cylinder used in oxyacetylene welding systems is shown in *Figure 27*. While colors are not standardized, the color of an oxygen cylinder used for industrial purposes is usually green. Oxygen cylinders vary in size. The most common size used in oxyacetylene welding and cutting has 227 cubic feet of oxygen capacity. It measures 9 inches in diameter, stands 51 inches high, weighs about 116 pounds, and is typically charged to a pressure of 2,000 psi at an ambient temperature of 70°F.

As stated in the previous warning, pure acetylene becomes volatile at 15 psi. For this reason, hollow steel cylinders cannot be charged with pure acetylene at pressures above 15 psi. In order to provide a sufficient acetylene supply for oxyacetylene welding, acetylene and its associated cylinders must be uniquely adapted to allow safe storage of acetylene at pressures high enough to be used as a supply gas.

Because the acetylene working pressure required to perform oxyacetylene welding is much lower than the oxygen work pressure, acetylene does not need to be stored in its cylinder at the high pressures required in oxygen storage (2,000 psi). However, because it must maintain a working pressure of less than 15 psi, its storage pressure must be sufficient to continually supply a constant working pressure. It has been found that acetylene can be safely compressed up to 275 psi when dissolved in acetone and stored in specially designed cylinders filled with porous material. Examples of such porous material include balsa wood, charcoal, and Portland cement. These porous filler materials help prevent the buildup of high-pressure gas pockets in the cylinder. Because acetone is a liquid chemical that dissolves large portions of acetylene under pressure without changing the nature of the gas, it does not affect the fuel properties of acetylene. Acetylene is stored in cubic feet. The most common cylinder sizes have 130, 290, and 330 cubic feet capacities. *Figure 28* shows a typical acetylene cylinder. Safety (blowout) plugs at the top and bottom of the cylinder melt at 220°F to release the gas in the event of a fire in order to prevent an explosion.

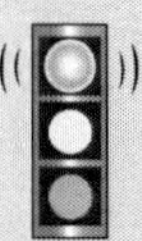

WARNING!

Acetone is an extremely volatile liquid chemical and is very explosive. Acetylene cylinders must not be stored on their sides. In unforeseen situations where acetylene cylinders have been turned or stored on their sides, the cylinders must stand upright for a minimum of 30 minutes before use. This allows the acetone to settle to the bottom of the cylinder. Remember: without acetone in the cylinder, acetylene becomes extremely volatile at pressures of 15 psi or greater. Proper storage and handling of this type of cylinder is of the utmost importance.

Care must always be taken to withdraw acetylene gas at pressures less than 15 psi and at rates that do not exceed 1/10 of the cylinder capacity. Consumption rates for various size tips can be determined from the torch manufacturer's data.

CAUTION

If acetone is not allowed to settle to the bottom of acetylene storage cylinders for a minimum of 30 minutes after a cylinder has been stored or turned on its side, acetone may be drawn from the cylinder during the welding process. Acetone is a solvent that contaminates the hoses, regulators, and torch. It disrupts the flame.

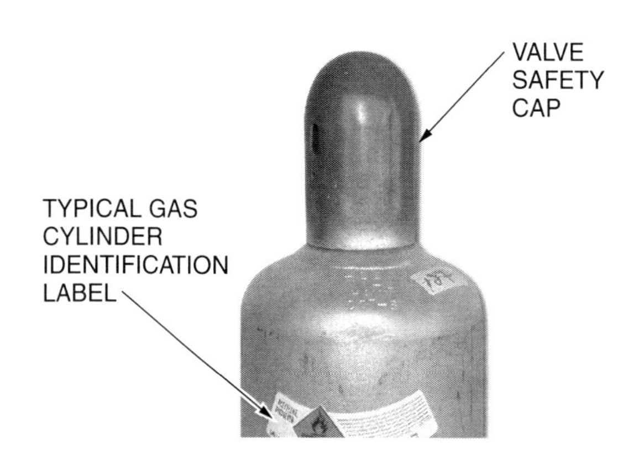

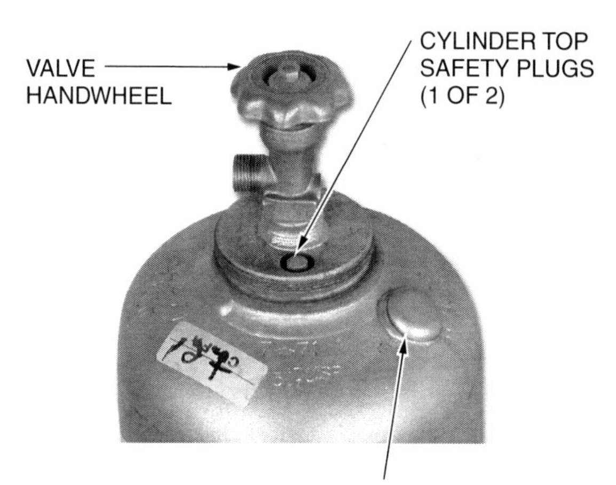

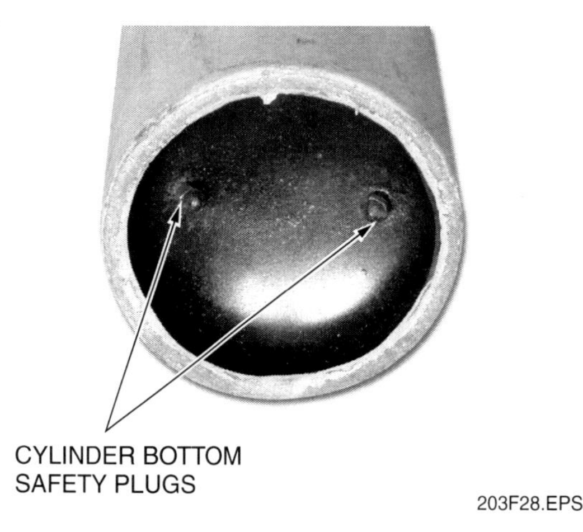

Figure 28 ◆ Typical acetylene cylinder, valve, safety cap, and safety plugs.

8.0.0 TYPES OF WELDED FITTINGS

Whenever field instruments must be connected directly to the process, the piping or tubing connecting the instrument to the process often must be welded and make use of welded fittings. Piping specifications for many processes that can flow through steel (including stainless steel) piping will frequently call for either socket or butt weld fittings to be used when connecting the process to the instrument. In both the socket and butt weld fitting installations, the fittings must be welded to the pipe or tubing that connects to them. Socket weld fittings do not need precise alignment because the pipe or tubing actually fits into a socket. Butt weld fittings, on the other hand, must have their edges perfectly aligned with the prepared edges of the pipe or tubing. This calls for tedious and time consuming alignment procedures that only experienced pipe welders should perform.

Whenever a process requires copper pipe or tubing as its carrier, the fittings used are typically solder-type fittings. These are normally installed using a torch system in conjunction with an approved solder that is compatible with the type of copper pipe, tubing, or fittings. These fittings may be butt- or socket-type fittings. *Figure 29* illustrates how socket and butt weld fittings are mated with the pipe or tubing.

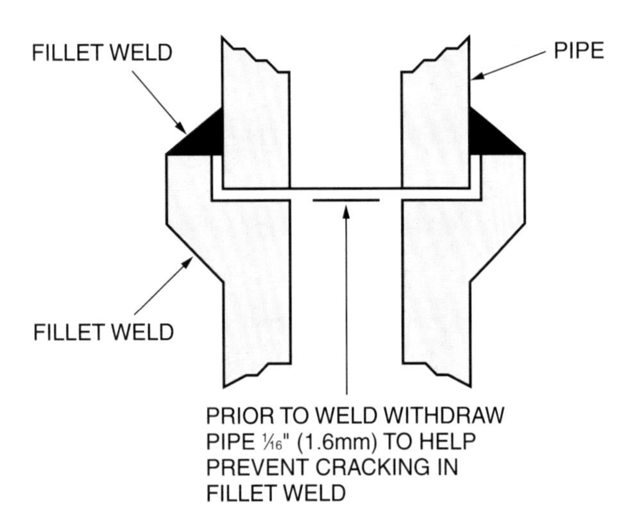

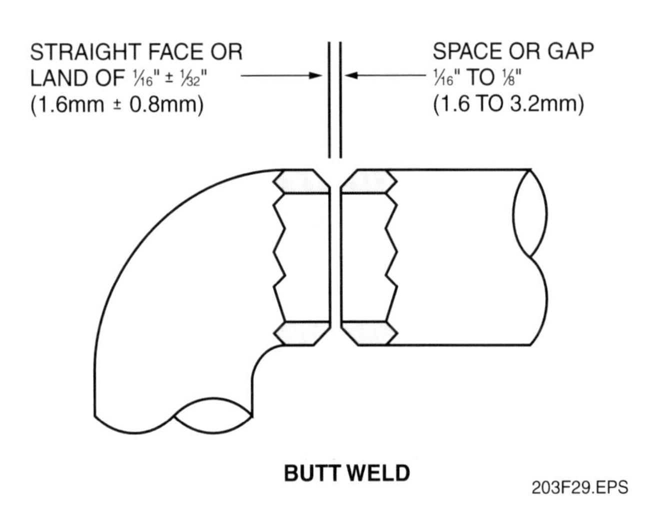

Figure 29 ◆ Typical socket and butt weld fittings.

Job specifications sometimes call for fittings to be back-welded. This term refers to running a weld bead around a threaded joint to ensure that it does not leak. Back-welding is used on process lines for volatile processes.

8.1.0 Socket Weld Fittings

Socket weld fittings are used to join smaller sizes of pipe, usually 2" (50 mm) and under, which require the strength and security of a welded joint.

The socket-welded joint is made by fitting pipe into the socket of the fitting and fillet welding around the pipe and the top of the fitting. Dimensions for standard forged steel socket welded fittings, pressure classes 3000, 6000, and 9000, are given in *Figure 30* and *Tables 5* and *6*.

8.2.0 Butt Weld Fittings

The most common style of welding fitting used in welded pipe systems (primarily in pipe sizes over 2") is the butt weld fitting.

Figure 31 shows a typical butt weld fitting and explains the identification markings used on fittings. Pressure/temperature ratings for the fittings duplicate those of seamless pipe of the same material, size, and wall thickness. Standard sizes for butt weld fittings are available in wall thickness and schedule numbers paralleling those of steel pipe.

Specific dimensions for butt welded fittings that are displayed in *Figure 32* are provided in *Tables 7* and *8*.

8.3.0 Copper Solder Joint Fittings

Both wrought copper and cast copper alloy solder joint pressure fittings are produced for use with copper water tube. Pressure and temperature ratings for the fittings are equal to those of **Type L copper tube**. However, in most cases the solder used to join the tube and fitting will determine the safe working pressure and temperature of the system. Recommended solder joint pressure and temperature ratings are given in *Table 9*.

Dimensions for common cast copper alloy solder joints are given in *Figure 33* and *Table 10*. The term laying length in reference to copper solder fittings refers to the distance from the center of the fitting to the shoulder or stop at the bottom end of the socket. American National Standards Institute (*ANSI B16.22-1980*) has established laying length and sizing designations for cast fittings only. Because of the various forming methods, no standardized dimensions are established for wrought copper solder fittings.

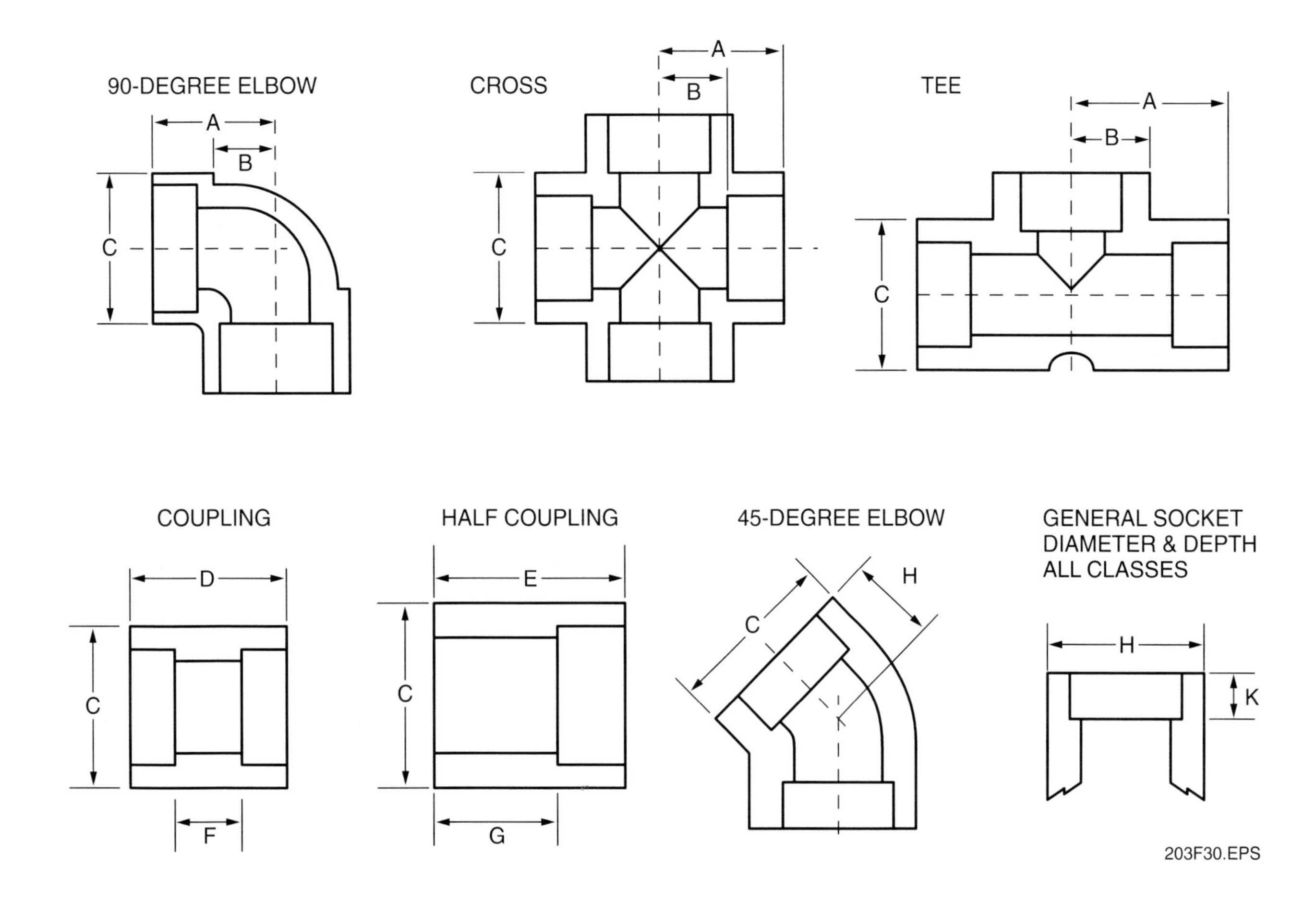

Figure 30 ◆ Dimensions for forged steel socket weld fittings.

Table 5 Socket Weld Fitting Chart (Imperial)

STANDARD FORGED STEEL SOCKET WELDED FITTINGS																						
Nominal Pipe Size (Inches)	Dimensions – Inches																					
	A			B			C			D	E	F	G	H			I			J		K
	3000	6000	9000	3000	6000	9000	3000	6000	9000	All Classes				3000	6000	9000	3000	6000	9000	Max.	Min.	Min.
1/8	0.82	0.82	—	0.44	0.44	—	0.68	0.74	—	1.01	1.00	0.25	0.62	0.69	0.69	—	0.31	0.31	—	0.430	0.420	0.38
1/4	0.82	0.91	—	0.44	0.53	—	0.86	0.92	—	1.01	1.00	0.25	0.62	0.69	0.69	—	0.31	0.31	—	0.565	0.555	0.38
3/8	0.91	1.00	—	0.53	0.62	—	1.01	1.09	—	1.01	1.07	0.25	0.69	0.69	0.82	—	0.31	0.44	—	0.700	0.690	0.38
1/2	1.00	1.13	1.38	0.62	0.75	1.00	1.23	1.33	1.60	1.14	1.26	0.38	0.88	0.82	0.88	1.00	0.44	0.50	0.62	0.865	0.855	0.38
3/4	1.25	1.38	1.62	0.75	0.88	1.12	1.46	1.62	1.84	1.38	1.44	0.38	0.94	1.00	1.06	1.25	0.50	0.56	0.75	1.075	1.065	0.50
1	1.33	1.56	1.75	0.88	1.06	1.25	1.78	1.96	2.23	1.50	1.62	0.50	1.12	1.06	1.19	1.31	0.56	0.69	0.81	1.340	1.330	0.50
1 1/4	1.56	1.75	1.88	1.06	1.25	1.38	2.16	2.30	2.64	1.50	1.69	0.50	1.19	1.19	1.31	1.38	0.69	0.81	0.88	1.685	1.675	0.50
1 1/2	1.75	2.00	2.00	1.25	1.50	1.50	2.42	2.62	2.92	1.50	1.75	0.50	1.25	1.31	1.50	1.50	0.81	1.00	1.00	1.925	1.915	0.50
2	2.12	2.24	2.74	1.50	1.62	2.12	2.96	3.27	3.50	1.99	2.24	0.75	1.62	1.62	1.74	1.74	1.00	1.12	1.12	2.416	2.406	0.62
2 1/2	2.24	—	—	1.62	—	—	3.60	—	—	1.99	2.31	0.75	1.69	1.72	—	—	1.12	—	—	2.721	2.906	0.62
3	2.87	—	—	2.25	—	—	4.29	—	—	1.99	2.37	0.75	1.75	1.87	—	—	1.25	—	—	3.550	3.535	0.62
4	3.37	—	—	2.62	—	—	5.39	—	—	2.25	2.63	0.75	1.88	2.37	—	—	1.62	—	—	4.560	4.545	0.75

Note:
- Dimensions for table are based on fittings manufactured to *ANSI B16.11* standard.
- Sight variations between inch and millimeter dimensions are due to rounding factors and permitted tolerances ± in standard.

203T05.EPS

Table 6 Socket Weld Fitting Chart (Metric)

STANDARD FORGED STEEL SOCKET WELDED FITTINGS

Nominal Pipe Size (mm)	Dimensions – Millimeters																					
	A			B			C			D	E	F	G	H			I			J		K
	3000	6000	9000	3000	6000	9000	3000	6000	9000	All Classes				3000	6000	9000	3000	6000	9000	Max.	Min.	Min.
6	20.83	20.83	—	11.18	11.18	—	17.27	18.80	—	25.65	25.40	6.35	15.75	18.00	18.00	—	8.00	8.00	—	10.90	10.65	10
8	20.83	23.11	—	11.18	13.46	—	21.84	23.37	—	25.65	25.40	6.35	15.75	18.00	18.00	—	8.00	8.00	—	14.35	14.10	10
10	23.11	25.40	—	13.46	15.75	—	25.65	27.69	—	25.65	27.18	6.35	17.53	18.00	21.50	—	8.00	11.50	—	17.80	17.55	10
15	25.40	28.70	35.05	15.75	19.05	25.40	31.24	33.78	40.64	28.96	32.00	9.65	22.35	21.50	22.50	25.50	11.50	12.50	15.50	21.95	21.70	10
20	31.75	35.05	41.15	19.05	22.35	28.45	37.08	41.15	46.74	35.05	36.58	9.65	23.88	25.00	27.50	32.00	12.00	14.50	19.00	27.30	27.05	13
25	33.78	39.62	44.45	22.35	26.92	31.75	45.21	49.78	56.64	38.10	41.15	12.70	28.45	27.00	30.00	34.00	14.00	17.00	21.00	34.05	33.80	13
32	39.62	44.45	47.75	26.92	31.75	35.05	54.86	58.42	67.06	38.10	42.93	12.70	30.23	30.00	34.00	35.00	17.00	21.00	22.00	42.80	42.55	13
40	44.45	50.80	50.80	31.75	38.10	38.10	61.47	66.55	74.17	38.10	44.45	12.70	31.75	34.00	38.00	38.50	21.00	25.00	25.50	48.90	48.65	13
50	53.85	56.90	69.60	38.10	41.15	53.85	75.18	83.06	88.90	50.55	56.90	19.05	41.15	41.00	45.00	44.50	25.00	29.00	28.50	61.35	61.10	16
65	56.90	—	—	41.15	—	—	91.44	—	—	50.55	58.67	19.05	42.93	45.00	—	—	29.00	—	—	74.20	73.80	16
80	72.90	—	—	57.15	—	—	108.97	—	—	50.55	60.20	19.05	44.45	47.50	—	—	31.50	—	—	90.15	89.90	16
100	85.60	—	—	66.55	—	—	136.91	—	—	57.15	66.80	19.05	47.75	60.50	—	—	41.50	—	—	115.80	115.45	19

Note:
- Dimensions for table are based on fittings manufactured to *ANSI B16.11* standard.
- Sight variations between inch and millimeter dimensions are due to rounding factors and permitted ± tolerances in the standard.

203T06.EPS

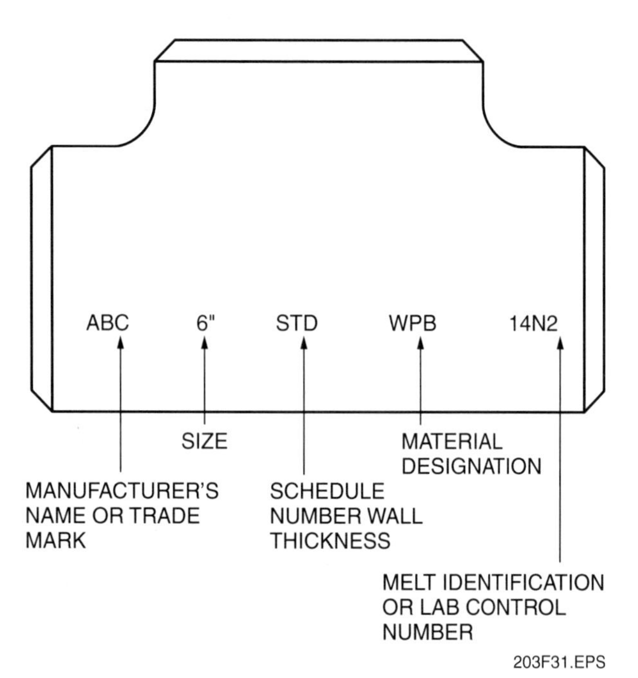

Figure 31 ◆ Butt weld fitting identification.

9.0.0 ◆ WELDING APPLICATIONS FOR INSTRUMENTATION

The primary environment for an instrumentation fitter or technician is in industrial facilities, such as petrochemical plants, power plants, mills, pipelines, compressor stations, and some factories. Most of the welding that takes place in these settings must be performed by a professional welder who holds certification both in the work being welded (steel, piping etc.) and the type of welding that is being performed on the work such as SMAW, GTAW, SMAW, orbital, and in some cases, **brazing** and oxyacetylene welding.

As an instrument fitter or technician, you should be familiar with all of these certifications and can become certified should you choose to do so. Benefits of certification include being able to do your own welding instead of having to wait for a welder to do it for you. However, if you are not comfortable in the welding environment, these tasks are best left to those who have chosen welding as their primary career.

Many instrument fitters have mastered the art of SMAW, or stick welding, and are able to use it comfortably along with their primary duties of installing instrumentation. The ability to stick weld proficiently as a fitter is very beneficial in that the fitter is able to fabricate his or her own instrument stands and supports without waiting for others to make them.

There are some instrument fitters who are able to perform fairly simple brazing or oxyacetylene tasks. These include using a cutting torch to cut plate and piping for instrument stand fabrication and brazing brackets to specialty metals. However, most piping brazing or soldering is best left to those certified in these procedures.

The arrival of orbital welding technology enabled many fitters and technicians to perform high quality piping welds with little expertise, often without certification. However, the project specifications and applicable company policies must be reviewed before making any welds with any type of welding equipment, especially when it is associated with process or utility piping.

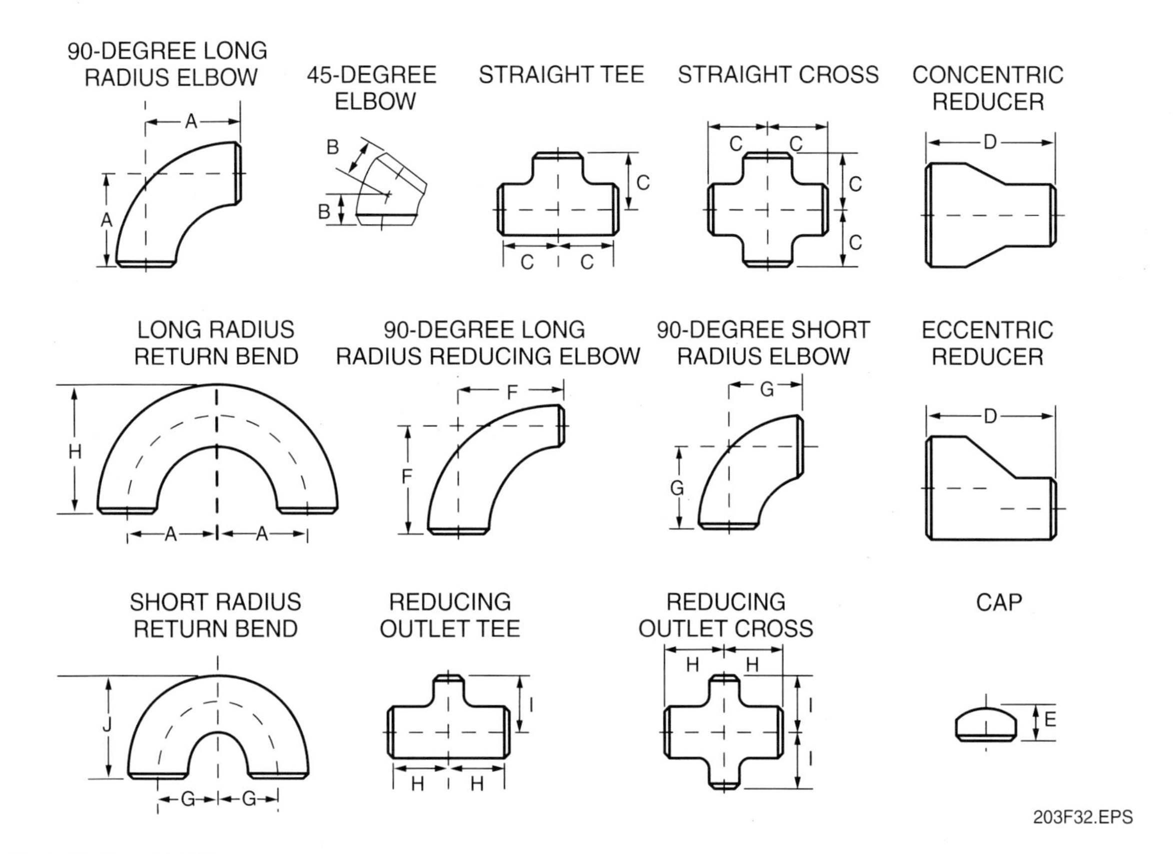

Figure 32 ◆ Butt weld fittings.

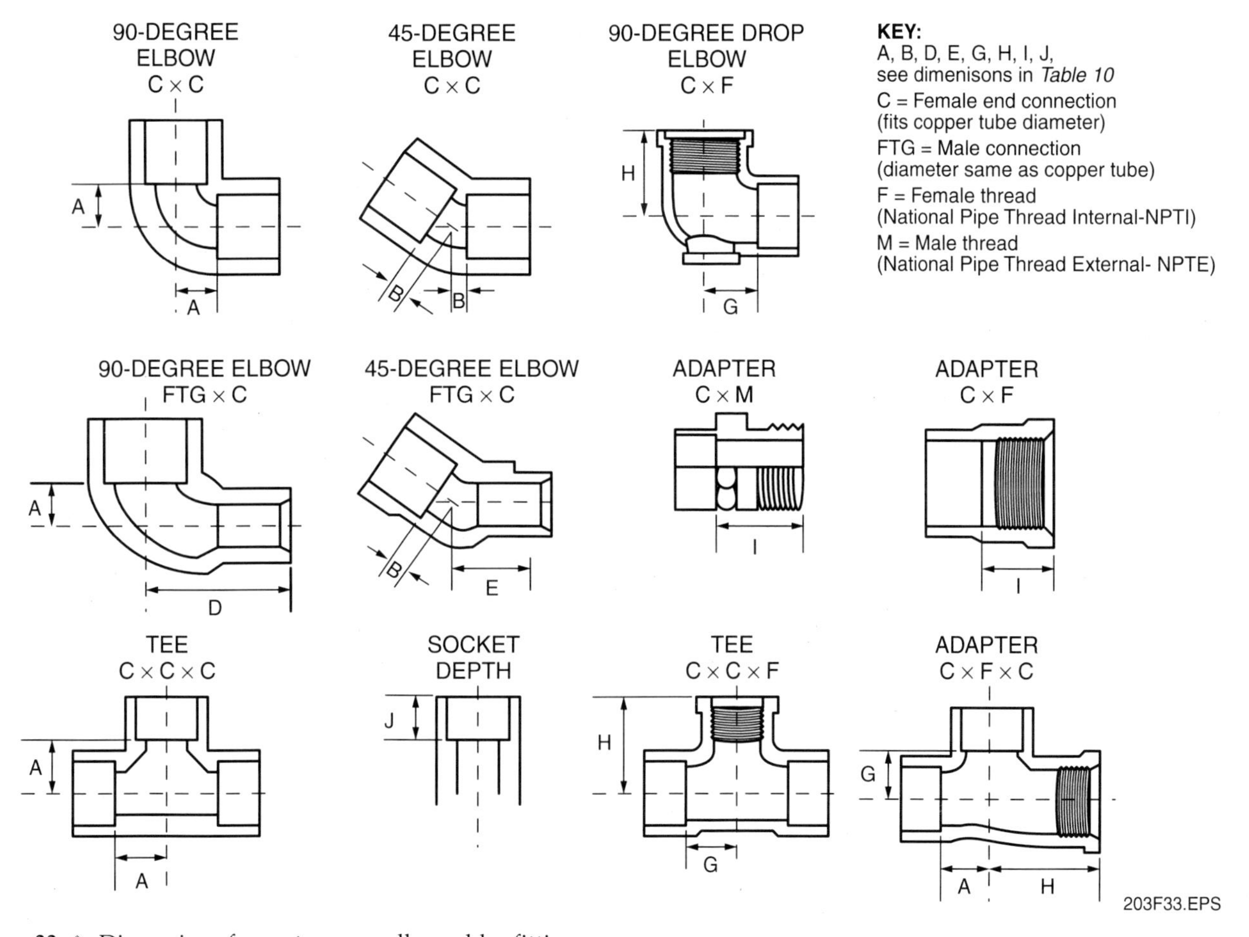

Figure 33 ◆ Dimensions for cast copper alloy solder fittings.

Table 7 Butt Weld Fitting Chart (Imperial)

DIMENSIONS FOR BUTT WELD FITTINGS – INCHES									
Nominal Pipe Size (Inches)	90° Long Radius Elbows	45° Elbows	Tees and Crosses	Reducing Couplings CON/ECC	Caps	Reducing 90° Elbows	Short Radius 90° Elbows	Long Radius Return Bends	Short Radius Return Bends
	A	B	C	D	E	F	G	H	J
½	1.50	0.62	1.00	—	1.00	—	—	1.88	—
¾	1.12	0.44	1.12	1.50	1.00	—	—	1.69	—
1	1.50	0.88	1.50	2.00	1.50	—	1.00	2.19	1.62
1¼	1.88	1.00	1.88	2.00	1.50	—	1.25	2.75	2.06
1½	2.25	1.12	2.25	2.50	1.50	—	1.50	3.25	2.44
2	3.00	1.38	2.50	3.00	1.50	3.00	2.00	4.19	3.19
2½	3.75	1.75	3.00	3.50	1.50	3.75	2.50	5.19	3.94
3	4.50	2.00	3.38	3.50	2.00	4.50	3.00	6.25	4.75
3½	5.25	2.25	3.75	4.00	2.50	5.25	3.50	7.25	5.50
4	6.00	2.50	4.12	4.00	2.50	6.00	4.00	8.25	6.25
5	7.50	3.12	4.88	5.00	3.00	7.50	5.00	10.31	7.75
6	9.00	3.75	5.62	5.50	3.50	9.00	6.00	12.31	9.31
8	12.00	5.00	7.00	6.00	4.00	12.00	8.00	16.31	12.31
10	15.00	6.25	8.50	7.00	5.00	15.00	10.00	20.38	15.38
12	18.00	7.50	10.00	8.00	6.00	18.00	12.00	24.38	18.38
14	21.00	8.75	11.00	13.00	6.50	21.00	14.00	28.00	21.00
16	24.00	10.00	12.00	14.00	7.00	24.00	16.00	32.00	24.00
18	27.00	11.25	13.50	15.00	8.00	27.00	18.00	36.00	27.00
20	30.00	12.50	15.00	20.00	9.00	30.00	20.00	40.00	30.00
22	33.00	13.50	16.50	20.00	10.00	—	22.00	44.00	—
24	36.00	15.00	17.00	20.00	10.50	36.00	24.00	48.00	36.00
26	39.00	16.00	19.50	24.00	10.50	—	—	—	—
28	42.00	17.25	20.50	24.00	10.50	—	—	—	—
30	45.00	18.50	22.00	24.00	10.50	—	—	—	—
32	48.00	19.75	23.50	24.00	10.50	—	—	—	—
34	51.00	21.00	25.00	24.00	10.50	—	—	—	—
36	54.00	22.25	26.50	24.00	10.50	—	—	—	—
38	57.00	23.62	28.00	24.00	12.00	—	—	—	—
40	60.00	24.88	29.50	24.00	12.00	—	—	—	—
42	63.00	26.00	30.00	24.00	12.00	—	—	—	—
44	66.00	27.38	32.00	24.00	13.50	—	—	—	—
46	69.00	28.62	33.50	28.00	13.50	—	—	—	—
48	72.00	29.88	35.00	28.00	13.50	—	—	—	—

203T07.EPS

Table 8 Butt Weld Fitting Chart (Metric)

DIMENSIONS FOR BUTT WELD FITTINGS – MILLIMETERS									
Nominal Pipe Size (mm)	90° Long Radius Elbows	45° Elbows	Tees and Crosses	Reducing Couplings CON/ECC	Caps	Reducing 90° Elbows	Short Radius 90° Elbows	Long Radius Return Bends	Short Radius Return Bends
	A	B	C	D	E	F	G	H	J
15	38	16	25	—	25	—	—	48	—
20	29	11	29	38	25	—	—	43	—
25	38	22	38	51	38	—	25	56	41
32	48	25	48	51	38	—	32	70	52
40	57	29	57	64	38	—	38	83	62
50	76	35	64	76	38	76	51	106	81
65	95	44	76	89	38	95	64	132	100
80	114	51	86	89	51	114	76	159	121
90	133	57	95	102	64	133	89	184	140
100	152	64	105	102	64	152	102	210	159
125	190	79	124	127	76	190	127	262	197
150	229	95	143	140	89	229	152	313	237
200	305	127	178	152	102	305	203	414	313
250	381	159	216	176	127	381	254	518	391
300	457	190	254	203	152	457	305	619	467
350	533	222	279	330	165	533	356	711	533
400	610	254	305	356	178	610	406	813	610
450	686	286	343	381	203	686	457	914	686
500	762	318	381	508	229	762	508	1,016	762
550	838	343	419	508	254	—	559	1,118	—
600	914	381	432	508	267	914	610	1,119	914
650	991	406	495	610	267	—	—	—	—
700	1,067	438	521	610	267	—	—	—	—
750	1,143	470	559	610	267	—	—	—	—
800	1,219	502	597	610	267	—	—	—	—
850	1,295	533	635	610	267	—	—	—	—
900	1,372	565	673	610	267	—	—	—	—
950	1,448	600	711	610	305	—	—	—	—
1,000	1,524	632	749	610	305	—	—	—	—
1,050	1,600	660	762	610	305	—	—	—	—
1,100	1,676	695	813	610	343	—	—	—	—
1,150	1,723	727	851	711	343	—	—	—	—
1,200	1,829	759	889	711	343	—	—	—	—

203T08.EPS

Table 9 Pressure and Temperature Chart for Solder and Brazed Joints

RATED PRESSURE/TEMPERATURE FOR SOLDERED AND BRAZED JOINTS														
Solder or Brazing Alloy	Working Temperature		Copper Tube Nominal Sizes – Inches and Millimeters											
			⅛" – 1" 6mm – 25mm		1¼" – 2" 32mm – 50mm		2½" – 4" 65mm – 100mm		5" – 8" 125mm – 200mm		10" – 12" 250mm – 300mm		Saturated Steam	
	°F	°C	psi	kPa	psi	kPa	psi	kPa	psi	kPa	psi	kPa	psi	kPa
50-50 Tin-Lead Solder	100	38	200	1,379.0	175	1,206.6	150	1,034.2	135	930.8	100	689.5	—	—
	150	66	150	1,034.2	125	861.8	100	689.5	90	620.5	70	482.6	—	—
	200	93	100	689.5	90	620.5	75	517.1	70	482.6	50	344.7	—	—
	250	120	85	586.0	75	517.1	50	344.7	45	310.2	40	275.8	15	103.4
95-5 Tin-Antimony Solder	100	38	500	3,447.5	400	2,758.0	300	2,068.5	270	1,861.6	150	1,034.2	—	—
	150	66	400	2,758.0	350	2,413.2	275	1,896.1	250	1,723.7	150	1,034.2	—	—
	200	93	300	2,068.5	250	1,723.7	200	1,379.0	180	1,241.1	140	965.3	—	—
	250	120	200	1,379.0	175	1,206.6	150	1,034.2	135	930.8	110	758.4	15	103.4
Brazing Alloys Melting at or Above 1,000°F/540°C	250	120	300	2,068.5	210	1,447.9	170	1,172.1	150	1,034.2	150	1,034.2	—	—
	350	176	270	1,861.6	190	1,310.0	150	1,034.2	150	1,034.2	150	1,034.2	120	827.4

Note:

- The pressure unit of one bar is equal to 14.5 psi or 100,000 newtons per square meter.
- Saturated steam at 15 psi (103.4 kPa) is produced at 250°F (121°C).
- Saturated steam at 120 psi (827.4 kPa) is produced at 350°F (176°C).
- Brazing alloys are recommended for low-temperature service between 0°F to –100°F (–18°C to –73°C).

203T09.EPS

Table 10 Cast Copper Alloy Solder Fitting Chart

DIMENSIONS FOR COMMON CAST COPPER ALLOY SOLDER FITTINGS																	
Tube Size	Dimensions – Inches								Tube Size	Dimensions – Millimeters							
	A	B	D	E	G	H	I	J		A	B	D	E	G	H	I	J
¼	0.25	—	0.75	—	0.38	0.56	0.62	0.31	8	6.5	—	19.0	—	9.5	14.5	16.0	7.87
⅜	0.31	0.19	0.88	0.75	0.44	0.69	0.62	0.38	10	8.0	5.0	22.0	19.0	11.0	17.5	16.0	9.65
½	0.44	0.19	1.12	0.88	0.56	0.88	0.75	0.50	15	11.0	5.0	28.5	22.0	14.5	22.0	18.0	12.70
¾	0.56	0.25	1.50	1.19	0.69	1.00	0.88	0.75	20	14.5	6.5	38.0	30.0	17.5	25.5	22.0	19.05
1	0.75	0.31	1.84	1.31	0.88	1.25	1.00	0.91	25	19.0	8.0	47.0	33.5	22.0	32.0	25.5	23.11
1¼	0.88	0.44	2.03	1.56	1.00	1.50	1.06	0.97	32	22.0	11.0	51.5	39.5	25.5	38.0	27.0	24.64
1½	1.00	0.50	2.28	1.75	1.12	1.62	1.06	1.09	40	25.5	12.5	58.0	44.5	28.5	41.5	27.0	27.69
2	1.25	0.56	2.78	2.12	1.38	1.94	1.12	1.34	50	32.0	14.5	70.5	54.0	35.0	49.0	28.5	34.04

Note:

- Dimensions apply to cast fittings only. Dimensions for wrought fittings have not been standardized.
- Dimensions for table based on fittings manufactured to ANSI standards.
- Tube size dimensions are nominal sizes.

203T10.EPS

Summary

Welding tasks associated with instrumentation installations are frequently performed by the technician because of the specifications related to the installation. Newer welding technologies also make these welding tasks easier to perform. Orbital welding has opened the door for many instrumentation personnel to perform welding tasks that previously could only be performed by certified welders. However, not all installations that require welding tasks can be performed through orbital welding, because of the configuration of the fittings and other obstacles that prevent the weld head of an orbital system from rotating around the weld area. For this reason, you should acquaint yourself with as many welding techniques as possible. In situations that require specialized welding tasks, such as SMAW, TIG, MIG, and oxyacetylene welding, technicians skilled in these techniques may be able to perform the tasks, or at least assist a certified welder in accomplishing them. In addition to developing the ability to perform these welding tasks, technicians should develop a working knowledge of the fittings and accessories associated with these installations.

Review Questions

1. A process in which two pieces of metal melt sufficiently for a thorough blending of the material in both pieces, and harden as they cool to form a single continuous mass is known as _____.
   a. overlapping
   b. fusion
   c. brazing
   d. arcing

2. All of the following types of clothing my be worn when welding *except* _____.
   a. leather
   b. wool
   c. polyester
   d. cotton

3. Gauntlets are usually found on _____.
   a. eye protection
   b. gloves
   c. aprons
   d. ear protection

4. Another name commonly used to describe shielded metal arc welding (SMAW) is _____ welding.
   a. TIG
   b. MIG
   c. braze
   d. stick

5. SMAW welding can be used to weld metal thicknesses greater than _____ gauge.
   a. 8
   b. 10
   c. 18
   d. 24

6. Direction of current or _____ is important in SMAW when direct current is used.
   a. polarity
   b. voltage
   c. resistance
   d. AC

7. In orbital welding, an orbital _____ rotates around the weld joint to make the required weld.
   a. power source
   b. weld head
   c. electrode holder
   d. shield gas

8. Many orbital system power units come equipped with computer-controlled welding _____ that allow the operator to select and set them based on the type and size of material to be welded.
   a. electrodes
   b. parameters
   c. instructions
   d. temperatures

9. Enclosed orbital weld heads are practical for welding tube sizes from _____.

a. 1/16" to 4"
b. 1/32" to 6"
c. 1/2" to 8"
d. 1" to 10"

10. In gas tungsten arc welding, shielding is obtained from a(n) _____.

a. oxygen gas
b. protective blanket
c. inert gas
d. darkened plate of glass

11. Tungsten has the highest melting point of any metal. In gas tungsten arc welding, tungsten is considered _____.

a. consumable
b. a poor electrode
c. nonconsumable
d. an alloy

12. Gas metal arc welding can be used in _____ to produce weld deposits with little or no spatter.

a. inclined positions only
b. horizontal positions only
c. vertical positions only
d. all positions

13. In oxyacetylene welding, a carburizing flame has _____ and adds carbon to the weld.

a. excess oxygen
b. excess acetylene
c. added acetone
d. no acetylene

14. Each of the following is an accessory that may be used with an oxyacetylene welding system *except* _____.

a. a striker
b. a cylinder wrench
c. a torch lubricator
d. soapstone

15. The color of a standard oxygen cylinder used for industrial purposes is usually _____.

a. blue
b. red
c. green
d. black

16. Pure acetylene must never become pressurized at or above _____ psi.

a. 15
b. 275
c. 321
d. 2,000

17. Acetylene cylinders must never be stored _____.

a. in ambient temperature exceeding 70°F
b. on their sides
c. standing up
d. next to oxygen cylinders

18. Before welding, _____ weld fittings must have their edges perfectly aligned with the prepared edges of the pipe or tubing.

a. socket
b. braze
c. slip
d. butt

19. Socket weld fittings are usually used to join sizes of pipe _____.

a. 2 inches and under
b. 2 inches and over
c. 4 inches to 6 inches
d. that cannot be welded

20. Many instrumentation technicians are proficient in _____, which is beneficial because they are able to fabricate their own instrument stands and supports.

a. stick welding
b. TIG welding
c. plasma welding
d. MIG welding

GLOSSARY

Trade Terms Introduced in This Module

Arc blow: The deflection of an electric arc from its normal path because of magnetic forces.

Arc length: The distance between the tip of the electrode and the weld puddle.

Arc voltage: The voltage across the welding arc.

Brazing: A means of joining metal using an alloy with a melting point higher than that of solder, but lower than that of the metals being joined.

Cylinder: A portable container used for transportation and storage of a compressed gas.

Ferrous: A term used to describe metal that contains iron.

Filler metal: The metal (material) added in making a welded, brazed, or soldered joint.

Flux: Material used to prevent, dissolve, or facilitate removal of oxides and other undesirable surface substances or to shield the arc from the atmosphere.

Gauntlet: A glove cuff designed for extra protection for the forearm. It is usually a 4" cuff. It slides on and off easily and allows for maximum forearm movement.

Inert gas: This is used to shield the electric arc from outside contaminants and gases that may react with the weld. An inert chemical is one with a full outer shell of electrons that do not normally react with other substances. Inert gases include argon and helium. Some other non-inert gases are used for welding, such as CO_2.

Left-hand threads: Threads that are machined to require a counterclockwise rotation to tighten them and a clockwise rotation to loosen them.

Nonconsumable: A metal or other material that is introduced at the weld area that does not melt and become part of the finished weld. A consumable does melt and become part of the finished weld.

Nonferrous: A term used to describe a metal, such as aluminum, that does not contain iron.

Overlapping: The protrusion of weld metal beyond the toe, face, or root of the weld. In resistance seam welding, this is the area in the preceding weld remelted by the succeeding weld.

Pass: A single progression of a welding or surfacing operation along a joint, weld deposit, or substrate. The result of a pass is a weld bead, layer, or spray deposit.

Purge gas: An inert gas that is distributed through the inside of piping or tubing to be welded, to prevent oxidation caused by combining oxygen in the ambient atmosphere with the extreme heat of the welding arc.

Right-hand threads: Threads that are machined to require a clockwise rotation to tighten them and a counterclockwise rotation to loosen them.

Shield gas: An inert gas that is distributed around the outside of piping or tubing to be welded, to prevent oxidation caused by combining oxygen in the ambient atmosphere with the extreme heat of the welding arc.

Soapstone: A tool used to permanently mark metal surfaces so that they may be fitted properly in preparation for welding.

Spatter: In arc and gas welding, the metal particles expelled during welding that do not form part of the weld.

Striker: A hand tool used to ignite a welding or cutting torch using a flint and a rasp-like wheel rubbed against each other.

Type L copper tube: An industry standard for copper tubing, referring to the tube wall thickness and identified by a blue stripe.

Additional Resources

This module is intended to present thorough resources for task training. The following reference works are suggested for further study. These are optional materials for continued education rather than for task training.

Metals and How to Weld Them, 1976. Cleveland, Ohio: James F. Lincoln Arc Welding Foundation.

Modern Welding, 1980. South Holland, IL: The Goodheart-Willcox Company, Inc.

Technical Guide for Gas Tungsten Arc Welding, EW-740, ITW/Hobart Welders, Troy, OH

Acknowledgments

Gerald Shannon	203F03, 203F04, 203F05, 203F06, 203F09, 203F10, 203F11, 203F22, 203F23, 203F24, 203F27, 203F28
Charles Rogers	203F07 203F13, 203F25, 203F26
©2002 Swagelok Company	203F12
Miller Electric Mfg. Co.	203F14

NCCER CURRICULA — USER UPDATE

NCCER makes every effort to keep its textbooks up-to-date and free of technical errors. We appreciate your help in this process. If you find an error, a typographical mistake, or an inaccuracy in NCCER's curricula, please fill out this form (or a photocopy), or complete the online form at **www.nccer.org/olf**. Be sure to include the exact module ID number, page number, a detailed description, and your recommended correction. Your input will be brought to the attention of the Authoring Team. Thank you for your assistance.

Instructors – If you have an idea for improving this textbook, or have found that additional materials were necessary to teach this module effectively, please let us know so that we may present your suggestions to the Authoring Team.

NCCER Product Development and Revision

13614 Progress Blvd., Alachua, FL 32615

Email: curriculum@nccer.org
Online: www.nccer.org/olf

❑ Trainee Guide ❑ AIG ❑ Exam ❑ PowerPoints Other ______________________________

Craft / Level: Copyright Date:

Module ID Number / Title:

Section Number(s):

Description:

Recommended Correction:

Your Name:

Address:

Email: Phone:

Module 12204-03

Process Control Theory

COURSE MAP

This course map shows all of the modules in the second level of the Instrumentation curriculum. The suggested training order begins at the bottom and proceeds up. Skill levels increase as you advance on the course map. The local Training Program Sponsor may adjust the training order.

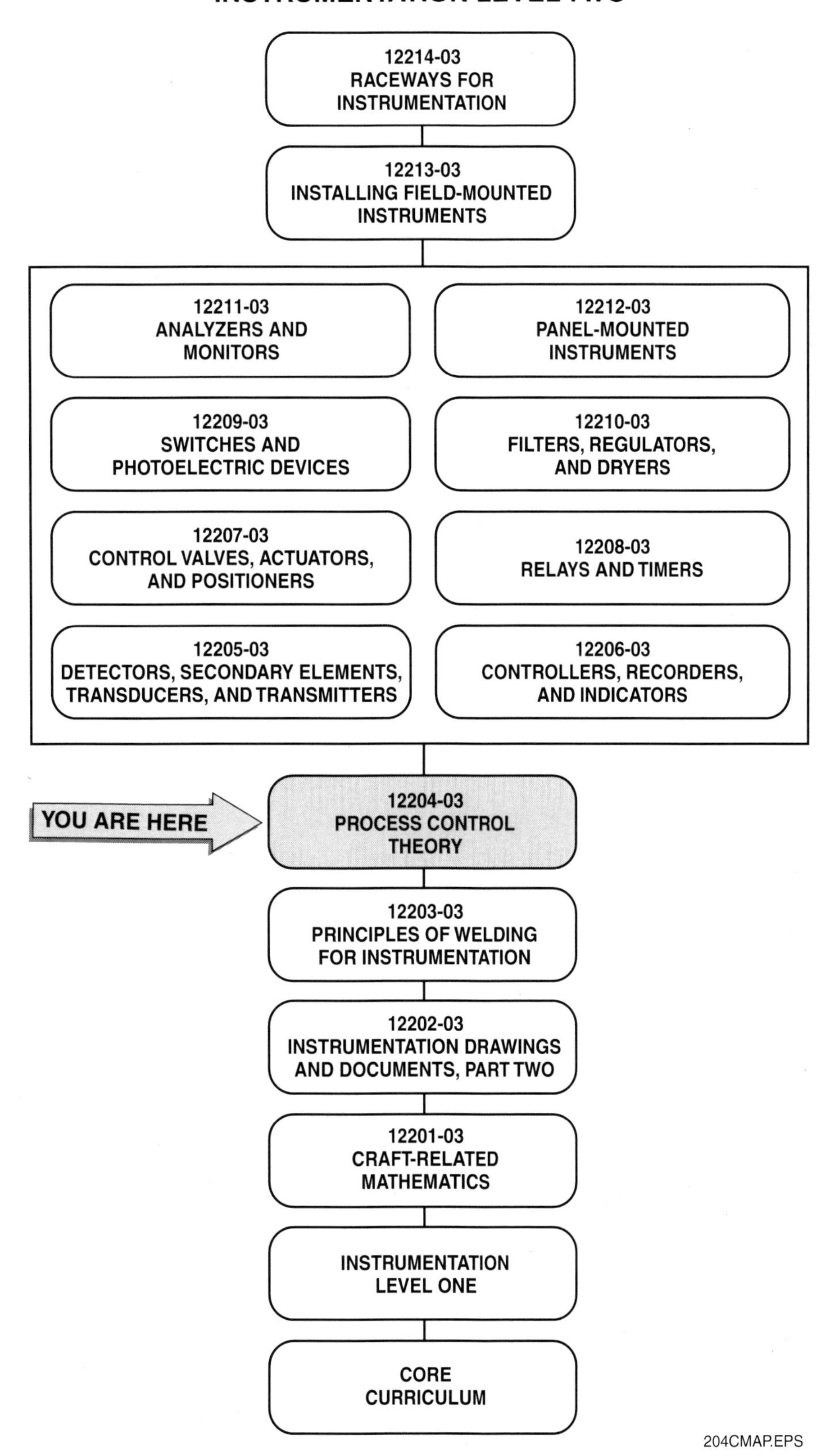

MODULE 12204-03 CONTENTS

1.0.0 INTRODUCTION 4.1

2.0.0 PROCESS CHARACTERISTICS 4.2

3.0.0 THE PROCESS CONTROL SYSTEM 4.3

4.0.0 COMPONENTS OF AN INSTRUMENT CHANNEL 4.4

4.1.0 Detector/Sensor 4.4

4.1.1 Direct vs. Inferred Measurements 4.4

4.2.0 Transducer 4.7

4.3.0 Amplifier/Signal Conditioner 4.7

4.4.0 Transmitter 4.7

4.5.0 Controller 4.8

4.6.0 Final Control Element (Control Valve) 4.8

4.6.1 Pneumatic Control Valve Actuators 4.9

4.6.2 Manual Actuators 4.10

4.6.3 Valve Positioners 4.10

4.6.4 Electric Proportional Valve Actuators 4.11

4.6.5 Solenoid Actuators 4.12

5.0.0 CONTROL LOOPS 4.13

5.1.0 Feedforward Control (Open Loop) 4.13

5.2.0 Feedback Control (Closed Loop) 4.13

5.2.1 Operation of Closed-Loop Control 4.14

5.2.2 Performance of a Closed-Loop System 4.15

5.2.3 Criteria for Closed-Loop Control Quality 4.17

5.3.0 Cascade Control 4.18

5.4.0 Ratio Control 4.20

6.0.0 CONTROL MODES 4.22

6.1.0 On-Off Control (Two-Position Control) 4.22

6.1.1 On-Off Control Characteristics 4.22

6.2.0 Modulating Control 4.26

6.2.1 Proportional (Gain or P) Control 4.27

6.2.2 Integral (Reset or I) Control 4.30

6.2.3 Derivative (Rate or D) Control 4.31

6.2.4 Proportional Plus Integral (PI) Control 4.32

6.2.5 Proportional Plus Derivative (PD) Controllers 4.35

6.2.6 Proportional Plus Integral Plus Derivative (PID) Controllers 4.38

7.0.0 TYPES OF CONTROL APPLICATIONS 4.39

7.1.0 Typical Temperature Control Loops 4.39

7.1.1 Pneumatic Temperature Control Loops 4.39

7.1.2 Electronic Temperature Control Loops 4.41

7.2.0 Typical Pressure Control Loops 4.41

7.2.1 Pneumatic Pressure Control Loops 4.42
7.2.2 Electronic Pressure Control Loops 4.42
7.3.0 Pressure- Sensing Control Loops 4.43
7.3.1 Pneumatic Flow Control Loops 4.43
7.3.2 Electronic Flow Control Loops 4.44
7.4.0 Typical Level Control Loops 4.45
7.4.1 Pneumatic Level Control Loops 4.45
7.4.2 Electronic Level Control Loops 4.45
SUMMARY 4.46
REVIEW QUESTIONS 4.46
GLOSSARY 4.49
REFERENCES & ACKNOWLEDGMENTS 4.51

Figures

Figure 1 Heat transfer process 4.2
Figure 2 Heat exchanger equipped with a process control system 4.3
Figure 3 Rosemount® Model 1151 Alphaline® pressure transmitter 4.8
Figure 4 Valve characteristic curves .4.9
Figure 5 Diaphragm actuators 4.9
Figure 6 Double-acting piston actuator 4.10
Figure 7 Current-to-air valve positioner 4.11
Figure 8 Typical electric valve actuator schematic diagram . . .4.12
Figure 9 Electric valve actuator 4.12
Figure 10 Solenoid valve construction 4.13
Figure 11 Heat exchanger with open-loop control 4.14
Figure 12 Closed-loop control system schematic diagram 4.15
Figure 13 Simple three-element closed-loop control system . . .4.16
Figure 14 Closed-loop responses 4.16
Figure 15 Possible stable responses to a disturbance 4.17
Figure 16 Three criteria for control quality 4.17
Figure 17 Pneumatic cascade control loop 4.19
Figure 18 Cascade level control 4.20
Figure 19 Ratio combustion control loop 4.21
Figure 20 Ratio control of a blending process 4.21
Figure 21 On-Off (two-position) control system 4.22
Figure 22 Response of an On-Off (two-position) control system 4.23
Figure 23 On-Off (two-position) control system with a neutral zone 4.23
Figure 24 Response of an On-Off (two-position) control system with a neutral zone 4.24
Figure 25 On-Off (two-position) tank level control system 4.24
Figure 26 Theoretical tank-level control system characteristics 4.25
Figure 27 Tank level system overshoots and undershoots 4.25
Figure 28 Process demand disturbance 4.26
Figure 29 Setpoint change 4.27
Figure 30 Proportional control system (gain of 1.0) 4.27
Figure 31 Valve position vs. float level (gain of 1.0) 4.28
Figure 32 Proportional control system (gain of 2.0) 4.29
Figure 33 Valve position vs. float level (gain of 2.0) 4.29
Figure 34 Proportional control system (gain of 0.5) 4.30
Figure 35 Valve position vs. float level (gain of 0.5) 4.30

Figure 36 Fixed input with an integral output 4.31
Figure 37 Derivative response to an input signal 4.32
Figure 38 Heat exchanger process system with a proportional plus reset (PI) controller 4.33
Figure 39 Effects of demand disturbance on a proportional plus reset (PI) control mode 4.33
Figure 40 Effect of a setpoint change .4.34
Figure 41 Proportional only (A) and proportional plus integral (B) startup comparison 4.35
Figure 42 Heat exchanger process with a proportional plus derivative (PD) controller 4.36
Figure 43 Effects of demand disturbance (PD) 4.37
Figure 44 Effects of a setpoint change (PD) 4.37
Figure 45 Effects of demand disturbance (PD) 4.39
Figure 46 Effects of a setpoint change on different control modes 4.40
Figure 47 Pneumatic temperature control loop 4.40
Figure 48 Tube-type heat exchanger 4.41
Figure 49 Electronic temperature control loop 4.41
Figure 50 Split-range pneumatic pressure control loop 4.42
Figure 51 Electronic pressure control loop 4.43
Figure 52 Split-range electronic pressure control loop 4.43
Figure 53 Pneumatic flow control loop 4.44
Figure 54 Electronic flow control loop 4.44
Figure 55 Pneumatic level control loop 4.45
Figure 56 Electronic level control loop 4.45

Tables

Table 1 Common Detectors and Their Principles of Operation 4.5–4.7
Table 2 Common Transducers 4.7

MODULE 12204-03

Process Control Theory

Objectives

When you have completed this module, you will be able to do the following:

1. Define process measurement and control.
2. Explain process characteristics that demand process control.
3. Describe the elements of an instrumentation channel, including:
   - Detector (sensor)
   - Transducer
   - Amplifier or signal conditioner
   - Transmitter
   - Controller
   - Final element (control valve)
4. Define and describe process control loop types, including:
   - Feedforward
   - Feedback
   - Cascade
   - Ratio
5. Define and describe process controller modes, including:
   - On-off control (two-position control)
   - Modulating control
     - Proportional (P)
     - Integral (I)
     - Derivative (D)
     - Proportional plus integral (PI)
     - Proportional plus derivative (PD)
     - Proportional plus integral plus derivative (PID)
6. Discuss various types of process control applications and loops.

Prerequisites

Before you begin this module, it is recommended that you successfully complete the following: Core Curriculum; Instrumentation Level One; Instrumentation Level Two, Modules 12201-03 through 12203-03.

Required Trainee Materials

1. Pencil and paper
2. Appropriate personal protective equipment

1.0.0 ◆ INTRODUCTION

The industrial plants of today require the measurement and control of numerous parameters to operate in the most efficient and reliable way possible. Conditions must be constantly monitored to provide a safe and comfortable atmosphere for the workers in the plant and to limit emissions from the plant that could have adverse effects on the environment. Because there is such a wide range of parameters monitored in a typical plant, the instrumentation industry has become a highly diversified field. Quality technical personnel are required to install, operate, calibrate, and maintain the devices used to measure and control these various process parameters. To perform these tasks, trainees must understand the meaning of process measurement and be familiar with the types and modes of control. Trainees must also be familiar with the terminology used and the basic principles involved in detecting and controlling these parameters. Many methods can be used to measure and control each parameter in a process. The more commonly used methods and the principles involved in measuring and controlling a parameter are covered in this section.

Process control is the study of manipulating material to produce a desired product. More specifically, it is the application of a control system to a process to provide the most efficient and reliable operation possible.

The goal of process control is to measure, compare, compute, and correct. The control system implements this concept. To produce the best control possible, the control system must be matched to the process, which means that the process must be understood. Understanding the process requires the use of new terms and concepts.

The process itself presents the greatest mystery in the application and implementation of control systems. The reaction of the process to a supply or demand **disturbance** is difficult to predict.

The key characteristics of a process must be identified, and the complexity of the process must be simplified. Processes are generally very complex in that they contain many smaller processes connected together in series.

This module introduces the fundamental concepts of process control theory and discusses the major principles of operation of process and control systems.

2.0.0 ◆ PROCESS CHARACTERISTICS

A major portion of any process control system is the process itself. A process may be defined simply as the operation or operations used in the treatment of material. In *Figure 1*, the operation of adding thermal energy to the oil flowing through the heat exchanger is the process. This type of process is called heat transfer.

A process is normally implemented by the use of equipment and hardware. The shell-and-tube heat exchanger and the piping comprise the physical system within which the thermal energy transfer occurs. In this example, the actual process is the heating of the oil flowing through the tube side of the heat exchanger. The hot water entering the shell side of the heat exchanger gives up some of its thermal energy to the tube side oil, raising its temperature. After the hot water transfers its thermal energy, it leaves the heat exchanger at a lower temperature.

In *Figure 1*, the oil-heating process shown can be considered an energy exchange. Many processes transfer material only, or both material and energy. In the process shown in *Figure 1*, energy enters the process as an input. It is then transferred and leaves the process as an output. The energy output equals the energy input minus the sum of the energy lost and the energy stored in the process.

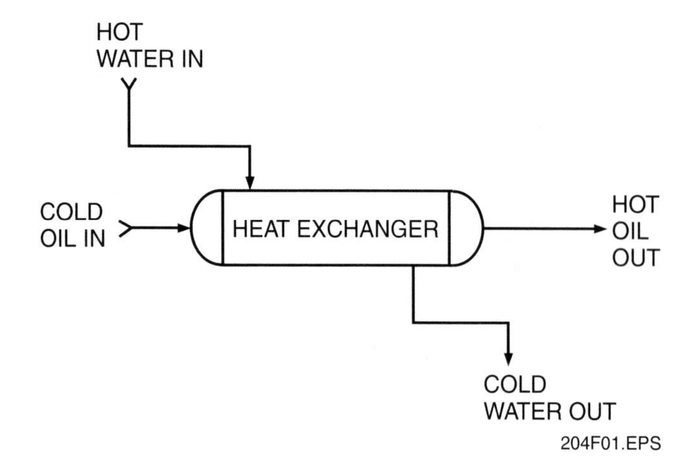

Figure 1 ◆ Heat transfer process.

In the heat exchanger, the energy input depends on the flow of shell side water and the temperature of the entering hot water. The energy output depends on the flow of oil and the temperature of the hot oil out. The energy lost is that lost through the walls of the heat exchanger to the surrounding environment. The energy stored is the amount which keeps the mass of metal of the heat exchanger at its operating temperature. Therefore, whether the individual parameters of the process are constant or changing in value is determined by the energy balance equation:

$$E_{out} = E_{in} - (E_{lost} + E_{stored})$$

Therefore, if E_{lost} increases because of ambient temperature changes, and E_{stored} and E_{in} are constant, then E_{out} must decrease, resulting in a drop in the outlet oil temperature. Once the balance is made and the energy terms are not changed, the process parameters will be constant.

If left to itself, the outlet oil temperature would ultimately come to a steady value. The output energy would then equal the input energy minus the sum of the energy lost and stored. When this condition exists, the process is in a steady-state condition, or in balance. Any disturbance to the energy in, energy out, energy lost, or energy stored will upset this balance and consequently cause a change in the values of the process parameters. In this particular example, as long as the energy out equals the energy in minus the sum of the energy lost and stored, the temperature of the outlet oil remains constant.

As an example, allow the hot water entering the shell side of the heat exchanger to increase in temperature. Due to this temperature increase, the energy input to the heat exchanger is increased. This increases the energy transfer to the oil, raising its temperature. This is the energy output.

However, at the same time, more energy is being lost to the surroundings because of the increased temperature differential. Also, the internal components of the heat exchanger are absorbing energy and storing it. These three actions continue until equilibrium is reached. At that time, the increased energy input is balanced by an increase in the energy output, the energy lost, and the energy stored. The net result is that the temperature of the oil out has increased along with a greater loss to the surroundings and an increase in stored energy within the heat exchanger.

Whether they are controlled manually or automatically, most processes perform efficiently and/or safely only when the process parameter values are held within specified limits. As described above, process parameters are subject to change if the energy balance is upset. Therefore, a fundamental function of process control is to manipulate the energy balance of a system in order to maintain the process parameters within predetermined limits.

3.0.0 ◆ THE PROCESS CONTROL SYSTEM

A process control system can be defined as a mechanism that measures the value of a selected process parameter and operates to limit the change in this parameter from a desired value. Its implementation is usually accomplished by an operator and/or a hardware system. *Figure 2* shows the heat exchanger after equipping it with a simple control system.

The process parameter that is selected for control is called the **controlled variable**. The controlled variable for this heat exchanger example is the outlet oil temperature. The measurement of the controlled variable is performed by a measuring element. In the case of the heat exchanger, the measuring element is a thermocouple connected to a temperature transmitter. The temperature transmitter detects the change in temperature and converts it to a signal suitable for transmission. Typical transmission signals are 4–20 mA DC current, 1–5 volts, 0–10 volts, and 3–15 psi air pressure. The transmission signal from the measuring element is termed the measured variable. The measured variable is a signal that represents the controlled variable. *Figure 2* shows the location of the thermocouple and the temperature transmitter within the control system.

The transmission signal received from the measuring element is utilized by the controller. A controller is a device that operates to ultimately regulate the controlled variable. The controlled variable is normally required to be maintained at a desired value. This desired value is termed the reference or **setpoint**. In the case of the heat exchanger, the setpoint is the desired temperature of the hot oil flowing out of the heat exchanger. The difference between the setpoint and the measured

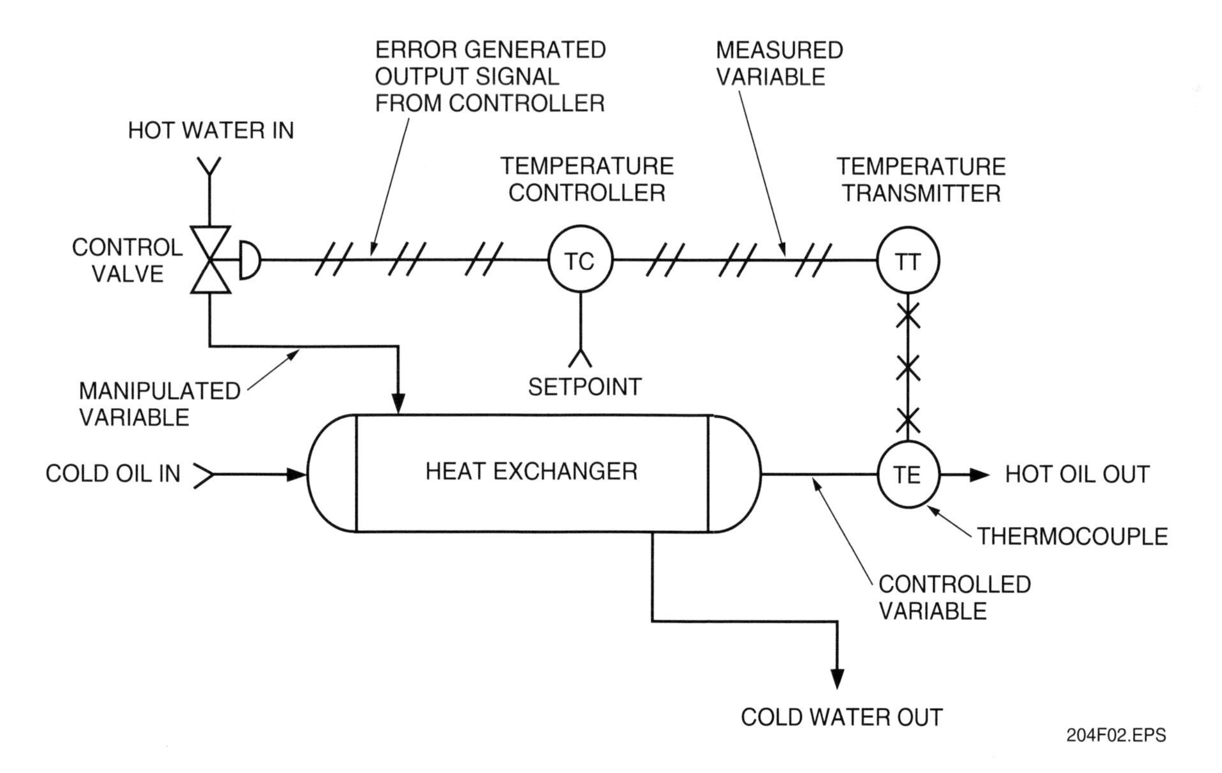

Figure 2 ◆ Heat exchanger equipped with a process control system.

variable is the error signal. The controller acts on the error signal and produces an output signal. The controller may act on the error in many different modes. The modes of control action are presented later in detail. The location of the controller within the control system is shown in *Figure 2*.

The output signal from the controller is received by a device that converts this signal into an action that is applied to the process. The device that performs this operation is the final control element. The final control element regulates a parameter of the process so that the controller can regulate the controlled variable. The process parameter that is regulated by the final control element is called the manipulated variable. The manipulated variable for the heat exchanger is the rate of hot water flow into the shell side of the heat exchanger. *Figure 2* shows the location of the final control element.

When a control system is applied to a process, two parameters of the process are chosen to interface with the control system. These parameters are the controlled variable and the manipulated variable. The manipulated variable is used by the control system to supply energy or material to the process to regulate the deviation of the controlled variable from the desired value. The process can be affected by many other variables. These parameters are not directly sensed by the control system and can also cause a change in the energy balance of the process. This unbalancing is seen by the control system as a change in the controlled variable. Generally, these parameters are divided into two large groups: supply and demand. The changing of the supply or demand parameter is known as a supply disturbance or a demand disturbance. A supply disturbance is a change in the energy or material into the process. In the heat exchanger in *Figure 2*, changes in the temperature of the entering hot water and/or the flow rate of the entering hot water are supply disturbances. A demand disturbance is the change in the energy or material output of the process. Some examples for the heat exchanger in *Figure 2* are a change in the temperature of the entering cold oil and/or the rate of oil flow through the tube side of the heat exchanger. A major objective of process control is to maintain the controlled variable at the setpoint regardless of either supply or demand disturbances. This is the reason for using a process control system.

4.0.0 ◆ COMPONENTS OF AN INSTRUMENT CHANNEL

All of the components that are necessary to detect, condition, transmit, control, and regulate the process make up the instrumentation **channel**. How they are interconnected determines the instrument control loop. In a typical instrumentation channel, these elements may include one or more detectors or sensors, transducers, signal amplifiers, transmitters, controllers, indicators, recorders, and/or control valves. The following sections will discuss all of these elements except indicators and recorders, since these two components only serve as observation elements.

4.1.0 Detector/Sensor

To sense the process parameters, the detector receives energy from the process and produces an output. It is important to realize that the sensing element always extracts some energy from the process. The measured quantity is always disturbed by the act of measurement. A good instrument is designed to minimize this disturbance.

Detectors are selected for measurement systems on the basis of the parameter being sensed, the desired accuracy, the range of measurement, and the particular type of output needed. *Table 1* lists some of the common detectors used, values involved, and detection principles.

4.1.1 Direct vs. Inferred Measurements

The wide variety of measurements made in industrial plants and the varying environmental conditions under which these measurements must be made require mention of the basic nature of measurement. Measurements for this discussion fall into two general categories: those measurements made directly and those that are inferred.

If a tank is filled to some level with water, how can the level be measured? One way is to measure the level with a dipstick. Another way is to look at a gauge glass on the side of the tank and measure the level in the glass with a ruler. These methods are examples of direct measurement.

Looking again at *Table 1*, the list of detection principles indicates that some of these parameters are not being measured directly. The parameter of interest is affecting a characteristic or property of the material that is actually measured. The change in this property or characteristic is what is actually being sensed. For example, when a difference in temperature exists between the junctions of a thermocouple, a voltage is produced. Therefore, voltage is the actual parameter being measured, and the temperature is inferred, or derived from this measurement.

Table 1 Common Detectors and Their Principles of Operation (1 of 3)

Parameter	Detector	Principle of Operation
Temperature	Bimetallic Strip	Different metals experience thermal expansion with changes in temperature.
	Thermocouple	When two wires composed of dissimilar metals are joined at one end and this junction is heated, a voltage is created.
	Resistance Temperature Detector (RTD)	The resistance of certain metals and semi-conductors changes significantly and in a repeatable manner as a function of temperature. An electrical excitation current is passed through the RTD, and the output voltage is an indication of the RTD s resistance based on temperature.
	Thermistor	Same as RTD but by far the largest value of change with temperature as compared with the other three sensors.
Pressure	Diaphragm/ Potentiometric	Converts pressure into a variable resistance. A mechanical device such as a diaphragm is used to move the wiper arm of a potentiometer as the input pressure changes.
	Diaphragm/ Piezoelectric	A certain class of crystals called *piezoelectric* produce an electrical signal when they are mechanically deformed. The voltage level of the signal is proportional to the amount of deformation. The crystal is mechanically attached to a metal diaphragm, with one side of the diaphragm connected to the process fluid to sense pressure. A mechanical linkage connects the diaphragm to the crystal.
	Diaphragm/ Capacitance	A differential pressure is applied to a diaphragm, which causes a filling fluid to move between the isolating diaphragm and a sensing diaphragm. As a result, the sensing diaphragm moves toward one of the capacitor plates and away from the other causing the capacitance to change.
	Diaphragm/ Variable Sensing (Inductance)	Inductance is the ability of a conductor to produce induced voltage when the current in the circuit varies. A longer wire has more inductance than a shorter wire. The diaphragm moves along the wire, changing the inductance with any pressure change.
	Diaphragm or Bellows/Strain Gauge	A strain gauge is a device that changes resistance when stretched. It consists of multiple runs of very fine wire that are mounted to a stationary frame on one end and a movable armature on the other. The movable arm is connected to the diaphragm or bellows.
	Bourdon Tube	A curved, oval tube will attempt to achieve a straight cylindrical shape when internal pressure is applied.
	Manometer	This is used to check low vapor pressure.
Level	Differential Pressure	There is a direct relationship between the hydrostatic pressure caused by a column of liquid, the specific gravity of the liquid, and the height of the vertical column of liquid. In applying differential pressure to level sensing, the hydrostatic pressure of the liquid in a tank or vessel is compared to atmospheric pressure to derive the difference between the two pressures and determine the level of the liquid.
	Bubbler	A dip tube is installed in a tank with its open end just off the bottom of the tank. A fluid is forced through the tube; when the bubbles escape from the open end of the tube, the pressure in the tube equals the hydrostatic head of the liquid. As the liquid level varies, the pressure in the dip tube changes. This pressure is detected by a pressure-sensitive instrument.
	Diaphragm	The process material exerts pressure against a diaphragm. This pressure is detected by a pressure-sensitive instrument.

204T01A.EPS

Table 1 Common Detectors and Their Principles of Operation (2 of 3)

Parameter	Detector	Principle of Operation
Flow	Orifice Plate	A plate with a calibrated hole in it is inserted in a process line and the differential pressure developed across the orifice plate is measured by tapping off the high and low sides to determine the flow rate.
	Venturi Tube	A converging cone-shaped inlet section in which the cross section of the process stream decreases and the velocity increases with concurrent increases of velocity head and decreases of pressure head. The differential pressure is measured by tapping off the high and low sides to determine flow rate.
	Flow Nozzle	A restriction with an elliptical contour inlet section and a cylindrical throat section. The differential pressure is measured by tapping off the high and low sides to determine flow rate.
	Wedge Flow Element	The wedge element is also a differential pressure sensing element with no sudden changes in shape and no sharp corners. It is often used to measure slurries or thick fluids.
	Pitot Tube	The pitot tube is also a differential pressure element. It is a cylindrical probe that is inserted into a stream. The velocity of fluid flow at the upstream face of the probe is reduced to almost zero. Velocity head pressure is converted into impact pressure, which is sensed through a small hole in the upstream face of the probe. A corresponding small hole in the side of the probe senses static pressure. A pressure-sensitive instrument measures the difference between the two pressures.
	Annubars	Two-chamber flow tubes that have several pressure openings distributed across the area of the process stream are inserted into the piping. They are high- and low-pressure holes with fixed separations. The differential pressure is measured between the high- and low-pressure holes.
	Turbine	A meter that is inserted in the piping that provides a frequency or pulse output signal. The fluid being measured enters the meter, then passes through a rotor. This causes the rotor to turn in proportion to the fluid velocity. Each turn of the rotor produces magnetic pulses through the use of a magnetic tip on the rotor and a magnetic pickup coil mounted outside the meter.
	Vortex Shedding	As fluid flows past a blunt body, or shedder, at low velocity, the fluid pattern remains streamlined. However, as velocity increases, the fluid separates from each side of the shedder and swirls to form vortices downstream of the shedder. A pressure sensor that is mounted on the upstream side of the flow shedder detects the pressure that is exerted on the shedder by the formation of the vortices.
	Magnetic	A nonmagnetic tube carries the process flow, which must have a degree of conductivity. Surrounding the tube are magnetic coils and cores that provide a magnetic field across the full width of the flow when electric current is applied. The fluid flowing through the tube is the conductor, and as the conductor moves through the magnetic field, a voltage is generated that is proportional to the volume of the flow rate.
	Ultrasonic	Piezoelectric crystals transmit sound across the stream of flow, and receivers (which are also made from piezoelectric crystals) receive the sound. The process flow across the acoustic signals alters the signal, and the alteration is measured to determine the flow.
	Positive Displacement	A meter that continuously entraps a known quantity of fluid as it passes through the meter. Since the number of times the fluid is entrapped and the volume of the entrapped fluid are both known, the amount of fluid is easily calculated.

204T01B.EPS

Table 1 Common Detectors and Their Principles of Operation (3 of 3)

Parameter	Detector	Principle of Operation
Flow (continued)	Mass	Most mass flowmeters operate on the principle of Coriolis, in which a tube is designed and built to have a predictable vibration characteristic whenever exposed to the forces exerted by the acceleration of fluid. A drive assembly connected to the center of the tube causes the tube to twist. This vibrates the tube, and position-sensing coils on each side of the flow tube sense this twisting.
	Rotameter (not a detector)	A type of variable-area flowmeter that consists of a tapered metering tube, a float, and a graduated scale. This device is not considered a detecting device, but a measuring device.

204T01C.EPS

4.2.0 Transducer

In an instrumentation channel, the output signal from one device is often incompatible with the required input signal of another device that is located in the same channel. In order to provide an interface between the two elements, a device called a transducer must be installed between the two incompatible devices. A transducer receives one form of energy and converts it into another form of energy. *Table 2* lists some of the more common instrumentation transducers, their commonly used abbreviations, and their typical functions. The values of their function may vary, either to a higher or lower value.

Table 2 only lists some of the more common transducers available. Keep in mind that any of these transducers can be installed in series with one another, thus allowing through their combination practically any signal conversion required in the system.

Table 2 Common Transducers

Transducer	Abbreviation	Typical Function
Analog signal (I) to pneumatic signal (P)	I/P	4–20 mA signal to 3–15 psi signal
Pneumatic signal (P) to analog signal (I)	P/I	3–15 psi signal to 4–20 mA signal
Voltage (E) to analog signal (I)	E/I	±0–5 volt signal to 4–20 mA signal
Analog signal (I) to voltage signal (E)	I/E	4–20 mA signal to ±0–5 volt signal
Analog signal (I or A) to digital signal (D)	I/D or A/D	4–20 mA signal to a discrete digital signal (12-bit resolution)
Digital signal (D) to analog signal (I or A)	D/A or D/I	Discrete digital signal (12-bit resolution) to a 4–20 mA signal
Voltage signal to digital signal	E/D	±0–5 volt signal to a discrete digital signal (12-bit resolution)

204T02.EPS

4.3.0 Amplifier/Signal Conditioner

The instrumentation signal must often travel great distances or, in the case of electronic signals, may be subjected to frequency interference or **noise**. In either case, the actual originating signal may become weak or may be altered to an impure state. Signal conditions such as these often warrant the use of signal amplifiers or signal conditioners whose function is to either boost the signal to a proportionately higher level for long transmission lengths or to condition it in a manner such that it is not severely affected by noise interference.

In the case of pneumatic signals, devices called pneumatic boosters may be used to amplify the air signals to compensate for leakage and/or pressure drop. In electronic or digital signals, devices are often incorporated that filter out noise and proportionately boost the signal. In most cases, amplifiers and signal conditioners are built into the transducer or transmitter as a single unit and don't require separate installation.

4.4.0 Transmitter

The function of the field transmitter is to receive the measurement from the primary element or sensor and send that signal on its way to an indicator, recorder, or controller. In many transmitter designs, the transmitter contains both the transducer and amplifier if these components are required in the specific application. *Figure 3* shows a photograph and a block diagram of a Rosemount® Model 1151 Alphaline® pressure transmitter that may be used for flow, liquid level, and

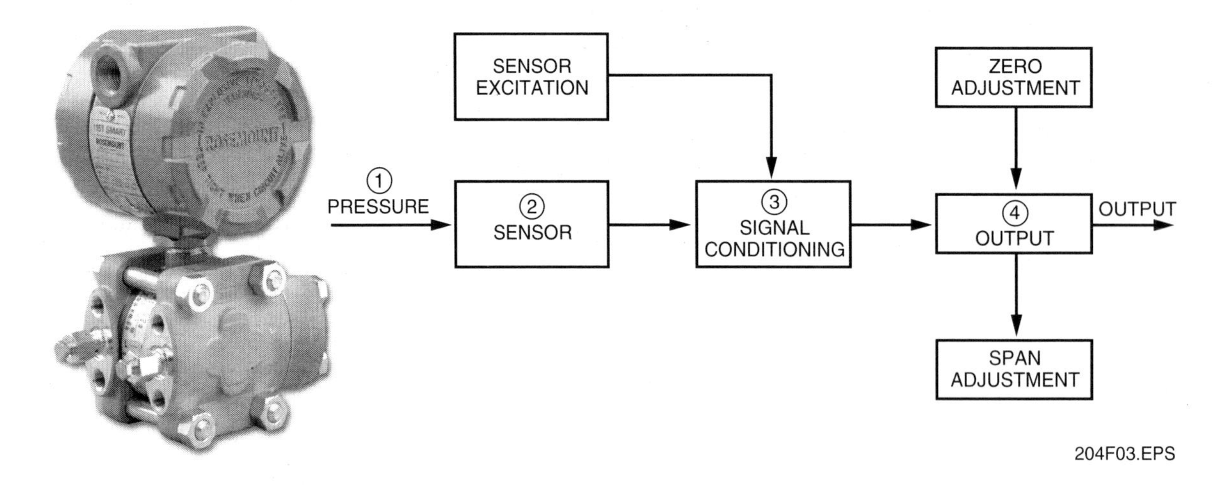

Figure 3 ◆ Rosemount® Model 1151 Alphaline® pressure transmitter.

other applications requiring accurate measurement of differential pressure. Notice that this transmitter contains a signal-conditioning component that is an integral part of the transmitter. In this figure, the signal flow can be summarized in four basic steps:

1. Pressure is applied to the sensor.
2. A change in pressure is measured by a change in the sensor output.
3. The sensor signal is conditioned for various parameters.
4. The conditioned signal is converted to an appropriate analog output.

4.5.0 Controller

A controller is a device that operates manually or automatically to regulate a controlled variable. There are three basic types of controllers: mechanical, pneumatic, and electronic. All three types serve the same purpose, which is to compare the process variable (received as a signal) with the setpoint value and generate an output signal that manipulates or changes the process to make the process variable equal to the controller's setpoint.

The controller is made up of a **feedback** transmission circuit, a comparator with a setpoint input, controller functions, and an output transmission system. The feedback transmission circuit makes sure that the sensor signal is converted into a usable controller signal. For example, if the input signal is 4-20 mA DC, the feedback circuit in the controller will convert the signal to 0 to 10 mV, which is the requirement for the controller. The output transmission circuit converts the signal from the feedback circuit into the form required by the final control element. More on controllers is found in the sections discussing types of control loops and methods of control modes.

4.6.0 Final Control Element (Control Valve)

A control valve is simply a variable orifice that is used to regulate the flow of a process fluid according to the requirements of the process. In a control valve, an actuator that is connected to the **valve plug stem** moves the valve between open and closed positions to regulate flow in the process. The valve body is mounted in the process fluid line and is used to control the flow of fluid in the process. There are two basic categories of control valves: gate valves and globe valves.

Gate valves are used either to allow full flow or to stop flow. They are designed to seat well and allow very little or no leakage. This characteristic of their design makes them useful for applications requiring the isolation of a system. They are either opened or closed, which makes them excellent for use with two-position controllers. They are not used to control flow rate, usually because the flow would wear the edges of the valve, eventually causing erosion and leakage.

The second type of valve, the globe or throttle valve, is used in a multitude of applications in the instrumentation industry. It provides a variable resistance or restriction to flow that is dependent on the plug's position in its seat. The **valve plug** is the component in the valve that forms to the valve seat to stop or restrict flow. It is generally constructed with rounded edges to reduce erosion by the process fluid stream. The gap between the plug and the seat, with the plug at its extreme open position, determines the maximum flow allowed. There are numerous variations of this valve available.

Trainees should be familiar with these two types of valves and also be familiar with the flow characteristic curve of a typical control valve. The flow characteristic curve of a control valve is the relationship between the flow rate through the valve and the percentage of valve travel. As a means of comparing and discussing the flow characteristics of a valve, a curve is usually plotted on a percentage of travel axis versus a percentage of flow axis. The three most common types of characteristics are linear, quick opening, and equal percentage, as illustrated in *Figure 4*.

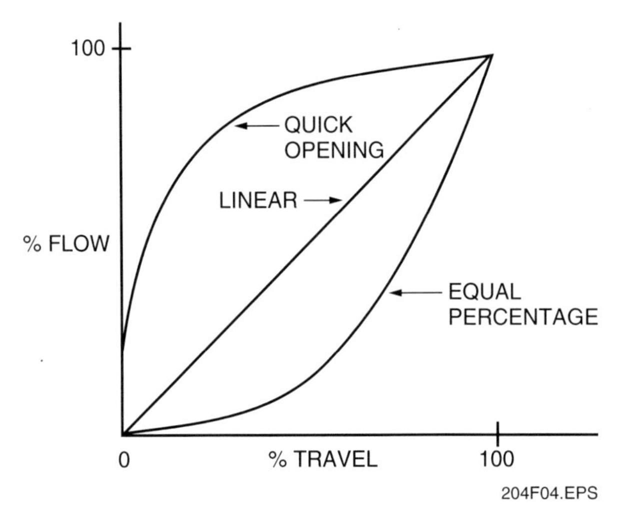

Figure 4 ◆ Valve characteristic curves.

The linear flow curve shows that the flow rate is directly proportional to the valve position. With this type of flow characteristic, a change in valve plug position will cause a similar change in flow rate at any given point in the valve travel.

A valve classified as quick opening produces a large linear change in flow rate at low valve travels to obtain about 90% of the total flow at only about 60% of total valve travel. This feature accounts for its name. Any additional increases in valve travel give sharply reduced changes in flow rate until the valve approaches wide open and the flow rate change is essentially zero.

An equal-percentage flow characteristic exhibits flow rates that increase exponentially with valve travel. The name is derived from the fact that a percentage increase in valve travel produces the same percentage increase in flow rate. The exponential curve means that any change in flow rate is always proportional to the flow rate just before the change in valve plug position is made. In other words, when the valve is lightly open and flow through the valve is small, the change in flow rate obtained by opening the valve a given amount is small. On the other hand, when the valve allows a large flow, the change in flow rate obtained by opening the valve the same given amount is large.

These characteristics are inherent characteristics of the valve design; they are observed with a constant pressure drop across the valve. In a process where the pressure drop may vary, the installed characteristic may be somewhat different. These inherent characteristics allow system designers to select valves that provide the proper process stability under expected operating conditions of the process. A throttle valve can have any of these characteristics.

4.6.1 Pneumatic Control Valve Actuators

The stem of a control valve can be moved by a mechanism called an actuator. There are two types of pneumatic actuators: the diaphragm actuator and the piston actuator. In control systems, the most commonly used of these types is the diaphragm actuator (*Figure 5*).

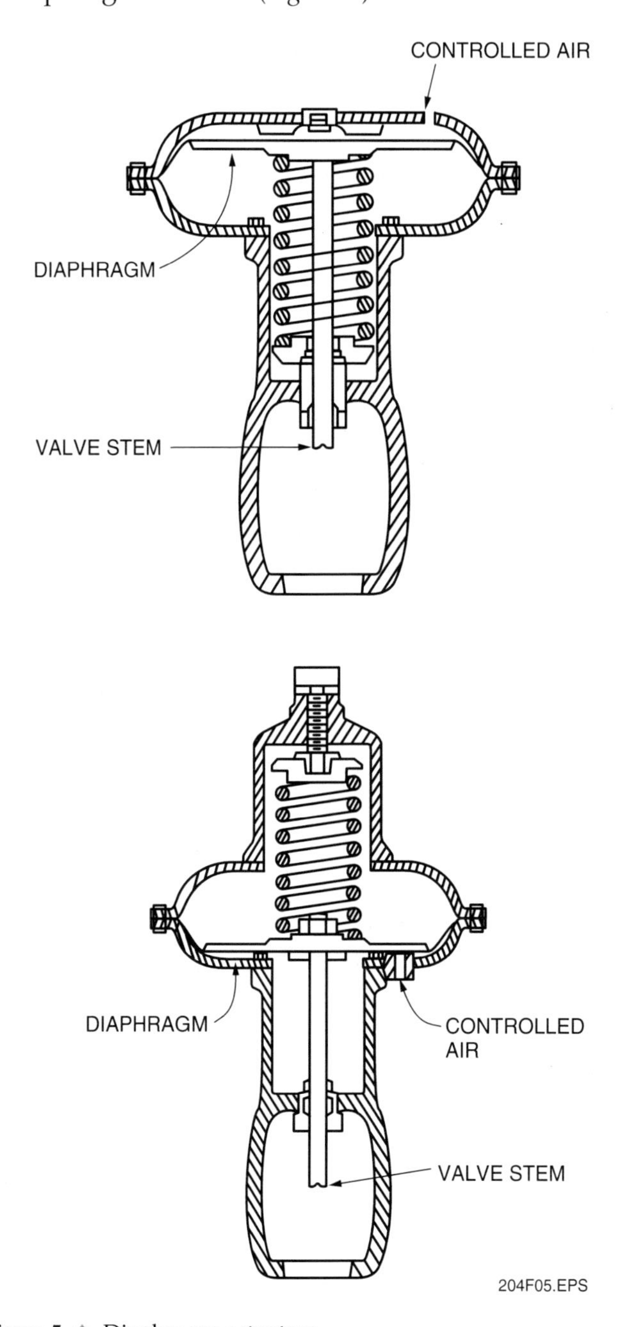

Figure 5 ◆ Diaphragm actuators.

The principle of the diaphragm actuator is very simple. A diaphragm is bolted to a dished metal head, forming a pressure-tight compartment. The controlled air is connected to the compartment. The motion of the diaphragm is opposed by a spring. Attached to the diaphragm is the valve stem, such that any movement of the diaphragm results in the same valve plug movement.

Both the diaphragm and the plug may be direct or indirect acting. The action of the actuator might be such that either an increase or a decrease in air pressure will lift the stem. The design of some actuators permits the action to be reversed mechanically.

The plug may be attached to the stem such that a lifting stem closes or opens the valve. Some plugs, or disks, are reversible on the stem.

The determination of the valve action desired is usually based on what valve position the valve should take if the supply fails. In one application, it may be desirable to have the valve go wide open when the air fails. On other applications, the system may be better served or safer if the valve fails shut.

Most piston actuators used in control valve applications are double acting (*Figure 6*). This means that a controlled air pressure is used on both sides of the piston to cause the desired valve stroke.

The pneumatic piston valve works on a similar principle to that of the diaphragm valve. The difference is that that the valve stem is connected to a piston-cylinder apparatus rather than to a flexible diaphragm.

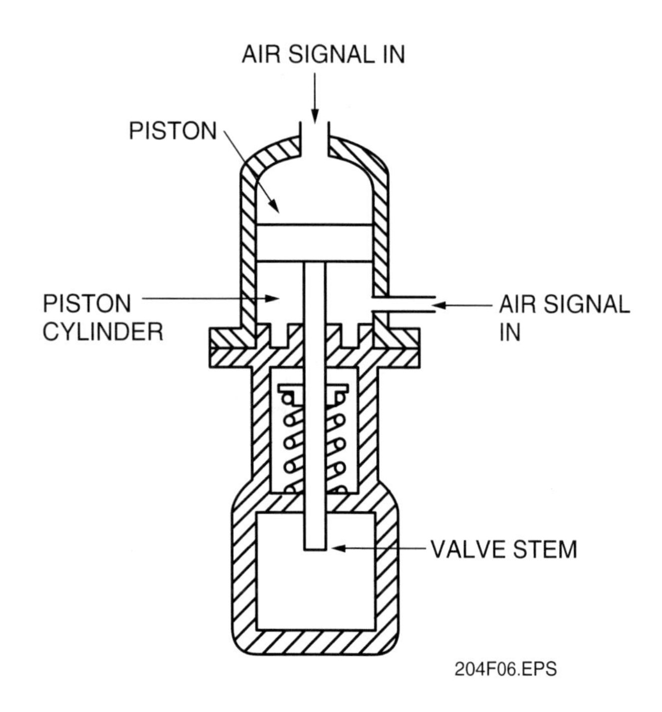

Figure 6 ◆ Double-acting piston actuator.

As a sensed pressure or a controller output signal varies, the differential pressure between it and reference pressure causes the piston to stroke within the cylinder. This design does not normally allow for a valve failure position. The piston- actuated valve fails as-is. A means of achieving a failed position is to install either a backup air volume or an internal spring mechanism to cause the valve to fail in either a closed or open position as preferred.

4.6.2 Manual Actuators

Manual actuators on automatic control valves are essentially handwheels. In almost every case, the manual actuator overrides the automatic actuator. Therefore, unless local operator control is desired, care should be taken to ensure that the manual actuator is not unintentionally engaged because it may restrict the normal movement of the valve stem.

4.6.3 Valve Positioners

A valve positioner is a device used primarily to maintain the control valve plug at a position that is directly proportional to its controller output. To perform this function, the stem motion of the valve actuator is compared to the signal from the controller. Any deviation of the valve stem from the desired position produces an error signal that activates a pneumatic relay. Air pressure to the valve actuator is then either increased or decreased to drive the valve stem to the desired position.

A positioner can also be used to reverse the signal to a valve, to overcome frictional forces within a valve on high pressure drop applications, and to split-range the output of a controller.

Split-ranging is necessary when one controller is used to control two final elements, each of which must be actuated at different values of controller output. For example, the temperature in a tank must be maintained at 70°F. When the temperature drops below 70°F, a steam valve opens to heat the tank. When the temperature rises above 70°F, a cooling water valve opens to reduce the temperature in the tank. Both the steam valve and cooling water valve have positioners. The steam valve positioner operates when the controller output signal varies between 12 mA DC and 4 mA DC. The cooling water valve positioner operates when the controller output signal varies between 12 mA DC and 20 mA DC.

An example of a current-to-air (I/P) valve positioner is shown in *Figure 7*. The operation of this instrument is relatively simple. When the output of the controller increases, the 4–20 mA DC input

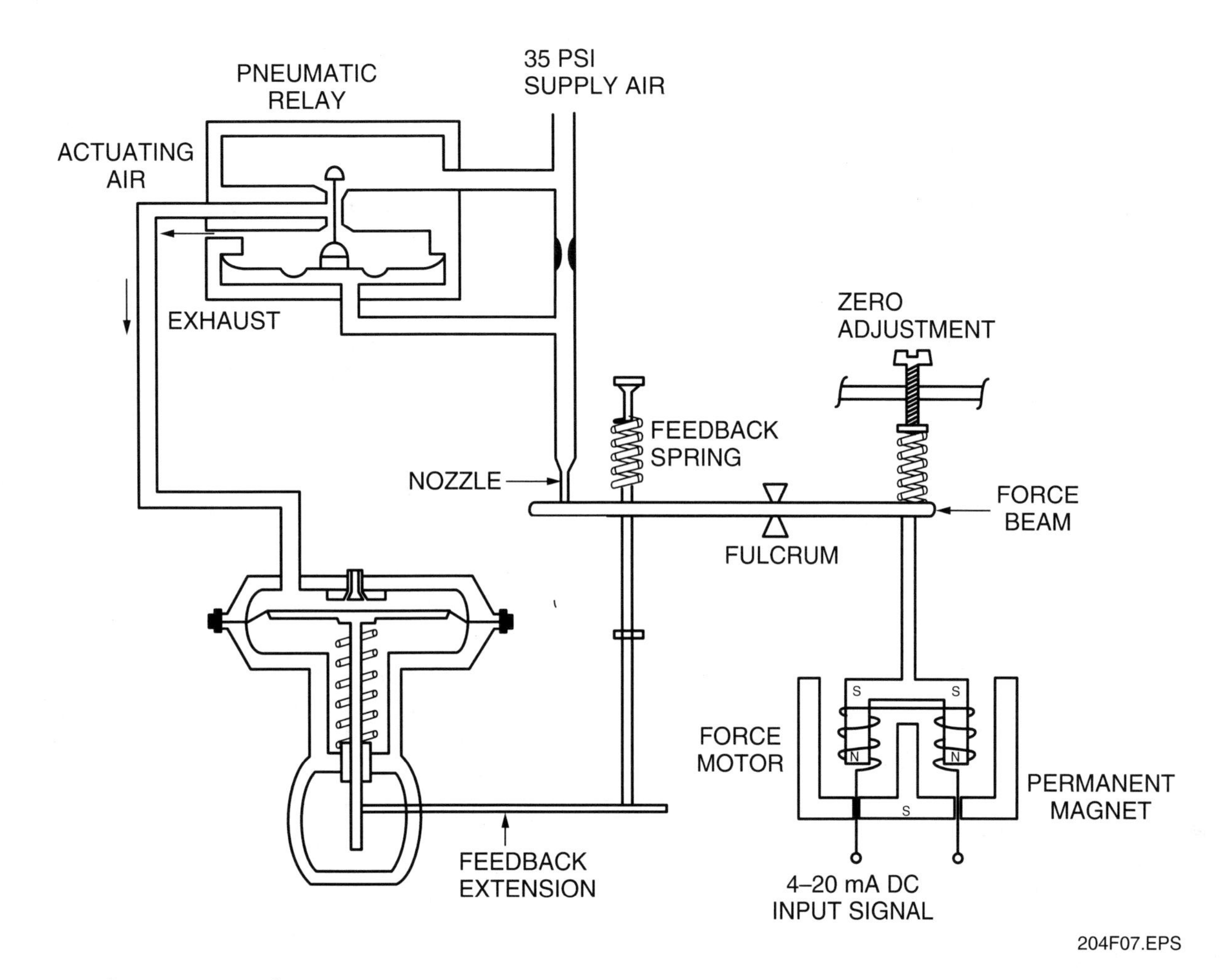

Figure 7 ◆ Current-to-air valve positioner.

signal to the positioner increases. Current flowing through the force motor coil develops a magnetic field as shown in *Figure 7*. Because opposite poles attract, a force is developed that moves the force motor coil downward. The coil is connected to a force beam, so the downward motion of the coil causes the right side of the force beam to pivot down. The pivoting of the force beam moves the left side of the beam closer to the nozzle. As a result of this action, less air can pass through the nozzle, so pressure builds downstream of it. The increase in pressure applies a force to a small diaphragm within the pneumatic relay. This force moves a pilot valve downward. The pilot valve exhaust plug seals off the exhaust port and opens the supply ball valve, allowing air to flow from the supply to the diaphragm actuator. The increased air flow to the diaphragm allows pressure above the diaphragm to increase. This increased pressure above the diaphragm develops a force that overcomes the force of the spring and causes the valve stem to move downward. The motion of the valve stem is transmitted to the positioner by the feedback extension. When both stem and feedback extension move downward, the feedback spring is compressed, applying a balancing force to the force beam. The balancing force pivots the beam to restore the balanced clearance between nozzle and force beam. The positioner and valve remain in this condition until further changes in controller output are produced.

4.6.4 Electric Proportional Valve Actuators

Electric operators that provide valve positioning proportional to an electronic control signal have several disadvantages, so they have found limited use in the process industries. Their primary use has been to operate valves located in remote areas where actuating air is unavailable. The primary disadvantage of electric valve actuators is slow operating speed. In addition, they are more expensive than pneumatic actuators.

Several different electric actuators are commercially available. A typical electric actuator operates in a manner very similar to a simple controller.

Recall that when the input signal changed, an error signal (the difference between the input signal and the reference signal) was produced (*Figure 8*).

The error signal was amplified by an error signal amplifier and converted to an AC signal by a chopper. The signal was then amplified again so that it had sufficient power to operate the servomotor. In the valve actuator, the output of the power amp produces the phase relationship necessary to cause rotation of the valve actuating motor.

The motor is connected to the valve stem through a gear train, shown in *Figure 9*, in order to increase the force applied to the valve stem by the electric motor. Movement of the valve stem positions the wiper arm of the balancing potentiometer, which reduces the error signal to zero. When the error signal is reduced to zero, motor rotation stops, and the valve remains in this new position until further changes in the control signal occur.

4.6.5 Solenoid Actuators

If the controller of a system is a two-position, or on-off, controller, an electrical solenoid is often used either to actuate the control valve (if it is small) or to actuate an air signal, which actuates the control valve. There is usually no type of feedback to the solenoid.

The solenoid consists of a valve body, a magnetic core attached to a valve stem, and a solenoid coil (*Figure 10*).

As current flows through the solenoid coil, a magnetic field is produced that develops an upward force on the movable magnetic core. When the magnetic force is large enough to overcome the force applied to the valve stem by the spring, the valve opens. When current decreases, the spring forces the valve shut.

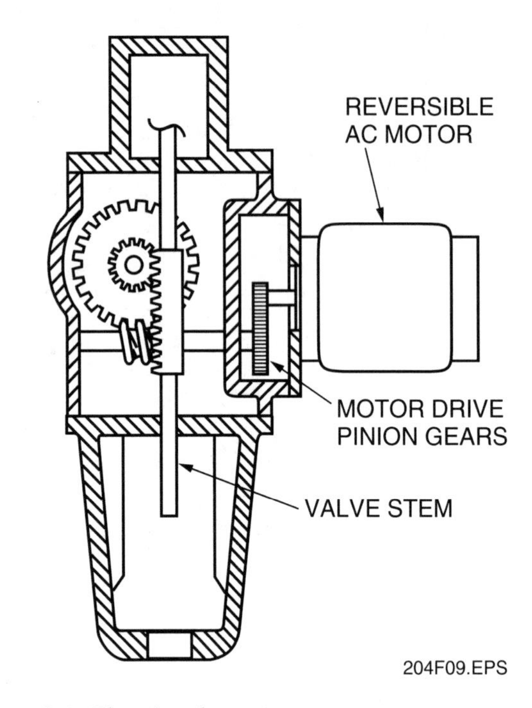

Figure 9 ◆ Electric valve actuator.

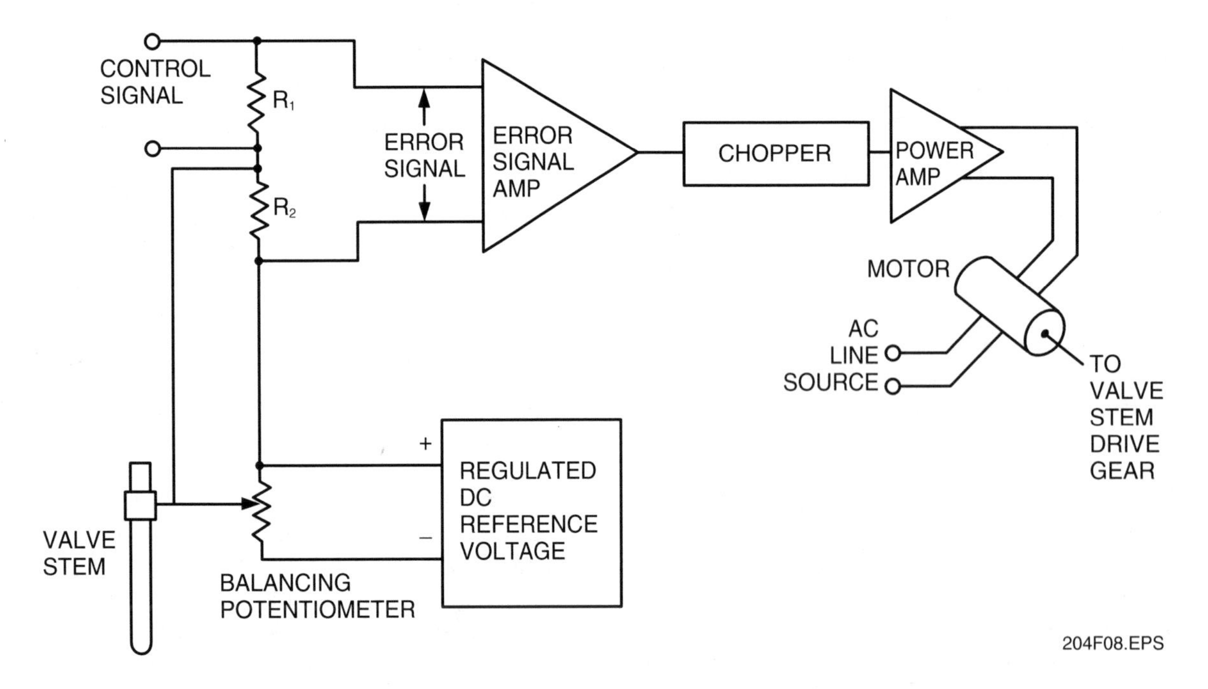

Figure 8 ◆ Typical electric valve actuator schematic diagram.

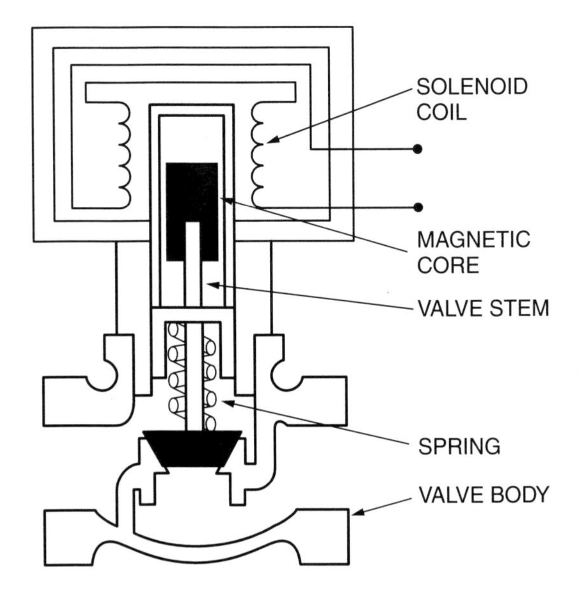

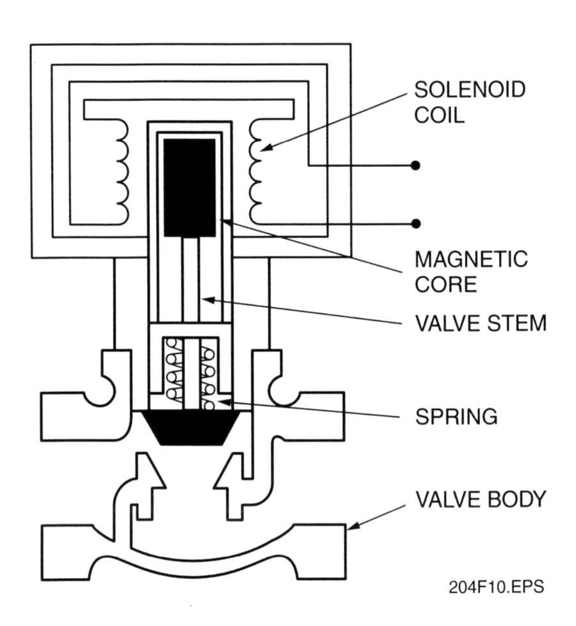

Figure 10 ◆ Solenoid valve construction.

5.0.0 ◆ CONTROL LOOPS

In a control loop, some parameter is measured, and then a controller takes action. This action manipulates a variable to bring the measured parameter back to a setpoint or desired value. While the devices that are included in the instrumentation channel are vital to the operation of the process control system, the way that they are selected and interconnected really determines the stability of the process. This is the function of the instrumentation control loop, which ultimately determines process control stability.

Four basic types of process control loops are common in the instrumentation industry: feedforward control (commonly referred to as **open loop**); feedback control (**closed loop**); cascade control; and ratio control. We will examine each of these more closely in the following sections.

5.1.0 Feedforward Control (Open Loop)

If the control variable measurement is not used as a signal to supply a controller that adjusts the inputs to the process, the loop is considered to be an open-loop control system. When open-loop control is used in industrial applications, it is often referred to as feedforward control. *Figure 11* shows a heat exchanger with an open-loop control system. The flow of oil through the tube side of the heat exchanger is held constant. The flow of water through the shell side of the heat exchanger is also held constant. If the two flow rates remain constant and are properly proportioned, the temperature of the oil leaving the tube side should achieve a stable value and is considered to be controlled. This is considered an open-loop control system because the exit oil temperature, which is the controlled variable, is not used to adjust any of the other inputs. With an open-loop control system, every input variable must be kept constant in order to keep the controlled variable constant. Some inputs, such as ambient temperature and the temperature of the hot water entering the shell side of the exchanger, may not be measured or controlled, resulting in the need for manual adjustment of the system.

5.2.0 Feedback Control (Closed Loop)

Another term that is used in closed-loop control systems is feedback control. This term receives its name from the fact that information concerning the controlled variable is fed back to the controller through the measuring element. There are two types of feedback possible: positive feedback and negative feedback. Positive feedback operates to aid an imbalance and increase the effect of a disturbance. Negative feedback works toward restoring balance or decreasing the effect of a disturbance. The result of positive feedback is that the process becomes unstable. If a temperature controller with positive feedback were employed on the heat exchanger, it would increase the heat transfer when the temperature of the outlet oil was above the setpoint and decrease the heat transfer when the temperature of the outlet oil was below

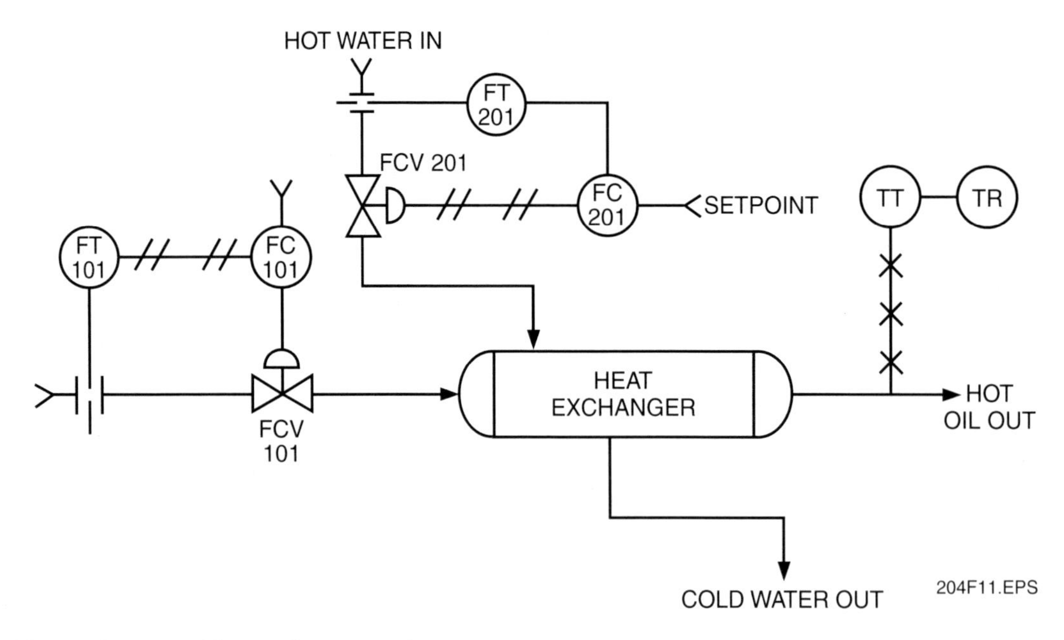

Figure 11 ◆ Heat exchanger with open-loop control.

the setpoint. Control loops that employ positive feedback eventually lock at one extreme or the other. For this reason, they are not employed in closed-loop control.

Negative feedback, when used in closed-loop control, results in a stable process. If the temperature of the outlet oil is above the setpoint, the heat transfer is reduced. The change in the heat transfer is in the opposite direction of, or negative to, the change in the outlet oil temperature. Negative feedback is implemented in the controller. The setpoint and the measured variable are compared in the controller in a manner that produces the difference between them. The result is that an increase in the setpoint has the opposite effect of an increase in the measured variable. This can be considered as a reversal of polarity taking place at the summing junction. All negative feedback controllers exhibit this characteristic, which is a polarity reversal giving the feedback its negative sense. A polarity reversal gives the same results as a 180-degree phase shift.

5.2.1 Operation of Closed-Loop Control

Figure 12 shows a schematic diagram of a closed-loop control system. This system has been applied to a heat exchanger. Assuming a decrease in the value of the controlled variable (outlet oil temperature), the general operation of the closed-loop control system is as follows:

1. The controlled variable change is sensed by the thermocouple element and sent to a temperature transmitter, which sends the measured variable to the controller.
2. The summing junction compares the measured variable to the setpoint and generates an error signal.
3. Since the measured variable is less than the setpoint, the controller sends a signal to the final control element (a flow control valve), which opens the valve.
4. The final control element regulates the manipulated variable (shell side inlet water flow) by increasing the water flow.
5. The heat exchanger reacts to the increase in hot water on the shell side by increasing heat transfer.
6. The increase in the controlled variable caused by the greater heat transfer is detected by the measuring element.

The control system response continues to repeat itself until the controlled variable has again achieved the same value as the setpoint. The same type of response occurs if the value of the controlled variable is increased or if the setpoint value is increased or decreased. This type of response is typical of closed-loop control systems that utilize negative feedback. An analysis of the response highlights one of the major flaws of closed-loop control; a change in the controlled variable has to occur before any response can begin. This is an inherent characteristic of closed-loop control. Proper tuning of the control system may minimize the effect of this flaw.

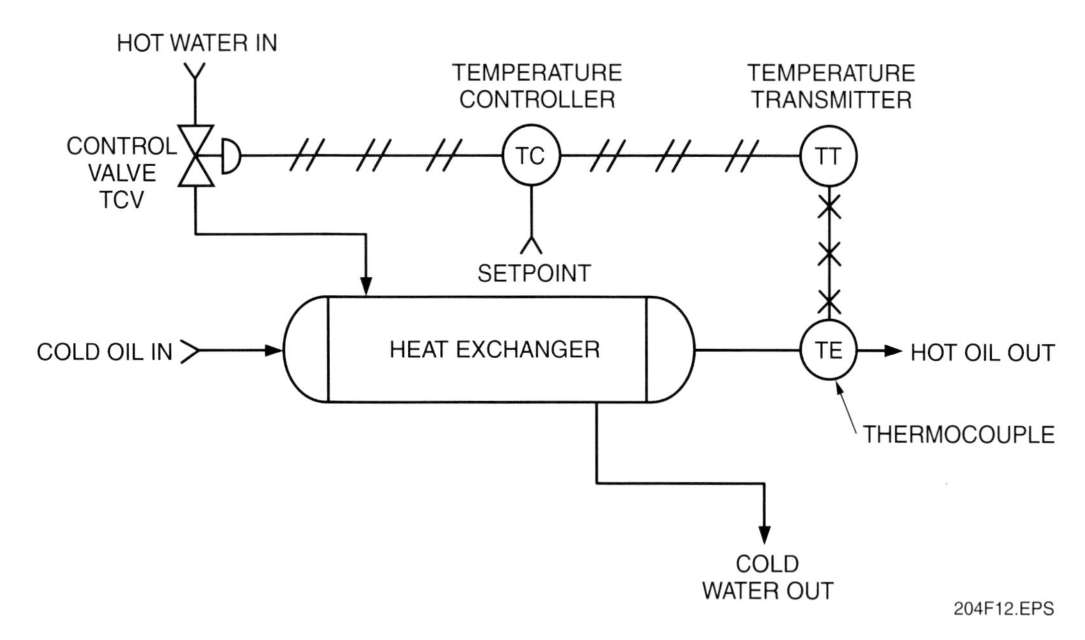

Figure 12 ◆ Closed-loop control system schematic diagram.

5.2.2 Performance of Closed-Loop Control

The primary performance criterion for any closed-loop control system is stability. If a control system of any design cannot reasonably maintain the controlled variable at a relatively constant value, it is not worth the investment. Stability is essentially the desired result of all applications of process control systems.

Before a process with a control system can be maintained in a stable condition, the inherent stability of the process itself must be determined. A closed-loop control system is considered stable if oscillations or transients caused by a disturbance eventually die out or at least remain constant. Preferably, the oscillations will be **dampened** rather quickly. As previously discussed, closed-loop control action consists of energy transfer, measurement, comparison, correction, and reaction. These events go around and around in a closed loop, through the control loop and the process. The simplest explanation of closed-loop control stability is in terms of time delays and energy loss or gain around the loop. A simple three-element schematic diagram is shown in *Figure 13*. The components of this diagram are the process, the controller, and the transmitter. As shown in *Figure 13*, a signal from the process output is fed back via the controller into the process as an input.

For example, if the controlled variable is increasing in temperature, the controller output must be in a direction that opposes the increase. In other words, the signal that is fed back must be in a direction that changes the process input negatively with respect to a change in the process output. As presented before, this is negative feedback and is essential to stability. A controller is normally made with a built-in 180-degree phase shift or polarity reversal. This is negative feedback. It is normally accomplished by the summing junction.

The controller is not the only component in the control loop that can introduce a phase shift. The process itself can introduce a phase shift, as can the final control element (TCV) and the measuring element. These phase shifts add to the 180-degree phase shift of the controller. If the phase shifts are large enough, the result can be that the total phase shift for the entire loop may be 360 degrees, or positive feedback. This results in instability.

In addition to the phase shift, the overall system gain must be considered. Looking strictly at the system input and output, the process transfers energy. If the energies lost and stored are ignored, and a comparison is made of energy in to energy out, the energy out will either be greater than, less than, or equal to the energy in. The term gain is used to describe this relationship. A gain of one exists when the energy out equals the energy in, a gain of greater than one exists when the energy out is more than the energy in, and a gain of less than one exists when the energy out is less than the energy in.

These two characteristics, phase shift and gain, provide a simple yardstick for measuring and defining closed-loop control stability. A control loop will be stable if a 360-degree phase shift exists in the control loop and the gain around the loop is less than or equal to one. However, if a 360-degree phase shift exists and the gain around the loop is greater than one, the loop will be unstable.

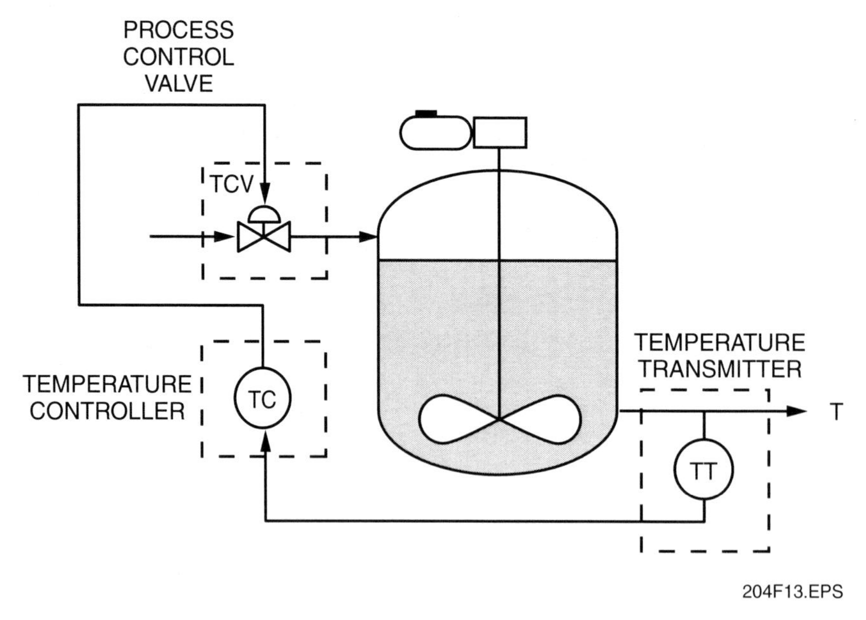

Figure 13 ◆ Simple three-element closed-loop control system.

In this condition, any disturbance will cause the loop to break into continuous oscillations of increasing **amplitude**. These oscillations increase in amplitude because of a loop gain greater than one. With the output increasing and this increase being fed back in phase with itself, the loop eventually locks at an extreme condition. The tendency for increasing amplitude oscillations can be a major limitation of closed-loop control.

A closed-loop control system can oscillate or cycle in response to a disturbance in three ways: increasing amplitude, constant amplitude, and decreasing amplitude. *Figure 14* graphically shows these three responses. *Figure 14(A)* shows the condition of increasing amplitude. If a 360-degree phase shift and a loop gain of greater than one exist, the amplitude of the oscillations increases. The condition of a 360-degree phase shift and a loop gain of exactly one is shown in *Figure 14(B)*. In this condition, the oscillations are continuous and equal in amplitude. *Figure 14(C)* shows the third possible response. In this condition, the phase shift is 360 degrees, but the loop gain is less than one. The amplitude of the oscillations becomes increasingly smaller until they disappear. Of the three possible responses, the decreasing amplitude response is the most desirable because the process regains its original condition.

Following a disturbance, there are three degrees of stability possible during a return to steady-state conditions. These responses are shown in *Figure 15*. An important point shown here is that all these waveforms do return to a steady-state condition. The waveform shown in *Figure 15(A)* is called

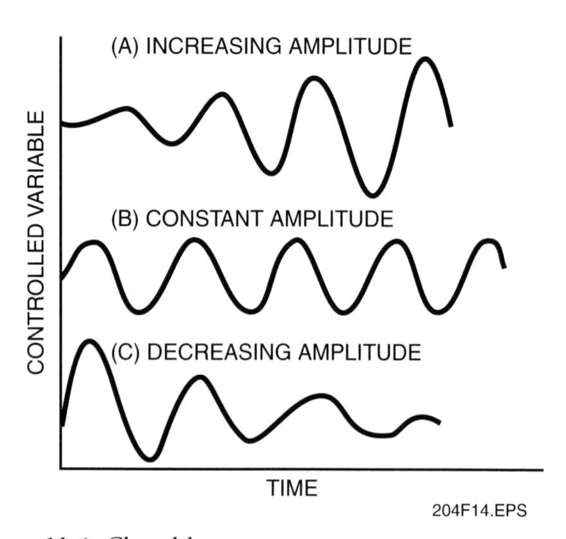

Figure 14 ◆ Closed-loop responses.

cyclic because of the oscillating but decreasing nature of the wave before stability it reached. The aperiodic wave, *Figure 15(B)*, shows the controlled variable returning to a steady-state condition in the shortest period of time with minimal overshoot. *Figure 15(C)* displays the overdampened recovery. In this waveform, the controlled variable returns to a steady-state condition without overshoot. However, the time period to recovery is longer than the minimum.

The question of which one of these stable responses is the best form of control requires further discussion. Each of the waveforms in *Figure 15* may be appropriate. The final determination depends on the quality criteria that are being applied.

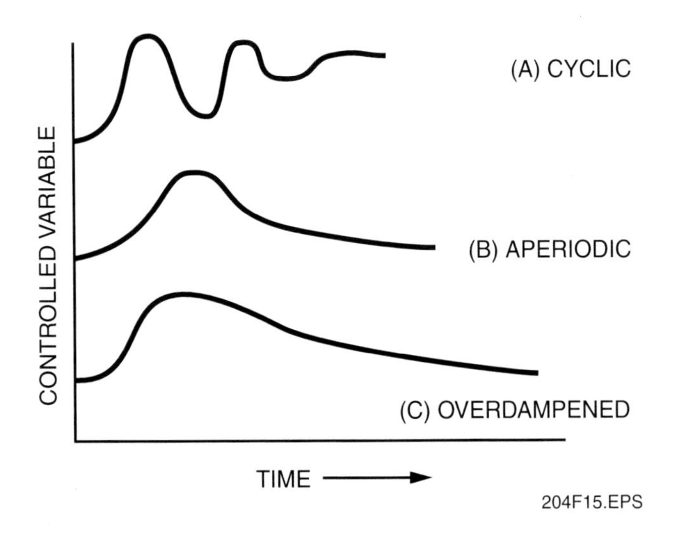

Figure 15 ◆ Possible stable responses to a disturbance.

5.2.3 Criteria for Closed-Loop Control Quality

How is good control measured? Essentially, there are three criteria for judging the quality of closed-loop control. Which of the criteria applies is directly dependent on the conditions of the process. A criterion used in one case may provide unsatisfactory control in another case. All these criteria apply to the shape of the waveform following a disturbance. *Figure 16* shows the waveforms for the three criteria.

The first criterion, shown in *Figure 16(A)*, is the minimum area criterion. This criterion requires that the area under the response curve be the smallest possible. It has been found that when the area is at its minimum, the deviation averages the smallest amplitude for the shortest time.

Experience has shown that the area is at a minimum when the amplitude ratio between the peaks of succeeding cycles is 0.25. This means that each wave cycle is only one-fourth as large as the cycle before it. This is the most widely applied control quality criterion. Its main application is a process where the duration of the deviation is as important as the size. A good application for this criterion is a process that cannot tolerate the controlled variable being out of a narrow band for a long period of time. Here, the best control is that which permits deviations from this band for the minimum time. Plastic and chemical processes require the application of this criterion.

The minimum disturbance criterion is displayed in *Figure 16(B)*. According to this criterion, the control action should cause the minimum disturbance to both the manipulated variable and the controlled variable. The non-cyclic curve shown in *Figure 16(B)* is an acceptable response because the number of disturbances is minimized. This criterion generally applies to control loops whose output feeds other connected processes. Normally, the common application of this criterion is in cascade control. Cascade control occurs when the output of one control loop becomes the input into another control loop. A sudden or cyclic response of the first loop may create oscillations in the second loop as well.

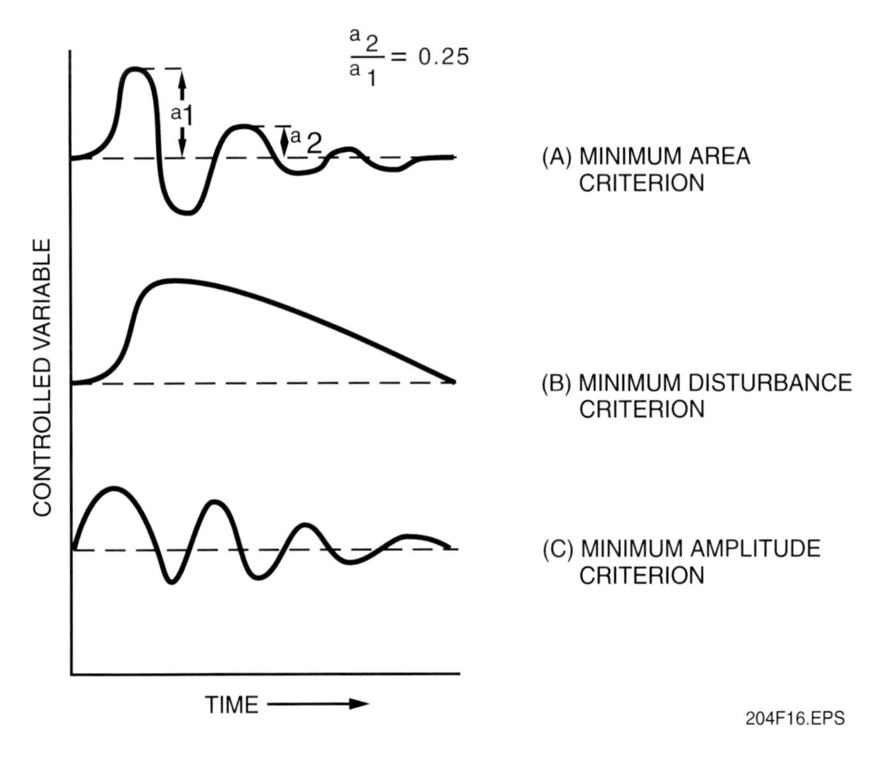

Figure 16 ◆ Three criteria for control quality.

The final quality criterion, shown in *Figure 16(C)*, is the minimum amplitude criterion. This criterion requires that the amplitude of the deviation be minimized. The application of this criterion is for processes where equipment damage or product loss may result if there are even momentary excessive deviations. In this application, the change in amplitude of the response curve is held to a minimum, but the deviation of the curve is allowed to be as long as necessary to achieve stable conditions. An example is the melting of metallic alloys. Even a short overshoot of temperature beyond an upper limit can ruin the metal by overheating it. This results in a marked decrease in the value of the metal due to an undesirable metallic structure.

Of these three criteria, the minimum area criterion is more commonly applied than the others.

5.3.0 Cascade Control

The control loops discussed so far are examples of feedforward and feedback control loops. In a feedback (closed-loop) control loop, a measured variable is compared to its desired value to produce a control action that reduces the magnitude of the deviation between the actual value and setpoint value. A feedback control system requires only one control loop to perform this function.

Cascade control is a modified form of feedback control. A single cascade control system has two control loops. The task of one of these control loops is to maintain the process variable at a desired value. This loop is called the primary or master control loop. The task of the second loop, the secondary or slave loop, is to manipulate a secondary variable in order to maintain the primary process variable at the setpoint. The value of the secondary variable is important only as it affects the primary variable.

In a cascade control loop, the master control loop determines the setpoint of the slave control loop. The slave loop is essentially the final control element of the master loop.

The overall function of a cascade control loop is the same as the simple feedback loops previously discussed. It is to maintain the value of the controlled variable at the setpoint under both steady-state and transient conditions. The second loop is included simply to make the control system more responsive and accurate.

Figure 17 provides an example of a pneumatic cascade control loop. In this system, the temperature in a chemical injection tank is maintained by controlling the temperature of a water jacket surrounding the tank vessel. The primary controlled variable of this system is chemical injection tank temperature. It is measured by a filled system temperature transmitter (TT). The output of the transmitter is sent to the master controller (TC).

If the chemical injection tank temperature drops below the setpoint that the master controller is adjusted to maintain, the output of the master controller increases. This signal appears to the slave controller (TC) as an increase in its setpoint for water jacket temperature. Therefore, the slave controller produces an increasing output signal.

Two valve postioners are used in this control loop because the output of the slave controller is split ranged. As the slave controller output increases from 9 to 15 psi, the output pressure of the steam valve positioner increases from its minimum value to its maximum value. When the slave controller output decreases from 9 to 3 psi, the output pressure of the cold water valve positioner increases from its minimum value to its maximum value. When slave controller output is midrange (9 psi), both positioners produce minimum output pressures.

Therefore, when the slave controller output increases above 9 psi, the steam valve positioner output increases, stroking the steam valve open. As a result, steam flow through the heat exchanger shell increases, heating the water jacket fluid passing through the heat exchanger tubes.

Temperature in the water jacket now increases. The heat added to the water jacket is transferred into the chemical injection tank, and the temperature of the material within the tank increases. Therefore, by controlling the temperature of the control loop, we can maintain the temperature in the chemical injection tank.

The same function could be performed by a single temperature control loop. The reason for using cascade control is obvious when you consider a disturbance in the secondary loop.

Temperature in the chemical injection tank is maintained by controlling the temperature in the water jacket. If water jacket temperature were to decrease, eventually tank temperature could also decrease if it were not controlled. Therefore, by controlling the water jacket temperature in addition to the tank temperature, we can minimize disturbances to the primary process.

Water jacket temperature is sensed by another filled system transmitter (TT). When this temperature decreases, the output of the transmitter decreases. The slave controller compares the temperature signal to the setpoint, which is reactor temperature. When the temperature falls below the setpoint, the slave controller output

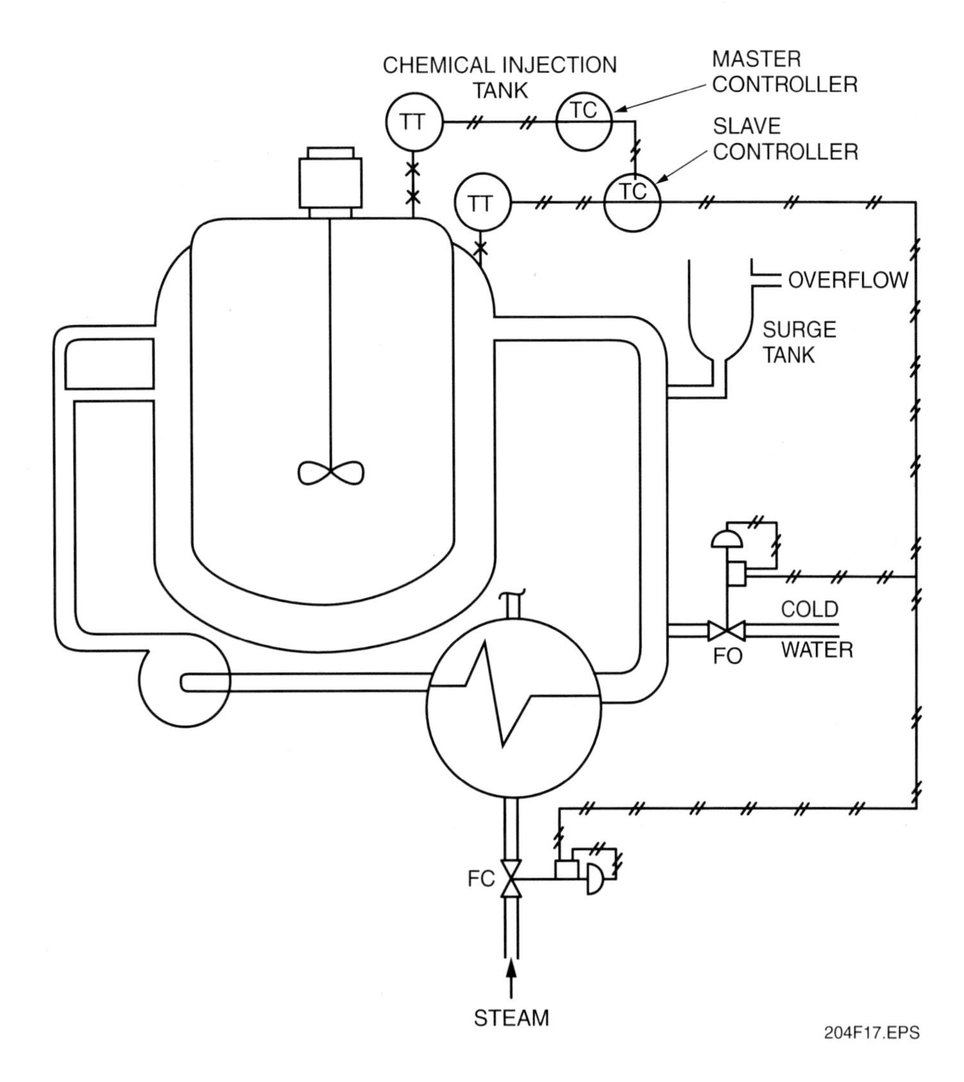

Figure 17 ◆ Pneumatic cascade control loop.

increases from 9 psi. This increase in pressure produces an increased output from the steam valve positioner, and the steam valve opens to allow more thermal energy into the system. As a result, water jacket temperature increases.

When either chemical injection tank or water jacket temperature increases, the output of the slave controller decreases below 9 psi. In this case, the cold water valve positioner output begins to increase. The cold water valve opens to allow cold water to mix with water already in the system. This action results in decreasing the water jacket temperature. The additional water added to the system causes water level in the surge tank to increase.

The function of the cascade control loop in *Figure 18* is to maintain the level in a large volume head tank within a narrow band. The tank supplies the suction head for a large pump that has a low suction pressure trip. Control of the level within this narrow band is made difficult because the pressure upstream of the control valve varies erratically. A rise in upstream pressure increases flow through the control valve and, as a result, causes level to increase. A single feedback loop could be used if upstream pressure were constant, but since the pressure varies erratically, unstable control could result. For example, when the tank level begins to increase due to increase in pressure, the corrective action would be to reduce the valve opening. If the pressure decreased while the valve was closing, tank inlet flow would be reduced. The resulting rate of decrease of the tank level could be great enough that the controller could not prevent an excessive undershoot. This, in turn, would cause the pump suction head to decrease and trip the pump.

Using cascade control, this excessive undershoot could be prevented. In the cascade level control loop illustrated in *Figure 18*, the level does not have to change before a corrective action can occur. Because the flow into the tank affects the level, if the flow can be kept constant during constant

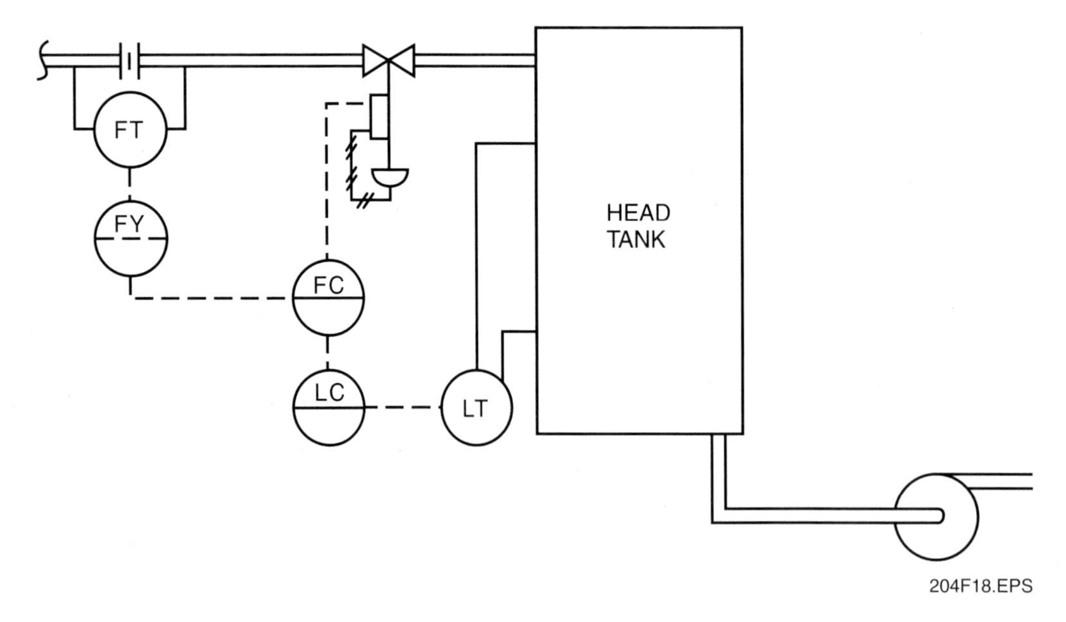

Figure 18 ◆ Cascade level control.

demand conditions, the level should not change. The increase in the pressure upstream of the valve mentioned earlier would cause an increase in flow through the orifice flow element. A differential pressure flow transmitter (FT) senses the increase in the differential pressure across the orifice and transmits the information to a square-root extractor (FY). The square-root extractor converts the differential pressure signal into a linear flow rate signal that is then sent to the flow controller (FC). The change in flow rate has not yet caused a change in the tank level, so the output of the master controller, which is the setpoint for the slave controller, remains constant. The change in flow rate, therefore, produces a mismatch between the actual flow rate and the flow rate desired by the master controller. As a result, the slave controller produces a control signal that reduces the opening of the control valve and returns inlet flow to the previous value. A slight change in level may occur, but it is not nearly as severe as the change that would occur if only a single loop were used. If pressure were to decrease again, the resulting change in flow would immediately be detected and the valve repositioned to prevent excessive fluctuations in tank level.

The advantage of cascade control, therefore, is that the control system does not have to wait for the primary controlled variable to change before it initiates a corrective action. A change in the secondary variable will produce a corrective action and, as a result, fluctuations of the primary controlled variable can be minimized.

5.4.0 Ratio Control

As the name implies, ratio control is the maintaining of a fixed ratio between two or more variables. The most common application of ratio control is the maintaining of a fixed relationship between two flows, such as a mixture of air and fuel for combustion control. Control is normally affected by adjusting one of the flows, the controlled flow, to follow in proportion to a second flow, the wild or uncontrolled flow.

The function of the control loop illustrated in *Figure 19* is to maintain a proper ratio between fuel flow and air flow to ensure that complete combustion of the fuel occurs. By ensuring that complete combustion takes place, efficient, economical operation of the plant can be realized, and emissions of harmful waste products can be minimized.

In this combustion control system, the fuel flow is the uncontrolled variable. Actually, fuel flow is controlled, but by another control loop. Air flow is the controlled variable in this system. Fuel flow and air flow are measured by variable head flow meters. Flow in the fuel line develops a differential pressure across an orifice. A differential pressure-type flow transmitter (FT) senses the flow proportional head through **vena contracta** or **radius taps**. The differential pressure signal is converted to a linear flow signal by a square-root extractor (FY) and transmitted to pen 2 of a two-pen flow recorder (FR 1-2) and to a flow ratio relay (FFY). The ratio relay is nothing more than a manually adjusted gain device. Most ratio relays have calibrated ratio scales that allow the direct setting of the required ratio or fraction.

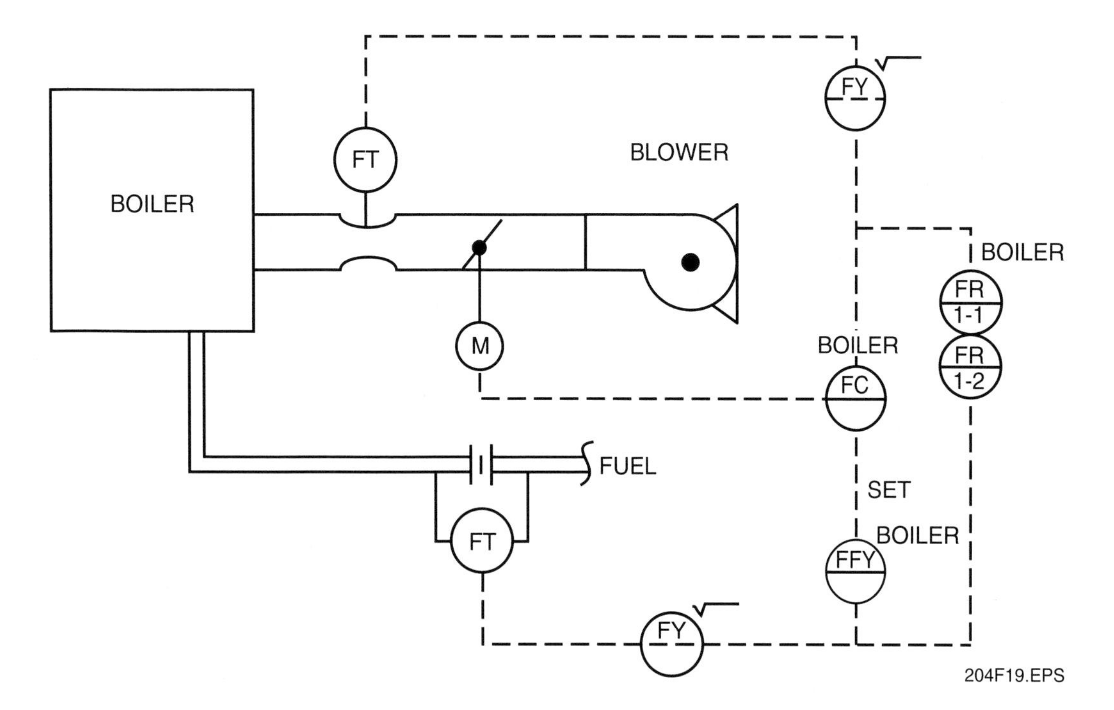

Figure 19 ◆ Ratio combustion control loop.

The output of the ratio relay is the setpoint of the flow controller (FC). Many manufacturers combine the functions of ratio relay and controller into a ratio controller. Such a device for flow control would be designated FFC, ratio flow controller. Flow through the air duct develops a differential pressure across a rectangular flow nozzle built into the blower duct. A differential pressure-type flow transmitter (FT) senses the pressure drop and transmits this information to a square-root extractor (FY) The square-root extractor converts the differential pressure signal into a linear flow signal and sends the information to pen 1 of the two-pen flow recorder (FR 1-1) and to the flow controller (FC). The air flow signal is the process variable input to the controller. The flow controller compares the biased fuel flow signal (the setpoint) to the air flow signal. A mismatch between the two produces an output from the controller that causes the motor-controlled damper to be repositioned in a direction that reduces the error.

Another common application of ratio control is maintaining a fixed relationship between two flows, such as the mixture of two materials in a blending operation (*Figure 20*).

In this system, the primary loop is uncontrolled and the secondary loop is controlled to maintain the proper mixture. Flow in both loops is measured by a variable head flow meter. Orifice plates produce a differential pressure that is sensed by differential pressure transmitters (FT).

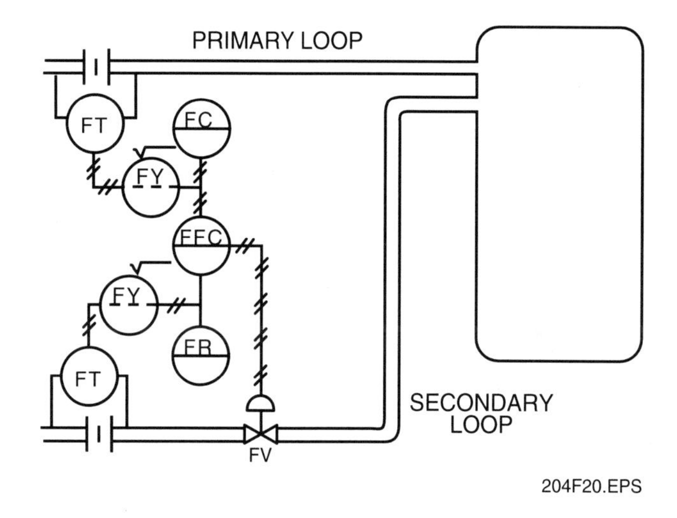

Figure 20 ◆ Ratio control of a blending process.

Both loops use square-root extractors (FY) to convert the differential pressure signal to a linear flow signal. The flow information is then transmitted to a flow recorder for each loop and finally to the ratio flow controller (FFC).

The signal from the primary loop is multiplied by an adjustable ratio factor within the controller and used as the setpoint. Secondary loop flow is then compared to this setpoint. If there is a difference between setpoint and secondary flow, an error signal is produced. The error signal, through the action of the controller, produces an output that positions a diaphragm actuated valve. This action returns the flow in the secondary loop to the desired ratio.

6.0.0 ◆ CONTROL MODES

Selecting the proper instruments for the channel and building the control loop do not complete the design of a process control system. We must decide what type of control mode (parameter of response) is needed in order to best achieve a stable process, which is the goal of process control design.

There are two basic modes of control, or parameters of response, that are applied to control loops: on-off control (two position) and modulating control. We will examine these modes more closely, but briefly stated, on-off control is as simple as it sounds. The final element is either on (open), or off (closed). However, modulating control, which is the more common application of the two modes, has many more elements to it. Modulating control includes the functions of proportional (gain-P), integral (reset-I), and derivative (rate-D), or a combination of any or all three. This section will discuss on-off control and examine the functions and combinations of modulating control.

6.1.0 On-Off Control (Two-Position Control)

A controller is a device that generates an output signal based on the input signal it receives. The input signal is actually an error signal, which is the difference between the measured variable and the desired value, or setpoint. This input error signal represents the amount of deviation between where the process system is actually operating and where it is supposed to be operating. The controller provides an output signal to the final control element, which adjusts the process to reduce this deviation. The characteristic of this output signal is dependent on the type, or mode, of the controller. This section describes the simplest type of controller, which is a two-position, or on-off, controller.

6.1.1 On-Off Control

A system using a two-position controller is shown in *Figure 21*. The process parameter being controlled is the volume of water in the tank. The controlled variable is the level in the tank. It is measured by a level transmitter that sends information to the controller. The output of the controller is sent to the final control element, which is a solenoid valve that controls the flow of water into the tank.

Within the controller, the measured variable is compared to the setpoint and an error signal is generated. The controller acts on the error signal. *Figure 22* shows the response of this control system.

As the water level decreases initially, a point is reached at which the measured variable drops below the setpoint. This creates a positive error signal. The controller opens the final control element fully. Water is injected into the tank, and the water level rises. As soon as the water level goes above the setpoint, a negative error signal is developed. The negative error signal causes the controller to shut the final control element. This opening and closing of the final control element result in a cycling characteristic of the measured variable. Notice in *Figure 22* that the frequency of these oscillations is very high. This can cause added wear of the system equipment. Therefore, a means is needed to reduce the frequency.

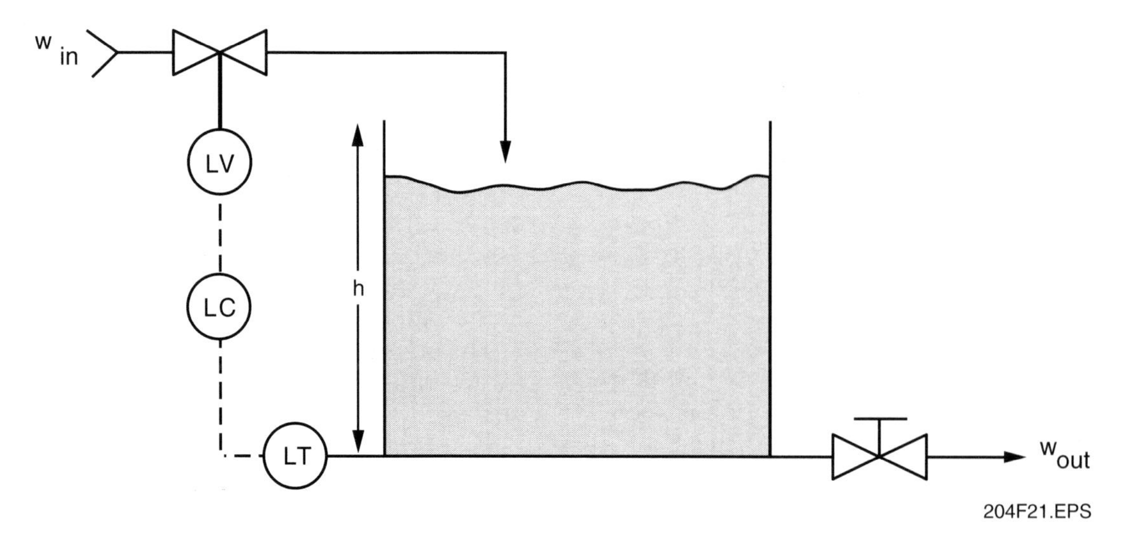

Figure 21 ◆ On-Off (two-position) control system.

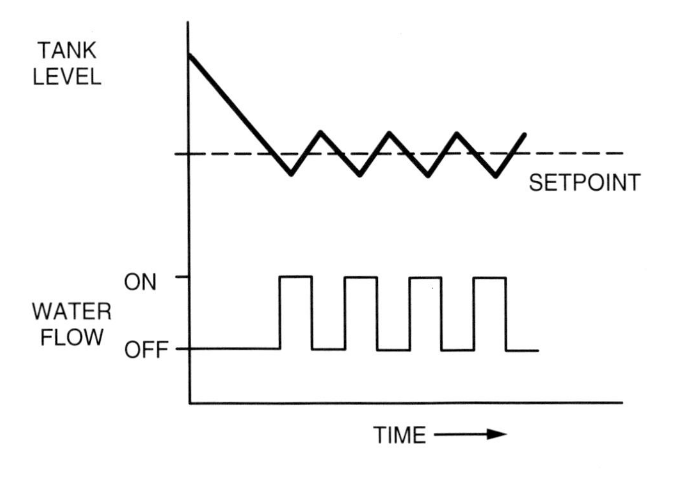

Figure 22 ◆ Response of an On-Off (two-position) control system.

The simplest means of reducing the frequency is to make the ON state and the OFF state of the controller occur at different points. This causes the valve to open at a lower level and to shut at a higher level. Thus, the level must change by a larger amount before the controller changes state. This additional level change requires more time; therefore, there is more time between changes of state. The type of controller that provides this mode of control is called a two-position controller with a neutral zone. Its characteristic open-loop waveform is shown in *Figure 23*.

Assuming that the controller output is in the ON state, an increase in the measured variable to the upper zone limit value causes the controller output to shift to the OFF state. The measured variable now decreases. It must decrease to the lower zone limit value for the controller output to shift back to the ON state. The neutral zone represents the region of measured variable overlap within which the controller output can be either on or off, depending on the state of controller output prior to the measured variable entering the neutral zone. When the measured variable is below the neutral zone, the controller output is always on. When the measured variable is above the neutral zone, the controller output is always off. Notice also that the controller output does not change state within the neutral zone. As such, the controller causes the process to cycle above and below the setpoint. The measured variable oscillates within a band defined by the width of the neutral zone.

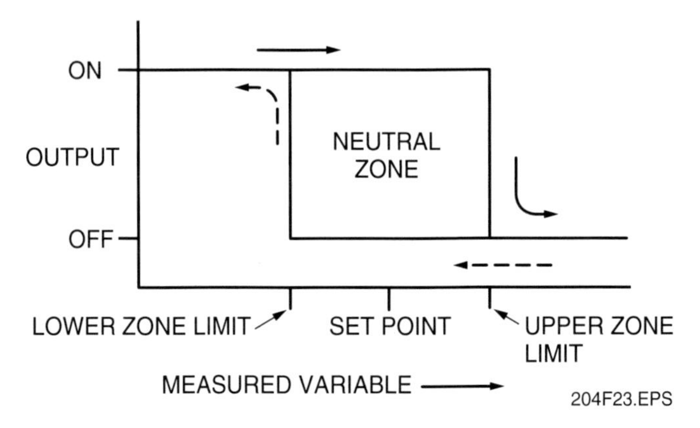

Figure 23 ◆ On-Off (two-position) control system with a neutral zone.

The neutral zone describes the region of overlap in which the controller does not change state. The controller of *Figure 23* provides a turn-off point that is higher than the turn-on point. The process variable must deviate above or below the setpoint by a specified amount to cause the controller to change state.

The use of a two-position controller with a neutral zone in the system shown in *Figure 21* would result in the response curve shown in *Figure 24*.

Comparing *Figure 24* with *Figure 22*, notice that the frequency of the oscillations has been reduced. Notice also, however, that the larger amplitude of the oscillations represents larger deviations from the setpoint.

Almost all switch-type devices that provide an on-off function have some amount of neutral zone. This is due to such phenomena as friction of moving parts, clearances between moving parts, and electrical component **hysteresis**.

Some systems need a wide neutral zone. For example, consider the wall thermostat found in most homes. With the thermostat set to 70°F, it would appear that the objective of the heating system would be to maintain a room temperature of 70°F. But recall that the thermostat turns the heating system on and off. Imagine a thermostat with no neutral zone; that is, with the turn-on and turn-off points occurring at the 70°F setpoint. The heating system would attempt to turn on and turn off at the same time. Imagine now that the thermostat has a neutral zone of only 1°F. This means that the heating system is turned on at 69.5°F and turned off at 70.5°F. This narrow neutral zone would cause the heating system to cycle between on and off many times each hour. This high frequency of turning on and off could cause added wear on the system, leading to early failure.

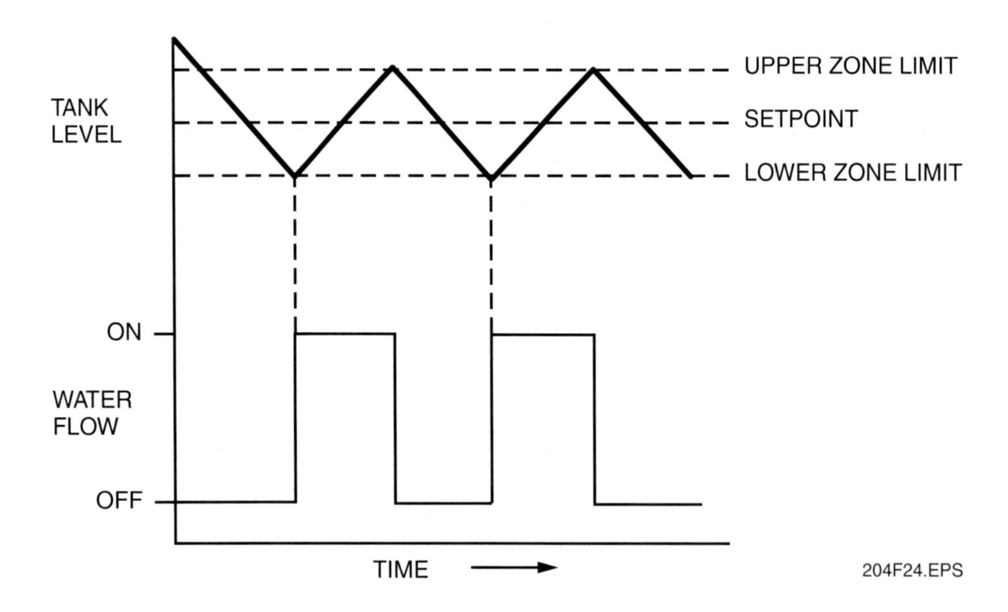

Figure 24 ◆ Response of an On-Off (two-position) control system with a neutral zone.

In other applications, however, narrow neutral zones are required. A test equipment calibration laboratory, for example, might require that room temperature be maintained at 70 ±2°F. The heating system must be designed to withstand frequent cycles of operation if two-position control is used.

Many two-position controllers are designed with an adjustable neutral zone to allow adjustment for ideal system operation. Note, however, that both the system process requirements and the system equipment limitations must be considered when establishing or changing the width of the neutral zone in a two-position controller.

Figure 25 shows a diagram for a tank level control system with upper and lower zone limits specified.

The final control element is a solenoid valve that, when open, supplies water flow to the tank at a rate of 10 gpm. The process is the tank of water, which must supply water at various demands to other systems. The measuring element senses the tank level and sends a signal to the controller. The controller setpoint signal is 55 gallons of water in the tank, and the neutral zone width is ±15 gallons.

Assume initially that the tank level is at 70 gallons and the solenoid valve is shut. Now a demand is placed on the system that causes a level decrease of 2.5 gallons per minute. Note that even though level is decreasing, the solenoid valve doesn't open until the level decreases by 30 gallons. The time required is:

$$\frac{30 \text{ gal.}}{2.5 \text{ gpm}} = 12 \text{ minutes}$$

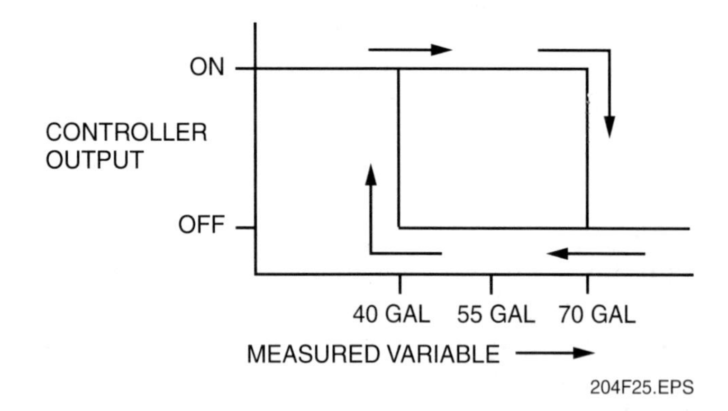

Figure 25 ◆ On-Off (two-position) tank level control system.

At the 40-gallon level, the controller opens the solenoid valve, and the tank level increases at a rate of:

$$10 \text{ gpm} - 2.5 \text{ gpm} = 7.5 \text{ gpm}$$

The time required to reach the 70-gallon level is:

$$\frac{30 \text{ gal.}}{7.5 \text{ gpm}} = 4 \text{ minutes}$$

The resultant theoretical characteristics are shown in *Figure 26*. As shown in the figure, the frequency of the waveform is dependent on the width of the neutral zone and the magnitudes of the supply rate and demand rate.

The characteristic waveforms also show that the level in the tank makes a full cycle every 16 minutes. In addition, the average value of the

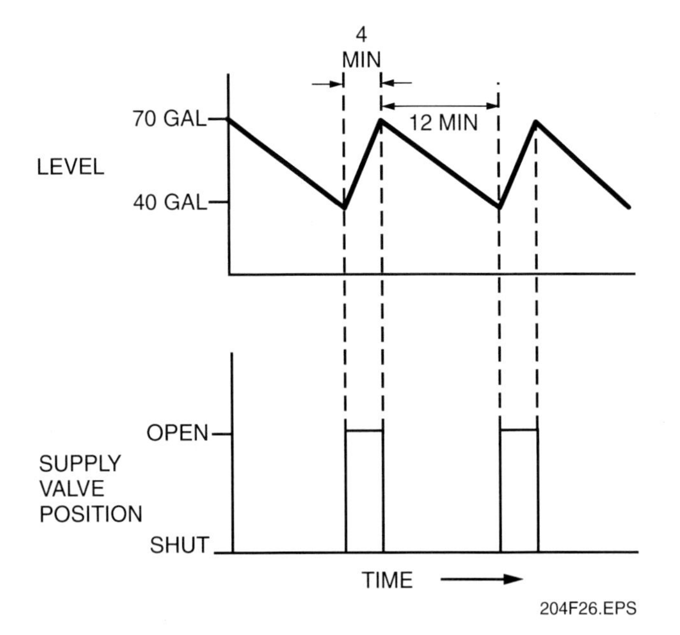

Figure 26 ◆ Theoretical tank level control system characteristics.

level equals the 55-gallon setpoint. The waveforms do not, however, account for system delay times. For example, if the pipe between the solenoid valve and the tank is long, a large volume of water continues to flow into the tank after the solenoid valve is shut. Also, when the solenoid valve is opened, the water must flow into the empty pipe length before reaching the tank. The effect is that tank level continues to rise above the 70-gallon level after the solenoid valve shuts and the volume of water flows out of the pipe length. Also, the tank level continues to decrease below the 40-gallon level after the solenoid valve opens and water flows into the empty pipe. These overshoots and undershoots are shown in *Figure 27*.

The magnitude of the overshoots and undershoots in a system is dependent on many factors. For this example, looking only at the delay caused by the length of the pipe, the overshoots are larger in magnitude than the undershoots. This is because the fill rate is higher than the drain rate. This causes the average value of the water level to be higher than the 55-gallon setpoint. The overshoots and undershoots can be large enough to cause problems such as overflow of the tank due to overshoot or loss of drain flow due to undershoot. The possibility of this happening is greater if the system is of small capacity. This would be the case in the example given if the tank volume were small compared to the flow rates.

Additional system delay can be caused by the delay times of the system equipment. For example, the time between when the level is detected to be at 70 gallons and the time that the solenoid valve reaches its fully shut position can be significant.

Ideally, in the example given, the solenoid valve is in phase with the process. That is, the solenoid valve follows the demand of the process exactly. Because of equipment delays, the solenoid valve lags behind the process.

The combined delays of the time needed for the process to change by a value larger than the neutral zone and the time needed for the system to respond to the process result in limited application of the two-position control mode in rapidly changing systems.

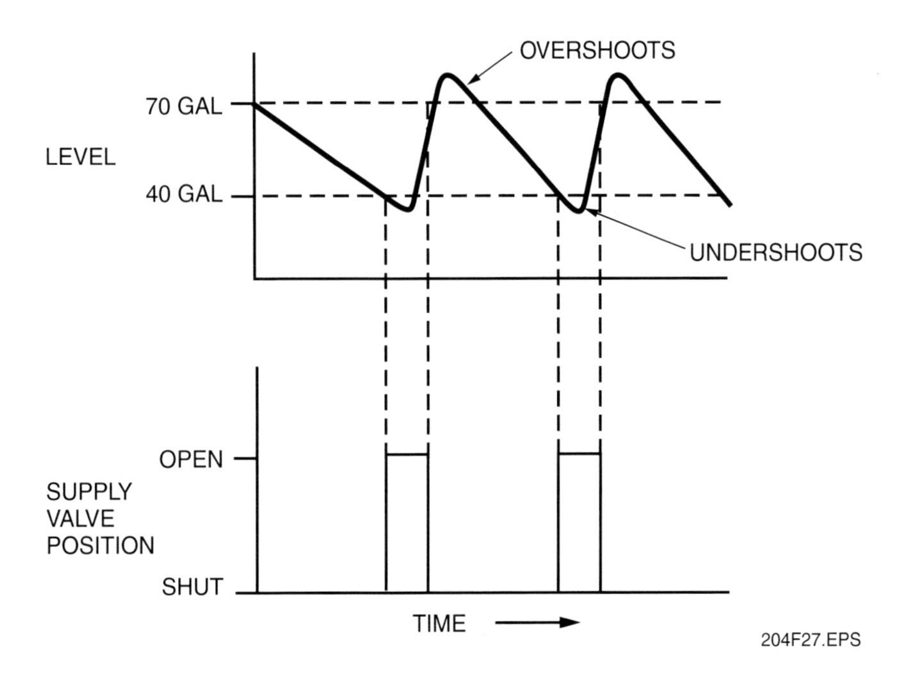

Figure 27 ◆ Tank level system overshoots and undershoots.

A demand disturbance can also adversely affect the process. Assume in *Figure 25* that the demand on the system is increased such that the level decreases at a rate of 7.5 gpm. The resultant characteristic curve is shown in *Figure 28.* The new value of system demand causes the tank to drain three times faster than before. This results in the undershoots being larger than the overshoots and causes the average tank level to be lower. The result is a varying average value determined by the process demand. This is not desirable in some systems.

One method of reducing the harmful effects of overshoots, undershoots, and system delays is to reduce the width of the neutral zone. In the example given, this would cause the solenoid valve to shut sooner and at a lower level and also open sooner and at a higher level. Recall, however, that reducing the neutral zone causes the system to cycle more frequently.

Increasing the setpoint causes the neutral zone of the controller to increase. This causes the supply valve to open and shut at higher tank levels. The width of the neutral zone has not been changed. Therefore, the peak-to-peak amplitude of the tank level oscillation is not changed, nor is the frequency of the oscillations. The supply and drain rates have not changed. Therefore, the amplitudes of the overshoots and undershoots remain the same. However, notice in *Figure 29* that the average value of the tank level is increased, which compensates for the previous undesired decrease in average value from the demand increase. The tank level now peaks at a higher actual level. Therefore, the amount that the setpoint can be increased may be limited by such factors as possible tank overflow.

Although the two-position control system has some disadvantages, there is one very good application for it. It is very suitable for use as a low pass filter. Assuming that the demand rate is random in both rate and duration, instead of a controller responding to each small change in level, it filters out these small changes and only responds to the level when it exceeds the neutral zone. It thereby reduces the frequency of valve cycles.

6.2.0 Modulating Control

In modulating control, the feedback controller operates in two steps. The first step is to compute the error between the controlled variable (the process feedback) and the setpoint. The second step is to produce an output signal to the final control element (control valve) in an effort to reduce the measured error to zero. This is a continuous operation, with constantly changing analog output values. Modulating or feedback controllers contain three basic functioning modes: proportional, integral, and derivative. Most modern controllers include all three functions and are referred to as PID controllers. However, loop operation and tuning parameters may activate only a single mode, a combination of two modes, or a combination of all three modes. The following sections will examine each of these functions individually, as well as the combinations of these functions.

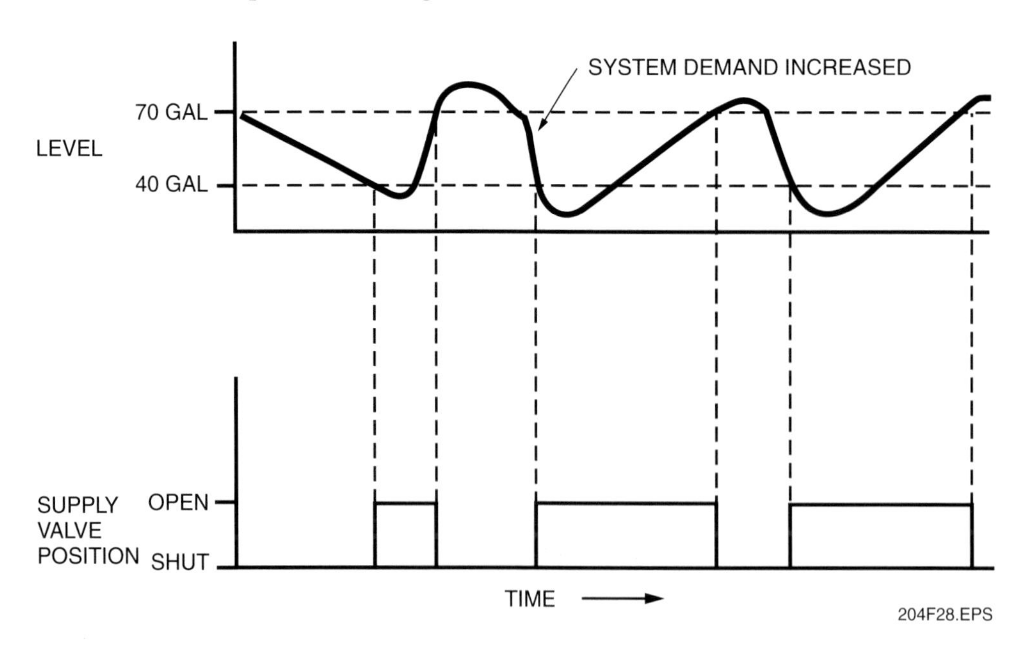

Figure 28 ◆ Process demand disturbance.

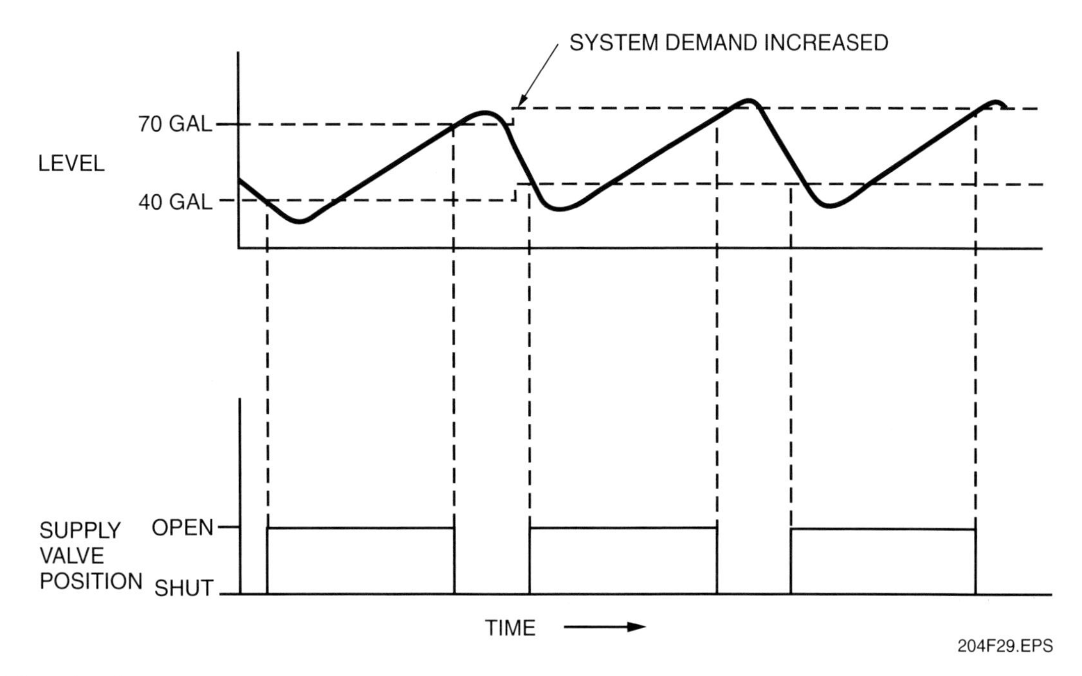

Figure 29 ◆ Setpoint change.

6.2.1 Proportional (Gain or P) Control

The section covering two-position controllers described the output of the controller as either a fully off or fully on signal. This can be viewed as a device with a very high gain at the point where it changes state. For example, when the input error signal to the controller is near the turn-on point, a small percentage change in the input causes the output to step change from 0% to 100%.

This section describes another type of control in which the final control element is throttled to various positions that are dependent on process system conditions. A proportional controller, also known as a gain controller, provides a stepless output that can position a control valve at intermediate positions, as well as full open and full shut. This is accomplished by a lower gain with the controller. With lower gain, a larger change in the input signal is required to cause the output to change from 0% to 100%. The controller operates within the band that is between the 0% output point and the 100% output point. The value of the controller gain determines how much of a change in the input signal is necessary to cause the output to change 100%. Within this band, the controller output is proportional to the input signal. The gain factor determines the proportional relationship.

With proportional control, the final control element has a definite position for each value of the measured variable. Another way to describe proportional control would be that it gives a linear output-to-input relationship.

Proportional band is the range of error signals that will give a controller output in this linear range. *Figure 30* shows a process that uses a very simple form of proportional control.

In the example system in *Figure 30*, the flow of supply water into the tank is controlled to maintain the tank water level within prescribed levels. The demand disturbances placed on the process system are such that the actual flow rates cannot

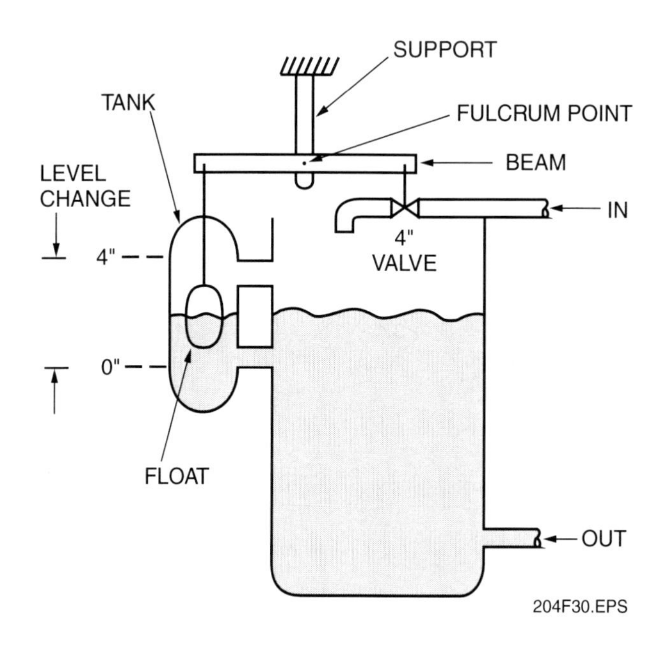

Figure 30 ◆ Proportional control system (gain of 1.0).

be predicted. Therefore, the system is designed to control tank level within a narrow band in order to minimize the chance of a large demand disturbance causing overflow or runout.

A **fulcrum** and **lever** assembly is used as the proportional controller. A float chamber is the level measuring element. A 4" stroke valve is the final control element. The fulcrum point is set such that a level change of 4" causes a full 4" stroke of the valve. Therefore, a 100% change in measured variable equals 4", and a 100% change in controller output equals 4". The gain of the controller is:

$$\text{Gain} = \frac{\text{Output change}}{\text{Input change}}$$

or

$$\text{Gain} = \frac{\text{4" valve stroke}}{\text{4" level change}}$$

Therefore:

$$\text{Gain} = 1.0$$

The input band over which the controller provides a proportional output is called the proportional band. It is defined as the change in input required to produce a full range change in output due to proportional control action, or:

$$\text{PB} = \frac{\text{100\% change in input}}{\text{100\% change in output}} \times 100\%$$

For the example given, the fulcrum point is such that a full 4" change in float height causes a full 4" stroke of the valve:

$$\text{PB} = \frac{\text{100\% Change in input}}{\text{100\% Change in output}} \times 100\%$$

Therefore:

$$\text{PB} = 100\%$$

Figure 31 shows the proportional relationship of the valve position versus the float level. As shown in the figure, the controller with a gain of 1 produces a 4" change in the output for a 4" change in input. Also, the controller can be described as having a proportional band of 100%, which means the input must change 100% to cause a 100% change in the output.

Notice that gain and proportional band are inversely related, or:

$$\text{Gain} = \frac{100\%}{\text{PB}(\%)} \quad \textit{and} \quad \text{PB}(\%) = \frac{100\%}{\text{Gain}}$$

This inverse relationship is important to remember. Some actual proportional controllers have an adjustment that is expressed in units of gain, whereas other proportional controllers have the adjustment expressed in units of percent proportional band. Regardless of the units used, the adjustment performs the function of determining the input-to-output proportional relationship.

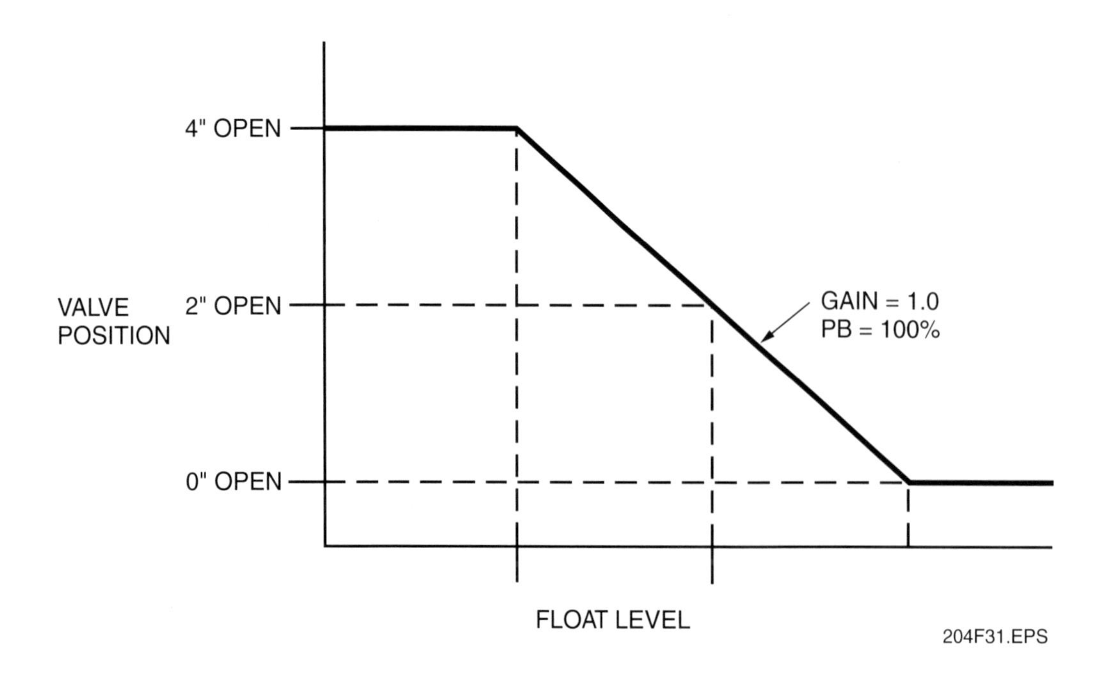

Figure 31 ◆ Valve position vs. float level (gain of 1.0).

Figure 32 shows the same level control system with the exception that now the fulcrum point has been changed to provide 100% stroke of the valve for a 50% change in float level.

Now, the example in *Figure 32* results in a gain of:

$$\text{Gain} = \frac{\text{Output change}}{\text{Input change}}$$

or

$$\text{Gain} = \frac{\text{4" valve stroke}}{\text{2" level change}}$$

Therefore:

$$\text{Gain} = 2.0$$

The proportional band of the controller is now:

$$\text{PB} = \frac{\text{\% change in input}}{\text{\% change in output}} \times 100\%$$

or

$$\text{PB} = \frac{\text{50\% change in input}}{\text{100\% change in output}} \times 100\%$$

Therefore:

$$\text{PB} = 50\%$$

The resultant characteristic curve is shown in *Figure 33*. As shown in the figure, the controller with a gain of 2 produces a 4" change in output for a 2" change in input. Also, the controller is described as having a 50% proportional band, which means that the input must change only 50% to produce a 100% change in the output.

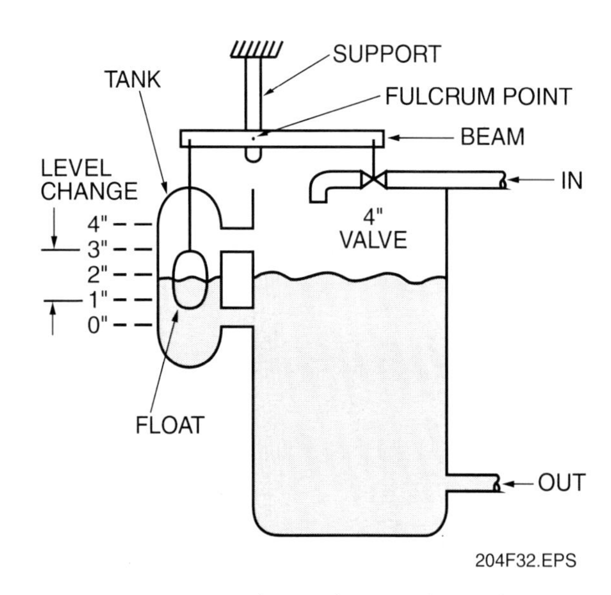

Figure 32 ◆ Proportional control system (gain of 2.0).

Figure 34 shows the same level control system, except that now the fulcrum point has been changed to provide 50% stroke of the valve for a 100% change in float level.

The example now results in a controller gain of:

$$\text{Gain} = \frac{\text{Output change}}{\text{Input change}}$$

or

$$\text{Gain} = \frac{\text{2" valve stroke}}{\text{4" level change}}$$

Therefore:

$$\text{Gain} = 0.5$$

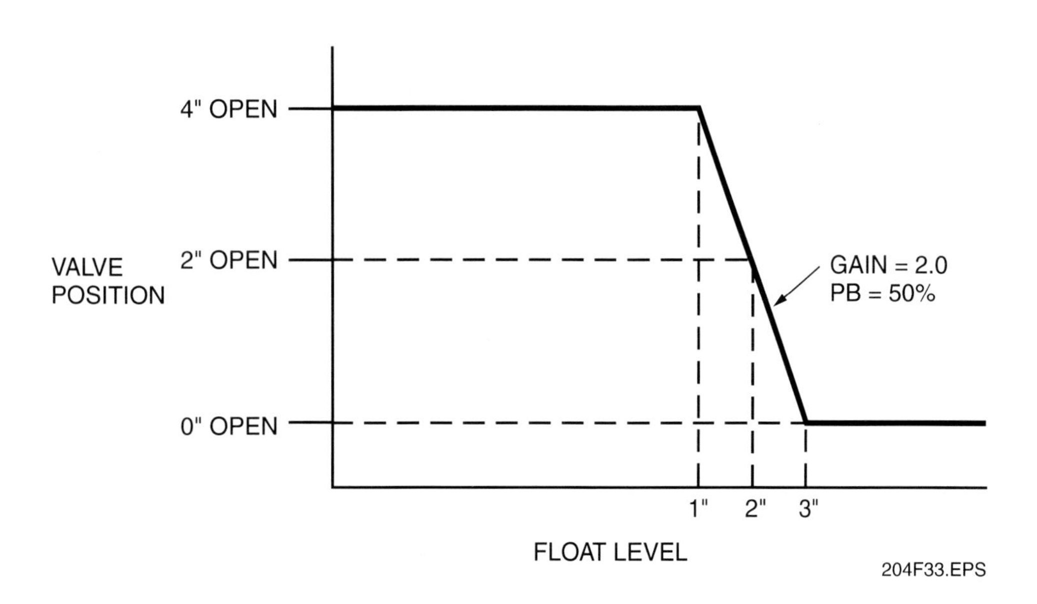

Figure 33 ◆ Valve position vs. float level (gain of 2.0).

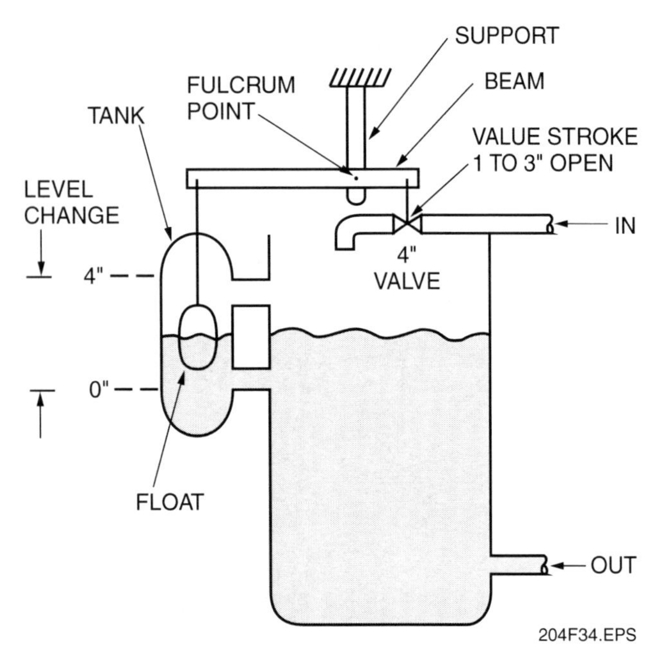

Figure 34 ◆ Proportional control system (gain of 0.5).

The proportional band of the controller now is:

$$PB = \frac{\%\ \text{change in input}}{\%\ \text{change in output}} \times 100\%$$

or:

$$PB = \frac{100\%\ \text{change in input}}{50\%\ \text{change in output}} \times 100\%$$

Therefore:

$$PB = 200\%$$

The resultant characteristic curve is shown in *Figure 35*. As shown in the figure, the controller with a gain of 0.5 produces a 2" change in output for a 4" change in input. Also, the controller is described as having a 200% proportional band, which, in theory, means that a 200% change in the input produces a 100% change in the output. However, the maximum possible output change cannot exceed 100%. Therefore, a 200% input change is not possible. As such, a proportional band of 200% means that an input change of 100% produces a 50% change in the output.

In general, with a gain greater than 1, which is a proportional band less than 100%, a process variable change of less than 100% causes a full 100% change in controller output. Conversely, with a gain less than 1, which is a proportional band greater than 100%, a full 100% change in process variable causes less than a 100% change in controller output.

6.2.2 Integral (Reset or I) Control

Integral control, also known as reset, or I, describes a controller in which the output rate of change is dependent on the magnitude of the input. Specifically, a smaller amplitude input causes a slower rate of change of the output. This controller is called an integral controller because it approximates the mathematical function of integration.

Since an integral is a calculus function that often appears complex, this section will address the integral function in terms of reset. Keep in mind, however, that the integral is calculated through the application of a unique formula that is performed within the logic of the PID controller.

The integral function of the controller allows the output of the controller to continue to change as long as an error between input signal and setpoint exists. The integral mode of the controller acts only when the error exists between the controlled variable and the setpoint for a period of time. That period of time is the reset value.

The purpose of the integral mode is to gradually eliminate offset, which is the difference between setpoint or desired value and the actual stable

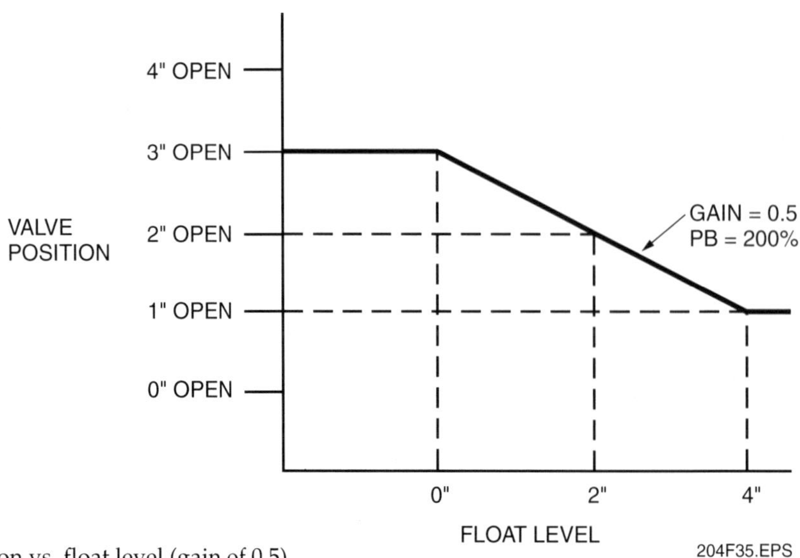

Figure 35 ◆ Valve position vs. float level (gain of 0.5).

value. Control loops having a low gain setting without integral can provide a stable performance but will often generate a large offset. The integral mode is slower in response than the proportional mode because it acts over a period of time. Another drawback with the integral function is integral windup, or reset windup, which occurs whenever the offset or deviation between the setpoint and the actual stable value cannot be eliminated, such as in open loops. In conditions such as an open loop, where the control variable measurement is not used as a signal to supply a controller that adjusts the inputs to the process, the controller is driven into its extreme output, and a loss of control is created for a period of time, usually followed by extreme cycling.

A controller with reset is equipped with a device (integrator) that performs the mathematical function of integration. The mathematical result of integration is called the integral. The integrator provides a linear output with a rate of change that is directly related to the amplitude of the step change input and a constant that specifies the function of integration. In the example shown in *Figure 36*, the input has a fixed step change with a 10% amplitude, while the constant of the controller integrator is set to change the output 0.2% per second for each 1% of the input.

The integrator responds to transform the step change into a gradually changing output signal. As you can see, the input amplitude is repeated in the output every five seconds. As long as the input remains constant at 10%, the output will continue to ramp up every five seconds until the integrator becomes saturated.

The integral function of a PID controller does not integrate the error signal in a true mathematical sense. In mathematics, an integral measures the area under a curve over a finite period of time. PID controllers integrate the error over an infinite period of time. This is known in mathematics as an improper integral. For this reason, if an error signal is not restored to zero, the PID integrator signal continues to increase over time until the signal saturates. This is called "windup." If the PID controller operated in a true integral function, windup would not occur because the area under the curve would only be measured over a finite time period, thus limiting the signal.

6.2.3 Derivative (Rate or D) Control

Derivative control is also known as rate, or D. It provides an output that is a derivative of the rate of change of error, which means that it is proportional to the rate of change of error. The derivative only acts when the error between measurement and setpoint is changing with time. The derivative

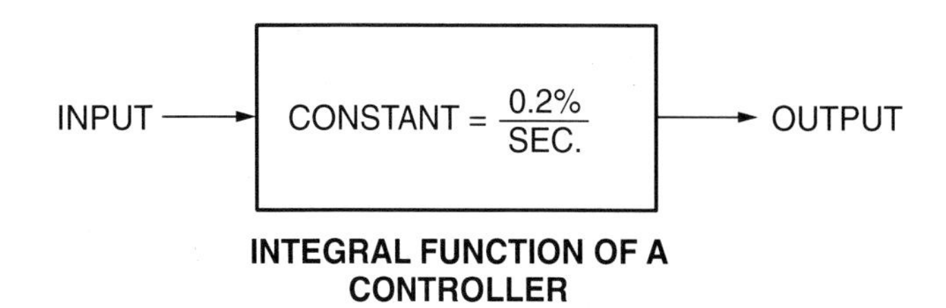

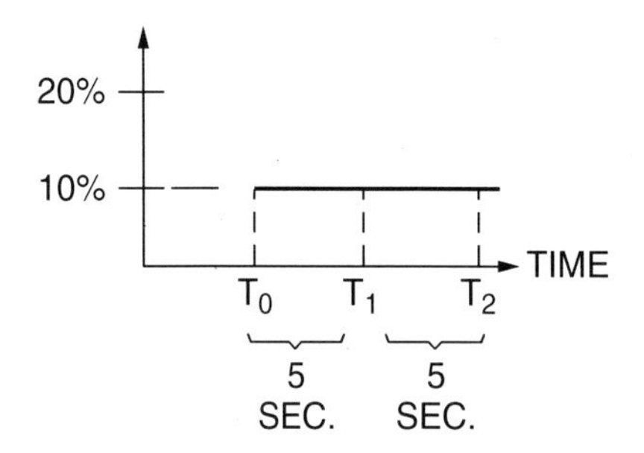

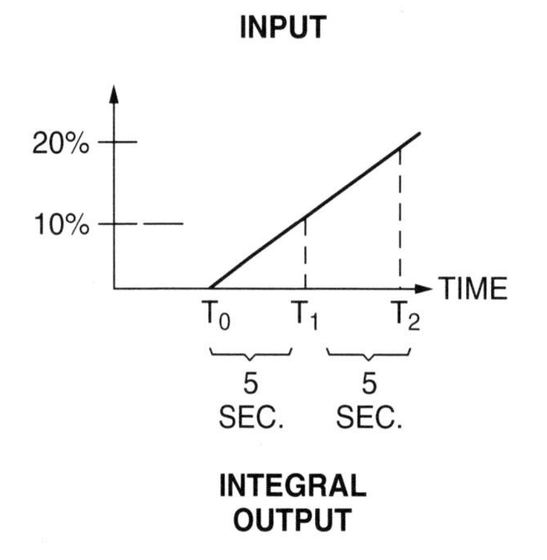

204F36.EPS

Figure 36 ◆ Fixed input with an integral output.

function speeds up the action of the controller in response to the error. It is used in control loops to provide quick stability to sudden upsets in the process variable. Derivative only comes into play while the error is changing. It disappears when the error stops changing, even though there may still be a large error.

The device in the controller that produces the derivative signal is called the differentiator. The differentiator acts to transform a changing signal to a constant magnitude signal (*Figure 37*). As long as the input rate of change is constant, the magnitude of the output is constant. A new input rate of change would give a new output magnitude. The derivative constant is expressed in units of seconds and defines the differential controller output.

Derivative cannot be used alone as a control mode, as can gain and reset. This is because a steady-state input produces a zero output in a differentiator. If only the derivative function were used in a controller, the input signal it would receive would be the error signal. The derivative function must receive its input signal from the proportional function of a controller.

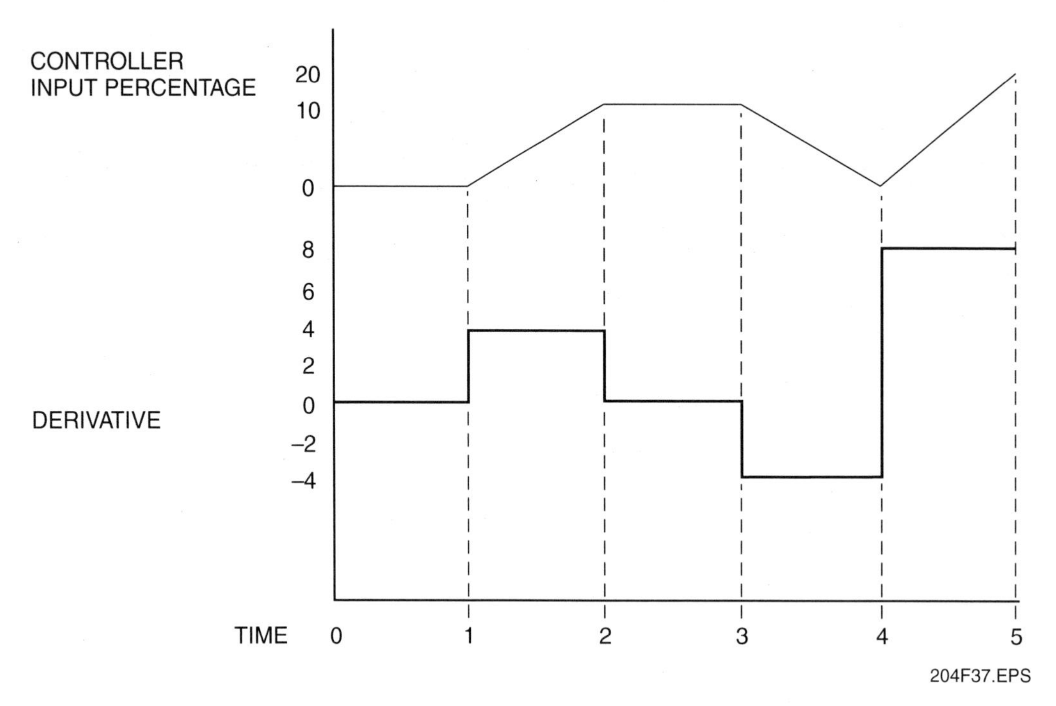

Figure 37 ◆ Derivative response to an input signal.

6.2.4 Proportional Plus Integral (PI) Control

The purpose of this section is to describe a control mode that is actually a combination of two previously discussed control modes: proportional and integral. Combining the two modes results in gaining the advantages and compensating for the disadvantages of the two individual modes.

The main advantage of the proportional control mode is that an immediate proportional output is produced as soon as an error signal exists at the controller. In reality, there is a signal delay time within the controller, but it is usually small, and the proportional controller is therefore generally considered a fast-acting device. This immediate output change enables the proportional controller to reposition the final control element within a relatively short period of time in response to the error.

The main disadvantage of the proportional control mode is that a residual offset error exists between the measured variable and the setpoint for all but one set of system conditions. This offset can be reduced by increasing the controller gain, but this increase leads to process instability. The offset can be eliminated by manually resetting the controller output after each demand disturbance, but this is rare because of the time involved.

The main advantage of the integral control mode is that the controller output continues to reposition the final control element until the error is reduced to zero (within system design limitations). This eliminates the residual offset error and, therefore, the need for gain setting changes or manual resetting of the controller output.

The main disadvantage of the integral control mode is that the controller output does not immediately direct the final control element to a new position in response to an error signal. The controller output changes at a defined rate of change, and time is needed for the final control element to be gradually repositioned.

The combination of the two control modes, the proportional plus integral control mode, combines the immediate output characteristic of the proportional mode with the zero residual offset characteristic of the integral mode. The addition of integral action to proportional control automatically performs the gain resetting that was manually accomplished. For this reason, proportional plus integral controllers are sometimes referred to as proportional plus automatic reset, or proportional plus reset, controllers.

To fully describe the effects produced by adding integral control mode to proportional control mode, a heat exchanger process system is used as an example (*Figure 38*). *Figure 39* shows the effects of demand disturbance on proportional plus reset control mode.

In *Figure 39*, waveform A shows the proportional only control mode response to the decrease in demand. A residual offset error remains.

Setting the integral constant to a low (but not the lowest) integral time results in waveform B, which peaks at a higher value and at a later time. This is because the addition of integral action causes a larger change in the controller output for a given error signal. Because the controller output is larger, more time passes until the peak value is reached.

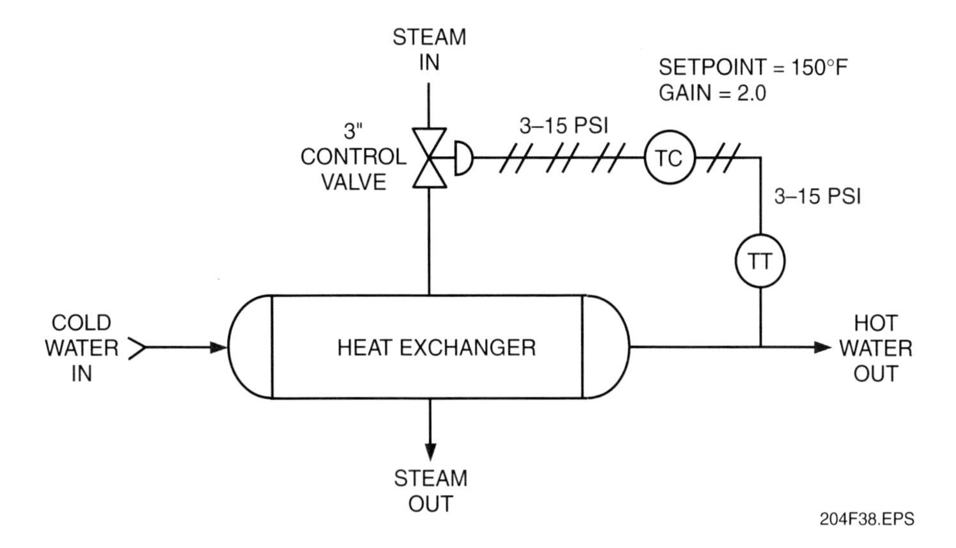

Figure 38 ◆ Heat exchanger process system with a proportional plus reset (PI) controller.

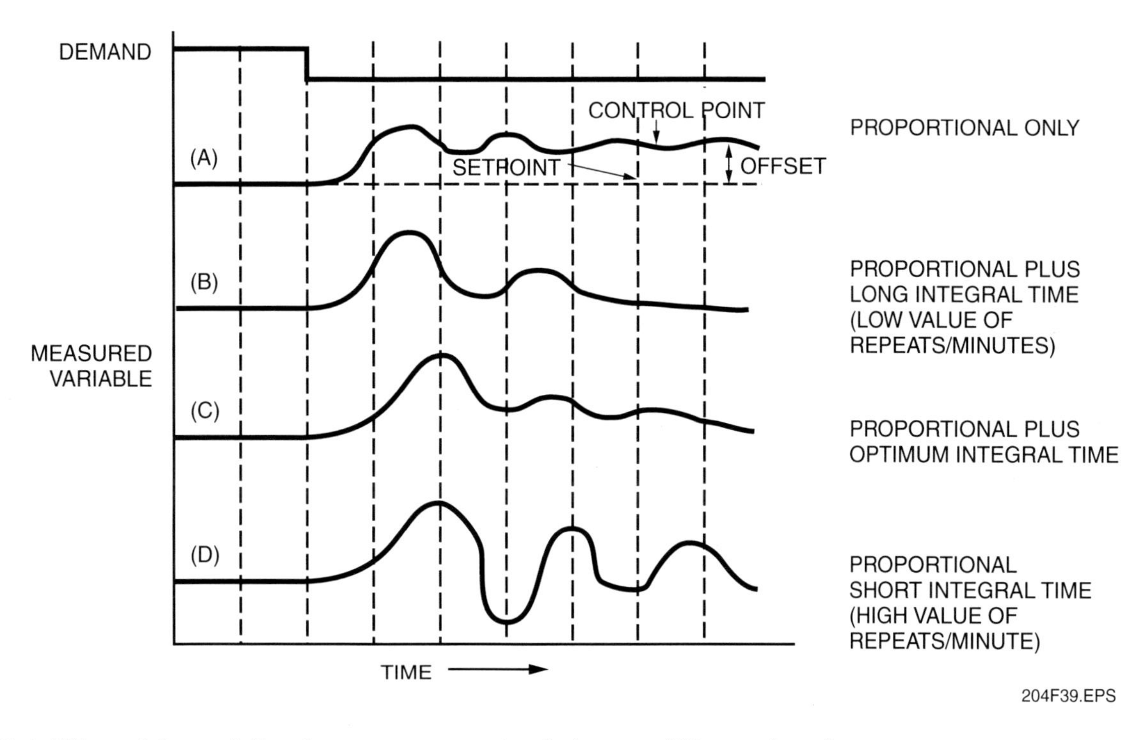

Figure 39 ◆ Effects of demand disturbance on a proportional plus reset (PI) control mode.

The advantage is that the addition of integral action eliminates the offset error after a period of time. This is an improvement, but there still is a relatively long amount of time during which the system control point is away from the setpoint.

Further reduction of the integral time results in waveform C, which is characterized by a higher peak value. This peak occurs at a still later time. Again, the offset error is eliminated, but now in less time. For most processes that employ proportional plus integral control mode, waveform C represents the optimum response characteristics. The increase in peak amplitudes and the period of the cycles is a small price to pay for the distinct advantage of offset elimination.

Setting the integral constant for a short integral time, as in waveform D, results in even larger peak amplitudes and periods of oscillation. Notice that the average value of the control point is returned to the setpoint relatively soon after the demand disturbance, but many processes cannot tolerate the large overshoots above and below the setpoint.

As shown by the waveforms, the addition of integral action to proportional action does not improve the stability of the process system. To the contrary, it actually makes the system less stable. This is because the integral action, after a period of time, causes the controller output to be larger for a given error signal than would be obtained by proportional only action. This can be viewed as the integral action causing an increase in controller total gain. The increase in total gain is desirable. It provides a larger change in position of the final control element in order to eliminate offset error. Recall, though, that increasing the gain leads to process system instability. To compensate for the resulting instability due to the addition of integral action to proportional action, the proportional constant of the controller is reduced to a lower gain value. This results in a higher proportional band value.

Figure 40 shows the waveforms that result from making a setpoint change to the process system. The response of the proportional only control mode to a setpoint change results in an offset between the new control point and the new setpoint. The response of the proportional plus integral control mode to a setpoint change results in larger peaks with a longer period, but the integral action causes the new control point to be at the new setpoint.

The proportional plus integral control mode can be adversely affected by sudden large error signals. The large error can be caused by a large demand deviation or a large setpoint change, or it can occur during initial system startup. A large sustained error signal eventually causes the controller output to drive to its limit. The result is called reset windup.

The effect of reset windup can best be seen by viewing the startup of a system operated in the proportional only control mode and comparing it with the startup of a system operated in the proportional plus integral control mode.

Referring again to the heat exchanger system in *Figure 38*, assume that the steam supply to the control valve has been shut down, and also assume that the controller has been left in automatic. The outlet water temperature decreases to a low value. Because of the large deviation from the setpoint, the controller output is at maximum and the control valve is fully open.

Figure 41 illustrates the resultant waveforms associated with a system startup using the two different control modes. In *Figure 41(A)*, at time t_0 steam is again supplied to the system. Because the controller output is 100%, the control valve is fully open, allowing steam to enter the heat exchanger. This causes an increase in the outlet water temperature. As soon as the value of the measured variable reaches the bottom of the 50% proportional band, which is equal to 100°F, the controller output decreases the control valve opening. After some amount of cycling, the process stabilizes at a control point.

In *Figure 41(B)*, initially the proportional band lower edge is at the 150°F setpoint. This is because the integral action of the controller has shifted the proportional band in its response to the large error signal. At time t_0 when the steam is supplied to the system, the measured variable increases, but the controller output cannot be decreased until the value of the measured variable reaches the proportional band. Notice that the bottom of the proportional band, due to the integral action, is initially 50°F higher than in the previous proportional only control mode of *Figure 41(A)*. Eventually, the measured variable does reach the proportional band at the 150°F point. Notice, however, that this is also

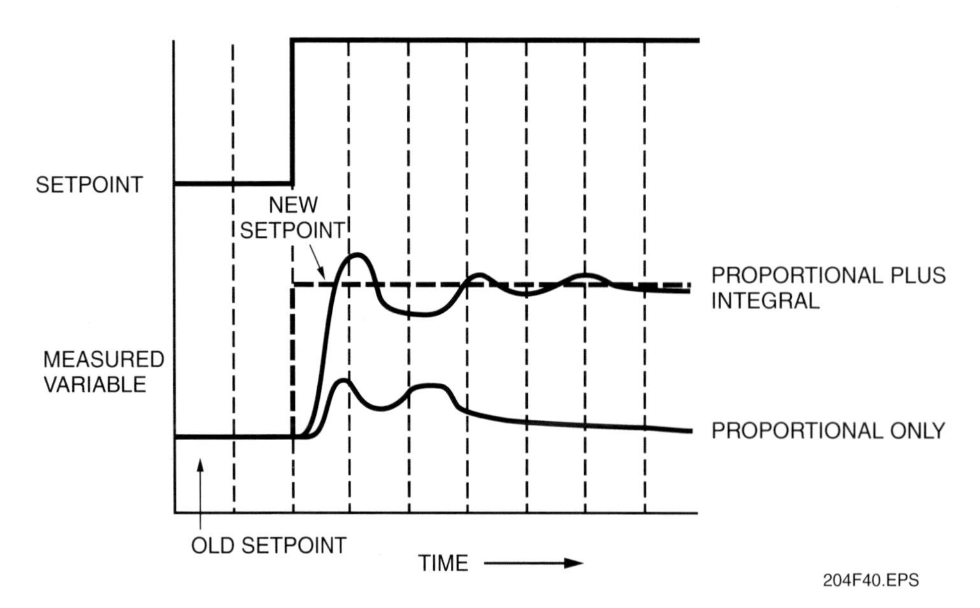

Figure 40 ◆ Effect of a setpoint change.

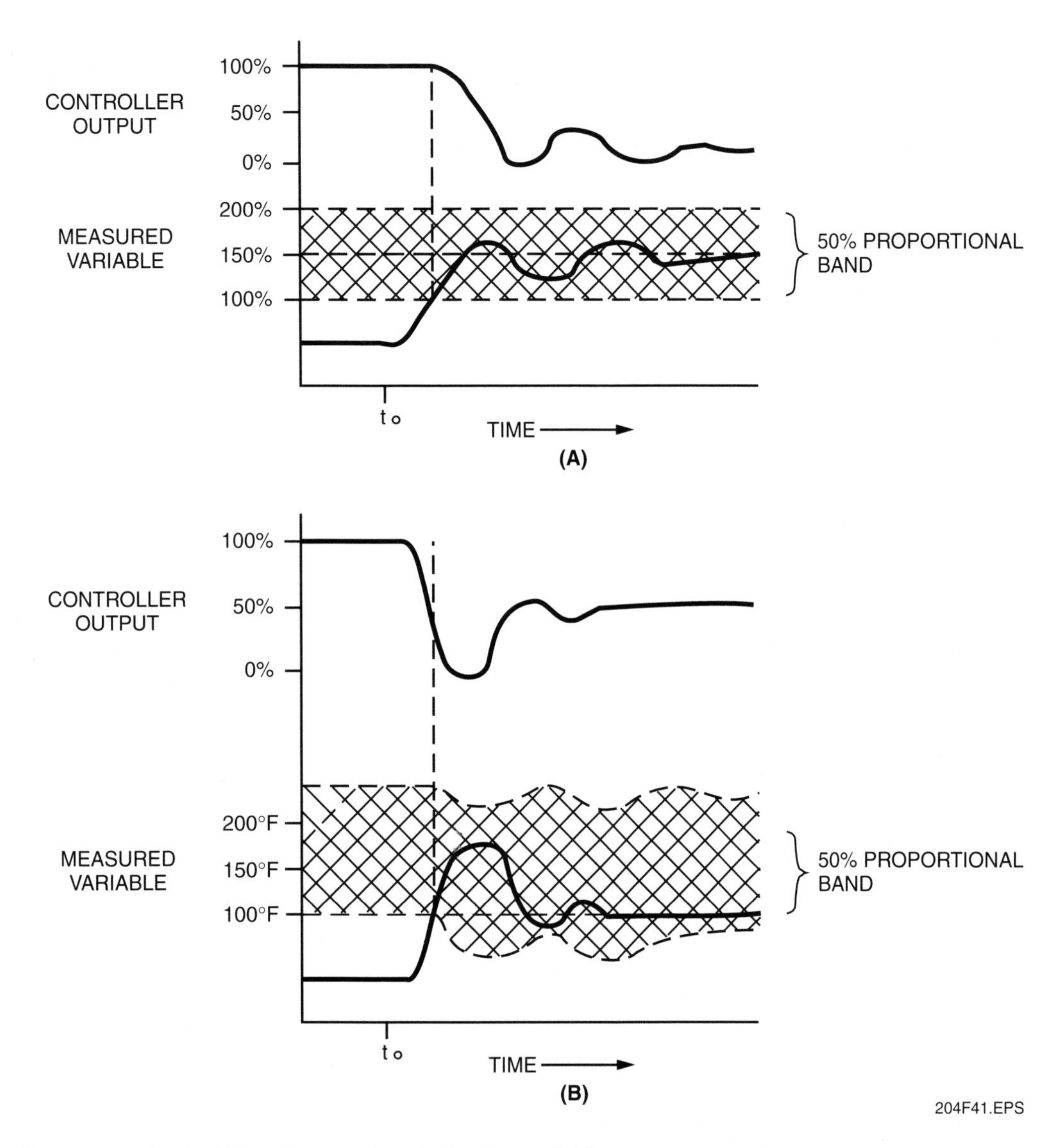

Figure 41 ◆ Proportional only (A) and proportional plus integral (B) startup comparison.

the setpoint, thereby producing a zero error signal. It would seem that the system would now stabilize due to the zero error, but the control valve is wide open, and the measured variable is rapidly changing. There is no way to avoid a very large overshoot, even though the proportional action now begins to decrease the controller output.

Because of the reset windup, in which large overshoots result, proportional plus integral controllers are not well-suited for process systems that must be frequently shut down and restarted or for process systems that are routinely subjected to large or sustained deviations between the setpoint and the measured variable.

Proportional plus integral controllers are often designed to minimize the effects of reset windup. This is accomplished in some controllers by disabling the integral section once the proportional section output reaches 100% so that no further proportional band shifting occurs. In other controllers, a more complicated method is used that involves the use of derivative control mode. This is discussed in a later section.

6.2.5 Proportional Plus Derivative (PD) Controllers

Previous discussion about the control of process systems has shown that the amplitude of the error signal, which is applied to a controller, is directly determined by the amount of deviation that exists between the measured variable and the setpoint. It is logical to assume, then, that the faster the measured variable is departing from the setpoint, the greater the eventual error will be.

A proportional controller acts to reduce, but not eliminate, the amount of deviation by reposition-

the final control element. A proportional controller with added integral action acts to reduce the deviation to zero after a period of time. However, both proportional and proportional plus integral control modes have inherent delay times that result in the controller output lagging behind the actual error condition in the process system. This lagging can be especially critical when the error signal is rapidly becoming larger, because neither the proportional nor the proportional plus integral controller can provide an output to the final control element soon enough to prevent a large overshoot from the desired setpoint.

The purpose of this section is to describe a control mode in which a derivative action is added to a proportional controller. This derivative action responds to the rate of change of the error signal, not the amplitude. It responds to the rate of change at the instant it starts. This causes the controller output to be initially larger in direct relation to the error signal rate of change. The higher the error signal rate of change, the sooner the final control element is positioned to the desired value. The added derivative action reduces initial overshoot of the measured variable. It therefore aids in stabilizing the process sooner.

This control mode is called proportional plus derivative. Because the derivative action responds to the rate of change of the error signal, it is also referred to as proportional plus rate control mode.

Figure 42 shows the same heat exchanger process that has been analyzed in previous chapters. For this example, however, the temperature controller used is a proportional plus derivative controller.

Figure 43 shows the resulting waveforms produced from a demand disturbance. Waveform A shows that the proportional only control mode responds to the decrease in demand, but because there is no integral action in the controller, a residual offset error remains. Proportional only action can be approximated with a proportional plus derivative controller by setting the derivative constant to its minimum or shortest value in terms of time. Some controllers state this adjustment in units of rate, so the setting would be at the lowest rate value.

Setting the derivative constant to a short derivative time results in waveform B, which has a lower peak and stabilizes sooner. This is because the addition of derivative action causes a larger initial controller output for a given error signal. This causes the final control element to be positioned faster for more rapid corrective action to the process. Notice that an offset error remains because of the lack of integral action.

Making the derivative time even longer results in waveform C, which is characterized by just one small overshoot and rapid stabilization to the new control point. Again, however, an offset error remains. For most process systems that employ proportional plus derivative control mode, waveform C represents the optimum characteristics. Recall that the main objective of using a controller that has derivative action is to achieve increased stability.

Setting the derivative constant for a long derivative time, as shown in waveform D, results in no overshoot, but there are many oscillations, and a longer period of time elapses until stabilization. This is because a long derivative time causes a large initial controller output for even small rate of change error inputs. This can be viewed as provid-

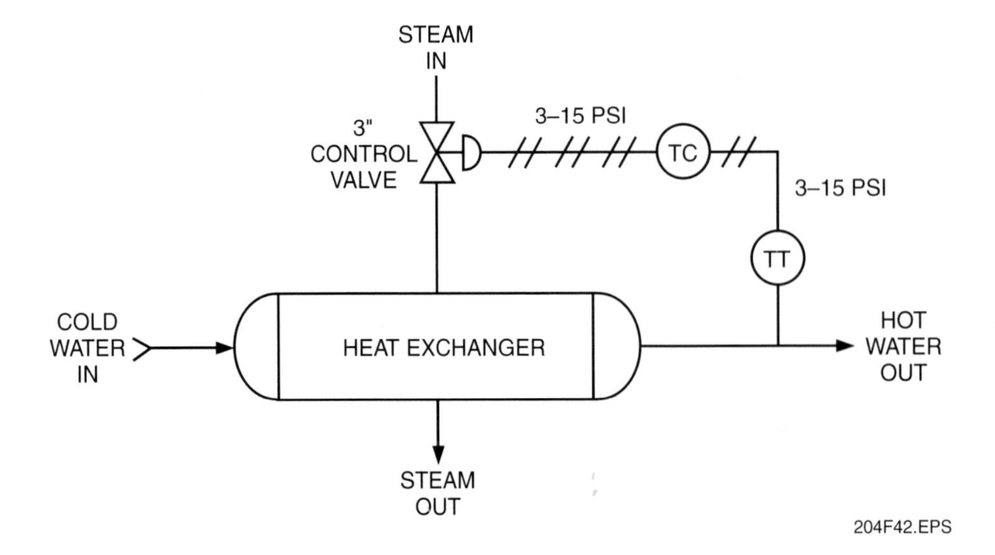

Figure 42 ◆ Heat exchanger process with a proportional plus derivative (PD) controller.

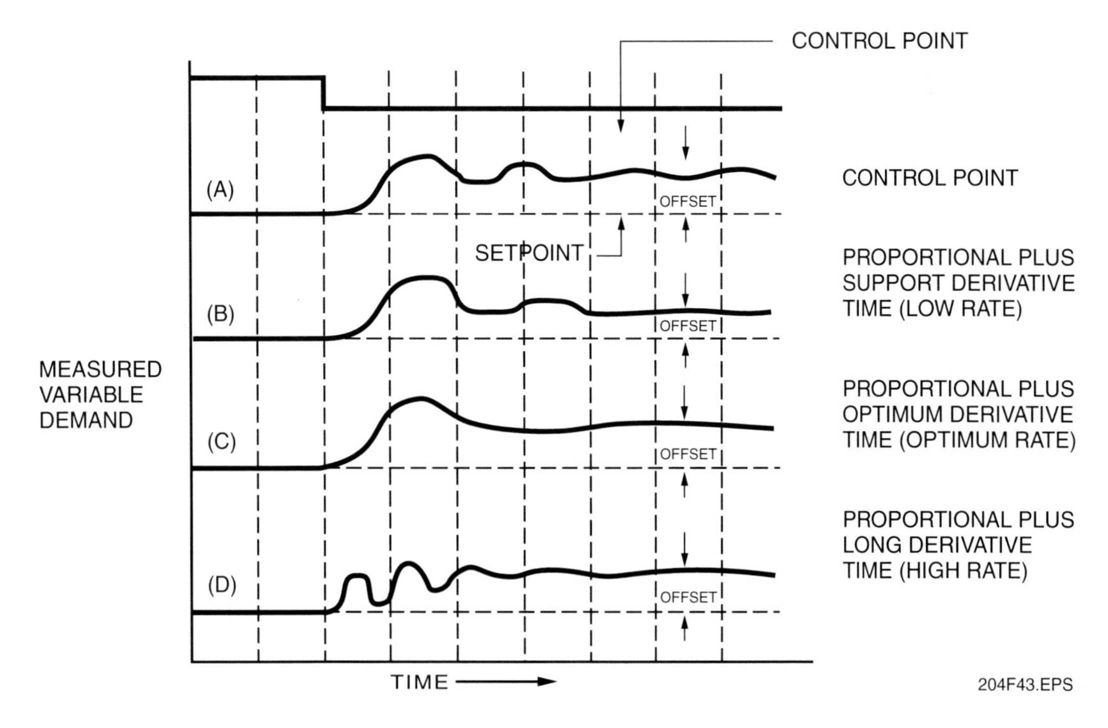

Figure 43 ◆ Effects of demand disturbance (PD).

ing the controller with an initial high gain, which, as described earlier, results in process instability. Many process systems cannot tolerate this rapid cycling.

As shown by the waveforms, the addition of derivative action provides increased stability. Because of the added stability, the gain of the controller can be set to a higher value, which is a lower proportional band setting, to reduce the amplitude of the offset error. The relationship of gain setting and offset error was discussed in the previous section on proportional control.

Figure 44 shows the waveforms that result from making a setpoint change. The response of the proportional plus derivative control mode to a setpoint change results in a smaller overshoot and faster stabilization as compared to proportional only control mode. Note that without integral action, both control modes show a residual

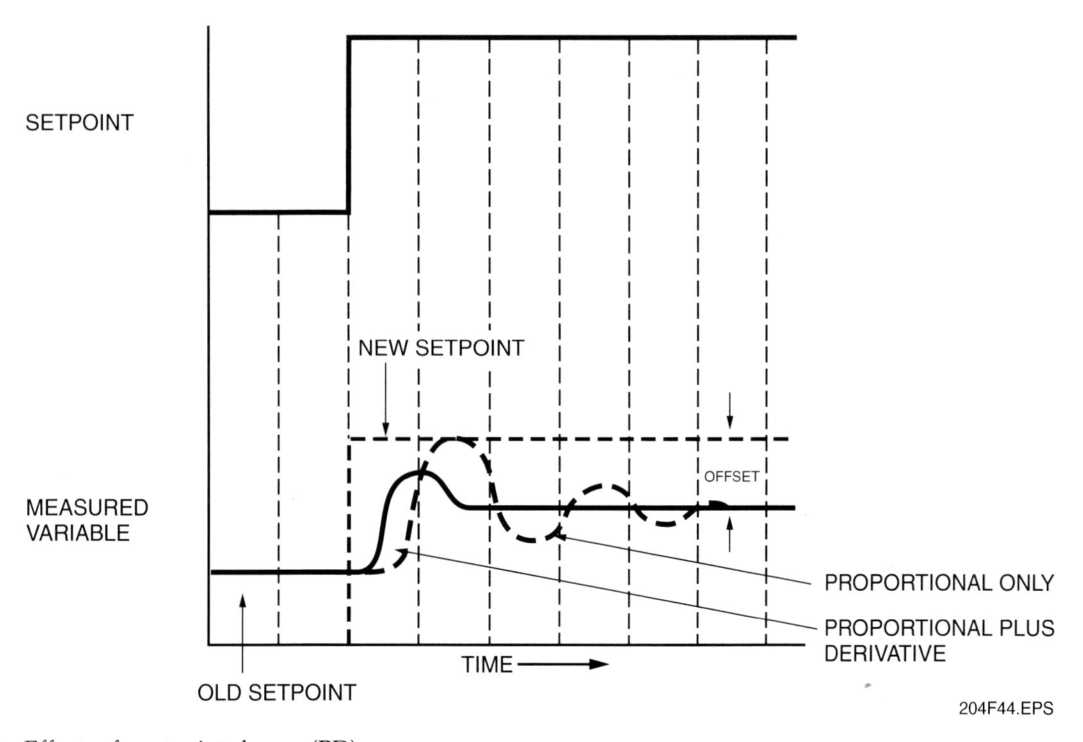

Figure 44 ◆ Effects of a setpoint change (PD).

offset error. As described previously, the addition of derivative action only provides for quicker application of the controller input to the final control element. Derivative action does not reduce offset error.

6.2.6 Proportional Plus Integral Plus Derivative (PID) Controllers

Earlier, we discussed the two-position control mode in which the controller output is in either the fully on or the fully off state. When properly applied, the two-position control mode can provide very good process control. However, continuous cycling is an inherent disadvantage. The two-position controller can be viewed as having a 0% proportional band, which is a very high gain. High gain is very useful for controlling the process accurately, but it is the high gain that results in the cycling nature of this control mode because the controller output changes state (fully on to fully off) in response to small changes in the input error signal.

Next, we discussed the proportional control mode in which the gain of the controller is reduced to a moderate level. This provides a fundamental means of returning stability to the control of the process. With a proportional controller, the output signal places the final control element at different positions corresponding to the error signal amplitude. For the final control element to remain at a desired position, an undesirable offset error must exist between the control point and the setpoint.

With the proportional plus integral control mode, the proportional section of the controller provides a moderate gain for better control stability, and the integral section increases the gain over a period of time to eliminate offset error. The addition of integral action, however, does reduce stability to some degree.

A natural extension of the possible control modes is the proportional plus derivative control mode. With this mode, the change will advance the output signal. The controller gain then returns to the moderate level of the proportional action. The result is that the measured variable overshoots are smaller and these overshoots have a reduced period. However, an offset error remains, which explains why this arrangement is not typically used.

For processes that can operate with continuous cycling, the relatively inexpensive two-position controller is adequate. For processes that cannot tolerate continuous cycling, a proportional controller is often employed. For processes that can tolerate neither continuous cycling nor offset error, a proportional plus integral controller can be used. For processes that need improved stability and can tolerate an offset error, a proportional plus derivative controller is employed.

There are some processes that cannot tolerate offset error and need good stability. The logical conclusion is to use a control mode that combines the advantages of proportional, integral, and derivative action. This section discusses just such a three-term control mode called proportional plus integral plus derivative. This mode is also identified as proportional plus reset plus rate.

The proportional plus integral control mode can be used to control most processes, including those that are by nature difficult to control. Derivative is added to decrease the initial overshoot amplitude and to reduce the period of the cycles.

Figure 45 shows the closed-loop characteristic waveforms resulting from a demand disturbance in a process controlled by a proportional plus integral plus derivative controller. Waveform A shows that the proportional action stabilizes the process and the integral action causes the control point to return to the setpoint. The derivative action does reduce the initial overshoot amplitude and reduce the cyclic period, but, due to the low derivative time set in the controller, the waveform is similar to that of proportional plus integral control.

Increasing the derivative time setting results in waveform B in which the overshoot amplitude is reduced further and the period of time between the overshoot peaks is shortened further. For most processes that employ three-mode control, waveform B represents the optimum characteristics.

Waveform C shows the result of a long derivative time. Notice that the initial overshoot amplitude is low and the period of the cycles is short. However, the time needed for the control point to return to the setpoint is longer. Recall that increasing the derivative time setting in a controller causes a reduction in the effective integral action. The waveforms of *Figure 45* show that as the derivative time is increased, a longer time is needed for integral action to return the control point to the setpoint.

The optimum characteristics shown by waveform B actually represent a compromise between waveform A and waveform C. Too much and too little derivative action both can reduce process stability.

In general, the addition of derivative action results in adjusting the controller for a higher gain, which is a lower proportional band. The higher gain is desirable to help make sure the controller is sensitive to even smaller error signals. The added stability resulting from the derivative

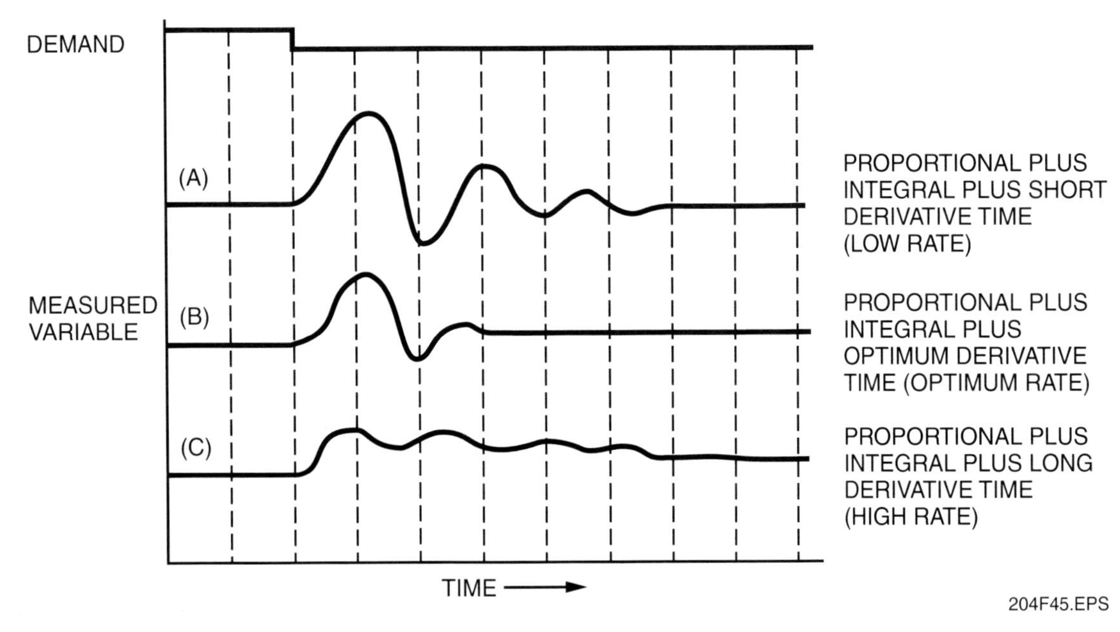

Figure 45 ◆ Effects of demand disturbance (PD).

action compensates for the lower stability resulting from the higher gain.

Also, the addition of derivative action allows adjusting the controller for a lower integral time, which equates to a great number of repeats per minute. This is desirable to enable a return of the control point to the setpoint in a shorter time. Again, the added stability resulting from the derivative action compensates for the lower stability resulting from the increased integral action.

Figure 46 shows the effects of a setpoint change on a process, as well as a comparison of the different results obtained by using different control modes.Waveform A shows the proportional control mode, which is characterized by several cycles before stabilization and residual offset error at the new control point.

Waveform B shows the proportional plus integral mode, which is characterized by a zero offset error, a higher initial overshoot amplitude, and a longer period until stabilization is reached.

Waveform C shows the proportional plus derivative mode, which is characterized by very good stability. Note that the addition of derivative allows the use of higher gain to help minimize the offset error.

Waveform D shows the proportional plus integral plus derivative mode, which is characterized by the advantages shown in both waveform B with its zero offset error, and waveform C, with its good stability. Note that the addition of integral caused some decrease in stability from waveform C due to the integral-derivative interaction, but in terms of overall response, waveform D is the best.

7.0.0 ◆ TYPES OF CONTROL APPLICATIONS

The primary purpose of process instrumentation is to provide automatic control. With automatic process control, products can be manufactured more economically and with higher quality. High-speed reactions that occur in many modern industries could not be successfully controlled without the rapid response possible from control systems.

A process can be expected to operate well only when certain process variables are held within specified limits. These variables cannot be held within these limits unless the energy or material input of the process equals the energy or material output of the process. Therefore, the basic function of a process control system is to maintain a balanced energy or material input-output relationship. Both pneumatic and electronic instruments play a large role in accomplishing this task.

7.1.0 Typical Temperature Control Loops

The following two examples of pneumatic and electronic temperature control loop applications are provided for discussion. Trainees are encouraged to seek out actual examples from a local facility to review.

7.1.1 Pneumatic Temperature Control Loop

The first control loop to be discussed is a simple pneumatic temperature control loop (*Figure 47*). This control loop is designed to maintain process temperature at some value by controlling the flow

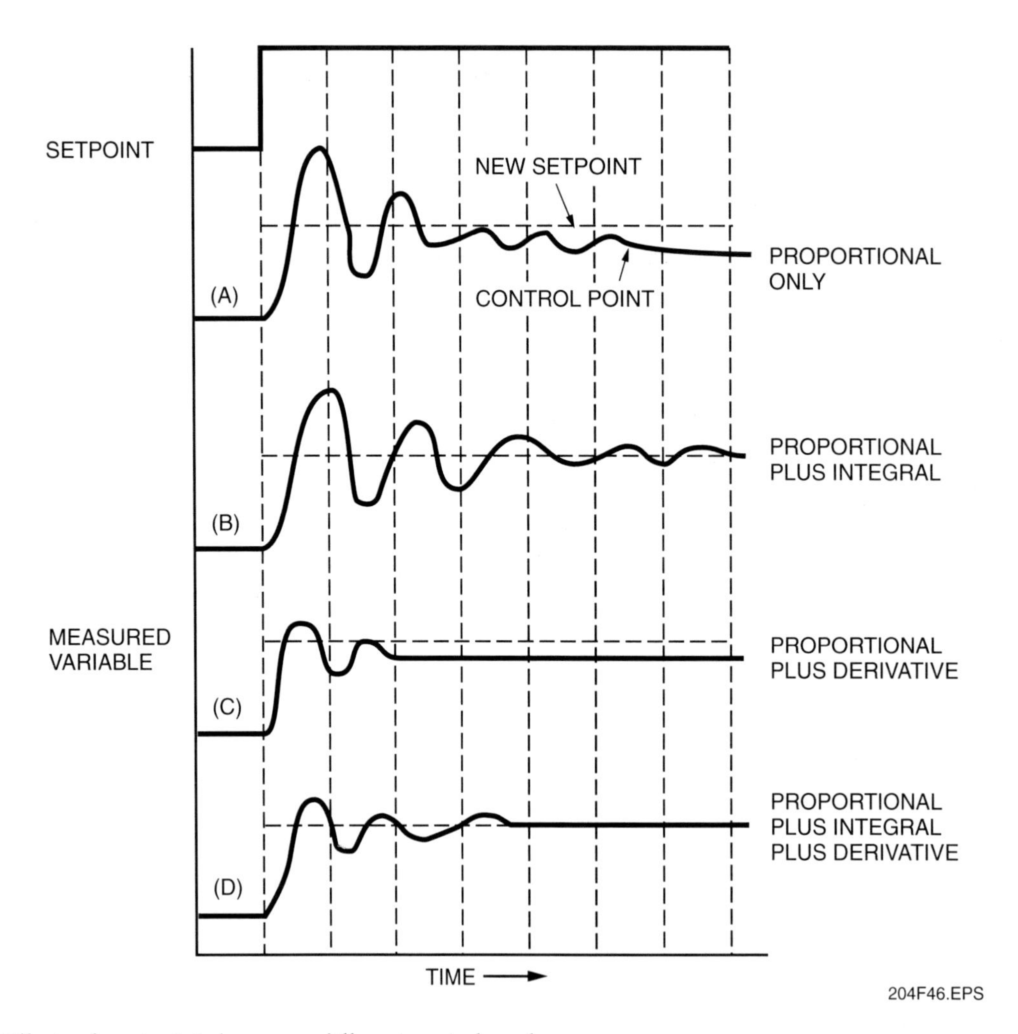

Figure 46 ◆ Effects of a setpoint change on different control modes.

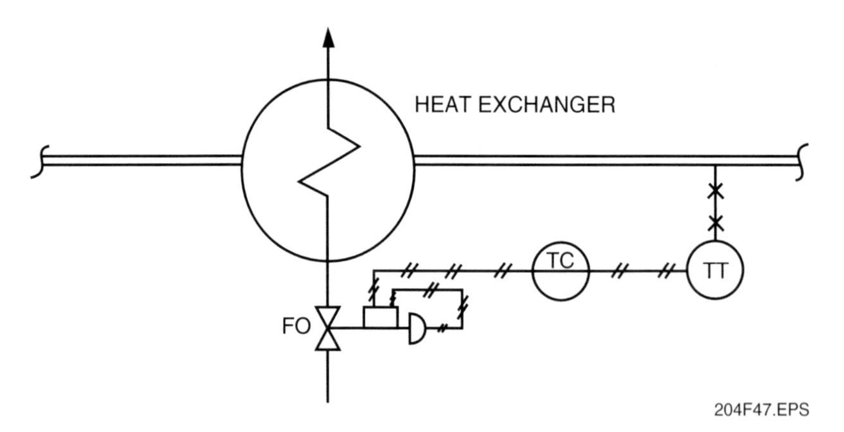

Figure 47 ◆ Pneumatic temperature control loop.

of cooling water to a heat exchanger. Temperature is sensed downstream of the heat exchanger with a locally mounted filled-system temperature transmitter (TT). The transmitter sends a pneumatic signal that is proportional to process temperature. The signal is received by a temperature controller (TC). The controller compares the actual process temperature to the setpoint. If there is a difference between the two, a pneumatic control signal is developed and sent to a valve positioner. The positioner then produces a change in air pressure to position a diaphragm-actuated globe valve. The resulting position of the valve throttles cooling water flow through the heat exchanger. In the heat exchanger, cooling water passes through the small tubes (as shown in *Figure 48*). The process fluid passes over the tubes as it flows through the heat exchanger.

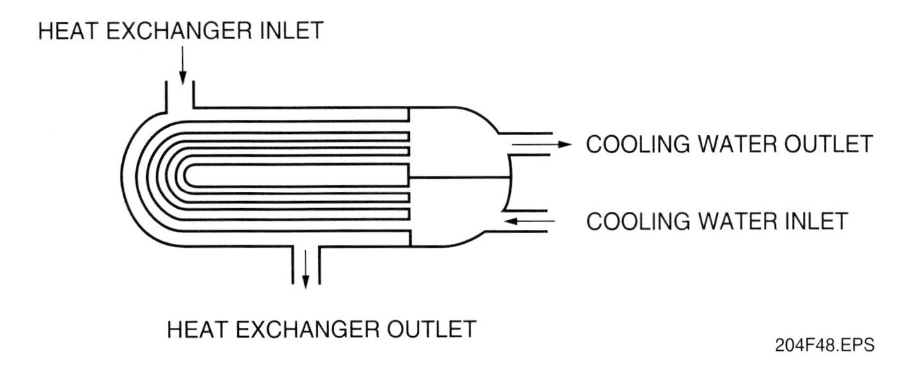

Figure 48 ◆ Tube-type heat exchanger.

If cooling water flow increases, more heat is transferred from the process fluid, so process temperature decreases. If cooling water flow decreases, less heat is transferred from the process fluid, resulting in a process temperature increase.

The control valve in this system typically fails open if a failure in the supply air system occurs. This feature prevents the process from overheating.

7.1.2 Electronic Temperature Control Loops

A simple electronic temperature control loop is illustrated in *Figure 49*. The function of this system is to maintain the temperature of the process fluid within a storage tank at some value by controlling a bank of electric heaters. Temperature in the tank is sensed by a resistance temperature detector (RTD). The symbol indicates that the sensing element is enclosed by a thermowell. The temperature transmitter (TT) performs the functions of the transducer and the transmitter of the basic control loop. It converts the temperature-dependent resistance value of the sensor into an electrical signal proportional to measured temperature. It transmits the information to the local temperature indicator (TI) and to the indicating temperature controller (TIC) located at the main control panel.

The temperature controller compares the actual process temperature to the desired temperature, the setpoint. When there is no difference between the two, the controller is normally adjusted to produce a midrange output signal, such as 12 mA DC for a 4–20 mA DC range. The 12 mA DC signal causes the heaters to be supplied with sufficient power to make up for heat loss from the process during steady-state conditions. If temperature were to decrease, perhaps because of a decrease in supply temperature, the resulting difference between actual process temperature and desired temperature would cause the controller output signal to increase from the midrange value. The increase in control signal current would increase the conduction angle of the **triac** power control circuit, thereby increasing the average power supplied to the heaters. When temperature was restored to the setpoint, the power supplied to the heaters would just be sufficient to maintain the proper temperature at the new steady-state conditions.

7.2.0 Typical Pressure Control Loops

The following two examples of pneumatic and electronic pressure control loop applications are provided for discussion. Trainees are encouraged to seek out actual examples from a local facility to review.

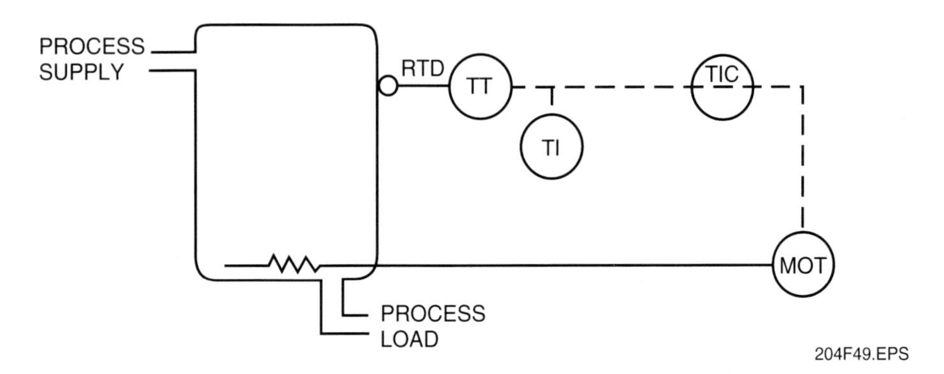

Figure 49 ◆ Electronic temperature control loop.

7.2.1 Pneumatic Pressure Control Loops

The pneumatic pressure control loop shown in *Figure 50* is designed to maintain a constant pressure within the closed tank by controlling a nitrogen supply valve and a vent valve. A locally mounted pressure transmitter (PT) senses pressure. The transmitter sends a signal to a pressure recorder (PR) and a pressure controller (PC). The output of the controller is transmitted to two valve positioners. Valve positioners are used in this control loop because each of the control valves cycles during only half of the controller output range. In other words, this control loop application requires that the control output be split-ranged.

The nitrogen supply valve positioner is calibrated to position the nitrogen supply valve in the lower half of the controller output range. The vent valve positioner is calibrated to operate the vent valve in the upper half of the controller output range. The result of this split-range configuration is that when the process pressure is at the setpoint, both the nitrogen supply valve and the vent valve are closed. When process pressure drops below the setpoint, the vent valve remains closed and the nitrogen supply valve opens to increase process pressure. When process pressure rises above the setpoint, the supply valve remains closed and the vent valve opens to reduce system pressure. In this application, both the supply valve and the vent valve typically fail closed. The feature prevents the system from overpressurizing or completely depressurizing.

7.2.2 Electronic Pressure Control Loops

The electronic pressure control loop shown in *Figure 51* provides an example of a simple control application that also requires a split-ranged output from the controller. The function of this control system is to maintain a constant pressure within a closed storage tank by controlling the position of a nitrogen supply valve and an atmospheric vent valve.

A locally mounted pressure transmitter (PT) senses pressure. An electric signal proportional to storage tank pressure is sent from the transmitter to an indicating pressure controller (PIC). The controller compares the actual tank pressure to the storage tank pressure setpoint and produces a control signal proportional to the difference between them. When there is no difference between setpoint and actual pressure, a mid-range signal value is produced and transmitted to two electric-to-air valve positioners, a nitrogen supply valve positioner, and an atmospheric vent valve positioner. The nitrogen supply valve positioner is calibrated to position the nitrogen supply valve in the lower half of the output signal range of the controller. When the controller output is midrange, the atmospheric vent valve positioner keeps the vent valve in the closed position. Both control valves are diaphragm-actuated globe valves.

As a result of the split-range calibration of the valve positioners, when storage tank pressure decreases below the setpoint, the vent valve remains shut and the nitrogen supply valve opens to admit more nitrogen into the tank, as shown in *Figure 52*. The addition of nitrogen into the tank increases the tank pressure to the desired value. When process pressure increases above the setpoint, the supply valve shuts, and the atmospheric vent valve opens. Opening of the vent valve allows nitrogen to bleed off to the atmosphere, which causes storage tank pressure to decrease to the setpoint. The position of either valve is dependent on the magnitude of the control signal. The further the signal varies from the mid-range value, the greater the movement of the valve stem.

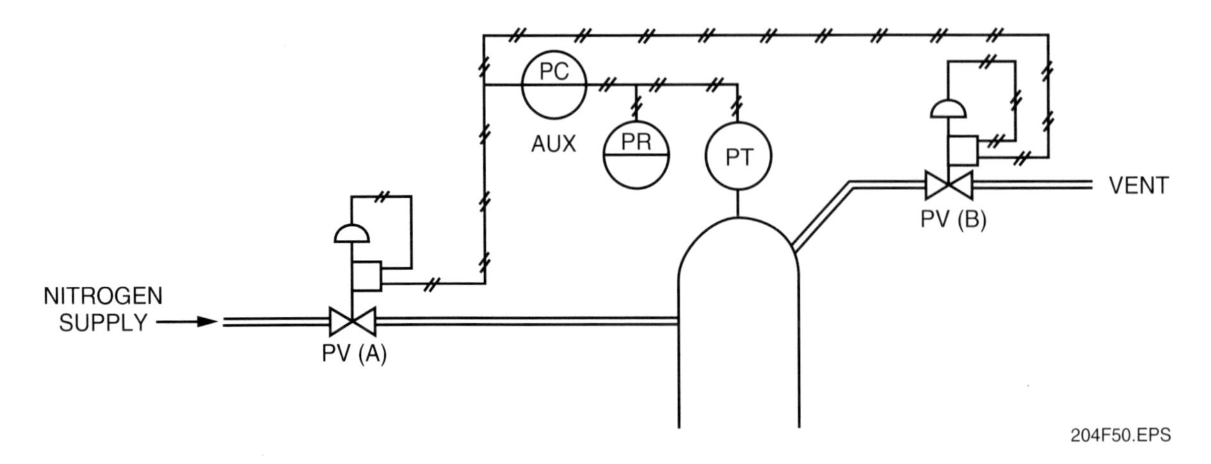

Figure 50 ◆ Split-range pneumatic pressure control loop.

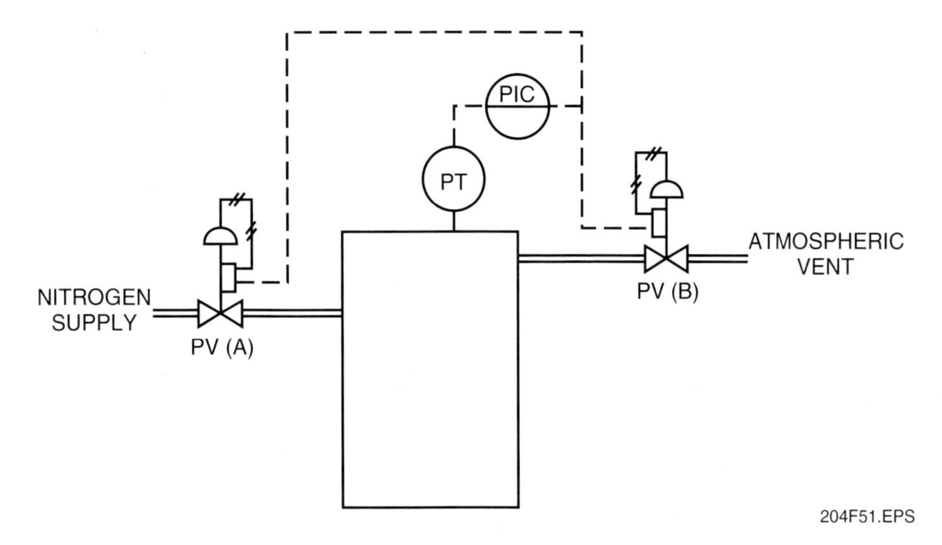

Figure 51 ◆ Electronic pressure control loop.

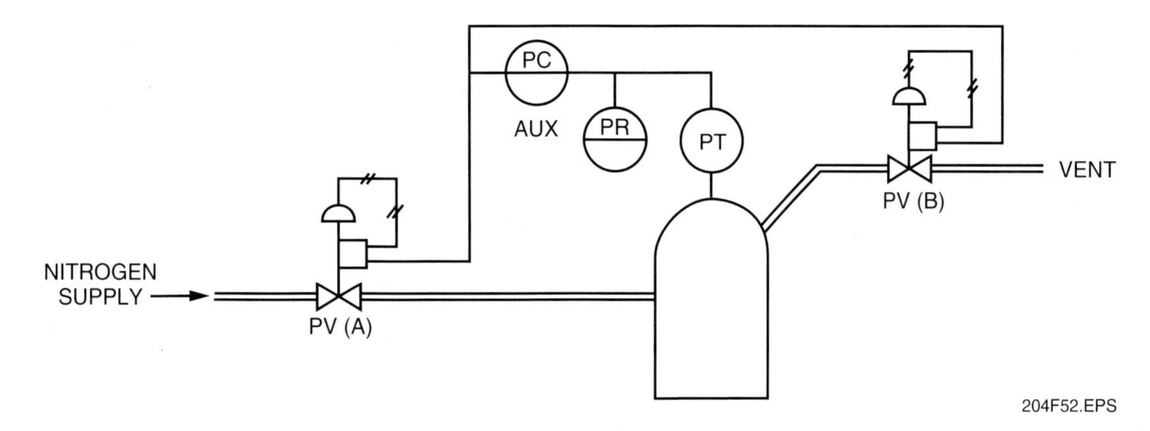

Figure 52 ◆ Split-range electronic pressure control loop.

In this control application, the supply valve typically fails shut on loss of signal, and the vent valve fails open. This feature prevents overpressurization of the tank.

7.3.0 Pressure-Sensing Control Loops

The following two examples of pneumatic and electronic flow control loop applications are provided for discussion. Trainees are encouraged to seek out actual examples from a local facility to review.

The nitrogen supply valve positioner is calibrated to position the nitrogen supply valve in the lower half of the controller output range. The vent valve positioner is calibrated to operate the vent valve in the upper half of the controller output range. The result of this split-range configuration is that when the process pressure is at the setpoint, both the nitrogen supply valve and the vent valve are closed. When process pressure drops below the setpoint, the vent valve remains closed and the nitrogen supply valve opens to increase process pressure. When process pressure rises above the setpoint, the supply valve remains closed, and the vent valve opens to reduce system pressure. In this application, both the supply valve and the vent valve typically fail closed. This feature prevents the system from overpressurizing or completely depressurizing.

7.3.1 Pneumatic Flow Control Loop

The pneumatic flow control loop illustrated in *Figure 53* is designed to maintain sufficient flow through the centrifugal pump to prevent the pump from overheating. This minimum flow must be maintained whenever the pump is operating, whether the pump discharge valve is open or fully closed.

Pump discharge flow is measured by a variable head flow meter. The differential pressure is created by a flow nozzle. A locally mounted flow transmitter (FT) senses the differential pressure

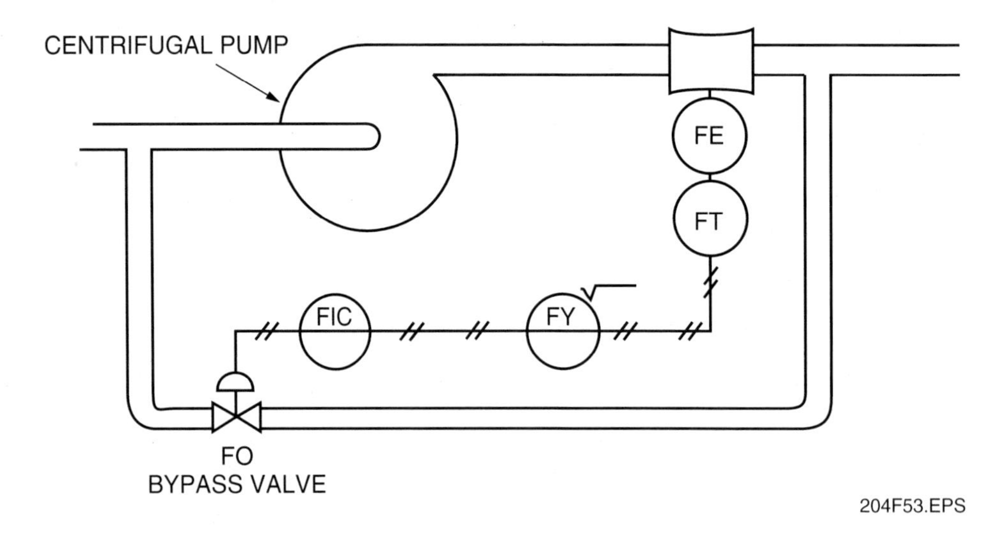

Figure 53 ◆ Pneumatic flow control loop.

and sends a pneumatic signal to a square-root extractor (FY). This converts the differential pressure signal to a linear flow signal and transmits the flow information to an indicating flow controller (FIC). The controller compares the actual flow signal to the desired flow or setpoint. If a difference exists, a change in control pressure is produced that actuates the bypass valve.

During most operating conditions, the pump discharge valve is sufficiently open to allow adequate flow for pump cooling. In this condition, the bypass valve is closed. When the pump discharge valve receives a signal to reduce flow, the flow controller output increases, and the bypass valve is positioned to increase pump bypass flow. If the pump discharge valve closes fully, the controller output increases to maximum, causing the bypass valve to open fully.

The bypass valve is typically designed to fail open so that in the event of a supply air failure the minimum pump discharge flow would always exist. This prevents the pump from overheating.

7.3.2 Electronic Flow Control Loop

The electronic flow loop illustrated in *Figure 54* regulates process flow by varying the speed of a centrifugal pump. Flow in the process is measured by a variable-head flow meter. A differential pressure-type flow transmitter (FT) senses the head produced across a flow nozzle and transmits a signal proportional to the differential pressure to a square-root extractor (FY). The square-root extractor converts the differential pressure information into a linear flow rate signal and transmits this information to a flow recorder (FR) and an indicating flow controller (FIC). These are both located at the auxiliary control board. The indicating flow controller compares the actual flow in the process to the desired flow. It produces a control signal to the triac power control circuit that is proportional to the difference between actual flow and setpoint.

The triac power control circuit varies the average power to control the pump speed, which ultimately controls pump discharge pressure. As a

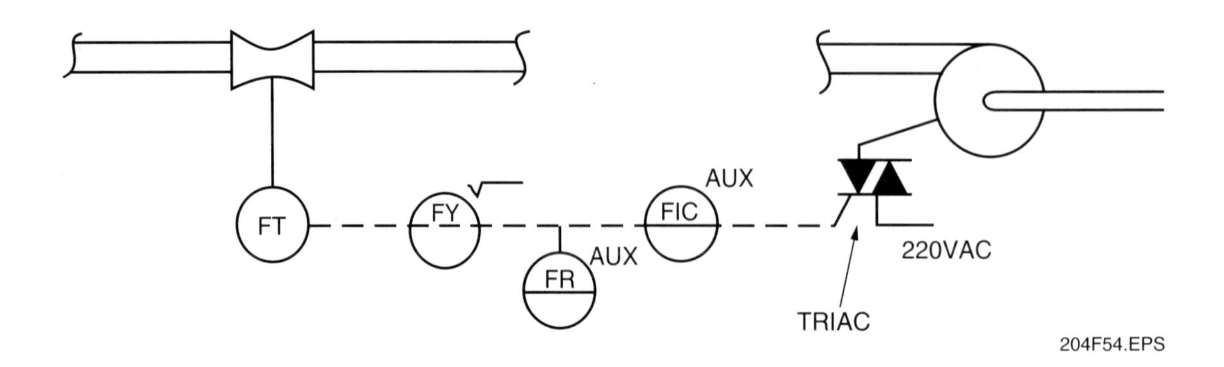

Figure 54 ◆ Electronic flow control loop.

result, the flow rate sensed at the flow nozzle can be varied by the pump in order to return the process flow rate to the setpoint.

7.4.0 Typical Level Control Loops

The following two examples of pneumatic and electronic level control loop applications are provided for discussion. Trainees are encouraged to seek out actual examples from a local facility to review.

7.4.1 Pneumatic Level Control Loop

The pneumatic level control loop illustrated in *Figure 55* is used to maintain the level of slurry within the closed, pressurized vessel. The level of slurry, a mixture of liquid and solid particles, is measured by a flange-mounted, diaphragm-type differential pressure transmitter (LT). The pneumatic output of the transmitter is sent to a level controller (LC) and a level recorder (LR). The controller compares the actual slurry level to the setpoint and produces an output signal proportional to the difference between the two.

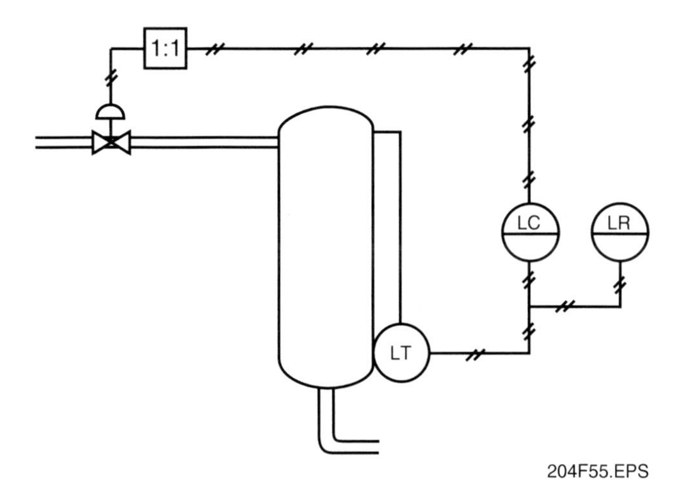

Figure 55 ◆ Pneumatic level control loop.

The volume of air transmitted by the controller is insufficient to operate the control valve, so a 1:1 booster relay is installed in the vicinity of the valve actuator in order to increase the controller output air signal. The control valve is in a pinch valve. The pinch valve has poor control characteristics, but because the slurry does not contact the working parts of the valve, it does not fail as easily as other types of control valves.

7.4.2 Electronic Level Control Loop

The function of the electronic level control loop illustrated in *Figure 56* is to control the supply of fluid into the tank in order to maintain tank level at the desired value. The level in the tank is measured by a differential pressure-type level transmitter. The sensing element of the transmitter detects the difference between the pressure exerted by the column of water within the reference leg and the pressure exerted by the column of water within the tank. The pressure difference is converted to an electrical signal by the transducing circuits within the transmitter (LT).

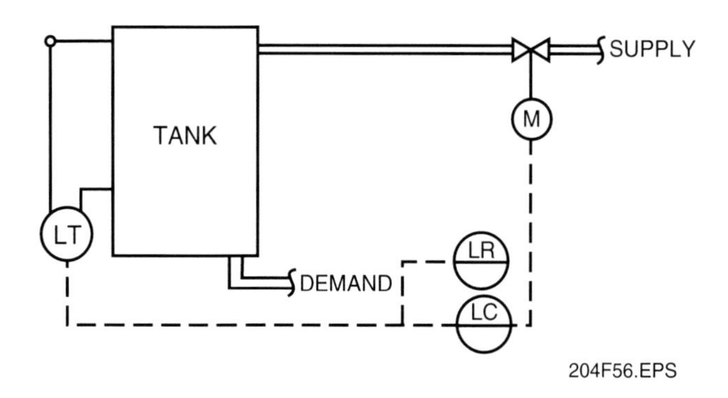

Figure 56 ◆ Electronic level control loop.

The tank level signal is then sent to a level recorder (LR) and a level controller (LC). The level controller compares the actual tank level to the setpoint and produces an electric control signal proportional to the difference between them. The final control element in this control loop is a motor-operated globe valve.

During alignment, the control loop is adjusted so that when the actual tank level is at the setpoint, the motor-operated control valve opens just enough to allow supply flow into the tank to equal demand flow from the tank. An increase in demand causes tank level to decrease. The decrease in level is sensed by the level transmitter and transmitted to the controller. At the controller, a difference signal is produced that results in an increase in the output from the controller. The motor-operated globe valve then opens to restore level to the setpoint.

Summary

The purpose of this module is to introduce the trainee to the factors that must be considered in designing process control systems. In process control, the primary objective is to control a process in a stable manner while meeting or exceeding the requirements of the process. The process is performed on material, while the control of the process is performed by hardware that makes up the instrumentation loop. Variable process characteristics demand process control. These characteristics include temperature, pressure, level, and flow.

The instrumentation loop hardware required to control the process includes devices such as detectors, transducers, signal conditioners, transmitters, controllers, and final elements such as control valves. Each of these devices must be selected based on criteria in the process variables. In this module, we learned that the detector is usually the first element in the control loop. Its function is to monitor or detect the state or change of a particular characteristic of the process. The transducer is used to convert one form of signal to another to make it usable to the next element in the line in the control loop. Transmitters are used to send signals from one location to another, usually from the field to a remote location like a control room. The controller is the device that takes the transmitted signal and compares it to a preset value, sending an output signal to the final element that brings the process back to the desired setpoint.

There are two fundamental types of control loops: open and closed. The closed loop involves the detection of a change in the controlled variable, while the open loop regulates an independent variable for the purpose of controlling the controlled variable. One variation of open-loop control is the feedforward loop. Some examples of closed-loop designs are the feedback, cascade, and ratio control types. Most control loops are closed.

In a modulating control loop, the feedback control determines the amount of error and produces a corrective signal proportional to that error in order to reduce the error to zero. The corrective signal may be generated by logic based on proportional (P), integral (I), or derivative (D) algorithms, or a combination of these. A proportional plus integral algorithm can be used to control most functions. Proportional plus integral plus derivative (PID) control is used when it is necessary to minimize overshoot.

Before designing and installing an instrumentation loop, each process application must be scrutinized to determine the type of loop, devices in the loop, and the type of control that is necessary to keep the process in control while maintaining process quality and quantity. There are many types of process control systems to serve many different purposes.

Review Questions

1. The concept of measuring, comparing, computing, and correcting is implemented by the _____.

a. channel
b. process
c. control system
d. operator

2. The operation or operations used in the treatment of materials comprise the _____.

a. feedback
b. process
c. instrumentation
d. control

3. Process parameters are subject to change if the energy balance is _____.

a. constant
b. high
c. low
d. upset

4. The process parameter that is selected for control is the _____ variable.

a. controlled
b. manipulated
c. constant
d. input

5. The controlled variable is ultimately regulated by the _____ loop.
   a. transmitter
   b. primary element
   c. feedback
   d. controller

6. Which parameter supplies energy or material to the process to regulate the deviation of a process output?
   a. Controlled variable
   b. Manipulated variable
   c. Constant variable
   d. Primary element

7. Which of the following flow measuring detectors applies a converging cone-shaped inlet section?
   a. Orifice plate
   b. Flow nozzle
   c. Venturi tube
   d. Pitot tube

8. A detector or sensor is considered good if it is designed to minimize _____.
   a. signal loss
   b. disturbance
   c. differential pressure
   d. continuous flow

9. If a tank's level is measured using a dipstick, this is considered _____ measurement.
   a. inferred
   b. referenced
   c. assumed
   d. direct

10. A thermocouple uses _____ measurement.
    a. direct
    b. transducer
    c. inferred
    d. pyrometer

11. If the output from one device is not compatible with the input it supplies to another device, a(n) _____ must be applied.
    a. transducer
    b. amplifier
    c. transmitter
    d. rotameter

12. Of the following, _____ transducers may convert a 0-5-volt signal to a 4–20 mA signal.
    a. I/P
    b. I/E
    c. E/I
    d. E/D

13. The function of the feedback transmission circuit in a controller is to _____.
    a. convert the signal from the feedback transmission circuit into a form required by the final element
    b. convert the sensor signal into a usable controller signal
    c. supply feedback to the transmitter from the primary element
    d. generate a voltage requirement for all feedback components in the control system

14. A variable orifice that is used to regulate the flow of process fluid according to the requirements of the process is known as a(n) _____.
    a. orifice plate
    b. variable transducer
    c. pitot tube
    d. control valve

15. The component attached to a control valve that compares the stem motion of the valve to the signal from the controller is the _____.
    a. actuator
    b. plug stem
    c. diaphragm
    d. positioner

16. The selection and interconnection of devices in an instrumentation channel that ultimately determine the stability of the process are referred to as the _____.
    a. control loop
    b. process system
    c. selective design
    d. instrumentation trade

17. Another name for an open loop is a _____ loop.
    a. feedback
    b. feedforward
    c. cascade
    d. ratio

18. Another name for a feedback loop is a(n) _____ loop.
   a. feedforward
   b. ratio
   c. closed
   d. open

19. The primary performance criteria for any closed-loop system is _____.
   a. accurate feedback
   b. gain less than 100%
   c. stability
   d. rate more than 1

20. Another name for a two-position control mode is a(n) _____ control.
   a. modulating
   b. proportional
   c. integral
   d. On-Off

21. In what type of control mode is the final control element throttled to various positions that are dependent on process system conditions?
   a. Derivative
   b. Proportional
   c. On-Off
   d. Integral

22. In what type of control mode is the output rate of change dependent on the magnitude of the input?
   a. Proportional
   b. Integral
   c. On-Off
   d. Open loop

23. The basic function of a process control system is to _____.
   a. maintain a balanced energy or material input-to-output relationship
   b. convert the input signal into a usable output signal
   c. monitor process errors and report their degree of error
   d. maintain the proper supply versus demand process ratio

24. You would most likely find an RTD in a _____ process control loop.
   a. flow
   b. temperature
   c. level
   d. pressure

25. The function of an FY is _____.
   a. accurate feedback
   b. for gain less than 100%
   c. to provide stability
   d. to convert a differential pressure signal to a linear flow signal

GLOSSARY

Trade Terms Introduced in This Module

***Amplitude*:** The height of a waveform. In process control, it refers to the severity of overshoots and undershoots of the setpoint.

***Channel*:** In process control, a channel consists of a group of equipment components that act together to perform one function.

***Closed loop*:** A process control configuration in which the process variable is measured and compared to the desired value or setpoint.

***Controlled variable*:** The variable that the control system attempts to keep at the setpoint value. The setpoint may be constant or variable.

***Dampened*:** Decreased in amplitude of an oscillation with time.

***Disturbance*:** A general term for a disruption of the desired value or setpoint.

***Feedback*:** Part of a closed-loop system that provides information about the current status or parameter for comparison with the desired condition of that parameter.

***Fulcrum*:** The pivot point on a lever.

***Hysteresis*:** A state or effect that causes changes to a system not to take immediate effect. For example, as a parameter is increased, the behavior makes a sudden jump at a particular value of the parameter. But as the parameter is then decreased, the jump back to the original behavior does not occur until a much lower value.

***Lever*:** A simple machine consisting of a rigid bar that is free to pivot on a fulcrum.

***Noise*:** Signals that interfere with output signals you are trying to measure. These signals may come from the product itself or from other electronic equipment.

***Open loop*:** A process control configuration in which no comparison is made between the actual value and the desired value or setpoint of a process variable.

***Radius taps*:** Similar to vena contracta taps, except the downstream tap is fixed at half of the pipe diameter from the orifice plate. See *vena contracta taps*.

***Setpoint*:** A variable, expressed in the same units as the measurement, which sets either the desired target for a controller or the condition at which alarms or safety interlocks are to be energized.

***Triac*:** An electronic, three-terminal device that is similar to two silicone-controlled rectifiers back-to-back with a common gate and common terminals.

***Valve plug*:** That part of a control valve that physically restricts or regulates the flow through the body of the control valve.

***Valve plug stem*:** An attachment point that is constructed onto the portion of a control valve that physically restricts the flow through the body of the control valve.

***Vena contracta taps*:** When a fluid flows through an abrupt contraction like an orifice plate, the flow stream does not follow the shape of the pipe walls, but takes a curved route. The flow stream continues to contract for a short distance downstream of the contraction and then expands to fill the entire area available for flow. The point of smallest cross-sectional area in the flow stream is the vena contracta. Vena contracta taps are located one pipe diameter upstream from the plate, and downstream at the point of vena contracta. This location varies with the orifice plate calculation. The vena contracta taps provide the maximum pressure differential, but also the most flow noise. Additionally, if the plate is changed, it may require a change in the tap location. Also, in small pipes, the vena contracta might lie under a flange. Therefore, vena contracta taps normally are used only in pipe sizes exceeding six inches.

Additional Resources

This module is intended to present thorough resources for task training. The following reference works are suggested for further study. These are optional materials for continued education rather than for task training.

Instrumentation, 1975. F.W. Kirk and N.R. Rimboi. American Technical Society.

Latest Standards on Terminology and Symbols, Instrument Society of America.

The Condensed Handbook of Measurement and Control, 1976. N.E. Battikha. Instrument Society of America.

Measurement and Control Basics, 2002. T.A. Hughes. Instrumentation Society of America.

Figure Credit

Rosemount, Inc. 204F03

NCCER CURRICULA — USER UPDATE

NCCER makes every effort to keep its textbooks up-to-date and free of technical errors. We appreciate your help in this process. If you find an error, a typographical mistake, or an inaccuracy in NCCER's curricula, please fill out this form (or a photocopy), or complete the on-line form at **www.nccer.org/olf**. Be sure to include the exact module ID number, page number, a detailed description, and your recommended correction. Your input will be brought to the attention of the Authoring Team. Thank you for your assistance.

Instructors – If you have an idea for improving this textbook, or have found that additional materials were necessary to teach this module effectively, please let us know so that we may present your suggestions to the Authoring Team.

NCCER Product Development and Revision

13614 Progress Blvd., Alachua, FL 32615

Email: curriculum@nccer.org
Online: www.nccer.org/olf

❑ Trainee Guide ❑ AIG ❑ Exam ❑ PowerPoints Other ______________________________

Craft / Level: Copyright Date:

Module ID Number / Title:

Section Number(s):

Description:

Recommended Correction:

Your Name:

Address:

Email: Phone:

Module 12205-03

Detectors, Secondary Elements, Transducers, and Transmitters

COURSE MAP

This course map shows all of the modules in the second level of the Instrumentation curriculum. The suggested training order begins at the bottom and proceeds up. Skill levels increase as you advance on the course map. The local Training Program Sponsor may adjust the training order.

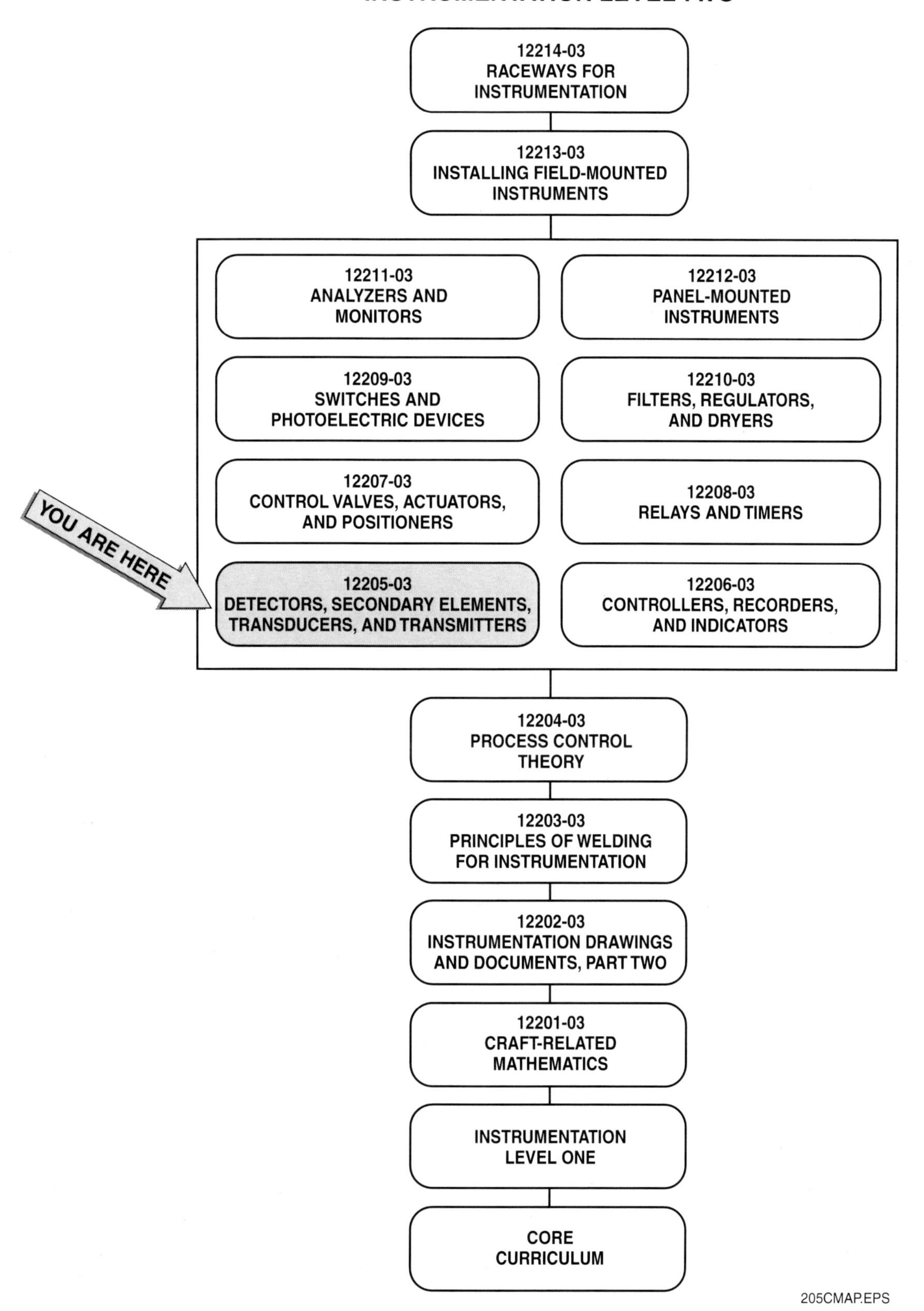

205CMAP.EPS

MODULE 12205-03 CONTENTS

1.0.0 INTRODUCTION 5.1
1.1.0 Review of Basic Instrument Control Channels 5.1
1.1.1 Detector (Sensor) 5.3
1.1.2 Transducer/Converter 5.4
1.1.3 Amplifier (Signal Conditioner) 5.4
1.1.4 Transmitter 5.4
1.2.0 Review of Measurement Terminology 5.4
1.2.1 Accuracy 5.4
1.2.2 Precision Versus Accuracy 5.5
1.2.3 Measurement Errors 5.6
1.2.4 Reproducibility and Drift 5.7
1.2.5 Sensitivity and Responsiveness 5.8
1.3.0 Standards and Elements of Measurement 5.8
1.3.1 Direct Versus Inferred Measurements 5.8
1.3.2 Measurement Standards 5.9
1.3.3 Primary Standards 5.9
1.3.4 Secondary Standards 5.9
1.3.5 Working Standards 5.9
1.3.6 Primary and Secondary Elements 5.10
1.3.7 Calibration 5.10
1.3.8 Significant Figures 5.10

2.0.0 DETECTORS 5.12
2.1.0 Orifice Plates 5.12
2.2.0 Venturi Tubes 5.15
2.3.0 Pitot Tubes 5.16
2.4.0 Annubar Tubes 5.17
2.5.0 Magnetic Flowmeters 5.17
2.6.0 Ultrasonic Flowmeters 5.18
2.7.0 Capacitance-Type Level Detectors 5.21
2.8.0 Ultrasonic Level Measurement 5.25
2.9.0 Nuclear Level Detection 5.26
2.10.0 Bimetallic Strip Thermometers 5.26
2.11.0 Thermocouples 5.28
2.11.1 Thermocouple Metals 5.28
2.11.2 Designations for Thermocouple Wire 5.28
2.11.3 Thermocouple Construction 5.29

3.0.0 SECONDARY ELEMENTS 5.30
3.1.0 Bourdon Tubes 5.31
3.1.1 C-Type Bourdon Tubes 5.31
3.1.2 Spiral-Type Bourdon Tubes 5.32

3.1.3 Helical-Type Bourdon Tubes . . . 5.32
3.2.0 Diaphragm Pressure Devices . . . 5.32
3.3.0 Pressure Capsules . . . 5.33
3.4.0 Bellows Pressure Devices . . . 5.33
3.5.0 Capacitance-Type Pressure Sensors . . . 5.34
3.6.0 Secondary Element Protection . . . 5.34
3.6.1 Diaphragm Seals . . . 5.34
3.6.2 Pulsation Dampeners . . . 5.35
3.6.3 Pressure Sensor Positioning . . . 5.35
4.0.0 TRANSDUCERS . . . 5.35
4.1.0 Transducer Functions . . . 5.35
4.2.0 Transducer Types . . . 5.35
4.3.0 I/P Transducers . . . 5.35
4.4.0 P/I Transducers . . . 5.36
4.5.0 Transducer Operation . . . 5.36
4.5.1 I/P Transducer Operation . . . 5.36
4.5.2 P/I Transducer Operation . . . 5.36
4.6.0 Metallic Strain Gauges . . . 5.38
4.6.1 Semiconductor Strain Gauges . . . 5.39
4.7.0 Pressure Strain Gauges . . . 5.39
4.7.1 Voltage-Divider Pressure Transducers . . . 5.40
4.8.0 Piezoelectric Transducers . . . 5.40
4.9.0 Linear-Variable Differential Transformers . . . 5.40
4.10.0 Accelerometers . . . 5.41
5.0.0 TRANSMITTERS . . . 5.41
5.1.0 Pneumatic Transmitters . . . 5.41
5.1.1 Force Balance Differential Pressure Pneumatic Transmitters . . . 5.41
5.1.2 Process Measuring Section . . . 5.42
5.1.3 Force Bar Section . . . 5.42
5.1.4 Balancing Section . . . 5.42
5.1.5 Pneumatic Input/Output Section . . . 5.44
5.1.6 Pneumatic Force Balance Transmitter Applications . . . 5.44
5.1.7 DP Cell Flow Measurement . . . 5.44
5.1.8 DP Cell Liquid Level Measurement . . . 5.45
5.1.9 DP Cell Pressure Measurement . . . 5.46
5.1.10 Pneumatic Force Balance Temperature Measurement . . . 5.46
5.1.11 Motion Balance Pneumatic Transmitters . . . 5.46
5.1.12 Measuring Section . . . 5.46
5.1.13 Link and Flapper-Nozzle Section . . . 5.46

5.1.14 Bellows and Relay Section 5.47
5.1.15 Applications of Motion Balance Transmitters 5.48
5.2.0 Electronic Transmitters 5.48
5.2.1 Force Balance Differential Pressure Electronic Transmitters 5.48
5.2.2 Variable Capacitance Cell Differential Pressure Electronic Transmitters 5.49
SUMMARY 5.50
REVIEW QUESTIONS 5.51
GLOSSARY 5.53
REFERENCES & ACKNOWLEDGMENTS 5.55

Figures

Figure 1 Basic instrument and process control channels 5.2
Figure 2 Expressing accuracy 5.5
Figure 3 Error due to parallax 5.6
Figure 4 Typical hysteresis loop 5.7
Figure 5 Typical hysteresis loop plot 5.8
Figure 6 Orifice plates 5.12
Figure 7 Concentric orifice plate with vent and drain holes 5.13
Figure 8 Flange and corner tap locations 5.14
Figure 9 Vena contracta and radius tap locations 5.14
Figure 10 Pipe tap locations 5.15
Figure 11 Orifice tap locations 5.15
Figure 12 Venturi tube 5.15
Figure 13 Eccentric venturi 5.16
Figure 14 Pitot tube 5.16
Figure 15 Pitot-venturi tube 5.16
Figure 16 Annubar tube 5.17
Figure 17 Right-hand rule for induced EMF 5.17
Figure 18 Magnetic flowmeter components 5.18
Figure 19 Piezoelectric crystal deformation 5.19
Figure 20 Ultrasonic flowmeters 5.20
Figure 21 Ultrasonic flow measurement 5.21
Figure 22 Frequency shift ultrasonic flow measurement 5.21
Figure 23 Capacitors 5.21
Figure 24 Bare capacitance probe 5.23

Figure 25 Insulated capacitance probe 5.23
Figure 26 Capacitance level detector circuit 5.24
Figure 27 Concentrically shielded probe 5.25
Figure 28 Ultrasonic level measurement 5.25
Figure 29 Subsurface ultrasonic level detection 5.26
Figure 30 Bimetal sensing elements 5.27
Figure 31 Thermostatic switches 5.27
Figure 32 Industrial bimetallic thermometer using helical element 5.27
Figure 33 Seebeck's thermocouple circuit 5.28
Figure 34 Various thermocouple joints 5.30
Figure 35 Sheathed thermocouple 5.30
Figure 36 C-type Bourdon tube 5.31
Figure 37 Pressure gauge 5.31
Figure 38 Spiral-type Bourdon tube 5.32
Figure 39 Helical-type Bourdon tube 5.32
Figure 40 Basic diaphragm gauge 5.33
Figure 41 Stacked pressure capsule gauge 5.33
Figure 42 Basic bellows 5.34
Figure 43 Capacitance pressure sensor 5.34
Figure 44 Diaphragm seal 5.35
Figure 45 Typical I/P transducer mounting 5.36
Figure 46 Typical P/I transducer mounting 5.36
Figure 47 Relationship between coil movement and flapper movement 5.37
Figure 48 Operation of I/P transducer 5.37
Figure 49 Arrangement of typical P/I transducer 5.38
Figure 50 Relationship of bellows and coils 5.38
Figure 51 Location of regulator in a P/I transducer 5.38
Figure 52 Application of a strain gauge 5.39
Figure 53 A resistance bridge circuit for measuring strain 5.39
Figure 54 Application of the strain gauge for pressure measurement 5.40
Figure 55 Voltage-generating (piezoelectric) transducer in a pressure-measuring system 5.40
Figure 56 Linear-variable differential transformer schematic and linear function 5.41
Figure 57 Linear-variable differential transformer used to measure acceleration 5.41
Figure 58 Pneumatic DP cell 5.42
Figure 59 Transmitter measuring section 5.42
Figure 60 Force bar section 5.43

Figure 61 Effect of diaphragm movement on flapper-nozzle position 5.43
Figure 62 Balancing section 5.43
Figure 63 Pneumatic input/output section 5.44
Figure 64 Pneumatic relay 5.44
Figure 65 DP cell flow installation 5.45
Figure 66 DP cell liquid level applications 5.45
Figure 67 DP cell used for pressure measurement 5.46
Figure 68 Force balance temperature transmitter 5.46
Figure 69 Motion balance transmitter .5.47
Figure 70 Common motion balance measuring elements 5.47
Figure 71 Link and flapper-nozzle assembly 5.47
Figure 72 Bellows and relay assembly 5.48
Figure 73 Motion balance temperature measurement installation 5.48
Figure 74 Motion balance pressure measurement installation 5.48
Figure 75 Electronic force balance DP transmitter 5.49
Figure 76 Sensor assembly 5.49
Figure 77 Output section 5.50
Figure 78 Variable capacitance cell differential pressure electronic transmitters 5.50

Tables

Table 1 List of Parameters Monitored and Typical Detectors 5.3
Table 2 Mechanical and Electrical Measuring Elements 5.11
Table 3 Dielectric Constants for Common Materials 5.22
Table 4 Symbols for Indicating Positive and Negative 5.29
Table 5 Color Coding of Thermocouple Wire 5.29
Table 6 Common Transducer Types .5.35

MODULE 12205-03

Detectors, Secondary Elements, Transducers, and Transmitters

Objectives

When you have completed this module, you will be able to do the following:

1. Identify the following primary elements (detectors) and describe their operation:
   - Orifice plate
   - Pitot tube
   - Bimetallic strip device
   - Thermocouple
2. Identify the following secondary elements and describe their operation:
   - Bourdon tube
   - Diaphragm device
   - Pressure capsule
   - Bellows device
3. Define an I/P and a P/I transducer and describe their operation.
4. Describe the operation of a strain gauge.
5. Identify a pneumatic DP transmitter and an electronic DP transmitter and describe their operation.
6. Identify the following primary components in a DP cell transmitter:
   - Process measuring section (hi and lo sides)
   - Force bar section
   - Flapper-nozzle (pneumatic only)
   - Pneumatic relay (pneumatic only)
   - Input/output sections (pneumatic and electronic)
7. Draw a one-line diagram including a measuring element, transducer, and transmitter.

Prerequisites

Before you begin this module, it is recommended that you successfully complete the following: Core Curriculum; Instrumentation Level One; Instrumentation Level Two, Modules 12201-03 through 12204-03.

Required Trainee Materials

1. Pencil and paper
2. Appropriate personal protective equipment

1.0.0 ◆ INTRODUCTION

In the module *Process Control Theory*, the basic instrument and process control channels were introduced and described in general terms. They are shown in *Figure 1*.

This module focuses on the first four blocks in each of these channels. These blocks are common to almost all instrumentation applications. It is very important that trainees have a good understanding of them and the variety of ways they can be applied.

1.1.0 Review of Basic Instrument Control Channels

The functions of each block are reviewed as follows:

- *Detector* or *sensor* – Senses the parameter being monitored (the process variable or the controlled

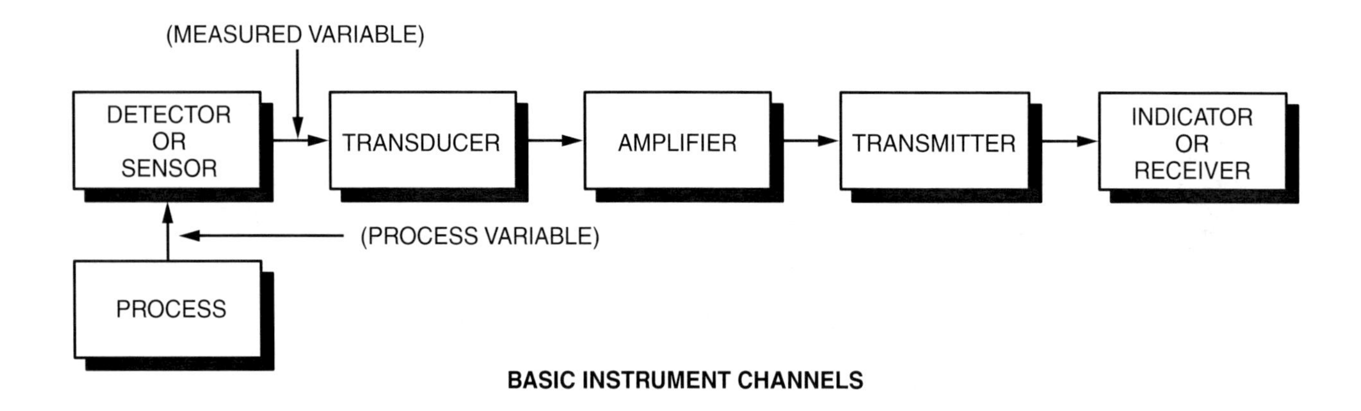

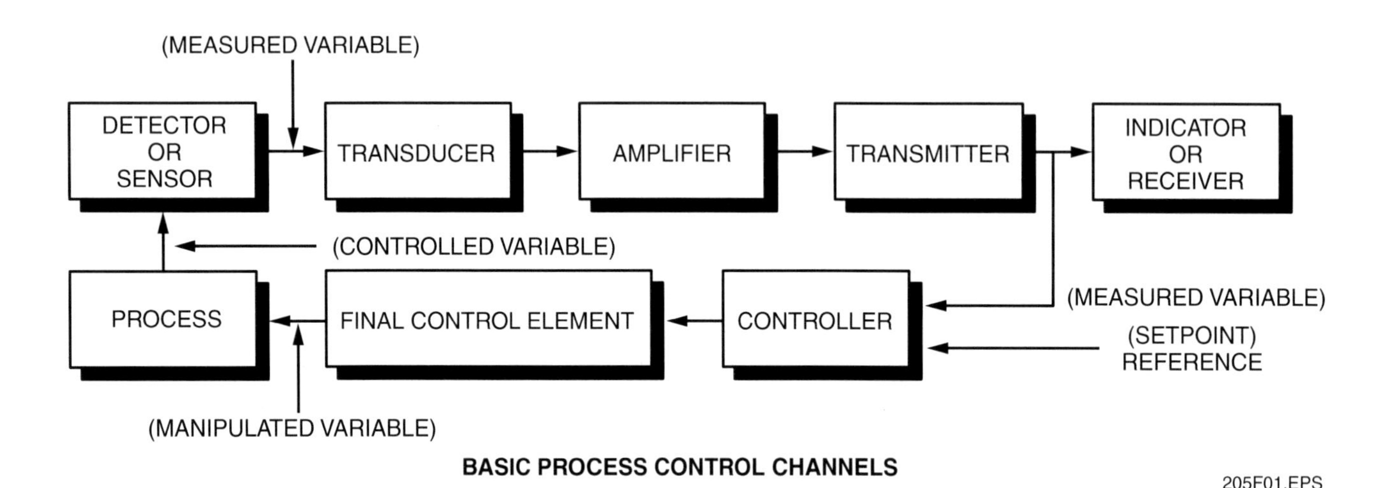

Figure 1 ◆ Basic instrument and process control channels.

variable) and changes that parameter to a mechanical or electrical signal that is proportionally related to the measured variable. This type of device is often referred to as the **primary element**, or measuring element, of the instrument channel.

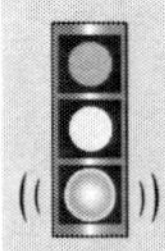

NOTE

Proper handling, storage, and protection of detectors and sensors are critical.

- *Transducer* – Converts the output signal of the detector to a signal that can be used easily. If the detector signal can be used directly, this conversion step is not needed. When it is used, it is most often found in the field in close proximity to the measuring element. In many cases, it is in the same housing or case as the measuring element and appears to be one device. Depending on the application, a transducer can be part of the primary element, part of the **transmitter**, or a stand-alone device.
- *Amplifier* – Increases the process signal to a usable **magnitude**. In many cases, signal conditioning occurs with amplification. It is common for instrument manufacturers to combine the amplification and signal conditioning functions with one of the other blocks. Amplifiers are often found as a part of the transducer or the transmitter.
- *Transmitter* – Sends data from one instrument component to another when components are physically separated. It may contain the detector, transducer, and amplifier (signal conditioning) functions.

A specific instrument channel may involve these basic components in any number and any combination. They need not appear in the order of *Figure 1*, and not all of the components described may be required. The reason for such a variety of potential variations is that instrumentation manufacturers produce (and name) devices in unique ways with regard to these four basic functions. Most of the time all four functions are performed, but it is possible with modern instruments to have one or more devices do them all. The examples in this module show these variations. It is important for you to know what to expect when reading the specifications, data sheets, or vendor manuals for these instruments.

1.1.1 Detector (Sensor)

The first contact that a measurement channel of instrumentation has with the process parameter to be measured is through the action of the detector or sensing device.

To sense the process parameter, the detector receives energy from the process and produces an output that is dependent on the measured quantity. It is important to realize that the sensing element always extracts some energy from the process. The process is always disturbed by the act of measurement. This effect is referred to as loading. A good instrument is designed to minimize this loading or disturbance of the parameter being measured. In measuring systems made up of mostly electric or electronic components, the loading of the signal source (process variable) is almost exclusively a function of the detector. Other components in the electronic instrument channel receive most of the energy or power they need from power supplies independent of the process itself. This is one major advantage of electronic measurement and control channels.

Detectors are also selected for measurement systems on the basis of the parameter being sensed, the desired accuracy, the range of measurement, and the particular type of output supplied. *Table 1* lists some typical detectors, parameters monitored, and detection principles. These detection principles and associated detectors are discussed in greater detail in the following sections of this module.

Detectors measure process variables such as pressure, temperature, fluid level, and liquid flow. As *Table 1* suggests, the most common detector output is a very small **displacement** or distance moved that is proportional to the measure of the process variable. The detector's output is usually not directly usable in the control or instrument channel. Often it must be converted, amplified, or conditioned in some way before it can be used to indicate or control the process parameter.

Table 1 List of Parameters Monitored and Typical Detectors

Parameter	Detector Detection	Principle
Temperature	Resistance Temperature Detector (RTD)	Resistance of certain metals varies linearly with temperature
	Thermocouple	Two dissimilar metals produce a voltage proportional to their temperature when joined
Pressure	Expansion of a Liquid	Liquid expands proportional to an increase in temperature
	Differential Pressure Cell	The difference of pressure is measured either across a restriction added into the process or between the process and atmospheric pressure
	Bourdon Tube	A hollow, arc-shaped tube that is mechanically connected to some type of linkage expands or retracts when internal pressure is applied or taken away, causing the linkage to follow
Level	Differential Pressure Cell	The difference of pressure is measured either across a restriction added into the process or between the process and atmospheric pressure
	Float	A material that is less dense than the material being monitored will float on the surface and follow changes in the level
Flow	Flow Restrictor Combined with a Differential Pressure Cell	The difference of pressure is measured across a restriction added into the process

1.1.2 Transducer/Converter

The function of the transducer is to convert the detector output to a form that can be processed by the instrument channel. The transducer is usually connected to the primary detecting element. Sometimes it is part of the primary element. Once the measured variable output of the transducer is converted to some usable form, it can be manipulated by the instrument channel components as necessary without loading the process variable that produced it.

The purpose of the transducer is to convert any input it receives to another type of output signal more readily usable by the next component or portion of the instrument or control channel. While this conversion is obviously necessary at the output of the detector (few detectors provide outputs that are directly usable), it is also possible that a second or a third transducer may be found in some instrument and control channels. Wherever a signal conversion must occur, there is a transducer.

It is possible to have one transducer convert the detector output to a form that can be amplified while another transducer converts the amplified signal to a form that can easily be transmitted to an indicator or a controller elsewhere in the plant (in the control room, for example). It is just as likely that the controller's output might need conversion or transducing again for transmission back into the plant to operate a control element, such as a valve or heater.

It is important that trainees understand that a transducer functions to convert signals (such as mechanical, electrical, or pneumatic) from one form to another that is needed at that point in the channel. A transducer can be found anywhere in a channel, even as part of another block like the transmitter. Detector outputs may be converted for use by the channel, so in many diagrams a transducing element will also be directly connected to the primary sensing element or detector.

1.1.3 Amplifier (Signal Conditioner)

Most measured variable signals may be increased in either amplitude or magnitude so they can be used by indicators or controllers directly (or transmitted to them). The amplifier block indicates that this usually happens somewhere in the channel. Actually, it may occur several times before the signal can be used.

Separate instruments performed this function in the past. Instrument manufacturers now often include the amplification stage as part of other elements in the channel. The amplification may take place in the transducer, the transmitter, or wherever the instrument manufacturer finds it most feasible. Often, if other conditioning or modification of the signal is required, amplification is also performed.

An amplifier block, therefore, may not appear on instrument channel diagrams, but it is likely that amplification is performed somewhere in the channel.

1.1.4 Transmitter

The transmitter is a device that originally was very specific in function. It created a signal to be sent from one location to another, usually involving some distance. The definition of the term *transmitter* has since grown to encompass that primary function and all of the others listed so far. It usually does not include the detector, but it could.

Again, this has evolved due to new technology in the instrument industry. Often, a detector is connected to the input side of a transmitter, with the output side connected to the final indicator or controller.

The transmitter very often contains the transducing element, amplifiers, and a signal-conditioning element. It frequently provides the output in a form ready for direct transmission to a remote location.

1.2.0 Review of Measurement Terminology

The following terms apply to the blocks in the basic instrument channel. Some of the terms have been presented in previous modules. Many of these terms are used in conjunction with each other in order to describe the properties of a process or an instrument in a comparative manner, such as **precision** versus accuracy.

1.2.1 Accuracy

Accuracy is the degree to which the output of an instrument approaches an accepted standard or true value. There is no absolute accuracy. However, a given detector may be said to be accurate to within ±0.1%. As the definition states, the output of a device is compared or referenced to some value or standard to determine whether the instrument is performing as required. Therefore, when used as a performance specification for an instrument, accuracy means reference accuracy.

Reference accuracy is a number or quantity that defines the limits that errors will not exceed when the device is used under referenced conditions. Reference accuracy can be expressed in a number of ways:

- It can be expressed in terms of the measured variable. For a temperature measuring device, the reference accuracy may be expressed as ±1°F.
- Reference accuracy can be expressed in percent of span. This can be explained by using the following example: A meter is used to indicate the water level in a tank between 50 inches and 150 inches. The reference accuracy of the indicator is ±0.5% of span. Therefore, the reference accuracy of the indicator is ½ inch of level because the span is 100 inches.
- Reference accuracy can be expressed in percent of the upper range value. If the upper range value of a pressure gauge is 100 psi, and the reference accuracy is ±0.1% of upper range value, the reference accuracy of the gauge would be 0.1 psi.
- It can be expressed in percent of scale length. For an indicating meter with a 6-inch scale length and a reference accuracy of ±1.2% of scale length, the reference accuracy would be ±0.072 inches (about ±5⁄64 inch).
- Reference accuracy can be expressed in percent of actual output reading. If the 50-inch to 150-inch level indicator discussed previously has a reference accuracy of ±1% of actual reading, and the tank presently has 125 inches of water, the indicator should be reading 125 inches ±1.25 inches.

When stating the accuracy of an instrument, it is very important to express the quantity to which the accuracy is referenced. To say that a component is accurate to within 0.1% is meaningless. The percent specification must be related to some specific magnitude. See *Figure 2* for an example of expressing accuracy in various ways.

1.2.2 Precision Versus Accuracy

The precision of an instrument is its measurement of repeatability. If a detector or sensor monitoring an unchanging process continues to produce results that are within close tolerances of one another, this demonstrates precision. However, it does not necessarily indicate the accuracy of the instrument.

Accuracy is the measure of how close a result is to the true value. Accuracy is usually expressed as the percent of a reference. An instrument can have precision without accuracy, but it cannot generally be accurate without having precision. Repeatability is one element of accuracy.

If two seemingly identical detectors are installed side-by-side in a process, it is possible that each will provide a different value for the measured variable. In fact, it is very common. There are a number of reasons why this may happen. The two detectors may have been installed in a slightly different manner, they may have been adjusted or **calibrated** in different ways, or one may be exhibiting more wear due to internal friction. Both instruments may be precise, but one may be more accurate than the other. Accurate measurement is the combined effort of the human input to the instrument, such as installation and **calibration**, with the built-in elements of accuracy within the instrument.

Example: A temperature-measuring instrument produces an output with a range of 180°F to 320°F, which is displayed on a meter with a scale length of 4 inches. The accuracy of this meter is expressed below in each of the five ways.

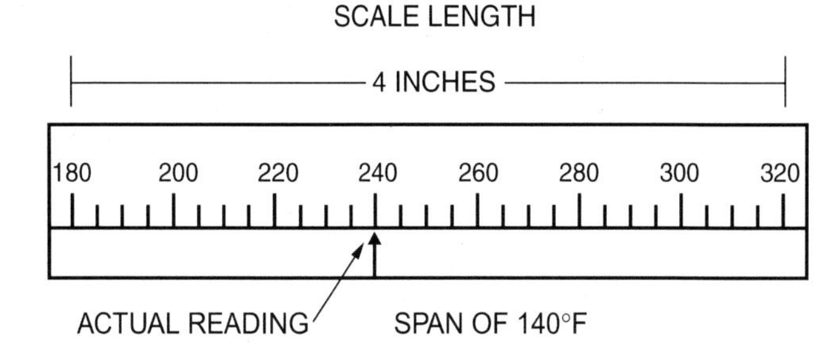

1. Expressed in terms of measured variable:
   Accuracy = ±2°F
2. Expressed in terms of percent of span:
   Accuracy = ±2% of span
   = (0.02 × 140°F)
   = ±2.8°F
3. Expressed in terms of percent of upper-range value:
   Accuracy = 0.5% of upper-range value
   = (0.005 × 320°F)
   = ±1.60°F
4. Expressed in terms of percent of scale length:
   Accuracy = ±2% of scale length
   = (0.02 × 4 inches)
   = ±0.08 inches
   = approximately ±5⁄64 inch
5. Expressed in terms of percent of actual reading:
   Accuracy = ±0.5% of actual reading
   = (0.005 × 240°F)
   = ±1.2°F

205F02.EPS

Figure 2 ◆ Expressing accuracy.

When monitoring and recording measurements to determine precision and accuracy, it is standard procedure to make and record a series of observations rather than relying on only one value or reading. This is the reason a technician should perform instrument verifications frequently. It is one reason why it is necessary to check and record the performance values of a device when it is first used and again when work with it has been completed. This is the only way to ensure that the accuracy of a device is acceptable within the limits specified.

1.2.3 Measurement Errors

A **measurement error** is the numerical difference between the measured value and the true value. In instrumentation terminology, the error of a detector or sensor is the difference between the measured variable and the actual process variable. For example, the error of a barometric pressure measurement might be –1 mm if the barometer reads 758 mm, while a more accurate measure indicates the true pressure is represented by 759 mm. The actual error is given by:

$$\begin{aligned}\text{Error (E)} &= \text{measured value} - \text{true value}\\ &= 758\text{ mm} - 759\text{ mm}\\ &= -1\text{ mm}\end{aligned}$$

While is it necessary to know the actual error, it is often more useful to convert it into an indication of how accurate the measurement is, or the accuracy of the instrument. To do this, we convert the actual numerical error value into a **relative error** value, which more closely expresses the degree of accuracy of the instrument. Relative error is defined as the ratio of the actual error value to the true value. In this instance it would be $\frac{1}{759}$. Multiplied by 100%, this expresses the relative error as a percentage of the true value:

$$\text{Relative error} = \frac{|\text{measured value} - \text{true value}|^*}{|\text{true value}|^*} \times 100\%$$

$$\text{Relative error} = \frac{|\text{error}|^*}{|\text{true value}|^*} \times 100\%$$

$$\begin{aligned}\text{Relative error} &= \frac{1}{759} \times 100\%\\ &= .00132 \times 100\%\\ &= .132\%\end{aligned}$$

* Bars denote the absolute value of the number.

The relative error is always expressed as a positive number, as the absolute-value notation signifies. Accuracy, then, is determined by the magnitude of the relative error. To obtain the various reference forms of the expression, the true value must be expressed in those reference values.

Errors in general measurement work are classified as:

- Personal errors
- Random errors
- Systematic errors
- Application errors

Personal errors are those caused by carelessness, lack of experience, and/or bias on the part of the worker. Carelessness is a common factor in many errors. Misreading a measurement can be due to carelessness or the lack of experience of the worker. A common error made as a result of inexperience is **parallax** error. Parallax occurs when a worker is not experienced enough to know that the measured value will change with the relative position of the eye when reading an analog meter. It is important to be careful in order to avoid this type of error. Parallax is demonstrated in *Figure 3*.

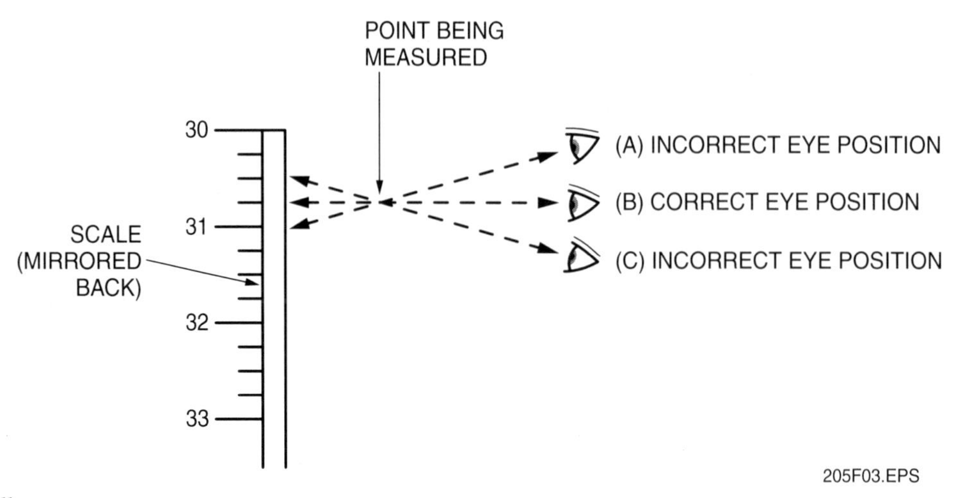

Figure 3 ◆ Error due to parallax.

Other errors due to carelessness or inexperience might be mathematical errors. They could also be caused by not knowing where to find or verify something in a procedure, technical manual, or specification sheet.

Bias results in a more subjective type of error. It usually is caused by having some preconceived notion or expectation of the magnitude or quality of the variable under measurement or of the device measuring it. Data are selected to substantiate the results or to make the job easier, rather than being accepted with equal confidence or objectivity. A common example of bias occurs when a technician determines that a borderline measurement value on a calibration check is within specifications because he or she would have to go back and recalibrate the device if one measurement point is out of line. The instrument will **drift** further out of tolerance and will need to be recalibrated anyway. It is also possible that a lot of wasted product could be produced in the meantime, or a plant could be operating less efficiently for the next year, possibly contaminating the environment. Bias must be controlled by the individual worker.

A common method of overcoming personal errors is to have readings made by more than one worker, but this can be time consuming and expensive.

Random errors occur when repeated measurements of the same quantity result in different values. The errors probably exist, but they are considered indeterminate.

Systematic errors are considered built-in errors that result from the characteristics of the materials used in manufacturing the instrumentation systems. Systematic errors are caused by such things as the natural inertia of moving parts, hysteresis, friction, and backlash in gearing. Inaccuracies arising from such causes are generally uniform or consistent in character. These errors are repeatable and result in the typical hysteresis loop of an instrument, which can be demonstrated by plotting incremental increases in first ascending and then descending steps. *Figure 4* shows a typical hysteresis loop. *Figure 5* shows this same loop plotted as a statement of continuous percent accuracy.

These errors are normally small enough to tolerate and actually make up the majority of the tolerance specified by the manufacturer as the percent accuracy of the instrument.

Application errors occur through the improper use or faulty installation of an instrument. Application errors can be minimized by following the manufacturer's and design specifications for installation and use and general rules of industrial good practice standards, such as those of the ISA.

1.2.4 Reproducibility and Drift

Reproducibility is the degree of closeness with which the same value can be measured at different times. It is usually expressed as a percentage of span of the instrument. Perfect reproducibility indicates that an instrument or instrument channel has no drift.

Drift is a gradual separation of the measured value from the calibrated value. This usually occurs over a long period of time during which the value of the variable is assumed not to change. Drift can be caused by permanent setting of the mechanical or physical components of the detector or instrument, stress on the equipment parts,

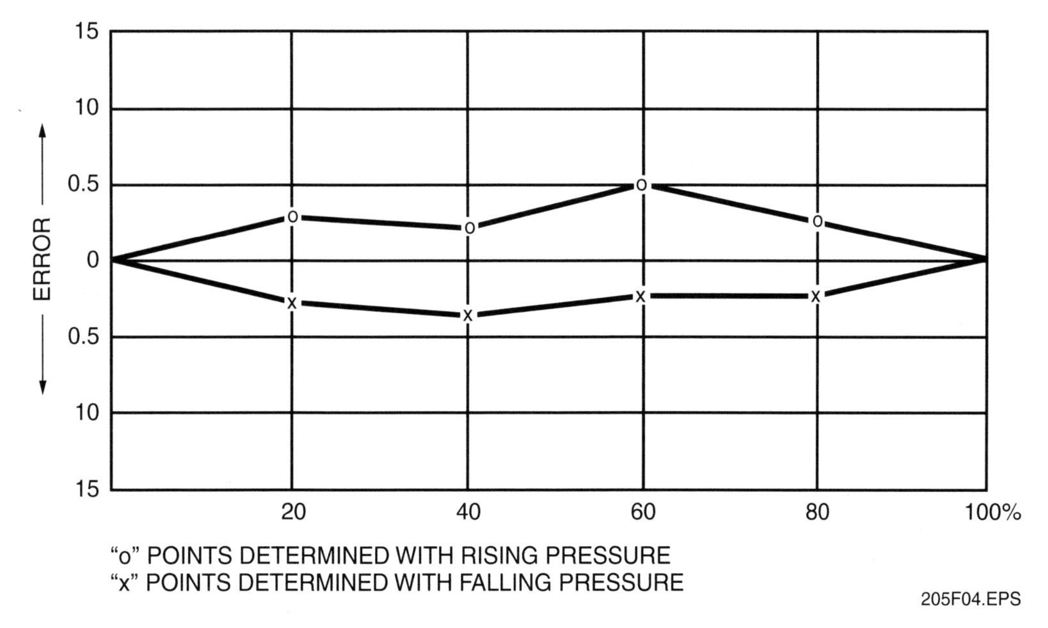

Figure 4 ◆ Typical hysteresis loop.

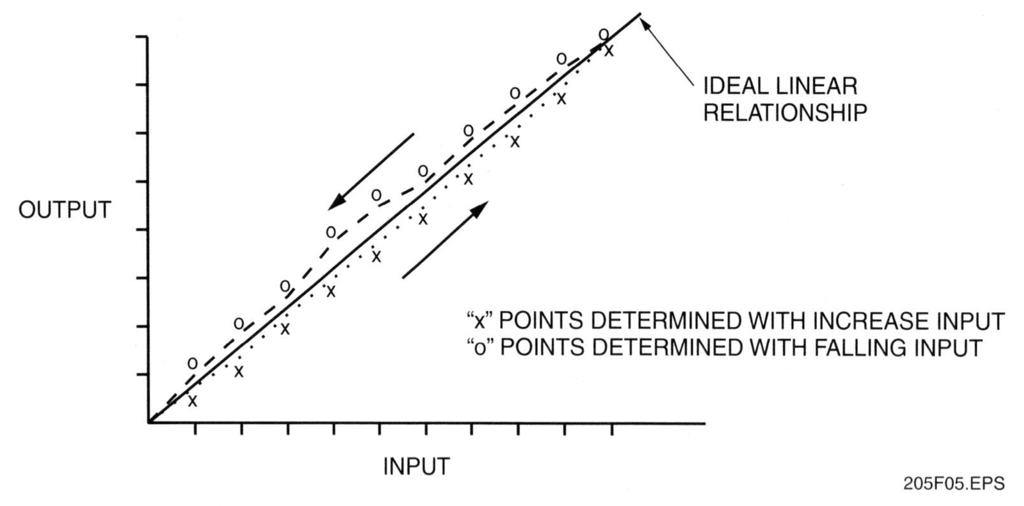

Figure 5 ◆ Typical hysteresis loop plot.

or fatigue in the metals or other construction materials. Alternatively, drift can be due to wear, erosion, or general deterioration due to age.

Drift is a primary cause for the need to recalibrate instruments. When maintaining instrumentation, a good way to plan normal calibration cycles of instrument channels is to keep a record of the drift accruing with respect to time. In this way, the instrument channel most likely to need recalibration can be anticipated.

1.2.5 Sensitivity and Responsiveness

The sensitivity of a device is the ratio of a change in output magnitude to the change of input that causes it, after steady state has been reached. It is a ratio that describes how much the input variable must change to produce some change in output magnitude. The sensitivity of a device is an important property that is determined or set by the designer based on the requirements of the application. The required sensitivity is decided by the design engineer based on the smallest change needed to be measured in the given process variable.

Sensitivity and **responsiveness** are frequently confused. Responsiveness is the amount of change in the process variable needed to cause a perceptible change or movement in the measured variable, or the amount of input change that can cause the output to start to change in any device.

A thermometer is said to be sensitive to 0.1°C when the thermometer will respond to a change of ±0.1°C.

For example, a pressure detector at 200 psi requires a change of ±2 psi to cause a change to be perceived at its output.

An evaluation of the percent responsiveness of the pressure detector would be:

$$\text{Responsiveness} = \frac{\text{change in input (pressure)}}{\text{value of output (pressure)}} \times 100$$

$$\text{Responsiveness} = \frac{2\text{ psi}}{200\text{ psi}} \times 100 = 1.0\% \text{ at } 200\text{ psi}$$

The value of responsiveness may vary throughout the range of the detector, just as accuracy can. Responsiveness may be improved through proper lubrication and adjustment of the instrument.

1.3.0 Standards and Elements of Measurement

The purpose of measurement is to determine the value of a quantity, condition, physical parameter, or phenomenon. A measuring instrument is simply a device used to sense and relay that value to another device that processes the information. The value determined by the instrument is generally, but not necessarily, **quantitative**. For the measurement to be really useful it must be reliable and accurate. To ensure this, the measuring instruments must always be functioning to compare the process variable being sensed to the equivalent of a known measure or standard.

1.3.1 Direct Versus Inferred Measurements

The wide variety of measurements made in industrial plants, as well as the varying environmental conditions under which these measurements must be made, make an understanding of the basic nature of measurement important. Measurements, for this discussion, fall into two general categories: those measurements that are directly made and those that are inferred.

The length of this page could be measured using a ruler or metric rule. The length of the page would be compared to the measurement marks or increments on the rule. Using the English system ruler, the measurement would probably be accurate to within ⅛ of an inch. Using the metric system rule (or meter stick) it would probably be accurate to within 1–3 mm. This is a direct measurement. The length of the paper has been determined by direct comparison of that parameter to a known or acceptable standard.

Although measuring by direct comparison is the simplest method, direct measurement is not always adequate or possible. For one thing, the human senses are not prepared to make direct comparisons of all quantities with equal facility. In many cases they are not sensitive enough. Human beings can make direct comparisons of small distances using a rule, with an accuracy of about 1 mm (approximately 0.04 in). Often, however, measurements require greater accuracy. Human senses just don't detect, in a quantitative way, what must be measured. Examples of such measurements include pressure, temperature, and flow. To measure these, a more complex form of measurement system is necessary. Direct measurement is much less common than one might think.

Inferred measurement occurs when there is an indirect comparison. The change in the process variable affects a characteristic or property of the material of the measuring system or detector. It is the change in the detector that is actually being measured. For example, when a temperature exists at the junction of a thermocouple, a voltage is produced. The voltage is the actual parameter being measured, and the temperature is inferred or derived from the characteristic voltage measured. Again, it is important that inferred relationships be referenced to known standards.

1.3.2 Measurement Standards

A standard is an accurate known quantity used for calibration of measurement instruments. A standard can also be an instrument of high accuracy.

Standards exist for every type of measurement. The ultimate set of standards is maintained by the U.S. National Bureau of Standards (NBS). On occasion, standards used for calibration purposes are sent to the NBS. The NBS checks these standards against their own standards for accuracy. These standards are then said to be traceable back to the NBS.

The three basic levels of measurement standards are:

- Primary or absolute standards
- Secondary reference standards
- Working standards

1.3.3 Primary Standards

Primary or absolute standards are constructed to conform to the legal definitions of different fundamental units of measurement. Examples of primary standards are the standard meter and a set of precision weights for ounces and pounds. Another example is a set of containers that hold precise amounts of liquids for liters, pints, quarts, and gallons. The term *absolute* is used to indicate that these measures are finite and are, therefore, independently accurate and correct. Other types of standard measures are ultimately traceable to these. Traceability is an important quality of instrumentation in high-tech and hazardous industrial applications. Primary standards are traceable to national standards, such as NBS.

1.3.4 Secondary Standards

Secondary reference standards are devices that are copied from existing primary or absolute standards. These standards are sometimes referred to as prototype standards. Often these standards are maintained accurate and traceable to the NBS primary standards by special companies that set up regional standards laboratories. Instead of sending measuring equipment to the NBS in Washington, D.C., for example, technicians can send it periodically to a regional laboratory. There the equipment is checked for accuracy against the regional laboratory's secondary standards.

Some construction and operating companies maintain their own secondary standards laboratory on site.

1.3.5 Working Standards

Working standards are the standards that are used to calibrate the instruments installed in actual systems. A detector's or sensor's accuracy is thereby traceable through local working standards to secondary reference standards and ultimately to the primary standards at the NBS.

This system of traceability is what ensures the reliable, safe functioning of industrial plant instrumentation. The paperwork associated with the calibration and adjustment of instrumentation in the field is the documentation of proper installation and setup of plant systems. In many industries, this paperwork is just as important as the actual installation and calibration of the equipment. Without it, technicians would be unable to assure government regulators, plant operators and owners, and the public that the plant is being operated safely and in compliance with environmental guidelines or regulations.

1.3.6 Primary and Secondary Elements

All instrument channels contain various component parts or elements that perform the prescribed measurement, conversion (transducing), conditioning (amplification) and transmitting functions described previously.

The primary sensing element is that part of the instrument or channel that first uses energy from the measured medium to produce a condition or signal that represents the value of the measured medium (process variable). In most cases, an industrial primary element converts the measured variable into a displacement. Often this mechanical displacement is converted by a **secondary element** into an electrical or electronic signal. *Table 2* lists common mechanical and electrical primary elements and the operations they typically perform.

The primary element may be very simple, consisting of no more than a mechanical spindle, arm, or contacting member used to provide movement or force to the secondary element. It may also be much more complex.

The purpose of the primary element is to sense the quantity of interest and to process the sensed information into a form that is usable by the instrument channel. It usually does not present a usable output that can be directly applied as indication or control of equipment.

Table 2 is only a partial listing. Many of these elements will be described in more detail later. It is significant, however, to note that many of the mechanical elements produce an output that is a physical displacement, which is easily converted (or transduced) by a secondary element to an electrical signal. Many electrical elements can provide a usable electrical signal output. Electrical elements have several important advantages:

- Amplification and signal conditioning is easily performed.
- **Mass-inertia effects** are minimized.
- Friction is minimized.
- An output power signal of almost any magnitude can be provided.
- Electronic transducer/transmitter combinations can often be miniaturized.

Displacement and force can also be converted to pneumatic signals for processing and transmission. Although this requires more components, it has a major advantage. A loss of electrical power will not immediately incapacitate the instruments, due to the storage of air pressure.

1.3.7 Calibration

Every measuring system must be provable. That is, it must prove its ability to measure reliably. The procedure for establishing an instrument's accuracy is called calibration.

Calibration is the testing of the validity of the measurements performed by an instrument in normal operation. This is done through comparison with measurements made by primary, secondary, or working standards. It is essential that the instrument or device function at a high degree of accuracy and reliability in order to achieve quality performance.

At some point after the installation of instruments and systems, known magnitudes of the process variable or input parameter must be applied to the detector as well as the entire instrument channel. The operation must be observed and recorded. The instrument manufacturer will provide sample procedures for calibrating the devices. This information must often be combined with reliable industrial practices and facility procedures for testing and calibrating system installations.

Application of a tag or sticker to an instrument or detector means that it has been tested and adjusted in accordance with accepted procedures and practices. It is certified as calibrated to the degree of accuracy recorded in accompanying documentation. This means that the accuracy of the instrument is valid and traceable to a known standard.

1.3.8 Significant Figures

The documentation of calibration is very important. It is important that the values or numbers used in the calibration records be consistent with the sensitivity range of the instrument and testing equipment being used.

In writing a measured value as a series of digits, some of these digits will have an element of doubt associated with them. The total number of significant figures used to describe the result depends on several factors. If the reading is interpreted by the observer as 3.6834, the accuracy of the detector is stated as ±0.05, and the test equipment is ±0.001, the reading should be described as the least accurate limit involved (in this case, 3.68 ±0.05).

Table 2 Mechanical and Electrical Measuring Elements

ELEMENT	OPERATION
I. Mechanical	
A. Contacting spindle, pin, or finger	Displacement to displacement
B. Elastic member	
1. Load cells	
a. Tension/compression	Force to linear displacement
b. Bending	Force to linear displacement
c. Torsion	Torque to angular displacement
2. Proving ring	Force to linear displacement
3. Bourdon tube	Pressure to displacement
4. Bellows	Pressure to displacement
5. Diaphragm	Force to linear displacement
6. Helical spring	Pressure to displacement
7. Liquid column	
C. Mass	
1. Seismic mass	Forcing function to relative displacement
2. Pendulum	Gravitational acceleration to frequency or period
3. Pendulum	Force to displacement
4. Liquid column	Pressure to displacement
D. Thermal	
1. Thermocouple	Temperature to electric current
2. Bimaterial (includes mercury in glass)	Temperature to displacement
3. Thermistor	Temperature to resistance change
E. Hydropneumatic	
1. Static	
a. Float	Fluid level to displacement
b. Hydrometer	Specific gravity to relative dsplacement
2. Dynamic	
a. Orifice	Fluid velocity to pressure change
b. Venturi	Fluid velocity to pressure change
c. Pitot	Fluid velocity to pressure change
d. Vane	Velocity to force
e. Turbine	Linear to angular velocity
II. Electrical	
A. Resistive	
1. Contacting	Displacement to resistance change
2. Variable-length conductor	Displacement to resistance change
3. Variable-area conductor	Displacement to resistance change
4. Variable dimensions of conductor	Strain to resistance change
5. Variable resistivity of conductor	Temperature to resistance change
B. Inductive	
1. Variable coil dimensions	Displacement to change in inductance
2. Variable air gap	Displacement to change in inductance
3. Changing core material	Displacement to change in inductance
4. Changing core positions	Displacement to change in inductance
5. Changing coil positions	Displacement to change in inductance
6. Moving coil	Velocity to change in inductance
7. Moving permanent magnet	Velocity to change in inductance
8. Moving core	Velocity to change in inductance
C. Capacitive	
1. Changing air gap	Displacement to change in capacitance
2. Changing plate areas	Displacement to change in capacitance
3. Changing dielectric	Displacement to change in capacitance
D. Piezoelectric	Displacement to voltage and/or voltage to displacement
E. Photoelectric	
1. Photovoltaic	Light intensity to voltage
2. Photoresistive	Light intensity to resistance change
3. Photoemissive	Light intensity to current

205T02.EPS

2.0.0 ◆ DETECTORS

Detectors and sensors are synonymous. Both are primary elements in the instrument channel. A detector is the first system element that responds quantitatively to the measured variable and performs the initial measurement operation. It performs the initial conversion of measurement energy.

The effect produced by the primary element may be a change of pressure, force, position, electrical potential, or resistance. Detectors are components of a measurement or control system that first uses or transforms energy from a given medium to produce an effect. That effect is a function of the value of the measured variable.

2.1.0 Orifice Plates

Orifice plates are the most common type of flow-measuring element. The orifice plate is a thin, circular, metal plate with a sharp-edged hole. The orifice plate is usually mounted between two flanges. Three kinds of orifice plates are used: the concentric, the eccentric, and the segmental (*Figure 6*).

The concentric orifice plate is the most commonly used of the three types. It is usually made of stainless steel from ⅛ to ½ inch thick, depending primarily on the diameter of the pipe for which it is manufactured. Other materials such as **Monel®** or **Hastelloy®** are used for fluids corrosive to stainless steel. The plate is usually manufactured with a tab or handle on which pertinent orifice plate data is stamped, such as orifice bore or hole size.

The ratio of the orifice bore to the internal pipe diameter is called the Beta.

$$B = d/D$$

Where:

B = orifice bore to internal pipe diameter ratio
d = orifice bore (inches)
D = internal diameter of pipe (inches)

The flow pattern and the sharp leading edge of the orifice plate that produces it are of major importance to the accuracy of the flow measurement when using a concentric orifice plate. Any nicks or rounding of the sharp edge change the flow pattern significantly and affect the accuracy of the measurement. The type of flow, as reflected by the **Reynolds number**, also has a considerable influence on the flow pattern. At a low Reynolds number, when **laminar flow** occurs, the velocity profile of the fluid reveals that the greatest flow rates occur at the center of the pipe. As the fluid passes through the orifice, only a small energy conversion is required to constrict the flow. A relatively small differential pressure results. At a high Reynolds number, the velocity profile is relatively flat, so a large energy conversion takes place. A relatively large differential pressure results. The pattern obtained at high Reynolds numbers is desirable.

Concentric orifice plates should be used for clean, vapor-free liquids and condensate-free vapors and gases. In liquid flow applications, as the fluid converges in order to pass through the orifice, particles tend to drop out and collect at the bottom of the upstream face of the orifice plate. Gases and vapors tend to collect at the top of the upstream face. In gas or vapor flow applications, condensate tends to form a puddle at the bottom of the horizontal line upstream of the orifice plate. Any of these conditions changes the area of the upstream fluid stream and causes inaccurate flow measurement.

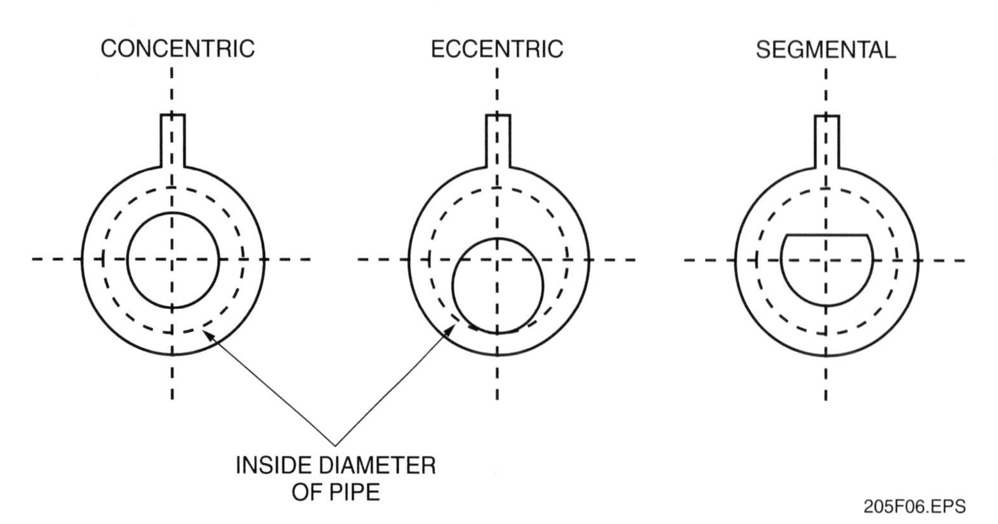

Figure 6 ◆ Orifice plates.

The collection of particles and condensate can be alleviated by drilling a small drain hole nearly flush with the inside diameter of the pipe at the bottom of the orifice plate (*Figure 7*). This small drain hole also permits drainage of a horizontal pipe that contains the orifice. The collection of gases or vapors can be eliminated by drilling a small vent hole nearly flush with the inside diameter of the pipe at the top of the orifice plate. This is also indicated in *Figure 7*.

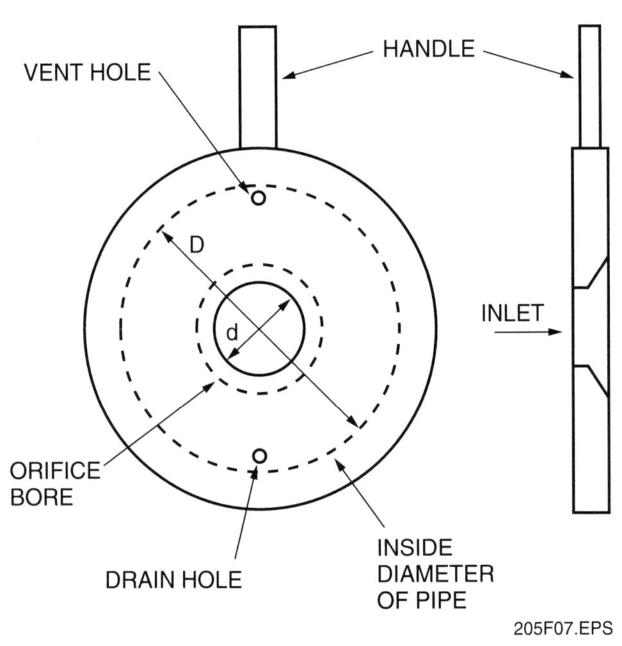

Figure 7 ◆ Concentric orifice plate with vent and drain holes.

All orifice plates, regardless of type, must be properly installed in the process. The designed upstream side of the plate must face the upstream side of the process, and the designed downstream side of the plate must face the downstream side of the process. As shown in side view in *Figure 7*, the downstream side of the orifice plate is **chamfered** to allow the process to immediately fill the downstream side of the pipe before the process contacts the downstream tap. If the orifice plate is installed in reverse, the reading will not be accurate.

The vent and drain holes have little effect on the flow measurement because of their size. For example, if the diameter of these holes is less than one tenth of the orifice bore diameter, the maximum flow through these small holes is less than 1% of the total flow. Drain and vent holes are inadequate for liquid flow applications where large quantities of solids or gases are present. They are also inadequate for gas or vapor flow applications where large quantities of condensate are present. For these applications, segmental or eccentric orifices plate are normally used.

The opening in a segmental orifice is a segment of a circle (*Figure 6*). The diameter of the opening is 98% of the inside diameter of the pipe. The circular section of the segment should be concentric with the pipe. The segmental orifice plate is useful because it eliminates damming of foreign materials on the upstream side of the orifice when the orifice plate is mounted in a horizontal pipe.

When the flow of a liquid containing solids or of a gas containing moisture is to be measured in a horizontal pipe, the segmental orifice is used with the circular section at the bottom of the pipe. When the flow of liquid containing gases is to be measured in a horizontal pipe, the segmental orifice is used with the circular section at the top of the pipe. When flow is measured in vertical runs, however, the concentric orifice should always be used.

The eccentric orifice has a circular hole bored tangent to the inside diameter of the pipe. The diameter of the opening is 98% of the inside diameter of the pipe. The eccentric orifice plate is used in the same way as the segmental orifice plate.

There are five commonly used tap locations for measuring the differential pressure across an orifice plate. They are the flange taps, corner taps, vena contracta taps, radius taps, and pipe taps.

Flange taps are the ones most often used in the U.S. for pipe sizes of 2 inches or greater. These taps are drilled through the orifice flanges 1 inch from the surface of the orifice plate. This arrangement is shown in *Figure 8*. Flange taps are not recommended for pipe sizes less than 2 inches, because the vena contracta may be less than 1 inch from the orifice plate.

Corner taps are in common use in Europe. These taps are drilled through the flange so that they sense the pressures at the edge of the orifice plate (*Figure 8*).

Vena contracta taps have an upstream or high-pressure tap located one pipe diameter upstream of the orifice plate and a downstream or low-pressure tap located at the vena contracta, the point of minimum pressure. These taps are shown in *Figure 9*. Theoretically, this is the optimum location for orifice plate taps because the greatest differential pressure is available between these points. However, the point of minimum pressure varies with the d/D ratio. Therefore, errors are introduced if the orifice plate bore is changed due to erosion.

Radius taps, as shown in *Figure 9*, are an approximation of the vena contracta taps. The upstream or high-pressure tap is located one pipe diameter upstream of the orifice plate. The downstream or low-pressure tap is located ½ pipe diameter downstream of the orifice plate.

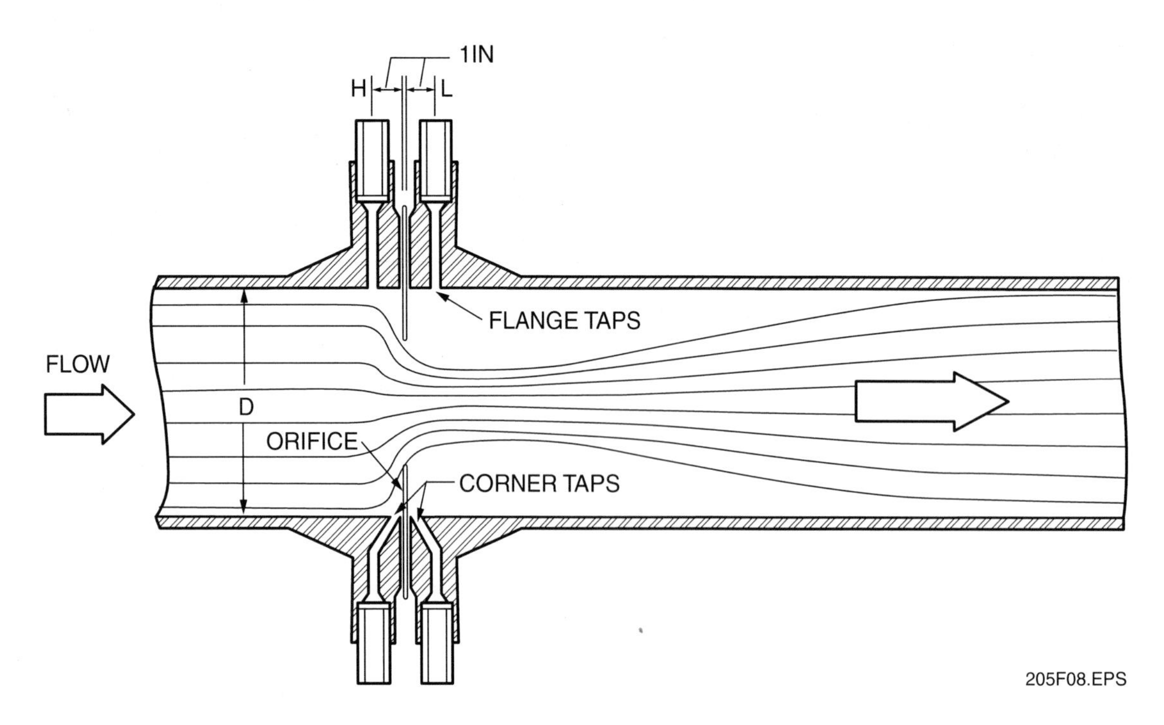

Figure 8 ◆ Flange and corner tap locations.

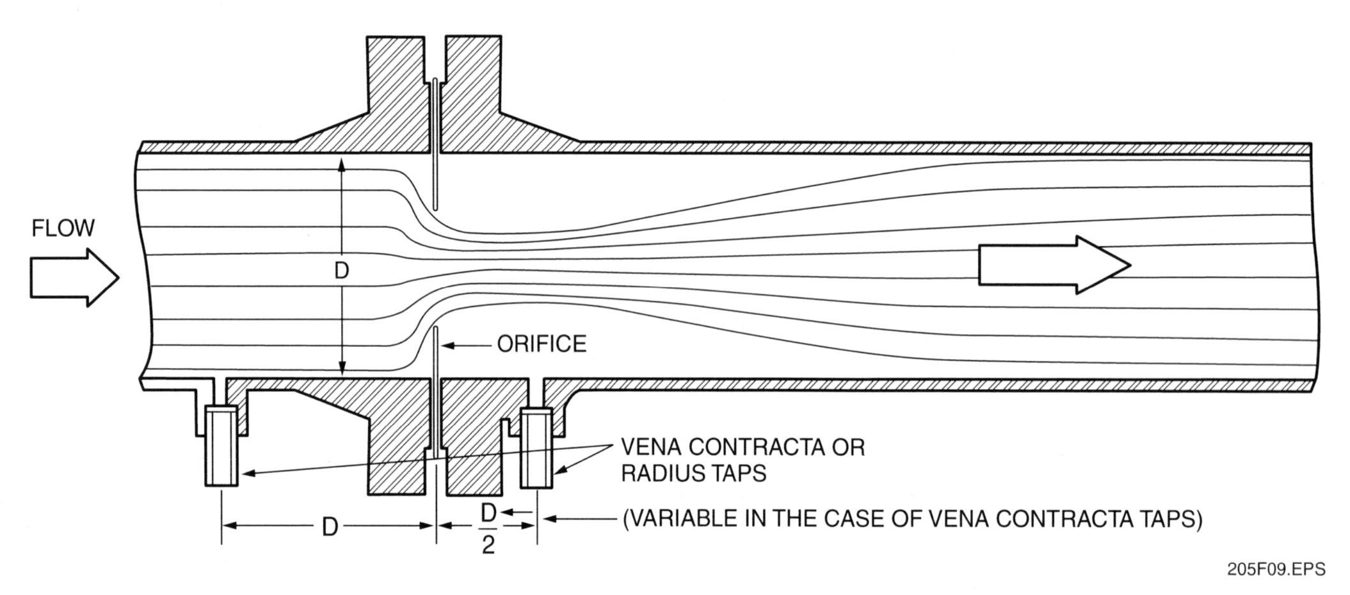

Figure 9 ◆ Vena contracta and radius tap locations.

Pipe taps or full flow taps measure the permanent pressure drop across an orifice plate. The taps are drilled into the pipe 2½ pipe diameters upstream of the orifice plate and 8 pipe diameters downstream of the orifice plate. This tap arrangement is shown in *Figure 10*. Because of the distance from the orifice, exact location of the taps is not critical. However, there is some likelihood of measurement errors created by head loss in the long length of the pipe.

The pressure taps for liquid flow are generally located along the horizontal centerline of the pipe. Location of the taps along the side of the pipe prevents trapped gas bubbles from interfering with the measurement and prevents sludge or particles from fouling sensing lines. For gas flow, the taps are generally located at the top vertical centerline of the pipe. This tap location allows condensate to drain from the sensing lines. For best accuracy, the pressure taps for segmental orifice plates must be at 180° from the center of tangency of the opening (*Figure 11*). The pressure taps for eccentric orifice plates must be located at 180° or 90° to the eccentric opening (*Figure 11*).

Orifice plates are the most widely used of the primary flow elements because they are inexpensive and easy to install. In addition, more **empirical data** has been collected on this device than has been

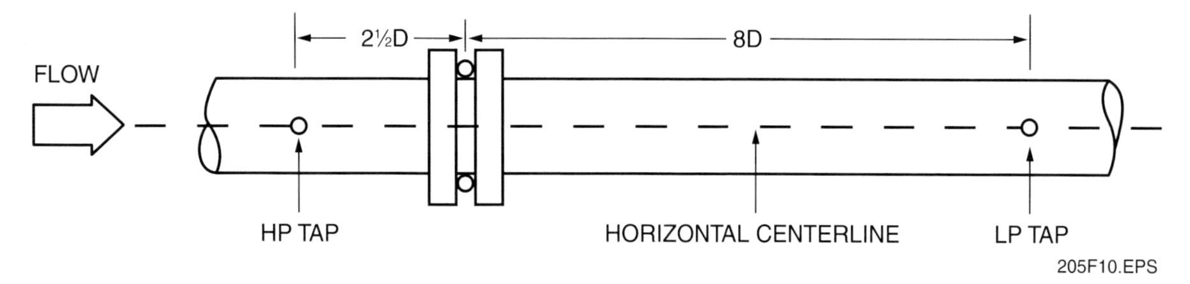

Figure 10 ◆ Pipe tap locations.

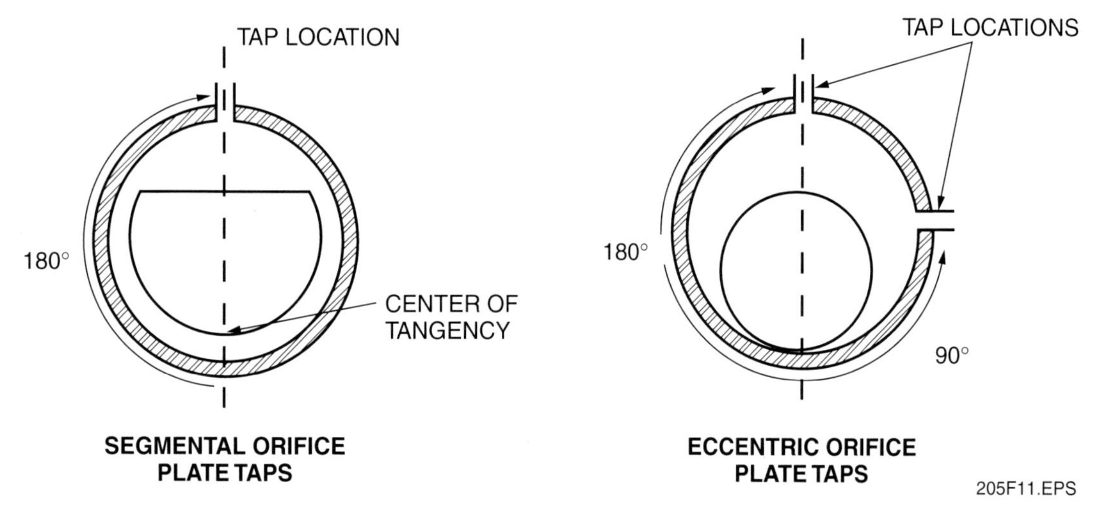

Figure 11 ◆ Orifice tap locations.

collected for any of the other primary elements. Orifice plates, however, have two serious disadvantages. First, as previously mentioned, they cause a high permanent pressure drop. Second, they are highly susceptible to erosion because of the sharp edges at the opening. Erosion of the sharp edges can cause serious inaccuracies in flow measurement.

2.2.0 Venturi Tubes

The venturi tube is the most accurate of all primary elements when it is properly calibrated. Fluids that contain large amounts of suspended solids, such as slurries, or those that are very **viscous**, can be measured by venturi tubes with maximum accuracy. Venturi tubes consist of a converging conical inlet section, a cylindrical throat, and a diverging recovery cone. A typical venturi tube is shown in *Figure 12*. The inlet section decreases the area of the fluid stream, causing the velocity to increase and the pressure to decrease. The low pressure is measured in the center of the cylindrical throat. At this point the velocity and pressure are neither increasing nor decreasing. In the diverging recovery cone, the velocity decreases and the pressure is recovered. The recovery cone allows a relatively large pressure recovery, such that the permanent pressure loss is only 10% to 25% of the differential pressure developed by the device. The differential pressure

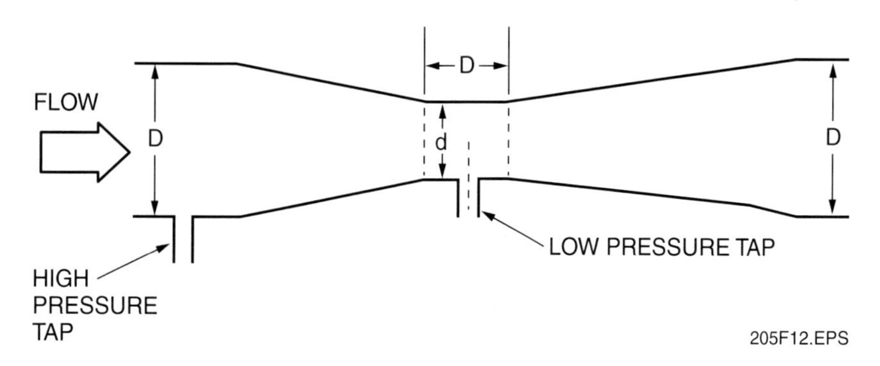

Figure 12 ◆ Venturi tube.

developed by the venturi is sensed between an upstream or high-pressure tap located one-half pipe diameter upstream of the inlet cone and a low-pressure tap located at the center of the throat.

Eccentric venturi tubes are occasionally used in systems where the flow of slurries is measured. In this type of venturi, the throat is flush with the bottom of the pipe. This design, shown in *Figure 13*, further ensures that a buildup of solids does not occur. It permits complete drainage of horizontal pipes.

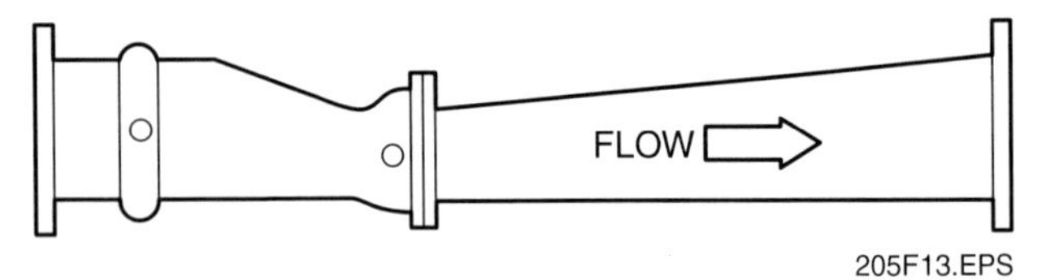

Figure 13 ◆ Eccentric venturi.

2.3.0 Pitot Tubes

The pitot tube measures fluid velocity at one point within the pipe. It is a type of head flowmeter. Since the velocity of a fluid passing through a pipe varies with its distance from the pipe wall, the flow indication obtained from a pitot tube can be highly inaccurate, particularly in laminar flow conditions. For this reason, the pitot tube has limited industrial application. Nevertheless, for velocity measurements, spot measurements and laboratory measurements, the pitot tube is the best and most frequently used device. It is also commonly used for flow measurement in large pipes and ducts, such as in ventilation systems.

A simple pitot tube consists of a cylindrical probe that is inserted into the flow stream. It has two openings. The first, called the impact opening, faces into the stream. The second, called the static opening, faces perpendicular to the flow stream. Refer to *Figure 14*.

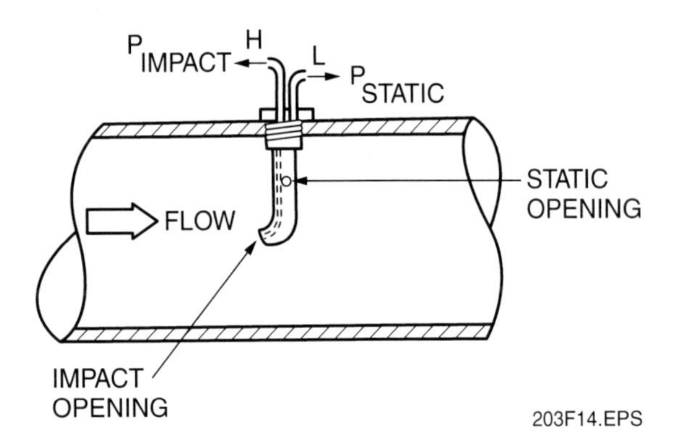

Figure 14 ◆ Pitot tube.

The differential pressure produced by the device is measured by a conventional differential pressure-measuring device, such as a bellows. It should be noted, however, that the high-pressure connection senses the impact pressure, while the low-pressure connection senses static pressure.

To obtain a true measurement of flow in a pipe, it is necessary to know the average velocity of the fluid. Velocity readings from the pitot tube positioned at several different distances from the pipe wall have to be taken, weighted in accordance with a factor based on distance from the wall, and averaged in order to obtain an accurate flow measurement. This is frequently done in test work, but it is hardly practical for industrial process flow measurement.

Several different variations of the pitot tube have been designed in order to provide a higher differential pressure than that produced by impact pressure alone. One such variation is the pitot-venturi tube shown in *Figure 15*.

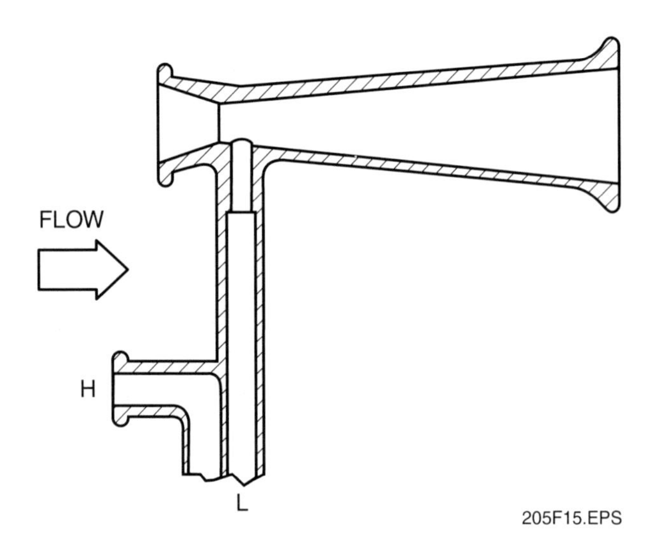

Figure 15 ◆ Pitot-venturi tube.

The pressure at the impact opening, which is the sum of impact pressure and static pressure, is developed the same way as in the conventional pitot tube. The pressure at the impact opening is compared to the reduced pressure at the throat of a small venturi that is also suspended in the fluid stream. Remember that the pressure at the throat of a venturi drops because of a velocity increase. The differential pressure created by this device is measured in the same manner as with the conventional pitot tube.

The pitot tube has two major disadvantages. One disadvantage is that it can measure velocity at only one point within a pipe or duct. The second disadvantage is that the impact opening is easily blocked if the pitot tube is used for the measurement of dirty or sticky fluids. Advantages include the fact that it produces no appreciable pressure drop and is easy to install. It is also relatively inexpensive compared to other differential pressure detectors.

2.4.0 Annubar Tubes

The annubar tube is an improvement over the conventional pitot tube in that it nearly eliminates the major disadvantages of conventional pitot tubes. It consists of two probes: one that senses fluid velocity and one that senses static pressure. The probes are suspended in the fluid line in much the same way as the conventional pitot tube.

The velocity-sensing probe (HP tap) has four openings or ports that face upstream into the flow stream. Each of the ports is located at positions representing equal cross-sectional areas to the flow stream. These ports are shown in *Figure 16*.

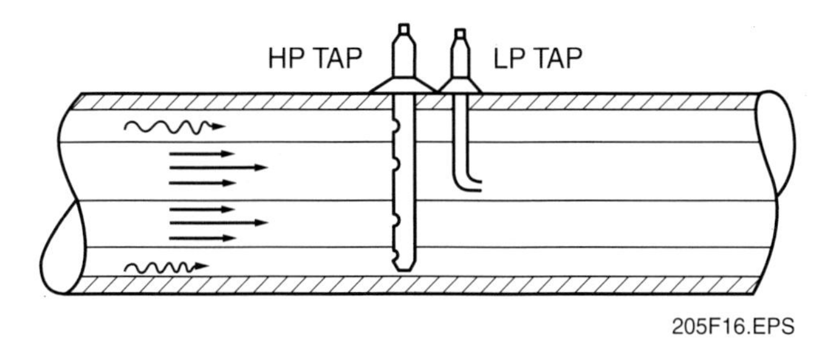

Figure 16 ◆ Annubar tube.

A line inserted into the upstream probe senses the average of the impact pressures present at the four openings. This average pressure is a result of the average velocity in the pipe. Therefore, the differential between average impact pressure and the static pressure sensed by the downstream probe gives an accurate indication of the fluid flow rate.

In addition to the fact that this device provides average velocity measurement, its other advantages are similar to the conventional pitot tube. It is easy to install, and it is inexpensive.

2.5.0 Magnetic Flowmeters

The magnetic flowmeter operates on the principle of Faraday's Law of Induction. This law states that any time a conductor is moved through a magnetic field at right angles, an electrical potential is developed. The magnetic flowmeter was developed to measure the volumetric flow rate of electrically conductive fluids. It is particularly useful for measuring the flow rate of fluids that present very difficult handling problems, such as corrosive acids, sewage, slurries, and paper pulp stock.

As was stated previously, when an electrical conductor moves through a magnetic field in a direction perpendicular to the **magnetic lines of flux**, an **electromagnetic field (EMF)** is induced in the conductor. The magnitude of the induced EMF is proportional to the magnetic flux density, the length of the conductor in the magnetic field, and the speed or velocity of the conductor.

The magnetic flux density is a term that describes the strength of the magnetic field. It is the number of magnetic lines of force per unit area. The units normally used for magnetic flux density, B, are webers per square meter, where a weber is the measure of the number of magnetic lines of force.

To determine the EMF induced in a conductor, the following equation is used:

$$\text{EMF induced} = lBv$$

Where:

l = length of the conductor

B = magnetic flux density

v = velocity of the conductor

The direction or polarity of the induced EMF is determined by using the right-hand rule, as illustrated in *Figure 17*. If the index finger points in the direction of the magnetic lines of force, from the north pole to the south pole of the magnet, and the thumb points in the direction of the motion of the conductor with respect to the field, the middle finger points toward the negative potential.

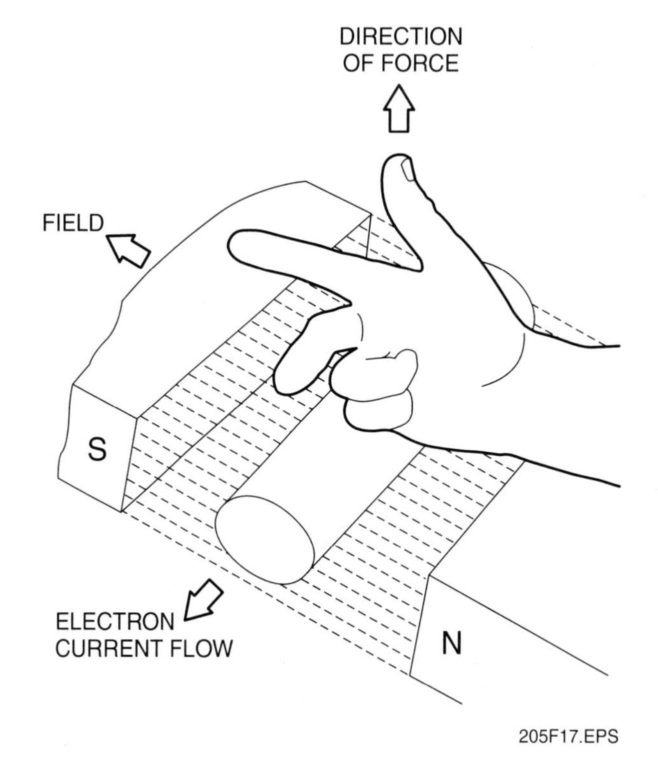

Figure 17 ◆ Right-hand rule for induced EMF.

A typical magnetic flowmeter consists of a tube, a set of coils, a laminated iron core (armature), and the EMF electrode assembly. These components and others are shown in *Figure 18*.

Current flow through the coil produces a magnetic field. The laminated core concentrates the magnetic lines of flux of this field around the tube. By concentrating these lines of flux, the magnetic flux density is increased. The tube directs the process fluid through the concentrated magnetic field. The process fluid is the moving electrical conductor. As the fluid passes through the magnetic field, an EMF is induced into it that is sensed by the electrode assembly. If the magnetic field strength and the distance between the electrodes is constant, the EMF sensed by the electrodes is proportional to the velocity of the fluid.

Previously, the characteristics of fluid flow were discussed. Remember from that discussion that the velocity of the fluid close to the pipe wall is less than the velocity of the fluid flowing at the center of the pipe. These variations in the velocity profile do not affect the flow measurement accuracy of the magnetic flowmeter. The electrodes sense the average EMF of the fluid. Therefore, the magnetic flowmeter measures the average velocity of the fluid regardless of whether the flow is laminar or turbulent. This also allows bidirectional flow measurement.

The tube carrying the process fluid must be made of a non-magnetic material to allow the field to penetrate to the liquid. In *Figure 18*, the flowmeter has an insulating pipe liner to prevent potential induced in the fluid from being short-circuited by the pipe. Typical materials used in the construction of the flowmeter tube are fiberglass reinforced plastic or nonmagnetic stainless steel. When stainless steel is used, the interior of the tube is lined with a nonconducting material, such as Teflon®, polyurethane, or glass.

The electrodes are flush with the tube interior. They actually come in contact with the process fluid. The electrodes must be made of good conducting material. They are usually made of type 316 stainless steel, but for highly corrosive service, platinum electrodes are often used. It is essential that these electrodes remain free of dirt. Dirt acts as an electrical insulator and reduces the accuracy of measurement.

2.6.0 Ultrasonic Flowmeters

Ultrasonic flowmeters are actually a group of devices based on well-documented theory. Their application to field installations has awaited the availability of cost-effective electronics capable of accurately measuring small changes in time or frequency. This section will discuss the two most common applications of ultrasonic flow measurement: the time difference method and the frequency shift method.

The determination of fluid flow through the use of ultrasonics implies the transmission and reception of sound waves. Ultrasonic sound waves are above the range of frequencies audible by the human ear (20,000 Hz). The typical frequencies used by ultrasonic flowmeters are in the range of 1 MHz to 10 MHz. However, before the use of sound waves to determine flow can be explained, some characteristics of sound and the method of detecting and transmitting it must be understood.

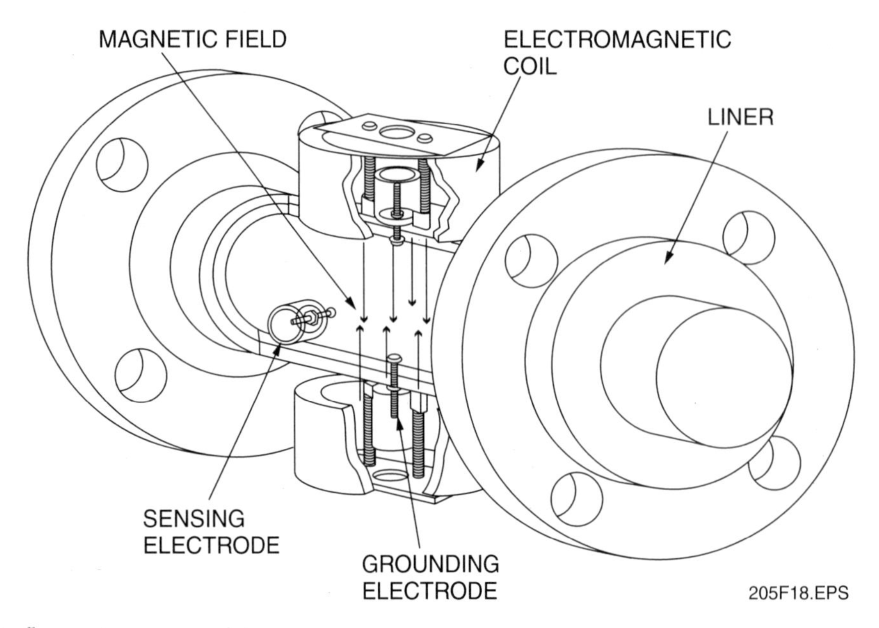

Figure 18 ◆ Magnetic flowmeter components.

Sound is the transmission of very small pressure variations through a medium to a receiving device. For instrument applications, the transmitter must convert an electrical signal to a mechanical motion, and the receiver must convert this motion back to an electrical signal. A device capable of making both of these conversions is the **piezoelectric crystal**. When a crystal is held between two flat metal plates and the plates are pressed together, a small EMF is developed between the two plates. It is as if the crystal becomes a battery for an instant. When the plates are released, the crystal springs back to its original shape, and an opposite-polarity EMF is developed between the two plates. In this way, the crystal converts mechanical energy into electrical energy. Furthermore, when an EMF is applied across the two plates on either side of a crystal, the normal shape of the crystal is distorted. When an EMF of opposite polarity is applied, the physical distortion of the crystal is reversed. In this way, the crystal converts electric energy into mechanical energy. The two reciprocal effects that occur in a crystal are known as the piezoelectric effect. Typical deformations that can occur in a crystal are shown in *Figure 19*.

Every crystal has a resonant frequency that depends on its structure, size, and shape. The electronic signal applied to the transmitting crystal should be of the frequency at which the crystal resonates in order for the crystal to sustain its mechanical oscillations. Similarly, the receiving crystal should be matched to the transmitting crystal.

As previously mentioned, sound is the transmission of small pressure variations through a medium. What actually occurs is that the molecules of the transmission medium are alternately compressed and rarefied (spread out). The molecules do not travel an appreciable distance. Only the variations in pressure actually move.

It was previously explained that flow rate is proportional to the density of the fluid, the cross-sectional area of the pipe through which the fluid flows, and the velocity of the flowing fluid. Using the following equations, the flow rate can be determined as follows:

$$\dot{V} = v \times A \text{ or } \dot{m} = \rho \times A \times v$$

Where:

$\dot{V}$ = volumetric flow rate (ft^3/min)
$\dot{m}$ = mass flow rate (lb_m/min)
v = fluid velocity (ft/min)
A = flow area (ft^2)
ρ = fluid density (lb_m/ft^3)

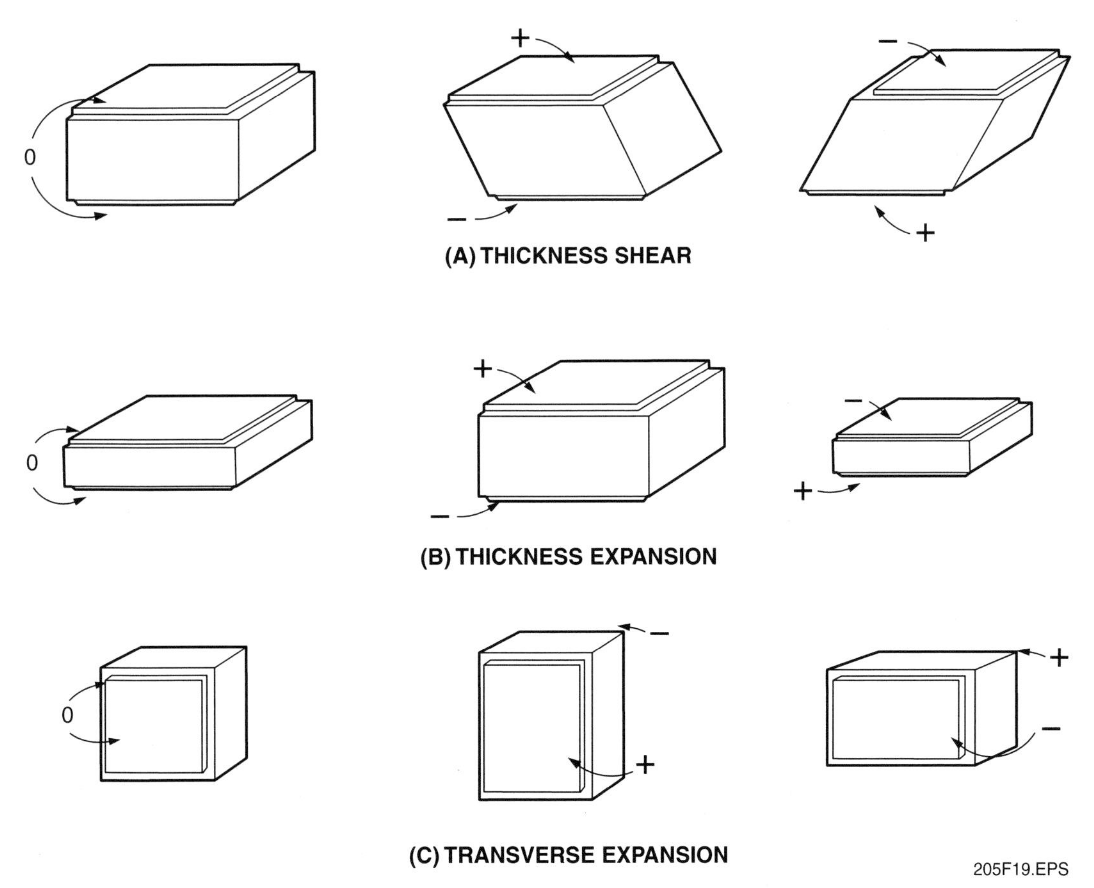

Figure 19 ◆ Piezoelectric crystal deformation.

If the flow area and the fluid density are known constants, the flow rate can be determined by measuring the velocity of the flowing fluid. Ultrasonic flowmeters are used to measure velocity. The method employed to measure the velocity of the fluid is rather simple. The rate at which sound is propagated through a given medium at rest is constant. For water it is 5,000 ft/s. If the fluid also has a velocity, the absolute velocity of the pressure-disturbance propagation is the algebraic sum of the two. This means that if sound is transmitted in the direction of the fluid flow, the actual velocity of the sound is the sum of the sound's velocity when the fluid is at rest plus the velocity of the fluid. Conversely, if the sound is transmitted in a direction opposite the fluid flow, the actual velocity will be the difference between the two velocities.

Figure 20(A) is a simple sketch of the detector placement for ultrasonic measurement of flow. The detectors used are piezoelectric crystals identical to those discussed previously. The operation of this instrument requires that the velocity of sound in the fluid, at rest, be accurately known. When flow exists, the sound will travel from transmitter to receiver at a greater velocity (v_m).

The distance from the transmitter to the receiver (L) and the speed of sound with no flow (C) are constants. Since the time it takes for the fluid to travel from the transmitter to the receiver (t_x) can be measured, the velocity of the fluid and thus the flow rate can be determined.

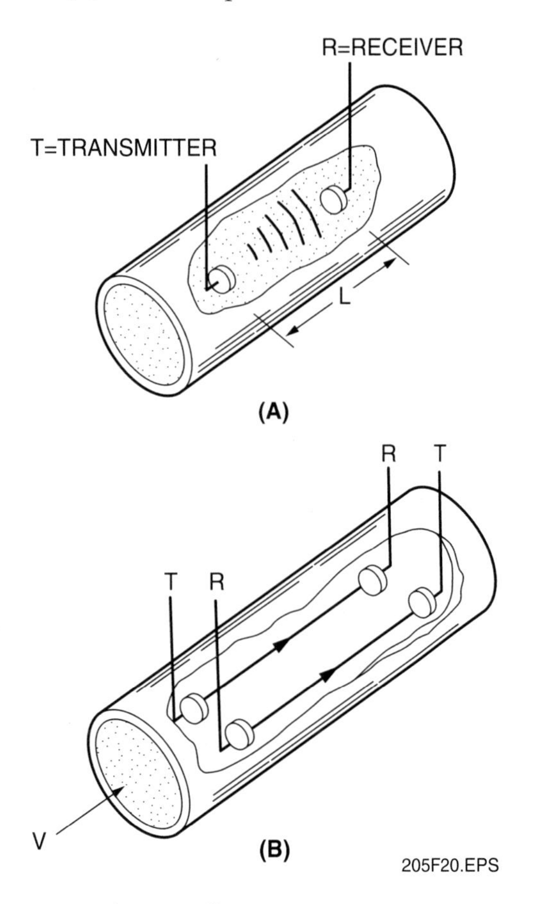

Figure 20 ◆ Ultrasonic flowmeters.

A similar method of determining the flow rate is illustrated in *Figure 20(B)*. In this application the difference in time between transmission with the flow and transmission against the flow is measured. Now the increment of time can be measured directly and is larger than the difference between v_m and C in the first example. The fluid velocity is calculated as shown below:

$$t_w = \frac{L}{C + v} \text{ and } t_a = \frac{L}{C - v}$$

and: $t = t_a - t_w$

so: $t = \frac{L}{C - v} - \frac{L}{C + v}$

or: $t = \frac{2Lv}{C^2 - v^2}$

if: $v^2 < C^2$

then: $t = \frac{2Lv}{C^2}$

or: $v = \frac{tC^2}{2L}$

Where:

L = distance between sensors
C = speed of sound with no flow
v = velocity of fluid
t_w = sound transit time with flow
t_a = sound transit time against flow

This equation shows that fluid velocity is changing linearly with the measured time difference. As the transit time increases, the fluid flow rate is also increasing.

The two devices analyzed so far offer little resistance to fluid flow, and therefore cause a negligible permanent head loss. However, both devices require that the detectors be placed within the fluid stream. An example of an ultrasonic flow measuring device that requires no piping penetrations is illustrated in *Figure 21*. Here the detectors are mounted externally to the pipe and function as both the transmitter and the receiver (transceiver) for the sound. The external mounts require the sound to be transmitted through the walls of the pipe. This adds another variable for which an adjustment must be made. In most cases the device is calibrated for only one schedule of pipe, so the transit time of the sound through the pipe is known. Also, the sound is no longer traveling in a path parallel to the fluid. The

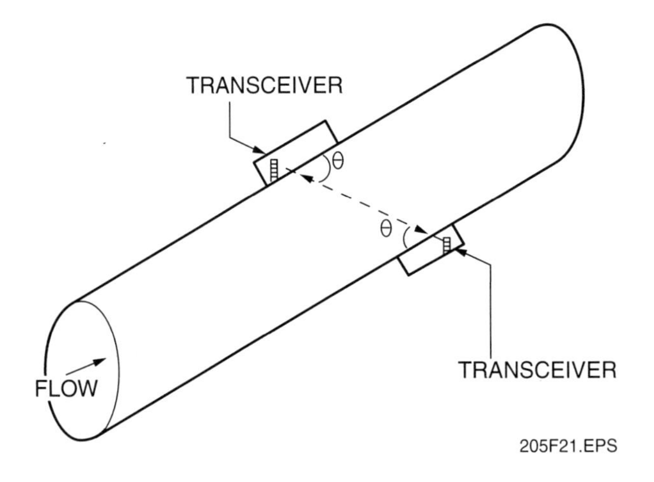

Figure 21 ◆ Ultrasonic flow measurement.

previously developed equation is still valid, however, if the fluid velocity is multiplied by the cosine of the angle θ.

The frequency shift method of ultrasonic flow measurement uses detectors placed as shown in *Figure 22*. The amplifiers are actually self-excited oscillators. Following the initial pulse, each succeeding pulse is triggered by the receipt of the previous pulse. Therefore, the frequency of oscillation is a function of the signal transient time through the fluid. The transient time, in turn, is a function of the magnitude of the fluid flow. The advantage of this system over the time difference method is that it can be made independently of the speed of sound in the fluid at rest.

The operation of this system is based on measuring a difference in frequency. Amplifier A oscillates at a frequency that is greater than the frequency of amplifier B. The frequency of either oscillator is equal to the inverse of its signal's transient time through the fluid.

2.7.0 Capacitance-Type Level Detectors

A capacitor consists of two conductors separated by an insulator. The insulator is referred to as the dielectric, and the conductors are called the plates of the capacitor. Capacitance is measured in **farads**. A one farad capacitor is capable of storing one **coulomb** of charge for each volt applied. *Figure 23* is a simplified sketch of a plate and a cylindrical capacitor.

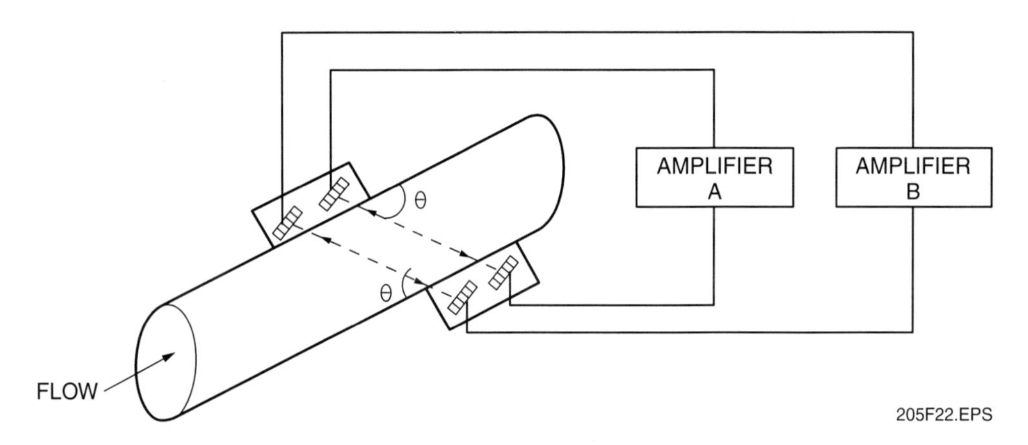

Figure 22 ◆ Frequency shift ultrasonic flow measurement.

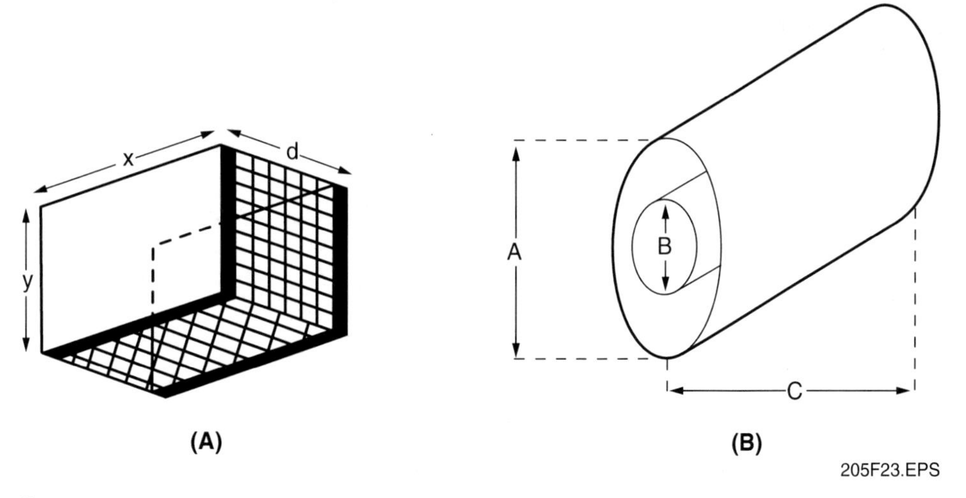

Figure 23 ◆ Capacitors.

For the plate capacitor, the capacitance is calculated using the following equation:

$$C = 0.2249\ (K)\ \frac{A}{d}$$

Where:

C = capacitance in **picofarads**
A = plate area (xy) in square inches
d = distance between plates in inches
K = dielectric constant
0.2249 = conversion factor

The dielectric constant is a factor that compares any material to a perfect vacuum. It is a numerical value on a scale of 1 to 100 that relates to the ability of the dielectric to store an electrostatic charge. A capacitor with a vacuum dielectric has a dielectric constant of one. Replacing vacuum with paper doubles the capacitance; paper has a dielectric constant of two. *Table 3* lists the dielectric constants of some common materials.

Table 3 Dielectric Constants for Common Materials

Material	Dielectric Constant
Vacuum	1.00
Air	1.006
Paper	2.00
Quartz	4.3
Teflon®	2.0
Water (32°)	88.0
Water (68°)	80.0
Water (212°)	48.0

For the measurement of level in non-conductive materials, a bare metal probe is used. The probe serves as one plate of the capacitor, while the tank walls serve as the other plate. Refer to *Figure 24* for a diagram of this type of installation. The actual measured capacitance is the sum of C_1, C_v, and C_L. Capacitance C_1 is unaffected by tank level. It represents the capacitance of the leads and measuring system. Capacitor C_v represents the capacitance of the portion of the probe exposed to the vapor. Its value is largely determined by the tank level and dielectric constant of the vapor, K_v. Capacitor C_L represents the capacitance of the portion of the probe exposed to the liquid. Its value is a function of the tank level and the liquid's dielectric constant, K_L. The dielectric constant of all gases is nearly unity, so K_v is smaller than K_L. Refer to the previous equations as necessary during the following discussion. Assuming the tank is empty, C_L is zero and C_v is maximum. It is still small because K_v is small. At this level, measured capacitance is minimum, $C_1 + C_{vmax}$. As tank level rises, C_L starts to increase and C_v begins to decrease. However, the rate at which C_L increases is greater than the rate at which C_v decreases because of the difference in their dielectric constants. When the tank is full, C_v is zero and C_L is maximum. Measured capacitance is at its maximum value of $C_1 + C_{L\,max}$. The total span of measured capacitance is the difference between C_1 when the tank is full and C_v when the tank is empty, as shown in the equation below:

$$\text{Span} = \text{upper range value} - \text{lower value}$$
$$\text{Span} = C_1 + C_{L\,max} - (C_1 + C_{v\,max})$$
$$\text{Span} = C_{L\,max} - C_{v\,max}$$

The change in capacitance is a linear function of tank level. Converting from a span given in picofarads to one given in inches of tank level requires knowing only the length of the probe. Refer again to *Figure 24*. Some value of resistance is also present between ground and the probe. The actual value will vary with tank level and the type of non-conductive material in the tank. This resistance must be high when compared to the impedance of the capacitors to make its shunt path for current insignificant. In this regard, there is a definite advantage to making the measurement at high frequencies.

The bare capacitance probe can only be used on non-conductive fluids. If the fluid is conductive, the shunt resistance will be small, making an accurate measurement of capacitance changes impossible.

The insulated probe has been developed for use with conductive liquids. *Figure 25* is a diagram of an insulated capacitance probe level measurement system. The most commonly chosen insulating material is Teflon®. Teflon® has a dielectric constant of 2. The capacitor formed by the insulator is referred to as C_a above the liquid and C_b below the liquid surface. When the tank is empty, the measured capacitance will be as shown in the simplified sketch in *Figure 26(A)*. Both C_v and C_a are at their maximum values at this time. Capacitor C_v is small due to the low dielectric constant of the vapor. Therefore, the effective capacitance of C_v and Ca in series is less than C_v. As tank level increases, both C_v and C_a decrease (due to the decreasing plate area) at a rate largely determined by C_v. However, as shown in *Figure 26(B)*, the series combination of RL and C_b is now in parallel with C_v and C_a. The value of RL is small because the fluid is conductive. The value of RL can be made insignificant compared to X_{cb} if the proper frequency is chosen. The value of Cb increases

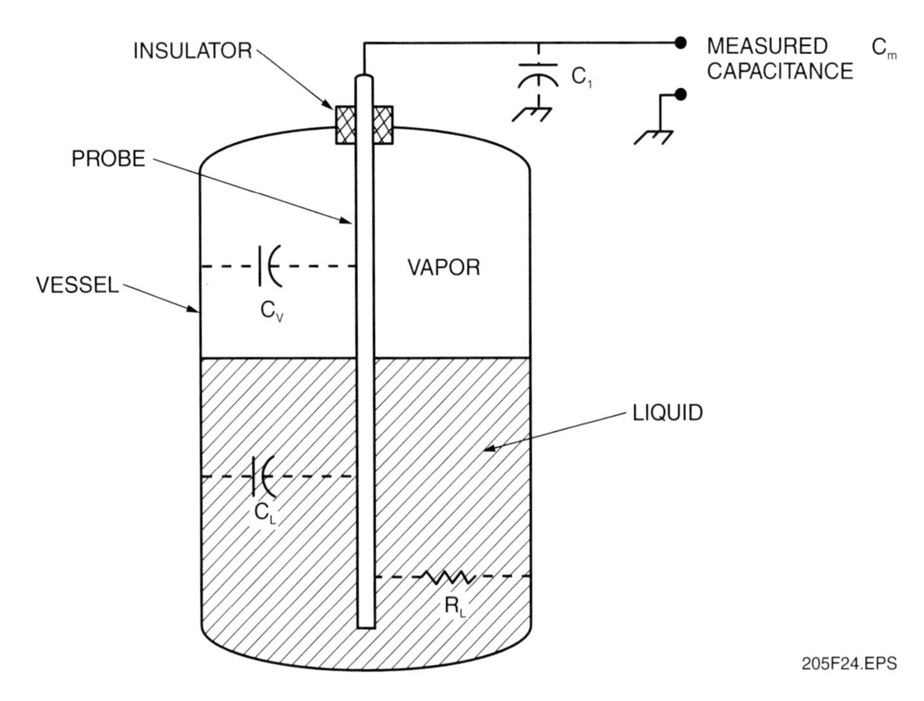

Figure 24 ◆ Bare capacitance probe.

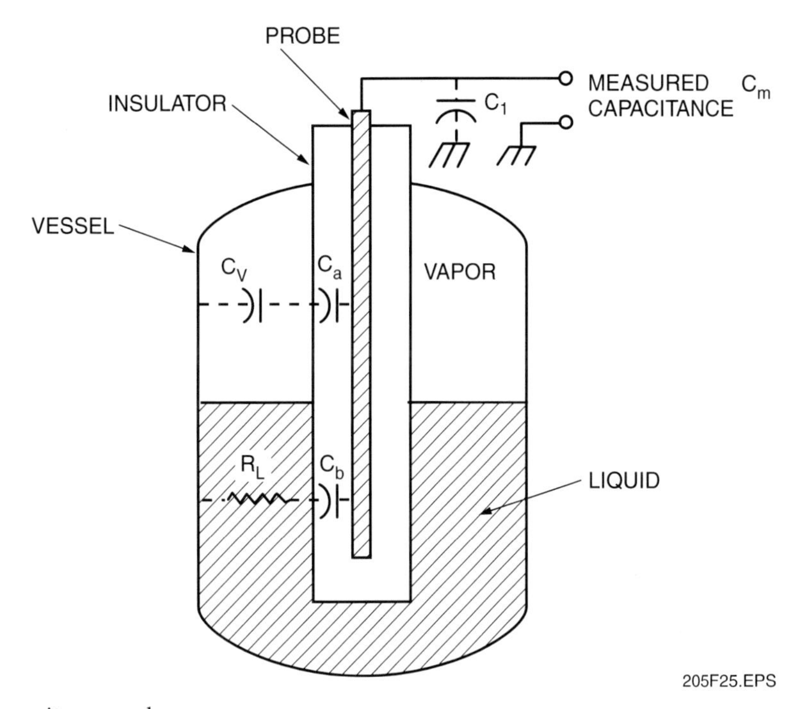

Figure 25 ◆ Insulated capacitance probe.

linearly with level at a rate much greater than C_v decreases, so the net measured capacitance increases linearly. When the tank is full, as shown in *Figure 26(C)*, the measured capacitance, C_b, is at its maximum value. The insulated probe can be used on nonconductive and slightly conductive fluids as well. However, the resistance of the fluids has to be taken into account. Although this complicates the measurement, it does not alter the result, which is a measurable change in capacitance as a linear function of liquid level. The level and measurement system capacitance, C_1, will represent a static capacitance value unaffected by liquid level. The value of C_1 will be added to the capacitances calculated in *Figure 26* and must be subtracted by the measurement system.

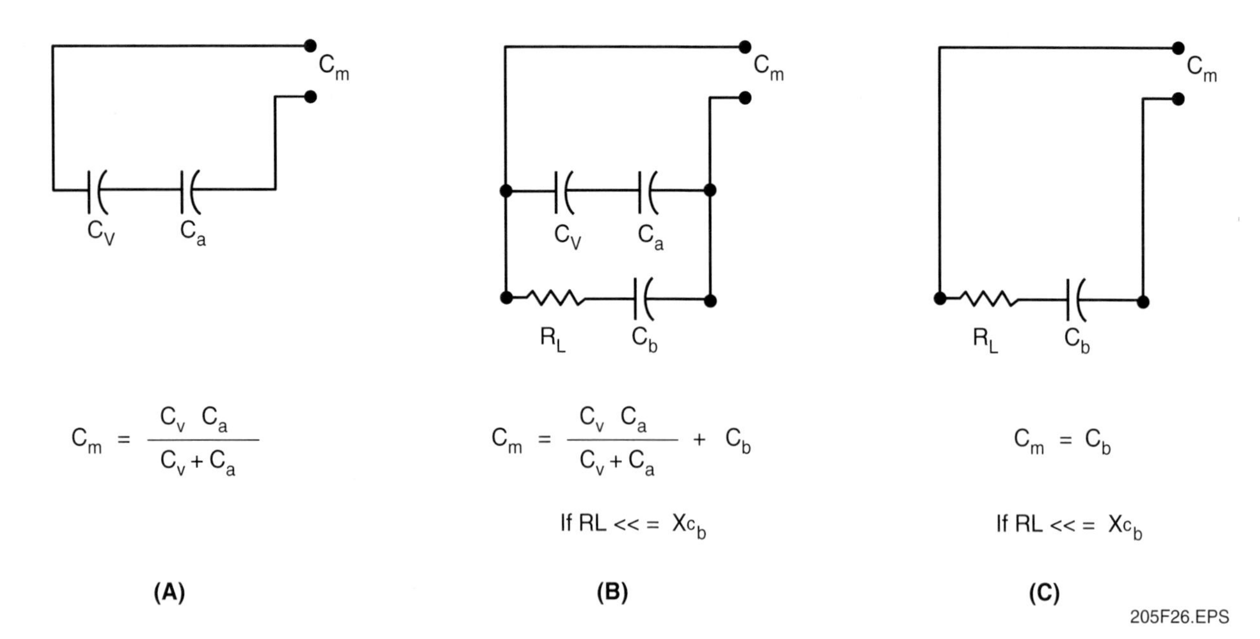

Figure 26 ◆ Capacitance level detector circuit.

Bare and insulated capacitive probes can use the walls of the vessel as one plate of the capacitor. However, this is not necessarily the only method. If the vessel is of non-uniform size, the second capacitive plate can be part of the detector. To accomplish this, the existing probe is surrounded by another cylindrical plate. The outer plate is insulated from the inner plate, and space is allowed for process liquid and vapor to exist between them. The outer plate is then grounded to the vessel. Thus, a capacitor is formed between the probe and outer plate with a dielectric that varies with process liquid level.

The accuracy of capacitive level detection systems is affected by two significant errors. The first is that coating or wetting of the probe by a conductive liquid can change the detected surface level. This explains the common use of Teflon®, since few materials will adhere to it. The second is that anything that can change the dielectric constant of the liquid will also affect the measured capacitance. As can be shown in *Table 3*, the dielectric constant of water varies greatly with temperature. This can be compensated for through the use of an RTD or thermistor as another input to the measurement system. Chemical and physical composition can also influence the value of the dielectric constant. For these reasons, careful consideration should be given before selecting capacitive level measurement.

Clean, non-coating nonconductive liquids (such as many hydrocarbons) in metallic tanks require special consideration. A bare (uninsulated) probe can be used, but one must ensure that the probe does not come in contact with any conductive liquid that may contaminate the nonconductive liquid (such as water in oil). If this occurs, the output will be driven to full scale, regardless of the actual level in the tank. Note that an insulated probe can also be used. The probe's metal fitting must be grounded to the metal tank wall, and the distance from the tank wall to the probe must be constant along the entire length of the probe to provide a linear change in analog output per change in fluid height. If this is not the case (for example, if the tank is irregular in shape), or if the tank is greater than 15 feet in diameter, a concentrically shielded probe should be used. The capacitance change per foot depends on the dielectric constant of the material as well as the tank diameter. Tank diameter does not affect the concentrically shielded probe.

The concentrically positioned metal shield provides a linear ground reference for applications in which nonconductive liquids are contained in irregularly shaped and/or plastic vessels. The electrode inside the shield is insulated with TFE Teflon®, PVDF (Kynar®), or polyethylene. Due to the close proximity of the shield to the measuring electrode, these level sensing elements produce a high capacitance change with level variation. The concentric shield also makes it possible to pre-calibrate the electronics outside the main vessel by using the same material to be measured as the reference medium. *Figure 27* shows the basic construction of a concentrically shielded capacitance probe. Note that the probe is centered concentrically within the shield and held in position by standoffs or spacers.

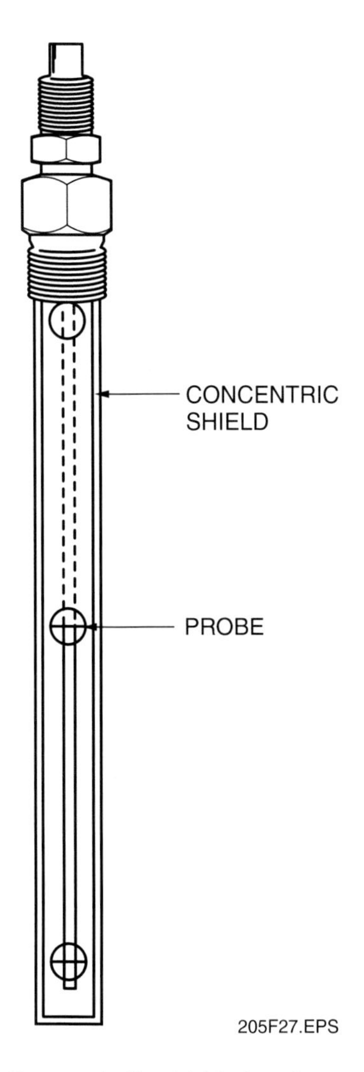

Figure 27 ◆ Concentrically shielded probe.

2.8.0 Ultrasonic Level Measurement

The determination of liquid level by the use of ultrasonics implies the transmission and reception of sound waves. Recall that sound is actually the transmission of small magnitude pressure variations through a medium. Without a vibrating transmitter, a medium of transfer, and a receiver sensitive to small pressure variations, there can be no sound. Sound, like any wave, is reflected whenever it encounters an interface of materials. An echo is an example of sound being reflected by the interface of air and a solid material. The amount of sound reflected varies with the combination of materials at the interface. The velocity of sound in a given material is constant at a constant temperature. However, the velocity will vary with chemical composition and temperature.

Ultrasonic liquid level measuring instruments normally utilize piezoelectric crystals identical to those used for ultrasonic flow measurement. *Figure 28* shows an illustration of an ultrasonic installation for the continuous measurement of liquid level. An oscillator sends an AC signal of short duration to the transmitter. The piezoelectric crystal converts the electrical signal to mechanical motion, causing sound waves to be produced at an ultrasonic frequency. These sound waves are directed at the liquid surface. Some of the waves are reflected from the surface back to a receiver. The receiver converts the sound waves to an electrical signal and determines the time interval from transmission to reception of the sound. Once the time is known, the distance to the liquid surface can be determined using the following equation:

$$\text{Distance to Surface} = \frac{\text{Velocity of sound} \times \text{time interval}}{2}$$

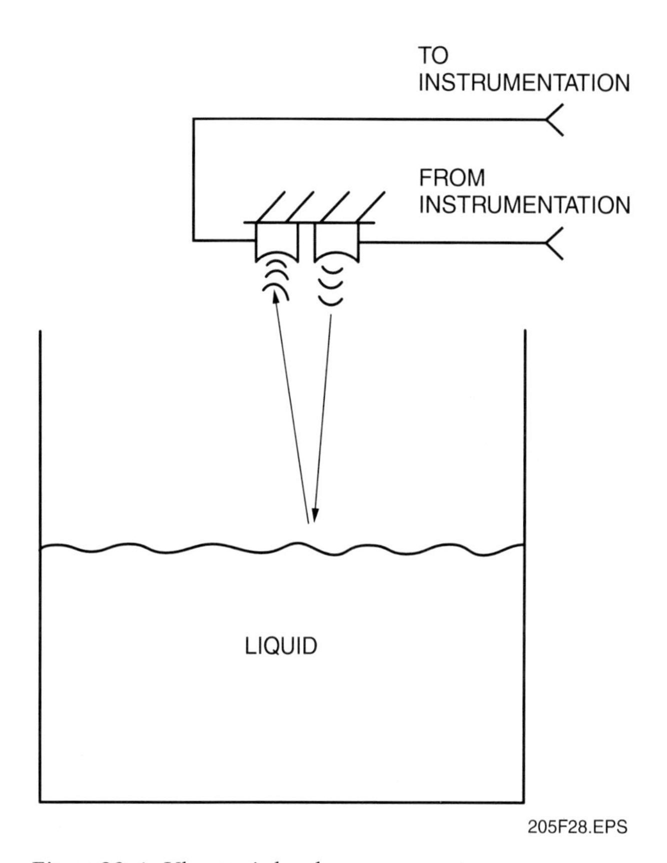

Figure 28 ◆ Ultrasonic level measurement.

If the distance to the surface is known with respect to a reference point, normally the bottom of the tank, the height of the fluid can be found.

Figure 29 illustrates two other possible installations of ultrasonic level detectors. In *Figure 29(B)*, the detectors are located at the bottom and inside the tank. The sound is still reflected off the air-liquid interface, but now the velocity of the sound wave in the liquid is used in determining the distance traveled. In *Figure 29(A)*, the detectors are located externally. The operation is identical to the circuit in *Figure 29(B)* except that the time it takes for the sound to travel through the tank wall is a constant value that is not related to

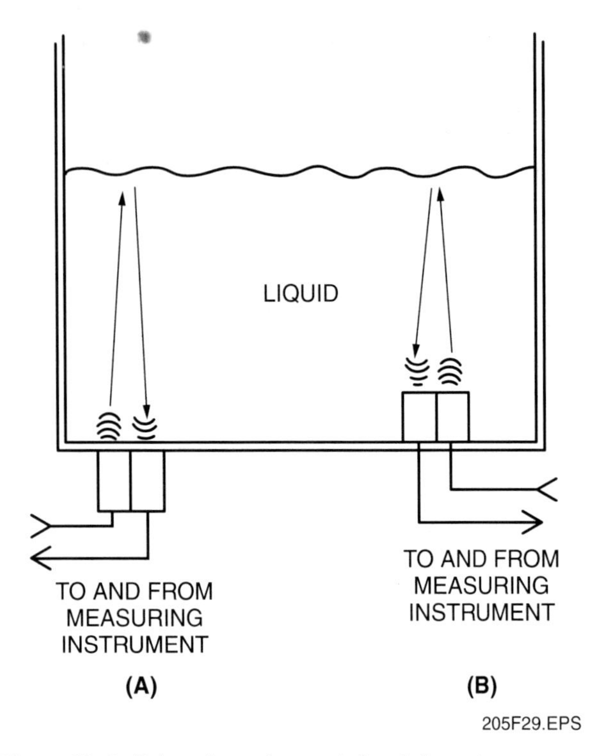

Figure 29 ◆ Subsurface ultrasonic level detection.

the actual level. Therefore, this time interval must be subtracted from the measured time before level can be accurately determined.

The reliable determination of liquid level using ultrasonics is limited by two factors. The first is that the placement and aiming of the detectors is critical. Care must be taken to ensure that the transmitted sound is reflected back to the receiver with sufficient intensity to reliably stop the timer. Secondly, and more significantly, the velocity of sound varies with temperature, atmospheric conditions, and the composition of the liquid. Since the measurement is based on accurately knowing this velocity, ultrasonic level detection is suitable only for installations where these conditions are relatively constant.

2.9.0 Nuclear Level Detection

The Geiger-Muller tube is the oldest and simplest type of radiation detector. It is often identified with the Geiger counter, which makes a loud and dramatic clicking sound when exposed to radiation.

The Geiger-Muller tube is a detector resembling a metal cylinder that is filled with an inert gas. It acts as one of the electrodes. A thin wire, which is insulated from the tube using glass caps, is passed down the center of the cylinder and acts as the other electrode. A high voltage in a range of 700 to 1,000 volts is applied to the thin wire. When the tube is exposed to gamma radiation, the inert gas inside the tube ionizes, and the ionized particles carry the current from one electrode to the other in the form of a pulse. As more gamma radiation reaches the gas in the tube, more pulses are generated. The pulse rate is counted by exterior electronic circuitry, which has an output in the form of pulses per second. This detector can be used as a level switch if it is calibrated to energize or de-energize a relay when radiation intensity indicates a high- or low-level condition.

Another type of nuclear level device is the ion chamber detector, which is a continuous level device. It is typically a 4- to 6-inch diameter tube, which may be up to twenty feet long, filled with pressurized inert gas. A very low voltage is applied to a large electrode that is inserted down the center of the ion chamber. As gamma radiation strikes the chamber, a very small signal is detected as the inert gas is ionized. This current, which is proportional to the amount of gamma radiation received by the detector, is amplified and transmitted as the level measurement signal.

When used in level measurement applications, the ion chamber will receive the most radiation and have the highest output when the measured process level is at its lowest. When the level rises and the greater quantity of process absorbs more gamma radiation, the output current of the detector decreases proportionally with the increase in level.

Radiation level detectors are used when nothing else will work. They are used when process penetrations required by a traditional level sensor present a risk to human life or the environment, or could do major damage to property. The liquids and bulk solids measured by nuclear gauges are among the most dangerous, highly pressurized, toxic, corrosive, explosive, and carcinogenic materials present. Because the nuclear gauge "sees" through tank walls, it can be installed and modified while the process is running, without expensive downtime or accidental release.

Because the installation of nuclear detectors requires a Nuclear Regulatory Commission (NRC) license, associated procedures are designed to guarantee that the installation will be safe. As detectors become more sensitive and are aided by computers, the size of the radiation sources and the resulting radiation levels drop. Therefore, the safety of these instruments is likely to continue to improve with time.

2.10.0 Bimetallic Strip Thermometers

A bimetal element (*Figure 30*) is composed of two different metals bonded together. One is usually copper or brass. The other, a special metal called Invar®, contains 36% nickel. When heated, the copper or brass has a more rapid expansion rate than the Invar® and changes the shape of the element.

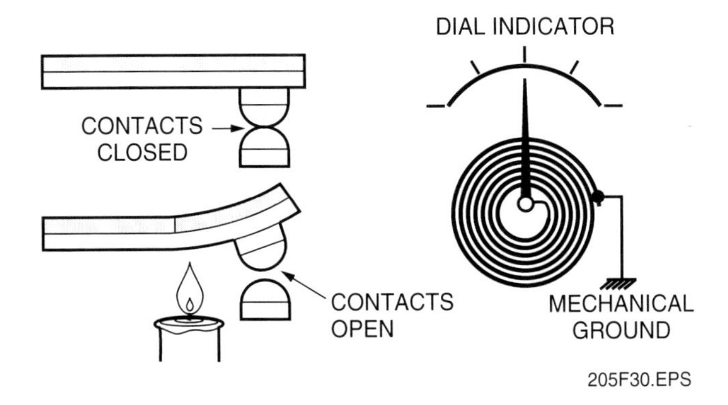

Figure 30 ◆ Bimetal sensing elements.

The movement that occurs when the bimetal changes shape is used to open or close switch contacts in a thermostat.

Bimetal elements are constructed in various shapes. The spiral-wound element is the most compact in construction and is widely used in thermostatic switches. In *Figure 31*, for example, a glass bulb containing mercury (a conductor) is attached to a coiled bimetal strip. When the bulb is tipped in one direction, the mercury makes an electrical connection between the contacts, and the switch is closed. When the bulb is tipped in the opposite direction, the mercury moves to the other end of the bulb, and the switch is opened. Thermostat switching action should take place rapidly to prevent arcing, which could cause damage to the switch contacts. A magnet is used to provide rapid action to help eliminate the arcing potential in some bimetal thermostats.

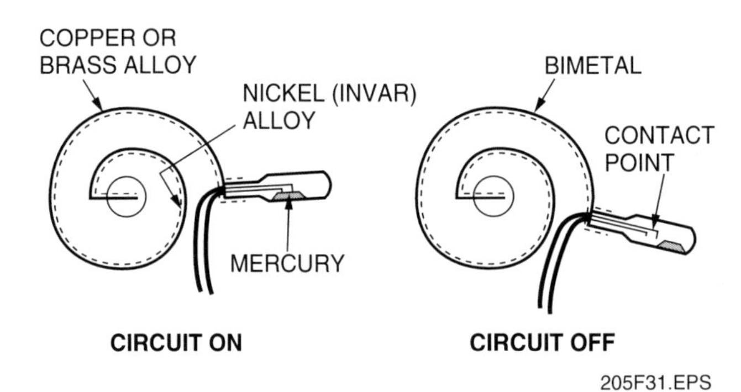

Figure 31 ◆ Thermostatic switches.

WARNING!

Mercury is toxic. Even short-term exposure may result in damage to the lungs and central nervous system. Mercury is also an environmental hazard. Do not dispose of thermostats containing mercury in regular trash. Contact your local waste management or environmental authority for disposal/recycling instructions.

In a bimetallic thermometer, as illustrated in *Figure 32*, one end of a spiral or helical element is fastened to the inside of the thermometer case. The other end of the element is attached to a shaft. The shaft and bimetallic element are supported and centered in the case by guide bushings and bearings. A pointer is fastened to the top of the shaft. When heat is transferred to the thermometer, the element expands, causing the shaft to rotate. The resultant motion moves the pointer over a circular scale to indicate the temperature. Because the deflection caused by a change in temperature is linear over only a small range of temperatures, the scale of a bimetallic thermometer is not linear over its entire range. Generally, the range of a bimetallic element is from -200°F to 1,000°F.

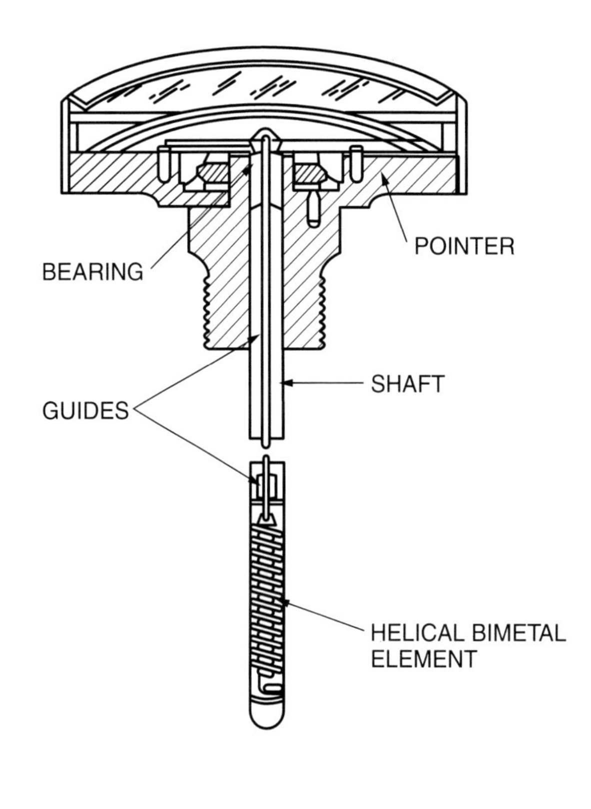

Figure 32 ◆ Industrial bimetallic thermometer using helical element.

The reason for winding the bimetallic element into a helix or spiral is to make the length of the bimetal longer than the available containing space. Increasing the length of bimetal produces a greater movement.

2.11.0 Thermocouples

A thermocouple is a detector that can convert a difference in temperature directly to electrical energy.

The first thermocouple was formed by Thomas Seebeck, a German scientist, in 1821. Seebeck's thermocouple circuit is shown in *Figure 33*. He observed a current flow that was directly related to the difference in temperature between the hot and cold junctions of two dissimilar metals. When heat was applied to one junction of the iron and copper wire, electric current began to flow in the circuit. The larger the difference in temperature, the larger the observed current flow.

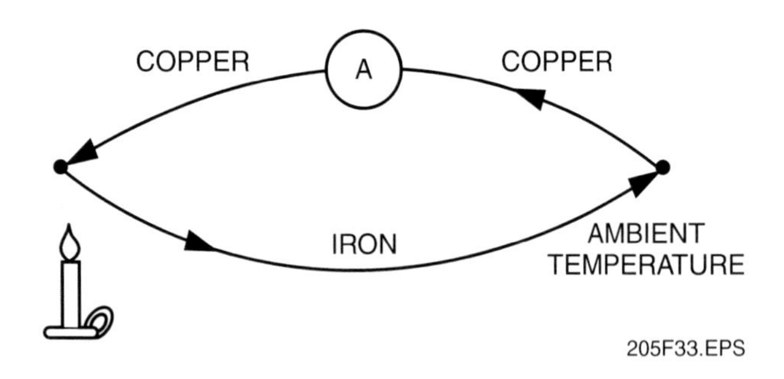

Figure 33 ◆ Seebeck's thermocouple circuit.

2.11.1 Thermocouple Metals

Thermocouples are available for use within the approximate limits of -300 to +3,200°F (-185 to +1,760°C). No single thermocouple meets all application requirements, but each possesses characteristics desirable for selected applications.

Thermocouples are classified into two groups identified as **noble metals** and base metals:

Noble metals:

- *10% rhodium versus platinum (Type S)* – Used for defining the international temperature scale from 630.5°C (1,166°F) to 1,063°C (1,945.4°F). It is chemically inert and stable at high temperatures in oxidizing atmospheres. It is widely used as a standard for calibration of base-metal thermocouples.
- *13% rhodium versus platinum (Type R)* – Similar to Type S and produces a slightly greater EMF for a given temperature.
- *30% rhodium versus platinum, 6% rhodium (Type B)* – Similar to Types S and R, but yields somewhat greater physical strength and stability and can withstand somewhat higher temperatures.

Base metals:

- *Copper versus constantan (Type T)* – Used over the temperature range of -300 to +700°F (-184 to +370°C). Constantan is an alloy of approximately 55% copper and 45% nickel. Its primary uses are in measuring subzero temperatures. It has superior corrosion resistance in moist atmospheres.
- *Iron versus constantan (Type J)* – Used over the temperature range of -200 to +1,400°F (-130 to +760°C). It exhibits good stability at 1,400°F (760°C) in nonoxidizing atmospheres.
- *Chromel® versus Alumel® (Type K)* – Used over the temperature range of -200 to 2,300°F and is more resistant to oxidation than other base-metal combinations. Chromel® is an alloy of approximately 90% nickel and 10% chromium. Alumel® is approximately 94% nickel, 3% manganese, 2% aluminum, and 1% silicon. This combination must be protected against reducing atmospheres. Alternate cycling between oxidizing and reducing atmospheres is particularly destructive.
- *Chromel versus constantan (Type E)* – Produces the highest thermoelectric output of any conventional thermocouples. It is used up to 1,400°F (760°C) and exhibits a high degree of calibration stability at temperatures not exceeding 1,000°F (538°C).

2.11.2 Designations for Thermocouple Wire

The technician must be able to recognize the type of thermocouple and the polarity of thermocouple wires in order to ensure that thermocouple circuits are properly installed. Technicians should be familiar with both the letter designations and the color codes used to identify thermocouple leads and thermocouple extension wires. To make this task easier, the Instrument Society of America (ISA) has developed a standard, *MC96.1 Temperature Measurement Thermocouples*, to establish uniformity in the designation of thermocouple wire.

Letter designations have been used for many years to identify thermocouple types. Although the letter designation is often associated with a particular thermocouple material, the letter designation applies only to the temperature-EMF relationship and not to the material. Other materials having the same temperature-EMF relationship and better physical properties for a particular application may be given the same letter designation. Earlier in this module, the types and the materials used to define the temperature-EMF relationships were listed.

Polarity is also indicated by the thermocouple letter designations. In the letter designation, P represents the positive wire and N represents the negative wire.

Extension wires are frequently used with thermocouple circuits. They are designated by the letter X. *Table 4* lists the lead designations for thermocouples.

Table 4 Symbols for Indicating Positive and Negative

THERMOCOUPLE WIRE		
Type	**Positive**	**Negative**
B	BP	BN
E	EP	EN
J	JP	JN
K	KP	KN
R	RP	RN
S	SP	SN
T	TP	TN

The use of color codes to indicate the value and polarity of electronic components has been an accepted practice for many years. The manufacturers of thermocouples also use color coding to designate both the type and the polarity of thermocouples and thermocouple extension wires. For thermocouples and thermocouple extension wires, the color red designates the negative wire. This red-negative designation is endorsed by the ISA for all thermocouple types. *Table 5* summarizes the practice of color coding leads of thermocouples and thermocouple extension leads.

Another common method of designating the positive and negative polarity of thermocouple wire is by the order in which the material of the leads is written or spoken. The first wire named is positive, and the second is negative. For example, in the iron versus constantan thermocouple (Type J), the iron wire is positive and the constantan wire is negative. The indicated polarity of the thermocouple leads apply for conditions where the measuring junction is at a higher temperature than the reference junction.

2.11.3 Thermocouple Construction

There are various fabrication methods used in the manufacture of thermocouples. The factors that are considered in determining the correct fabrication process include the types of material used, the thermal and chemical environment to which the thermocouple is to be subjected, and the electrical insulation and mechanical strength requirements. Analysis of these factors determines how to join the dissimilar metals and what type of insulation and protection tube to use.

The primary requirement when joining thermocouple wires is to maintain good thermal and electrical conductivity without damaging the metallurgical properties of the wires. Several joining methods are illustrated in *Figure 34*. Method (A) is employed in the manufacture of Type J and K thermocouples using 8 AWG and 14 AWG gauge wire. It uses the resistance-welding technique. Method (B) is known as a butt weld. It is used to join 8 AWG through 20 AWG gauge wires in Type E, J, and K thermocouples. This joint is also made with resistance welding. The twisted wires shown in method (C) are either arc welded or gas welded. This manufacturing method can be used in the fabrication of Types E, J, K and T thermocouples using 8 AWG through 28 AWG gauge

Table 5 Color Coding of Thermocouple Wire

DUPLEX INSULATED THERMOCOUPLE WIRE					
Thermocouple			Color of Insulation		
Type	**Positive**	**Negative**	**Overall**	**Positive**	**Negative**
E	EP	EN	Brown	Purple	Red
J	JP	JN	Brown	White	Red
K	KP	KN	Brown	Yellow	Red
T	TP	TN	Brown	Blue	Red

SINGLE CONDUCTOR EXTENSION WIRE				
Extension Wire Type			Color of Insulation	
Type	**Positive**	**Negative**	**Positive**	**Negative**
B	BPX	BNX	Gray	Red-Gray Trace
E	EPX	ENX	Purple	Red-Purple Trace
J	JPX	JNX	White	Red-White Trace
K	KPX	KNX	Yellow	Red-Yellow Trace
R or S	SPX	SNX	Black	Red-Black Trace
T	TPX	TNX	Blue	Red-Blue Trace

SINGLE CONDUCTOR INSULATOR EXTENSION WIRE				
Extension Wire Type			Color of Insulation	
Type	**Positive**	**Negative**	**Positive**	**Negative**
B	BPX	BNX	Gray	Red-Gray Trace
E	EPX	ENX	Purple	Red-Purple Trace
J	JPX	JNX	White	Red-White Trace
K	KPX	KNX	Yellow	Red-Yellow Trace
R or S	SPX	SNX	Black	Red-Black Trace
T	TPX	TNX	Blue	Red-Blue Trace

205T05.EPS

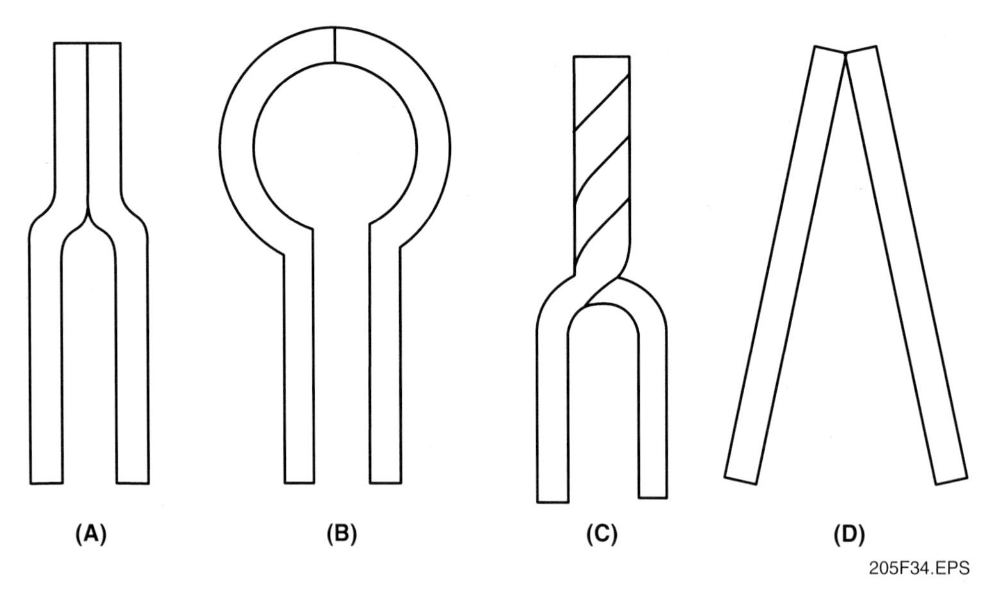

Figure 34 ◆ Various thermocouple joints.

wire. Finally, method (D) illustrates the V-joint, which may be arc, gas, or TIG welded. It is used to join Types B, R, and S thermocouples, as well as all types previously discussed.

There are several insulating materials that can be used to cover the sensing wires in order to prevent electrical shorts and to provide abrasion protection. Some materials, such as Teflon® and polyvinyl chloride, are used because of their natural resistance to absorbing moisture. Certain fiber materials may also be used when impregnated with wax, resin, or a similar substance. These impregnated fibers should not be used in applications where temperature exceeds 400°F, as this would vaporize the wax or resin and leave the thermocouple wire unprotected. For high-temperature applications, materials like fibrous silica and fiberglass are among the best of the non-ceramic types. These insulators are capable of withstanding operating temperatures up to 1,200°F. If temperatures greater than 1,200°F will be measured, a ceramic insulator must be used. Ceramic insulators can operate in environments exceeding 2,500°F.

The insulator is typically covered by a protective tube used to prevent mechanical damage to the sensing element. The tube also provides a layer of separation between the element and any harmful environmental conditions. Materials used for tube construction are: various metals, such as carbon steel, stainless steel, and high nickel alloys; ceramics; and combinations of metals and ceramics, called cermets. While metals provide adequate protection against mechanical damage, they are sometimes unsuitable due to their susceptibility to corrosion and because of their tendency to become porous at high temperatures. Ceramic or cermet tubes are used at temperatures exceeding approximately 2,100°F because of their high strength, high density, and resistance to corrosion.

Another type of construction that has become increasingly popular is the sheathed thermocouple. This device consists of a matched pair of thermocouple wires surrounded by a metal case. The inside of the case is filled with a non-compacted ceramic insulating material. The metal is then reduced in diameter by some mechanical process that tightly compacts the insulation around the thermocouple wires. This type of construction has the most desirable characteristics of the previously discussed types. In addition, it is easier and less costly to fabricate. A sheathed thermocouple is shown in *Figure 35*.

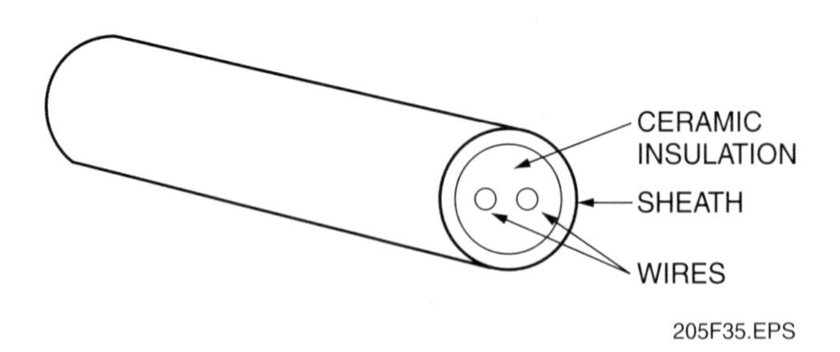

Figure 35 ◆ Sheathed thermocouple.

3.0.0 ◆ SECONDARY ELEMENTS

A secondary element is often used with a primary element to obtain a usable output from a detector. For example, an orifice plate causes a pressure drop, which can be measured to calculate flow. The orifice plate is the primary element because it causes the pressure drop. The device that measures the resulting pressure drop is called the secondary element. There are many types of secondary elements from which to choose.

3.1.0 Bourdon Tubes

Another device that uses displacement distance to measure pressure is the Bourdon tube. In this case, however, it is the device itself that displaces. Bourdon tubes come in three basic types: C-type, spiral type, and helical type. *Figure 36* shows a C-type Bourdon tube.

3.1.1 C-Type Bourdon Tubes

A C-type Bourdon tube is fabricated by flattening the side of a hollow tube, then bending the tube into the shape of a C. One end of the tube is sealed, and the open end is fixed to a support base. The material of the tube is selected for its elastic properties. It deforms under pressure, then returns to its original shape when pressure is removed.

The principle under which the C-type Bourdon tube works is that a bent tube with a cross-section that is not a perfect circle straightens out as pressure is applied.

The actual amount of tip displacement is relatively small. It is generally about ¼ inch. The angular displacement does not change in a linear manner. The small and nonlinear tip motion characteristics are compensated for mechanically, as shown with the direct indicating pressure gauge in *Figure 37*.

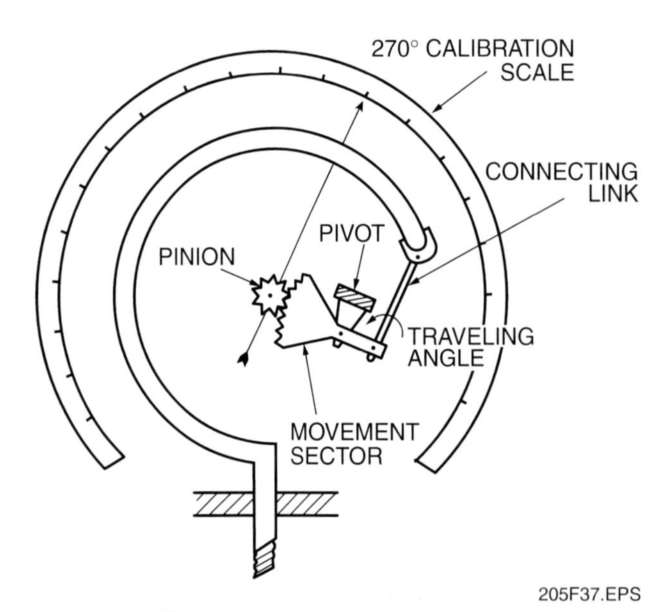

Figure 37 ◆ Pressure gauge.

Bourdon tubes are made of different materials. Each of these materials is appropriate for certain applications.

C-type Bourdon tubes have some disadvantages. The first is the large overhang of the tube, which makes it susceptible to shock or vibration. Another disadvantage is that using the gauge to monitor liquid pressure might yield a different indication than monitoring an equal gas pressure, due to the weight of the liquid in the tube. A third disadvantage is that the internal gears and links can become dirty or worn and cause severe loss of accuracy.

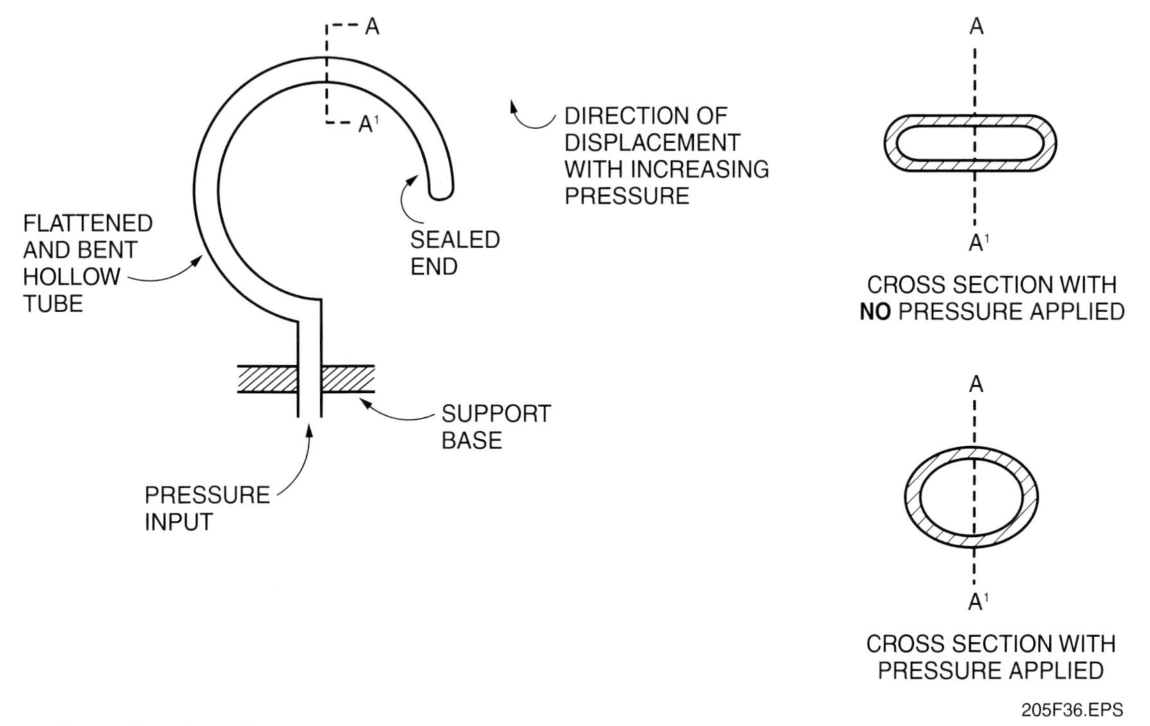

Figure 36 ◆ C-type Bourdon tube.

3.1.2 Spiral-Type Bourdon Tubes

As described earlier, one limitation of the C-type Bourdon tube is the relatively small amount of tip movement. *Figure 38* shows a spiral-type Bourdon tube that provides more tip movement. Because the spiral length is ¾ times longer than C-type length, the tip movement is ¾ times greater.

The spiral type works under the same observed principles as the C-type: as the applied pressure increases, the spiral uncoils.

Because of the increased tip movement, mechanical amplification is not normally needed. This results in an increase in sensitivity and accuracy because there is no lost motion from loose or sticking links, levers, or gears.

The same materials that are used for C-type Bourdon tubes are also used for spiral-type Bourdon tubes.

Spiral types are normally used in the range of 0 to 4,000 psig; however, some unflattened, heavy-wall spirals are available that can measure pressure up to 100,000 psig.

The effects of nonlinear tip motion, if not mechanically compensated for, can be minimized by operating the spiral type over only a portion of its range.

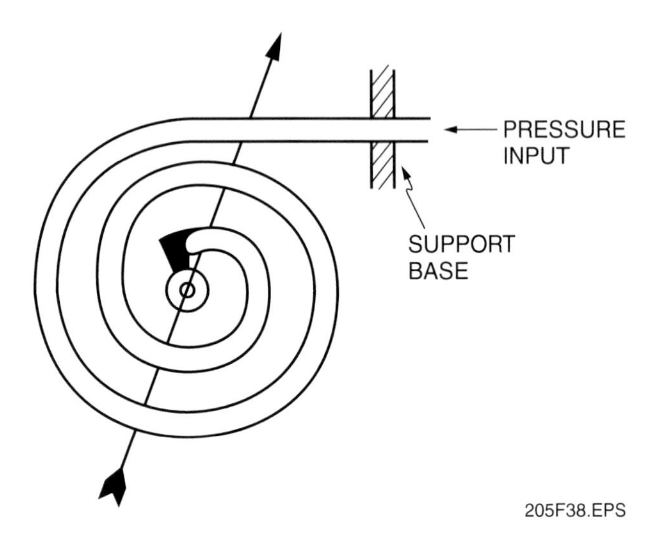

Figure 38 ◆ Spiral-type Bourdon tube.

3.1.3 Helical-Type Bourdon Tubes

Figure 39 shows a helical-type Bourdon tube which provides even more tip movement.

The observed principles of operation are the same as for the C-type.

High-pressure helical types might have as many as twenty coils, while low-pressure helical types might have only two or three coils. Recall that the change in tip motion decreases as the applied pressure becomes larger. Adding more coils compensates for this motion decrease.

Helical types are generally used for pressure ranges above 4,000 psig, up to a maximum range of 80,000 psig.

For a given pressure gauge application, the decision regarding the type of Bourdon tube (C-type, spiral type, or helical type) to use is generally made by the manufacturer.

3.2.0 Diaphragm Pressure Devices

The metallic diaphragm gauge consists of a metal disc built into a housing with one side of the disc exposed to the pressure to be measured and the other side exposed to atmospheric pressure. A basic diaphragm gauge is shown in *Figure 40*.

The distortion of the diaphragm under pressure is transmitted to the gauge dial by a linkage connected to the center of the diaphragm. Under pressure, a circular metallic diaphragm will exhibit a deflection curve shaped roughly like the letter S. The proportional limits occurs at approximately 0.5% of the bursting load. Therefore, the deflection of a flat metallic diaphragm is proportional to pressure in a linear fashion only for a small range of low pressures and low vacuums.

The deflection characteristics of a metal diaphragm can be somewhat improved by corrugating the surface of the disc. A corrugated diaphragm produces approximately four times the deflection of a flat diaphragm subjected to the same pressure. In addition, the deflection curve is slightly more linear.

The actual deflection depends on the diameter of the diaphragm, the thickness of the diaphragm,

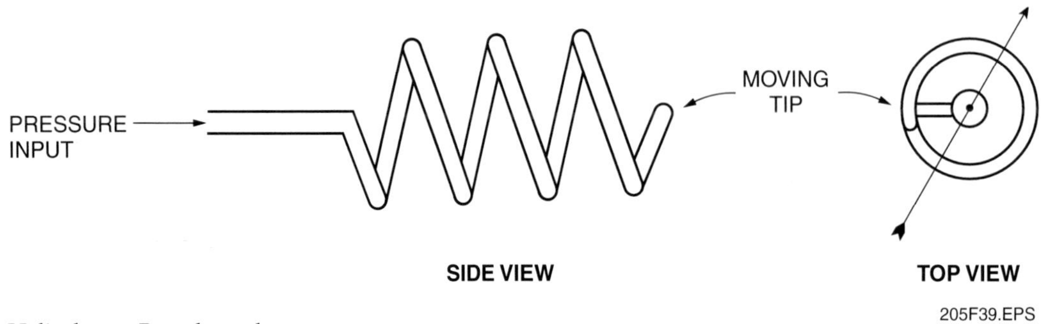

Figure 39 ◆ Helical-type Bourdon tube.

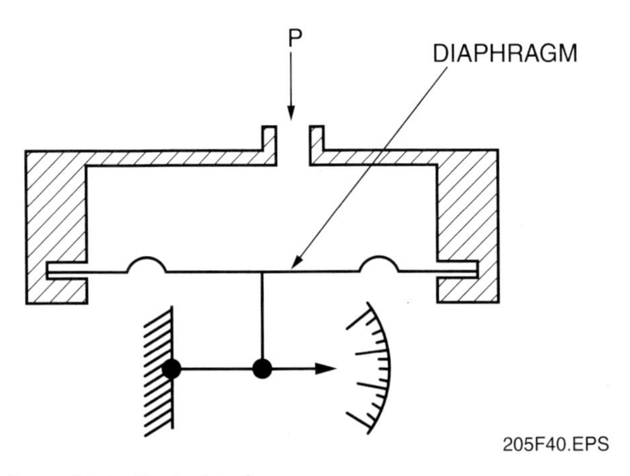

Figure 40 ◆ Basic diaphragm gauge.

the shape of the corrugations, the number of corrugations, the elasticity of the diaphragm, and the amount of pressure applied.

3.3.0 Pressure Capsules

A technology commonly used to measure low pressure is the pressure capsule. The capsule is usually incorporated as an internal component of either a pressure gauge or a transducer, which converts input pressure to another form of output (such as 4–20 mA). A pressure gauge or transducer incorporating a pressure capsule is typically applied in lower pressures that fall in the range of 0–30 psi. Specialty gauges or transducers containing pressure capsules are available for higher pressure ranges up to 200 psi.

Pressure capsules resemble two metal diaphragms that are soldered or welded together, then stacked together with other capsules (*Figure 41*). They are available in a variety of metals in order to provide compatibility with the chemical composition of the processes, as well as the process pressure and temperature.

Aluminum pressure capsules provide chemical, corrosion, and abrasion resistance in chemical processes. Stainless steel pressure capsules are ideal for medical, beverage, and food processing applications where stringent sanitation requirements are necessary. Ni-Span C®, which is a nickel-chromium alloy, is used for many tasks. These include monitoring liquid levels in bulk storage tanks and closed pressure vessels, monitoring pressure in climate control and energy management, monitoring air/fuel ratio measurement in industrial furnaces, and performing leakage measurement in natural gas meters. Phosphor-bronze capsules, due to their flexibility or elasticity, are usually used in low-temperature applications that are subjected to extreme vibration or pulsations.

The Ni-Span C® and bronze materials are predictably elastic, meaning they respond proportionally and predictably to mechanical pressure changes. They are generally rated to withstand maximum pressures approximately 1½ times their rated range without being permanently deformed. In most cases, this overpressure capacity is sufficient to prevent deformation. In some applications, however, such as industrial process machinery, the pressure may peak beyond the rated overpressure and cause a dent or bubble in the body of the capsule assembly. Once this happens, the capsule assembly is out of calibration and must be replaced.

3.4.0 Bellows Pressure Devices

It was the need for a pressure-sensing element that was more sensitive to low pressures than the Bourdon tube or the basic metallic diaphragm and capable of providing greater power for actuating recording and indicating mechanisms that resulted in the development of the metallic bellows. The use of metallic bellows has been most successful on pressures ranging from 0.5 to 75 psig. It is important not to overpressurize a bellows. *Figure 42* illustrates a basic bellows sensing element.

The bellows-type pressure gauge is usually built as a one-piece, collapsible, seamless metallic unit with deep folds formed from very thin-walled tubing. The moving end of the bellows is usually connected with a simple linkage to an indicator

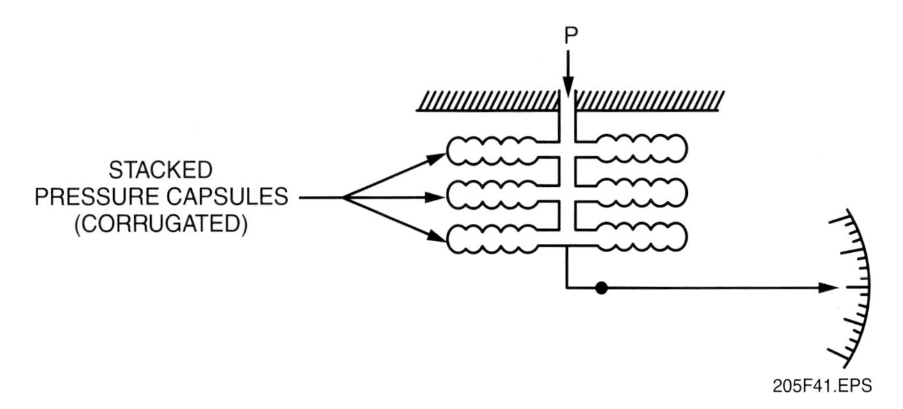

Figure 41 ◆ Stacked pressure capsule gauge.

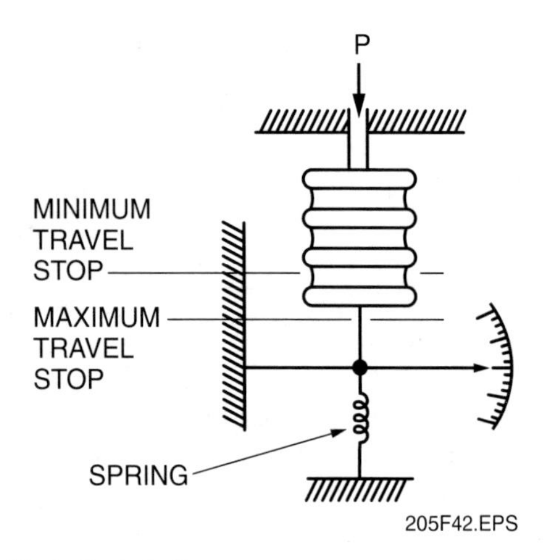

Figure 42 ◆ Basic bellows.

pointer. The flexibility of a metallic bellows is similar to that of a helical, coiled compression spring. The relationship between increments of load and deflection is linear up to the elastic limit. However, this linear relationship exists only when the travel of the bellows occurs under the influence of a minimum compressive force. All travel of the bellows must be made on the compressive side of the point of pressure equilibrium. For this reason, the spring in *Figure 42* is exerting on the bellows a compressive force that the movement or measuring action of the bellows must overcome. In practice, the bellows is always opposed by a spring, and the deflection characteristics of the unit are the net result of the spring and the bellows.

3.5.0 Capacitance-Type Pressure Sensors

As described earlier, pressure applied to a metallic diaphragm causes the diaphragm to deflect. As the diaphragm deflects, it moves away from the stationary metal housing. If the diaphragm and housing were viewed as the two plates of a capacitor, a change in applied pressure would cause proportional change in capacitance.

Figure 43 shows a diaphragm assembly arranged as a capacitance pressure sensor.

The pressure to be measured is applied to isolating diaphragm A. Isolating diaphragm B can be referenced to atmosphere for psig readings, vacuum for psia readings, or another pressure for differential pressure readings. These two diaphragms are called *isolating* because they prevent the fluid being measured from coming into contact with the sensing diaphragm.

The applied pressure is transmitted by the isolating diaphragm to the silicone oil. This oil serves as a non-corrosive pressure transmitting fluid and also as the capacitor dielectric.

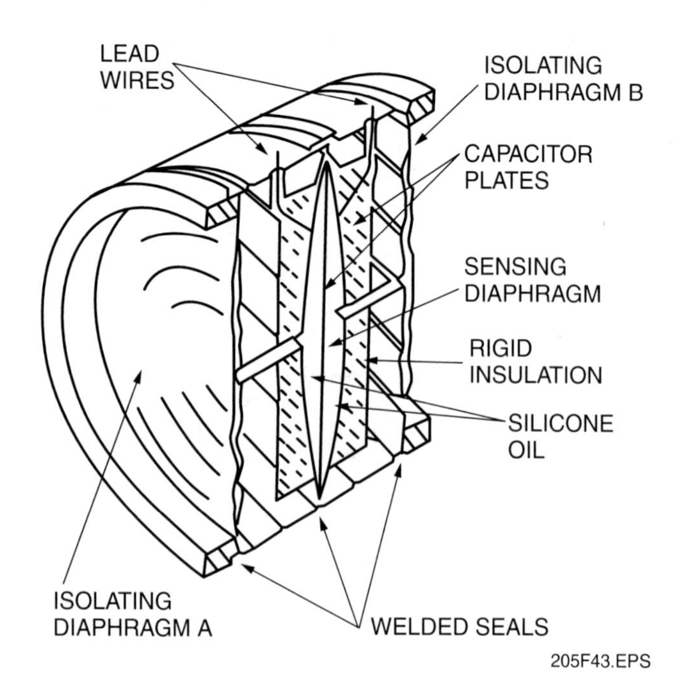

Figure 43 ◆ Capacitance pressure sensor.

The sensing diaphragm deflects by an amount and in a direction as determined by the net applied pressure. If the pressure that is applied to isolating diaphragm A is larger than that applied to isolating diaphragm B, the sensing diaphragm will deflect to the right.

Using different types of material to make the sensing diaphragm results in capacitance pressure sensors of different pressure ranges.

3.6.0 Secondary Element Protection

Secondary element protection includes diaphragm seals, pulsation dampeners, and pressure sensor positioning.

3.6.1 Diaphragm Seals

Stainless steel diaphragm seals can be used when a sensor is not corrosion resistant or is subject to possible contamination. The space between the sealing diaphragm and the sensor is filled with a suitable liquid whose pressure duplicates that of the process side of the diaphragm. The seals are usually assembled in the factory to ensure complete filling, which is very important. Diaphragm seals are commonly used on sensing instrument lines where the movement of the sensor is minimal. Excessive displacement may involve an error arising from the straining of the diaphragm seal.

Figure 44 shows a diaphragm seal used with a pressure gauge to prevent the corrosive liquid of the process from coming into contact with the Bourdon tube.

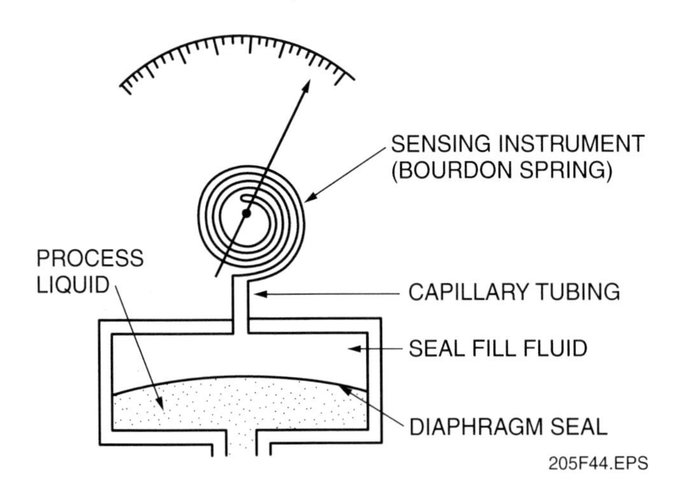

Figure 44 ◆ Diaphragm seal.

3.6.2 Pulsation Dampeners

In some process systems, the pressure can vary rapidly or pulsate. Pulsations can cause excessive wear of the internal moving parts of a pressure sensor. Dampeners that block the pulsations and pass the steady-state pressure can be included in sensing lines. One design consists of a **sintered** stainless steel disc or cylinder held in a stainless steel body. Another is a captive stainless steel pin in an orifice opening. Because plugging may present a problem, periodic cleaning of pulsation dampeners may be required.

3.6.3 Pressure Sensor Positioning

The pressure detector should be positioned on a stable, shock-mounted housing adjacent to the process line. The detector should be installed in such a way as to reduce the transmission of piping or vessel expansion strains, process heat, and system vibration to the sensor mechanism.

The response of most pressure sensors to a full-scale pressure change is extremely rapid. The response will, however, be affected by the length and diameter of the pipe or tubing used to connect the sensor to the process takeoff point. The longer the sensing line, the slower the response. For low-pressure measurement, the length of the sensing line should be short and the diameter small.

4.0.0 ◆ TRANSDUCERS

Transducers are used in instrument systems to ensure proper signal transmission to a system's final element or controller. In most cases, the final element is a pneumatically operated control valve. Transducers convert input signals of one form into output signals of another form. Simply stated, a transducer is a device that receives an input signal in some form and transforms that input into an output of another, usable form.

4.1.0 Transducer Functions

An industrial plant contains many types of instrument systems. The transducers used in these instrument systems must accept a signal generated by a controller or transmitter and convert it into a usable signal for another component in the loop. For example, an instrument system that has an electronic controller and a pneumatic final element requires a transducer that will convert an electrical signal into a proportional pneumatic signal.

4.2.0 Transducer Types

Many types of transducers are available for different applications. Transmitters are available that incorporate a combination of transmitting capabilities along with signal conversion in the form of a transducer. Pressure transmitters such as resistor-type strain transmitters are equipped with internal transmitters that have a voltage (0–5) or current (4–20 mA) output. Other sensors, such as low-pressure capacitance sensors, are also equipped with internal transducers.

Regardless of the type or location of transducer, its main function is to convert one signal form into another usable form. *Table 6* lists some of the more common signal conversions used in the instrumentation industry.

Table 6 Common Transducer Types

Type	Input Signal	Output Signal
I/P	Current	Pneumatic
P/I	Pneumatic	Current
I/R	Current	Resistance
R/I	Resistance	Current
MV/I	Resistance	Current
pH/I	pH	Current

4.3.0 I/P Transducers

Most I/P, or current-to-pneumatic, transducers are field-mounted instruments. A transducer is usually mounted close to its receiving instrument to avoid long, sluggish pneumatic signal transmission. In many cases, a transducer is mounted directly onto a receiving instrument such as a control valve actuator. *Figure 45* shows a typical I/P transducer mounting.

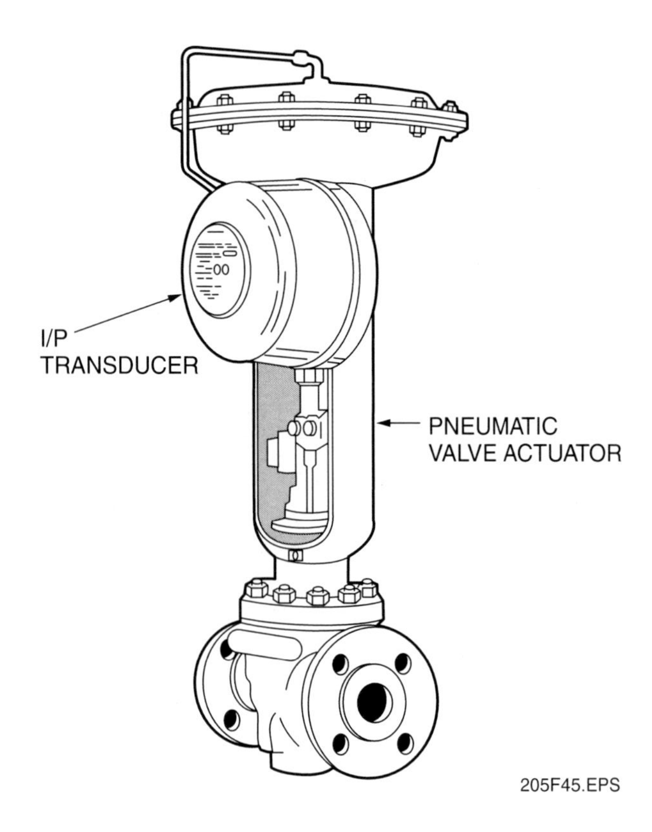

Figure 45 ◆ Typical I/P transducer mounting.

4.4.0 P/I Transducers

P/I, or pneumatic-to-current, transducers are generally used in chemical areas or humid spaces where pneumatic measuring instruments are used. Electrical measuring instruments are rarely used in these areas because of high corrosion. In order for the pneumatic signals to reach their controllers with minimal lag time, these signals must be converted to proportional mA signals. P/I transducers are usually mounted in a cabinet on a wall just outside these contaminated areas. *Figure 46* shows a typical P/I transducer mounting.

4.5.0 Transducer Operation

Both P/I and I/P transducers operate like typical electro-pneumatic instruments. The transducer circuitry relies on electronic and pneumatic components. Most P/I transducers receive a 3 to 15 psi signal from a controller or transmitter. Signals received outside of this range cause an incorrect output signal or no output signal at all. Most I/P transducers require either a 4 to 20 mA or a 10 to 50 mA input signal for proportional pneumatic signal conversion.

The following sections explain the operation of P/I and I/P transducers.

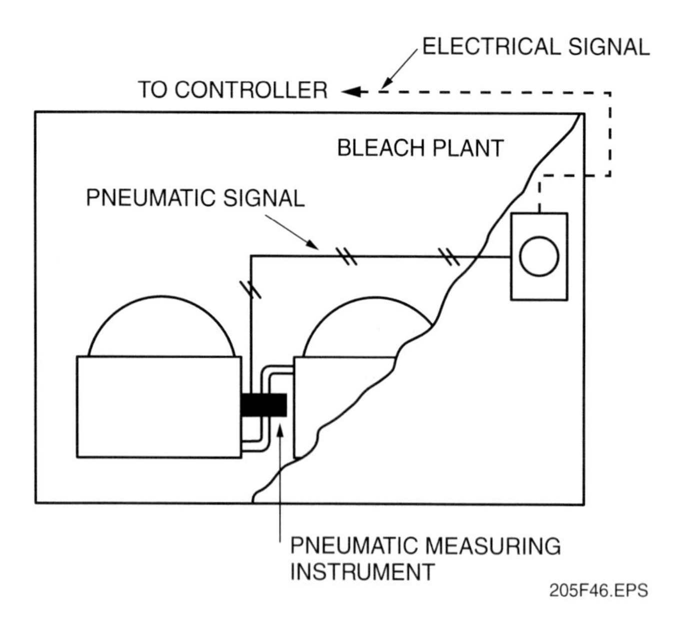

Figure 46 ◆ Typical P/I transducer mounting.

4.5.1 I/P Transducer Operation

An I/P transducer receives a DC milliampere input signal in the range of 4 to 20 mA or 10 to 50 mA. This signal is received by a coil positioned in the magnetic field of a permanent magnet. The coil reacts to the mA signal by producing a thrust parallel to the shape of the magnet. This thrust varies the gap between a flapper nozzle and flapper.

Figure 47 shows the relationship between coil movement and flapper movement.

The change in distance between the flapper and flapper nozzle changes the output pressure of the relay. This relay output pressure, also called the transducer output pressure, is fed to a feedback bellows. The bellows pivots on an adjusting fulcrum and moves the flapper nozzle. As the flapper nozzle moves, a throttling relationship is established between the flapper and flapper nozzle. *Figure 48* shows the operation of an I/P transducer.

4.5.2 P/I Transducer Operation

P/I transducers convert a pneumatic signal to a proportional mA output signal. This type of transducer usually consists of a pneumatic receiver and solid-state circuitry. The solid-state circuitry includes the following components:

- *Magnetic coils* – Induce an electromagnetic force (EMF)
- *Oscillator* – Generates a constantly changing electrical signal
- *Detector* – Receives a constantly changing electrical signal

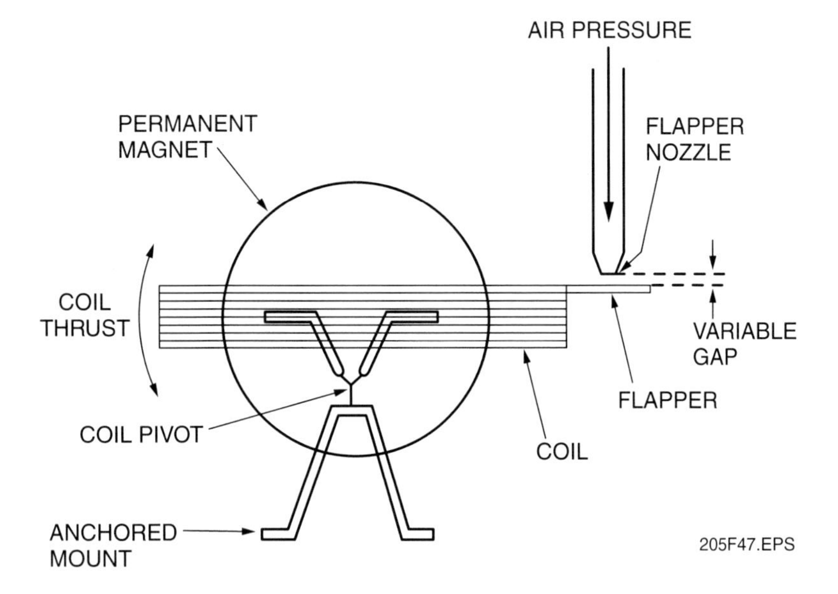

Figure 47 ◆ Relationship between coil movement and flapper movement.

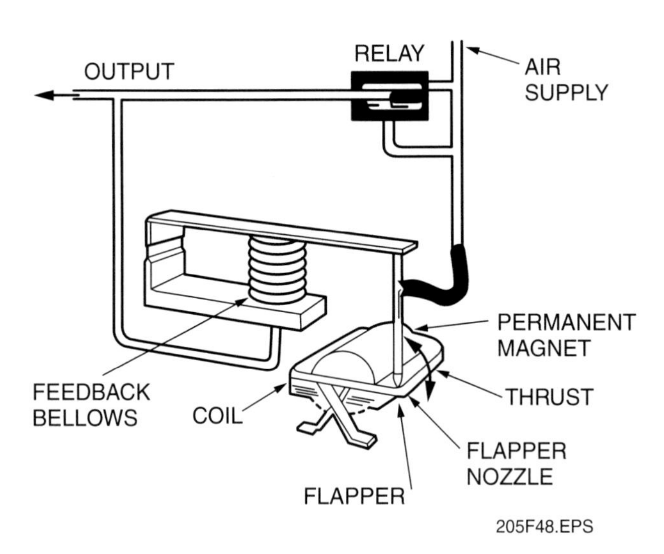

Figure 48 ◆ Operation of I/P transducer.

- *Amplifier* – Receives a constantly changing signal from an oscillator and responds to a voltage drop across a resistor
- *Driver* – Energizes the output transistor

Figure 49 shows the arrangement of a typical P/I transducer.

When a P/I transducer receives an input signal, the signal is applied to a bellows. As the bellows expands and contracts, a connecting linkage changes the inductance of the coils. *Figure 50* shows the relationship of the bellows and coils.

The change in inductance alters the signal amplitude of an oscillator inside the transducer. A detector receives the amplitude change and produces a proportional voltage drop across a resistor. This voltage drop is fed to an amplifier that drives the output transistor to produce an output current.

When a properly operating P/I transducer with a 3–15 psi input and a 4–20 mA output receives a pneumatic input signal of 9 psi, it should respond with a current output of 12 mA and is referred to as being in a balanced state. This can be verified in that an input range of 3–15 psi has a span of 12 psi (15 – 3 = 12). The median or middle of that span is 9 psi (½ of 12 = 6, and 6 + 3 = 9). We add the 6 psi to 3 because the range doesn't begin at 0 but at 3. Likewise, on the output side the range is 4–20 mA, with a span of 16 mA (20 – 4 = 16). The median or middle of that span is 12 mA (½ of 16 = 8, and 8 + 4 = 12). We add the 8 mA to 12 because the output range doesn't begin at 0 but at 4. So if the transducer has a 9 psi input and a 12 mA output, it is in balance.

A regulator is often incorporated in the electronic circuitry of a P/I transducer to maintain a balanced state within the transducer (*Figure 51*). By adding a regulator to the circuitry of the transducer, the output signal of the oscillator, which supplies current to the coils, is more accurately controlled by the regulator. The signal received by the detector from the oscillator (through the regulator) remains in balance with the pneumatic input signal.

A change in the 9 psi input signal disrupts the transducer's balanced mode. This causes a different current to flow through the coils and changes the oscillator signal amplitude.

A positive change in current flow in a P/I transducer produces an increase in output current. A negative change in current flow results in a decrease in output current.

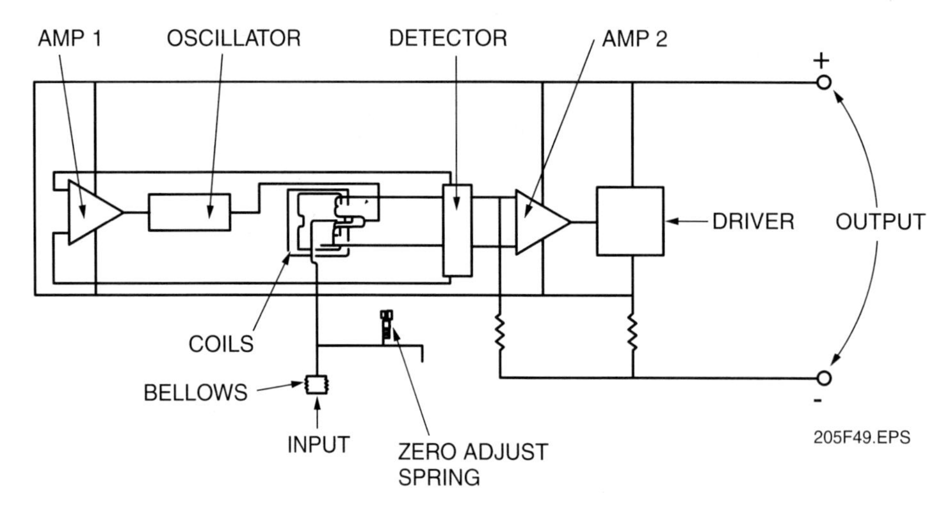

Figure 49 ◆ Arrangement of typical P/I transducer.

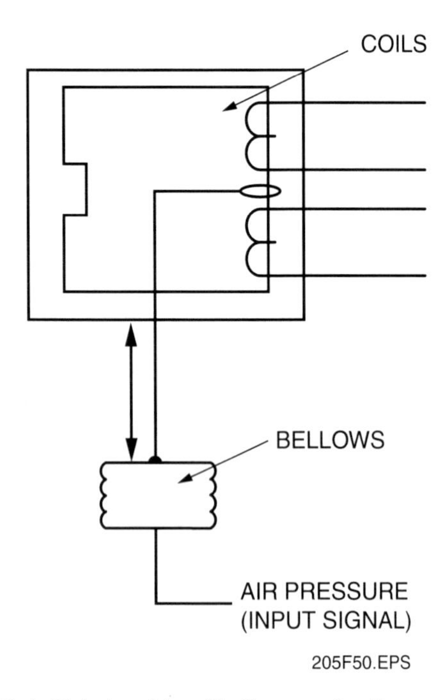

Figure 50 ◆ Relationship of bellows and coils.

4.6.0 Metallic Strain Gauges

The strain gauge is a transducer employing electrical resistance variation to sense the strain or other results of force. It is a very versatile detector for measuring weight, pressure, mechanical force, and displacement. A bonded strain gauge, as seen in *Figure 52*, is an electrical conductor of fine wire looped back and forth on a flexible mounting plate. The plate is usually bonded or cemented to the test piece member undergoing stress. The extension in length of the hairpin loops increases the effect of a stress applied in the direction of length. In short, a tensile stress would elongate the wire and thereby increase its length and decrease its cross-sectional area. The result would be an increase in its resistance, because the resistance (R) of a metallic conductor, at a constant temperature, varies directly with length (L) and inversely with cross-sectional area (A). Symbolically, where k is a constant depending on the wire:

$$R = \frac{kL}{A}$$

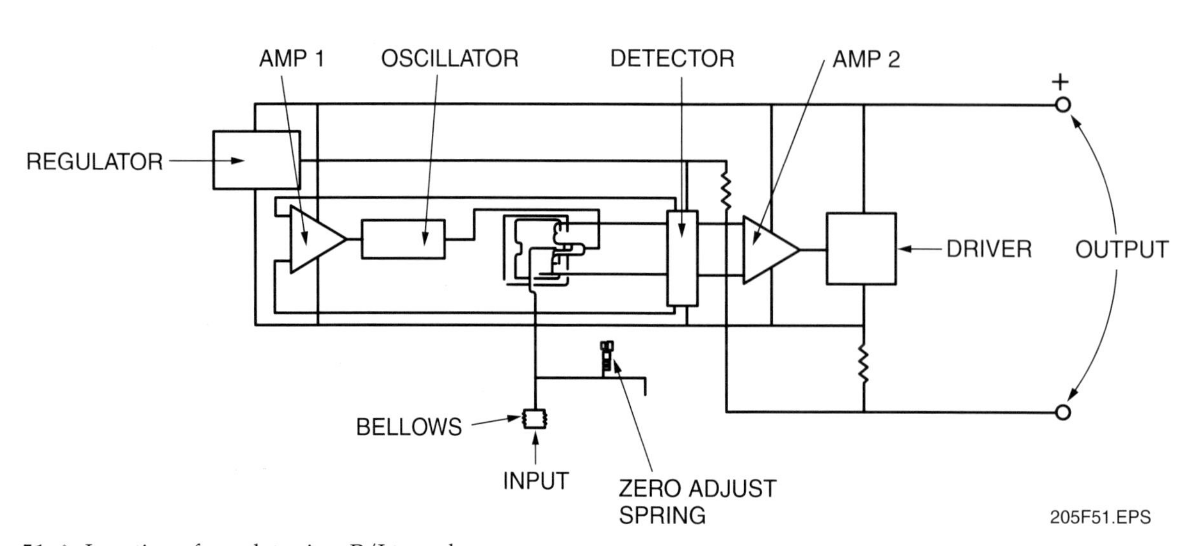

Figure 51 ◆ Location of regulator in a P/I transducer.

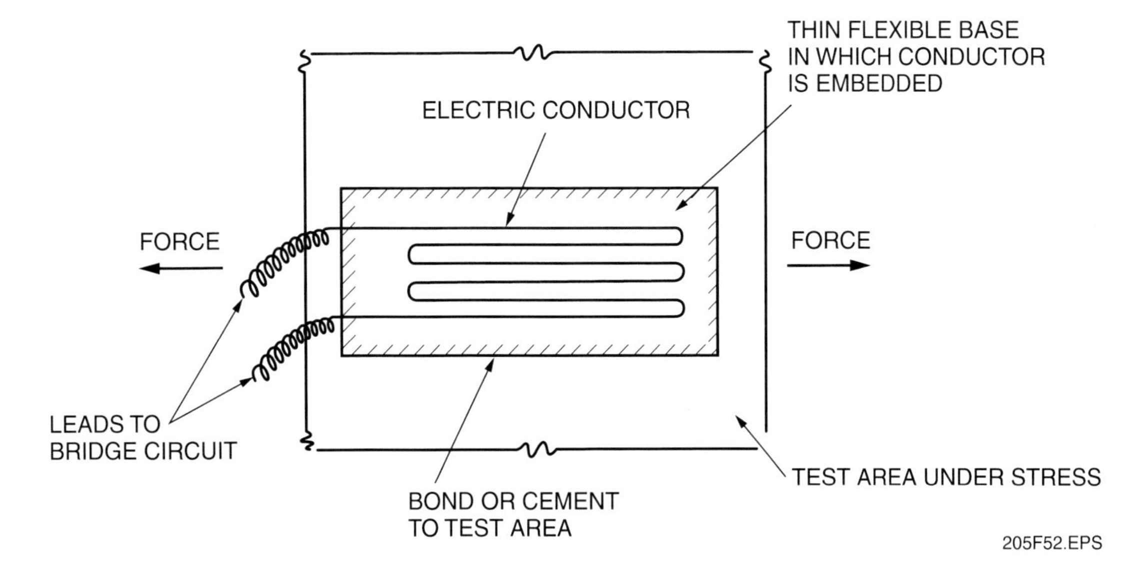

Figure 52 ◆ Application of a strain gauge.

With a good bond between the strain gauge and the test piece, either tensile or compressive strains can be measured. Generally, the current passed through the gauge is about 25 mA, the diameter of the conductor is about 0.001 inch, and the resistance of the conductor is about 100 ohms.

Because a current flow through the strain-gauge element has a heating effect proportional to the square of the current, any resulting change in resistance will have to be applied as a correction. One way to correct for the heating effect is to use a dummy gauge in the opposite arm of the bridge circuit, which has the same current flowing through it. The schematic arrangement is shown in *Figure 53*.

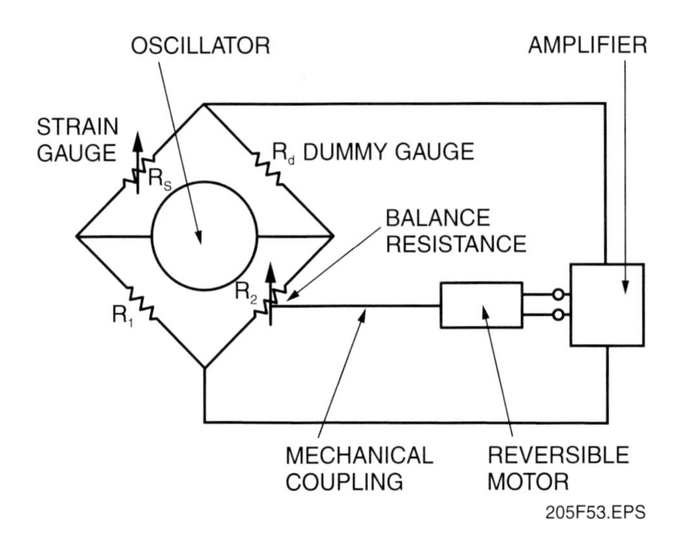

Figure 53 ◆ A resistance bridge circuit for measuring strain.

The type of measuring device used to balance the bridge circuit or to measure its unbalance depends on the required speed of response. For high-speed analysis, as used in research laboratories, an oscilloscope is used to measure the degree of unbalance. In industrial applications, a potentiometer is sufficient. Strain gauges are available in lengths from 1⁄16 to 1½ inch, with a resistance range of 60 to 6,000 ohms. They are calibrated to determine strain due to tensile and compressive loads.

4.6.1 Semiconductor Strain Gauges

Strain-gauge sensitivity has been improved by using semiconductors. The flexible silicon strain gauge is a very practical device. It is as stable as the metallic type and has a higher output level. Semiconductor strain gauges can detect microinches of change in length per unit inch of length.

4.7.0 Pressure Strain Gauges

A means of detecting very small variations in pressure incorporates strain gauges cemented or bonded on both sides of a flexible base (*Figure 54*). Each gauge will be in either compression or tension depending on the vertical motion of the bellows. In turn, the motion of the bellows depends on the pressure. Any slight pressure variation can be detected by the strain gauge bridge circuit and read on a suitable meter. This type of strain gauge, known as a bellows-cantilever beam strain gauge, is suitable for measuring pressures from 0 to 50,000 psi.

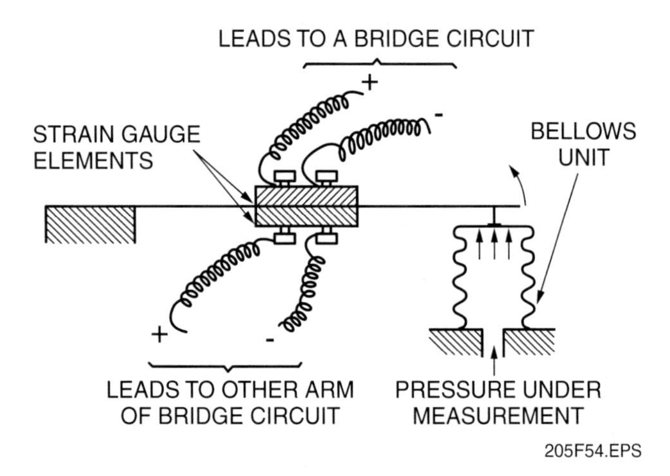

Figure 54 ◆ Application of the strain gauge for pressure measurement.

4.7.1 Voltage-Divider Pressure Transducers

A simple method of using the motion of a pressure-sensitive element, such as a bellows or diaphragm, is to actuate the arm of a potentiometer voltage divider. The moving arm of the voltage divider produces a voltage output proportional to the pressure variable, instead of a simple resistance variation. The voltage-divider potentiometer is suited to a DC indicating and recording system, which may not require amplification.

4.8.0 Piezoelectric Transducers

Deforming various crystals such as quartz, Rochelle salt, tourmaline, and barium titanate creates electrical potential proportional to the applied force or pressure. This action of generating a voltage by the application of a force or pressure is the principle used in the piezoelectric crystal transducer. A sectional view of a voltage-generating (piezoelectric) pressure transducer is shown in *Figure 55*. Piezoelectric crystals are found in such devices as phonograph pickup cartridges and ceramic microphones. Industry uses them in accelerometers, which are transducers that convert acceleration characteristics of mechanical vibration into a proportional electrical signal.

As mentioned, the piezoelectric crystal is inherently a dynamic responding sensor and is not really suitable for steady-state conditions. The crystal sensing element has a high output impedance and low current output. This type of transducer is used with an AC amplifier to increase the crystal output signal for readout purposes. Because it is a high-impedance measuring system, careful shielding is required.

The main advantages of piezoelectric transducers are self-generating power, dynamic response, small size, and rugged construction. The disadvantages are the electronic-signal-conditioning systems caused by the high-impedance output, sensitivity to temperature variations, and, with long leads, noise generation.

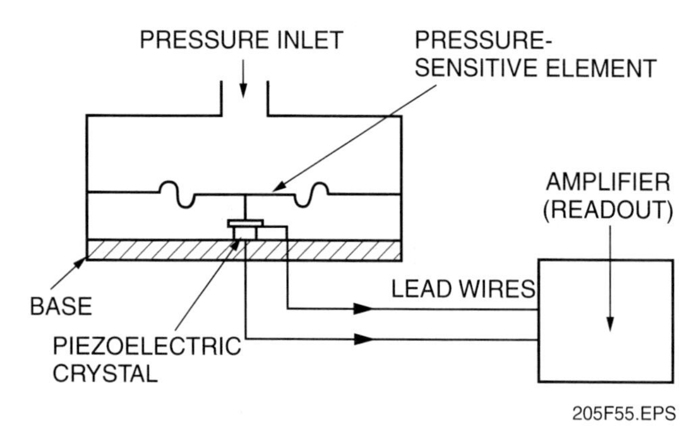

Figure 55 ◆ Voltage-generating (piezoelectric) transducer in a pressure-measuring system.

4.9.0 Linear-Variable Differential Transformers

The differential transformer may be designed to provide an electrical output that is linearly proportional to a mechanical displacement. The transducer is generally referred to as the linear-variable differential transformer (LVDT). The LVDT is an electromechanical transducer of the variable-inductance type (*Figure 56*).

Three coils are wound on a cylindrical coil form or tube. The center coil is the primary, which induces a voltage in each of two secondary coils, wound in opposite directions, on either side of it. The magnetic core is free to move axially inside the assembly as a result of a displacement.

When the primary coil is energized by alternating current, voltages are induced in the two secondary coils. Because these coils are connected in series opposition, the two voltages in the secondary circuit are opposite in phase. Thus, the resulting output of the transformer circuit is the difference between the two voltages. In the null, or balanced, position the movable magnetic core is at a central point. If the core is moved from the null position, owing to displacement variation, the voltage induced in the secondary coil will either increase or decrease, according to the direction of the displacement change. In measuring systems, the transformer is designed to produce a differential-voltage output which varies linearly as seen in *Figure 56*.

The linear-variable differential transformer can be used as a transducer to measure other variables such as pressure, weight, and acceleration. The accelerometer is a good example of an LVDT application.

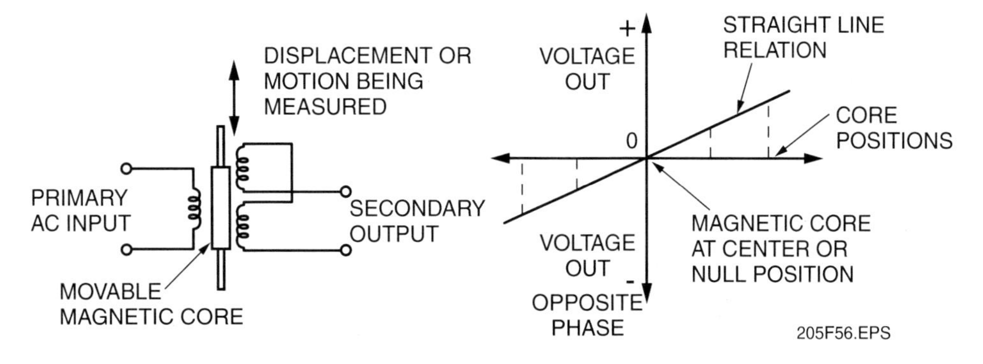

Figure 56 ◆ Linear-variable differential transformer schematic and linear function.

4.10.0 Accelerometers

In accordance with Newton's basic law of motion,

Force (F) = mass (m) × acceleration (a)

The elastic deformation of the sensitive measuring element, instead of being produced directly by an external force, may be a function of acceleration. An accelerometer is shown in *Figure 57*.

The secondary-circuit output signal, which can be picked up by any of the conventional measuring circuits described in this and preceding sections, is a function of the displacement of the moving magnetic core caused by acceleration. Accelerometers of the base-mounting type, capable of withstanding the extreme conditions of missile testing, operate on a similar principle. Accelerometers can be applied to detect vibration and convert it into a usable signal.

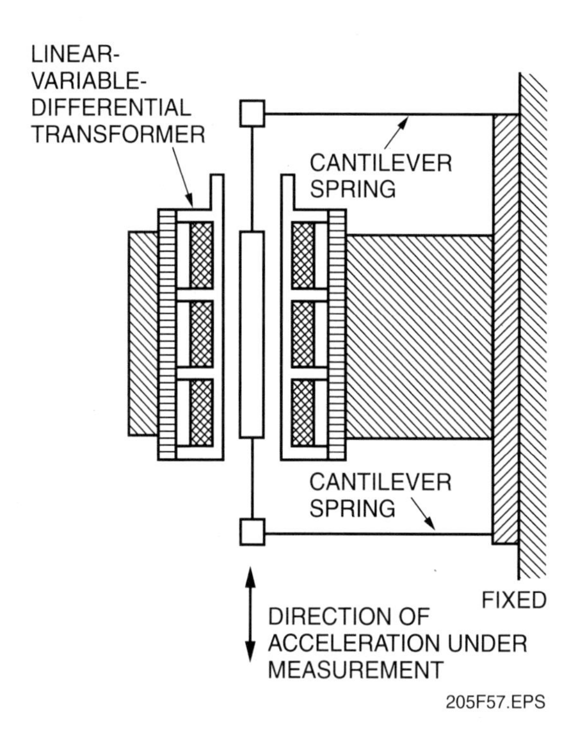

Figure 57 ◆ Linear-variable differential transformer used to measure acceleration.

5.0.0 ◆ TRANSMITTERS

A transmitter usually consists of a sensor and an output device. The sensor detects a signal from a primary element that is measuring a process variable. The output device converts this initial signal to a transmittable signal that is proportional to the measurement. This signal can then be used to indicate, record, or control the process variable.

Transmitters can be pneumatic or electronic. The trainee should understand transmitters and how they function.

5.1.0 Pneumatic Transmitters

The two main operating types of pneumatic transmitters are force balance transmitters and motion balance transmitters. These units differ in their basic operating principles. Most pneumatic signals use 3 psi to correspond to a zero signal so that, should a leak occur (or if the air supply is lost), a loss of pressure will be seen. The 3 psi baseline also allows calibrations above and below the calibration point.

5.1.1 Force Balance Differential Pressure Pneumatic Transmitters

The most common type of pneumatic transmitter is the force balance differential pressure transmitter. This type of transmitter is also called a DP cell. The DP cell usually operates on a regulated 20 psi air supply. The output signal of the DP cell is usually 3–15 psi.

A pneumatic DP cell can be divided into four basic sections:

- Process measuring section
- Force bar section
- Balancing section
- Pneumatic input/output section

Figure 58 shows a pneumatic DP cell.

5.1.2 Process Measuring Section

The process measuring section of a DP cell consists of a transmitter body and a diaphragm capsule. The diaphragm capsule splits the transmitter body into a high-pressure chamber and a low-pressure chamber. *Figure 59* shows a transmitter measuring section.

Process fluid is piped to the high- and low-pressure sides of the transmitter body. Pressure acts on the sealed metallic diaphragm and causes a small movement of the diaphragm link.

5.1.3 Force Bar Section

The bottom of a DP cell force bar section is connected to the diaphragm link. A force bar seal prevents process fluid from leaking out of the transmitter body. The top of the force bar is connected to the pneumatic flapper-nozzle. *Figure 60* shows a force bar section.

Any difference in pressure in the transmitter body causes a movement of the diaphragm link. This movement is transferred to the force bar at the link connection. An increase in the high-pressure side causes the base of the force bar to move toward the low-pressure side.

This action draws the flapper closer to the nozzle. An increase in pressure on the low-pressure side moves the flapper away from the nozzle. *Figure 61* shows the effect of diaphragm movement on flapper-nozzle position.

5.1.4 Balancing Section

The balancing section of a DP cell consists of a feedback bellows, range rod, range wheel, and zero adjustment. *Figure 62* shows this balancing section.

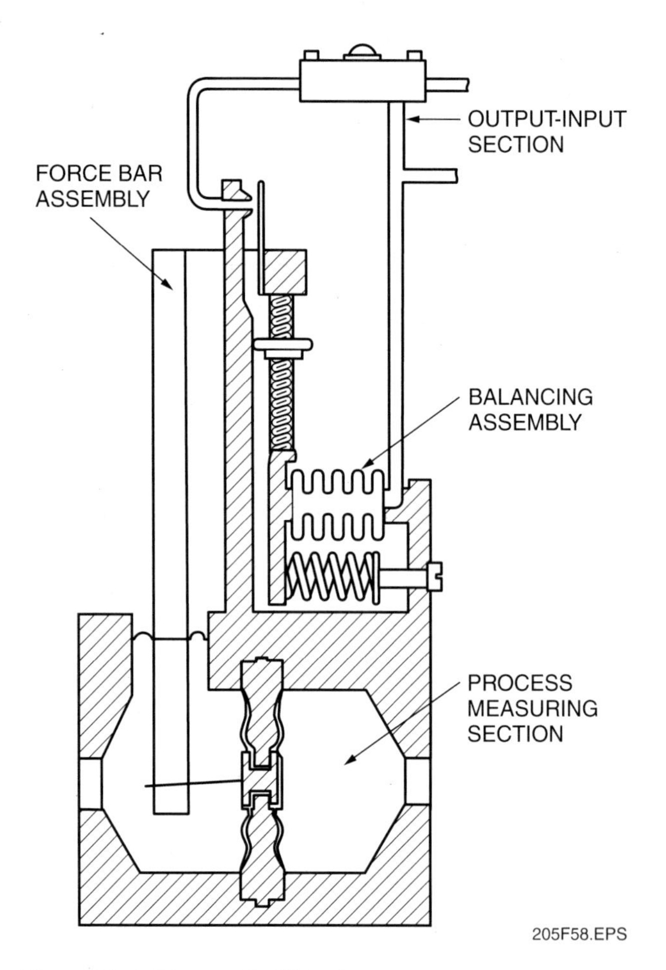

Figure 58 ◆ Pneumatic DP cell.

Movement of the flapper-nozzle by the force bar causes changes in the amount of air supplied to the feedback bellows. The bellows air supply is changed until the bellows exerts enough pressure on the range rod to balance the original movement of the force bar. The term *force balance* comes from these counterbalancing actions.

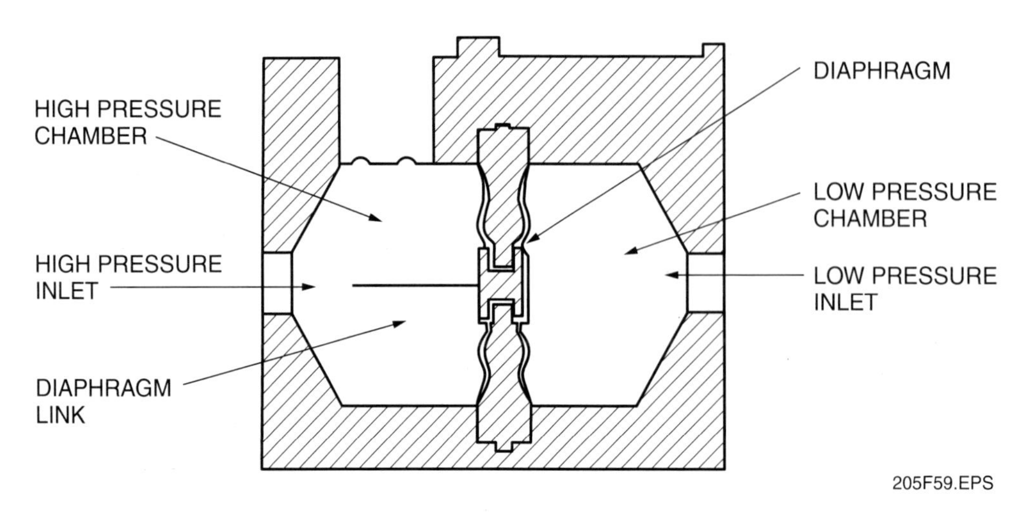

Figure 59 ◆ Transmitter measuring section.

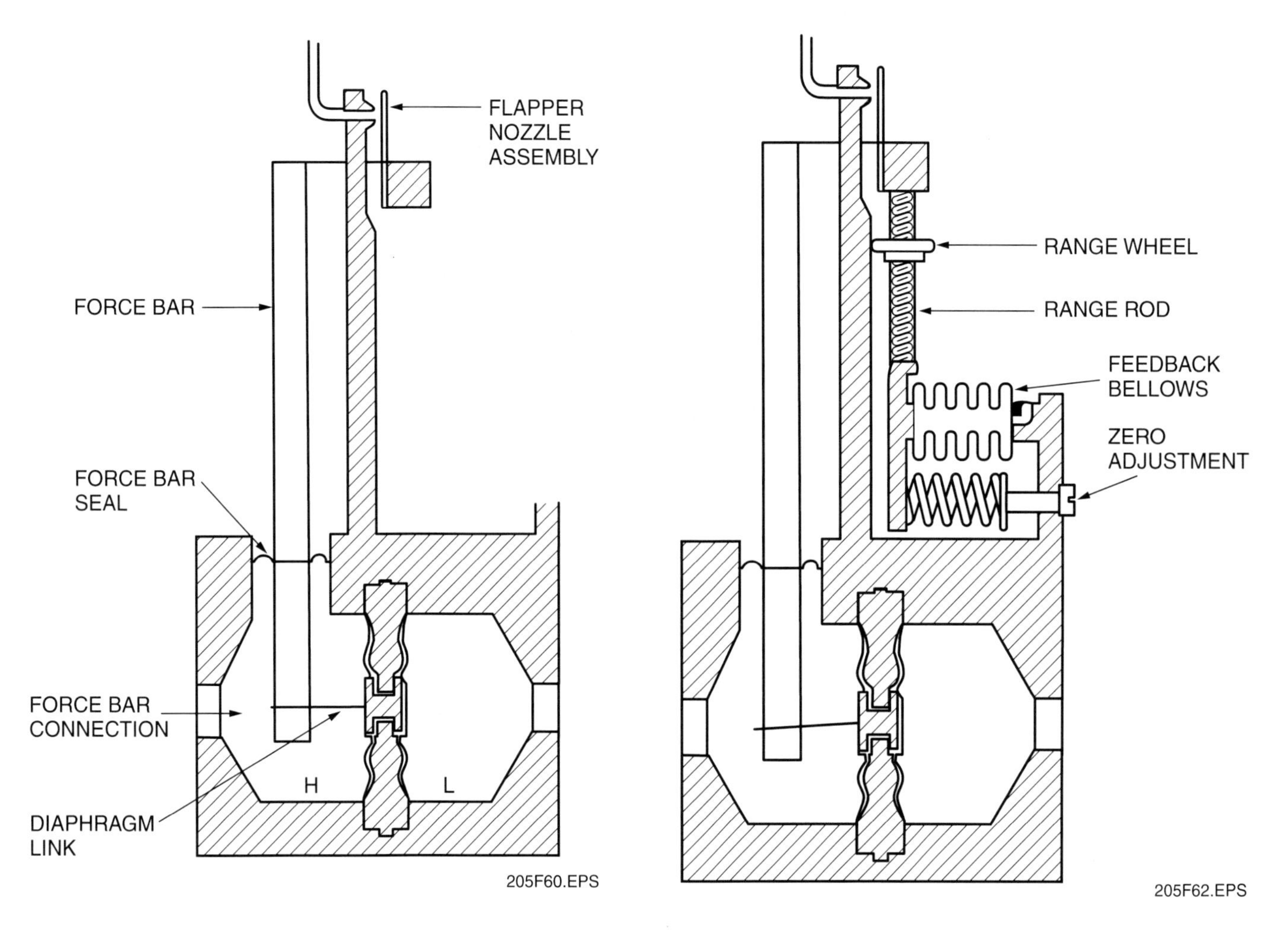

Figure 60 ◆ Force bar section.

Figure 62 ◆ Balancing section.

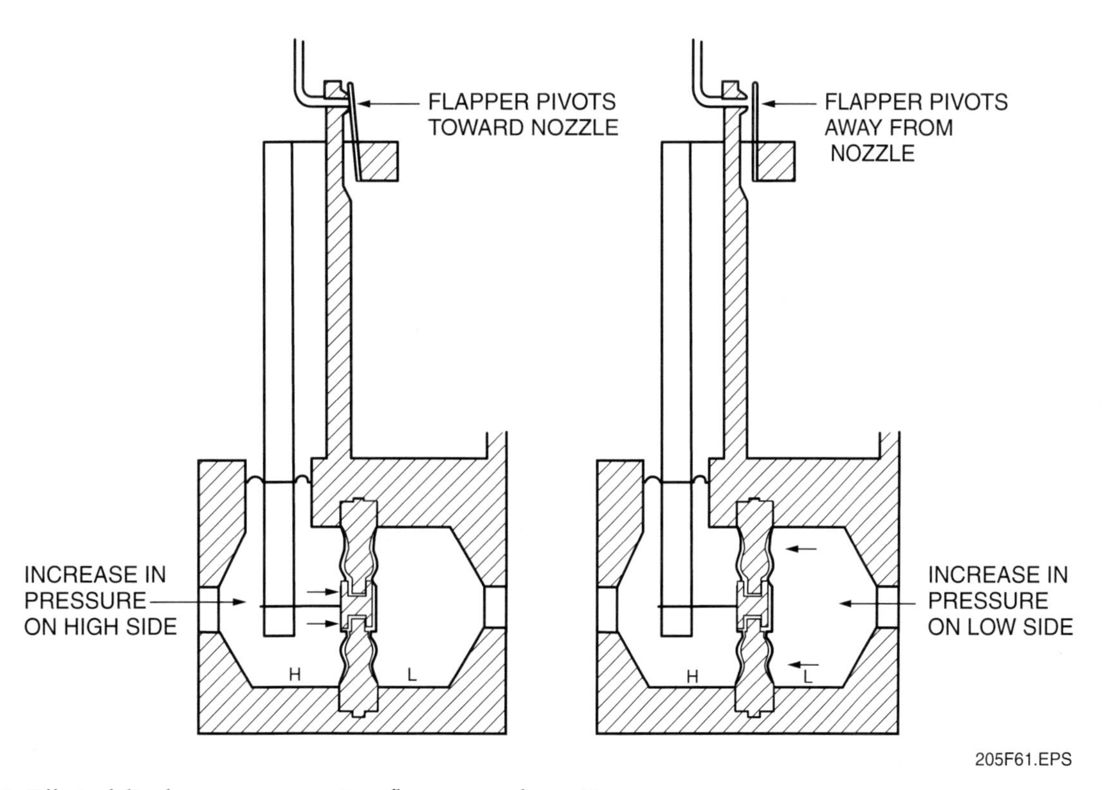

Figure 61 ◆ Effect of diaphragm movement on flapper-nozzle position.

The zero adjustment is used to establish a zero reference, usually 3 psi, by allowing a manual setting of the flapper-nozzle position. The range wheel is the balancing point for the motion of the range rod. By adjusting the range wheel up or down on the range rod, the total measuring range of the instrument can be changed.

5.1.5 Pneumatic Input/Output Section

The pneumatic input/output section of a DP cell consists of a regulated 20 psi air supply, a pneumatic relay, a nozzle, and an output connection. *Figure 63* shows a pneumatic input/output section.

The pneumatic relay (*Figure 64*) operates from a 20 psi air supply. A fixed restrictor within the relay body provides a constant supply of air to the nozzle. The nozzle pressure acts on a diaphragm inside the relay. The diaphragm, in turn, acts on a ball check valve that supplies air to the feedback bellows and output signal.

An increase in process pressure on the high-pressure side of the capsule causes the flapper to cover the nozzle. This causes air pressure to build up on the nozzle supply side of the diaphragm. The downward movement of the diaphragm causes the ball check valve to open and increase the air pressure to the feedback bellows and output signal. This continues until the pressure of the bellows against the range rod counterbalances the original movement of the force bar.

If the flapper moves away from the nozzle, the diaphragm within the relay reduces the air pressure to the feedback bellows and output supply. This action occurs until the unit is again in balance.

5.1.6 Pneumatic Force Balance Transmitter Applications

Pneumatic force balance transmitters can be used in most process measurement applications where differential pressure measurement devices are used.

Common process measurements made with pneumatic force balance transmitters are:

- Flow measurement
- Liquid level measurement
- Pressure measurement
- Temperature measurement

5.1.7 DP Cell Flow Measurement

For flow measurement, a DP cell is most commonly used with an orifice plate. *Figure 65* shows a DP cell flow installation without the air supply or pneumatic output signal lines connected to it.

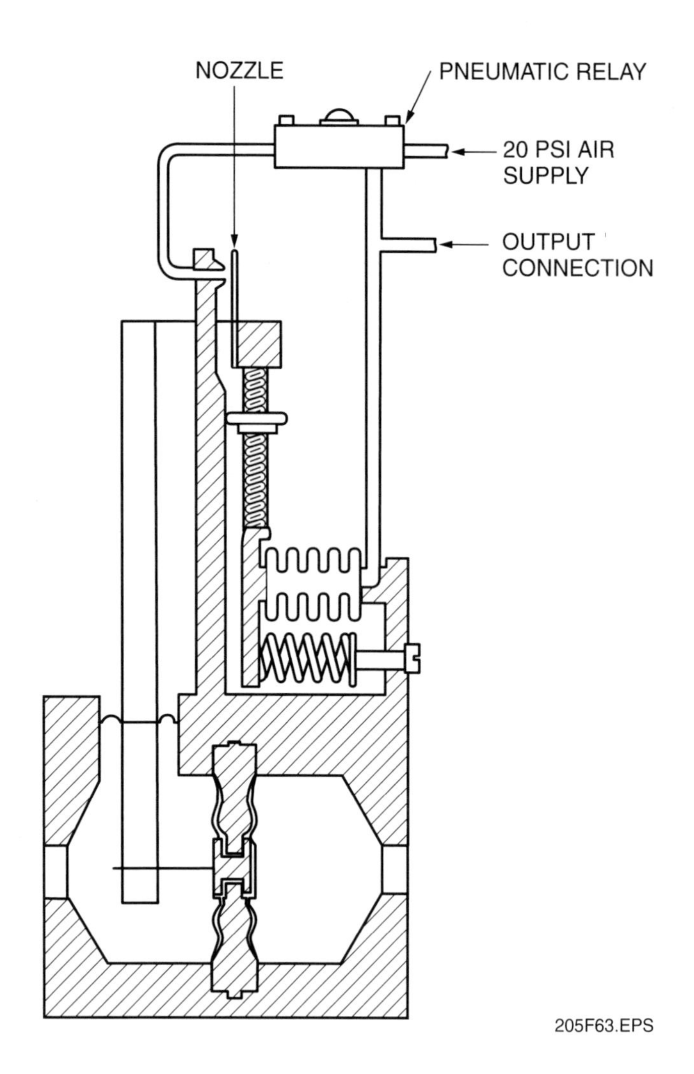

Figure 63 ◆ Pneumatic input/output section.

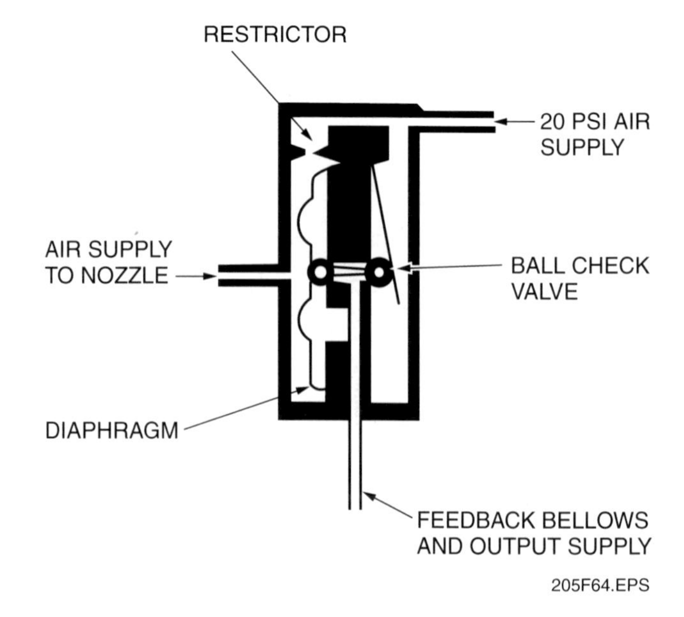

Figure 64 ◆ Pneumatic relay.

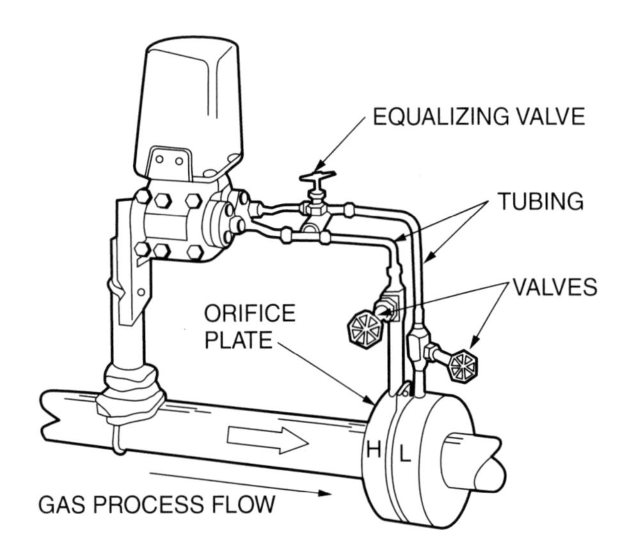

Figure 65 ◆ DP cell flow installation.

The DP cell can be mounted on the process piping or at a location as close as possible to the orifice plate. The valves located near the orifice plate can be closed to make repair or removal possible without shutting off the process flow.

The equalizing valve on the process tubing is used in replacement, removal, and calibration procedures. When the valve is open, it permits equal pressure to be applied to both sides of the diaphragm capsule. High pressure applied to only one side of the diaphragm capsule can cause damage.

5.1.8 DP Cell Liquid Level Measurement

With a slight modification in tubing connections, a DP cell can be used to measure the liquid level in both open, nonpressurized tanks and closed, pressurized tanks. *Figure 66* shows DP cell liquid level applications.

Tank A in *Figure 66* is an open, nonpressurized vessel. The high-pressure side of the DP cell is connected near the bottom of the tank. The low-pressure side is vented to atmospheric pressure.

Increasing the level in the tank causes more pressure to be applied to the high side of the capsule. This causes an increase in the output signal. The low side of the DP cell is open to the atmosphere. Increasing or decreasing liquid pressure on the high side is the only pressure change that affects the signal from the DP cell.

Tank B in *Figure 66* is a closed, pressurized vessel. The high side of the DP is connected near the bottom and the low side at the top. A rise in level increases the pressure to the high side due to the increase of head pressure of the liquid. A rise in level also increases the pressure to the low side due to compression of the air or vapor gap above the liquid. As would be expected, a fall in liquid level decreases pressure to both sides.

The DP cell measures the difference between the actual liquid pressure (high side) and the compressed vapor pressure in this application.

Tank C in *Figure 66* is also a closed, pressurized vessel. Due to some process conditions, the low leg can fill up with the process fluid or condensate and cause measurement errors. For example, if steam from a container of hot water fills the low-pressure piping leg, this steam will condense into a liquid as it cools off. This will create a false liquid level pressure on the low side and cause errors in measurement.

To eliminate the problem of liquid collecting in the low leg, the low leg is intentionally filled with liquid. The DP cell can then be mechanically adjusted to balance out the liquid pressure on the low leg. This is referred to as a wet leg installation.

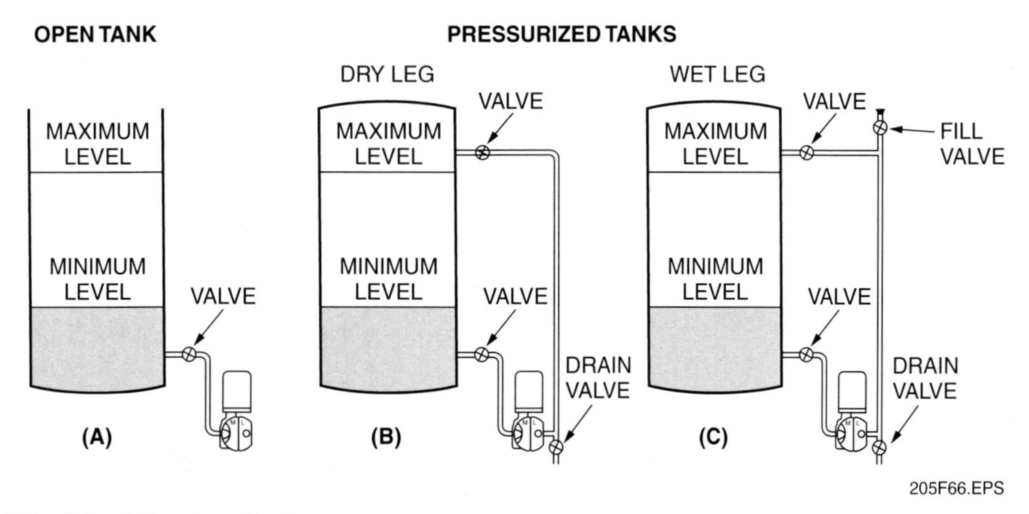

Figure 66 ◆ DP cell liquid level applications.

5.1.9 DP Cell Pressure Measurement

For pressure measurement, the DP cell is constructed with only a high side connection. The low side is usually a metal blank with a small vent for atmospheric pressure. *Figure 67* shows a DP cell used for pressure measurement. A standard DP transmitter can also be used, as long as the low side is vented to the atmosphere.

The DP cell can be mounted directly on a vessel or pipeline to measure pressure. The valve permits removal or replacement of the DP cell without process shutdown.

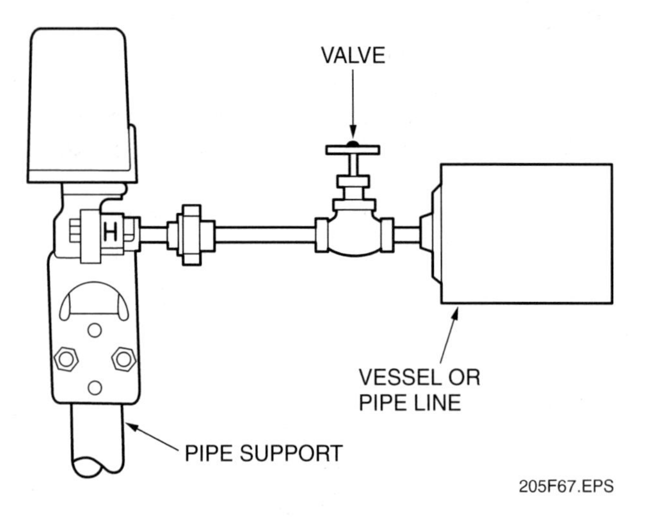

Figure 67 ◆ DP cell used for pressure measurement.

5.1.10 Pneumatic Force Balance Temperature Measurement

The body of a pneumatic force balance temperature transmitter is different from that of a normal DP cell. *Figure 68* shows a force balance temperature transmitter.

A pneumatic force balance temperature transmitter has a thermal element that senses temperature changes. The bellows on the end of this thermal element responds to temperature changes by expanding and contracting. This force is transferred to the force bar, causing flapper-nozzle position changes. The change in the flapper-nozzle position is sensed by the relay, and balancing pressures are sent to the feedback bellows. A compensation bellows unit on the force bar eliminates the effects of ambient temperature.

5.1.11 Motion Balance Pneumatic Transmitters

Motion balance transmitters convert the input motion of basic measuring elements into a pneumatic output that is directly proportional to the measured variable.

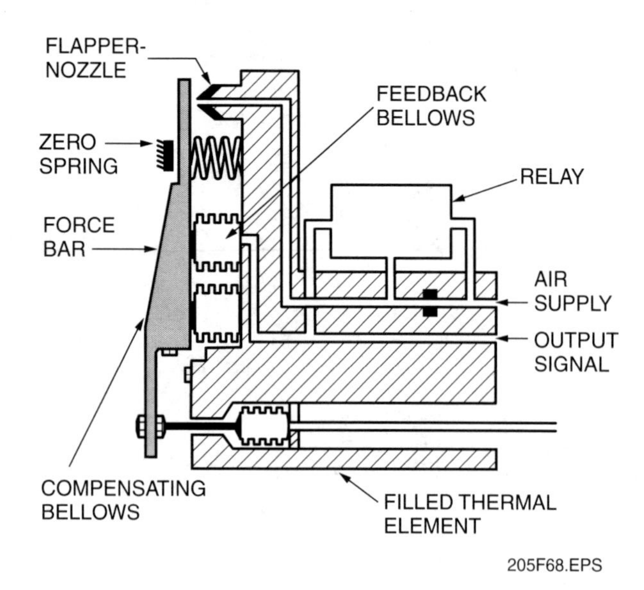

Figure 68 ◆ Force balance temperature transmitter.

Basic measuring elements such as Bourdon tubes, bellows, and diaphragm assemblies can move distances of up to ½ inch. Pneumatic motion balance transmitters convert this input motion into an output pressure. *Figure 69* shows a motion balance transmitter.

A motion balance transmitter receives an air supply input of 20 psi. Its output of 3 to 15 psi is proportional to the measured value. The three sections of a motion balance transmitter are:

- Measuring section
- Link and flapper-nozzle section
- Bellows and relay section

5.1.12 Measuring Section

The measuring section of a motion balance transmitter is the basic measuring element. In most motion balance transmitters, this basic element is either a Bourdon tube, a diaphragm assembly, or a bellows unit. These measuring elements rely on a link to transfer the measurement motion. *Figure 70* shows some common measuring elements used in motion balance transmitters.

The measuring element section provides the initial motion to the transmitters.

5.1.13 Link and Flapper-Nozzle Section

Changes in the measured variable move the connecting link on the measuring element. This motion is transferred to the link assembly. The link assembly moves up or down on its pivot point. This movement causes the flapper-nozzle relationship to change. *Figure 71* shows the link and flapper-nozzle assembly.

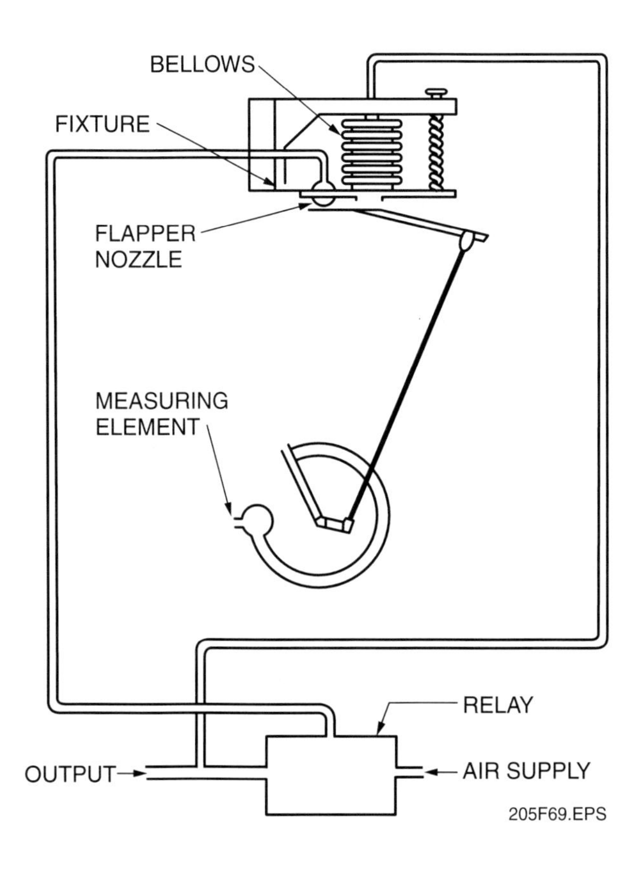

Figure 69 ◆ Motion balance transmitter.

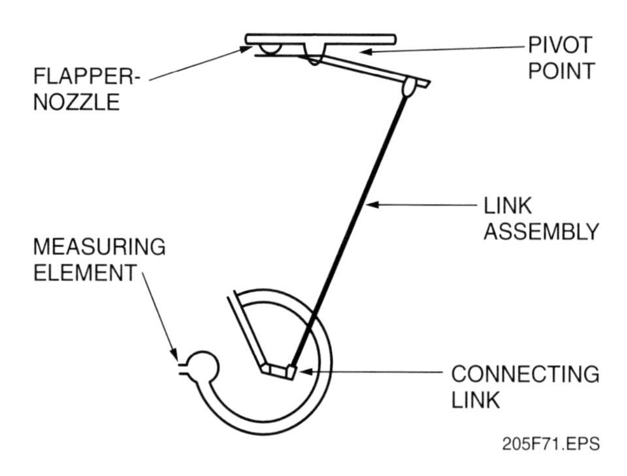

Figure 71 ◆ Link and flapper-nozzle assembly.

5.1.14 Bellows and Relay Section

Changes in the flapper-nozzle position affect the output of the pneumatic relay. At this point, the action of the motion balance transmitter is the same as the action of a force balance transmitter.

As the nozzle is covered or uncovered, the back pressure changes in the pneumatic relay assembly. These changes are fed back to the bellows. Any movement of the flapper toward or away from the nozzle causes a pressure change

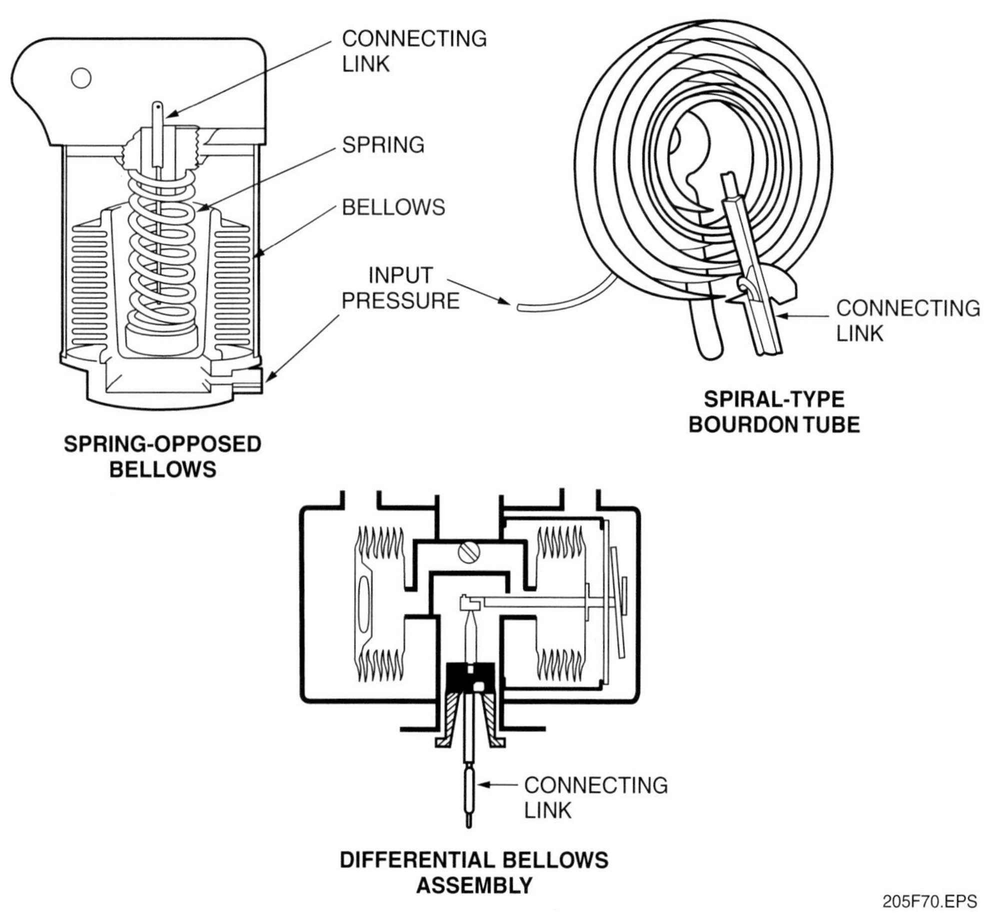

Figure 70 ◆ Common motion balance measuring elements.

in the bellows. The pressure change continues until the unit is in balance. As in a force balance transmitter, the feedback pressure is the output signal of the transmitter. A zero adjustment is provided for an initial 3 psi output. *Figure 72* shows the bellows and relay assembly.

5.1.15 Applications of Motion Balance Transmitters

Motion balance transmitters are most commonly used in temperature and pressure measurement.

In pneumatic motion balance temperature measurement, a thermal expansion system is used as the basic measuring element. The filled thermal piping is connected to a Bourdon tube inside the transmitter housing. As the temperature of the process rises and falls, the thermal fluid expands and contracts within the Bourdon tube, causing a motion at its connecting link. This motion activates the transmitter. *Figure 73* shows a motion balance temperature measurement installation.

In pneumatic motion balance pressure measurement, the transmitter is connected directly into the process piping or vessel. A Bourdon tube or bellows assembly is located in the transmitter housing and is used as the basic measuring element. A change in process pressure moves the connecting link of the measuring element. This motion activates the transmitter. *Figure 74* shows a motion balance pressure measurement installation.

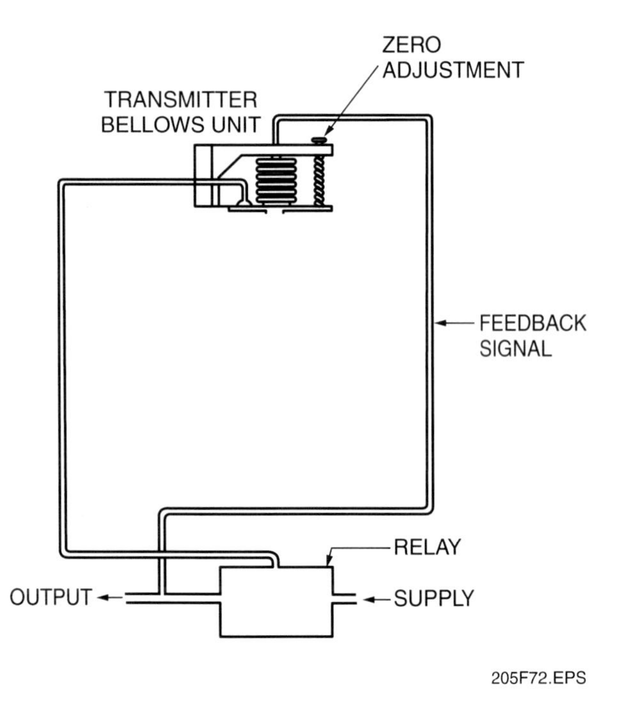

Figure 72 ◆ Bellows and relay assembly.

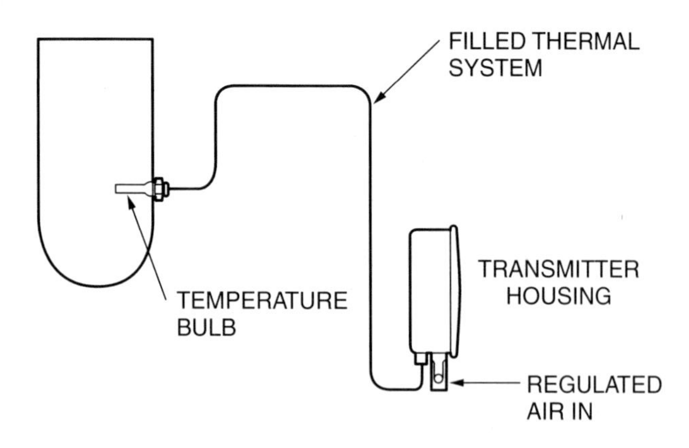

Figure 73 ◆ Motion balance temperature measurement installation.

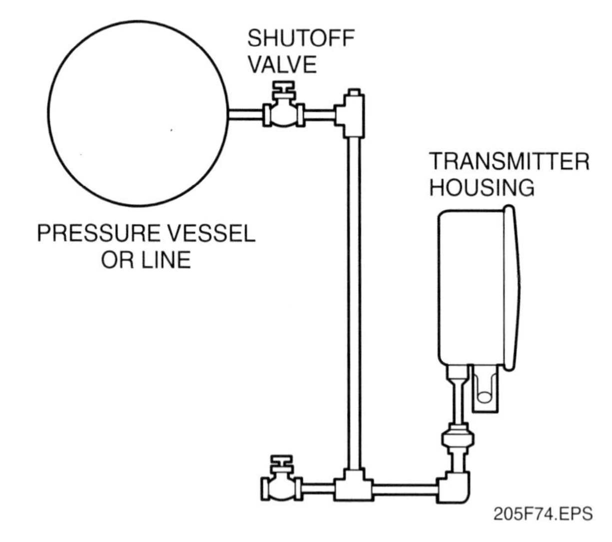

Figure 74 ◆ Motion balance pressure measurement installation.

5.2.0 Electronic Transmitters

Electronic transmitters offer a higher degree of accuracy than pneumatic transmitters and are used extensively. They also have a faster and longer-range transmission system. Their operation varies with the manufacturer.

Two common electronic transmitter systems are the force balance system and variable capacitance cell.

5.2.1 Force Balance Differential Pressure Electronic Transmitters

An electronic force balance transmitter has a supply voltage of up to 95 VDC. Its output can be 10 to 50 mA or 4 to 20 mA. Current systems are commonly 4 to 20 mA.

The sensor system of an electronic force balance transmitter is similar to that of a pneumatic DP cell. The major difference is the output device. *Figure 75* shows an electronic force balance DP transmitter.

A sensor assembly is a sealed metallic diaphragm. Pressure on this diaphragm causes a small movement of the diaphragm link. This link movement is transferred to a force bar that transmits it to the range rod. The range rod transmits the force to the armature, causing it to move. *Figure 76* shows a sensor assembly.

The armature portion of a range rod can move back and forth within the transformer. For example, if the armature is moved toward the secondary winding, the voltage at that winding will increase. *Figure 77* shows an output section.

An increase in voltage causes an increase in the output signal of an oscillator-amplifier. If the armature moves in the other direction, the output signal decreases.

5.2.2 Variable Capacitance Cell Differential Pressure Electronic Transmitters

Variable capacitance cell differential pressure electronic transmitters are designed to accurately measure very low differential pressures below 0.5 psi (less than 25 millibars).

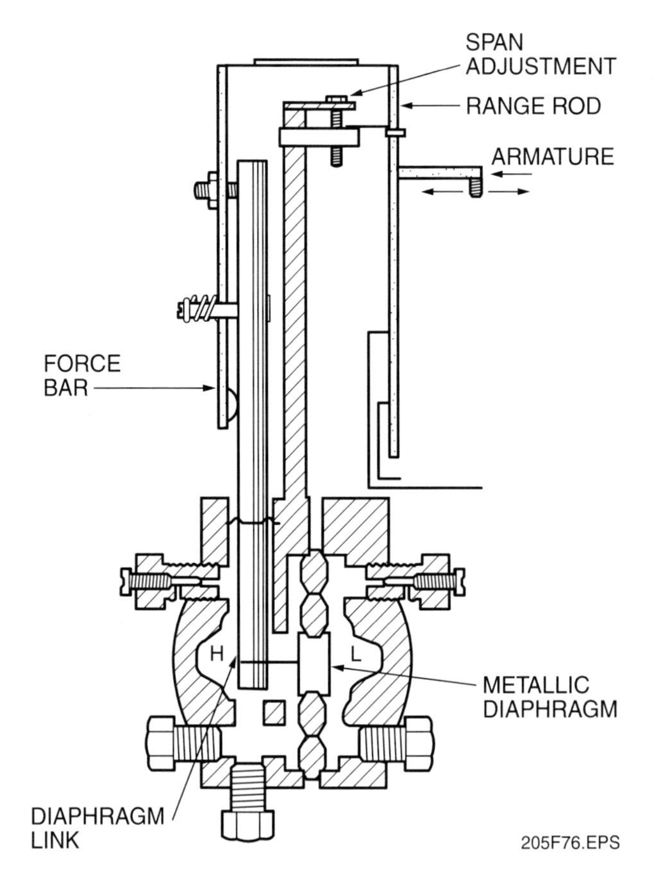

Figure 76 ◆ Sensor assembly.

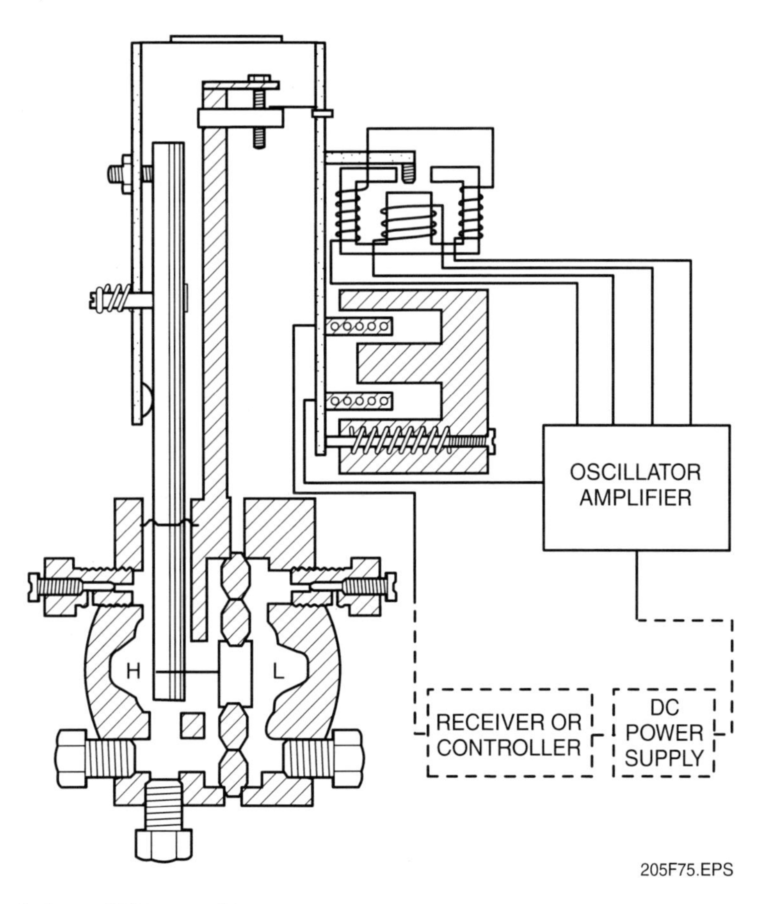

Figure 75 ◆ Electronic force balance DP transmitter.

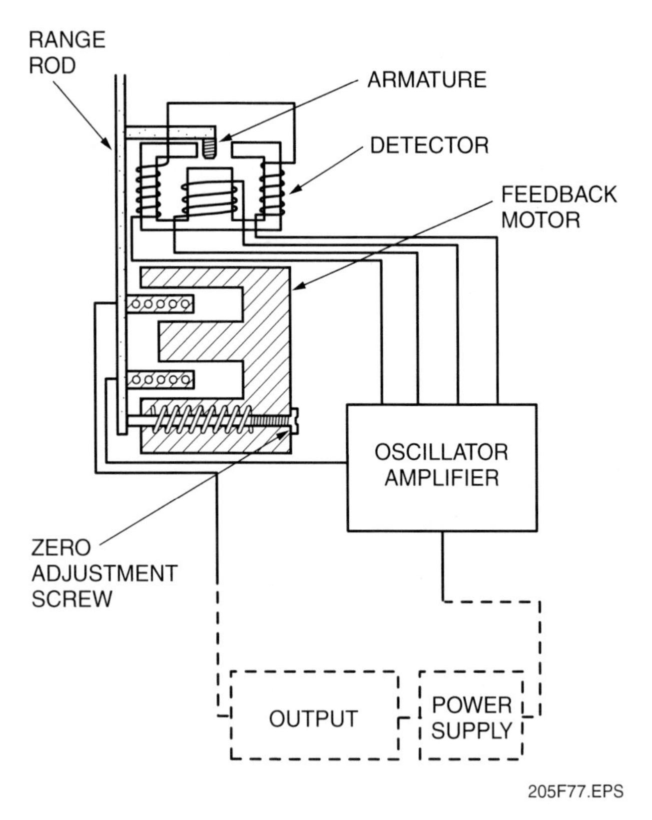

Figure 77 ◆ Output section.

Figure 78 shows the operation of a variable capacitance cell differential pressure transmitter. Small lead wires from the circuit assembly are attached to the capacitor plates located inside the diaphragm assembly. A regulated current is passed through the wires into each of the capacitor plates. Pressure changes within the diaphragm assembly cause a ceramic membrane to deflect toward or against the plates, changing the capacitance between the two plates. These changes are converted in the circuit assembly to a 4–20 mA output signal that is directly proportional to the pressure changes.

Summary

This module covered the basics of elements, transducers, and transmitters. Accuracy, sensitivity, range, and similar concepts were covered as they apply to measuring devices.

Major components involved in measuring pressure, flow, temperature, and level were covered. The principles of operation, as well as typical applications, were discussed.

Transducers were discussed, as was the role they play in processes. The various types of transducers were shown, along with applications and explanations of their operation.

Transmitters were shown to be found in both electrical as well as pneumatic types. The typical range in which transmitters operate was given, as well as the theory of operation of each type of transmitter.

The interrelationship of all these different devices was stressed. Various trade tips were given to help the trainee better utilize the information discussed.

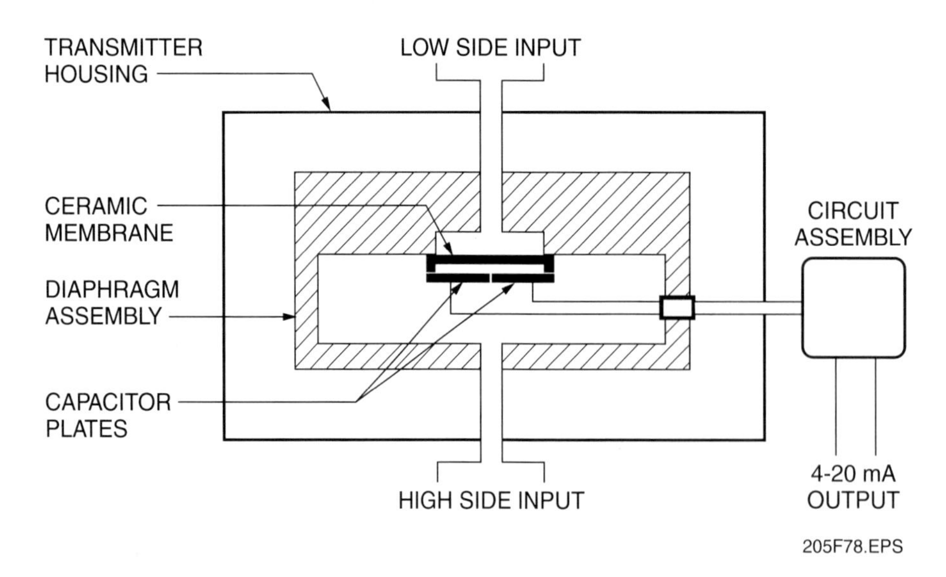

Figure 78 ◆ Variable capacitance cell differential pressure electronic transmitters.

Review Questions

1. Another name for a sensor is a _____.

a. transducer
b. detector
c. transmitter
d. secondary element

2. The detection principle of an RTD is that resistance varies linearly with _____.

a. pressure
b. flow
c. temperature
d. level

3. A device that converts one form of input signal to another form of output signal is called a _____.

a. transducer
b. transceiver
c. DP cell
d. pneumatic relay

4. Sending a signal from one location to another, usually involving some distance, is the job of a _____.

a. transducer
b. primary element
c. transmitter
d. DP cell

5. The measurement of repeatability of an instrument is referred to as _____.

a. accuracy
b. validity
c. precision
d. span

6. The first system element that responds quantitatively to the measured variable and performs the initial measurement operation is a _____.

a. secondary element
b. transducer
c. transmitter
d. detector

7. An orifice plate is normally mounted _____.

a. between two flanges
b. inside the measuring section of a DP cell
c. on the wall of a tank
d. in the control room

8. For clean, vapor-free liquids, a(n) _____ orifice plate should be used.

a. eccentric
b. segmental
c. concentric
d. ventless

9. A(n) _____ provides maximum accuracy when detecting flow in fluids that contain large amounts of suspended solids, such as slurries.

a. orifice plate
b. pitot tube
c. venturi tube
d. annubar tube

10. In a magnetic flowmeter, the process fluid _____.

a. acts as the moving conductor
b. produces the magnetic field
c. acts as the electrodes
d. induces the EMF into the coil

11. A Bourdon tube must be fabricated from a material that has elastic properties so that it _____.

a. is not affected by temperature
b. can provide a suitable sealing surface
c. will not return to its original shape
d. can deform under pressure

12. The deflection characteristics of a metal diaphragm pressure device can be somewhat improved by _____.

a. tempering the metal
b. corrugating the surface
c. polishing the surface
d. layering the surface

13. The use of metallic bellows has been most successful in the _____ pressure range.

a. 5.0 to 75 psig
b. 3 to 15 psig
c. 0.5 to 75 psig
d. 4–20 mA

14. When measuring differential pressure with an instrument that is equipped with a capacitance pressure sensor, _____ is applied to isolating diaphragm A.

a. a vacuum
b. another pressure
c. the pressure to be measured
d. the atmosphere

15. In a diaphragm seal detector, the space between the sealing diaphragm and the sensor are filled with a suitable liquid whose pressure duplicates _____.
    a. the pressure of the process side
    b. atmospheric pressure
    c. water
    d. mercury

16. An I/P transducer converts _____ signals.
    a. pneumatic to current
    b. digital to analog
    c. current to pneumatic
    d. resistance to pneumatic

17. In a chemical area or humid space where pneumatic measuring instruments are used, a(n) _____ transducer may be found.
    a. I/P
    b. P/P
    c. P/I
    d. E/P

18. The component that receives the input signal in an I/P transducer is the _____.
    a. pneumatic relay
    b. transformer
    c. flapper and nozzle
    d. coil

19. The input signal in a P/I transducer is first applied to _____.
    a. a bellows
    b. a coil
    c. a DP cell
    d. flapper and nozzle

20. What type of transducer employs electrical resistance variation to sense strain or other results of force?
    a. I/P
    b. Metallic strain gauge
    c. Metallic bellows
    d. Corrugated diaphragm

21. The zero point on pneumatic transmitters is typically set to 3 psi instead of 0 psi _____.
    a. to indicate the presence of a leak
    b. to allow for a shorter range
    c. because pneumatic transmitters cannot be calibrated to 0 psi as the zero point
    d. to maintain a positive pressure on the low side

22. A pneumatic force balance differential pressure transmitter is also referred to as a(n) _____.
    a. motion balance transmitter
    b. accelerometer
    c. DP cell
    d. strain gauge

23. The feedback bellows of a pneumatic DP cell is located in the _____.
    a. process measuring section
    b. force bar section
    c. balancing section
    d. input/output section

24. Whenever condensate or process fluid may fill the low leg on a DP cell liquid measurement transmitter, the low leg may be filled with a liquid that is compatible to the process, and is referred to as a _____.
    a. drip leg
    b. dry leg
    c. wet leg
    d. open tank

25. The major difference between an electronic force balance transmitter and a pneumatic DP cell is the _____.
    a. force bar
    b. metallic diaphragm
    c. high and low sides
    d. output device

GLOSSARY

Trade Terms Introduced in This Module

Calibrate: To check the readings of an instrument against a known standard and adjust the instrument to correct any errors.

Calibration: The procedure laid down for determining, correcting, or verifying the absolute values corresponding to graduations on a measuring instrument.

Chamfered: Constructed with a slanted surface that connects two external surfaces, forming two angles.

Coulomb: The amount of electricity passing a given point in 1 second at a current of 1 ampere.

Displacement: The measured distance traveled by a point from its position at rest.

Drift: The gradual change of an instrument output from the correct value.

Electromagnetic field (EMF): The force fields around an electrical conductor that carries current.

Empirical data: Information or data derived from observation or experiment.

Farad: A measurement of capacitance that is equal to one coulomb of charge for each volt applied.

Hastelloy®: A nickel alloy composed of nickel, chromium and molybdenum that is highly resistant to corrosion.

Laminar flow: Smooth, uniform, non-turbulent flow of a fluid in parallel layers, with little mixing between layers. This type of flow is characterized by small values of the Reynolds number.

Magnetic lines of flux: Imaginary lines representing a magnetic field. These lines have the direction of the magnetic flux at each point. Flux is a vector quantity having both magnitude and direction.

Magnitude: In pneumatic or other non-electrical signals, the maximum measurable level of the signal, such as psi, psig, or kPag.

Mass-inertia effect: A tendency of a mass (object) to resist any change in its present motion.

Measurement error: The difference between a measured value and an actual value.

Monel®: A high-nickel alloy resistant to corrosion and widely used to resist the action of acids.

Noble metal: A metal that is so inert that it is usually found as uncombined metal in nature. Platinum, gold, and silver are noble metals. Also, metal that does not corrode.

Parallax: An optical illusion that occurs in analog meters and causes reading errors. It occurs when the viewing eye is not in the same plane, perpendicular to the meter face, as the indicating needle.

Picofarad: One-millionth of one-millionth of one farad, abbreviated *pf*.

Piezoelectric (crystal): Any material that provides a conversion between mechanical and electrical energy. For a piezoelectric crystal, if mechanical stresses are applied on two opposite faces, electrical charges appear on another pair of faces.

Precision: The degree of reproducibility of measurement by an instrument.

Primary element: The element in a measurement device that is acted on directly by the process.

Quantitative: Describes a property that can be measured and described as a number or quantity.

Relative error: The expression of an error as a percent of the value being measured.

Responsiveness: The ability of an instrument to follow changes.

Reynolds number: A dimensionless parameter that indicates transition from laminar to turbulent flow in pipes.

Secondary element: The element of a measurement device that takes the cascade output from the primary element and sends a signal proportional to it to the controller.

Sintered: Heated and compressed.

Transducer: A device that functions primarily to convert its input signal to an output signal of a different form.

Transmitter: A device that functions primarily to prepare and send input information to a remote location.

Viscous: A liquid or slurry that is thick or resistant to flow.

Additional Resources

This module is intended to present thorough resources for task training. The following reference works are suggested for further study. These are optional materials for continued education rather than for task training.

Process Measurement Fundamentals, 1997. GPS.

Basic Instrumentation, 1966. Patrick J. Higgins. McGraw-Hill Book Company.

Instrumentation, 1975. Franklin Kirk and N.R. Rimboi. American Technical Society.

Figure Credits

Dwyer Instruments, Inc. 205F32 (photo)

NCCER CURRICULA — USER UPDATE

NCCER makes every effort to keep its textbooks up-to-date and free of technical errors. We appreciate your help in this process. If you find an error, a typographical mistake, or an inaccuracy in NCCER's curricula, please fill out this form (or a photocopy), or complete the online form at **www.nccer.org/olf**. Be sure to include the exact module ID number, page number, a detailed description, and your recommended correction. Your input will be brought to the attention of the Authoring Team. Thank you for your assistance.

Instructors – If you have an idea for improving this textbook, or have found that additional materials were necessary to teach this module effectively, please let us know so that we may present your suggestions to the Authoring Team.

NCCER Product Development and Revision

13614 Progress Blvd., Alachua, FL 32615

Email: curriculum@nccer.org
Online: www.nccer.org/olf

❑ Trainee Guide ❑ AIG ❑ Exam ❑ PowerPoints Other ______________________________

Craft / Level: Copyright Date:

Module ID Number / Title:

Section Number(s):

Description:

Recommended Correction:

Your Name:

Address:

Email: Phone:

Module 12206-03

Controllers, Recorders, and Indicators

COURSE MAP

This course map shows all of the modules in the second level of the Instrumentation curriculum. The suggested training order begins at the bottom and proceeds up. Skill levels increase as you advance on the course map. The local Training Program Sponsor may adjust the training order.

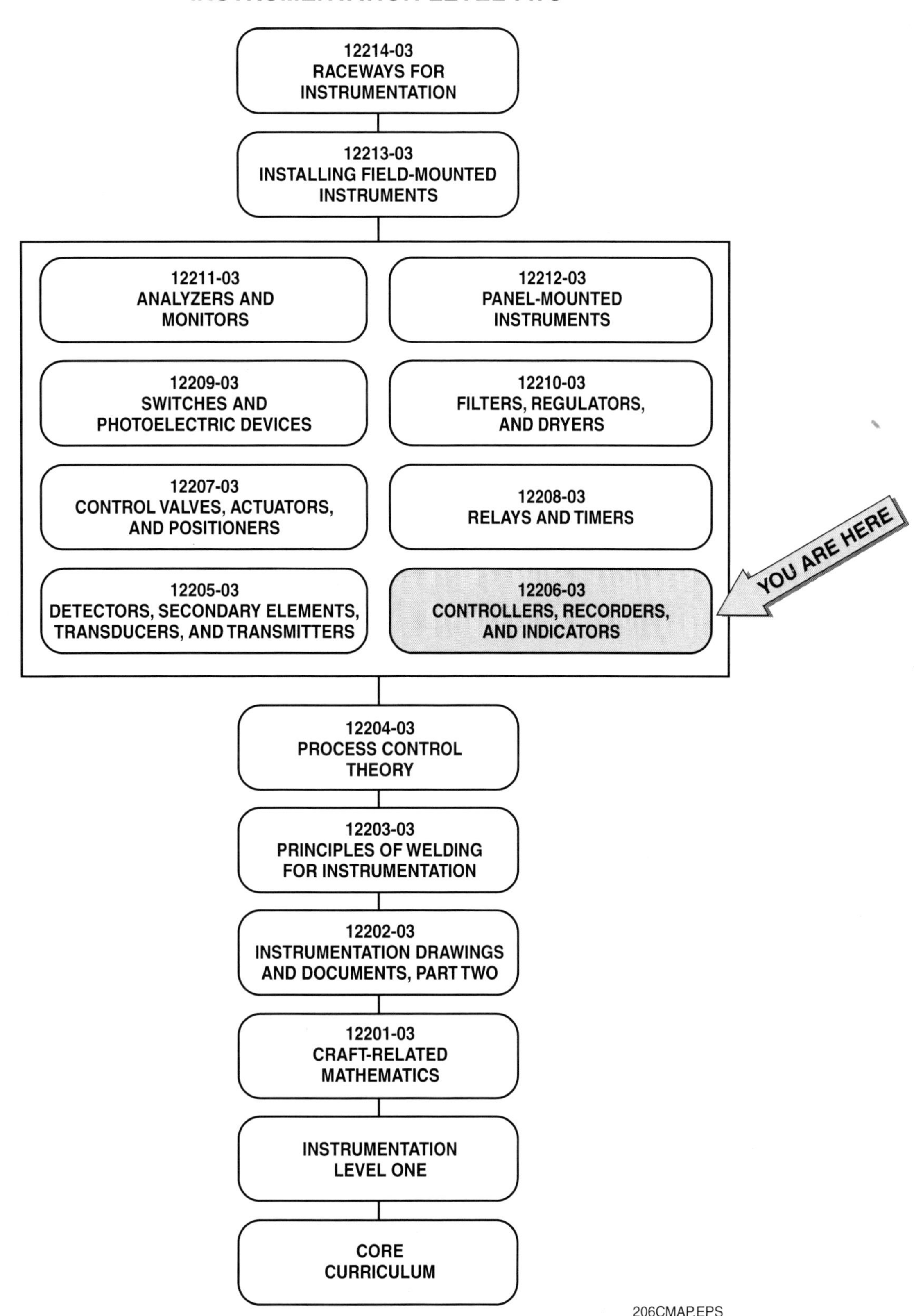

MODULE 12206-03 CONTENTS

1.0.0 INTRODUCTION 6.1

2.0.0 PNEUMATIC CONTROLLERS 6.2

2.1.0 On-Off Control 6.2

2.2.0 Proportional Control 6.3

2.3.0 Proportional Controllers with Reset (Integral) 6.4

2.4.0 Proportional Controllers with Integral and Derivative (PID) 6.7

2.4.1 Masoneilan® Controllers 6.7

2.4.2 Taylor® Controllers 6.9

2.4.3 Foxboro-Eckardt Controllers 6.16

3.0.0 ELECTRONIC CONTROLLERS 6.16

3.1.0 On-Off Control 6.17

3.2.0 Proportional Control 6.17

3.3.0 Proportional Controllers with Reset (Integral) 6.18

3.4.0 Proportional Controllers with Reset (Integral) and Derivative (PID) 6.18

4.0.0 RECORDERS AND INDICATORS 6.20

4.1.0 Recorders 6.20

4.1.1 Mechanical Recorders 6.20

4.1.2 Electronic Recorders 6.21

4.1.3 Newer Electronic Recorders 6.22

4.1.4 Pneumatic Recorders 6.23

4.2.0 Indicators 6.24

4.2.1 Gauges 6.24

4.2.2 Electrical Indicators 6.25

4.2.3 Electronic Indicators 6.26

4.2.4 Pneumatic Indicators 6.26

4.2.5 Thermal Indicators 6.27

4.2.6 Magnetic Indicators 6.27

SUMMARY 6.28

REVIEW QUESTIONS 6.29

GLOSSARY 6.31

REFERENCES & ACKNOWLEDGMENTS 6.32

Figures

Figure 1 Components of a pneumatic controller 6.2
Figure 2 Pneumatic controller in the On-Off mode with a signal to the final element 6.3
Figure 3 Pneumatic controller in the On-Off mode with no signal to the final element 6.3
Figure 4 Nozzle moving away from flapper 6.4
Figure 5 Reduction in output pressure in a pneumatic proportional controller 6.5
Figure 6 Increased output pressure in a pneumatic proportional controller 6.5
Figure 7 Graph of a 5% proportional band 6.6
Figure 8 Controller with proportional and reset mode 6.6
Figure 9 Proportional controller with reset and derivative 6.7
Figure 10 Masoneilan® 12000 Level Controller 6.8
Figure 11 Feedforward and feedforward/feedback 6.8
Figure 12 Taylor® Model 1414R Controller 6.9
Figure 13 Schematic diagram of the Taylor® Controller 6.9
Figure 14 Input comparison lever for the Taylor® Controller . . .6.10
Figure 15 Proportional control mechanism 6.10
Figure 16 Proportional and reset mechanisms 6.11
Figure 17 Derivative mechanism 6.12
Figure 18 Taylor® Controller simplified schematic 6.13
Figure 19 Manual control mechanism .6.15
Figure 20 Basic electronic controller flow graph 6.17
Figure 21 On-Off control devices 6.17
Figure 22 Electronic controller with proportional control mode 6.18
Figure 23 Electronic controller with proportional and reset control mode 6.18
Figure 24 Electronic controller with proportional, reset, and derivative control 6.19
Figure 25 Block diagram of a typical electronic controller 6.19
Figure 26 ABB Series 53MC5000A Process Control Station .6.20
Figure 27 A level control process using the ABB Series 53MC5000A 6.20
Figure 28 Mechanical recorder 6.21
Figure 29 Helix-type Bourdon measuring element 6.21
Figure 30 Mechanical recorder operation 6.21
Figure 31 Circular-chart paper for recording temperature 6.22
Figure 32 ABB Commander™ SR100A strip-chart recorder . . .6.23

Figure 33 ABB Commander™ 1900 circular-chart recorder . . .6.23
Figure 34 ABB Screenmaster™ 300 with large color display 6.23
Figure 35 Three-pen pneumatic recorder 6.24
Figure 36 Recorder bellows assembly with link arm 6.24
Figure 37 Ammeter circuit 6.25
Figure 38 Simple voltmeter circuit 6.25
Figure 39 Ohmmeter with internal resistance adjustment 6.26
Figure 40 Digital indicators with seven-segment display 6.26
Figure 41 Seven-segment display 0 through 9 . . . . .6.27
Figure 42 Measuring indicating section 6.27
Figure 43 Optical pyrometer 6.27
Figure 44 The appearance of the filament during optical pyrometer adjustment 6.28

MODULE 12206-03

Controllers, Recorders, and Indicators

Objectives

1. Describe the operation of a controller.
2. Describe the operation of a recorder.
3. Describe the operation of an indicator.
4. Using samples, pictures, or specification sheets, identify common types of controllers, recorders, and indicators.
5. Identify the common parts of a pneumatic controller.
6. Describe the functions of an electronic controller.
7. Identify the common parts of an electronic controller.
8. Identify the three main sections of a recorder.
9. Connect and use a short recorder.

Prerequisites

Before you begin this module, it is recommended that you successfully complete the following: Core Curriculum; Instrumentation Level One; Instrumentation Level Two, Modules 12201-03 through 12204-03.

Required Trainee Materials

1. Pencil and paper
2. Appropriate personal protective equipment

1.0.0 ◆ INTRODUCTION

A **controller** in an instrumentation system receives a signal from a process measuring element and sends a signal to a final element. This signal manipulates the final element to maintain a desired process condition such as specific flow rate, pressure, temperature, level, or consistency.

Most controllers function by comparing the input signal from the measuring element to a predetermined setpoint in the control. If the input signal and the setpoint do not match, the controller must change the position of the final element. Other controllers take action on a predetermined schedule. The following sections explain pneumatic controllers and electronic controllers. Most controllers can operate in a number of different modes. The mode selected by the operator depends on the control requirements of the process being controlled.

- *Automatic control* – Automatic control is the most frequent control method used. The error signal is developed and processed through the PID controller. The resulting control signal is then sent to the final control device. While in automatic control, the final control element will respond automatically to a disturbance or change in the process.
- *Manual control* – Manual control is used when the operator wants to take direct control over the final control element. In this mode of operation, the final control element responds only to manual changes made by the operator. The PID output is disabled in this mode and has no effect on the final control.
- *Local control* – When a controller is placed in local control, the setpoint to the controller comes from the setpoint adjustment on the controller itself. The only way to change the setpoint is to manually change it with the setpoint adjustment. Do not confuse local control with manual control. Manual control deals with the output of the controller, while local control deals with where and how the controller gets its setpoint.
- *Remote control* – When a controller receives its setpoint from a remote source, it is operating in remote control. If the source of the setpoint comes from another controller, this mode is called cascade control.

2.0.0 ◆ PNEUMATIC CONTROLLERS

Pneumatic controllers rely on a pneumatic input signal for operation. This signal can come from a measuring element, an **I/P transducer** (which converts an electronic signal into a pneumatic signal), or a pneumatic transmitter. All pneumatic controllers have common components (*see Figure 1*):

- *Air supply* – This provides the controller with the operating medium.
- *Flapper and flapper-nozzle* – This is the mechanism that determines the output signal. It uses compressed air for response.
- *Bellows* – This is a device that expands and contracts to activate the air control relay.
- *Air control relay* – This is a device that directs the pneumatic signal toward or away from the final element.

Pneumatic controllers can operate in one of four control configurations. These are:

- On-Off
- Proportional
- Proportional with reset (integral)
- Proportional with reset (integral) and derivative (PID)

2.1.0 On-Off Control

On-Off control is often referred to as discrete control or two-position control. The output of the controller changes from one fixed position to another fixed position. Control adjustments are made by changing the setpoints or the **differential gap**. The differential gap is determined by the difference between two setpoints: the high (maximum) and the low (minimum) settings.

This type of control is the simplest of all controls and is the least expensive. It provides reasonable flexibility because the valve size or travel (final element) is adjustable. It is limited, however, to processes that permit cycling control. Consistent measured values cannot be obtained with this type of control.

Pneumatic controllers in the On-Off control can adjust the final element to one of two positions. When the controller detects that the measured variable has deviated from its setpoint, the flapper moves.

When the flapper moves away from the nozzle, a minimum pneumatic signal is sent to the air control relay. This causes a maximum pneumatic signal to be sent to the final element, moving it to one extreme. *Figure 2* shows a pneumatic controller configured as an On-Off controller with the signal going to a final element because the flapper is not sealing off the nozzle. The air used to expand the bellows is escaping out the nozzle.

When the flapper moves against the nozzle, a maximum signal is sent to the air control relay bellows. This causes a minimum signal to be sent to the final element. The final element moves to its other extreme position because the bellows expands and shuts off the air to the final element.

Figure 3 shows a pneumatic On-Off controller with no signal to a final element.

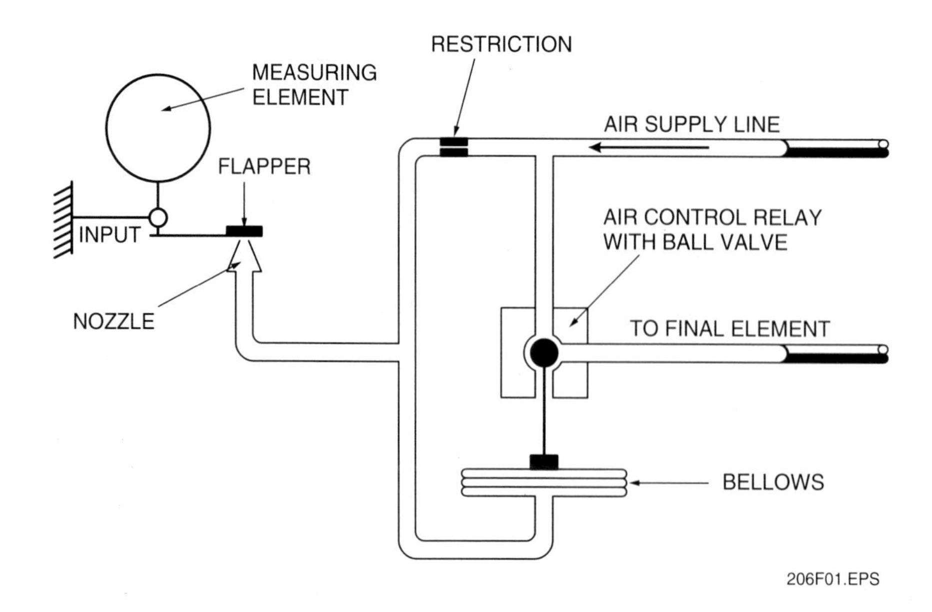

Figure 1 ◆ Components of a pneumatic controller.

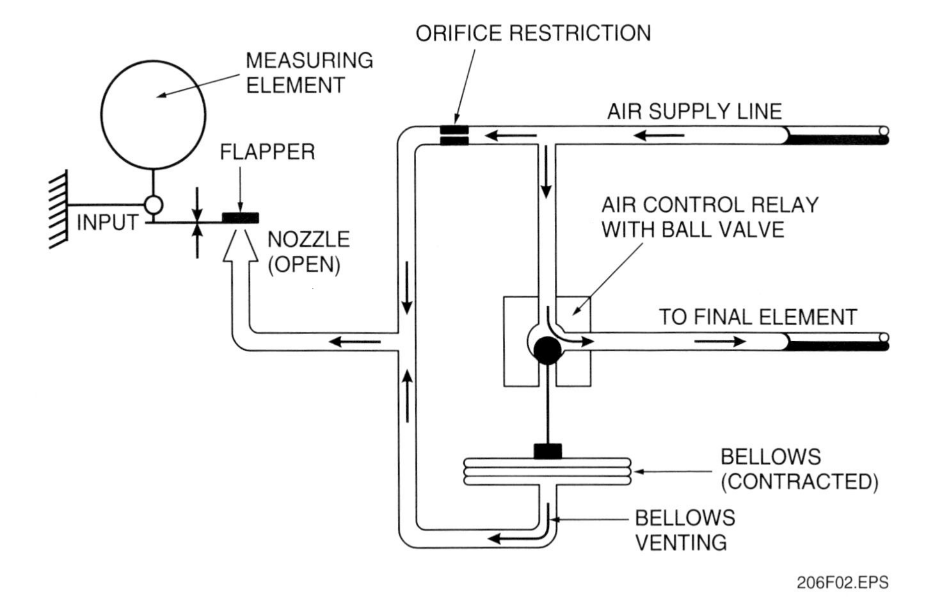

Figure 2 ◆ Pneumatic controller in the On-Off mode with a signal to the final element.

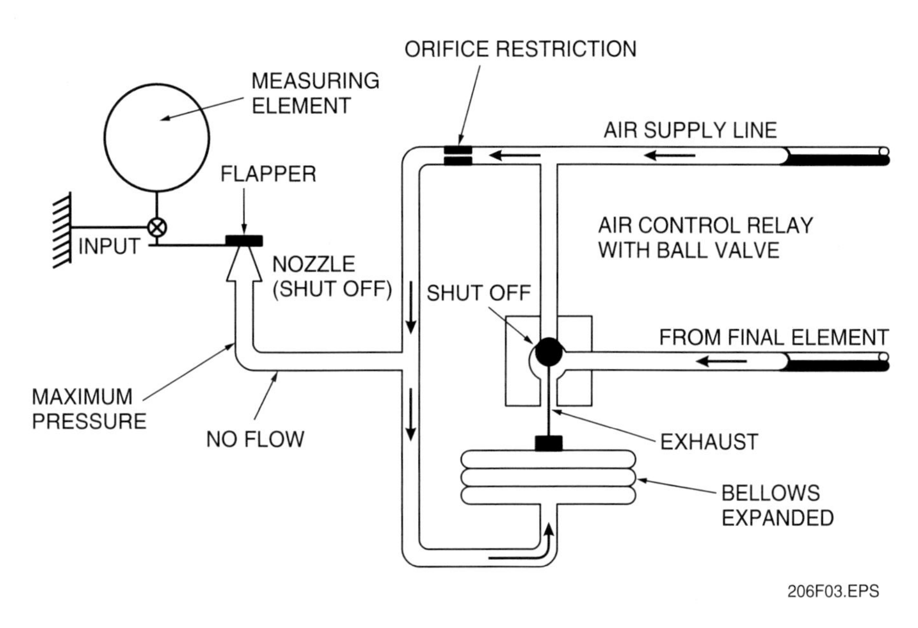

Figure 3 ◆ Pneumatic controller in the On-Off mode with no signal to the final element.

2.2.0 Proportional Control

This mode, also known as gain, provides an output that is proportional to, or has a linear relation to, the direction (increase or decrease) and magnitude of the error in the signal. The larger the gain or proportional setting, the larger the change in the controller output caused by a given error between setpoint and controlled variable.

Pneumatic controllers that provide a proportional output use a response mechanism in addition to the other common components. This response mechanism consists of the following components:

- *Proportioning bellows* – expands or contracts as the response line pressure changes. It moves the nozzle.
- *Opposing spring* – maintains the tension on the proportioning bellows.
- *Positioning link* – transfers the movement of the bellows to the nozzle.
- *Response line* – carries the portion of the controller output signal to the bellows.

Figure 4 shows the nozzle moving downward, away from the flapper. *Figure 5* shows reduced output pressure in a pneumatic proportional controller.

If the controller output signal to the final element is increased, the proportioning bellows contracts. This causes the nozzle to move upward toward the flapper. *Figure 6* shows increased output pressure in a pneumatic proportional controller.

All controllers with a proportional mode have a **proportional band** adjustment. Proportional band is the range of values of the measured variable that the final element moves through as the controller output changes from a minimum value.

For example, a 5% proportional band means the controller output signal (position of the final element) changes from maximum to minimum with a 5% change in the measured variable. *Figure 7* shows a graph of a 5% proportional band.

2.3.0 Proportional Controllers with Reset (Integral)

Another function that can be added to a controller is an integral reset function. Reset, or integral, provides an output that is proportional to the time integral of the input. In other words, the output continues to change as long as an error between setpoint and variable exists. The integral mode acts only when the error exists for a period of time between setpoint and the controlled variable.

One problem that must be observed closely when using reset is the integral (reset) windup that occurs when the deviation cannot be eliminated, such as in open loops. The controller is driven into its extreme output. This condition can cause a loss of control for a period of time, followed by extreme cycling.

In addition to the components of the proportional controller, a controller with reset action uses the following components:

- *Reset capacity tank* – This is a pressurized container that discharges according to a set schedule.
- *Reset bellows* – This opposes the proportioning bellows and allows balance between proportional action and reset action.
- *Adjustable reset* – This restriction provides a range of pressures to the reset bellows.

Figure 8 shows a controller with proportional and reset mode.

In this mode, the air signal from the air control relay enters the proportional bellows. The signal also travels through the adjustable reset restriction and reset capacity tank to the reset bellows. The reset bellows opposes the motion of the proportional bellows.

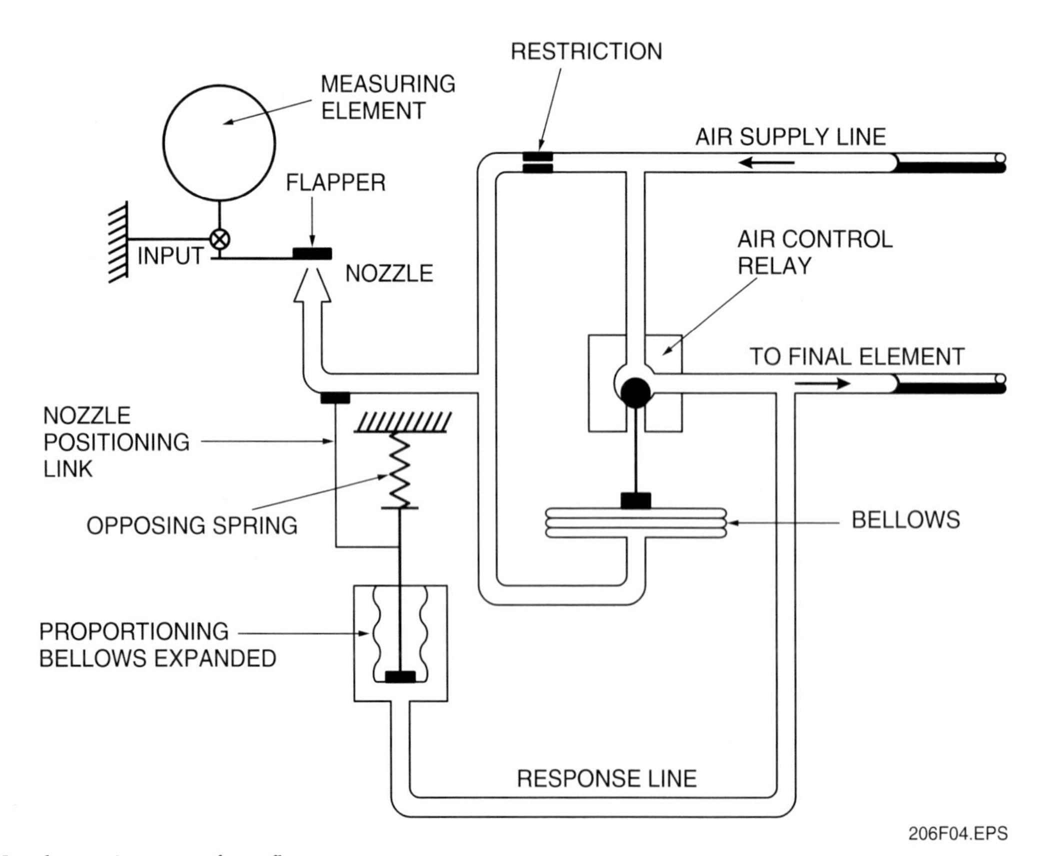

Figure 4 ◆ Nozzle moving away from flapper.

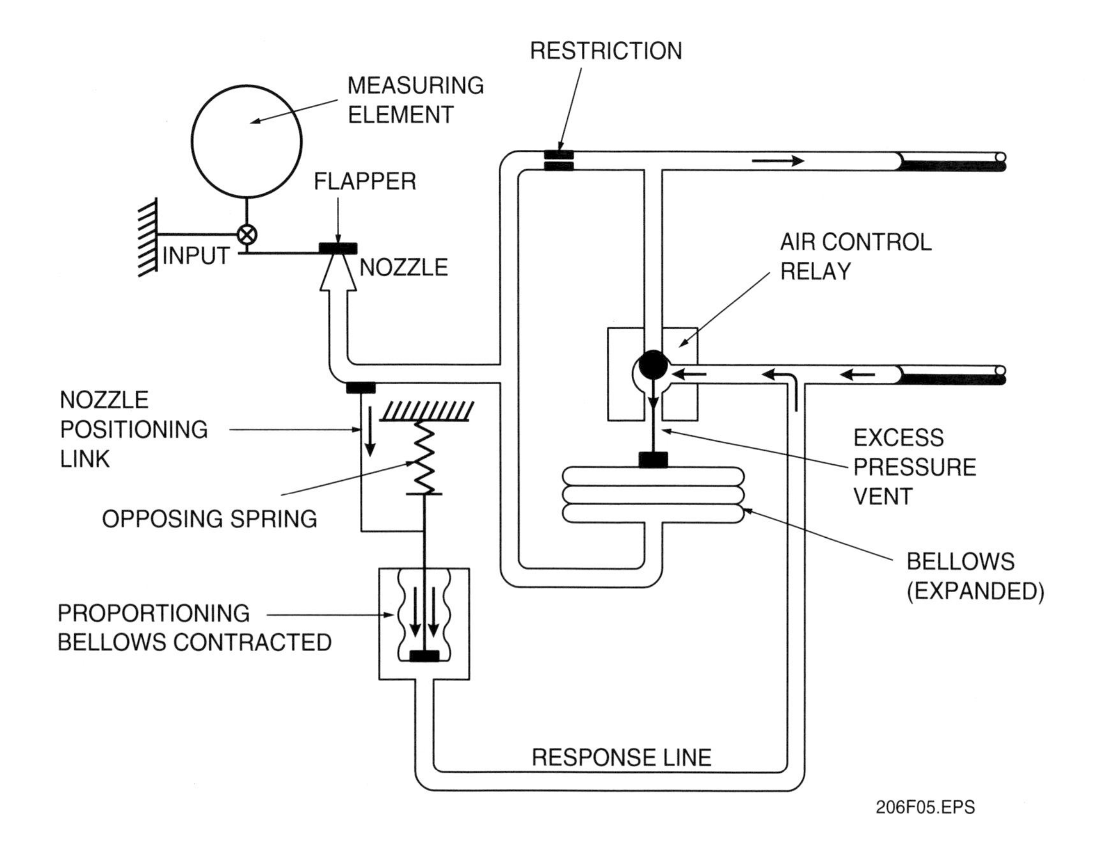

Figure 5 ◆ Reduction in output pressure in a pneumatic proportional controller.

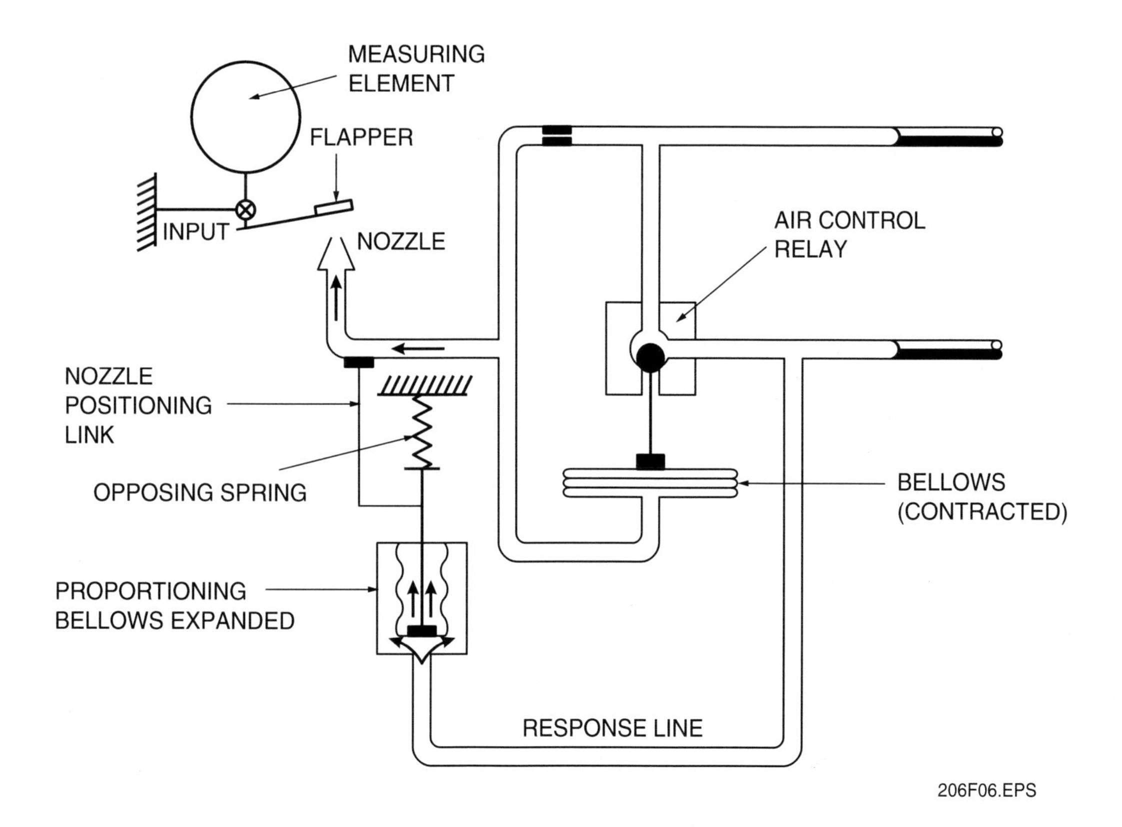

Figure 6 ◆ Increased output pressure in a pneumatic proportional controller.

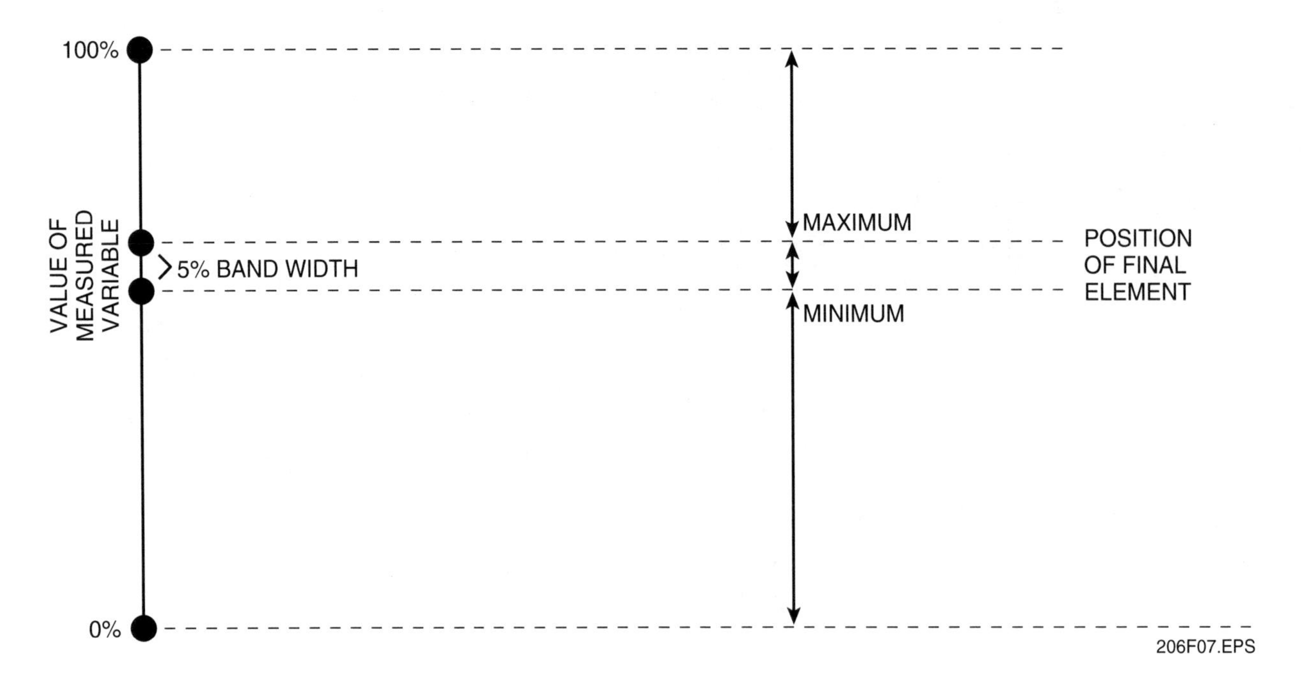

Figure 7 ◆ Graph of a 5% proportional band.

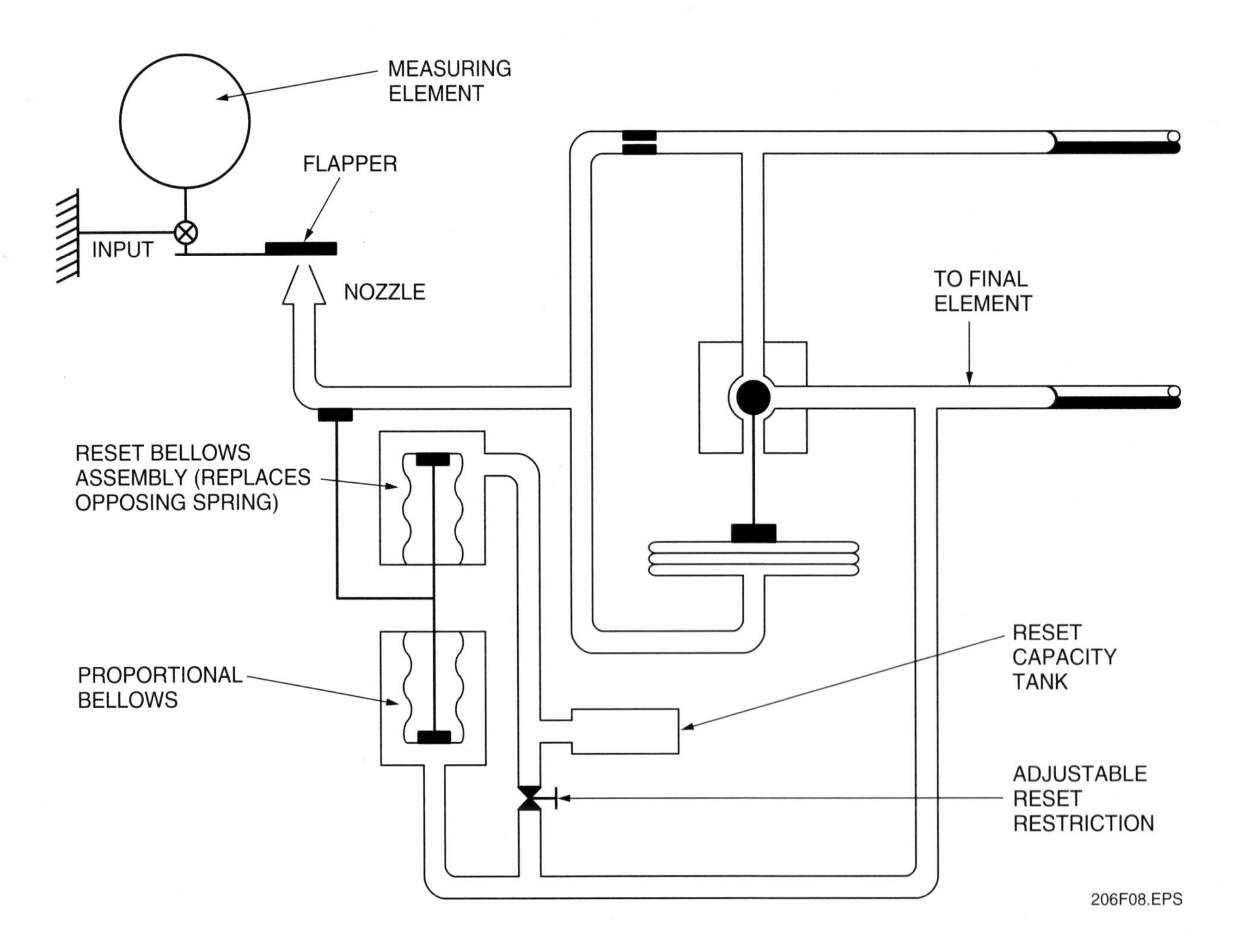

Figure 8 ◆ Controller with proportional and reset mode.

2.4.0 Proportional Controllers with Integral and Derivative (PID)

Figure 9 shows a proportional controller with reset (integral) and derivative, often called a PID controller. In this type of controller, the adjustment derivative restriction and derivative capacity tank delay the corrective motion of the proportional bellows. The amount of delay is determined by the rate at which the difference between setpoint and control point increases or decreases. A derivative control mode is necessary for controlling processes that respond slowly to changes.

2.4.1 Masoneilan® Controllers

One example of a pneumatic controller is the Masoneilan® 12000 Level Controller (*Figure 10*). This controller measures the process level through a displacer unit, compares the actual level with the setpoint, and produces a 3–15 psig control signal proportional to the difference between the two.

Three adjustments are normally made to align the Masoneilan® Level Controller: the specific gravity setting, the proportional band adjustment, and the setpoint adjustment.

The specific gravity adjustment knob, shown in *Figure 10*, corrects for specific gravity so that the controller is applicable for all ranges and specific gravities.

The specific gravity adjustment knob connects to the adjustment slot on either the direct or reverse side of the reversing arc shown in *Figure 10*, depending on whether the controller is configured for direct response or is reverse acting. This configuration depends on factors such as the action of the final element (control valve), the process, and the location of the control valve.

To adjust the specific gravity setting, loosen the specific gravity adjustment knob and slide it along the slot until the index is aligned with the correct gravity on the specific gravity scale located on the reversing arc.

> **NOTE**
> While the specific gravity scale is graduated only for specific gravities as low as 0.5, the standard controller may be used for gravities as low as 0.2.

For general usage and convenience, the values indicated on the proportional scale are in percent of the level change. 100% proportional band indicates a change in output pressure of 12 psi for full level change.

The proportional band should be set as narrow as the process will permit without cycling. If the controlled level cycles, widen the proportional

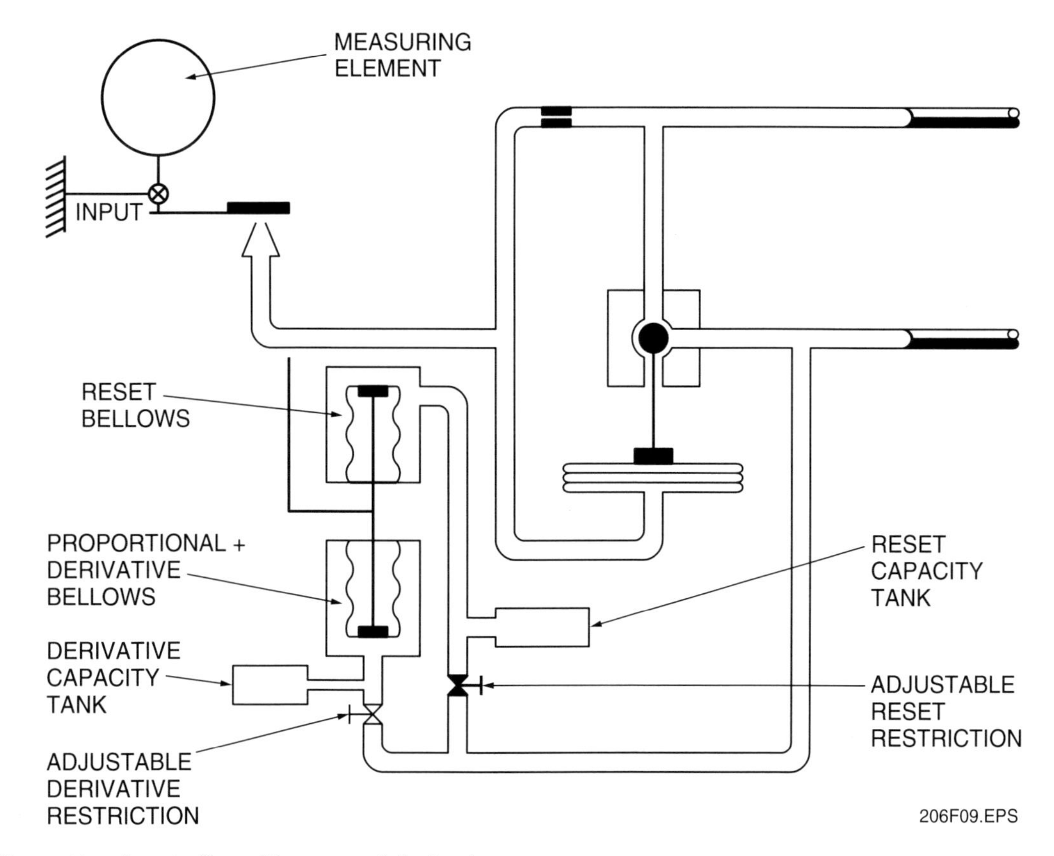

Figure 9 ◆ Proportional controller with reset and derivative.

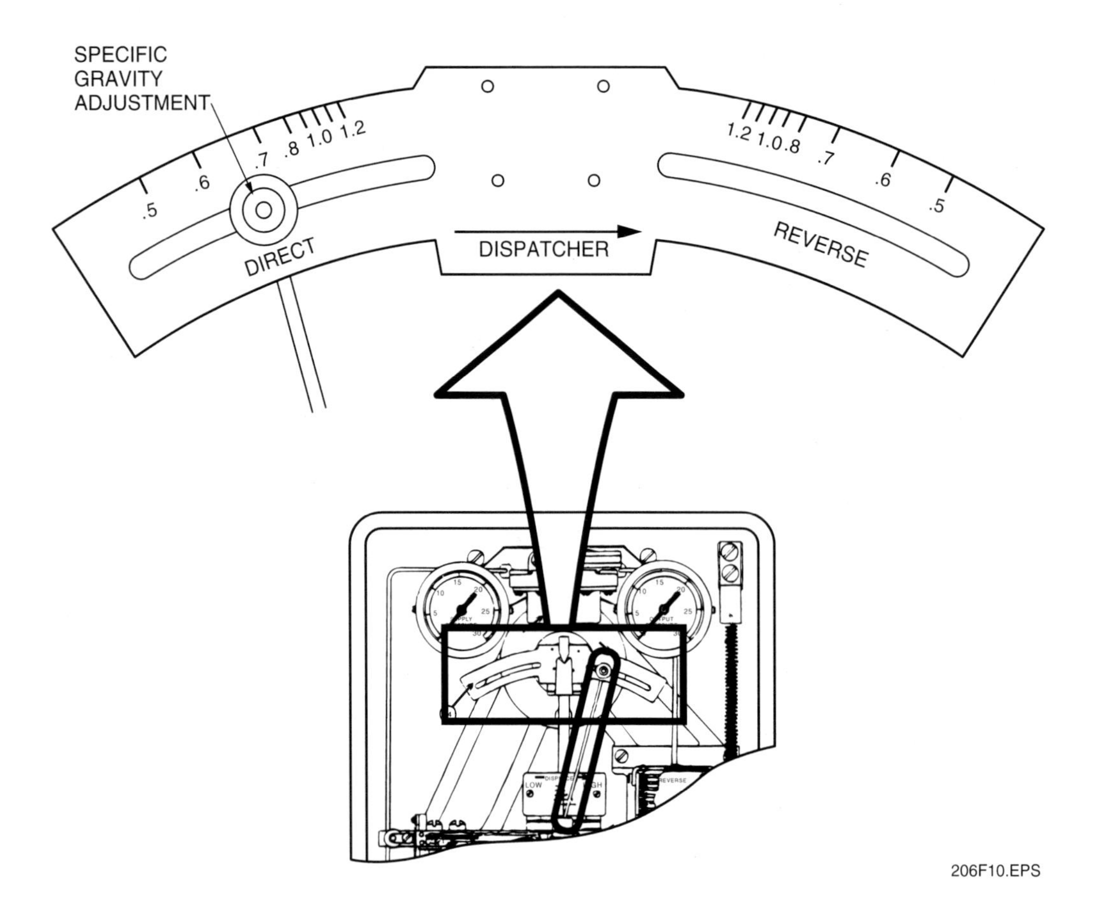

Figure 10 ◆ Masoneilan® 12000 Level Controller.

band. If the controlled level drifts around the control point, narrow the proportional band until the level shows a distinct tendency to cycle, then widen it sufficiently to eliminate this cycling.

If the level stabilizes at a point away from the setting of the index, determine whether an increase or decrease of output pressure will correct the condition, and adjust accordingly. Small adjustments will correct for extreme conditions.

This instrument uses an angle motion feedback mechanism to provide the proportional control action. Feedback is provided by a feedback bellows in conjunction with a feedback spring. The bellows expands, causing more spring tension to balance the control arm. This balancing motion restores the instrument to equilibrium. The resulting increase in the control signal is proportional to the change in the process level.

Keep in mind that feedback measurement comes directly from the affected process and is usually altered by a disturbance or change in an element coming into the process. **Feedforward** measurement comes directly from the disturbance. *Figure 11* better illustrates these differences. The operator shown in *Figure 11* could be replaced with a controller. Feedforward could be represented by the disturbance indication and the

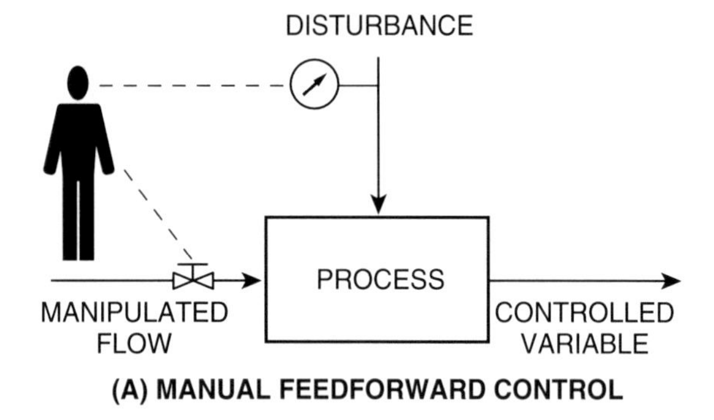

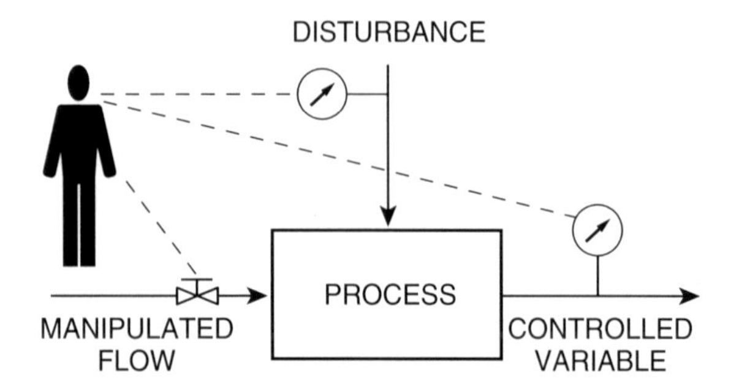

206F11.EPS

Figure 11 ◆ Feedforward and feedforward/feedback.

feedback represented by the indication of the controlled variable. The manipulated flow would be the final element or control valve. Feedforward and feedback systems independently adjust the control valve (final element).

2.4.2 Taylor® Controllers

The Taylor® Model 1414R Controller is one example of a proportional-plug-rate controller (*Figure 12*). In this controller, a **moment**-balance feedback mechanism is used to produce the three-mode action.

The schematic for this controller (*Figure 13*) might make it look complicated, but it is actually a rather simple instrument. As the operation of the controller is described, only those portions of this schematic necessary for the present discussions are introduced. This should make the discussion easier to follow.

This controller, like all controlling instruments, has a comparison and detection mechanism at which errors between the measured variable signal

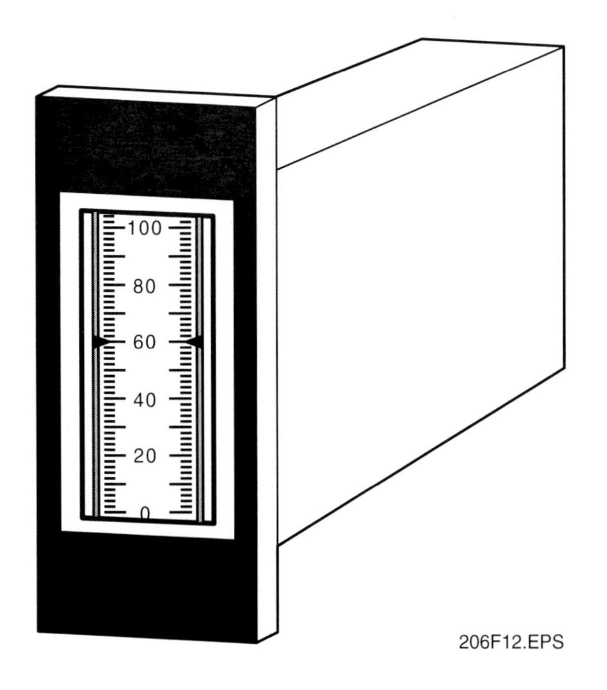

Figure 12 ◆ Taylor® Model 1414R Controller.

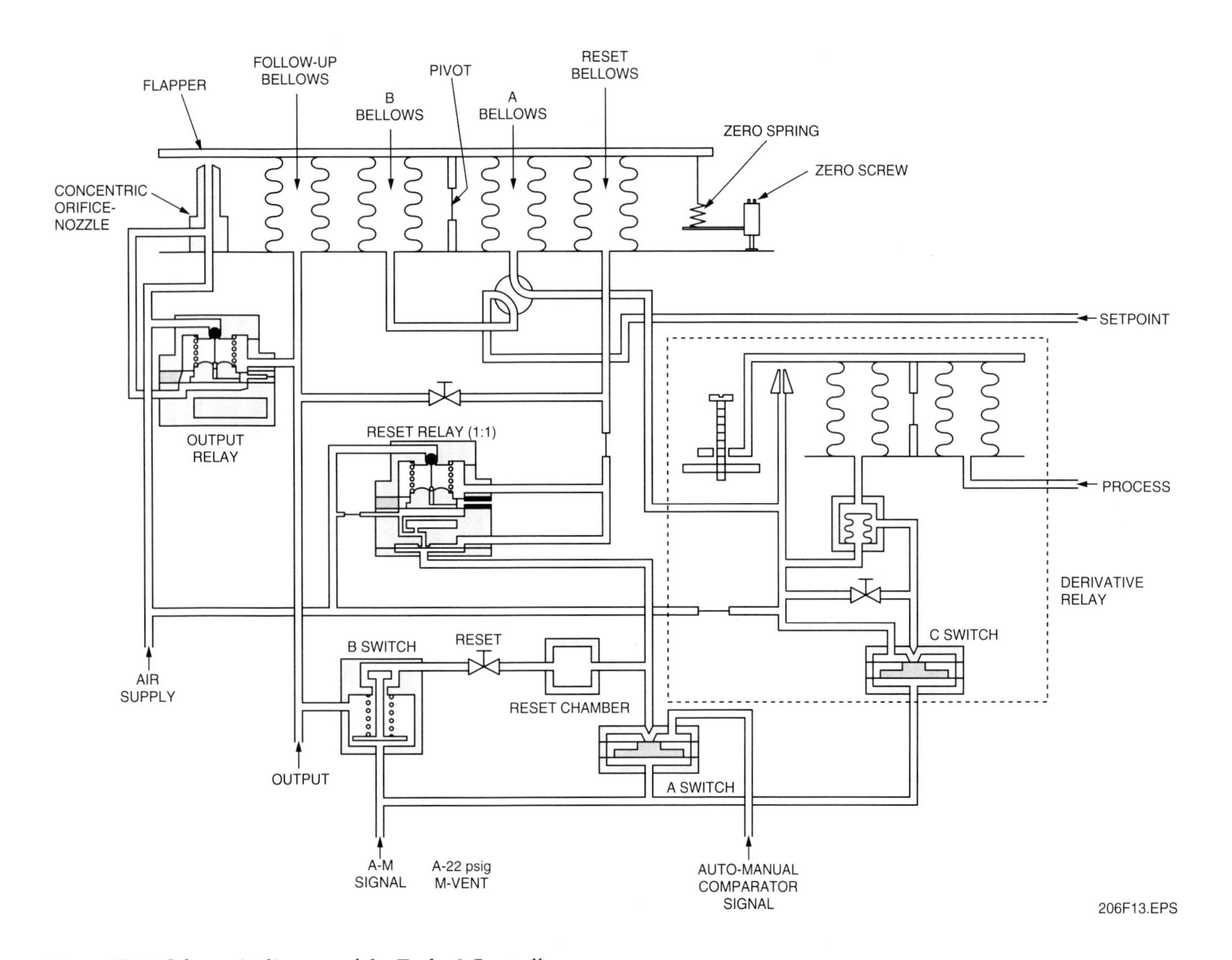

Figure 13 ◆ Schematic diagram of the Taylor® Controller.

and setpoint signal produce a change in pneumatic signal pressure. In the Taylor® Controller, the measurement signal and setpoint signal pressures are applied to bellows positioned at opposite ends of a first-class lever (*Figure 14*).

A force is created when pressure is applied to either of the bellows. This force is applied to the force beam to produce a moment. If the measurement moment and the setpoint moment are not equal, the zero spring is displaced. The resulting movement of the force beam changes flapper/nozzle clearance, causing a change in nozzle backpressure. The change in nozzle backpressure actuates an output relay, causing the output pressure to change. This is the feedforward path of the controller.

The output pressure of the instrument is sensed by two feedback bellows: the follow-up bellows, located to the left of the pivot, and the reset bellows, located to the right of the pivot (*Figure 15*). Output pressure is applied directly to the follow-up bellows. It is also applied through the gain adjustment to the reset bellows. A portion of the pressure applied to the reset bellows is bled through a fixed restrictor to atmosphere. The gain adjustment controls the flow into the reset bellows. If the gain restrictor allows more flow into the bellows than can vent through the fixed restrictor, pressure in the reset bellows increases.

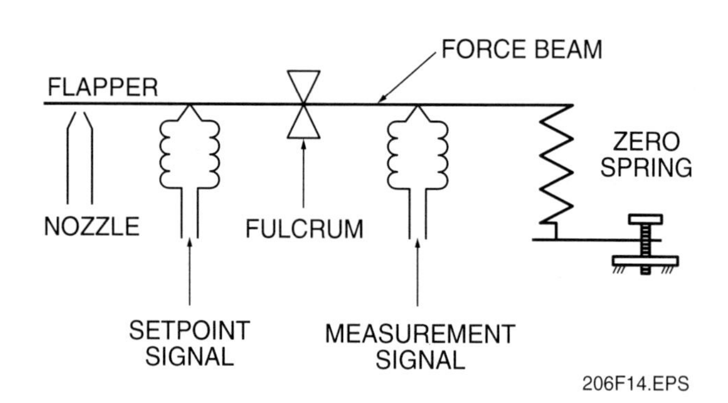

Figure 14 ◆ Input comparison lever for the Taylor® Controller.

The pressure applied to these bellows produces a force. When this force is applied to the beam, a moment is produced. The difference between the follow-up and reset moments is the proportional response of the controller. When the gain valve is closed, pressure within the reset bellows remains constant. An error between the measured variable and setpoint produces a change in the output, which is sensed by the follow-up bellows alone. The force produced by the follow-up bellows produces a balancing moment that restores the system to equilibrium. The follow-up bellows is larger than the setpoint or measurement bellows. Therefore, to balance the change in input pressure, only a relatively small change in the output is required.

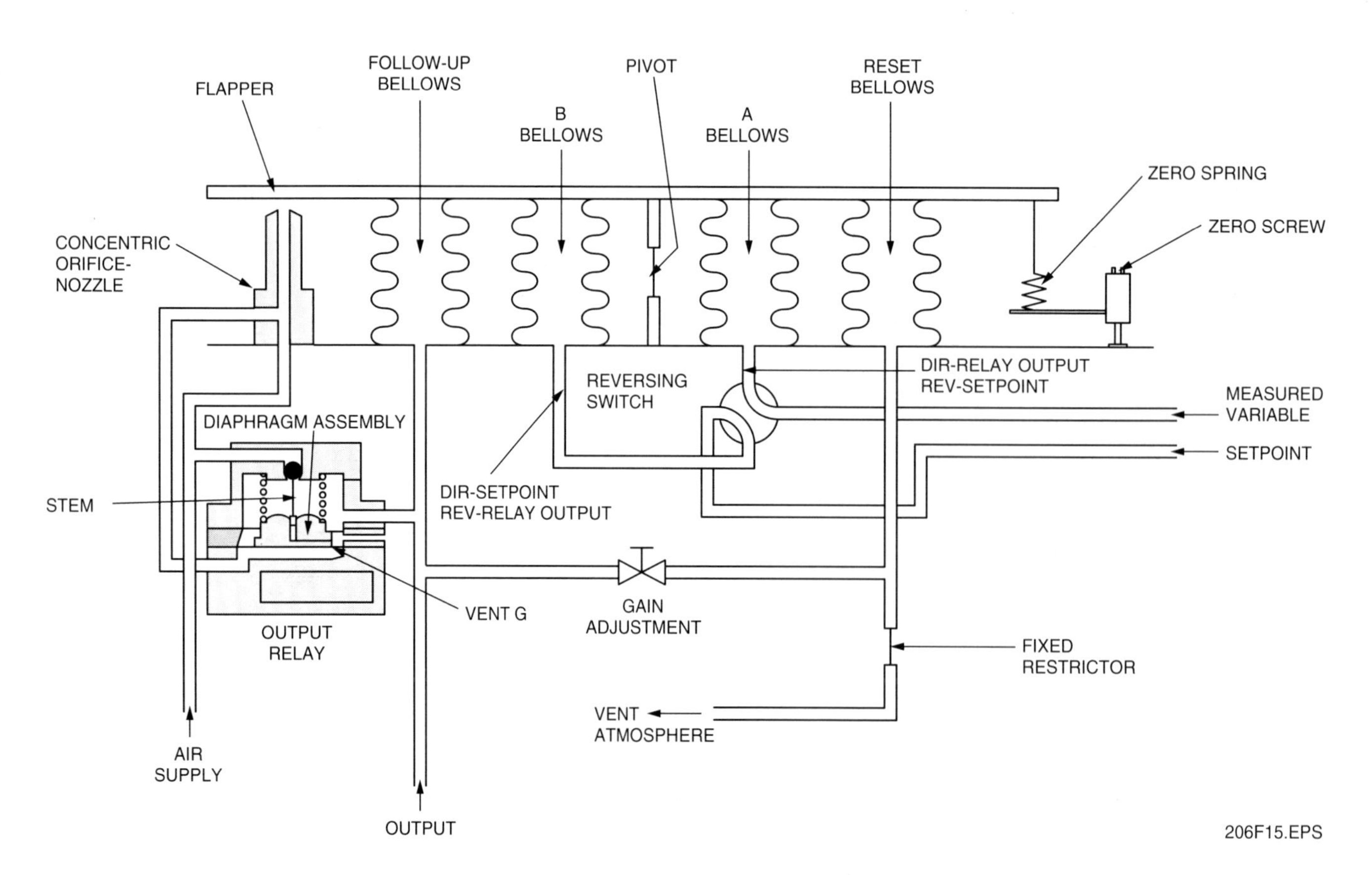

Figure 15 ◆ Proportional control mechanism.

The gain with the gain valve closed is less than 1, making the proportional band very wide.

When the gain valve is fully open, the pressure within the reset bellows can increase in response to a change in output. An error between the measured variable and setpoint produces a change in the output, which is now sensed by both the follow-up bellows and the reset bellows. The force produced by the reset bellows nearly equals the force produced by the follow-up bellows. Therefore, for the follow-up bellows to produce a moment that balances the moment created by the error signal, there must be a large change in the output signal. With the gain valve fully open, the gain of the instrument is high, and the proportional band is very narrow.

The reset mechanism of this controller involves a reset relay, a reset adjustment valve, and a reset capacity (*Figure 16*).

The proportional action of this controller is not changed by the addition of the reset components shown in *Figure 16*. Notice, though, that pressure from the reset bellows is now vented to the atmosphere through the reset relays.

The relay controls the bleed of air to the atmosphere. The reset relay is controlled by output pressure applied through the reset valve and the reset chamber. When the output pressure increases, air flows through the reset valve and the chamber to the reset relay and causes the vent to close. This action allows pressure within the reset bellows to increase. When the output pressure decreases, air flows from the reset relay through the reset valve and reset chamber. This action causes the vent to open, bleeding off pressure from the reset bellows. The reset valve and chamber causes the response of the reset relay to lag behind the change in output pressure.

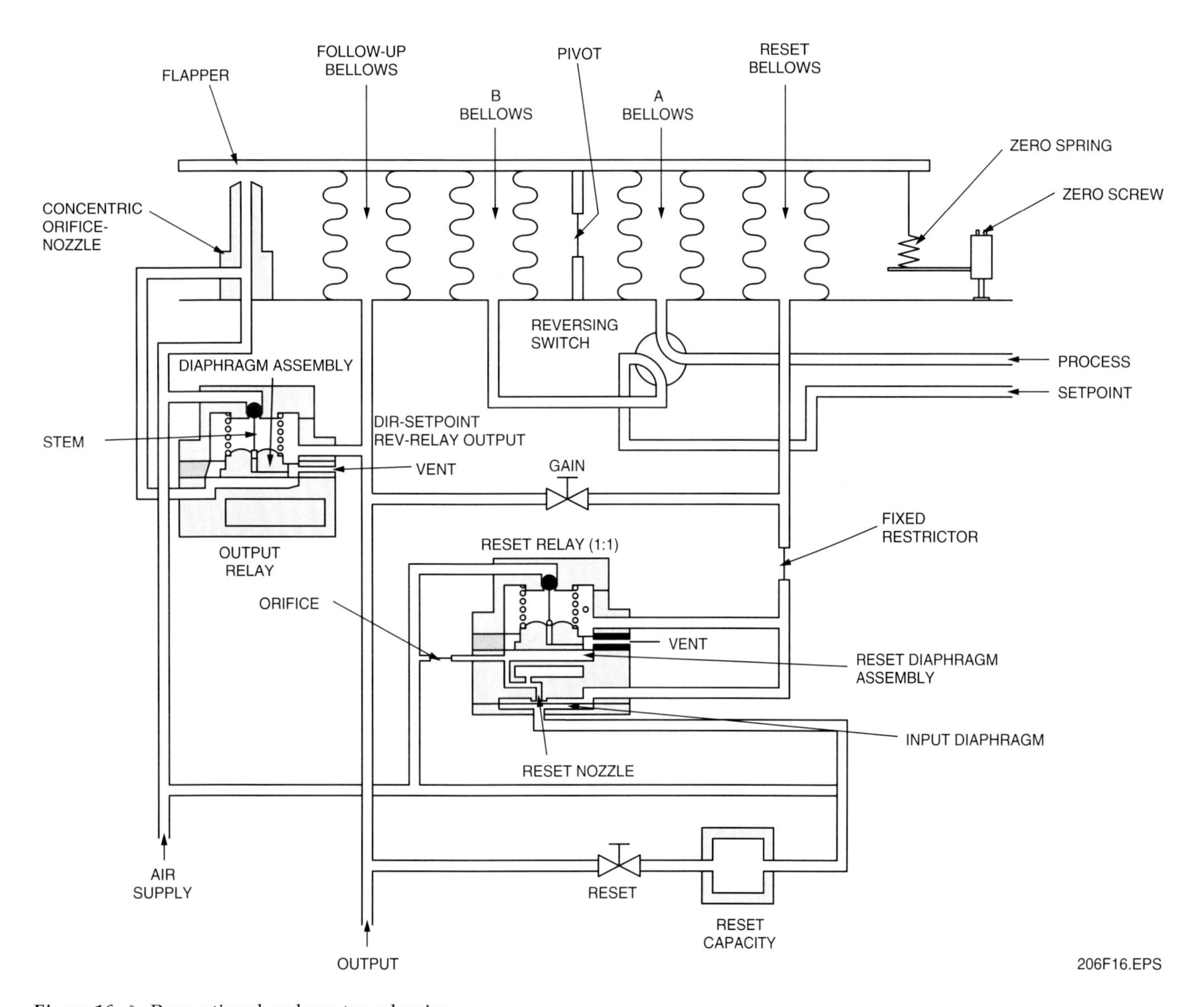

Figure 16 ◆ Proportional and reset mechanisms.

The components necessary to provide **derivative action** are shown in *Figure 17*. Derivative action in the Taylor® Controller is provided by a derivative relay that Taylor® calls the Pre-Act™ relay. The process variable signal is applied to a bellows. The process bellows converts the pressure to a force that acts on the derivative flapper beam to produce an input moment. A change in the input moment displaces the zero screw. The resulting rotation of the flapper beam changes the flapper/nozzle clearance, causing a change in the nozzle backpressure.

The flapper beam in the derivative relay is a first-class lever. The resistance arm of this lever is the length of the beam between the pivot and the nozzle. The effort arm is the length of the beam between the derivative bellows and the pivot. The resistance arm is much longer than the effort arm. Therefore, a relatively small displacement of the effort arm causes a large displacement of the resistance arm. A small change in the measured variable causes a relatively large change in nozzle backpressure, provided no balancing force is produced.

The output of the derivative nozzle can be fed back to produce a balancing force through the derivative adjustment valve. When the derivative valve is fully open, a change in nozzle backpressure is immediately sensed by the derivative follow-up bellows. The pressure within the follow-up bellows produces a force that balances the force produced by the measurement variable signal. Therefore, with the derivative valve fully open, the derivative mechanism is simply a 1:1 relay.

With the derivative valve only partially open, the pressure sensed by the follow-up bellows lags behind nozzle backpressure. Initially, the gain of the derivative unit is very high, and a small change in input pressure causes a large change in nozzle backpressure. Over time, the pressure across the valve equalizes, and the nozzle backpressure decreases to the value of the input pressure.

Derivative nozzle backpressure is also sensed by the stabilizer bellows. The stabilizer bellows changes the volume of the follow-up chamber when nozzle backpressure changes. The resulting pressure change in the follow-up bellows provides a small instantaneous balancing force to limit the initial gain of the derivative relay. This makes the derivative relay less sensitive to noise in the process variable signal.

In *Figure 18*, three components have been added to the schematic of the Taylor® Controller. Each is a pneumatic switch actuated by an automatic-manual select signal. When the controller is switched to the automatic mode, a 22 psig control signal opens the B-switch to apply output pressure to the reset circuit. The control signal closes the A-switch to block the manual control pressure. The

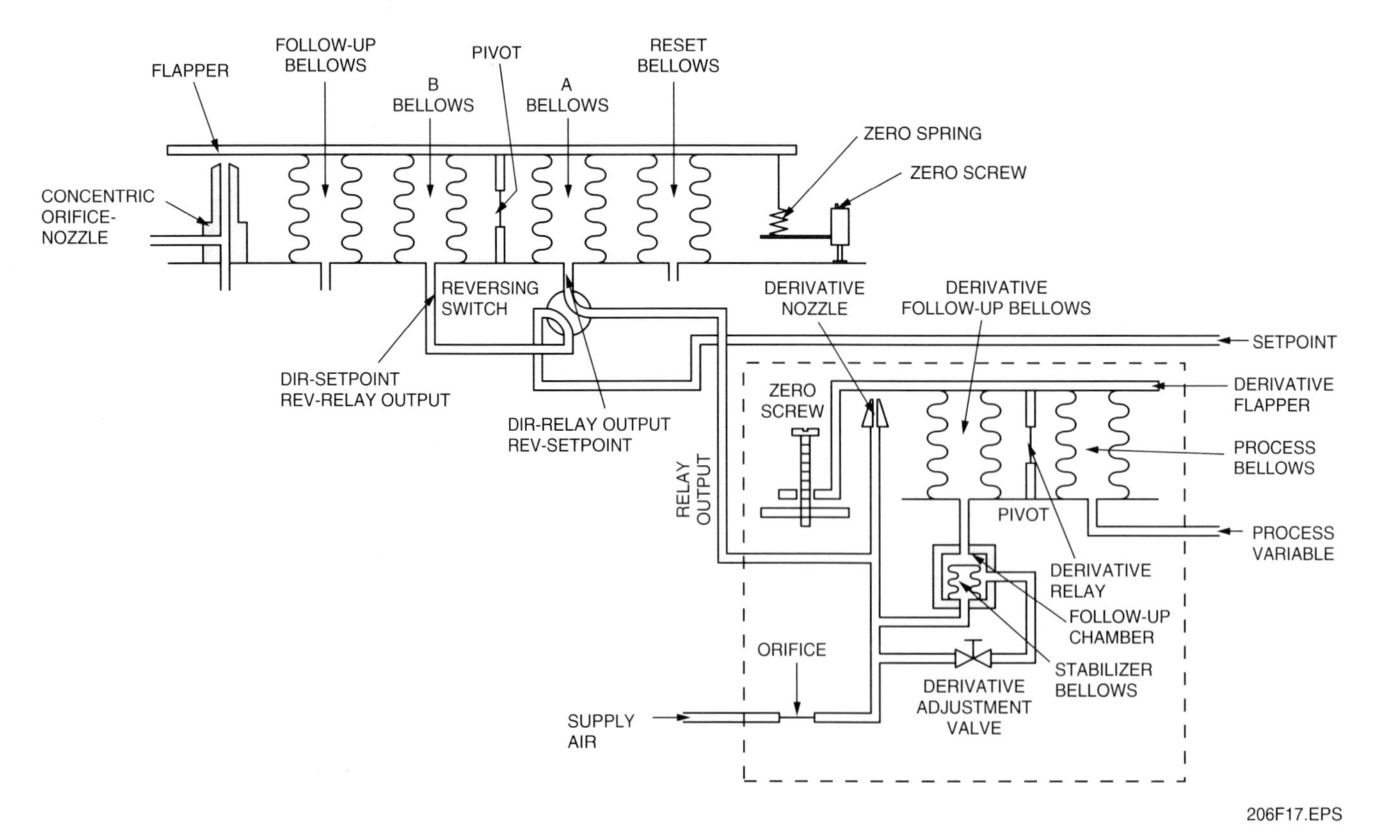

Figure 17 ◆ Derivative mechanism.

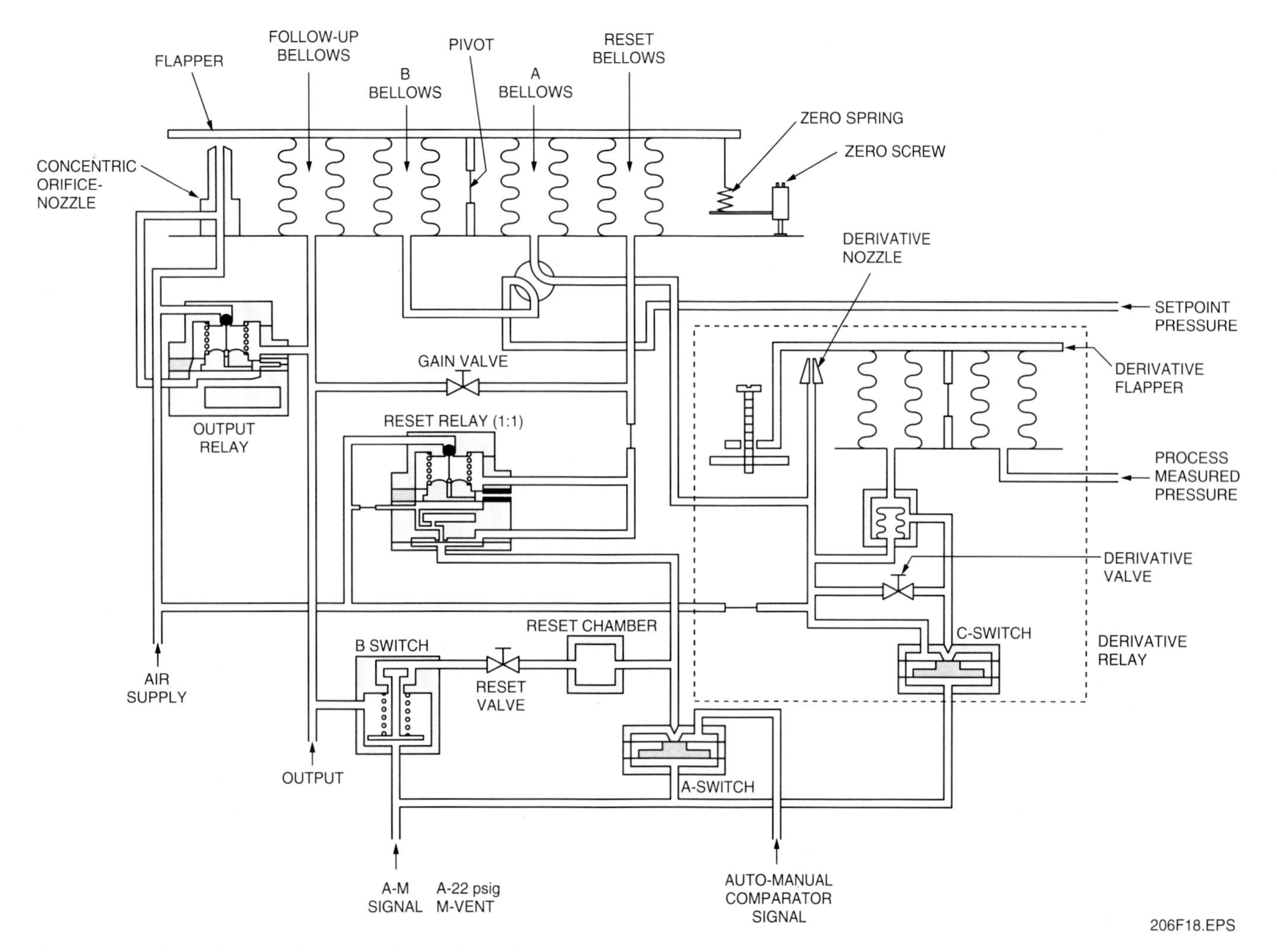

Figure 18 ◆ Taylor® Controller simplified schematic.

control signal also closes the C-switch, preventing the derivative relay from bypassing. If the C-switch were open, the derivative relay would be a 1:1 relay, just as if the derivative valve were fully open.

All of the major components of the Taylor® Controller work together to produce the three-mode control action in response to an increase in the measured variable.

Assume that the controller is balanced and that the setpoint pressure and measured variable pressure are equal. The output pressure is 9 psig. The reset valve, gain valve, and derivative valve are all partially open.

When the measured variable signal increases, the process bellows expands, applying a force to the derivative flapper lever (*Figure 18*). This force causes the lever to pivot counterclockwise about the pivot, reducing the flapper/nozzle clearance. This action causes nozzle backpressure to increase. Initially, this increase is very large due to the amplifying action of the lever.

This large change in derivative nozzle backpressure expands the A-bellows of the comparison mechanism, exerting a force on the comparison beam. A large difference now exists between the measurement force and the setpoint force. As a result of this difference between the forces, the comparison beam pivots counterclockwise, reducing the flapper/nozzle clearance. Nozzle backpressure increases because of this action.

The increasing nozzle backpressure is applied to the underside of the output relay diaphragm and forces the diaphragm assembly upward. As the assembly moves upward, the supply port of the relay opens, and the exhaust ports close. The output pressure then increases. This is the feedforward path of the controller.

Over time, the pressure across the derivative valve equalizes. As this pressure is equalizing, the output from the derivative relay decreases. The force exerted on the beam by the A-bellows decreases, causing the flapper/nozzle clearance to increase. A decrease in nozzle backpressure and therefore, controller output, results from the increase in the flapper/nozzle clearance.

When the output pressure of the controller increases, a force is exerted on the upper side of the output relay diaphragm. This force produces a slight downward movement of the diaphragm assembly, which throttles the supply port of the relay. The output pressure also provides feedback to produce the proportional and reset actions.

To provide proportional response, increasing output pressure is applied to the follow-up bellows and through the gain valve to the reset bellows. Because the gain valve is partially open, pressure in the reset bellows increases in response to the change in output pressure. Therefore, the force created by the follow-up bellows is partially counteracted by the force produced in the reset bellows. The output must increase more in order to balance the force produced by the A-bellows. The resulting change in the output pressure is proportionally larger than the change in the input signal.

When the output pressure increases, a differential pressure is created across the reset valve. Air begins flowing through the valve in order to equalize the pressure. The airflow through the reset valve causes pressure beneath the reset relay input diaphragm to increase. The increasing pressure forces the input diaphragm upward, causing the diaphragm to block the exhaust nozzle. Normally, supply air pressure flows through the exhaust nozzle, into the chamber above the reset diaphragm assembly, and through the relay exhaust port before venting to the atmosphere. Because the exhaust nozzle is blocked by the input diaphragm, the pressure builds up in the chamber below the reset diaphragm assembly. The increasing pressure below the reset diaphragm forces the reset diaphragm upward. This action closes the exhaust port of the relay. With the exhaust port closed, pressure within the reset bellows begins to increase. The resulting force created by the reset bellows reinforces the initial unbalancing force produced by the A-bellows. The force beam, therefore, moves closer to the nozzle, causing the output pressure to increase further. The reset cycle continues to increase the output pressure until the setpoint and measurement pressures are equal. When this condition is reached, equilibrium within the controller is restored.

Manual control is provided by the manual regulator (*Figure 19*). In addition to the manual regulator, an auto-manual comparator is included within the controller so that bumpless transfer between automatic and manual control modes can be accomplished.

When the controller is in the manual control mode, the output of the manual relay provides controller output. In addition, a valve that vents the auto/manual select signal to the atmosphere is actuated. With no pressure applied to the E-switch diaphragm, the clutch is engaged, connecting the manual control adjustment to the regulator input arm. Venting of the auto/manual select signal also actuates the D-switch. Without the select signal pressure, the D-switch diaphragm moves the pilot valve to a position that keeps the tracking bellows vented.

When the manual control adjustment is rotated to increase controller output pressure, the input arm rotates clockwise. As a result of this rotation, spring tension increases. Spring tension exerts a force on the regulator flapper level that moves the flapper closer to the regulator nozzle. The reduction in flapper/nozzle clearance causes nozzle backpressure to increase. In the manual relay, the increasing nozzle backpressure exerts a force on the relay diaphragm. The force moves the relay diaphragm downward, closing off the relay exhaust port and opening the supply port. This action within the relay causes an increase in the manual output pressure. The output pressure increases until it produces an upward force on the diaphragm assembly sufficient to balance the downward force produced when the manual control adjustment is rotated. A follow-up bellows within the manual regulator senses the increased output pressure and applies a force to the regulator flapper lever to balance the regulator. The resulting change in regulator output pressure is proportional to the rotation of the manual control adjustment.

The output of the manual regulator is also applied to the manual bellows in the auto-manual comparator. The increase in pressure produces an upward force on the comparator flapper lever. The upward force rotates the lever about the pivot, causing a reduction in the clearance between the comparator nozzle and flapper. This reduction increases nozzle backpressure. Comparator nozzle backpressure is the output of the auto-manual comparator.

The manual signal from the comparator is applied through the open A-switch to the input diaphragm of the reset relay (*Figure 18*). The reset relay is actuated by the manual signal in the same way that it was actuated by the reset signal. When the reset relay is actuated, the pressure within the reset bellows increases, producing an upward force. The upward force is applied to the force beam. The force beam is rotated counterclockwise, causing the flapper/nozzle clearance to be reduced. As before, the reduction in the flapper/nozzle clearance causes nozzle backpressure to increase, which actuates the output relay, causing automatic output pressure to increase. The output from the automatic output relay is fed back to the manual-auto comparator auto bellows to restore

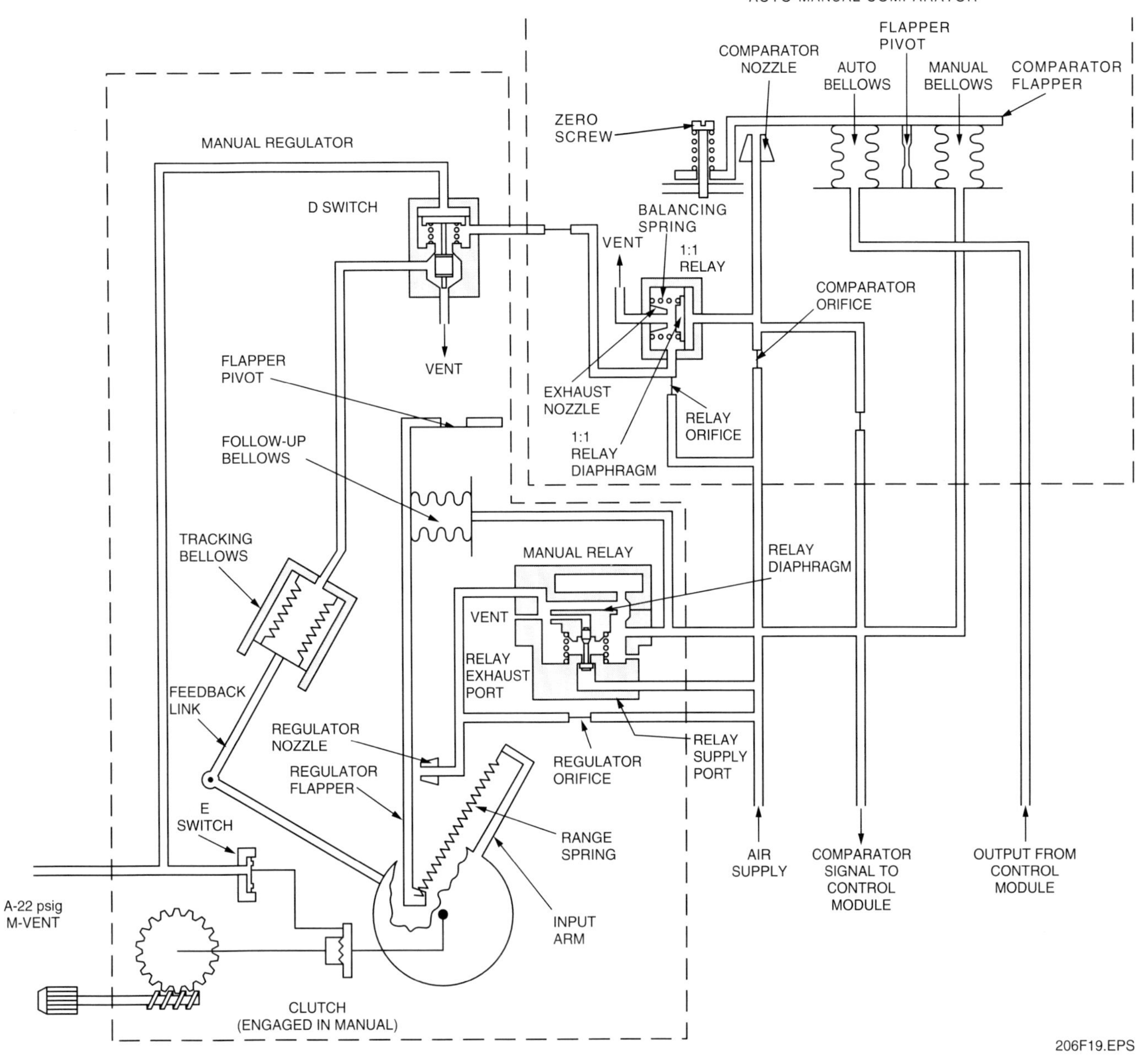

Figure 19 ◆ Manual control mechanism.

the balance. As a result of the comparator action, the automatic output pressure is forced to track with the manual output pressure. Thus, when switching from manual to automatic, the process is not disturbed.

How then is the manual output forced to track the automatic output? When the controller is in automatic, the auto-manual select signal is 22 psig. The select signal causes the E-switch to disengage the manual control adjustment from the regulator input arm. The select signal also actuates the D-switch. With the 22 psig applied to the D-switch diaphragm, the vent valve is closed and supply air is provided to the tracking bellows. The amount of pressure within the tracking bellows is regulated by the comparator 1:1 relay.

Automatic output pressure is sensed by the auto bellows of the auto-manual comparator. If controller output pressure increases, an upward force is applied by the auto bellows to the comparator flapper lever. This upward force moves the flapper away from the nozzle, causing the nozzle backpressure to decrease. The decreasing nozzle backpressure is sensed by the comparator 1:1 relay. The force applied to the 1:1 relay diaphragm decreases. The diaphragm balancing spring moves the diaphragm assembly to the right, uncovering the exhaust nozzle. As a result, a path is opened that vents pressure within the tracking bellows to the atmosphere. The reduction in pressure within the tracking bellows causes the bellows to contract. As the bellows contracts, the feedback link moves to the right. This movement rotates the input arm, causing spring tension to increase. Spring tension applies a force to the regulator flapper lever. This force moves the flapper lever closer to the regulator nozzle. The manual output pressure is applied to the manual bellows. A force is produced by the manual bellows as a result of the increased manual output pressure that balances the comparator mechanism. The action of the auto-manual comparator forces the manual output pressure to track the change in automatic output pressure. When the controller is shifted from automatic to manual control, a bumpless transfer occurs.

2.4.3 Foxboro-Eckardt Controllers

Although new technology, such as electronic and digital control, has begun to replace pneumatic control technology, many older industrial facilities still rely on pneumatic technology in process control.

Foxboro-Eckardt's Pneumatic Controller CP 600/400 Compact P and the Manual Control Unit CP 600/400/200 are examples. They offer a flat scale indication, using a bellows measuring system, with modular construction. They can also be mounted in individual housings. These controllers also come equipped with a convenient I/P internal transducer.

The controllers offer any combination of control, including only P, a combination of P and D, a combination of P and I, and PID.

The Manual Control module can be linked to an extra CP 600 module and then linked with the master controller section, which is designed with pneumatic tubing to the back panel for easy plug-in installation.

3.0.0 ◆ ELECTRONIC CONTROLLERS

Electronic controllers have the ability to receive a process variable signal, compare that signal to one that the operator selects, and adjust an output signal used to operate a final control element.

Electronic controllers have many advantages compared to pneumatic controllers:

- *Location* – The controller can be placed in a room with other controllers, thereby allowing an operator to monitor and control many control loops or systems at one location.
- *Signal isolation* – Incoming signals can be isolated from others to minimize interference in the controller circuitry.
- *Signal conversion* – Controllers can accept a variety of electronic input signals, convert them to voltages acceptable to the controller, and, after processing, convert the final signal to a current or voltage that the next controller (cascade) or final element can utilize.
- *Signal comparison* – The operator may adjust a voltage in the setpoint section of the controller that will provide a standard to which the incoming process signal is compared. The controller will also accept a cascade signal from another source as the standard for comparison. Any difference in the two signals will be seen as an error or differential voltage, which the controller will process to provide a corrective output signal.
- *Process indication* – The controller may be equipped with a meter, lights, or other indicators to visually represent the present condition of the process. The indications may also include the setpoint and a visual comparison between the process measurement and the setpoint.
- *Power supply* – The controller may provide a safe, low noise DC voltage to the transmitter or other peripheral equipment.
- *Signal connection* – The controller requires electrical wiring only for the input and output signals and the supply. It does not require any piping or tubing connections.
- *Control modes* – Circuitry within the controller can be used to provide different amounts of output change for a given input change and a different output response rate for a given input change.
- *Direct manual control* – The controller allows the operator to manually change the output signal by adjusting the setpoint or by overriding the automatic signal and manually adjusting the output. The controller circuitry can be modified to **annunciate**, visually or audibly, any controlled variable or output changes that go beyond predetermined limits.

The most basic type of electronic controller consists of the following:

- *Measuring slide wire* – provides a variable resistance that is adjusted by moving a contacting slider.
- *Unbalance detector* – compares the input signal to a setpoint.
- *Amplifier* – increases the unbalance detector signal to a usable size.
- *Control slide wire* – transmits the amplified signal to the unbalance detector.
- *Control motor* – positions a final element.

Figure 20 shows an electronic controller flow graph.

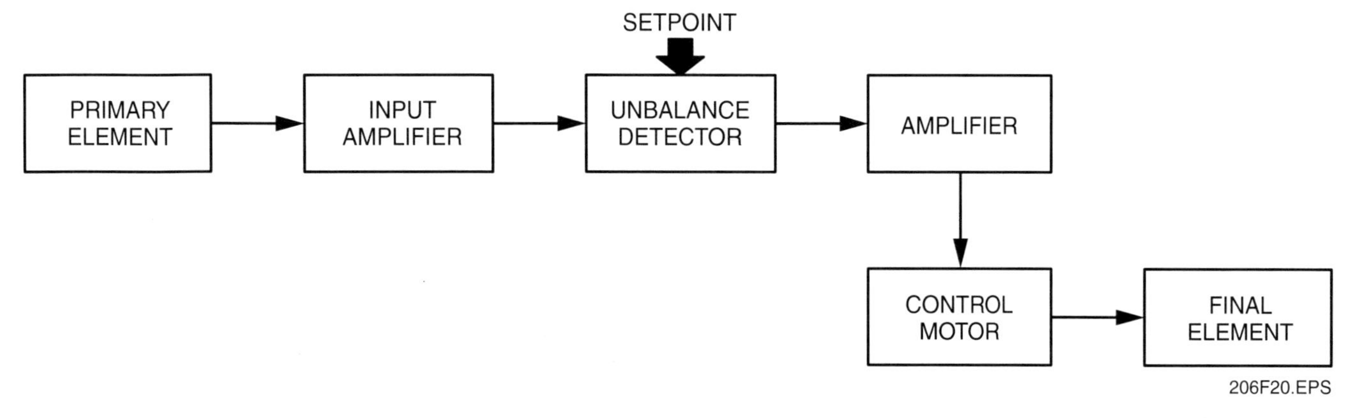

Figure 20 ◆ Basic electronic controller flow graph.

Electronic controllers are available with the same control characteristics as pneumatic controllers:

- On-Off
- Proportional
- Proportional with reset (integral)
- Proportional with reset (integral) and derivative (PID)

3.1.0 On-Off Control

On-Off control can be achieved by any device that opens and closes an electrical circuit when the measured variable moves from the setpoint. Examples of these devices are thermostats and switches. These devices can be found in flow-related, pressure-related, level-related, and temperature-related processes. Some examples are shown in *Figure 21*.

3.2.0 Proportional Control

Electronic controllers with a proportional control provide an electrical output signal that changes proportionally as the measured variable moves from the setpoint. The input signal from a primary element is received by a measuring slide wire that changes the measurement into a form usable by a controller. The signal then travels to an unbalance detector that compares the measurement to a setpoint. The difference between these two values is amplified and transmitted to a control motor. The control motor repositions the final element.

As the control motor is manipulating the final element, the same signal goes to a control slide wire. The control slide wire sends a signal to the unbalance detector, which cancels the original unbalance. *Figure 22* shows an electronic controller with a proportional control mode.

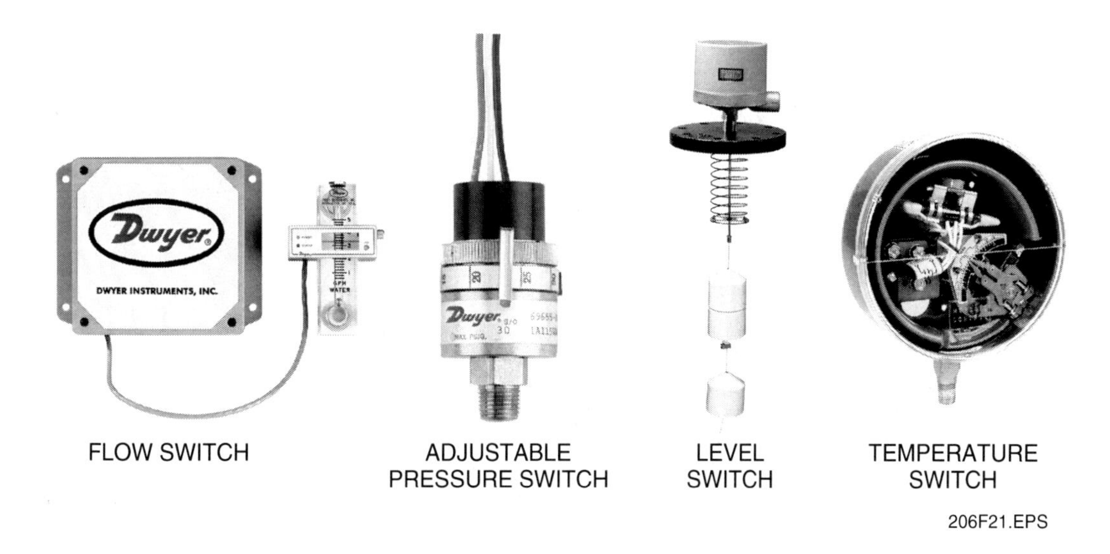

Figure 21 ◆ On-Off control devices.

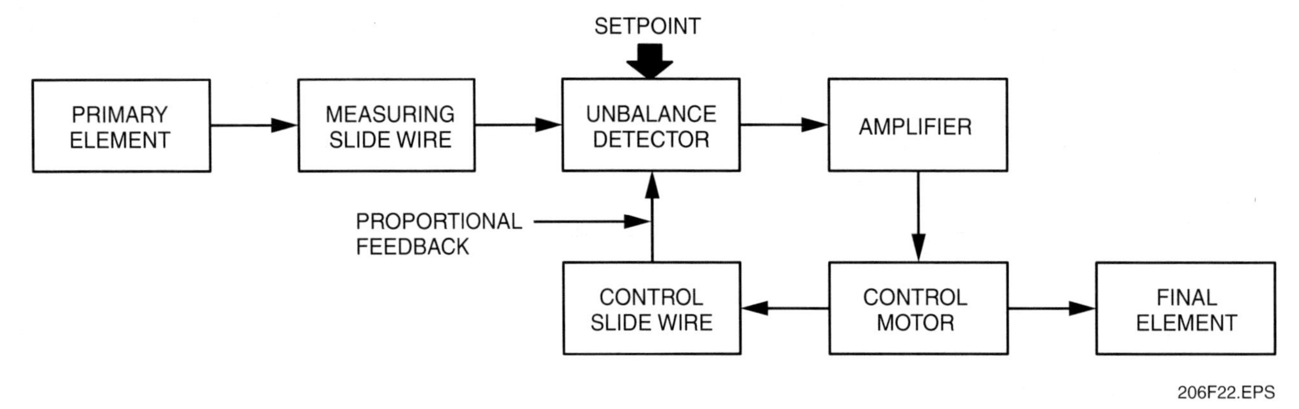

Figure 22 ◆ Electronic controller with proportional control mode.

A narrow or wide proportional band can be achieved with minor adjustments. With a narrow proportional band, a small movement from the setpoint causes a large unbalance. The control motor must drive a large amount to restore controller balance.

With a wide proportional band, a large movement from setpoint causes a small unbalance. This causes the control motor to drive only a small amount to restore balance. *Figure 22* shows the location of a proportional band adjustment.

3.3.0 Proportional Controllers with Reset (Integral)

An electronic controller with a proportional and reset control alters the feedback signal from the control slide wire. This controller also uses a reset unit that allows a shift in the control slide wire and measuring slide wire while an unbalance exists. *Figure 23* shows an electronic controller with a proportional and reset control mode.

If the position of the control motor or final element does not balance the controller, the reset unit sends a signal to the unbalance detector until the controller is in balance.

3.4.0 Proportional Controllers with Reset (Integral) and Derivative (PID)

Electronic controllers with proportional, reset (integral), and derivative control use a derivative unit in addition to the other components explained. The derivative unit receives the proportional feedback signal from the control slide wire and delays the signal to the unbalance detector.

This delay varies with the rate at which the unbalance occurs. The more rapid the unbalance, the longer the delay. The delay results in a greater unbalance signal to the control motor than if the feedback signal were to immediately restore controller balance. This allows the control motor to move a greater amount than with a proportional control mode alone. *Figure 24* shows an electronic controller with proportional, reset, and derivative control modes.

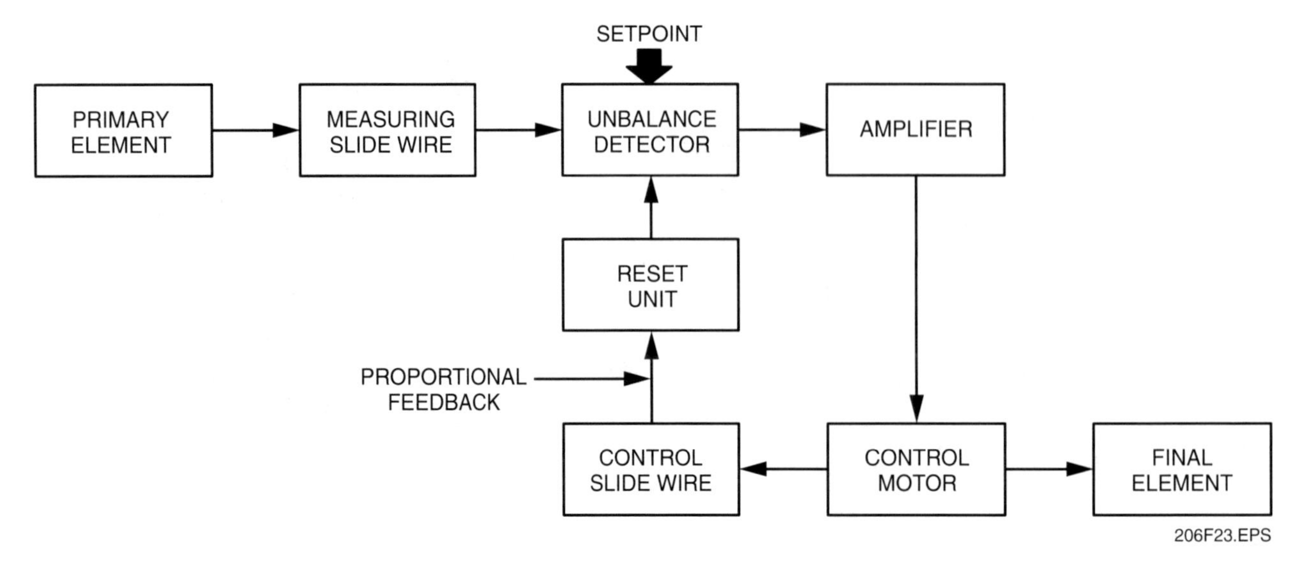

Figure 23 ◆ Electronic controller with proportional and reset control mode.

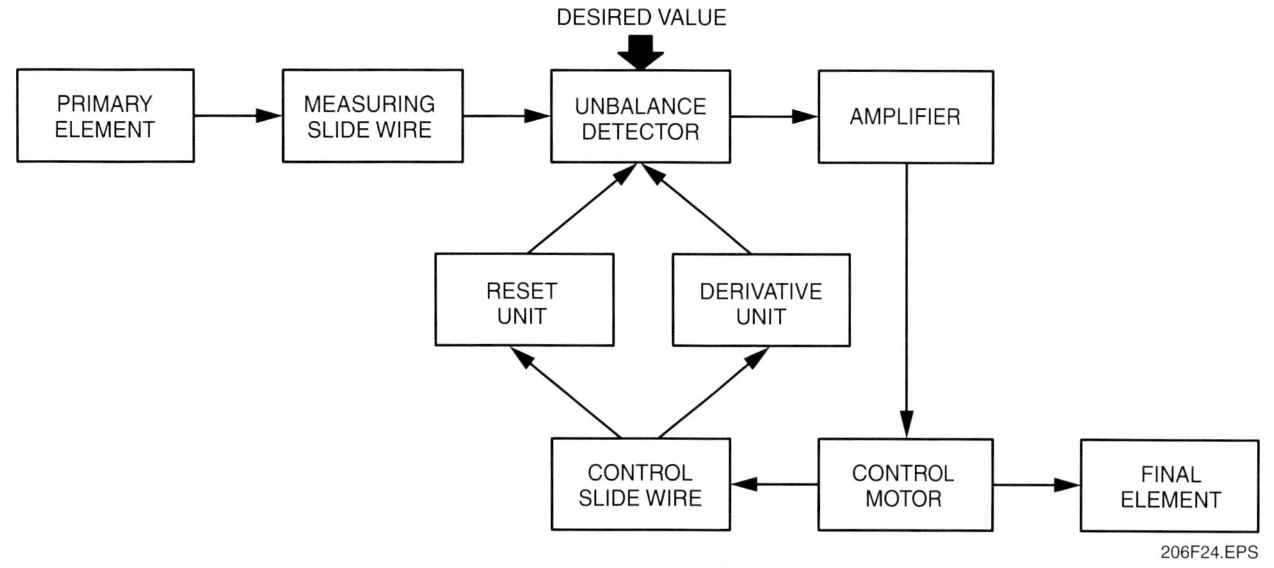

Figure 24 ◆ Electronic controller with proportional, reset, and derivative control.

Through the use of high gain operational amplifiers, the electronic controller has become more compact and more powerful. Controllers that use this technology can receive a 4–20 mA input signal. Through the use of the operational amplifiers acting as differential amplifiers, these controllers develop an **error signal burst** on the setpoint and other control parameters set into the controller. This can be done without moving parts, which can become dirty or worn out. An operational amplifier can be more accurate than a slide resistor. A block diagram of an operational burst controller can be seen in *Figure 25*.

An example of this type of controller is the ABB Automation MICRO-DCI™, Series 53MC5000A Process Control Station (*Figure 26*). The display is a highly visible dot matrix. The control station is available in a one-, two-, three-, or four-loop design, with easy PID self-tuning control.

The 53MC5000A can perform any process application, from simple PID to the most complex control design. It incorporates an embedded and programmable microprocessor.

The 53MC5000A can handle a wide variety of control applications. At the operator interface level, this is achieved through multiple screens.

The Flexible Control Strategy is a sequence of control modules that can be configured to fit most process applications. A one-loop 53MC5000A has one set of these control modules. A two-loop version has a second, independent set of control modules. The four-loop version has four sets. A library of commonly used control configurations can be found in the 53MC5000A, including PID, ratio, two-loop, and cascade. The user simply inputs specific process parameters to implement a control strategy.

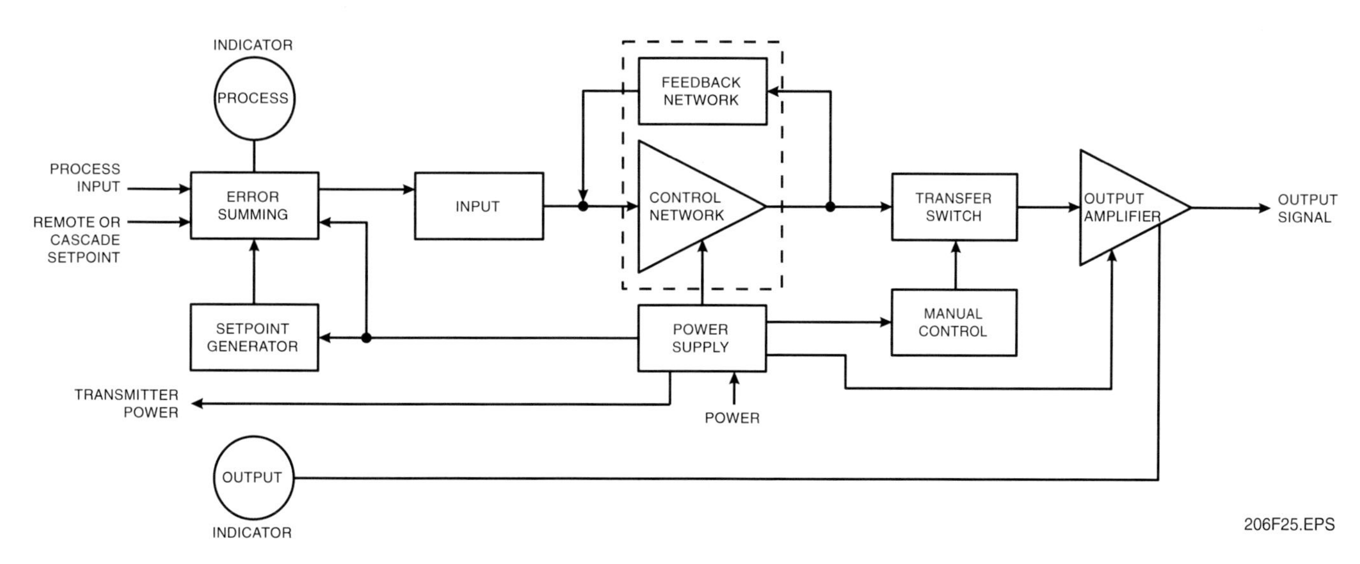

Figure 25 ◆ Block diagram of a typical electronic controller.

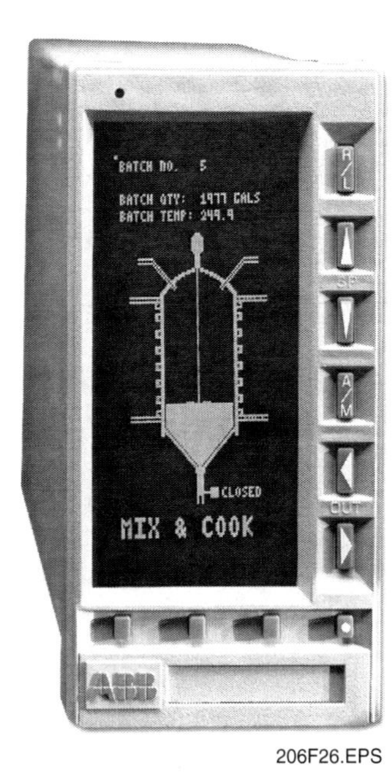

206F26.EPS

Figure 26 ◆ ABB Series 53MC5000A Process Control Station.

Figure 27 shows an example using the 53MC5000A, in which a chemical processing facility surge tank is normally maintained at a specific level using a variable speed pump. If the influent flow becomes excessive, and the level rises to an unacceptable point, a constant-speed pump is started. After a safe level is reached, the constant-speed pump is stopped, and normal control resumes.

4.0.0 ◆ RECORDERS AND INDICATORS

Recorders and indicators are used in process plants to provide readouts of process conditions. Recorders and indicators may be field mounted to give readouts at precise locations. They may also be located in central control rooms to monitor many process activities in a central area.

4.1.0 Recorders

Recorders are instruments that provide permanent records of changes in a process variable. There are three basic categories of recorders:

- Mechanical
- Electronic
- Pneumatic

Recorders can be either chart recorders, which leave a continuous record on chart paper, or database recorders, which log all recorded data in a database for future use. By far the most common is the chart recorder.

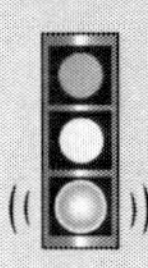

NOTE

Because of their physical similarity, it is sometimes easy to confuse a controller and a recorder. Always check the device to avoid making this mistake.

4.1.1 Mechanical Recorders

Mechanical recorders, also known as direct process recorders, are used primarily in remote locations. They generally record only one or two process variables. *Figure 28* shows an older single-pen mechanical recorder.

A mechanical recorder can be divided into three main sections:

- *Measuring section* – The measuring section of a mechanical recorder is usually self actuating. Pressure and temperature are measured by this type of recorder. Bourdon tubes and filled thermal systems are common measuring elements

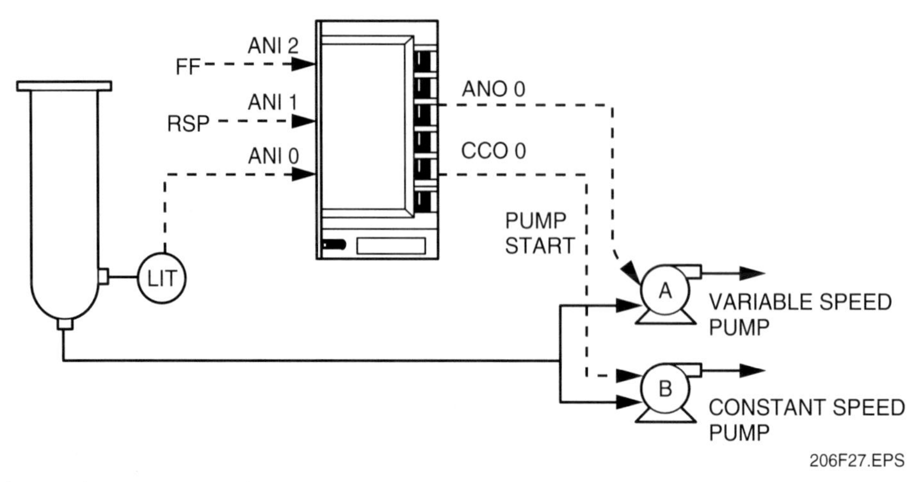

206F27.EPS

Figure 27 ◆ A level control process using the ABB Series 53MC5000A.

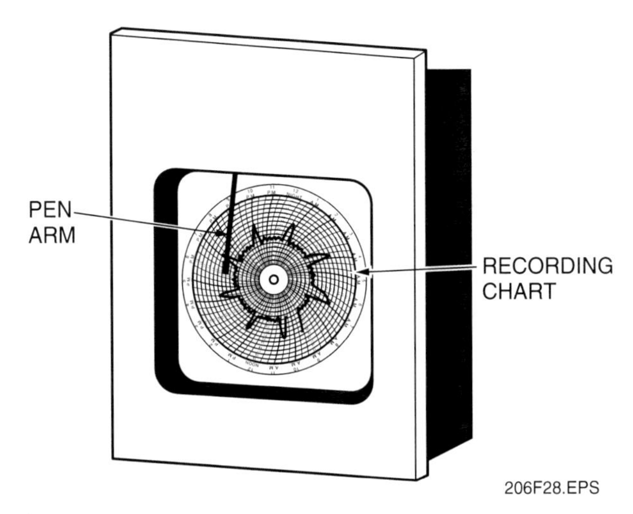

Figure 28 ◆ Mechanical recorder.

used in the measuring section. Pressure applied to the Bourdon tube causes it to move. The movement is transferred to the recording section through the link assembly. *Figure 29* shows a helix-type Bourdon measuring element.

- *Recording section* – The recording section consists of a link arm, pivot point, and pen arm. The link arm is connected to a link assembly on the measuring element. *Figure 30* shows the operation of a mechanical recorder.

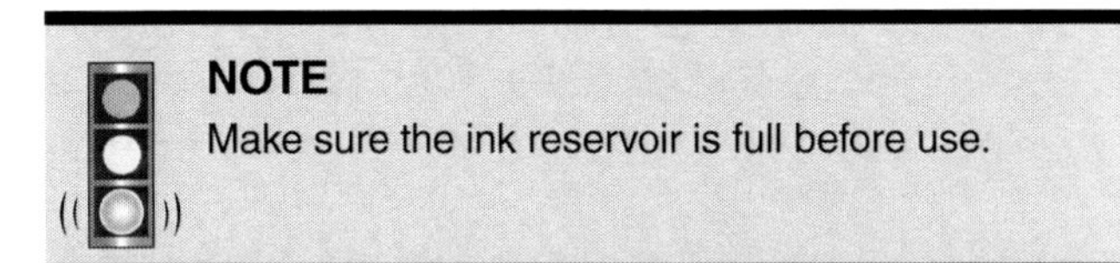

NOTE
Make sure the ink reservoir is full before use.

- *Chart drive section* – The chart drive section consists of a chart drive motor and chart paper. Mechanical chart drive assemblies are usually spring driven. A spring-driven motor can be used anywhere, because an outside power source is not required. These motors generally operate for one to seven days without being rewound. A spring clip on the face of the motor holds the chart paper in place. *Figure 31* shows the circular chart paper used to record temperature. The outer ring of the chart paper indicates the time of the recording. The degrees of temperature are indicated in the graph section. The dark uneven line shown in *Figure 31* is a sample process temperature recording.

4.1.2 Electronic Recorders

Electronic recorders are usually located in a control room. Electronic recorders that record more than one process variable are called multi-pen recorders.

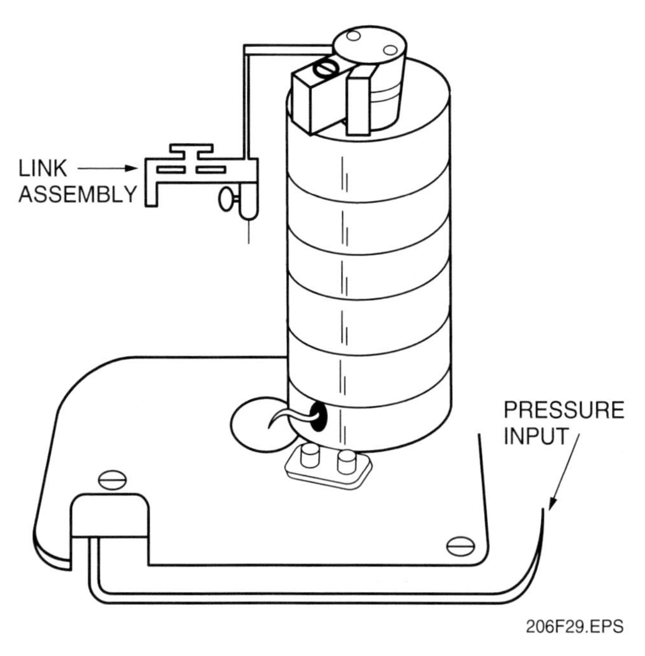

Figure 29 ◆ Helix-type Bourdon measuring element.

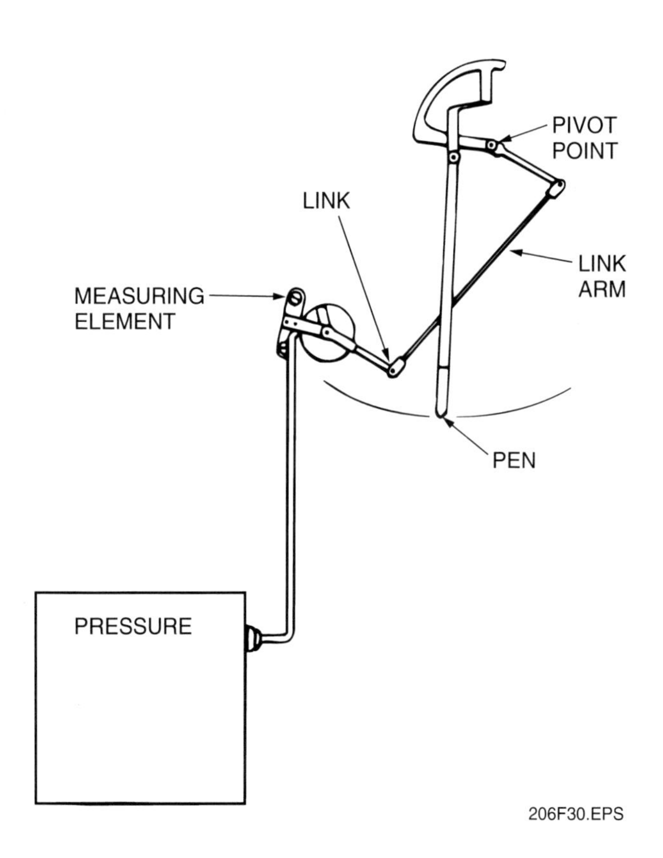

Figure 30 ◆ Mechanical recorder operation.

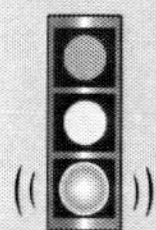

NOTE

The following paragraphs explain each section of an older type of panel-mounted electronic three-pen recorder.

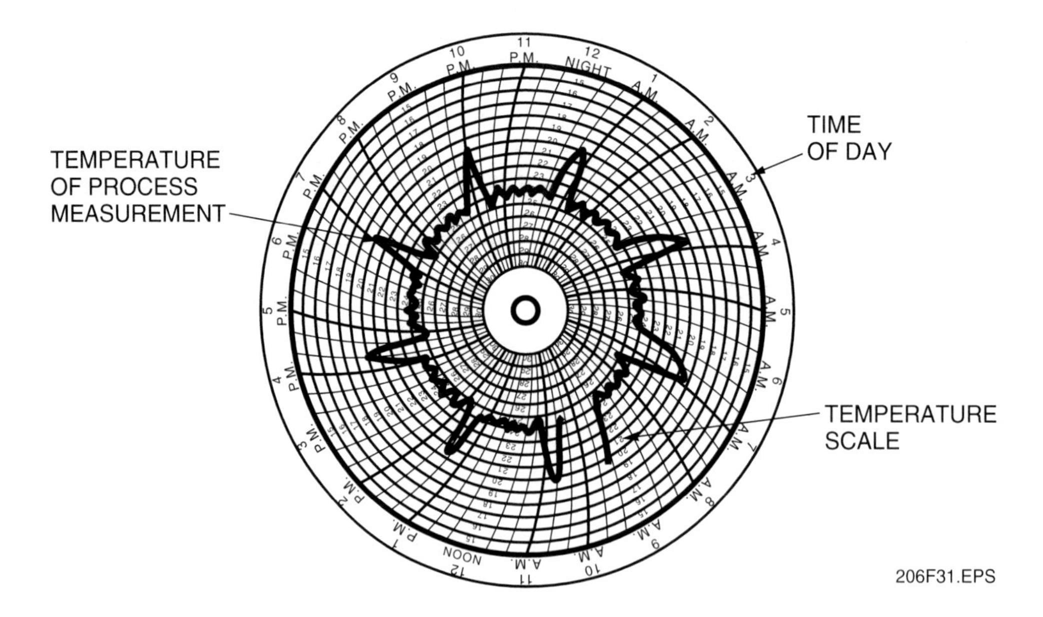

Figure 31 ◆ Circular-chart paper for recording temperature.

An electronic recorder can be divided into three main sections:

- *Measuring section* – The measuring section receives up to three different electronic process measurement signals. The signals generally are 1–5 VDC, or 4–20 mA converted to a 1–5 VDC signal. The measurement signals are received by a signal cord set located at the rear of the unit. The signal is converted into a physical movement by an electronic servomotor. The servomotor rotates to an exact position for each change in the input signal. A belt attaches the servomotor to the recording section.
- *Recording section* – The recording section consists of a pen cartridge assembly that travels up and down on a rod. A belt is attached to the back of the pen holder and the servomotor. The connecting belt drives the pen assembly up and down on the rod and positions the pen on the chart paper. Newer units have a disposable cartridge pen assembly.
- *Chart drive section* – The chart drive section consists of a gear assembly driven by a small electric motor. The chart paper is packaged in a roll that is designed to last up to 30 days.

The time of day is indicated on the bottom and top of the chart paper. Actual process measurements are indicated on the graph line. The measuring element transfers movement to the link arm. The link arm then transfers movement to the pen arm through the pivot point. The pivot point changes the straight back-and-forth movement of the link arm into an arc-type motion. On this type of unit, the pen tip on the end of the pen arm has a small well that holds a liquid ink supply.

4.1.3 Newer Electronic Recorders

Most newer recorders are microprocessor controlled and are fully programmable for a variety of functions. For example, ABB Automation manufactures the following electronic recorders:

- *Electronic six-channel strip-chart recorder – Figure 32* shows a 100mm strip chart recorder that provides accurate and reliable recording of up to six channels. The SR100A shown also provides a range of advanced processing capabilities, such as flow totalization, math blocks, logic equations, configurable displays, and full message printing, which can be configured using the front panel or with PC configuration software. The SR100A can be supplied either for panel-mounting or for portable use.
- *Electronic four-channel circular-chart recorder* – Like the six-channel strip-chart recorder, the electronic circular-chart recorder shown in *Figure 33* is microprocessor controlled and can be programmed from the front panel. The unit has math functions, alarm outputs, analog or digital inputs, digital outputs, a serial communications link for a PC, and pneumatic inputs/outputs.
- *Electronic nine-channel paperless recorder* – The recorder shown in *Figure 34* uses a 10-inch video color display instead of paper to show the data being recorded. All data is recorded on a diskette and can be transmitted via a serial data link to a PC. Like the other electronic recorders described, this unit is fully programmable using a screen display. It can accept eight analog or digital inputs and send alarms through four or eight relay contacts.

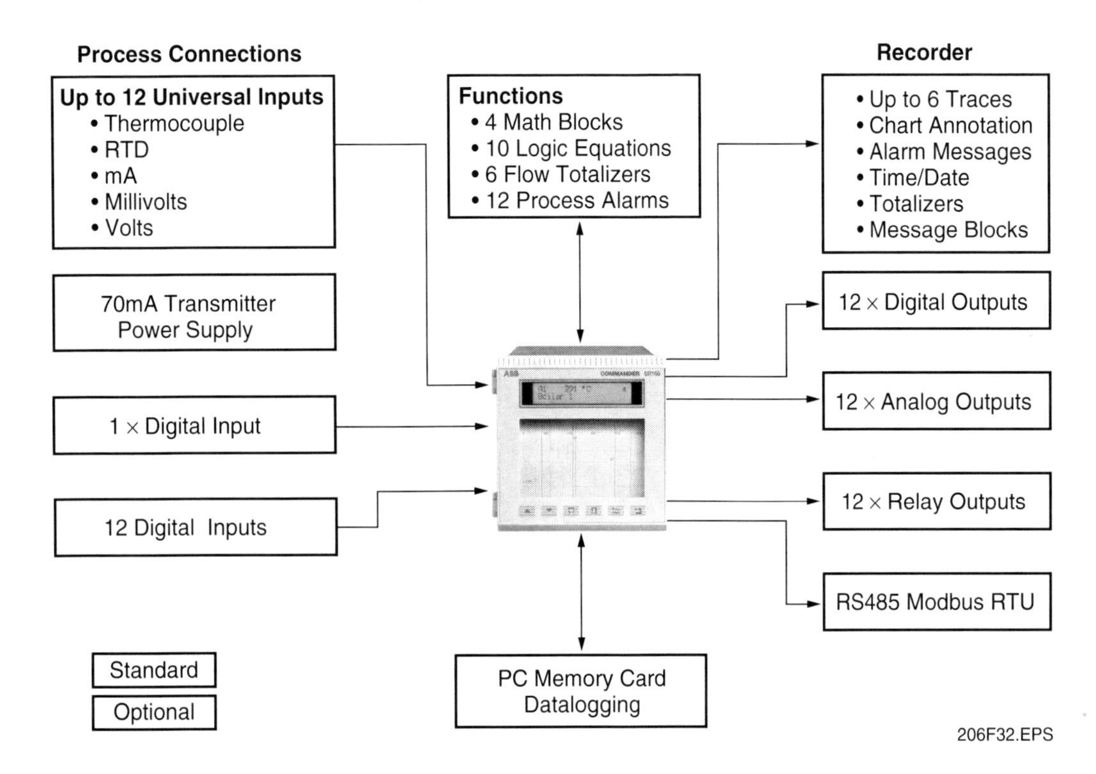

Figure 32 ◆ ABB Commander™ SR100A strip-chart recorder.

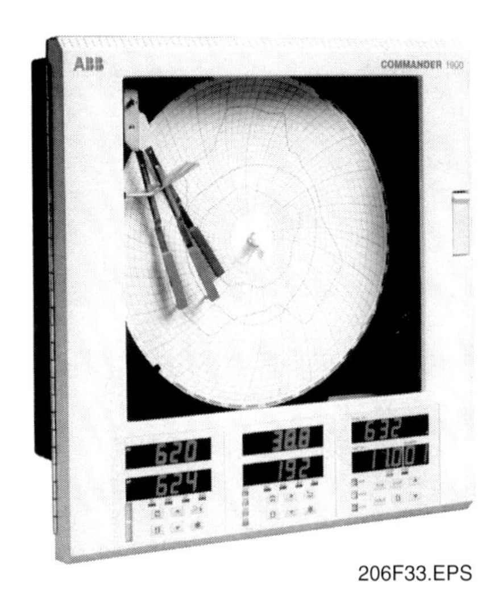

Figure 33 ◆ ABB Commander™ 1900 circular-chart recorder.

Figure 34 ◆ ABB Screenmaster™ 300 with large color display.

4.1.4 Pneumatic Recorders

Pneumatic recorders usually operate from a 3 to 15 psi input signal. These signals can be sent from a local or remote transmitter location. Depending on the construction of the unit, one or more variables may be recorded. *Figure 35* shows a three-pen pneumatic recorder.

A three-pen pneumatic recorder can be divided into three main sections:

- *Measuring section* – The measuring section receives up to three different pneumatic process measurement signals. The signals generally range between 3 and 15 psi. The signals are connected to the back of the unit through a flexible tubing cord set. The cord set may be either male or female. Be sure to get the proper connector for the type of cord set used. A bellows assembly converts an air signal into a physical movement. *Figure 36* shows a recorder bellows assembly with a link arm. Changes in the air signal cause the bellows to expand or contract. This movement is transferred to the attached link arm. The link arm then transfers movement to the pivot point of the recording section.

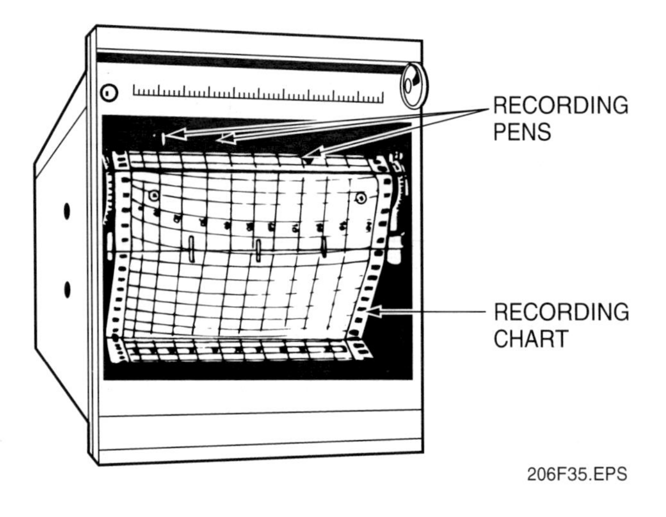

Figure 35 ◆ Three-pen pneumatic recorder.

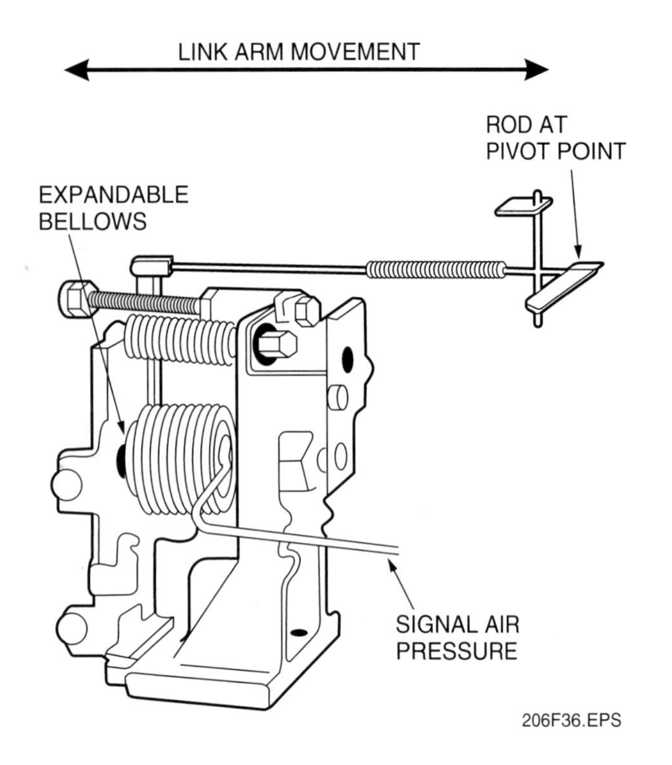

Figure 36 ◆ Recorder bellows assembly with link arm.

- *Recording section* – The recording section consists of a pivot point, pen arm, and capillary-type inking system. The back-and-forth motion of the bellows is transferred by the link arm to the pivot point. The back-and-forth motion is converted into an arc-type motion at the pivot point. This arc-type motion is then transferred to the pen arm assembly, where it positions the pen arm on the chart paper. Ink is supplied to the pen tip through a capillary tubing inking system. A supply reservoir is located on the side of the instrument case. The ink is in liquid form, and the colors are manually loaded from small plastic bottles into the ink reservoir. Lengths of small plastic capillary tubing carry the ink to the pen tips and onto the chart paper.
- *Chart drive system* – The chart drive system consists of a gear assembly driven by a small electric or pneumatic motor. The chart paper is a long continuous strip that lasts approximately two weeks. The chart drive assembly moves the paper at a steady speed under the pen tips. The time of day is marked on the outer edges of the chart paper, and the process indicators are marked on the graph divisions.

4.2.0 Indicators

Indicators are the parts of a measuring device that allows the operator to see the status of the variable being measured. Indicators come in many forms, from the simple dipstick or gauge to the digital readout on a meter.

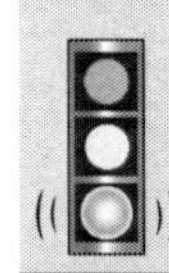

NOTE

Indicators are by far the largest and most varied family of instrumentation devices.

4.2.1 Gauges

The most common measuring element used in gauges is a C-type Bourdon tube. The measuring element is a C-shaped hollow tube, sealed at one end. Pressure causes the tube to attempt to straighten, creating a small linear movement. This movement is transferred to the link at the end of the tube. The link then transfers the motion to the pointer movement.

The pointer movement magnifies the small linear movement it receives from the link. The pointer movement converts the magnified movement into a rotary motion that moves the pointer across the graduated scale. A return spring in the movement causes the pointer to return to zero when pressure is relieved from the gauge. The knife-edge pointer is usually used on precision gauges with many fine graduations. The mirrored strip helps align the pointer edge on the graduation to avoid a parallax error. Parallax error is caused by viewing a dial from an angle. A good example of the parallax effect is viewing a speedometer from the passenger seat of a car. The reading the passenger sees is several miles per hour different from the reading the driver sees.

The dial is a circular metal or plastic plate with a scale. The scale on the dial indicates the value of an indicated variable. The individual marks on the scale are graduations. Index graduations are the heaviest black lines aligned with the numbers on the scale. Major graduations are intermediate points between the index graduations. Minor graduations are the smallest values or subdivisions of the scale.

4.2.2 Electrical Indicators

The most common indicators used in electrical measurement are electrical meters. Electrical meters measure the magnetic effects of a small current. The basic electric meter or **galvanometer** is composed of three basic elements:

- *Magnet* – The magnets used in meters are permanent magnets. A permanent magnet is one that retains magnetic strength after a magnetizing force is removed.
- *Core assembly* – A stationary iron core is located between the two poles of a permanent magnet. A moveable coil assembly pivots inside the core. A pointer is attached to the moveable coil. The coil becomes an electromagnet when current passes through it. The electromagnetic coil balances within the magnetic forces of the permanent magnet. As the current in the coil increases, the magnetic strength also increases to drive the pointer upscale. When current is not passed through the meter, a spring keeps the pointer at its zero position.
- *Scale* – The scale of the meter indicates the variable being read. The scale of the meter also indicates the maximum strength of the variable the meter can safely read.

The basic galvanometer is modified to create the three most commonly used electrical indicators:

- *Ammeter* – The ammeter is used to determine the amount of current flowing in an electrical device. An ammeter is a basic galvanometer with an electrical shunt circuit. The shunt circuit is a resistor that allows the meter to bypass the majority of the current to prevent damage to the meter. The scale on the meter is marked to indicate the total current of the circuit. *Figure 37* shows an ammeter circuit.
- *Voltmeter* – A voltmeter is used to measure the amount of voltage in the circuit. A voltmeter is a basic galvanometer with a high value resistor in series with the coil circuit. *Figure 38* shows a voltmeter circuit.
- *Ohmmeter* – An ohmmeter is used to measure the resistance in ohms in a circuit, conductor, or load. The ohmmeter is a basic galvanometer with a battery, fixed resistor, and rheostat. The battery provides power for the circuit. The

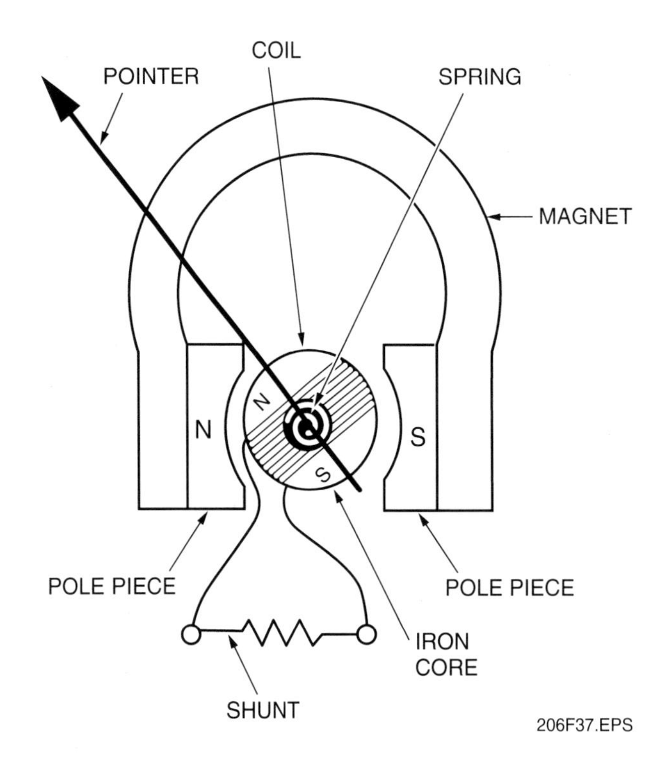

Figure 37 ◆ Ammeter circuit.

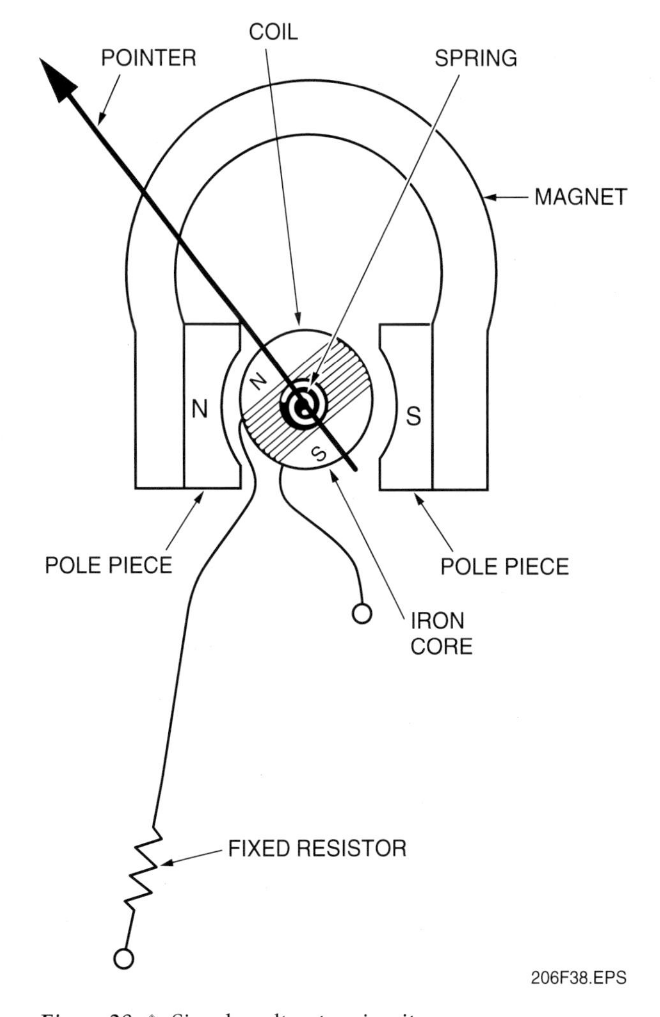

Figure 38 ◆ Simple voltmeter circuit.

rheostat allows the circuit to be adjusted to zero resistance. A zero setting avoids error in measuring an unknown resistance by canceling out the meters and resistance. *Figure 39* shows an ohmmeter circuit.

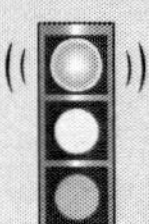

WARNING!

Never use an ohmmeter on any energized circuit. Excessive voltage can cause damage to the meter that may result in injury to the user.

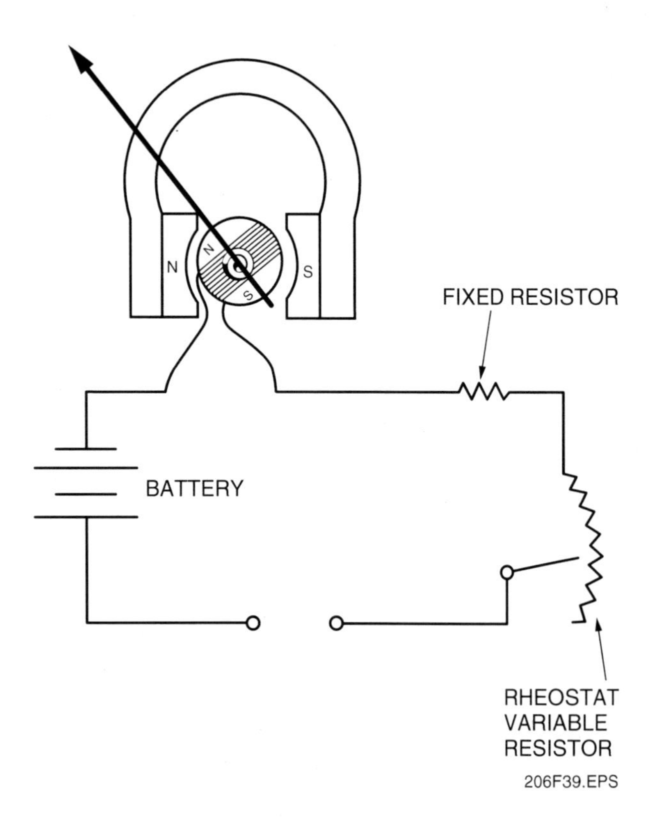

Figure 39 ◆ Ohmmeter with internal resistance adjustment.

4.2.3 Electronic Indicators

The most common types of electronic indicators are servomotors and digital displays. The servomotor indicator functions like an electronic recorder but does not contain a recording section.

A digital indicator receives a transmitted signal and processes the signal into a lighted visual digital display. This lighted digital display can represent any process variable, such as pounds of pressure or a **pH** scale.

Most of these indicators use a digital circuit called a seven-segment display. *Figure 40* shows digital indicators with a seven-segment display.

A seven-segment display is an electronic component used to display the numbers 0 through 9. Illuminating certain segments in sequence creates individual numbers. *Figure 41* shows digits 0 through 9 on a seven-segment display.

4.2.4 Pneumatic Indicators

Most pneumatic indicators operate from a 3 to 15 psi (pneumatic) input signal. Pneumatic panel-mounted indicators operate in a manner much like pneumatic recorders. This section explains the measuring indicating section of a pneumatic panel indicator.

Refer to *Figure 42*. The measuring indicating section consists of a spring-opposed bellows, link assembly, pivot point, and indicator arm. The bellows assembly receives a pneumatic input and expands accordingly. This expanding motion is transferred to a link assembly at the base of the bellows unit. The link assembly receives the signal and transfers it in a straight back-and-forth motion to a pivot point.

The indicator arm is attached to the pivot point. When the bellows expands, the indicator arm is driven upscale. The spring at the base of the bellows returns the indicator to the zero position when there is no input signal.

206F40.EPS

Figure 40 ◆ Digital indicators with seven-segment display.

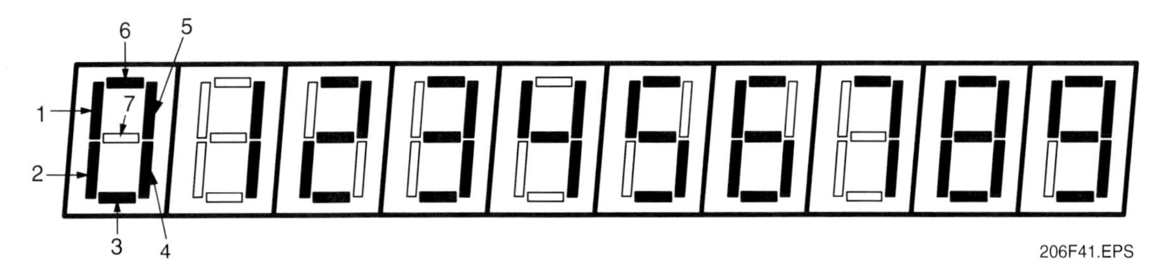

Figure 41 ◆ Seven-segment display 0 through 9.

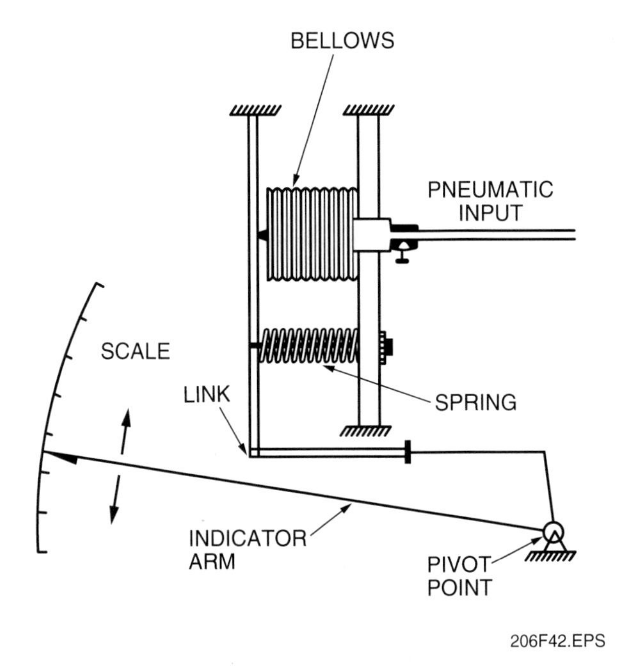

Figure 42 ◆ Measuring indicating section.

4.2.5 Thermal Indicators

A thermal indicator is one which changes its visual indication in response to a change in temperature. A good example of a thermal indicator is an optical or brightness pyrometer (*Figure 43*).

The optical pyrometer consists of a tube that houses a lens arrangement and a platinum filament. The filament is connected to a battery.

When a very hot object is viewed through the pyrometer, that object has a certain brightness, depending on the heat coming from the object.

The battery can be adjusted on the pyrometer so that the filament glows with approximately the same intensity as the object. When the brightness of the filament and the brightness of the object match, the temperature of the object can be found. This is shown in *Figure 44* and is done by reading the setting of the battery adjustment control.

4.2.6 Magnetic Indicators

Magnetic indicators use magnetism to provide an indication. There are many types of magnetic indicators, from reed switches, which will indicate in a window, to syncros and servo-driven indicators.

A very simple type of magnetic indicator is shown in *Figure 45*. This is basically a float type of device, which couples its position through a nonmagnetic dip tube to an inner magnet. As the float moves up and down, the inner magnet follows. The position of the inner magnet is indicated by an indicating pointer.

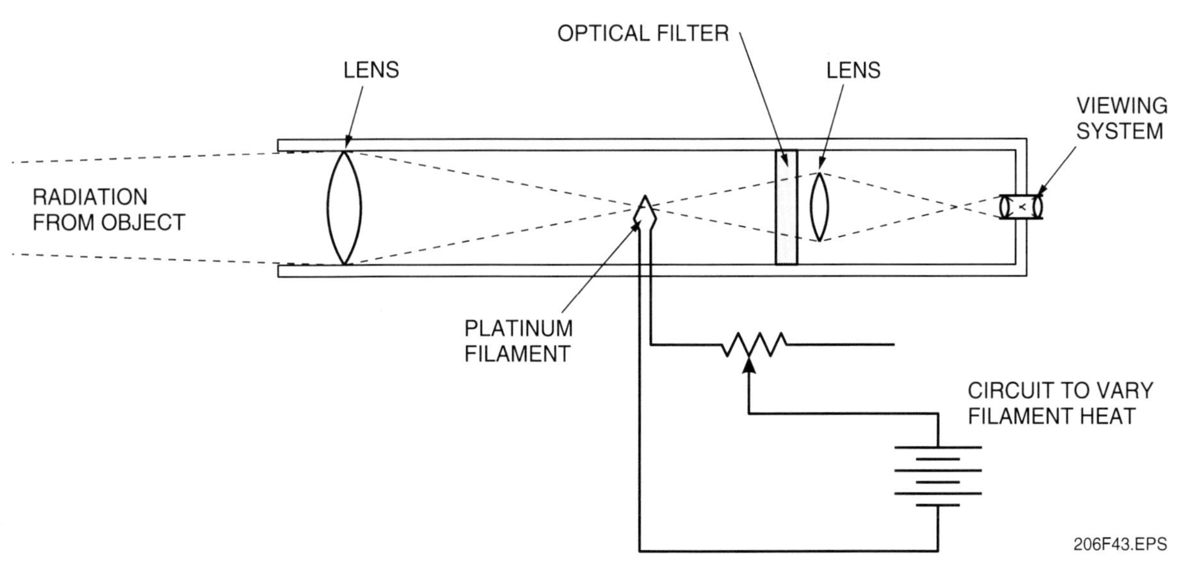

Figure 43 ◆ Optical pyrometer.

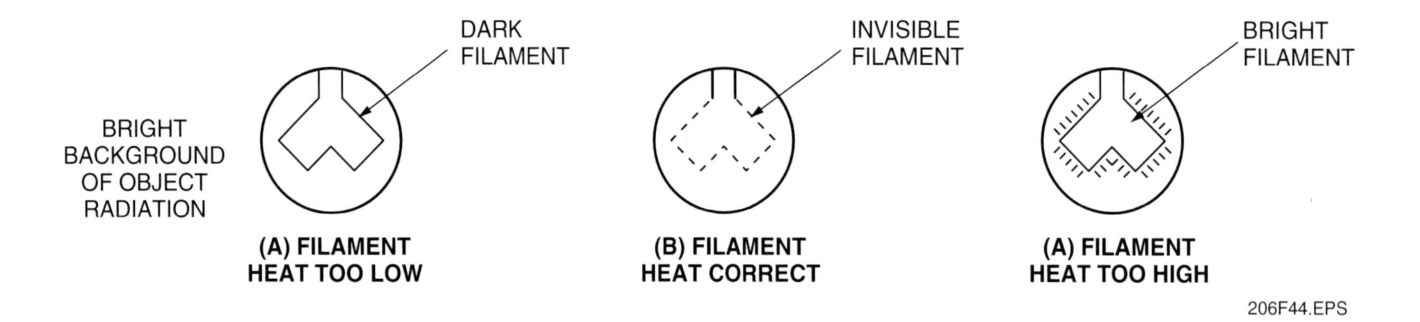

Figure 44 ◆ The appearance of the filament during optical pyrometer adjustment.

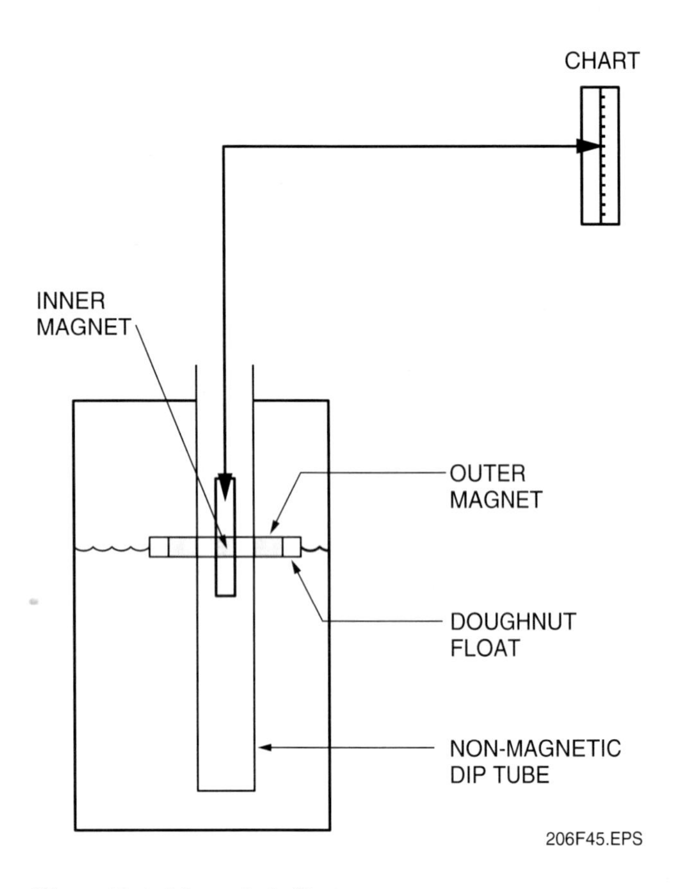

Figure 45 ◆ Magnetic indicator.

Summary

A controller in an instrumentation system receives a process signal from a measuring element and sends a signal to a final element. The final element uses this signal to maintain a desired process condition, such as flow rate, temperature, level, or consistency. Two types of controllers discussed in this module are pneumatic and electronic.

Pneumatic controllers rely on a pneumatic input signal for operation. This signal can come from a measuring element, an I/P transducer, or a pneumatic transmitter.

Electronic controllers can accept a variety of electronic input signals, convert them to voltages acceptable to the controller and, after processing, convert the final signal to a current or voltage that the next controller or final element can use.

Both pneumatic and electronic controllers are available with different control characteristics, such as On-Off, proportional, proportional with reset, and proportional with reset and derivative.

Recorders and indicators are used in process plants to provide readouts of process conditions.

There are three basic categories of recorders: mechanical, electronic and pneumatic.

Each of these recorders is divided into three basic sections: measuring, recording, and chart drive. Recorders may be field mounted to give readouts at precise locations or located in control rooms to monitor many process activities in a central area.

Indicating instruments are generally equipped with scales graduated in units of the measured variable. The shape and size of the scale will vary, depending on the needs of the specific application. There are many different types of indicators, such as gauges, analog meters, and digital meters.

Review Questions

1. In automatic control, the error signal is developed and processed through the _____.

a. final element
b. recorder
c. On-Off device
d. PID controller

2. An I/P transducer converts a(n) _____ signal to a pneumatic signal.

a. mechanical
b. electronic
c. air
d. hydraulic

3. Another name for On-Off control is _____ control.

a. automatic
b. manual
c. discrete
d. switching

4. The proportional function of a controller is also known as _____.

a. gain
b. integral
c. derivative
d. algorithm

5. The reset function of a controller is also known as _____.

a. derivative
b. algorithm
c. gain
d. integral

6. A controller that has the functions of proportional, reset, and derivative is often referred to as a(n) _____ controller.

a. PID
b. PRD
c. multifunctional
d. automatic

7. In an electronic controller with a _____ proportional band setting, a small movement from the setpoint causes a large unbalance.

a. wide
b. narrow
c. non-linear
d. derivative

8. In an electronic controller with proportional and reset control, the reset control unit sends a signal to the _____ until the controller is in balance.

a. final element
b. control valve
c. sensor
d. unbalance detector

9. The ABB Automation Series 53MC5000A Controller incorporates a(n) _____.

a. hand-held configurer
b. infrared sensor
c. programmable microprocessor
d. database downloader

10. Instruments that provide permanent records of changes in a process variable are known as _____.

a. indicators
b. controllers
c. recorders
d. databases

11. Common measuring elements used in the measuring system of mechanical recorders include _____ and filled thermal systems.

a. thermometers
b. I/P transducers
c. infrared detectors
d. Bourdon tubes

12. Electronic recorders that record more than one process variable are called _____ recorders.

a. multipurpose
b. process-variable
c. multi-pen
d. multifunctional

13. Of the following, a _____ is a type of indicator.

a. multi-pen
b. PID
c. chart drive
d. gauge

14. _____ causes the tube of a C-type Bourdon tube to attempt to straighten.

a. Pressure
b. Temperature
c. Flow
d. Level

15. Most electronic indicators that use a digital display are _____-segment displays.

a. nine
b. seven
c. five
d. one

GLOSSARY

Trade Terms Introduced in This Module

***Annunciate*:** To indicate, either visually or audibly, as determined by a preset value.

***Controller*:** An instrument that receives a process signal in a loop from a measuring element and sends a signal to a final element.

***Derivative action*:** A controller mode in which there is a continuous linear relationship between the controller output and the derivative of the error signal.

***Differential gap*:** The difference between a high and low setpoint.

***Error signal burst*:** Amplified bits of data that are transmitted in rapid succession from a single device.

***Feedforward*:** Feedforward measures a process disturbance, predicts its effect on the process, and applies a corrective action. It is an open loop.

***Galvanometer*:** An instrument for detecting the existence of, and determining the strength of, small electrical currents.

***I/P transducer*:** A converting device that proportionally changes or converts an electronic signal into a pneumatic signal.

***Moment*:** A bending effect caused by an applied force.

***pH*:** The measurement of acid or alkaline content.

***Proportional band*:** The range of values of the measured variable through which the final element moves.

***Recorder*:** An instrument that provides a permanent record of changes in a process variable.

Additional Resources

This module is intended to present thorough resources for task training. The following reference works are suggested for further study. These are optional materials for continued education rather than for task training.

Process Instruments and Controls Handbook, Second Edition, D.M. Considine, McGraw-Hill Book Company.

Instrument Engineers Handbook, Volume II, Process Control, Bela Liptak, Chilton Book Company.

Figure Credits

Dwyer	206F21, 206F40
ABB	206F26, 206F32, 206F33, 206F34

NCCER CURRICULA — USER UPDATE

NCCER makes every effort to keep its textbooks up-to-date and free of technical errors. We appreciate your help in this process. If you find an error, a typographical mistake, or an inaccuracy in NCCER's curricula, please fill out this form (or a photocopy), or complete the online form at **www.nccer.org/olf**. Be sure to include the exact module ID number, page number, a detailed description, and your recommended correction. Your input will be brought to the attention of the Authoring Team. Thank you for your assistance.

Instructors – If you have an idea for improving this textbook, or have found that additional materials were necessary to teach this module effectively, please let us know so that we may present your suggestions to the Authoring Team.

NCCER Product Development and Revision

13614 Progress Blvd., Alachua, FL 32615

Email: curriculum@nccer.org
Online: www.nccer.org/olf

❑ Trainee Guide ❑ AIG ❑ Exam ❑ PowerPoints Other ______________________________

Craft / Level: Copyright Date:

Module ID Number / Title:

Section Number(s):

Description:

Recommended Correction:

Your Name:

Address:

Email: Phone:

Module 12207-03

Control Valves, Actuators, and Positioners

COURSE MAP

This course map shows all of the modules in the second level of the Instrumentation curriculum. The suggested training order begins at the bottom and proceeds up. Skill levels increase as you advance on the course map. The local Training Program Sponsor may adjust the training order.

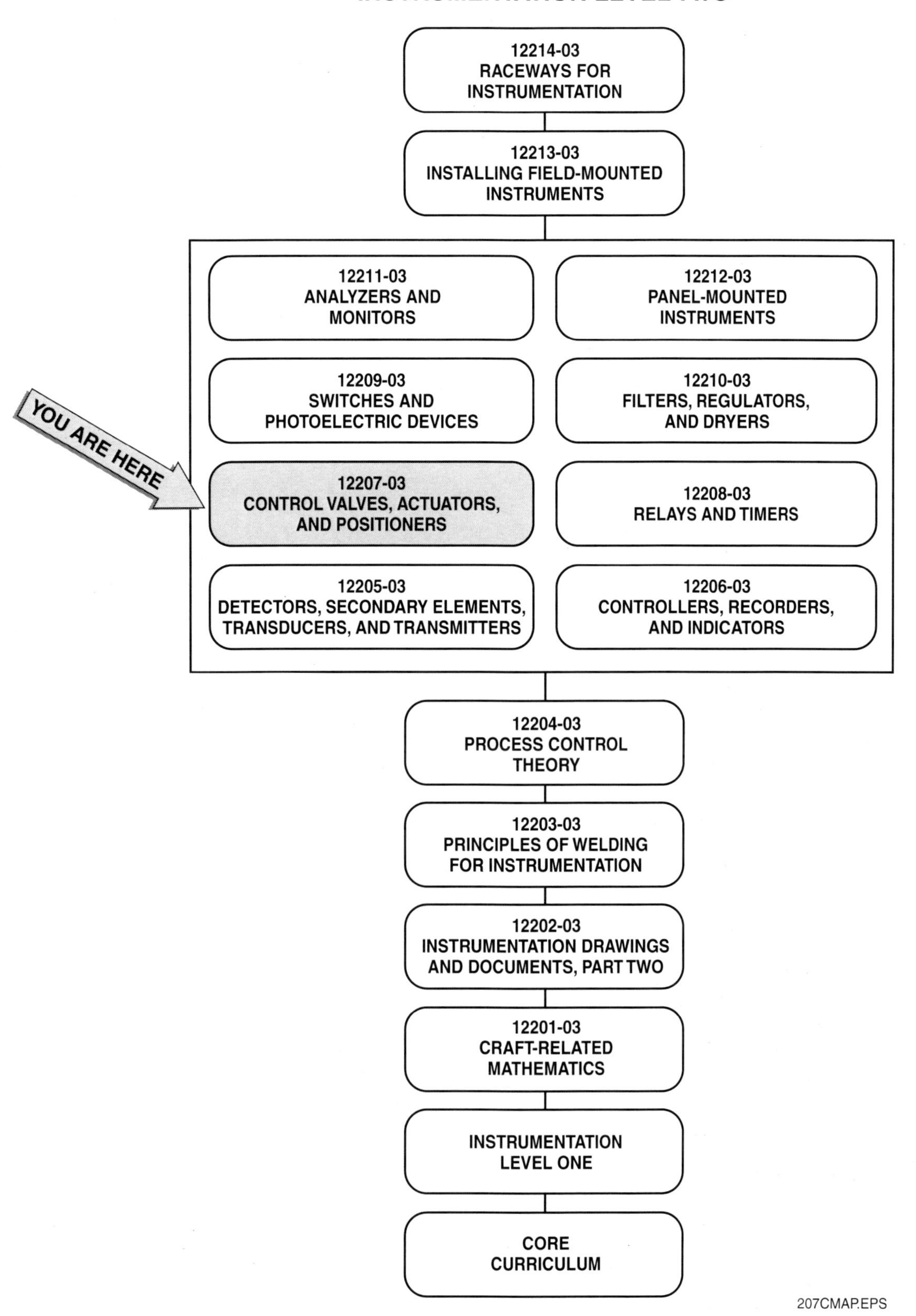

207CMAP.EPS

MODULE 12207-03 CONTENTS

1.0.0 INTRODUCTION 7.1

2.0.0 PRINCIPLES OF OPERATION AND CONSTRUCTION OF VARIOUS CONTROL VALVES 7.1

2.1.0 Globe Valves 7.1

2.2.0 Gate Valves 7.5

2.3.0 Knife Valves 7.8

2.4.0 Ball Valves 7.8

2.5.0 Plug Valves 7.10

2.6.0 Butterfly Valves 7.13

2.7.0 Needle Valves 7.16

3.0.0 PRINCIPLES OF OPERATION AND CONSTRUCTION OF VARIOUS ACTUATORS 7.16

3.1.0 Valve Actuator Terms 7.17

3.1.1 Fail Open 7.17

3.1.2 Fail Closed 7.17

3.1.3 Fail-as-Positioned 7.18

3.1.4 Spring-to-Open 7.18

3.1.5 Spring-to-Close 7.18

3.1.6 Air-to-Open 7.18

3.1.7 Air-to-Close 7.18

3.1.8 Upstream 7.18

3.1.9 Downstream 7.18

4.0.0 PRINCIPLES OF OPERATION AND CONSTRUCTION OF VARIOUS POSITIONERS 7.19

4.1.0 Direct- or Reverse-Acting Positioners 7.20

4.2.0 Pneumatic Positioners 7.20

4.3.0 Analog Positioners 7.20

4.4.0 Modular Assembly Positioners 7.21

4.5.0 Digital Positioners 7.21

4.6.0 Process Precautions for Positioners 7.21

5.0.0 VALVE SELECTION, TYPES, AND APPLICATIONS 7.22

5.1.0 Valve Selection 7.22

5.2.0 Valve Types and Applications 7.23

5.3.0 Valve Markings and Nameplate Information 7.23

5.3.1 Rating Designation 7.24

5.3.2 Trim Identification 7.24

5.3.3 Size Designation 7.25

5.3.4 Thread Markings 7.25

5.3.5 Valve Schematic Symbols 7.25

SUMMARY 7.27
REVIEW QUESTIONS 7.27
GLOSSARY 7.31
REFERENCES & ACKNOWLEDGMENTS 7.33

Figures

Figure 1 Valve functions 7.2
Figure 2 Globe valve 7.3
Figure 3 Angle globe valve 7.3
Figure 4 Seat/disc arrangements 7.4
Figure 5 Globe valve plugs 7.4
Figure 6 Gate valve 7.5
Figure 7 Wedge types 7.6
Figure 8 Solid or single wedge 7.6
Figure 9 Flexible wedge 7.6
Figure 10 Split wedge 7.7
Figure 11 Parallel disc gate valve 7.7
Figure 12 Parallel disc 7.7
Figure 13 Spring-loaded parallel disc gate valve (cutaway) 7.8
Figure 14 Parallel disc gate valve 7.8
Figure 15 Relieving disc gate valve 7.8
Figure 16 Knife gate valve 7.9
Figure 17 Ball valve components 7.9
Figure 18 Ball valve with slanted seals 7.10
Figure 19 Venturi type ball valve 7.10
Figure 20 Top-entry ball valve 7.10
Figure 21 Split-body ball valve 7.10
Figure 22 Plug valve 7.11
Figure 23 Plug valve – port types 7.11
Figure 24 Lubricated taper plug valve 7.12
Figure 25 Cam-operated and nonlubricated plug valve 7.12
Figure 26 Typical nonlubricated plug valve with elastomer sleeve 7.13
Figure 27 Butterfly valve 7.13
Figure 28 Circulating water pump discharge valve 7.14

Figure 29 Butterfly valve seat arrangements 7.14
Figure 30 Seat ring designs 7.15
Figure 31 Flanged and screwed butterfly valves 7.15
Figure 32 Stem/disc arrangement 7.15
Figure 33 Needle valve 7.16
Figure 34 Bar-stock instrument valve 7.16
Figure 35 Spring and diaphragm actuator 7.17
Figure 36 Spring-to-open valve operator (travels up to open) 7.18
Figure 37 Spring-to-close valve operator (travels down to close) 7.18
Figure 38 Air-to-open valve operator (travels down to open) . . .7.19
Figure 39 Air-to-close valve operator (travels up to close) 7.19
Figure 40 Effects of direct-acting positioners 7.20
Figure 41 Typical pneumatic positioner 7.20
Figure 42 Analog (I/P) positioner 7.20
Figure 43 Masoneilan® modular design positioner 7.21
Figure 44 Moore Series 760D digital valve controller (positioner) 7.21
Figure 45 Valve markings 7.23
Figure 46 The meaning of bridgewall markings on valves 7.23
Figure 47 Actuator nameplate 7.24
Figure 48 Typical piping system schematic symbols 7.26

Tables

Table 1 Maximum Operating Temperature for Seat Materials 7.11
Table 2 Valve Rating Designations .7.24
Table 3 Examples of Threaded-Type Symbols 7.25

MODULE 12207-03

Control Valves, Actuators, and Positioners

Objectives

When you have completed this module, you will be able to do the following:

1. Describe the construction principles of operation of various control valves.
2. Describe the construction principles of operation of various actuators.
3. Describe the principles of operation of various positioners.
4. Describe the variables measured and used as inputs for various types of positioners.
5. Discuss valve selection criteria and identify various control valves, actuators, and positioners using specification sheets, pictures, or samples.

Prerequisites

Before you begin this module, it is recommended that you successfully complete the following: Core Curriculum; Instrumentation Level One; Instrumentation Level Two, Modules 12201-03 through 12204-03.

Required Trainee Materials

1. Pencil and paper
2. Appropriate personal protective equipment

1.0.0 ◆ INTRODUCTION

Instrumentation workers encounter many different types of fluid transport systems. These systems may be as simple as potable water plumbing or as complex as petrochemical process piping systems. Regardless of the design, they all have at least one thing in common: there is some type of control valve, valve **actuator**, or **positioner** in the system. Often, there are many.

Valves are necessary in these systems to start, stop, and control or throttle flow. They are essential to isolate high-energy or hazardous fluids from the atmosphere. Actuators for these valves provide a means of automatic operation controlled by a system input, such as pressure, flow, temperature, or level.

Any discussion of valves and actuators would be incomplete without considering the use and operation of valve positioners. These devices are essential when remote operability of a valve is required due to inaccessibility for manual operation. They also provide a means of operating extremely large valves against high differential pressures using pneumatics or hydraulics as an energy source.

2.0.0 ◆ PRINCIPLES OF OPERATION AND CONSTRUCTION OF VARIOUS CONTROL VALVES

A valve is a controlled restriction or orifice in a pipe system, with some means of operating it. Valves, as shown in *Figure 1*, are used for three purposes:

- To control the volume or amount of flow
- To control the direction of flow
- To control the pressure at which a system operates

2.1.0 Globe Valves

Globe valves are used to stop, start, and regulate fluid flow. They are important elements in power plant systems and are commonly used as the standard against which other valve types are judged.

As shown in *Figure 2*, the globe valve **disc** can be totally removed from the flow path or it can completely close the flow path. The essential principle of globe valve operation is the perpendicular

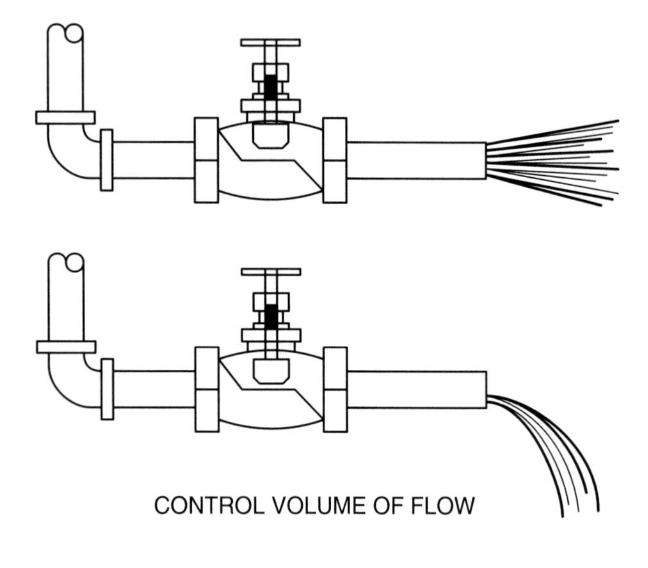

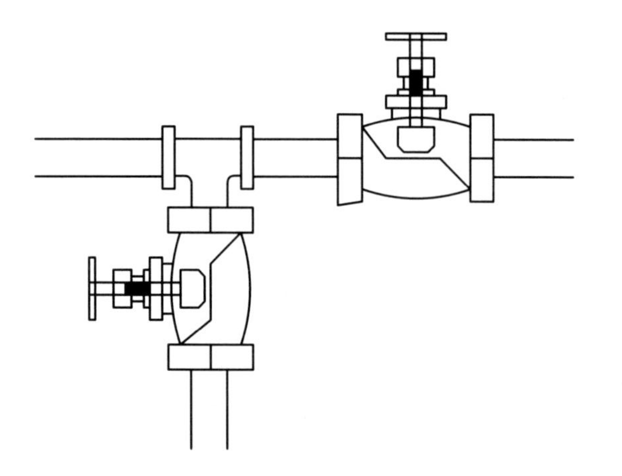

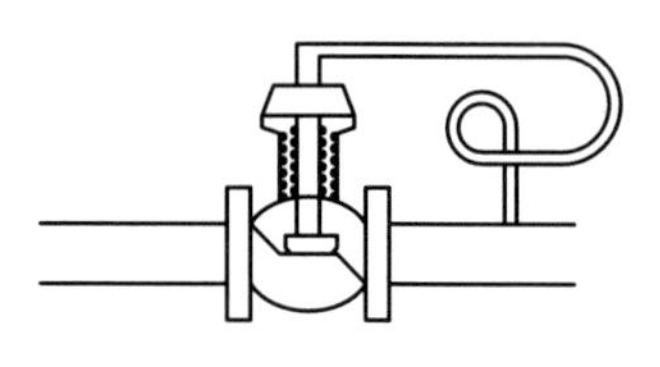

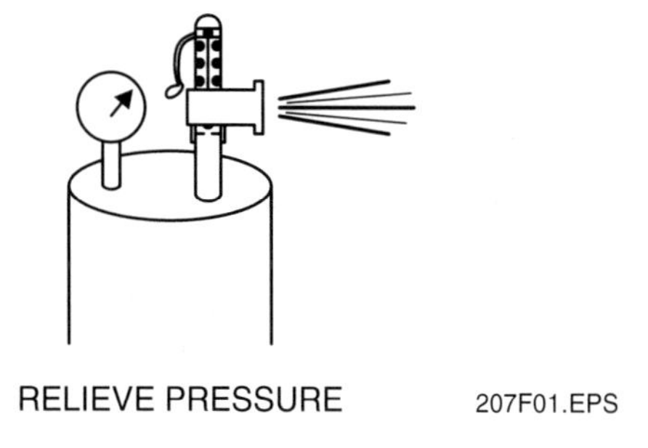

Figure 1 ◆ Valve functions.

movement of the disc away from the **seat**. This causes the annular space between disc and seat ring to gradually close as the valve is closed. This characteristic gives the globe valve good **throttling** ability, which permits its use in regulating flow. Therefore, the globe valve is used not only for start-stop functions, but also for flow-regulating functions.

It is generally easier to obtain very low seat leakage with a globe valve as compared to a **gate valve**. This is because the disc to seat ring contact is more at right angles, which permits the force of closing to tightly seat the disc.

Globe valves can be arranged so that the disc closes against the direction of fluid flow (flow-to-open) or so that the disc closes in the same direction as fluid flow (flow-to-close). When the disc closes against the direction of flow, the **kinetic energy** of the fluid impedes closing but aids opening of the valve. When the disc closes in the same direction as flow, the kinetic energy of the fluid aids closing but impedes opening of the valve. This characteristic makes the globe valve preferable to the gate valve when quick-acting stop valves are necessary.

Along with its advantages, the globe valve has a few drawbacks. Although valve designers can eliminate any or all of the drawbacks for specific services, the corrective measures are expensive and often narrow the valve's scope of service. The most evident shortcoming of the simple globe valve is the high head loss from the two or more right-angle turns of flowing fluid. Obstructions and breaks in the flow path add to the loss.

High-pressure losses in the globe valve can cost thousands of dollars a year for large high-pressure lines. The fluid-dynamic effects from the pulsation, impacts, and pressure drops in traditional globe valves can damage **valve trim**, stem **packing**, and actuators. Troublesome noise can also result. In addition, large sizes require considerable power to operate, which may make gearing or levers necessary.

Another drawback is the large opening needed for assembly of the disc. Globe valves are often heavier than other valves of the same flow rating. The cantilever mounting of the disc on its stem is also a potential trouble source. Each of these shortcomings can be overcome, but only at costs in dollars, space, and weight.

The **angle valve** of *Figure 3* is a simpler modification of the basic globe form. With the ports at right angles, the diaphragm can be a simple flat plate. Fluid can flow through with only a single 90-degree turn, discharging more symmetrically than the discharge from an ordinary globe. Installation advantages also may suggest the angle valve. It can replace an elbow, for example.

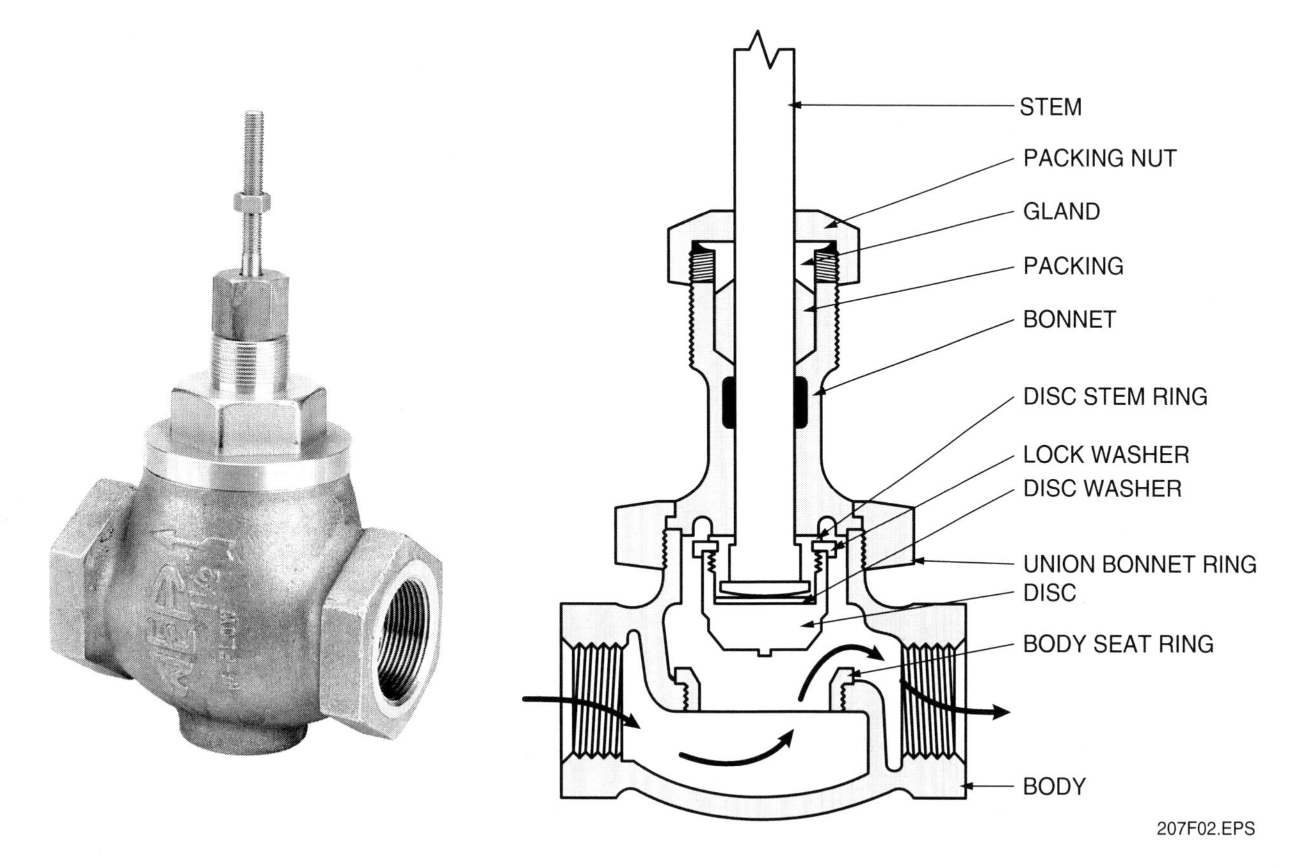

Figure 2 ◆ Globe valve.

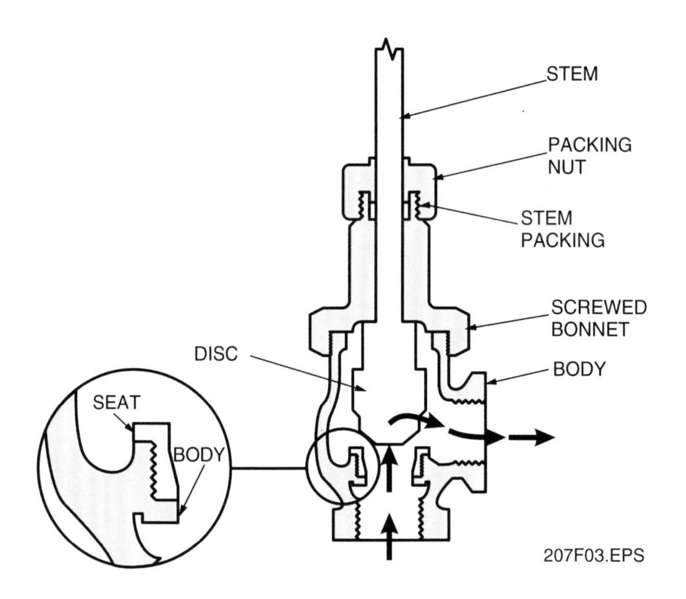

Figure 3 ◆ Angle globe valve.

For moderate conditions of pressure, temperature, and flow, the angle valve closely resembles the ordinary globe. Many manufacturers have interchangeable trim and **bonnets** for the two body styles, with the body differing only in the outlet end. The angle valve's discharge conditions are so favorable that many high-technology control valves use this configuration. They are self-draining and tend to prevent solid buildup inside the **valve body**, like straight flow-through globe valves.

Like valve bodies, there are also many variations of disc and seat arrangements for globe valves. The three basic types are shown in *Figure 4*.

The ball-shaped disc shown in *Figure 4(A)*, fits on a tapered, flat-surfaced seat and is generally used on relatively low-pressure, low-temperature systems. It is generally used in a fully open or shut position, but it may be employed for moderate throttling of a flow.

Figure 4(B) shows one of the proven modifications of seat/disc design, a hard nonmetallic insert ring on the disc to make closure tighter on steam and hot water. The composition disc is resistant to erosion and is sufficiently resilient and cut resistant to close on solid particles without serious permanent damage.

The composition disc is renewable. It is available in a variety of materials that are designed for different types of service, such as high- and low-temperature water, air, or steam. The seating surface is often formed by a rubber O-ring or washer.

The **plug**-type disc, *Figure 4(C)*, provides the best throttling service because of its configuration. It also offers maximum resistance to **galling**, **wire drawing**, and erosion. Plug-type discs are available in a variety of specific configurations, but in

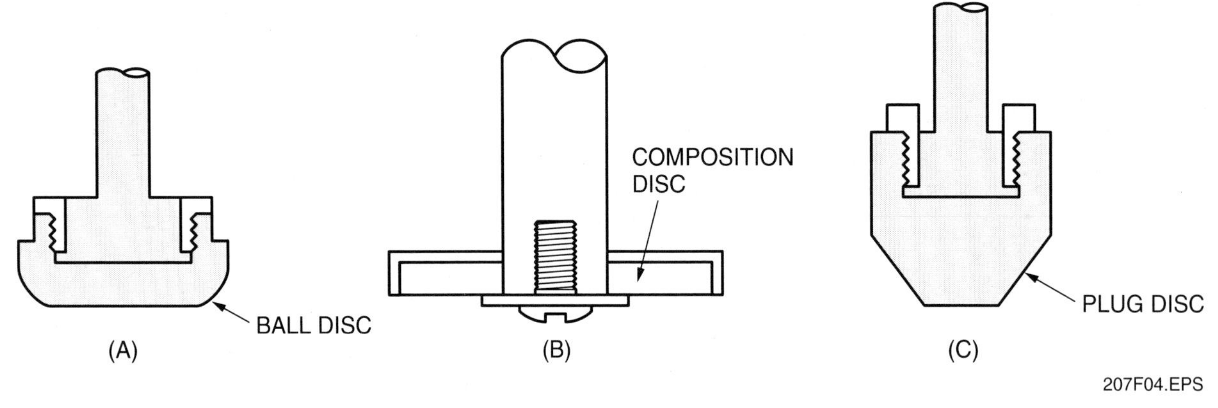

Figure 4 ◆ Seat/disc arrangements.

general they all have a relatively long tapered configuration. Each of the variations has specific types of applications and certain fundamental characteristics. *Figure 5* shows the various types.

The equal percentage plug, as its name indicates, is used for equal percentage flow characteristic for predetermined valve performance. Equal increments of valve **lift** give equal percentage increases in flow.

Linear flow plugs are used for linear flow characteristics with high-pressure drops.

V-port plugs provide linear flow characteristics with medium- and low-pressure drops. This design also prevents wire drawing during low flow periods by restricting the flow through the orifices in the V-port plug when the valve is only partially open.

Needle plugs are used primarily for instrumentation applications and are seldom available in valves over 1" in size. These plugs provide high-pressure drops and low flows. The threads on the stem are usually very fine. Consequently, the opening between the disc and seat does not change rapidly with stem rise. This permits closer regulation of flow.

All of the plug configurations are available in either a conventional globe valve design or the angle valve design. When the needle plug is used, the valve name changes to needle valve. In all other cases the valves are still referred to as globe valves with a specific type of disc.

Globe and angle valves should be installed so that the pressure is under the disc (flow-to-open). This promotes easy operation. It also

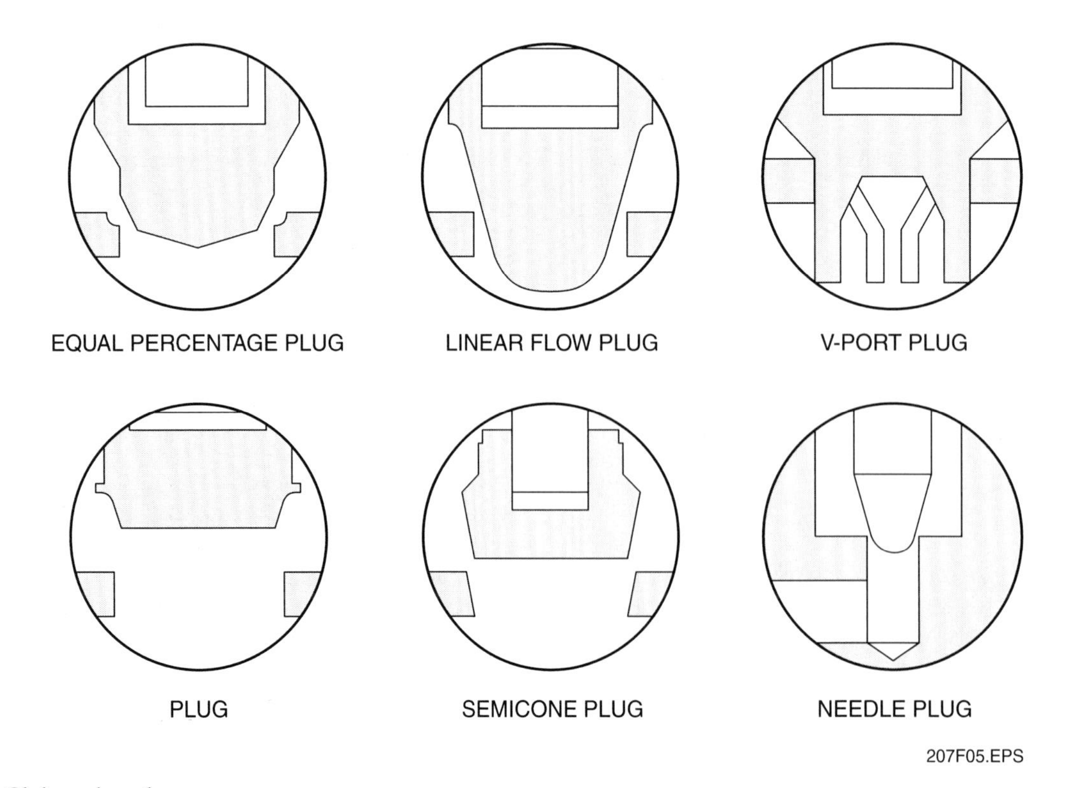

Figure 5 ◆ Globe valve plugs.

helps to protect the packing and eliminates a certain amount of erosive action on the seat and disc faces. However, when high temperature steam is the medium being controlled, and the valve is closed with the pressure under the disc, the **valve stem**, which is now out of the fluid, contracts on cooling. This action tends to lift the disc off the seat, causing leaks that eventually result in wire drawing on seat and disc faces. Therefore, in high-temperature steam service, globe valves may be installed so that the pressure is above the disc (flow-to-close).

2.2.0 Gate Valves

Gate valves are used to start or stop fluid flow, but not to regulate or throttle flow. The term gate is derived from the appearance of the disc in the flow stream, which is similar to a gate. *Figure 6* shows a gate valve. The disc is completely removed from the flow stream when a gate valve is fully open. The disc offers virtually no resistance to flow when the valve is open, so there is little pressure drop across an open gate valve. When the valve is fully closed, a disc to seal ring contact surface exists for 360° and good sealing is provided. With proper mating of disc to seal ring, very little or no leakage occurs across the disc when the gate valve is closed.

Gate valves are not used to regulate flow because the relationship between valve stem movement and flow rate is nonlinear. Operating with the valve in a partially open position can cause disc and seat wear, which will eventually lead to valve leakage.

The primary consideration in the application of a gate valve with relation to a globe valve is that the gate valve represents much less flow restriction than the globe valve. This reduced flow restriction is the result of straight-through body construction and the design of the disc. A gate valve can be used for a wide variety of fluids and provides a tight seal when closed. The major disadvantages of using a gate valve as opposed to a globe valve are as follows:

- It is not good for throttling applications.
- It is prone to vibration in the partially open state.
- It is more subject to seat and disc wear than a globe valve.
- Repairs, such as lapping and grinding, are generally more difficult to accomplish.

Gate valves are available with a variety of fluid-control elements. Classification of gate valves is usually made by the type of fluid-control element used.

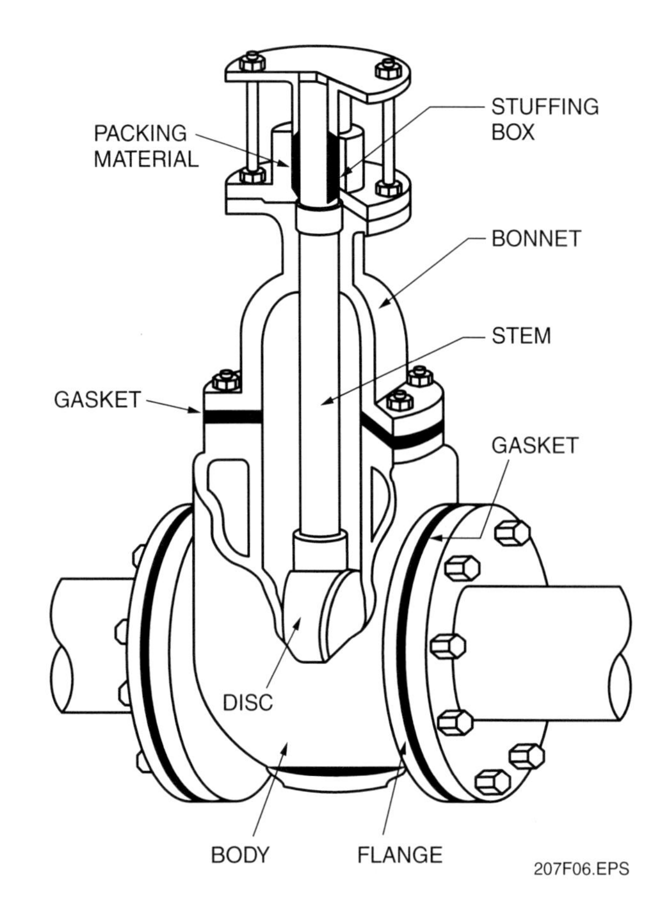

Figure 6 ◆ Gate valve.

The fluid-control elements are available as:

- Solid **wedge**
- Flexible wedge
- Split wedge
- Double disc (parallel disc)

Solid, flexible, and split wedges (*Figure 7*), are employed in valves with inclined seats, while the double discs are used in valves with parallel seats. Regardless of the style of wedge or disc used, they are all replaceable. In services where solids or high velocity may cause rapid erosion of the seat or disc, these components should have a high surface hardness and replaceable seats as well as discs. If the seats are not replaceable, any damage would require refacing of the seat. Valves being used in corrosive service should always be specified with renewable seats.

The solid or single wedge, shown in *Figure 8* is the most commonly used disc because of its simplicity and strength. A valve with this type of wedge may be installed in any position. It is suitable for almost all fluids. It is most practical for turbulent flow.

The flexible wedge (*Figure 9*) is a one-piece disc with a cut around the perimeter to improve ability to match error or change in angle between the

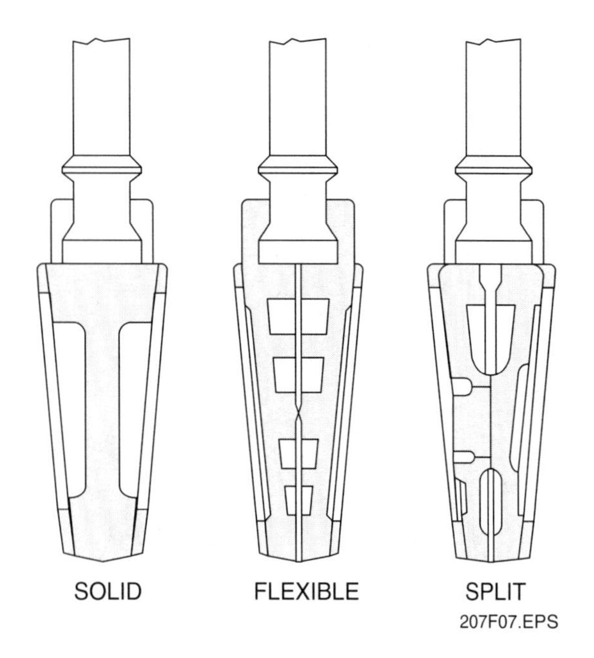

Figure 7 ◆ Wedge types.

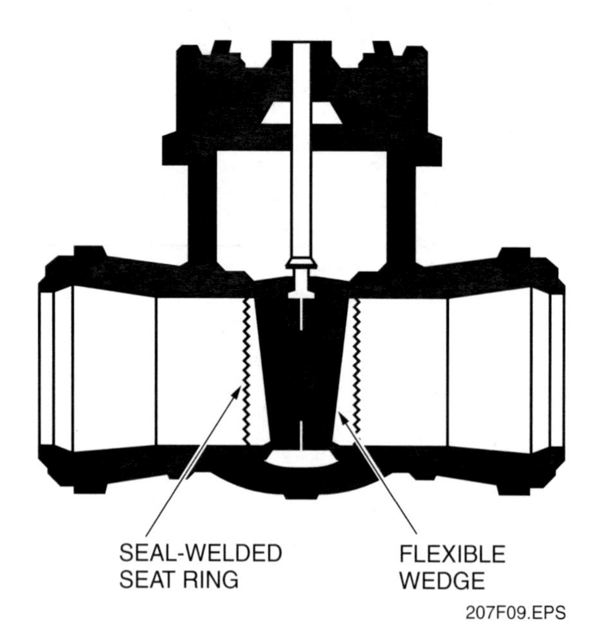

Figure 9 ◆ Flexible wedge.

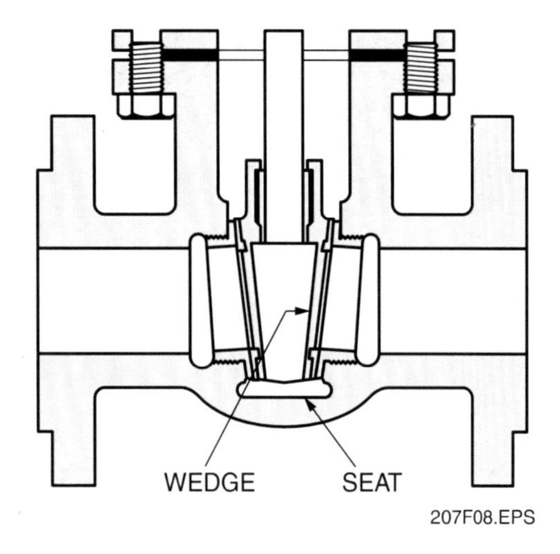

Figure 8 ◆ Solid or single wedge.

seats. The cut varies in size, shape, and depth. A shallow, narrow cut gives little flexibility but retains strength. A deeper and wider cut, or a cast-in-recess, leaves little material at the center, allowing more flexibility but potentially compromising strength and risking permanent set. A correct profile of the disc half in the flexible wedge design can provide uniform deflection properties at the disc edge, so that the wedging force applied in seating will force the disc seating surface uniformly and tightly against the seat.

Gate valves used in steam systems have flexible gates. This prevents binding of the gate within the valve when the valve is in the closed position. When steam lines are heated, they will expand, causing some distortion of valve bodies. If a solid gate fits snugly between the seat of a valve in a cold steam system, when the system is heated and pipes elongate, the seats will compress against the gate, wedging the gate between them and clamping the valve shut. A flexible gate, by contrast, flexes as the valve seat compresses it. This prevents clamping.

The major problem associated with flexible gates is that water tends to collect in the body neck. Under certain conditions, the admission of steam may cause the valve body neck to rupture, the bonnet to lift off, or the seat ring to collapse. It is essential that correct warming-up procedures be followed to prevent this. Some very large gate valves also have a three-position vent and bypass valve installed. This valve allows venting of the bonnet either upstream or downstream of the valve and has a position for bypassing the valve.

Split wedges, as shown in *Figure 10*, are of the ball and socket design, which are self-adjusting and self-aligning to both seating surfaces. The disc is free to adjust itself to the seating surface if one half of the disc is slightly out of alignment because of foreign matter lodged between the disc half and the seat ring. This type of wedge (disc) is suitable for handling noncondensing gases and liquids, particularly corrosive liquids, at normal temperatures. Freedom of movement of the discs in the carrier prevents binding, even though the valve may be closed when hot (later contracting as it cools). This type of valve should be installed with the stem in the vertical position.

The parallel disc (*Figure 11*), was also designed to prevent valve binding due to **thermal transients**. Both low-pressure iron valves and high-pressure steel types have this disc. The principle of

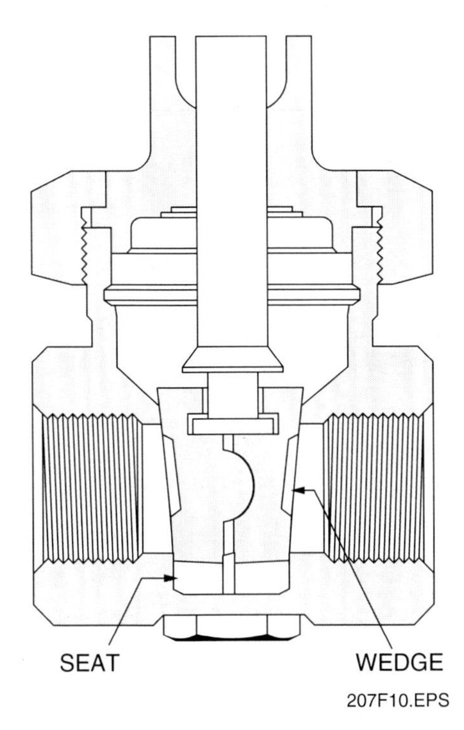

Figure 10 ◆ Split wedge.

operation is that wedge surfaces between the parallel-faced disc halves press together under stem thrust and spread the discs to seal against the seats. The tapered wedges may be part of the disc halves or may be separate elements. The lower wedge must bottom out on a rib at the valve bottom so that the stem can develop seating force. In one version (*Figure 12*), the wedge contact surfaces are curved to keep the point of contact close to optimal.

In other parallel disc gates (*Figures 13* and *14*), the two halves do not move apart under wedge action. Instead, the upstream pressure holds the downstream disc against the seat. A carrier ring lifts the discs. A spring or springs hold the discs apart and seated when there is no upstream pressure.

Another design found on parallel gate discs provides for sealing only one port. In these designs, the high-pressure side pushes the disc open (relieving disc) on the high-pressure side, but forces the disc closed on the low-pressure side (*Figure 15*). With such designs, the amount of seat leakage tends to decrease as differential pressure across the seat increases.

These valves usually have a flow direction marking to show which side is the high-pressure (relieving) side. Take care to make sure these valves are not installed backwards in the system.

Some parallel-disc gate valves used in high-pressure systems are equipped with an integral bonnet vent/bypass line. A three-way valve is used to position the line to bypass in order to

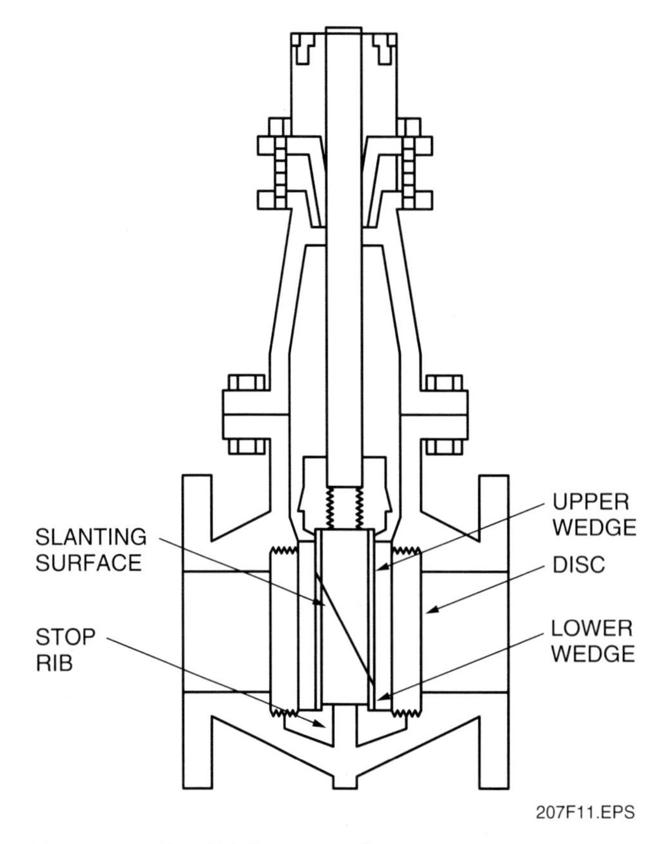

Figure 11 ◆ Parallel disc gate valve.

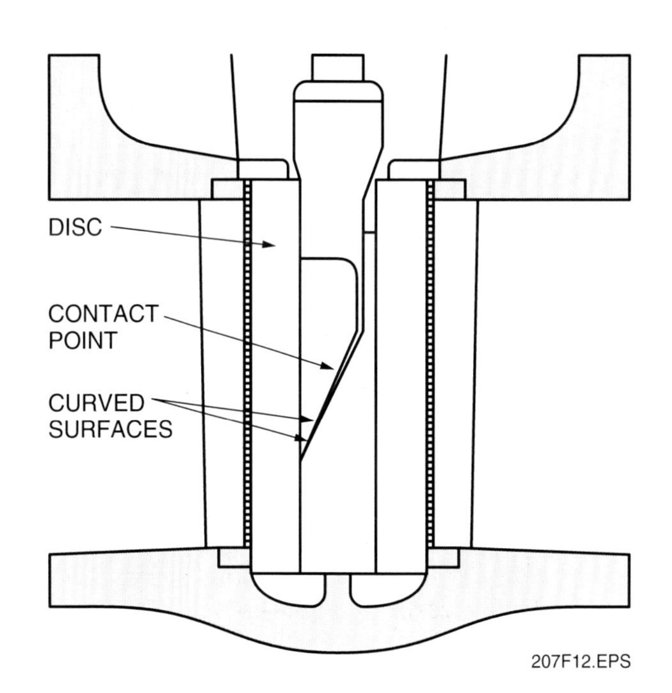

Figure 12 ◆ Parallel disc.

equalize pressure across the discs prior to opening. When the gate valve is closed, the three-way valve is positioned to vent the bonnet to one side or the other. This prevents moisture from accumulating in the bonnet. The three-way valve is positioned to the high-pressure side of the gate valve when the

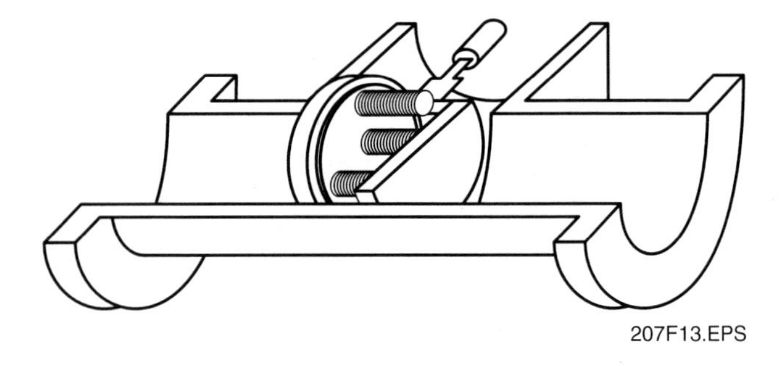

Figure 13 ◆ Spring-loaded parallel disc gate valve (cutaway).

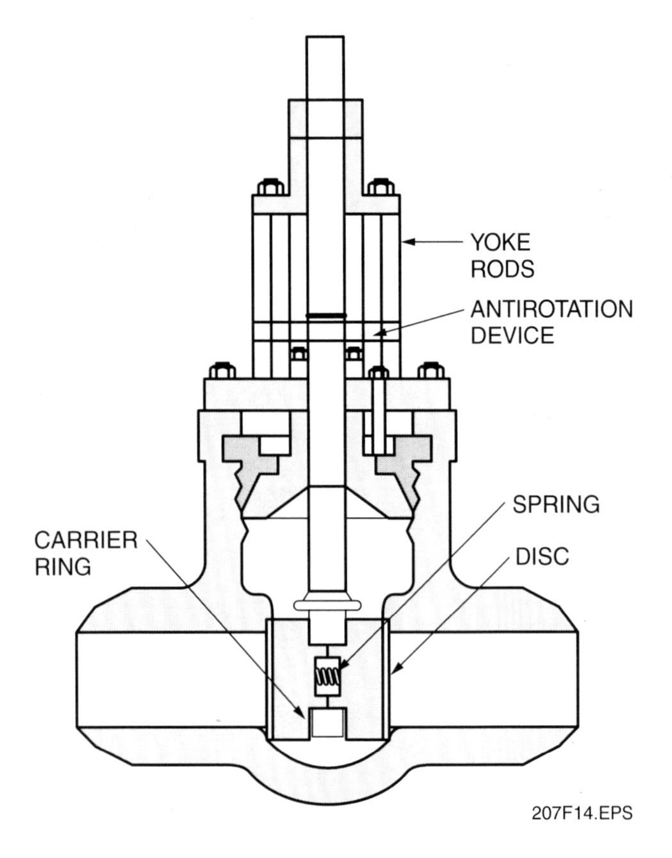

Figure 14 ◆ Parallel disc gate valve.

gate valve is closed to ensure that flow does not bypass the isolation valve. The high-pressure acts against spring compression and forces one gate off its seat. The three-way valve vents this flow back to the pressure source.

2.3.0 Knife Valves

A special type of gate valve is the knife gate valve (*Figure 16*). This valve serves in slurry and waste lines and in other low-pressure applications. The sharp edge of the disc bottom is easily forced closed in contact with a metal or **elastomeric** seat. When open, the disc is in the air after passing through a full-width packing box.

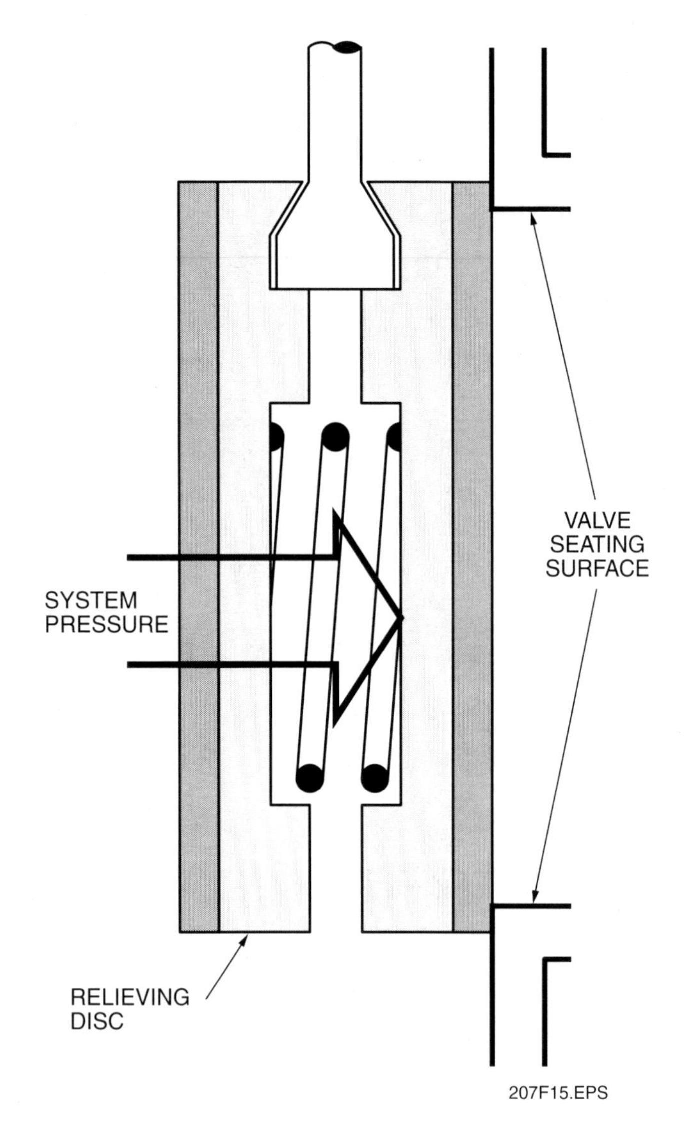

Figure 15 ◆ Relieving disc gate valve.

Knife valves have a disc can cut through deposits and flow-stream solids such as resin slurry. These values, like most gate valves, can be positioned by manual, electrical, pneumatic, and hydraulic actuation.

2.4.0 Ball Valves

Ball valves, as the name implies, are stop valves that use a ball to stop or start the flow of fluid. They are rotary action valves. The ball performs the same function as the disc in the globe valve. However, it turns instead of traveling up and down. When the motor is operated to open the valve, the ball rotates to a point where the hole through the ball is in line with the valve body inlet and outlet. When the valve is shut, which requires only a 90° rotation for most valves, the ball is rotated so that the hole is perpendicular to the flow openings of the valve body, and flow is stopped.

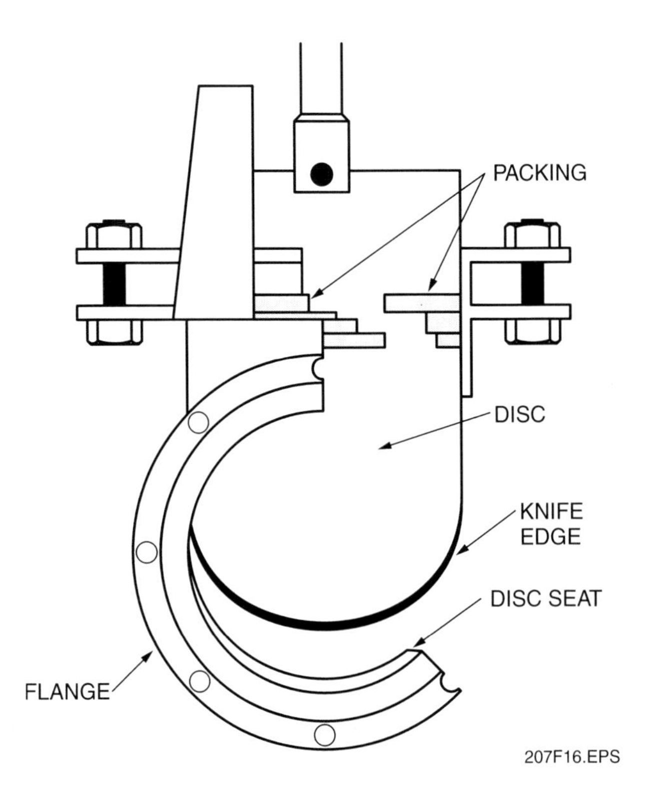

Figure 16 ◆ Knife gate valve.

Most ball valves are quick-acting, requiring only a 90° turn to operate the valve either completely open or closed. *Figure 17* shows the major components of a ball valve. The ball valve, in general, is the least expensive of any valve configuration. In early designs having metal-to-metal seating, the valves could not provide bubble-tight sealing and were not fire-safe. With the development of elastomeric materials and with advances in plastics, the original metallic seats have been replaced with materials such as fluorinated polymers, nylon, neoprene, and Buna-N. Ball valves also have low maintenance costs.

In addition to quick on-off operation, ball valves are compact, require no lubrication, and provide tight sealing with low **torque**. With a soft seat on both sides of the ball, most ball valves provide equally effective sealing of flow in either direction. Many designs permit adjustment for wear.

Because conventional ball valves have relatively poor throttling characteristics, they are not generally satisfactory for this service. In a throttling position, the partially exposed seat rapidly erodes because of the high-velocity flow. However, a ball valve has been developed with a spherical surface-coated plug off to one side in the open position. This rotates into the flow passage until it blocks the flow path completely. Seating is accomplished by the eccentric movement of the plug. The valve requires no lubrication and can be used for throttling service.

Ball valves are designed on the simple principle of floating a polished ball between two plastic seating surfaces, permitting free turning of the ball. Because the plastic is subject to **deformation** under load, some means must be provided to hold the ball against at least one seat. This is normally accomplished through spring pressure, differential line pressure, or a combination of both. *Figure 18* shows a combination of line pressure and a spring on the ball.

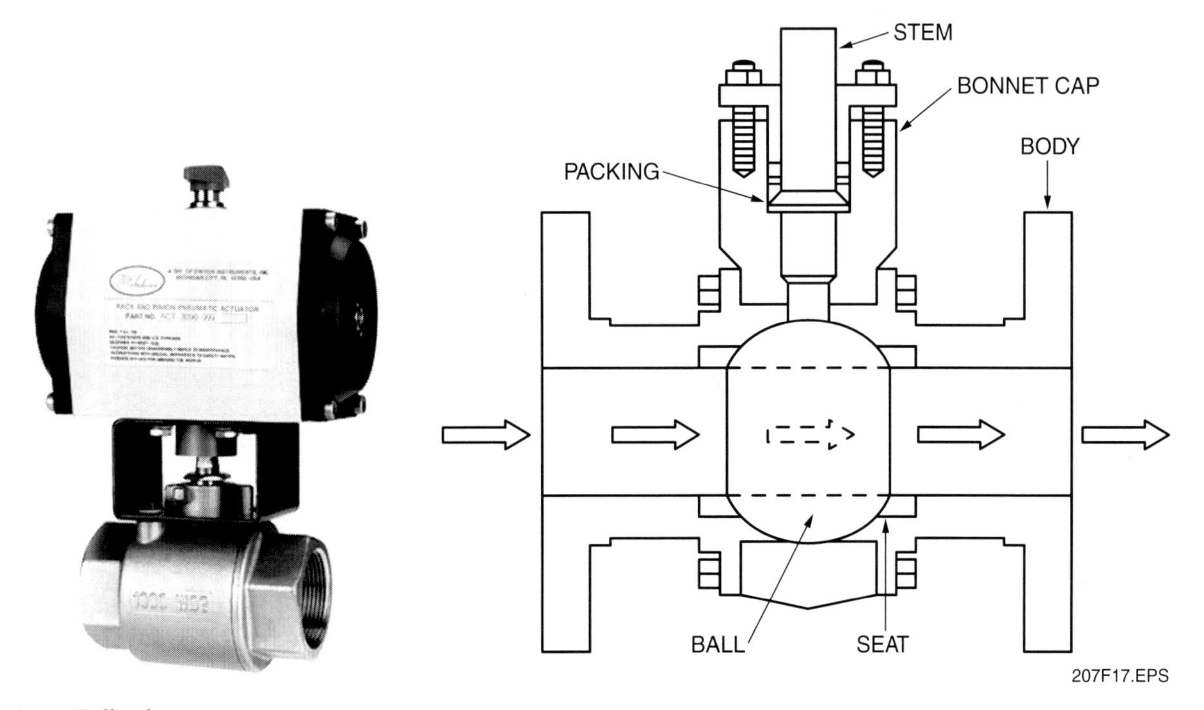

Figure 17 ◆ Ball valve components.

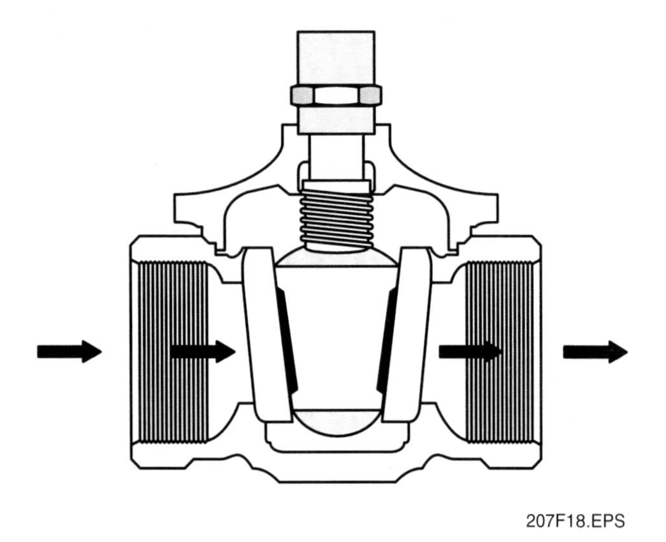

Figure 18 ◆ Ball valve with slanted seals.

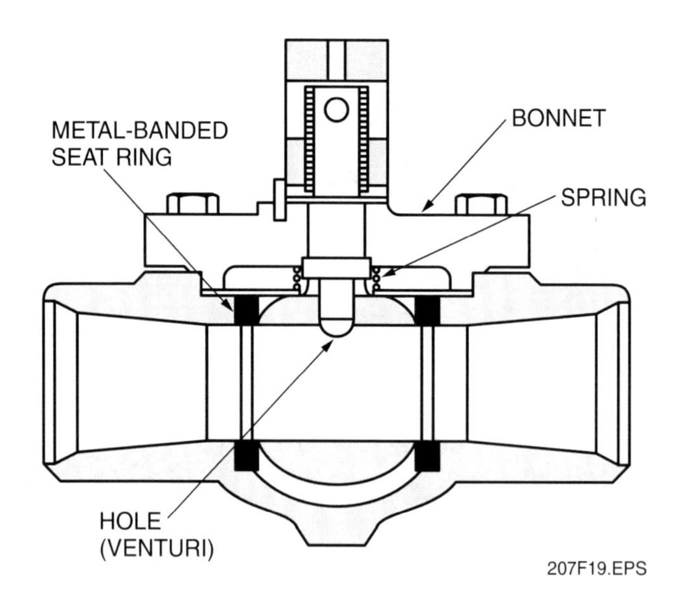

Figure 19 ◆ Venturi type ball valve.

Ball valves are available in the venturi (*Figure 19*) and full-port patterns. The latter has a ball with a bore equal to the inside diameter of the pipe. Balls are usually metallic in metallic bodies with trim (seats) produced from elastomeric materials. All-plastic designs are also available. Seats and balls are replaceable.

Ball valves are available in top-entry (*Figure 20*) and split-body (end-entry) types (*Figure 21*). In the former, the ball and seats are inserted through the top, while in the latter, the ball and seats are inserted from the ends.

The resilient seats for ball valves are made from various elastomeric materials. The most common seat materials are TFE (virgin), filled TFE, nylon, Buna-N, neoprene, and combinations of these materials. Because of the presence of the elastomeric materials, these valves cannot be used at elevated temperatures. To overcome this disadvantage, a graphite seat has been developed that will permit operation up to 1,000°F. Typical maximum operating temperatures of the valves at their full pressure ratings are shown in *Table 1* for various seat materials.

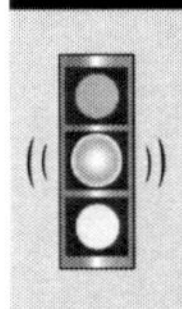

CAUTION

Use care when selecting the seat material to make sure it is compatible with the materials being handled by the valve and suitable according to its temperature limitations.

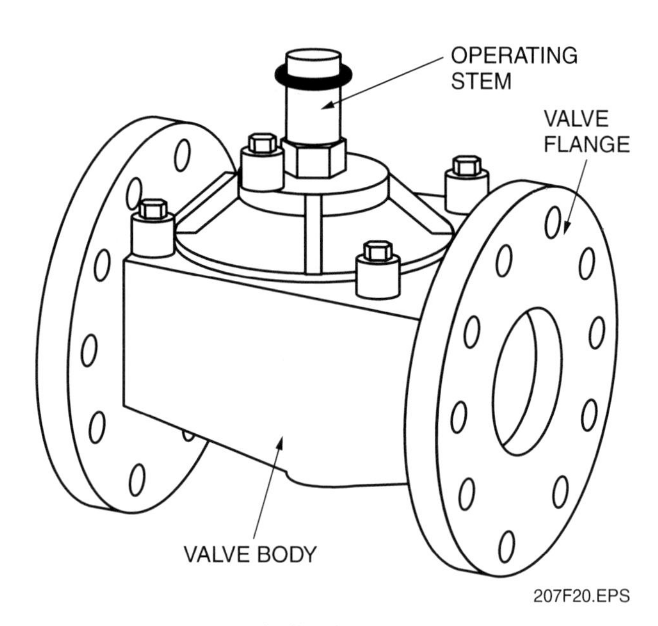

Figure 20 ◆ Top-entry ball valve.

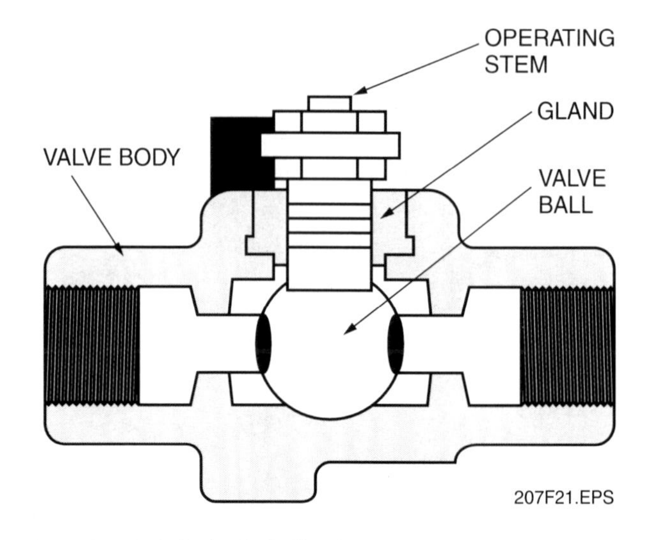

Figure 21 ◆ Split-body ball valve.

2.5.0 Plug Valves

Plug valves are used to stop or start fluid flow. They are rotary-action valves. The name is derived from the shape of the disc, which resembles a plug. A plug valve is shown in *Figure 22*.

Table 1 Maximum Operating Temperature for Seat Materials

Seat Material	Operating Temperatures
TFE (Virgin)	450°F
Filled TFE	400°F
Buna-N	180°F
Neoprene	180°F

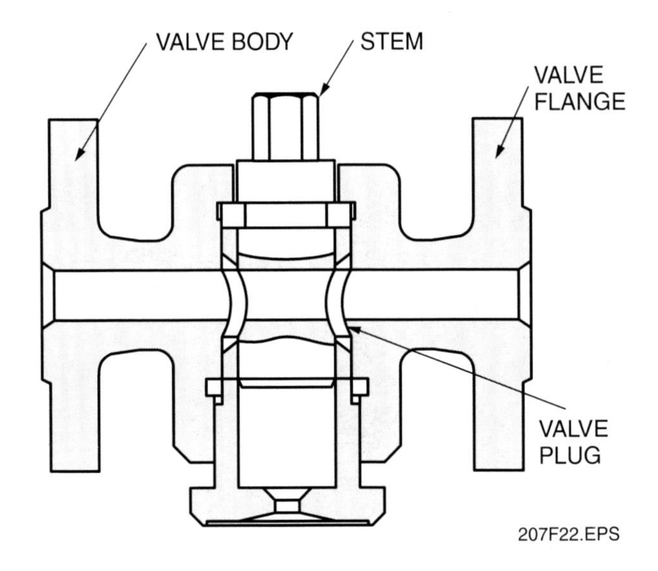

Figure 22 ◆ Plug valve.

The body of a plug valve is machined to receive the tapered or cylindrical plug. The disc is a solid plug with a bored passage at a right angle to the longitudinal axis of the plug. In the open position, the passage in the plug lines up with the inlet and outlet ports of the valve body. When the plug is turned 90° from the open position, the solid part of the plug blocks the ports and stops fluid flow.

Plug valves are available in either a lubricated or nonlubricated design and with a variety of styles of port openings through the plug. There are numerous plug designs.

An important characteristic of the plug valve is its easy adaptation to multiport construction. Multiport valves are widely used. Their installation simplifies piping. They provide a much more convenient operation than multiple gate valves do. They also eliminate pipe fittings. The use of a multiport valve eliminates the need for two, three, or even four conventional shutoff valves, depending on the number of ports in the plug valve.

Plug valves are normally used only for on-off operations, particularly where frequent operation of the valve is necessary. These valves are not normally recommended for throttling service because, as with the gate valve, a great percentage of flow change occurs near shutoff at high velocity. However, a diamond-shaped port has been developed for throttling service.

Multiport valves are particularly good on transfer lines and for diverting services. A single multiport valve may be installed in lieu of three or four gate valves or other types of shutoff valves. Many of the multiport configurations do not permit complete shutoff of flow, however. In most cases one flow path is always open. These valves are intended to divert the flow to one line while shutting off flow from the other lines. If complete shutoff of flow is required, a suitable multiport valve should be used, or a secondary valve should be installed on the main feed line ahead of the multiport valve to permit complete flow shutoff.

It should also be noted that in some multiport configurations, flow to more than one port simultaneously is also possible. Great care should be taken in specifying the port arrangement to guarantee proper operation.

Plugs are either round or cylindrical with a taper. They may have various types of port openings, each with a varying degree of free area relative to the corresponding pipe size (*Figure 23*).

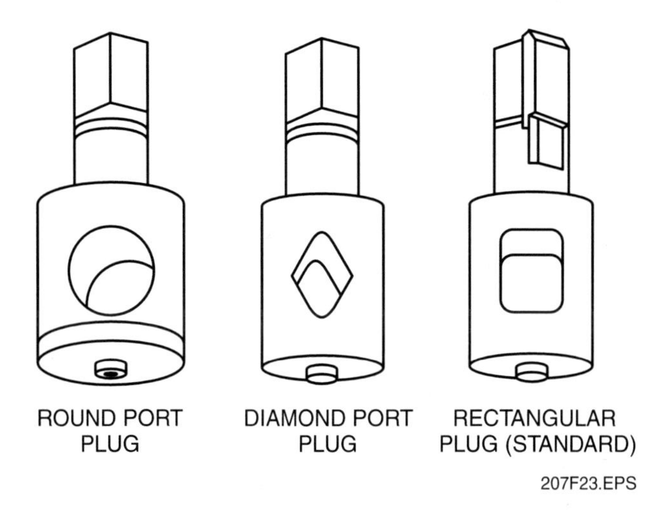

Figure 23 ◆ Plug valve – port types.

- *Rectangular port* is the standard-shaped port with a minimum of 70% of the area of the corresponding size of standard pipe.
- *Round port* means that the valve has a full round opening through the plug and body, of the same shape as standard pipe.
  - *Full port* means that the area through the valve is equal to or greater than the area of standard pipe.
  - *Standard opening* means that the area through the valve is less than the area of standard pipe. These valves should be used only where restriction of flow is unimportant.

- *Diamond port* means that the opening through the plug is diamond-shaped. This has been designed for throttling service. All diamond port valves are venturi, restricted-flow type.

Clearances and leakage prevention are the chief considerations in plug valves. Many plug valves are of all-metal construction. In these versions, the narrow gap around the plug can permit leakage. If the gap is reduced by sinking the taper plug deeper into the body, actuation torque will climb rapidly, and galling can occur.

Lubrication remedies this. A series of grooves around the port openings in the plug or body is supplied with grease prior to actuation, not only to lubricate the plug motion but also to seal the gap (*Figure 24*). Grease injected into a fitting at the stem top travels down through a **check valve** in the passageway and then past the plug top to the grooves on the plug and down to a well below the plug.

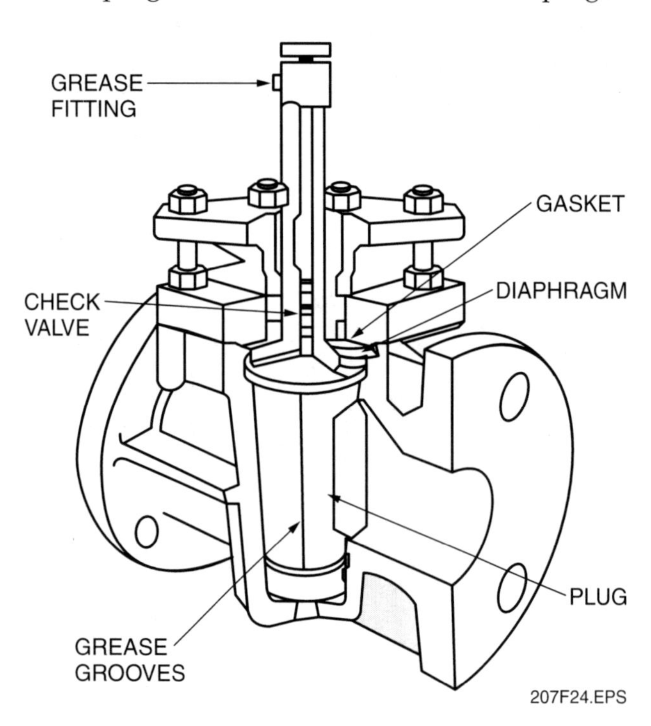

Figure 24 ◆ Lubricated taper plug valve.

The lubricant must be compatible with the temperature and nature of the fluid. The most common substances controlled by plug valves are gases and liquid hydrocarbons. Some water lines have these valves too, if lubricant contamination is not a serious danger. This type can go to 24-inch size, with pressure capability of 6,000 psig. Steel and iron bodies are available. The plug can be cylindrical or tapered.

The correct choice of lubricant is extremely important for successful lubricated plug valve performance. In addition to providing adequate lubrication to the valve, the lubricant must not react chemically with the material passing through the valve. The lubricant must not contaminate the material passing through the valve, either. All manufacturers of lubricated plug valves have developed a series of lubricants that are compatible with a wide range of media. Their recommendations should be followed regarding which lubricant is best suited for the service.

To overcome the disadvantages of lubricated plug valves, two basic types of nonlubricated plug valves were developed. A nonlubricated valve may be a lift-type, or it may have an elastomer sleeve or plug coating that eliminates the need to lubricate the space between the plug and seat.

Lift-type valves provide a means of mechanically lifting the tapered plug slightly from its seating surface to permit easy rotation. The mechanical lifting can be accomplished through either a cam (*Figure 25*) or an external lever.

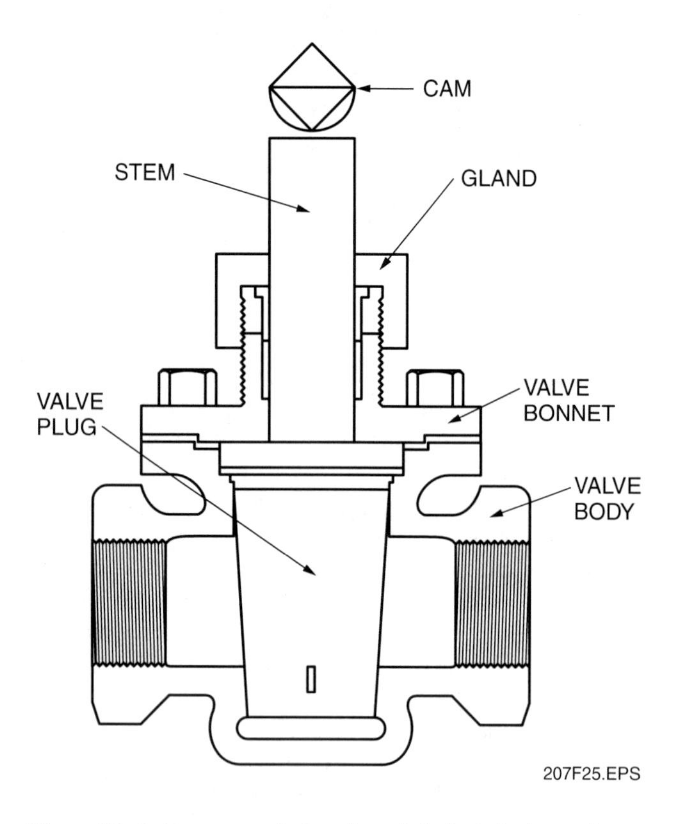

Figure 25 ◆ Cam-operated and nonlubricated plug valve.

A typical nonlubricated plug valve with an elastomer sleeve is shown in *Figure 26*. In this particular valve, a sleeve of TFE completely surrounds the plug. It is retained and locked in place by the metal body. This results in a continuous primary seal between the sleeve and the plug, both while the plug is rotated and when the valve is in either the open or closed position. The TFE sleeve is durable and essentially inert to all but a few rarely encountered chemicals. It also has a low coefficient of friction and therefore is selflubricating.

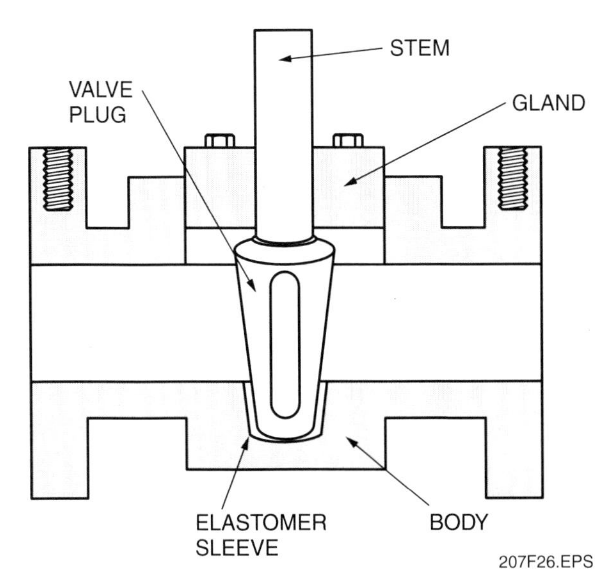

Figure 26 ◆ Typical nonlubricated plug valve with elastomer sleeve.

Lubricants are available in stick form and in bulk. Stick lubrication is usually employed when a small number of valves are in service or when they are widely scattered throughout the plant. However, for a large number of valves, gun lubrication is the most convenient and economical solution.

Valves are usually shipped with an assembly lubricant. This assembly lubricant should be removed and the valve completely relubricated with the proper lubricant before the valve is put into service.

Regular lubrication is critical for best results. Extreme care should be taken to prevent any foreign matter from entering the plug when inserting new lubricant.

2.6.0 Butterfly Valves

Butterfly valves (*Figure 27*) have many advantages over gate, globe, plug, and ball valves in a variety of installations, particularly in the larger sizes. Savings in weight, space, and cost are the most obvious advantages. Maintenance costs are low, because there is a minimum number of moving parts and there are no pockets to trap fluids. Butterfly valves are especially well-suited for the handling of large flows of liquids or gases at relatively low pressures and for the handling of slurries or liquids with large amounts of suspended solids.

The butterfly valve is suitable for throttling as well as open-closed applications. It is a rotary-action valve. Operation is quick and easy because a 90° rotation of the stem moves the flow control element from the fully closed to the fully opened position. Butterfly valves are built on the principle of a pipe damper. The flow control element is a

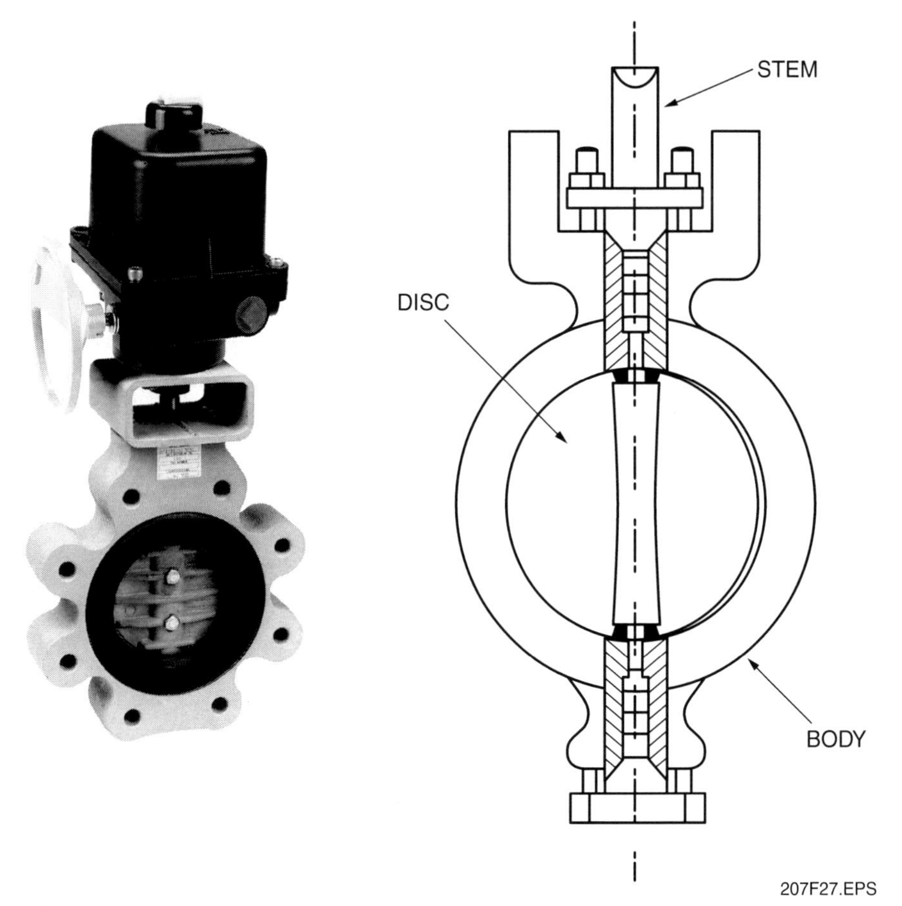

Figure 27 ◆ Butterfly valve.

disc of approximately the same diameter as the inside diameter of the adjoining pipe. It rotates on either a vertical or horizontal axis. When the disc lies parallel to the piping run, the valve is fully opened. When the disc approaches the perpendicular position, the valve is shut. Intermediate positions, for throttling purposes, can be secured in place by handle-locking devices.

Flow is stopped when the valve disc seals against a seat on the inside of the valve body. Originally, a metal disc was used to seal against a metal seat. This arrangement did not provide a leak-tight closure, but it did provide sufficient closure in such applications as power plant water distribution lines. Valves of this design are still available. With the advent of newer elastomeric materials, most butterfly valves are now produced with an elastomeric seat against which the disc seals. An example of this is the circulating water pump discharge valve shown in *Figure 28*. This arrangement provides a leak-tight closure.

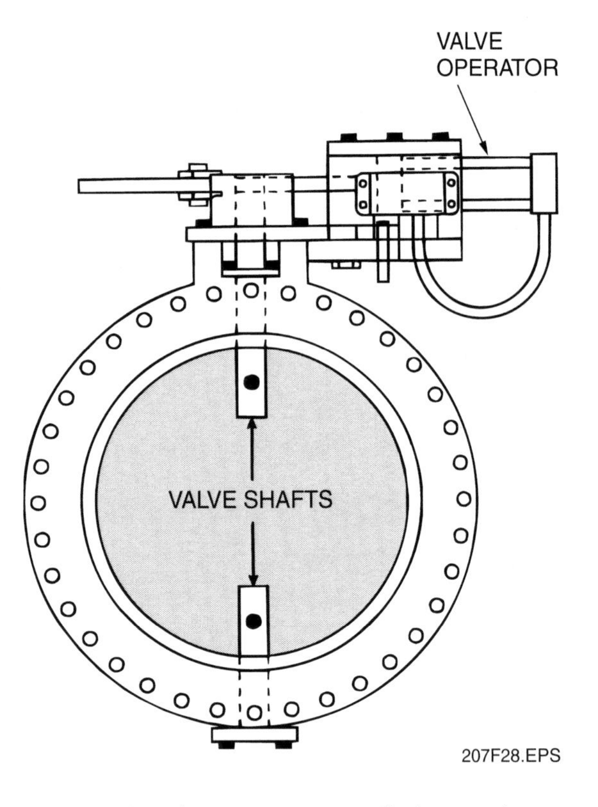

Figure 28 ◆ Circulating water pump discharge valve.

Figure 29 shows a seat ring arrangement using a clamp ring *(A)* and a backing ring used on a serrated-edge rubber ring *(B)*. These designs prevent the extrusion of the O-rings. *Figure 30* shows additional seat ring designs.

Body construction varies. The most economical is the wafer type, which simply fits between two pipeline flanges. Another type of lug wafer valve

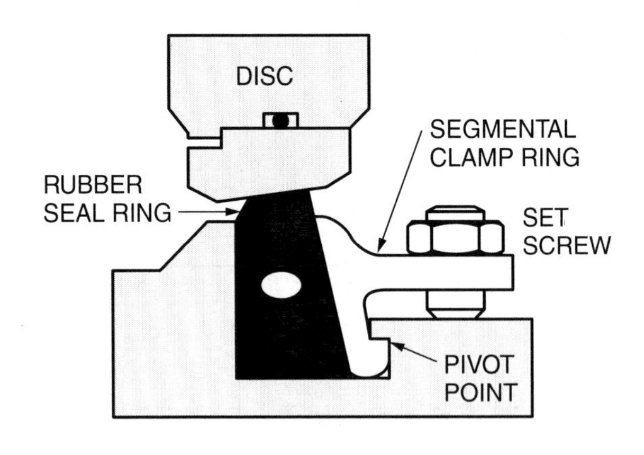

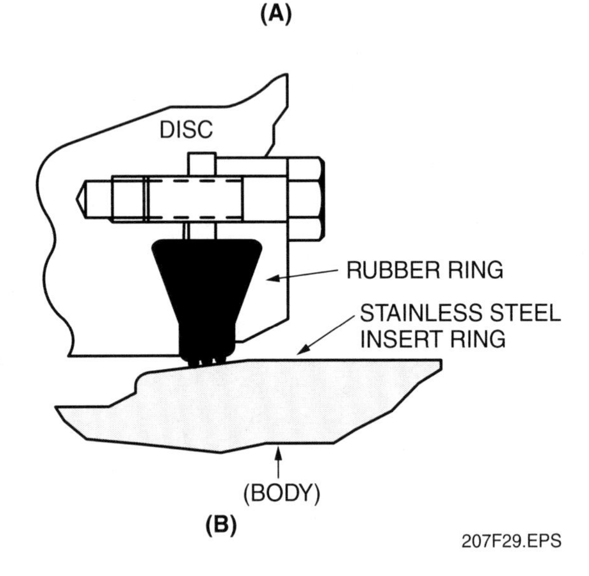

Figure 29 ◆ Butterfly valve seat arrangements.

is held in place between two pipe flanges by bolts that join the two flanges and pass through holes in the valve's outer casing. Valves are also available with conventional flanged ends for bolting to pipe flanges and in a screwed end construction. *Figure 31* shows these valve body designs.

The stem and disc for a butterfly valve are separate pieces. The disc is bored to receive the stem. Two methods are used to secure the disc to the stem so that the disc rotates as the stem is turned.

In the first method, the disc is simply bored through and the disc secured to the stem by means of bolts or pins that pass through the stem and disc, as shown in *Figure 32(A)*. The alternate method, shown in *Figure 32(B)*, involves boring the disc as before, then broaching the upper stem bore to fit a squared or hex-shaped stem. This method allows the disc to "float" and seek its center in the seat. Uniform sealing is accomplished, and external stem fasteners are eliminated. This is advantageous for covered discs and corrosive applications.

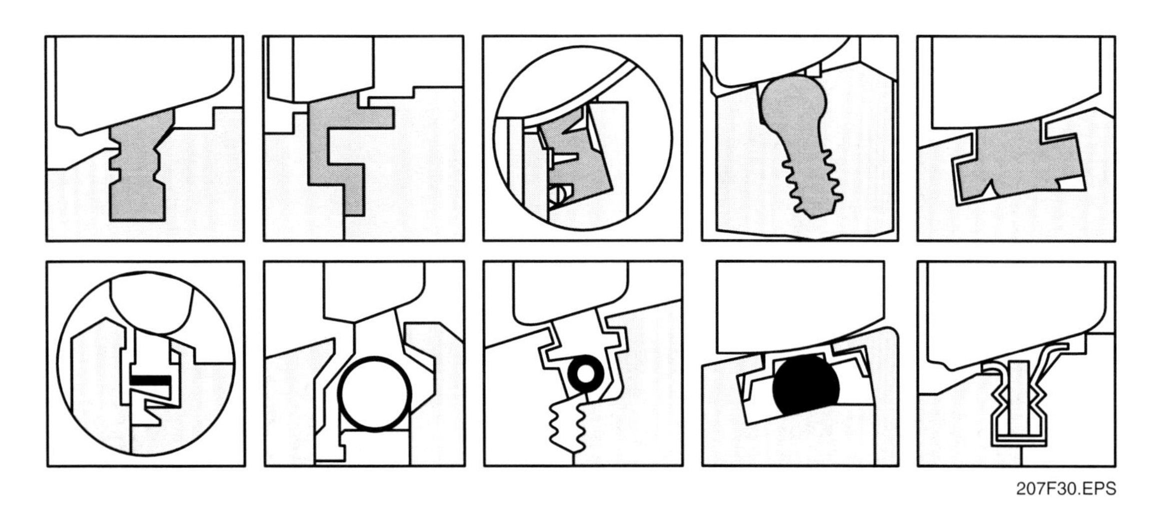

Figure 30 ◆ Seat ring designs.

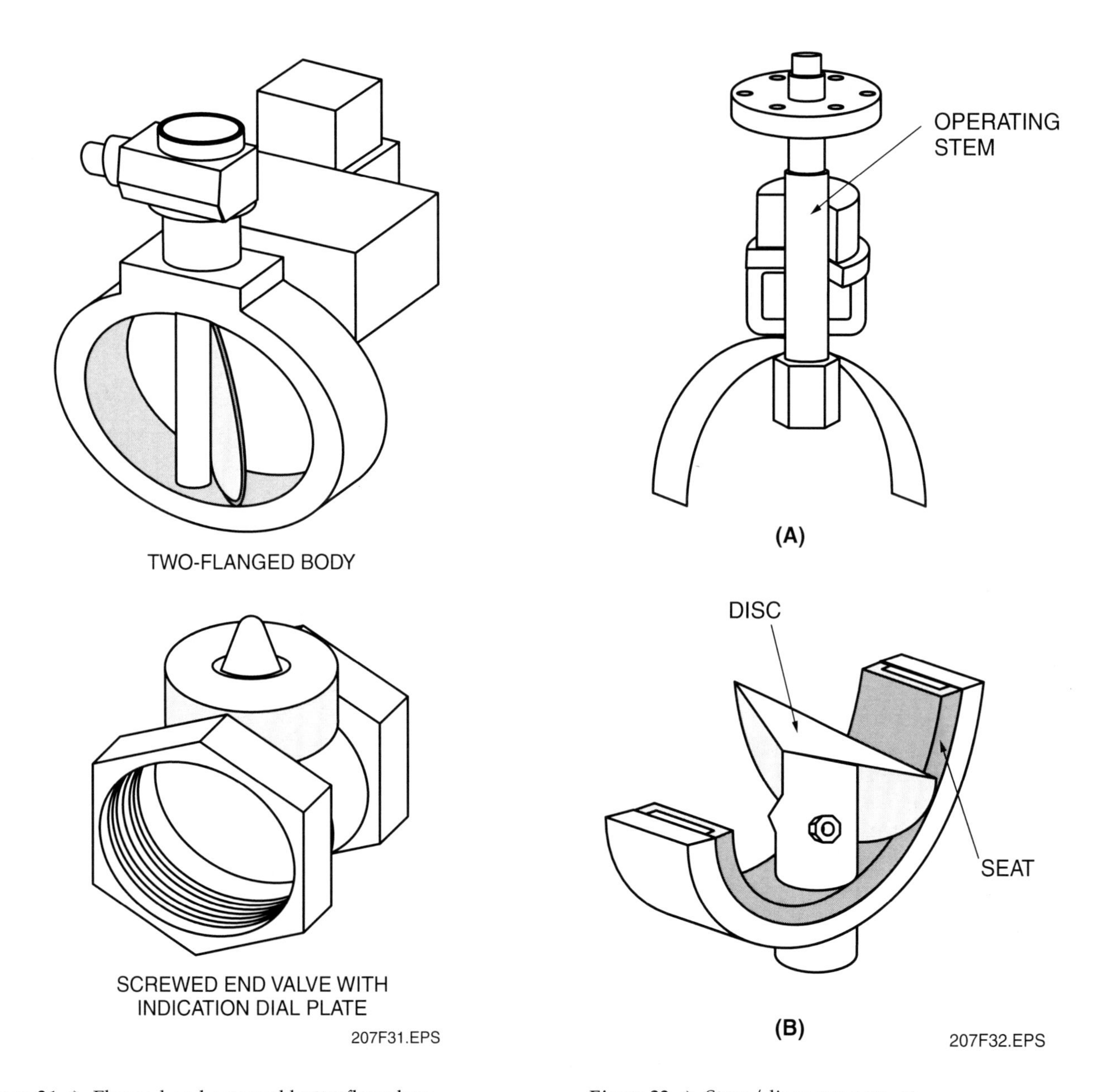

Figure 31 ◆ Flanged and screwed butterfly valves.

Figure 32 ◆ Stem/disc arrangement.

In order for the disc to be held in the proper position, the stem must extend beyond the bottom of the disc and be fitted into a bushing in the bottom of the valve body. One or two similar bushings are necessary along the upper portion of the stem as well. These bushings must either be resistant to the media being handled, or they must be sealed so that the corrosive media cannot come into contact with them. Both methods are employed, depending on the valve manufacturer.

Stem seals are accomplished either with packing in a conventional stuffing box or by means of O-ring seals. Some valve manufacturers, particularly those specializing in the handling of corrosive materials, use a stem seal on the inside of the valve so that no material being handled by the valve can come into contact with the valve stem. If a stuffing box or external O-ring seal is employed, the material passing through the valve will come into contact with the valve stem.

Butterfly valves may be operated by means of air or fluid power and electricity. Cylinder actuators, **diaphragm actuators**, piston actuators, and electric actuators are available for this type of operation.

Operating torques required for specific valves and services should always be checked with the valve manufacturer before sizing the automatic operating accessory. Most manufacturers will supply the valve with the automatic operation if specified.

2.7.0 Needle Valves

Needle valves, as shown in *Figure 33*, are used for making relatively fine adjustments in the amount of fluid allowed to pass through an opening. The needle valve has a long, tapering, needle-like point on the end of the valve stem. This needle acts as a disc. The longer part of the needle is smaller than the orifice in the valve seat. It therefore passes through it before the needle seats. This arrangement permits a very gradual increase or decrease in the size of the opening and thus allows a more precise control of flow than could be obtained with an ordinary globe valve. Needle valves are often used as component parts of other, more complicated valves. For example, they are used in some types of reducing valves. Most constant-pressure pump governors have needle valves to minimize the effects of fluctuations in pump discharge pressure. Needle valves are also used in some components of automatic combustion control systems where very precise flow regulation is necessary.

Bar-stock bodies are common. In globe types, a ball swiveling in the stem end will give the necessary rotatability for seating without damage (*see Figure 34*).

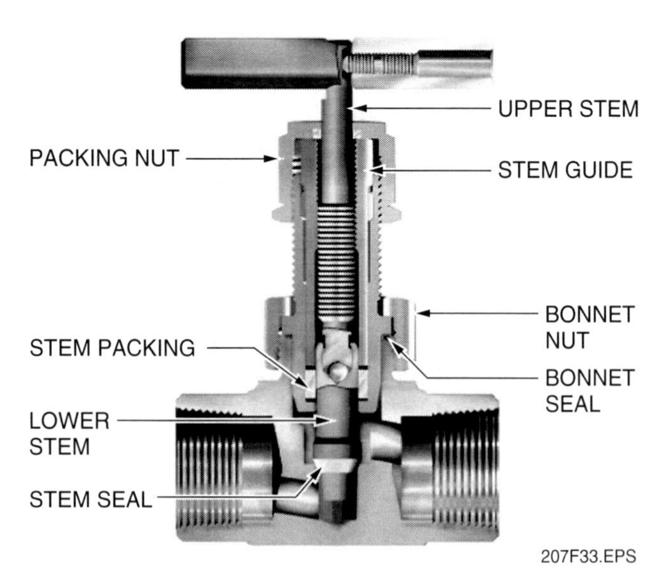

Figure 33 ◆ Needle valve.

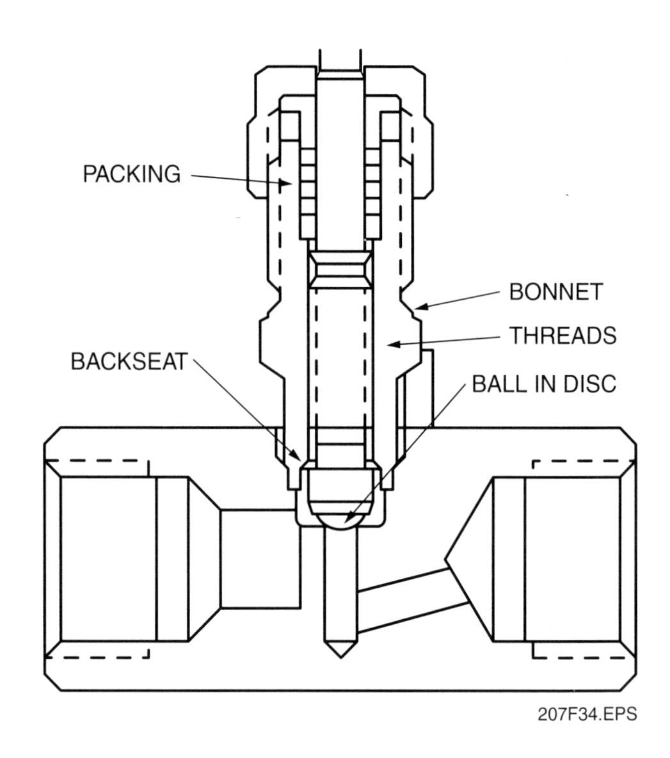

Figure 34 ◆ Bar-stock instrument valve.

3.0.0 ◆ PRINCIPLES OF OPERATION AND CONSTRUCTION OF VARIOUS ACTUATORS

The actuator provides the power to vary the orifice area of the valve in response to a signal received. The stem carries the load from the actuator to the plug.

The two most common types of actuators are the spring/diaphragm assembly and the electric motor. The spring and diaphragm actuators shown in *Figure 35* are inexpensive, simple, and reliable, with

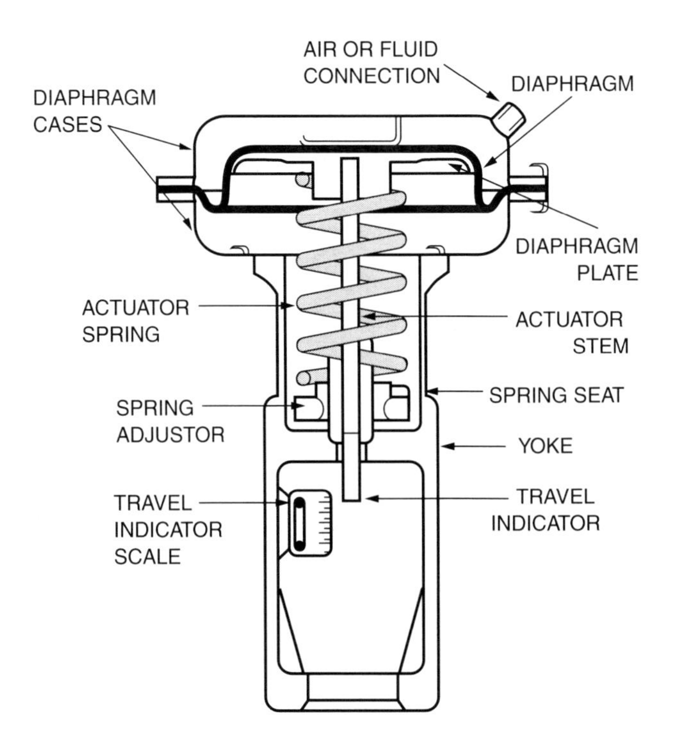

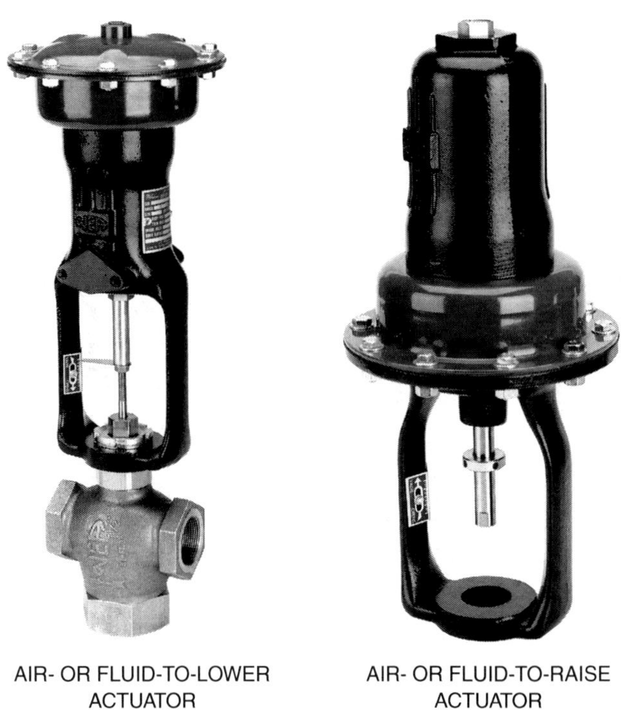

Figure 35 ◆ Spring and diaphragm actuator.

few moving parts. Some disadvantages are that they have limited power, limited seat shutoff capabilities, and are slow in operating. They will not handle high variations in the load applied to the stem. They are available as air- (or fluid)-to-lower and air- (or fluid)-to-raise types. The difference is that the diaphragm is located at the top of the spring in the lowering type and forces the stem downward against spring pressure. The diaphragm in the raising type is located under the spring and forces the stem upward against spring pressure. Depending on the valve, either type can be used to open or close the valve described in the following sections.

The electric motor is generally used with a gear box and does not need air or an I/P converter, but it may require a P/E converter. The motor provides high power and can be reversed. It is often used for rotary-action applications.

3.1.0 Valve Actuator Terms

As stated earlier, valves serve to stop, start, or regulate fluid flow. To facilitate these basic functions, various actuators are used in conjunction with valves. For the purpose of describing how valves and their actuators operate during normal or abnormal conditions, various general terms are used. Other terms are used to depict the orientation of a valve with relation to other components. This section will explains these terms.

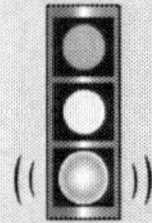

NOTE

When reviewing the actuator illustrations shown in this section, pay close attention to the taper of the plug in the valve. If the taper goes from wide to narrow from top to bottom, the plug must travel down to close and up to open. If the taper goes from narrow to wide from top to bottom, the plug must travel up to close and down to open.

3.1.1 Fail Open

The term *fail open*, when used to describe a valve or valve operator, means that when the controlling medium is lost, the valve will automatically open. A common valve operator that provides this feature is a pneumatic spring and diaphragm operator. Air supplied on one side of the diaphragm serves to move the valve in the closed direction. On loss of the controlling air supply, a spring, which is an integral part of the operator, forces the valve to move to the open position.

3.1.2 Fail Closed

The term *fail closed*, when used to describe a valve or valve operator, is similar to fail open, but the spring is positioned to close the valve on loss of control air. The term fail closed means that on loss of the controlling medium, the valve is moved automatically to the closed position. This is a common application.

3.1.3 Fail-as-Positioned

Fail-as-positioned simply means that, on loss of the controlling medium, the valve remains in the same position it was in before the loss. A common example is a motor-operated valve. Loss of the electrical power supply would cause the valve to be left in the position it was in at the time of the power loss.

3.1.4 Spring-to-Open

Spring-to-open is a phrase that describes the effect that spring force has on a valve operator. The spring is positioned to open the valve. Some other operating force must be used to shut the valve. *Figure 36* shows a spring-to-open valve operator.

In this valve, the spring exerts an upward force on the stem to open the valve. Air pressure is used to shut the valve.

3.1.5 Spring-to-Close

In a spring-to-close valve operator, the spring is used to move the valve to the closed position. The spring exerts a force that moves the stem down to close the valve. Air pressure is used to open the valve. *Figure 37* shows a spring-to-close valve operator.

3.1.6 Air-to-Open

Air-to-open valve operators use air pressure to open the valve and spring force to shut the valve. *Figure 37* shows a spring-to-close valve operator, but it is also an air-to-open operator. *Figure 38* is an air-to-open valve operator that uses the spring force to exert an upward force to close the valve. It could also be referred to as a spring-to-close valve operator.

3.1.7 Air-to-Close

Air-to-close valve operators use air to move the valve to its closed position. *Figure 39* is an example of an air-to-close operator. The air pressure exerts an upward force on the diaphragm to close the valve.

Spring force opposes the force induced by the air pressure. This operator could also be referred to as a spring-to-open operator.

3.1.8 Upstream

The term *upstream* is used to describe the location of a component with regard to another component within a fluid system. To say a component is located upstream means the component is located in the direction opposite the flow with relation to another component.

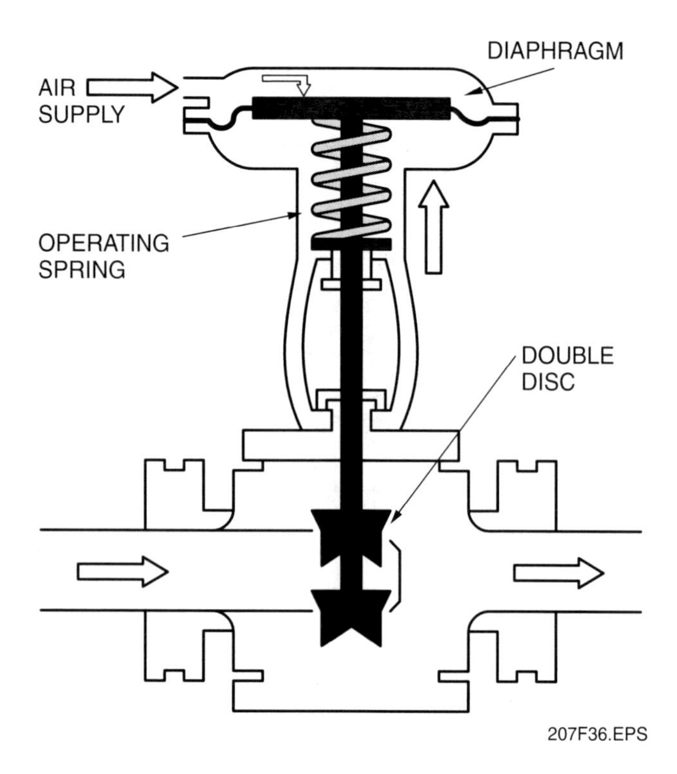

Figure 36 ◆ Spring-to-open valve operator (travels up to open).

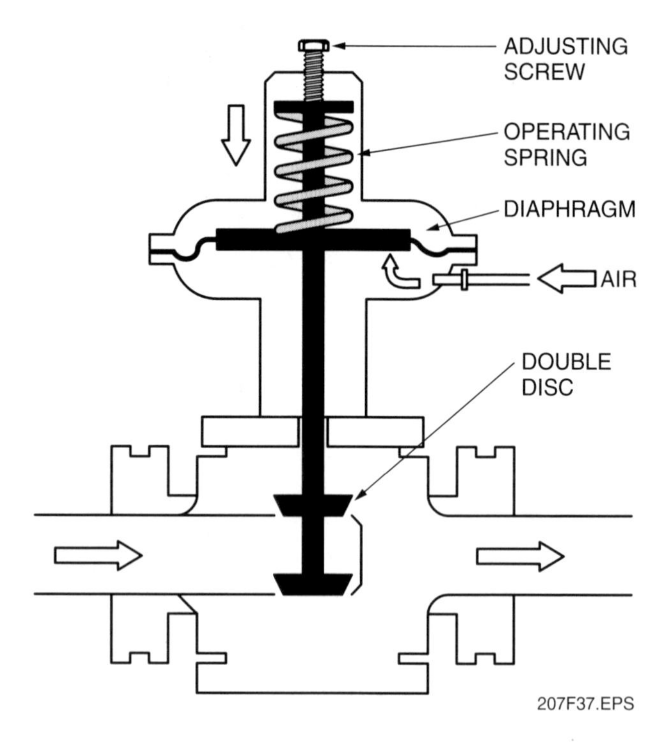

Figure 37 ◆ Spring-to-close valve operator (travels down to close).

3.1.9 Downstream

Downstream is exactly the opposite of upstream. A downstream component is located in the direction of flow with relation to a given component.

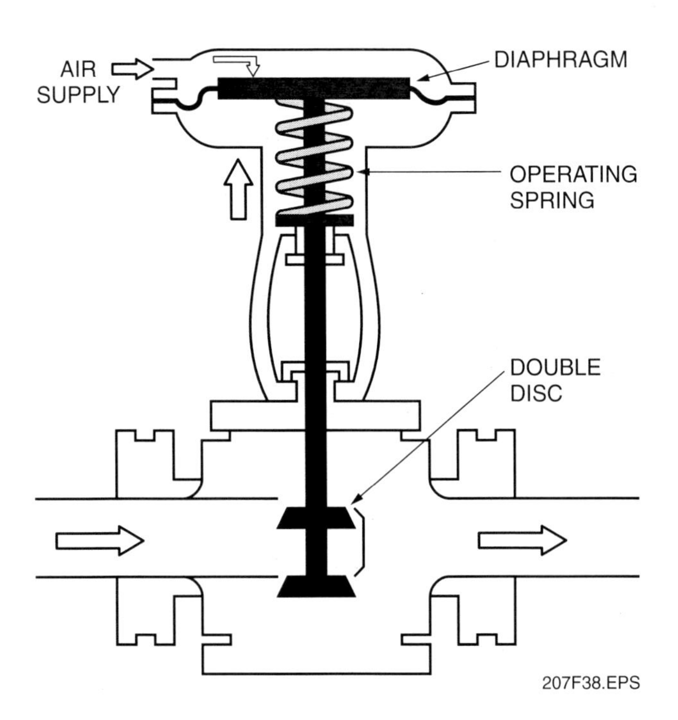

Figure 38 ◆ Air-to-open valve operator (travels down to open).

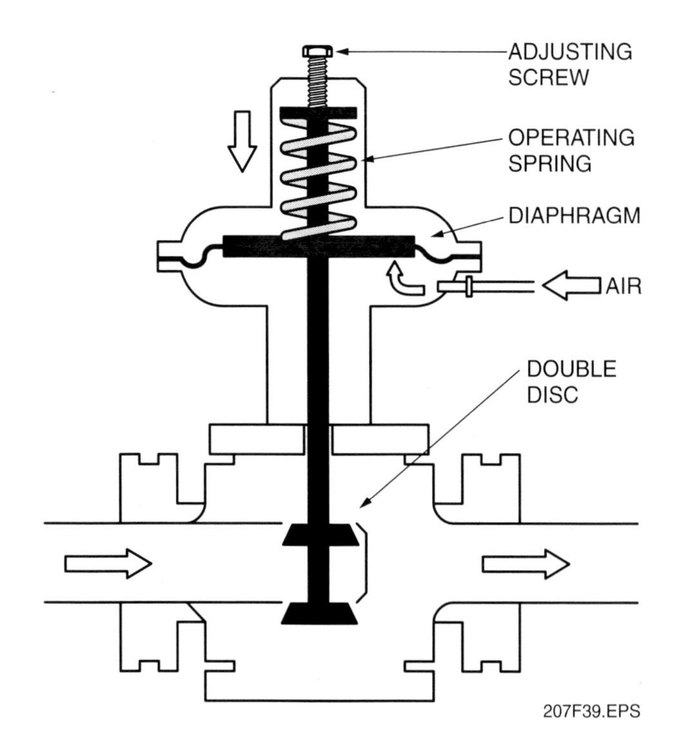

Figure 39 ◆ Air-to-close valve operator (travels up to close).

4.0.0 ◆ PRINCIPLES OF OPERATION AND CONSTRUCTION OF VARIOUS POSITIONERS

The primary function of a valve positioner is to maintain the control valve orifice at a position that is directly proportional to its controller output pressure. Another function of the positioner may be to supply a proportionally larger pneumatic signal to the valve in order to move the valve. A typical control pneumatic pressure of 8–15 psig may not be adequate. To accomplish this, the positioner compares the stem position of the valve actuator to the signal from the controller. Any deviation from the desired position produces an error signal that activates a pneumatic relay in the positioner. The positioner then either increases or decreases pressure to the valve to drive the valve stem to the desired position.

A positioner can also reverse the signal to a valve and overcome frictional forces within a valve on high-pressure drop applications. Positioners are typically mounted on the side of a diaphragm actuator and on top of a piston and rotary actuator. Several designs are available in which the positioner is an integral part of the actuator. Because of the large volume of pneumatic pressure required to operate the valve, the positioner must have an independent, regulated air supply. In many cases, a valve positioner greatly improves the performance of a process control loop. The control loop would be more stable without the positioner in other cases. A properly sized spring and diaphragm actuator provides satisfactory process control in these cases.

Studies have shown that the use of positioners is clearly beneficial in slow processes and clearly detrimental in fast processes. Applications in which positioners should be used include temperature control, liquid level control, and gas flow control mixing and blending. Positioners are also required where it is necessary to split-range the controller output to more than one valve. An example is in a temperature control loop where the output of the controller (3–9 psi) operates a valve controlling cooling liquid flow to the process, while the 9–15 psi signal span operates a valve controlling steam flow that heats the process. A positioner is also required when the controller output signal pressure of 3–15 psi is insufficient to actuate the control valve. In this situation, the positioner would be used to amplify the controller output to a range of higher pressure to provide increased actuator thrust. Finally, the use of a positioner should be considered for systems where it is necessary to provide control of the process with minimum overshoot and the fastest possible recovery following a disturbance. This is especially critical for process control loops with long transmission lines between the controller and the actuator.

Even though positioners are referred to as *valve positioners* in the instrumentation industry, they are really actuator positioners.

4.1.0 Direct- or Reverse-Acting Positioners

The terms *direct* and *reverse* are used frequently when discussing control valves, positioners, and controllers.

While control bodies and control valve actuators can be described as being direct-acting or reverse-acting, thinking about such things when working through a system problem only adds to the confusion. Fortunately, when it comes to positioners, 99 percent of the time they will mimic the input signal from the controller. That is, they will be direct-acting. *Figure 40* illustrates this.

4.2.0 Pneumatic Positioners

Pneumatic positioners receive a pneumatic signal from the controller and supply a pneumatic signal to the valve that is proportional to the signal received from the controller. These positioners are typically supplied with three pressure gauges that monitor supply, signal, and output pressures. *Figure 41* is an example of a typical pneumatic positioner.

4.3.0 Analog Positioners

Analog positioners receive a 4–20 mA signal from the controller and convert it to mechanical or electromechanical methods to position the valve. The main difference between an analog positioner and a pneumatic positioner is the addition of an I/P transducer, which converts the 4–20 mA signal from the controller into a pneumatic signal. However, it operates in the same fashion as the pneumatic positioner in that it monitors the valve actuator for response and compares its location to the input signal from the controller. An example of an analog (I/P) positioner is shown in *Figure 42*. As noted, there are only two pressure gauges present on the face of the analog positioner, instead of three

207F41.EPS

Figure 41 ◆ Typical pneumatic positioner.

207F42.EPS

Figure 42 ◆ Analog (I/P) positioner.

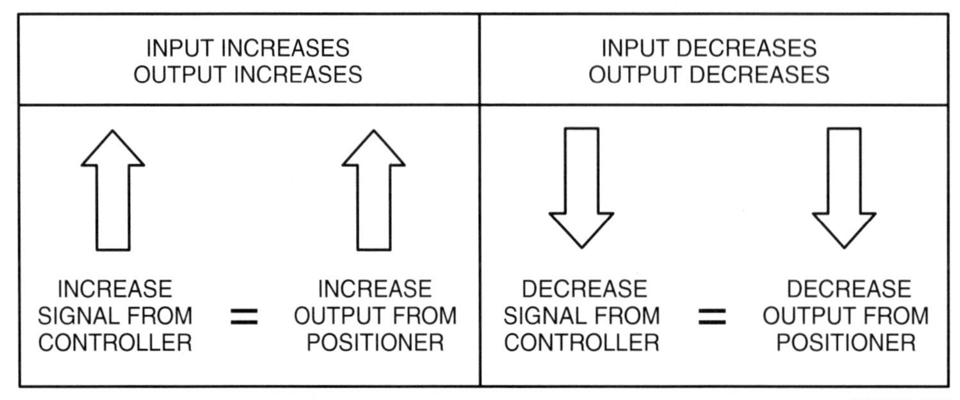

207F40.EPS

Figure 40 ◆ Effects of direct-acting positioners.

as typically found on a pneumatic positioner. One of the gauges reads the supply air, while the other reads the output to the valve. The third gauge on a pneumatic positioner reads the input from the controller. In this case, the input is a 4–20 mA signal.

In this particular model positioner, the mounting to the valve actuator is the same as its counterpart pneumatic model positioner. Thus, it is unnecessary to change control valves if the controller input is changed from a pneumatic to an analog signal, or in a less likely scenario, from an analog to a pneumatic signal.

4.4.0 Modular Assembly Positioners

Manufacturers of instrumentation valves, controllers, actuators, and positioners understand that the transition from pneumatic to analog signal and now to intelligent, or computerized, control is not accomplished overnight due to budget limitations. They are continually developing products that consider the cost in making a total changeover from one medium to another. One such product example is the positioner shown in *Figure 43*, a Masoneilan® Model 4700 positioner. It is a modular design assembly that can be configured as either a pneumatic or electro-pneumatic device. To switch from pneumatic input and pneumatic output to analog input and pneumatic output (electromechanical), the input gauge is removed and the electronic component bolted in place. These positioners can be used with either reciprocating or rotary actuators.

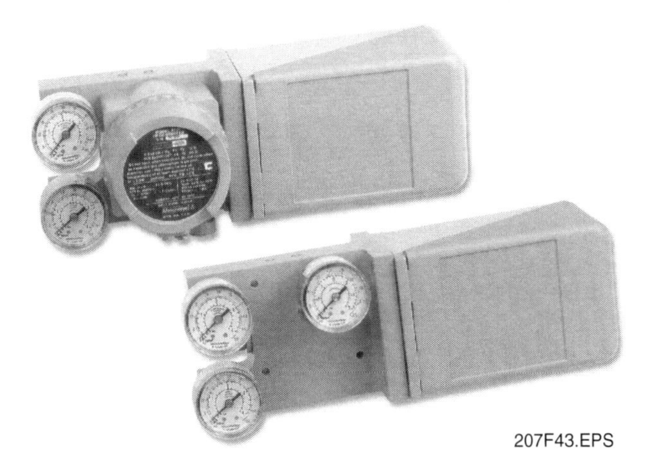

207F43.EPS

Figure 43 ◆ Masoneilan® modular design positioner.

4.5.0 Digital Positioners

Manufacturers in the process industries are realizing the many benefits offered through the integration of digital positioners with **plant asset management** (PAM) systems. Through the integration of digital positioners with PAM systems, manufacturers are establishing a predictive maintenance environment that helps eliminate costly scheduled repairs. Digital valve positioners are seen as the key to creating that predictive maintenance environment.

The recent flood of digital valve positioner products is the leading force driving growth in the worldwide control valve market, according to one study. Demand for more information from valves is influencing users to purchase digital valve positioners for their data storage and remote communication capabilities. Digital positioners are also being applied on valves in remote locations, reducing the need for maintenance personnel to visit the site.

Figure 44 shows one example of a digital positioner, often referred to as a smart valve control. It is a Moore Series 760D ValvePAC®. The positioner can be configured, calibrated, and operated locally or from a remote location. An optional software product is also available that allows you to configure, calibrate, and perform valve diagnostics from a remote personal computer.

207F44.EPS

Figure 44 ◆ Moore Series 760D digital valve controller (positioner).

4.6.0 Process Precautions for Positioners

The following precautions pertain to the use and operation of valve (actuator) positioners:

- Pneumatic or electric power supplies should not be isolated or disconnected from an operating positioner unless the technician understands the failure modes beforehand.
- When returning a positioner to service, observe the valve and system indications to make sure the associated control signal is properly controlling the process. Large or rapid valve oscillations should not occur.

- When restoring pneumatic or electric power to a valve positioner, the positioner and valve response should be understood beforehand. The technician must determine if the valve should open, close, or remain as is.

5.0.0 ◆ VALVE SELECTION, TYPES, AND APPLICATIONS

Because of the diverse nature of valves, with valve types overlapping each other in both design and application, the valve selection process must be examined. This section discusses valve selection, valve types, and valve applications.

5.1.0 Valve Selection

With valve selection, there are many factors that must be taken into consideration. Cost is often an overriding factor, although experience has shown that sparing expense now may result in additional expense later. When selecting a valve during system design, overall system performance must be taken into consideration. Questions that must be asked include these:

- At what temperature will the system be operating? Are there any internal parts that would be adversely affected by the temperature? Valves designed for high temperature steam systems are not necessarily suited for the extreme low temperatures that may be found in a liquid nitrogen system.
- At what pressure (or vacuum) will this valve be operating? How does the temperature affect the valve's pressure rating? System integrity is a major concern on any system. The valve must be rated at or above the maximum system pressure anticipated. Due to factors such as valve design, packing construction, and end attachments, the valve is often considered a weak point in the system.
- Are there any sizing constraints? It seems obvious that a 2" valve would not be installed in a 10" pipe, but what may not be obvious is how the yoke size, actuator, or positioner figures in the scenario. Valve manufacturers provide dimensional tables to aid in valve selections.
- Will this valve be used for on-off or throttle application? Throttle valves are generally globe valves, although in some applications a ball valve or butterfly valve may be used.
- To what type of erosion will the valve be exposed? Will it require hardened seats and discs? Will it be throttled close to its seat and need a special pressure drop valve?
- What kind of pressure drop is allowed? Globe valves exhibit the largest pressure drop or head loss characteristics, whereas ball valves exhibit the least.
- What kind of differential pressure will this valve be operated against? Will this differential pressure be used to seat or unseat the valve? Will the high differential pressure deform the body or disc and bind the valve? Will this also require a bypass valve?
- How will this valve be connected to the system? Will it be welded, screwed or flanged? Should it be butt-welded or socket welded? Will it be union threaded or pipe threaded? Should the flanges be raised, flat, **phonographic**, male/female, or tongue-and-groove?
- In what type of environment will the valve be installed? Is it a dirty environment where an exposed stem would score the **yoke bushing** and cause premature failure? Is it a clean environment where different stem lubricant should be used?
- What kind of fluid is being handled? Is it hazardous in such a way that packing leakage may be detrimental? Is it corrosive to the packing or to the valve itself?
- What is the life expectancy required? Will it require frequent maintenance? If so, is it easily repairable, or does the cost of labor justify replacement instead of repair?

Of course, there are many other questions which may be required. After these are answered, a suitable valve may be selected.

If an installed valve is to be replaced, a valve identical to the one removed should be installed. If that valve is no longer manufactured, valve selection should be made in the same fashion as for a new application, except that the valve dimensions are the limiting factor unless piping alterations can be made. Several questions should be asked:

- Are the system parameters the same as when the system was designed, or has the system intent changed?
- Have any problems been noted since system fabrication that could be remedied by installing a different valve design at this time?
- With what type of operator should the new valve be fitted? Is the new valve compatible with the installed operator?

5.2.0 Valve Types and Applications

As we have noted in valve selection, there are many factors that determine the application and/or type of valve to be used. Some of the factors include these:

- The temperature at which the system will be operating
- The sizing constraints or the pressure at which the system will be operating
- The kind of fluid that is in the system
- The type of environment in which the valve will have to operate
- The type of actuator the valve will use

5.3.0 Valve Markings and Nameplate Information

Before the present system of valve and flange coding, manufacturers had their own systems. With the development of components rated at higher temperatures and pressures, in conjunction with more stringent regulations, a standard was needed. The Manufacturers Standardization Society (MSS) first developed SP-25 in 1934. In 1978, SP-25 was revised to incorporate all the changes that had developed since 1934. To preclude errors in cross-referencing, the American National Standards Institute (ANSI) and the **American Society for Testing Materials (ASTM)** have adopted the MSS marking system.

Two markings that are frequently used on valves are the flow direction arrow, indicating which way the flow is going, and the bridgewall marking, shown in *Figure 45*. The bridgewall marking is usually found on globe valves and is an indication of how the seat walls are angled in relation to inlet and outlet ports of the valve. Specifically, it shows whether the wall of the seat on the inlet side angles up or down. The wall of the seat on the outlet side of the globe valve will always be angled opposite to the angle on the inlet side, as indicated in *Figure 46*.

Not all globe valves are designed with angled bridgewalls. However, some applications may specifically require the process to enter either on the top side or the bottom side of the disc in a globe valve.

If the process enters on the top side (bridgewall angled up on the inlet), the force of the process will assist in the closing of the valve. However, if the process enters on the bottom side (bridgewall angled down on the inlet), the force of the process will assist in the opening of the valve.

No markings for flow are normally used on gate, plug, butterfly, or ball valves. If a gate valve has a flow arrow, it is because the gate valve has a double gate. Double gate valves are capable of relieving fluid pressure in the event that a high pressure difference exists across the shut gate. Standard practice is for the outlet-side gate to relieve to the inlet side. This type of valve is used for specific applications. Therefore, system plans should be consulted for correct valve orientation.

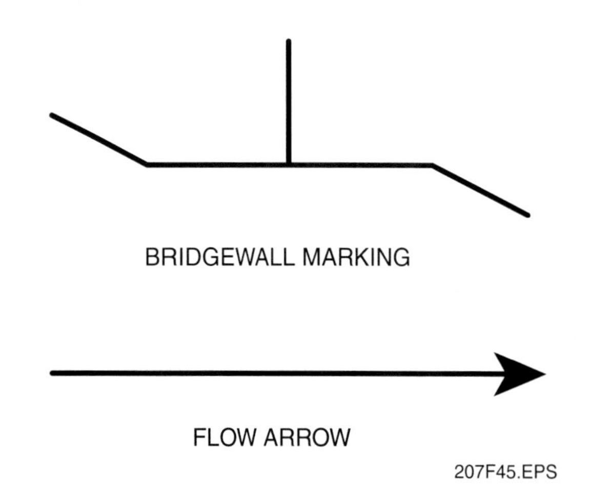

Figure 45 ◆ Valve markings.

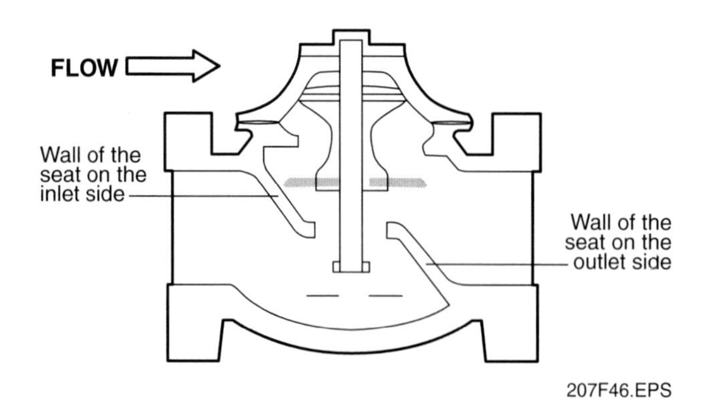

Figure 46 ◆ The meaning of bridgewall markings on valves.

There are normally two identification sets: one permanently embossed, welded, or cast into the valve body, and the other a valve identification plate (*Figure 47*). Typically, as a minimum, the following information will be included within the two sets:

- Rating designation markings
- Material designation markings
- Melt identification markings
- Trim identification markings (if applicable)
- Size markings
- Thread identification markings (if applicable)

The omission of markings and component marking requirements must also be discussed.

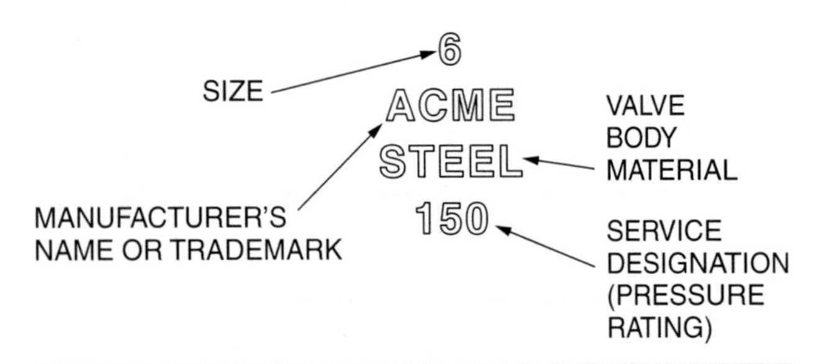

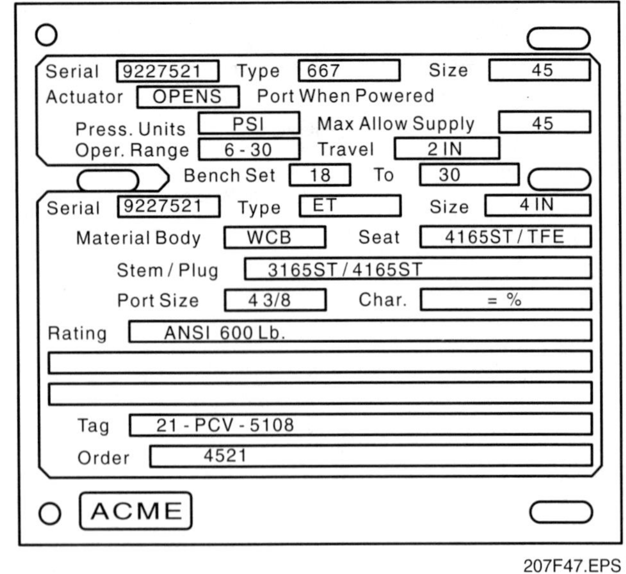

Figure 47 ◆ Actuator nameplate.

5.3.1 Rating Designation

The rating designation of a valve gives the pressure and temperature rating as well as the type of service. *Table 2* shows commonly used service designations.

The product rating may be designated by the class numbers alone, as with a steam pressure rating or pressure class designation. The ratings for products that conform to recognized standards but are not suitable for the full range of pressures or temperatures of those standards may be marked as appropriate. The numbers and letters representing the pressure rating at the limiting conditions may also be shown.

Table 2 Valve Rating Designations

Correspond to Steam Working Pressure (SWP)	Correspond to Cold Working Pressure (CWP)
Steam pressure (SP)	Water, oil pressure (WO)
Working steam pressure (WSP)	Oil, water, gas pressure (OWG)
Steam (S)	Water, oil, gas pressure (WOG)
	Gas, liquid pressure (GLP)
	Working water pressure (WWP)
	Water pressure (WP)

The rating designation for products that do not conform to recognized national product standards may be shown by numbers and letters representing the pressure ratings at maximum and minimum temperatures. If desired, the rating designation may be shown as the maximum pressure followed by **CWP** and the allowed pressure at the maximum temperature (for example, 2,000 CWP 725/925°F).

Other typical designations are given as the first letter of the system for which they are designated:

- A – Air service
- G – Gas service
- L – Liquid service
- O – Oil service
- W – Water service
- DWV – Drainage waste and vent service

5.3.2 Trim Identification

Trim identification marking is required on the identification plate for all flanged-end and butt welding end steel or flanged-end ductile iron body valves with trim material that is different than the body material. Symbols for materials are the same. The identification plate may be marked with the word *trim* followed by the appropriate material symbol.

Trim identification marking for gate, globe, angle, and cross valves, or valves with similar design characteristics, consists of three symbols. The first indicates the material of the stem. The second indicates the material of the disc or wedge face. The third indicates the material of the seat face. The symbol may be preceded by the words *stem*, *disc*, or *seat*, or it may be used alone. If used alone, the symbols must appear in the order given.

Plug valves, ball valves, butterfly valves, and other quarter-turn valves require no trim identification marking unless the plug, disc, closure member, or stem is of different material than the body. In such cases, trim identification symbols on the nameplate first indicate the material of the stem and then the material of the plug, ball, disc, or closure member.

Those valves with seating or sealing material different than the body material must have a third symbol to indicate the material of the seat. In these cases, symbol identification must be preceded by the words *stem*, *disc* (or *plug*, *ball*, or *gate*, as appropriate) and the word *seat*. If used alone, the symbols must appear in the order given.

5.3.3 Size Designation

Size markings are in accordance with the product-referenced marking requirements. For size designation for products with a single nominal pipe size of the connecting ends, the word *nominal* indicates the numerical identification associated with the pipe size. It does not necessarily correspond to the valve, pipe, or fitting inside diameter.

Products with internal elements that are the equivalent of one pipe size or are different than the end size may have dual markings unless otherwise specified in a product standard. Unless these exceptions exist, the first number indicates the connecting end pipe size. The second indicates the minimum bore diameter, or the pipe size corresponding to the closure size (for example, 6 × 4, 4 × 2½, 30 × 24).

At the manufacturer's option, triple marking size designation may be used for valves. If triple size designation is used, the first number must indicate the connecting end size at the other end. For example, 24 × 20 × 30 on a valve designates a size 24 connection, a size 20 nominal center section, and a size 30 connection.

Fittings with multiple outlets may be designated at the manufacturer's option in a run×run×outlet size method. For example, 30 × 30 × 24 on a fitting designates a product with size 30 end connections and a nominal size 24 connection between.

5.3.4 Thread Markings

Fittings, flanges, and valve bodies with threaded connecting ends other than American National Standard Pipe Thread or American National Standard Hose Thread will be marked to indicate the type. The style or marking may be the manufacturer's own symbol provided confusion with standard symbols is avoided. Fittings with left-hand threads must be marked with the letters LH on the outside wall of the appropriate opening.

Marking of products with ends threaded for API casing, tubing, or drill pipe must include the nominal size, the letters *API*, and the thread type symbol as listed in *Table 3*.

Table 3 Examples of Threaded-Type Symbols

Name/Description	Symbol
Casing – Short round thread	CSG
Casing – Long round thread	LCSG
Casing – Buttress thread	BCSG
Casing – Extreme-line	XCSG
Line pipe	LP
Tubing – Non-upset	TBG
Tubing – External-upset CSG	UP TBG

Marking of products using other pipe threads must include the following:

- Nominal pipe, tubing, drill pipe, or casing size
- Outside diameter of pipe, tubing, drill pipe, or casing
- Name of thread
- Number of threads per inch

5.3.5 Valve Schematic Symbols

The last and most important aspect of valve identification is the ability to identify different types of valves from blueprints and schematics. In general, the symbols that denote various control valves, actuators, and positioners are standard symbols as shown in *Figure 48*. However, in certain cases these symbols vary depending on site-specific prints. The legend of a typical system print or schematic will show the symbols that represent all components on the drawing.

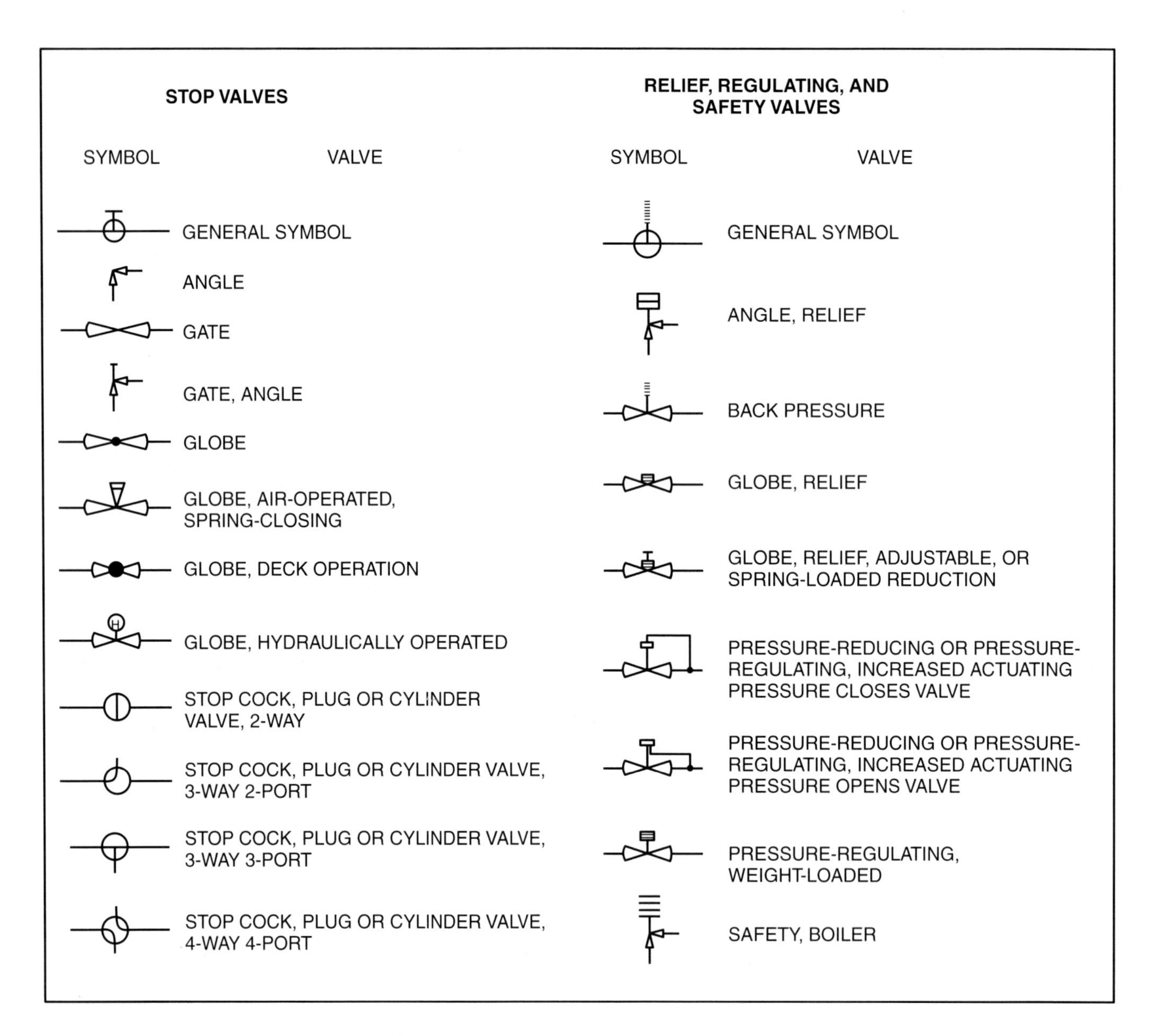

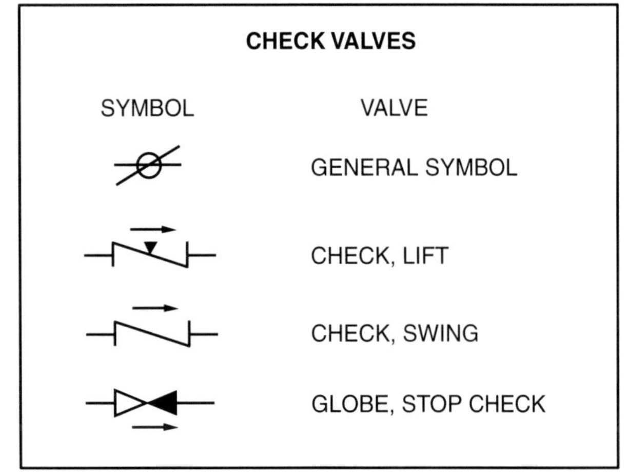

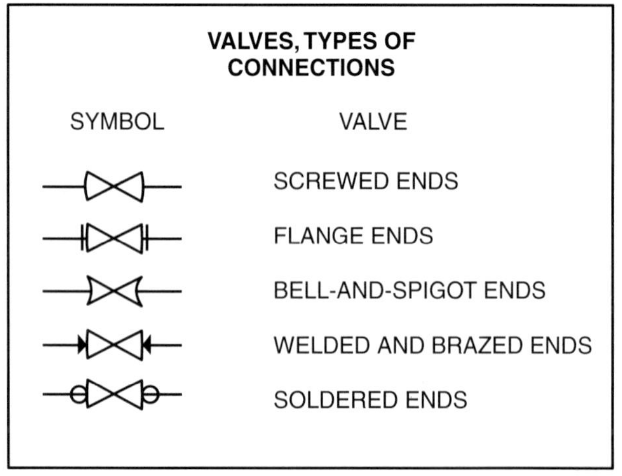

Figure 48 ◆ Typical piping system schematic symbols.

Summary

Control valves, actuators, and positioners are integral to the operation and control of fluid flow systems. Each device has characteristics that make it specially suited for certain applications. The applications for which they are used vary widely and include fine control of flow, temperature regulation, pressure regulation, and flow isolation.

Proper selection of the control valve for a specific system is determined by several factors such as system design pressure and temperature, piping size, and system flow conditions.

Actuators and positioners serve as energy transmission devices that cause valve stem movement. They are designed to use several different energy sources to perform their functions. These devices provide a means through which valves may be operated remotely or against extremely high differential pressures.

The information presented in this module provides detailed descriptions of the construction and operation of various control valves, actuators, and positioners. This information will prove invaluable in the proper selection and correct installation of any of these valves, actuators, and positioners in a fluid system.

Review Questions

1. The three functions of a control valve are to _____ flow.

a. start, stop, and monitor
b. throttle, start, and monitor
c. start, stop, and control
d. control, throttle, and monitor

2. One disadvantage of the globe valve is _____.

a. high head loss
b. inability to regulate flow
c. high seat leakage
d. inability to serve in quick-acting start-stop applications

3. A globe valve that closes against the direction of fluid flow is said to be _____.

a. flow-to-close
b. flow-to-open
c. fail-to-open
d. fail-to-close

4. Gate valves should not be used to _____.

a. start flow
b. stop flow
c. throttle flow
d. control an on/off process

5. A knife gate valve is likely to be used in which of the following applications?

a. High pressure
b. Fine liquids only
c. Slurry or waste
d. Throttling steam

6. Which of the following valves is a rotary-action valve?

a. Gate
b. Needle
c. Globe
d. Ball

7. Which of the following valves is easily adapted to multiport construction?

a. Gate
b. Plug
c. Globe
d. Knife

8. Which one of the following valves is built on the principle of a pipe damper?

a. Globe
b. Plug
c. Knife
d. Butterfly

9. A needle valve may be used when an application requires _____.

a. fine adjustment flow
b. coarse flow
c. on-off control
d. slurry flow

10. One disadvantage of the spring and diaphragm actuator is _____.

a. high cost
b. complex construction
c. lack of reliability
d. limited power

11. When looking at an illustration of valve plug and seat installation, if the taper of the plug goes from wide to narrow from top to bottom, the plug must travel _____.
    a. up to close, down to open
    b. down to close, up to open
    c. down to close, down to open
    d. up to close, up to open

12. A valve that automatically opens whenever the controlling medium is lost is referred to as a _____.
    a. fail closed
    b. fail-as-positioned
    c. fail open
    d. spring close

13. A fail closed valve is closed by _____.
    a. a spring
    b. a diaphragm
    c. pneumatic pressure
    d. a motor actuator

14. If a fail-as-positioned valve is 50% open and pneumatic pressure is lost to the valve, the valve will _____.
    a. close
    b. open
    c. remain 50% open
    d. fail in the close position

15. An air-to-open valve is the same thing as a _____ valve.
    a. fail open
    b. fail-as-positioned
    c. spring-to-open
    d. spring-to-close

16. An air-to-close valve is the same as a(n) _____ valve.
    a. spring-to-close
    b. air-to-open
    c. fail-as-positioned
    d. spring-to-open

17. A valve positioner compares the stem position of the valve actuator to the _____.
    a. signal from the controller
    b. signal from the sensing element
    c. input to the valve
    d. upstream pressure or flow

18. On a direct-acting positioner, if the output from the controller decreases, the positioner's output will _____.
    a. increase
    b. remain the same
    c. decrease
    d. be 0 psi

19. A typical pneumatic positioner has _____ pressure gauges.
    a. three
    b. two
    c. one
    d. zero

20. A device that converts a 4–20 mA signal into a pneumatic signal is called a(n) _____.
    a. P/I transducer
    b. I/P transducer
    c. mA/psi converter
    d. current-to-air regulator

21. Which one of the following positioners may interface with a PAM program?
    a. Pneumatic
    b. Analog
    c. Intelligent
    d. I/P

22. Pneumatic or electric power supplies should not be isolated or disconnected from operating positioners unless the technician understands the _____ beforehand.
    a. oscillation
    b. operating medium
    c. controller response
    d. failure modes

23. When choosing among valves, the _____ valve exhibits the largest pressure drop.

a. globe
b. gate
c. ball
d. butterfly

24. If a gate valve has a flow arrow indicated on it, it is because the gate valve has a _____.

a. single gate
b. double gate
c. bridgewall marking
d. vent port

25. Fittings that are marked LH on the outside wall indicate _____.

a. low heat
b. liquid hydrogen only
c. left-hand threads
d. low hardening material

GLOSSARY

Trade Terms Introduced in This Module

Actuator: The part of a regulating valve that converts electrical or fluid energy to mechanical energy to position the valve.

Angle valve: A type of globe valve in which the piping connections are at right angles.

American Society for Testing Materials (ASTM): Founded in 1898, a scientific and technical organization, formed for the development of standards on the characteristics and performance of materials, products, systems, and services.

Ball valve: A type of plug valve with a spherical disc.

Bonnet: The part of a valve containing the valve stem and packing.

Butterfly valve: A quarter-turn valve with a plate-like disc that stops flow when the outside area of the disc seals against the inside of the valve body.

Check valve: A valve that allows flow in one direction only.

CWP: Cold working pressure.

Deformation: A change in the shape of a material or component due to an applied force or temperature.

Diaphragm actuator: A valve actuator in which pressure exerted on the diaphragm is used to position the valve stem.

Disc: Part of a valve used to control the flow of system fluid.

Elastomeric: Elastic or rubberlike. Flexible, pliable.

Galling: An uneven wear pattern between trim and seat that causes friction between the moving parts.

Gate valve: A valve with a straight-through flow design that exhibits very little resistance to flow. It is normally used for open/shut applications.

Globe valve: A valve in which flow is always parallel to the stem as it goes past the seat.

Kinetic energy: Energy of motion.

Lapping: The removal of mating surface defects using an abrasive compound.

Lift: The actual travel of the disc away from the closed position when a valve is relieving.

Linear flow: Flow in which the output is directly proportional to the input.

Packing: Material used to make a dynamic seal, preventing system fluid leakage around a valve stem.

Phonographic: When referring to the facing of a pipe flange, serrated grooves cut into the facing, resembling those on a phonograph record.

Plant asset management (PAM): Interfacing plant maintenance through computerized programs designed to maximize the use and effectiveness of plant assets.

Plug: The moving part of a valve trim (plug and seat) that either opens or restricts the flow through a valve in accordance with its position relative to the valve seat, which is the stationary part of a valve trim.

Plug valve: A quarter-turn valve with a ported disc.

Positioner: A field-based device that takes a signal from a control system and ensures that the control device is at the setting required by the control system.

Seat: The part of a valve against which the disc presses to stop flow through the valve.

Thermal transients: Short-lived temperature spikes.

Throttling: The regulation of flow through a valve.

Torque: A twisting force used to apply a clamping force to a mechanical joint.

Valve body: The part of a valve containing the passages for fluid flow, valve seat, and inlet and outlet connections.

Valve stem: The part of a valve that raises, lowers, or turns the valve disc.

Valve trim: The combination of the valve plug and the valve seat.

Wedge: The disc in a gate valve.

Wire drawing: The erosion of a valve seat under high velocity flow through which thin, wire-like gullies are eroded away.

Yoke bushing: The bearing between the valve stem and the valve yoke.

Additional Resources

This module is intended to present thorough resources for task training. The following reference work is suggested for further study. This is optional material for continued education rather than for task training.

Valve Handbook, 1997. Philip L. Skousen. Columbus, OH: McGraw Hill Professional Publishing, Inc.

Figure Credits

Dwyer Instruments, Inc.	207F02, 207F17, 207F27, 207F35, 207F41, 207F42
Parker Hannifin Corp.	207F33
Masoneilan, Division of Dresser, Inc.	207F43
Siemens Energy and Automation, Inc.	207F44

NCCER CURRICULA — USER UPDATE

NCCER makes every effort to keep its textbooks up-to-date and free of technical errors. We appreciate your help in this process. If you find an error, a typographical mistake, or an inaccuracy in NCCER's curricula, please fill out this form (or a photocopy), or complete the online form at **www.nccer.org/olf**. Be sure to include the exact module ID number, page number, a detailed description, and your recommended correction. Your input will be brought to the attention of the Authoring Team. Thank you for your assistance.

Instructors – If you have an idea for improving this textbook, or have found that additional materials were necessary to teach this module effectively, please let us know so that we may present your suggestions to the Authoring Team.

NCCER Product Development and Revision

13614 Progress Blvd., Alachua, FL 32615

Email: curriculum@nccer.org
Online: www.nccer.org/olf

❑ Trainee Guide ❑ AIG ❑ Exam ❑ PowerPoints Other ______________________________

Craft / Level: Copyright Date:

Module ID Number / Title:

Section Number(s):

Description:

Recommended Correction:

Your Name:

Address:

Email: Phone:

Module 12208-03

Relays and Timers

COURSE MAP

This course map shows all of the modules in the second level of the Instrumentation curriculum. The suggested training order begins at the bottom and proceeds up. Skill levels increase as you advance on the course map. The local Training Program Sponsor may adjust the training order.

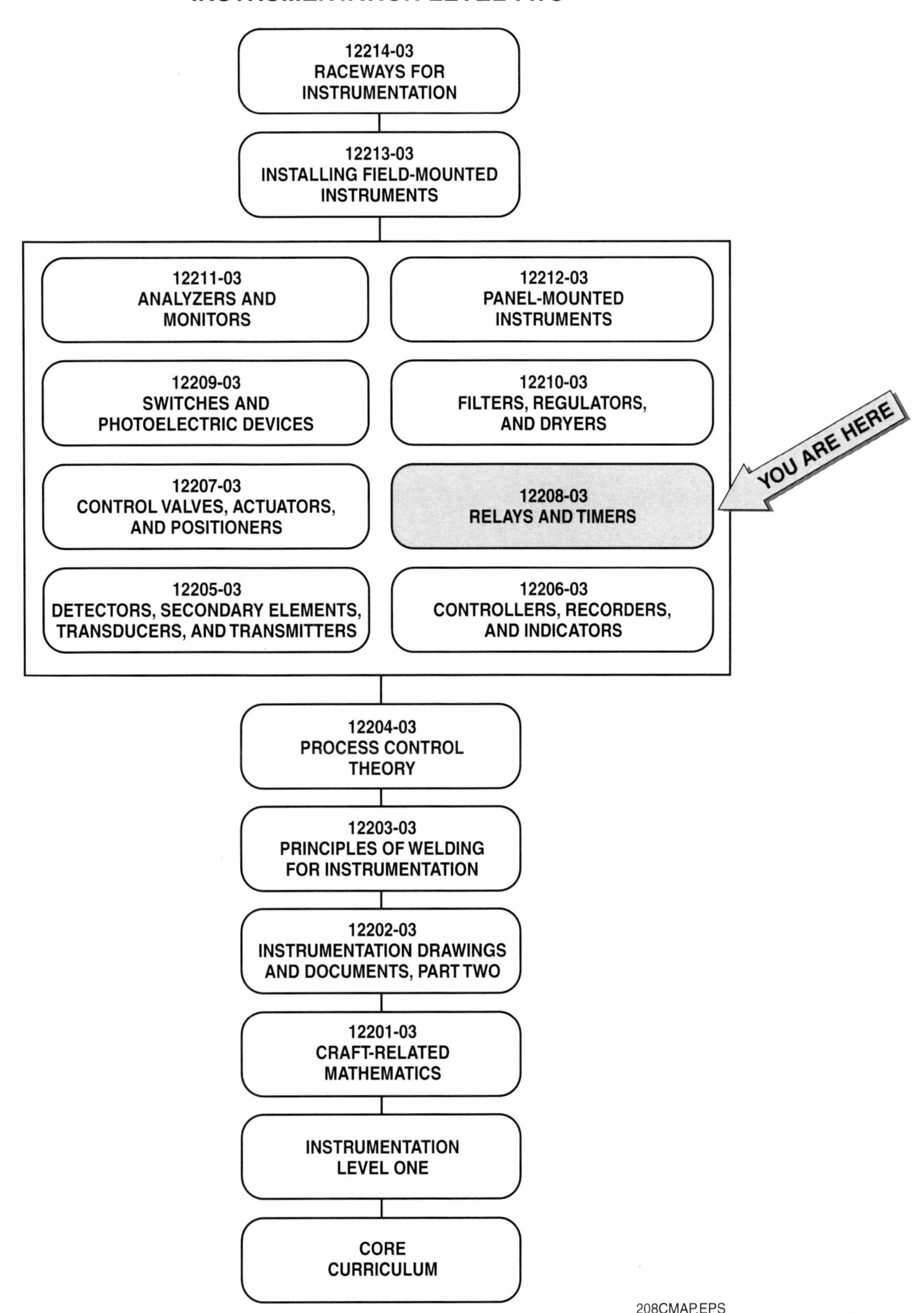

208CMAP.EPS

MODULE 12208-03 CONTENTS

1.0.0 INTRODUCTION 8.1

2.0.0 ELECTRICAL RELAYS 8.2

2.1.0 Electromechanical Relays 8.2

2.1.1 Reed Relays and Switches 8.2

2.1.2 General Purpose Relays 8.4

2.1.3 Control Relays in Instrumentation 8.7

2.2.0 Solid-State Relays 8.7

2.2.1 Comparison of Electromechanical Relays to Solid-State Relays 8.7

3.0.0 PNEUMATIC RELAYS, REPEATERS, AND BOOSTERS 8.10

3.1.0 Force-Balance Transmitter Relays 8.10

3.2.0 Computing Relays 8.11

3.2.1 Pneumatic Multiplying and Dividing Relays 8.11

3.2.2 Pneumatic Adding, Subtracting, and Inverting Relays 8.11

3.2.3 Pneumatic Scaling and Proportioning Relays 8.12

3.2.4 High- and Low-Pressure Selector and High-Pressure Limiter Relays 8.12

3.2.5 Booster Relays 8.14

4.0.0 TIMERS AND TIME CLOCKS 8.15

4.1.0 Dashpot Timer Relays 8.15

4.1.1 Pneumatic Timers 8.15

4.2.0 Synchronous Time Switches 8.17

4.3.0 Solid-State Timers 8.18

SUMMARY 8.20

REVIEW QUESTIONS 8.20

GLOSSARY 8.23

REFERENCES & ACKNOWLEDGMENTS 8.24

Figures

Figure 1 Typical electromechanical relays 8.3
Figure 2 Reed switch 8.3
Figure 3 Reed relay and microminiature relay with terminal schematics 8.4
Figure 4 Examples of general purpose relays 8.4
Figure 5 General purpose relay 8.5
Figure 6 Typical control relay shading coil, contact-pressure spring, and inflexible moving contact arm 8.5
Figure 7 Various contact arrangements of relays 8.6
Figure 8 Basic contact forms accepted by NARM 8.8
Figure 9 Typical solid-state relay 8.8
Figure 10 Comparison between an EMR and an SSR 8.9
Figure 11 Simplified 1:1 force-balance pressure transmitter relay 8.10
Figure 12 1:1 force-balance pressure transmitter with volume booster 8.11
Figure 13 Pneumatic force-bridge multiplying and dividing relay 8.12
Figure 14 Pneumatic adding, subtracting, and inverting relay 8.13
Figure 15 Pneumatic fixed ratio amplifying relay 8.13
Figure 16 High-pressure selector relay 8.14
Figure 17 Low-pressure selector relay 8.14
Figure 18 High-pressure limit relay 8.14
Figure 19 Volume booster relay 8.15
Figure 20 Fluid dashpot timing relay .8.16
Figure 21 Synchronous time switch 8.17
Figure 22 Clock gear train, motor, and switch mechanism 8.17
Figure 23 Single-pole switch mechanism 8.18
Figure 24 Terminal board and control dial 8.18
Figure 25 Typical solid-state timer specification sheet 8.19
Figure 26 Octal socket pin numbers .8.19

Tables

Table 1 Relay Contact and Pole Designations 8.6
Table 2 Relay Switching Arrangement Code Number 8.7

MODULE 12208-03

Relays and Timers

Objectives

When you have completed this module, you will be able to do the following:

1. Describe the basic functions of relays.
2. Describe and identify electromechanical relays and explain how they operate.
3. Install and connect relays in sockets.
4. Describe and identify solid state relays and explain how they operate.
5. Describe and identify pneumatic relays and repeaters. Explain how these operate.
6. Describe and identify hydraulic relays and explain how they operate.
7. Describe and identify timers and time delay relays, including:
   - Dashpot
   - Synchronous time clock
   - Solid state
8. Describe the operation of a volume booster.
9. Install various types of timers.

Prerequisites

Before you begin this module, it is recommended that you successfully complete the following: Core Curriculum; Instrumentation Level One; Instrumentation Level Two, Modules 12201-03 through 12204-03.

Required Trainee Materials

1. Pencil and paper
2. Appropriate personal protective equipment

1.0.0 ◆ INTRODUCTION

A relay is defined as an electrical or mechanical device used for remote or automatic control of other devices or systems. Relays can be electrical, pneumatic, or hydraulic.

Electrical relays are typically used to control contacts and switches in electrical control circuits. Pneumatic and hydraulic relays are used when a small air or hydraulic signal needs to control a much larger pressure or volume.

A low-power signal can operate a relay regardless of the power level the relay is controlling. The signal does not have to be the same type of power. For example, the operating mechanism of a relay may be pneumatic or hydraulic, while the energy that is being controlled may be electrical. This arrangement could be reversed in that an electrical signal may operate a relay that controls pneumatic or hydraulic energy. These configurations require much smaller or lower-capacity supply systems to be installed to the operating mechanism of the relay. They only require the larger or higher-capacity systems to be installed on the energy system controlled by the relay.

There are many relay types, sizes, and applications. The aim of this module is to describe the basic operating characteristics and applications of some of the more common types of relays used in industry. This information helps the technician to make an educated choice when selecting and installing such relays.

2.0.0 ◆ ELECTRICAL RELAYS

Most electrical relays can be classified as either electromechanical or solid state. Electromechanical relays (EMRs) consist of devices with contacts that are closed by some type of magnetic effect. **Solid-state relays (SSR)**, by contrast, have no mechanical contacts and switch entirely by electronic means. A third category that is sometimes recognized is a type of solid-state relay called the hybrid relay. Hybrid relays are a combination of electromechanical and solid state technology. The advantages, limitations, and applications of each relay category are discussed in this module.

2.1.0 Electromechanical Relays

Electromechanical relays (EMRs) generally can be subdivided into three classifications: reed relays, control relays, and general purpose relays. The major differences between these three types of relays are their costs, life expectancies, and intended uses in circuits.

Reed relays are used in applications requiring low and stable contact resistance, low capacitance, high insulation resistance, long life, and small size. These include automatic test equipment and instrumentation. Reed relays can be fitted with coaxial shielding for high-frequency applications. They are available with the very high isolation voltage ratings typically required for medical applications. Also, their low cost and versatility make them suitable for many security and general purpose applications. Reed relays offer several advantages over electromechanical relays, one of which is switching speed.

Reed switches are often misidentified as reed relays. The difference between the two reed devices is discussed in Section 2.1.1, *Reed Relays and Switches*.

Refer to *Figure 1(A)*. Control relays are generally used to control machinery, tools, motors, and other equipment, including valves and lighting. Other terms used to identify control relays are **contactors**, starters, and power relays. Often, the conductors associated with the circuits being controlled by control relays are relatively large, in order to handle the amperage levels associated with these loads. The conductors supplying the coils of these relays are relatively small in size, because the load is electrically isolated from the coil. The parts of these relays are explained later in this module.

Refer to *Figure 1(B)*. General purpose relays are plug-in relays available with relatively low coil voltage ratings such as 24VAC and 120VAC. DC coil voltage relays are also available. The greatest benefit of this type of relay is that all the wiring stays in place with the socket if the relay requires replacement. It is relatively inexpensive and is often referred to as a throw-away relay.

2.1.1 Reed Relays and Switches

Reed switches are magnetically operated components with contacts hermetically sealed in a glass capsule. Positioning a permanent magnet next to the switch or placing the switch in or near a DC electromagnet causes the contact reeds to flex and touch, completing a circuit. Protective inert gas, or a vacuum within the capsule, keeps the contacts clean and protected for the life of the device. These very small, inexpensive devices provide millions of switching cycles in normal circuits.

Applications include alarms, relays, and proximity sensors. Any time an electric current must be turned on or off without mechanically touching the switch, the reed switch is a good choice. Reed switches are often referred to as proximity switches because the switch is not required to actually contact the magnetic device or to be a component of it in order to operate.

Figure 2 illustrates a reed switch with a single set of normally open contacts. Reed switches are available with combinations of normally open (NO) and normally closed (NC) sets of contacts.

Reed relays are similar in design to reed switches with the exception that they contain their own coil or electromagnet. Like all electromechanical relays, they require an electrical signal to activate the coil. Reed relays extend the contact terminals and the coil terminals to the outside of the sealed unit for connections.

Figure 3 shows a typical reed relay and a comparable sealed, electromechanical, microminiature relay with terminal schematics for each. An M.S. on a schematic denotes a magnetic shield, while an E.S.S. denotes an electrostatic shield. As with the reed switches, reed and microminiature relays are also available with combinations of normally open (NO) and normally closed (NC) sets of contacts. Reed relays are typically rated by the following factors:

- Minimum coil operating voltage
- Release voltage, which is the level of voltage at which the coil drops out and returns to its rest state
- Contact voltage and amperage ratings

Reed relay applications in the instrumentation industry include control of valves and solenoid operation, where contact arcing must be isolated for hazardous locations. Reed relays are also used in applications in which repetitive and rapid switching is required.

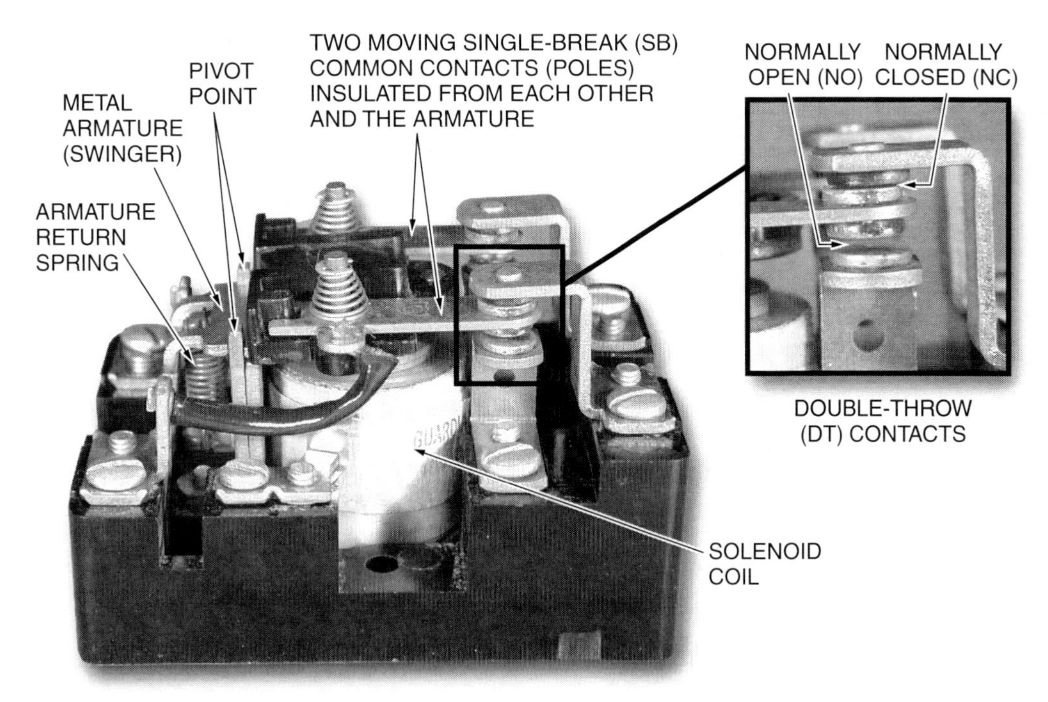

(A) DOUBLE-POLE (DP), DOUBLE-THROW (DT), OPEN-FRAME CONTROL RELAY

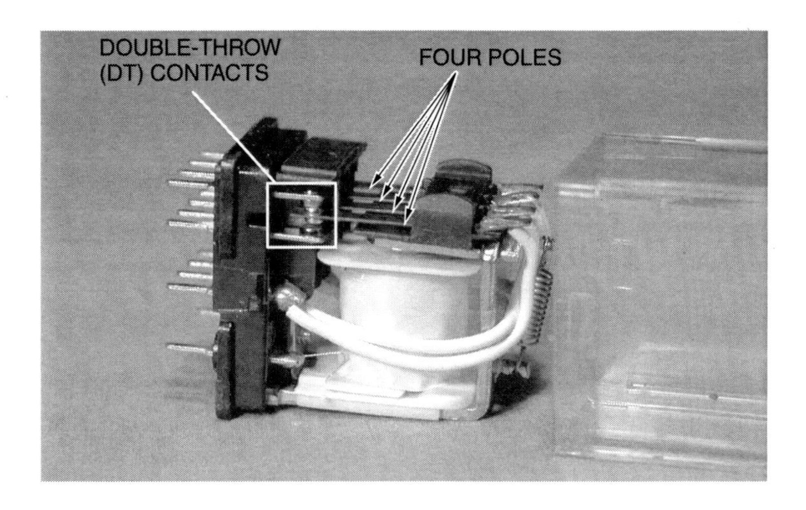

(B) FOUR-POLE (4P), DOUBLE-THROW (DT), GENERAL PURPOSE RELAY

208F01.EPS

Figure 1 ◆ Typical electromechanical relays.

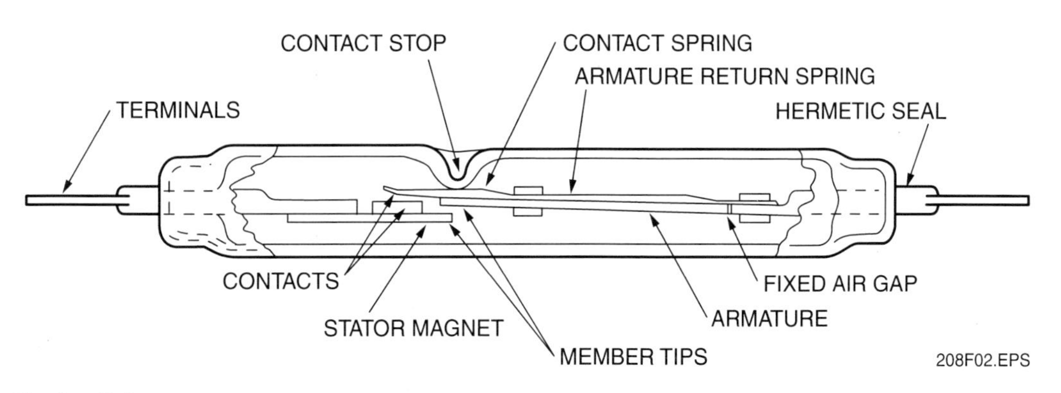

Figure 2 ◆ Reed switch.

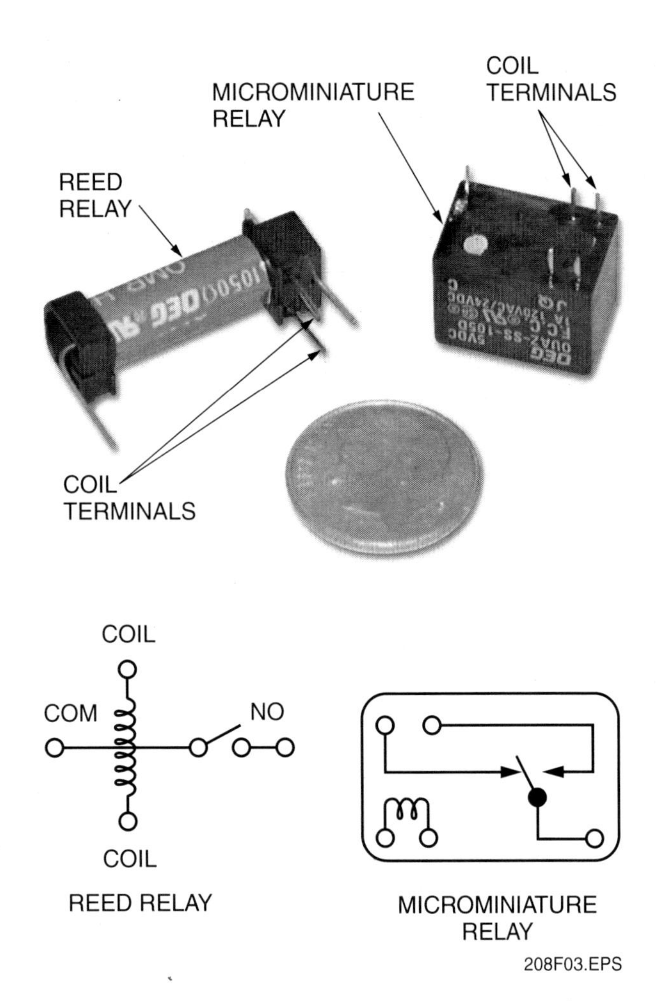

Figure 3 ◆ Reed relay and microminiature relay with terminal schematics.

2.1.2 General Purpose Relays

Figure 4 illustrates several different styles of general purpose relays. These relays are designed for commercial and industrial application where economy and fast replacement are high priorities. Most general purpose relays have a plug-in feature that makes for quick replacement and simple troubleshooting. General purpose relays can be **latching relays**, which means they remain actuated in the absence of an input signal. A reset signal or mechanism is used to reset the relay. These types of latching relays typically have two coils: one to latch and another to release. Both require a momentary signal.

Refer to *Figure 5*. A general purpose relay consists of an electromagnet or coil, contact springs, a contact **armature**, and the mounting. When the coil is energized, the flow of current through the coil creates a magnetic field, which pulls the armature downward to close or open a set of contacts.

The general purpose relay is available for both AC and DC operations. These relays are available with coils that can open or close the contacts by applying voltages from millivolts to several hundred volts. Relays with coil voltages of 6V, 12V, 24V, 48V, 115V, and 230V designs are the most common. Today, manufacturers offer a number of general purpose relays that require as little as 4 milliamperes at 5VDC or 22 milliamperes at 12VDC for use in controlling solid state circuits.

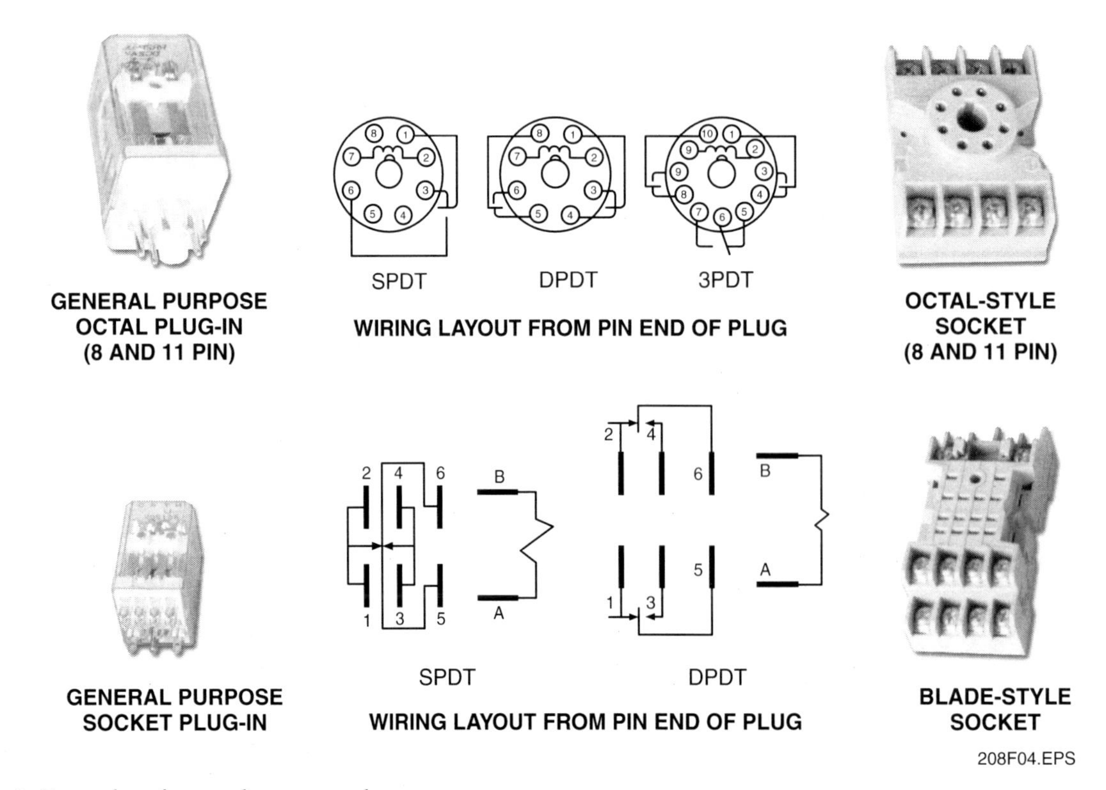

Figure 4 ◆ Examples of general purpose relays.

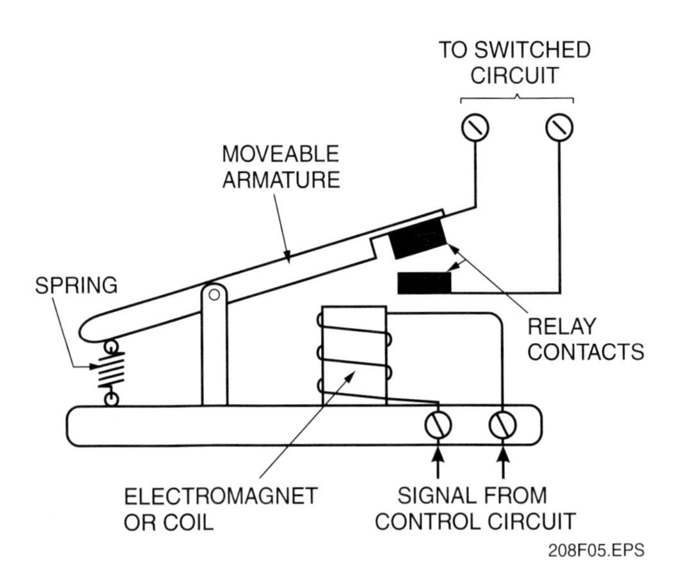

Figure 5 ◆ General purpose relay.

When the solenoid coil is specified for use with an AC control signal, a shading coil (aluminum or copper) at the top of the core (*Figure 6*) is used to create a weak, out-of-phase, auxiliary magnetic field. As the main field collapses when the AC current periodically drops to zero, the weak field generated by the shading coil is strong enough to keep the armature in contact with the core and prevent the relay from chattering at a 120Hz rate. If the shading coil is loose or missing, the relay will produce excessive noise and be subject to abnormal wear and coil heat buildup. Light-duty relays use a flexible, thin-gauge copper spring stock for the moving contact arm. The armature is designed to swing over a slightly greater distance than necessary in both directions. This overswing, known as armature over-travel, causes the flexible moving arm to bend slightly in the normally open and closed positions. This eliminates any contact-pressure mating or bounce problems. In control relays that use an inflexible moving contact arm (*Figure 6*), a contact-pressure spring is used to allow flexible positioning of the moving contact arm to eliminate contact-pressure mating and bounce problems. As in light-duty relays, this is accomplished through armature over-travel. A degree of contact cleaning is also accomplished by armature over-travel. As the contacts mate, the over-travel causes the contact surfaces to wipe slightly. this action removes oxides from the mating surfaces of the contacts and helps keep contact resistance low.

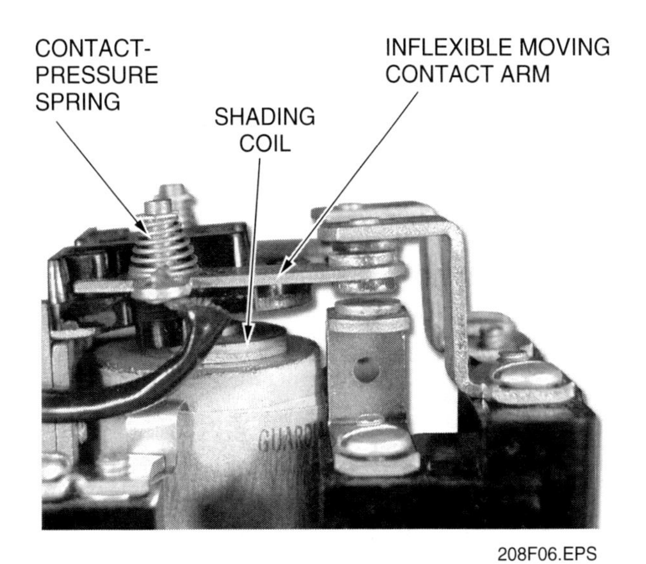

Figure 6 ◆ Typical control relay shading coil, contact-pressure spring, and inflexible moving contact arm.

Like all relays, general purpose relays are rated and selected according to their maximum voltage rating, contact current rating, and the coil voltage rating. The maximum voltage rating is the maximum voltage that the relay can handle safely. The insulation, contacts, and other design elements of the relay are rated according to a safe maximum voltage. This is the maximum voltage rating of the relay. The contact current rating represents the maximum current that the contacts may handle without severely arcing, burning, melting, or otherwise damaging the contact surfaces. The coil voltage represents the nominal voltage level needed to operate the coil properly. In latching relays that are constructed with two individual coils, the latching voltage represents the minimum momentary voltage required to operate the relay, while the release voltage is the minimum momentary voltage required to return the relay to its rest state. These are usually low DC voltage levels and are found in low-voltage electronic circuits.

Relays are available in a wide range of contact configurations. Terms typically associated with relay contacts include poles, throws, and breaks. *Figure 7* and *Table 1* help define these terms. Pole describes the number of completely isolated circuits that can pass through the switch at one time. Isolated means that maximum rated voltage of the same polarity may be applied to each pole of a given switch without danger of shorting between poles or contacts. A double-pole switch can carry current through two circuits simultaneously, with each circuit isolated from the other. With double-pole switches, the two circuits are mechanically connected so that they open or close at the same time, while still being electrically insulated from each other. This mechanical connection is represented in the symbol by a dashed line connecting the poles.

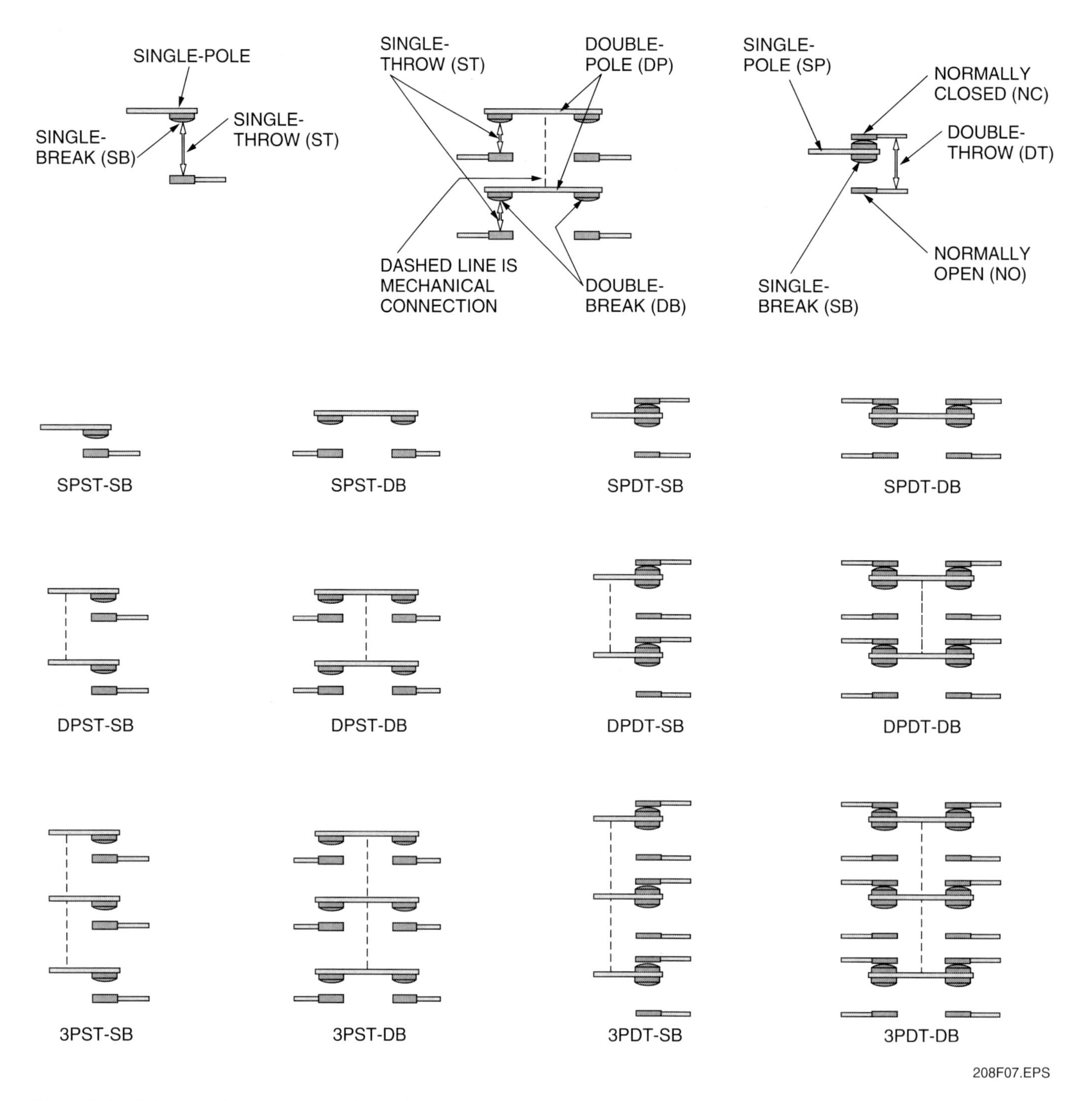

Figure 7 ◆ Various contact arrangements of relays.

Table 1 Relay Contact and Pole Designations

Designation	Meaning
ST	Single-Throw
DT	Double-Throw
NO	Normally Open
NC	Normally Closed
SB	Single-Break
DB	Double-Break (industrial relays)
SP	Single-Pole

Designation	Meaning
DP	Double-Throw
3P	Three-Pole
4P	Four-Pole
5P	Five-Pole
6P	Six-Pole
(N)P	N = numeric number of poles

Throws are the number of different closed contact positions per pole that are available on the switch. In other words, throw denotes the total number of different circuits that each individual pole is capable of controlling. The number of throws is independent of the number of poles. It is possible to have a single-throw switch with one, two, or more poles, as shown in *Figure 7*.

Breaks are the number of separate contacts the switch uses to open or close each individual circuit. If the switch breaks the electrical circuit in one place, then it is a single-break switch. If the switch breaks the electrical circuit in two places, then it is a double-break switch.

To simplify the listing of contact and switching arrangements, general purpose relays carry **NARM** (National Association of Relay Manufacturers) code numbers. The numerals are used as abbreviations of the switching arrangements and are listed in *Table 2*. *Figure 8* illustrates the basic contact forms that are available. These forms are also designated by NARM.

Table 2 Relay Switching Arrangement Code Number

Contact Code and NARM Designator	
1-SPST-NO	15-4PST-NO 16-4PST-NC
2-SPST-NC	17-4PDT
3-SPST-NO-DM	18-5PST-NO
4-SPST-NC-DB	19-5PST-NC
5-SPDT	20-5PDT
6-SPDT-DB	21-6PST-NO
7-DPST-NO	22-6PST-NC
8-DPST-NC	23-6PDT
9-DPST-NO-DB	24-7PST-NO
10-DPST-NC-DB	25-7PST-NC
11-DPDT	26-7PDT
12-3PST-NO	27-8PST-NO
13-3PST-NC	28-8PST-NC
14-3PDT	29-8PDT

2.1.3 Control Relays in Instrumentation

Control relays are used in many different areas of instrumentation, including process control and power monitoring. Older process control systems use programmable logic control (PLC) technology that receives input from field devices and transmits the output to relay banks. The control relays in these banks are used to control motor valves, solenoids and other electrical equipment. Newer PLC technology does not require control relays, as the output signals from the PLC are sent directly to the controlled devices. However, many of the older systems are still in place and functioning properly.

Power monitoring, which is considered an instrumentation function, is often accomplished with control relays. Current and voltage monitors often use control relays to de-energize power systems in the event of overcurrent conditions and over- or under-voltage conditions. Phase-failure control relays are designed to drop out when one or more of the phases on a three-phase power system fails. Under- and over-speed monitors use control relays to control power to conveyor and other systems that depend on constant or uniform speed control. Liquid level control relays are used to control the level in conductive liquids by passing a low voltage through the liquid from a suitable probe to an earth return, which can either be a conductive container or another probe. Special control relays can be used as alarm modules that accept several standard current or voltage instrument signals from the process. The modules have trip or alarm points that can be set directly on the control relay, allowing flexibility and multiple alarm functions from one device.

2.2.0 Solid-State Relays

Solid-state relays (SSRs) are becoming more popular because, in addition to their size advantage, they offer long operating life and no contact bounce. They offer higher reliability than EMRs because there are no moving parts. Current leakage and the inability to carry higher loads are some disadvantages associated with SSRs. The devices are also sensitive to over voltage spikes. SSRs are used in circuit designs that aren't sensitive to minor leakages. In addition to instrumentation applications, SSRs are used in automatic test equipment, data acquisition systems, mobile communications, and security systems, which require high reliability and long life.

The SSR business is growing at the expense of mechanical relays, particularly for applications that need millions of cycles of repetitive operation. SSRs are a good choice for customers seeking to control relays with less power. *Figure 9* shows an example of an SSR, but it only represents one type.

2.2.1 Comparison of Electromechanical Relays to Solid-State Relays

Both EMRs and SSRs are designed to provide a common switching function, but they accomplish the final results in very different ways. As previously noted, EMRs provide electrical switching by using electromagnetic energy to move a set or sets of contacts, creating or breaking the path for the electrical circuit or circuits passing through them.

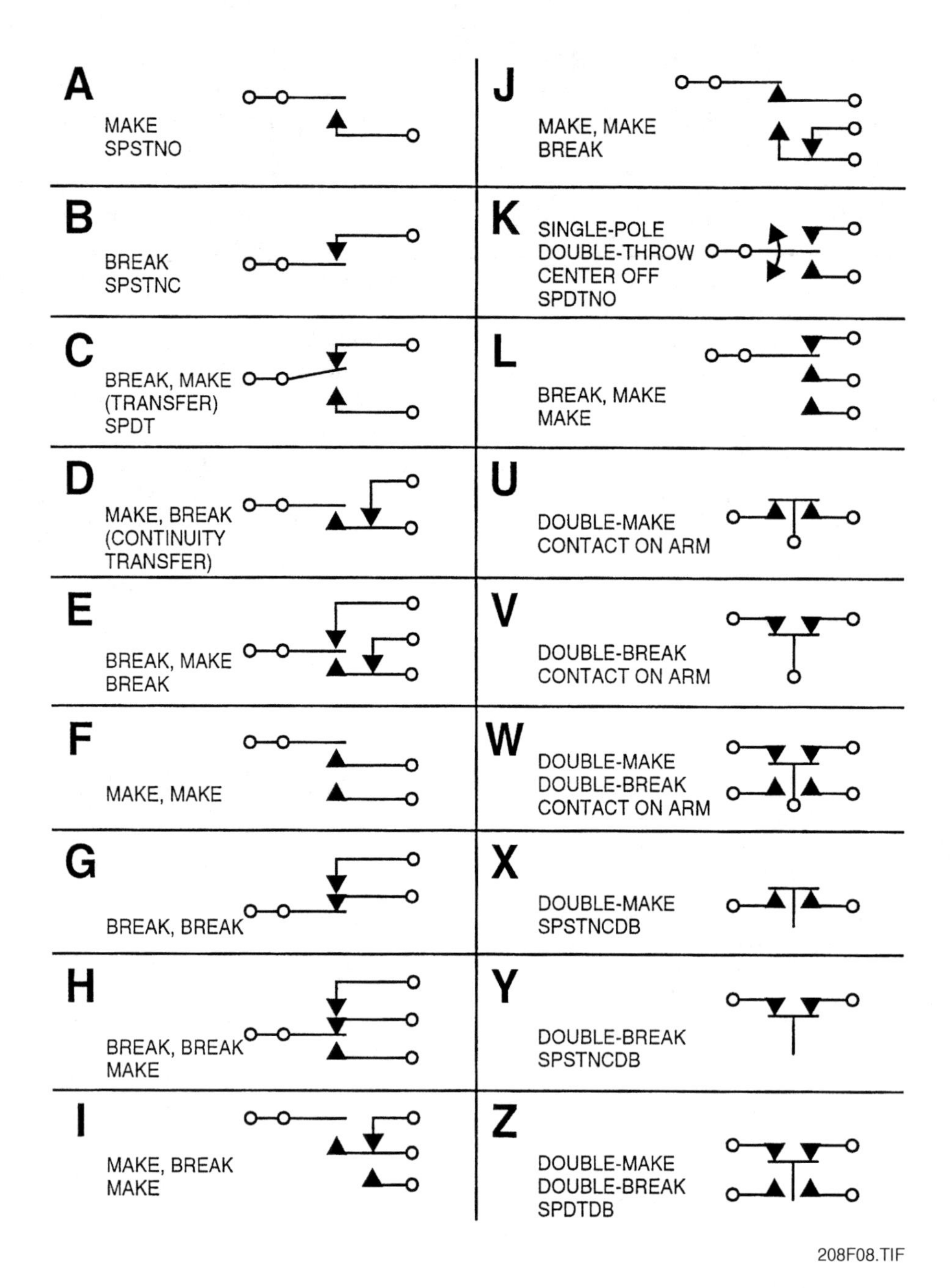

Figure 8 ◆ Basic contact forms accepted by NARM.

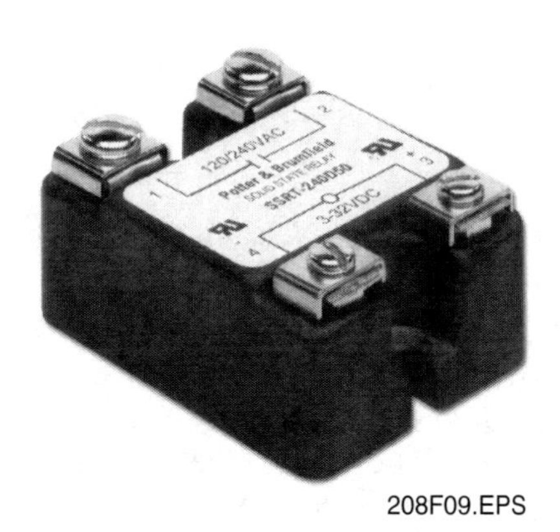

Figure 9 ◆ Typical solid-state relay.

The SSR, on the other hand, depends on electronic devices such as silicon-controlled rectifiers (**SCRs**) and triacs to create switching by triggering solid state, semiconductor devices to pass or block the flow of electrical current without using mechanical contacts. Simply put, the EMR provides or breaks a mechanical path for the flow of current, while the SSR allows or prevents current flow through it based on an electrical trigger or exciter that it receives. *Figure 10* shows a basic schematic for both an EMR and an SSR.

In *Figure 10A*, the electromagnet is controlled by the control signal applied to it. When this signal is at the appropriate level to cause the electromagnet to operate, the magnetic field created by

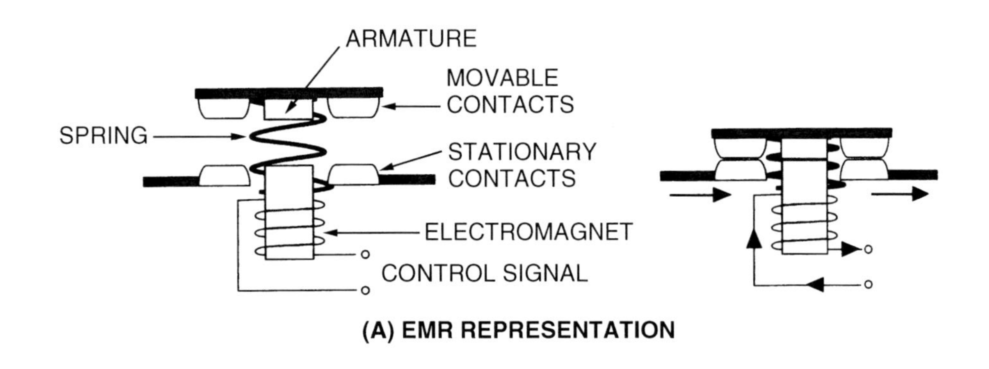

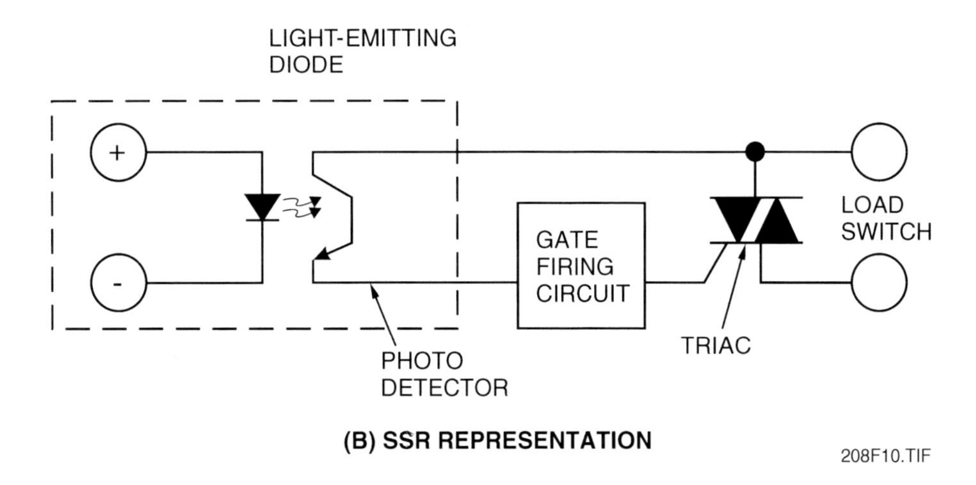

Figure 10 ◆ Comparison between an EMR and an SSR.

the electromagnet will attract the opposite pole of the armature, pulling it down until it rests against the core of the electromagnet. This movement causes the movable contacts to mate with the stationary contacts. In this example, two breaks are used to make the circuit. The function of the spring is to cause the contacts to separate whenever the signal is removed from the electromagnet and the magnetic field ceases to exist.

In *Figure 10B*, the control voltage (signal) is connected to the positive (+) and negative (–) terminals. The **LED** (light-emitting diode) accepts the relay control voltage and converts this energy to light energy. This light is collected by a photo detector that controls the gate firing circuit. When the control voltage or signal specified for this relay is reached, the gate circuit fires and in turn triggers the triac, allowing current to flow through the triac. Removal or reduction of control voltage reduces light output by the LED and stops the triggering of the gate firing circuit. This particular SSR requires a DC signal or control voltage that usually falls in the range of 3 to 32 volts DC.

Some disadvantages of using SSRs instead of EMRs were discussed earlier. They include the tendency of an SSR to leak current, a limited ability to carry heavy loads, sensitivity to over-voltage spikes, and higher cost compared to electromagnetic relays. However, the advantages of using SSRs instead of EMRs exceed the disadvantages:

- *Improved reliability* – The usual failure modes of conventional relays are contacts welding together or fatigue failure of the contact supports or springs. Because the SSR has no moving parts, these types of failures are eliminated.
- *Noiseless operation* – With no moving parts, operation is noiseless.
- *Voltage drop constant with time* – The voltage drop across the closed contacts of a conventional relay will vary depending on contact force, corrosion, or number of cycles. The SSR does not suffer from these problems.
- *Low input requirements* – The SSR requires much less power to turn on than a conventional relay of similar current-carrying capability.
- *Self-protection* – The thermal shutdown designed into the SSR provides internal protection against overload. Short-circuited loads may result in contact welding in a conventional relay, while the SSR simply turns itself off.

- *Sealed housing and weatherproof connector* – This feature allows the SSR to be installed in locations that would be detrimental to ordinary relays.

3.0.0 PNEUMATIC RELAYS, REPEATERS, AND BOOSTERS

Pneumatic relays, **repeaters** (transmitters) and boosters were first developed as a means of transmitting process signals located in hazardous locations to remote locations, such as control rooms, without the extreme costs associated with explosion-proof enclosures. Pneumatic relays and repeaters also eliminate the hazards of directly connecting volatile or potentially dangerous processes to the control room. However, pneumatic transmission requires a myriad of accessory items in order to assemble and maintain a properly operating system. Items like amplifying, reducing, boosting, and computing relays, repeaters, and a number of primary measuring elements are generally required. This section examines some of these pneumatic devices.

3.1.0 Force-Balance Transmitter Relays

The force-balance principle is most commonly used in pneumatic transmitter relays. A simplified 1:1 force-balance transmitter relay is shown in *Figure 11*. Process pressure acts downward on the flexible diaphragm. The resulting force is counterbalanced by the force of nozzle back pressure acting upward on the diaphragm. Air, at a supply pressure slightly higher than the maximum process pressure to be measured, flows through the restriction to the underside of the diaphragm and bleeds through the nozzle to atmosphere. At **equilibrium**, the nozzle seat clearance is such that the flow of air in the nozzle is equal to the continuous flow through the restriction. If the process pressure increases, the diaphragm **baffles off the nozzle**, causing the back pressure to increase until a new equilibrium is achieved. If the process pressure drops, the diaphragm moves away from the nozzle, causing the back pressure to drop until equilibrium is reestablished. Because nozzle back pressure is directly related to process pressure, the signal can be used remotely for indication, recording, or control.

Though such a transmitter is accurate, it has some limitations. Because all the air must flow through the restriction, the speed of response is slow, particularly if there is much volume in the output side of the circuit. Also, a leak in the output side would cause an error or even make the unit completely inoperative. *Figure 12* shows essentially the same repeater, but with refinements. This system employs a volume booster, which has considerable air-handling capacity, speeding up the response, minimizing transmission lag, and coping with leakage. In addition, it provides a constant pressure drop across the nozzle, regardless of the level of operation. This improves accuracy because the total diaphragm travel is lessened. The relationship of nozzle seat position to nozzle back pressure is more linear than that of the direct nozzle circuit in *Figure 11*.

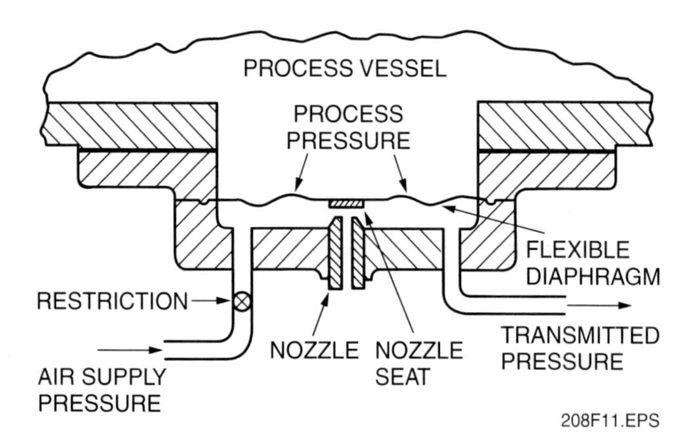

Figure 11 ◆ Simplified 1:1 force-balance pressure transmitter relay.

The booster contains an exhaust diaphragm assembly consisting of two diaphragms that move integrally and provide a bleed to atmosphere. On the underside of the diaphragm is a differential spring, which at balance exerts a force equivalent to 3 psi acting on the diaphragm. Therefore, the nozzle back pressure that acts above the exhaust diaphragm is always nominally 3 psi higher than the transmitted pressure. Because the nozzle bleeds into the transmitted pressure, the pressure drop across the nozzle is always constant regardless of output pressure. As in *Figure 11*, the air pressure on the underside of the transmitter diaphragm counterbalances the process pressure. If the process pressure increases, the diaphragm moves the seat closer to the nozzle, increasing the nozzle back pressure. This moves the exhaust diaphragm downward, closing off the exhaust port as it contacts the pilot valve and moving the pilot valve downward to open the supply port. The result is an increase in transmitted pressure until the transmitted pressure fed back to the underside of the diaphragm equals the process pressure. A decrease in process pressure causes the nozzle back pressure to drop, and the exhaust diaphragm moves upward. The pilot valve closes off the supply port while the diaphragm opens the exhaust port, causing the transmitted pressure to

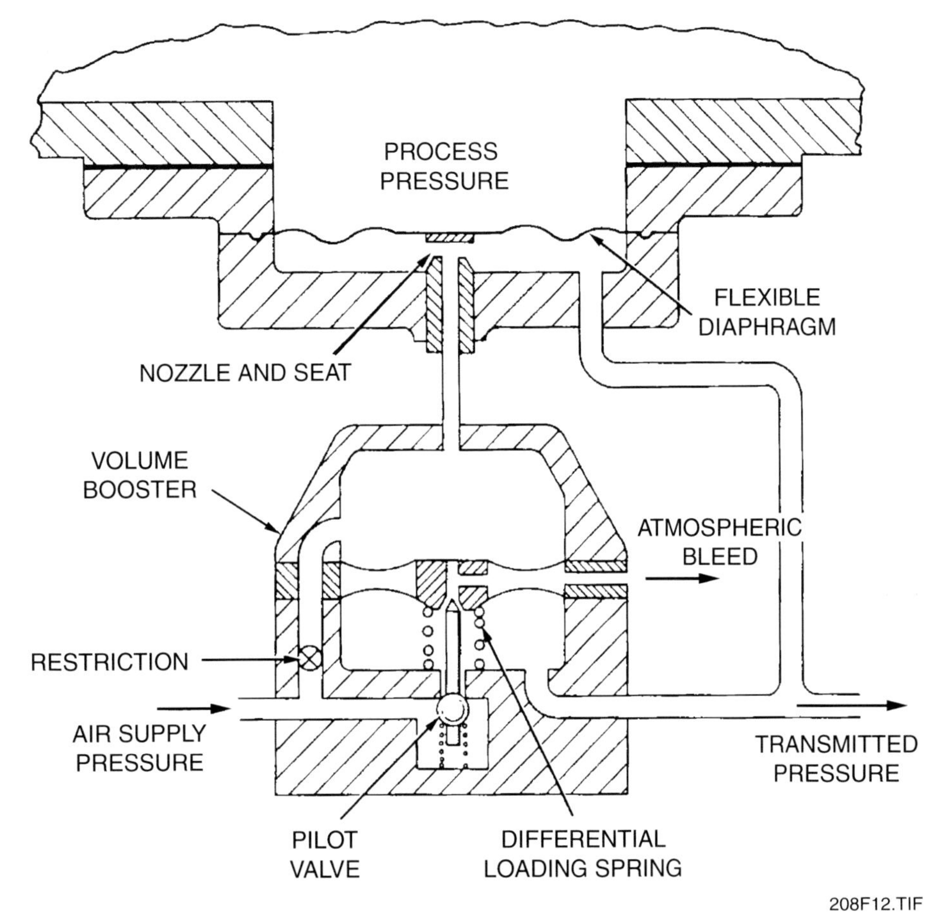

Figure 12 ◆ 1:1 force-balance pressure transmitter with volume booster.

drop until equilibrium is established. At balance, there is a continuous flow of air passing through the restriction, detection nozzle, and out to atmosphere via the exhaust diaphragm.

3.2.0 Computing Relays

Computing relays perform mathematical functions using pneumatic input signals as **operands**. The output signal is the result. Adding, subtracting, multiplying, and dividing are the most common functions of computing relays.

3.2.1 Pneumatic Multiplying and Dividing Relays

In the force-bridge multiplier-divider shown in *Figure 13*, input pressures act on bellows in Chambers A, B, and D. The output is a feedback pressure in Chamber C. The bridge consists of two weight beams, which pivot on a common movable fulcrum. Each beam operates a separate feedback loop. Any unbalance in moments on the left-hand beam causes a movement of the fulcrum position until a moment balance is restored. An unbalance in moments on the right-hand beam results in a change in output pressure until balance is restored. Equations that characterize the operation of the force-bridge are:

$$A \times a = B \times b \text{ and } D \times a = C \times b$$

The equations reduce to:

$$A \times C = B \times D$$

or

$$C = \frac{B \times D}{A}$$

Multiplication results when the two input variables are connected to Chambers B and D. Division results when the dividend is connected to either Chamber B or D, with the divisor connected to A. Simultaneous multiplication and division results when B, D, and A chambers are used.

3.2.2 Pneumatic Adding, Subtracting, and Inverting Relays

In the force-balance, arithmetic-computing relay in *Figure 14*, a signal pressure in chamber A acts downward in a diaphragm with **unit effective area**. A signal in chamber B also acts downward on

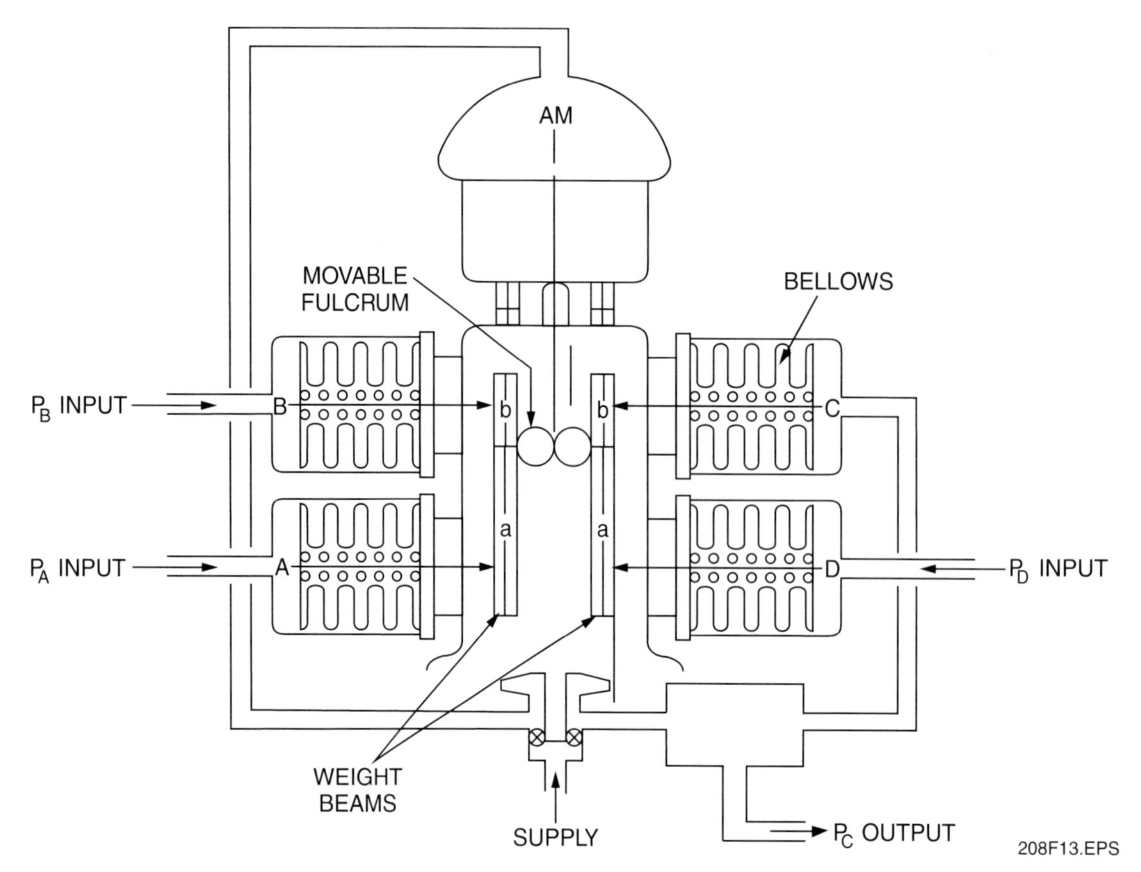

Figure 13 ◆ Pneumatic force-bridge multiplying and dividing relay.

an **annular** diaphragm configuration, which also has an effective area of unity. Signal pressures in chambers C and D similarly act upward on unit effective diaphragm areas. Any unbalance in forces moves the diaphragm assembly with its integral nozzle seat. The change in nozzle seat clearance changes the nozzle back pressure and also changes the output pressure, which is fed back into chamber D until force-balance is restored. The basic equation that describes the operation of the relay is:

$$T = A + B - C + K$$

K is the spring **constant**. It is adjustable to give an equivalent bias of +18 psi.

3.2.3 Pneumatic Scaling and Proportioning Relays

Scaling, or **proportioning**, involves multiplication by a constant. Several approaches are available: special fixed-ratio relays, pressure transmitters, proportional controllers, and adjustable ratio relays.

The fixed-ratio scaler is the simplest if the correct ratio is available and if adjustability and exact ratio are unnecessary. *Figure 15* shows such a relay. The input pressure is connected to the top chamber and acts on the upper diaphragm. The output acts upward on the small bottom diaphragm. The **gain** (increase) is a function of the relative effective areas of the large and small diaphragm as determined by the dimensions of the diaphragm ring. The bottom spring applies a negative bias to the input, and the adjustable top spring allows exact zero setting. The operating equation is:

$$T = AP1 + K$$

Where A is the gain constant and K is the spring setting.

Where the scaling must be exact and does not have to be adjusted periodically, pressure transmitters are an economical, reliable, and accurate choice. Where the scaling factor must be modified occasionally, conventional ratio relays, which often consist of the proportioning section of a controller, are commonly used.

3.2.4 High- and Low-Pressure Selector and High-Pressure Limiter Relays

Pneumatic selector and limiter relays work on the principle of satisfying a pressure setting that is either manually set on the relay or, in the case of limiters, is preset in the manufacturing of the relay.

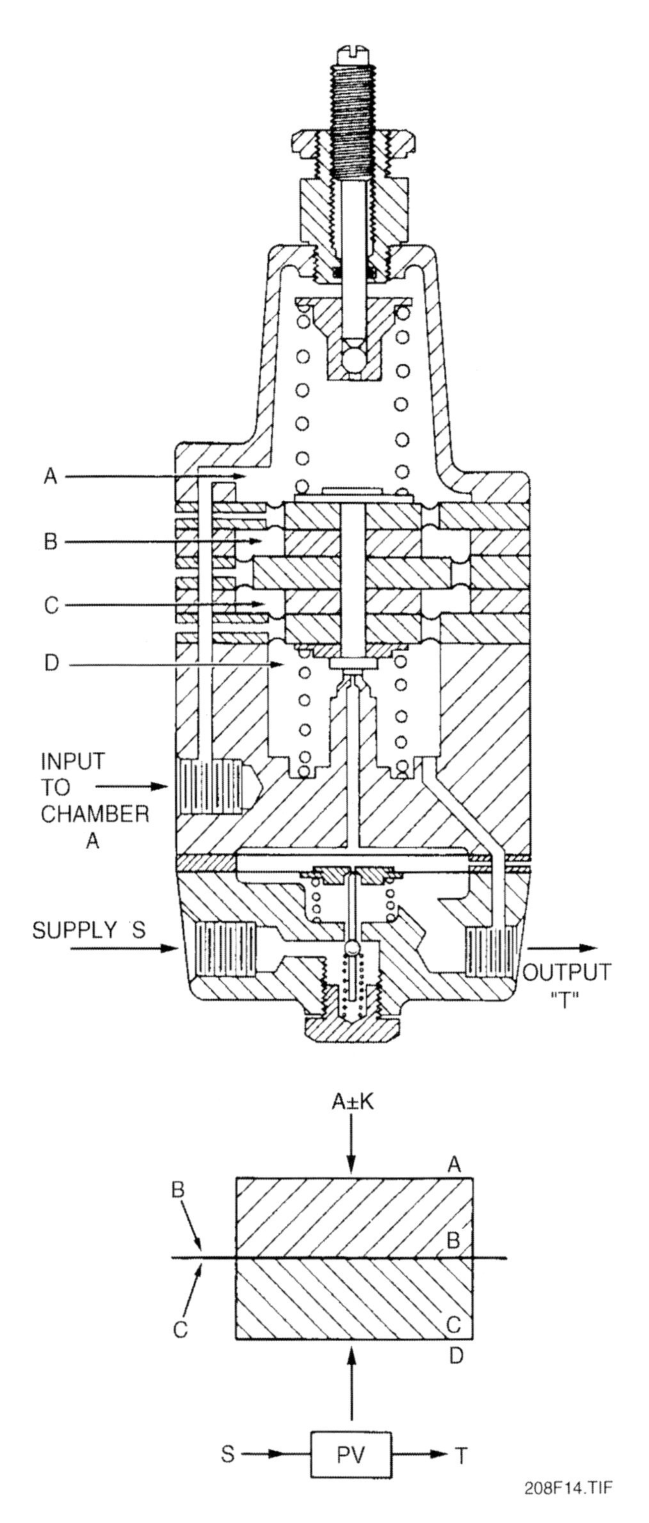

Figure 14 ◆ Pneumatic adding, subtracting, and inverting relay.

A pneumatic selector generally has more than one input and selects the highest input pressure as its output signal. A high-pressure pneumatic limiter, on the other hand, is typically used to transmit an input pressure as an output at whatever pressure is received until it reaches the high setting of the relay setpoint. At that time, the relay will only transmit that setpoint level of pressure. It limits the output based on the input's comparative level relative to the high setting. Examples and more in-depth discussion follow regarding the pneumatic selector and limiter relays.

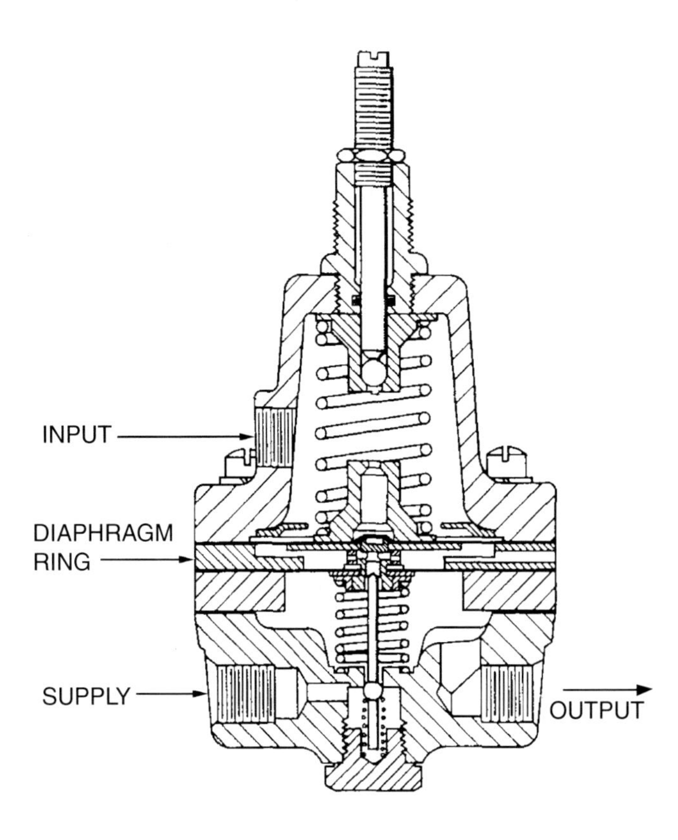

Figure 15 ◆ Pneumatic fixed ratio amplifying relay.

Selector – The high-pressure selector relay compares two pressures and transmits the higher of the two in its full value. In *Figure 16*, the two input pressures act against a free-floating flapper disc. The differential pressure across the flapper always results in closure of the low-pressure port.

In the low-pressure selector in *Figure 17*, if input A is less than input B, the diaphragm assembly throttles the pilot plunger to make the output equal to input A, the conventional action of a 1:1 booster relay. If input B is less than A, the supply seat of the pilot plunger is wide open so that pressure B is transmitted in its full value.

Limiter – A high-pressure limit relay, like the example shown in *Figure 18*, maintains an output pressure equal to the input pressure except when the input pressure increases above the high limit setting. At that time, the output pressure will be equal to the high limit setting. The high limit relay consists of a pressure regulator and a volume booster with a ratio of 1:1. The pressure regulator provides the high limit setting pressure to the 1:1 volume booster. The booster serves as a low-signal selector to choose either the input pressure or the reproduced high limit setting pressure as the high-limit relay output. In this sense it is functioning as a pneumatic selector switch.

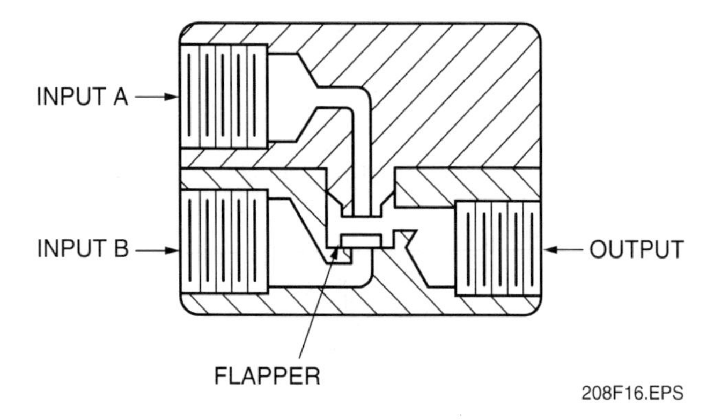

Figure 16 ◆ High-pressure selector relay.

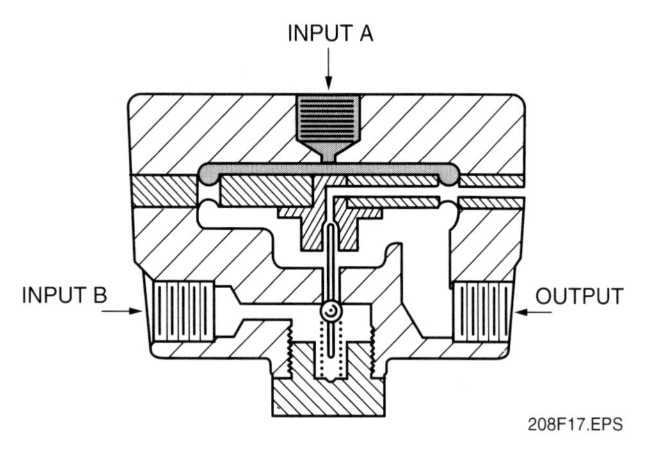

Figure 17 ◆ Low-pressure selector relay.

The high limit setting pressure is applied to the top of the exhaust diaphragm and forces the valve plunger open. This allows the input pressure to pass to the output and to the bottom of the exhaust diaphragm. As long as the input pressure to the bottom of the exhaust diaphragm is less than the high limit setting pressure on top of the exhaust diaphragm, the relay output will be equal to its input. If the input pressure exceeds the high limit setting pressure, the exhaust diaphragm will move upward, the valve plunger will close, and the input pressure will exhaust until it equals the high limit setting pressure. Thus, the high limit relay maintains an output pressure equal to the input pressure except when the input pressure increases above the high limit setting; at which time the output pressure will be equal to the high limit setting.

3.2.5 Booster Relays

Booster relays are essentially air-loaded, self-contained pressure regulators. They can be classified in three broad groups:

- Volume boosters used to multiply the available volume of the air signal
- Ratio relays used to multiply or divide the pressure of an input signal

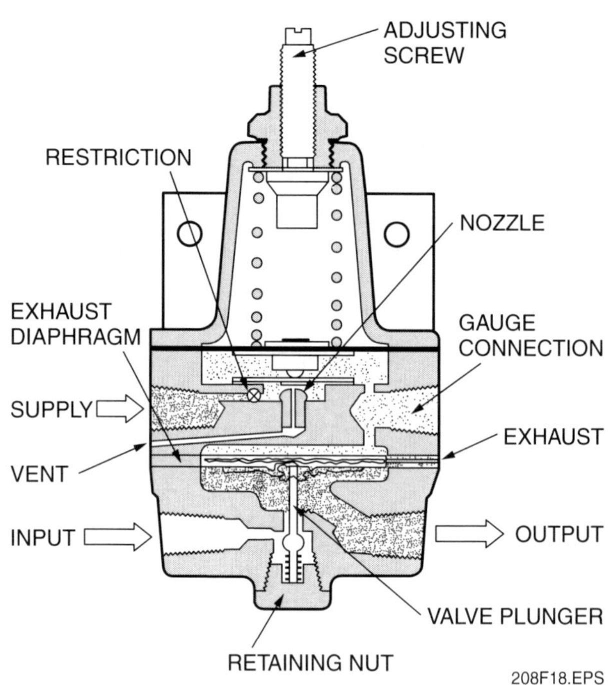

Figure 18 ◆ High-pressure limit relay.

- Reversing relays that produce a decreasing output signal for an increasing input signal

One of the main characteristics of boosters is their high exhaust capacity. That is, the bleed rate matches closely the maximum output capacity. *Figure 19* shows a cross-section of a typical volume booster relay. A sensing tube connecting the outlet to the inner diaphragm chamber provides an aspirating effect under high flow conditions to provide additional valve lift. Under equilibrium conditions, the signal air acting on the top diaphragm is balanced by the outlet pressure pushing the lower diaphragm. Any imbalance of these forces causes a change in position of the valve plug.

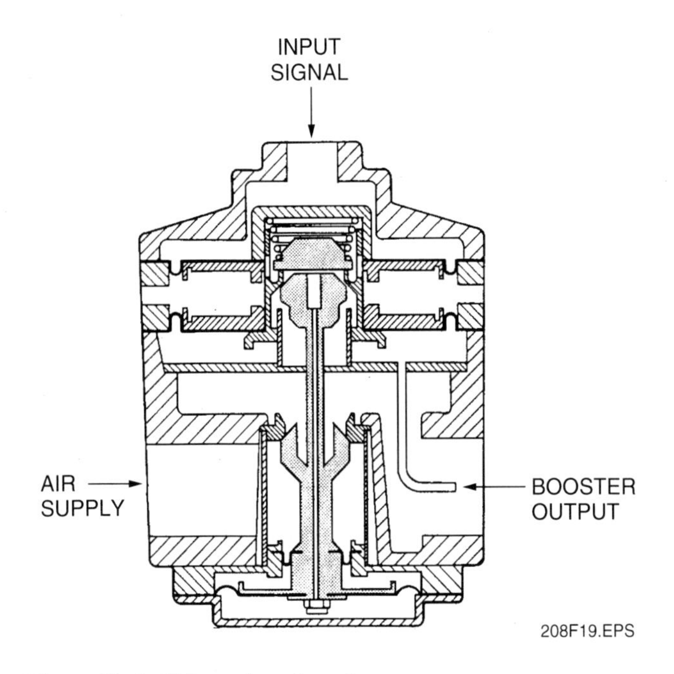

Figure 19 ◆ Volume booster relay.

In the case of ratio relays, there is a difference between the effective areas of the signal and feedback diaphragm. Typical ratios are: 1:2, 1:3, 2:1, 3:1, and 5:1, where the first number refers to the signal and the second to the output pressure. These ratios can be maintained within approximately one percent. Booster relays are usually made of die cast aluminum to withstand up to 250 psig supply pressure. They are generally provided with ¼ inch pipe tap connections except on high-capacity models, which may be ½ inch or ¾ inch. They are highly accurate devices and respond to signal variations as low as 0.01 psi. Applications include the use of volume relays to increase the frequency response of a control valve. This is sometimes preferable to the use of positioners on fast control loops.

Reversing relays are employed when two control valves, one air-to-open and the other air-to-close, are operated from the same controller. They might also be used to reverse part of the output pressure from a single-acting positioner to a double-acting piston actuator.

Ratio relays can provide for split-ranging. For example, a 1:2 ratio relay could change a 3 to 9 controller signal to a 3 to 15 psig output signal.

4.0.0 ◆ TIMERS AND TIME CLOCKS

Many industrial control applications require timing relays that can provide dependable service and are easily adjustable over the timing ranges. The proper selection of timing relays for a particular application can be made after a study of the service requirements and with a knowledge of the operating characteristics of each available device.

Basically, there are three categories of timers: **dashpot timers**, synchronous clock timers, and solid-state electronic timers. Although each device works in a different way, all timers have the ability to introduce some degree of time delay into a control circuit.

4.1.0 Dashpot Timer Relays

Magnetically operated, oil dashpot timing relays (*Figure 20*) may be used on voltages up to 600 volts AC or DC. The contacts are operated by the movement of an iron core. The magnetic field of a solenoid coil lifts the iron core against the retarding force of a piston moving in an oil-filled dashpot. This type of relay is not very accurate. The piston must be allowed to settle back down to the bottom of the dashpot between successive timing periods. If the piston is not allowed to make a full return, the timing is erratic. The dashpot timing relay provides time delay after the magnet is energized. The contact may be normally open or normally closed for different applications.

Fluid dashpot timing relays are used for a number of applications:

- To control the accelerating contactors of motor starters
- To time the closing or opening of valves on refrigeration equipment
- For any application where the operating sequence requires a delay

Multi-contact dashpot timer relays are used for DC motor starting. When the coil is energized on this type of timing relay, the contacts close in succession with a time lag between each closing.

4.1.1 Pneumatic Timers

One type of dashpot timer is a pneumatic timer. The construction and performance features of the pneumatic timer make it suitable for the majority of industrial applications. Pneumatic timers have the following characteristics:

- They are unaffected by normal variations in ambient temperature or atmospheric pressure.
- They are adjustable over a wide range of timing periods.
- They have good repeat accuracy.
- They are available with a variety of contact and timing arrangements.

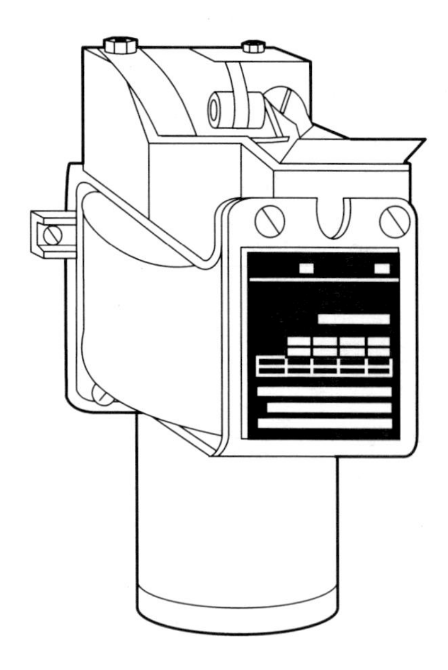

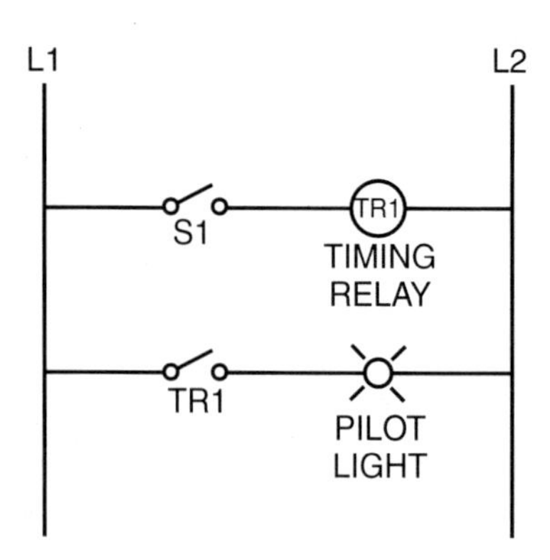

When S1 is closed, the timing relay is energized. After the timing cycle is complete, switch TR1 closes and the pilot light is energized.

208F20.EPS

Figure 20 ◆ Fluid dashpot timing relay.

This type of relay has a pneumatic time-delay unit that is mechanically operated by a magnet structure. The time-delay function depends on the transfer of air through a restricted orifice by use of a reinforced synthetic rubber bellows or diaphragm. The timing range is adjusted by positioning a needle valve to vary the amount of orifice or vent restriction.

The process of energizing or de-energizing pneumatic timing relays can be controlled by pilot devices such as pushbuttons, limit switches, or thermostatic relays. Because the power drawn by a timing relay coil is small, sensitive control devices may be used to control the operating sequence.

Pneumatic timing relays are used for motor acceleration and in automatic control circuits. Automatic control is necessary in applications where repetitive accuracy is required, such as controls for machine tools, sequence operations, industrial process operation, and conveyor lines. To aid in selecting and installing timers properly, manufacturers provide catalog information and specification sheets to explain their product.

A typical specification sheet for a pneumatic timer used in industrial applications contains certain information. The next few paragraphs examine examples of these specifications to determine how they should be used in practical situations.

Special features – When reading specifications, look for the special features of a device. In certain cases, the special features will make the device more usable and attractive. In other cases, the special features may drive up the price of the device and may be unnecessary.

A given specification sheet might describe features like these:

- Two timed contact cartridges that are individually convertible from normally open to normally closed simply by removing the cartridge, turning it over, and remounting the unit. The advantage of this conversion is that it permits changing contact arrangements without changing the timer.
- The ability to convert from on-delay to off-delay on the job site. This interchangeability is one major advantage of pneumatic timers and synchronous motor timers over solid-state timers. Solid-state timers generally are either on-delay or off-delay and are not usually convertible. There are, however, solid-state timers that are convertible and adjustable.
- A push-to-test operator for manual operation of the timing contacts, allowing the unit to be time adjusted without energizing the relay in a line circuit.

A specification sheet might also list the following information:

- *Adjustment* – Each timer must have some type of adjustment to establish its timing range. In a given unit, adjustment might be made by a front-mounted thumbwheel screw. As the timing range is increased or decreased, a timing indicator such as a thumbwheel will move to establish a reference point in relation to a graduated scale. This scale indicates the percentage of time delay.

- *General specifications* – Manufacturers' timer specifications will generally state the time range for the device. For example, a timer could have a range from 0.2 seconds to 60 seconds with a reset time of 0.075 seconds and an accuracy of plus or minus 15% of the setting. What this means is that the timer can provide a time delay as short as 0.2 seconds or as long as 60 seconds but may vary as much as 15% from this value on a repeat basis. The device will take 0.075 seconds to reset.

The specifications for a timer could indicate that input coil voltages are available up to 300 volts at either 50 or 60Hz. These voltages make such a timer more adaptable to a variety of industrial circuits in the United States and abroad.

The operating temperature range for a timer could be substantial, provided there is no freezing moisture or water.

The current rating of the contacts for a typical timer depends on the voltage applied to the contacts. For 120V, for example, the contact make rating could be 60 amperes, with a break rating of 6 amperes and a continuous current of 10 amperes. The difference is that any time a contact breaks a circuit, the arc formed can have a destructive burning effect on the contacts. Because the arc is not as prevalent when the contacts make, they can handle much more current on making than on breaking. In some circuits with special loads, this must be taken into consideration. However, in most cases the continuous current rating is used. The DC rating on a typical timer's contacts is much lower than the same voltage rating of AC. This is because DC is continuous and delivers more power to the contacts in a given moment than an equivalent AC.

4.2.0 Synchronous Time Switches

A time switch, also called a time clock, is an electrically operated switch used to control the amount of time and the time of day that power is switched on and off to a piece of equipment. The term synchronous is used to describe such time switches because AC synchronous motors are used for the clock motors. A typical synchronous time switch is shown in *Figure 21*.

To help understand how these controls operate, the technician must first become acquainted with the interior components of a time switch. Descriptions of these components follow.

The clock gear train (*Figure 22*) is heavy and rugged enough to withstand prolonged normal usage. It is designed and built to function properly even under adverse conditions.

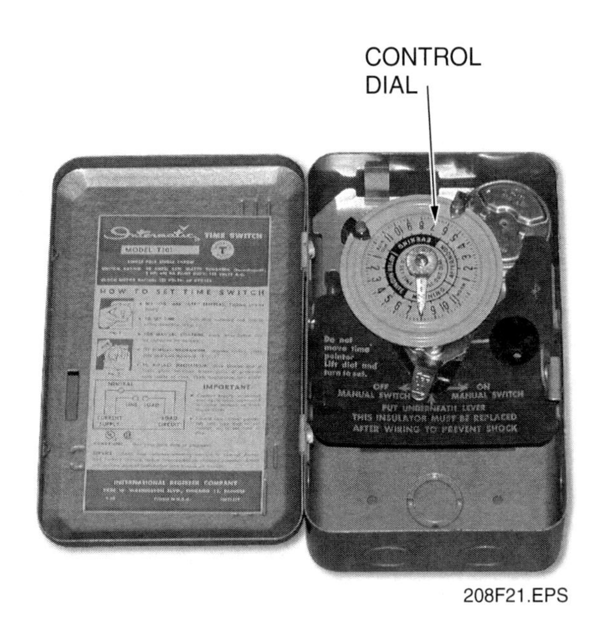

Figure 21 ◆ Synchronous time switch.

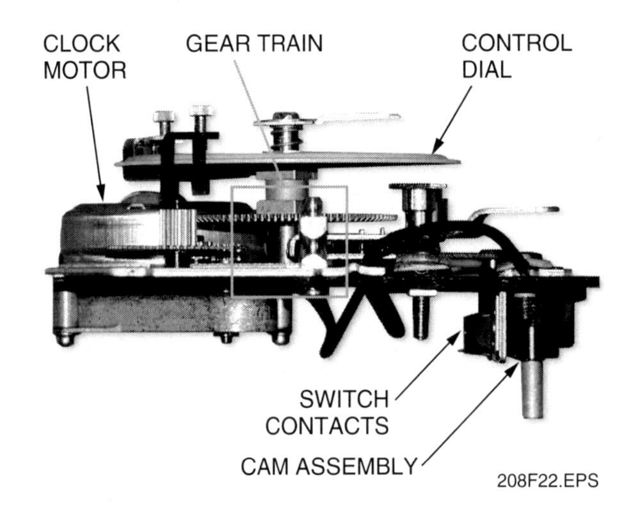

Figure 22 ◆ Clock gear train, motor, and switch mechanism.

The switch mechanism (*Figures 22* and *23*) is made of channeled spring brass U-beam blades operated by a rugged cam to give instant and positive make and break action. The contacts are usually self-cleaning and made of a special alloy to prevent pitting. They are rated to carry an inrush current ten times their normal amperage rating without arcing or sticking. They are supplied as single- or double-pole switches.

Time switches are usually powered by heavy-duty industrial-type motors. These motors are a synchronous type and are self-lubricating and extremely quiet. They never need service or attention and are practically immune to adverse temperature and humidity conditions.

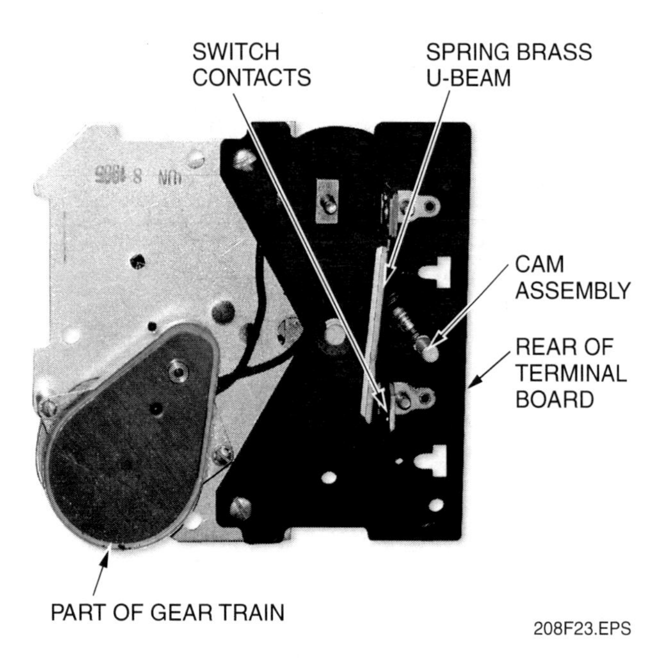

Figure 23 ◆ Single-pole switch mechanism.

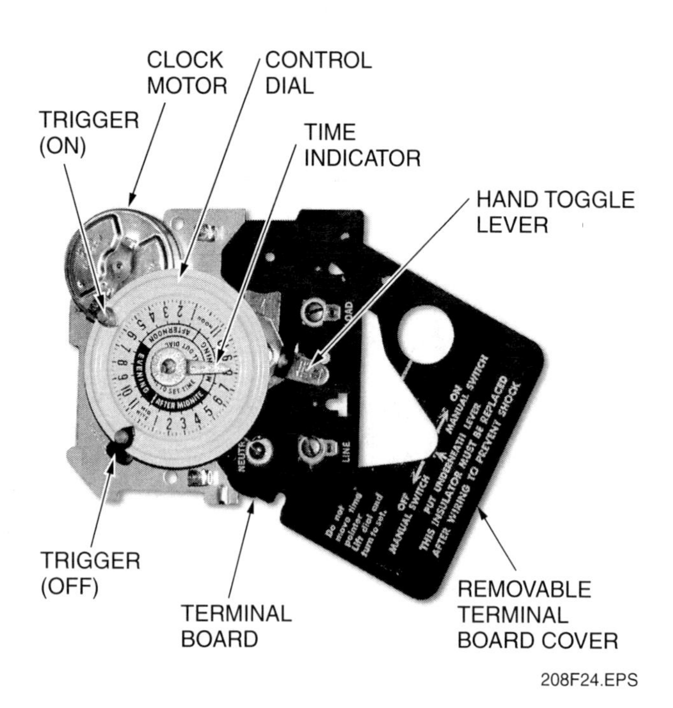

Figure 24 ◆ Terminal board and control dial.

The terminal board (*Figure 24*) allows easy connection of equipment wiring. Note the connections for the line and load wires. On most two-pole time switches, the line terminals of both poles are located on the left side of the enclosure and the load terminals are located on the right side of the enclosure.

The control dial is divided accurately into units of time. Most are painted with contrasting colors to facilitate the placement of trippers. The on and off trippers are adjustable to provide the desired on and off periods. These dials provide 24-hour coverage. The trippers actuate a toggle mechanism that turns the contact cam slightly to open and close the contact(s). The trippers are secured in position on the rotating dial by thumbscrews. The dial is usually set at the correct time by pulling the spring-loaded dial toward the front of the control and rotating until the correct time corresponds to the time indicator. Consult the manufacturer's recommendations before attempting to set the dial.

The toggle lever allows operation of the switch without disturbing the dial settings. The toggle lever can be moved to the on position to determine if the system is functioning properly.

In operation, the time switch motor is wired in parallel with the switching contacts across two phases or from a neutral terminal to a line terminal. The toggle lever also moves from OFF to ON and back when the toggle mechanism is actuated by the on or off trippers. As the motor rotates the dial and trippers, the contacts are opened or closed, depending on the desired operation. Timers provide automatic control over a great variety of functions and equipment. Some of the more common uses include the following:

- Chemical injection
- Water heaters
- Defrost control

4.3.0 Solid-State Timers

Solid-state timers derive their name from the fact that the time delay is provided by solid-state electronic devices enclosed within the timing device. Because solid-state timing devices are almost always replaced and never repaired, this module does not elaborate on how the device electronically provides time delay. Briefly, however, most solid-state devices use a resistance/capacitance (RC) network with a transistor and silicon-controlled rectifier (SCR) to provide the time delay and switching characteristics.

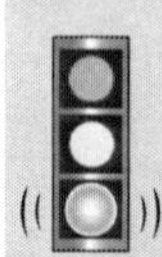

NOTE

Make sure that solid-state devices are connected properly before applying power. An improper connection could damage the solid-state device.

Figure 25 illustrates a typical solid-state timer with the manufacturer's specifications, wiring diagram (pin arrangement) and operation chart. This

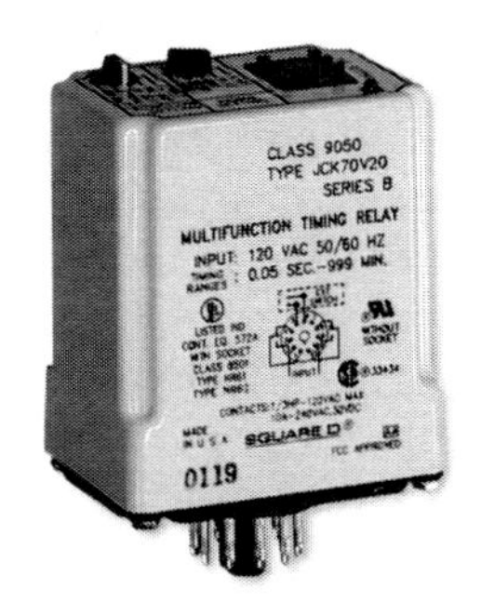

Economical Solid-State DPDT operate delay relay for new or replacement applications. For use in automatic control circuits, machine tool programming, sequencing controls, heating & cooling operations, warm-up delays, etc. A complete range of delay times are available as shown below. Models with fixed delay and models with remote adjustment are also available.

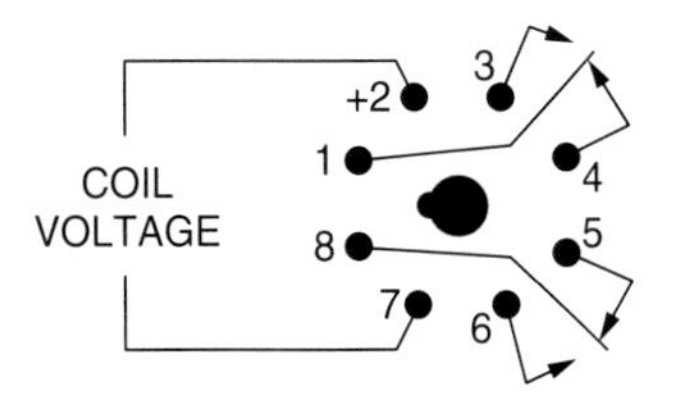

PIN ARRANGEMENT

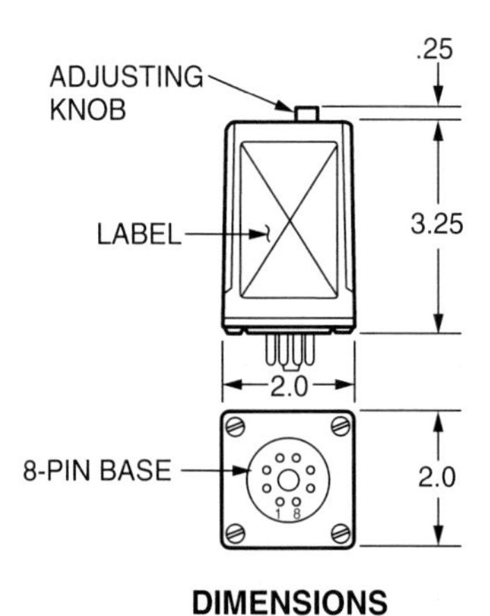

DIMENSIONS

SPECIFICATIONS

- 4 ADJUSTABLE TIMING RANGES
- 100ms RESET TIME
- TEMPERATURE RANGE = -20 TO +60° C
- TRANSIENT PROTECTION = 2,500 VOLTS
- FALSE TRIGGER PROTECTION
- ± 2% REPEATABILITY
- MAX. TIME ACCURACY = ± 5%
- CONTACTS = DPDT
  10 AMPS @ 115V
  6 AMPS @ 230V

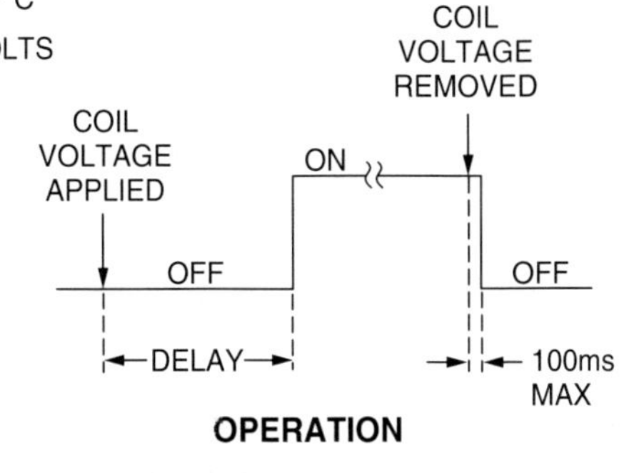

OPERATION

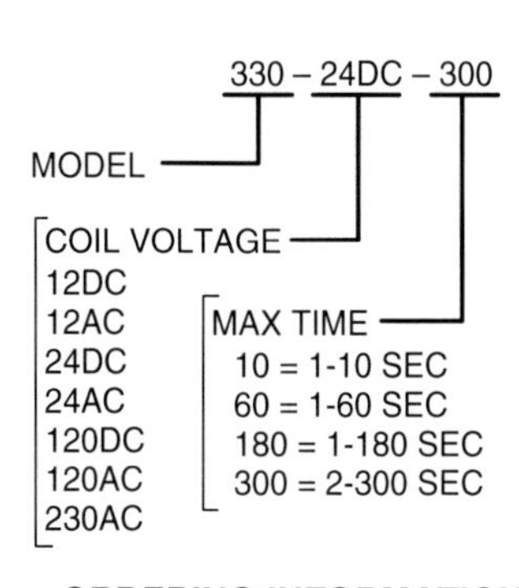

ORDERING INFORMATION

208F25.EPS

Figure 25 ◆ Typical solid-state timer specification sheet.

timer uses an eight-pin plug-in relay socket. This enables the timer to be replaced without disconnecting any wires once the socket has been wired. To read the pin numbers, find the identification notch (*Figure 26*) and count clockwise starting with one. Not all manufacturers mark the numbers on the pins, so remembering this is important.

Using the pin arrangement diagram, it is evident that pin 1 is common to both pins 3 and 4. Pins 1 and 3 are NO contacts, and pins 1 and 4 are NC contacts. This is called a single-pole, single-throw (SPST) switch. Likewise, pins 8 and 6 are NO, and pins 8 and 5 are NC. Pins 8, 5 and 6 also make up an SPST switch. Together, these two switches (pins 1, 3, 4, and 8, 5, 6) would make a double-pole, double-throw (DPDT) switch, as listed under the specifications for this timer.

Refer to the specifications in *Figure 25* and find the current rating for the contacts, 10 amps at 115VAC and 6 amps at 230VAC. This is the amount of current that each contact can safely handle. It determines the maximum size load the timer can turn on or off. It is important not to confuse the rating of the contacts with the coil voltage needed to run the timer. The coil voltage for this model can be anywhere from 12V to 230V and may be AC or DC according to the ordering information, but the con-

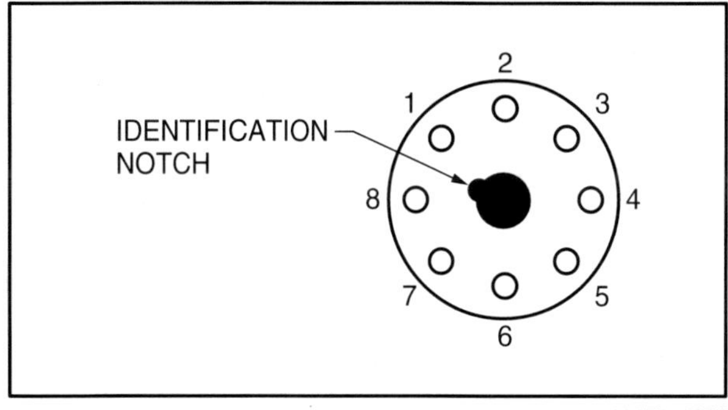

208F26.EPS

Figure 26 ◆ Octal socket pin numbers.

tact rating does not change. Thus it is possible that a coil voltage of 12VDC (pins 2 and 7) can control a 120VAC load (through pins 1 and 3).

To explain the logic of a timer, most manufacturers of solid-state timers provide an operating graph. The operating graph tells the technician that this model timer is an on-delay timer. To understand this, refer to the operating graph in *Figure 25*. The graph refers to the condition of the contacts, normal condition or switched, at reset, during timing, and after timing. On the graph, the *off* level is on the bottom and represents the contacts in their normal condition. The *on* level is on the top and represents the contacts in switched condition. For example, refer to pins 1 and 3 (NO) in *Figure 25*. When the coil voltage is applied (pins 2 and 7), contacts 1 and 3 remain open for the set delay period. After the delay, pins 1 and 3 close, turning on the load, and remain closed for as long as the voltage is applied to the coil. When the coil voltage is removed from the coil, pins 1 and 3 instantly open. Thus the logic of this timer is on-delay and can be used in any on-delay circuit as long as it meets specifications, such as for contact rating and time duration.

The timer in *Figure 25*, like most solid-state timers, has a knob mounted on top to adjust the time delay. This knob can be set between 0 and 10. The numbers 0 to 10 on the timer do not indicate the time delay directly. They represent the percentage of the total time for which the timer can be set. If the total range of the timer is 1 to 60 seconds, a setting of 5 on the timer dial would indicate a 30-second delay. If the total range of the timer was 2 to 300 seconds, a setting of 5 on the timer dial would indicate a 150-second delay. Manufacturers use this system of making and setting the time range when repeatability can be about +0.1%. If greater repeatability is required, a digital adjustable switch can be used. The repeatability of this type of timer is generally about +0.005%.

Summary

Relays and timers play an important role in the operation of today's highly sophisticated manufacturing and industrial plants. Relays and timers make the remote or automatic operation of electrical or mechanical equipment possible. They may be electromechanical, electronic, or pneumatic. The application usually determines the type of relay or timer. However, the technician must often decide on such factors as compatibility, accuracy, and operating characteristics.

Electrical relays usually control one or more sets of contacts. These contacts in turn control or send electrical signals to other electrical devices. Electrical relays generally fall into two categories: electromechanical and solid state. Electromechanical relays use a magnetic effect to open or close mechanical contacts, whereas solid-state relays use electronic devices such as SCRs and triacs to simulate open or closed contacts.

Pneumatic relays use air signals to control other devices or to amplify and compute other air signals. Pneumatic repeaters or amplifiers boost or maintain the air pressure and volume at the required value. Computing relays perform mathematical functions such as adding, subtracting, or multiplication of air signals.

Timers are used to introduce delays into controllers and control circuits or to allow for timed operations to occur accurately. Timers are usually dashpot, synchronous clocks, or solid state.

Dashpot timers provide a time delay for closing or opening of contacts by allowing liquid or air to slowly escape from a dashpot through an orifice.

Electric time clocks are normally used to provide automatic operation of electrical equipment by using a synchronous AC motor to turn the dial of a clock. The dial then opens and closes the mechanical contacts, which supply power to the equipment at the proper time of day.

Solid-state timers perform their timing function using solid-state electronics. This method has the advantage of having no moving parts to wear out or replace. Theoretically, solid-state timers should never wear out or need maintenance.

Review Questions

1. Regardless of the level of power being controlled by a relay, the activation of the relay is accomplished by a(n) _____ power level signal.

a. higher
b. inverse
c. identical
d. lower

2. Relays that are a combination of electromechanical and solid-state technology are known as _____.

a. synchronous relays
b. hybrid relays
c. EMRs
d. SSRs

3. Reed relays are used in applications requiring all of the following characteristics *except* _____.
   a. low contact resistance
   b. low capacitance
   c. high insulation resistance
   d. high current

4. Relays that are often referred to as throw-away relays are _____ relays.
   a. general purpose plug-in
   b. solid-state
   c. electromechanical
   d. dashpot

5. The primary difference between a reed switch and a reed relay is _____.
   a. reed relays contain their own coil and reed switches depend on an external magnetic field to operate
   b. reed switches contain their own coil and reed relays depend on an external magnetic field to operate
   c. a reed relay operates only on AC power
   d. a reed switch cannot be used to switch DC

6. Latching types of general purpose relays typically have _____ coils.
   a. DC
   b. interlocking
   c. two
   d. plug-in

7. Electromechanical control relays are used in all of the following functions except _____.
   a. power monitoring
   b. current and voltage monitoring
   c. phase failure control
   d. high-speed power switching

8. A disadvantage associated with solid-state relays is _____.
   a. current leakage
   b. contact bounce
   c. a long operating life
   d. a lack of moving parts

9. In a solid-state relay, an LED converts electrical energy into _____.
   a. pneumatic energy
   b. light energy
   c. current
   d. capacitance

10. The force-balance principle is most commonly used in pneumatic _____.
   a. receivers
   b. transmitters
   c. valves
   d. controllers

11. Using pneumatic input signals as operands, computing relays perform _____ functions.
   a. transmitting
   b. relaying
   c. mathematical
   d. switching

12. Pneumatic selector and limiter relays work on the principle of satisfying a preset _____.
   a. voltage level
   b. equilibrium
   c. force of balance
   d. pressure setting

13. One disadvantage of an oil dashpot timing relay is that the timing may be erratic if _____.
   a. it is used on voltages above 300VAC
   b. it is equipped with normally open contacts
   c. the dashpot is used to control magnetic starters
   d. the piston is not allowed to make a full return to the bottom of the dashpot

14. Clock motors on time switches are typically _____.
   a. DC motors
   b. protected by thermal overload relays
   c. rated at a higher voltage rating than the contacts in the clock
   d. AC synchronous motors

15. Switching functions in a solid state timer are provided by _____.
   a. SCRs
   b. SSRs
   c. EMRs
   d. LEDs

GLOSSARY

Trade Terms Introduced in This Module

Annular: Related to or forming a ring.

Armature: The part or windings of an electric device at which the voltage is induced.

Baffles off the nozzle: Forces pressure through a small orifice in an attempt to reach equilibrium.

Constant: A known, unchanging value that is generally used in calculations as a multiplier or divisor (factor).

Contactor: A heavy-duty relay used to switch heavy current, high voltage, or both.

Dashpot timer: This timer uses air or liquid, which passes into or out of a contained space through an opening with either a fixed or variable diameter. The smaller the opening, the longer the time delay.

Equilibrium: Balance between forces.

Gain: The magnitude of the output signal divided by the magnitude of the input signal.

Latching relay: A general purpose relay that remains actuated in the absence of an input signal. A reset signal or mechanism is required to reset the relay.

LED: Light-emitting diode.

NARM: National Association of Relay Manufacturers

Operands: In all computing expressions, two types of components are present: operands and operators. Operands are the objects that are manipulated, and operators are the symbols that represent specific actions. For example, in the expression 5 + 6, 5 and 6 are operands and + is an operator. All expressions have at least one operand.

Proportioning: See scaling.

Repeater: An amplifier or transmitter.

Scaling: Uniformly increasing or decreasing an output value of an instrument by some known factor.

SCR: Silicon-controlled rectifier. A solid state electronic switching device.

Solid state relay (SSR): A relay that uses solid state electronic devices in place of mechanical contacts to perform switching operations.

Unit effective area: In instrumentation, the area of a diaphragm that is exposed to the flow of air or fluid. In most cases, the unit of measurement is square inches.

Additional Resources

This module is intended to present thorough resources for task training. The following reference works are suggested for further study. These are optional materials for continued education rather than for task training.

Process Instruments and Controls Handbook, Second Edition. D. M. Considine. McGraw-Hill Book Company.

Instrument Engineers Handbook, Volume II, Process Control. Bela Liptak. Chilton Book Company.

Electric Motor Controls Automated Industrial Systems, Second Edition. Gary Rockis and Glen Mazur. American Technical Publishers.

Instrumentation, Third Edition. Franklyn W. Kirk and Nicholas R. Rimboi. American Technical Society.

Standards and Practices for Instrumentation, Instrumentation Society of America.

Figure Credits

Rosemount, Inc.	204F03
Gerald Shannon	208F01, 208F03, 208F06, 208F21, 208F22, 208F23, 208F24, 208F25
Veronica Westfall	208F04
Tyco Electronics	208F09
Siemens Energy and Automation, Inc.	208F10(A), 208F18

NCCER CURRICULA — USER UPDATE

NCCER makes every effort to keep its textbooks up-to-date and free of technical errors. We appreciate your help in this process. If you find an error, a typographical mistake, or an inaccuracy in NCCER's curricula, please fill out this form (or a photocopy), or complete the online form at **www.nccer.org/olf**. Be sure to include the exact module ID number, page number, a detailed description, and your recommended correction. Your input will be brought to the attention of the Authoring Team. Thank you for your assistance.

Instructors – If you have an idea for improving this textbook, or have found that additional materials were necessary to teach this module effectively, please let us know so that we may present your suggestions to the Authoring Team.

NCCER Product Development and Revision

13614 Progress Blvd., Alachua, FL 32615

Email: curriculum@nccer.org

Online: www.nccer.org/olf

❑ Trainee Guide ❑ AIG ❑ Exam ❑ PowerPoints Other ______________________________

Craft / Level: Copyright Date:

Module ID Number / Title:

Section Number(s):

Description:

Recommended Correction:

Your Name:

Address:

Email: Phone:

Module 12209-03

Switches and Photoelectric Devices

COURSE MAP

This course map shows all of the modules in the second level of the Instrumentation curriculum. The suggested training order begins at the bottom and proceeds up. Skill levels increase as you advance on the course map. The local Training Program Sponsor may adjust the training order.

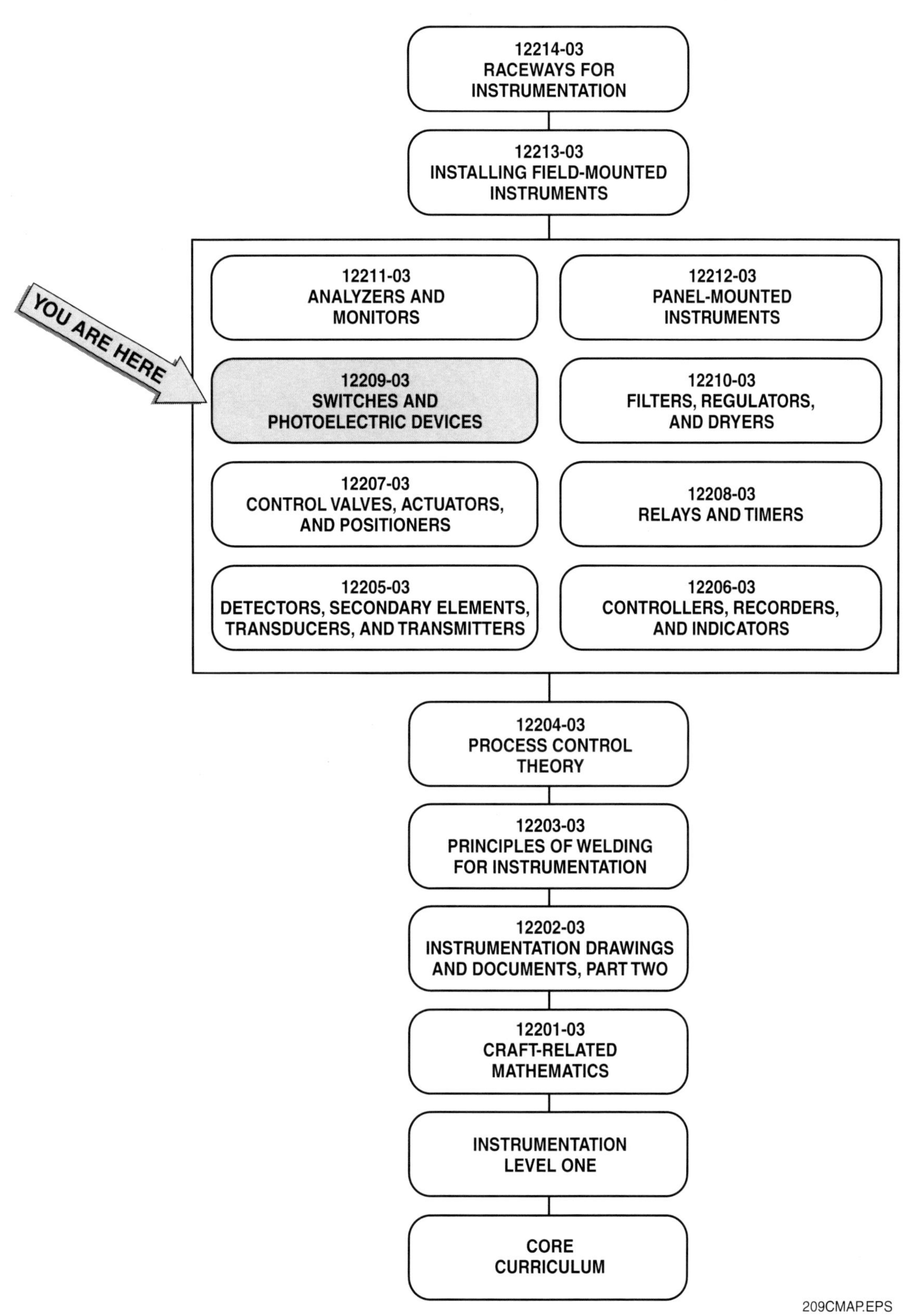

209CMAP.EPS

MODULE 12209-03 CONTENTS

1.0.0 INTRODUCTION 9.1

2.0.0 SWITCH DEFINITION, PROPERTIES, AND DESCRIPTION 9.1

2.1.0 Switch Definition 9.1

2.2.0 Switch Classifications 9.2

2.2.1 Switch Contacts 9.2

2.2.2 Pole of a Switch 9.2

2.2.3 Closed Positions or Throws of a Switch 9.3

2.2.4 Typical Switch Wiring 9.3

2.3.0 Switch Descriptions 9.3

2.3.1 Panel-Mounted Switches 9.4

2.3.2 Float Level Switches 9.4

2.3.3 Pressure Switches 9.6

2.3.4 Limit Switches 9.8

2.3.5 Electronic Switches 9.9

3.0.0 PHOTOELECTRIC DEVICES 9.10

3.1.0 Photocell Switches 9.10

3.2.0 Solar Cells 9.12

3.3.0 Infrared Devices 9.13

3.3.1 Motion Detectors 9.13

3.3.2 Industrial Process IR Sensors 9.13

3.4.0 Fiber Optics 9.13

4.0.0 PROXIMITY SENSORS 9.14

SUMMARY 9.15

REVIEW QUESTIONS 9.15

GLOSSARY 9.17

REFERENCES & ACKNOWLEDGMENTS 9.18

Figures

Figure 1 Knife blade contacts and butt contacts 9.2
Figure 2 Poles of a switch 9.2
Figure 3 Single- and double-throw switches 9.3
Figure 4 Single-pole, single-throw switch 9.3
Figure 5 Three-way switch (single-pole, double-throw) 9.4
Figure 6 Double-pole, single-throw switch 9.4
Figure 7 Typical panel-mounted switches 9.5
Figure 8 Rod-operated float valve with wiring symbol 9.5
Figure 9 Isolated or hermetically sealed level switches 9.5
Figure 10 Explosionproof, pressure-activated liquid or dry material level switch 9.6
Figure 11 Typical pressure switches 9.6
Figure 12 Pressure switch with calibrated adjustments 9.7
Figure 13 Differential pressure switch 9.8
Figure 14 Differential pressure switch application 9.9
Figure 15 Machine limit switch with wiring symbols 9.9
Figure 16 Limit switch protection 9.9
Figure 17 Typical microswitches used as limit switches and interlock switches 9.9
Figure 18 Silicon-controlled rectifier (SCR) 9.10
Figure 19 Cutaway view of a photocell or light-dependent resistor (LDR) .9.10
Figure 20 Photocell switch and typical application 9.11
Figure 21 Direct scanning methods 9.11
Figure 22 Typical instrumentation photo sensors used for direct scanning .9.12
Figure 23 Reflective scanning method used to count and cap bottles 9.12
Figure 24 Typical solar cell 9.12
Figure 25 Photovoltaic array components 9.12
Figure 26 Motion detector floodlights .9.13
Figure 27 Single-line diagram of a motion detector with a photocell light controller 9.13
Figure 28 Typical IR instrumentation equipment 9.14
Figure 29 Fiber-optic applications 9.14
Figure 30 Proximity sensor 9.15
Figure 31 Line-powered inductive proximity sensor 9.15

MODULE 12209-03

Switches and Photoelectric Devices

Objectives

When you have completed this module, you will be able to do the following:

1. State the purpose of a switch.
2. Identify commonly used switches.
3. Describe the operation of various types of switches.
4. Classify switches, using wiring symbols, according to the number of poles and the number of throws.
5. State the purpose of an SCR.
6. Describe the operation of photoelectric devices.
7. Identify commonly used photoelectric devices.
8. State the electrical characteristics of a solar cell.

Prerequisites

Before you begin this module, it is recommended that you successfully complete the following: Core Curriculum; Instrumentation Level One; Instrumentation Level Two, Modules 12201-03 through 12204-03.

Required Trainee Materials

1. Pencil and paper
2. Appropriate personal protective equipment

1.0.0 ◆ INTRODUCTION

One of the most common electrical devices in commercial and industrial installations is the switch. Virtually every system and piece of electrical equipment has some type of switch. Photoelectric devices, such as motion sensors and photocells, are also becoming very popular.

There are many types and name brands of switches and photoelectric devices. You are encouraged to learn the principles of operation of the most popular types and apply this knowledge to other devices encountered in the field.

The aim of this module is to describe the characteristics and operation of the most popular types of switches and photoelectric devices so that proper equipment can be selected and installed in commercial and industrial applications.

2.0.0 ◆ SWITCH DEFINITION, PROPERTIES, AND DESCRIPTION

The following sections provide a switch definition, along with switch descriptions, classifications, and properties.

2.1.0 Switch Definition

An electrical switch is a device that is manually or automatically operated to make or break an electrical **circuit** operating within the rated current and voltage of the switch. Standard electrical switches should not be used to interrupt a circuit under **short circuit** or **overload** conditions. Fuses or circuit breakers should be used to prevent such conditions.

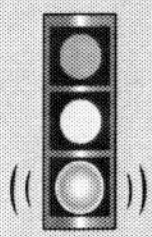

NOTE

When selecting and installing switches, make sure that each switch is rated for the voltage and current of its circuit.

2.2.0 Switch Classifications

Switches can be classified in many different ways:

- According to the number of poles:
  - Single-pole
  - Double-pole
  - Multiple-pole
- According to the number of positions (throws):
  - Single-throw
  - Double-throw
  - Multiple-throw
- According to the method of insulation (arc control):
  - Air
  - Oil
  - Gas
- According to the method of operation:
  - Electrical
  - Mechanical
  - Motion
  - Light
  - Pneumatic
  - Hydraulic
  - Manual
  - Magnetic
- According to function:
  - Break a circuit
  - Make a circuit

2.2.1 Switch Contacts

Switches contain contacts that are made of conducting material, such as copper. Current flow is established when the contacts touch each other. Current flow is shut off when the contacts are separated. As previously explained, there are many methods used to make and break electrical contact. A spring mechanism is often used to hold contacts securely in place when the switch is closed. Switches incorporating solid-state technology do not have moving parts, but this module focuses mainly on mechanical-type switches.

Two common types of contacts are knife blade and butt contacts (*Figure 1*). Many times the contacts are coated with a conductive material to prevent **oxidation** and to ensure a low-resistance contact point.

Switch contacts that are burnt or charred will need to be replaced to maintain a low resistance contact point.

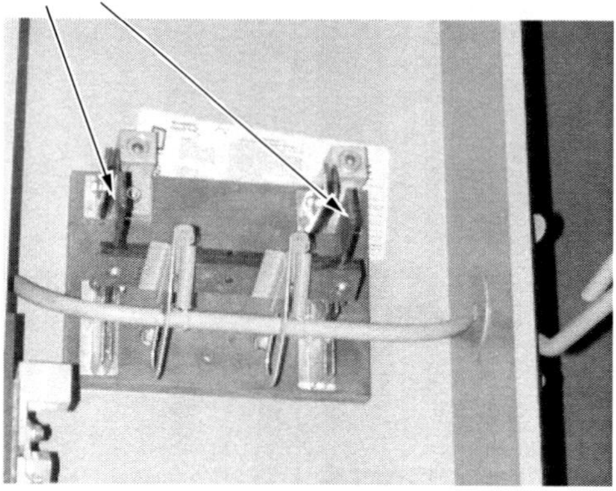

Figure 1 ◆ Knife blade contacts and butt contacts.

2.2.2 Pole of a Switch

The pole of a switch is where the movable contacts are attached. The poles, in conjunction with the contacts, are used to make and break the electrical circuit. The poles are electrically insulated and isolated from one another.

A single-pole switch will make and break only one conductor or leg of a circuit, whereas a two-pole switch will make and break two legs of a circuit, and so on (*Figure 2*). The dotted line between

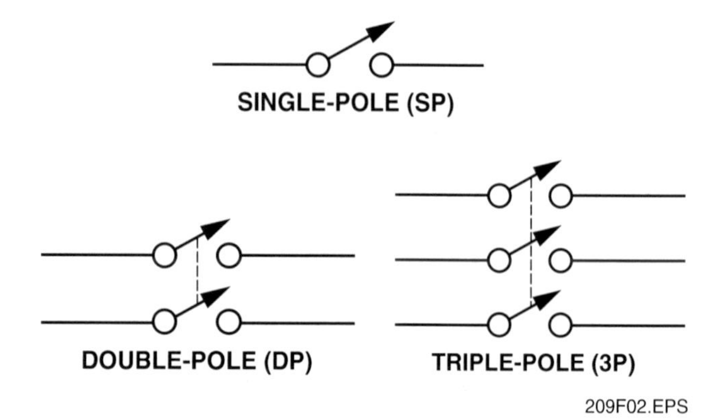

Figure 2 ◆ Poles of a switch.

the double-pole and triple-pole contacts indicates that the poles operate at the same time. Switch poles are often abbreviated as shown in *Figure 2*.

2.2.3 Closed Positions or Throws of a Switch

As shown in *Figure 3*, a single-throw switch will only make a closed circuit when the switch is in one position. A double-throw switch will make a closed circuit when placed in either of two positions. The poles and throws of a switch are often abbreviated as shown in *Figure 3*. When more than two closed positions are needed, special multiple throw switches should be employed.

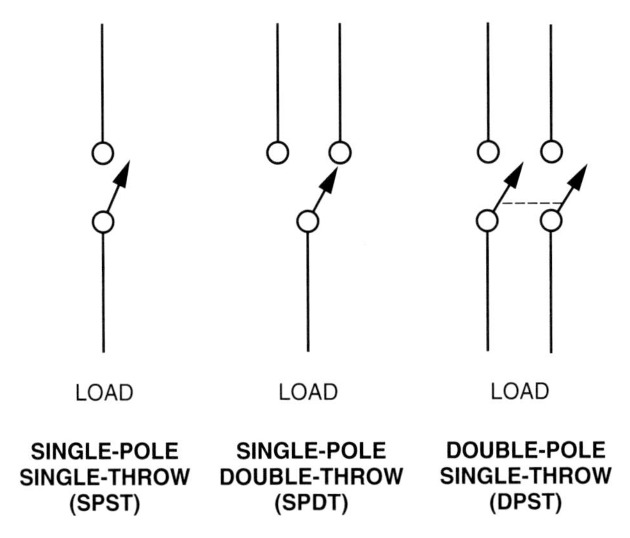

Figure 3 ◆ Single- and double-throw switches.

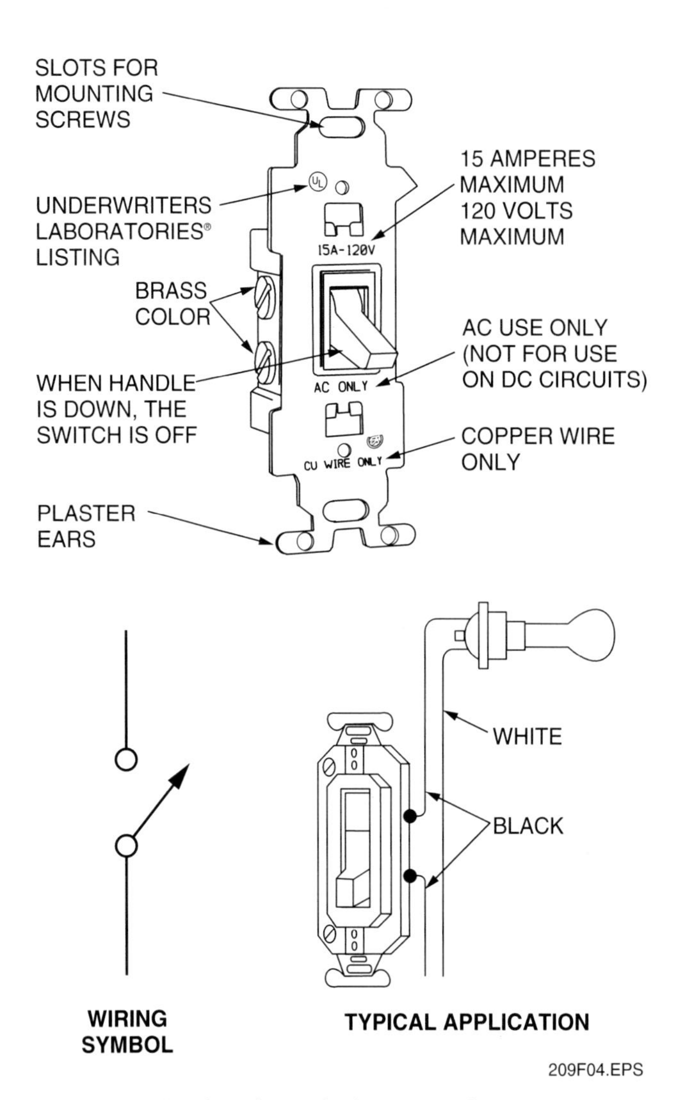

Figure 4 ◆ Single-pole single-throw switch.

2.2.4 Typical Switch Wiring

The following explanations use an electrical light switch and light bulb analogy to illustrate the wiring of various switch poles and throw configurations when applying a signal or power to a load:

- *Single-pole, single-throw switch* – Toggle switches or wall switches can be used to control lights and fractional horsepower motors. The most common switch is the single-pole, single-throw switch used to control lights or equipment from one location. This switch has two brass-colored screw terminals for wire connections (*Figure 4*).
- *Single-pole, double-throw switch* – The three-way switch is a single-pole, double-throw switch that is commonly used in pairs to control a light or piece of equipment from two locations. The common toggle type used in lighting has three screw terminals for wire connections: one black or copper, which is the common, or pole, and two brass- or silver-colored, which are the throws, as shown in *Figure 5*.
- *Double-pole, single-throw switch* – The double-pole, single-throw switch shown in *Figure 6* is normally used to control 240-volt equipment. This switch has four brass-colored screw terminals for wire connections. The terminals on the left side of the switch are connected through one set of contacts, and the terminals on the right side of the switch are connected through another set of contacts.

2.3.0 Switch Descriptions

Switches used for instrumentation are available in a variety of types, styles, current/voltage ratings, and pole/contact configurations. Some of the more common are covered in the following sections.

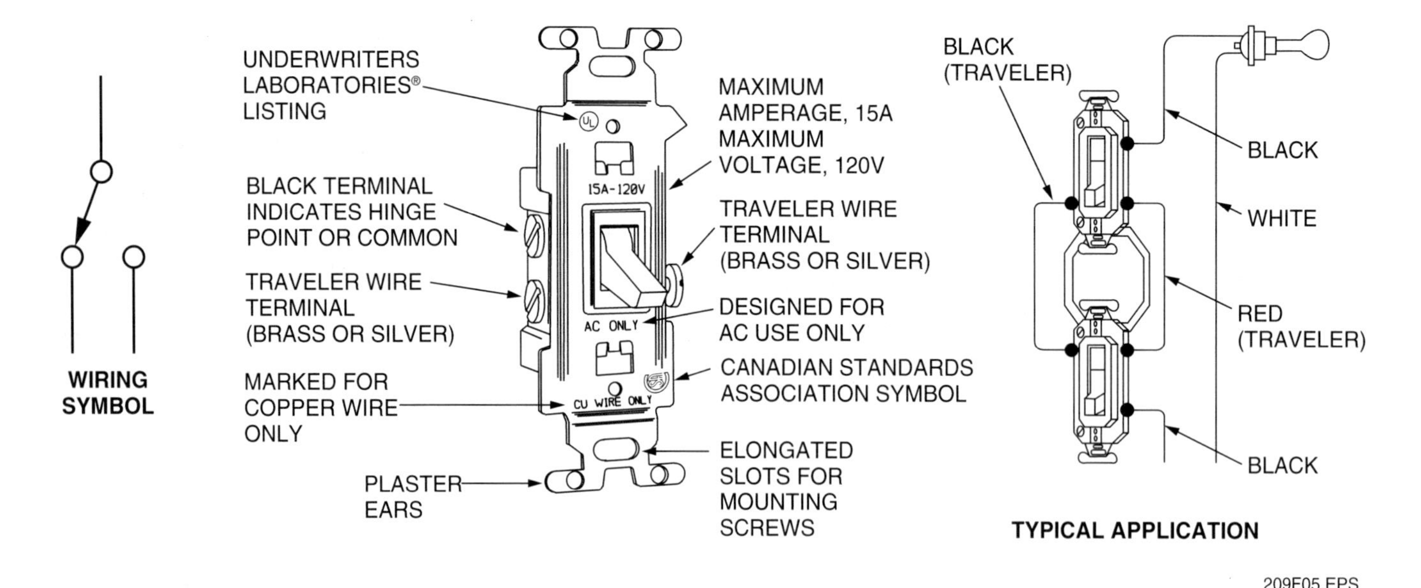

Figure 5 ◆ Three-way switch (single-pole, double-throw).

2.3.1 Panel-Mounted Switches

Figure 7 shows some of the most common types of switches that may be installed on instrument panels. Except for rotary or joystick switches that can have many poles and multiple throws, most push-button, rocker, and snap-action switches have the same pole and throw designations as toggle switches. Switches are also available with center-off positions, momentary contact positions, and make-before-break contacts.

2.3.2 Float Level Switches

A float switch contains a floating device that is placed in contact with the fluid to be controlled. Float switches are often used to start and stop motor-driven pumps used to empty and fill tanks. Float switch contacts may be either normally open (NO) or normally closed (NC), and they cannot contact the process fluid.

Rod-operated float switches (*Figure 8*) are commonly used to open or close solenoid valves to control fluids.

Level switches (*Figure 9*) are hermetically sealed or isolated from the fluid being monitored and are designed to be mounted inside or through the exterior of fluid tanks. *Figure 10* shows an explosion-proof, exterior-mounted switch that uses pressure against a diaphragm to detect the level of liquids or dry bin materials in a storage vessel. Two or more switches must be positioned at various levels on a tank to detect high, low, and (if desired) one or more intermediate levels.

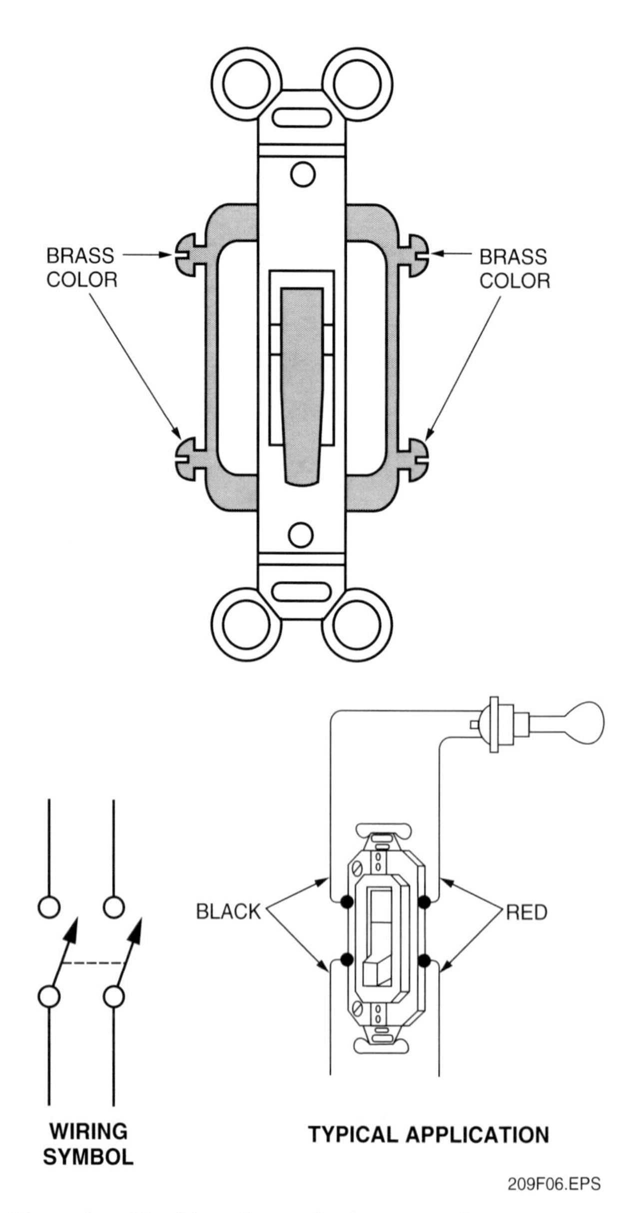

Figure 6 ◆ Double-pole, single-throw switch.

Figure 7 ◆ Typical panel-mounted switches.

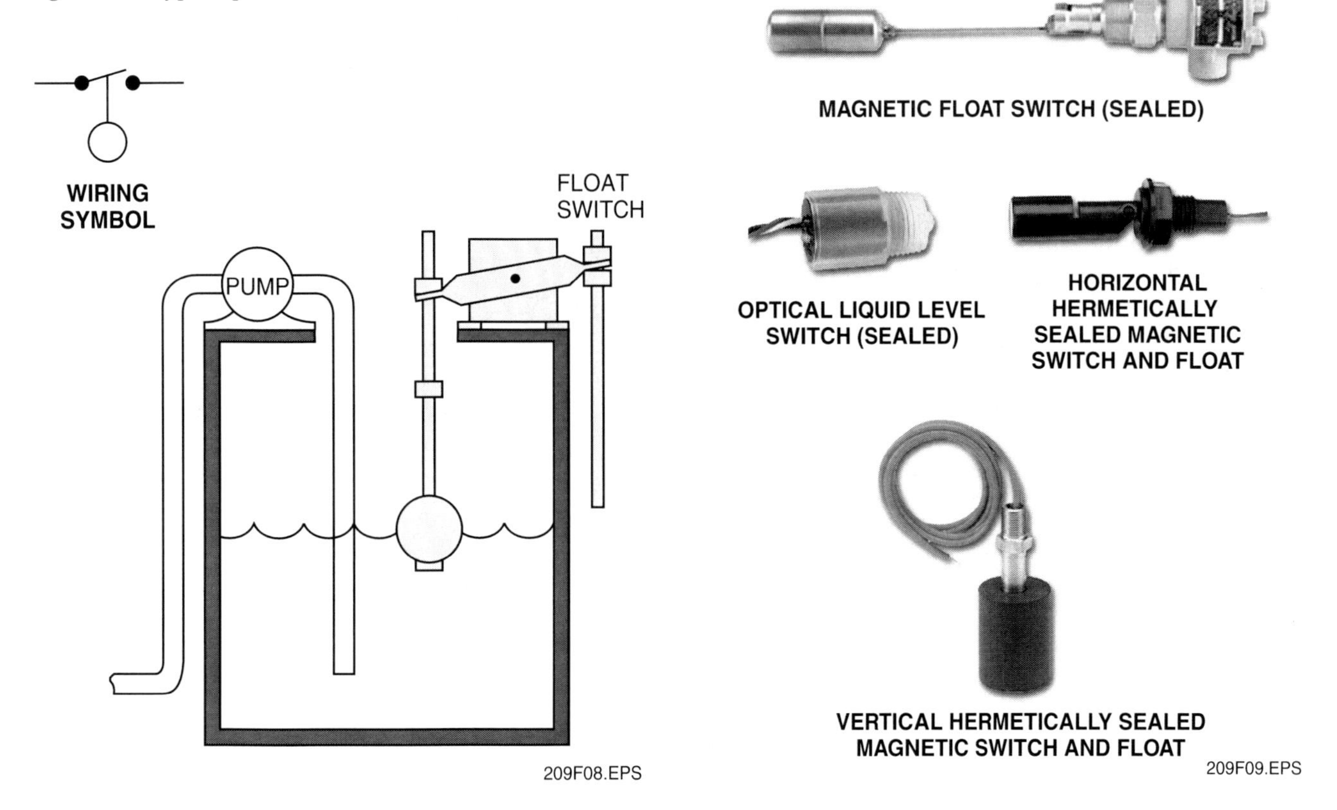

Figure 8 ◆ Rod-operated float valve with wiring symbol.

Figure 9 ◆ Isolated or hermetically sealed level switches.

Figure 10 ◆ Explosionproof, pressure-activated liquid or dry material level switch.

2.3.3 Pressure Switches

Pressure switches are used to control the pressure of liquids and gases by starting and stopping motors and compressors when the pressure in the system reaches a preset level. Air compressors, for example, are started directly or indirectly on a call for more air by a pressure switch (*Figure 11*).

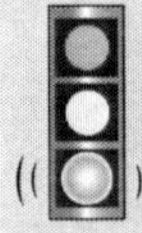

NOTE

When selecting pressure switches, pay particular attention to the pressure rating of the switch and the operating pressure of the system. Never exceed the switch ratings.

Pressure switches use mechanical motion from pressure changes to operate one or more sets of contacts. A typical pressure switch may use a bellows, diaphragm, or Bourdon tube as the pressure-sensing element. Most pressure switches have an adjustment screw to adjust the high-pressure setpoint, while other, more complex switches have an adjustable **deadband** range that allows the low-pressure setpoint to be adjusted, as shown in *Figures 11* and *12*.

Many pressure switches come with two sets of contacts: one NO set and one NC set. When the pressure reaches the setpoint, one set of contacts will open, and the other will close. In selecting a pressure switch for a particular application, each of the following should be taken into consideration:

- *Setpoint range* – This is the span of pressures within which the pressure sensing element can be set to actuate the contacts of the switch. For example, a pressure switch may have a setpoint range of 20 to 100 psi.
- *Deadband adjustment* – This is the span of pressure between the setpoint limit (which actuates the contacts) and the **reset point** (which resets the contacts to their normal position). For example, a pressure switch may have a setpoint range between 20 and 100 psi with a deadband of only 5 to 15 psi between a high-pressure setpoint and a low-pressure setpoint (reset).
- *Rating of the switch and contacts* – The rating includes the type of switch used, such as single-pole, single-throw, or single-pole, double-throw. It also includes the amount of current and voltage that the contacts can safely switch.
- *Accuracy* – The accuracy of the switch refers to the ability of the switch to repeatedly actuate at the setpoint. This value is typically stated as a percentage of the maximum operating pressure of the switch.

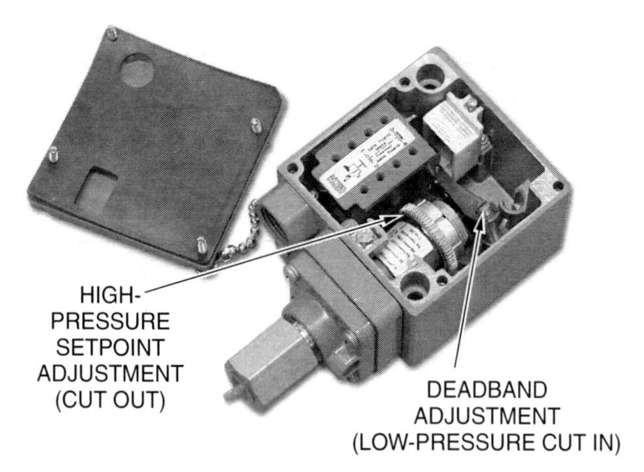

MACHINE PRESSURE SWITCH

HIGH-PRESSURE SETPOINT ADJUSTMENT (CUT OUT)

DEADBAND ADJUSTMENT (LOW-PRESSURE CUT IN)

AIR COMPRESSOR PRESSURE SWITCH

209F11.EPS

Figure 11 ◆ Typical pressure switches.

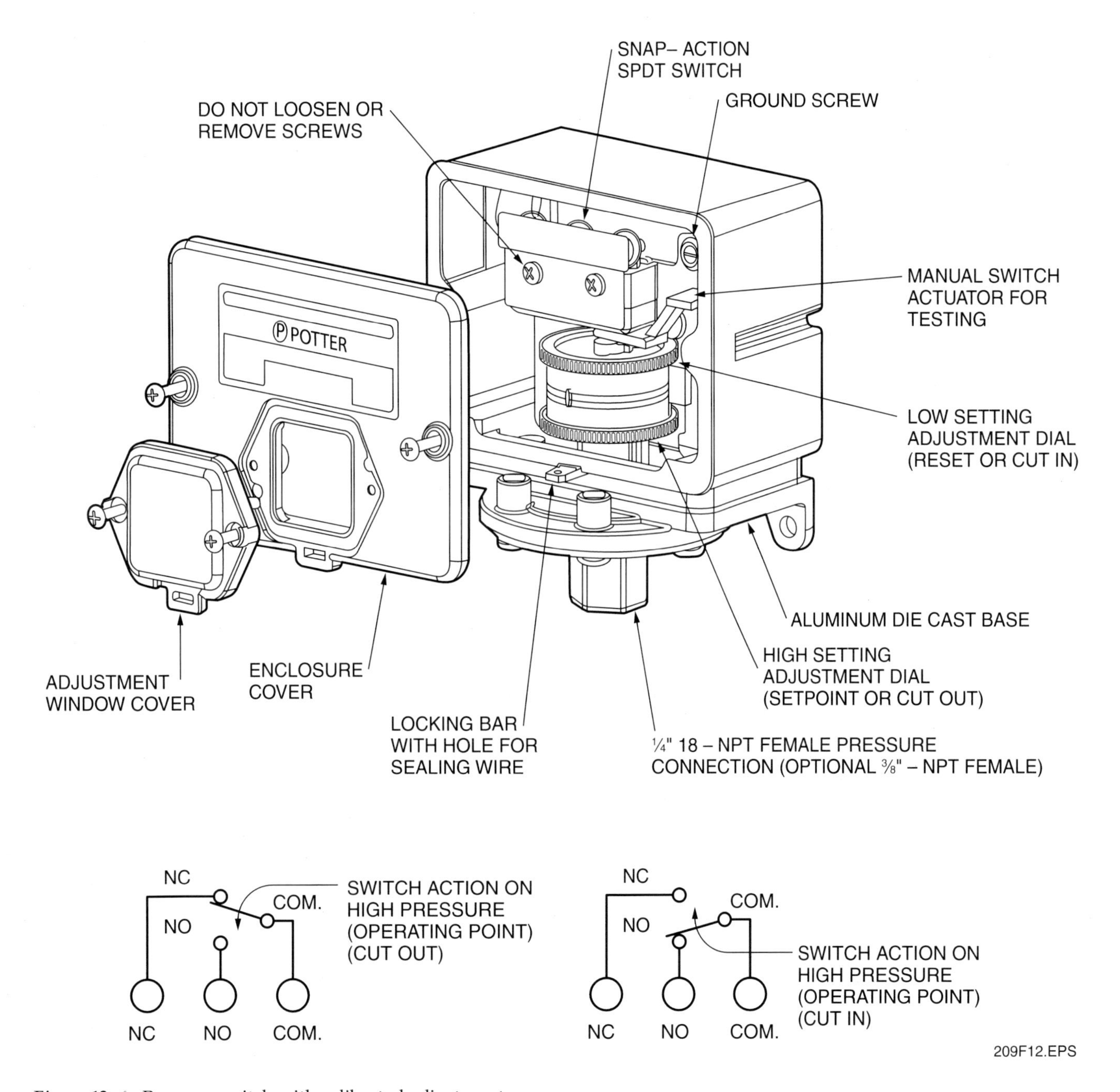

Figure 12 ◆ Pressure switch with calibrated adjustments.

Another application of a pressure switch is the detection of high and low liquid or gas flow in pressurized lines. This is accomplished with a differential pressure switch (*Figure 13*). As shown in *Figure 14*, the pressure drop across a calibrated orifice in the pressure line provides an indication of the flow rate. The high-pressure setpoint and deadband of the pressure switch are set so that an alarm sounds or the flow is shut down if the flow rises above (or falls below) the desired rate.

2.3.4 Limit Switches

Limit switches react to physical changes. There are many types of limit switches. One type contains a metal element that reacts mechanically to changes in heat. It will open a circuit if temperatures exceed preset limits. Another common type of limit switch contains a mechanical actuator that activates a switch when it reaches a preset travel limit. Another type, similar to the limit switch that stops a garage door at either end of its travel, is actuated when its actuator is touched by a *dog* mounted on a chain or rod.

In general, the operation of a limit switch begins when the moving machine or part strikes an operating lever, which activates the limit switch. *Figure 15* shows a limit switch with its operating lever.

Limit switches can be used either as control devices for regular operation or as emergency switches to prevent the improper functioning of equipment. The contacts may be normally open (NO) or normally closed (NC), and they may be momentary (spring-returned) or maintained-contact types.

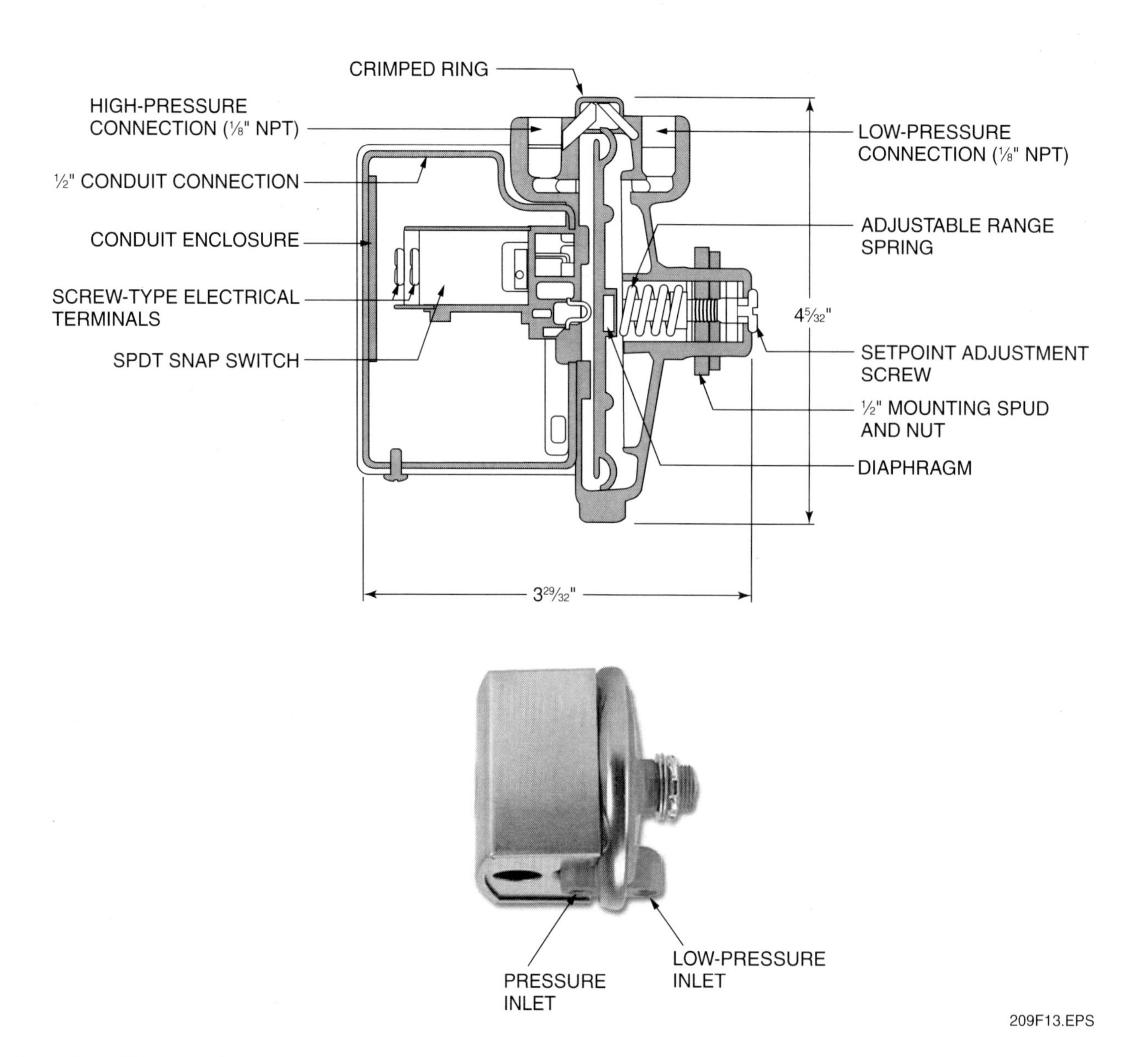

Figure 13 ◆ Differential pressure switch.

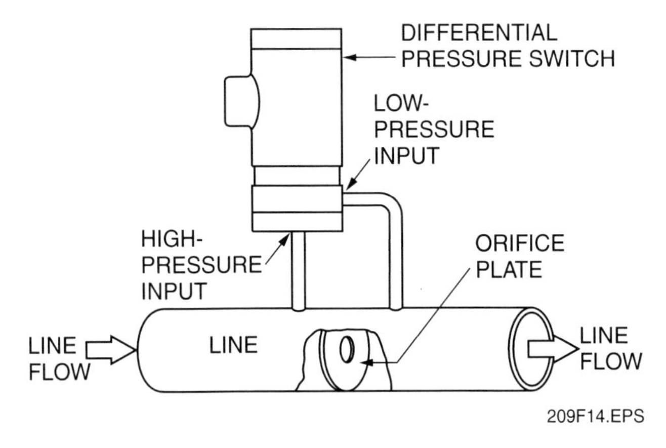

Figure 14 ◆ Differential pressure switch application.

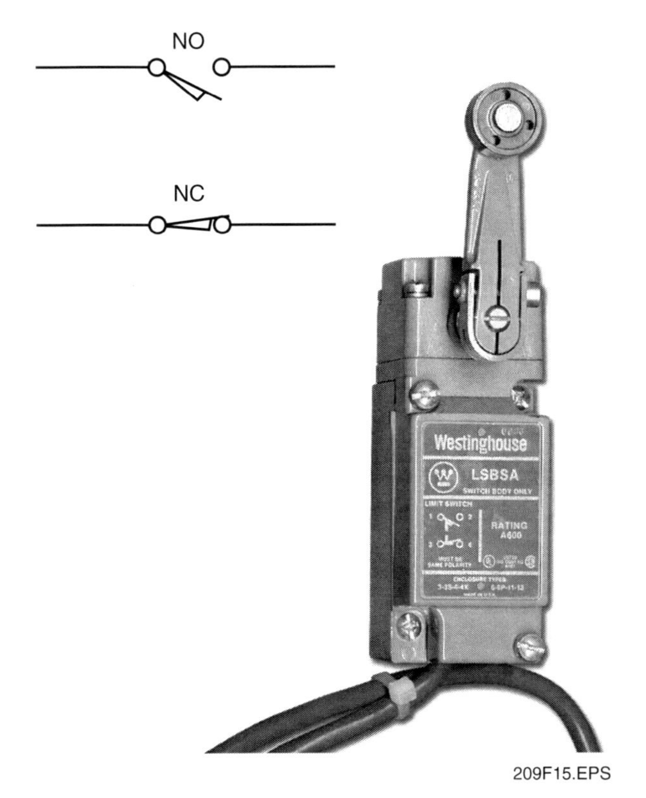

Figure 15 ◆ Machine limit switch with wiring symbols.

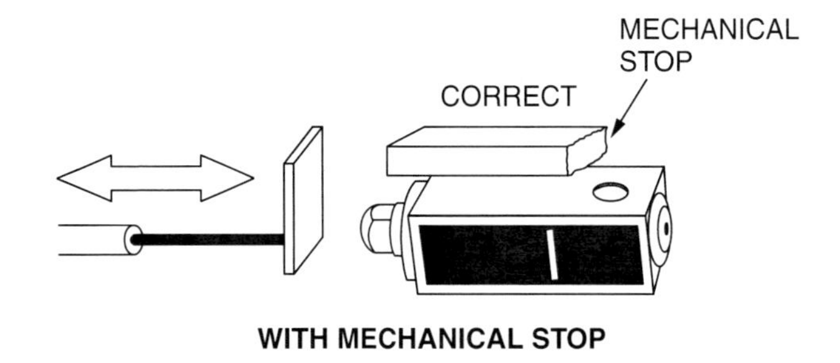

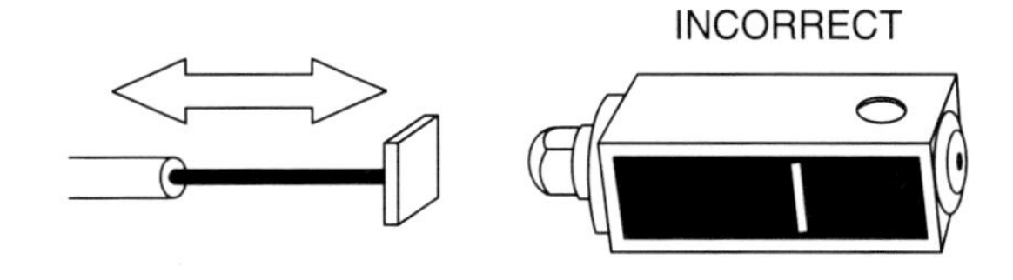

Figure 16 ◆ Limit switch protection.

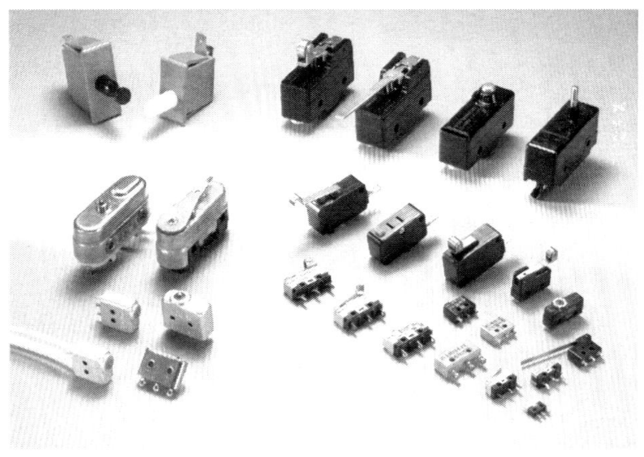

Figure 17 ◆ Typical microswitches used as limit switches and interlock switches.

The installation of limit switches involves selecting the best actuator, or operating lever, and then mounting the limit switch in the correct position in relationship to the moving part. It is important that the limit switch not be operated beyond the manufacturer's recommended specifications of travel. If overtravel of the moving part will affect the switch or cause the switch to operate beyond its travel range, an adequate mechanical stop, as shown in *Figure 16*, should be used to prevent damage to the switch.

Various types of microswitches (*Figure 17*) are used in or as limit switches. They are also used as equipment and instrument panel access-safety interlocks.

2.3.5 Electronic Switches

Silicon-controlled rectifiers, or SCRs, are electronic devices that normally operate in the ON or OFF state very much like a switch does. SCRs are used extensively in motor controls and power supplies (*Figure 18*).

The basic purpose of the SCR is to switch on and off large amounts of power. It performs this function with no moving parts to wear out and no contacts that require replacing or maintaining.

The SCR can be used in AC and DC circuits, but it will only allow current to pass in the forward direction (from **anode** to **cathode**). It has an anode (positive) lead, a cathode (negative) lead, and a gate lead. No current will flow through the SCR until the gate lead receives an electrical signal. Once the gate receives a signal, the SCR is like a

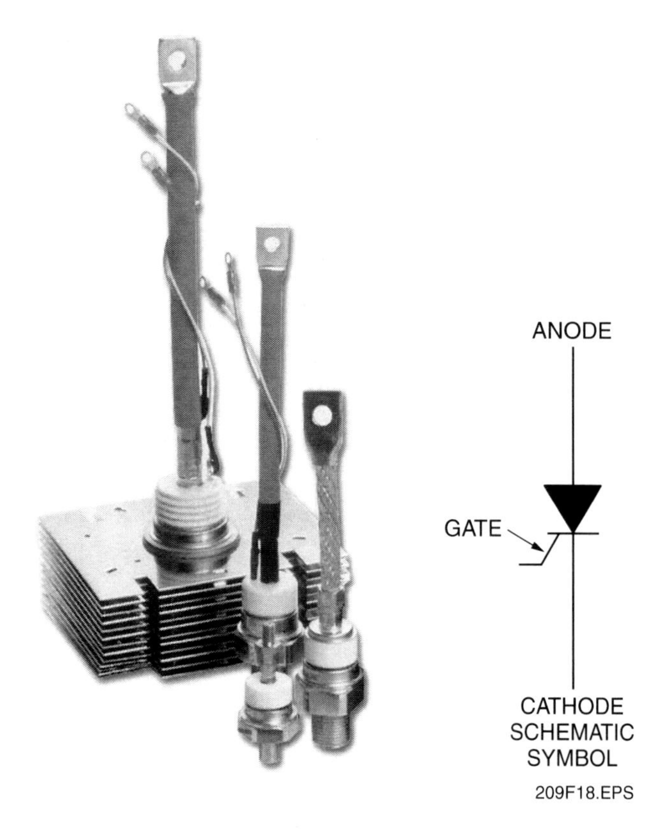

Figure 18 ◆ Silicon-controlled rectifier (SCR).

closed switch, and it conducts current easily in the forward direction. The gate signal can now be removed, but conduction will not stop until the current tries to reverse direction or the current drops to near zero.

Once conduction has stopped, the SCR will be like an open switch, and no current will flow. A new gate signal must be applied before the SCR will conduct again. The SCR is similar to a **diode** in that it will only allow current to flow in the forward direction and never in the reverse direction (cathode to anode). A diode, however, is not a switch, has no gate, and acts as an electrical check valve that continually allows forward current flow only.

3.0.0 ◆ PHOTOELECTRIC DEVICES

Light-sensitive devices, sometimes called photoelectric transducers, alter their electric characteristics when exposed to visible or **infrared radiation (IR)** light. Light-sensitive devices include photocells, solar cells, motion detectors, and phototransistors.

Such light-sensitive devices can trigger many different kinds of circuits for the control of alarms, lights, motors, relays, and other actuators.

Some of the more common photoelectric devices and their principles of operation are discussed in this section.

3.1.0 Photocell Switches

Photocells are also called many other names, including photoconductive cells, light-dependent resistors (LDRs), and photoresistors. Basically, photocells are variable resistors in which the resistance values depend on the amount of light that falls on the device.

The resistance of the photoconductive material shown in *Figure 19* varies inversely with the amount of light that shines on it. In other words, as more light hits the photocell, the lower the resistance of the photoconductive material becomes. The photoconductive material is usually made of cadmium sulfide (CdS) or cadmium selenide (CdSe) and is placed directly between the two terminals of the photocell. In bright light, the resistance between the terminals can be as low as 50 ohms; in darkness, it can be as high as 5 megohms.

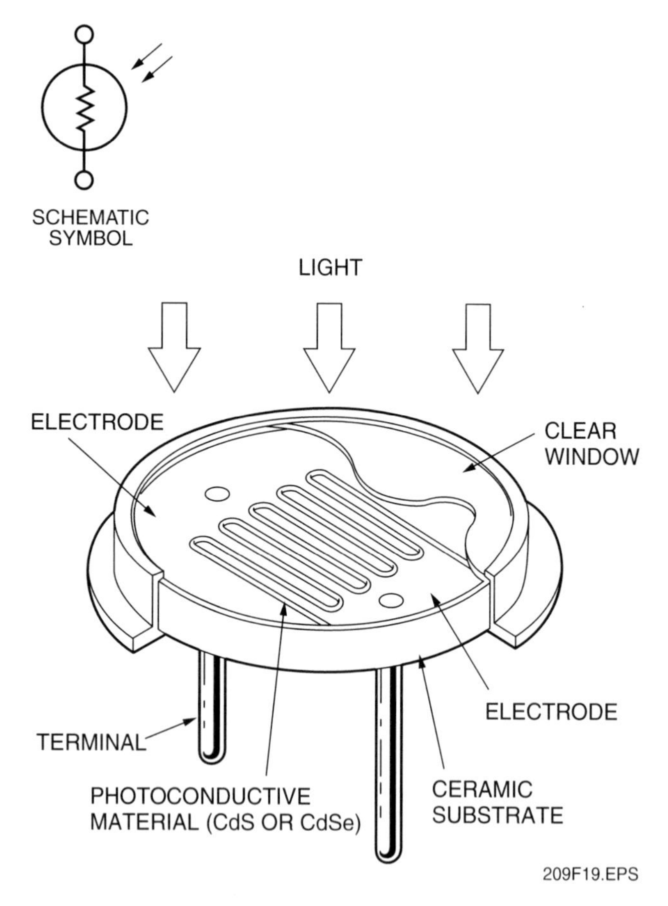

Figure 19 ◆ Cutaway view of a photocell or light-dependent resistor (LDR).

The most common use for photocells is the photocell switch used to control lights between dusk and dawn. Essentially, the photocell switch, shown in *Figure 20*, acts as a single-pole, single-throw (SPST) switch that closes when it gets dark and opens when it gets light. Unlike the common SPST

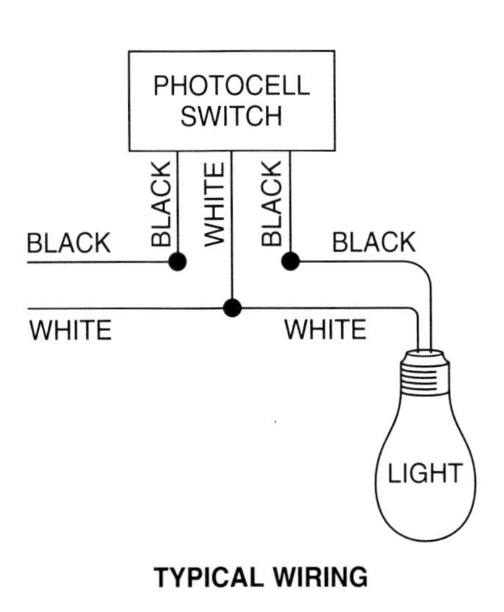

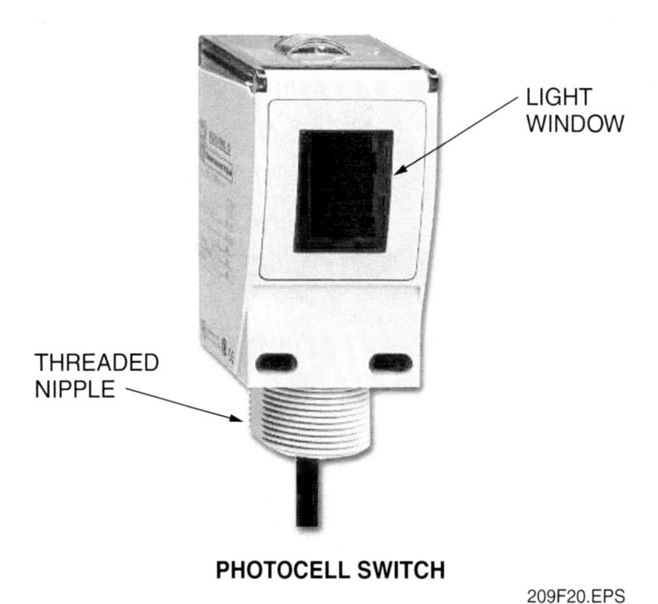

Figure 20 ◆ Photocell switch and typical application.

toggle switch previously discussed, the photocell must receive power to operate, which means that it requires both conductors of the circuit (L1 and L2).

Photocell switches like the one in *Figure 20* can be rated for up to 300 watts and can directly control a set of floor lights rated 300 watts or less.

Photocell switches can also be used in control circuits. A light source forms a light beam, and a photoreceiver scans across an area to detect the presence of an object. When the light beam is broken, the photocell or photoreceiver responds by sending a DC signal to the control circuit. *Figure 21* shows the direct scan method of photoelectric detection. The light source and the photoreceiver are positioned directly opposite each other, such that light from the source shines directly at the receiver. In *Figure 21(A)*, the photocell is being used to count the products on the conveyor belt. In *Figure 21(B)*, when the roll of paper gets big enough to block the light beam, the photoreceiver sends a signal to the controller, which will stop the winding until a new spool is loaded onto the paper spindle.

Typical instrumentation photo sensors used for direct scanning are shown in *Figure 22*. Most of these sensors use IR only or visible red light beams to avoid interference from ambient area lighting.

Reflective scan is another popular scanning method used to detect moving objects. In reflective scan, the light source and the photoreceiver are mounted in the same housing and on the same side of the object to be detected (*Figure 23*). This method is used when there is limited space or when mounting restrictions prevent aiming the light beam directly at the photoreceiver. As shown in *Figure 23*, the reflected light is absorbed by the scanner/receiver. When a bottle breaks the light beam, there is no reflection and the photodetector signals the capper to apply the bottle cap.

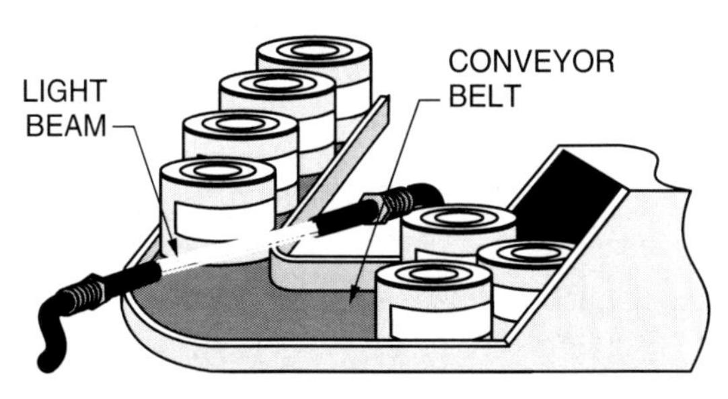

(A) Counting products is a common application of photoelectric controls. Counting batches or groups of cans or other items prior to packaging or group processing is also common.

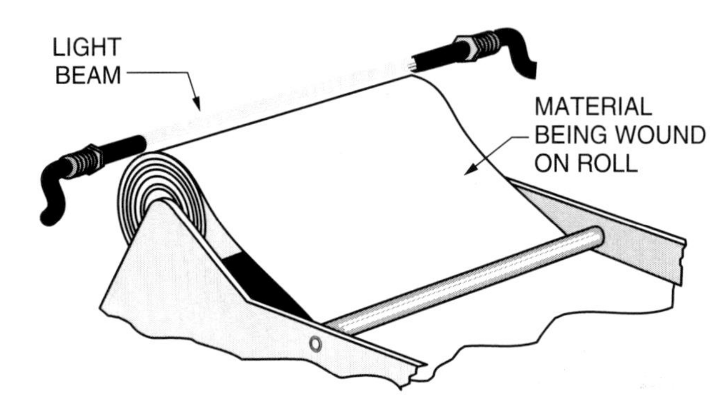

(B) The size of a paper or fabric roll can be controlled by positioning a light source and a photoreceiver so the roll diameter blocks the beam.

209F21.EPS

Figure 21 ◆ Direct scanning methods.

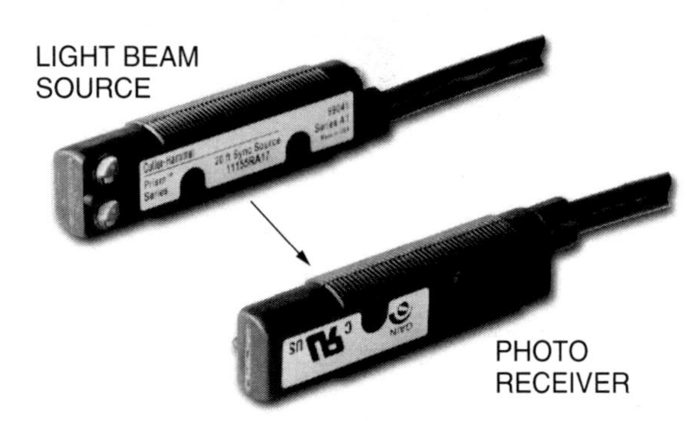

Figure 22 ◆ Typical instrumentation photo sensors used for direct scanning.

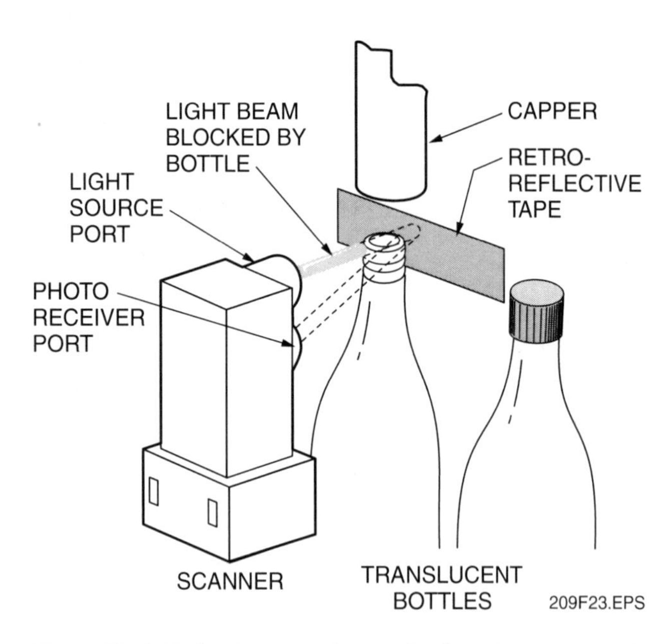

Figure 23 ◆ Reflective scanning method used to count and cap bottles.

3.2.0 Solar Cells

Solar cells, also called photovoltaic detectors, generate a DC voltage proportional to the amount of light that is absorbed by the cell.

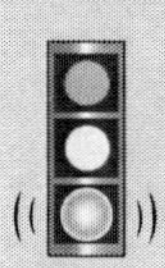

NOTE

Keep sensor and reflector lenses clean to ensure proper operation of detectors. Avoid abrasive cleaners or cloth that may scratch the lenses.

A typical cell, as shown in *Figure 24*, only generates about ½ volt DC. Therefore, cells are often arranged in series and parallel combinations to obtain the desired output voltage and current levels. Cells can be joined together to make a solar module, a solar panel, or a solar array. *Figure 25* shows the components of a photovoltaic array system that could be used to charge a battery bank or power a piece of DC equipment.

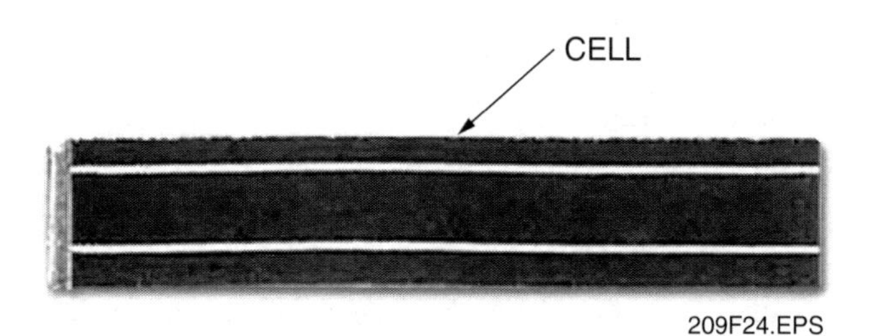

Figure 24 ◆ Typical solar cell.

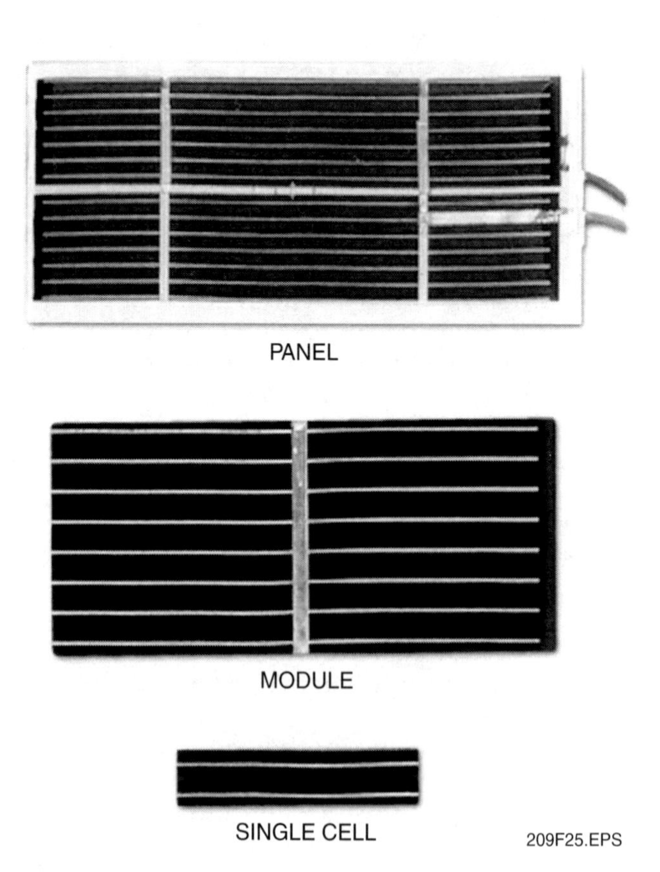

Figure 25 ◆ Photovoltaic array components.

3.3.0 Infrared Devices

Any object that is not at a temperature of **absolute zero** (–273°C, or –460°F) will emit invisible electromagnetic energy in the form of IR. A detector can pick up this transmitted energy and produce an electrical signal proportional to the amount of IR detected. This signal can then be used to control other devices.

IR detectors are commonly used for motion/people detectors, TV remote control detectors, and IR scanners that detect hot spots in electrical equipment. The most common application is the motion/people detector.

3.3.1 Motion Detectors

Motion detectors are typically used in security systems. The security system may be an elaborate zone-controlled system that incorporates motion detectors, door switches, smoke detectors, and moisture sensors, or it may be as simple as a set of motion detector floodlights mounted above a garage. The basic operation of the detector is the same in all cases.

Figure 26 shows a set of floodlights controlled by an IR motion detector switch. Many times, a photocell switch is added to the light controller, which automatically turns the lights off during daylight hours. Also, the light controller usually comes equipped with a manual override method that can be used to turn the lights on at any time.

Figure 26 ◆ Motion detector floodlights.

When the light controller contains both the photocell switch and an IR detector switch, it operates as if there were three SPST (single-pole, single-throw) switches involved (*Figure 27*). The photocell switch and the motion detector switch are in series, so both must be closed (there must be motion, and it must be dark) in order for the light to come on. The third switch is the manual bypass switch, which is in parallel with the series motion detector and photocell switches.

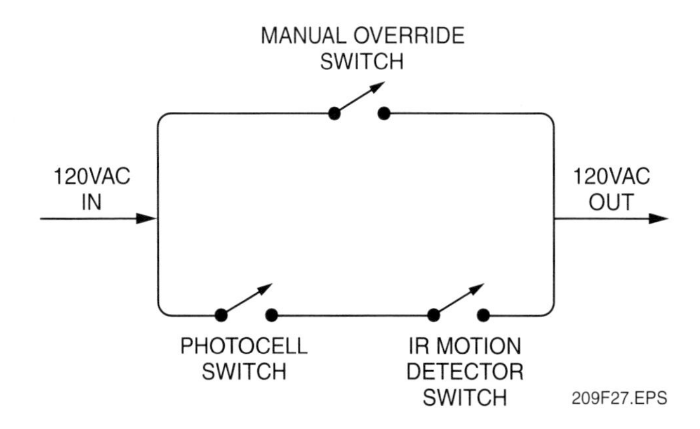

Figure 27 ◆ Single-line diagram of a motion detector with a photocell light controller.

3.3.2 Industrial Process IR Sensors

IR sensors are used in the instrumentation of temperature-sensitive industrial processes, including paper production and stack temperature monitoring in cement and petrochemical facilities. They are particularly useful when the use of temperature probes is impossible or impractical. *Figure 28* shows a typical fixed IR monitoring system that includes controller functions for industrial processes. Also shown is a calibrated IR handheld thermometer used to perform random temperature checks. A visible laser pointing beam is used to identify the area measured by the instrument's IR sensor.

3.4.0 Fiber Optics

Fiber optics transmit light and can be used in all types of photoelectric devices. In one application, fiber-optic sensors use an emitter to send the light, a receiver to receive the light sent by the emitter, and a flexible cable packed with tiny fibers to transmit the light. There may be separate fiber cables for the emitter and the receiver, depending on the sensor, or there may be a single cable with two internal fiber cables. When a single cable is used, the emitter and receiver use various methods to distribute emitter and receiver fibers within a cable. Glass fibers are used when the emitter source is infrared light, while plastic fibers are used when the emitter source is visible light.

Figure 29 illustrates three different methods for using fiber optics in photoelectric sensing. In *Figure 29(A)*, the method illustrated is referred to as thru-beam, in which light is emitted and received

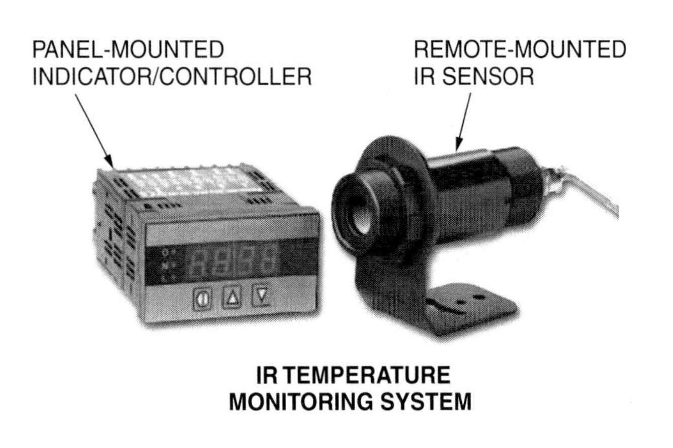

VISIBLE LASER POINTING BEAM

HANDHELD THERMOMETER

209F28.EPS

Figure 28 ◆ Typical IR instrumentation equipment.

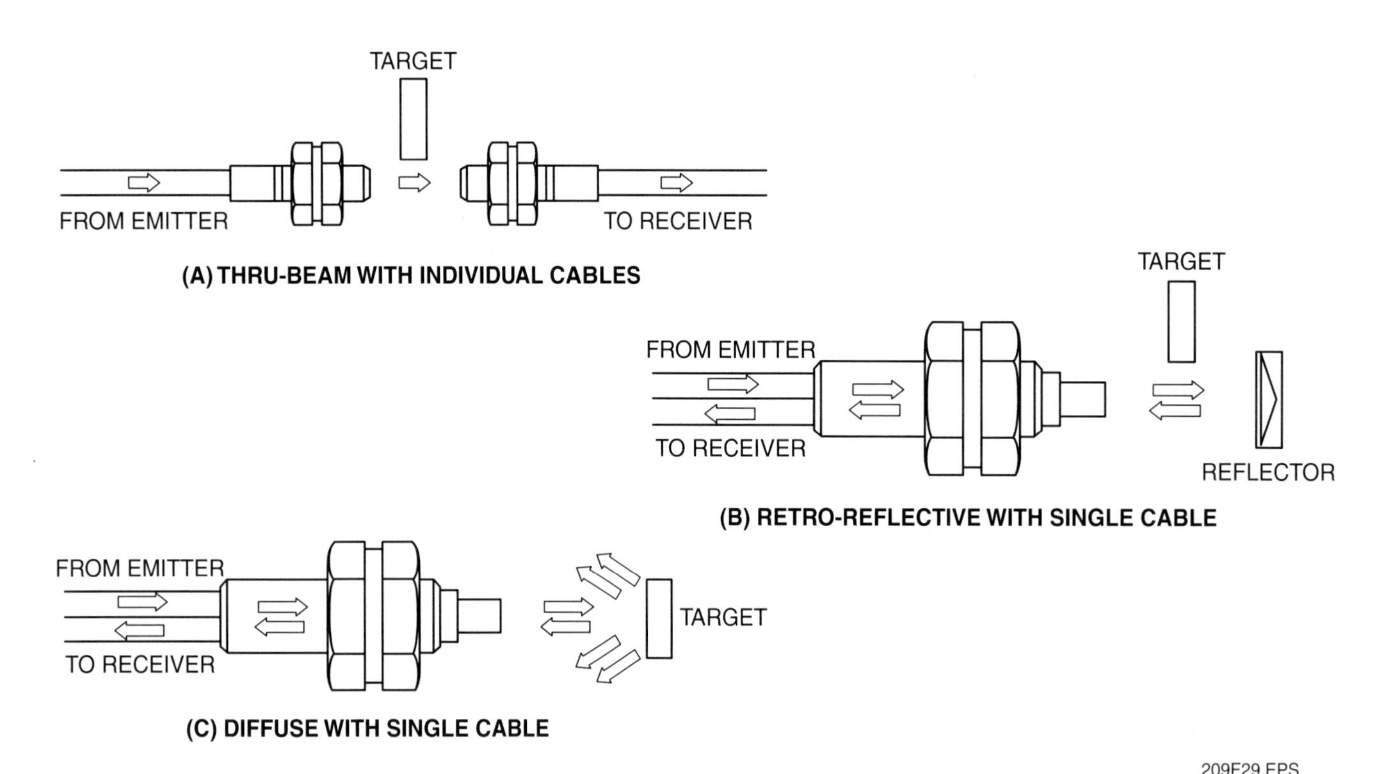

Figure 29 ◆ Fiber-optic applications.

with individual cables. The target either breaks this light transmission or allows it to pass from emitter to receiver. *Figure 29(B)* and *Figure 29(C)* show the retro-reflective and the diffuse methods in which the light is emitted and received with the same cable. In the retro-reflective method, a reflective material is installed to bounce the emitted light back to the receiver, as was the case in the infrared installations previously discussed. The diffuse method uses the target to diffuse or reflect the emitted light from the cable.

4.0.0 ◆ PROXIMITY SENSORS

Proximity sensors are used to detect the presence or absence of an object without ever touching it. They are becoming increasingly popular for industrial applications because they are versatile, safe, reliable, and can be used in place of mechanical limit switches. A typical proximity sensor is shown in *Figure 30*.

The most popular type of proximity sensor is the line-powered inductive type (*Figure 31*), which operates on the **eddy current** killed **oscillator**

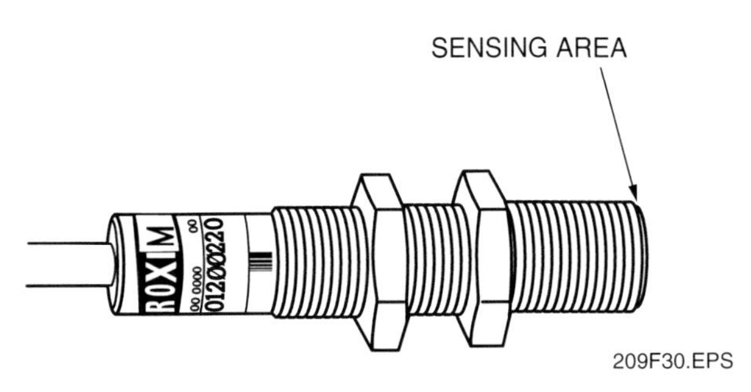

Figure 30 ◆ Proximity sensor.

(ECKO) principle. The oscillator consists of an inductor/capacitor (LC) tuned tank circuit which generates an AC magnetic flux as it oscillates and emits this field from the face of the switch. When a metal object is placed near the magnetic lines of flux, eddy currents are induced in the metal object, which in turn disrupt the tuning of the LC tank circuit. The oscillations die out or are killed. When the oscillations die, the integrator signals the trigger to switch the output of the proximity sensor.

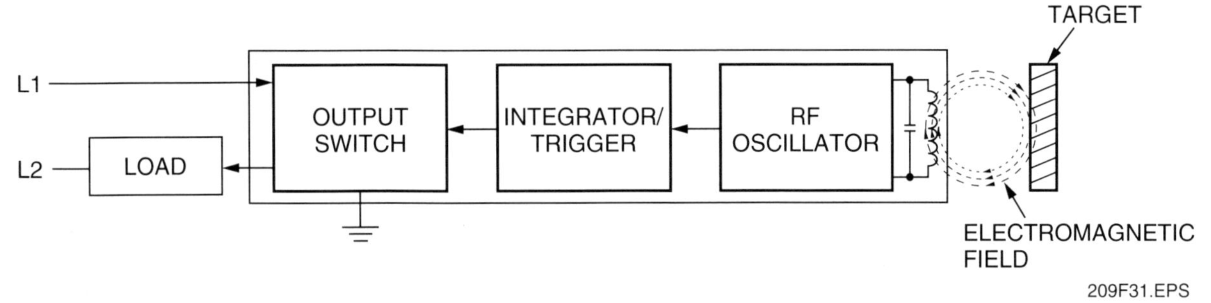

Figure 31 ◆ Line-powered inductive proximity sensor.

Summary

Understanding the classifications and operation of switches and photoelectric devices is very important to anyone working in the instrumentation and electrical fields. Without such knowledge, the selection, installation, and maintenance of electrical equipment and circuits would be difficult, if not impossible.

Switches are classified according to the number of poles, the number of throws, and according to their function (such as pressure or temperature). Their operation depends mostly on the type of contacts involved and the method of operating the contacts. Overloading is the most frequent cause of switch failure.

Photoelectric devices are generally classified only according to their function. The operation of these devices usually depends on complicated electronics that are not readily accessible to the user. For photoelectric devices, overloading and improper wiring connections are the most common causes failure.

Review Questions

1. Switches are normally used to make or break an electrical circuit that is operating _____.

a. within the rated voltage of the switch only
b. under an overload condition
c. within the rated current of the switch only
d. within the voltage and current ratings of the switch

2. Switch contacts are often coated with a material to _____.

a. provide lubrication
b. increase circuit resistance
c. decrease contact pressure
d. prevent oxidation

3. The poles of a switch define the number of _____ in a switch.

a. contacts
b. isolated circuits
c. toggles
d. throws

4. A double-pole, single-throw switch with four brass-colored screw terminals for wire connections is normally used to control _____ circuits or equipment.

a. 100V
b. 240V
c. 480V
d. three-phase

Refer to *Figure 1* to answer Question 5.

5. Label the switches according to the number of poles and the number of throws:

Switch A is _____
Switch B is _____
Switch C is _____

a. SPST
b. DPDT
c. SPDT
d. DPST

6. Switches designed to be activated mechanically by the motion of machinery are called _____ switches.

a. pressure
b. limit
c. float
d. photoelectric

7. An SCR is an electronic device that _____.

a. is triggered by infrared radiation
b. can switch on and off large amounts of power
c. uses four leads during normal operation
d. will conduct current in the forward and reverse direction

8. An increase in the he amount of light that shines on a photocell _____.

a. decreases its resistance
b. increases its resistance
c. increases its operating voltage
d. decreases its operating voltage

9. Light is converted into electrical power by a(n) _____.

a. photocell
b. proximity sensor
c. solar cell
d. SCR

10. A motion detector switch typically uses a(n) _____ sensor to detect motion.

a. vibration
b. temperature
c. infrared
d. pressure

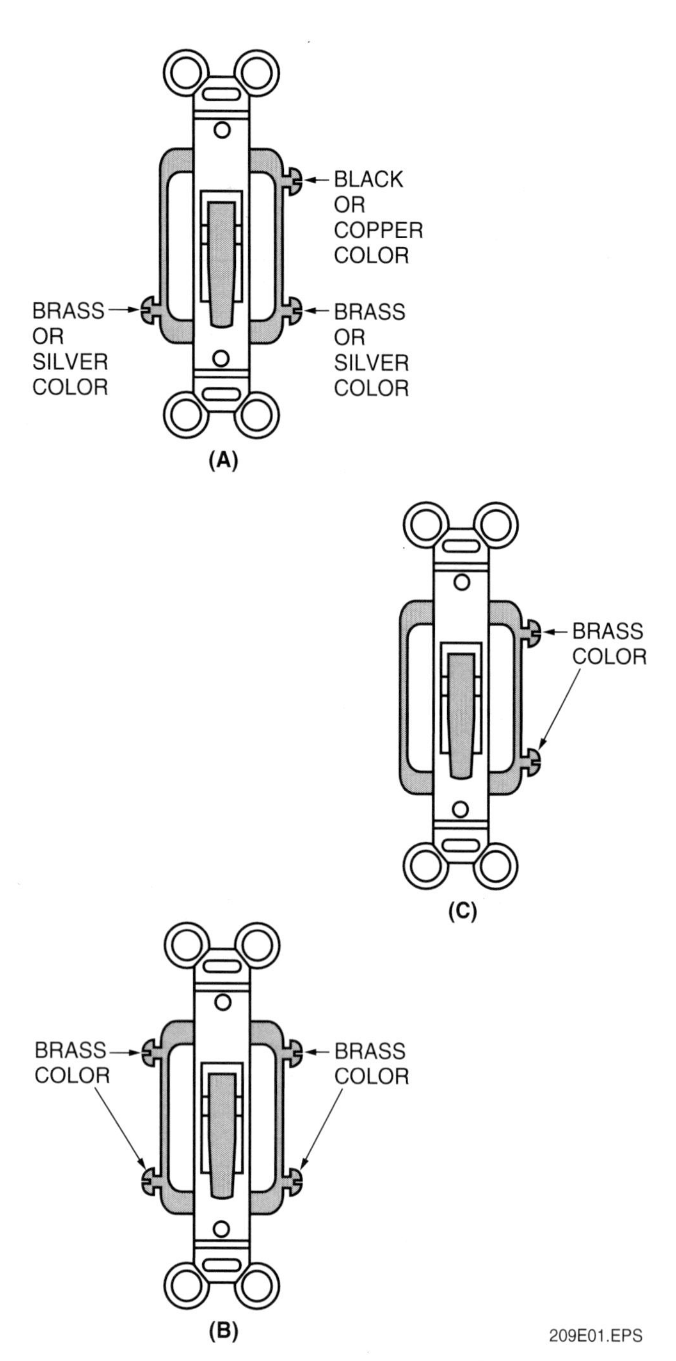

Figure 1

GLOSSARY

Trade Terms Introduced in This Module

***Absolute zero*:** A hypothetical temperature at which all molecular movement stops (–273°C, or –460°F).

***Anode*:** The positive terminal of an electrical device.

***Cathode*:** The negative terminal of an electrical device.

***Circuit*:** The complete path of an electric current.

***Deadband*:** The difference between the setpoint and the reset point in calibration.

***Diode*:** An electronic device that allows current to flow in one direction only.

***Eddy currents*:** Currents induced into the core by transformer action.

***Infrared radiation (IR)*:** Describes invisible heat waves having wavelengths longer than those of red light.

***Oscillator*:** An electrical device that generates an electrical frequency.

***Overload*:** A condition in which a circuit operates at greater than rated current.

***Oxidation*:** The chemical reaction in which oxidizers, typically oxygen, combine with other elements.

***Reset point*:** The pressure that resets the contacts in a pressure switch after the setpoint has been achieved.

***Short circuit*:** A circuit condition in which a path for current flow is provided that is not the intended path through the load.

Additional Resources

This module is intended to present thorough resources for task training. The following reference works are suggested for further study. These are optional materials for continued education rather than for task training.

American Electricians' Handbook, Twelfth Edition. Terrell Croft and Wilford I. Summers. NY; McGraw-Hill.

National Electrical Code Handbook, National Fire Protection Association.

Figure Credits

Veronica Westfall	209F01 (top)
Gerald Shannon	209F01 (bottom), 209F11, 209F15, 209F26
RadioShack	209F07
Dwyer Instruments, Inc.	209F09, 209F10, 209F13 (photos), 209F14, 209F28 (top), 209F30
Potter Electrical Signal Co.	209F12
Courtesy of Honeywell Control Products Business	209F17
Powerex, Inc.	209F18
Reprinted courtesy of Square D/ Schneider Electric	209F20 (photo)
Cutler Hammer	209F22
Solar World www.solarworld.com	209F24, 209F25
Raytek Corporation, Santa Cruz, CA	209F28 (bottom)

NCCER CURRICULA — USER UPDATE

NCCER makes every effort to keep its textbooks up-to-date and free of technical errors. We appreciate your help in this process. If you find an error, a typographical mistake, or an inaccuracy in NCCER's curricula, please fill out this form (or a photocopy), or complete the online form at **www.nccer.org/olf**. Be sure to include the exact module ID number, page number, a detailed description, and your recommended correction. Your input will be brought to the attention of the Authoring Team. Thank you for your assistance.

Instructors – If you have an idea for improving this textbook, or have found that additional materials were necessary to teach this module effectively, please let us know so that we may present your suggestions to the Authoring Team.

NCCER Product Development and Revision

13614 Progress Blvd., Alachua, FL 32615

Email: curriculum@nccer.org
Online: www.nccer.org/olf

❏ Trainee Guide ❏ AIG ❏ Exam ❏ PowerPoints Other ______________________________

Craft / Level: Copyright Date:

Module ID Number / Title:

Section Number(s):

Description:

Recommended Correction:

Your Name:

Address:

Email: Phone:

Module 12210-03

Filters, Regulators, and Dryers

COURSE MAP

This course map shows all of the modules in the second level of the Instrumentation curriculum. The suggested training order begins at the bottom and proceeds up. Skill levels increase as you advance on the course map. The local Training Program Sponsor may adjust the training order.

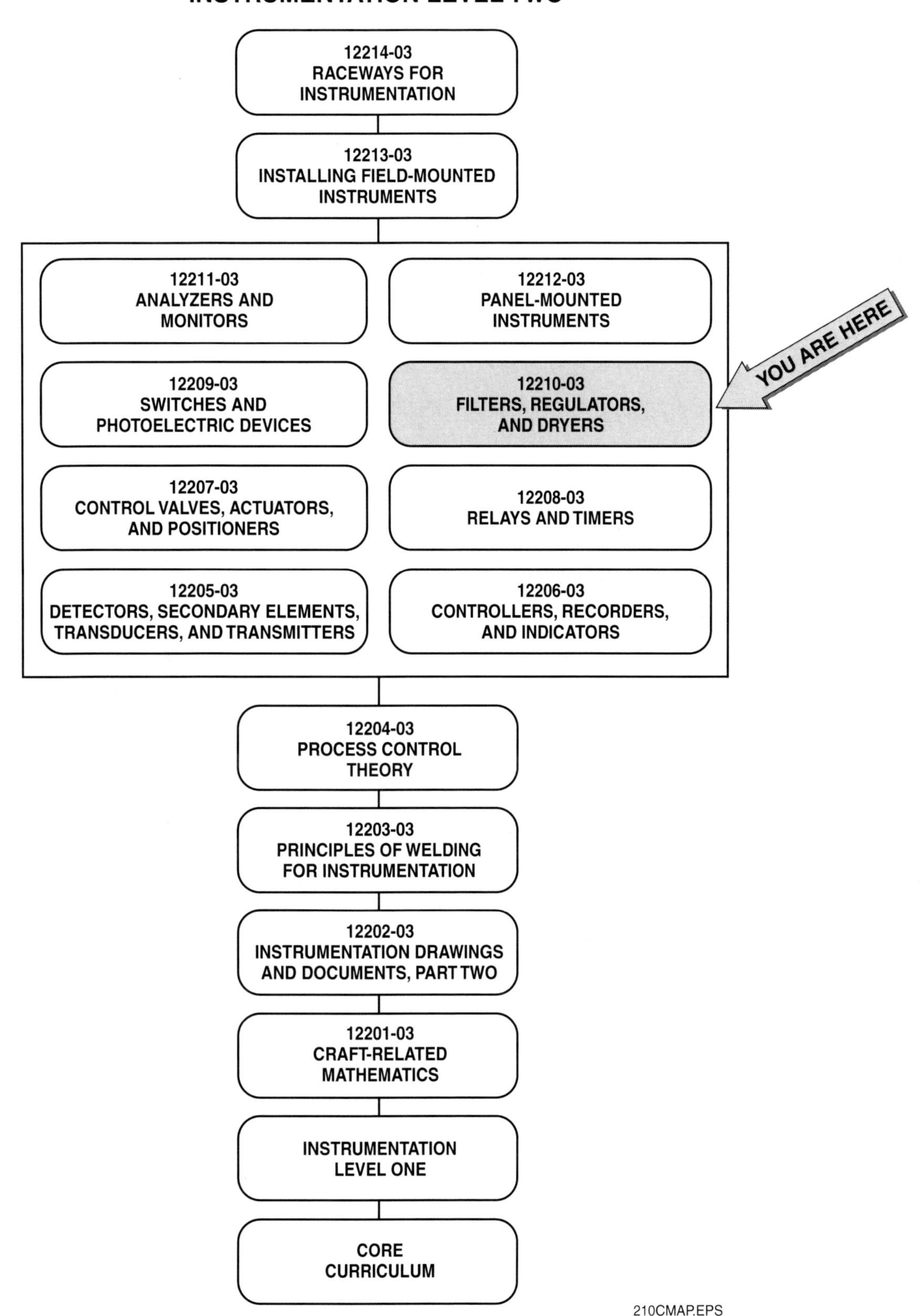

MODULE 12210-03 CONTENTS

1.0.0 INTRODUCTION 10.1

2.0.0 PRINCIPLES OF OPERATION OF VARIOUS FILTERS 10.1

2.1.0 Electronic/Electrostatic Filters 10.1

2.2.0 Compressed Air Filters. 10.3

3.0.0 PRINCIPLES OF OPERATION OF VARIOUS REGULATORS 10.4

3.1.0 Electrical/Electropneumatic Regulators 10.5

3.2.0 Pneumatic Regulators 10.6

3.2.1 Direct-Operated Regulators 10.7

3.2.2 Pilot-Operated Regulators 10.8

4.0.0 PRINCIPLES OF OPERATIONOF VARIOUS DRYERS 10.9

4.1.0 Absorbent (Deliquescent) Dryers 10.10

4.2.0 Refrigerated Dryers 10.10

4.3.0 Adsorptive Desiccant (Heat-Reactivated) Dryers. 10.11

4.4.0 Adsorptive Desiccant (Heatless) Dryer 10.11

5.0.0 FILTER, DRYER, AND REGULATOR SELECTION AND IDENTIFICATION 10.12

5.1.0 General Guidelines for Pressure Regulator Selection 10.13

5.2.0 Guidelines for Selecting a Dryer System 10.13

5.3.0 Component Schematic Symbols 10.14

SUMMARY. 10.16

REVIEW QUESTIONS. 10.16

GLOSSARY. 10.19

ACKNOWLEDGMENTS 10.20

Figures

Figure 1 Electrostatic precipitator 10.2
Figure 2 Electronic air cleaning process 10.2
Figure 3 Screen-type filter 10.3
Figure 4 Mechanical-type filter 10.4
Figure 5 Impregnated filter elements .10.4
Figure 6 Filter unit 10.4
Figure 7 Automatic-drain air filter with a 40-micron filter element 10.5
Figure 8 Charts indicating pressure drop across the filter versus air flow 10.5
Figure 9 Transducer converts low-level electrical signal to proportional air signal for valve actuation 10.6
Figure 10 Force-balance electropneumatic positioner 10.6
Figure 11 Typical adjustable air regulator 10.7
Figure 12 Self-operated regulator with a two-way stabilizer vent valve 10.7
Figure 13 Direct-operated regulators .10.8
Figure 14 Externally connected regulators 10.9
Figure 15 Pressure regulator with a pilot-operated reducing valve 10.9
Figure 16 Pressure regulator with a reducing valve operated by an instrument pilot 10.9
Figure 17 Typical closed-vessel pressure control diagram . . .10.10
Figure 18 An absorbent (deliquescent) dryer 10.10
Figure 19 Refrigerated dryer 10.11
Figure 20 Blower-purged, heat-activated dryer and diagram . .10.12
Figure 21 Absorptive desiccant (heatless) dryer 10.12
Figure 22 Heatless dryer system 10.13
Figure 23 Typical pneumatic system schematic symbols 10.15

Tables

Table 1 Comparison of Kilowatts and Costs Based on Gas Movement 10.2
Table 2 Comparison of Microns to Inches 10.3
Table 3 Listing of Commercially Available Filters 10.5

Filters, Regulators, and Dryers

Objectives

When you have completed this module, you will be able to do the following:

1. Define and discuss principles of operation of various filters.
2. Define and discuss principles of operation of various regulators.
3. Define and discuss principles of operation of dryers.
4. Define and discuss variables measured and used as inputs to various types of regulators.
5. Discuss selection criteria, and identify various filters, regulators, and dryers using specification sheets, pictures, and samples.

Prerequisites

Before you begin this module, it is recommended that you successfully complete the following: Core Curriculum; Instrumentation Level One; Instrumentation Level Two, Modules 12201-03 through 12204-03.

Required Trainee Materials

1. Pencil and paper
2. Appropriate personal protective equipment

1.0.0 ◆ INTRODUCTION

Fluid system designs vary widely in industry, but there are several components common to all of them. In particular, pneumatic systems will usually have some type of regulator to control various system parameters. Filters are also used to prevent system damage from unwanted **contaminants**.

In pneumatic systems, air dryers are necessary to minimize the moisture in an air system. This module explains construction, operation, and selection criteria for various types of filters, regulators, and dryers used in pneumatic systems.

2.0.0 ◆ PRINCIPLES OF OPERATION OF VARIOUS FILTERS

Any type of system with air flowing to intricate, delicate components is susceptible to damage from contaminants or debris. Filters are used to remove these unwanted elements.

This section discusses a variety of filters, including their construction and operational characteristics.

2.1.0 Electronic/Electrostatic Filters

An electrostatic precipitator (*Figures 1* and *2*) is an extremely efficient air pollution control device that can remove more than 99% of the undesirable particulates in a gas stream. This high efficiency is possible because, unlike other pollution control devices, the precipitator applies the collecting force only to the particles to be collected, not to the entire gas stream.

A typical precipitator operates on 480 volts AC, with the assistance of a step-up transformer and a rectifier to convert the AC to DC. The electrodes have a discharge voltage in a range of 55 to 70 kilovolts DC. If no particles are present, there is no electrode discharge. The combination of this feature along with a very low pressure drop (0.5" **water column**) across the system results in power requirements approximately 50% of comparable wet filter systems and 25% of equivalent bag filter

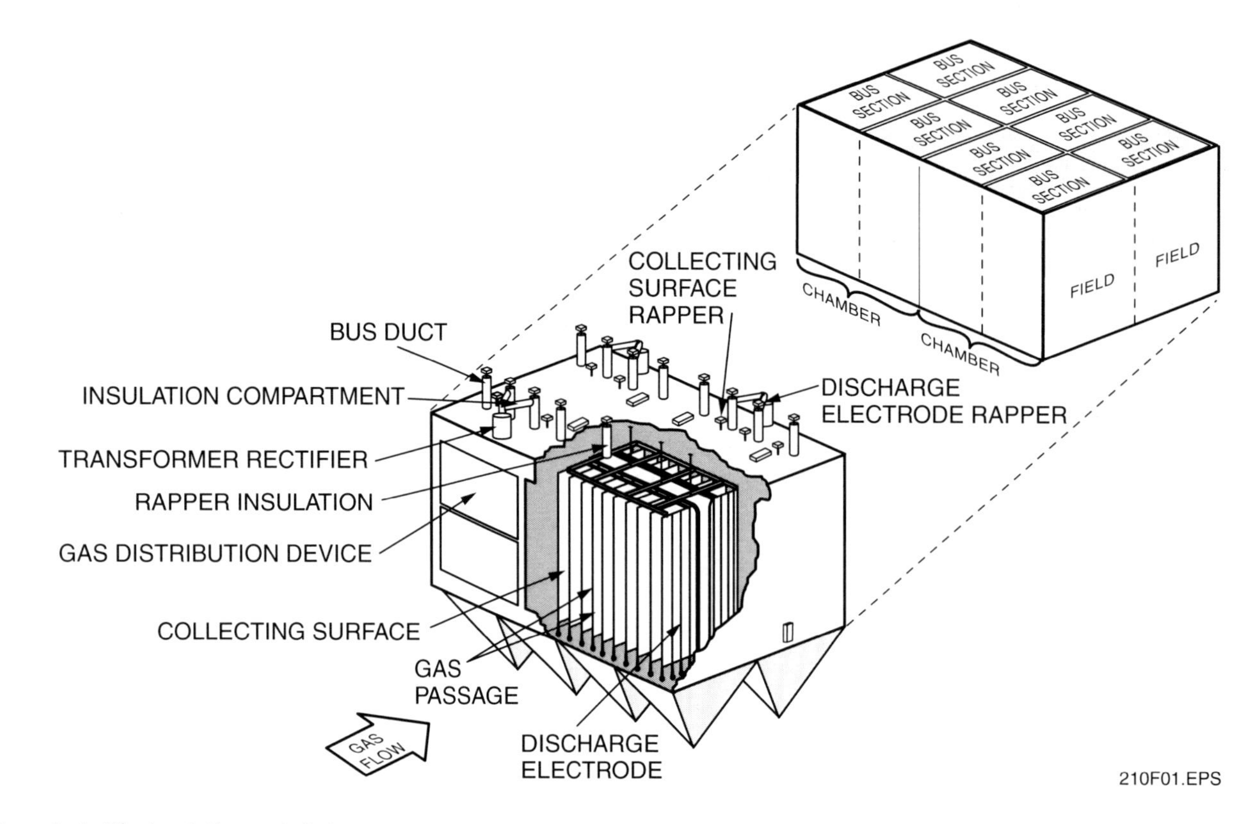

Figure 1 ◆ Electrostatic precipitator.

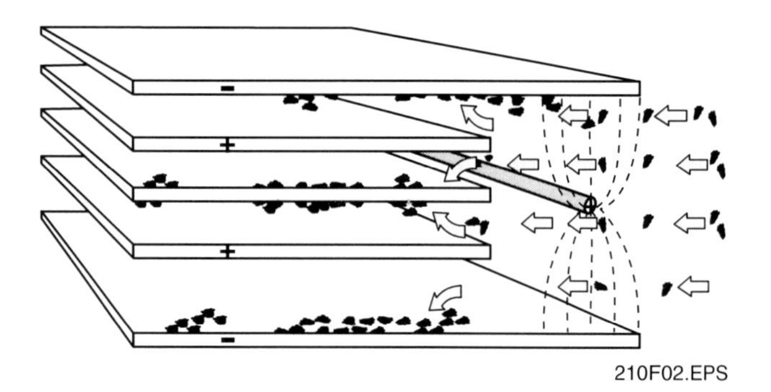

Figure 2 ◆ Electronic air cleaning process.

Table 1 Comparison of Kilowatts and Costs Based on Gas Movement

ACFM	kW	Hourly Operating Cost
20,000	10	$0.50
50,000	21	$1.05
120,000	62	$3.10

systems. *Table 1* shows a comparison of kilowatts and approximate costs of operating various sizes of precipitator systems based on actual cubic feet per minute (ACFM) air or gas movement.

In operation, a voltage source creates a negatively charged area, usually by means of wires suspended in the gas-flow path. On either side of this charged area are grounded collecting plates. The high potential difference between these plates and the discharge wires creates a powerful electric field. As the polluted gas passes through this field, particles suspended in the gas become electrically charged and are drawn out of the gas flow by the collecting plates, adhering to the plates until removed for storage or disposal. Removal is accomplished mechanically through periodic vibration, rapping, or rinsing. The gas stream, free of particulate pollution, continues on for release to the atmosphere.

Precipitators may be designed as either plate or pipe types. The plate type just described is widely used for dry dust. The pipe type, used for removal of liquid or sludge particles and particulate fumes, uses the same principles with a different mechanical setup. The discharge electrodes are suspended within a series of pipes (collecting plates) contained in a cylindrical shell under a header plate.

As noted earlier, the precipitator applies the separating force directly to the particles, regardless of gas-stream velocity, thereby requiring less power than other control devices. Because it can operate completely dry, it is ideal for use in the following situations:

Precipitators work well where water availability and disposal are problems.

Precipitators work well where very high efficiency on fine materials is required. For example, electrostatic precipitators remove very fine particulate matter from zinc oxide fumes with 97 to 98% efficiency.

Precipitators work well where large volumes of gas must be treated. In nonferrous metals production, gas flows of 600,000 ft^3/min (250 m^3/s) at temperatures up to 800°F (425°C) are treated with up to 99.6% efficiency.

Precipitators work well where valuable dry materials must be recovered, as in rotary-kiln or spray-drying operations and **calcining**.

2.2.0 Compressed Air Filters

After compression, contaminants can combine in the piping system with condensed moisture, pipe scale, and rust, creating a damaging, abrasive mixture. Instruments and tools can malfunction; spray painting and breathing air can be rendered unusable; maintenance costs will increase, and products will be spoiled. Selecting the proper type of compressed air filter depends on the class and size of the contaminants that must be removed from the system.

Compressed air filters of various designs are available for whatever application is required. The type of filter installed depends on the kind and size of the particles to be removed. The size of foreign particles is measured in microns. One micron is equal to approximately 1⁄25,400 of one inch. *Table 2* shows conversion values from microns to inches. It is not uncommon to have compressed air requirements that specify filtration as low as 0.01 microns. Filters are available that can support those filtration requirements. Some applications may have much more lenient requirements, such as shop air or other applications, and only require filtration in ranges up to 50 microns (0.0020").

Table 2 Comparison of Microns to Inches

Opening in Microns	Opening in Inches
0.01	0.000000393
0.1	0.00000393
1.0	0.0000393
5.0	0.0001965
10.0	0.000393
20.0	0.000786
30.0	0.001179
40.0	0.001572
50.0	0.001965

NOTE
The lower limit of visibility for the naked human eye is about 40 microns (0.001572"). An average human white blood cell measures 25 microns, while an average human red blood cell measures 8 microns.

Filtration methods used to remove foreign particles from compressed air may include various materials used as **filtration elements**, such as porous metal or stone, felt material, resin-impregnated paper, or resin-impregnated wool fiber. Foreign particles may also be removed by a **centrifuge** or cyclone action.

In the screen-type filter shown in *Figure 3*, air enters at the top, passes through the screen, and then passes out of the filter. The various parts are identified in the illustration.

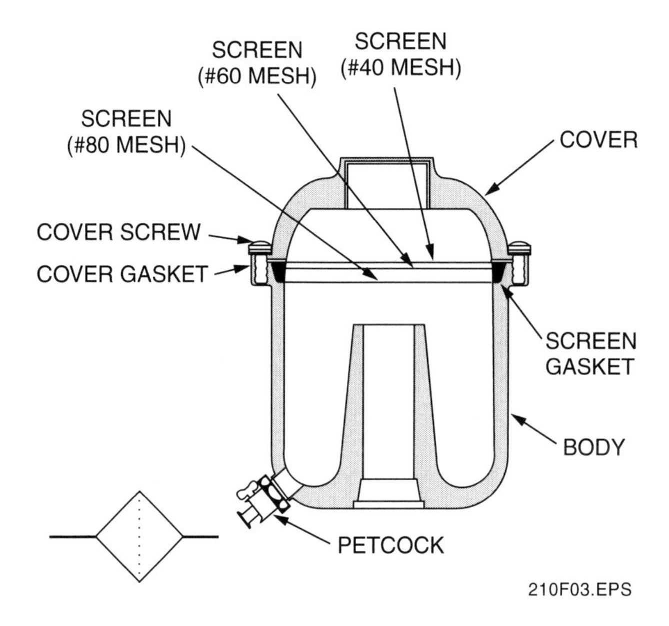

Figure 3 ◆ Screen-type filter.

In the mechanical-type filter illustrated in *Figure 4*, air enters at the bottom. As the air passes through the housing, it rotates four rotors at high speed. This action is similar to that of a centrifuge. The foreign particles, which are heavier than the air, are thrown against the outer walls of the housing by centrifugal force. The foreign particles then drop downward to the bottom of the housing and into a trap. Two rotors revolve in the same direction, while the other two rotors revolve in the opposite direction. As the air stream passes from one rotor to another, the sudden reversal of the air stream provides a cleaning action for removing the foreign particles.

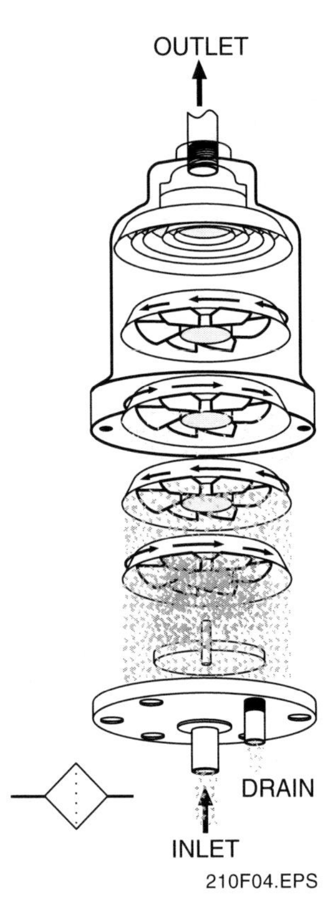

Figure 4 ◆ Mechanical-type filter.

Filtering elements that are used with various types of filter housings and devices are sometimes fabricated into a **phenolic-impregnated cellulose** cartridge that looks like a tube. The material is fused and **polymerized** for **cohesiveness** between the layers of impregnated cellulose, making the element **impervious** to gases, moisture, and common solvents. The ribbon-like impregnated cellulose is wound edgewise on a mandrel (core) to form a cylindrical element. When the element is installed in its matching housing or filter device, the air passes through the ribbons, enters at the top of the housing, passes through the filter element, and finally passes outward through the outlet. This action is sometimes referred to as *edge filtration* because the foreign particles only cling to the outer surface of the element. The particles can be removed from the element surface by reversing the flow through the filter. *Figure 5* shows various impregnated elements that can be used for a variety of filtration needs.

Various combinations of edge and cyclonic filtration are possible, as shown in *Figure 6*. Air enters at the left-hand port. A deflector plate provides a swirling or cyclonic action to the air flow. The action is similar to that of a centrifuge. The larger foreign particles that are heavier than the air are thrown against the walls of the bowl by centrifugal force. The smaller or finer particles are removed by the ribbon-type filter element.

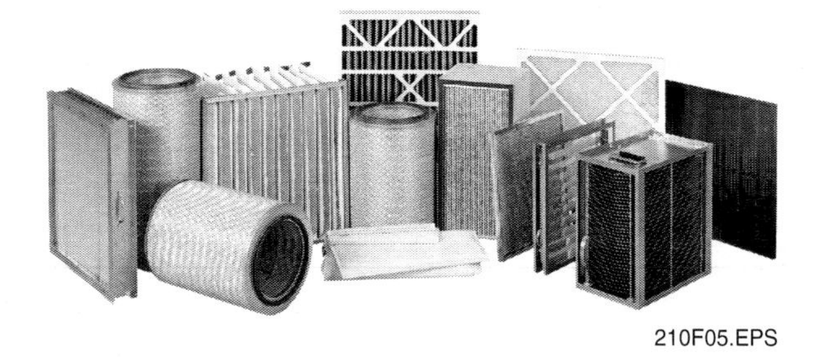

Figure 5 ◆ Impregnated filter elements.

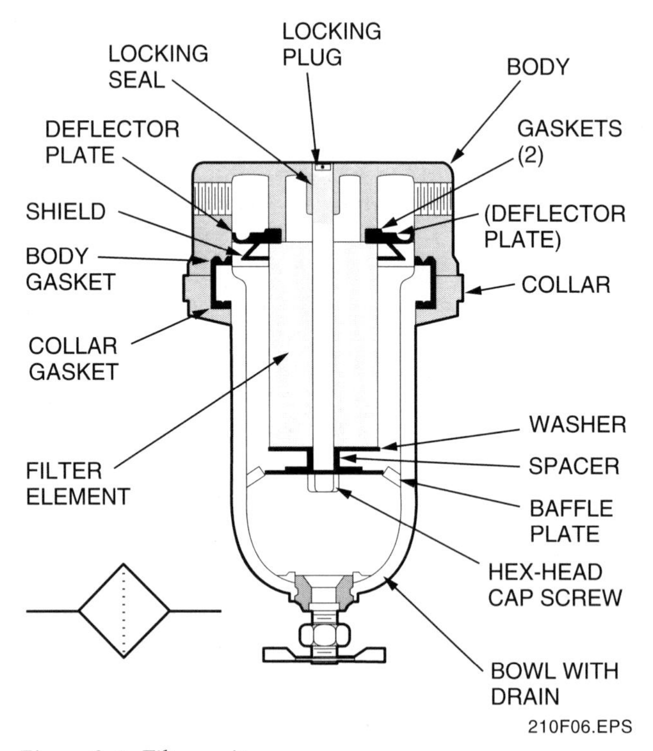

Figure 6 ◆ Filter unit.

An automatic-drain air filter is illustrated in *Figure 7*. As the liquid accumulation level rises in the bowl, the float opens the pilot valve, admitting air to the pilot chamber. As the pressure increases in the chamber, the diaphragm extends and opens the scavenger valve. Liquid and other impurities are blown out the drain opening. *Figure 8* provides charts that show pressure drop across the filter versus air flow (cubic feet per minute) through the filter. *Table 3* lists some commercially available types of air filters that can be used for pneumatic systems. Always follow the manufacturer's recommendations regarding application and installation.

3.0.0 ◆ PRINCIPLES OF OPERATION OF VARIOUS REGULATORS

A regulator is a device designed to maintain or control the pressure or flow output of a system. These devices vary in design and construction, but all have a common operational feature: they

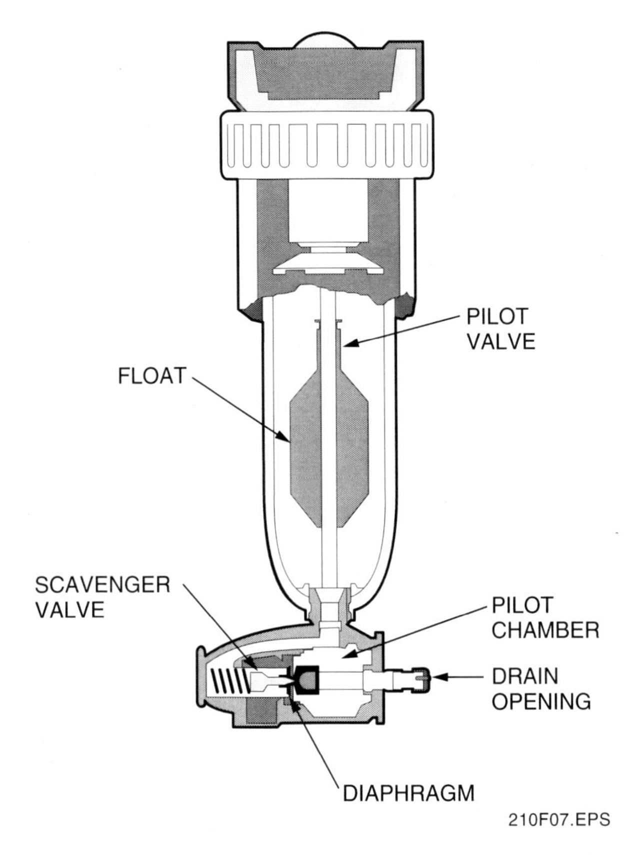

Figure 7 ◆ Automatic-drain air filter with a 40-micron filter element.

use some type of input or preset value to provide the desired range of output value. The following topics will describe a variety of different regulating devices and their operation.

3.1.0 Electrical/Electropneumatic Regulators

Electropneumatic transducers, which convert analog electronic controller signals into proportional air signals, are widely used in measurement and control applications. Two of these applications are converting the millivolt output of a thermocouple to a pneumatic signal, and converting the voltage output of a **tachometer** to a pneumatic signal.

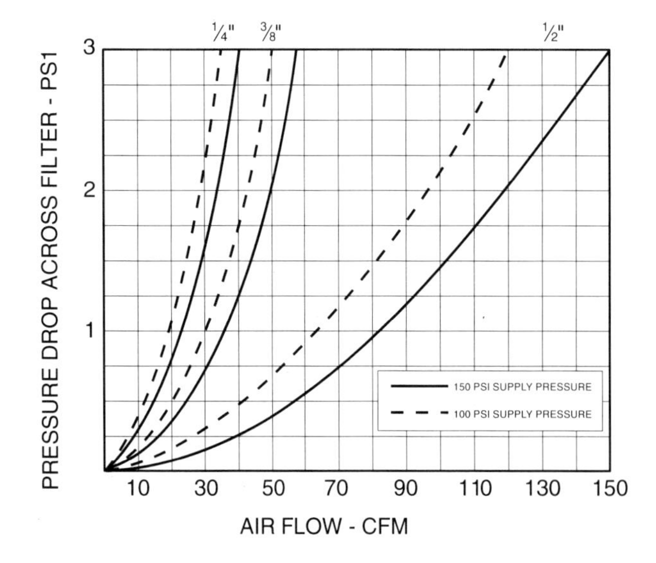

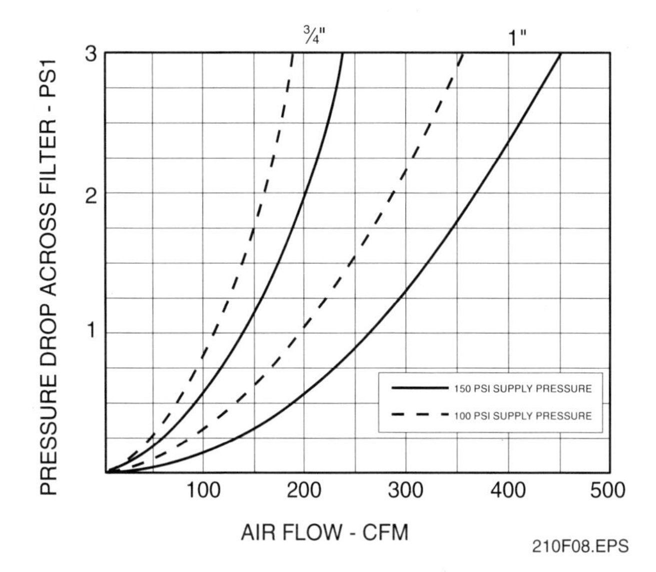

Figure 8 ◆ Charts indicating pressure drop across the filter versus air flow.

The most frequent use of the electronic to pneumatic converter is to convert the output of an electronic controller to the pneumatic signal necessary to operate some type of diaphragm-actuated valve or other regulator.

Table 3 Listing of Commercially Available Filters

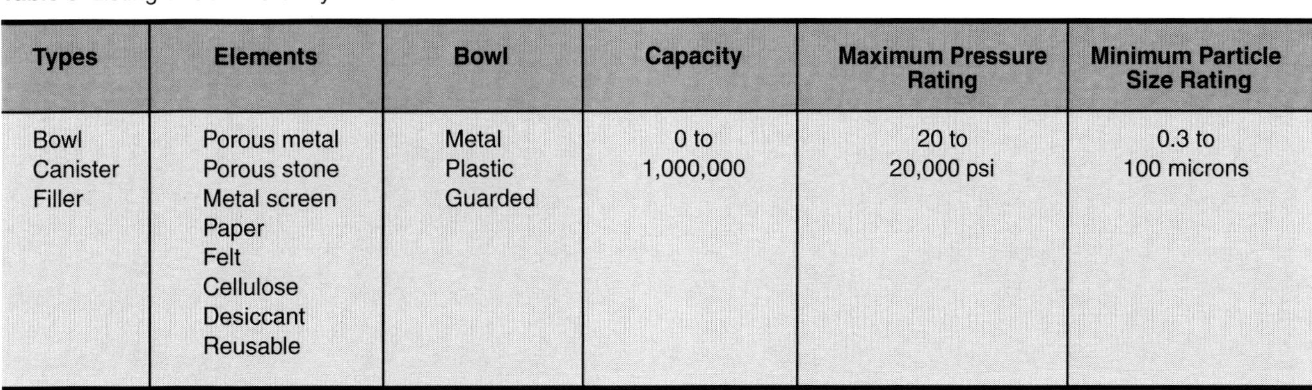

Types	Elements	Bowl	Capacity	Maximum Pressure Rating	Minimum Particle Size Rating
Bowl Canister Filler	Porous metal Porous stone Metal screen Paper Felt Cellulose Desiccant Reusable	Metal Plastic Guarded	0 to 1,000,000	20 to 20,000 psi	0.3 to 100 microns

Figure 9 shows the component parts of a transducer. The output from the transducer may be fed directly to the control valve, or it may serve as the input to a pneumatic positioner, as shown in *Figure 10*.

3.2.0 Pneumatic Regulators

Pneumatic regulators are simple devices that use energy from a pneumatic system to operate or regulate other devices. Pneumatic regulators fall into two main categories: direct-operated, also called self-operated, and pilot-operated.

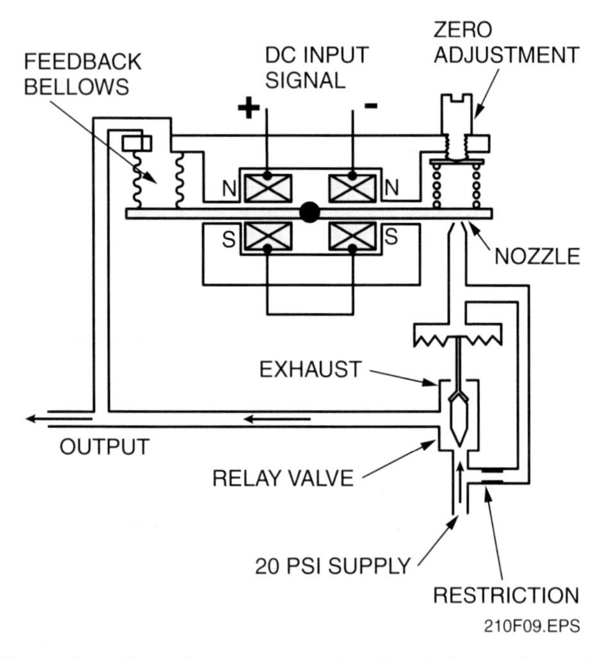

Figure 9 ◆ Transducer converts low-level electrical signal to proportional air signal for valve actuation.

These two categories of regulators may be further subdivided into pressure-reducing regulators, back-pressure regulators, and pressure relief valves. In order to select the proper pneumatic regulator for the application, it is necessary to define what must be accomplished in the application.

The majority of applications require a pressure-reducing regulator, such as in the regulation of supply air and the operation of process control valves. Assuming the application calls for a pressure-reducing regulator, one or more of the following parameters must be determined before a regulator can be selected correctly:

- Outlet pressure to be maintained
- Inlet pressure to the regulator
- Capacity required
- Shutoff capability required
- Process fluid
- Process fluid temperature
- Accuracy required
- Pipe size required
- End connection style
- Material requirements
- Control lines
- Stroking speed
- Over-pressure protection

This section examines the two categories of regulators and their variations. Because most of the regulators are applied in pressure-reducing applications, the regulators discussed are all pressure-reducing regulators.

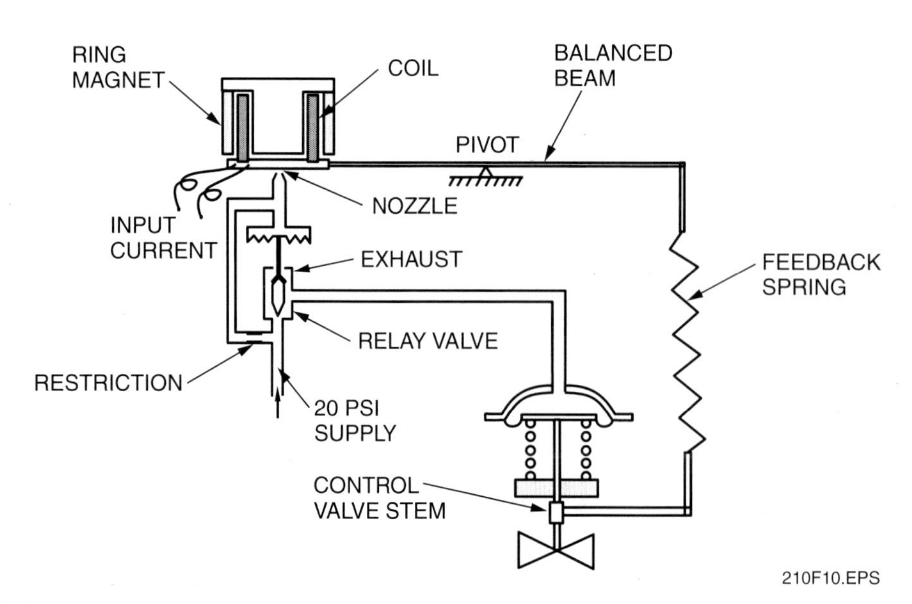

Figure 10 ◆ Force-balance electropneumatic positioner.

3.2.1 Direct-Operated Regulators

A direct-operated pressure-reducing regulator uses the downstream pressure as its controlling element and responds directly to it. A simple diaphragm-actuated direct-operated regulator is shown in *Figure 11*. This type of regulator is typically installed to regulate supply air or various gases. It may be of the adjustable type or it may be preset at the factory and lack an adjustment knob. All direct-operated regulators operate by monitoring the downstream pressure through an internal pressure monitoring process in the sensing chamber, causing the diaphragm to respond either upward or downward. The diaphragm, valve pin, and valve move upward, compressing the regulating spring. Upward movement stops when the forces below the diaphragm balance the forces above the diaphragm. When there is no downstream flow demand, the balance of forces occurs with the valve closed. When there is downstream flow demand, the balance of forces occurs when the valve opens sufficiently to compensate for demand, maintaining the desired outlet pressure.

As indicated previously, some direct-operated regulators may be non-adjustable and are preset at the factory during the manufacturing process. These are generally installed in applications where a constant pressure setting is critical. Such applications include the regulation of patient oxygen supplies or the regulation of gas pressure to furnaces, burners, and other appliances, as illustrated in the gas regulator shown in *Figure 12*.

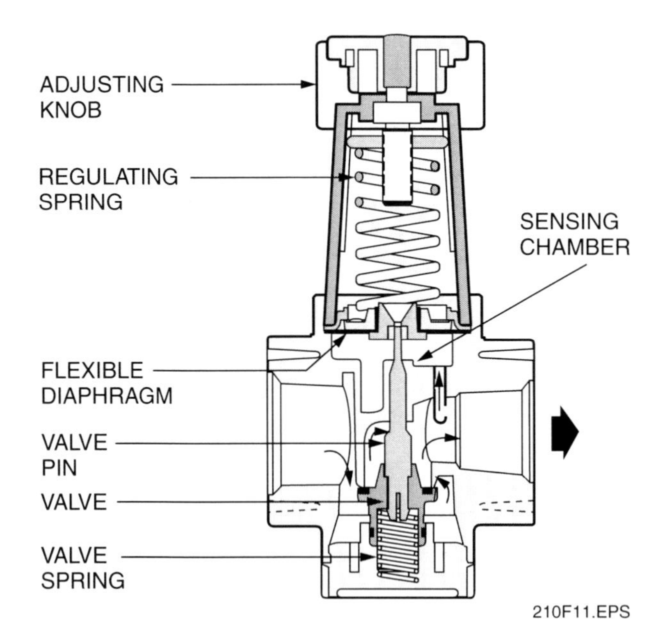

Figure 11 ◆ Typical adjustable air regulator.

In this type of regulator, downstream pressure is registered under the diaphragm via the external control line and is used as the operating medium. Increased demand lowers the downstream pressure and allows the spring to move the diaphragm and stem assembly down, opening the valve disc and supplying more gas to the downstream system. Decreased demand increases the downstream pressure and moves the diaphragm and stem assembly up, closing the valve disc and decreasing the gas supply to the downstream system.

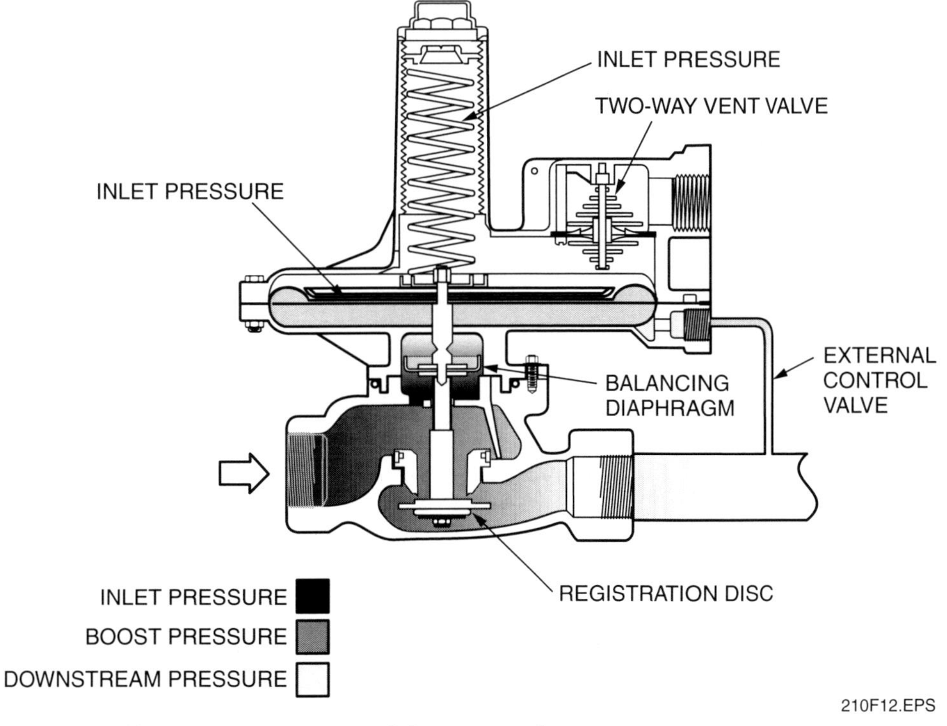

Figure 12 ◆ Self-operated regulator with a two-way stabilizer vent valve.

When the regulator responds to an increase in downstream pressure, the diaphragm moves upward. As the diaphragm rises, movement of air forces the lower vent flapper upward, carrying the upper flapper with it. This allows the air above the diaphragm to vent to atmosphere rapidly enough to minimize lag in diaphragm movement.

Pneumatic, diaphragm-actuated, direct-operated regulators may be self-contained, meaning that they have no external connection to the pressure pipeline, or they may be connected externally to the pressure pipeline. Direct-operated regulators may be weight loaded, pressure loaded, or spring loaded. *Figure 13* illustrates examples of self-contained regulators, each with a different form of loading, while *Figure 14* shows examples of externally connected regulators having two different forms of loading. Regardless of which type of loading is used, the pressure entering the regulator acts against the loading force. When this pressure increases, the loading force changes the valve opening so the desired outlet pressure is maintained.

3.2.2 Pilot-Operated Regulators

One of the disadvantages of the direct-operated regulator is poor sensitivity, which eventually led to the development of the pilot-operated regulator. In this type of regulator, a small controlling regulator operates the main regulator. Because the pilot is the controlling device of the main regulator, many of the controlling performance criteria apply to the pilot device. For example, **droop** is determined mainly by the pilot. By using very small pilot orifices and light springs, droop can be minimized. The **lockup pressure** of the pilot regulator also determines the lockup characteristics for the system. The main regulator spring provides tight shutoff whenever the pilot is locked up.

Pilot-operated regulators are preferred for high flow rates or where precise pressure control is required. Pilot-operated regulators can have internal or external pilots. *Figure 15* shows an illustration of an external pilot operating a main regulator.

Another pilot method sometimes used to control main pressure regulators is the installation of an instrument as the pilot regulator. In general, the pilot regulator is a pneumatic controller with a bellows or pressure spring as the sensing element. The controller usually includes proportional band adjustment for varying the sensitivity. The regulation is actually performed by the control valve. *Figure 16* demonstrates an instrument pilot-operated regulator.

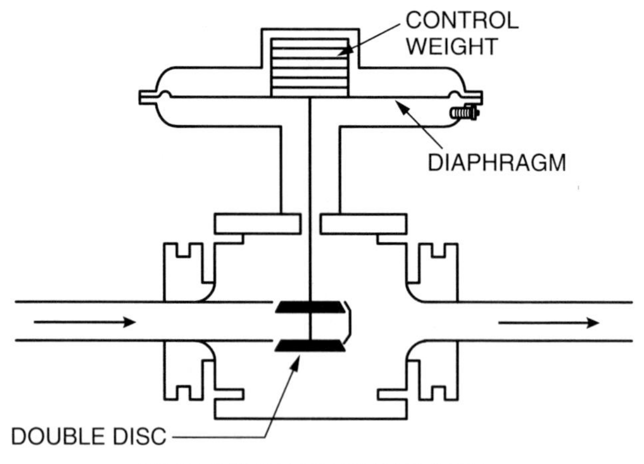

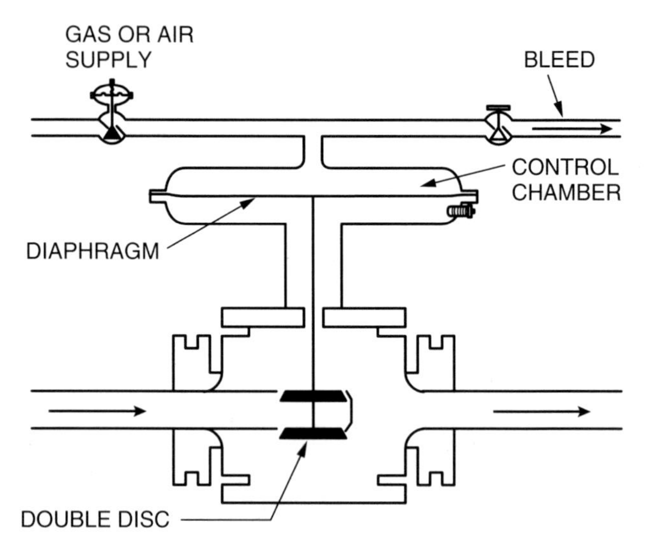

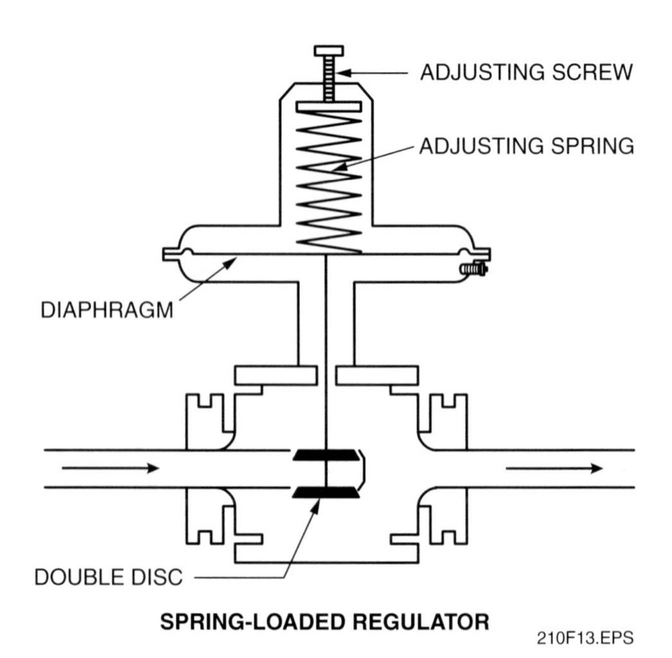

Figure 13 ◆ Direct-operated regulators.

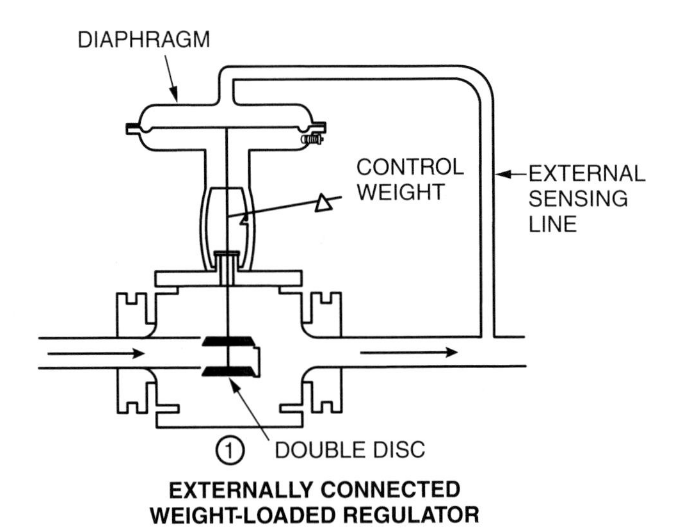

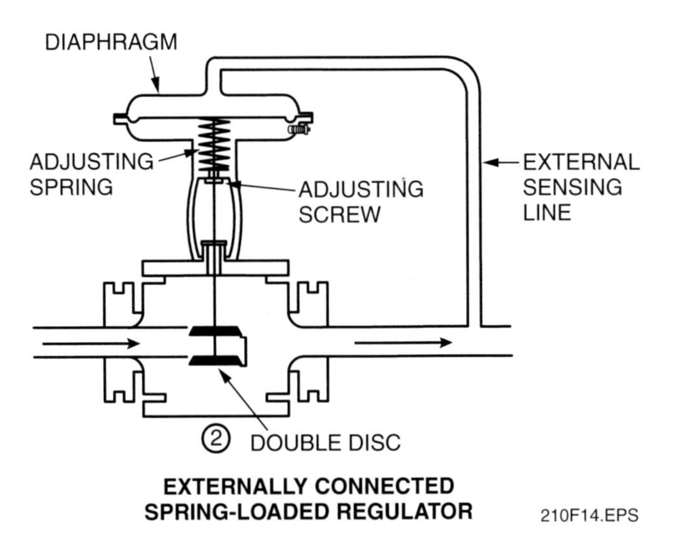

Figure 14 ◆ Externally connected regulators.

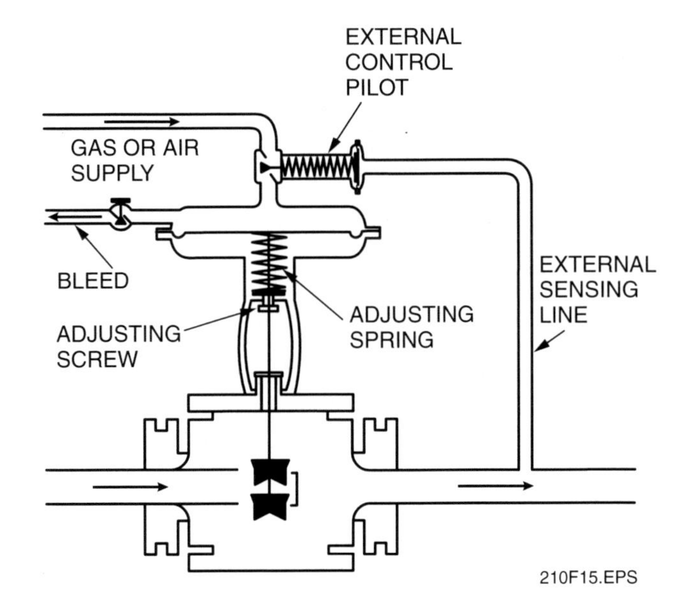

Figure 15 ◆ Pressure regulator with a pilot-operated reducing valve.

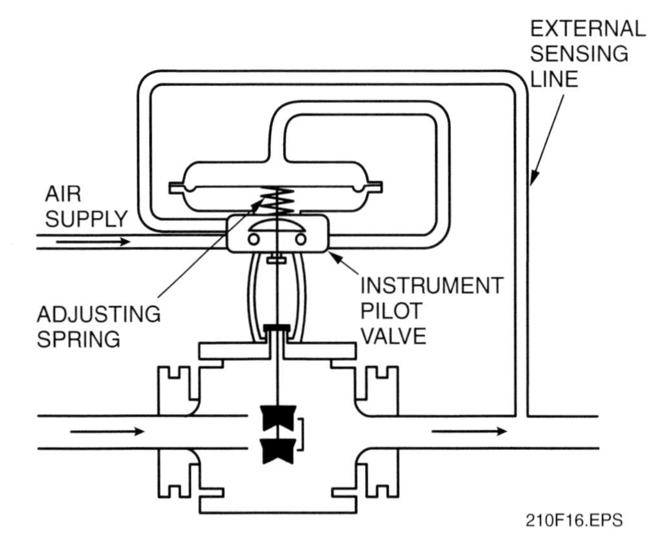

Figure 16 ◆ Pressure regulator with a reducing valve operated by an instrument pilot.

The regulation of pressure inside a closed vessel is a common control application. Gas holders (tanks for gas supply) require such control. Liquid oxygen, nitrogen, and hydrogen are commonly stored in vessels in which pressure must be closely regulated. In these cases, an instrument pilot and a diaphragm control valve are commonly used, as shown in *Figure 17*. The instrument pilot may be a pneumatic pressure controller providing a pneumatic signal to a control valve, or a pressure controller with an electrical output. The electrical output may be sent to an electro-pneumatic relay that provides the pneumatic signal by converting the electrical signal received from the controller. In some applications, the pneumatic pressure pilot and a diaphragm control valve provide more precise pressure control. This type of system is often used to achieve faster response time when there is a great distance between the sensing point and the control point.

4.0.0 ◆ PRINCIPLES OF OPERATION OF VARIOUS DRYERS

Anyone responsible for maintaining and operating plant compressed-air systems will be aware of problems caused by water in the air system. The problems are very apparent to those who operate pneumatic tools, rock drills, automatic air-powered machines, sandblasting equipment, pneumatic logic devices and controls, air gauging equipment, and paint spray equipment.

The following problems are among the many that can be caused by water in an air system:

- Washing away necessary lubricants
- Causing rust and scale to form within pipelines

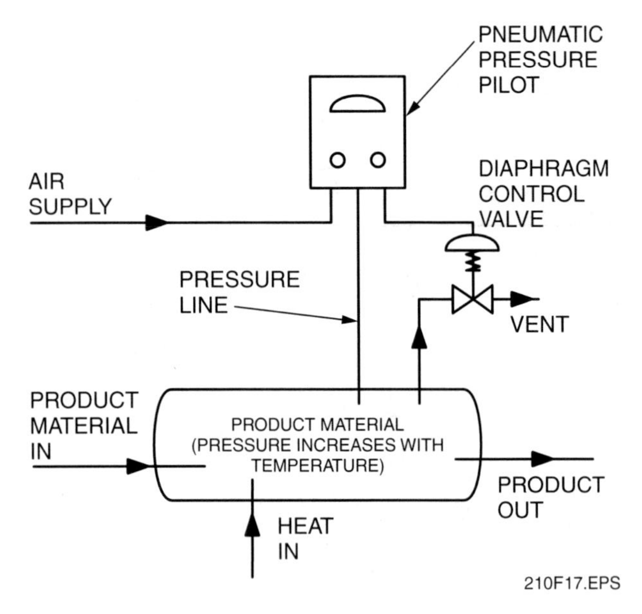

Figure 17 ◆ Typical closed-vessel pressure control diagram.

- Increased wear and maintenance of pneumatic devices
- Sluggish and inconsistent operation of air valves and cylinders
- Malfunction and high maintenance of control instruments and air logic devices
- Product spoilage by spotting in paint and other types of spraying
- Rusting of parts that have been sandblasted
- Freezing in exposed lines during cold weather
- Further condensation and possible freezing of moisture at the exhaust whenever air devices are rapidly exhausted

The increased use of control systems and automatic machinery has made these problems more common and has caused increased awareness of the need for drier compressed air. Air-drying equipment is now a standard part of any well-designed compressed-air system.

4.1.0 Absorbent (Deliquescent) Dryers

Water vapor in compressed air can be removed by an **absorption** process. Absorption is a process in which moisture or water vapor is pulled into a chemical substance in much the same way that a sponge absorbs liquids. There are two basic absorption methods. In the first, water vapor is absorbed in a solid block of chemicals, called a **desiccant**, without liquefying the solid. The chemicals used in the solid insoluble type are typically dehydrated chalk and magnesium perchlorate. Another type (*Figure 18*) uses **deliquescent dryers** like lithium chloride and calcium chloride salts, which react chemically with water vapor and liquefy as the absorption proceeds. Periodic replenishment of some sort for both types of dryers must be performed.

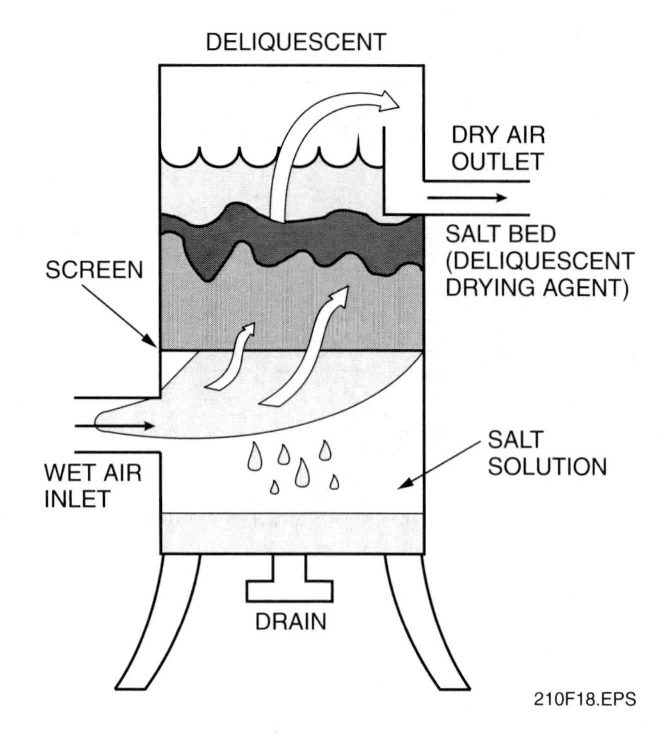

Figure 18 ◆ An absorbent (deliquescent) dryer.

Problems exist with the deliquescent drying process. One problem is that most of the drying agents are highly corrosive. Another problem is that the desiccant pellets can soften and bake at temperatures exceeding 90°F (32°C). This may cause an increased pressure drop. In addition, a fine corrosive mist may be carried downstream with the air, causing corrosion to the system components. However, this type of air dryer has the lowest initial and operating cost of the more common air dryers. Maintenance is simple, requiring only periodic draining and replacement of the deliquescent drying agent.

4.2.0 Refrigerated Dryers

When the temperature of air is lowered, its ability to hold gaseous water is reduced. This is what takes place in an aftercooler. However, the typical minimum air temperature attainable in an aftercooler is limited by the temperature of the cooling water or air. If extremely dry air is needed, a refrigerated dryer is employed, as shown in *Figure 19*.

In these devices, hot incoming air is allowed to exchange heat with the cold outgoing air in a heat exchanger. The lowest temperature to which the air is cooled is between 32.5°F and 33°F, which prevents frost from forming. This type of air-drying equipment has relatively low initial and operating costs.

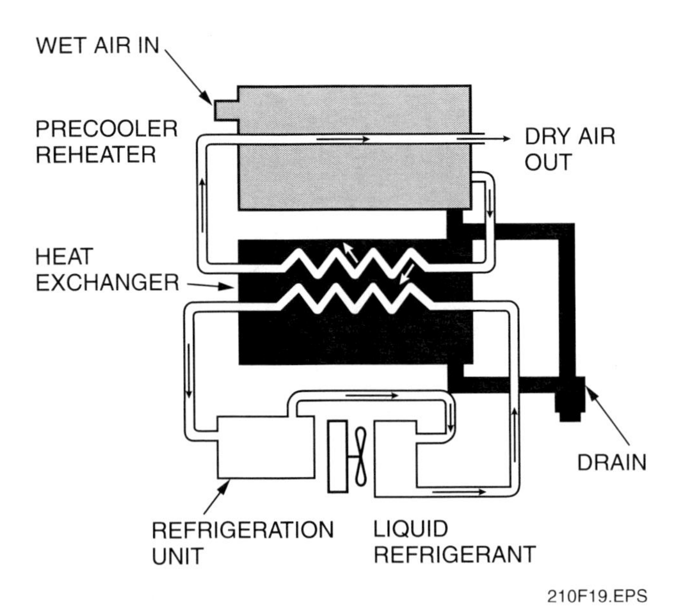

Figure 19 ◆ Refrigerated dryer.

4.3.0 Adsorptive Desiccant (Heat-Reactivated) Dryers

Heat-reactivated, adsorptive, desiccant dryers are used for drying many common gases such as carbon monoxide, carbon dioxide, argon, oxygen, and nitrogen, as well as hydrocarbon gases such as ethylene, ethane, methane, and natural gases. **Adsorption** differs from absorption. In absorption, chemical dryers soak up the moisture in the air or gas being dried. In adsorption, a chemical desiccant gathers moisture on its surface. Heated air is used to remove the gathered moisture from the surface, and the air is exhausted into the atmosphere. Most of these drying applications require an extremely low water content of 1 to 25 ppm by weight. Some common heat-reactivated compressed-gas dryers are discussed in the following paragraphs.

Some heat-reactivated dryer models are fully automatic, with their drying tower cycles controlled by a preset cam timer or optional programmable logic controller (PLC). Ambient air is pulled through the intake filter silencer on the blower and forced into a heater that has a temperature-controlled thermostat preset to heat the air to 400°F before it enters the regeneration chamber. The heated air flows through the desiccant in a reverse direction to the drying circuit, adsorbing the moisture from the surface of the desiccant, and exhausts into the atmosphere.

A standard drying and reactivation time cycle is 8 hours. While one tower is drying the incoming air for 4 hours, the other tower is regenerating for 3 hours, cooling for 50 minutes, and re-pressurizing to the line pressure for 10 minutes before switchover to the other chamber. One such model is shown in *Figure 20*.

This type of dryer provides pressure dew points of as low as –40 to –100°F (–40 to –74°C) and reduces the dew point and temperature elevation spikes commonly experienced with internal heat-reactivated dryers.

In the split-stream type of adsorption dryers, a throttling valve diverts some gas from the inlet side of the dryer. This gas is heated and then strips moisture from the adsorbent during the regeneration. The gas is then cooled. Water condenses and is drained from the unit. The gas is then returned to the inlet without purge losses. Split-stream reactivation is commonly used to dry natural gas and many other hydrocarbon gases, or other gases that cannot be purged to the atmosphere. The distinct advantage of the split-stream reactivation versus a closed-loop reactivation is obviously the absence of the reactivation blower. However, pressure loss is incurred to operate the throttling valve and the regeneration system.

4.4.0 Adsorptive Desiccant (Heatless) Dryer

Another type of desiccant dehydrator in use is the heatless dryer. These units require no electric heaters or external source of purge air to operate. In *Figure 21(A)*, compressed air enters at the bottom of the left tower and passes upward through the desiccant, where it is dried to a very low moisture content. The dry air passes through the check valve to the dry air outlet. At the same time, a small percentage of the dry air is passed through the orifice between the towers and flows down through the right tower, reactivating the desiccant and passing out through the purge exhaust. At the end of the cycle, the towers are automatically reversed, as shown in *Figure 21(B)*.

Some models are available that combine a filtration system with a drying system, like the example shown in *Figure 22*. In the illustration shown, the system is manufactured in a single enclosure, completely assembled and ready for installation. The filters have automatic drains for discharge of collected liquids. In this installation, easily accessible filter cartridges remove solids to 5 microns in the first stage and liquids and solids to .03 microns in the second stage. In most of these types of systems, the dryer works best if the compressed air provided is first pre-dried in a refrigeration-type air dryer (35°F to 50°F pressure dew point). This system can be continuously operated.

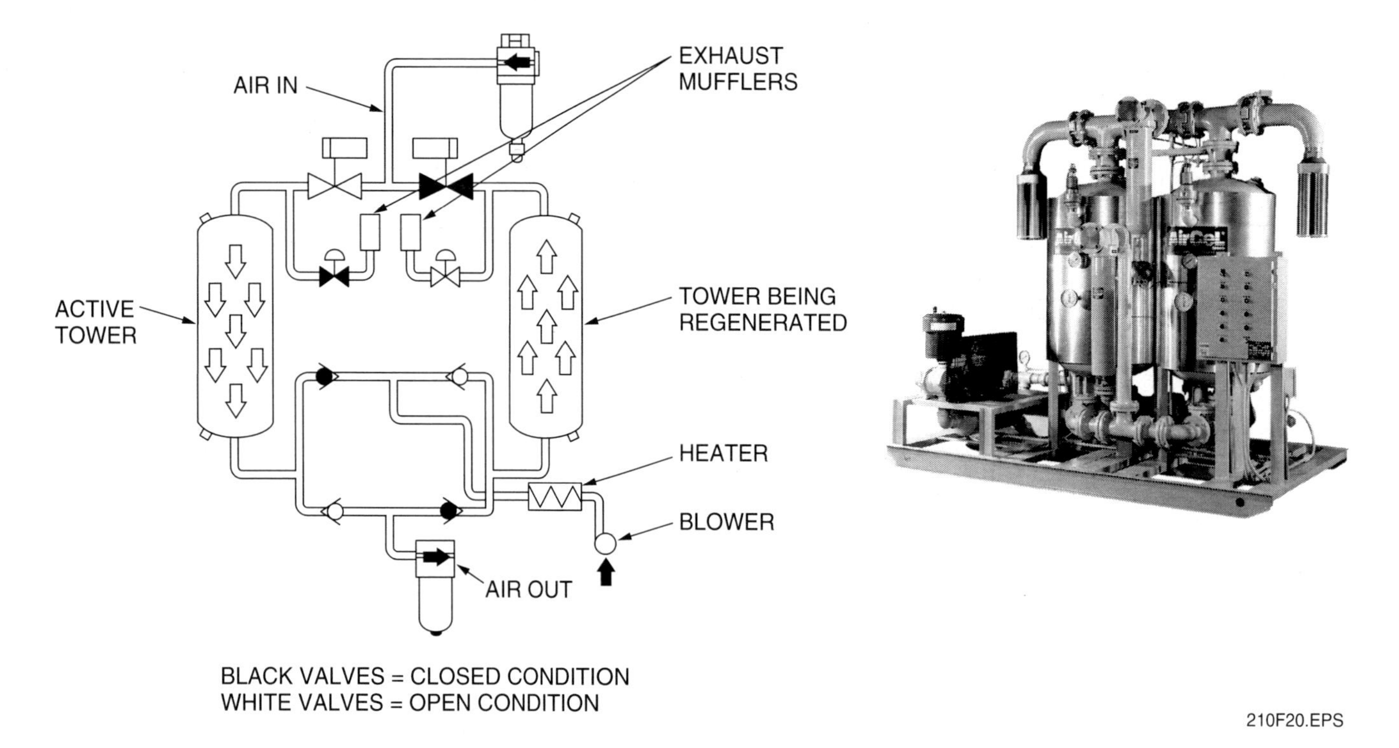

Figure 20 ◆ Blower-purged, heat-activated dryer and diagram.

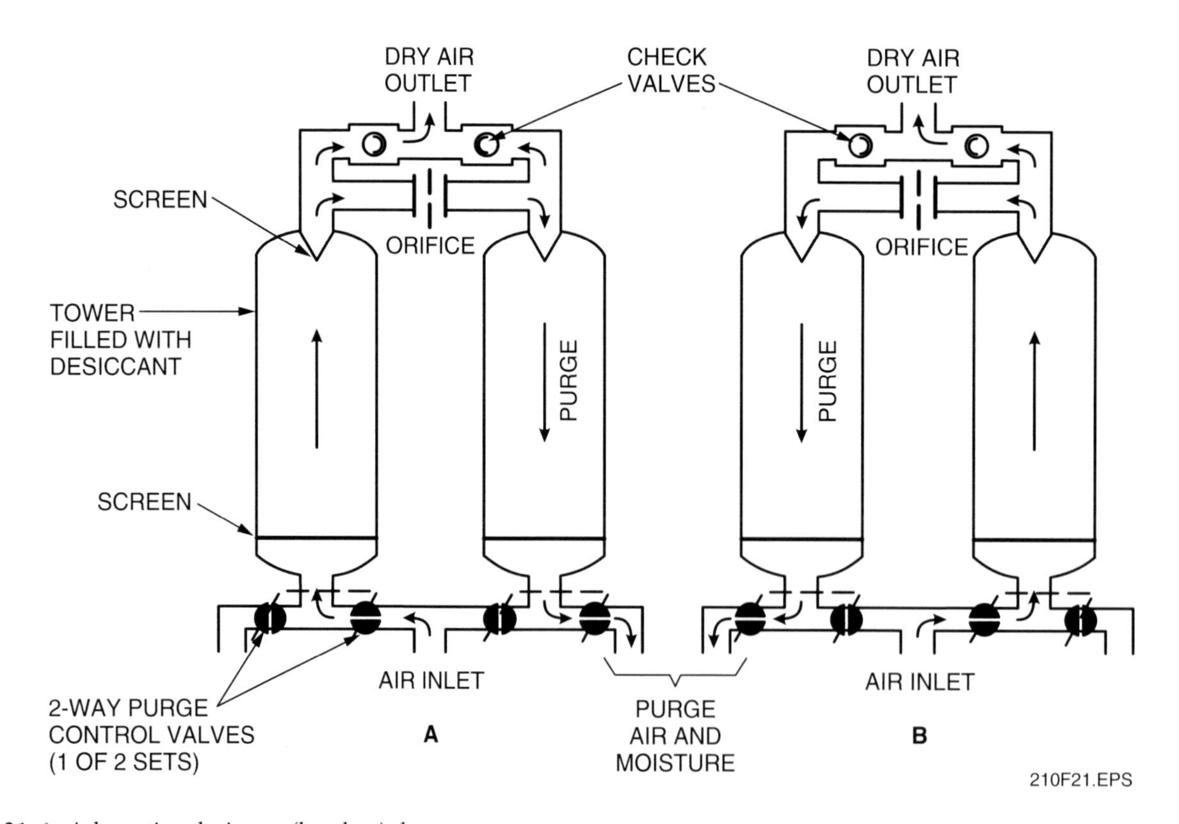

Figure 21 ◆ Adsorptive desiccant (heatless) dryer.

5.0.0 ◆ FILTER, DRYER, AND REGULATOR SELECTION AND IDENTIFICATION

There are literally hundreds of variations of designs, sizes, and capabilities of filters, dryers, and regulators. Each type has a specific application. This makes the selection process as important as correct installation in the system. The following sections provide some guidelines for identifying and selecting the right component for the system.

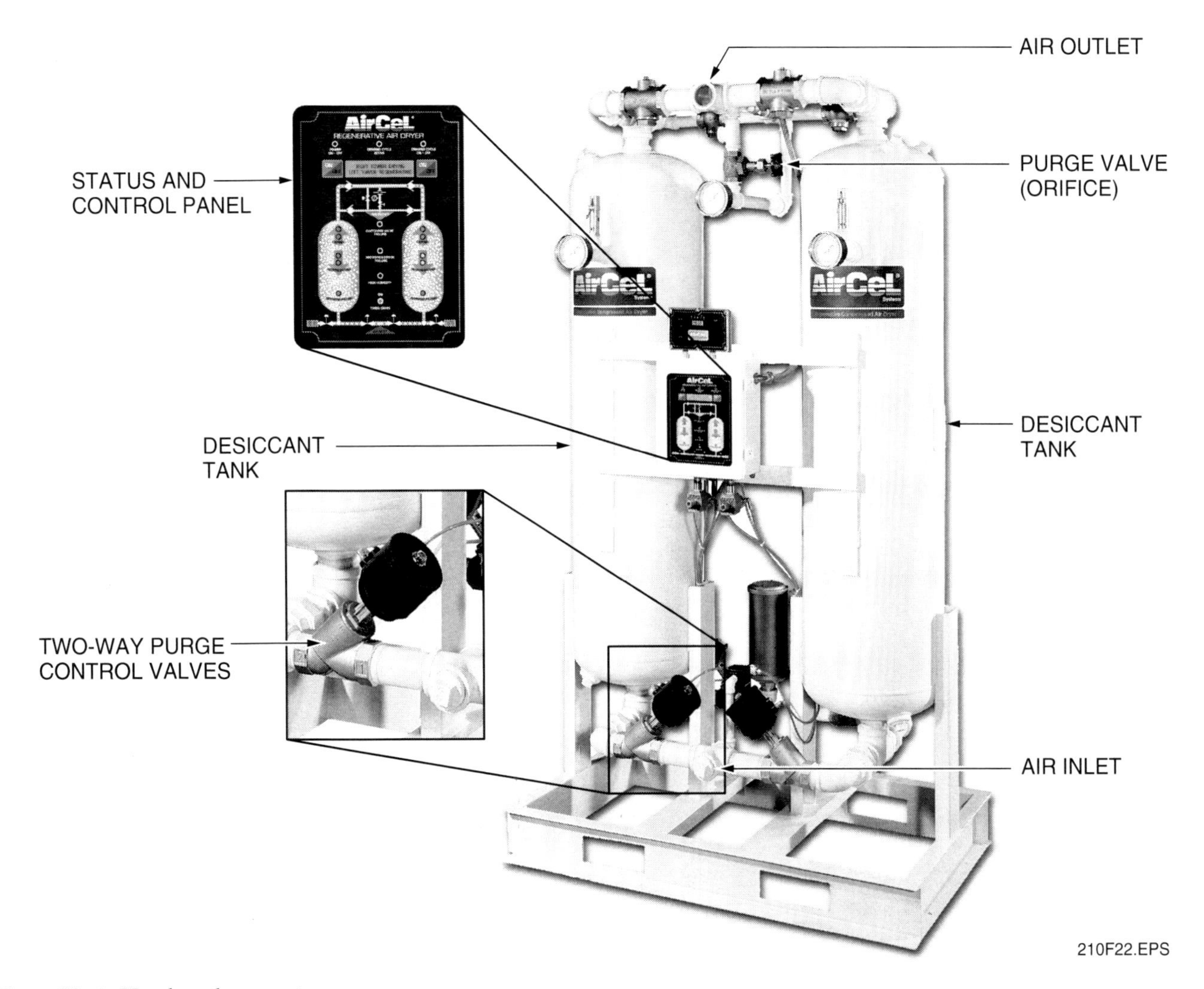

Figure 22 ◆ Heatless dryer system.

5.1.0 General Guidelines for Pressure Regulator Selection

This section provides guidance for selecting regulators. These guidelines assist in selecting the appropriate components for the intended application.

All regulators should be installed and used in accordance with federal, state, and local codes and regulations.

Adequate overpressure protection should be installed to protect the regular and all downstream equipment (in the event of regulator failure).

The capacity of a regulator when it has failed full open is usually greater than the regulating capacity. For that reason, use the wide-open flow rates when sizing a relief valve. These full-open capacities can be larger than normal service flow rates.

If two or more available springs have published pressure ranges that include the desired pressure setting, use the spring with the lower range for finer control.

The recommended selection for port diameters is the smallest port diameter that will handle the flow.

Always check the product bulletin to determine materials and temperature capabilities available. Use stainless steel diaphragms and seats for higher temperatures, such as in steam service.

Regulator body size should never be larger than pipe size. In many cases, the regulator body is one size smaller than the pipe size.

Always refer to the product bulletins to make the final selection.

5.2.0 Guidelines for Selecting a Dryer System

Moisture, either liquid or vapor, is present in compressed air as it exits the compressor system. If this moisture is not properly removed, a compressed air system can lose efficiency and require dramatically increased maintenance, which can result in costly downtime.

To avoid these problems, compressed air system designers have made a number of purification devices available to remove remaining water vapor and other contaminants. The proper selection of these devices becomes more critical as pneumatic applications and compressed air systems become increasingly sophisticated.

Ambient air, which includes atmospheric humidity (water vapor), is drawn into the compressed air system, where it is compressed to a desired discharge pressure. Once the compressed air is discharged, its temperature is elevated and its moisture content is high. If a lubricated compressor is used, a small quantity of compressor lubricant, in both liquid and vapor form, is discharged with the compressed air.

Because the majority of pneumatic instruments and processes cannot tolerate hot compressed air, compressors are normally supplied with aftercoolers and moisture separators.

Aftercoolers are heat exchangers that use either water or ambient air to cool the compressed air. As the water and lubricant vapors within the compressed air cool, a significant amount of them condenses into liquid. The amount of condensation depends on the temperature of the air when it leaves the aftercooler. Usually, this temperature ranges from 10°F to 15°F above the temperature of the cooling air or water.

Following this stage, the condensed water is collected and removed by the moisture separator and discharged through a drain valve. However, the compressed air is still saturated with water vapor at the discharge of the aftercooler/moisture separator. It may require some form of dryer system if additional condensation is generated downstream when the compressed air cools further.

Plant maintenance personnel and system designers must determine the air quality requirements for their specific compressed air applications.

The three areas that should be addressed are these:

- Review the air quality requirements of instrumentation, tools, and other air-powered equipment; these are available from the manufacturer.
- Determine the air quality required for use in processes using compressed air; these can be obtained from the process designer.
- Estimate the expected ambient conditions for all pneumatic equipment, processes and piping; for example, outdoor locations during the winter months require compressed air to be dried to a lower dew point than indoor, heated locations.

Any oversight regarding these issues can result in misapplied purification equipment, inefficient system operation, high operating and maintenance costs, and even unnecessary capital expenditures.

Specifying a dew point that is much lower than the requirements may add considerably to both the initial cost and the operating costs of the system. Trying to operate on dew points that are well above plant requirements will result in higher costs both in labor and equipment replacement.

A much lower dew point may be required in the winter than in the summer. In the winter, the cooling water temperature and ambient temperature are generally lower than in the summer, resulting in a variation in temperature of the air supplied to the dryer. These considerations must also be addressed when selecting air supply drying systems.

5.3.0 Component Schematic Symbols

One very important aspect of component identification is the ability to identify filters, regulators, and dryers from blueprints and schematics. *Figure 23* illustrates some common symbology used in the trade. The legend of any system print or schematic will show symbols that represent all components on the drawing.

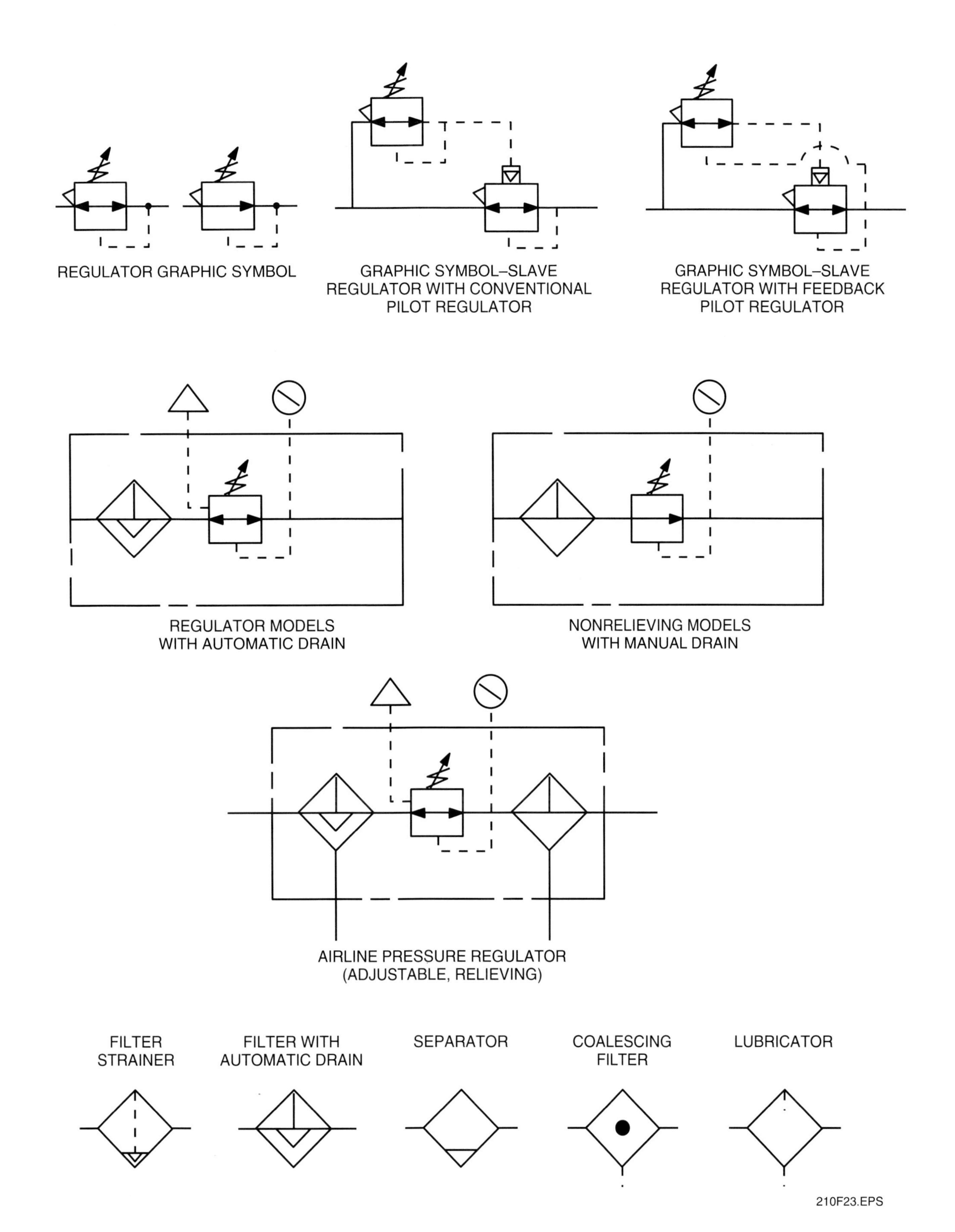

Figure 23 ◆ Typical pneumatic system schematic symbols.

Summary

Filters, regulators, and dryers share one common element; they are not directly related to the process, but they all act as elements of quality in the supply of usable, maintenance-free instrument air and some gases. It is very common to see all three devices used in conjunction with one another in a hierarchical fashion, with one serving the other in a predetermined sequence.

Filters need regulators to function properly, and vise versa. If a filtering system receives unregulated air or gas pressure, it cannot function properly and may even be damaged or destroyed by extreme pressures. On the other hand, if the pressure is below the working range of the filter, the filter cannot work efficiently. It depends on the regulator to provide the ideal working pressure.

Regulators also depend on filters to do their job before the regulator receives its input pressure. Contaminants and foreign matter in unfiltered air can cause regulators to plug or become restricted, which could result in improper regulation or control.

Dryers are typically the last step in the sequence of delivering quality air to a system. They must receive both regulated and filtered air in order to operate efficiently and maintenance free.

Because all three devices depend on one another for proper operation, it is of the utmost importance that air quality supply systems be designed and selected according to certain guidelines.

Review Questions

1. An advantage of using an electrostatic precipitator is its _____.

a. application to the entire gas stream
b. low electrical cost
c. high pressure drop
d. discharge even when no particles are present

2. One micron is equivalent to approximately _____ inches.

a. 0.0000393
b. 0.00000393
c. 0.000393
d. 0.000000393

3. A recommended way to remove foreign particles from an edge filtration filter element is to _____.

a. rinse the element in solvent
b. use a scraping device on the element
c. reverse the flow through the element
d. replace the element

4. A device designed to maintain a constant pressure or flow in the output of a system is a _____.

a. filter
b. diaphragm
c. control valve
d. regulator

5. The most frequent application of an electronic to pneumatic converter is to convert the signals necessary to _____.

a. operate panel indicators
b. operate electronic recorders
c. operate pneumatic valves or regulators
d. supply digital devices

6. Another name for direct-operated pneumatic regulators is _____ regulators.

a. pilot-operated
b. self-operated
c. internally controlled
d. feedback

7. Regardless of the type of loading used in regulators, the pressure entering the regulator acts against the _____.

a. process pressure
b. upstream pressure
c. loading force
d. downstream pressure

8. The controlling force in a direct-operated regulator is _____.

a. a pilot regulator output
b. upstream pressure
c. valve feedback
d. downstream pressure

9. A major disadvantage of direct-operated regulators is _____.
   a. poor sensitivity
   b. high equipment cost
   c. proportional band drift
   d. high sensitivity

10. The primary element in an absorption drying method that uses a process to liquefy the solids is _____.
    a. a solid block of chemicals
    b. a deliquescent drying agent
    c. refrigerated air
    d. a heating element

11. The condition of the output air in a typical refrigerant-type drying system is _____.
    a. extremely wet
    b. below 32.5°F to 33°F
    c. between 32.5°F to 33°F
    d. slightly damp

12. Standard drying and reactivation time cycle for a heat-reactivated automatic dryer system is _____.
    a. 50 minutes
    b. 3 hours
    c. 8 hours
    d. 24 hours

13. In most combination filter/dryer systems, the system works best if the compressed air provided is first _____.
    a. increased to 50 psig
    b. filtered to 5 microns
    c. cooled to below 32.4°F
    d. dried in a refrigeration-type air dryer

14. The recommended port diameter when selecting pressure regulators is _____.
    a. the largest diameter that will handle the flow
    b. the smallest diameter that will handle the flow
    c. one size larger than the pipe diameter
    d. 1½ times the calibrated flow requirement

15. Because the majority of pneumatic instruments and processes cannot tolerate wet, hot compressed air, compressors are normally supplied with moisture separators and _____.
    a. filters
    b. regulators
    c. aftercoolers
    d. desiccants

GLOSSARY

Trade Terms Introduced in This Module

Absorption: The process of taking in moisture through the pores of a moisture-attracting substance. A sponge absorbs moisture.

Adsorption: A process in which moisture is condensed and held on a chemical desiccant applied to the surface of a substance.

Calcining: A process that uses heat and calcium salts (crushed limestone) to reduce liquid and acidic high-level wastes to dry, powdered, nonacidic waste.

Centrifuge: A device that rotates at various speeds around a fixed, central point; it can separate liquids from solids or liquids of different densities by way of the force resulting from its rotation.

Cohesiveness: The ability to stimulate the strength of the internal bonding properties of two or more elements.

Contaminant: Any material or substance that is unwanted or adversely affects the fluid power system its components, or both.

Deliquescent: The process of melting or becoming liquid by absorbing moisture from the air.

Deliquescent dryer: A single-stage, fixed-bed chemical reactor that reduces the dew point of the compressed air passed through it by adsorption.

Desiccant: A chemical used to attract and remove moisture from air or gas.

Droop: The amount a regulator deviates below its setpoint as flow increases. Some regulators exhibit boost instead of droop.

Edge: A filter medium in which the passages are formed by the adjacent surfaces of stacked discs, edge-wound ribbons, or single-layer filaments.

Filtration element: The porous device that performs the actual process of filtration.

Impervious: Having properties of anti-absorption; does not take in, but repels.

Lockup pressure: Increase over setpoint when the regulator is at no-flow condition.

Phenolic-impregnated cellulose: A fibrous material of vegetable origin (cellulose) that has its voids filled (impregnated) with phenol, a petroleum distillate, making it into a substance that can be used as a filtering medium.

Polymerize: To form a complex compound (polymer) by linking together many smaller elements (residues).

Tachometer: A speed measuring device that is calibrated in revolutions per minute (rpm).

Water column: A method of pressure measurement that refers to how far a column of water in a calibrated glass tube is moved by an applied pressure.

Figure Credits

Donaldson Company, Inc.	210F05
Aircel International	210F20, 210F22

NCCER CURRICULA — USER UPDATE

NCCER makes every effort to keep its textbooks up-to-date and free of technical errors. We appreciate your help in this process. If you find an error, a typographical mistake, or an inaccuracy in NCCER's curricula, please fill out this form (or a photocopy), or complete the online form at **www.nccer.org/olf**. Be sure to include the exact module ID number, page number, a detailed description, and your recommended correction. Your input will be brought to the attention of the Authoring Team. Thank you for your assistance.

Instructors – If you have an idea for improving this textbook, or have found that additional materials were necessary to teach this module effectively, please let us know so that we may present your suggestions to the Authoring Team.

NCCER Product Development and Revision
13614 Progress Blvd., Alachua, FL 32615
Email: curriculum@nccer.org
Online: www.nccer.org/olf

❑ Trainee Guide ❑ AIG ❑ Exam ❑ PowerPoints Other ______________________________

Craft / Level: Copyright Date:

Module ID Number / Title:

Section Number(s):

Description:

Recommended Correction:

Your Name:

Address:

Email: Phone:

Module 12211-03

Analyzers and Monitors

COURSE MAP

This course map shows all of the modules in the second level of the Instrumentation curriculum. The suggested training order begins at the bottom and proceeds up. Skill levels increase as you advance on the course map. The local Training Program Sponsor may adjust the training order.

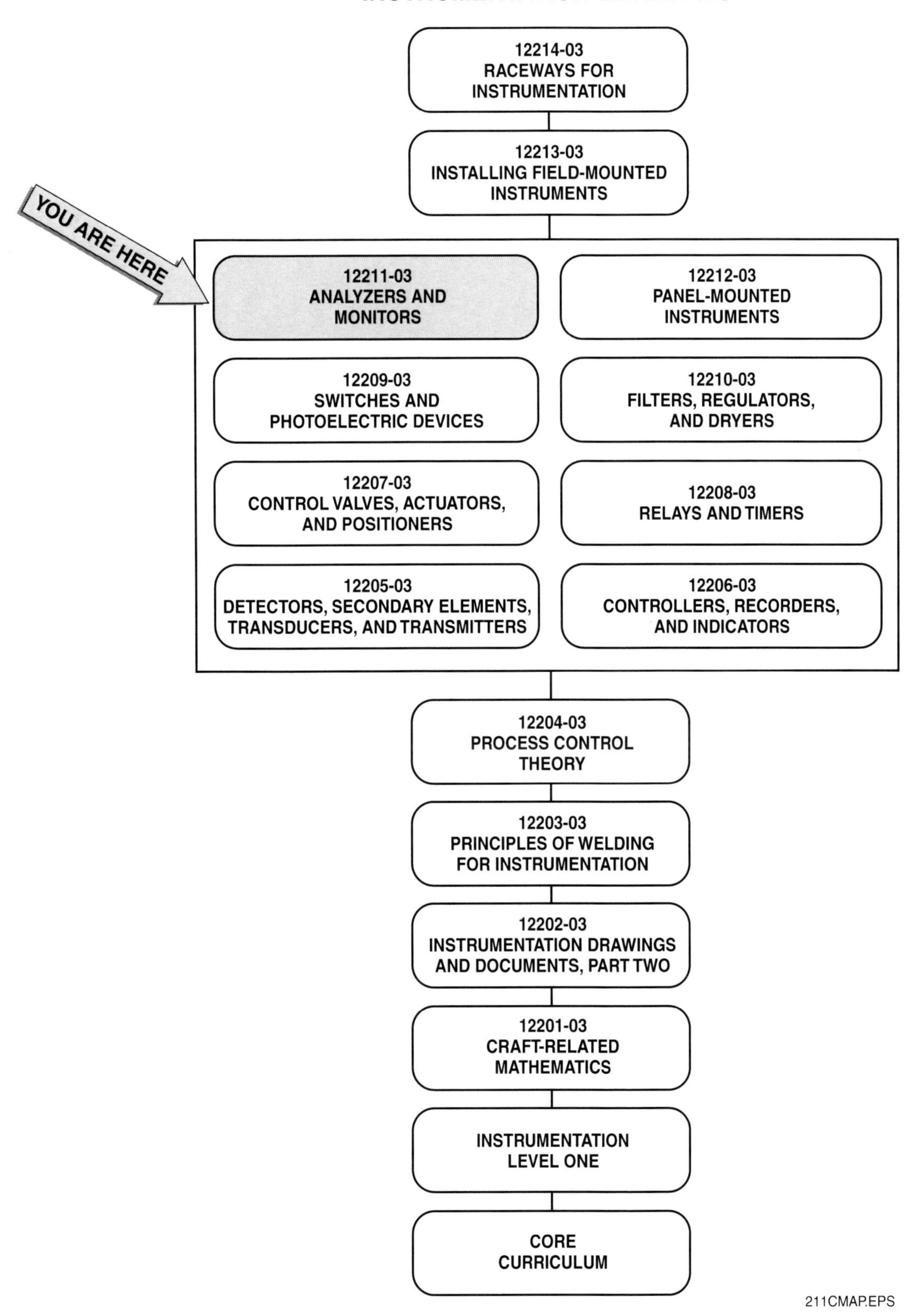

MODULE 12211-03 CONTENTS

1.0.0 INTRODUCTION 11.1

2.0.0 CHEMISTRY 11.1

2.1.0 Chemical Reactivity 11.2

2.2.0 Atomic Mass 11.4

2.3.0 Concentration 11.5

2.4.0 Electrolytes 11.6

2.5.0 pH 11.7

3.0.0 pH MEASUREMENT 11.7

3.1.0 pH Probes 11.9

3.2.0 ph Analyzer/Controller 11.10

4.0.0 CONDUCTIVITY MEASUREMENT 11.10

5.0.0 DENSITY AND SPECIFIC GRAVITY 11.13

5.1.0 Measurement of Density and Specific Gravity 11.14

5.1.1 Hydrometer 11.14

5.1.2 Air Bubbler Method 11.15

5.1.3 Differential Pressure Method 11.15

5.1.4 Specific Gravity and Density Values 11.16

6.0.0 VISCOSITY MEASUREMENT 11.16

6.1.0 Viscometers 11.16

6.2.0 Rotating Spindle Viscometers 11.17

6.3.0 Vibrating Reed Viscometers 11.18

6.4.0 Temperature Effects on Viscosity 11.18

7.0.0 CHEMICAL COMPOSITION MEASUREMENT 11.18

7.1.0 Gas Chromatography 11.19

7.2.0 Thermal Conductivity Gas Analysis 11.20

7.3.0 Smoke Density Measurement 11.21

7.4.0 Hydrocarbons 11.21

7.4.1 Flame Ionization Methods 11.21

7.4.2 Infrared Methods 11.22

SUMMARY 11.22

REVIEW QUESTIONS 11.22

GLOSSARY 11.25

REFERENCES & ACKNOWLEDGEMENTS 11.27

Figures

Figure 1 Modern periodic table 11.3
Figure 2 Ionic bonding 11.4
Figure 3 Covalent bonding 11.4
Figure 4 pH scale range 11.8
Figure 5 pH indicator scale 11.8
Figure 6 In-line pH probe and meter 11.9
Figure 7 Simplified diagram of the essential parts of a pH measuring instrument 11.9
Figure 8 Analyzer/controller 11.10
Figure 9 Simple conductivity meter 11.11
Figure 10 Basic types of conductivity cells 11.12
Figure 11 Comparison of heaviness of metals and liquids ...11.13
Figure 12 Hydrometer 11.14
Figure 13 Arrangement of bubbler system to measure specific gravity 11.15
Figure 14 Differential pressure density meter 11.16
Figure 15 Basic method of determining viscosity 11.17
Figure 16 Rotating spindle viscometer 11.17
Figure 17 Vibrating reed viscometer 11.18
Figure 18 Viscosity variation of an oil with temperature 11.18
Figure 19 Viscosity variation of water with temperature 11.18
Figure 20 Chromatograph block diagram 11.19
Figure 21 Thermal conductivity detector 11.20
Figure 22 Wheatstone bridge and filament circuitry for thermal conductivity detector 11.20
Figure 23 Opacity measurement of smoke 11.21
Figure 24 Flame ionization detector for hydrocarbon measurement 11.21

Tables

Table 1 pH - pOH Relationship 11.8
Table 2 Weight Density of Some Commonly Used Liquids 11.13

MODULE 12211-03

Analyzers and Monitors

Objectives

When you have completed this module, you will be able to do the following:

1. Describe the purpose of the periodic table.
2. Describe the scale used for measuring pH.
3. Define conductivity, and describe the method used to measure conductivity.
4. Define specific gravity, and list two methods used for measuring specific gravity.
5. Define viscosity, and list two methods used for measuring viscosity.
6. Describe the purpose of gas chromatography.
7. Describe the purpose of thermal conductivity gas analysis.
8. List two methods commonly used for measuring hydrocarbons.

Prerequisites

Before you begin this module, it is recommended that you successfully complete the following: Core Curriculum; Instrumentation Level One; Instrumentation Level Two, Modules 12201-03 through 12204-03.

Required Trainee Materials

1. Pencil and paper
2. Required safety equipment

1.0.0 ◆ INTRODUCTION

Progress in the field of industrial instrumentation has been extraordinary in the past few decades. The next few decades will doubtlessly provide even greater expansion in the technology of instrumentation, process analyzers, and process monitors. For this reason, there will be a great demand for trained workers to design, install, service, and operate the wide variety of industrial analyzers and monitors important in the industrial field and the process industries.

The objective of this module is to introduce several of the popular types of industrial analyzers and monitors. It describes typical uses and the basic principles of operation of each.

In order to understand and appreciate the methods used in analyzers and monitors, several scientific topics must be covered and understood. The first and most important topic is basic chemistry. An introduction to some basic chemistry concepts is essential to a complete understanding of analytical measurements.

2.0.0 ◆ CHEMISTRY

Matter is anything that occupies space and has mass. The smallest particle of matter that can enter into a chemical reaction is an **atom**. An atom consists of **protons**, **neutrons**, and **electrons**. The protons are positively charged, the neutrons have no charge, and the electrons are negatively charged. Under normal conditions, atoms are electrically neutral because the positive charge from the protons equals the negative charge from the electrons. Since the neutrons have no charge, they only contribute to the mass of the atom. The combined mass of the protons, neutrons and electrons of an atom is called the atomic mass number.

Atoms are named by the number of protons they have. All atoms with the same number of protons have the same chemical properties and are referred to as a particular **element**. There are 92 elements that occur naturally. The first element is hydrogen, which has the chemical symbol H. It has one proton, one electron, and no neutrons.

Hydrogen is also the lightest (meaning that it has the smallest mass) of all elements. The heaviest naturally occurring element is uranium, which has the symbol U. It has 92 protons, 92 electrons, and 146 neutrons. By use of nuclear reactors and particle accelerators, elements with as many as 116 protons have been identified.

Many elements are familiar to all of us. The charcoal used in outdoor grills is nearly pure carbon. Electrical wiring, jewelry, and water pipes are often made from copper, a metallic element. Another such element, aluminum, is used in many household utensils. The shiny silver-colored liquid in older thermometers is yet another metallic element, mercury.

In chemistry, an element is identified by its symbol. The symbol consists of one to three letters, usually derived from the name of the element. Thus the symbol for carbon is C, and the symbol for aluminum is Al. Sometimes the symbol comes from the Latin name of the element or one of its compounds. For example, the elements copper and mercury, which were known in ancient times, have the symbols Cu (cuprum) and Hg (hydrargyrum). Other examples include:

Antimony	**Sb**	Lead	**Pb**	Sodium	**Na**
Gold	**Au**	Potassium	**K**	Tin	**Sn**
Iron	**Fe**	Silver	**Ag**		

The periodic table shown in *Figure 1* is used to organize the properties of the elements.

Elements that are similar chemically fall directly beneath one another in the table. Later sections of this module will discuss the rationale for this and show how the table is used for a variety of purposes. Trainees need only be concerned with two general features of the periodic table.

1. The horizontal rows are referred to as periods. Thus, the first period consists of the two elements hydrogen (H) and helium (He). The second period starts with lithium (Li) and ends with neon (Ne). Each period ends with a very unreactive element called a **noble gas**.
2. The vertical columns are known as groups. Three different notation conventions are used for designating the groups, as shown in *Figure 1*. The three conventions are 1 - 18, IA – VIIIA or 1A - 8A, and IB - VIIB or 1B - 8B. Using the second notation convention for example, IA and IIA are located at the far left, and groups IIIA through VIIIA are located at the far right of the table. Elements that fall into these groups are often referred to as main, or representative, groups, as indicated on the table. The elements in the center of the table, IB through VIIIB, are called transition elements. For example, Sc through Zn in period 4 are transition elements.

Certain main groups are given special names. The elements in Group IA are called alkali metals. Those in Group IIA are referred to as alkaline earth metals. The Group VIIA elements are called halogens. The noble gases comprise Group VIIIA. Elements in the same main group show very similar chemical reactions. For example, sodium (Na) and potassium (K) in Group IA both react violently with water to produce hydrogen gas. However, the noble gases in Group VIIIA are very stable and very seldomly react with other chemicals.

2.1.0 Chemical Reactivity

Only the electrons in the outermost, or **valence shell** affect the chemical properties of an element. The most stable arrangement of electrons is one in which all shells, up to and including the valence shell, have their maximum number of electrons. When elements interact chemically, they tend to gain, lose, or share electrons with other elements to complete their valence shells. The valence of an element is the number of electrons it gains or loses or the number of pairs of electrons it shares when it chemically interacts with other elements. Sodium has one electron in its outermost shell. When sodium interacts with other elements, it loses this electron. Because losing an electron leaves an excess positive charge, sodium has a valence of +1. Chlorine has seven electrons in its outermost shell. When chlorine interacts with other elements, it gains an electron to complete its valence shell. Because gaining an electron leaves an excess negative charge, chlorine has a valence of –1. Some elements naturally have complete valence shells. These are the elements referred to as inert, or noble gases.

A molecule is the smallest particle of a substance that can have a stable, independent existence. A molecule may be composed of one or more atoms of the same or different elements. An example of a molecule composed of only one atom is helium gas. Helium is an inert gas that does not chemically react with other elements. An example of a molecule composed of more than one atom of the same element is oxygen gas. A molecule of oxygen gas contains two oxygen atoms. The two atoms of oxygen share electrons, giving each a stable configuration of valence shell electrons. This is why oxygen gas is normally referred to as O_2.

A compound is a special type of molecule. Compounds are composed of atoms of two or more different elements. An example of a compound is water. A molecule of water is composed of two atoms of hydrogen and one atom of oxygen. The chemical formula for water is H_2O. The three atoms share electrons such that all three have stable configurations of valence shell electrons.

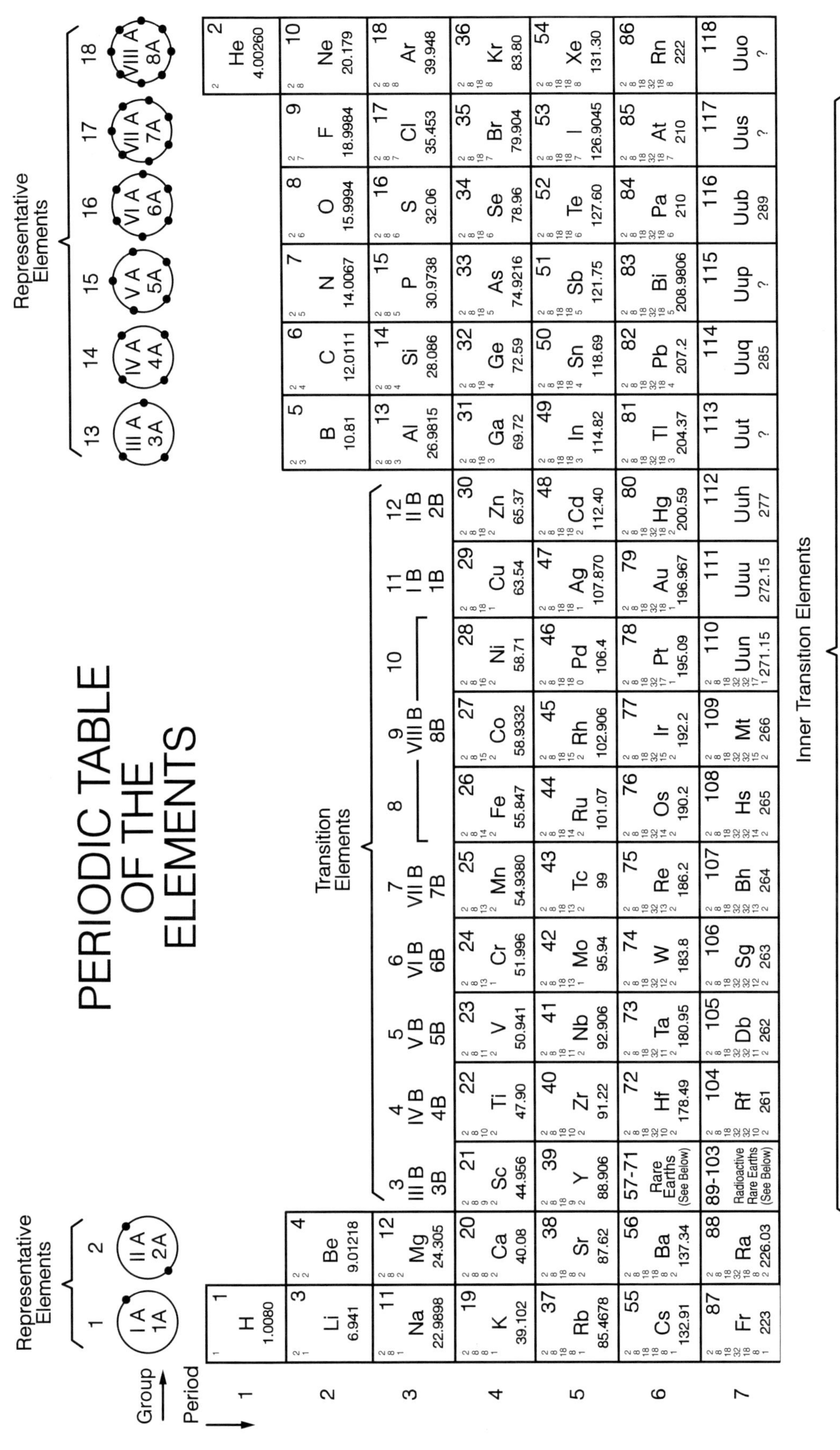

Figure 1 ◆ Modern periodic table.

When atoms react chemically to form molecules, either a transfer or sharing of valence electrons occurs. Chemical bonds formed by the transfer of electrons are referred to as **ionic bonds**. An ion is an atom, or group of atoms, with a net positive (more protons than electrons) or net negative (more electrons than protons) charge. Recall that protons and electrons have the same magnitude but opposite polarity charges and that a neutral atom has as many protons as electrons. A good example of an ionic bond is the chemical reaction that occurs between sodium and chlorine. Sodium has one electron in its outer shell. To reach a stable configuration, it must lose that electron. Chlorine has seven electrons in its valence shell and must gain an electron to achieve a stable configuration. When sodium and chlorine react chemically, a transfer of one electron from sodium to chlorine occurs. That is, the valence electron of sodium now orbits the chlorine **nucleus** exclusively. The sodium atom, minus its electron, is now an ion with a net charge of +1. The chlorine atom, with the electron it gained, is also an ion with a net charge of –1. The two atoms are electrostatically attracted to each other and form a molecule of sodium chloride, or NaCl. *Figure 2* shows the electron structure of the sodium (Na) and the chlorine (Cl) atoms before and after the chemical reaction.

Many chemical reactions are based on the sharing, rather than transfer, of electrons by atoms. Atoms that form molecules in this way are **covalently bonded**. In a covalent bond, the valence electrons may be in orbit around either atom at any instant. The shared electrons allow both atoms to appear to have filled their valence electron shell, thus achieving a stable configuration. A good example of a covalent bond is water. *Figure 3* shows the electron configuration of each hydrogen and oxygen atom before and after covalent bonding. Note that after covalent bonding, each atom appears to have a complete valence electron shell structure.

Regardless of the type of chemical bond, the resulting molecule has properties completely different from the elements that formed it. Sodium metal reacts violently in water, and chlorine is a pungent, poisonous gas; following ionic bonding, sodium and chlorine form common table salt. The reaction of hydrogen and oxygen gas to form water is another example. There are over 100 known elements, but there are hundreds of thousands of identified compounds.

2.2.0 Atomic Mass

Compounds are composed of atoms of different elements. The atoms combine in definite proportions to form each molecule of the compound. For

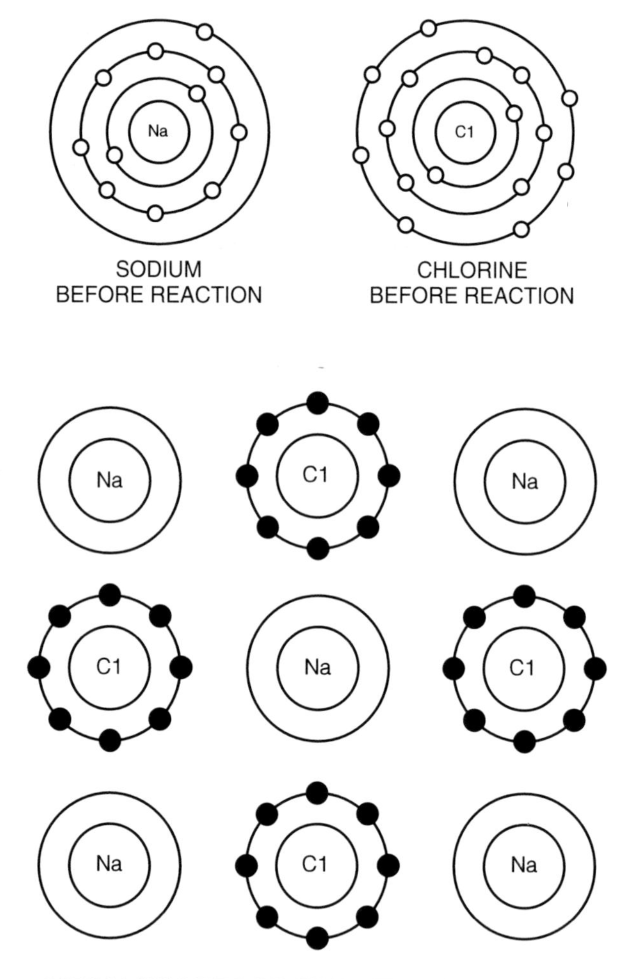

Figure 2 ◆ Ionic bonding.

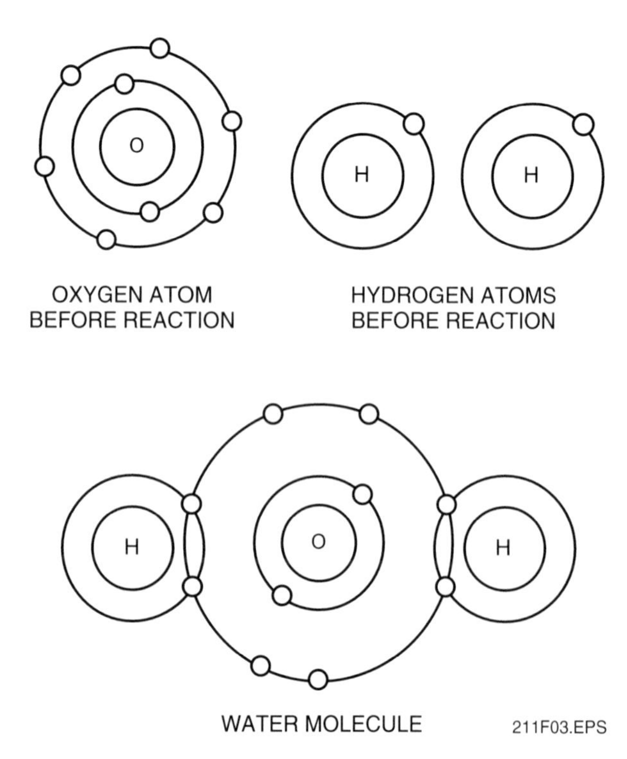

Figure 3 ◆ Covalent bonding.

example, the combining of oxygen and hydrogen atoms at a ratio of two hydrogen atoms for each oxygen atom forms molecules of water. In order for a chemical process to efficiently use the atoms of the elements necessary to form a compound, the atoms must be available in the proper proportions. To make salt, there must be as many sodium atoms as chlorine atoms available for the reaction. Otherwise, an amount of chlorine will remain following the reaction. It is impossible to individually count the number of atoms present to ensure a complete reaction. Normally, chemists determine the mass of the element necessary for a complete reaction. Therefore, a method must exist to relate the number of atoms present to the mass of atoms.

The mass of an individual atom is extremely small, on the order of 10^{-25} kg. Rather than use such small numbers when dealing with atomic mass, it is simpler to establish a new unit of measurement of approximately the same size as an atom. The carbon atom with six protons, six neutrons, and six electrons was chosen as the reference. It has a mass of 1.9926×10^{-26} kg. Since the majority of the mass of any atom is concentrated in the central nucleus, composed of protons and neutrons (each of which have about the same mass), the actual unit of measure is one twelfth the mass of carbon, or 1.6605×10^{-27} kg. This new unit of measure is given the name **atomic mass unit (AMU)**. One AMU is approximately the mass of one proton or neutron. Therefore, if the number of protons and neutrons in an atom is known, the mass of the atom can be approximated easily. The sum of the protons and neutrons in an atom is the atomic mass number. The atomic mass number is the mass of an atom expressed in AMUs and rounded off to the nearest integer value.

In most practical applications, it is easier to deal with **atomic weight** in grams than with AMUs. The **gram atomic weight (GAW)** of an element is just the atomic mass number of the element expressed in grams. In the early 1800s, the Italian chemist Amedeo Avogadro proved the relationship between the mass of an element and the number of atoms present. Avogadro proved that one gram atomic weight of any element contains 6.02×10^{23} atoms of that element. This constant (6.02×10^{23}) is known as **Avogadro's number**. Another term frequently encountered is the **mole**. A mole is the amount of a substance that contains 6.02×10^{23} atoms or molecules.

Example 1:

How many grams of carbon and oxygen are required to form 1 mole of carbon dioxide (CO_2)?

Since each atom of carbon will react with two atoms of oxygen to form one molecule of carbon dioxide, 1 mole of carbon and 2 moles of oxygen will be required. Refer to *Figure 1* to obtain the atomic weight.

$$1 \text{ mole of C} \times \frac{6.02 \times 10^{23}}{\text{mole}} \times \frac{1 \text{ GAW}}{6.02 \times 10^{23}} \times \frac{12.011 \text{ grams}}{1 \text{ GAW}}$$

$$= 12.011 \text{ grams of carbon}$$

$$2 \text{ moles of O} \times \frac{6.02 \times 10^{23}}{\text{mole}} \times \frac{1 \text{ GAW}}{6.02 \times 10^{23}} \times \frac{15.9994 \text{ grams}}{1 \text{ GAW}}$$

$$= 31.9988 \text{ grams of oxygen}$$

The results indicate that to form 1 mole of carbon dioxide (6.02×10^{23} molecules) 12.011 grams of carbon and 31.9988 grams of oxygen are required. The resulting carbon dioxide will weigh 44.098 grams. This mass could also be referred to as one **gram molecular weight (GMW)** of carbon dioxide. A gram molecular weight is the sum of the gram atomic weights of the molecules' component elements times the number of atoms of each element present.

2.3.0 Concentration

The fundamental reason for most analytical measurements is to determine how much. For this reason, it is necessary to understand the meaning of the terms frequently used to express the concentration of solutions.

When crystals of salt are stirred into a sufficient quantity of water, the salt disappears and a clear mixture of salt and water is formed. The salt is said to have dissolved or gone into solution in the water. The solution consists of two components, the **solute** (the dissolved salt) and the solvent (the water). In this solution, the solute is uniformly distributed among the molecules of the solvent. A solution, then, is a uniform mixture of solute and solvent. A solution containing a small amount of solute compared to the amount of solution present is said to be dilute. As more solute is added to the solution, it becomes more concentrated. The two terms we will use to describe the concentration of a solution are **molarity** and **parts per million (ppm)**.

Molarity is the ratio of the number of moles of solute present per liter of the solution.

Example 2:

What is the molarity of a 2-liter solution of salt water containing 40 grams of salt (NaCl)?

1 mole of salt = 1 gram molecular weight of salt

1 GMW salt = 1 gram atomic weight of sodium +1 GAW of chlorine

1 mole salt = 22.990 grams of sodium + 35.453 grams of chlorine

1 mole salt = 58.443 grams of salt

$$\# \text{ of moles of salt present} = \frac{\text{grams of salt per mole}}{\text{grams of salt present}}$$

$$\# \text{ of moles of salt present} = \frac{40 \text{ grams}}{58.443 \text{ grams/mole}}$$

$$\text{Molarity} = \frac{\text{No. of grams of salt per mole}}{\text{No. of grams of salt present}}$$

$$M = \frac{.684 \text{ moles}}{2 \text{ liters}} = .342 \text{ moles/liter}$$

The result of Example 2 is expressed as .342 moles per liter. It could also be given as a .342 M solution or a .342 molar solution.

The term parts per million is frequently used to report small concentrations of impurities in solutions. Parts per million (ppm) is a ratio of solute to solution. As with most ratios, both the solute and solution should be expressed in the same units. The size of the solution term, however, should be a factor of one million larger. A concentration of 10 ppm chlorine means that for every 10 grams of chlorine there are 1×10^6 grams of solution.

2.4.0 Electrolytes

Chemists in the 19th century noted that certain substances in solution produced abnormal changes in the properties of the solvents. These substances produced elevations in the boiling points and depressions in the freezing points considerably greater than predicted on the basis of normal molecular effects. In his experiments on electricity, Michael Faraday, an English physicist, observed that, while pure water was a very poor conductor of electricity, water solutions of certain substances, which he called electrolytes, were very good conductors. In 1883, Svante August Arrhenius, a Swedish chemist, noted that Faraday's electrolytes included all of the substances that, in solution, produced abnormal changes in the boiling and freezing points. Arrhenius proposed that if each solute molecule separated or dissociated on dissolving into two or more particles, each particle might act as an individual molecule, and the abnormal effects noted on the boiling and freezing points could be explained. Also, if the dissociated particles were electrically charged, they could conduct electricity by motion through the solution. Arrhenius published what is called the *Theory of Ionization*. Its principle points are these:

1. Certain substances dissociate into ions when dissolved in water. These substances are called electrolytes. Acids, bases, and salts are electrolytes.
2. Dissociation takes place during the process of dissolving. It results from the action of the solvent on the solute.
3. Ions are atoms or groups of atoms that have an electric charge. Metallic ions have positive charges; positive ions are called cations. Nonmetallic ions have negative charges; negative ions are called anions.
4. The total number of positive and negative ions in solution are equal. The solution as a whole, therefore, is electrically neutral.
5. Ions can wander freely throughout the solution. When an electric current is passed through a solution containing an electrolyte, the cations migrate to the cathode, the negative pole, and pick up electrons. The anions migrate to the anode, the positive pole, and give up electrons.

Acids, bases, and salts are among the most important of all chemical compounds. They are important because they are electrolytes and because they dissociate into ions when dissolved in water.

An acid is a compound that produces hydrogen ions (H^+) in a water solution. In water solutions, all ions are surrounded by the polar water molecules. Thus, it is common to refer to either the hydronium ion (H_3O^+) concentration or the hydrogen ion (H^+) concentration in a water solution. A hydronium ion (H_3O) is simply a hydrogen ion (H^+) attached to a water molecule. An example of a common acid is sulfuric acid (H_2SO_4). In solution, it dissociates to form hydrogen and sulfate ions according to the following equation:

$$H_2SO_4 = 2H^+ + SO_4^=$$

A base is a compound that produces hydroxyl ions (OH^-) in water solution. An example of a common base is sodium hydroxide (NaOH). In solution, it dissociates to form sodium ions and hydroxyl ions according to the following equation:

$$NaOH = Na^+ + OH^-$$

A salt is a compound that dissociates in solution to form positive ions other than hydrogen or hydronium ions and negative ions other than hydroxyl ions. An example of a common salt is sodium chloride (NaCl). In solution, it dissociates to form sodium ions and chloride ions according to the following equation:

$$NaCl = Na^+ + Cl^-$$

Acids, bases, and salts are all electrolytes. Their water solutions conduct electricity. Acids turn **litmus** red. In very weak solutions, they taste sour. Bases turn litmus blue. In very weak solutions, they taste bitter.

The strength of an acid or base is measured by the extent to which it ionizes in solution to form the hydrogen or hydronium ions and hydroxyl ions, respectively. The measure of the acidity or basicity (which is also called **alkalinity**) of a solution depends on the concentrations of hydrogen ions/ hydronium ions and hydroxyl ions in solution. If the concentration of hydrogen ions or hydronium ions exceeds the concentration of hydroxyl ions, the solution is acidic. If the concentration of hydroxyl ions exceeds the concentration of hydrogen ions or hydronium ions, the solution is basic, or alkaline. If the concentrations of the two ions are equal, the solution is said to be neutral. In this application, concentrations are given as gram moles per liter molarity, where 1 gram mole of hydrogen ions weighs 1 gram, 1 gram mole of hydronium ions weighs 19 grams, and 1 gram mole of hydroxyl ions weighs 17 grams. In all cases, one gram mole contains the same number of ions, 6.023×10^{23}.

2.5.0 pH

The quantitative measure of the acidity or alkalinity of a solution is the pH of the solution. Sulfuric acid and boric acid are considered acids, just as caustic soda and lime are recognized as alkalies or bases. However, these compounds vary markedly in their strength, or *activity*. A 5% by weight solution of boric acid can be used as an eye wash with no danger. Similar use of a 5% by weight solution of sulfuric acid would be disastrous.

Thus, we must consider something besides concentration of an acid or a base to understand its quality or effectiveness for a given function. In the following discussion, references to strong or weak acids could also apply to strong or weak bases.

One of the most important identifying factors of an acid is the activity of the hydrogen it contains. Sulfuric acid and boric acid both contain hydrogen, but the hydrogen in sulfuric acid dissociates in the presence of water to become free hydrogen ions. When boric acid is added to water, very little of the hydrogen is liberated as free hydrogen ions. Instead, it remains relatively inactive in the undissociated molecules of the acid. The true measure of acidity concerns the measure of the dissociated or free hydrogen ion concentration of a given solution. The expression of such a value would be very awkward if stated in terms of the number of hydrogen ions per cubic centimeter ounce, or some unit of weight or volume.

In 1909, the Danish chemist Sorenson proposed that the initial letter of the French word potenz, meaning potency or power, be called the hydrogen ion **exponent** and expressed as pH. Technically, the pH of a solution is defined as the negative **logarithm** of the hydrogen ion concentration, written (H^+), when expressed in gram moles per liter.

$$pH = -\log (H^+)$$
$$(H^+) = 10^{-pH}$$

Pure water is neither an acid nor a base. It is neutral. At 25°C, an equilibrium exists in pure water such that the hydrogen ion concentration equals the hydroxyl ion concentration and both have values of 1.0×10^{-7} grams moles per liter. Using the definition of pH, this means that the pH of pure water at 25°C is 7.

If the pH of any solution is 7, the solution is neutral. If the pH is less than 7, the solution is acidic. If the pH is greater than 7, the solution is basic. A solution with a pH of 4 to 6 is weakly acidic, with a pH less than 4 is strongly acidic, with a pH of 8 to 10 is weakly basic, and with a pH greater than 10 is strongly basic.

An important relationship in water solutions is the relationship between the hydrogen or hydronium ion concentration and the hydroxyl ion concentration. The term used to express the hydroxyl ion concentration of a solution is **pOH**. The pOH of a solution is defined as the negative logarithm of the hydroxyl ion concentration, written (OH-), when expressed in gram moles per liter.

$$pOH = -\log (OH^-)$$
$$(OH^-) = 10^{-pOH}$$

For water solutions, the product of the hydrogen ion concentration (H^+) and the hydroxyl ion concentration (OH^-) is always 1×10^{-14}. This means that the sum of the pH and the pOH is always 14.

$$(H^+) \times (OH^-) = 1 \times 10^{-14}$$
$$pH + pOH = 14$$

pH is a measure of the acidity or basicity of water solutions. It is a measure of the strength of acids and bases. *Table 1* gives pH-pOH relationships. Referring to the previous example, a 5% by weight sulfuric acid solution has a pH of approximately 0.3. A 5% by weight boric acid solution would have a pH value of about 5.0. Thus, two acids having approximately the same hydrogen concentration have very different pH values, due to the degree of dissociation of the hydrogen present in each.

3.0.0 pH MEASUREMENT

The numbering system for a pH scale ranges from 0 to 14. The number 7 is at the center of this span and indicates a neutral solution. Acid levels occupy the positions between 7 and 0, with the

Table 1 pH - pOH Relationship

Hydrogen Ion Concentration (H+) (gm moles/ℓ)	Hydroxyl Ion Concentration (OH-) (gm moles/ℓ)	pH	pOH Description
1	10^{-14}	0	14 Strongly Acidic
10^{-1}	10^{-13}	1	13
10^{-3}	10^{-11}	3	11
10^{-4}	10^{-9}	5	9 Weakly Acidic
10^{-7}	10^{-7}	7	7 Neutral
10^{-9}	10^{-5}	9	5 Weakly Basic
10^{-11}	10^{-3}	11	3
10^{-13}	10^{-1}	13	1
10^{-14}	1	14	0 Strongly Basic

211T01.EPS

smaller numbers indicating the highest acid levels. The numbers between 7 and 14 represent the base scale, with the larger numbers indicating the highest base levels. Refer to *Figure 4*.

The pH level of a solution can be determined by direct measurement of the DC voltage developed between two electrodes immersed in the solution under test. The electronics in the instrument convert the voltage developed by the electrodes into an indication of pH. Hand-deflecting instruments, chart recorders, and digital readout displays are common today. A number of different display techniques are used in pH measurement. In this module, we will assume that the display device is a hand-deflection meter or a recorder. Recorders provide a permanent record of pH levels for a variety of different time spans.

Figure 5 shows a typical pH indicator scale for a hand-deflection instrument. There are ten small graduations or divisions between each two numbers. Each division, therefore, represents 0.1 pH unit. If the indicating hand is deflected to the third small graduation to the right of the number 5, the

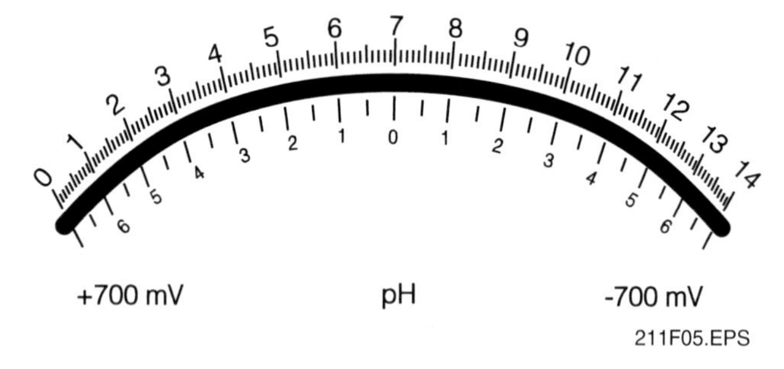

Figure 5 ◆ pH indicator scale.

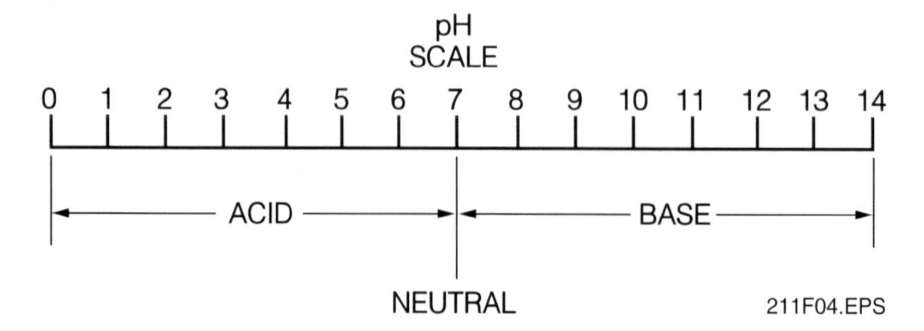

Figure 4 ◆ pH scale range.

pH level would be 5.3. This indicates an acid level of approximately 20%. A pH meter can also be used to indicate positive and negative voltage values. In this particular indicator, full-scale deflection is +700 to -700 millivolts. In practice, the deflecting hand should come to rest at zero when measurements are not being taken.

The curved black strip of the pH meter scale shown in *Figure 5* represents a mirror finish. This part of the meter scale is designed to produce a reflection of the indicating hand. In practice, an operator looks at the scale in such a way that the hand and its reflection are exactly in line. As a result, the reflection cannot be seen. When this occurs, the indication should be quite accurate because the operator is not looking at the scale from an angle. There should be no parallax effect.

3.1.0 pH Probes

The probe or electrode part of a pH instrument is often thought of as a battery in which the voltage varies with the pH level of the solution. It contains either two separate electrodes in a single probe housing, or two distinct probes. In either case, the electrode or part of the common probe is sensitive to hydrogen. A special glass bulb or membrane material is used that has the ability to pass H^+ ions inside of the sensitive bulb. When the electrode is placed in the solution, a voltage proportional to the hydrogen ion concentration is developed between an inner electrode and the outer electrode or glass bulb material. This pH-sensitive electrode is often called a half-cell.

A second discrete electrode, or the alternate part of the common probe, is used to develop a reference voltage. The reference electrode is primarily responsible for producing a stable voltage that is independent of solution properties. This electrode, or half-cell, develops a fixed voltage value when placed in the solution. When the reference half-cell and the pH glass-bulb half-cell are combined, they form a complete probe. *Figure 6* shows a typical pH probe and meter.

Figure 7 shows a general diagram of the essential parts of a pH measuring instrument. In this case, pH is observed on a hand-deflection meter scale. Quantitative measurements of pH are produced by this type of instrument. A majority of the

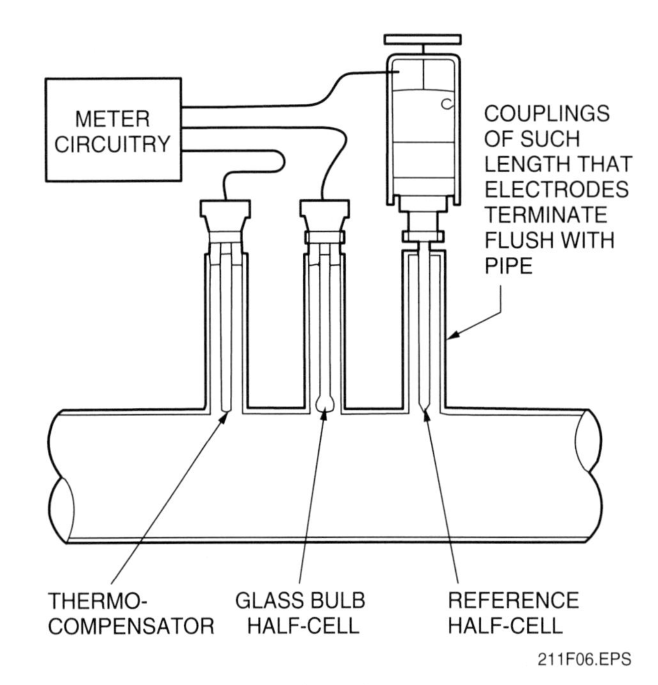

Figure 6 ◆ In-line pH probe and meter.

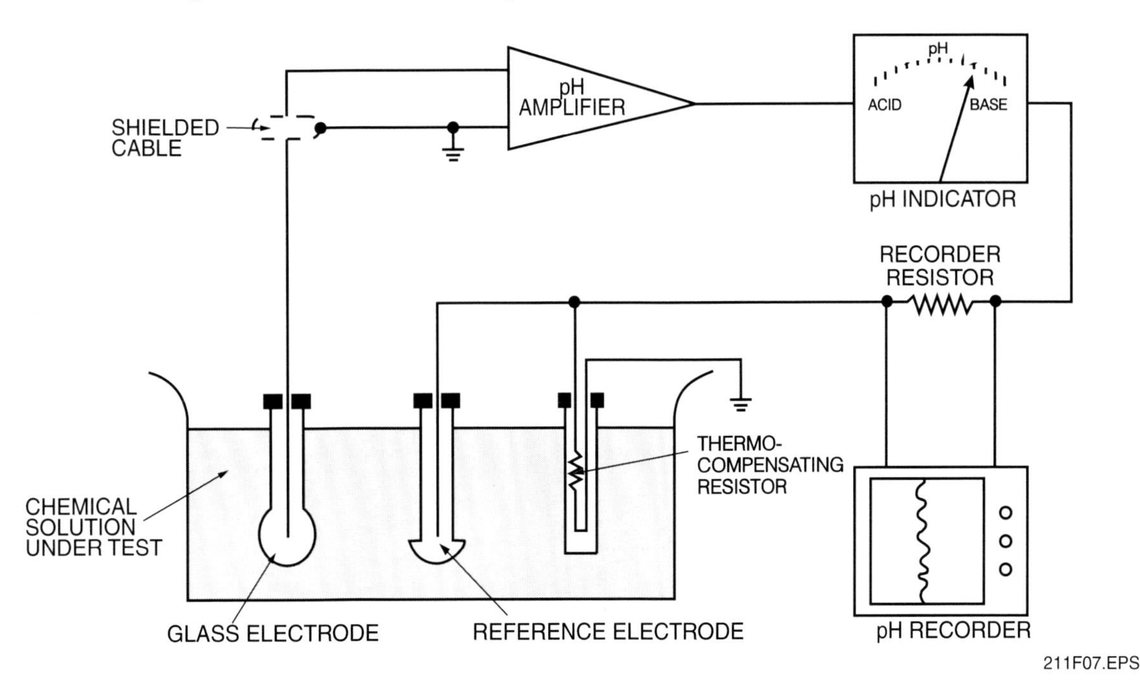

Figure 7 ◆ Simplified diagram of the essential parts of a pH measuring instrument.

pH instruments used in industry today are this type. They range from small portable units, housed in convenient carrying cases, to larger stationary units.

The essential parts of a pH instrument, regardless of its type or style, are the measuring half-cell electrode, the reference half-cell electrode, a high-impedance amplifier, and an indicator. This type of instrument is normally classified as a direct-reading unit because it responds to voltage values produced directly from the solution under test. Nearly all industrial pH instruments are this type as opposed to indirect-reading units, which produce an indication due to color changes in a material sample.

3.2.0 pH Analyzer/Controller

Many of the new pH analyzers also function as controllers and are microprocessor-based instruments. An example is shown in *Figure 8*. This analyzer/controller still requires an input from an electrode or probe system, but calibration is made easy through a dedicated calibration key. The key permits rapid calibration by reading a standardized **buffer solution**. The liquid crystal display (LCD) shows the pH value of the buffer solution and allows that value to be changed from the keyboard to the correct value of the buffer solution. A hold key is provided to maintain analyzer output level and alarm status during calibration procedures. The display will show the calculated value of pH to within .01 pH, at an update rate of 0.5 second per reading. The useful limits for this display are -2 pH and 14 pH, although higher displays are available.

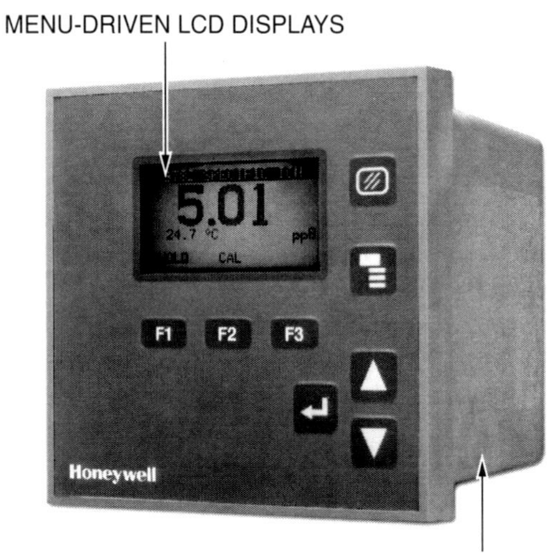

Figure 8 ◆ Analyzer/controller.

This type of analyzer also features extensive on-line diagnostics, as well as keyboard-controlled test functions. A keyboard-entered security function prevents access to instrument settings as desired. It is designed for panel, pipe, or wall mounting and is typically enclosed in a watertight and corrosion-resistant industrial case.

4.0.0 ◆ CONDUCTIVITY MEASUREMENT

The pH measurement system was designed to be sensitive to just one ion, the hydrogen ion. Similar systems can be designed to produce outputs proportional to other selected ions using other types of glass or plastics. However, there are applications where it is necessary to determine the total ionic concentration of all ions. This is accomplished using a conductivity bridge.

In a solution of pure water, the only ions in solution would be those produced by the dissociation of water into hydroxyl (OH^-) and hydrogen (H^+) ions. Because the passage of current flow through a solution relies on ions as the charge-carrying conductors, very little current flow could exist in a solution of pure water. As impurities are added to the water, the number of available ions increases, reducing the resistance to current flow. Conductivity is a measure of how easily a material or solution conducts electricity. Therefore, the conductivity of a solution is directly proportional to the total number of ions in the solution.

Conductivity (k) is the inverse of resistivity (ρ). Recall that resistivity is the opposition to current flow offered by a material independent of its external size. The relationship between conductivity and resistivity is shown in the equation below:

$$k = \frac{1}{\rho(\text{ohm–meters})}$$

$$k = \frac{1}{\text{ohm–meters}} = \frac{\text{ohms}^{-1}}{\text{meter}} \text{ or } \frac{\text{mhos}}{\text{meter}}$$

Although conductivity could be expressed as ohms^{-1}/meter, it is more frequently given its own unit of measure. The mho is the standard unit of conductivity. It is ohm spelled backwards, because the mho is the inverse of the ohm.

Resistivity is a property of a material. The resistance of an actual sample of a material can be found using the following equation:

$$\text{Resistance (R)} = \rho\frac{\ell}{a}$$

Where:

ρ = the resistivity in ohm-meters

ℓ = the length of the sample in meters

a = the area of the sample in meters^2

Similarly, conductivity is a property of a material. If the resistance of a sample were known, the conductivity could be determined using the following equation:

$$k = \frac{1}{R} \times \frac{\ell}{a}$$

Where:

k = conductivity in mhos/cm
R = resistance in ohms
ℓ = the length of the sample in cm
a = the area of the sample in cm^2

Although resistivity and conductivity are very similar terms, resistivity is normally used in relation to solid materials. Conductivity is reserved for liquid solutions.

A typical primary element used to measure the conductivity of a solution is two parallel metallic electrode plates immersed in the solution. The external circuitry must provide a source of voltage to the electrodes and a method of determining the current flow between them. A simplified instrument is shown in *Figure 9.*

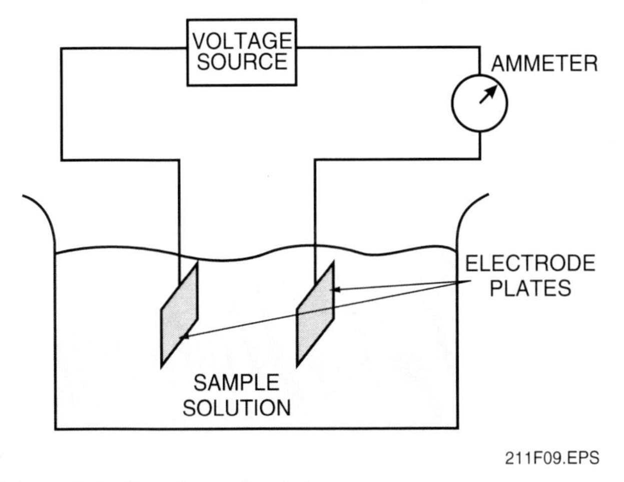

Figure 9 ◆ Simple conductivity meter.

If the voltage and current flow in Figure 9 are accurately known, the resistance between the electrodes can be determined using Ohm's law. Once the resistance of the sample is known, its conductivity (k) can be calculated using the following equation:

$$k = \frac{1}{R} \times \frac{\ell}{a}$$

In the equation, the length would be the distance between the plates, and the area would be that of a plate. Essentially, we have measured the resistance of a column of the sample and multiplied it by the ratio ℓ/a, making the measurement independent of the dimensions of the sample.

Rather than attempting to measure the distance between, and the area of, the plates in the field, the manufacturer of the electrodes will provide a cell constant. The cell constant (K) is:

$$K = \frac{\ell}{a}$$

Where length and area are in centimeters.

Electrodes can be purchased with cell constants ranging from 0.001 cm^{-1} to 50 cm^{-1}. Cells with high constants will usually have small, widely spaced electrodes and are used to measure solutions with high concentrations of ions. Conversely, low cell-constant electrodes are used to measure low concentrations of ions and have large, closely spaced electrodes.

The cell constant of a conductivity cell may be checked, or that of an unknown cell determined, by using a buffer solution. An acceptable method of determining a cell constant is given in the American Society for Testing and Materials (ASTM) *Standard D 1125-64, Standard Methods of Test for Electrical Conductivity of Water and Industrial Wastewater.* Briefly, the standard states that if the conductivity of a solution and the resistance measured with the cell under test are known, the cell constant could be determined using the following equation:

$$k = \frac{1}{R} \times \frac{\ell}{a}$$

Where K replaces the term.

Solving for K yields the cell constant:

$$K = \frac{k}{1/R} = Rk$$

To obtain a solution of known conductivity, a precise amount of potassium chloride is dissolved in distilled water. To obtain an accurate constant, three measurements should be made using different batches of the same solution. The average reading is the cell constant.

The use of a DC voltage to drive the conductivity cell would cause significant errors. First, because of the mobility of ions in solution, continued application of a DC current would cause changes in the electrolytic concentration near the electrodes. The cell anode would attract the negative ions, and the cathode would attract positive ions. These ions would shield the electrode voltage source from the rest of the solution, causing reduced current flow. Second, the chemical reaction between the electrolyte and the cell plates could cause oxidation-reduction *half cells* with sufficient voltage to cause measurable errors. For both these reasons, most conductivity bridges use

AC potentials. Both effects, commonly referred to as polarization, are avoided in the AC bridge.

The conductivity of any solution is strongly influenced by temperature, so manual or automatic temperature compensation must be provided. The actual change in conductivity with respect to temperature is determined by the type of ions in the solution. Therefore, automatic compensation must rely on assumptions about the composition of the electrolyte. The actual compensation is often done with a **thermistor**. The nonlinear characteristics of the thermistor are matched to the nonlinear changes in conductivity that occur in the sample solution. Instruments that may be used on many different solutions and that provide direct conductivity readouts are most often manually compensated using tables for different electrolytes. These are available in *The Handbook of Chemistry and Physics* or *The International Critical Tables*, published by McGraw-Hill.

The mechanical features of conductivity cells are illustrated in *Figure 10*. There are four basic types:

- *Dip cells* – Designed for permanent installation in pipelines and tanks

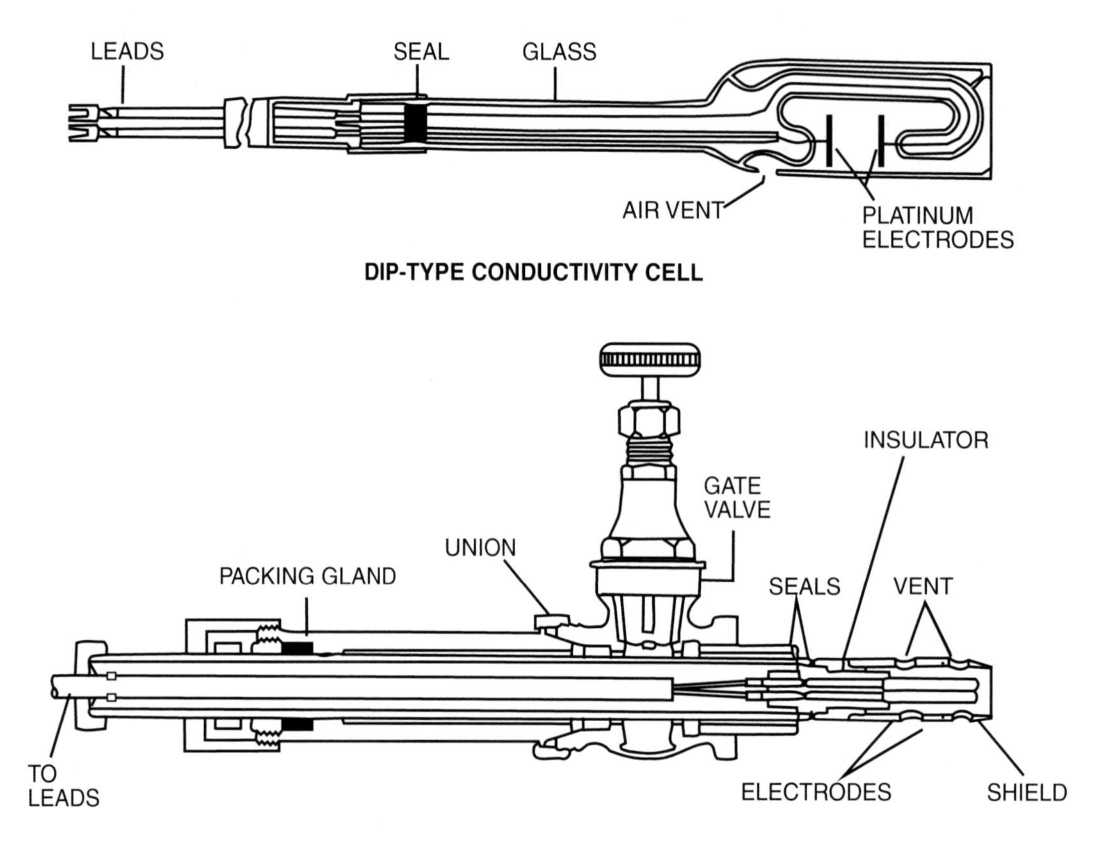

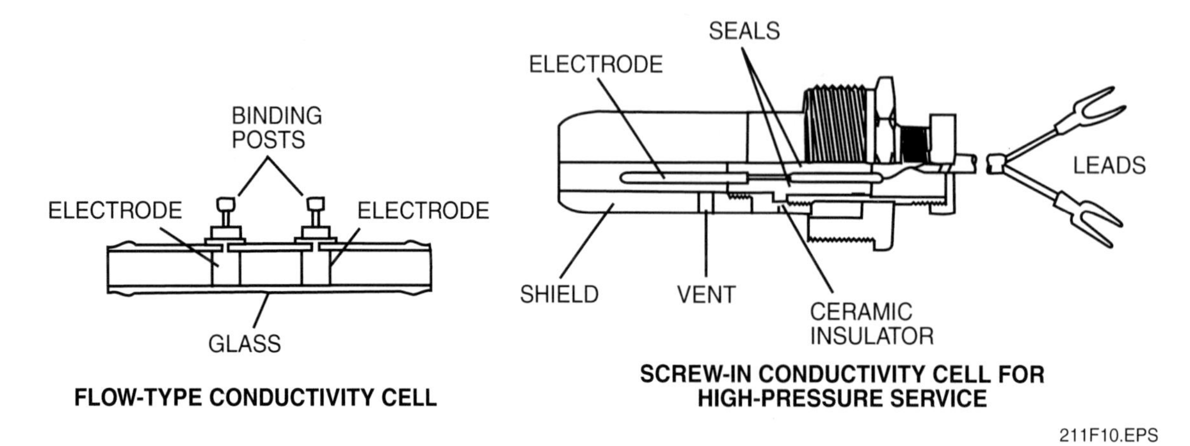

Figure 10 ◆ Basic types of conductivity cells.

- *Insertion cells* – With removal devices, designed to permit removal of an element without closing down the line in which it is installed
- *Flow cells* – Glass or plastic, with internal electrodes close to the wall to offer little resistance to the flowing medium (in small sizes, the flow-cell tubes are connected to the system with rubber or plastic tubing; in large sizes, standard pipe flanges are used)
- *Screw-in cells* – Used for high-pressure service (the system must be shut down before replacing the cells)

5.0.0 ◆ DENSITY AND SPECIFIC GRAVITY

In industrial processes and scientific work, it is frequently necessary to know the mass density, weight density, or specific gravity of a material. Materials are commonly classified as solids, liquids, and gases. The term *fluid* is applicable to both liquids and gases.

If small cubes of different metals having the same volume are examined, they will be found to have different weights. Referring to *Figure 11*, aluminum is one of the lightest metals, and lead is one of the heaviest. In order to compare the relative heaviness of different substances, the term density is used. Because of its common occurrence, water is generally used as the basis for comparison. The temperature must also be specified because of **volumetric** changes. Water has its greatest density at 39°F. Some comparison scales refer to that temperature.

In scientific work, mass density means the mass per unit volume. In the metric system, mass density is expressed in grams per cubic centimeter. In the British system, it is expressed as slugs per cubic foot.

$$\text{Mass density D} = \frac{\text{mass m}}{\text{volume V}}$$

When a body is weighed on a beam balance or scale, it is directly balanced by a known, calibrated mass. This weighing operation is, in fact, a measure of the mass of the body, because the attraction of gravity on both the unknown and known masses is the same. On the other hand, when a pound mass is weighed on a spring scale, the deflection of the scale will be influenced by the local value of the earth's gravitational attraction. The weight of a body depends on this attraction of gravity, whereas the mass of the body is not dependent on it. *Table 2* shows approximate weight densities of some commonly used liquids.

Table 2 Weight Density of Some Commonly Used Liquids

Liquid	Temperature	Density (lbs/ft³) (rounded off)
Gasoline	60	46.8
Acetone	60	49.0
Kerosene	60	51.0
SAE 30 Lube Oil	60	53.0
Water	60	62.4
Mercury	60	847.0

Specific implies a ratio. Weight is the measure of the earth's attraction on a mass or body, and is called *gravity*. The ratio of the weight of a unit volume of some substance to the weight of an equal volume of a standard substance, measured under standard pressure and temperature conditions is called specific gravity. Thus, the specific gravity of a liquid is the ratio of its density to the density of water.

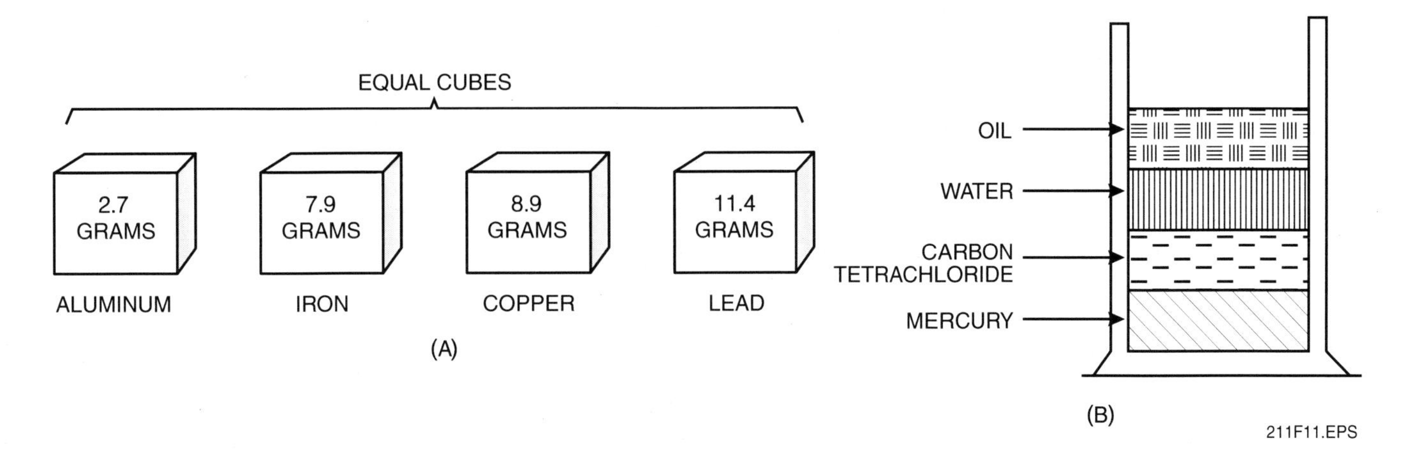

Figure 11 ◆ Comparison of heaviness of metals and liquids.

5.1.0 Measurement of Density and Specific Gravity

To find the density of a substance, its weight and volume must be known. The weight is then divided by the volume to arrive at the weight per unit volume. The equation for this is:

$$D = W \div V$$

A container that is 4 feet long, 3 feet wide and 2 feet deep has a volume of 24 cubic feet ($4 \times 3 \times 2 = 24$). If the contents of this container weigh 1,497.6 pounds, the density can be found by dividing the weight by the volume, or,

$$D = 1497.6 \div 24 = 62.4$$

This is the density of water at 4°C and is usually used as the standard for comparing densities of other substances. The temperature of 4°C was selected because water has its maximum density at this temperature. Changes in temperature will not change the weight of a substance, but they will change the volume of the substance through expansion or contraction. This changes the weight per unit volume, or density.

The following formulas are used to find the specific gravity (sp gr) of solids and liquids, with water used as the standard substance.

sp gr = weight of the substance ÷ weight of equal volume of water

or

sp gr = density of the substance ÷ density of water

The specific gravity of water is 1 ($62.4 \div 62.4 = 1$).

The same formulas are used to find the specific gravity of gases by substituting air, oxygen, or hydrogen for water. If a cubic foot of a certain liquid weighs 68.64 pounds, then its specific gravity is 1.1 ($68.64 \div 62.4 = 1.1$).

Specific gravity and density are independent of the size of the sample under consideration and depend only on the substance of which it is made. A device called a **hydrometer** is used for measuring the specific gravity of liquids.

5.1.1 Hydrometer

The common hydrometer consists of a glass float that is weighted at the bottom. The float has a hollow stem, inside of which is a graduated scale. The arrangement is shown in *Figure 12*. To find the density of a liquid, the hydrometer is floated in the liquid. The position of the surface of the liquid on the hydrometer scale indicates the liquid density.

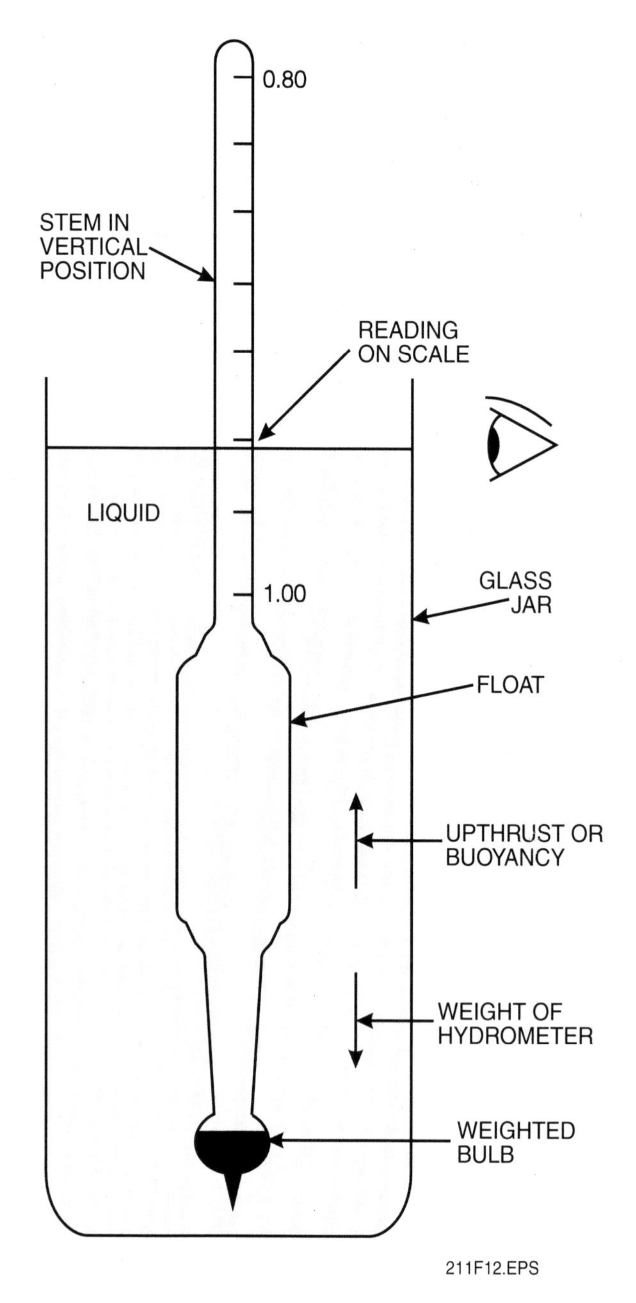

Figure 12 ◆ Hydrometer.

The hydrometer is based on the application of Archimedes' principle of flotation, which states that the weight of a floating object is equal to the weight of the fluid it displaces. The float increases the buoyancy of the hydrometer because of the liquid displaced. Hydrometers used for liquids that are lighter than water have a large float, and scale graduations start with a specific gravity of 1.00 at the bottom. A variety of hydrometers, whose operations are based on the principles discussed, are available on the market.

Hydrometers are used in the laboratory and in industry for testing liquids such as salt solutions, petroleum products, and acids. The storage-battery hydrometer is used to test the specific gravity of the acid solution and thereby estimate the degree to which the battery should be charged. Hydrometers are also used in the testing of antifreeze mixtures, of milk for possible dilution, and for estimating the strength of wines and beers.

5.1.2 Air Bubbler Method

The air bubbler arrangement shown in *Figure 13* gives a continuous indication or record of specific gravity. The method requires that a sample of the liquid be drawn from the process system. The container provides a fixed depth of liquid into which a standpipe is inserted at constant immersion. Air is fed past a regulating valve and a sight gauge into the standpipe. Bubbles are permitted to escape only from the bottom end of the standpipe. The pressure in the system and the standpipe equals the pressure due to the head of liquid above the lower end of the standpipe. This pressure will vary with the specific gravity of the liquid, and the receiving instrument is calibrated in terms of specific gravity. The air pressure, in pounds per square foot, equals the pressure head of liquid in the same units. Symbolically,

$$p = h \times 62.4 \times \text{sp gr}$$

Where:

p = pressure of the air in the system and standpipe in lb/ft^2

h = height of the liquid above the opening at the bottom of the standpipe in feet

sp gr = specific gravity of the liquid

5.1.3 Differential Pressure Method

Figure 14 shows the operating principle of a density-measuring system. The chamber contains the liquid for which the weight density (w_m) is under measurement. Two bubble tubes are immersed at distances d_1 and d_2, respectively, as indicated. Air is bled through a restrictor to each tube, and it builds up pressure until it equals the heads wd_1 and wd_2. The escape of air, which bubbles out through the holes at the bottoms of the tubes, holds the pressure constant at this value.

Naturally, there are further requirements to be added to the basic scheme in order to accommodate the range of densities from 0.9 to 1.0. An added pressure head is introduced in the low-pressure

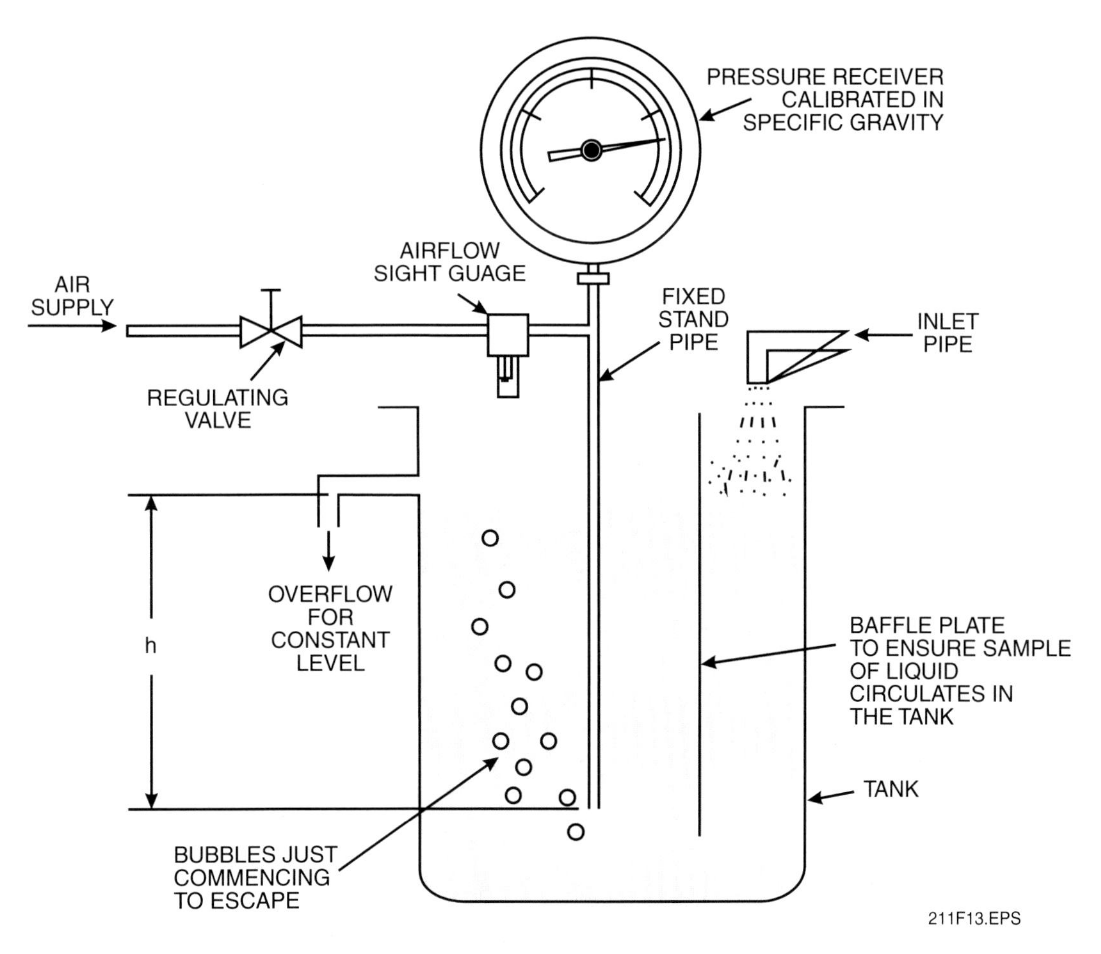

Figure 13 ◆ Arrangement of bubbler system to measure specific gravity.

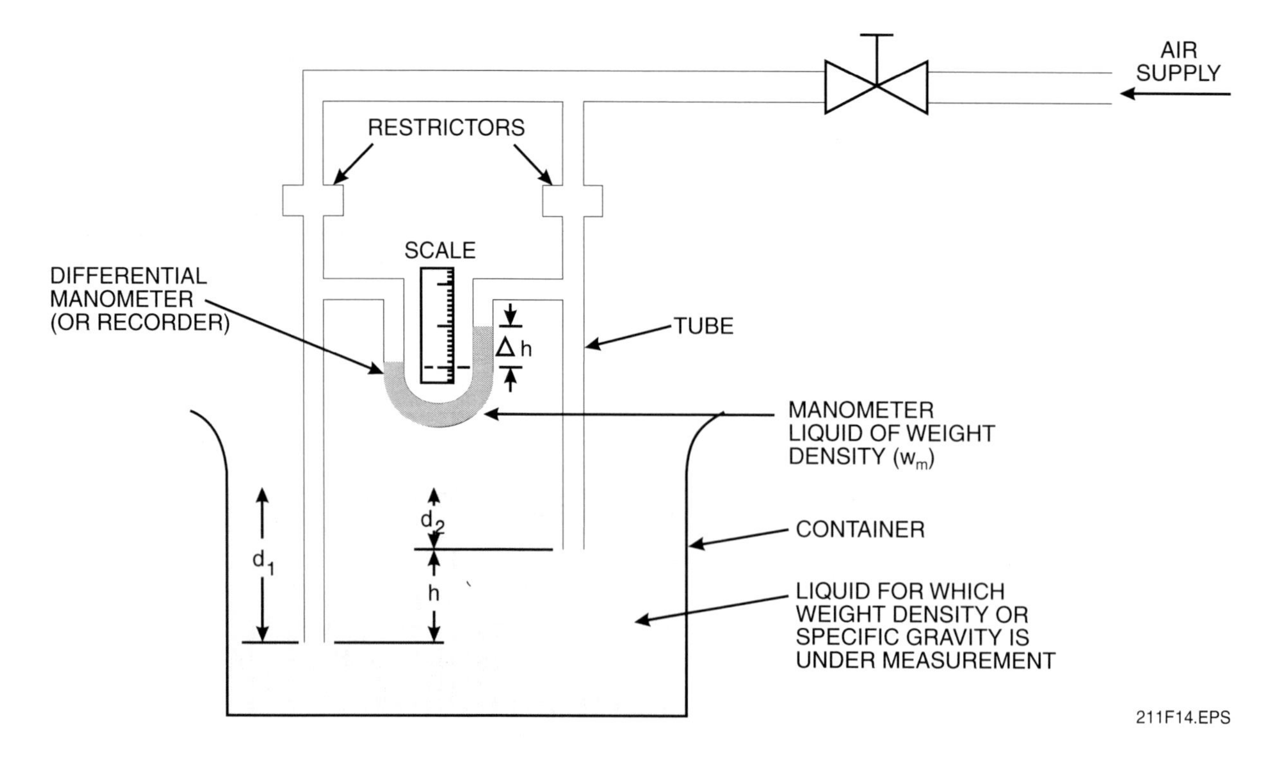

Figure 14 ◆ Differential pressure density meter.

side of the differential manometer by means of a reference chamber. This provides full-scale travel of the measuring device for the density range under measurement.

5.1.4 Specific Gravity and Density Values

The specific gravity values of certain solids and liquids are listed as follows:

- *Lead* – 11.34
- *Mercury* – 13.546
- *Tool steel* – 7.70 to 7.73
- *Oil* – 0.88 to 0.94
- *Water at 39°F = 4°C (max density)* – 1.00
- *Petroleum, gasoline* – 0.70 to 0.75

6.0.0 ◆ VISCOSITY MEASUREMENT

The internal friction, or resistance to flow, set up within a fluid (liquid or gas) is called viscosity. The water on the surface of a river moves more rapidly than the water near the sides or bottom. Water flow in a pipeline, or airflow in a shaft (or ducting) has greater velocity at the center than it has next to the metal surfaces. This difference in velocity is partly due to the friction between the fluid and the boundary surface, which causes the adjacent layers of fluid to move slowly. These slowly moving layers of fluid, in turn, tend to retard the motion of adjacent layers. Viscosity is responsible for most of the loss of energy in liquid and gas pipelines. For this reason, viscosity is a very important factor in fluid flow.

The property of viscosity is particularly significant in the study of oil. In industry, heavier oils have higher viscosities. This term does not bear any relation to their densities. The effectiveness of lubricating oils depends on, among other factors, the viscosity of the oils. Generally, lubricating oils should be sufficiently viscous that they will not be squeezed out of bearings and yet not so viscous as to increase the resistance to the motion of the moving parts being lubricated.

For these reasons, the study of viscosity and its measurement is very important. Viscosity is a comparative measure of the ease with which particles in a fluid can change their relative positions and yield to an external force. For example, a thick liquid like honey offers more resistance to flow than does water. More specifically, the viscosity of a liquid or gas is a physical property that determines the magnitude of the resistance of the fluid to a shearing force.

6.1.0 Viscometers

The viscosity of a fluid can be determined by several methods. One method is to measure the time required for a known quantity of the fluid to escape through a long tube of small diameter. For determining the viscosity of paints, varnishes, nail polish, and similar liquids, a common test is to time the flow of the liquid through a standard opening. Instruments used to determine viscosity values are called viscometers or viscosimeters.

A simple and basic method for measuring viscosity is the flow of a liquid through a capillary or small-diameter tube such as is used for thermometers. Provided that flow is truly laminar or viscous, all the necessary conditions for viscosity measurement are easily achieved. *Figure 15* shows the method for measuring viscosity by the flow of liquid through a small diameter tube.

The discharge rate Q can be found by using a graduated jar to measure the volume of liquid and a stopwatch to determine the time. The coefficient of absolute viscosity can then be computed:

$$Q = \frac{\text{volume collected}}{\text{time taken}}$$

The discharge rate Q is inversely proportional to the viscosity.

6.2.0 Rotating Spindle Viscometers

Another method of measuring viscosity is through the use of a rotating spindle viscometer. A rotating spindle viscometer continuously measures viscosity by drawing a sample of process liquid into a chamber. A motor rotates a spindle at a fixed speed in this chamber. The resistance of the fluid sample against the spindle determines the viscosity of the fluid. The viscosity of the fluid is in direct proportion to the force, or torque, needed to drive the spindle. *Figure 16* shows a rotating spindle viscometer.

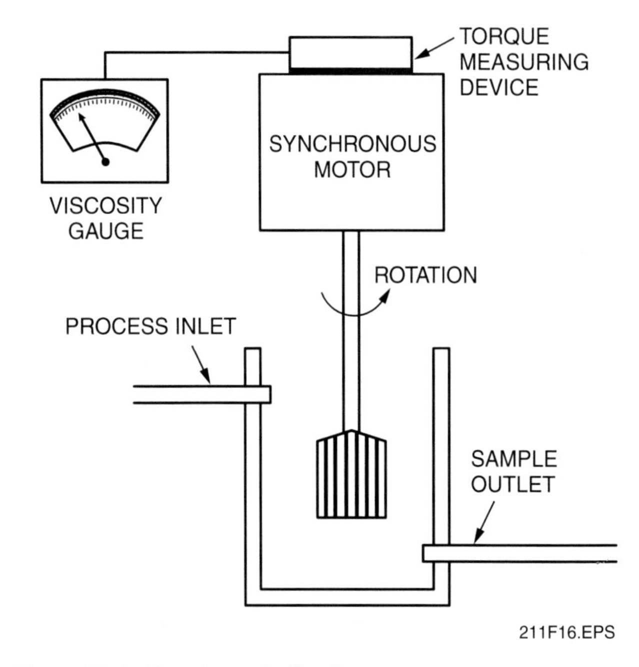

Figure 16 ◆ Rotating spindle viscometer.

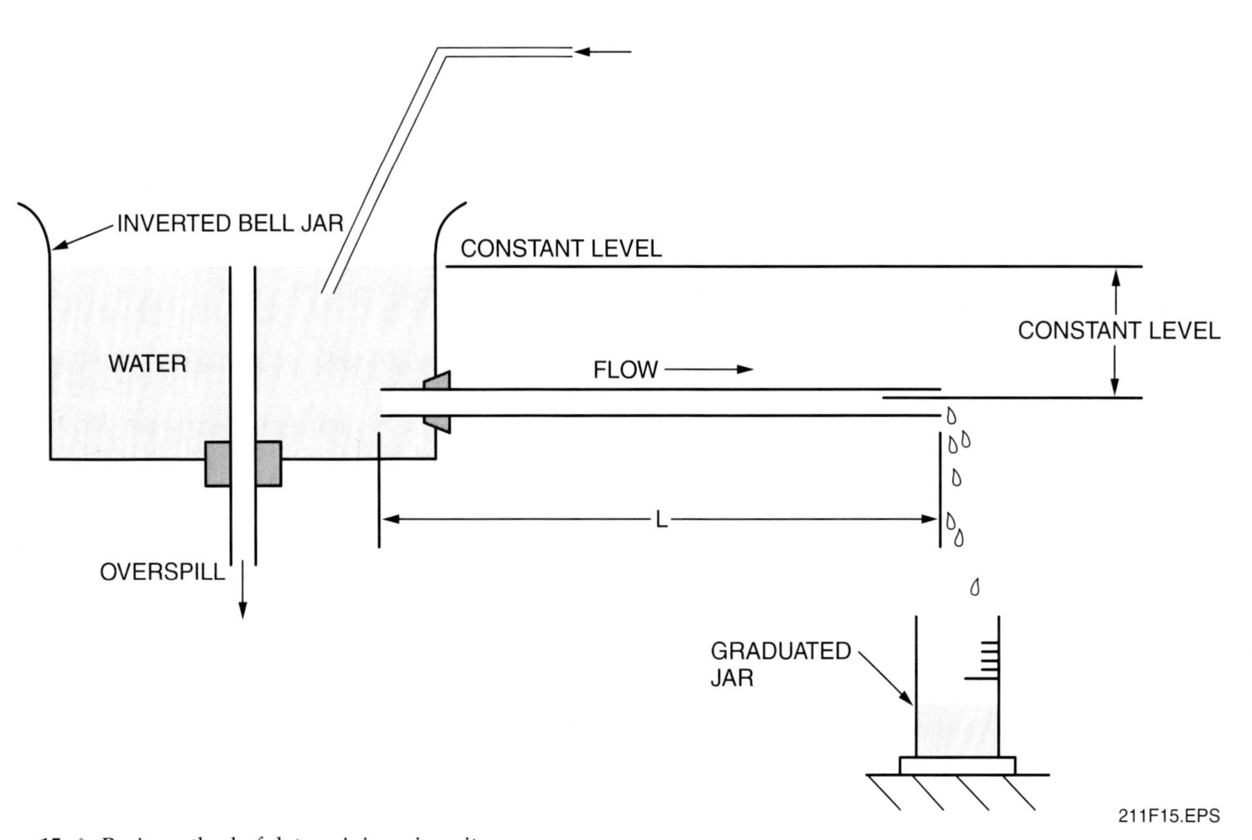

Figure 15 ◆ Basic method of determining viscosity.

6.3.0 Vibrating Reed Viscometers

A vibrating reed viscometer continuously measures liquid viscosity with a vibrating reed sensor. This sensor is a tuning fork made of stainless steel installed in the process piping. An electrical current causes the sensor to vibrate. The viscosity of the fluid causes these vibrations to die out. The amount of time it takes for the sensor to stop vibrating in the liquid is proportional to the viscosity of the liquid. Electronic circuits measure this time and convert the measurement to a signal that can be used to indicate the liquid's viscosity. *Figure 17* shows a vibrating reed viscometer.

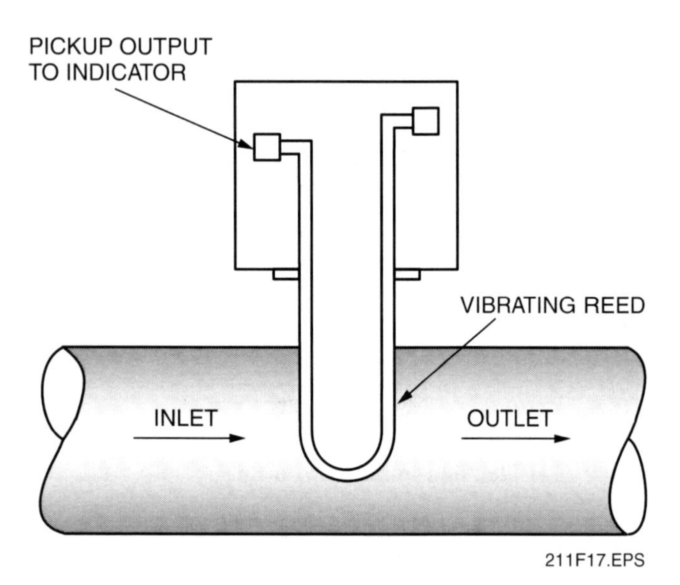

Figure 17 ◆ Vibrating reed viscometer.

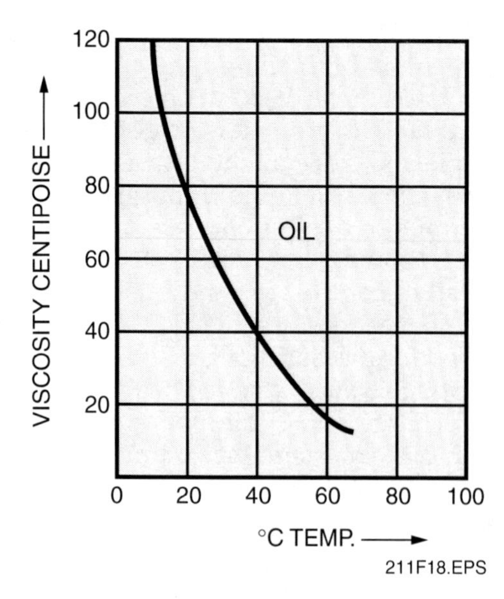

Figure 18 ◆ Viscosity variation of an oil with temperature.

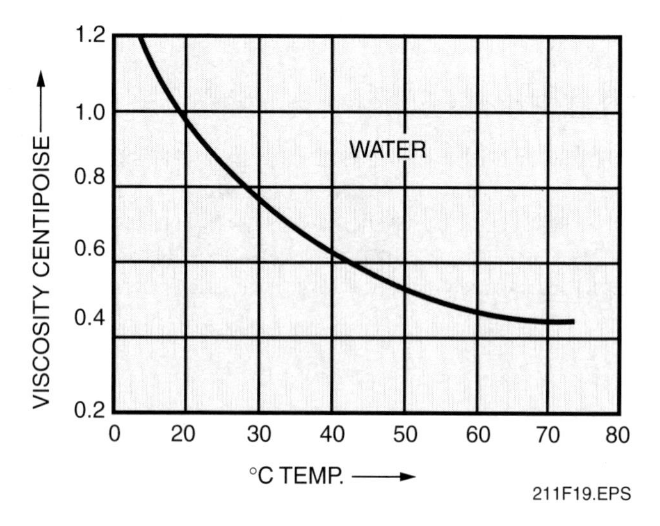

Figure 19 ◆ Viscosity variation of water with temperature.

6.4.0 Temperature Effects on Viscosity

Of all the physical conditions, temperature has the greatest effect on viscosity. An increase in temperature causes a greater separation among the liquid molecules, a reduction in the shear stress, and therefore a decrease in viscosity. Lubricating oils may fail to form a protective film at low temperatures. The large decrease in absolute viscosity with increasing temperature can be observed in *Figure 18* and *Figure 19*.

For example, tar at 15°C has a viscosity of 1.65 × 106 poises, whereas at 45°C its viscosity is 1.3 × 103 poises. For a temperature span of 30°C the viscosity changes by a factor of a thousand.

7.0.0 ◆ CHEMICAL COMPOSITION MEASUREMENT

Chemical composition analysis is vitally important in nearly all of the continuous-process industries. It is particularly important in the petrochemical industries. Incoming raw materials may be sampled to ensure compliance with required chemical specifications or to optimize the yield and minimize production costs. In-process analysis is necessary at various stages to ensure that there will not be excessive waste at the end of the production process because of a failure to meet the final-product specification. Many analyzing instruments are available to make continuous on-line measurements that permit a fast response to a control mechanism. The control mechanism adjusts the appropriate inputs or production stages to bring the process back into control. This type of continuous analysis permits much greater product throughput with less waste than is possible with the more conventional batch analysis technique.

Chemical composition analysis is also used to assure public health and safety, as in the case of municipal water treatment, sewage disposal plants, and industrial plants emitting toxic liquids or gases. The objective in all cases is to determine what chemical elements are present or how much of one or more of several known elements are present. The former is referred to as qualitative analysis and the latter as quantitative analysis.

7.1.0 Gas Chromatography

The name **chromatography** is somewhat misleading, but it originated with a discovery by Mikhail Tswett in 1906. In his experiments, he found that when a solution of mixed colored pigments was allowed to filter through a column of firmly packed pulverized calcium carbonate, the individual pigments passed down the column at different rates, and so could be separated into distinctive color bands. The result was called a chromatogram. A modern variation of this is found in paper chromatography, in which compounds in solution migrate at different speeds across a sheet of porous paper.

Gas chromatography is a method of separating mixtures of gases by passing a sample mixture and a carrier gas, usually helium, through a column packed with a suitable fractionating agent. During the passage through the column, the fractionating agent causes the various components to be retained for different periods of time. When there is a suitable detection device in the stream leaving the column, the times of emergence and quantities of the various components can be determined. For a given column and fractionating agent with constant carrier pressure, carrier flow rate, and column and detector temperature, the retention times are all held constant. Therefore, the time that elapses between sample injection and component detection will serve to identify that particular component.

The basic block diagram of a chromatograph is shown in *Figure 20*. It comprises four major components: the sample introduction units, the fractionating column, a detector, and a recording device. The fractionating column, the heart of the system, is usually constructed from a copper or stainless steel tube about ¼-inch in diameter and from 1 to 50 feet long. Adsorption columns are packed with a fine mesh material such as charcoal, silica gel, or activated **alumina**. Partition-type columns are packed with an inert material such as fire brick or **diatomaceous earth** and coated with a nonvolatile liquid called a partitioner.

The capillary column is quite different in structure. It comprises tubes with an inside diameter of 0.01 inch and a length of 150 to 1,000 feet coiled into a compact helix. The sample introduced into such a column must be kept small, but the column is highly selective and has a relatively high speed

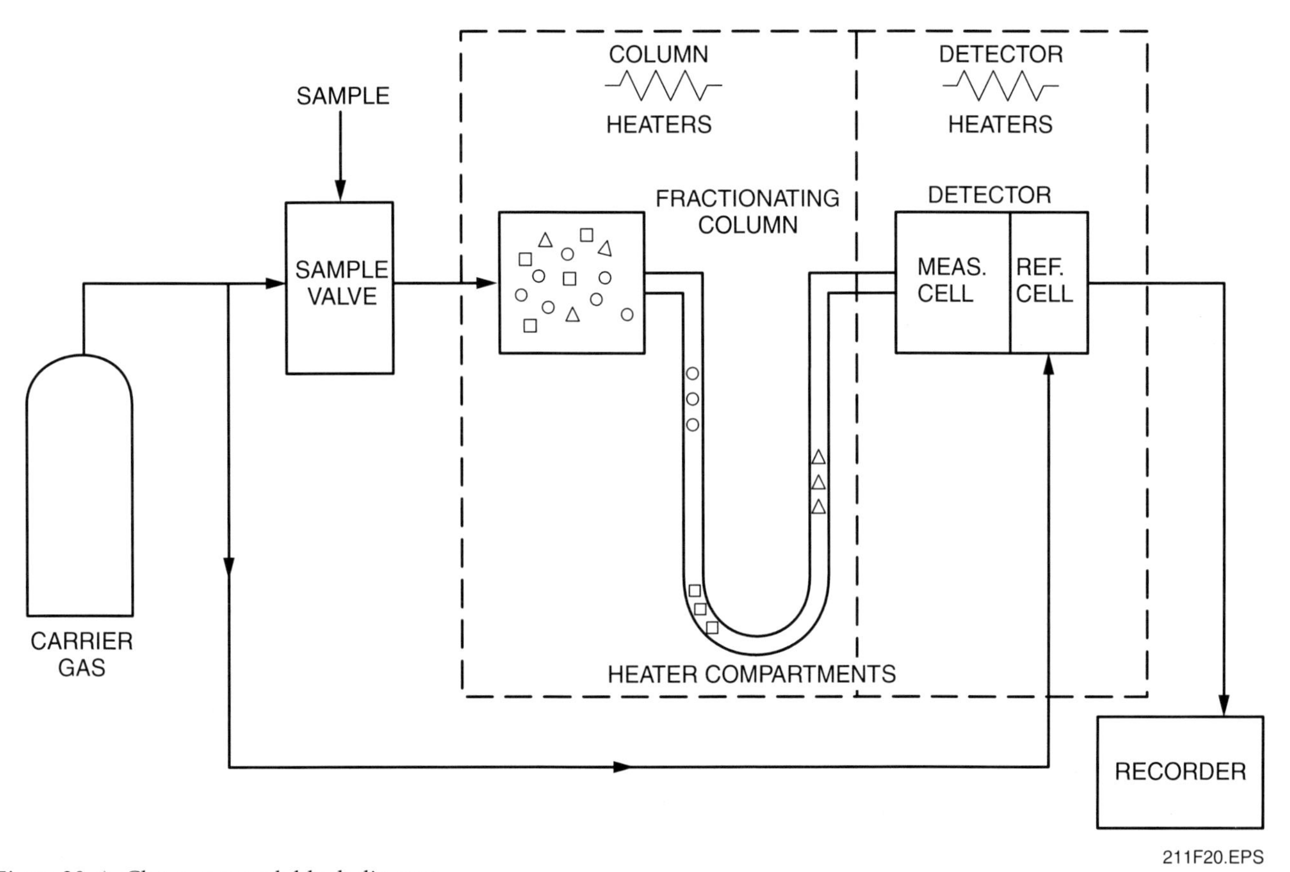

Figure 20 ◆ Chromatograph block diagram.

of response. This response is as fast as a few seconds compared to many minutes for adsorption and partition-type columns.

There are several types of detectors. Differential thermal conductivity cells are the most popular, because they are relatively inexpensive, sensitive, and stable. They may comprise either matched resistance wires or thermistors arranged in a bridge circuit. Other types are the hydrogen flame ionization detector, argon ionization detector, and catalytic combustion detector. The detector causes a recording instrument to indicate a concentration measurement based on the magnitude of the peak.

7.2.0 Thermal Conductivity Gas Analysis

This is the most common method of gas analysis because of its simplicity, reliability, relative speed, and easy adaptation to continuous recording and control. It is based on the fact that various gases differ considerably in their ability to conduct heat. The gas analyzer employs a hot wire mounted in a chamber containing the gas. The hot wire is maintained at an elevated temperature with respect to the cell walls by passing an electric current through it. An equilibrium temperature is attained by the wire when its electric power input is equalized by all thermal losses from it.

The two most commonly used detector transducers are resistive wires and thermistors. Their operation is similar, except that the resistance of the filament-type wires increases with temperature, while the resistance of the thermistor-type wires decreases with temperature increases. *Figure 21* shows an example of a thermal conductivity gas detector.

Figure 21 ◆ Thermal conductivity detector.

By proper design, it is possible to minimize all heat loss except that due to gaseous conduction. Under these circumstances, the temperature rise of the wire is inversely related to the thermal conductivity of the gas confined within the cell. The hot wire is made of material such as platinum having a high temperature coefficient of resistance. It is therefore possible to measure changes in the equilibrium temperature merely by measuring the change in resistance of the wire.

A Wheatstone bridge is normally used to measure this resistance change. In practice, it is preferable to use two hot-wire cells in adjacent arms of a Wheatstone bridge. One of the cells contains a reference gas and the other contains the gas to be analyzed. Thus the bridge responds to the difference in temperature rise of the two hot wires.

Figure 22 illustrates the circuitry for the Wheatstone bridge and a filament-type detector.

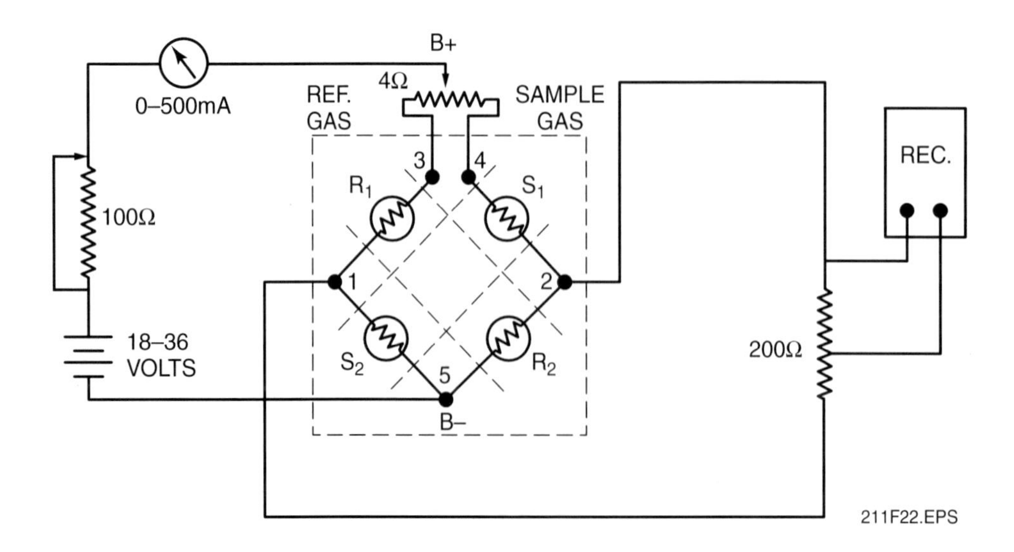

Figure 22 ◆ Wheatstone bridge and filament circuitry for thermal conductivity detector.

This type of analyzer is particularly suited to applications involving mixtures of two gases, such as air with oxygen, carbon dioxide, or hydrogen. It is possible, but often difficult, to use this technique to analyze complex mixtures of several gases. An important application is in the measurement of combustibles in the products of combustion. The thermal conductivity oxygen in the products of combustion can also be measured by this system, but the combustible gas hydrogen has to be added in order to turn the oxygen to water.

7.3.0 Smoke Density Measurement

The measurement of smoke density is important in all large heating plant installations because of the growing concern over air pollution. All large municipalities now have strict regulations governing the density of smoke expelled into the atmosphere and the time during which smoke of a given density can be expelled.

The simplest method of measurement is by visual comparison with the Ringelmann scale, which is nothing more than a series of six colored squares ranging from white (0) through increasingly darker shades of gray to solid black (5). This technique is obviously not satisfactory for continuous measurement, but it is in general use by air pollution inspection authorities.

For large plants, an automatic, continuous smoke density sampling system is necessary. Such a system generally utilizes a lamp and photocell located on opposite sides of the stack and is known as an opacity measurement system. Refer to *Figure 23*.

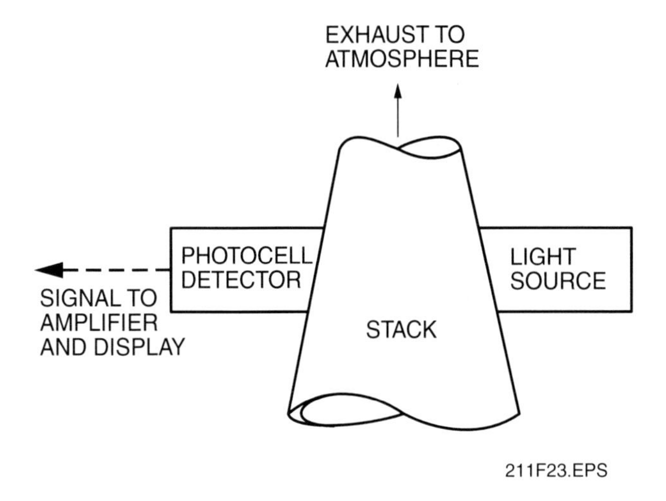

Figure 23 ◆ Opacity measurement of smoke.

As smoke flows up the stack, it obscures the light in proportion to its density. The photocell measures the intensity of the light being transmitted through the smoke and converts this to a proportional electrical signal that can be recorded or used to energize an alarm device.

7.4.0 Hydrocarbons

The instrument methods used for measuring hydrocarbons in air are the flame ionization and infrared methods.

7.4.1 Flame Ionization Methods

These methods involve measuring the ionization of carbon atoms in a hydrogen flame. A pure hydrogen flame contains very few ions. The presence of trace amounts of organic compounds will produce substantial numbers of charged carbon ions.

The sample is mixed with hydrogen and passed to the flame ionization detector (*Figure 24*). Air is supplied to maintain the flame. The electric potential across the flame jet to an ion collector placed above the flame is measured by an electrometer. The potential is proportional to the concentration of carbon ions in the sample.

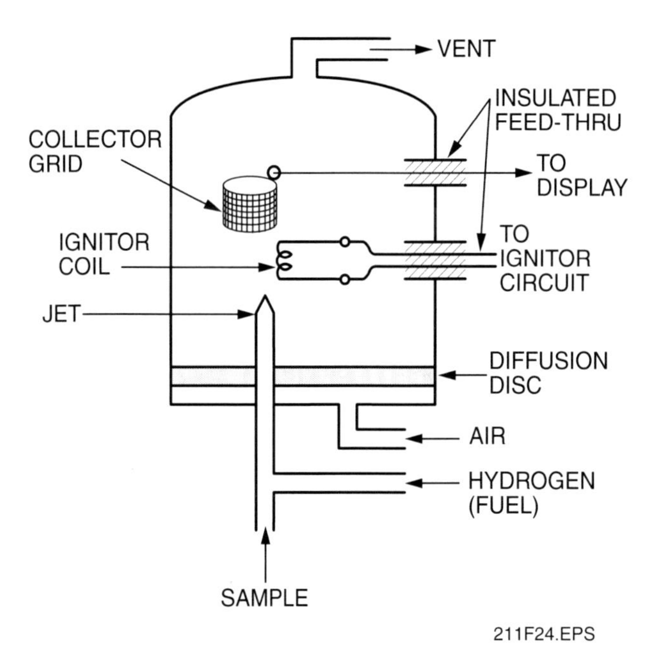

Figure 24 ◆ Flame ionization detector for hydrocarbon measurement.

7.4.2 Infrared Methods

These methods have been used for total hydrocarbon monitoring, particularly from automobile exhausts. The non-dispersive infrared (NDIR) analyzer with hexane-sensitized detectors has been particularly successful. In this technique, the energy in the IR absorption bands in the air sample is compared to in the energy in the bonds of normal hexane. The technique requires corrections for carbon dioxide, water vapor, and lower hydrocarbons. When infrared light is passed through the gas, some of the light is absorbed. The amount of absorption depends on the composition of the gas.

Summary

This module defines some of the basic chemical measurements and describes instruments used to monitor them. An understanding of basic chemistry is very important to a technician responsible for the care and calibration of analyzers and monitors. Chemistry is the study of the composition and properties of matter and the changes that matter undergoes. By continuously monitoring the quantity and quality of certain chemicals or elements during a continuous manufacturing process, strict quality standards can be maintained. A measuring device such as an analyzer or monitor is simply a device for determining the value of a quantity, condition, or physical variable. Generally, it is the magnitude or quantity of the variable being measured that is of interest. Many types of analyzers and monitors are available that continually analyze a critical parameter in a continuous manufacturing process. These devices provide a fast response to the control mechanism, which will adjust the appropriate input to bring the process back within specified limits.

Review Questions

1. An atom is made up of _____.

a. ions, periods, and groups
b. elements, neutrons, and mass
c. protons, electrons and neutrons
d. ions, electrons, and neutrons

2. The periodic table lists the following in groups and periods, based on the number of protons in the nucleus and the number of electrons in the valence shell *except* _____.

a. atoms
b. elements
c. noble gases
d. compounds

3. Inert or noble gases _____.

a. have incomplete valence shells
b. are in Group 7 on the periodic table
c. do not react chemically with other elements
d. are also known as transition elements

4. Molarity refers to the _____ of a solution.

a. viscosity
b. concentration
c. atomic mass
d. gram atomic weight

5. PPM is defined as the _____.

a. parts per mole of a solute per liter of solution
b. parts per million ratio of solute to solution
c. pints per million
d. ratio of moles to a metric measure of solution

6. Ions are atoms or groups of atoms that _____.

a. have an electrical charge
b. have more protons than neutrons
c. are generated by electromagnetic interference
d. have more neutrons than protons

7. pH is the measure of the _____ of a solution.

a. opacity
b. conductivity
c. concentration
d. acidity or alkalinity

8. True or False? A solution that has a pH of 8 is slightly acidic.

a. True
b. False

9. The numbering system for a pH scale ranges between _____.
   a. 0 and 7
   b. 0 and 10
   c. 0 and 14
   d. 0 and 100

10. Conductivity is a measure of the _____ in a solution.
    a. OH- ions
    b. cations
    c. anions
    d. ions

11. True or False? The conductivity of any solution is strongly influenced by temperature.
    a. True
    b. False

12. Conductivity meters read out in units of _____.
    a. ohms
    b. mhos
    c. impedance
    d. millivolts

13. Specific gravity is defined as the _____.
    a. density of the solution being measured divided by the density of water
    b. density of the solute divided by the density of the solution
    c. density of the solution being measured
    d. specific weight of the solution

14. Specific gravity can be measured using a _____.
    a. viscometer
    b. pH meter
    c. hydrometer
    d. opacity meter

15. Viscosity is measured by all of the following methods *except* _____.
    a. a vibrating reed submerged in the solution
    b. a rotating spindle submerged in the solution
    c. measuring the time required to fill a graduated cylinder
    d. the linear movement of a piston in a cylinder

16. An increase in temperature will cause _____.
    a. an increase in the SAE viscosity rating
    b. a decrease in the viscosity of a solution
    c. an increase in the viscosity of a solution
    d. an increase in the potential energy of a solution

17. Gas chromatography is used to _____.
    a. separate mixtures of gases for identification
    b. increase the thermal conductivity of a gas
    c. prevent chemical reactions in noble gases
    d. determine the amount of various gases

18. True or False? Thermal conductivity gas analysis can identify the composition of a gas mixture by measuring how well the gas mixture conducts heat and then comparing this to the ability of a reference gas to conduct heat.
    a. True
    b. False

19. Smoke density measurements are typically made with _____.
    a. gas hydrometers placed in the smoke stack
    b. gas chromatography as the gas exits the smoke stack
    c. a lamp and a photocell placed on opposite sides of the smoke stack
    d. a smoke viscometer

20. Hydrocarbons in the air can be analyzed by mixing an air sample with pure hydrogen and then burning the mixture. When the mixture burns, _____.
    a. the flame will turn orange
    b. the heat produced is proportional to the hydrocarbons present
    c. ions will be produced according to the amount of hydrocarbons present
    d. the flame produces different colors based on the hydrocarbons present

GLOSSARY

Trade Terms Introduced in This Module

Alkalinity: The measure of basicity of a solution.

Alumina: Aluminum oxide produced from bauxite through a complicated chemical process. It is a white powdery material that looks like granulated sugar.

Atomic mass unit (AMU): A unit of mass for expressing masses of atoms, molecules, and nuclear particles.

Atom: The smallest particle of a chemical element that can exist.

Atomic weight: The average relative weight of an element, referred to some element taken as a standard.

Avogadro's number: The constant 6.02 × 1023, which is the number of molecules present in a gram-molecule or the number of atoms in a gram-atomic weight of any element.

Buffer solution: A solution having a known pH value, used to calibrate pH analyzers.

Chromatography: A chemical analysis technique in which gases, liquids, or solids are separated from mixtures or solutions by selective adsorption.

Covalent bond: A chemical bond in which valence shell electrons are shared between atoms.

Diatomaceous earth: A chalklike, fossilized material used to filter out solid wastes in water. It is also used as an active ingredient in some pesticides.

Electron: The negatively charged part of an atom that orbits around the central nucleus.

Element: A substance that cannot be separated by ordinary chemical means.

Exponent: A number denoted by a small numeral placed above and to the right of a numerical quantity. An exponent indicates the number of times the quantity is multiplied by itself. In the case of X^n, X is multiplied by itself n times. It is said that X is raised to the power of n.

Gram atomic weight (GAW): The atomic mass number of an element expressed in grams.

Gram molecular weight (GMW): The sum of the gram atomic weights of a molecule's component elements multiplied by the number of atoms of each element present.

Hydrometer: A device used to measure the specific gravity of a solution.

Ionic bond: A chemical bond formed by the transfer of electrons.

Litmus: A mixture of pigments extracted from certain lichens that turns blue in basic solution and red in acidic solution.

Logarithm: The logarithm of a number N to a given base b is the power to which the base must be raised to produce the number N.

Logarithmic scale: A mathematical scale used for examining the rate of change. The units are based on the power of ten (0.1, 1, 10, 100, 1000, and so on). It is useful when the rates under study vary considerably.

Molarity: The concentration of a solution that equals the number of moles of solute divided by the number of liters of solution.

Mole: Mass in grams equal to the gram molecular weight of a substance.

***Neutron*:** An uncharged atomic particle that is similar in size to the proton and is present in the nucleus of all atoms except hydrogen.

***Noble gas*:** Any of the elements in the eighth group of the periodic table that naturally have completely full valence shells.

***Nucleus*:** The central part of an atom. The nucleus contains the protons and neutrons.

***pOH*:** The term used to express the hydroxyl ion concentration in a solution.

***Parts per million (ppm)*:** The concentration of solute in a solution, expressed as the parts of solute per million parts of solution.

***Proton*:** A positively charged atomic particle that is similar in size to the neutron and is present in the nucleus of an atom.

***Solute*:** A substance that is dissolved in another substance, forming a solution.

***Thermistor*:** A resistor that is sensitive to temperature change. The resistance of a thermistor changes proportionally with a temperature change.

***Valence shell*:** The outermost electron shell of an atom.

***Volumetric*:** Having to do with volume.

Additional Resources

This module is intended to present thorough resources for task training. The following reference works are suggested for further study. These are optional materials for continued education rather than for task training.

Instrumentation, 1975. F.W. Kirk and N.R. Rimboi. American Technical Society.

Basic Instrumentation, 1966. Patrick J. O'Higgens. McGraw-Hill.

Figure Credits

Honeywell Control Products	211F08
GOW-MAC® Instrument Company	211F21, 211F22

NCCER CURRICULA — USER UPDATE

NCCER makes every effort to keep its textbooks up-to-date and free of technical errors. We appreciate your help in this process. If you find an error, a typographical mistake, or an inaccuracy in NCCER's curricula, please fill out this form (or a photocopy), or complete the online form at **www.nccer.org/olf**. Be sure to include the exact module ID number, page number, a detailed description, and your recommended correction. Your input will be brought to the attention of the Authoring Team. Thank you for your assistance.

Instructors – If you have an idea for improving this textbook, or have found that additional materials were necessary to teach this module effectively, please let us know so that we may present your suggestions to the Authoring Team.

NCCER Product Development and Revision

13614 Progress Blvd., Alachua, FL 32615

Email: curriculum@nccer.org

Online: www.nccer.org/olf

❑ Trainee Guide ❑ AIG ❑ Exam ❑ PowerPoints Other ______________________________

Craft / Level: Copyright Date:

Module ID Number / Title:

Section Number(s):

Description:

Recommended Correction:

Your Name:

Address:

Email: Phone:

Module 12212-03

Panel-Mounted Instruments

COURSE MAP

This course map shows all of the modules in the second level of the Instrumentation curriculum. The suggested training order begins at the bottom and proceeds up. Skill levels increase as you advance on the course map. The local Training Program Sponsor may adjust the training order.

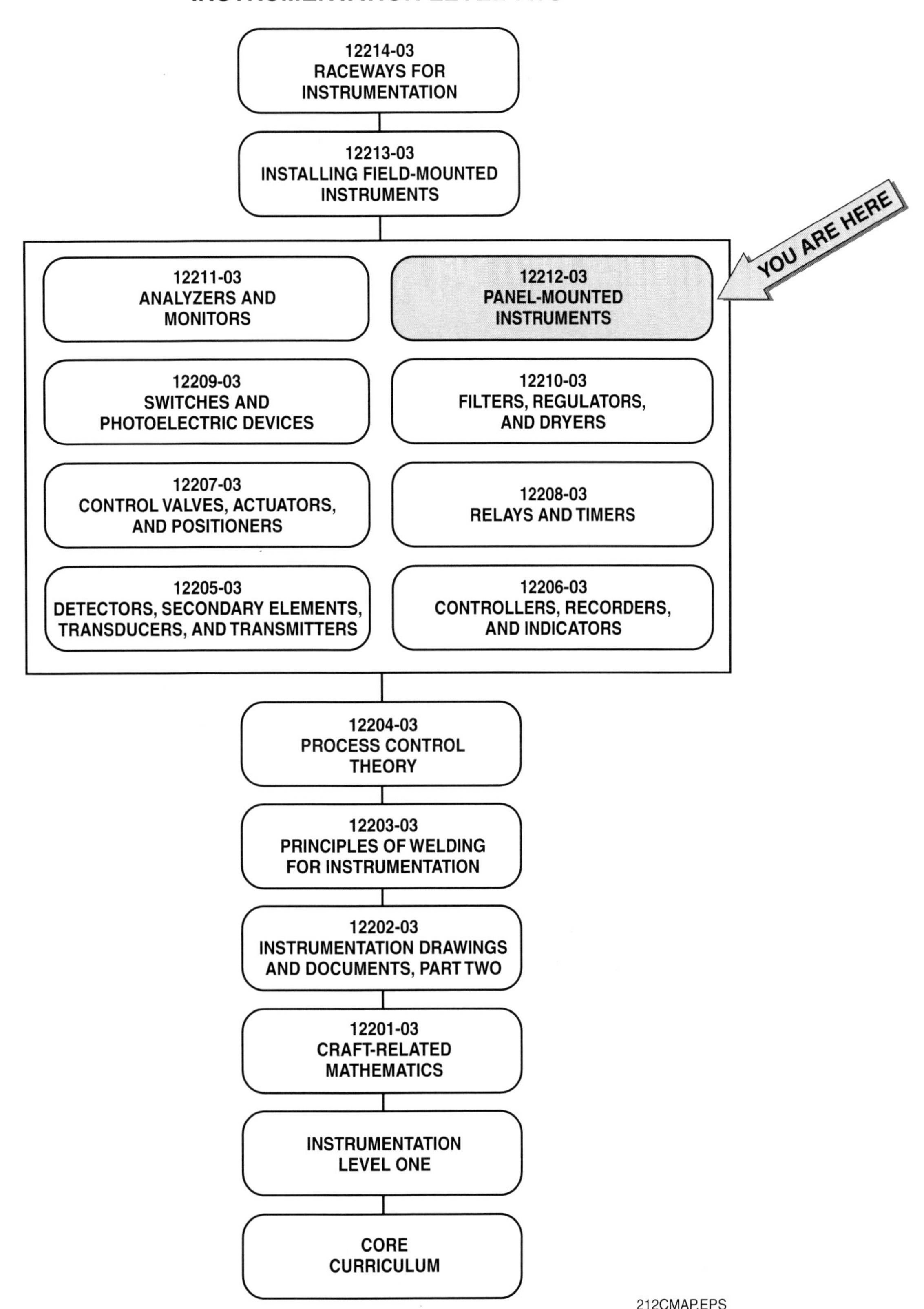

MODULE 12212-03 CONTENTS

1.0.0 INTRODUCTION 12.1

2.0.0 P&IDs RELATING TO PANEL-MOUNTED INSTRUMENTS 12.1

2.1.0 P&ID Symbology 12.2

3.0.0 LAYING OUT PANEL-MOUNTED INSTRUMENTS 12.3

3.1.0 Accessibility 12.3

3.2.0 Safety 12.3

3.3.0 Developing a Layout Template 12.3

3.4.0 Selecting the Proper Layout Tools 12.4

3.4.1 Scribers 12.5

3.4.2 Steel Rules 12.6

3.4.3 Steel Squares 12.6

3.4.4 Combination Sets 12.6

3.4.5 Dividers 12.6

3.4.6 Prick Punches 12.7

3.4.7 Center Punches 12.7

3.4.8 Toolmakers' Hammers 12.7

3.4.9 Straightedges 12.8

3.4.10 Layout Dye (Blueing) 12.8

3.5.0 Completing the Layout 12.8

3.5.1 Manufacturers' Templates 12.9

3.5.2 Creating a Template 12.10

4.0.0 SELECTING THE PROPER INSTALLATION TOOLS 12.10

4.1.0 Hydraulic Knockout Punches 12.10

4.2.0 Power Shears and Nibblers 12.11

5.0.0 MAKING THE PANEL CUTOUT AND INSTALLING THE INSTRUMENT 12.11

SUMMARY 12.13

REVIEW QUESTIONS 12.14

GLOSSARY 12.15

REFERENCES & ACKNOWLEDGMENTS 12.16

Figures

Figure 1 General instrument symbols (bubbles) 12.2
Figure 2 Instrument line symbols 12.4
Figure 3 Pressure instruments 12.5
Figure 4 Manufacturer's installation instructions 12.5
Figure 5 Examples of scribers 12.5
Figure 6 Steel rules 12.6
Figure 7 Steel squares 12.6
Figure 8 Combination set 12.6
Figure 9 Divider 12.7
Figure 10 Prick punch set 12.7
Figure 11 Center punches 12.7
Figure 12 Toolmaker's hammer 12.8
Figure 13 Straightedges 12.8
Figure 14 Layout dye (blueing) 12.8
Figure 15 Manufacturer's template 12.9
Figure 16 Instrument provided with trim collar 12.9
Figure 17 Template preparation 12.10
Figure 18 Hydraulic knockout drives 12.10
Figure 19 Power nibbler and power shear 12.11
Figure 20 How to scribe a line 12.11
Figure 21 Intersecting lines 12.12
Figure 22 Center punch marks 12.12
Figure 23 Holes drilled in panel 12.13
Figure 24 Saw cutting line 12.13

MODULE 12212-03

Panel-Mounted Instruments

Objectives

When you have completed this module, you will be able to do the following:

1. Identify panel-mounted instruments from piping and instrumentation drawings.
2. Lay out panel-mounted devices for installation.
3. Install various panel-mounted instruments.

Prerequisites

Before you begin this module, it is recommended that you successfully complete the following: Core Curriculum; Instrumentation Level One; Instrumentation Level Two, Modules 12201-03 through 12204-03.

Required Trainee Materials

1. Pencil and paper
2. Appropriate personal protective equipment

1.0.0 ◆ INTRODUCTION

Instrument panels provide a convenient method of centrally locating instruments. An instrument panel might contain anywhere from two or three instruments to a hundred or more instruments. The instruments on the panel can be electrically driven or direct driven. For example, a pressure gauge may be physically connected to a pressure source.

To install an instrument panel and install the instruments into a panel, trainees must be able to interpret system drawings, as well as instrument physical specifications and requirements.

Instruments come in various sizes and dimensions. They may be flush mounted or surface mounted. The instrument connection points may also affect how the instrument is connected to the panel.

These and other variables must be considered when constructing an instrument panel. This module covers the guidelines that are used to make reliable instrument panels.

The first step in fabricating or ordering the instrument panel is identifying what instrumentation is to be installed on the panel. In order to do this, you may have to identify the instrumentation on piping and instrumentation diagrams (P&IDs). A review of P&IDs relating to panel-mounted instruments is covered in the next section.

2.0.0 ◆ P&IDs RELATING TO PANEL-MOUNTED INSTRUMENTS

An instrumentation diagram, particularly a P&ID, is essentially a map of the process to be installed. These diagrams identify each instrument in a process, its function, and its relationship to other components in the system.

Instrument symbols, such as circles, letters, numbers, and lines, are used to provide information about the instruments. A symbol may represent a particular device, identify the function of a device, or show how a device is connected to other instruments in the process.

A symbol that resembles a circle on a P&ID is called a balloon or a bubble. These two terms mean the same thing. This module refers to the symbol as a bubble. A bubble contains lines, letters, and numbers that identify the location of the instrument, as well as its function in the process. It further specifies whether the instrument is used to measure, indicate, record, or control the process.

2.1.0 P&ID Symbology

- *Bubbles* – Some typical bubble symbols are shown in *Figure 1*, with a listing of what they normally represent. Discrete instruments are those that are considered *stand alone*. If the instrument has a shared display or a shared control, it is usually shown as a circle enclosed in a square. Hexagons are used to designate computer functions. Programmable logic controls (PLCs) are shown as diamonds enclosed in squares. Any of these shapes may indicate a panel-mounted device. Panel-mounted instruments, however, are considered accessible to the operator and are indicated by those symbols shown in the first and the last column of *Figure 1*. These bubbles have a solid line or lines drawn through the bubble to indicate that they are to be installed on the front of a panel. As the notes in *Figure 1* show, instruments mounted behind a panel are indicated with dashed or broken lines instead of solid lines through the bubble. The symbols or bubbles shown in the middle col-

	PRIMARY LOCATION ***NORMALLY ACCESSIBLE TO OPERATOR	FIELD MOUNTED	AUXILIARY LOCATION ***NORMALLY ACCESSIBLE TO OPERATOR
DISCRETE INSTRUMENTS	1 * IP1**	2	3
SHARED DISPLAY, SHARED CONTROL	4	5	6
COMPUTER FUNCTION	7	8	9
PROGRAMMABLE LOGIC CONTROL	10	11	12

* Symbol size may vary according to the user's needs and the type of document. A suggested square and circle size for large diagrams is shown. Consistency is recommended.

** Abbreviations of the user's choice, such as IP1 (Instrument Panel #1), IC2 (Instrument Console #2), and CC3 (Computer Console #3) may be used when it is necessary to specify instrument or function location.

*** Normally inaccessible or behind-the-panel devices or functions may be depicted by using the same symbol with dashed horizontal bars, such as:

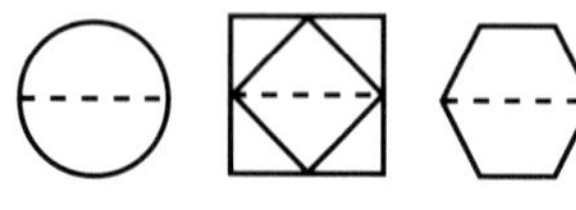

212F01.EPS

Figure 1 ◆ General instrument symbols (bubbles).

umn of *Figure 1* are considered field-mounted instruments. Their installations are covered in other training modules.

- *Line symbols* – P&IDs also use various line symbols to indicate how panel-mounted instruments are connected to each other and to the process. They also indicate the type of signal that is received from or transmitted to the field. Lines used as symbols may differ in several ways, including their relative thickness and whether they are solid lines or dashed/broken lines.

Signal lines on P&IDs running to and from panel-mounted instruments represent unique and critical information that must be referenced when installing these instruments. Because of the many instrument loops that can be present in large industrial facilities, many different types of power and communication methods between devices may be present in any one control room installation. Signal lines may be pneumatic, electric, digital, or optical. In order to correctly install these different types of panel-mounted instruments, it may be necessary for the installer to have the knowledge to physically mount the instrument in the panel and to make the necessary connections to the instrument. *Figure 2* shows common line symbols that may be found on P&IDs when referenced for the installation of panel-mounted instruments.

3.0.0 ◆ LAYING OUT PANEL-MOUNTED INSTRUMENTS

The actual location on the panel face where the instrument is to be installed is usually decided by engineering or operations, although this may not always be the case. The installer at times may be responsible for the design and layout of an instrument panel. When performing this task, the installer will have to arrange the panel in a logical and convenient layout. The instruments should be placed so they can be read easily. Access must also be provided for operation of the instrument panel controllers and adjustments. Other considerations that must be considered include these:

- Are there any other devices that will be located behind the panel?
- What type of input, output, and supply connections are there? Where are they located?

The connections can sometimes be located on the front, rear, top, or bottom. In some instances, there may be some on the side. *Figure 3* shows three typical pressure instruments. Note the different sizes, shapes, and mounting requirements.

3.1.0 Accessibility

The working space and entrance to the work area must be considered when locating a local instrument panel and the instruments within it. Enough space must be left around a panel so that normal maintenance and production activities can be performed. Many times, troubleshooting requires that readings be taken at the rear of the instrument. Adequate space is necessary for this.

The location of a local instrument panel should allow as much protection as possible from overhead piping and possible damage from dust, dirt, and process contamination.

3.2.0 Safety

When laying out an instrumentation panel or additions to a panel, there are safety issues that must be considered before beginning the installation.

If adding instruments to a panel, all attempts should be made to work on the panel in a de-energized state, following any and all facility lockout/tagout procedures.

If responsible for the location of a new panel, pay attention to congested work areas and allow for clearance both in front of and behind the panel for future maintenance and troubleshooting procedures.

Provide protection for existing equipment in the panel from metal shavings, dirt, and other debris that may be created during installation.

Instrumentation devices are sensitive to vibration and other installation-related effects. Limit vibration, hammering, and other installation-related effects both on the new devices being installed and on the existing equipment.

Be aware of safety equipment such as fire extinguishers, safety handrails, and other installed safety devices. Make sure the installation does not interfere with quick access or the proper operation of these safety components.

Be aware of sharp edges that may be created during the installation of instruments. Panel edges can become razor sharp, causing injury to the installer during installation and to the operator after installation.

3.3.0 Developing a Layout Template

Panel instruments vary in size, shape, and weight. They require different considerations when being positioned and supported. Most instrument manufacturers provide detailed instructions for mounting an instrument and using supporting brackets or racks.

Many panel instruments come with a rack assembly that is permanently attached to the

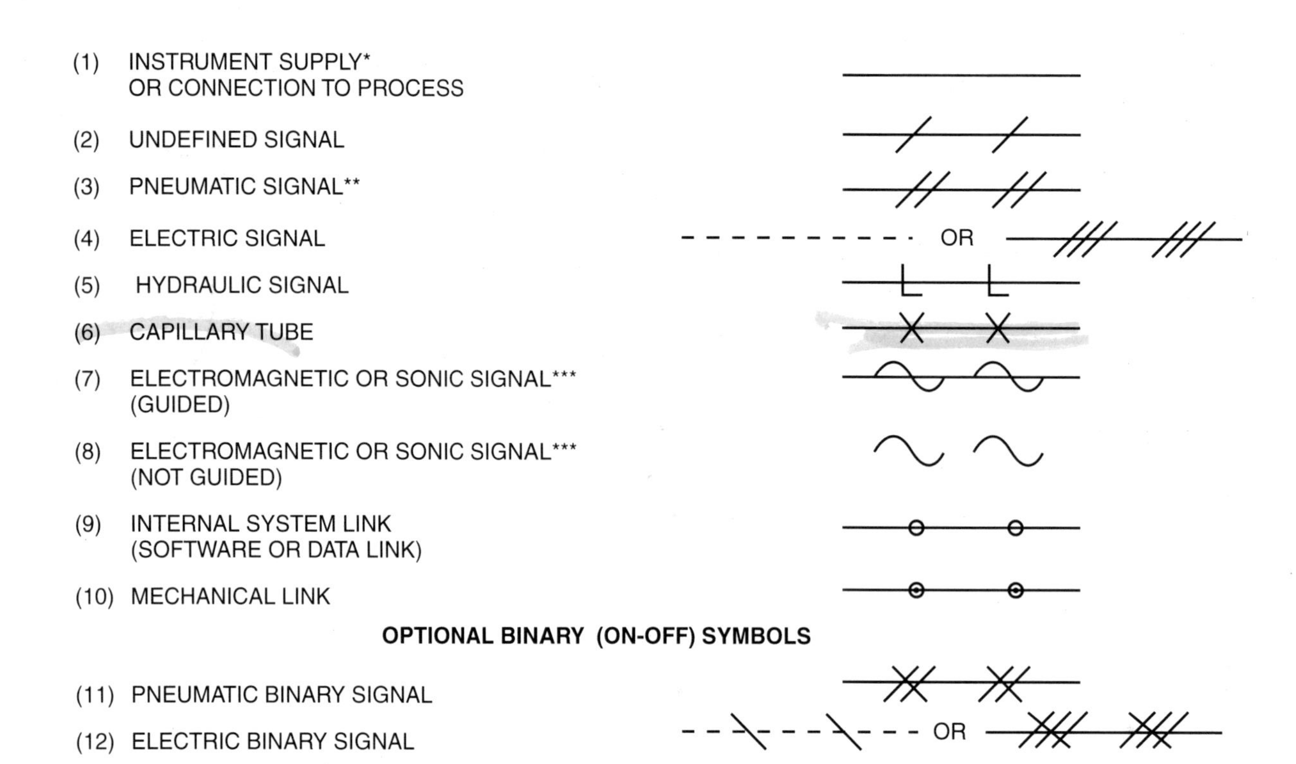

NOTE: "OR" means user's choice. Consistency is recommended.

* The following abbreviations are suggested to denote the types of power supplies. These designations may also be applied to purge fluid supplies.

AS – Air Supply
 IA – Instrument Air } Options
 PA – Plant Air
ES – Electric Supply
GS – Gas Supply

HS – Hydraulic Supply
NS – Nitrogen Supply
SS – Steam Supply
WS – Water Supply

The supply level may be added to the instrument supply line, such as AS-100, a 100 psig air supply, and ES-24DC, a 24volt direct current power supply.

** The pneumatic signal symbol applies to a signal using any gas as the signal medium. If a gas other than air is used, the gas may be identified by a note on the signal symbol or otherwise.

*** Electromagnetic phenomena include heat, radio waves, nuclear radiation, and light.

212F02.EPS

Figure 2 ◆ Instrument line symbols.

panel when the instrument is installed. This assembly permits easy removal and access to the instrument itself.

The following sections explain how to **lay out** an instrument in an existing panel, using a typical manufacturer's installation instructions. This includes:

- Selecting the proper layout tools
- Laying out the measurements

The following sections explain the basic measurement and power tools required to lay out an instrument in a panel.

3.4.0 Selecting the Proper Layout Tools

Proper tool selection is important in laying out instrument panels. Layout tools are used to measure specific sizes and shapes. The tools needed to

Figure 3 ◆ Pressure instruments.

fabricate parts vary with each job. A copy of the manufacturer's installation instructions should also be obtained to assist in proper layout tool selection. *Figure 4* shows an example of manufacturer's installation instructions.

Many different tools can be used to lay out and measure instrument panels. Some common layout tools are:

- Scribers
- Steel rules
- Steel squares
- Combination sets
- Dividers
- Prick punches
- Center punches
- Toolmakers' hammers
- Straightedges
- Layout dye (blueing)

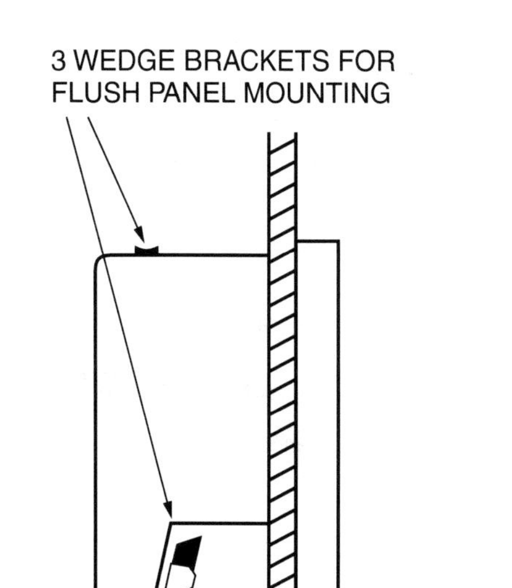

Figure 4 ◆ Manufacturer's installation instructions.

3.4.1 Scribers

Scribers are used to mark lines by scratching them on panel surfaces. They are made from tool steel and are available in different lengths. Some have a carbide tip for longer life. A scriber is ground to a needle point and is used with a handle or holder. The point of some scribers is bent at a 90-degree angle to mark the inside of cylindrical objects. An oil or diamond stone should be used to keep a scriber point sharp.

Figure 5 shows two types of scribers.

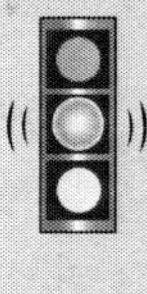

CAUTION

Do not carry or place a scriber in clothes pockets unless the point of the scriber can be retracted, removed, or adequately covered. This protects both the user and the needle point of the scriber.

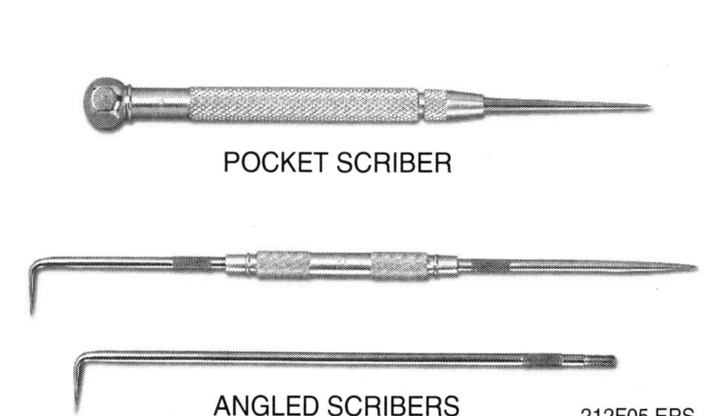

Figure 5 ◆ Examples of scribers.

3.4.2 Steel Rules

A steel rule or stainless steel machinist's rule is a tool marked on one or both sides to measure distances (*Figure 6*). It is also used as a straightedge to mark lines. One side of a steel rule may be marked in eighths and sixteenths of an inch, and the other side in thirty-seconds and sixty-fourths of an inch. Some are marked in both English and metric divisions. Steel or stainless steel rules are available as rigid, semiflexible, or fully flexible rules. They are commonly available in lengths from 6 to 36 inches. Some rules are available in lengths up to 12 feet.

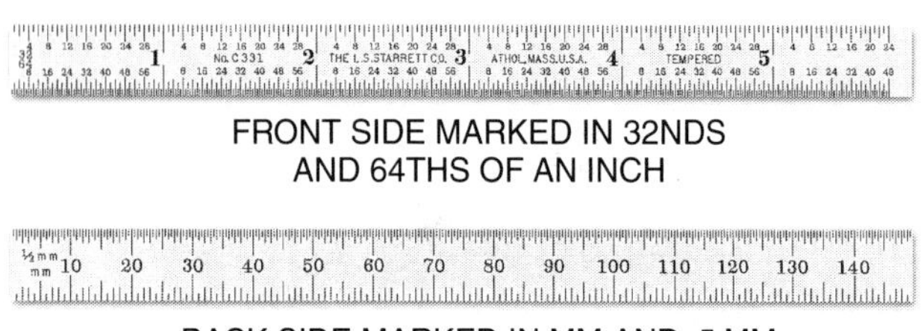

Figure 6 ◆ Steel rules.

Steel or stainless steel rules should be kept clean using solvent and cloth. After cleaning, a thin coating of lightweight machine oil may be applied to steel rules to protect them.

3.4.3 Steel Squares

Steel or stainless steel machinist, carpenter, or framing squares are used to square up stock and patterns (*Figure 7*). They are also used to lay out and check right angles and as rules or straightedges. The two parts of a steel square are the blade and the tongue. Depending on the type of square, the blade and tongue may be marked in eighths, sixteenths, or thirty-seconds of an inch.

Steel or stainless steel squares should also be cleaned with solvent and cloth. After cleaning, a thin coating of lightweight machine oil should be applied to steel squares to protect them.

3.4.4 Combination Sets

A combination set (*Figure 8*) is useful in layout work. It serves the following purposes:

- As a tool to lay out 45-degree angles, 90-degree angles, and parallel lines
- As a tool for measuring the height of components
- As a protractor to measure or lay out angles
- As a means of locating the centers of round or square bars
- As a steel rule, level, or marking gauge

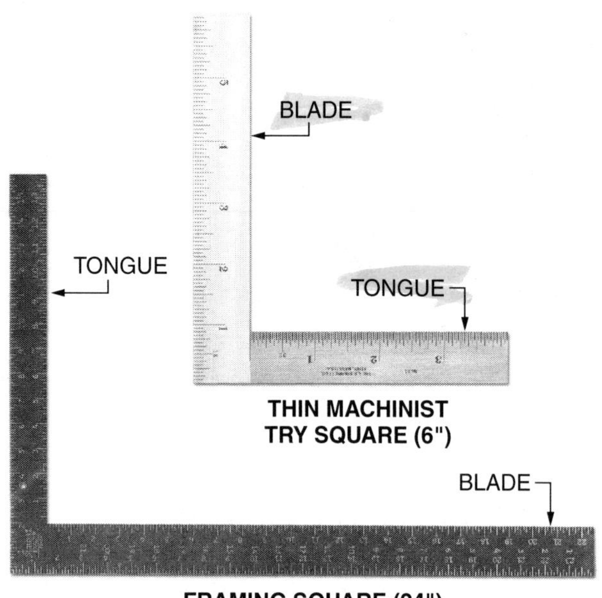

Figure 7 ◆ Steel squares.

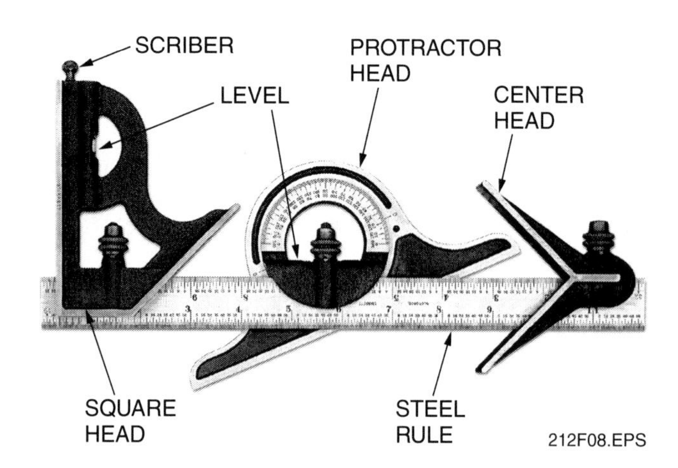

Figure 8 ◆ Combination set.

A combination set consists of the steel rule or blade marked in English or metric divisions, a square head, a center head, and a protractor head. The square head, and sometimes the protractor head, contains a spirit level.

A combination set should be cleaned with solvent and cloth. After cleaning, a thin coat of lightweight machine oil should be applied to protect it. When not in use, the set should be kept in a protective container or wrapped in a cloth.

3.4.5 Dividers

Dividers are used to transfer measurements, compare distances, and scribe circles and arcs. Dividers have hardened steel points attached to two legs that can be adjusted by a screw. Divider points should be kept sharp. When working with metal, dividers are used only to scribe arcs or circles that will later be removed by cutting. All other arcs or circles are drawn with pencil compasses to avoid scratching the material.

Figure 9 shows a divider.

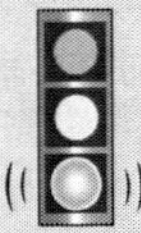

NOTE

Do not carry or place dividers in clothes pockets unless the points of the dividers are adequately covered.

212F09.EPS

Figure 9 ◆ Divider.

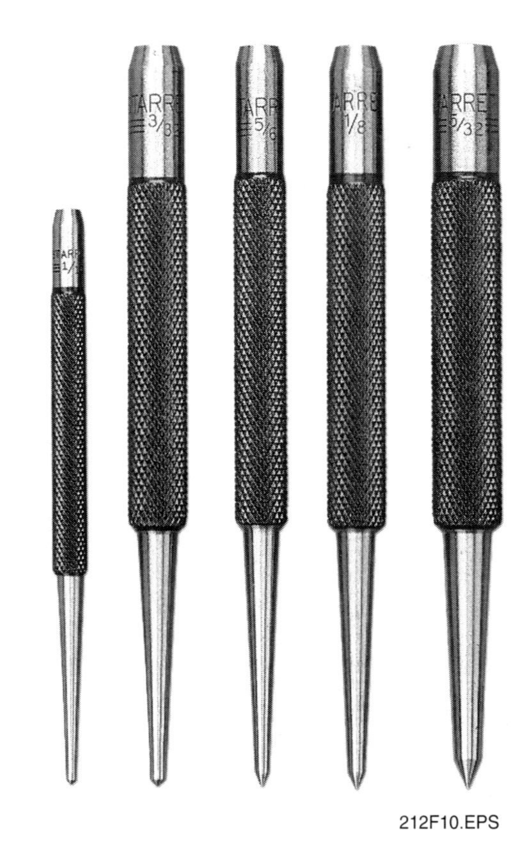

212F10.EPS

Figure 10 ◆ Prick punch set.

3.4.6 Prick Punches

Prick punches are used with a light hammer to mark centers and to make small indentations along layout lines. They are made of steel and are 4 to 6 inches long. A prick punch point is ground at an angle of 30 to 60 degrees. *Figure 10* shows a prick punch set.

The point of a prick punch should be kept sharp and maintained at the proper angle. The struck end should be ground flat, and the edge should be beveled to prevent **mushrooming**.

3.4.7 Center Punches

A center punch is used to enlarge center marks made with a prick punch. An awl can also be used. This is done so a drill will start easily when drilling holes. A center punch is similar to a prick punch but has a blunter point ground at a 45-degree angle. Center punches are usually square or hex-shaped to prevent rolling. Some center punches are spring-loaded automatic punches that do not require a hammer to make an indentation. Pushing down on the punch trips a hammer mechanism inside the punch. *Figure 11* shows center punches.

The point of the center punch should be kept sharpened to the proper angle. The struck end should be beveled to prevent mushrooming.

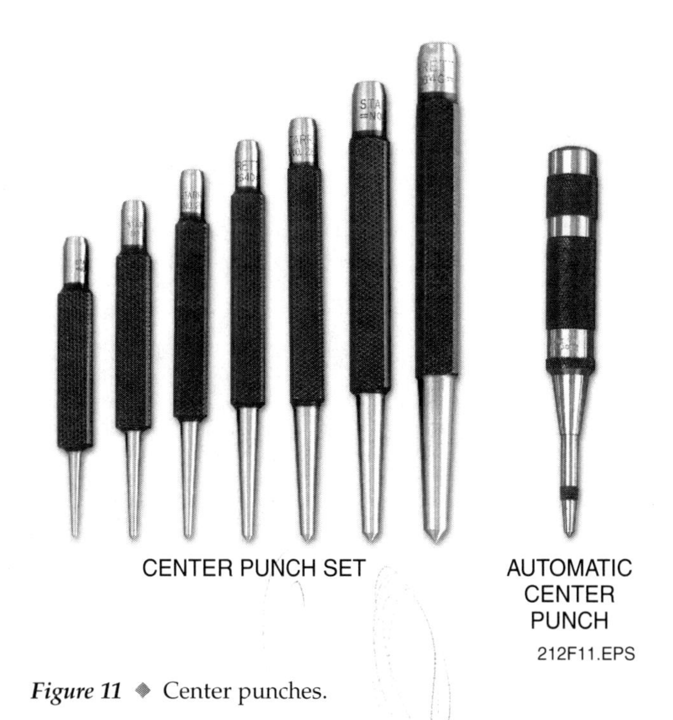

212F11.EPS

Figure 11 ◆ Center punches.

3.4.8 Toolmakers' Hammers

Although not a necessity, a lightweight toolmaker's hammer (*Figure 12*) provides faster and easier spotting and punching of hole locations. A magnifying lens, built into the head of the hammer, makes it very easy to observe layout lines

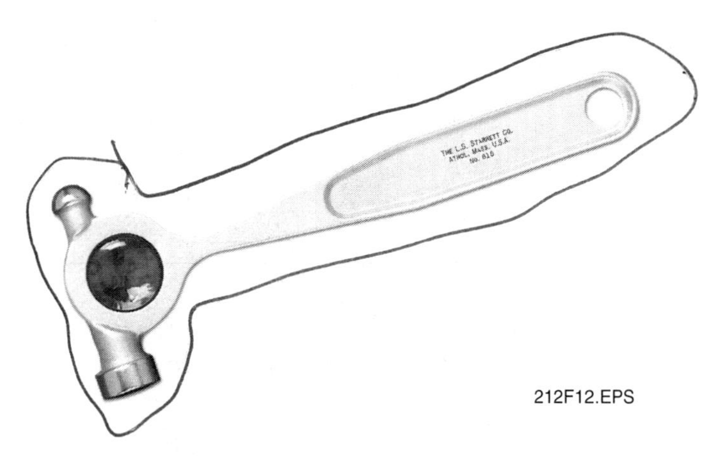

Figure 12 ◆ Toolmaker's hammer.

and accurately place prick and center punch marks at exact locations without taking one's eyes off the work.

3.4.9 Straightedges

A straightedge (*Figure 13*) is a piece of flat metal with the edges machined perfectly straight. It is used to check machined surfaces for flatness or straightness. It is also clamped along the edge of a piece of metal as a reference line from which other lines are scribed.

A straightedge should be cleaned with solvent and cloth. If made of steel, a thin coat of lightweight machine oil should be applied to protect it. A straightedge should be stored in a protective sleeve when not in use.

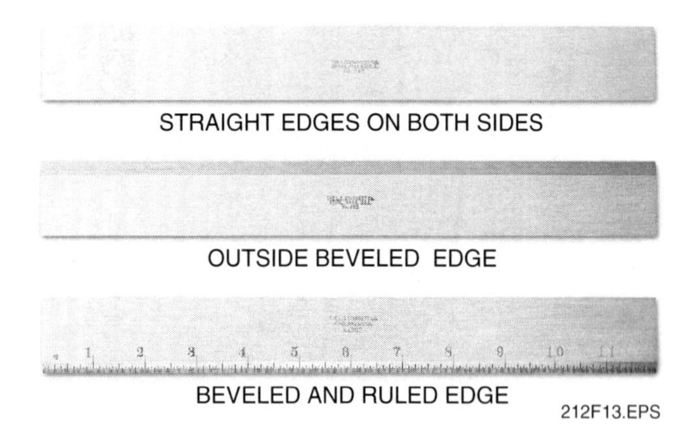

Figure 13 ◆ Straightedges.

3.4.10 Layout Dye (Blueing)

As an aid when scribing layout lines, a semi-opaque, dark Prussian-blue dye, often called blueing, can be applied to provide a temporary surface that is easily scribed. It makes scribe lines very visible. The dye is made of powdered ferrocyanides of iron dissolved in an alcohol base. It dries very quickly and is removed using alcohol and a rag. When dry, it resists rubbing, flaking, cutting lubricants, and machining/cutting heat buildup. If a mistake is made during scribing, the dye at the area of the error can be removed and reapplied, or the error can be overcoated with dye and the scribing repeated. When the final machining or cutting is accomplished, the remaining dye is easily removed. Layout dye is available in brush or spray containers as shown in *Figure 14*.

Figure 14 ◆ Layout dye (blueing).

3.5.0 Completing the Layout

Before any cutting, drilling, or other installation work is done, the panel must be laid out using a template. A template is a full-size pattern that shows where holes and cutouts should be made on the panel facing. In the fabrication of the metal panels, templates are also used to indicate where cuts or bends should be made. Templates are made from sketches or drawings and are usually constructed from heavy paper or cardboard. A template's outline can be transferred or traced directly onto the panel facing to be cut and drilled. It also can be taped or otherwise directly secured to the panel facing, with the cutters and drill bits following the template's lines directly.

3.5.1 Manufacturers' Templates

Most manufacturers include an installation template with their product. *Figure 15* shows an example of a simple template provided by one manufacturer. The installation instructions usually show the exact cutout size required to install the instrument. In *Figure 15* these dimensions are 8½ inches wide by 10⅛ inches long. As with any panel layout, be sure that structure doors do not interfere with the operation of the instrument and that the instrument is suited for the classification or environment in which it will be installed.

On any instrument, make sure never to lay out the panel cutout larger than the instrument can cover. Doing so may cause the instrument not to fit properly. Always note any flange that is part of the instrument and record the cutout measurements from the inside edge of the flange. Some instruments have a panel trim or trim collar that permits some adjustment in the size of the cutout, as shown in *Figure 16*. Make sure that any lines that are transferred using a scribe are located inside the cut line so the scribed scratches are not left on the panel facing after the instrument is installed. Also, when using manufacturers' templates, always check the measurements against the actual instrument to be installed. On rare occasions the wrong template may be included with an instrument. It is too late to discover this after cutting too large a hole in the panel facing. Double-checking these measurements before cutting requires little time and effort.

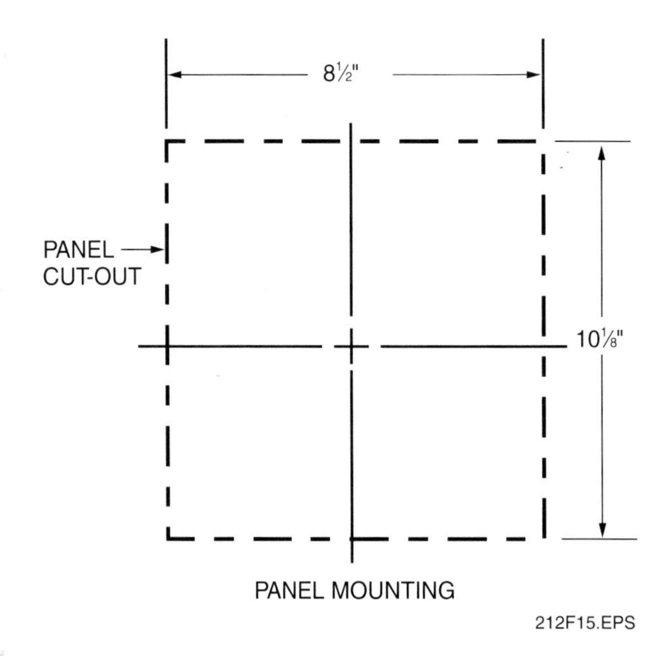

Figure 15 ◆ Manufacturer's template.

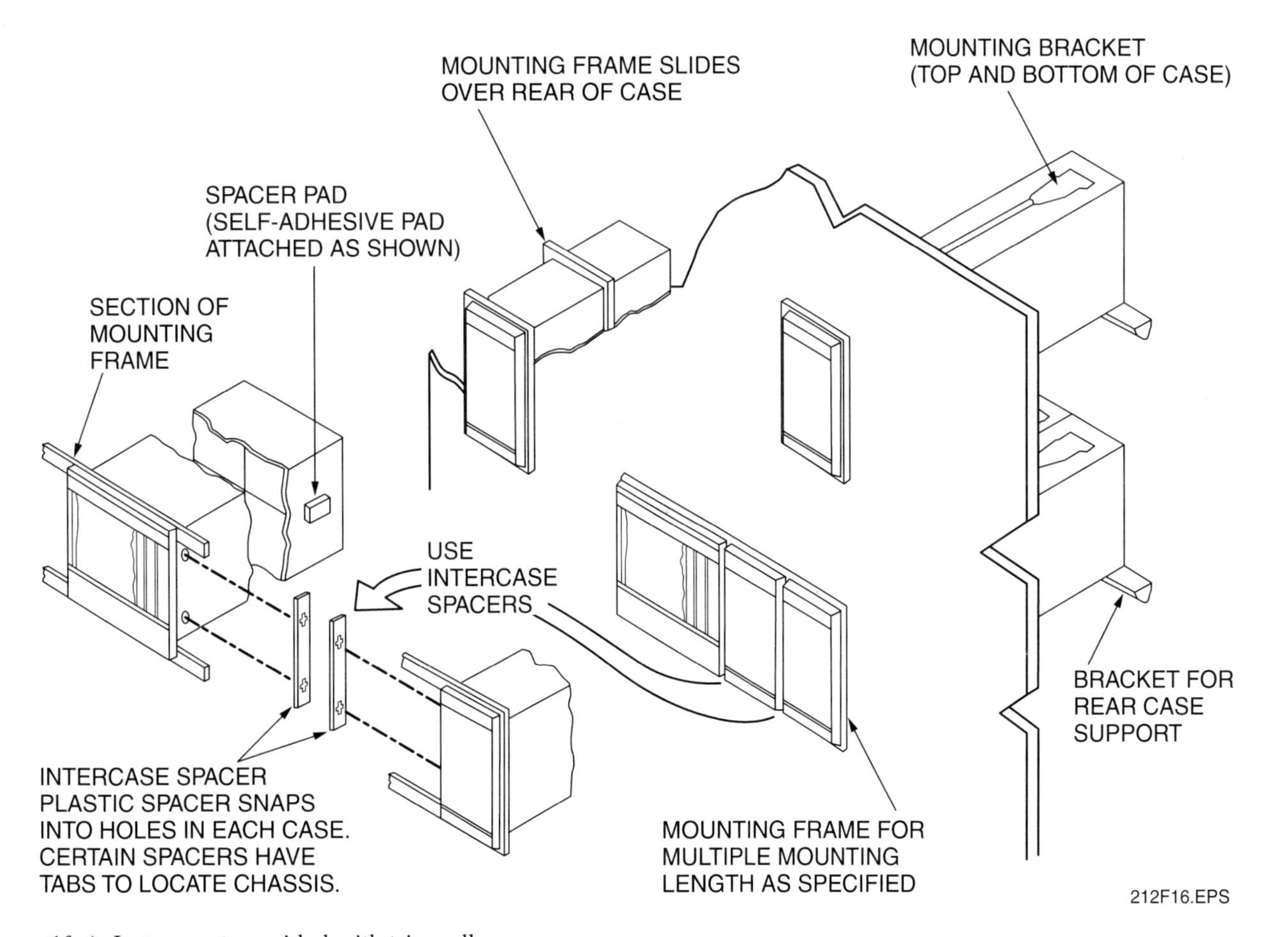

Figure 16 ◆ Instrument provided with trim collar.

3.5.2 Creating a Template

Follow these steps to prepare a template for the example installation, as shown in *Figure 17*.

Step 1 Using a metal square and a pencil, draw an 8½-inch by 10⅛-inch rectangle on a piece of cardboard or paper.

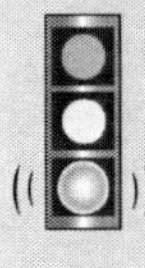

NOTE
These rectangle dimensions were provided in the example manufacturer's installation instructions shown in *Figure 15*.

Step 2 Cut away the excess paper or cardboard.

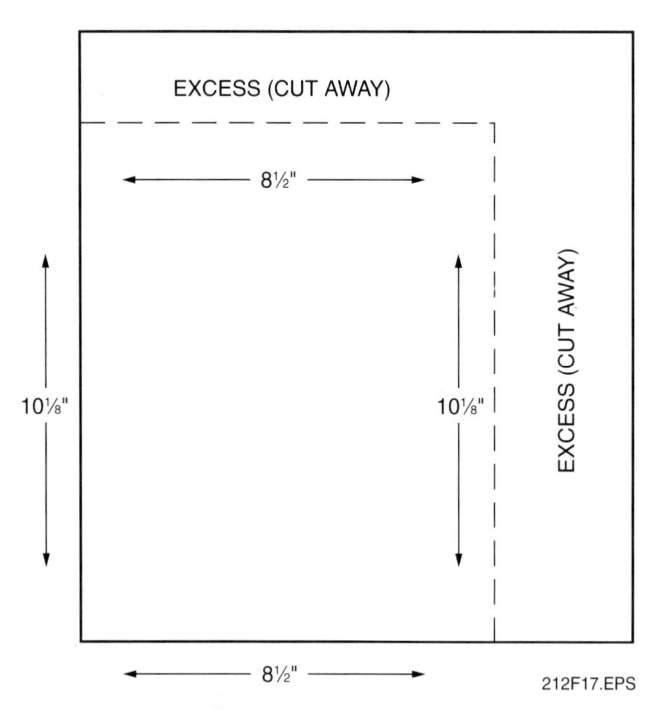

Figure 17 ◆ Template preparation.

4.0.0 ◆ SELECTING THE PROPER INSTALLATION TOOLS

Most instrument panels are made of steel or aluminum sheet metal. Modification or fabrication of these panels requires the use of tools that will cut this type of material. The most common tools used to cut out instrument panels are hole saws, power drills and drill bits, power shears, power nibblers, hydraulic punches, and reciprocating or saber saws.

4.1.0 Hydraulic Knockout Punches

Hydraulic knockout punches are powered by hydraulic pumps. These punches use the same dies and punches as manual knockout punches. Hydraulic knockout punches are usually used to make large holes and to punch through thick metal sheeting. Non-round punches can also be used. *Figure 18* shows hand-operated and foot-operated hydraulic knockout drives.

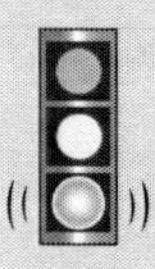

NOTE
When punching out stainless steel panels, always use appropriate dies designed for stainless steel.

FOOT-PUMP MODEL

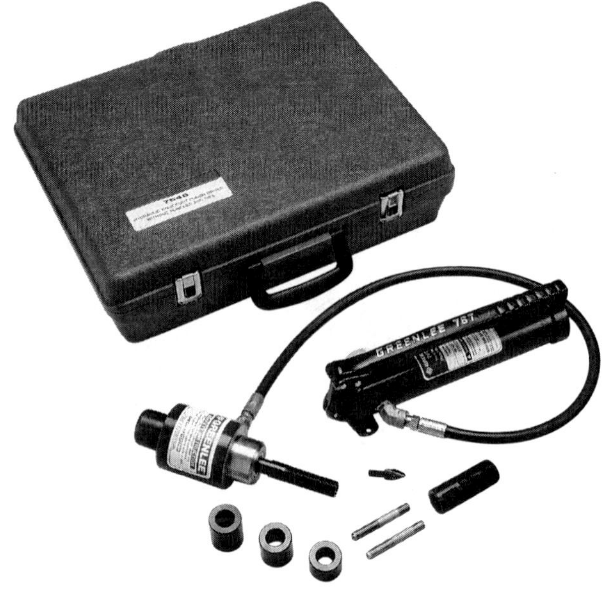

HAND-OPERATED MODEL

212F18.EPS

Figure 18 ◆ Hydraulic knockout drives.

4.2.0 Power Shears and Nibblers

Power shears and power nibblers are available as compressed air- or electric-powered tools with cutting capacities of up to 10-gauge steel or 12-gauge stainless steel. Nibblers are slow and create much fine debris, but they are especially useful when a shallow cutting depth is encountered or fine trimming and high accuracy are required. A power shear is faster than a reciprocating saw and does not create small debris when cutting. The narrow moving jaw of the power shear can be inserted in a relatively small starter hole. In most cases, depending on the capacity of the tool, larger starting holes will be required for power shears and nibblers than for reciprocating saws. *Figure 19* shows a typical power nibbler and power shear.

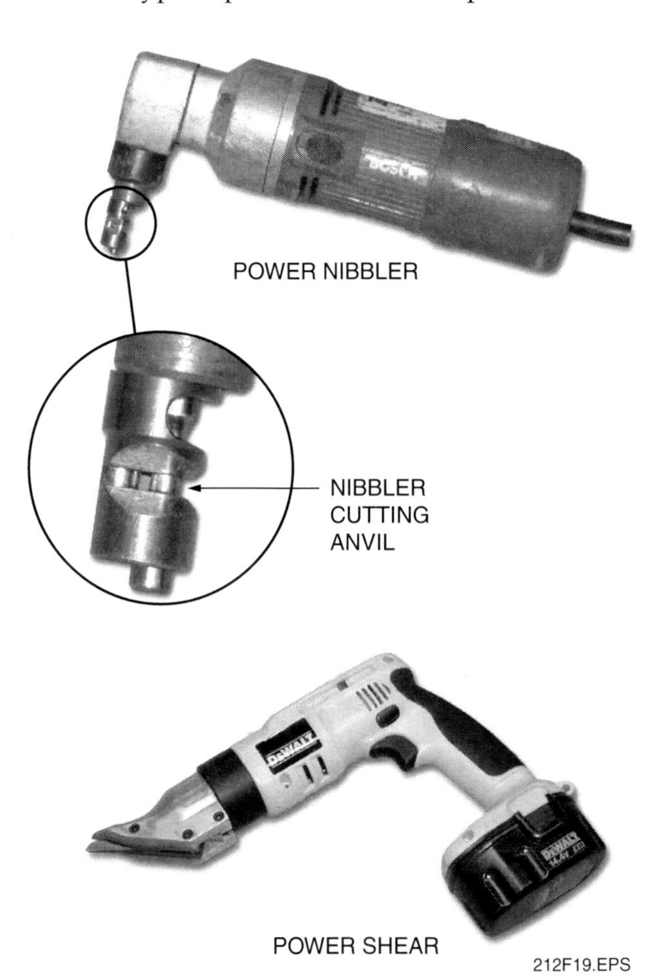

Figure 19 ◆ Power nibbler and power shear.

5.0.0 ◆ MAKING THE PANEL CUTOUT AND INSTALLING THE INSTRUMENT

Before making any cuts on the panel, make sure faceplates, instrument flanges, and panel trim collars will not interfere with adjoining instruments. The cutouts may indicate sufficient spacing between instruments, but keep in mind that some instruments require more trim or faceplate clearance than others. Always measure the trim plates and collars and allow a minimum of ½ inch between one trim plate or collar and the finished edge of any adjoining plates or collars on instruments.

Step 1 Apply blueing to the area being scribed. When the blueing is dry, place the template over the panel face and secure it to the panel with tape. Using a scribe, mark the outline of the template at the sides of corners and around curves, as shown in *Figure 20*. Do not cut the tape holding the template. The lines will be connected with a scribe and ruler after the template is removed.

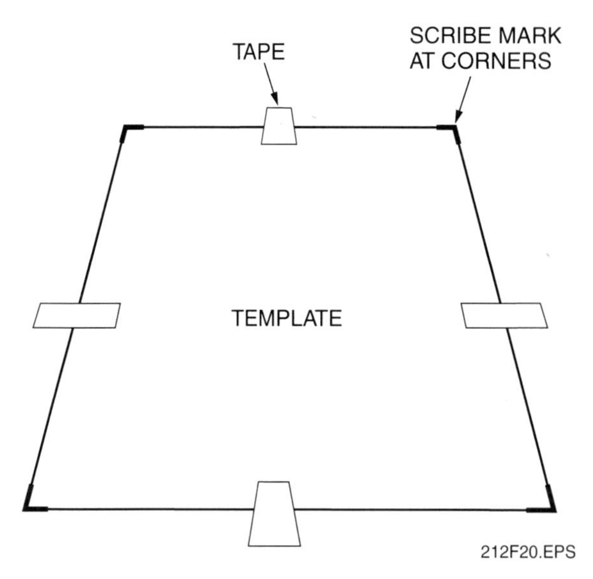

Figure 20 ◆ How to scribe a line.

Step 2 Remove the template from the panel. Using the marks established from the template, use the scriber and a metal rule to trace all straight lines.

Step 3 The size of the cutting tool or saw blade used will determine the size of hole to be drilled in the panel. These holes must be large enough for the tool or saw blade to be fully inserted into the hole. This allows the tool or power saw to be held flat against the panel surface. The saw manufacturer usually provides a guide identifying the blade needed to cut a particular material. Because a power saw will be used, the saw blade height in this example is ¼ inch. This will require a 5⁄16-inch hole. Measure out 3⁄16 inch in both directions along the scribed lines from the corner point of the layout pattern. Scribe lines into the inside of the pattern from these measured points until they intersect. The end result should look like *Figure 21*.

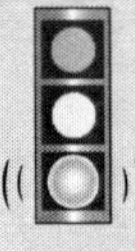

NOTE

On thin gauge metal, a heavy weight or backer should be positioned and held behind the panel at the area to be center-punched to prevent denting or distorting the metal.

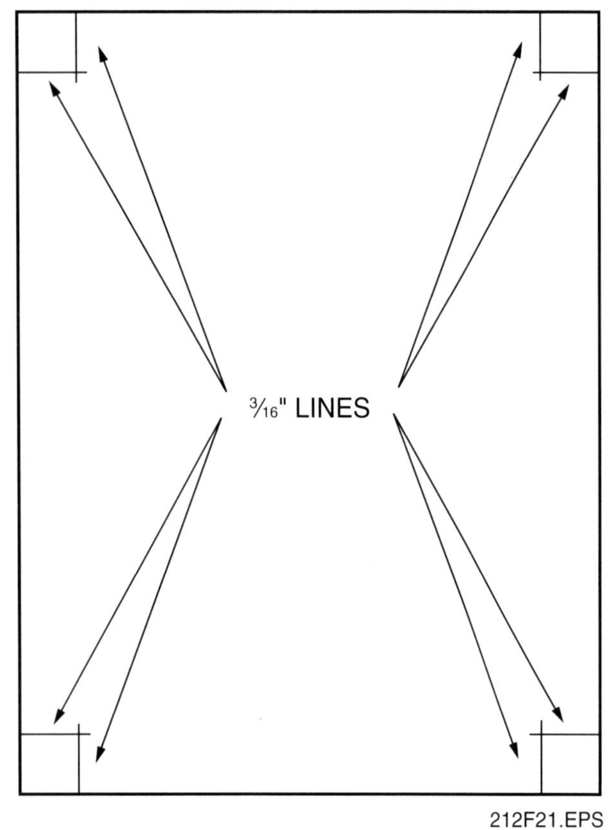

Figure 21 ◆ Intersecting lines.

Step 4 Center-punch at the point where the lines intersect in all four corners using the center punch and hammer. *Figure 22* shows the center punch marks. These will make drilling holes in the panel easier and more accurate.

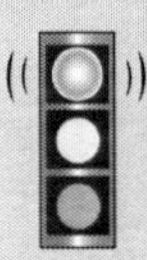

WARNING!

If power cannot be shut off, make sure the drill, saw, or file does not contact any equipment or wiring in the following steps. A severe electrical shock could occur.

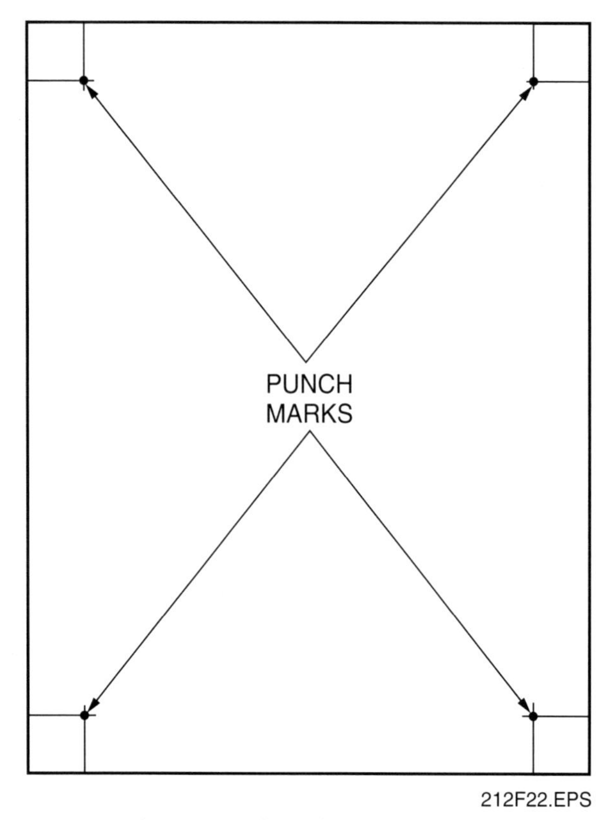

Figure 22 ◆ Center punch marks.

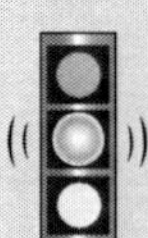

CAUTION

When drilling, cutting, or filing the template hole in the following steps, make sure that the end of the drill, saw, or file does not contact wiring and equipment behind the panel. Also, cover the area behind the panel with protective sheeting so that metal shavings from the drilling, cutting, and filing do not fall into the equipment.

Step 5 Drill a **pilot hole** at all four center punch marks with an electric drill and a ⅛-inch drill bit. After the initial tap holes are drilled, redrill all four holes with an electric drill and a 5⁄16-inch bit. *Figure 23* shows the holes drilled in the panel.

Step 6 The next step is to cut the panel to the correct size. Insert a reciprocating saw into the 5⁄16-inch drilled hole, and hold it flat against the panel. Press the saw trigger and cut to the outside corner adjacent to the hole. Then change the position of the saw in the hole and cut along the outside, or panel side, of the scribed line with a light, steady pressure in the direction of the cut. Cut all the way down the line to the corner by the next hole. *Figure 24* shows the saw cutting

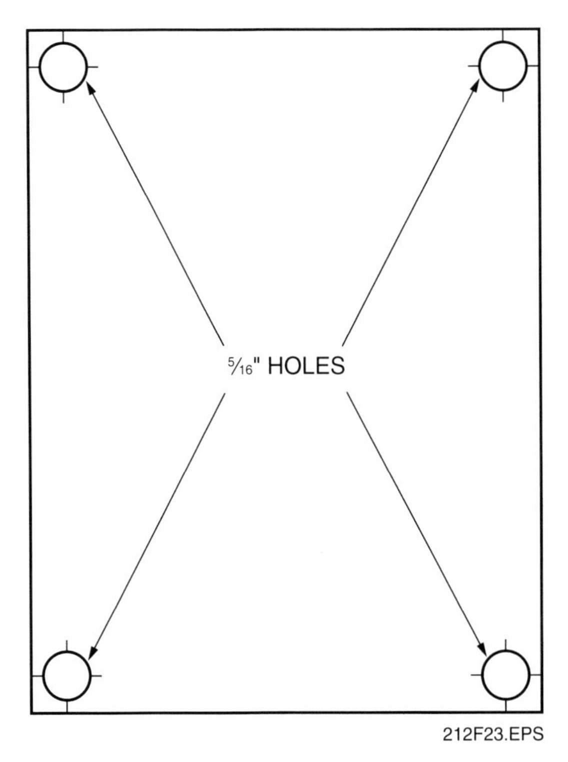

Figure 23 ◆ Holes drilled in panel.

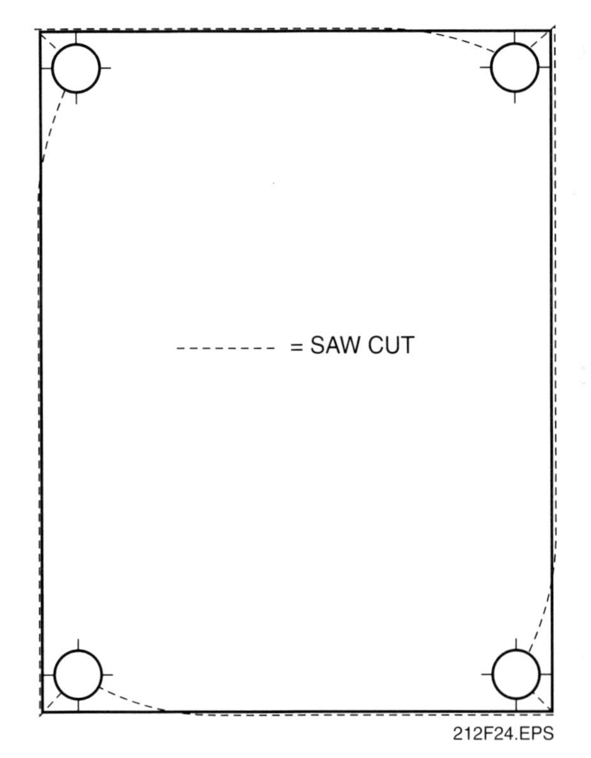

Figure 24 ◆ Saw cutting line.

lines. Do not try to pull the saw through the metal. This could cause the blades to heat up and break. Repeat the cutting procedure for each side of the panel until the hole is complete.

CAUTION

Be careful when nearing the completion of the cutout. The cut piece can fall on the installer or into the panel, possibly damaging other equipment. A piece of tape secured to the cutout piece and panel facing will prevent the piece from falling out.

Step 7 Smooth out the hole, if necessary, using a nibbler or a flat metal file. Remove any excess metal that may be in the corners of the hole. Clean up all metal particles in front of and behind the panel.

Step 8 With the hole properly sized, cleaned, and smooth, remove the mounting brackets that are supplied with the instrument. Slide the instrument into the hole and have another worker hold it flat and securely against the panel. If the instrument will not fit into the hole, use a metal file to remove excess metal and clean the area again.

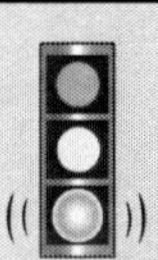

NOTE

If using carbon steel, protect any exposed metal with paint to prevent corrosion.

Step 9 Reattach the mounting brackets and secure them from behind the panel. Restore power as necessary.

Summary

This module covered panel-mounted instruments. Panels are used to place single or multiple instruments in a common area that allows for ease in reading or maintaining the instrument(s). As an instrument installer, you need the ability to read P&IDs and other drawings in order to determine which instruments are to be panel mounted.

When laying out an instrument panel, the first thing that should be determined is where the panel is to be located. Make sure the panel will be easily accessible for operation and maintenance. It should be positioned so that it will not be easily damaged. Potential problems caused by adjacent equipment, such as dust, debris, and heat, should be minimized.

Review Questions

1. In order to identify what instrumentation is to be installed on a panel, the technician should identify the instrumentation on the _____.
   a. P&ID
   b. layout drawing
   c. site plan
   d. panel template

2. Each of the following is usually represented by a symbol on a P&ID *except* _____.
   a. a particular device
   b. the function of a device
   c. the type of process (liquid or solid)
   d. how a device is connected to other devices in the process

3. Instruments that normally stand alone without sharing indicators and other functions with other devices are referred to as _____
   a. unique display
   b. discrete
   c. computer functional
   d. sole support

4. An instrument that has a shared display or a shared control is usually symbolized by a _____.
   a. solid line drawn through the bubble
   b. square enclosed in a circle
   c. diamond enclosed in a square
   d. circle enclosed in a square

5. Bubbles that are drawn with dashed or broken lines through them indicate an instrument _____.
   a. with a computer function
   b. that is a PLC
   c. located behind a panel
   d. located in the field

6. True or False? On a P&ID, a PLC is represented by a diamond enclosed in a square.
   a. True
   b. False

7. True or False? Hexagons are used to designate computer functions on a P&ID.
   a. True
   b. False

8. In selecting the proper tools for laying out a specific panel-mounted instrument, a copy of _____ should be obtained.
   a. a tool catalog
   b. the process flow
   c. the plant architectural plan
   d. the instrument manufacturer's installation instructions

9. True or False? A scriber is used to create an indentation to help start a drill bit.
   a. True
   b. False

10. A tool used to lay out and check right angles is called a _____.
    a. steel rule
    b. square
    c. divider
    d. scriber

11. True or False? Oil should not be used as a protective coating for stainless steel tools.
    a. True
    b. False

12. True or False? One of the things you can do with a combination set is make measurements.
    a. True
    b. False

13. The tool used to enlarge center marks made with a prick punch is the _____.
    a. combination set
    b. center punch
    c. scriber
    d. divider

14. All of the following tools are typically used in cutting out an instrument panel for an instrument installation *except* a _____.
    a. saber saw
    b. power drill
    c. trim saw
    d. punch

15. A minimum spacing of _____ inch must be maintained between the edge of one trim plate or collar and the edge of another trim place or collar.
    a. ¼
    b. ½
    c. ¾
    d. 1

Trade Terms Introduced in This Module

***Fabricate*:** To make a part from materials.

***Lay out*:** To arrange in a definite order or pattern.

***Mushrooming*:** When referring to a punch, the flattening out of the struck end caused by prolonged use.

***Pilot hole*:** A small hole drilled as a guide for a larger drill bit.

Additional Resources

This module is intended to present thorough resources for task training. The following reference work is suggested for further study. This is optional material for continued education rather than for task training.

Instrument Engineers Handbook, 1982. Pennsylvania: Chilton Book Company.

Figure Credits

Dwyer Instruments	212F03
The L.S. Starrett Company	212F05, 212F06, 212F07, 212F08, 212F10, 212F11, 212F12, 212F13, 212F14
ABB Inc.	212F16
Gerald Shannon	212F09, 212F19
Greenlee Textron, Inc.	212F18

NCCER CURRICULA — USER UPDATE

NCCER makes every effort to keep its textbooks up-to-date and free of technical errors. We appreciate your help in this process. If you find an error, a typographical mistake, or an inaccuracy in NCCER's curricula, please fill out this form (or a photocopy), or complete the online form at **www.nccer.org/olf**. Be sure to include the exact module ID number, page number, a detailed description, and your recommended correction. Your input will be brought to the attention of the Authoring Team. Thank you for your assistance.

Instructors – If you have an idea for improving this textbook, or have found that additional materials were necessary to teach this module effectively, please let us know so that we may present your suggestions to the Authoring Team.

NCCER Product Development and Revision

13614 Progress Blvd., Alachua, FL 32615

Email: curriculum@nccer.org
Online: www.nccer.org/olf

❑ Trainee Guide ❑ AIG ❑ Exam ❑ PowerPoints Other ______________________________

Craft / Level: Copyright Date:

Module ID Number / Title:

Section Number(s):

Description:

____

____

____

____

Recommended Correction:

____

____

____

____

Your Name:

Address:

____

Email: Phone:

Module 12213-03

Installing Field-Mounted Instruments

COURSE MAP

This course map shows all of the modules in the second level of the Instrumentation curriculum. The suggested training order begins at the bottom and proceeds up. Skill levels increase as you advance on the course map. The local Training Program Sponsor may adjust the training order.

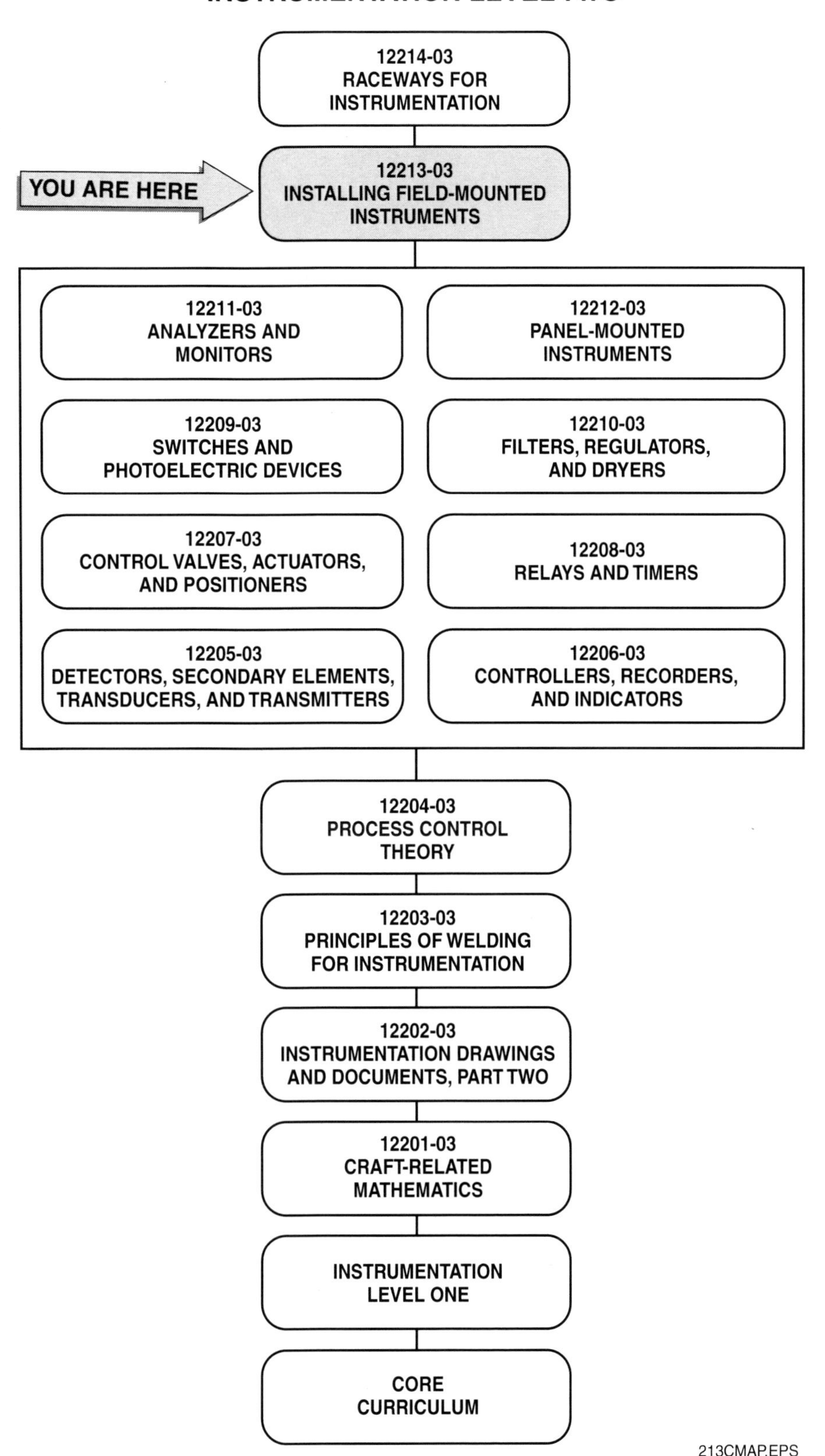

213CMAP.EPS

MODULE 12213-03 CONTENTS

1.0.0 INTRODUCTION 13.1

2.0.0 STAND-MOUNTED INSTRUMENTS 13.2

2.1.0 Floor-Mounted Stands 13.2

2.2.0 Wall-Mounted Stands 13.2

2.3.0 Fabricating the Stand 13.2

2.3.1 Tools and Materials Required 13.3

2.3.2 Measuring, Cutting, and Assembling the Pipe and Plate 13.5

2.4.0 Securing the Stand 13.9

2.4.1 Securing to Concrete Floors 13.10

2.4.2 Securing to Metal Grating Floors 13.11

2.5.0 Mounting Instruments on Stands 13.13

2.5.1 Instrument Locations 13.13

3.0.0 IN-LINE-MOUNTED INSTRUMENTS 13.13

3.1.0 Differential Pressure Flowmeters 13.14

3.1.1 Orifice Plates, Flow Nozzles, and Venturi Tubes 13.14

3.2.0 Velocity Flowmeters 13.17

3.2.1 Turbine Flowmeters 13.17

3.2.2 Vortex-Shedding Flowmeters 13.18

3.2.3 Magnetic Flowmeters 13.19

3.2.4 Ultrasonic Flowmeters 13.21

3.3.0 Volumetric Flowmeters 13.22

3.3.1 Rotary-Vane Flowmeters 13.22

3.3.2 Oval-Gear Flowmeters 13.23

3.3.3 Nutating-Disc Flowmeters 13.23

3.4.0 Mass Flowmeters 13.24

3.4.1 Coriolis Mass Flowmeters 13.24

3.5.0 Variable-Area Flowmeters (Rotameters) 13.25

3.5.1 Rotameter Installation 13.26

3.6.0 Density Meters 13.26

3.6.1 Angular Position Detectors 13.27

3.6.2 Hydrometers 13.27

3.6.3 Sound Velocity Meters 13.27

3.6.4 Vibrating Plate Detectors 13.28

3.6.5 Vibrating U-Tube Detectors 13.28

4.0.0 VESSEL-MOUNTED INSTRUMENTS 13.28

4.1.0 Probe-Type Level Instruments 13.28

4.1.1 Capacitance Probe (RF Probe) 13.29

4.1.2 pH Probes 13.30

4.2.0 Displacer-Type Level Instruments 13.32

4.2.1 Chamber-Installed Displacers 13.32

5.0.0 STRAP-MOUNTED INSTRUMENTS 13.32
5.1.0 Types of Supports 13.32
5.1.1 Cable-Mounted Instrument Supports 13.33
5.2.0 Thermostats and Heat Tracing 13.35
5.2.1 Thermostatically Controlled Tracing 13.35
5.3.0 Strapping 13.38
5.4.0 Radiation Meters 13.38
6.0.0 INSERTION-MOUNTED INSTRUMENTS 13.39
6.1.0 Thermowells 13.40
6.1.1 Material 13.41
6.1.2 Accuracy 13.41
6.1.3 Tapered or Straight Shank 13.41
6.1.4 Velocity Ratings 13.41
6.1.5 Thermowell Insertion 13.41
6.2.0 Connector Heads 13.41
6.3.0 Installation 13.43
6.4.0 Thermocouple Extension Wire 13.43
6.4.1 Thermocouple Insulation 13.43
6.5.0 Resistance Temperature Detectors 13.44
7.0.0 FLANGES 13.45
7.1.0 Flange Sizes 13.45
7.2.0 Flange Pressure Ratings 13.45
7.3.0 Flange Facings 13.45
7.4.0 Flange Gaskets 13.46
7.5.0 Methods of Joining Flanges 13.46
7.5.1 Socket-Welded Flanges 13.47
7.5.2 Screwed-Joint Flanges 13.47
7.5.3 Butt-Welded Flanges 13.47
8.0.0 MANIFOLD VALVE ASSEMBLIES 13.47
8.1.0 Single-Valve Manifolds 13.47
8.2.0 Two-Valve Manifolds 13.48
8.3.0 Three-Valve Equalizing Manifolds 13.48
8.4.0 Five-Valve Equalizing Manifolds 13.48
8.5.0 Five-Valve Blowdown Manifolds 13.49
8.6.0 Manifold Installation 13.49
SUMMARY 13.51
REVIEW QUESTIONS 13.52
GLOSSARY 13.55
REFERENCES & ACKNOWLEDGMENTS 13.56

Figures

Figure 1 Prefabricated floor-mounted stands 13.2
Figure 2 Pipe stands 13.3
Figure 3 Type 3 single wall-mounted stand 13.3
Figure 4 Fabrication tools 13.4
Figure 5 Quartering the pipe 13.5
Figure 6 Folding a paper strip 13.5
Figure 7 Creased paper strip 13.5
Figure 8 Marking the pipe 13.6
Figure 9 First 45-degree cut 13.6
Figure 10 Second 45-degree cut 13.7
Figure 11 Cutting the second upright 13.7
Figure 12 Saddle joint showing dimension X 13.7
Figure 13 Calculating support cutback 13.8
Figure 14 Marking the support cutback 13.8
Figure 15 Layout of the baseplate and pipe plug 13.8
Figure 16 Marking support location on baseplate 13.9
Figure 17 Concrete wedge anchors 13.10
Figure 18 Plumbing the stand 13.11
Figure 19 Metal grating flooring 13.12
Figure 20 Instrument mountings 13.13
Figure 21 Orifice plates 13.14
Figure 22 Typical orifice plate/nozzle installation 13.15
Figure 23 Centering an orifice plate or flow nozzle 13.16
Figure 24 Pressure taps 13.17
Figure 25 Turbine flowmeter with a magnetic sensor 13.18
Figure 26 Vortex-shedding flowmeter 13.18
Figure 27 Flanged magnetic flowmeters 13.19
Figure 28 Magmeter vertical installation with the flow going upward 13.20
Figure 29 Magmeter horizontal piping installation at a lower elevation 13.20
Figure 30 Ultrasonic flowmeter 13.21
Figure 31 Diagonal and reflective modes 13.22
Figure 32 Rotary-vane flowmeter 13.22
Figure 33 Oval-gear flowmeter 13.23
Figure 34 Nutating-disc flowmeter 13.23
Figure 35 Coriolis flowmeter tube 13.24
Figure 36 Rotameter 13.26
Figure 37 Angular position density sensor 13.27
Figure 38 In-line hydrometer indicator 13.27
Figure 39 In-line hydrometer in rotameter housing 13.27
Figure 40 Sound velocity meter 13.28
Figure 41 Vibrating plate density detector 13.29

Figure 42 Vibrating U-tube density detector 13.29
Figure 43 Use of capacitance probes 13.30
Figure 44 Batch control of pH 13.31
Figure 45 Continuous control of pH .13.31
Figure 46 Chamber-installed displacers 13.33
Figure 47 Cable-mounted base assembly 13.34
Figure 48 Cable assembly 13.34
Figure 49 Base with 2" NPT coupling for direct-mounting of support 13.34
Figure 50 Square-leg extension 13.34
Figure 51 Right-angle elbow extensions 13.35
Figure 52 U-bolt/cable interchangeability 13.35
Figure 53 Cable-mounted configurations 13.36
Figure 54 Line-sensing thermostat .13.37
Figure 55 Thermostat and sensor installation 13.37
Figure 56 Sensing bulb mounting 13.37
Figure 57 Sensing bulb position in reference to flow 13.38
Figure 58 Radiation density instrument 13.39
Figure 59 Installation for pipe diameters greater than 6" 13.39
Figure 60 Installation for pipe diameters smaller than 6" 13.40
Figure 61 Mounting a density instrument 13.40
Figure 62 Standard thermowell assembly 13.40
Figure 63 Standard NPT thermowell 13.42
Figure 64 Flange-type thermowell 13.42
Figure 65 Weatherproof head 13.42
Figure 66 Miniature weatherproof head 13.42
Figure 67 General purpose head 13.43
Figure 68 Explosion-proof head 13.43
Figure 69 Thermowell positioning 13.44
Figure 70 Industrial RTD/thermowell assembly 13.45
Figure 71 Determining flange size 13.45
Figure 72 Flange facings 13.46
Figure 73 Flange gaskets 13.47
Figure 74 Single-valve manifold 13.48
Figure 75 Two-valve manifold 13.48
Figure 76 Three-valve equalizing manifold 13.49
Figure 77 Five-valve equalizing manifold 13.49
Figure 78 Five-valve blowdown manifold 13.50
Figure 79 Manifold valve inlet/outlet configurations 13.50
Figure 80 Futbol assembly 13.51
Figure 81 Rod-out tool 13.51

Table

Table 1 NEC Requirements for Explosion-Proof Heads 13.51

MODULE 12213-03

Installing Field-Mounted Instruments

Objectives

When you have completed this module, you will be able to do the following:

1. Identify and describe various methods used in installing instruments in the field, including the following:
   - Stand mounted
   - In-line mounted
   - Structure mounted
   - Strap mounted
   - Insertion mounted
2. Determine and select the proper method of installation and location based on the instrument, environment and situation.
3. Plan and prepare support components for field-mounted instruments.
4. Install and describe the purpose of various valve manifold assemblies associated with the installation of field-mounted instruments.

Prerequisites

Before you begin this module, it is recommended that you successfully complete the following: Core Curriculum; Instrumentation Level One; Instrumentation Level Two, Modules 12201-03 through 12212-03.

Required Trainee Materials

1. Pencil and paper
2. Appropriate personal protective equipment

1.0.0 ◆ INTRODUCTION

Instrumentation personnel are required to perform many tasks. One of the tasks for which they are responsible is the installation of field process instrumentation. Instrumentation plays a major role in a system, whether it is used as an indicator or a control element. The mounting of this instrumentation also plays a major role. Proper instrument mounting ensures that any adverse conditions affecting the area or systems around the mounted equipment will have little or no effect on the instrumentation.

There are many ways that field instruments can be supported structurally. Stand mounting, in-line mounting, structure mounting, strap mounting, and insertion mounting are just a few of the major types of mounting.

Instrumentation personnel are not responsible for the design of the system, only the assembly of the system according to the design. They must be able to interpret the engineering drawings and specifications for the system and its components and assemble the system accordingly. However, some drawings show where the instrumentation must be located to sense a given parameter, but do not identify exactly where the instrument should be mounted. It is often left up to the installer to determine a logical and convenient location for mounting.

This module identifies the various types of field-mounted instruments and covers their installation. The module also reviews the selection and fabrication of mounting hardware and supports. Lastly, this module examines manifold systems used in these installations.

Note: The designations "National Electrical Code," "NE Code," and "NEC," where used in this document, refer to the *National Electrical Code®*, which is a registered trademark of the National Fire Protection Association, Quincy, MA. *All National Electrical Code (NEC) references in this module refer to the 2002 edition of the NEC.*

2.0.0 ◆ STAND-MOUNTED INSTRUMENTS

Most industrial process field instruments are mounted on instrument pipe stands. Some instrument pipe stands are mounted on the floor, and some are welded to columns. Almost all provide for the instrument to be mounted in the upright position. Usually, the instrument pipe stand is built to provide a 2" mounting. The instrument is attached by a bracket. The bracket is usually, but not always, provided by the manufacturer. Always check the specifications on the particular project for the exact instrument pipe stand sizes, types, and materials.

2.1.0 Floor-Mounted Stands

Examples of prefabricated floor-mounted pipe stands are shown in *Figure 1*. These modules can be used independently, or together with secondary modules, to construct a universal support package for any instrument in any position.

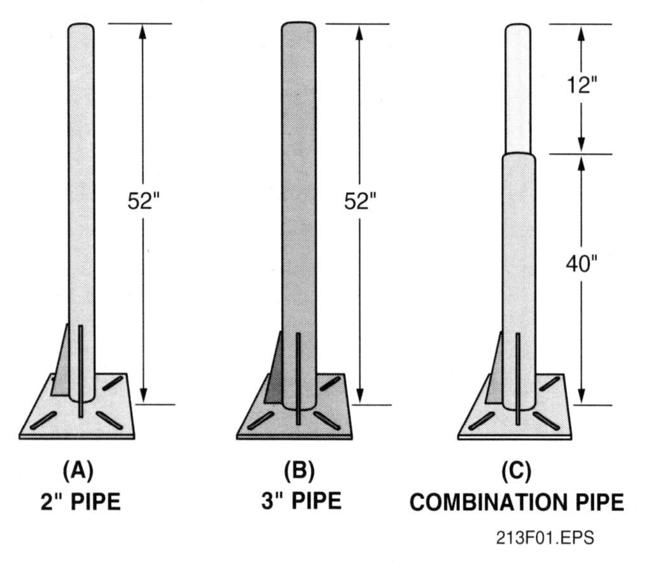

Figure 1 ◆ Prefabricated floor-mounted stands.

Figure 1(A) shows a 2" pipe extension 52" high used for supporting instruments of moderate size and weight. *Figure 1(B)* shows a 3" pipe extension 52" high used to support large and heavy instrument assemblies. The assembly in *Figure 1(C)* is a 3" pipe extension 40" long with a 12" section of 2" pipe. This design combines the rigidity of the 3" pipe stand with the mounting convenience of 2" pipe.

The mounting brackets are considered by many manufacturers to be accessories to the instrument purchased. The mounting brackets may be shipped already attached to the instrument or loose in the box with the instrument. They may also be purchased separately from the instrument.

Instrument stands can be purchased, or they can be fabricated with the assistance of a welder. Three typical instrument pipe stands are shown in *Figure 2*. Type 1 can be mounted on a concrete floor and supports a single instrument. Type 2 stands mount in the same manner and hold two instruments.

Actual dimensions of these stands will vary from project to project but are usually found in the instrument installation specifications.

The techniques used to fabricate pipe stands are covered later in this module.

2.2.0 Wall-Mounted Stands

Wall mounting of field instrumentation is often required instead of floor mounting in facilities where support bases cannot be mounted on floor surfaces. This situation may be caused by many factors, such as the presence of corrosive material, the absence of a floor or base support, or conditions in which the floor areas around specific equipment must be kept open. In addition, it is often more convenient to mount an instrument on a wall surface near the process being controlled or monitored. In general, lower maintenance instrumentation is wall mounted, while higher maintenance instrumentation is stand mounted, making these instruments easier to access.

Wall brackets may be built from various materials including carbon steel, stainless steel, PVC, or any material that is compatible with the environment in which it is to be installed. However, most projects include specification sheets that dictate the on-site fabrication of instrument stands. Always check these specifications before field mounting any instrument. *Figure 3* illustrates a simple Type 3 single wall-mounted stand. Tee brackets can be added to hold two or more instruments.

2.3.0 Fabricating the Stand

This section covers the procedures necessary to fabricate a typical two-instrument floor stand with a baseplate, constructed from 2" carbon steel pipe and ⅜" plate. It is similar to the Type 2 double floor mount illustration shown in *Figure 2*. Keep in mind that specifications on a particular project may or may not allow fabricated instrument stands. If fabricated stands are allowed, construction details such as type of material, pipe size, plate size, height, and other factors may be regulated by these specifications. Always check these specifications before mounting any instrument.

NOTE: Actual dimensions of these stands will vary from project to project but are usually found in the instrument installation specifications.

3/8" PLATE CAP TO EXCLUDE WATER AND DIRT
ABOUT 4'
2" PIPE
ABOUT 10"

TYPE 1
SINGLE FLOOR MOUNT

3/8" PLATE CAP (ON EACH UPRIGHT TO EXCLUDE WATER AND DIRT)
20"
CROSSPIECE
10"
45°
SADDLE
SUPPORT (2" PIPE)
54"
3/8" PLATE
1"

TYPE 2
DOUBLE FLOOR MOUNT

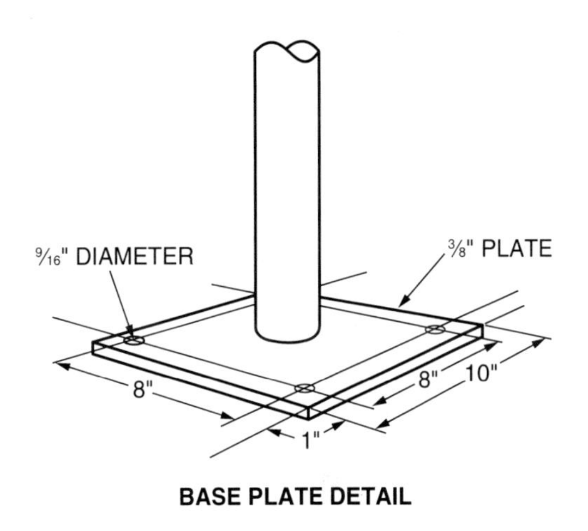

BASE PLATE DETAIL

213F02.EPS

Figure 2 ◆ Pipe stands.

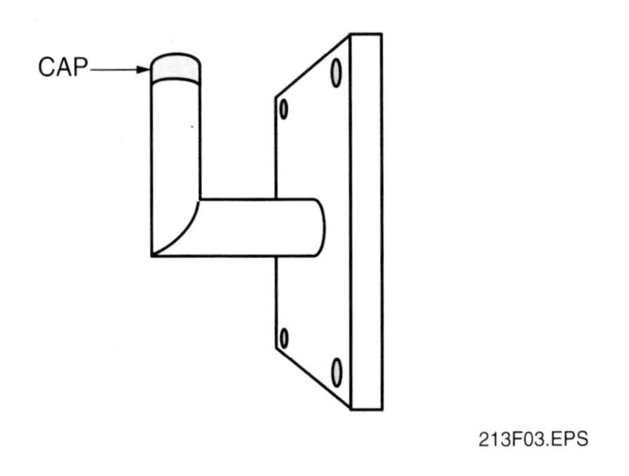

Figure 3 ◆ Type 3 single wall-mounted stand.

2.3.1 Tools and Materials Required

The suggested tools and materials required for fabricating the stand, some of which are shown in *Figure 4*, are listed here:

- Scribe
- Dividers
- Center punch
- Hammer
- Layout dye (optional)
- Electric hacksaw (porta-band) or metal chop saw to cut pipe
- Pipe cutter to cut pipe

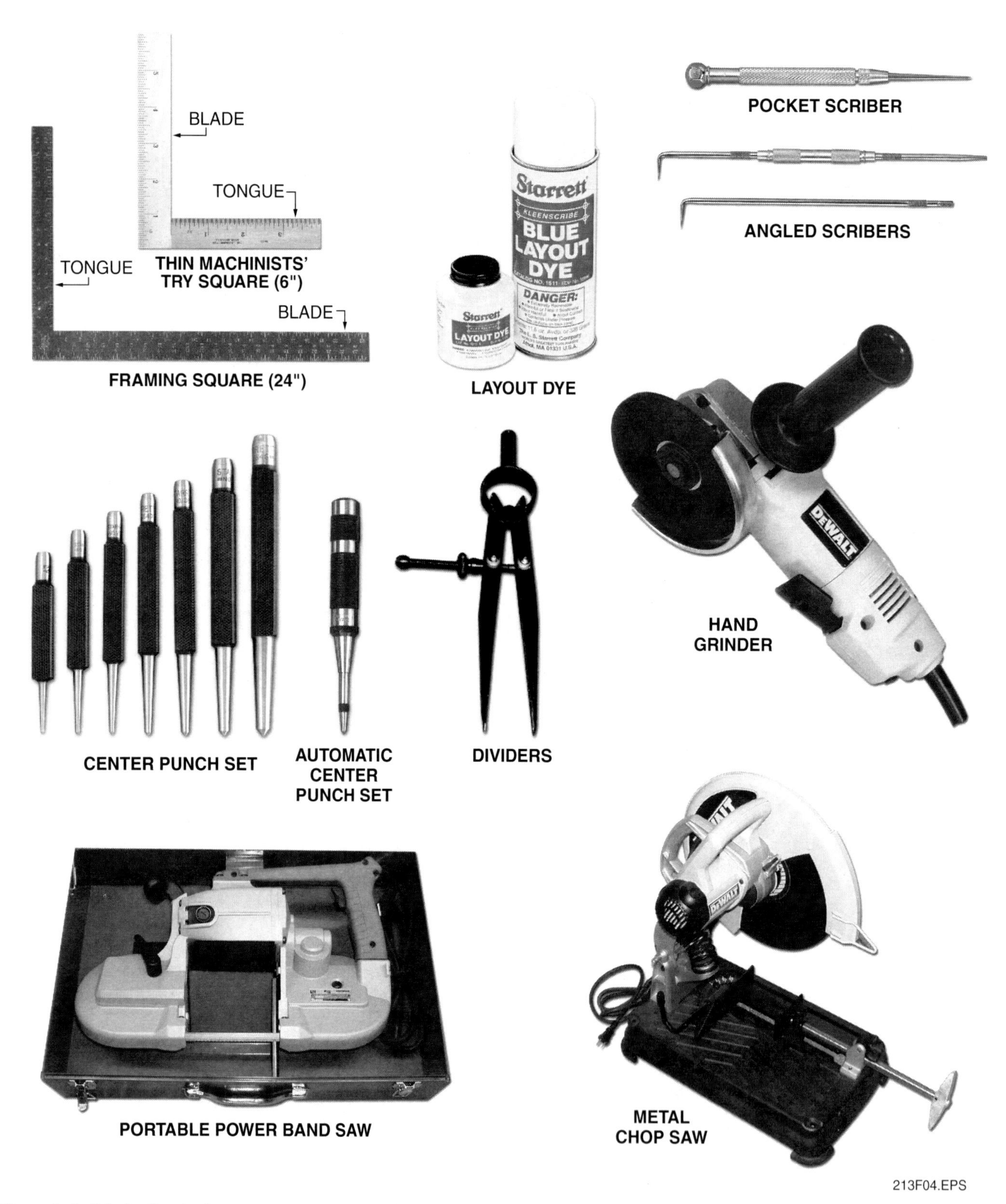

Figure 4 ◆ Fabrication tools.

- Acetylene cutting torch to cut plate
- Hand grinder
- Pencil or soapstone marker
- Measuring tape
- Try square or framing square
- Pipe wraparound
- 10' of 2" Schedule 40 carbon steel pipe
- ⅜" mild steel plate, minimum size 8" × 12"
- SMAW welding machine, or if not qualified to weld, a qualified stick welder

The proper and safe use of the tools listed above was covered in the Level One modules *Hand Tools for Instrumentation* and *Power Tools for Instrumentation* and the Level Two module *Principles of Welding for Instrumentation*. A complete understanding of these modules as well as a thorough knowledge of the skills taught is necessary in order to fabricate an instrument stand. Do not attempt to fabricate this stand without a thorough understanding of these skills.

2.3.2 Measuring, Cutting, and Assembling the Pipe and Plate

The first step of the procedure is marking and cutting the three sections of piping required for the top of the stand. To do this, identify the centerline of the pipe. This is done by quartering and marking the pipe.

One easy way to do this is to wrap measuring tape around the pipe and read the **circumference**. Then divide the measurement by four and mark the pipe at those increments, as shown in *Figure 5*.

Another method some instrument fitters find quicker is to wrap a piece of paper around the pipe and mark the point where it overlaps. Using the part of the paper that was wrapped around the pipe, fold the end and then mark together (*Figure 6*).

Fold the paper as shown in *Figure 6*. Flatten the paper so a crease is formed at point A. Then fold the paper in half again as shown. Flatten the paper so a crease is formed at point B. Unfold the paper. The paper now has creases at one quarter intervals around the circumference of the pipe as shown in *Figure 7*. Use this as a guide to mark the pipe every quarter of the way around the circumference.

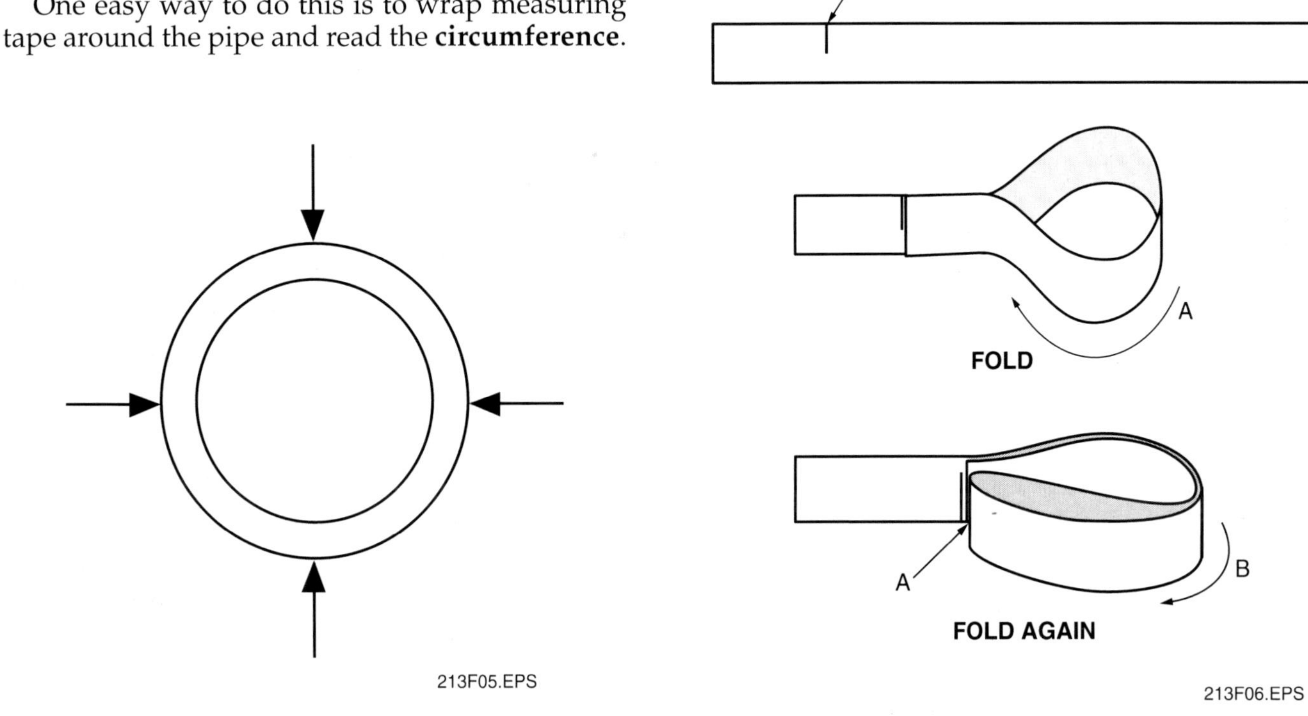

Figure 5 ◆ Quartering the pipe.

Figure 6 ◆ Folding a paper strip.

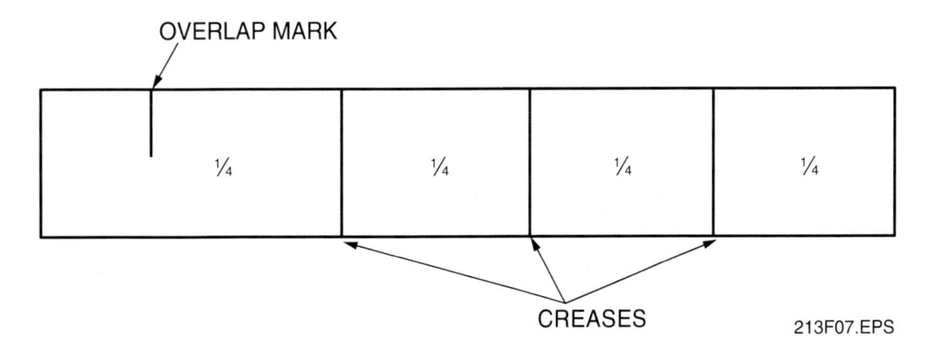

Figure 7 ◆ Creased paper strip.

Measure 10" from the end of the pipe and mark. Then measure 20" from the first mark and place the second mark. Measure 10" from the second mark and mark again. The total of the three measurements should use 40" of pipe. Roll paper tape around the pipe with the center crease on the 10" mark. Then mark the pipe at the quarter marks on either side of the center crease. Measure one-half the pipe OD in one direction from one of the side quarter marks, and one-half the OD from the other side quarter mark. Refer to *Figure 8*.

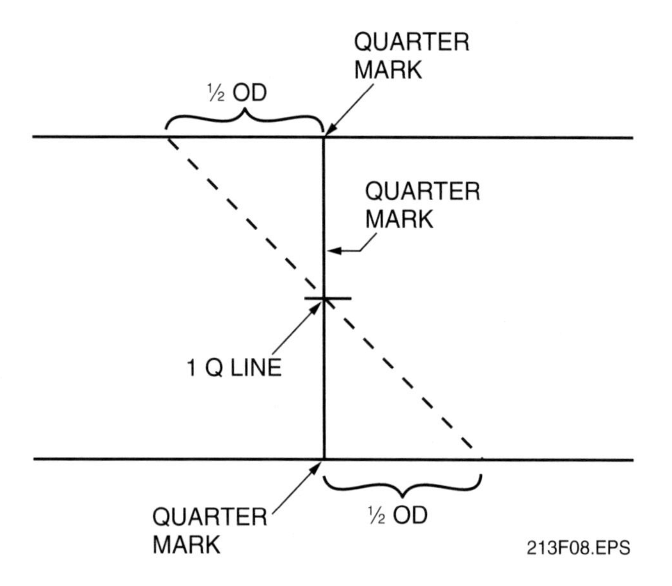

Figure 8 ◆ Marking the pipe.

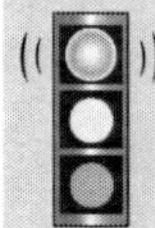

WARNING!

Follow all safety precautions outlined in the hand and power tool modules in *Instrumentation Level One*. Wear safety goggles at all times. Concrete or metal chips present a serious hazard to eyes, including the risk of blindness. Power tools present shock and electrocution hazards. Use properly grounded equipment and GFCI protection. Wear gloves to protect against sharp edges on metal plate and pipe ends.

Draw a line between the two marks you just made on the pipe. This line will provide the saw blade path for a 45-degree cut on the end of the pipe. Now cut the pipe with the porta-band saw. The 10" piece that is cut off is one of the two uprights.

The end of the remaining pipe stock is now 45-degrees, as seen in line A of *Figure 9*. The next step is to cut the other end of the pipe at a 45-degree angle, 20" down the centerline of the pipe.

To do this, wrap the paper tape around the pipe at the 20" mark with the center crease at the mark. Then mark the pipe at the side quarter marks on the tape. This is shown in *Figure 10*.

Refer to *Figure 10*. Mark a 45-degree line at point B as previously explained with the exception of the following:

- The square must lie in the same plane as when the pipe was marked before.
- The 45-degree mark must slant in the opposite direction from the first one.
- Using the porta-band saw, cut the pipe on the new 45-degree line.

One end of the remaining pipe stock is also on a 45-degree angle. Because it was cut to form the horizontal crosspiece, this angle will be used to form the second upright. At the remaining 10" mark on the pipe stock (*Figure 11*), cut the pipe at a 90-degree angle with a pipe cutter or saw.

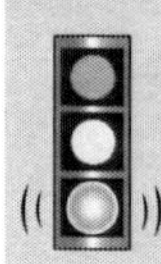

NOTE

The preceding instructions apply to pipe being cut freehand with a hand saw or porta-band saw. The 45-degree cuts could very easily be cut on a metal chop saw equipped with an adjustable vise, which can be set to specific angles.

Notice that in *Figure 2*, no dimension is given for the support itself. Dimensions are given, however, for the uprights and the entire stand. The instrument fitter can calculate the length of the support using these dimensions, the thickness of the baseplate, the distance between the plate and the floor, and dimension X, as shown in *Figure 12*.

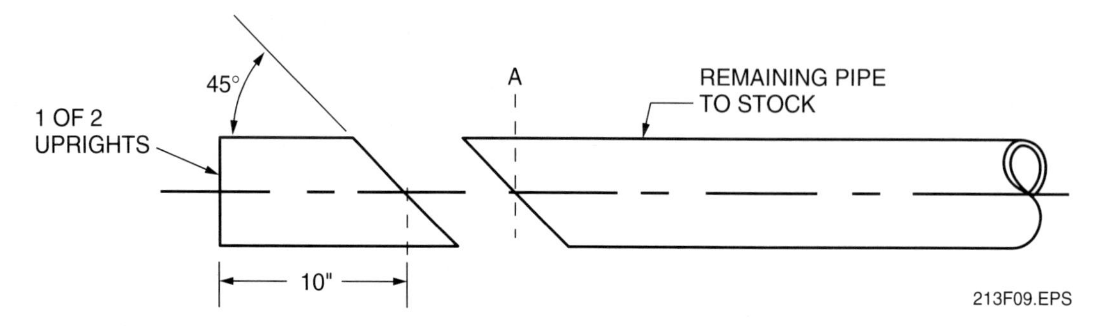

Figure 9 ◆ First 45-degree cut.

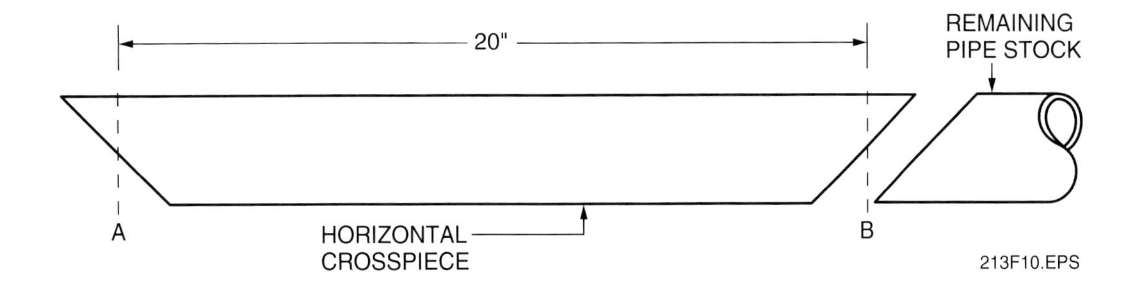

Figure 10 ◆ Second 45-degree cut.

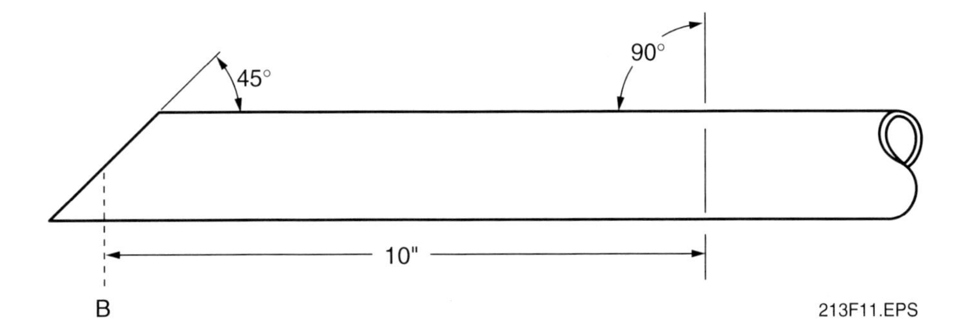

Figure 11 ◆ Cutting the second upright.

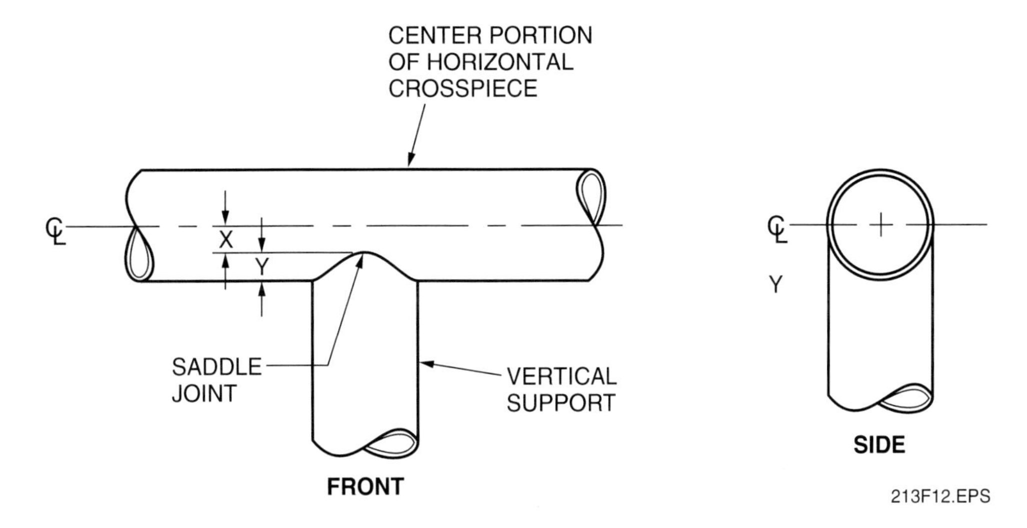

Figure 12 ◆ Saddle joint showing dimension X.

Figure 12 is an enlarged view of the joint between the support and the horizontal crosspiece. Notice that the support is saddled at the upper end.

Because the top of the upright stops below the centerline of the crosspiece, there is not enough information to calculate the height of the support. Dimension X must be determined. This involves calculating the depth of the saddle.

This pipe will not carry fluid, so the fabricator need not be as careful about saddle fabrication as for process pipe. Measure the inside diameter of the pipe. On 2" Schedule 40 pipe, this diameter is slightly over 2". Place a rule across the end of the pipe so that a 2" chord is produced on the OD (*Figure 13A*).

Hold another rule at right angles to the first one and measure the distance between the chord and the top of the pipe OD. This is the cutback distance Y as shown in *Figure 12*. On 2" Schedule 40 pipe, the distance is about $\frac{11}{16}$".

With the cutback distance Y, the length of the vertical support can now be calculated. To get the distance from the crosspiece centerline to the top of the saddle (dimension X), subtract the cutback distance from half the OD ($2\frac{3}{8}$" ÷ 2" = $1\frac{3}{16}$"), and $1\frac{3}{16}$" – $\frac{11}{16}$" = $\frac{1}{2}$ = X. Refer to *Figure 13B*.

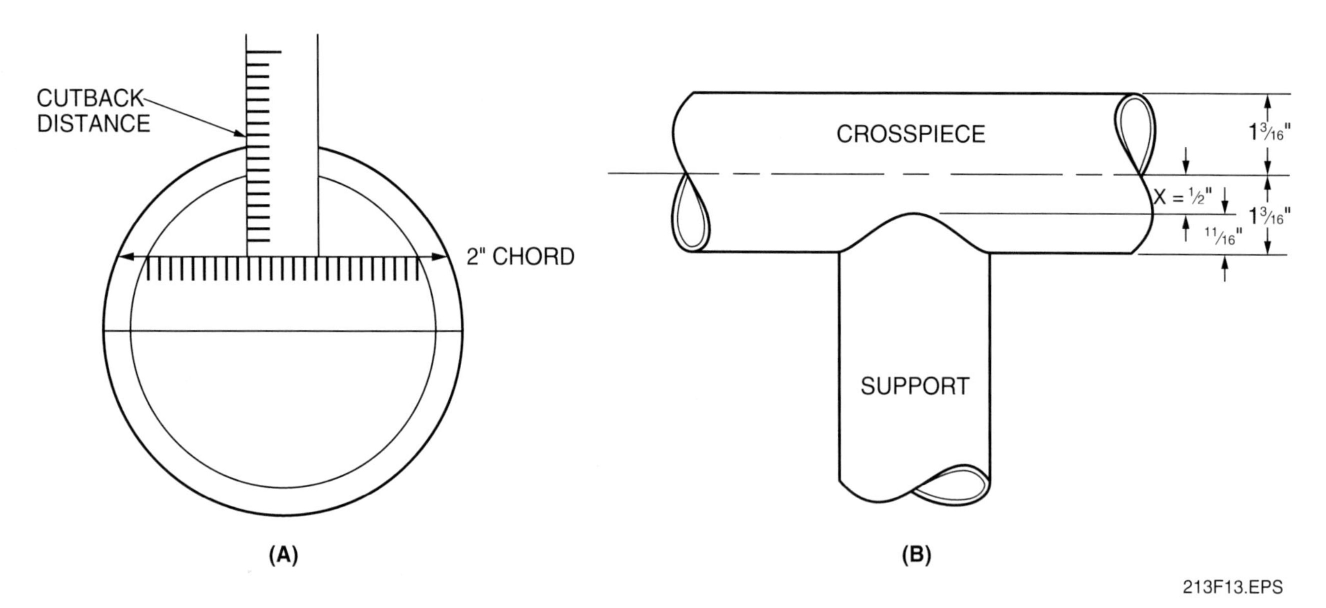

Figure 13 ◆ Calculating support cutback.

Find the distance from the top of the upright to the top of the support by adding the upright length to the distance from the crosspiece centerline to the top of the saddle (½" +10" = 10½"). This gives all the dimensions except the length of the support, which can be calculated as follows:

Bottom of baseplate from floor	=	1"
+ Baseplate thickness	=	⅜"
+ Top of upright to top of support	=	10½"
Total	=	11⅞"

Because the overall distance from the floor to the top of the stand is 54", the support length is 54" – 11⅞" = 42⅛".

The next step is to mark and cut the support. Mark the saddle by measuring the cutback distance (11/16") from one end of the pipe and make a semicircle on both sides, running from the OD to the cutback mark (*Figure 14*). The welder cuts the saddle with a torch.

The baseplate must now be made, as well as the two caps for the upright. These are made from ⅜" flat stock. Using a steel scribe, compass, center punch, and hammer, outline the baseplate and mark the holes. Outline two round weld plugs to fit in the ends of the uprights. Pipe caps can sometimes be purchased and used in place of the round plugs. Add 1/16" to the ID of the pipe. This will produce plugs of the right size. In this case, the plugs should have a diameter of 2⅛", as shown in *Figure 15*.

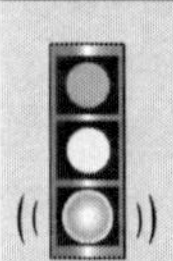

NOTE

Caps are used where possible to exclude water and dirt.

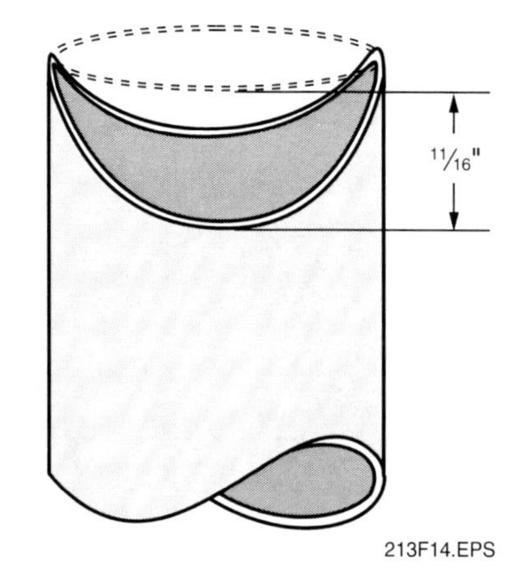

Figure 14 ◆ Marking the support cutback.

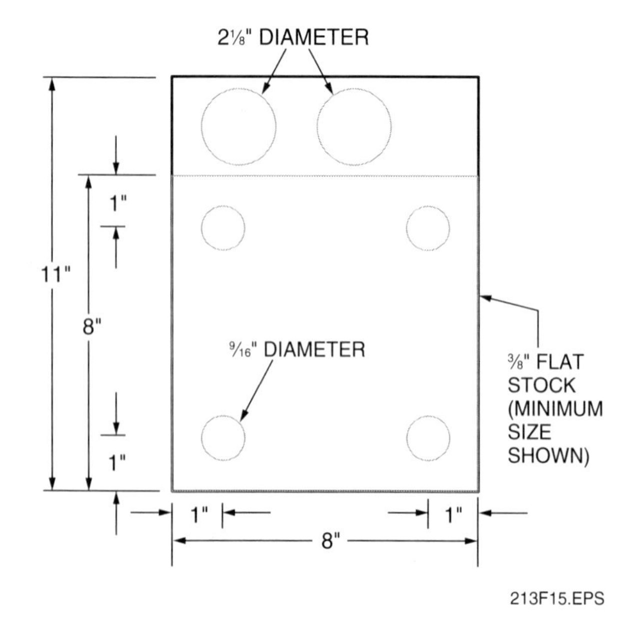

Figure 15 ◆ Layout of the baseplate and pipe plug.

Have the welder cut the saddle, baseplate, and plugs with the torch at right angles to the pipe. Now all the pieces are cut and the stand is ready to be assembled.

The first step in assembling the stand is to weld the support to the baseplate. Before welding the support to the baseplate, however, the center of the baseplate must be found in order to identify the proper location for the support. To do this, complete the following steps:

Step 1 Mark the center of each side of the baseplate, as shown in *Figure 16(A)*.

Step 2 Draw a straight line between the marks on opposite sides of the baseplate to find the center, as shown in *Figure 16(B)*.

Step 3 Using the OD of the support pipe as a guide, make a mark on each horizontal and vertical line at a distance of ½ of the OD from the center of the baseplate as shown in *Figure 16(C)*. For example, the support pipe OD is 2.125", so two marks are made on each line approximately 1⁄16" from the center.

Step 4 Using a try square, scribe a box on the plate using these marks as a guide, as shown in *Figure 16(D)*.

There is now a template on which to place the pipe. Complete the following steps on order to weld the support to the baseplate:

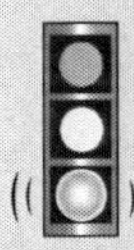

NOTE

The support pipe is usually mounted in the center of the baseplate unless this is physically impossible.

Step 1 Using the try square, hold the support perpendicular to the baseplate while it is being tacked in place. Tack weld at four equal points around the pipe, checking with a try square and hammer before an opposite weld is made to make sure the pipe back is vertical.

Step 2 After the pipe has been tacked at four points and aligned, complete the weld around the pipe. The heat will buckle the baseplate slightly, so the vertical pipe will not be exactly perpendicular. If a little extra heat is applied properly, the pipe can be drawn back to almost true. It should be as close to perpendicular as possible.

Step 3 Have the uprights welded to the horizontal crosspiece.

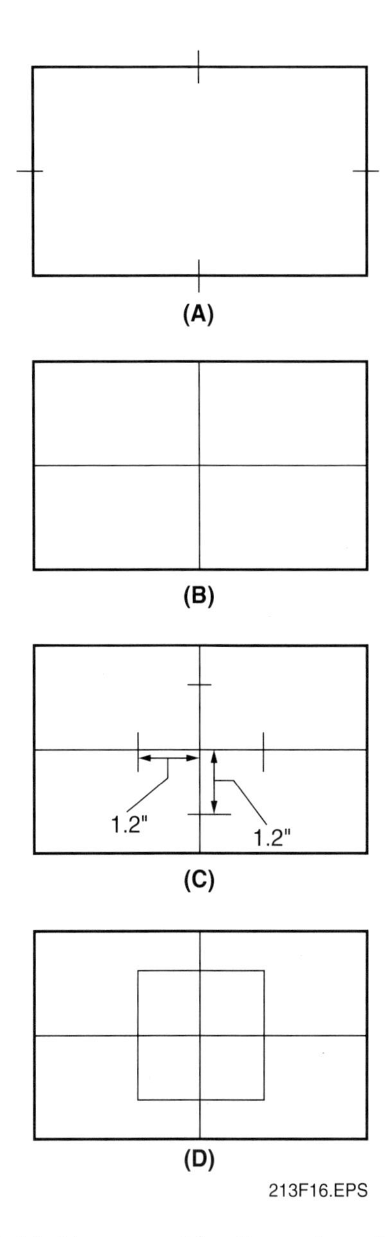

Figure 16 ◆ Marking support location on baseplate.

Step 4 Have the plugs welded into the upright ends.

Step 5 Have the horizontal crosspiece welded to the support. Use a square frequently.

Step 6 Use a grinder to smooth the work.

Step 7 Drill a small hole in the support at the baseplate as a drain for trapped moisture. On most projects, the final step is to galvanize or paint the completed stand.

2.4.0 Securing the Stand

After the stand is finished, the next task is to secure the stand to its resting place in a secure and level manner. It is important that all rough edges and welding slag be removed. The stand must be properly painted or galvanized to be compatible with the environment in which it is placed and to provide corrosion protection for years to come.

Most field floor stands are anchored to a concrete floor, although some instrument stands on other than grade level must be secured either to metal plate or metal grating flooring. Securing floor stands to both types of surfaces with concrete flooring is covered first in this module. A review of the Level One module *Fasteners* will help in selecting the proper fasteners. It also provides detailed instructions as to their proper installation.

2.4.1 Securing to Concrete Floors

If stand location is not specified in the drawings or other specifications, locate the stand so that it is convenient to the primary element point in or on the process, but not in a location that will cause interference with plant operation or personnel. Also provide sufficient room around the instrument stand so that routine calibration or instrument replacement is not a challenge. After the location is selected, verify that the concrete is level in that location and that nothing other than standard concrete reinforcement has been placed in the concrete. This includes conduit, cabling, or anything that may be damaged by contact with the drill bit.

Follow these steps when anchoring floor stand bases to concrete flooring:

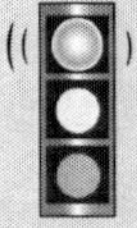

WARNING!

Follow all safety precautions as outlined in the hand and power tool modules in Level One. Wear safety goggles at all times. Concrete or metal chips present a serious hazard to eyes, including the risk of blindness. Power tools present shock and electrocution hazards. Use properly grounded equipment and Ground Fault Circuit Interrupter (GFCI) protection. Wear gloves to protect against sharp edges on metal plate and pipe ends.

Step 1 Clean floor area where the stand is to be mounted with low-pressure air or a broom.

Step 2 Set the stand in the selected position. Mark the four hole locations on the concrete by tracing the plate hole outlines on the concrete with a pencil or other appropriate marker.

Step 3 Remove the stand and select the proper anchor type and size, making sure the anchor stud passes through the plate holes freely, but with minimum clearance.

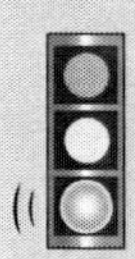

NOTE

Although there are several different types of concrete anchoring systems available, a wedge- or sleeve-type anchor is recommended (*Figure 17*). These anchors provide a permanent threaded stud protrusion, which allows ease of leveling and plumbing of the instrument stand. This is done by installing a leveling nut underneath the plate and an anchoring nut on top of the plate. Remember to select a wedge anchor that provides enough length so that a leveling nut and flat washer can be placed under the plate. The length must also be sufficient for another flat washer, a lock washer, and the fastening nut to be placed on top of the plate, with a minimum of thread length exposed above the final nut.

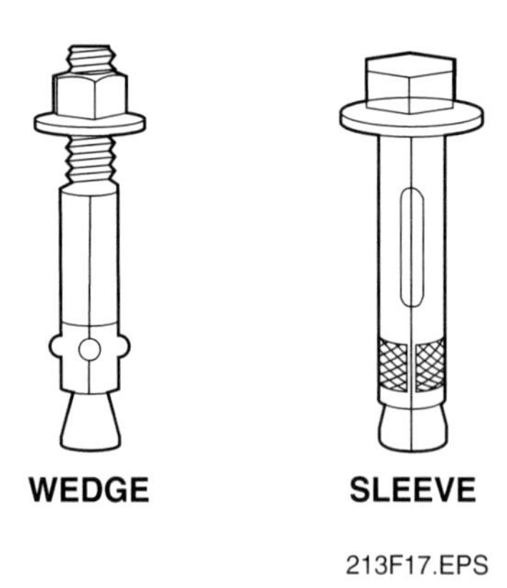

Figure 17 ◆ Concrete wedge anchors.

Step 4 Select a concrete masonry drill bit that is the same size as the anchor stud.

Step 5 Drill the four marked holes, making sure each hole is deep enough to allow for at least six threads below the surface of the concrete. Sufficient thread length must protrude as described in the note in Step 3.

Step 6 Clean out the holes and the area around the holes by using a squeeze bulb.

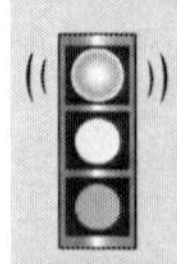

WARNING!

Do not use compressed air. Eye injury may occur.

Step 7 Install an extra, expendable stud nut on each of the stud bolts to protect the threads on the end of the stud. There should be no threads exposed above the nut. Drive the anchor bolts into the holes with a hammer, striking the flat surface of the nut only. When at least six threads on each of the studs are below the surface of the concrete, stop hammering.

Step 8 To set the anchors in place before installing the stand plate, remove and discard the hammered nut from each of the stud bolts. Place a flat washer first, then a new nut on the stud bolt. Thread the nut all the way down until it and the flat washer are snug with the concrete surface. Tighten the nut with the proper hand wrench to maximum resistance. The anchor is now set in the concrete. Follow the same procedure for all four stud bolts.

Step 9 Remove the nut and washer and check the finished work, making sure that all four studs are securely set. If any stud is loose in the hole, repeat Step 8.

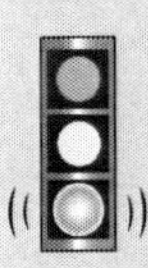

NOTE

It is not desirable to install the stand plate onto the studs directly against the concrete floor. This method does not provide any means of leveling or plumbing the stand.

Step 10 To plumb and level the stand, install a nut, then a flat washer on each of the protruding studs. Thread the nuts down until the tops of the flat washers are all at the specified or desired stand plate clearance from the concrete. Place the stand plate on the studs, making sure that it is all the way down at all four studs, against the tops of the flat washer. Place a magnetic torpedo level on the upright pipe support of the stand. While watching the level, adjust the leveling nuts installed underneath the flat washers until the bubble in the level is plumb. Relocate the level 90 degrees on the pipe support and repeat the process. Again, relocate the level 90 degrees and repeat the process. Continue this process until the level is back to the starting point and all positions indicate plumb (refer to *Figure 18*).

Step 11 Leave the stand on the studs, against the flat washers, and install another flat washer on top of the plate. Then install a lock washer and another nut. Tighten the nuts until all four are snug. Do not over tighten, which can sometimes cause the anchors to break loose. After all nuts are tightened, recheck the plumbness of the stand and repeat Step 10 if necessary.

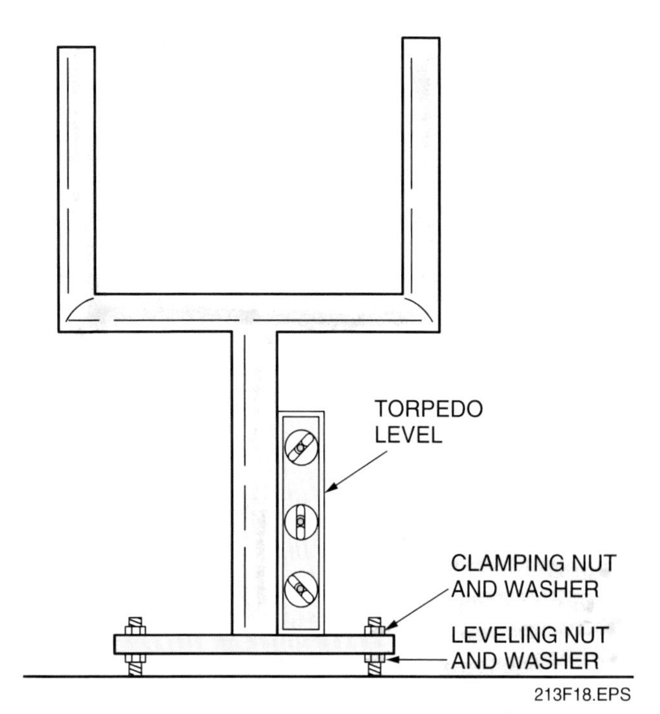

Figure 18 ◆ Plumbing the stand.

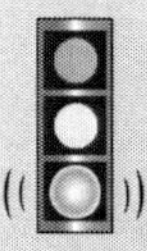

NOTE

Follow the same steps when installing instrument wall stands to concrete wall surfaces. Often, it is the policy of the plant or facility to install some type of concrete grout around the baseplate of the stand to fill in the gap between the concrete floor and the baseplate. This is for cosmetic purposes only and is not normally a task assigned to the instrument installer.

2.4.2 Securing to Metal Grating Floors

Not all instruments mounted on stands are located on grade level, especially in multi-level installations such as power plants and tower installations. These above-grade levels are typically designed with a metal grating flooring like the example shown in *Figure 19*. However, the same fabricated stand can be mounted to a metal grating using machine bolts, flat washers or **fender washers**, lock washers, and nuts.

Follow these steps when anchoring floor stand bases to metal grating flooring:

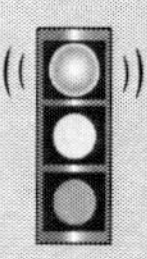

WARNING!

Follow all safety precautions as outlined in the hand and power tool modules in Level One. Wear safety glasses at all times. Power tools present shock and electrocution hazards. Use properly grounded equipment and GFCI protection. Wear gloves to protect against sharp edges on metal plate and pipe ends

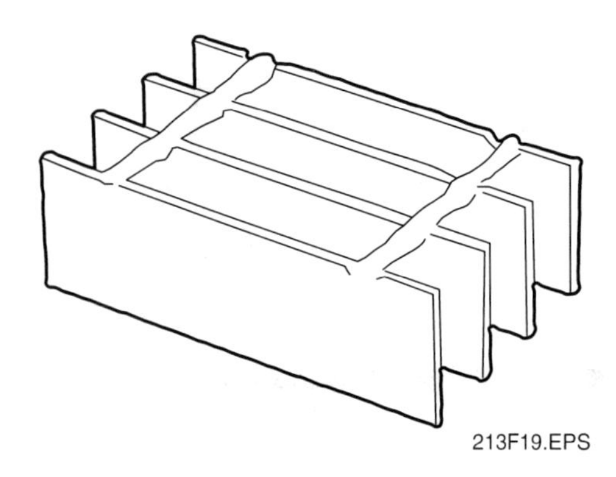

Figure 19 ◆ Metal grating flooring.

Step 1 Locate the stand in the position specified or selected.

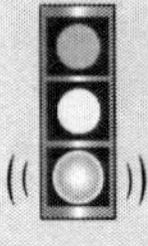

NOTE

The four holes in the baseplate must line up with open sections on the grating. If all four holes cannot be located so they are positioned over open portions of the grating, either the metal grating or the baseplate holes must be cut out so bolts may pass through the baseplate and the metal grating.

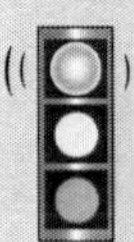

WARNING!

Unlike mounting to a concrete base, working on a metal grating presents hazards to equipment and personnel below the open grating. Proper plant safety policies must be followed. These include posting guards below the work area, installing signs or taping off the area immediately under the work area, covering with fireproof tarps any equipment susceptible to falling debris or welding slag, and constant observation of all activity beneath the work area during the mounting process.

Step 2 Select machine bolts that match the holes in the baseplate, allowing free passage with minimum clearance. They must be long enough to pass through two flat washers, the baseplate, the grating, two thick fender washers, two lock nuts, a lock washer, and a nut. The extra bolt length is not as critical, because the extra bolt protrusion under the grating is not normally likely to come in contact with personnel. However, it is still good practice to select a bolt length that doesn't leave excessive bolt protrusion after the final nut is installed.

Step 3 Do not place the stand baseplate in position. First, install a standard flat washer over each bolt. Insert the four bolts, with washers installed, through the baseplate holes from the stand side, so that the bolts are protruding out the bottom of the baseplate. Install a washer and lock nut on each bolt. Tighten the lock nuts and bolts.

Step 4 Do not place the stand baseplate in position. Loosely install another lock nut on each bolt protrusion under the plate. These nuts are the leveling nuts discussed in Step 10 of the concrete installation. The baseplate should now be positioned between two flat washers, making contact with the plate and two lock nuts on the bottom side.

Step 5 Do not place the stand baseplate in position. Because the lower lock nuts on the bottom of the plate will probably slip between the grating, they will not function as leveling nuts. A fender washer must be installed on each bolt. These must be large enough in diameter not to slip between the grating bars.

Step 6 After all four fender washers are in place, it's a good idea to tape them with masking tape to the nut or plate. This prevents them from falling off and through the grating when the baseplate is positioned on the grating. Grease or petroleum jelly can also be applied to hold them in place.

Step 7 Place the stand baseplate in position. If working alone, change position and go under the grating where the bolts are protruding. Place another sizable fender washer, lock washer, and nut over each of the four protruding bolt threads.

Step 8 Run the nut up by hand as much as possible and then slightly snug the nut using the proper hand wrench.

Step 9 Plumb and level the stand using the same techniques described earlier. Technicians working alone may have to change position several times from the bottom side of the grating to the upper side of the grating. It's best to find a helper in this situation.

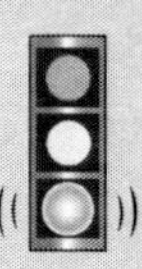

NOTE

Grouting or cosmetic dressing up of the gap between the plate and grating is not possible in this type of installation. Make sure all work is neat in appearance.

2.5.0 Mounting Instruments on Stands

Most instruments have provisions for several types of mountings. The majority are mounted on instrument pipe stands. Often, a single instrument is designed for wall, panel, valve yoke, and instrument pipe stand mounting. Hardware for each option may be furnished with the instrument, so there may be clamps and bolts left over after mounting an instrument to an instrument pipe stand.

Instruments are usually designed to mount to instrument pipe stands in several different ways, depending on the instrument type and manufacturer. Most use a mounting bracket with a U-bolt, two lock washers, and two nuts. Some use a bracket with two or four bolts holding a clamp around the pipe. Others just slip down over the top of the pipe and are held in place by setscrews, as shown in *Figure 20*.

2.5.1 Instrument Locations

Instruments must be accessible for repair, calibration, and adjustment from the ground, permanent ladder, or platform. This requirement is not intended to apply to transmitters close-coupled to the points of measurement.

Pressure taps serving pressure-actuated instruments must, whenever possible, be taken from the vessel concerned. Where the pressure tap is to be taken from the flow line for controllers, it may not be located less than eight pipe diameters from the associated control valve.

Wherever possible, locally mounted instruments must be mounted at approximately 4'-6" above the floor, ground, or platform, in an accessible and visible position.

Normally, an instrument is located as close as possible to the process line it monitors.

Unless otherwise specified, mount a D/P transmitter for an orifice meter run as follows:

- Mount liquid and steam service below the line.
- Mount gas and vapor service above the line.
- Mount liquid and steam in the horizontal line below the orifice.
- Mount gas and vapor in the vertical line above the orifice.

3.0.0 ◆ IN-LINE-MOUNTED INSTRUMENTS

Even though some in-line instrumentation components may relate to pressure or other aspects of process control, the majority of in-line applications deal directly with the measurement and control of flow. This section covers these types of elements. It

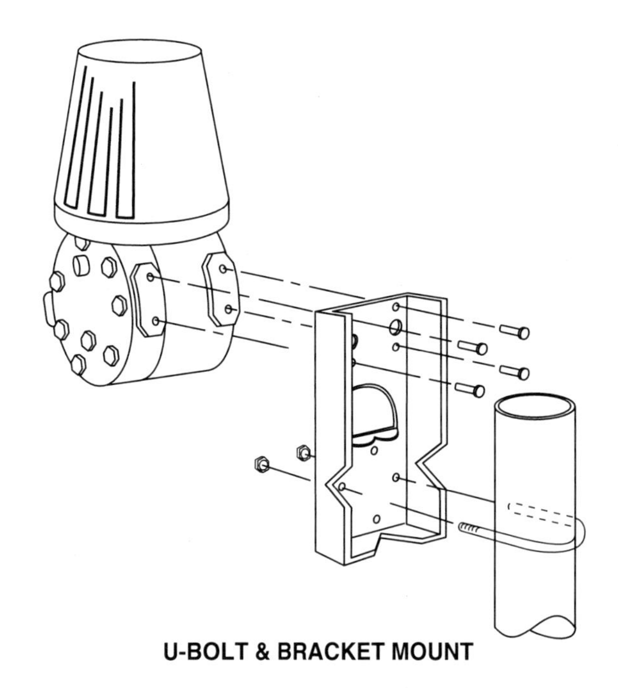

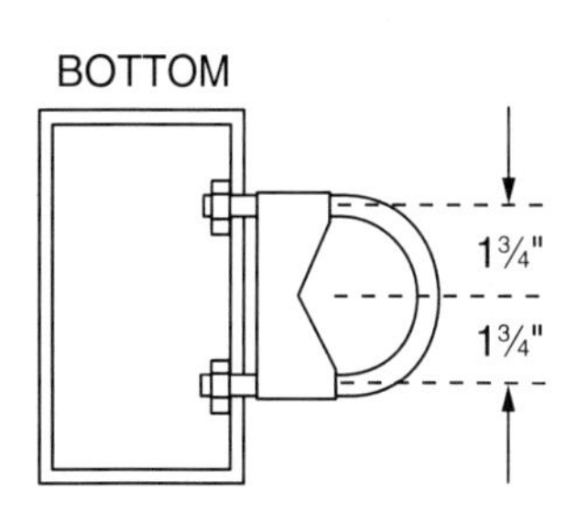

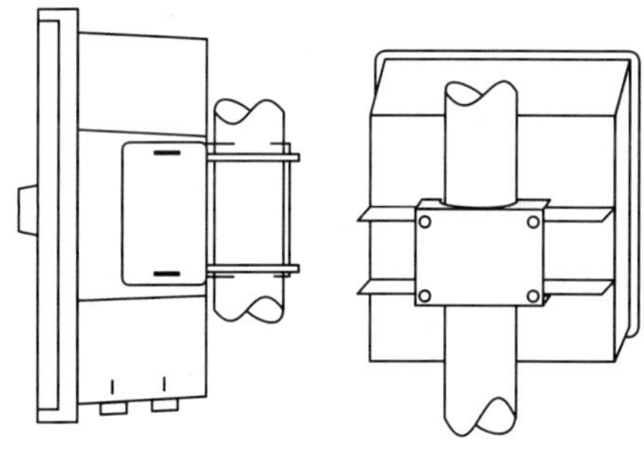

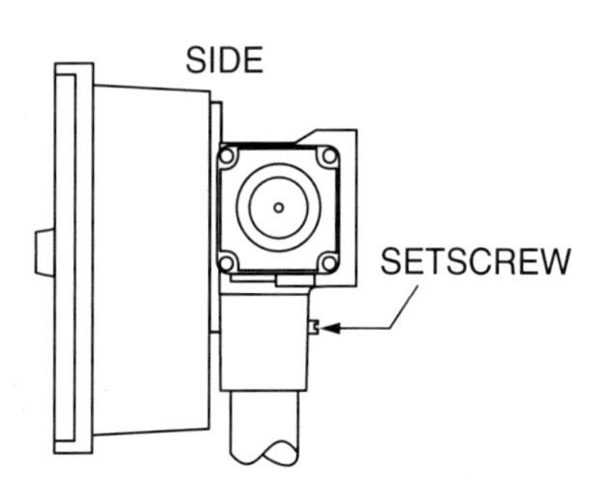

Figure 20 ◆ Instrument mountings.

also describes various liquid density meters and their installation. The installation of temperature control or monitoring instrumentation is also covered. It addresses the installation of insertion-type instrumentation, such as thermowells, thermocouples, and RTDs.

There are many types of flow-measuring devices used in the instrumentation industry. New technology constantly adds more to the list. The techniques used to measure flow fall into five general classes: differential pressure, velocity, volumetric, mass, and variable-area flowmeters. This section describes each of these, lists some examples, and follows up on installation methods for each.

3.1.0 Differential Pressure Flowmeters

Differential pressure flowmeters measure flow by inferring the flow rate from the differential pressure drop (dP or DP) across a restriction in the process pipe. This restriction causes an increase in flow velocity at the restriction and a corresponding pressure drop across the restriction. Examples of differential pressure restriction devices include orifice plates, flow nozzles, and venturi tubes. The equation that determines the results of the relationship between the pressure drop across this restriction and the rate of flow is:

$$Q = K\sqrt{DP}$$

Where:

Q = the volumetric flow rate

K = a constant for the pipe and liquid type

DP = the differential pressure drop across the restriction

The constant (K) depends on factors such as the type of liquid, the size of the process pipe, and the temperature of the liquid. The type and configuration of the restriction will also affect the constant (K). The following section discusses three of the most common restriction-type, differential pressure flow-measuring devices and their installation. These include orifice plates, flow nozzles, and venturi tubes.

3.1.1 Orifice Plates, Flow Nozzles, and Venturi Tubes

The orifice plate is the most common type of restriction used in process control applications to measure flow. Orifice plates are, in most cases, thin plate with openings concentric with the inner diameter of the pipe. The opening may also be eccentric or segmental, as shown in *Figure 21*.

The eccentric type of orifice plate is generally used when the downstream pressure connection cannot be located close enough to a concentric orifice for suitable accuracy. Segmental orifices might be considered extreme concentric orifices. They allow the downstream connection to be located even further downstream and offer even less restriction in the pipe to dirty fluids.

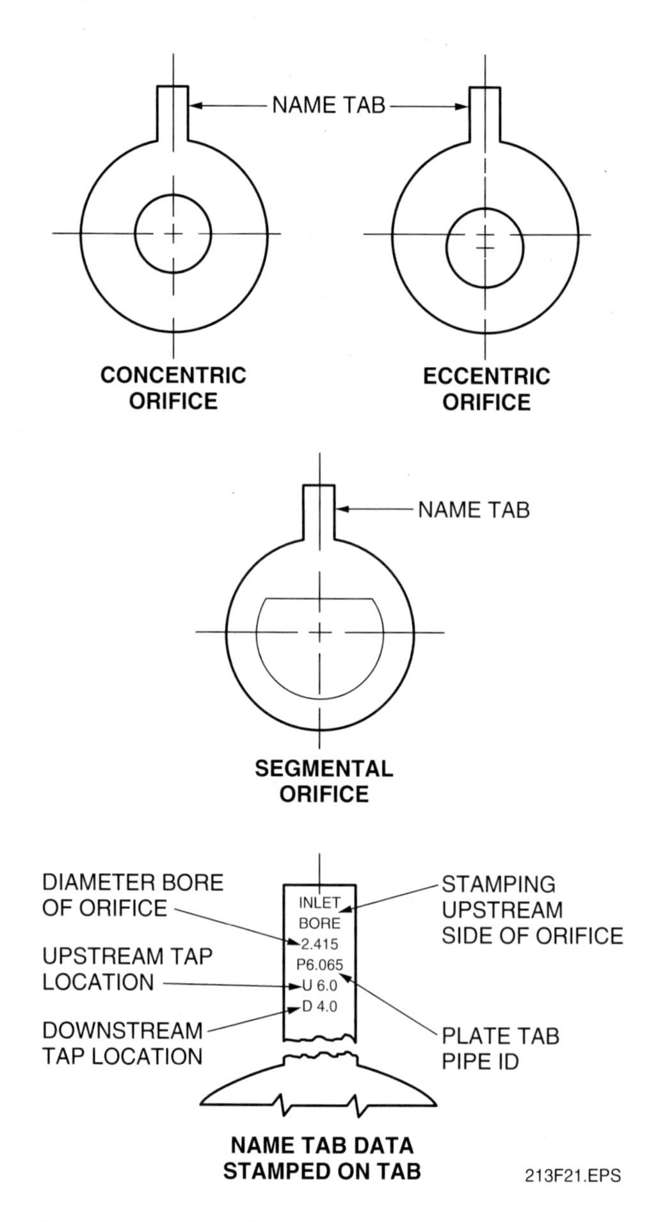

Figure 21 ◆ Orifice plates.

Orifice plates are usually installed between flanges in the pipe. Special orifice plate flanges are sometimes used. These are extra thick and have pressure taps radially through the flange. Orifice plates may be welded into the pipe itself, but for most weld-in installations, a flow nozzle is used.

Flow nozzles (and venturi tubes) have two important advantages. First, the permanent pressure loss is less for a flow nozzle than for an orifice plate with an opening of the same size. Second, flow nozzles and venturi tubes can handle approximately 60% more flow than the orifice plate.

Like orifice plates, flanged flow nozzles are installed between pipeline flanges. Quite often, however, nozzles are welded in place inside a pipeline or are pinned in place against a ring welded in the pipeline.

Typical orifice plate/nozzle installation between flanges is shown in *Figure 22*. The first consideration should be proper position of the primary element in relation to the direction of flow. Orifice plates must have the sharp unbeveled edge upstream and the beveled edge downstream. Orifice plates have the word inlet stamped on the name tab to indicate the side on which the fluid is to enter.

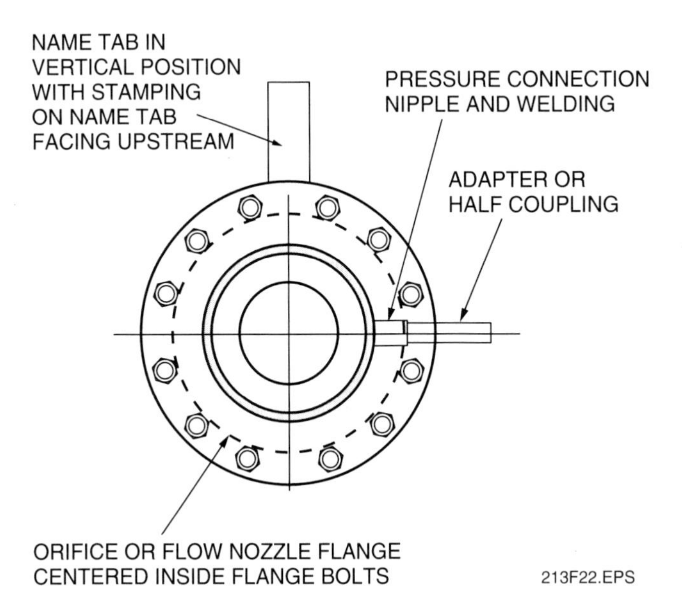

Figure 22 ◆ Typical orifice plate/nozzle installation.

Flow nozzles must be placed in the pipeline projecting downstream from the flanges with the large opening upstream. The flow nozzle also has a name tab indicating the upstream side.

Venturi tube sections usually have an arrow or other indication to show the direction of fluid flow through them.

The second consideration when installing the primary element is the position of the drain hole through the primary element. The name tab must be in a vertical position on horizontal installations. It must be 90 degrees away from the pressure tap connections on vertical installations.

The third important consideration is to make certain that the primary element is centered. A method of centering the primary element using the outside diameter of the flange as a guide is shown in *Figure 23*. First, the concentricity of the flange in relation to the pipe inside diameter is measured. Next, the primary element is placed in position and shifted until the difference between the two sets of measurements at each checkpoint is the same.

Segmental and concentric orifices often have centering pins to locate the opening in the orifice plates properly in relation to the pipe. Make certain the opening on such orifices is directly in line with, and on the opposite side of, the pipe from the pressure tap connections.

There are three principal methods of making pressure tap connections (*Figure 24*). The socket-weld adapter method is preferred and recommended over the welded half coupling and screwed-type methods because it ensures a good permanent connection with the least chance of introducing errors into the metering due to a poor pressure tap connection. The socket-weld (sock-o-let) adapter shown is furnished with ordered sets of primary element fittings for pressure and temperature conditions above 300 psig and 550°F.

The half coupling welded to the pipe shown in the welded half coupling method is the standard type of connection, used up to 300 psig and 550° F. Descriptions of the two most common methods of making pressure tap connections follow:

Socket-Weld Adapter Method

Step 1 Locate the pressure tap connections (U and D, dimensions from the installation data sheet), and note the size of the U and D connections.

Step 2 Drill pressure tap holes partially into the pipe wall, ensuring the drilling is perpendicular to the pipe wall.

Step 3 Counterbore each hole to receive each socket-weld adapter.

Step 4 Weld the socket-weld adapters in place so that their centerlines are perpendicular to the line pipe centerline. Bore or ream before proceeding with the next step.

Step 5 Drill through the pipe wall by using the socket-weld adapters as guides for the drill. Note that the inside diameter of the adapter is the same as the pressure tap hole in the pipe wall. It should be reamed, if necessary.

Step 6 Remove any burrs, projections, wires, or other irregularities from the inside edge of the drilled pressure tap holes.

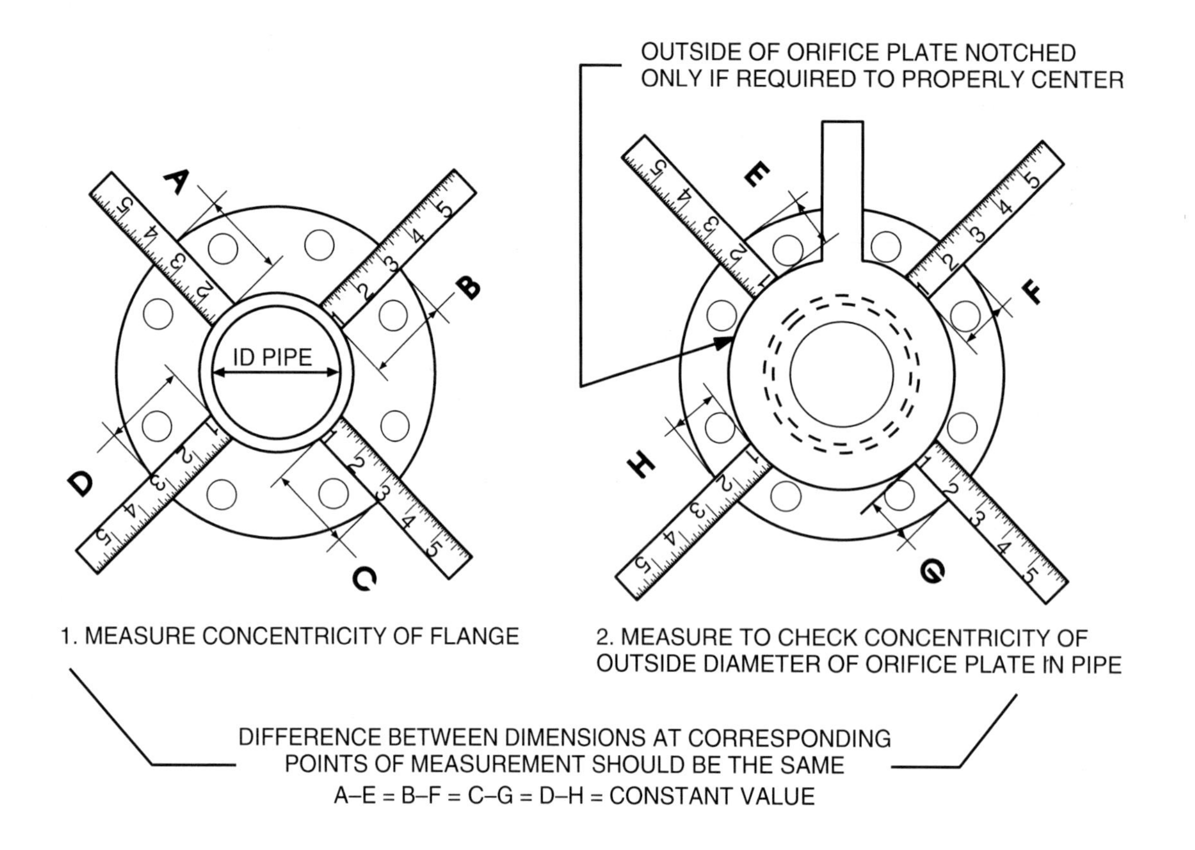

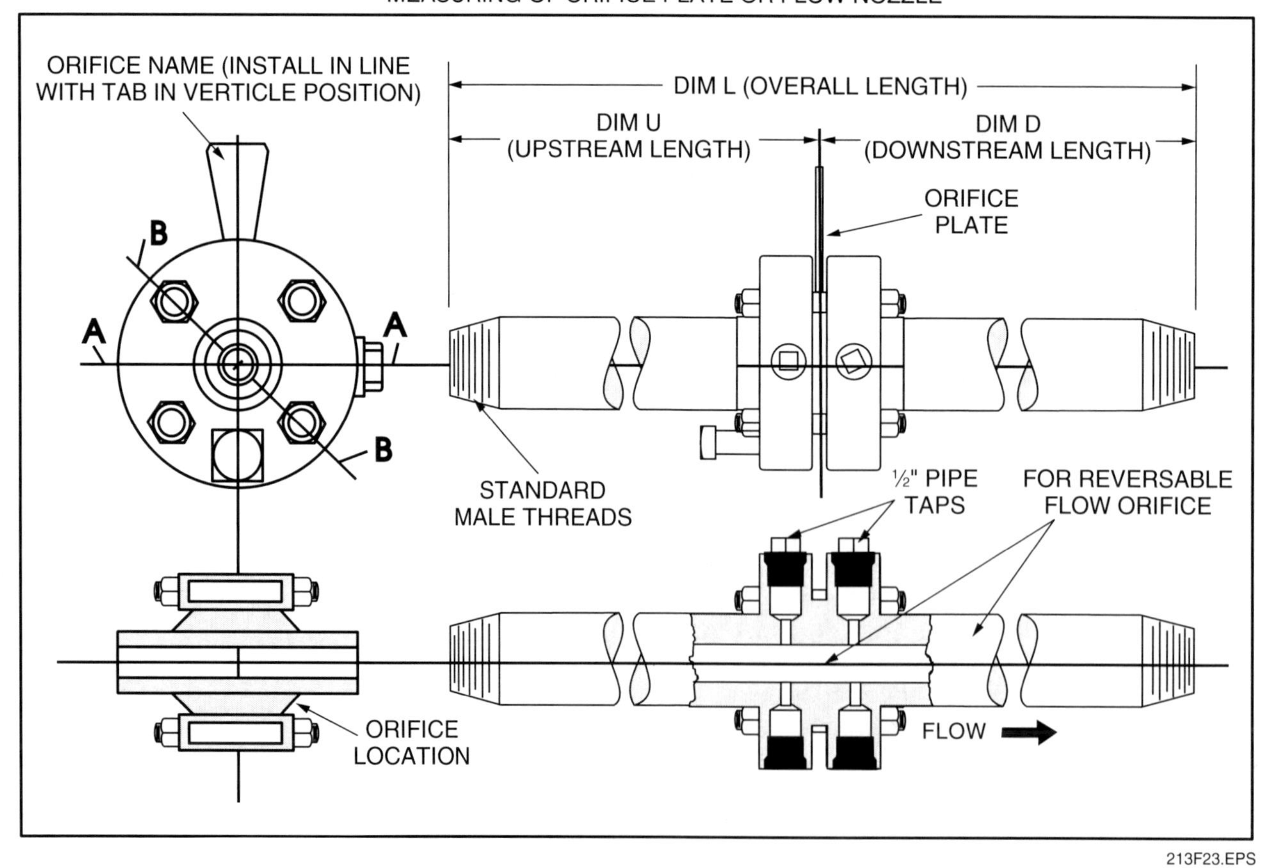

Figure 23 ◆ Centering an orifice plate or flow nozzle.

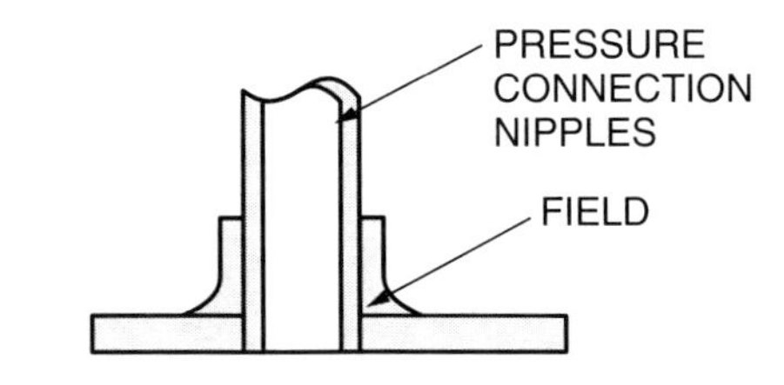

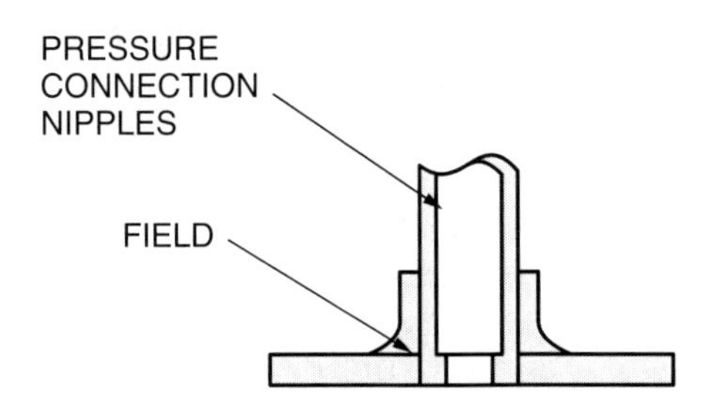

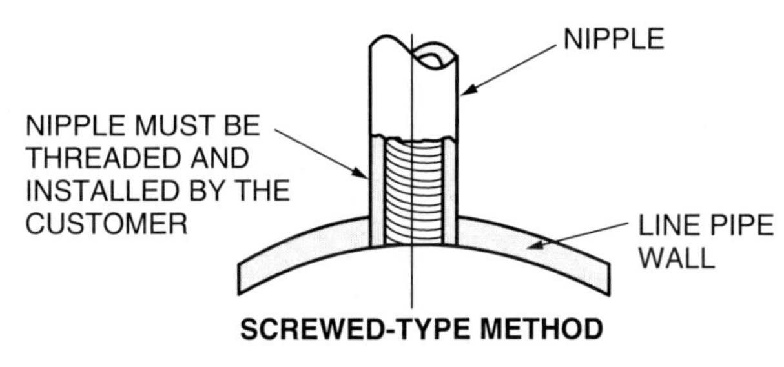

Figure 24 ◆ Pressure taps.

Welded Half Coupling Method

Step 1 Locate the pressure tap connections (U and D dimensions from the installation data sheet), and note the size of the U and D connections.

Step 2 Drill pressure tap holes part of the way into the pipe wall, making certain drilling is perpendicular to the pipe wall.

Step 3 Shape half couplings by grinding or machining to fit the half coupling to the line pipe. Chamfer for welding.

Step 4 Weld the half coupling in position, centering over the partially drilled pressure tap holes.

Step 5 Finish drilling the pressure tap holes through the pipe wall, being careful to drill perpendicular to the centerline of the line pipe.

Step 6 Remove any burrs, projections, wires, or other irregularities from the inside edge of the drilled pressure tap holes.

Step 7 Bore, smooth, or ream before proceeding with the next step.

For high temperature service (above 800°F) a special type of socket-weld adapter is used. It is made of 2.25% chrome and 1% molybdenum alloy steel for temperatures up to 1,050°F. It can be made of stainless steel for higher temperatures.

3.2.0 Velocity Flowmeters

Velocity flowmeters measure flow rate by measuring the velocity (speed) of the flow and multiplying the result by the area through which the fluid flows. Various methods are used to determine the velocity of the process flow. These include using a turbine that the process liquid turns; using a **vortex**-shedding device that measures on the principle of the process flow striking a **bluff** or solid object, using a magnetic flowmeter that can only measure flow of conductive processes that pass through a series of magnetic fields, and using an ultrasonic flowmeter in which electrodes within the flowmeter transmit **acoustic** signals to receivers that compare the transmitted signal to the flow-distorted signal. Each of these types of meters and their installation requirements are discussed in this section.

3.2.1 Turbine Flowmeters

In a turbine flowmeter, as liquid or gas flows through the turbine, it turns an impeller blade that is sensed by infrared beams, photo-electric sensors, or magnets. The example shown is *Figure 25* illustrates a turbine flowmeter with a magnetic sensor. An electrical pulse is generated and converted to a frequency output proportional to the flow rate. Turbine flowmeters are best used with clean, low-viscosity liquids. Some turbine flowmeters can be used with air. However, if there are air bubbles or vapor pockets in the liquid, the reading will be inaccurate. There should be a laminar (stable) flow through the cross-section of the pipe.

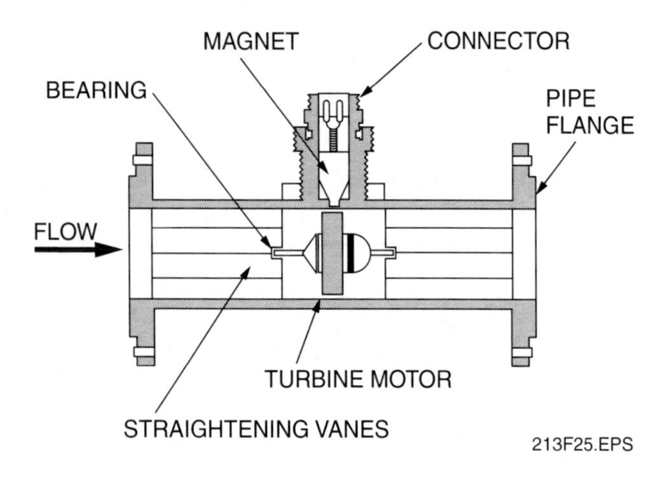

Figure 25 ◆ Turbine flowmeter with a magnetic sensor.

Most turbine flowmeters are designed to be mounted in the process piping run between two flanges. They are manufactured with various flange/piping sizes. In order for the turbine flowmeter to read accurately, it is recommended that there be a straight pipe length of at least 10 times the meter's inner diameter upstream and at least 5 times the meter's inner diameter downstream of the sensor. Check each flowmeter for specific requirements.

3.2.2 Vortex-Shedding Flowmeters

When a blunt object, also called a bluff body, is placed in the path of a fast-moving liquid flow, the fluid is unable to remain attached to the object on its downstream sides. It will alternately separate from one side and then the other. The fluid in the contact layer on the bluff body becomes detached and curls back on itself. This is illustrated in *Figure 26*. The result of this separation and backflow is the formation of vortices or eddies. Vortex formation causes fluid on that side of the bluff body to move with higher velocity than the fluid on the other side. Therefore, the fluid on the side of the vortex exerts less pressure than the fluid on the other side, because it has greater kinetic energy. Initially, a vortex is in a fixed position relative to the bluff body. The vortex grows in strength and size, eventually detaches itself, and sheds downstream. Then the process reverses itself, with a vortex being created on the other side of the bluff body. This process creates a vortex path that extends downstream of the bluff body, having alternating vortices spaced at equal distances. Using a strobe light, the vortex shedding can be seen in water flowing through clear plastic piping. The vortex looks like a miniature tornado. However, while a tornado is funnel shaped, the vortex is shaped more like a column of fluid, as wide at the base as at the top, stretching across the entire pipe diameter.

A vortex flowmeter typically consists of a flowmeter body mechanically connected to the electronics housing. The electronics may be remotely mounted for safety or convenience. The flowmeter body contains the bluff body and the vortex sensing assembly. The meter is typically made from 316 stainless steel or Hastelloy®. Vortex meters can theoretically be made of any material, but are only truly cost effective when made of these two materials.

Vortex meters can be installed vertically, horizontally, or at any angle. It is best to position the meter so that liquids flow against gravity to keep the pipe full. If that is not feasible, elevate the downstream piping above the meter installation level to maintain a full pipe. Install the meter to avoid standing liquid when the pipe is empty, and avoid installations that permit gas bubbles to form in the liquid flow. Check valves may be used when installing a vortex meter to keep it full of liquid when there is no active flow in the process. Many manufacturers recommend that the meter be installed a minimum of 30 times the pipe diameter downstream of any control valves.

The vortex meter requires an installation free from any distortions or swirl in order to achieve maximum performance. Sufficient lengths of **conditioning piping** are required upstream and downstream of the meter to condition the flow. This piping must have the same inside diameter as the meter itself. Each manufacturer supplies guidelines for piping requirements.

A component called a flow straightener is sometimes installed to condition or stabilize the process fluid before it enters the flowmeter, in order to minimize shifts and extend rangeability. However, traditional flow straightener designs require additional piping or straightening vanes, both upstream and downstream of the meter. Additional end fittings are also necessary to accommodate these straighteners. This results in increased hardware and installation costs. Even when flow straighteners are installed upstream

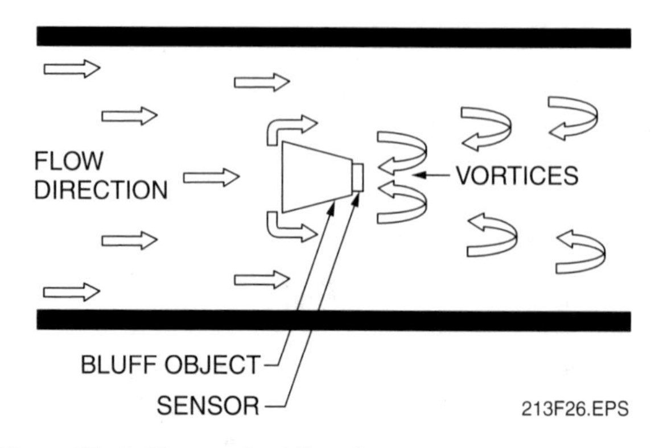

Figure 26 ◆ Vortex-shedding flowmeter.

from a vortex flowmeter, a certain amount of conditioning piping is still required. Nearly half of all vortex meter installations will require concentric reducers/expanders to neck down the process

Mating flanges on the process piping must be of the same nominal size as those on the flowmeter. Flanges with a smooth bore, similar to weld neck flanges, are preferred. Don't use reducing flanges. Most recommended specifications call for either schedule 40 or schedule 80 mating pipe. The mating pipe should be high quality and have an internal surface that is free from scale, pits, holes, reaming scores, and bumps for at least four times the pipe diameter upstream from the meter and two times the pipe diameter downstream from the meter. The bores of the adjacent piping, meter, and gaskets must be carefully aligned to prevent interference or distortion in the flow.

Some vortex meter designs allow the sensors to be replaced without shutting down the process. Those meters that don't allow this must be installed with bypass piping and block valves if unscheduled process shutdowns are a problem.

Because excessive pipe vibration or process noise can affect the accuracy of the meter measurement, proper piping supports on either side of the meter must be installed to prevent vibration, or the meter can be rotated in the process piping so that the sensor is located in a plane that is different than the vibration. Most electronics in vortex flowmeters are equipped with some type of noise filtering circuitry. Keep in mind, however, that increasing noise reduction by electronically filtering it out also decreases the low flow **sensitivity** of the meter. If vibration or process chatter is a problem, it may be necessary to relocate the meter.

3.2.3 Magnetic Flowmeters

The conventional magnetic flowmeter (magmeter) consists of a magnetic flow tube and a magnetic flow transmitter. The magnetic flow tube is generally a section of nonmagnetic pipe, such as stainless steel, which has been lined with a nonconducting material. In the center of the flow tube is a pair of electrodes. These electrodes protrude through the non-conductive liner and make contact with liquid flowing through the tube. Magnetic coils are mounted on the outside of the pipe and generate a magnetic field inside the tube when energized by a current. One of the laws of electricity, Faraday's *Law of Magnetic Induction* states that a voltage will develop in a conductive liquid moving through a magnetic field. This voltage is measured at the electrodes. The liner insulates the electrodes so that the voltage will not be conducted into the pipe.

Figure 27 shows an illustration of typical flanged magnetic flowmeters.

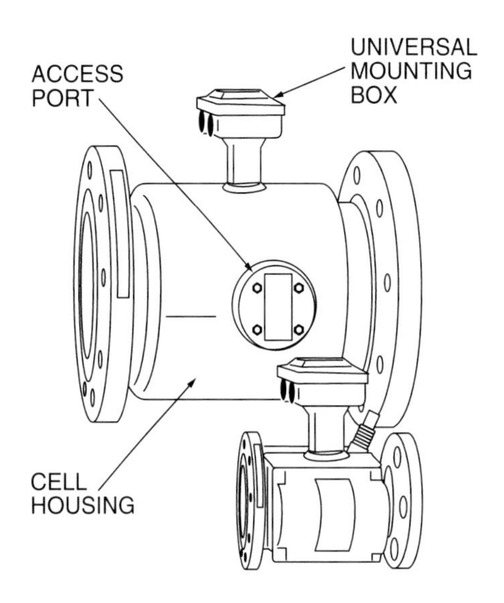

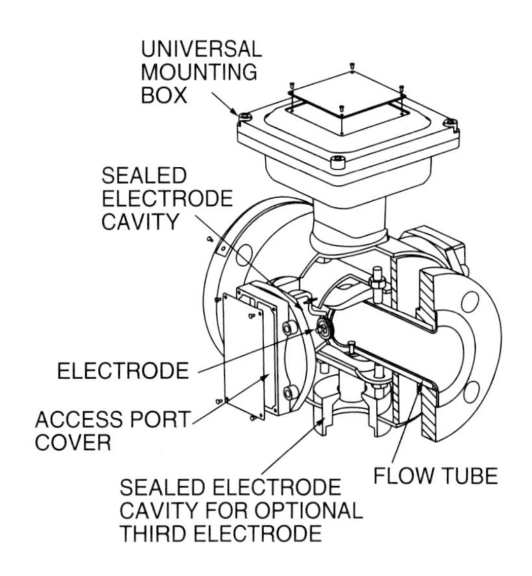

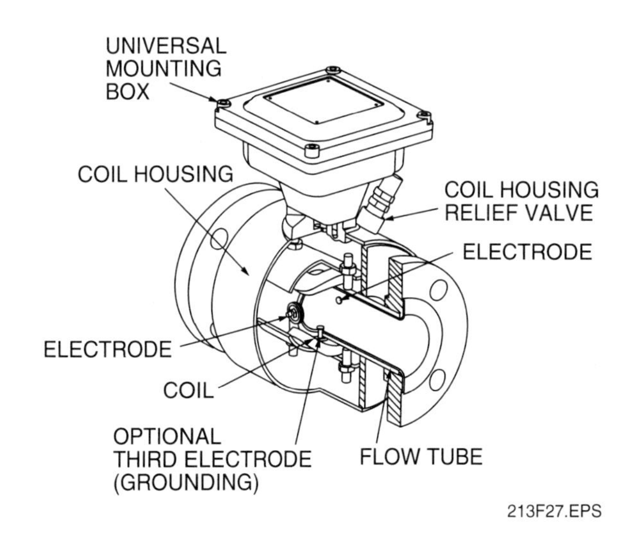

Figure 27 ◆ Flanged magnetic flowmeters.

Magmeters should always be filled with liquid. Empty or partially full flow tubes should be avoided. For this reason, the suggested installation positioning is in vertical lines with the flow going upward, as shown in *Figure 28*. Installation in horizontal lines is also acceptable, but the flow tube should be positioned so that the electrodes are not at the top of the piping. This prevents air from coming into contact with the electrodes, causing errors in measurement. The flow tube orientation should not trap any entrained air or cause a buildup of undissolved solids within the flow tube. The magmeter can be installed in a horizontal line at a lower elevation than adjacent piping by using piping elbows to help maintain full pipe conditions, as shown in *Figure 29*. Magmeters can also be reduced in size, as compared to the process piping, in order to ensure full pipe conditions or to achieve the proper fluid velocities. This can be accomplished by using reducers.

As mentioned previously and illustrated in *Figure 28*, you should attempt to install the magnetic flowmeter in vertical lines, with the flow going upward to ensure a full meter. Also, there may be times when the magmeter must be installed very close or adjacent to pipe fittings. Fittings such as elbows can create excessive wear along the leading edge of the liner where it flares out over the face of the flange. Liner protectors or orifice plates can protect the leading edge of the liner from the abrasive effects of the process fluid.

Some piping configurations will allow the liquid to drain during conditions of zero flow. Check valves can be used to prevent the flow tube from emptying under such conditions. Magmeters installed in horizontal pipes can become partially full of liquid at velocities below one foot per second unless they are initially sized and installed properly. A partially full magmeter will exhibit the same output problem as an empty magtube after the liquid level falls below electrode level. An empty or partially full flow tube will generate an unpredictable output from the transmitter.

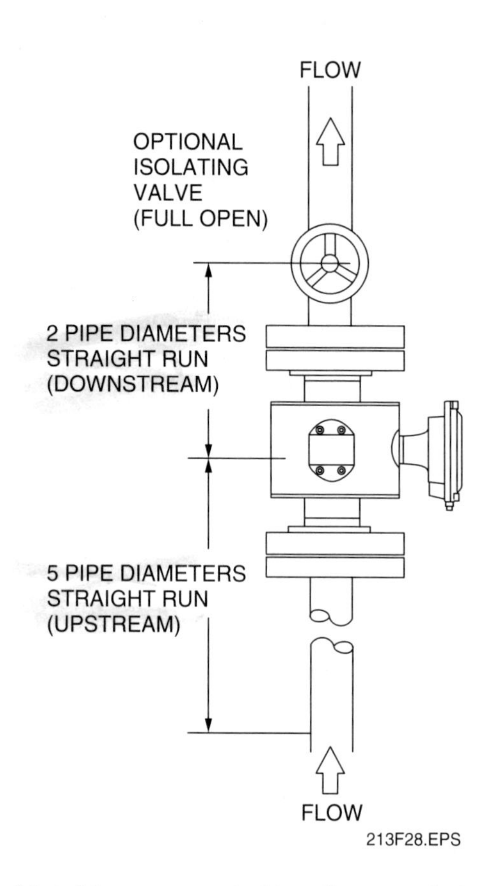

Figure 28 ◆ Magmeter vertical installation with the flow going upward.

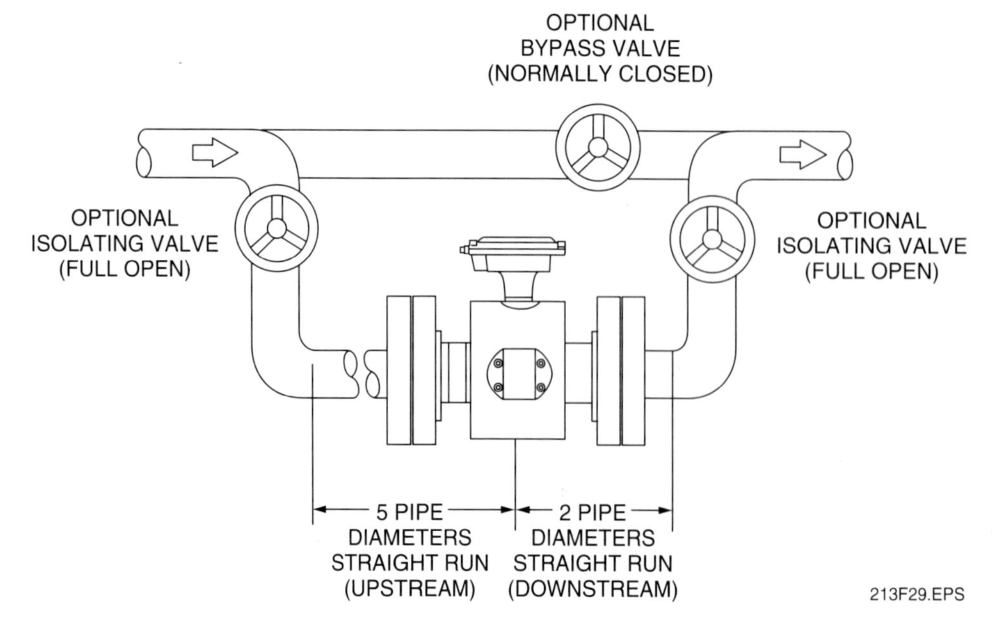

Figure 29 ◆ Magmeter horizontal piping installation at a lower elevation.

To prevent this, most manufacturers offer an empty tube zero feature. This feature requires a contact input to the magmeter transmitter (provided by the user) to indicate when an empty tube exists. This contact can come from a flow switch or pump relay. When such a contact is activated, the outputs of the transmitter will be locked at zero. They will remain locked until the contact is opened.

Some transmitters also have automatic empty-tube electronic circuitry that senses when the electrodes are uncovered or covered by the liquid. The transmitter automatically locks or releases the output as it senses the electrodes being uncovered or covered. Such transmitters will indicate zero flow when the tube is completely empty or less than 50% full of liquid.

As with most flowmeters, piping supports should be installed on each side of the meter to prevent damage due to vibration and to support the weight of the tube. Flow tubes should not be installed in long horizontal runs of 30' or more without proper pipe supports.

The magnetic flowmeter functions poorly if it is not properly grounded. Proper grounding involves bonding the flow tube to the process liquid, meaning that the meter body must be in mechanical contact with the process at both its ends. Transient electrical currents are common in magmeter installations. Metal pipelines form excellent conductors for these transient currents. After the magmeter is installed, it becomes part of the pipeline path for any transient currents traveling down the pipeline or through the process. These currents can create a zero shift in the magnetic flowmeter output.

Bonding provides a bypass electrical path in which stray currents are routed around the flow tube instead of through it. Bonding can be achieved by installing metal grounding straps from one of the meter inlet flange bolts, around the meter housing, to one of the meter outlet flange bolts.

In nonconductive piping systems or within metal piping lined with a non-conductive material, grounding rings must be used to install a magnetic flow tube. The grounding ring is simply a paddle-type orifice plate with a bore equal to the nominal diameter of the flow tube. A grounding ring is installed between the flow tube and adjacent process piping on the upstream and downstream sides. The purpose of grounding rings is to make contact with the process fluid. The flow tube is then bonded to the process fluid via its connections with the grounding rings.

Grounding rings can become expensive accessories for very large flow tubes, or in instances when exotic materials are required for chemical compatibility. Some manufacturers offer a grounding electrode option for such applications. A grounding electrode is a third electrode placed in the flow tube for bonding with a process fluid. Grounding electrodes have limitations because they provide a much smaller area of contact with the fluid than do pipe flanges and grounding rings. They are not satisfactory to provide adequate grounding in all instances. *Figure 27* shows an option for a third electrode, which would be the grounding electrode.

3.2.4 Ultrasonic Flowmeters

Ultrasonic flowmeters measure the velocity of sound as it passes though the process fluid flowing in a pipe. Crystals are used as sound transmitters to send acoustic signals through the piping and across the flow to receivers that are also crystals. The most common type of crystal used is piezoelectric. This process is illustrated simply in *Figure 30*.

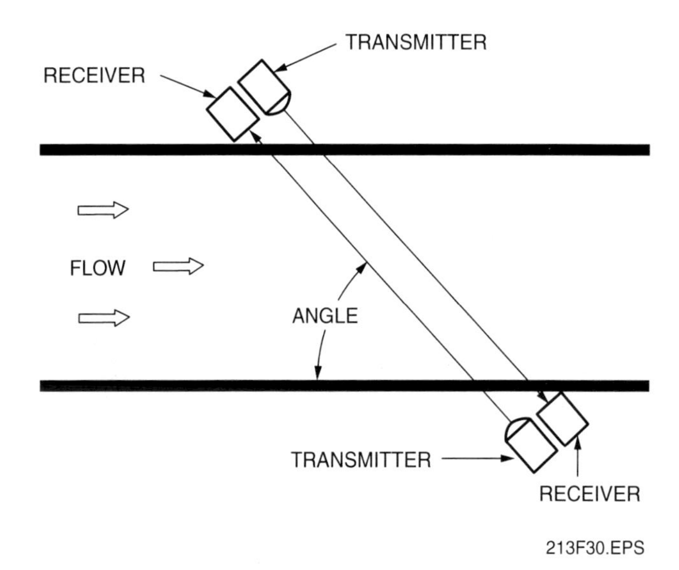

Figure 30 ◆ Ultrasonic flowmeter.

Ultrasonic flowmeters are normally installed on the outside of liquid-filled pipes. This is so the measuring element is nonintrusive and will not induce a pressure drop or disturbance in the process stream. Even though they cost more than other types of inline flowmeters, they can be easily attached to the outside of existing piping without shutting down the process or using additional piping sections or bypass valves.

Some models of ultrasonic flowmeters permit the transducers (transmitters and receivers) to be directly clamped onto the pipe. Model variations may have all the transducers mounted on the same side of the pipe and reflect the signal off the opposite side of the pipe back to the receiver. Others diagonally mount the transducers on opposite sides of the piping. These variations are referred to as diagonal mode and reflective mode, as shown in *Figure 31*. The transducers are either clamped on with chains or sit in a rail track. They require only minutes to install.

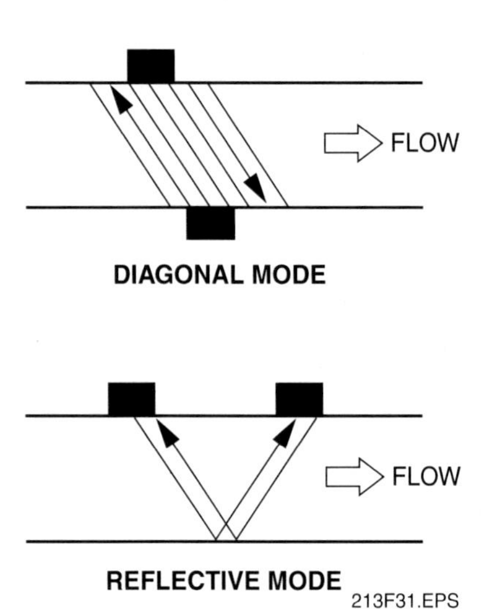

Figure 31 ◆ Diagonal and reflective modes.

3.3.0 Volumetric Flowmeters

Volumetric or positive displacement (PD) flowmeters measure flow by directly dividing a stream into distinct segments of known volume, counting segments, and multiplying by the volume of each segment. Measured over a specific period, the result is a value expressed in units of volume per unit of time. Positive displacement meters frequently report total flow directly on a counter, but they can also generate output pulses, with each pulse representing a discrete volume of fluid. The pulses can be read on a local display counter or transmitted to a control system.

These meters are frequently used in hydrocarbon batching, blending, and custody transfer operations. They are very accurate when used with relatively viscous fluids and have rangeability of up to 10:1. Generally, they are not good for gas flow, due to their inability to seal effectively.

As mechanical devices with many moving parts that are prone to wear, positive displacement meters are not suitable for use with dirty or gritty fluids. They also extract some energy from the fluid stream, producing a pressure loss. Because leakage around gears or vanes can cause erroneous readings, these devices are best suited to viscous fluids, which tend to seal the small clearances. However, if a heavy fluid residue coats the inner chambers of the device, a reduced volume will be delivered from each segment, producing an output higher than actual flow. Some examples of volumetric flowmeters include rotary-vane, oval-gear, and nutating-disc flowmeters.

3.3.1 Rotary-Vane Flowmeters

Rotary-vane flowmeters are widely used in liquid processes where accuracy is crucial. The rotary-vane flowmeter has spring-loaded vanes that are free to slide in and out of a channel in a rotor, thereby making constant contact with the device's eccentric cylinder wall. This type of flowmeter converts the trapped liquid into a rotational velocity that is proportional to the flow through the device. The forces exerted by the flowing fluid rotate blades in the meter device. The blades and body interior of the meter are machined to close tolerances during fabrication because they must form a tight seal with each other over the life of the flowmeter. A fixed volume of liquid is swept to the meter's outlet from each compartment as the impeller rotates. The revolutions of the drum are counted and registered. *Figure 32* illustrates a rotary-vane flowmeter.

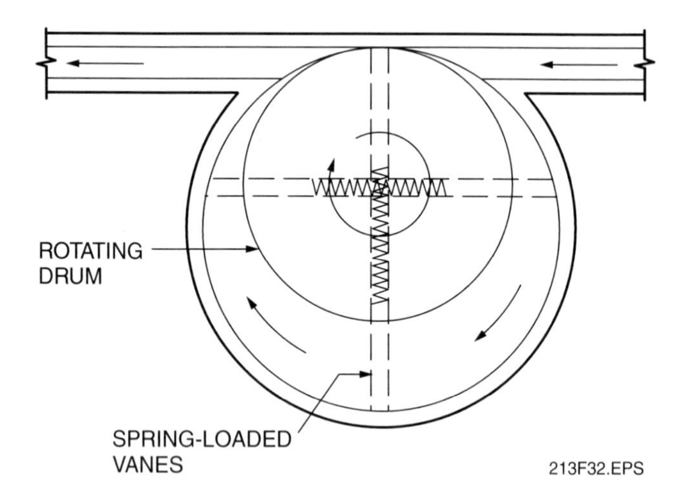

Figure 32 ◆ Rotary-vane flowmeter.

Because a rotary-vane flowmeter is a positive displacement device, it must maintain a full chamber or housing in order for the measurement to be accurate. Unlike a positive displacement pump that must only have a pump housing and a static (not moving) supply on the suction or inlet side, the rotary-vane flowmeter depends on the pressure of the flow to drive the rotating drum. This operates the vanes. Therefore, a full-flowing process stream is required.

Most rotary-vane flowmeters are made with flat-face flanges on both the inlet and outlet sides of the meter. Unlike those flowmeters that require critical upstream and downstream conditioning piping to avoid disturbances, the rotary-vane flowmeter does not require conditioning piping. However, for any flow measurement to be accurate, piping sizes must be maintained. The flanges should be aligned so as not to cause any interference in the flow.

3.3.2 Oval-Gear Flowmeters

Oval-gear flowmeters *(Figure 33)* are typically used on very viscous liquids, which are difficult to measure using flowmeters that are not positive displacement flowmeters. Oval-gear flowmeters use toothed rotors to sweep a known calculated volume of liquid passing through the measurement housing during each rotation. Magnets embedded within the rotors trigger a reed switch that is remotely mounted on a sealed plate that is isolated from the flow stream. Each contact closure of the reed switch represents one unit of measurement of liquid volume passing through the meter. The process liquid doesn't pass directly between the oval gears themselves. Instead, it is swept out in the pocket created by the oval shape of the gear and the inner chamber wall. The precisely meshed gear teeth prevent any passage of the flow between them through the center of the chamber.

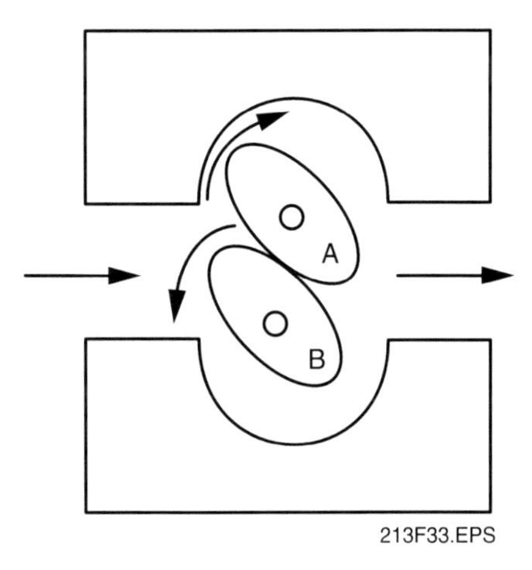

Figure 33 ◆ Oval-gear flowmeter.

Installation of all positive displacement flowmeters is similar because they all require the housing to remain full of process liquid during measurement. The measurement of flow in these positive displacement meters is calculated on the rate of pulses picked up by the magnetic pickup devices. Therefore, air pockets or process voids cannot be distinguished from normal rotational flow and may cause a degree of error in the measurement. Vertical installations, and the addition of check valves, are two methods often used to ensure a full meter housing and an accurate measurement.

3.3.3 Nutating-Disc Flowmeters

Nutating-disc flowmeters are often installed in water service processes as a low-cost flow measurement method where extreme accuracy is not required. As shown in *Figure 34*, the nutating disc uses a cylindrical measurement chamber in which the disc is allowed to wobble, or nutate, as the process fluid flows through the meter. This causes the spindle to rotate. The turning of the spindle creates the output signal of the flowmeter.

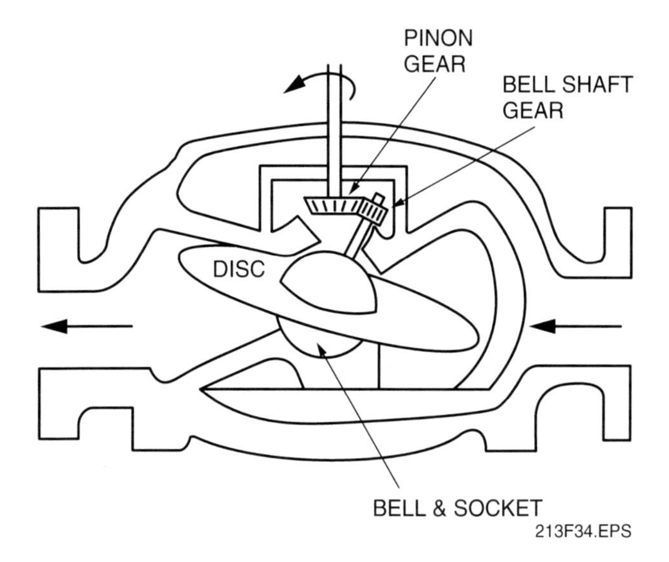

Figure 34 ◆ Nutating-disc flowmeter.

Nutating-disc flowmeters are installed to matching flanges in a piping system. These meters must follow the same basic guidelines that regulate the other types of positive displacement meters. That is, the meter body must be kept full, with a constant flow free of entrapped air in order for the measurement to be accurate. This is because the measurement is based on the rotational action of the spindle and not the amount of fluid in the system. One rotation of the disc is equivalent to some predetermined flow measurement, as determined by the manufacturer of the flowmeter. This information is provided in the technical literature supplied with the flowmeter. Unless this information is known, it is virtually impossible to relate spindle rotation to any measurement of flow.

3.4.0 Mass Flowmeters

Mass flow measurement is often a more accurate and meaningful measure of material transfer than volumetric measurement. For example, a pipeful of steam can condense to a significantly smaller volume of water, and a gallon of hot liquid may measure considerably less when cooled to room temperature. However, the mass remains constant in both examples. For this reason, mass measurement is preferred for critical applications such as batch mixing, billing, and material balances.

Until the more recent development of mass flowmeters, mass flow measurement of fluids was performed using volumetric flowmeters and other instruments interfaced with flow computers. The flow computer would calculate mass rate from the volumetric rate measured by the meter, along with a measurement/calculation of fluid density. Density would be measured directly by a densitometer or calculated in the flow computer using measurements for process temperature and pressure. Such interdependent volumetric systems have performance limitations:

- The relationship between the density of the fluid and the measured process conditions must be precisely known.
- More than one instrument is required. Each instrument adds its own inaccuracy to the overall measurement.
- The system responds slowly to step changes in flow due to the linkages between various components of the system.

Mass flow measurement of slurries was formerly handled using a volumetric meter and densitometer combination for continuous measurement. Slurries could also be measured batchwise, using a scale or load cells. Mass slurry measurement from tanks can be done by highly precise tank level measurement systems. Such a system must use pressure and differential pressure measurements to determine the mass of the fluid in the tank and the density of the fluid, respectively. Static pressure measurement is also required for closed tanks. These meters all have moving parts and complex mechanical designs, sometimes motor driven, to impart momentum to the fluid and measure the resulting forces. Their complex mechanics and high maintenance costs were influential in the development and popularity of mass flowmeters, such as the Coriolis mass flowmeter.

3.4.1 Coriolis Mass Flowmeters

Gustave Gaspara Coriolis was a French engineer who first described an apparent deflecting force in nature affecting tidal and weather systems. The Coriolis flowmeter was named after him because it operates on the principle of deflecting forces. These are generated whenever a particle in a rotating body moves relative to the body in a direction toward or away from the center of rotation. Measurement of the Coriolis force exerted by a flowing fluid on a rotating tube can provide a measure of mass flow rate. An example of a Coriolis flowmeter is shown in *Figure 35*.

The Coriolis flowmeter, like the magmeter, consists of a coupled flow tube and transmitter. The transmitter is generally remotely mounted from the flow tube, but designs with the transmitter integrally mounted to the flow tube are available.

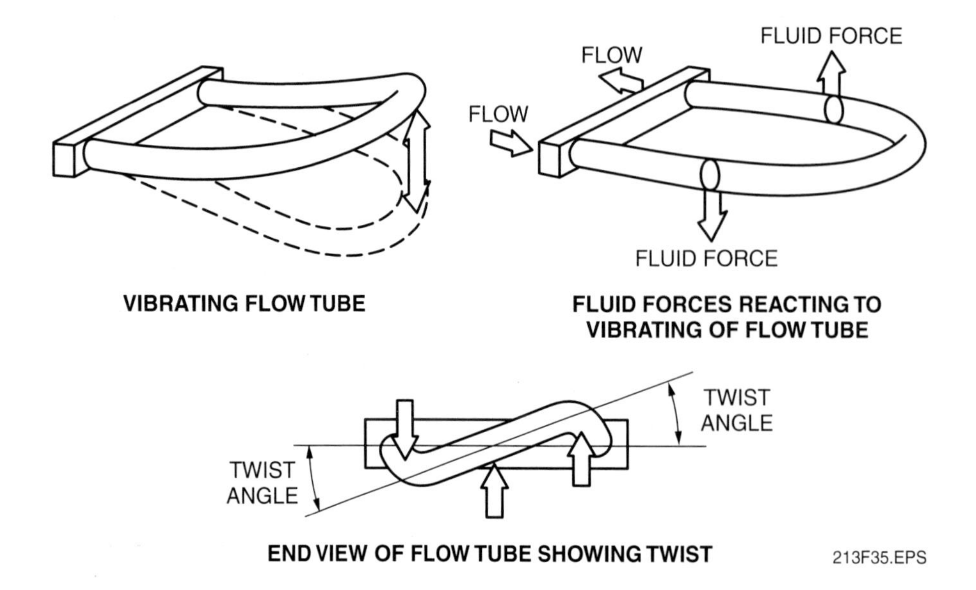

Figure 35 ◆ Coriolis flowmeter tube.

In most designs, the tube is anchored at two points and vibrated between the anchors. The tube can be curved or straight. When flow is present, Coriolis forces produce a secondary twisting vibration that results in a small phase difference in the relative motions at the sensing points. Deflection of the tubes due to Coriolis forces only exists when axial motion and tube vibrations are present. Vibration at zero flow, or flow with no tube vibration, does not produce output from the meter.

Coriolis flowmeters can measure flow through the tube in either the forward or reverse direction. If the design contains two parallel tubes, flow is divided into two separate streams by a simple splitter located near the meter inlet. After flowing through the tubes, the streams converge again near the meter exit. If the design contains a single continuous tube (or two tubes joined in series), the flow is not split inside the meter.

Electrical drivers cause these tubes to vibrate in response to flow. These drivers consist of a coil connected to one tube and a magnet connected to the other. The transmitter applies alternating current to the coil, which causes the magnet to be repelled and attracted, forcing the tubes away from and towards each other. The same technology is commonly used for sensors, although the flow sensor can be any type that will produce an output based on the position, velocity, or acceleration of the tubes.

The meter should be installed so that it will remain full of liquid and so air cannot be trapped inside the tubes. Generally, the best installation will be in a vertical pipe with the flow going upward. However, the meter can be installed in a horizontal line if the application permits. Always try to avoid a vertical flow downward. It may be more cost effective to prevent the entrance of air into the liquid than to eliminate it after the fact.

Strainers, filters, air and vapor eliminators, and other protective devices can be installed upstream of the meter to remove secondary phases of trapped particles that could cause premature wear or measurement error. An air eliminator decreases the velocity of the liquid, allowing time for entrained air to separate and rise through the liquid phase as free air. The rise and fall of the liquid level in the eliminator, due to the accumulation of free air, either closes or opens a valve that discharges the air. Consult the manufacturer before mounting two Coriolis meters near each other, because this type of installation can cause signal interference between the two units. Potential measurement errors may also occur.

Normal vibration in the process piping should not affect the performance of the Coriolis meter, especially in newer designs. The meter should be installed so that it is supported by the process piping and never in such a way that the Coriolis meter supports the piping. There should be no special supports or pads in contact with the flow tube. However, standard piping supports should be located on either side of the meter to support its weight. Read the installation instructions supplied by the manufacturer to determine the vibration sensitivity of the design. If the instructions appear complex and special hardware or supports are required for installation, the design is most likely sensitive to vibration.

3.5.0 Variable-Area Flowmeters (Rotameters)

The rotameter is a variable-area flowmeter. It consists of a tapered metering tube and a float that is free to move up and down within the tube. The metering tube is mounted vertically with the small end at the bottom. The fluid to be measured enters at the bottom of the tube, passes upward around the float, and flows out at the top. *Figure 36* is a representation of a rotameter.

When there is no flow through the rotameter, the float rests at the bottom of the metering tube where the maximum diameter of the float is approximately the same as the bore of the tube. When fluid enters the metering tube, the buoyant effect of the fluid lightens the float, but it has a greater density than the fluid, and the buoyant effect is not sufficient to raise it.

There is a small annular (ring-like) opening between the float and the tube. The pressure drop across the float increases and raises the float to increase the area between the float and tube until the upward hydraulic forces acting on it are balanced by its weight less the buoyant force. The metering float is floating in the fluid stream. The float moves up and down in the tube in proportion to the fluid flow rate and the annular area between the float and the tube. It reaches a stable position in the tube when the forces are in equilibrium.

With upward movement of the float toward the larger end of the tapered tube, the annular opening between the tube and the float increases. As the area increases, the pressure differential across the float decreases. The float will assume a position in dynamic equilibrium when the pressure differential across the float and the buoyancy effect balance the weight of the float. Any further increase in flow rate causes the float to rise higher in the tube. A decrease in flow causes the float to drop to a lower position. Every float position corresponds to one particular flow rate and no other for a fluid of a given density and viscosity. It is only necessary to provide a reading or calibration scale on the tube. Flow rate can be determined by observing the position of the float in the metering tube.

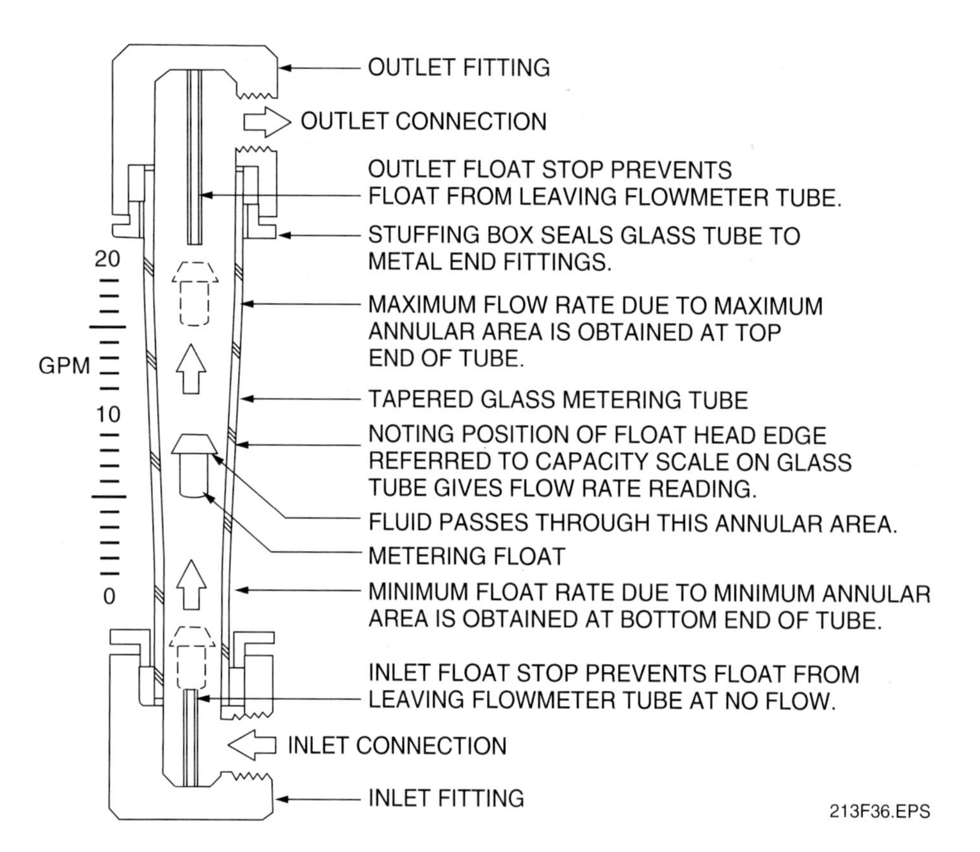

Figure 36 ◆ Rotameter.

Metal metering tubes are used in applications where glass is not satisfactory. In this case the float position must be indirectly determined by either magnetic or electrical techniques. The use of indirect float position sensors also provides functions other than direct visual indication. Rotameters are available that transmit pneumatic, electronic, or time pulse signals, or that provide recording, totalizing, or control functions.

3.5.1 Rotameter Installation

All rotameters must be installed in a vertical position, with the inlet of the flow at the bottom and the outlet at the top. There is a wide variety of rotameters available for all types of environments. Many of the general-purpose rotameters are manufactured with inexpensive plastic housings and tubes. More expensive stainless steel, brass and bronze units are available for those environments that require specific metal components due to corrosiveness or other incompatibilities. Accessories such as inlet filters, quick-connect fittings, and other supporting devices are available for most models and designs. Rotameters are typically installed using threaded fittings, which can be adapted to other types of piping systems.

3.6.0 Density Meters

The density of a substance is one of the important characteristics used to obtain information about its composition, concentration, mass flow, and, in fuels, caloric content. Direct density measurement at operating conditions eliminates the need for separate pressure, temperature, supercompressibility, or humidity measurements when the ultimate purpose is to determine mass flow. All mass flowmeters can be used as densitometers if the volumetric flow rate through them is kept constant.

Liquids are measured in terms of their density. The measurement is often expressed in terms of specific gravity, which is the ratio of the density of the measured substance to the density of water. The compressibility of most liquids is slight, and the effects of pressure on density measurement generally may be disregarded.

The effect of pressure on gas density cannot be disregarded. Because gases are compressible materials, the same detector will seldom be able to detect their specific gravity and their density at operating conditions.

3.6.1 Angular Position Detectors

Refer to *Figure 37*. The process fluid sample flows continuously through the detector at a rate under 30 gph. The gauge chamber contains three displacer floats, each of different density and volume. The solid displacers are spaced 90 to 100 degrees apart and are assembled to a common shaft. Each fluid density sample positions the shaft and displacers at a precise angular position. The displacer moments are a function of float position and buoyant force. With the three displacer moments in balance, the assembly is in equilibrium at all times. The angular position of the assembly is transmitted to the electrical components through magnetic coupling. The output signal to remote readout devices is available in either analog or digital form.

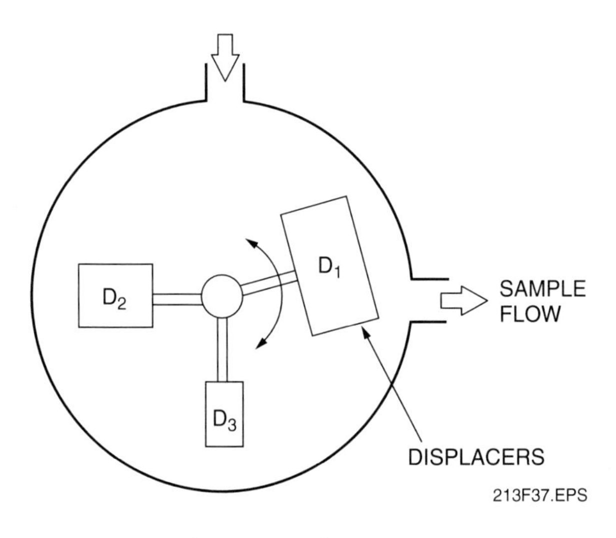

Figure 37 ◆ Angular position density sensor.

3.6.2 Hydrometers

One of the simplest in-line density indicators is illustrated in *Figure 38*. It consists of a transparent glass tee with a hydrometer element. The process fluid sample enters from the bottom and overflows to maintain constant level in the tee. The sample flow rate is maintained at a rating of less than 1 gpm to minimize velocity and turbulence effects. Where the process temperature is allowed to vary, a thermometer is added to the assembly for temperature correction capability.

The hydrometer element can also be mounted inside a rotameter housing, as shown in *Figure 39*.

3.6.3 Sound Velocity Meters

The sound velocity density detector consists of a device for measuring the speed of sound in the liquid while compensating for the effects of temperature and pressure.

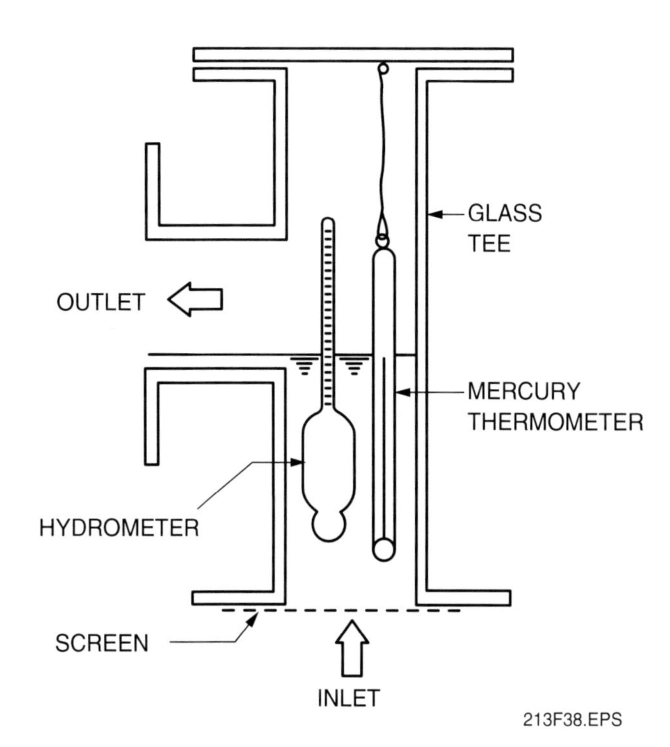

Figure 38 ◆ In-line hydrometer indicator.

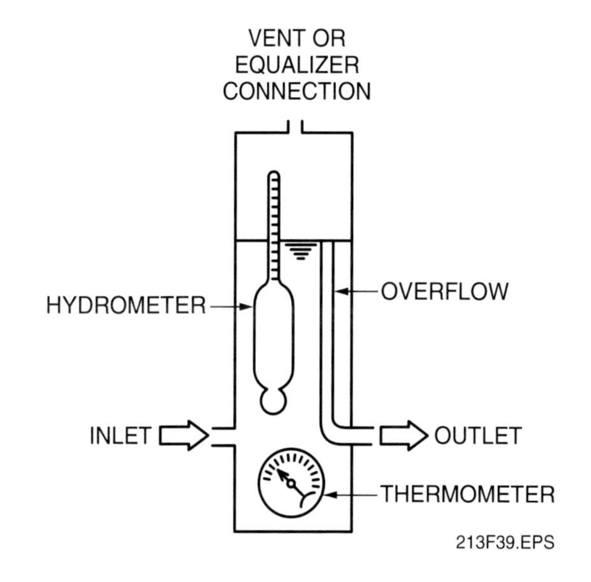

Figure 39 ◆ In-line hydrometer in rotameter housing.

Sound velocity meters (*Figure 40*) have several limitations. The liquid must be clear. Emulsions, dispersions, and slurries scatter the sound waves and cannot be monitored. Similarly, undissolved gases and bubbles cause errors and occasionally render instruments inoperable. Because the velocity of sound is a function of bulk modulus and density, variations in bulk modulus will cause significant measurement errors. The instruments are generally used in pipelines to detect the interface between two different hydrocarbon products.

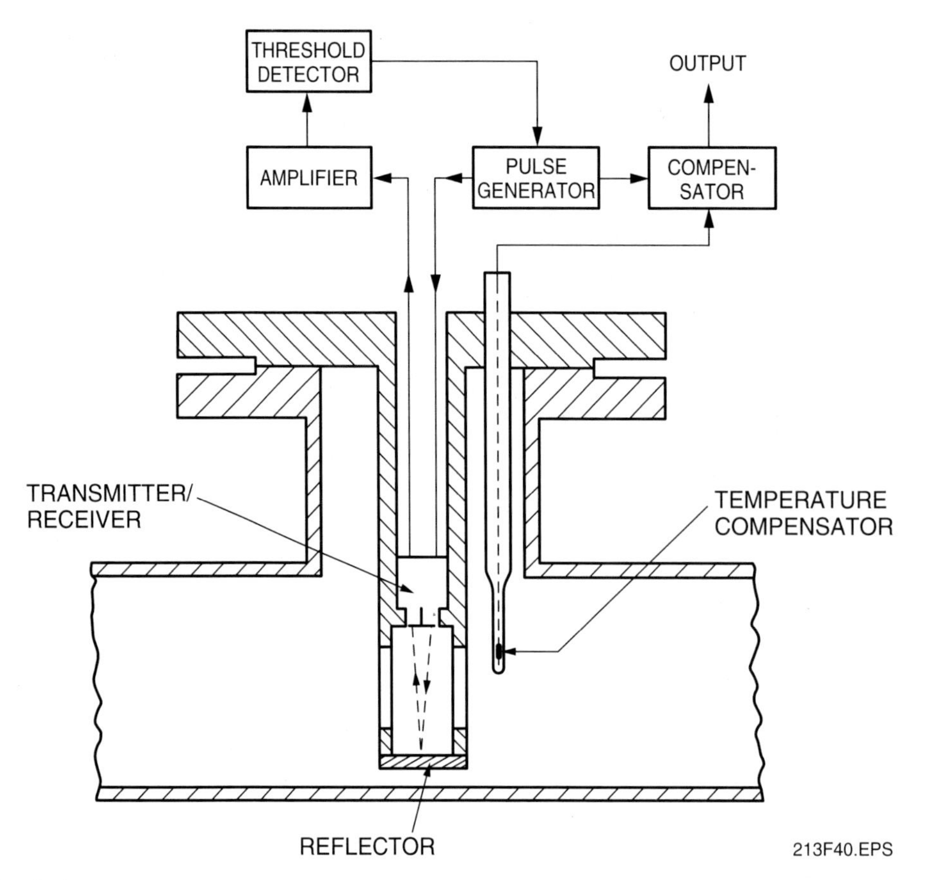

Figure 40 ◆ Sound velocity meter.

3.6.4 Vibrating Plate Detectors

The vibrating plate detector consists of a flexible rectangular plate fixed at both ends within a tube through which the liquid is flowing. A distributed mass of fluid of defined volume interacts with the plate, causing it to vibrate. The system is permitted to oscillate at its resonant frequency. A simple measurement of the period of oscillation is made and related to the density of fluid substance in which it is immersed. This type of detector is shown in *Figure 41*.

3.6.5 Vibrating U-Tube Detectors

If a body is excited into mechanical vibration by a pulsating drive, the amplitude of its vibration will be proportional to its mass. This concept is used in the detector illustrated in *Figure 42*. The total mass of the U-tube assembly is affected by the process fluid density. A pulsating current through the drive coil brings the U-tube into mechanical vibration. An increase in process density increases the effective mass of the U-tube and decreases the corresponding vibration amplitude. An armature and coil arrangement is provided to detect the vibration at the pickup end. The armature vibrates together with the U-tube and induces an AC voltage proportional to fluid density in the pickup coil. The AC voltage is then converted into DC millivolts, which are more compatible with remote recorders and controllers.

For installations in which the process temperature is expected to vary, an automatic temperature compensating circuit can be added. This circuit consists of a resistance-type temperature element in the process stream and solid state circuitry in the converter to perform the required temperature correction.

4.0.0 ◆ VESSEL-MOUNTED INSTRUMENTS

Instrumentation mounted on vessels usually consists of level sensors or analyzers, because the flow of process is not a factor in reservoir-type containment measurements. Of these two process variables, level measurement is the most common installation found on vessels.

4.1.0 Probe-Type Level Instruments

Level-measuring instrumentation can provide the most challenging installation procedures. No matter what continuous reading technology is chosen, chances are that an installer will have to climb to the top of a tank or bin to install the hardware and hook up the electronics. Working in an

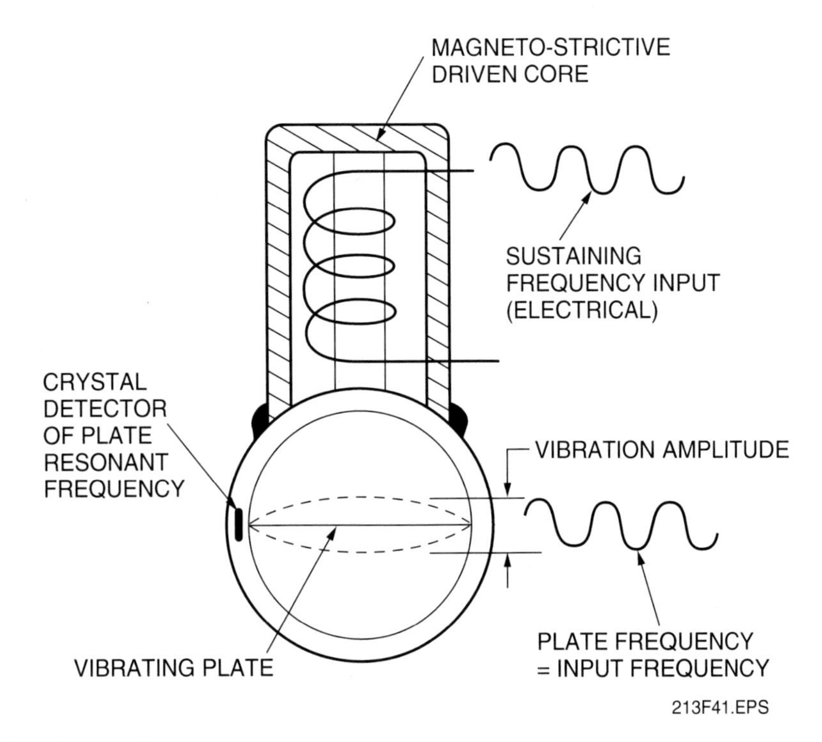

Figure 41 ◆ Vibrating plate density detector.

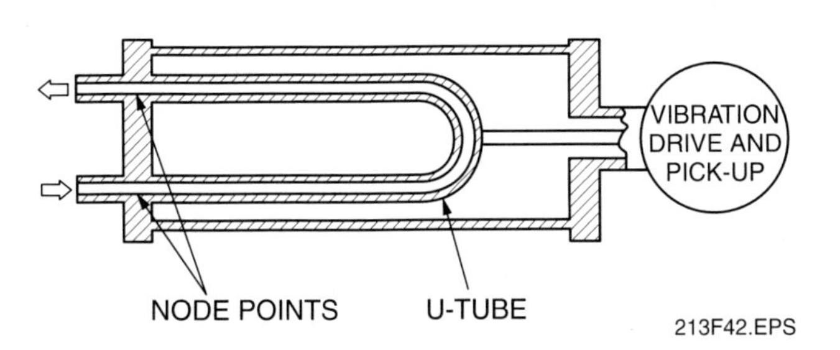

Figure 42 ◆ Vibrating U-tube density detector.

uncomfortable environment may tempt installers to take shortcuts or be less careful with installation procedures. However, when installing instruments in structures such as tanks and vessels, both local and national codes must be followed for signal cabling, flange installations, and vessel penetrations. Various types of flanges and their installation are covered later in this module.

Continuous-reading technologies such as radio frequency (RF) and ultrasonic require careful placement of the sensing elements. They should always be mounted perpendicular to the surface of the liquid being monitored, out of the way of the fill line. Although the exact amounts vary with manufacturer and technology, the distance from the vessel side must be maintained to ensure accurate readings. In continuous RF applications, vessel grounding may be necessary.

Level instrumentation probes must not be allowed to come into contact with vessel openings, or false readings will occur. Displacement of the probe can occur, especially during bulk-solid product discharge.

4.1.1 Capacitance Probe (RF Probe)

A capacitor consists of two conductors (plates) that are electrically isolated from one another by a nonconductor (dielectric). When the two conductors are at different potentials (voltages), the system is capable of storing an electric charge. The storage capability of a capacitor is measured in farads. This type of level measuring system is often referred to as an *RF* system or an admittance level sensor.

The capacitor plates have an area and are separated by a gap. This gap is filled with a nonconducting material (dielectric) measured in a dielectric constant. The dielectric constant of a substance is proportional to its conductance. The lower the dielectric constant, the less conductive the substance is. Capacitance is calculated as:

$$C = KA \div D$$

Where:

A = area of capacitor plates

D = distance between the capacitor plates

K = dielectric constant

C = capacitance

If the area of the plates and the distance between the plates of a capacitor remain constant, capacitance can only vary as a function of the dielectric constant of the substance filling the gap between the plates. If a change in level causes a change in the total quantity of dielectric in the capacitance system, because the lower part of the area is exposed to a conductive liquid while the upper part is in contact with a vapor, the capacitance measurement will be proportional to level. *Figure 43* shows how capacitance probes may be applied in either a conductive liquid or a nonconductive liquid by using or eliminating an insulated probe.

During installation, the capacitance probe should be mounted so that its operation is unaffected by incoming or outgoing material flow. Material impacts can cause false readings or damage to the probe and insulator. When measuring low-dielectric materials, it is important that the entire probe be covered, not just the tip. When rod or cable extensions are used, allow for 8" to 12" of active probe coverage.

Install the probe so that it does not contact the vessel wall or any structural elements of the vessel. If a cable extension is used, allow for swinging of the cable as the material level in the vessel rises, so that the plumb bob on the end of the cable does not touch the vessel wall. The probe should not be mounted where material can form a bridge between the active probe and the vessel wall. In addition, the probe should not be mounted at an upward angle to avoid material buildup.

If more than one capacitance level sensor is mounted in the vessel, a minimum distance of 18" should be provided between the probes. At closer distances, the electromagnetic fields might interfere with each other. If a capacitance probe is installed through the side wall of a vessel and the weight of the process material acting on the probe is sometimes excessive, a protective baffle should be installed above the sensor.

Typical insertion lengths of standard capacitance probes range from 7" to 16". These probes are typically side mounted. Vertical probes can be extended by solid rods up to a length of 4' to 5'. A steel cable with a weight can be used to suspend the probe up to 50'. Most capacitance level sensors have ¾" to 1½" NPT mounting connectors. The matching female coupling is usually welded to the vessel wall, and the capacitance probe is screwed into the mating connector. Low-profile capacitance sensors also are available and are flange mounted. Installing flanges is covered later in this module.

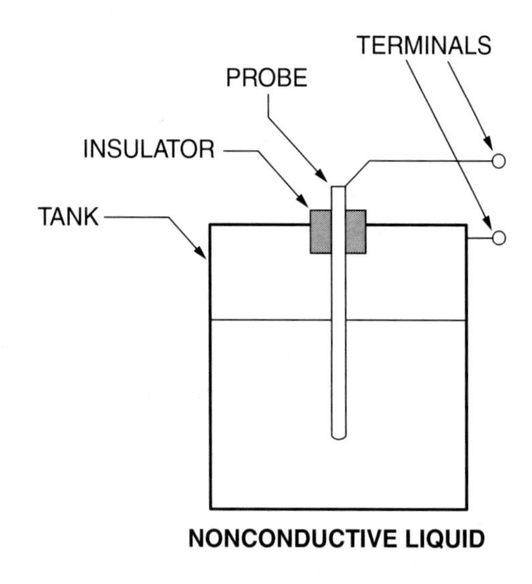

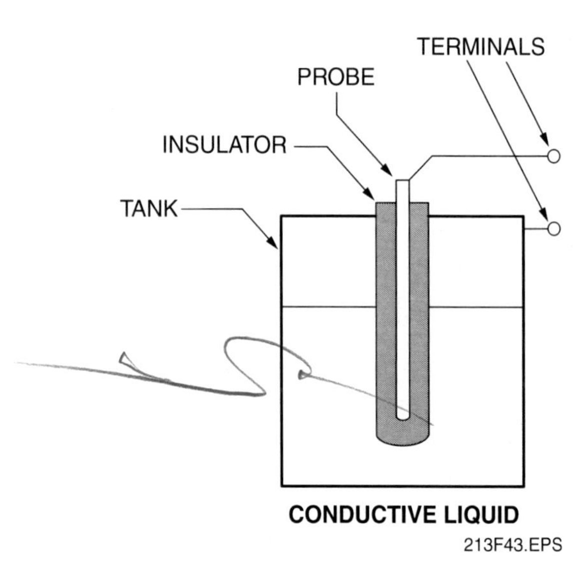

Figure 43 ◆ Use of capacitance probes.

4.1.2 pH Probes

Accurate measurement and control of pH is a common requirement in industrial plants. Applications include monitoring of cooling tower water, boiler feedwater, process steam, waste treatment, and plant **effluent**. Control of pH depends on measurement reliability.

Basically, pH control involves supplying the proper amount of reagent to bring the pH to the desired value. Control can be performed on either a batch or continuous basis.

Batch control is normally used when the total volume of the solution to be treated is relatively low, such as in waste treatment processes where liquids can be collected efficiently and treated in tanks. The amount of reagent required for neutralization can be determined from a titration curve of tank volume and reaction time. Slow reagent addition rates and good stirring permit more accurate control with less likelihood of the pH overshooting.

In the batch process shown in *Figure 44*, the tank inlet and reagent feed point are shown located away from the pH electrodes and effluent discharge pipe. This separation is needed to ensure proper mixing of reagent before measurements are made. Immersed electrodes near the outlet ensure rapid sensor response.

Continuous control is similar to batch control except that there is a continuous flow of **influent** and treated effluent (*Figure 45*).

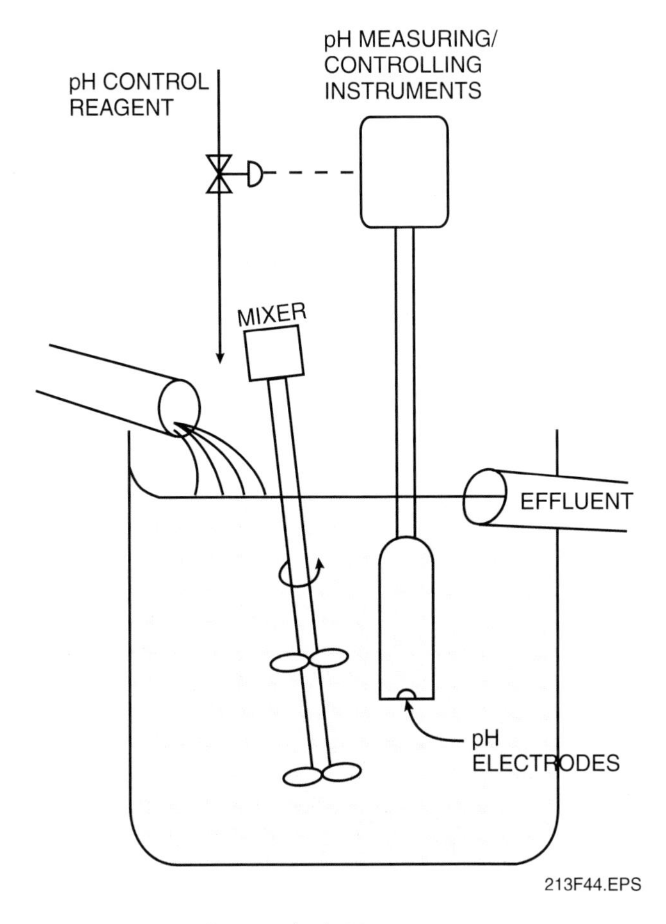

Figure 44 ◆ Batch control of pH.

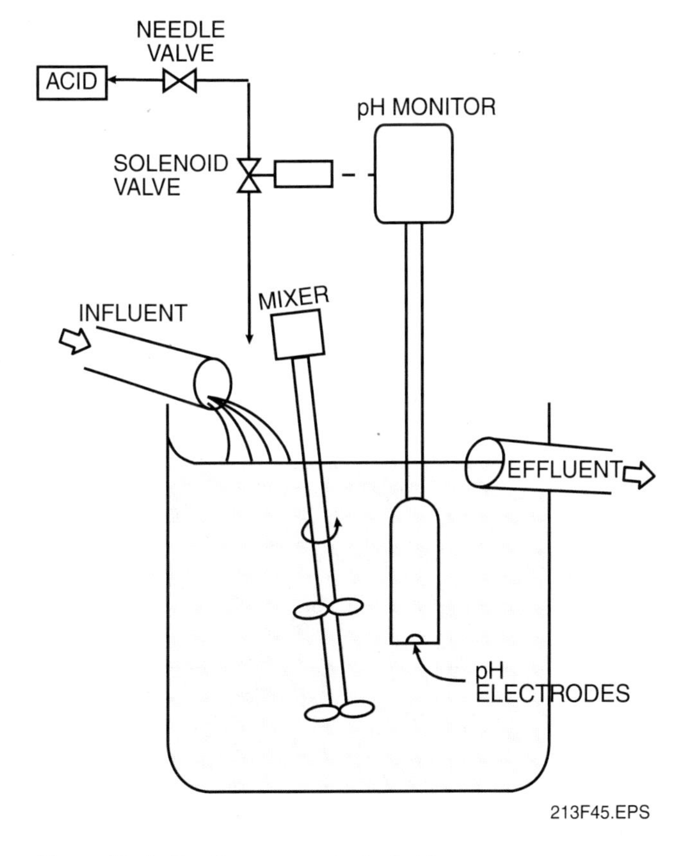

Figure 45 ◆ Continuous control of pH.

An improperly placed electrode assembly can cause excessive **deadtime** for control action and result in cyclic control and wasted reagent. Ideal deadtimes range from 5 to 30 seconds. Excessive deadtime often can be avoided by locating the electrode assembly close to where the reagent is added.

Vertical mounting of electrodes is preferred. They should always be exposed to a representative sample of the process solution. The entire assembly should remain wet at all times to keep the electrodes from drying out.

Because pH is a high-impedance measurement, it is best to use a controller, preamplifier, or signal conditioner as close to the electrode as possible. The preamplifier converts a high-impedance signal to a low-impedance signal-making it less susceptible to noise and signal loss on transmission back to the receiving instrument over unshielded wire.

A pH probe is always installed vertically. In any other position, an air bubble may prevent solution from entering the tiny pinhole at the end of the probe. Therefore, improper sensing can occur.

It is critical to prevent moisture at the cable connections to the probe and the controller. The place most susceptible to such moisture is the probe. Moisture could short out the probe. The probe holder, for immersion applications, is designed to mate to a female ¾" NPT fitting. Thread sealant tape should always be applied to the threads. A damp-proof connector should be used for the cable-to-controller connection, where the environment of the controller is such that spray or mist may be present in the air.

The location of probes in solution can assist in the success of a pH system. The probe is always placed downstream of the chemical injection point. The solution being measured should have proper mixing in order for proper pH levels to be sensed and controlled. The probe, however, should not be near the turbulent mixing point. Air bubbles affect the proper sensing of the solution and cause the readings from the controller to wander back and forth. However, it would be incorrect to locate the probe in an area with little to no circulation, as it would not yield a representative reading. A depth of between 1' and 2' from the bottom of the tank is suggested for immersion applications.

Cleaning of the probe is very important. Probes are coated even in very clean solutions. Each application will determine how frequently the probe must be cleaned. To have an effective pH control system, probe cleaning must be part of a preventive maintenance schedule. Some applications may need cleaning only once a month, while others may need it once a week. The pH of the solution should be monitored to verify that the probe controlling the solution is still working correctly. Litmus paper or a portable, battery-operated pH meter can verify the main probe's accuracy. The level of accuracy required by the system determines which method to use.

Design the plumbing for easy access to the probe. Inaccessible objects are forgotten too often.

Cable should be kept as short as possible, preferably no longer than 30'. Beyond this, test to see if signal strength is sufficient. For long runs, a preamplifier may be needed to boost the signal.

Proximity to mechanical devices emitting AC power and closeness to fluorescent lighting are other factors that can affect the signal passing through the cable, causing the pH readout to be erratic. If the cable is to run parallel to AC power, place it in a conduit at least 6" away from the AC power line. If crossing the AC power line, do so at 90 degrees. If the cable has to run close to a fluorescent light, a conduit might be required. A triaxial cable can be used as an alternative to the standard cable to provide increased shielding for the signal.

Cables can become damaged. Treat them accordingly. One indicator of a broken cable is that the pH readout drifts, even after the controller has been checked.

4.2.0 Displacer-Type Level Instruments

Archimedes was an Italian mathematician and scientist, born about 287 B.C.E. Archimedes' Principle states that a body immersed in a liquid will be buoyed up (forced up in the liquid) by a force equal to the weight of the liquid it displaces. This upward pressure acting on the area of the displacer creates the force called buoyancy.

A float will sink into liquid until a volume of liquid weighing as much as the float is displaced. When the specific gravity of the liquid and the cross-sectional area of the float remain constant, the float rises and falls with the level. Therefore, the float will assume a constant relative position with the level, and its position is a direct indication of level. The amount of liquid displaced by variable displacers depends on how deeply the device is submerged in the liquid. With variable displacement devices, the amount of displacement varies with the level of the liquid.

The span of the displacer is the distance at which the displacer responds to the forces of buoyancy. Buoyant force depends on the amount and density of the liquid displaced. It is important to note the relationship of specific gravity to the change in weight of the displacer as the level changes. Displacers used in liquids with lower specific gravity will not change weight as dramatically as those used in liquids with higher specific gravity. This is why displacer level measuring systems cannot be used in applications where they could be immersed in liquids of varying specific gravities.

A displacer must be connected to a measuring mechanism. When sensing the changes in buoyant force, this mechanism converts the force into an indication of level. A displacer body can be suspended directly in a tank or installed in a float chamber on the outside of the vessel.

4.2.1 Chamber-Installed Displacers

A displacer chamber is recommended if the liquid level can be disturbed by surges. Shutoff valves or blocking valves are also recommended so that the transmitter can be removed without having to shut down the process or drop the level in the vessel. A purge assembly is also recommended on the chamber so that it may be flushed out on a regular basis. *Figure 46* illustrates various chamber displacement installations.

5.0.0 ◆ STRAP-MOUNTED INSTRUMENTS

Strap-mounted instruments are strapped or similarly attached to the shell of a vessel, operating equipment housing, or piping. They are used to monitor some physical characteristic or factor such as flow, vibration, temperature, radiation, or expansion.

Examples of strap-mounted instruments include ultrasonic flowmeter transducers, vibration detectors, heat sensors, heat tracing instruments, and some types of load cells.

5.1.0 Types of Supports

A typical strap-installed instrument is supplied with its own strap-mounting installation kits because strap mounting is the only way it may be installed in order to function properly. However, other instruments that monitor or control process functions such as flow, level, pressure, and temperature may have their transmitters or other components of their system conveniently strapped to piping or other structures in order to achieve ideal location in relation to the process they are monitoring or controlling. These types of installations are accomplished by using universal strap-mounting kits available in a variety of configurations. This section of the module will discuss some of the strap-mounting kits that come with instruments, as well as some of the universal strap-mounting kits available for a variety of instruments.

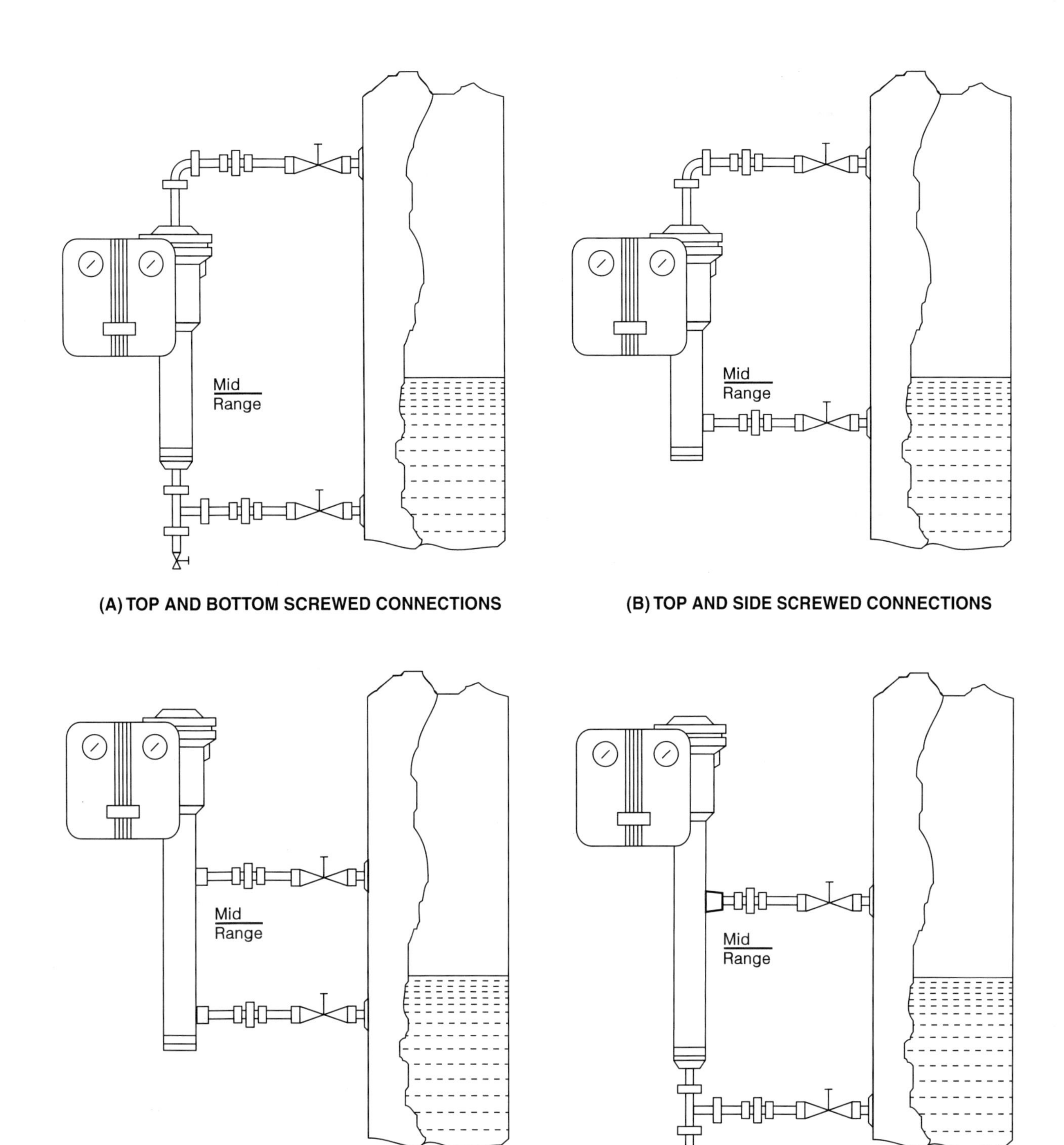

Figure 46 ◆ Chamber-installed displacers.

5.1.1 Cable-Mounted Instrument Supports

Support modules are used to mount instruments. A support module allows one or many instruments to be mounted locally. This allows an instrument to be mounted close to its sensing point, eliminating inaccuracies caused by line interference. This also allows the instrument to be positioned in a location that facilitates easy accessibility and maintenance.

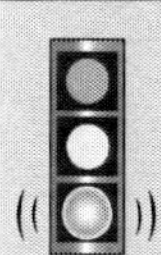

NOTE

The size and metallurgy of the mounting structure must be considered.

Cable-mounted assemblies usually come with a base and a cable assembly (*Figure 47*). The base as shown is usually made of ¼" steel plate. The plate is formed into a channel to provide gripping edges that prevent movement of the base when attached to the process line. Brackets are usually welded to the base to provide a connection to the cable assembly.

The cable assembly shown in *Figure 47* consists of two cables. The ones shown in the figure are 5⁄16" in diameter. This high tensile strength cable, with each steel strand individually galvanized, is available in various lengths depending on the size of the pipe to which the cable mount will be attached. One end of the cable is plain and terminates on the base bracket in a wire rope clip with a grooved seat. The other end of the cable has a ⅝" NC threaded connection (see *Figure 48*), which is also terminated on the base bracket by a ⅝" hex nut. The nut is loaded by a set of pre-calibrated compression washers to allow for fine adjustments.

Bases are available in many different configurations. One configuration is shown in *Figure 49*. This base has a threaded coupling for a 2" NPT pipe extension for direct-mounting of pipe supported on vertical or horizontal lines.

Another type of base is available with a square-leg extension. This is shown in *Figure 50*. This 2" square-leg extension is for mounting on a vertical or horizontal process line. The square leg eliminates rotation on the extension while providing horizontal adjustment. This support is ideal for auxiliary equipment such as junction boxes and air sets.

The right-angle elbow extension (*Figure 51*) is another common type of base. This extension has a 2" round pipe with a vertical leg. It is designed for mounting instruments on horizontal process lines. This mount has adjustable mounting capabilities in any position and on any plane on horizontal process lines.

The last type of base extension examined here is the right-angle extension for a vertical leg. It has a 2" round pipe and is designed for mounting on vertical process lines.

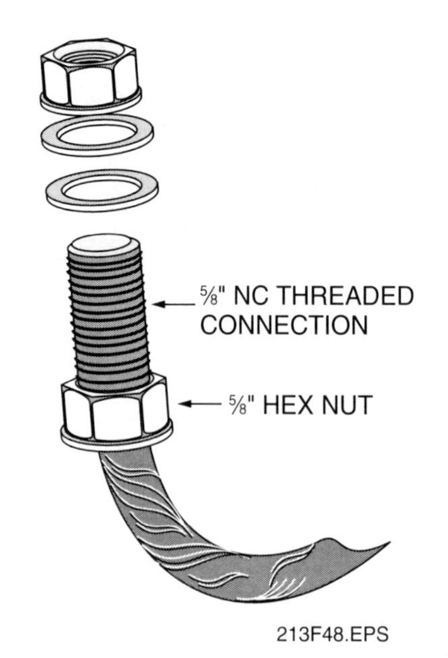

Figure 48 ◆ Cable assembly.

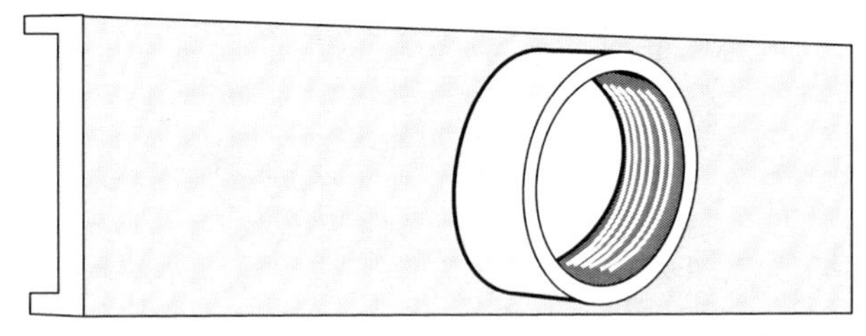

Figure 49 ◆ Base with 2" NPT coupling for direct-mounting of support.

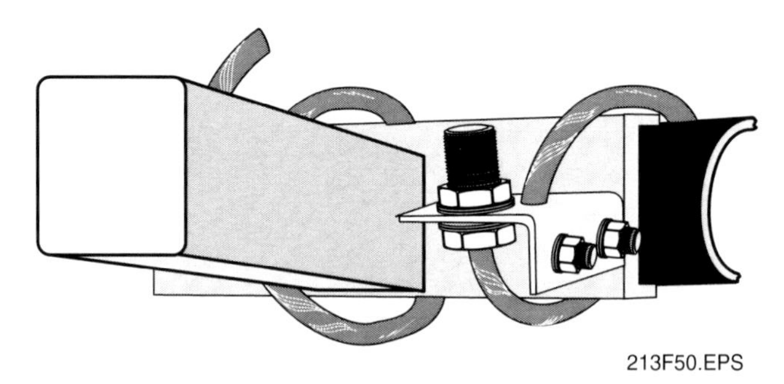

Figure 50 ◆ Square-leg extension.

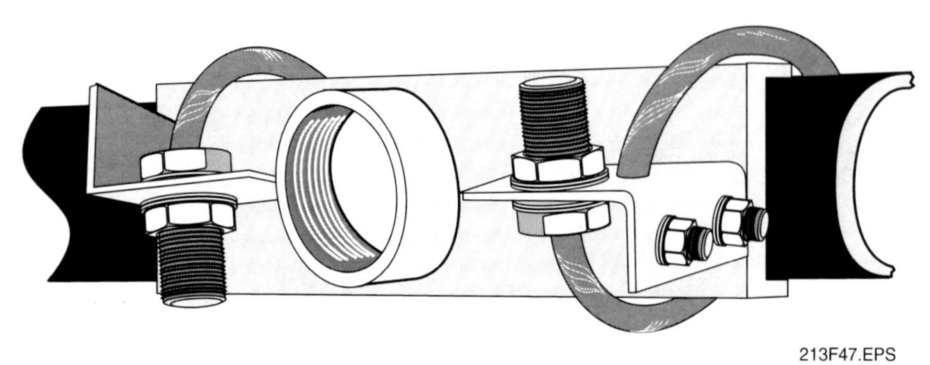

Figure 47 ◆ Cable-mounted base assembly.

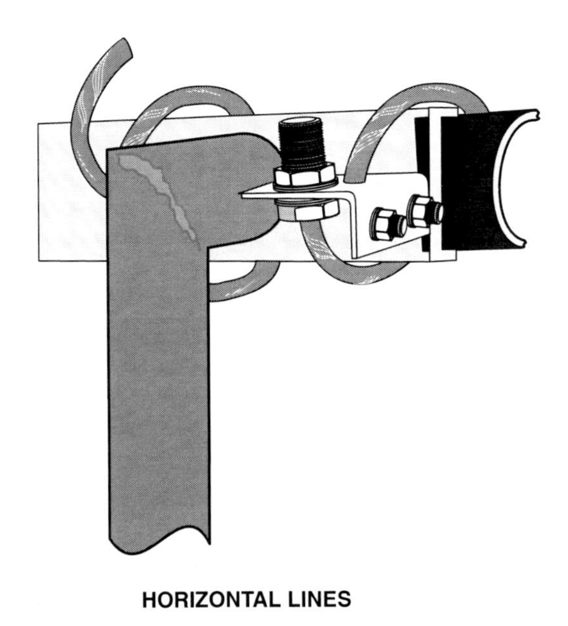

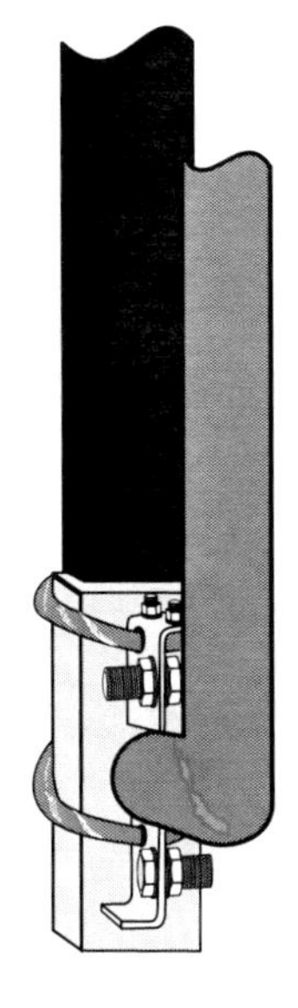

Figure 51 ◆ Right-angle elbow extensions.

Before installing a cable mount to a line, ensure that the line can withstand the weight of the instruments. Also ensure that the mount is capable of holding the instruments.

Most U-bolt mounts are interchangeable with cable mounts, as shown in *Figure 52*. Other cable-mount configurations are illustrated in *Figure 53*.

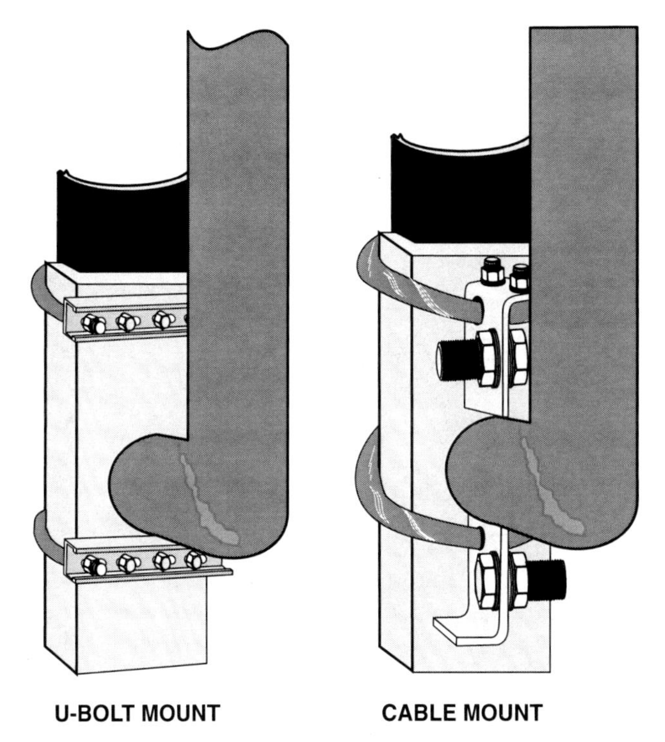

Figure 52 ◆ U-bolt/cable interchangeability.

5.2.0 Thermostats and Heat Tracing

Thermostats sense temperature and make necessary adjustments to maintain temperature within a specified band. Examples of thermostats are temperature controllers that maintain constant room temperatures and controllers used to maintain temperatures in heat trace systems.

5.2.1 Thermostatically Controlled Tracing

Most process control applications use line-sensing control in which the sensor of a thermostat controller is attached directly to a pipe. Line-sensing control uses a mechanical thermostat or electronic controller to measure the actual pipe temperature, not the ambient temperature surrounding the pipe. Line-sensing control is used in narrow temperature maintenance situations because it provides the highest degree of temperature control and the lowest energy use. *Figure 54* shows a line-sensing thermostat typical of those used in hazardous areas to directly control a single heat tracing circuit or as a pilot control of a contactor used to turn on and turn off multiple heat tracing circuits.

Junction boxes, usually secured to the object being traced, provide easy access to any controls or wiring splices. *Figure 55* shows a typical thermostat and sensor installation. The junction box is strapped to the pipe, and the thermostat is connected to the junction box. This junction box could also be attached by a clamp-mounted device.

The **sensing bulb** is strapped to the outside of the process pipe and is used to control a heat tape.

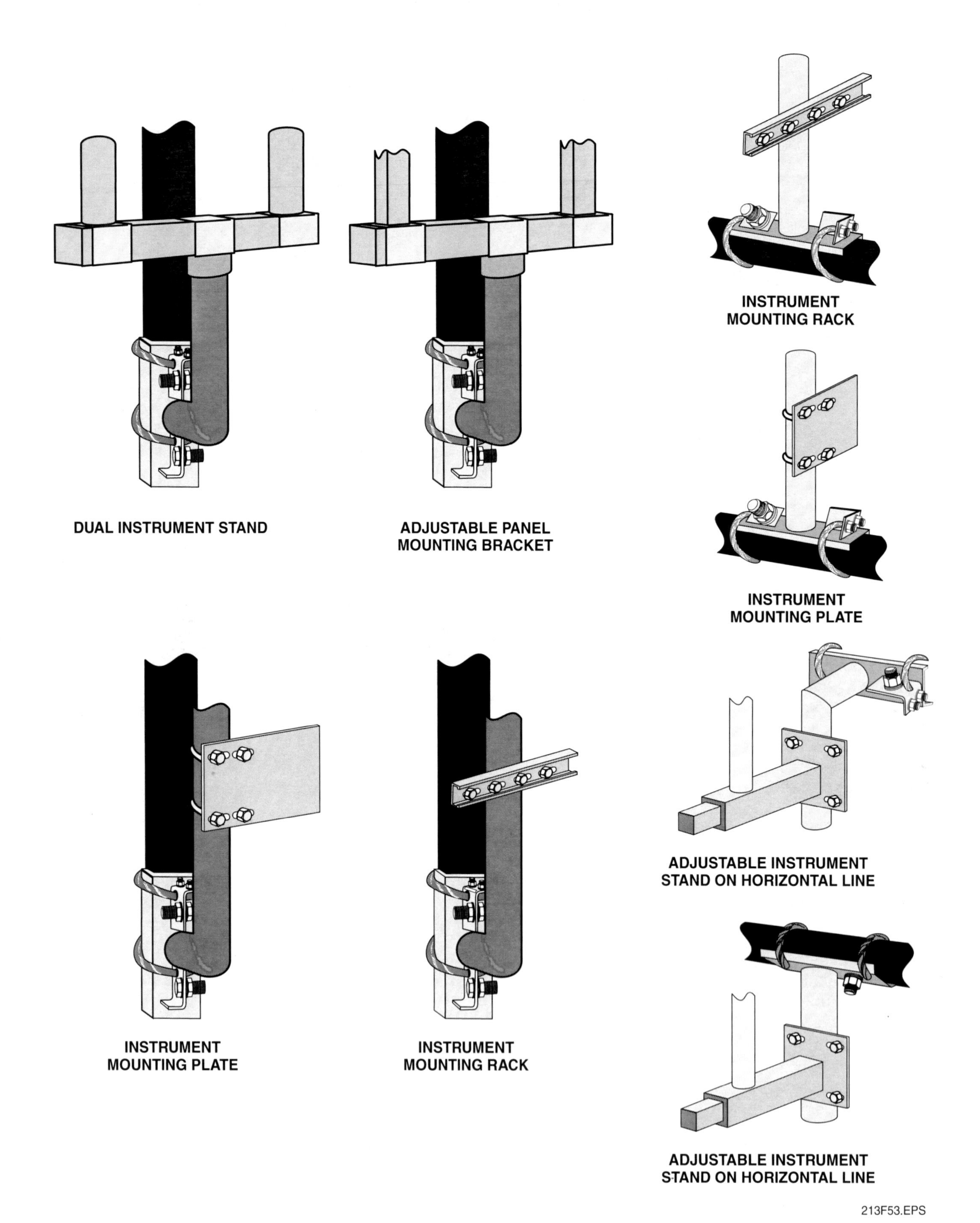

Figure 53 ◆ Cable-mounted configurations.

Figure 54 ◆ Line-sensing thermostat.

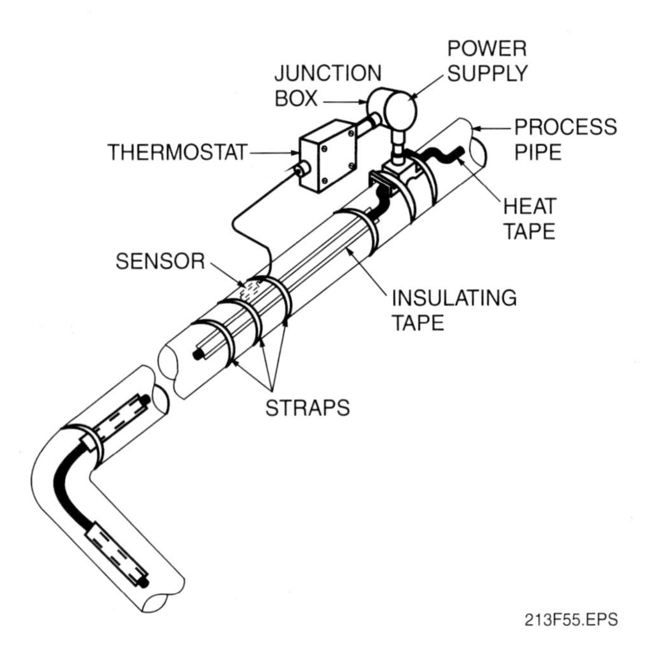

Figure 55 ◆ Thermostat and sensor installation.

As with all sensing bulbs, and particularly for this discussion, the following should be considered when strapping a sensing bulb to a pipe.

The location and mounting of the sensing bulb are very important. It must be in good **thermal** contact with the pipe. Refer to *Figure 56*. In most cases, the bulb should be mounted on the top of the pipe so the liquid in the bulb is close to the outlet of the pipe, as shown in *Figure 56(A)*. If it is necessary to mount the bulb on a vertical line, the capillary tube of the bulb should always enter from the top of the bulb, as shown in *Figure 56(B)*. It should never enter from the bottom.

To keep the bulb from being affected by external ambient temperature, it should be wrapped in insulation, as in *Figure 56(C)*, so that only line temperature affects the bulb. Special insulation forms are available. Plastic tape can also be used.

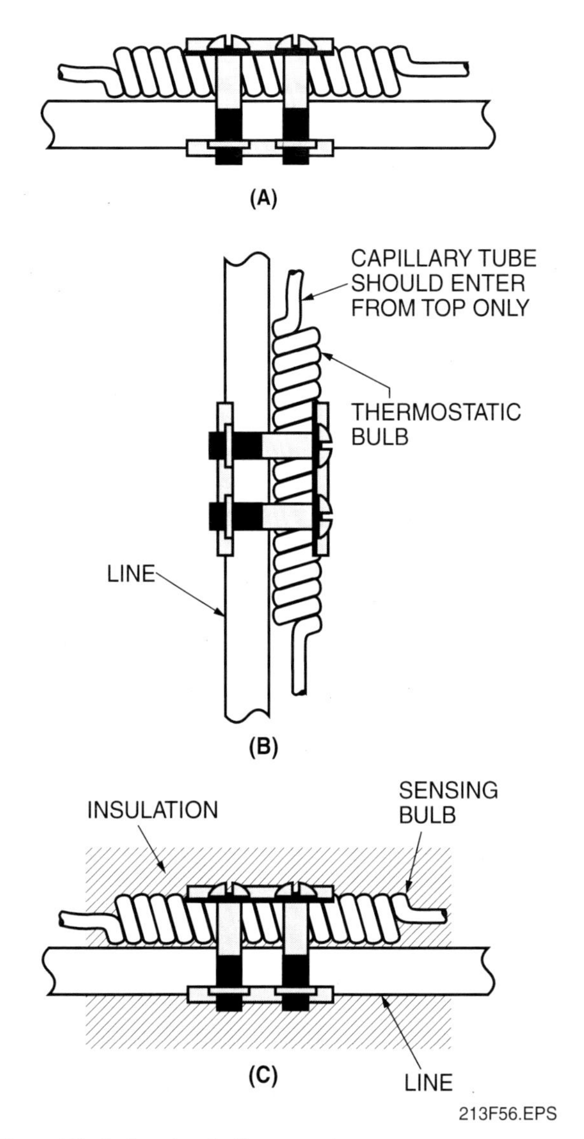

Figure 56 ◆ Sensing bulb mounting.

Copper straps and non-rusting machine screws and nuts should be used to fasten the bulb to the line. The bulb must have excellent thermal contact with the line. The connection must be clean and tight. Both the line and the bulb should be cleaned with a solvent before assembling.

Some lines may carry only a vapor. However, there may be some droplets of liquid as well.

Figure 57 illustrates conditions inside a line carrying primarily vapor, particularly on installations that require a large diameter line. At *Figure 57(A)*, vapor and some droplets of liquid are flowing through a large-diameter suction line. Due to the large diameter, the velocity of the vapor at times is quite slow. The droplets of liquid will settle at the bottom of the line. The line area at *Figure 57(B)* is smaller. As a result, the velocity of the vapor will be higher than in *Figure 57(A)*. This means less separation of the liquid from the flowing vapor. The inside of the tube will be uniformly coated with liquid. The line at

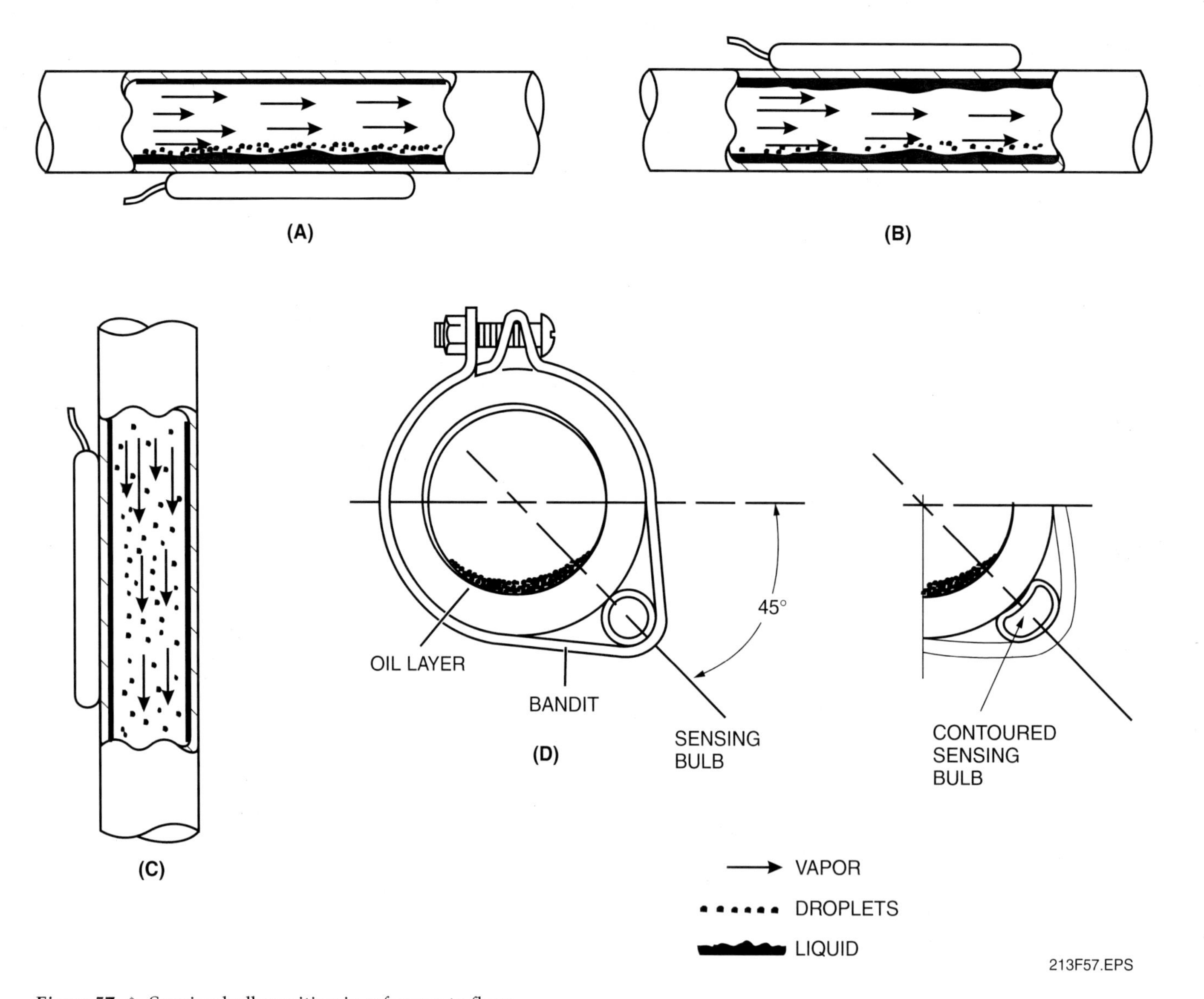

Figure 57 ◆ Sensing bulb position in reference to flow.

Figure 57(C) is shown in the vertical position. In this position, there will be no separation of the liquid from the vapor.

The sensing bulb on large lines should be located near the underside of the line rather than on top. This is better because droplets of liquid may separate from the flowing vapors. The recommended bulb position (on large suction lines) is shown in *Figure 57(D)*. Some sensing bulbs are crimped or creased lengthwise to provide double contact and to help align the sensing bulb with the surface of the suction line.

5.3.0 Strapping

Stainless steel strapping is available for the use of strapping instrumentation. A stainless steel strap, called a bandit, is shown in *Figure 57(D)*. Stainless steel straps are generally used in either ½" or ¾" widths and are 0.010" or of various thicknesses, depending on the size of the instrument to be mounted. Stainless steel bands are secured with matching seals or clips.

Before strapping an instrument to a support, ensure that the installation will meet the manufacturer's requirements, which may be found in the instrument's literature.

5.4.0 Radiation Meters

Radiation detectors can be used to measure levels in a tank and density of material. The basic components of a density gauge are a radioactive source beaming through a pipe and a detector to measure the amount of transmitted radiation (*Figure 58*).

When gamma rays pass through a process fluid, they are absorbed in proportion to the material density. An increase in process density results in a smaller output from the detector because the fluid absorbs more transmitted radiation.

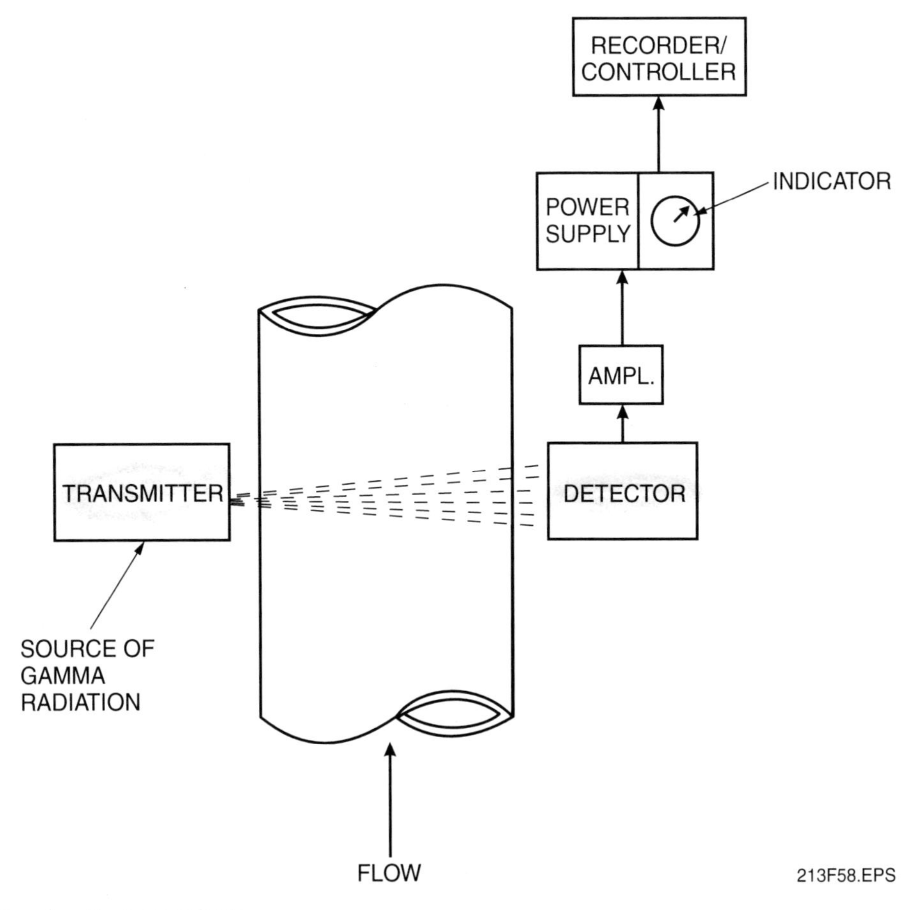

Figure 58 ◆ Radiation density instrument.

When density inside pipes or containers is measured, the radiation source and detector should be mounted as shown in *Figure 59* when the pipe is greater than 6" in diameter.

In smaller-diameter pipe of less than 6", the radiation path is not adequate to provide high accuracy and sensitivity. Therefore, the installation in *Figure 60* should be used to lengthen the radiation path and increase accuracy.

The detectors should be temperature-controlled to maintain accuracy. The process temperature is of no concern, but it is necessary to use thermal insulation between the cell and the process.

The transmitter and the detector can be mounted to the line by clamping with U-bolts or strapping to the line. Be sure to read the mounting requirements in the manufacturer's literature before installing. The density meter installed in a process line is shown in *Figure 61*.

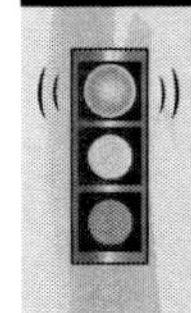

WARNING!

All personnel installing radiation detectors must be properly trained in handling radioactive material. All OSHA standards must be implemented to ensure proper personnel safety.

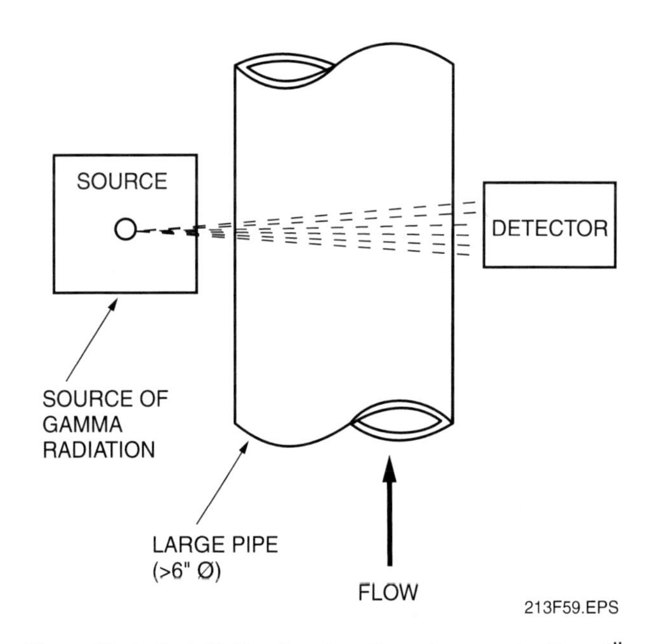

Figure 59 ◆ Installation for pipe diameters greater than 6".

6.0.0 ◆ INSERTION-MOUNTED INSTRUMENTS

Insertion-mounted instrumentation, for the purpose this module, are devices that require penetration into the stream of process flow or into the

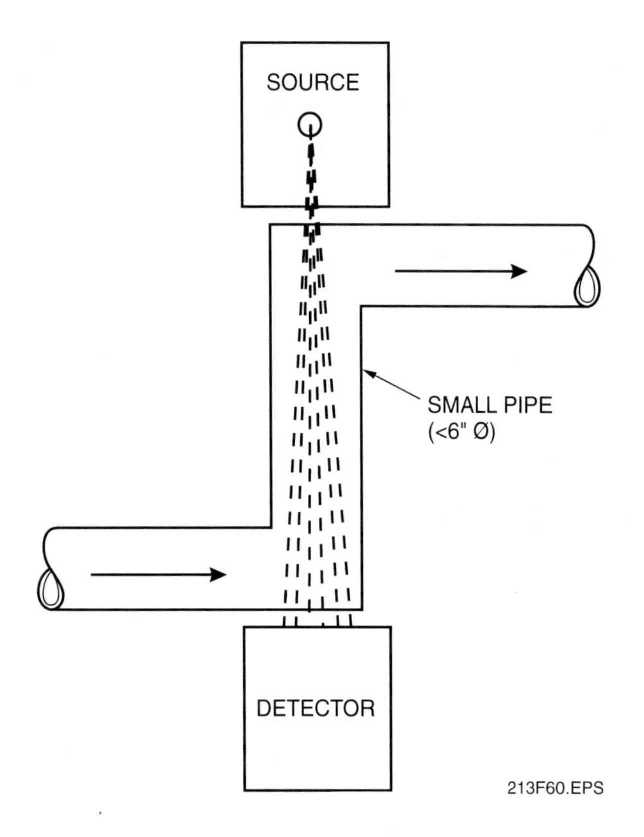

Figure 60 ◆ Installation for pipe diameters smaller than 6".

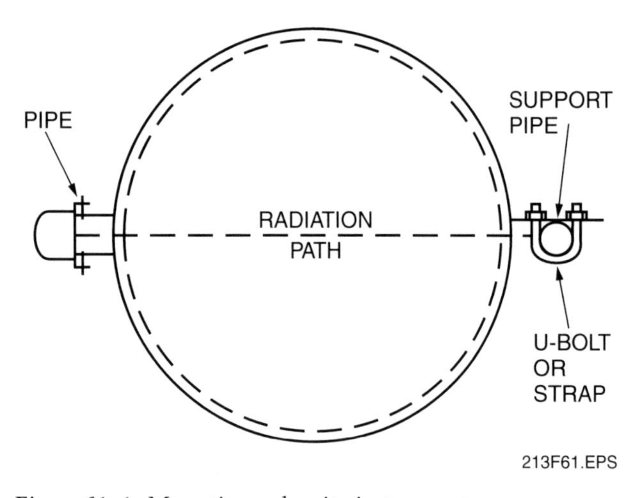

Figure 61 ◆ Mounting a density instrument.

process environment in the case of temperature-sensing in heating units or atmospheric analyzing in gases or vapors. Although capacitance probes may be considered insertion-mounted instruments in some situations, they were covered earlier in this module in the section discussing vessel-mounted instrumentation.

This section focuses only on thermowells, thermowell/thermocouple connector heads, thermocouple extension wire, and resistance temperature detectors (RTDs).

6.1.0 Thermowells

Most thermowell assemblies use female NPT threads for mating to male NPT pipe fittings. Because of normal tolerance and torque variations during assembly, variations as much as ⅝" in element lengths are allowed for adjustment. This will allow for thermostatic expansion of a thermocouple or RTD within the well without damaging the thermocouple, RTD, or thermowell. A standard thermowell is shown in *Figure 62*.

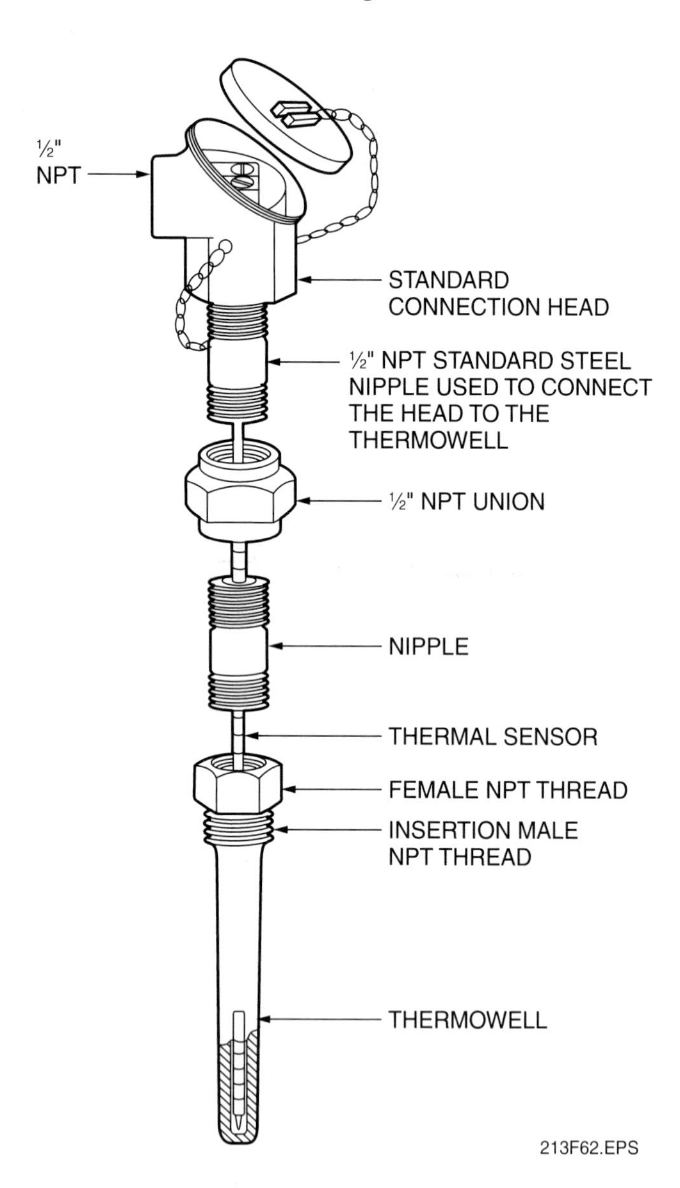

Figure 62 ◆ Standard thermowell assembly.

All metal thermowells have female NPT connection threads, and all connection heads have female NPT threads. Therefore, a nipple with male NPT threads on both ends is needed to attach the connection head to the thermowell. A secondary purpose of the nipple is to physically remove the connection head from a hot thermowell to a safe distance so that the connection head can be safely opened to make necessary replacements or repairs. A union is used to connect two nipples together. This combination allows easy removal of the head and element from the well.

The outlet port of the connection head contains a ½" NPT female thread. This is used to connect to conduit pipe containing the extension leads. Nipples and unions should also be used here to aid in connecting and disconnecting the conduit pipe. They may also be used to add extensions to the conduit pipe where required.

6.1.1 Material

Thermowells are made from drilled bar stock. Standard well materials include brass, carbon steel, and stainless steel. Wells are also available in special grades of stainless steel, chrome-molybdenum alloys, silicon bronze, Hastelloy® B and C, nickel, titanium, Inconel® and Monel®. The choice is usually determined by the corrosive conditions under which the thermowell must operate, consistent with the maximum service pressure and temperature. Before selecting the thermowell material, refer to the chart on recommended materials for various applications and the pressure-temperature ratings for each well type and material.

6.1.2 Accuracy

If all other variables are held constant, accuracy depends on the insertion length U, the distance between the tip of the well and the underside of the threads, and the attachment means. Where space and pressure-temperature ratings permit, accuracy can be improved by increasing the insertion length. The error in temperature measurement of short insertion length thermowells is created by the thermal conduction from the tip of the wall and to the outside atmosphere. This is commonly referred to as the error caused by the stem effect.

Another source of error is the stem effect caused by the thermal sensor itself. Commonly used thermal sensors include thermocouples, RTDs, thermistors, bi-metal thermometers, and liquid-filled thermometers. Generally, the shorter the insertion length of the sensors, the greater the error due to stem effect. This error can be reduced and even eliminated by the proper selection of the thermal sensor. Thermowells with insertion lengths less than 6" in liquids and 9" in gases can produce large errors due to the combined stem effect of the thermowell and the thermal sensor.

6.1.3 Tapered or Straight Shank

Tapered shank wells provide greater stiffness for the same sensitivity. The higher strength-to-weight ratio gives these wells a higher natural frequency than that of equivalent-length straight shank wells. This permits operation at higher fluid velocity.

6.1.4 Velocity Ratings

Well failures, in most cases, are not due to the effects of pressure and temperature. Fluid flowing by the well forms a turbulent wake that has a definite frequency based on the diameter of the well and the velocity of the fluid. It is important that the well have sufficient stiffness so that the wake frequency will never equal the natural frequency of the well itself. If the natural frequency of the well was to coincide with the wake frequency, the well would vibrate to destruction and break off in the piping.

6.1.5 Thermowell Insertion

Thermowells can be inserted into a process by a flanged connection or by standard NPT connections. A general-use standard NPT thermowell is shown in *Figure 63*.

The flange-type thermowell is illustrated in *Figure 64*. This well can come in many types of material, depending on the system requirements.

Typical flange faces that are available are raised-face (ASA serrated), flat-face, and ring-type flange.

6.2.0 Connector Heads

The first type of head used with thermocouples and RTDs is the weatherproof head. This type of head is a rugged cast aluminum or cast iron screw cover unit that is available in various NPT probe connection sizes and a choice of terminal blocks. This head is illustrated in *Figure 65*.

Another type of weatherproof head is shown in *Figure 66*. The miniature weatherproof connecting head is designed for the termination of single or dual thermocouples, RTDs, and strain gauges. Pipe threads are provided for probe mounting and extension wire fittings. Probe-mount thread accommodates ¼" NPT or ⅜" NPT. The cover includes a gasket to seal against moisture and a chain to prevent loss of the cover. Molded of reinforced thermoset compound, the standard head can be used in ambient temperatures to 350°F (175°C), and the Hi-Temp head can be used to 800°F (425°C).

1¾"
U INSERTION LENGTH
THREAD
ALLOWANCE
2½"
O DIA.
½" NPSM
4¼" DIA.
?
P NPT
A STEM LENGTH
¼" FROM
BOTTOM
213F63.EPS

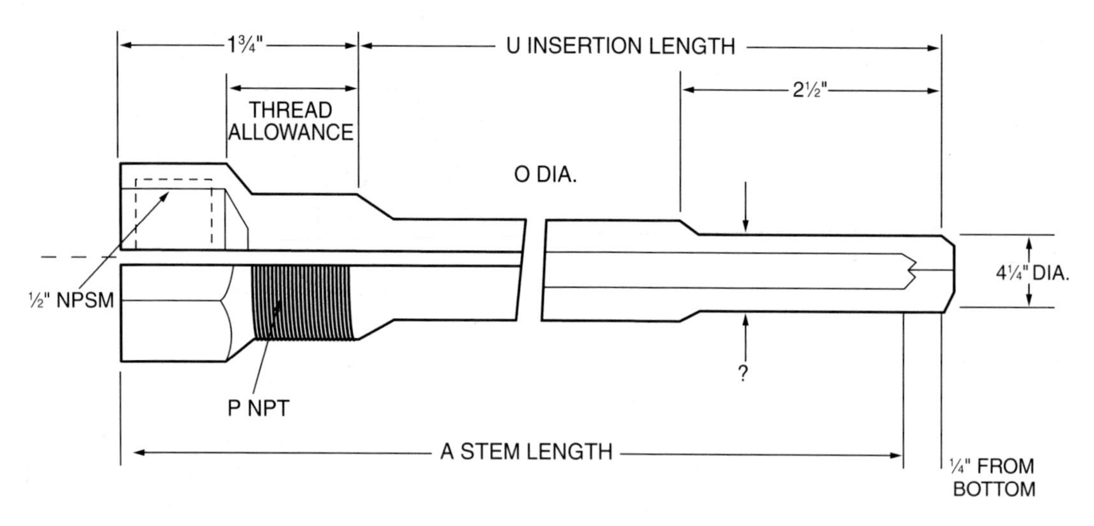

Figure 63 ◆ Standard NPT thermowell.

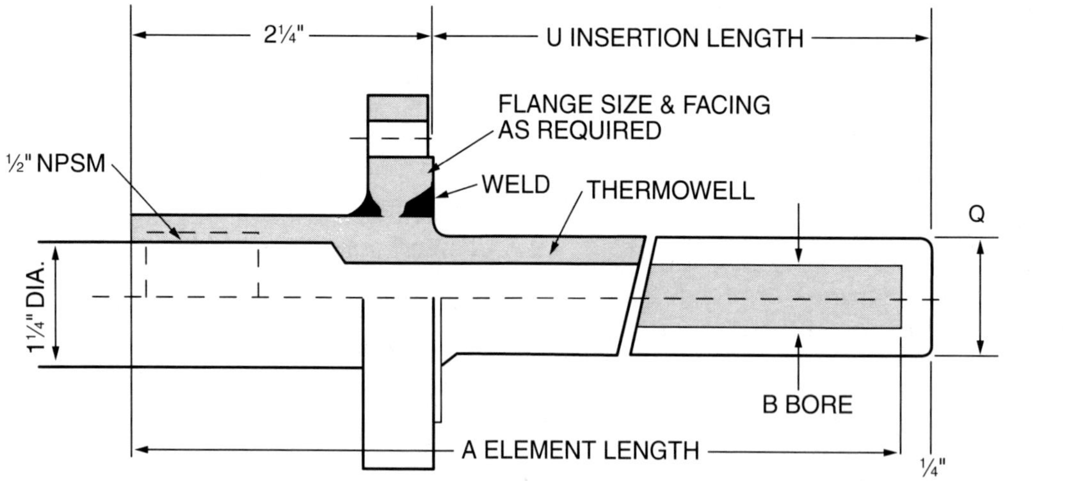

2¼"
U INSERTION LENGTH
FLANGE SIZE & FACING
AS REQUIRED
½" NPSM
WELD
THERMOWELL
Q
1¼" DIA.
B BORE
A ELEMENT LENGTH
¼"
FROM BOTTOM
213F64.EPS

Figure 64 ◆ Flange-type thermowell.

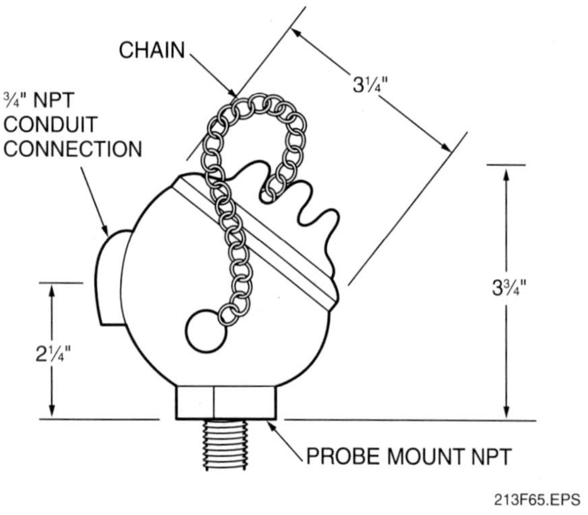

CHAIN
¾" NPT
CONDUIT
CONNECTION
3¼"
3¾"
2¼"
PROBE MOUNT NPT
213F65.EPS

Figure 65 ◆ Weatherproof head.

2⅛"
THERMOSET
PLASTIC
BODY & CAP
1¼"
¼" NPT WIRE
ENTRANCE
CHAIN
PROBE MOUNT THREAD
213F66.EPS

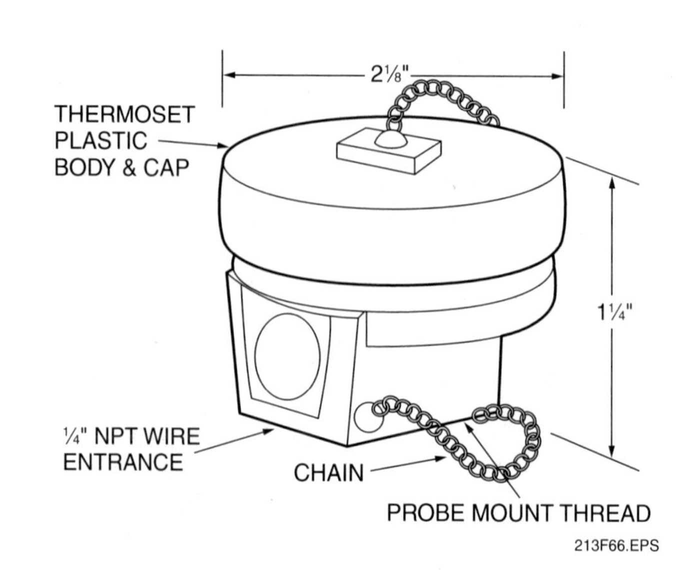

Figure 66 ◆ Miniature weatherproof head.

The general purpose heads are rugged cast aluminum units with stamped steel covers. The covers are available in various NPT probe connection sizes and terminal blocks. The general purpose head is shown in *Figure 67*.

The last of the heads is the explosion-proof head. These heads are cast aluminum, mainly with screw cover units that comply with the *National Electrical Code* (NEC) as shown in *Table 1*. They are designed for use in hazardous environments where extreme caution must be observed. *Figure 68* shows an explosion-proof head.

6.3.0 Installation

Industrial process plant thermowells installed in 4" and larger pipelines are generally mounted at a 90-degree angle to the pipe. In 3" pipelines, they are mounted at a 45-degree angle. Thermowells can also be installed in elbows, usually 3" or larger. *Figure 69* shows two typical installations of thermowells. The dotted lines in each drawing indicate provisions for mounting thermowells in lines smaller than 3". When mounting a thermowell at 90 or 45 degrees to the pipe walls, remember that the tip of the thermowell should be in the middle third of the pipe diameter. The thermowell's tip, when installed in an elbow or at an angle, should face the flow. Note the flow arrows in *Figure 69*.

6.4.0 Thermocouple Extension Wire

When thermocouple leads must be extended, special thermocouple extension wires must be used. Any extension wire not compatible with the materials used in the thermocouple will affect the performance of the thermocouple.

Thermocouple extension wire has approximately the same thermoelectric properties as thermocouple wire, but it is only guaranteed to be accurate within a limited temperature range. Thermocouple extension wire can be used to connect thermocouples to instruments at a lower cost or to provide improvement in the thermo-electrical circuit. Thermocouple extension wire used with base metal thermocouples is generally made from the same alloys. However, due to the high cost of noble metal thermocouple wire, such as platinum-rhodium and platinum combination, the extension wire for these applications is a copper/copper alloy combination. The X in the ANSI code denotes extension grade wire.

6.4.1 Thermocouple Insulation

Thermocouple insulation provides electrical insulation for thermocouple and thermocouple extension wire. If the insulation breaks down for any

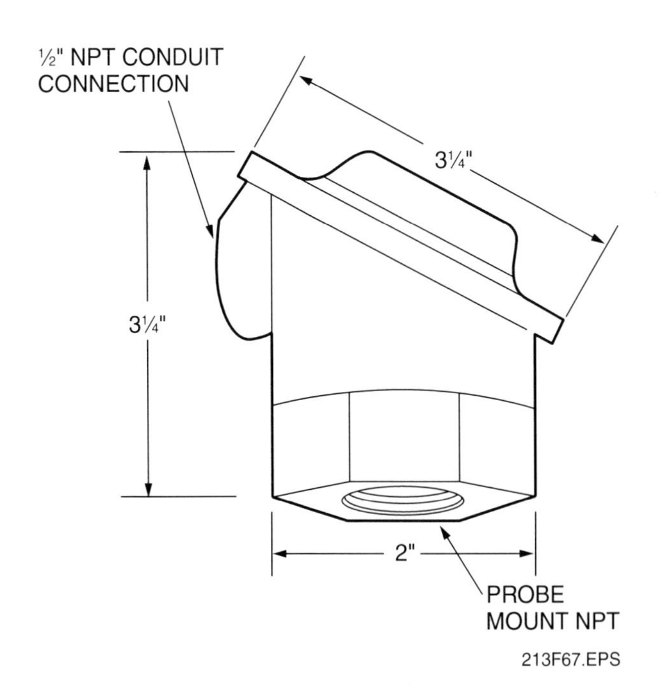

Figure 67 ◆ General purpose head.

Table 1 NEC Requirements for Explosion-Proof Heads

Class	Group	Maximum Temp.
Class I	Group C	180°C (356°F)
	Group D	280°C (536°F)
Class II	Group E	200°C (392°F)
	Group F	200°C (392°F)
	Group G	165°C (329°F)
Class III		Please see *UL Standard 866*

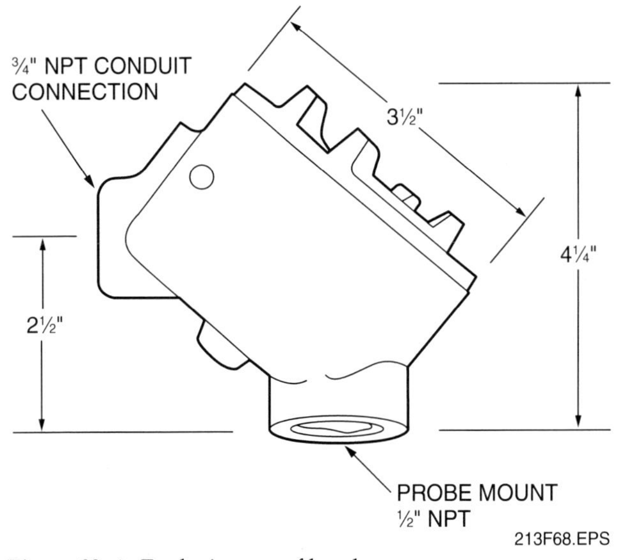

Figure 68 ◆ Explosion-proof head.

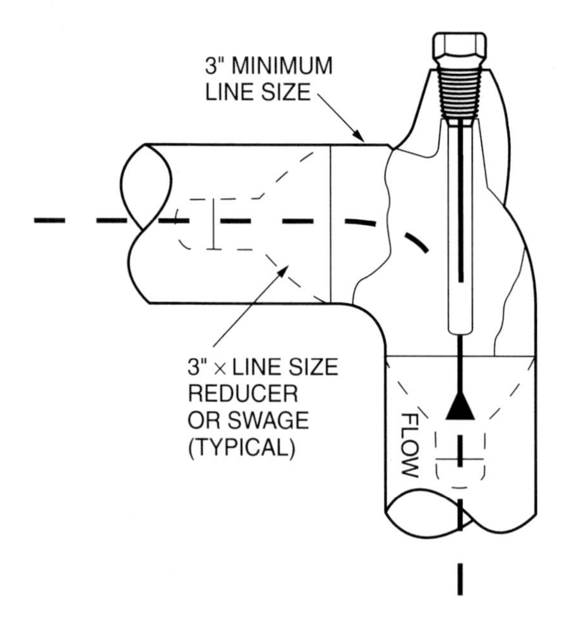

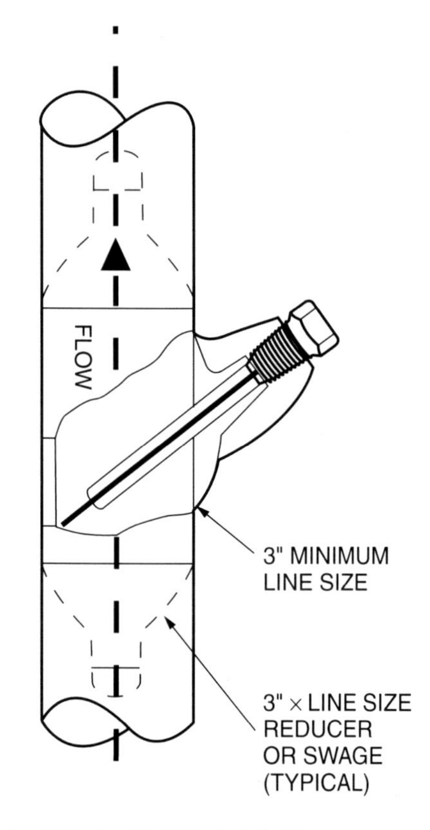

Figure 69 ◆ Thermowell installations.

reason, the indicated temperature may be in error. When selecting insulation, moisture, abrasion, flexing, chemical attack, temperature extremes, and any other adverse environmental considerations must be evaluated. Insulation is rated for a maximum continuous use temperature. After excessive temperatures are encountered, the insulation may become conductive or conductive residues may form, even though the insulation remains physically intact.

Fibrous insulation is either braided or corded on the conductors. In general, fibrous insulations are used where extreme moisture and abrasion resistance are not a concern. These insulations are available for upper temperatures of 900°F (482°C) for fiberglass, 1,000°F (538°C) for asbestos, and 1,600°F (780°C) for high-temperature silica fiber.

Plastic insulation is used in comparatively low-temperature applications and provides good moisture and abrasion resistance. It is available with typical upper temperatures of 220°F (104°C) for PVC and 500°F (260°C) for Teflon®.

6.5.0 Resistance Temperature Detectors

Resistance thermometry is based on the increasing electrical resistance of conductors with increasing temperature. The conductors used for RTDs include platinum, nickel of various purities, 70% nickel/30% iron (Balco®), and copper, listed in order of decreasing temperature range capability. These conductors are all available as fine wire for sensor winding. Platinum is also available as a deposited film sensor. Nickel and Balco® are available in foil-type sensors. All share, in varying degrees, the characteristics of repeatability, high-temperature coefficient, long-term stability, and linearity over a useful temperature range. While no sensor material surpasses platinum in overall performance, each has at least one characteristic that may encourage its selection.

Industrial platinum RTD offers ruggedness at a negligible loss in temperature coefficient compared to the standard SPRT, while retaining stability and repeatability.

Industrial RTDs are supplied in the same configurations as thermocouples. *Figure 70* illustrates the most common process temperature assembly. Two-, three-, and four-wire lead configurations are used in order of increasing possible precision in the temperature measurement. Two-wire sensors are common in applications where leads are short, such as machine installations.

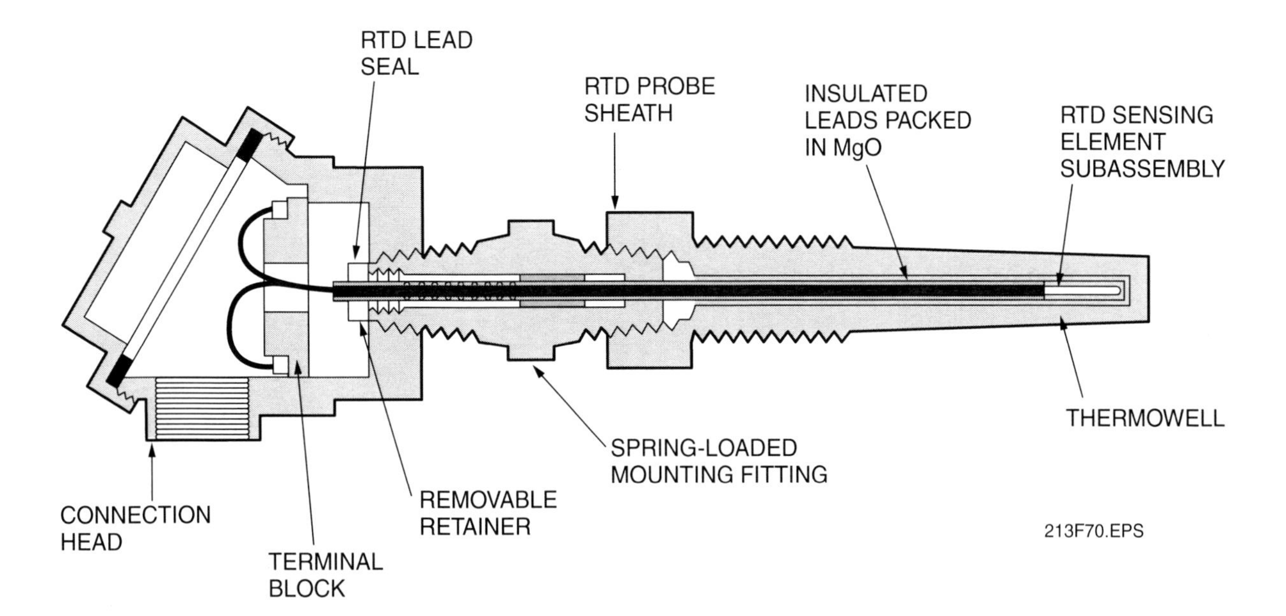

Figure 70 ◆ Industrial RTD/thermowell assembly.

7.0.0 ◆ FLANGES

Flanges are pipe fittings that allow one section of pipe to be bolted to another section of pipe or to instruments, vessels, or valves. Flanges provide a leakproof joint that can be taken apart easily for inspection or repair. The type of flange is identified by the design specifications as to size, type of pipe involved, and the type of fitting required. Normally, this is determined by the design engineer and conforms to the specifications and requirements of the fluid to be carried by the system. Flanges are normally separated by gaskets at their mating faces.

Flanges are normally identified by their size, rating, type of face, type of mounting, and composition. For example, a flange may be identified as a 3", 300 lb, raised-face, weld neck flange, 316 stainless steel.

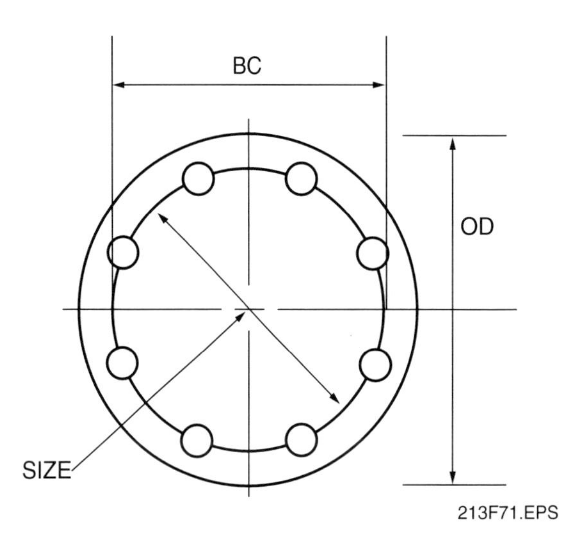

Figure 71 ◆ Determining flange size.

7.1.0 Flange Sizes

The size (in inches) of a flange refers to the NPS nominal bore (inside) diameter of the pipe with which the flange is to be used. *Figure 71* illustrates the dimensions used to determine flange size.

In *Figure 71*, BC stands for bolt center dimension. The size is the nominal bore diameter, and OD is the outside diameter of the flange itself.

7.2.0 Flange Pressure Ratings

The pressure rating of a flange is referred to as the primary pressure rating. This is a standard number used to describe the strength of the flange. It is not the working pressure of the flange under all conditions. Temperature has a great deal to do with the safe pressure limit of any given flange. As temperature increases, the pressure rating decreases.

The composition of a flange determines its strength, temperature range, and resistance to chemical effects. The flange used for a specific application should be made of the same material as the piping used in the process system.

7.3.0 Flange Facings

Figure 72 shows the more common flange facings. Each style of facing has specific applications. Trainees must be able to identify the various styles of flange facings. Study the different styles illustrated and the description of each.

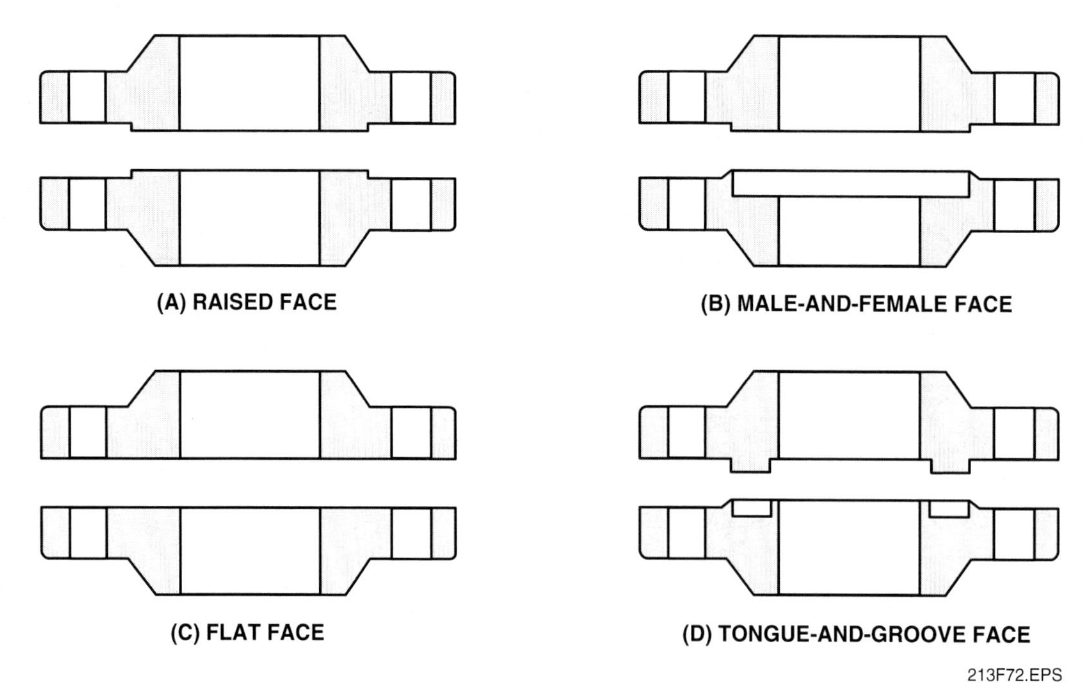

Figure 72 ◆ Flange facings.

The raised-face flange shown in *Figure 72(A)* is the most common facing used in piping systems. Because both halves of a flange pair are the same, there are no stocking or assembly problems.

Male-and-female facings are used for screwed flange assembly. The small male and female faces allow easy assembly of standard weight pipe. This is shown in *Figure 72(B)*.

Flat faces shown in *Figure 72(C)* are a variation of the raised face. The chief use of this style is for mating with 125-pound and 250-pound cast iron valves and fittings.

Tongue-and-groove facings shown in *Figure 72(D)* are made to provide minimum gasket area and high joint efficiency with flat gaskets.

7.4.0 Flange Gaskets

Most flanges and flange facings require a gasket between the mating surfaces in order to form a leakproof seal. The type of gasket required will vary according to the application and the type of flange material. Gaskets must be made of the proper material so as not to react with the process fluids. They must also have the correct temperature rating. The proper gasket for a joint is determined by the specifications. When in doubt as to the proper gasket to use in making up a joint, ask a supervisor for a specification reference. Using the wrong gasket is extremely dangerous.

Flat ring gaskets are made of various materials such as paper, cloth, rubber, iron, nickel, copper, aluminum, and other metals. The thickness will normally range between 1⁄64" and 1⁄8". Paper, cloth, and rubber gaskets should not be used in temperature applications over 250°F. Most ferrous or nickel-base metal gaskets are used for the maximum temperature the flange itself will withstand. *Figure 73* shows common types of flange gaskets.

Ring-joint gaskets are available in two types: octagonal and oval cross section. Both are standardized. The octagonal is considered superior. These rings are made of the softest carbon steel or iron available. For lower temperatures, the rings are made of plastic to resist corrosion or to insulate the joint from electric currents.

Serrated gaskets are flat metal gaskets with concentric grooves machined into their surfaces. With the contact area reduced to a few concentric lines, the required bolt load is reduced considerably, forming an efficient joint. Serrated gaskets are used with smooth-finished flange faces.

Corrugated gaskets are used when a stiffness between flat nonmetallic and metallic gaskets is needed. The ridges of the corrugations tend to concentrate the gasket loading along concentric rings.

Laminated gaskets are made of metal with a soft filler. The laminate can be parallel to the flange face or spiral wound. Laminated gaskets require less bolt load to compress them than solid metal gaskets and can be used in high-pressure, high-temperature applications.

7.5.0 Methods of Joining Flanges

The method of attaching flanges to piping will depend on the method used to join the piping. There are several methods of mounting flanges to piping systems, including butt-weld, socket-weld, and screwed-joint piping systems.

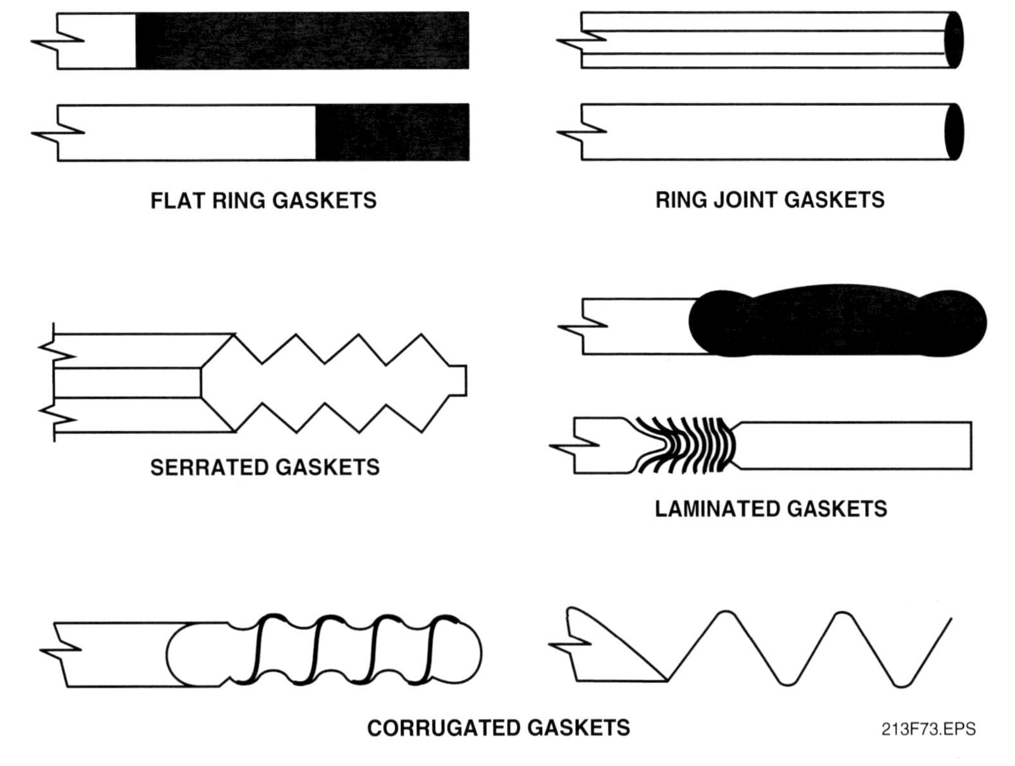

Figure 73 ◆ Flange gaskets.

7.5.1 Socket-Welded Flanges

Socket-weld piping systems normally accept only one type of flange mount. The raised-face flange is normally socket-welded to the pipe and then bolted to fittings, valves, or instrument.

7.5.2 Screwed-Joint Flanges

Screwed-joint piping systems are designed to accept only one type of flange mount. The flat-face flange is carefully threaded to the pipe and then bolted to a fitting, valve, or instrument.

7.5.3 Butt-Welded Flanges

Butt-welded piping systems can accept several different types of flange mounts. The type of flange mounting in these systems is specified by the design engineer.

8.0.0 ◆ MANIFOLD VALVE ASSEMBLIES

Manifold valve assemblies can consist of complex valve combinations that can direct, bypass, and shut off process flow in many directions. This module is concerned with the multi-valve assemblies or manifolds that are typically associated with instrumentation installation. These valve assemblies allow for the control and diversion of process liquids during calibration or instrument replacement. Most instrumentation that is directly connected to the process in one fashion or another, particularly in the case of differential pressure transmitters, must have some type of manifold valve assembly installed at the initial installation of the device.

Because differential pressure instruments operate on the basis of the difference in pressure between the high side (upstream from the primary element or sensor) and the low side (downstream from the primary element or sensor), many of these multiport valve assemblies installed on the instruments are designed with an **equalizer valve**. This valve, when opened, allows the high side and the low side to equalize in pressure. This convenient method allows for quick zeroing of the instrument, because if the high side pressure equals the low side pressure, no difference in pressure exists between the two sides. Therefore, the output should be zero (in the case of direct-acting transmitters). If the transmitter were set up to act in reverse, the output would be at its maximum if the two pressures equaled each other. This section of the module discusses some of the types of manifold valve assemblies.

8.1.0 Single-Valve Manifolds

Single-valve manifolds are used in simple systems where it is necessary to provide a positive shutoff of flow and to throttle flow during operation. They also provide for two instrument connections and a test connection. Any or all of these connections may be blocked according to desired use. *Figure 74* shows a cutaway view of a single-valve manifold along with its schematic diagram.

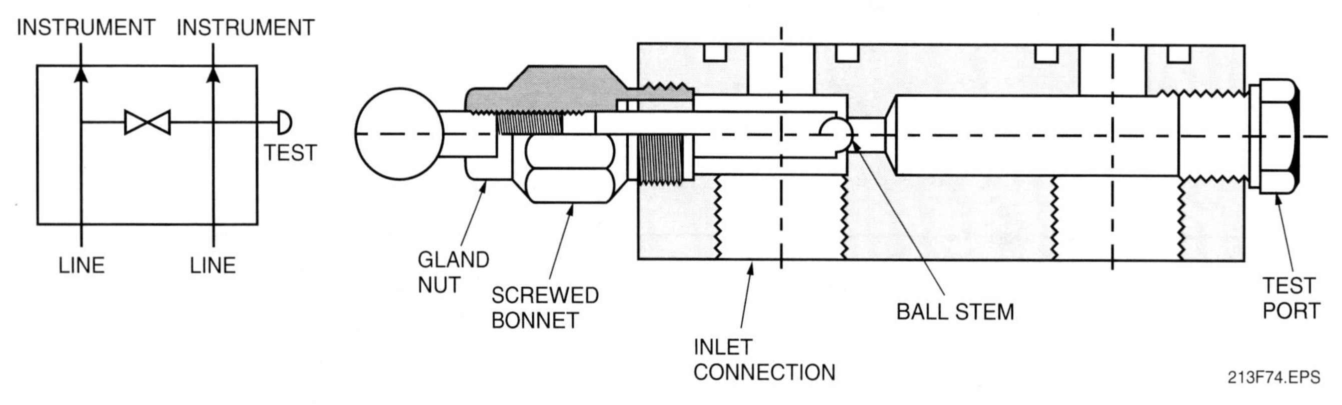

Figure 74 ◆ Single-valve manifold.

8.2.0 Two-Valve Manifolds

Two-valve manifolds are used as barriers between line pressure and instrumentation equipment. The second valve and line serve the dual purpose of providing a vent connection to prevent line pressure from being locked in the instrumentation. It could also serve as a calibration connection. *Figure 75* shows a cutaway view of a two-valve manifold with its schematic diagram.

8.3.0 Three-Valve Equalizing Manifolds

The three-valve equalizing manifold is most commonly used in conjunction with a differential pressure transmitter. It consists of two block valves and one equalizing valve. Each side of the equalizing valve body usually has been ported and plugged for either vents or test connections. *Figure 76* shows a cutaway view of a three-valve equalizing manifold with its schematic diagram.

8.4.0 Five-Valve Equalizing Manifolds

The five-valve equalizing manifold provides two mainline block valves and a double block-and-bleed arrangement for the equalizer line. Static pressure and calibration connections are also provided. This manifold is primarily intended for gas service and is used to connect differential pressure transmitters to system flow meters. *Figure 77* shows a simplified cutaway view of a five-valve equalizing manifold with its schematic diagram.

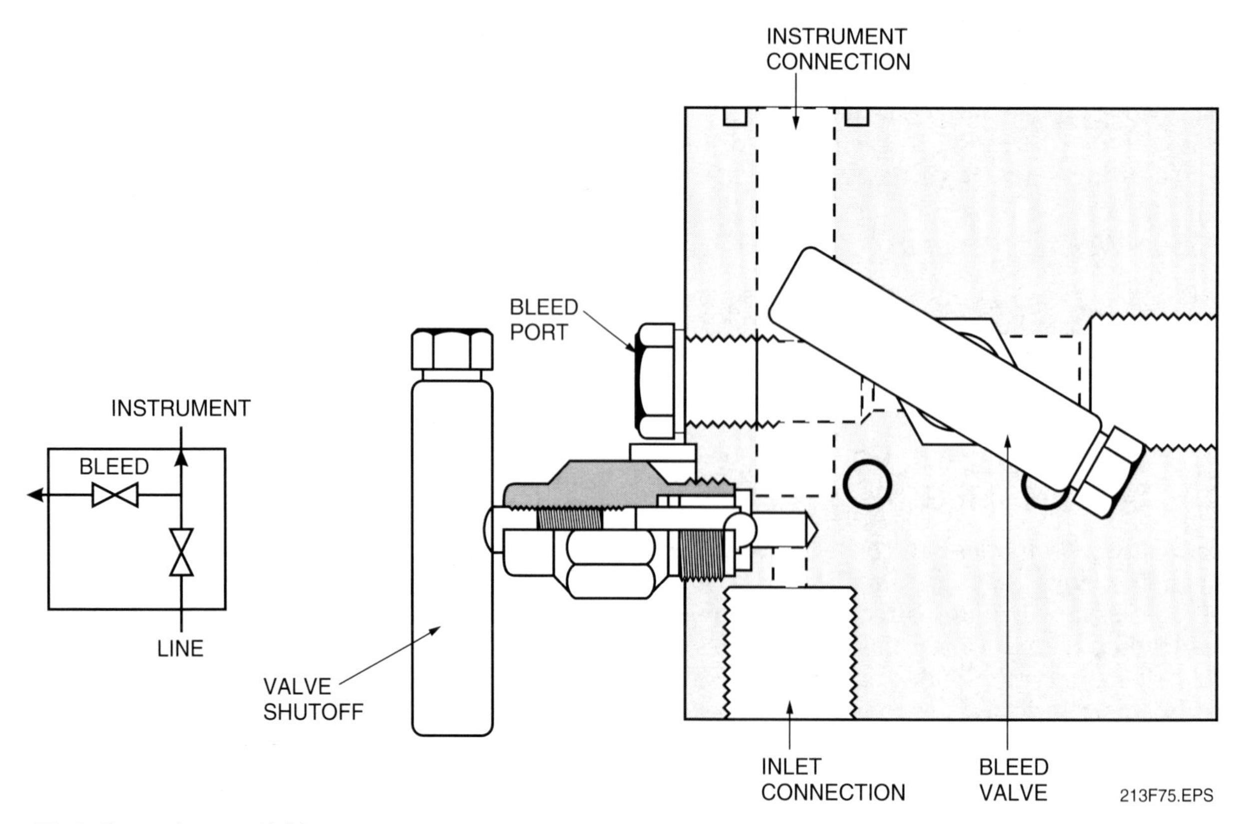

Figure 75 ◆ Two-valve manifold.

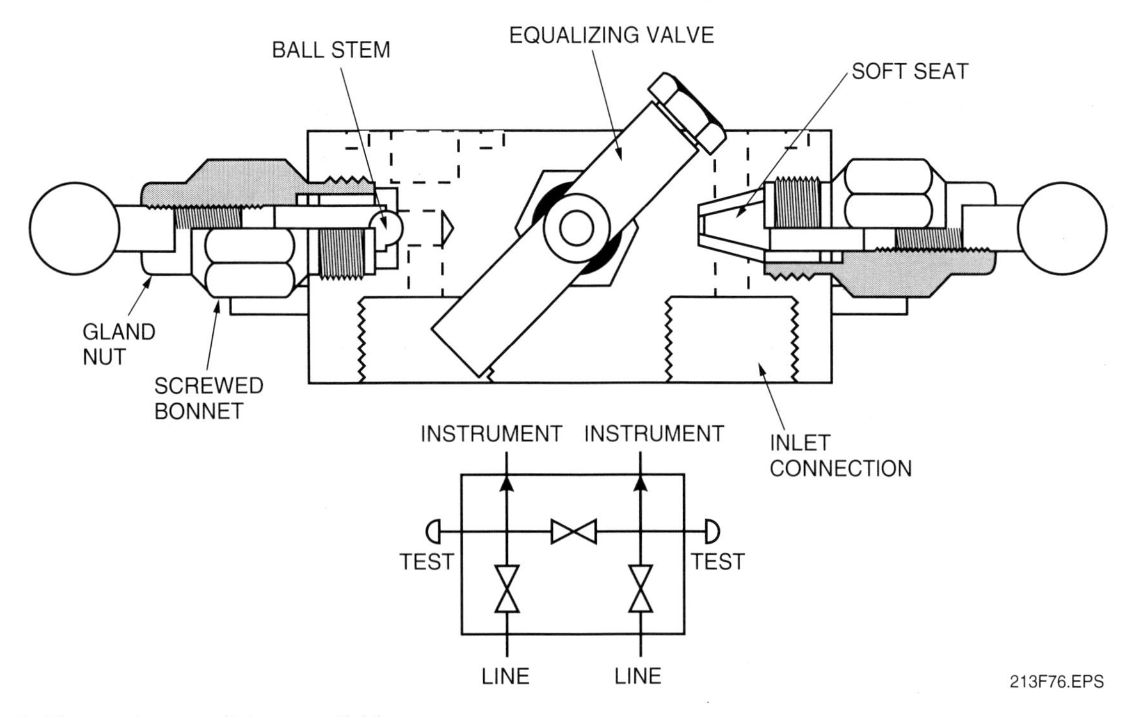

Figure 76 ◆ Three-valve equalizing manifold.

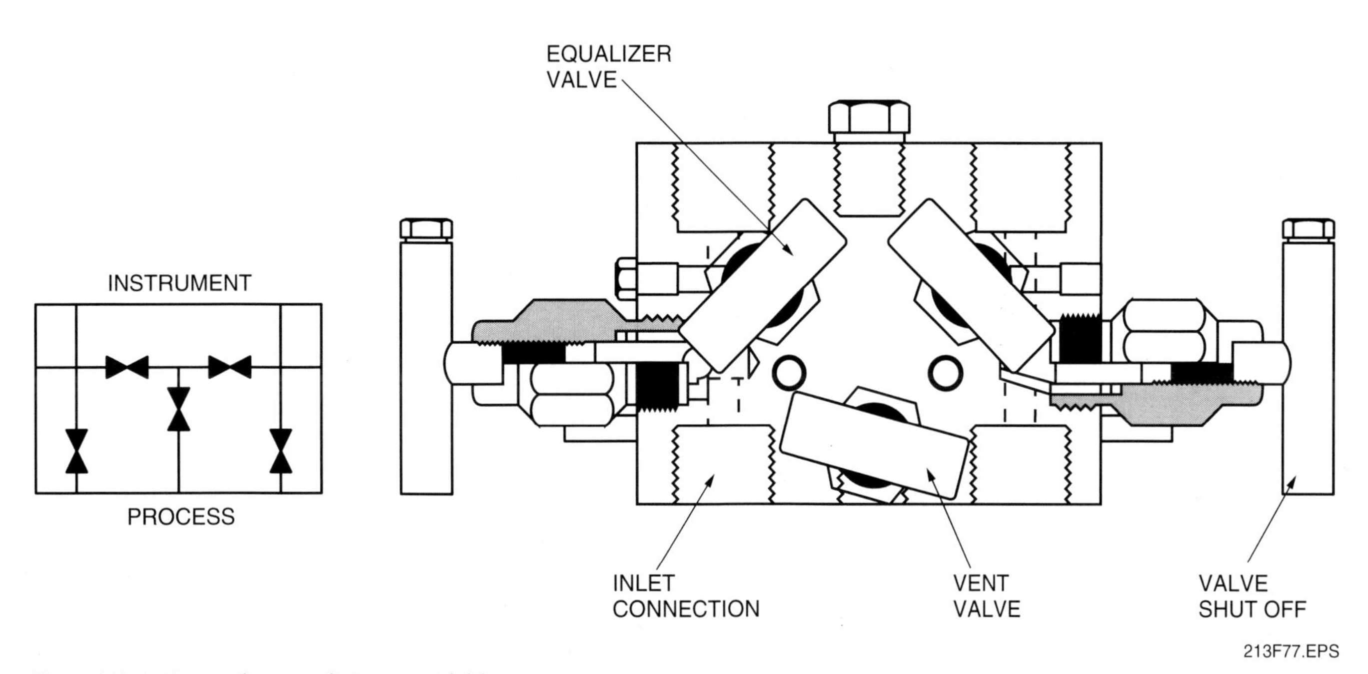

Figure 77 ◆ Five-valve equalizing manifold.

8.5.0 Five-Valve Blowdown Manifolds

The five-valve **blowdown** manifold combines a three-valve equalizing manifold with two special blowdown valves. This system provides an extremely compact, reliable, and economical unit eliminating eight nipples, four tees, and two shutoff valves. There are two block valves and one equalizing valve on the manifold. The two blowdown valves are designed with left- and right- hand block valves. The blowdown outlets are on the bottom of the two valves after the block valves. *Figure 78* shows a simplified cutaway view of a five-valve blowdown manifold with its schematic diagram.

8.6.0 Manifold Installation

Prior to actual manifold installation, several factors dealing with manifold construction and operational parameters must be verified. The following

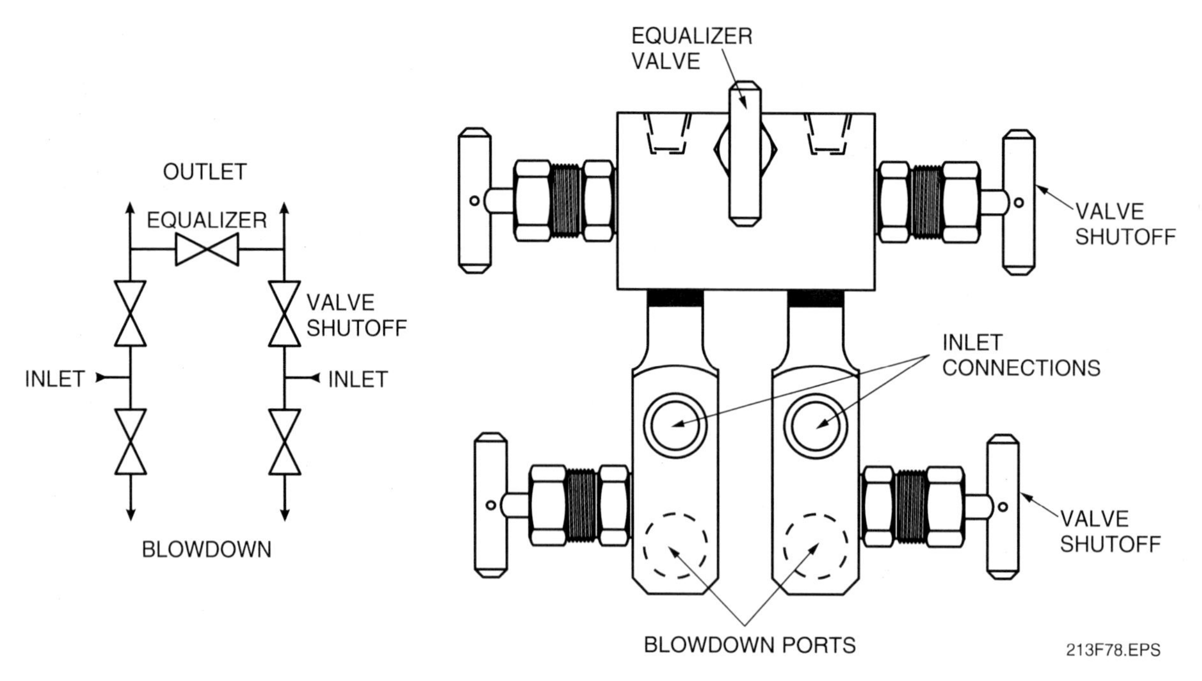

Figure 78 ◆ Five-valve blowdown manifold.

are considerations that can be checked using either a system schematic or system specifications/requirements:

- Inlet/outlet configuration
- Pressure/temperature ratings
- Packing type
- Disc configuration
- Manifold-to-system orientation

Figure 79 shows the common manifold valve inlet/outlet configurations. Selection and installation of the correct manifold is dependent on existing system configurations with the previous considerations.

The installation of various instruments has been made easy and efficient by the use of various manifold designs. Some design features include adjustable packing for leakproof extended service,

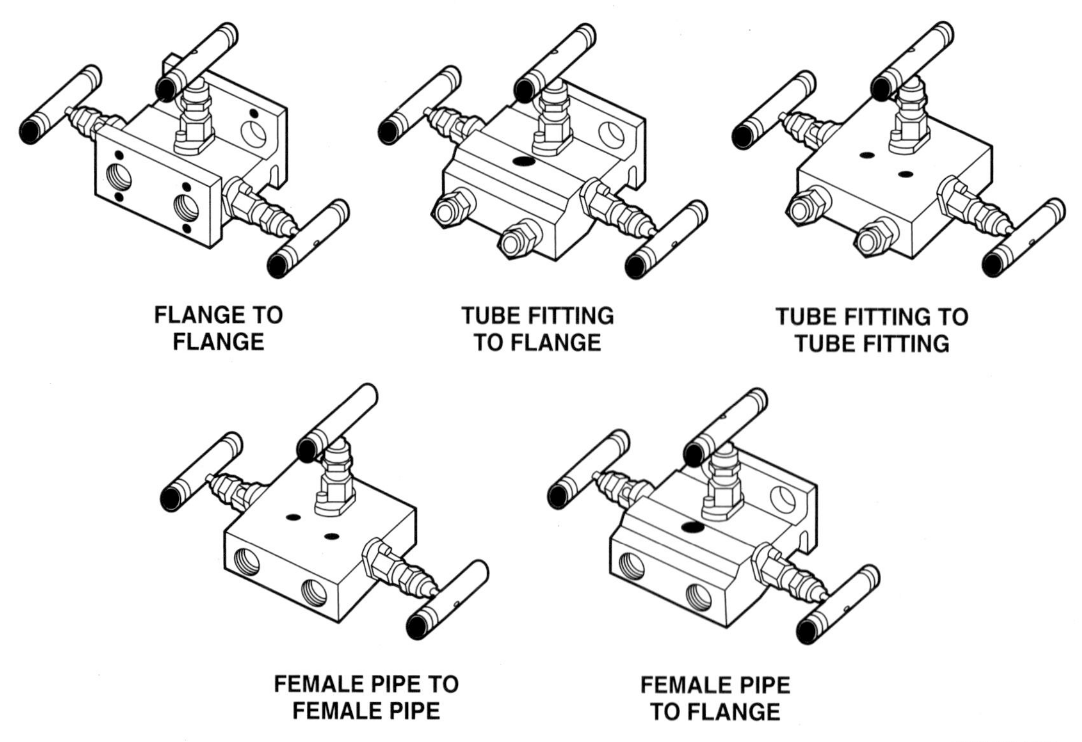

Figure 79 ◆ Manifold valve inlet/outlet configurations.

integral hard backseats to prevent stem blowout, and free swiveling ball-end stems that provide bubble-tight, metal-to-metal valve closure without seat galling.

Two manifold accessories that have proven invaluable for installing transmitters are the close-couple futbol and the rod-out device. The futbol (*Figure 80*) allows for the shortest possible distance between the orifice flange union and the instrument manifold. It comes complete from the manufacturer with bolts, washers, and a Teflon® O-ring. Two futbols are required per installation.

The rod-out is a device to clear orifice passages that have become restricted. It is designed to be used while the valve is open and under pressure. It is available in carbon steel, stainless steel, and Monel® with inlet connections to fit standard manifold instrument flanges. A type of rod-out tool is shown in *Figure 81*.

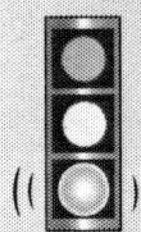

NOTE

Prior to manifold installation, make sure all O-rings are thoroughly lubricated with the manufacturer's recommended lubricant.

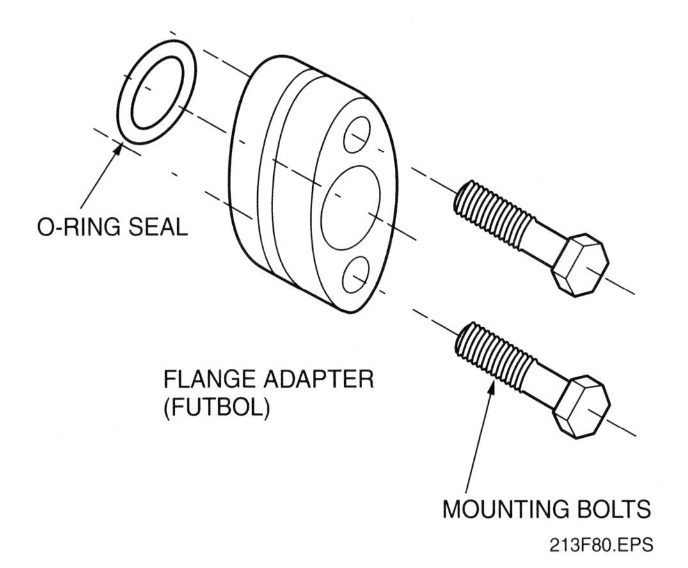

Figure 80 ◆ Futbol assembly.

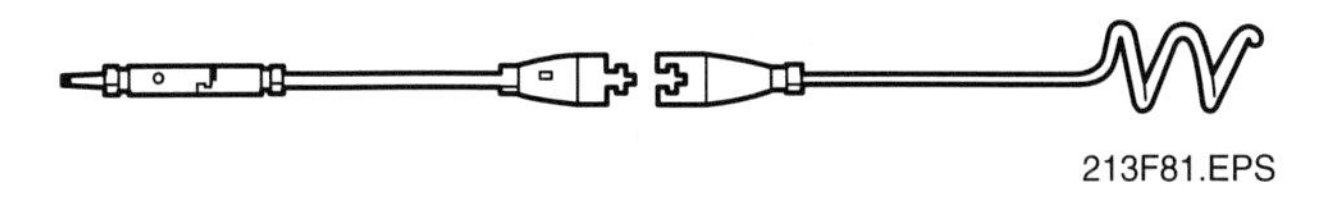

Figure 81 ◆ Rod-out tool.

Summary

The way in which an instrument is mounted often determines its accuracy and longevity. If stand-mounted differential pressure instruments are not mounted horizontally level and vertically plumb, they may operate at a degree of error because of the unevenness in height between the high and low sides.

It is also important that stand-mounted instruments be installed in locations that are relatively close to the process variable they are monitoring or controlling so that lengthy process lines to the instrument do not introduce error through process flow or pressure drops. Stand-mounted instruments should be mounted in locations that are not subjected to physical damage through contact with process elements such as corrosive material, dust, and water, unless these instruments are rated to withstand these types of elements.

In-line instruments must be mounted in the process so that there is little negative effect on the process itself. If incorrect fittings, piping, and other process-limiting components are incorrectly added in the installation of in-line instruments, the process may be severely affected. Erroneous instrument output is likely to occur. Any time instruments or elements of instrumentation are installed directly into a process, certain standards and specifications must be followed. It is important that the installer understand and follow these standards and specifications before and during installation of any in-line instrumentation.

Whenever structure-mounted instruments are installed, it is essential that the structure is not damaged or weakened in any fashion. Structure-mounted instruments must not be mounted where they impede safe personnel movement or where unsafe acts are required to maintain or calibrate these instruments.

Strap-mounted instruments must be mounted using straps that are the correct size and resistant to damage caused by harmful environmental exposure to corrosion, heat, or other damaging elements.

Insertion-mounted instruments must be mounted so that the element or element housing does not contact the process piping or vessel. The instrument cannot impede the flow of the process and must be mounted in accordance with applicable standards to ensure accuracy and longevity of the device.

Most field-mounted instrumentation requires some type of valve manifold system to allow for quick calibration and instrument replacement. Without valve manifolds, replacement of instrumentation would require a shutdown of the process or potentially unsafe personnel actions in order to block the process while working on the instrument.

Review Questions

1. Drawings and/or specifications will always specify where an instrument is to be mounted.
   a. True
   b. False

2. The instrument pipe stand is usually built to provide _____ mounting.
   a. ¾"
   b. 1"
   c. 2"
   d. 3"

3. Mounting brackets are always supplied with the instrument.
   a. True
   b. False

4. Which of the following is a factor that determines whether or not an instrument can be mounted on a floor stand?
   a. Available stand material
   b. Lack of power tools
   c. The absence of a floor or base support
   d. Noncorrosive materials present

5. Each of the following is required to fabricate an instrument stand *except* a _____.
   a. hand grinder
   b. pipe reamer
   c. measuring tape
   d. pipe wraparound

6. Wedge-type or sleeve-type concrete anchors work well for floor stands because they _____.
   a. are easier to install than other anchors
   b. require no installation tools
   c. make it easy to level the stand
   d. look better than other types of anchors

7. The holes drilled in concrete for the wedge anchors are cleaned out using _____.
   a. compressed air
   b. a squeeze bulb
   c. a water hose
   d. nitrogen

8. What additional hazard exists when mounting stands to metal gratings on upper levels as opposed to mounting stands on ground level?
   a. Shock hazard
   b. Stand imbalance
   c. Falling objects
   d. Tripping hazard

9. Which of the following sets of hardware are usually required when mounting an instrument to a stand with a bracket?
   a. U-bolt, two lock washers, and two nuts
   b. J-bolt, two flat washers, and one nut
   c. Cable strap, two lock washers, and two nuts
   d. Cable, strap, U-bolt, and four nuts

10. Which of the following symbols represents differential pressure?
    a. ΣP
    b. ΩP
    c. ΠP
    d. ΔP

11. The most common type of restriction used to measure flow in process control applications is a(n) _____.
    a. flow nozzle
    b. venturi tube
    c. pitot tube
    d. orifice plate

12. The recommended straight pipe length upstream from a turbine flowmeter is _____.
    a. 5 times the meter's inner diameter
    b. 10 times the meter's inner diameter
    c. 2 times the meter's inner diameter
    d. 10 times the meter's outer diameter

13. Another name for the blunt object located inside a vortex-shedding flowmeter is a _____.
    a. vortice
    b. diverter
    c. straightener
    d. bluff body

14. The rotameter is a _____ type of flowmeter.
    a. differential pressure
    b. mass
    c. volumetric
    d. variable-area

15. All rotameters must be installed in a(n) _____ position.

a. horizontal
b. vertical
c. angular
d. outlet-down

16. What is the most common measuring instrumentation found on vessels?

a. Flow
b. Temperature
c. Pressure
d. Level

17. A capacitance probe is a(n) _____ type of system.

a. RF
b. analog
c. digital
d. acoustic

18. Control of pH involves supplying the proper amount of _____ to bring the pH to the desired value.

a. reagent
b. sulfur
c. water
d. acid

19. The span of the displacer in a displacer-type level instrument is the distance at which the displacer responds to the forces of _____.

a. nature
b. gravity
c. buoyancy
d. atmospheric pressure

20. Cable-mounted instruments can be closely mounted to their sensing point, eliminating inaccuracies caused by _____.

a. temperature changes
b. flow restrictions
c. pressure drops
d. line interference

21. Thermostatically controlled heat tracing systems require the use of a thermostat and a(n) _____.

a. sensor
b. transmitter
c. orifice plate
d. gauge

22. Generally, industrial process plant thermowells installed in 4" pipelines or larger are mounted at a _____-degree angle in relationship to the pipe.

a. 12½
b. 33⅓
c. 45
d. 90

Research

23. Of the following materials used to make RTDs, which provides the smallest temperature range?

a. Copper
b. Platinum
c. Balco®
d. Nickel

650

24. Flat metal gaskets with concentric grooves machined into their surfaces are called _____ gaskets.

a. flat ring
b. serrated
c. ring-joint
d. flat facing

25. The manifold valve assembly usually installed on differential pressure transmitters is the _____ manifold assembly.

a. single-valve
b. two-valve
c. five-valve blowdown
d. three-valve equalizing

GLOSSARY

Trade Terms Introduced in This Module

Acoustic: Relating to sound.

Blowdown: The process of removing air, water, or sludge from an instrument line or the bottom of a vessel in order to drain or reduce concentration levels.

Bluff: A non-streamlined object or mass inserted in a vortex flowmeter to cause or create a vortex effect whenever a process fluid flow strikes the surface of the bluff.

Circumference: The distance around the outside of a circle or round object, such as pipe.

Conditioning piping: Piping that is installed ahead of and sometimes after an in-line instrument, having an interior surface condition that is free of flaws or burrs and the same ID as the ID of the in-line instrument.

Deadtime: The time between reagent addition and the first measurable change resulting from the addition.

Effluent: The out-flowing branch of the main stream.

Equalizer valve: A device that allows pressure to become equal from one side of a manifold to the other side.

Fender washer: A flat washer that is designed with a standard bolt hole inside diameter but a larger-than-normal outside diameter.

Influent: The in-flowing branch of the main stream.

Sensing bulb: An encapsulated bulb that goes through a physical change as surrounding conditions change.

Sensitivity: The ratio of a change in output magnitude to the change of input that causes the change, after steady state has been achieved.

Thermal: A state of matter dependent on temperature.

Thermostat: An automatic control device used to maintain temperature at a fixed or adjustable setpoint.

Vortex: A mass or masses (vortices) of fluid rotating about an object.

Additional Resources

This module is intended to present thorough resources for task training. The following reference work is suggested for further study. This is optional material for continued education rather than for task training.

The Condensed Handbook of Measurement and Control, 1997. N.E. Battikha. Research Triangle Park, NC: The Instrumentation Society of America.

Acknowledgments

Gerald Shannon	213F04 (divider, hand grinder, portable power band saw, metal chop saw)
The L.S. Starrett Company	213F04 (squares, scribers, dye, punch set)
Tyco Thermal Controls	212F54

NCCER CURRICULA — USER UPDATE

NCCER makes every effort to keep its textbooks up-to-date and free of technical errors. We appreciate your help in this process. If you find an error, a typographical mistake, or an inaccuracy in NCCER's curricula, please fill out this form (or a photocopy), or complete the online form at **www.nccer.org/olf**. Be sure to include the exact module ID number, page number, a detailed description, and your recommended correction. Your input will be brought to the attention of the Authoring Team. Thank you for your assistance.

Instructors – If you have an idea for improving this textbook, or have found that additional materials were necessary to teach this module effectively, please let us know so that we may present your suggestions to the Authoring Team.

NCCER Product Development and Revision

13614 Progress Blvd., Alachua, FL 32615

Email: curriculum@nccer.org

Online: www.nccer.org/olf

❑ Trainee Guide ❑ AIG ❑ Exam ❑ PowerPoints Other ______________________________

Craft / Level: Copyright Date:

Module ID Number / Title:

Section Number(s):

Description:

Recommended Correction:

Your Name:

Address:

Email: Phone:

Module 12214-03

Raceways for Instrumentation

COURSE MAP

This course map shows all of the modules in the second level of the Instrumentation curriculum. The suggested training order begins at the bottom and proceeds up. Skill levels increase as you advance on the course map. The local Training Program Sponsor may adjust the training order.

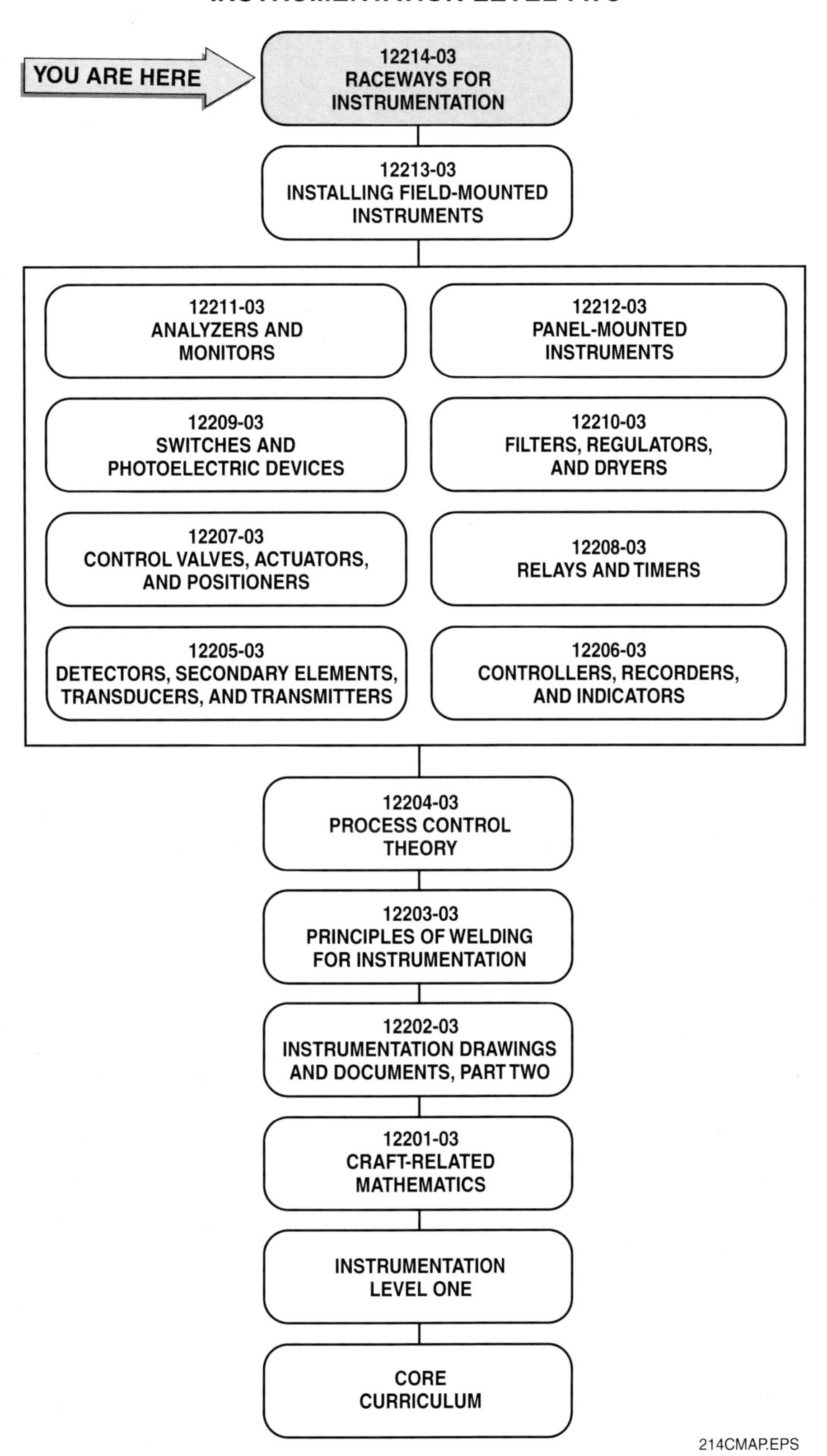

214CMAP.EPS

MODULE 12214-03 CONTENTS

1.0.0 INTRODUCTION 14.1

2.0.0 CONDUIT 14.2

2.1.0 Metal Conduit 14.2

2.1.1 Electrical Metallic Tubing (EMT) and Fittings 14.2

2.1.2 Rigid Metal Conduit (RMC) and Fittings 14.4

2.1.3 Intermediate Metal Conduit (IMC) 14.9

2.1.4 Flexible Metal Conduit (FMC), Liquidtight Flexible Metal Conduit (LFMC), and Fittings 14.9

2.1.5 Conduit Nipples 14.11

2.2.0 Nonmetallic Conduit 14.11

2.2.1 Rigid Nonmetallic Conduit and Fittings 14.11

2.2.2 Electrical Nonmetallic Tubing (ENT) 14.12

2.3.0 Boxes 14.12

2.3.1 Metal Boxes 14.12

2.3.2 Nonmetallic Boxes 14.13

2.4.0 Conduit Supports 14.13

2.4.1 One-Hole and Two-Hole Straps 14.13

2.4.2 Standoff Supports 14.14

2.4.3 Strut Supports 14.14

2.4.4 Beam Clamps 14.14

2.5.0 Installing Conduit 14.14

2.5.1 Size Requirements 14.15

2.5.2 Conduit Bend Requirements 14.15

2.6.0 Cutting Conduit 14.16

2.6.1 Cutting RMC and IMC 14.16

2.6.2 Cutting EMT 14.19

2.6.3 Cutting FMC and LFMC 14.20

2.6.4 Cutting RNC (PVC) and ENT 14.21

2.7.0 Deburring Conduit 14.21

2.7.1 Deburring RMC and IMC 14.21

2.7.2 Deburring EMT 14.23

2.7.3 Deburring FMC and LFMC 14.23

2.7.4 Deburring RNC (PVC) and ENT 14.24

2.8.0 Threading Conduit 14.24

2.9.0 Joining Conduit 14.26

2.9.1 Joining Metal Conduit 14.26

2.9.2 Joining PVC Conduit 14.27

3.0.0 METAL WIREWAYS 14.28

3.1.0 Wireway Fittings 14.30

3.1.1 Connectors 14.30

3.1.2 End Plates 14.30
3.1.3 Tees 14.30
3.1.4 Crosses 14.31
3.1.5 Elbows 14.31
3.1.6 Nipples 14.31
3.1.7 Telescopic Fittings 14.31
3.2.0 Wireway Supports 14.31
3.2.1 Suspended Hangers 14.31
3.2.2 Gusset Brackets 14.32
3.2.3 Standard Hangers 14.32
3.2.4 Trapeze Hangers 14.32
3.2.5 Nonmetallic Wireways 14.32
4.0.0 SURFACE METAL AND NONMETALLIC RACEWAYS 14.33
5.0.0 CABLE TRAYS 14.34
5.1.0 Cable Tray Fittings 14.34
5.2.0 Cable Tray Supports 14.35
5.2.1 Direct Rod Suspension 14.35
5.2.2 Trapeze Mounting 14.36
5.2.3 Wall Mounting 14.36
5.2.4 Pipe Rack or Bridge Mounting 14.36
6.0.0 STORING RACEWAYS 14.36
7.0.0 HANDLING RACEWAYS 14.37
SUMMARY 14.37
REVIEW QUESTIONS 14.37
GLOSSARY 14.41
REFERENCES & ACKNOWLEDGMENTS 14.42

Figures

Figure 1 Metal conduit wall thickness comparison 14.3
Figure 2 Compression fittings 14.3
Figure 3 Setscrew EMT couplings 14.4
Figure 4 Rigid metal conduit 14.4
Figure 5 Rigid metal conduit/coupling and I-/T-type conduit bodies 14.5
Figure 6 Type C conduit body 14.5
Figure 7 Typical Type L conduit bodies 14.5
Figure 8 Determining the type of L conduit body 14.6
Figure 9 Type T conduit body 14.6
Figure 10 Type X conduit body 14.6
Figure 11 RMC unions 14.7
Figure 12 Conduit hubs 14.7
Figure 13 Bushings and locknuts 14.7
Figure 14 Seal-off fittings 14.8
Figure 15 Flexible metal conduit 14.9
Figure 16 Liquidtight flexible metal conduit connectors 14.10
Figure 17 Adapter fittings 14.10
Figure 18 Offset nipples 14.11
Figure 19 PVC conduit 14.11
Figure 20 PVC expansion coupling 14.12
Figure 21 Boxes 14.12
Figure 22 Knockout removal 14.13
Figure 23 Pryout removal 14.13
Figure 24 Knockout punch 14.13
Figure 25 One-hole and two-hole straps ....14.14.
Figure 26 Standoff support ....14.14.
Figure 27 Strut supports ....14.14.
Figure 28 Beam clamp and hanger 14.15
Figure 29 Hand hacksaws and portable bandsaw 14.16
Figure 30 Cutting conduit with a hand hacksaw 14.17
Figure 31 Conduit ends after cutting 14.17
Figure 32 Portable pipe vise, portable power pipe threader, and power-driven vise 14.18
Figure 33 Cutter placement 14.18
Figure 34 Rotating the cutter 14.19
Figure 35 Hand-operated PVC conduit cutters 14.21
Figure 36 Using a half-round file 14.22
Figure 37 Reamers 14.22
Figure 38 Reamer use 14.23

Figure 39 EMT deburring tool 14.23
Figure 40 EMT deburring tool ready for use 14.23
Figure 41 Deburring PVC with a knife 14.24
Figure 42 Hand-operated ratchet threader 14.24
Figure 43 Lubricating the threading die 14.25
Figure 44 Handheld threader power head 14.25
Figure 45 Correct entrance angle for conduit 14.26
Figure 46 Assembly of conduit to a box 14.27
Figure 47 Applying solvent cement .14.28
Figure 48 Inserting conduit into the fitting 14.28
Figure 49 Wireway system layout 14.29
Figure 50 Assembly sections of wireway 14.30
Figure 51 Connector 14.30
Figure 52 End plate 14.30
Figure 53 Tee 14.31
Figure 54 Cross 14.31
Figure 55 90-degree inside elbow 14.31
Figure 56 Nipple 14.31
Figure 57 Suspended hanger 14.32
Figure 58 Gusset bracket 14.32
Figure 59 Standard hanger 14.32
Figure 60 Trapeze hanger 14.32
Figure 61 Smaller surface raceways 14.33
Figure 62 Pancake raceway 14.33
Figure 63 Dual-purpose surface metal raceway 14.33
Figure 64 Underfloor raceway duct .14.34
Figure 65 Cable tray system 14.35
Figure 66 Direct rod suspension 14.35
Figure 67 Trapeze mounting 14.36
Figure 68 Wall mounting 14.36

MODULE 12214-03

Raceways for Instrumentation

Objectives

When you have completed this module, you will be able to do the following:

1. Identify various types of metallic conduit and fittings and state their uses.
2. Identify types of nonmetallic conduit and fittings and state their uses.
3. Identify raceway supports and their uses.
4. Prepare various types of conduit for installation.
5. Describe wireways and cable trays and their associated fittings.

Prerequisites

Before you begin this module, it is recommended that you successfully complete the following: Core Curriculum; Instrumentation Level One; Instrumentation Level Two, Modules 12201-03 through 12213-03.

Required Trainee Materials

1. Pencil and paper
2. Appropriate personal protective equipment
3. A copy of the latest edition of the *National Electrical Code®*

1.0.0 ◆ INTRODUCTION

The primary function of a **raceway** system is to provide a carrier for the installation of electrical **conductors** used in instrumentation and electrical applications. It also provides a degree of protection for the conductors.

The selection of the proper raceway components depends on many factors, including the ***National Electrical Code*** **(NEC)** classification of the area in which the raceway is to be installed, the voltage or current levels of the conductors, the conductor application (power or control), physical and environmental exposure, and customer specifications or preferences. Based on these and other factors, the selection of a raceway system for a particular project may be made from a variety of systems. These include **conduit**, **cable tray**, **wireway**, other specialty raceway systems, and a combination of these systems. Some of these systems may be subdivided into groups that include metal, nonmetallic, rigid, and flexible types of raceways.

The uninterrupted connections of metal raceway system components may also provide a permanent, electrically conductive path that ensures electrical continuity and the capacity to safely conduct and direct any fault current to the grounding electrode system.

The various types of raceway systems and fittings, basic raceway installation skills, and NEC requirements applicable to raceway systems are covered in this training module. This module also covers raceway supports and environmental considerations for raceway systems, as well as general raceway information.

In studying this module, refer to the following NEC Articles:

- *NEC Article 250*, Grounding
- *NEC Article 342*, Intermediate Metal Conduit
- *NEC Article 344*, Rigid Metal Conduit
- *NEC Article 348*, Flexible Metal Conduit
- *NEC Article 350*, Liquidtight Flexible Metal Conduit

Note: The designations "National Electrical Code," "NE Code," and "NEC," where used in this document, refer to the *National Electrical Code®*, which is a registered trademark of the National Fire Protection Association, Quincy, MA. *All National Electrical Code (NEC) references in this module refer to the 2002 edition of the NEC.*

- *NEC Article 352*, Rigid Nonmetallic Conduit
- *NEC Article 358*, Electrical Metallic Tubing
- *NEC Article 376*, Metal Wireways
- *NEC Article 378*, Nonmetallic Wireways
- *NEC Article 392*, Cable Tray

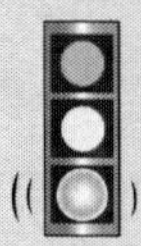

NOTE

Mandatory rules of the NEC are characterized by the use of the word shall. Explanatory material is in the form of Fine Print Notes (FPN). When referencing specific sections of the NEC, always check to see if any exceptions apply to that section.

2.0.0 ◆ CONDUIT

Conduit, as applied in the instrumentation trade, is typically used to provide a conductor carrier system. Conduit also provides protection for electrical conductors that transmit both signal and power to and from electrical or electronic instrumentation devices in field and control room environments. The type of conduit selected for a particular installation usually depends on such factors as environment, NEC regulations, potential exposure, and customer specifications. In order to properly select the type of conduit that is appropriate for an installation, all these factors, and possibly others that may affect the conductors, must be evaluated before installing any conduit to carry the instrumentation conductors.

There are two basic material classifications of conduit: metal and nonmetallic. These two classes of conduit may be available in rigid or flexible types. Each type is specifically designed and rated for a particular application.

Equally important to the proper selection of the conduit in a raceway system is the selection of the proper fittings, boxes, and accessories designed for a particular conduit system. Conduit fittings, boxes, and accessories are generally not interchangeable or compatible with other systems. However, this rule has exceptions. These exceptions will be examined in this module.

A conduit raceway system is only as good as the methods applied in its installation, including properly sizing, cutting, routing, and securing the conduit using the correct tools, fittings, and procedures. This module covers some of the more common installation methods.

2.1.0 Metal Conduit

Most exposed installations that require conduit for protection of the conductors will be run in some type of metal conduit. The type of conduit selected will normally be determined by the environment to which the conduit will be subjected. The majority of outdoor industrial installations, such as those located in petrochemical facilities and power generation stations, will require rigid metal conduit. Other outdoor commercial installations may only require electrical metallic tubing (**EMT**), using **raintight** connectors. Flexible metal conduit may be used in certain limited applications requiring flexible connections, such as installation where equipment vibration is a factor, or maintenance situations that call for easy removal and replacement of the equipment.

The following sections examine the types of metal conduits and their associated fittings more closely. Always refer to the appropriate NEC articles and sections previously listed before selecting or installing any raceway system. These sections list the uses permitted and not permitted for each type of conduit, as well as the minimum installation requirements associated with each type of conduit.

2.1.1 Electrical Metallic Tubing (EMT) and Fittings

Electrical metallic tubing is the lightest duty metal conduit available. It is widely used for residential, commercial, and limited industrial wiring systems, particularly indoors. It is lightweight, easily cut and bent, and is the least expensive of the metal conduits. Because the wall thickness of EMT is less than that of rigid metal conduit, it is often referred to as thinwall conduit. A comparison of inside and outside diameters (ID and OD) of EMT to rigid metal conduit and intermediate metal conduit (**IMC**) is shown in *Figure 1*. Other metal conduit types are discussed later in this module.

NEC Article 358 permits the installation of EMT for either exposed or concealed work where it will not be subjected to severe physical damage during or after installation. EMT is permitted in wet locations as long as the fittings are raintight and the associated support hardware is made of a corrosion-resistant material. Refer to ***NEC Section 358.12*** for restrictions that apply to the use of EMT.

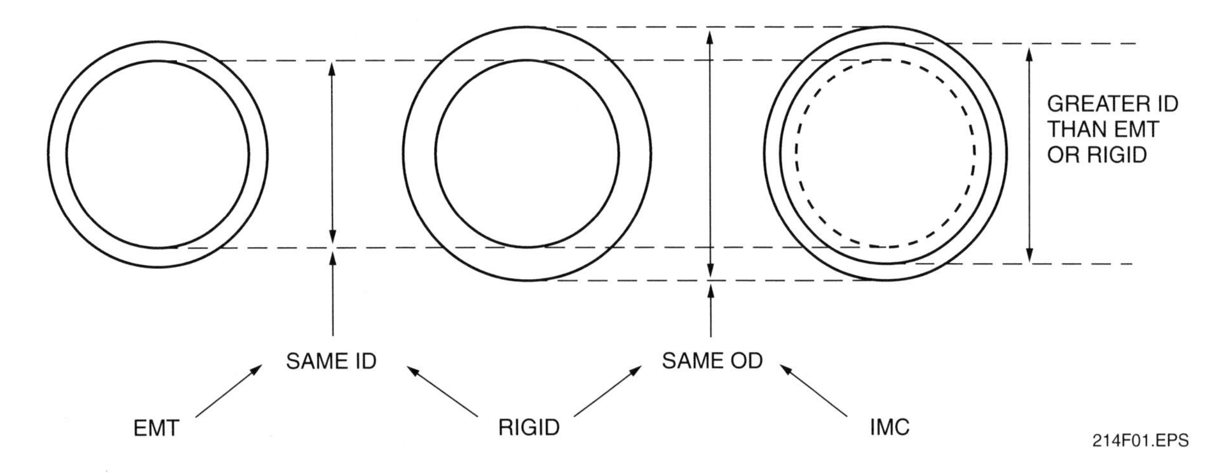

Figure 1 ◆ Metal conduit wall thickness comparison.

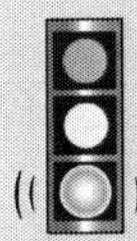

NOTE

EMT must be installed to prevent water from entering the conduit system through boxes and fittings. In locations where walls are frequently washed, such as in meat processing plants, poultry houses, and similar operations, boxes must be installed to provide a ¼" air space or gap between the box and the mounting or support surface to allow any water or moisture to drain down behind the box and not into it.

EMT may be used as a **bonding** path or grounding conductor to direct any undesirable fault current to ground, as long as the conduit system is properly bonded according to all NEC requirements as set forth in ***NEC Section 250.96(A)***. In order to qualify as an electrically continuous bonded system, the conduit system must be tightly connected at each joint and provide a continuous grounding path from each electrical load enclosure or housing to the service equipment grounding conductor. The EMT connectors are designed to provide such a path as long as they are installed correctly and match the conduit system in which they are applied.

Because the wall thickness of EMT is too thin to allow threads to be cut, specially designed fittings were developed to interconnect EMT conduit and to connect it to boxes and enclosures. Where buried in concrete, EMT couplings and connectors shall be concrete-tight fittings, according to ***NEC Section 358.42***. Likewise, this section states that EMT conduit installed in wet locations shall be raintight. Raintight EMT couplings and connectors are designed with a compression ring and compression nut that, when tightened, form a raintight seal between the fitting and the conduit. Examples of an EMT compression coupling and an EMT compression connector are shown in *Figure 2*.

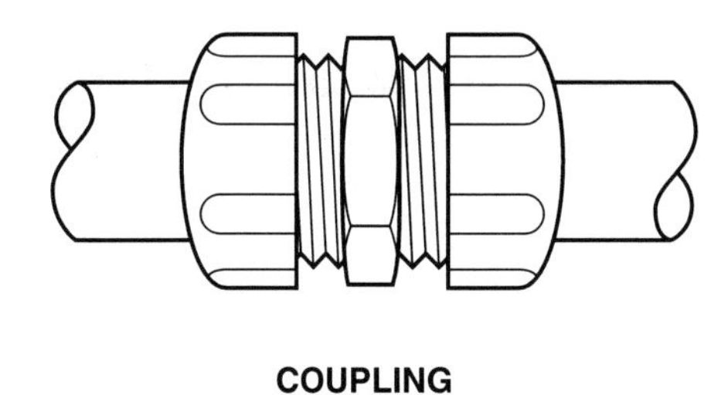

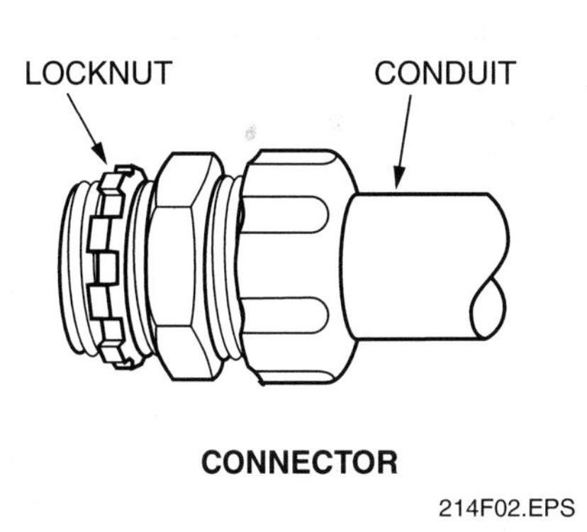

Figure 2 ◆ Compression fittings.

Whenever an EMT conduit system is installed in a dry location, usually indoors, setscrew-type fittings are used. The cut and deburred ends of the EMT conduit are inserted into the sleeve portion of a setscrew fitting until the end of the conduit bottoms out on the inside shoulder of the fitting, ensuring that the conduit is fully inserted in the fitting. The setscrews are then tightened evenly. Care must be taken not to over tighten the screws, which could cause an indention of the conduit into the conductors if they are already installed or a potential snagging point or sharp edge that may damage the conductors when installing them in the conduit. Many setscrew fitting screws are made with relatively soft metal material, which, when tightened to an extreme, will cause the screw head to sever from the screw shank. Several examples of setscrew couplings are shown in *Figure 3*.

214F03.EPS

Figure 3 ◆ Setscrew EMT couplings.

2.1.2 Rigid Metal Conduit (RMC) and Fittings

Rigid metal conduit has sufficient wall thickness to permit the cutting of threads for connection. Rigid metal conduit provides the best physical protection for conductors. It is available in 10' lengths and typically includes a threaded coupling attached to one end. RMC sizes range from ½" to 6" and are limited to sizes within this range by ***NEC Section 344.20***. Because of its threaded fittings, RMC provides an excellent bonding path for equipment grounding, as verified in ***NEC Sections 344.2*** and ***250.118***. More specific details about NEC requirements associated with rigid metal conduit may be found in other sections of ***NEC Article 344***.

Most RMC installed in industrial applications is made of steel with a galvanized coating. Galvanized rigid steel conduit is sometimes referred to as GRC (galvanized rigid conduit) instead of RMC, indicating that it is rigid steel conduit with a galvanized finish. Rigid metal conduit is also available in a variety that has a plastic coating applied to it. This type of conduit is installed in corrosive environments that require the physical conductor protection provided by rigid metal conduit, such as off shore drilling rigs, ships, corrosive petrochemical facilities, and other industrial environments. However, it would not be compatible with the exposed metal surfaces of uncoated metal conduit.

CAUTION

Special tools are required to avoid damaging the coating.

Aluminum rigid conduit is an alternative raceway in some corrosive environments; however, it does not offer the level of physical protection afforded by RMC. Whenever joining aluminum conduit to fittings, an anti-oxidizing compound should be applied to the aluminum threads. This guards against oxidation and subsequent seizing of the two parts caused by the formation of oxidation. It also helps maintain a conductive grounding path through the connection. Many such compounds are available that provide protection against the formation of oxidation in aluminum conduit connections.

NOTE

Do not bury aluminum conduit in soil or concrete that contains high levels of calcium chloride. Corrosion can be caused by calcium chloride.

Because aluminum conduit is nonferrous, it often acts to reduce voltage drop in the conductors installed in it. Magnetic fields, which create EMF in the conductors, are not induced in the conduit. In AC circuits, conductors installed in aluminum conduit may have their voltage drop reduced by as much as 20%. Aluminum conduit is much easier to install due to its lightweight characteristics as compared to rigid steel conduit. For example, a 10' section of 3" aluminum conduit weighs approximately 23 pounds, while its counterpart steel conduit weighs in the area of 68 pounds. However, aluminum conduit does not provide any shielding effects from outside transient voltages, as does rigid steel conduit. Examples of rigid metal conduit are shown in *Figure 4*.

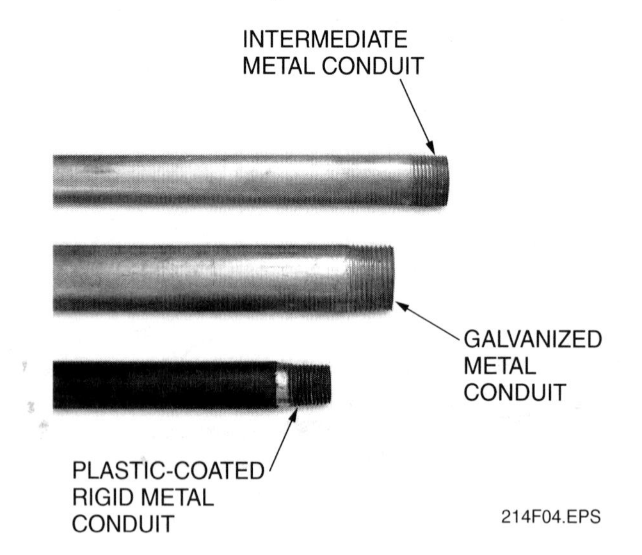

214F04.EPS

Figure 4 ◆ Rigid metal conduit.

In order for a conduit raceway system to maintain its physical protection properties, corrosion resistance, and specifications as designed, the properties of the fittings installed with the conduit must match the conduit properties. Rigid steel or IMC galvanized couplings should be installed on rigid steel or IMC galvanized conduit. Aluminum couplings should be installed on rigid aluminum conduit. Plastic-coated couplings should be installed on plastic-coated rigid steel conduit. Interchanging or mixing properties destroys or limits the integrity of the raceway system (except with RMC and IMC). Proper installation methods such as cutting, deburring, and threading required to maintain these properties are discussed in a later section.

In addition to couplings, **condulets** are used extensively in the installation of rigid metal conduit systems. *Figure 5* illustrates a rigid metal conduit and coupling, along with applications that use L- and T- type conduit bodies. Conduit bodies are fittings that provide access through a removable (usually gasketed) cover to the interior of the conduit system at a junction of two or more sections of the system or at a terminal point of the system.

Conduit bodies are used in rigid metal conduit systems when a change of direction is required in the conduit run or when a pull-point for the installation of conductors is needed on a relatively long run of conduit. The next few paragraphs and figures take a more in-depth look at some of the conduit bodies available:

- *Type C* – When installing a conduit body solely for the purpose of providing a conductor pull-point, a Type C conduit body is generally used. Its purpose is similar to that of a coupling in that it connects two sections of conduit together. However, it also provides an access into the conduit run itself. A Type C conduit body is shown in *Figure 6*.
- *Type L* – When a change of direction is needed in a rigid metal conduit run, it is not always feasible or cost effective to install a junction box, especially when all components of a raceway system must meet stringent specifications. A common practice is to install Type L conduit bodies in these situations. Examples of conduit bodies shown in *Figure 7* are LL, LB, and LR.

Figure 6 ◆ Type C conduit body.

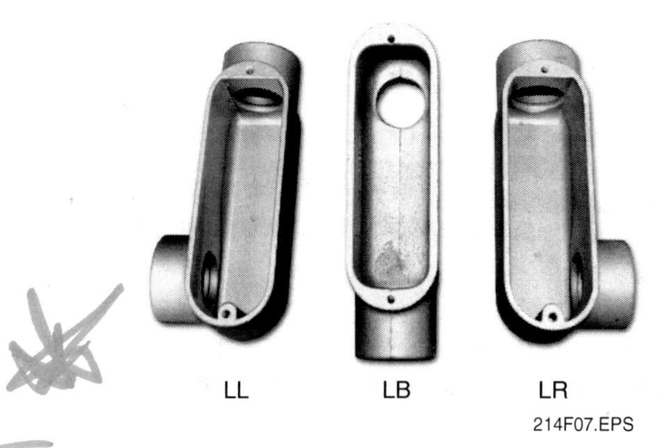

Figure 7 ◆ Typical Type L conduit bodies.

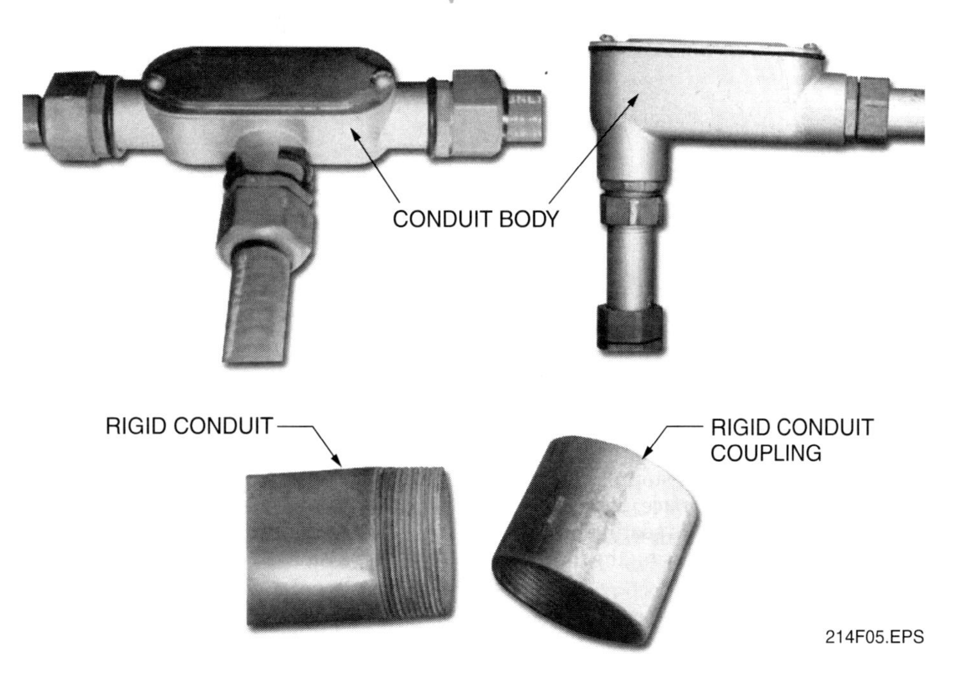

Figure 5 ◆ Rigid metal conduit/coupling and L-/T-type conduit bodies.

The letter L means that the fitting forms a 90-degree turn, or an L. The letter or letters designate where the opening of the conduit body is located. Knowledge of the methods used to describe a Type L conduit body helps the installer in determining what type of L conduit body is needed for a particular situation and in ordering the correct type of L conduit body. The location of the opening and how the conduit body is installed in the conduit run are important. The opening must always face outward from the mounting surface, and the long end of the conduit body must always be connected to the conduit end that is considered the pulling end. In other words, the conductors should be pulled into the conduit body from the long end and not the short end. This permits a straighter pull, and there is less danger of damaging the conductor insulation on the edge of the conduit body while pulling. To determine the type of L conduit body, hold the body like a pistol (*Figure 8*). If the opening is on the top (back), it is a Type LB. If the opening is on the left, it is a Type LL. If the opening is on the right, it is a Type LR. If the opening is on both sides, it is a Type LRL.

- *Type T* – Type T conduit bodies are used to provide a junction point for three intersecting conduits and are commonly found in RMC conduit systems. A Type T conduit body is shown in *Figure 9*.
- *Type X* – Type X conduit bodies are used to provide a junction point for four intersecting conduits. Though not as common as Ts in RMC conduit systems, they are still widely used. A Type X conduit body is shown in *Figure 10*.

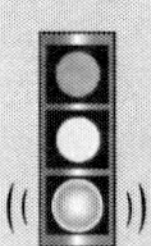

NOTE

NEC Article 314 covers requirements associated with the installation of conduit bodies and the conductors installed within them.

In addition to the more commonly applied couplings and conduit bodies, rigid metal conduit systems often require other fittings. These fittings accommodate the obstacles encountered in the proposed conduit path or satisfy NEC requirements or customer specifications. Although there are many special fittings available for specific applications, this module only examines those most frequently used in RMC installations:

- *RMC unions* – Unions are used to connect conduit to conduit, conduit to fittings, conduit to boxes, conduit to devices, and conduit to equipment when it is either impossible or very difficult to install a standard coupling because of the inability to turn (thread) one or more parts. Unions used with rigid metal conduit are shown in *Figure 11*.
- *Hubs* – Hubs are special fittings used to terminate conduit to a box or enclosure, usually in a wet or damp location. They have an O-ring that fits into a groove that rests on the surface of the box or enclosure, forming a watertight seal. One of the most common mistakes made by hub installers is over-tightening the hub, which causes the O-ring to squeeze out of the groove, voiding the sealing characteristics of the fitting. *Figure 12* shows examples of hubs commonly

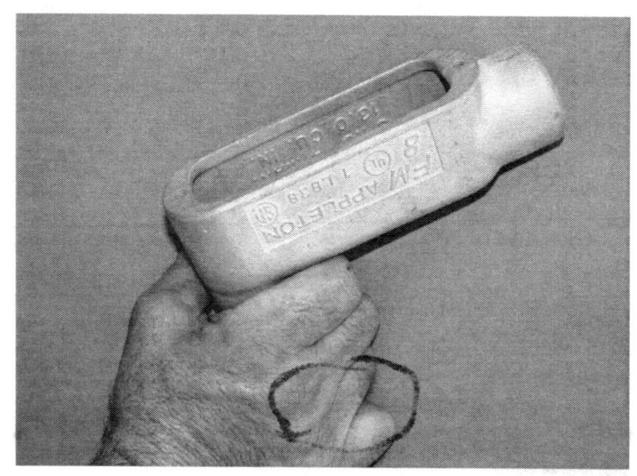

214F08.EPS

Figure 8 ◆ Determining the type of L conduit body.

214F09.EPS

Figure 9 ◆ Type T conduit body.

214F10.EPS

Figure 10 ◆ Type X conduit body.

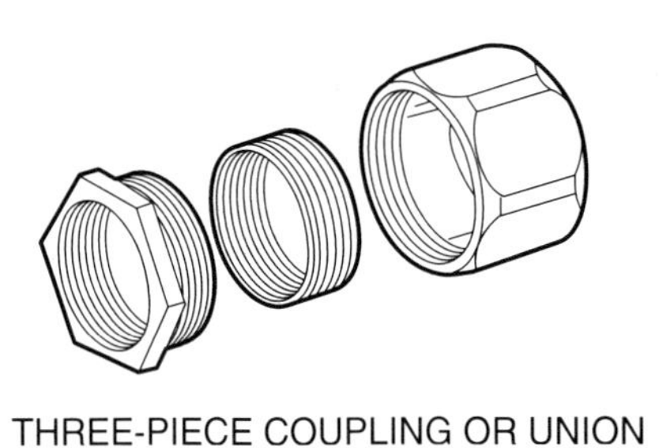

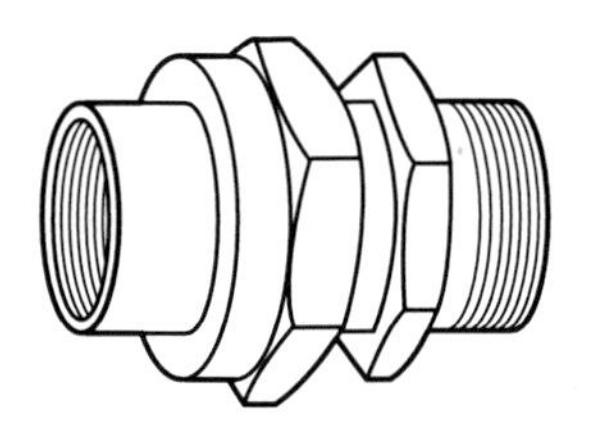

Figure 11 ◆ RMC unions.

Figure 12 ◆ Conduit hubs.

installed in RMC systems. Most have a screw on the locknut, providing a grounding lug. Some do not.

- *Bushings and locknuts* – Bushings and locknuts (*Figure 13*) often work together in the termination of conduit to an unthreaded box or enclosure. A bushing is usually installed inside the box or enclosure to protect the conductors from the sharp edges of the conduit. It is usually an

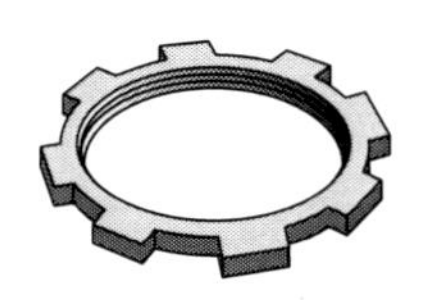

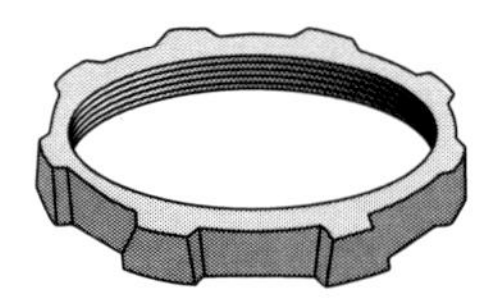

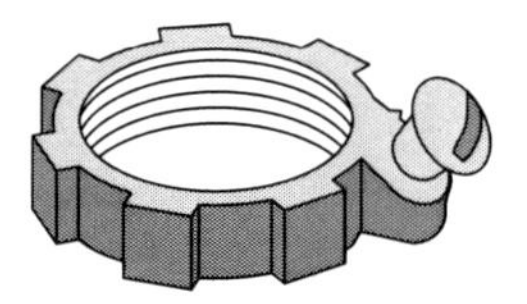

Figure 13 ◆ Bushings and locknuts.

insulating bushing. Like the hubs previously studied, a bushing may or may not provide a grounding lug. A bushing may also have a bonding screw, like the example shown in Figure 13. This screw may be tightened once the bushing is in place to ensure electrical bonding of the bushing and conduit system to its box or enclosure.

Locknuts are used on conduit, fittings, nipples, and other devices to help secure these devices to boxes or enclosures. When terminating a threaded rigid metal conduit end to a box or enclosure through a punched hole, a locknut is required on the outside and on the inside of the box. A bushing should not be used in place of the inside locknut, but should be installed after a locknut has been installed. The locknut should be used to tighten the conduit to the box. The bushing only provides conductor protection. Often, installers will install only a bushing on the inside of the box or enclosure. This is an unacceptable method of conduit termination. Like hubs and bushings, locknuts may or may not have a grounding lug. Some locknuts provide sealing capabilities in much the same manner as the weatherproof hub, using an O-ring on the facing of the locknut.

- *Seal-off fittings* – Seal-off fittings (*Figure 14*) are required in rigid conduit systems to minimize the passage of gases and vapors and to prevent the passage of flames from one portion of the electrical installation to another through the conduit. For Class I, Division 1 locations, ***NEC Section 501.5(A)(1)*** states that each conduit run entering an explosion-proof enclosure containing switches, circuit breakers, fuses, relays, resistors, or other apparatus that may produce arcs, sparks, or high temperatures must have seal-offs placed as close as practicable to, and in no case more than 18" from, such enclosures. There shall be no junction box or similar enclosure in the conduit run between the sealing fitting and the apparatus enclosure.

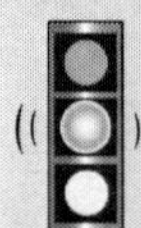

CAUTION

Do not fill seal-off fittings with sealing compound until all conductors are installed and checked for short circuits and open circuits. Conductors cannot be installed or removed from a conduit system once the seals are poured. Sealing compound hardens to a state similar to that of cured mortar or ceramic.

A seal-off fitting serves no purpose if it is not properly filled with sealing compound according to the project specifications and the manufacturer's recommendations. Sealing compound must be **approved** for the purpose, must not be affected by the surrounding atmosphere or liquids, and must not have a melting point of less than 200°F (93°C). Most sealing compound kits contain a powder in a polyethylene bag within an outer container. The following are steps in filling a seal-off fitting:

Step 1 To mix, remove the bag of powder, fill the outside container with water up to the marked line on the container, pour in the powder, and mix.

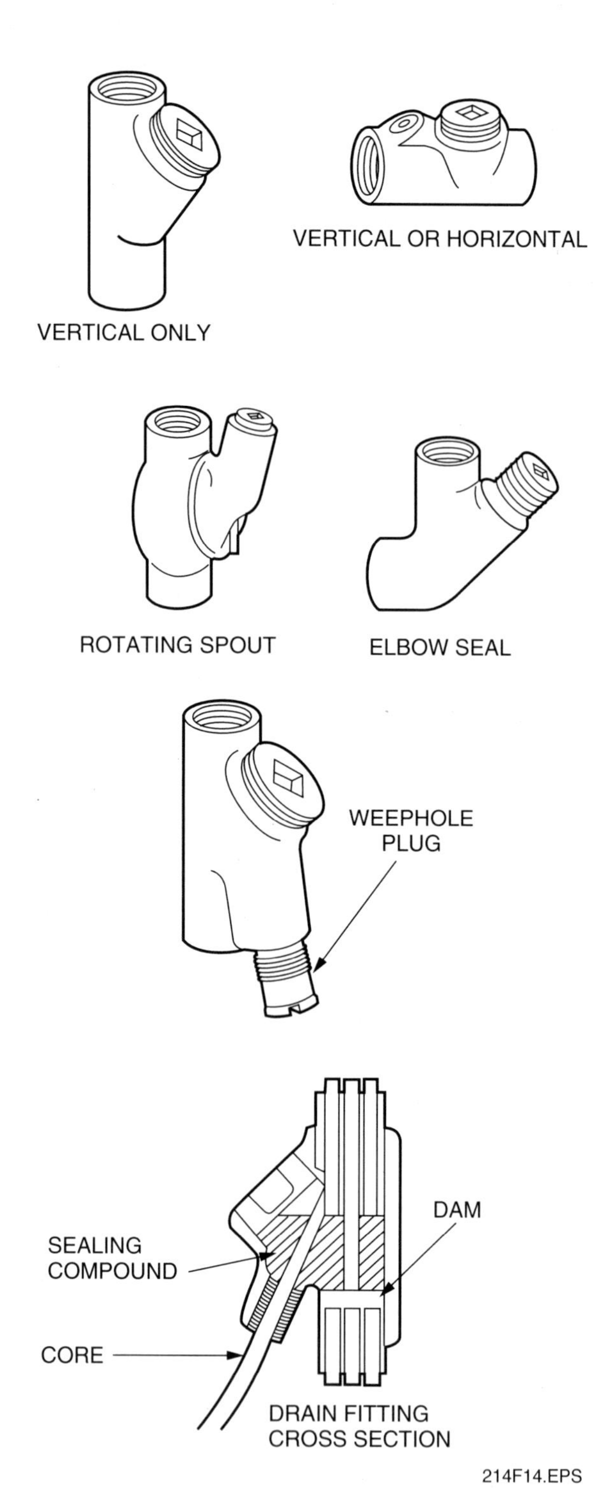

Figure 14 ◆ Seal-off fittings.

Step 2 To pack the seal off, remove the threaded plug or plugs from the seal-off fitting and insert the fiber supplied with the packing kit. Tamp the fiber between the conductors and the hub before pouring the sealing compound into the fitting.

Step 3 After tamping the fiber, pour in the sealing cement and tightly replace the threaded plug.

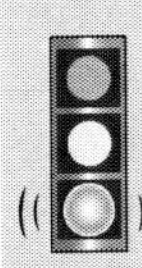

NOTE

The fiber packing prevents the liquid compound from running out and allows it to harden.

The seal-off fittings shown in *Figure 14* are typical of those used in the electrical and instrumentation industry. Seal-off fittings are available for both vertical and horizontal installations. Always refer to the appropriate NEC section before installing any conduit system that may require the installation of seal-off fittings.

2.1.3 Intermediate Metal Conduit (IMC)

Intermediate metal conduit has a wall thickness that is less than that of rigid metal conduit but greater than that of EMT. The weight of IMC is approximately two-thirds that of rigid metal conduit. Because of its lower purchase price, lighter weight, and thinner walls, IMC installations are generally less expensive than comparable rigid metal conduit installations. IMC installations have high strength ratings nonetheless. Specific information on intermediate metal conduit may be found in ***NEC Article 342***.

The outside diameter of a given size of IMC is the same as that of the comparable size of rigid metal conduit. Therefore, rigid metal conduit fittings may be used with IMC. Because the threads on IMC and rigid metal conduit are the same size, no special threading tools are needed to thread IMC. Some contractors feel that threading IMC is more difficult than threading rigid metal conduit because IMC is somewhat harder.

The internal diameter of a given size of IMC is a little larger than the internal diameter of the same size of rigid metal conduit because of the difference in wall thickness. However, the NEC rules for conductor fill for both IMC and rigid metal conduit are the same. The slightly larger internal diameter of IMC often makes pulling wire easier than pulling the same size and quantity of wire through rigid metal conduit. IMC has a maximum trade size of 4".

Bending IMC is considered easier than bending rigid metal conduit because of the reduced wall thickness. However, bending is sometimes complicated by kinking, which may be caused by the increased hardness of IMC.

NEC Section 342.120 requires that IMC be identified along its length at 5' intervals with the letters IMC. ***NEC Section 110.21*** describes this marking requirement.

Use of IMC may be restricted in some jurisdictions. It is important to investigate the requirements of each jurisdiction before selecting any materials.

2.1.4 Flexible Metal Conduit (FMC), Liquidtight Flexible Metal Conduit (LFMC), and Fittings

Flexible metal conduit, sometimes called flex, may be used for many kinds of wiring systems. Flexible metal conduit is made from a single strip of steel or aluminum, wound and interlocked. It is typically available in sizes from ⅜" to 4" in diameter. Flexible metal conduit is shown in *Figure 15*.

LIQUIDTIGHT CONDUIT

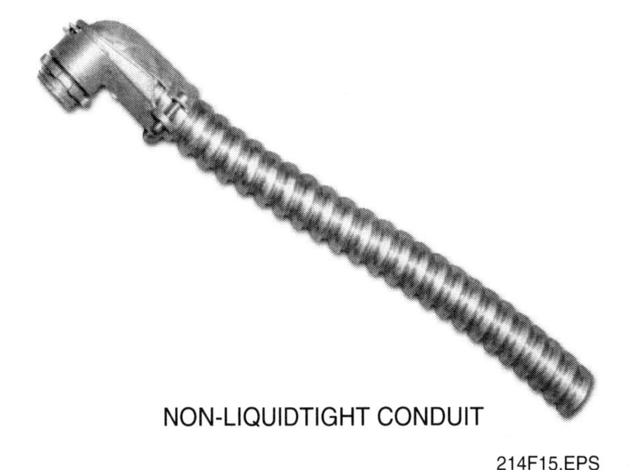

NON-LIQUIDTIGHT CONDUIT

214F15.EPS

Figure 15 ◆ Flexible metal conduit.

Flexible metal conduit is installed by some contractors for branch circuit wiring in walls and ceilings. The use of flexible metal conduit to connect equipment or machines that vibrate or move slightly during operation is considered good practice. Final connection to equipment having an electrical connection point that is not easily accessible is often accomplished with flexible metal conduit.

Flexible metal conduit is easily bent, but the minimum bend radius is the same as for other types of conduit. It should not be bent more than the equivalent of four quarter bends (360 degrees total) between pull points, such as conduit bodies and boxes.

Flexible metal conduit is generally available in two types: non-liquidtight and liquidtight.

NEC Articles 348 and ***350*** cover the uses of flexible metal conduit and liquid flexible metal conduit, respectively.

Non-liquidtight flexible metal conduit is only permitted in dry locations.

Liquidtight flexible metal conduit has an outer covering of liquidtight, sunlight-resistant, flexible material that acts as a moisture seal. It is intended for use in wet locations. It is used primarily for equipment and motor connections when movement or vibration of the equipment is likely to occur. Bend, size, and support requirements for liquidtight conduit are the same as for all flexible conduit. Fittings used with liquidtight conduit must also be liquidtight.

Support requirements for flexible metal conduit are found in ***NEC Section 348.30***. Straps or other means of securing the flexible metal conduit must be spaced closer together (every 4½' and within 12"of each end) than for rigid conduit. However, at terminals where flexibility is necessary, lengths of up to 36" without support are permitted.

Flexible metal conduit can be connected to boxes, enclosures, or equipment using specially designed connectors. These connectors are available in straight, 45-degree, and 90-degree configurations. Illustrations of flexible metal conduit fittings are shown in *Figure 16*.

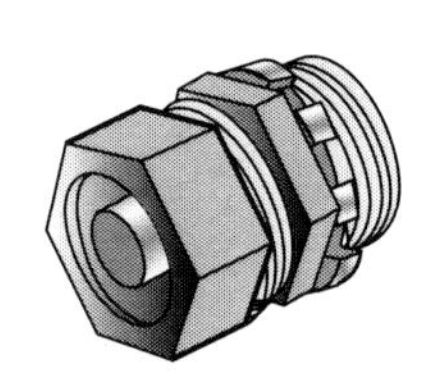

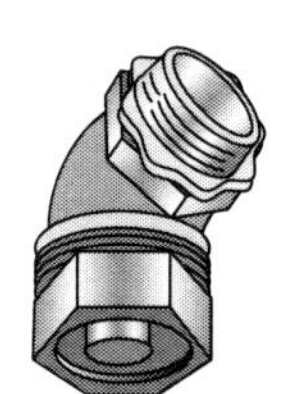

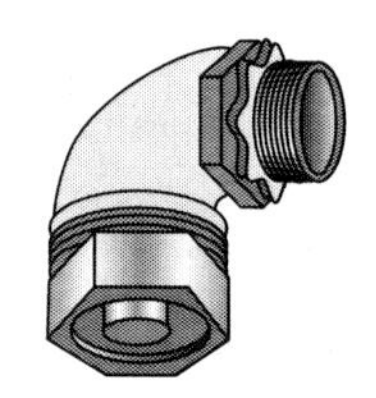

Figure 16 ◆ Liquidtight flexible metal conduit connectors.

Flexible metal conduit can also be made to connect to both EMT setscrew or raintight installations by using special adapter fittings such as those shown in *Figure 17*. When using one of these adapters, make sure that flexible conduit is pushed as far as possible into the fitting so that the sharp ends of the flex are not exposed to the conductors.

NOTE

Always use flexible metal conduit fittings with flexible metal conduit and liquidtight flexible metal conduit fittings with liquidtight flexible metal conduit. As shown in *Figure 16*, liquidtight connectors are similar to EMT raintight connectors in that a compression ring and compression nut are used to form a raintight seal.

CAUTION

Liquidtight flexible metal conduit connected to a setscrew flex connector does not provide protection from liquid entry.

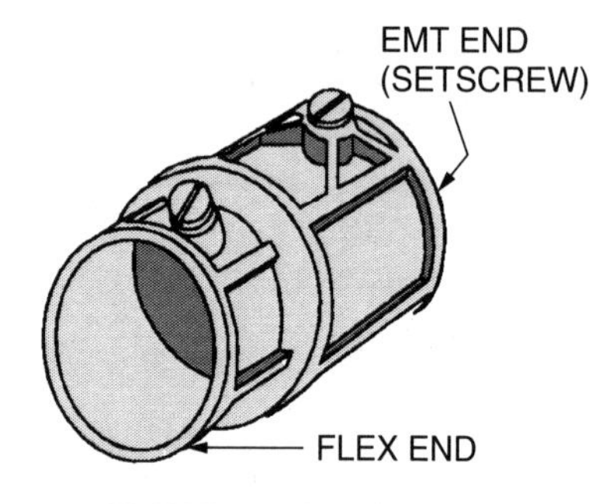

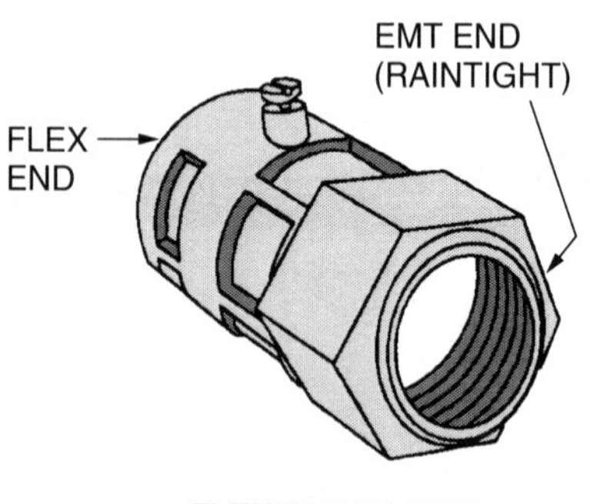

Figure 17 ◆ Adapter fittings.

2.1.5 Conduit Nipples

The NEC does not specifically define a conduit nipple. However, it does state in ***NEC Chapter 9, Table 1, Note 4***, that conduit or tubing nipples having a maximum length not to exceed 24" may be filled to 60% of total cross-sectional area instead of 40%, as required in ***NEC Table 1***. From this exception it is generally deduced that a conduit nipple is considered any section of conduit not greater than 24" in length.

Conduit nipples are available in various lengths from conduit manufacturers and are often handmade in the field. The NEC does not allow **running threads** on any conduit or nipple, which means the threads are cut from one end only and extend from one end to the other with no interruption. When an installation requires short nipples (6" or less), it is generally a good practice to purchase fabricated nipples. It is very difficult to properly chuck the nipple in any type of vise or clamp in order to thread both ends of the nipple without damaging the threads already cut.

Offset nipples can also be used to connect two pieces of electrical equipment that are in close proximity, where a slight offset is required. Offset nipples are available in various configurations, as shown in *Figure 18*.

Figure 18 ◆ Offset nipples.

2.2.0 Nonmetallic Conduit

Nonmetallic conduit is any conduit that does not conduct electricity. Based on that description, there are several types of conduit that may be classified as nonmetallic conduit, such as conduit made from concrete products, fiberglass, and other synthetic materials. Even though these types of nonmetallic conduit are appropriate in special applications, the most commonly used types of nonmetallic conduit in the electrical and instrumentation industries are rigid nonmetallic conduit (PVC) and electrical nonmetallic tubing (ENT).

Because nonmetallic conduit does not conduct electricity, it does not have the bonding characteristics of metal conduit. Therefore, a separate, bare grounding wire must be installed inside the conduit to provide a path to the service equipment grounding conductor. This conductor provides a path for any undesirable fault currents that may develop due to short circuits or faults. PVC and ENT are discussed in more detail in the following sections.

2.2.1 Rigid Nonmetallic Conduit and Fittings

The most common type of rigid nonmetallic conduit is made of polyvinyl chloride (PVC). ***NEC Article 352*** covers this type of conduit and refers to it as rigid nonmetallic conduit (RNC). Because PVC conduit is noncorrosive, chemically inert, and non-aging, it is often used in wet or corrosive environments. Corrosion problems found with steel and aluminum rigid metal conduit do not occur with PVC. However, PVC conduit may deteriorate under some conditions, such as extreme sunlight, unless rated for this environment.

All PVC conduit is marked according to standards established by the National Electrical Manufacturers Association (NEMA), the **Underwriters Laboratory (UL)**, and ***NEC Section 350.12***, which specifies that RNC be so marked at least every 10'. A section of PVC conduit is shown in *Figure 19*.

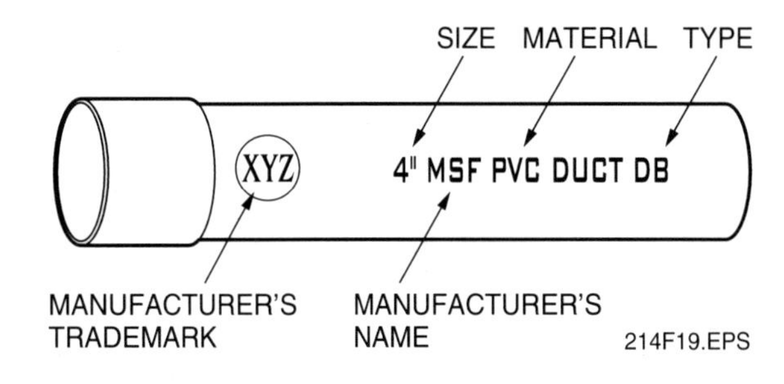

Figure 19 ◆ PVC conduit.

Because PVC conduit is lighter than steel and aluminum rigid conduit, IMC, and EMT, it is easier to handle. PVC conduit can usually be installed much faster than other types of conduit because the joints are made up with cement and require no threading.

PVC conduit contains no metal. This characteristic reduces the voltage drop of conductors carrying alternating current in PVC compared to identical conductors in steel conduit.

Because PVC is nonconducting, it cannot be used as an equipment grounding conductor. An equipment ground conductor sized from ***NEC Table 250.122*** must be pulled in each PVC conductor run.

PVC is available in lengths up to 20'. Some jurisdictions require it to be cut to 10' lengths prior to installation. PVC expands and contracts significantly when exposed to changing temperatures. Expansion and contraction are greater in long runs. To avoid damage to PVC conduit caused by temperature changes, expansion couplings such as the one shown in *Figure 20* are used. The inside of the coupling is sealed by one or more O-rings. This type of coupling may allow up to 6" of movement.

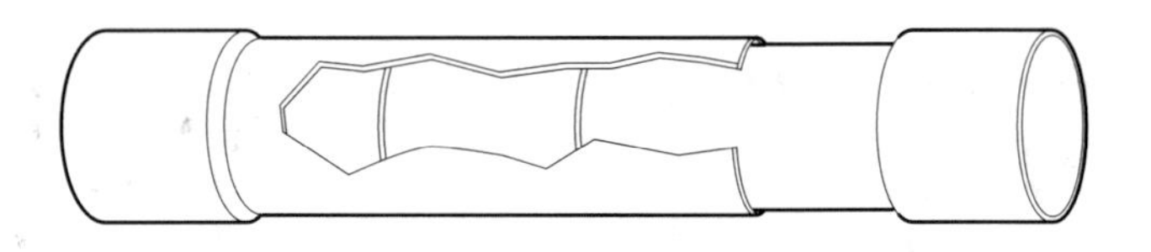

Figure 20 ◆ PVC expansion coupling

In some jurisdictions, PVC with expansion fittings cannot be installed if the conduit is exposed to direct sunlight. Check the requirements of the local jurisdiction prior to installing PVC.

Support requirements for PVC are found in ***NEC Table 350.30(B)***. As with other conduit, it must be supported within 3' of each termination. The maximum spacing between supports depends on the size of the conduit. Some of the regulations for maximum spacing of supports are presented here:

- *½" to 1" conduit* – every 3'
- *1¼" to 2" conduit* – every 5'
- *2½" to 3" conduit* – every 6'
- *3½" to 5" conduit* – every 7'
- *6" conduit* – every 8'

2.2.2 Electrical Nonmetallic Tubing (ENT)

Another nonmetallic conduit product that has become very common in residential and some commercial wiring applications is electrical nonmetallic tubing (ENT), known as Smurf. This slang term is typically used by those dealing with ENT, evidently associating its typical bright blue color with the blue cartoon characters, the Smurfs®. ENT, as defined by ***NEC Section 362.2,*** is a nonmetallic pliable corrugated raceway of circular cross section, with integral or associated couplings, connectors, and fittings for the installation of electrical conductors. In appearance, ENT looks and handles like plastic flexible conduit. It is very lightweight, easy to install, and its associated fittings snap directly onto the ends of the ENT. There are no other fittings that are compatible with ENT. Only those specifically designed for it will work. As would be expected due to its lightweight properties, ENT has limited usage. ***NEC Section 362.10*** lists the uses permitted, while ***NEC Section 362.12*** lists those uses not permitted for ENT. It has a limited maximum temperature exposure of 122°F, cannot be used for voltages over 600 volts, and cannot be exposed to direct sunlight, unless identified as sunlight resistant. One of its primary uses is as a concealed raceway for branch circuits in walls, floors, and ceilings. ENT must be marked every 10' with approved marking, according to ***NEC Sections 110.21*** and ***362.120***.

2.3.0 Boxes

A box is installed at each outlet, switch, or junction point for all wiring installations. Boxes are made from either metallic or nonmetallic material. Some boxes are shown in *Figure 21*.

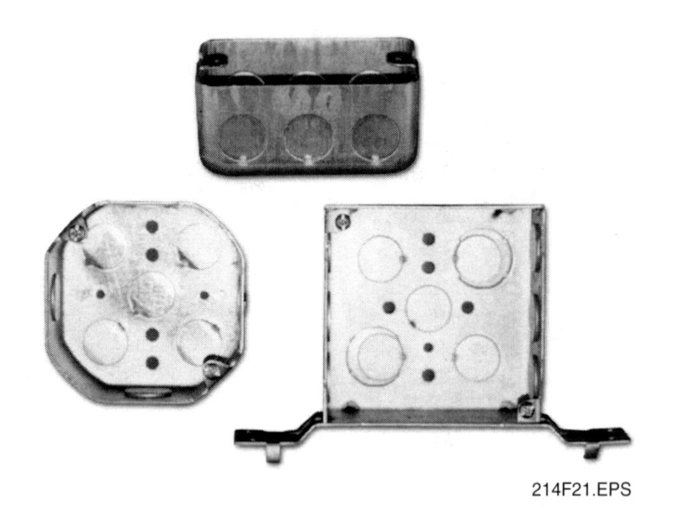

Figure 21 ◆ Boxes.

2.3.1 Metal Boxes

Metal boxes are normally made from sheet metal, except for those designed for hazardous locations, which are generally made from some type of cast metal. The surface is galvanized to resist corrosion and to provide a continuous ground.

Metal boxes for general use are made with removable circular sections called knockouts or pryouts. The circular sections are removed to make openings for conduit or cable connections. There are two basic types of knockouts:

One type is made by cutting into the metal until only a thin layer remains. This type of knockout is easily removed when sharply hit by a hammer and punch, as shown in *Figure 22*.

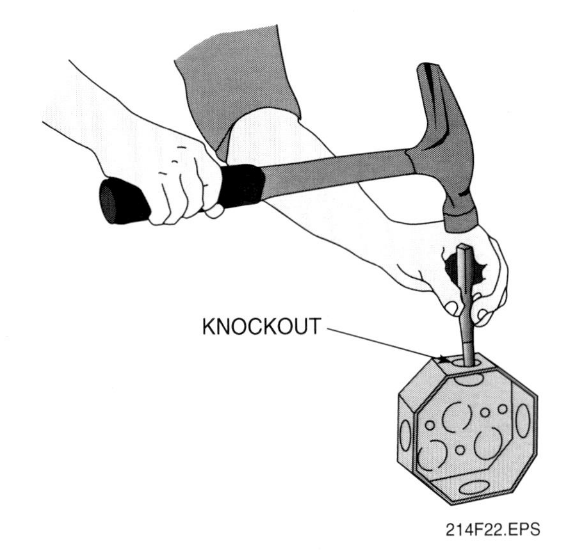

Figure 22 ◆ Knockout removal.

The other type of knockout is called a pryout. A section is cut completely through the metal but only part of the way around, leaving solid metal tabs at two points. A slot is cut in the center of the knockout. To remove the knockout, a screwdriver is inserted in the slot and twisted to break the solid tabs, as shown in *Figure 23*.

Conduit must often enter boxes, cabinets, or panels that do not have precut knockouts. In these cases, a knockout punch is used to make a hole for the conduit connection. A knockout punch is shown in *Figure 24*.

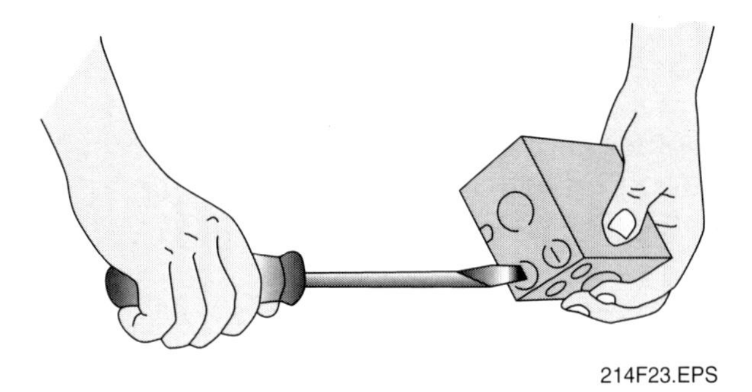

Figure 23 ◆ Pryout removal.

2.3.2 Nonmetallic Boxes

Nonmetallic boxes are made of PVC or Bakelite, which is a fiber-reinforced plastic. Nonmetallic boxes are often used in corrosive environments.

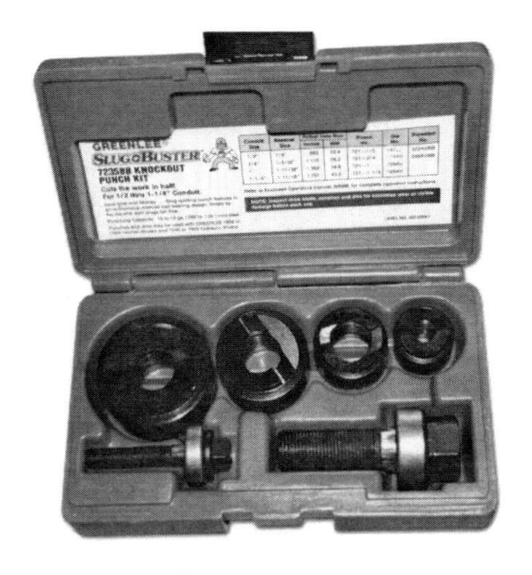

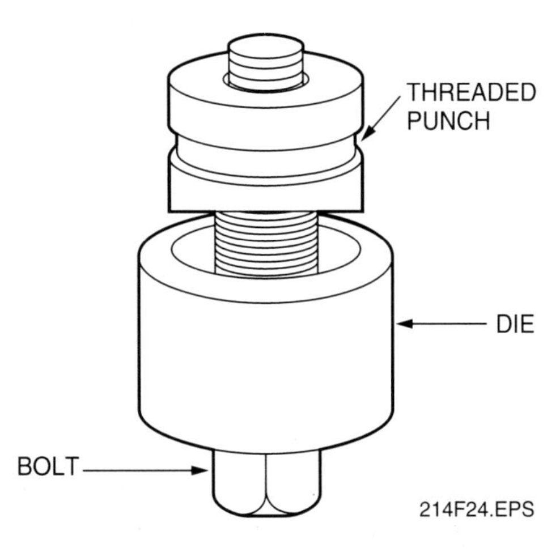

Figure 24 ◆ Knockout punch.

NEC Section 314.3 covers the use of nonmetallic boxes and the types of conduit, fittings, and grounding requirements for specific applications.

2.4.0 Conduit Supports

Conduit supports are available in many types and configurations. The conduit supports discussed in this section are standard devices often found in instrumentation installations.

2.4.1 One-Hole and Two-Hole Straps

One-hole and two-hole straps are used to attach and support conduit to a mounting surface. The spacing of these supports must conform to the minimum support spacing for each type of conduit as explained in the NEC. One-hole and two-hole straps are used for all types of conduit, including EMT, RMC, IMC, PVC, and flex. One-hole and two-hole straps are shown in *Figure 25*.

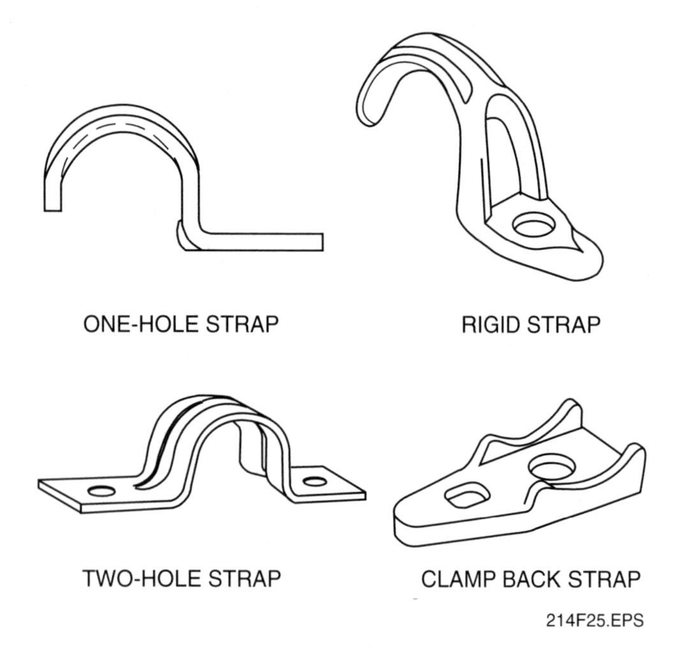

Figure 25 ◆ One-hole and two-hole straps.

The clamp back strap is used to space conduit ¼" away from surfaces in wet locations.

One-hole straps can also be used with a backplate to maintain a ¼" spacing from the surface as required for installations in wet locations.

2.4.2 Standoff Supports

A standoff support is used to support conduit away from the supporting structure. In the case of one-hole and two-hole straps, the conduit must be kicked up wherever a fitting occurs. If standoff supports are used, the conduit is held away from the supporting surface and no offsets (kicks) are required in the conduit at the fittings. Standoff supports may be used to support all types of conduit, including RMC, IMC, EMT, PVC, and flex, as well as tubing installations. A standoff support is shown in *Figure 26*.

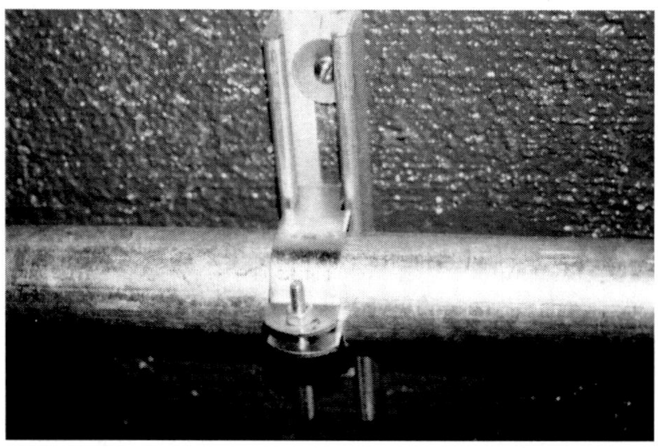

Figure 26 ◆ Standoff support.

2.4.3 Strut Supports

Strut supports and similar framing material are used in conjunction with strut-type conduit clamps to support conduit. A strut support may be attached to a ceiling, wall, or other surface, or it may be supported to form a trapeze. A strut support is shown in *Figure 27*.

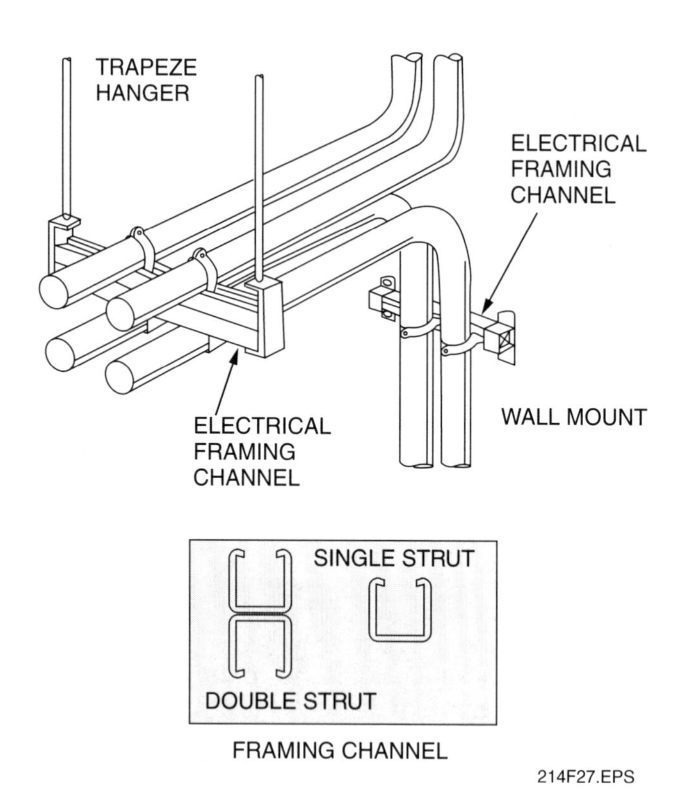

Figure 27 ◆ Strut supports.

2.4.4 Beam Clamps

Beam clamps are used with suspended hangers. The raceway is attached to or laid in the hanger. The hanger is suspended by a threaded rod. One end of the threaded rod is attached to the hanger. The other end is attached to a beam clamp. The beam clamp is then attached to a beam. A beam clamp with a wireway support assembly is shown in *Figure 28*.

2.5.0 INSTALLING CONDUIT

The types of conduit and their related NEC articles are listed below. They are arranged in order from least difficult and costly to most difficult and costly:

1. ENT (Electrical Nonmetallic Tubing) – ***NEC Article 362, Part II***
2. FMC (Flexible Metallic Conduit) – ***NEC Article 348, Part II***

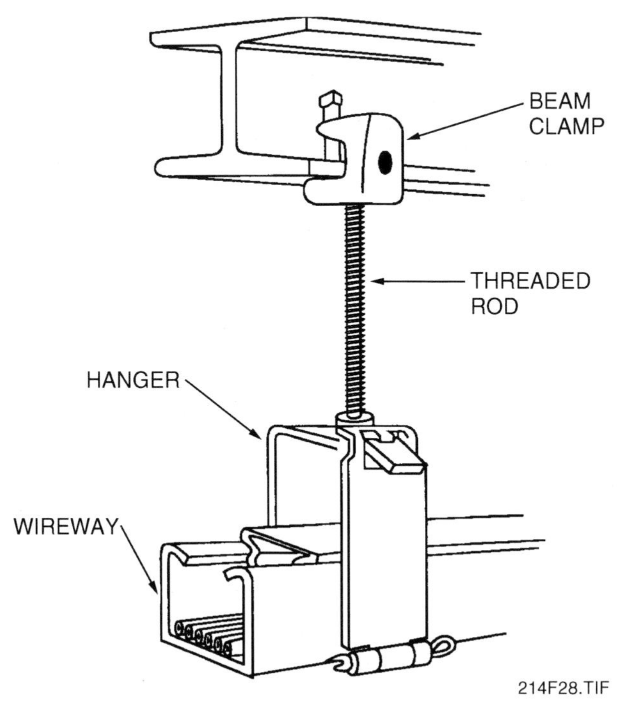

Figure 28 ◆ Beam clamp and hanger.

3. LFMC (Liquidtight Flexible Metal Conduit) – ***NEC Article 350, Part II***
4. RNC (Rigid Nonmetallic Conduit) – ***NEC Article 352, Part II***
5. EMT (Electrical Metallic Tubing) – ***NEC Article 358, Part II***
6. IMC (Intermediate Metal Conduit) – ***NEC Article 342, Part II***
7. RMC (Rigid Metal Conduit) – ***NEC Article 344, Part II***

Before installing a conduit system, the appropriate NEC article must be referenced in order to ensure a safe and compliant installation. Some general concerns regarding conduit installation include the following:

- ENT, FMC, and EMT with setscrew fittings may not be installed in wet or damp locations.
- ENT is not used in industrial facilities as a raceway in any outdoor installation.
- Each conduit fitting must match the conduit system for which it is designed. Only those conduit systems permitted in wet or damp locations may be installed in those environments.
- Rigid metal conduit and intermediate metal conduit may be installed in any environment or occupancy. This is not always feasible, however, due to the relatively high material and labor costs.
- Most commercial indoor raceway installations are done with EMT conduit systems with setscrew connectors and couplings.
- FMC is generally used as whips, which can be defined as short sections of flexible conduit (no longer than 6') that connect indoor commercial conduit branch circuits to lighting fixtures and other fixtures, devices, or equipment.
- LFMC may be found in industrial and commercial outdoor installations, connecting branch circuits to motors and other vibrating or replaceable equipment.

2.5.1 Size Requirements

In order to determine what size conduit to install, many factors must be considered. Do not assume that the engineering specifications on a project specify the conduit size based only on the NEC requirements and local codes. Because the NEC and local codes only reflect the minimum requirement, it is common for the conduit in a raceway system to be increased one or two sizes above the minimum to allow for ease in conductor installation and for any conductor additions due to future expansion.

In order to determine the minimum conduit size, it is necessary to know the size, type, and number of conductors to be installed in the conduit. The size and number of conductors necessary to carry the load can be determined by referring to ***NEC Tables 310.16*** through ***310.21*** to locate the correct conductor size based on the load in amperes, type of conductor insulation, and ambient temperature of the location. Once that information is obtained, refer to ***NEC Chapter 9, Tables 1, 4, 5, or 5A***, to determine the correct minimum size of conduit based on conductor dimensions and specific conduit properties.

NEC Chapter 9, Table 1, states that conduit holding more than two conductors can only be filled to 40% of the conduit's cross-sectional area. In order to determine how many conductors can be installed in a conduit, the cross-sectional area of the conduit must be determined. This can be found in the ***NEC Chapter 9, Table 4***. Then the total cross-sectional area of the conductors must be determined using ***NEC Chapter 9, Tables 5*** and ***5A***. The total cross-sectional area of the conductors cannot be more than 40% of the cross-sectional area of the conduit (3 or more conductors), according to ***NEC Chapter 9, Table 1***.

2.5.2 Conduit Bend Requirements

One of the main concerns when installing a conduit run is whether the conductors can be installed without damage to them. The type of insulation, the number of conductors, and the number of bends must also be considered.

NEC Section 344.26 states that a run of conduit between pull points (for example, outlet-to-outlet, fitting-to-fitting, or between outlet and fitting) shall not contain more than the equivalent of four 90-degrees bends, or 360 degrees total. This total bend requirement includes bends located immediately at an outlet or fitting.

2.6.0 Cutting Conduit

Properly cutting conduit, regardless of the type, is an important element of installing conduit systems. It is a common myth that no training or practice is needed in order to properly and safely cut conduit. This task is often assigned to inexperienced workers. This practice can result in wasted conduit, poor fit, and possible injuries. It is important to consider the cutting process of conduit to be as critical as any of the other processes involved in the installation of a conduit raceway system. Identifying, selecting, and safely using the proper tool or tools needed to cut a particular type of conduit are basic steps that every conduit installer should follow. This section presents the basic steps of cutting conduit.

2.6.1 Cutting RMC and IMC

Because of their similar properties, rigid metal conduit (RMC) and intermediate metal conduit (IMC) use the same cutting techniques. Because IMC is harder than RMC, it will consume blades and cutting wheels faster. Whether the tool of choice is a hacksaw or a pipe cutter, certain techniques must be applied to achieve a finished, quality cut.

A hacksaw or bandsaw is commonly used to cut RMC and IMC. *Figure 29* shows typical hand hacksaws and a portable bandsaw. The following steps should be followed when preparing and cutting RMC and IMC:

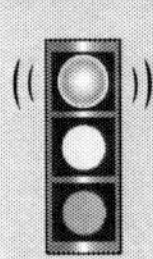

WARNING!

Always wear safety goggles and hand protection as needed when cutting any conduit. When using an electrically powered hacksaw for cutting, make sure the hacksaw (commonly called a bandsaw) is either double insulated or is connected to a GFCI-protected circuit.

Step 1 If using a hacksaw, a blade with 24 or 32 teeth per inch is usually recommended for cutting RMC and IMC. The teeth must point to the front of the saw when installed.

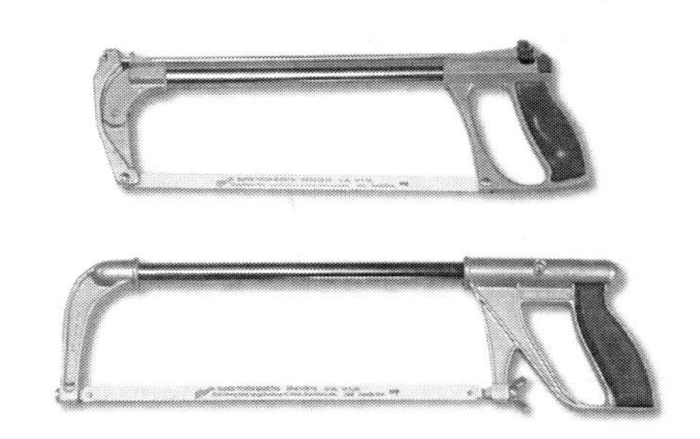

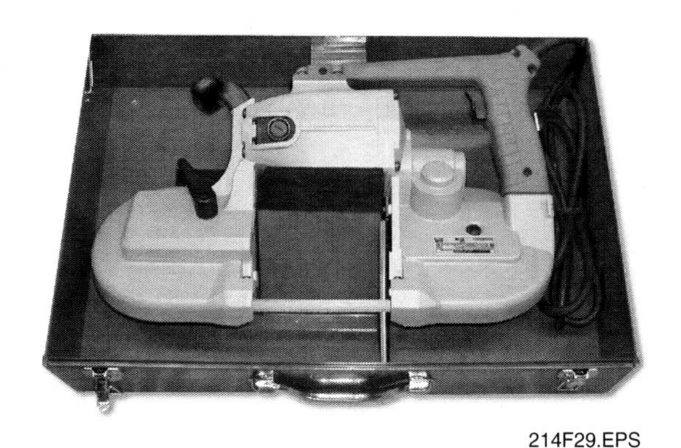

214F29.EPS

Figure 29 ◆ Hand hacksaws and portable bandsaw.

Step 2 Properly measure and mark the conduit, allowing enough conduit stub for fittings, bends, and other factors affecting the finished installation.

Step 3 Make sure the appropriate personal protective equipment is in place. If using a power bandsaw, make sure the saw is double-insulated or connected to a GFCI-protected circuit.

Step 4 Verify that the saw blade is installed correctly.

Step 5 Secure the conduit in an approved vise or clamping device, making sure that the vise or clamping device is also secured to prevent it from tipping over during cutting.

Step 6 If using a hand hacksaw, rest the blade on the conduit mark where the cut is to be made. Pull backward gently until a groove is made. Keep the hacksaw blade perpendicular to the conduit. Then make even strokes forward and backward until the cut is finished (*Figure 30*).

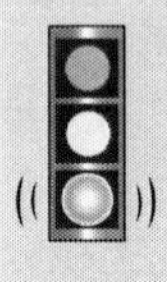

NOTE

Most of the cutting action takes place in the forward stroke of the saw. The backward stroke cleans the cut.

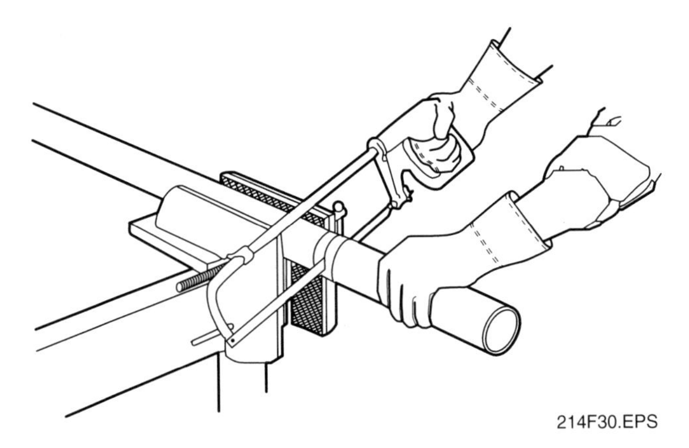

Figure 30 ◆ Cutting conduit with a hand hacksaw.

Step 7 If using a power bandsaw, use a fine-tooth blade and position the bandsaw blade on the conduit so the conduit rests firmly against the rear stop plate of the bandsaw. Position the saw at a right angle to the conduit, and maintain that position during the cutting process.

Step 8 Grip the trigger grip with one hand (without squeezing the trigger), grip the forward support knob with the other hand, and gently squeeze the trigger. Do not apply downward pressure on the saw. Allow the saw blade to perform the cutting process by using the saw's own weight as downward pressure. Cut all the way through the conduit, and do not allow the cut portion to break off from the conduit.

Step 9 Check the finished cut for straightness and smoothness. *Figure 31* shows correct and incorrect finished cuts. Metal conduits must be deburred or reamed following the cutting process. That procedure is covered in a later section.

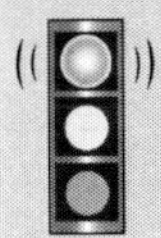

WARNING!
Cut conduit edges can be razor-sharp. Wear proper hand protection and avoid injuries.

A pipe cutter is also commonly used to cut RMC and IMC. Unlike a hacksaw or bandsaw, it uses a cutting wheel or a series of cutting wheels to cut into the pipe wall as pressure is applied by gradually turning a tee-handle in a clockwise direction on the cutter. In order to achieve a complete cut around the pipe wall, the cutter is either rotated around the pipe or the pipe is rotated in a fixed cutter.

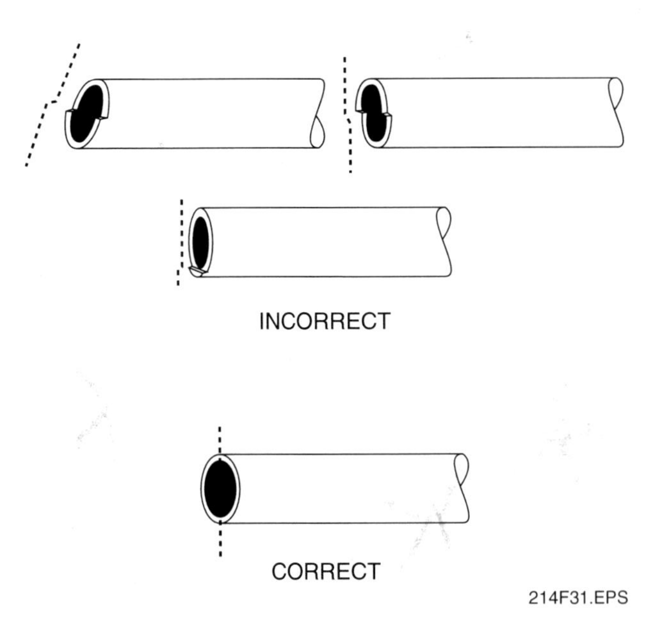

Figure 31 ◆ Conduit ends after cutting.

If the pipe is clamped in a portable pipe vise (*Figure 32*), the cutter is manually rotated around the pipe. If the pipe is secured, or chucked, in a portable power pipe threader, it is rotated in a pipe cutter held against a tool rest that is part of the machine.

A tool used to ease hand threading is the power drive, commonly referred to as a mule because of the gear reduction that provides tremendous torque at the pipe chuck.

The pipe chuck on the power drive is considered a speed chuck. The jaws of the chuck are opened by tightly gripping the large dimpled solid wheel with the hand and sharply jerking it in a downward clockwise direction. Once the pipe is in place, the jaws are tightened by gripping the wheel with the opposite hand and sharply snapping the wheel in a downward counterclockwise direction until the pipe or conduit is held tightly in place.

The mule normally comes equipped with a footswitch. It may be used in combination with a hand threader, keeping both hands free for threading. Another common accessory is a tripod stand on which the mule may be mounted, making it a portable power drive.

Follow these safety precautions when cutting RMC and IMC with a pipe cutter:

- Always wear safety goggles and hand protection when cutting any conduit.
- Keep all loose clothing free from rotating parts.
- Remember that power-driven vises are extremely high-torque machines and do not immediately stop when de-energized.

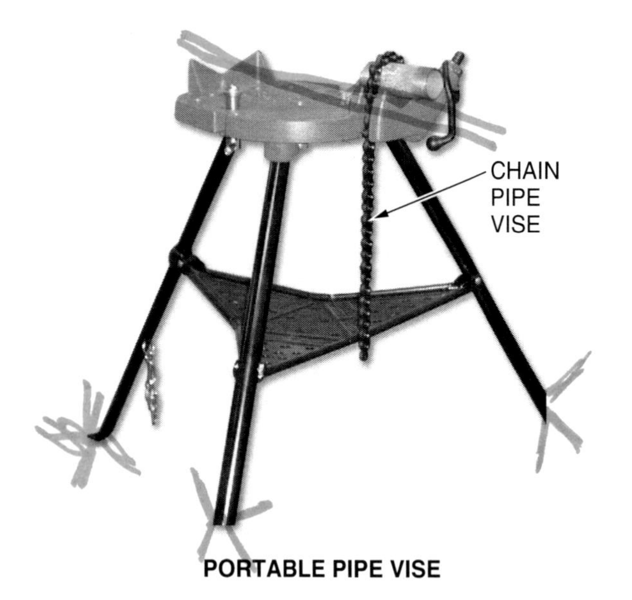

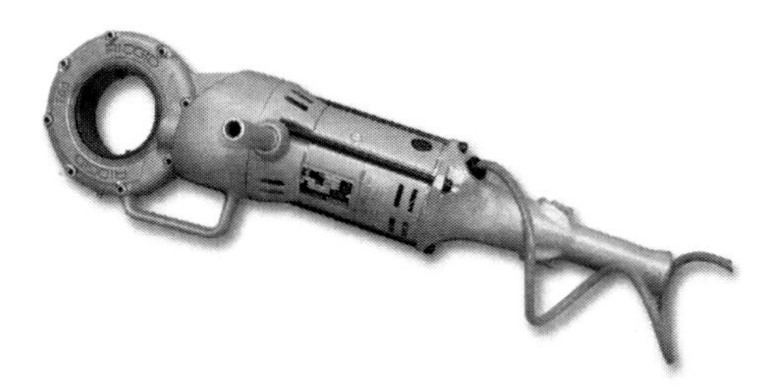

Figure 32 ◆ Portable pipe vise, portable power pipe threader, and power-driven vise.

- Should a pipe cutter bind or otherwise bite into the conduit, do not attempt to stop the action or hold it back. Severe injuries can result. De-energize the machine and allow the rotating action to stop before trying to remove the cutter.
- Make sure all electrically powered equipment is properly grounded.

The following steps are basic procedures used in cutting RMC and IMC using a pipe cutter:

Step 1 Inspect the cutting wheel or wheels for chips, cracks, dullness, or other deformities and replace as needed. Inspect the operation of the tee handle, lubricating as needed. Inspect the vise and electrical connections as applicable. Make sure the vise is secured to prevent it from tipping over during cutting.

Step 2 Properly measure and mark the conduit, allowing enough conduit stub for fittings, bends, and other factors affecting the finished installation.

Step 3 Make sure the appropriate personal protective equipment is in place and all loose clothing is secured if using a power vise or other rotating equipment.

Step 4 Secure the conduit in an approved vise or power pipe threader, making sure that the vise or pipe threader is secure.

Step 5 Open the cutter, and place it over the conduit with the cutter wheel on the mark. *Figure 33* shows the correct placement of the cutter wheel on the conduit for manual counterclockwise rotation of the cutter or clockwise rotation of the pipe in a power pipe threader.

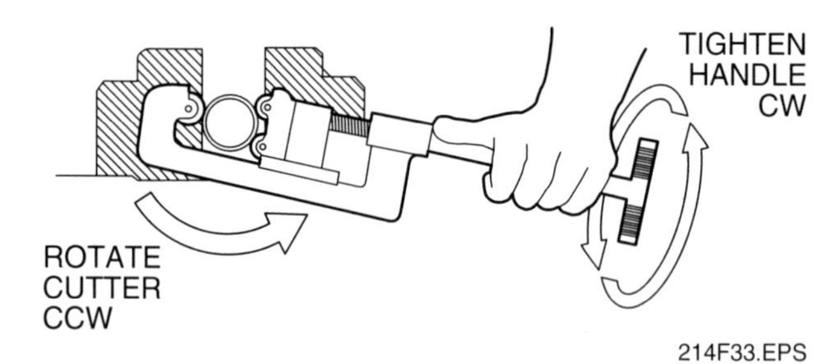

Figure 33 ◆ Cutter placement.

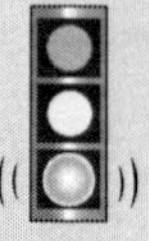

NOTE

Add a few drops of cutting oil to the wheel of the cutter, and continue to add cutting oil as needed as the cutting process progresses.

Step 6 Gradually apply cutter wheel pressure onto the conduit wall by tightening the tee handle in a clockwise direction, as shown in *Figure 33*. Do not attempt to rapidly cut the pipe by overtightening the tee handle. Turn the handle clockwise approximately one-quarter turn for each two or three rotations around the conduit. Overtightening can distort the conduit wall and can also break or damage the cutting wheel.

Step 7 If using a manual vise rather than a power-driven vise, rotate the entire pipe cutter in a counterclockwise direction around the pipe in a cutting motion two or three times before turning the tee handle another one-quarter turn. After every two or three rotations around the pipe, turn the tee handle another one-quarter turn inward. This is shown in *Figure 34*.

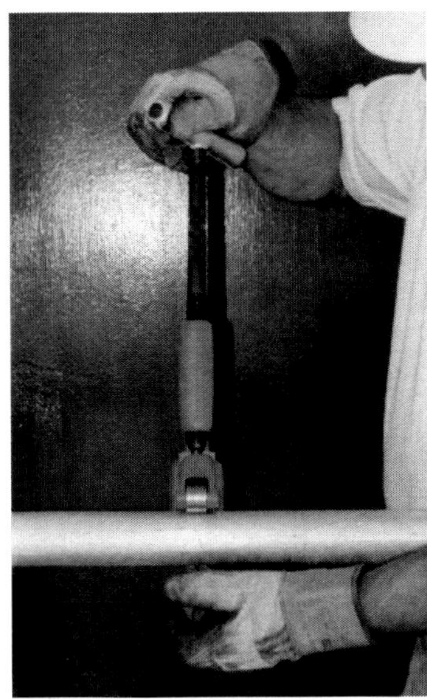

214F34.EPS

Figure 34 ◆ Rotating the cutter.

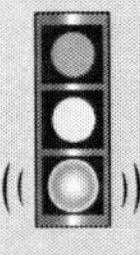

NOTE

When using a power-driven vise, the pipe cutter does not need to be rotated around the conduit, but the tee handle must be turned inward one-quarter turn for every two or three rotations of the conduit.

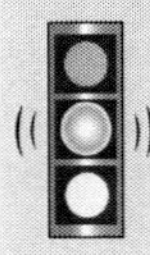

CAUTION

Proper lubrication must be used when cutting with a power-driven vise. Do not attempt to hold back a cutter that bites into the conduit and attempts to rotate with it.

Step 8 As the cut nears completion, make sure that the unchucked piece of conduit is supported to prevent it from falling in an unsafe location. Then, continue the action in Step 7 until the conduit is completely cut.

Step 9 Stop the power-driven vise if used. Unchuck the conduit. Clean the cutter and vise.

2.6.2 Cutting EMT

Electrical metallic tubing is fairly simple to cut. It is usually cut with a hand hacksaw, power hacksaw (bandsaw), or, in some cases, with a pipe/tubing cutter. The safety issues, warnings, and steps that apply to cutting RMC and IMC with a hand hacksaw or a power bandsaw also apply to EMT. The following steps are basic procedures used to cut EMT:

WARNING!

Always wear safety goggles and hand protection as needed when cutting any conduit. When using an electrically powered bandsaw for cutting, make sure the saw is either double-insulated or is connected to a GFCI-protected circuit.

Step 1 A blade with 24 or 32 teeth per inch is usually recommended for cutting EMT. The teeth must point to the front of the saw.

Step 2 Properly measure and mark the conduit, allowing enough conduit stub for fittings, bends, and other factors affecting the finished installation.

Step 3 Make sure the appropriate personal protective equipment is in place. If using a power bandsaw, make sure the saw is double-insulated or connected to a GFCI-protected circuit.

Step 4 Verify that the saw blade is installed correctly.

Step 5 Secure the conduit in an approved vise or clamping device, making sure that the vise or clamping device is also secured to prevent it from tipping over during cutting.

Step 6 If using a hand hacksaw, rest the blade on the conduit mark where the cut is to be made. Pull backward gently, until a groove is made. Keep the hacksaw blade perpendicular to the conduit. Then make even strokes forward and backward until the cut is finished.

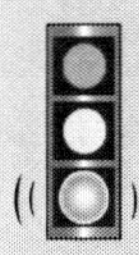

NOTE

Most of the cutting action takes place in the forward stroke of the saw. The backward stroke cleans the cut.

Step 7 If using a power bandsaw, position the bandsaw blade on the conduit so that the conduit rests firmly against the rear stop plate of the bandsaw. Position the saw at a right angle to the conduit, and maintain that position during the cutting process.

Step 8 Grip the trigger grip with one hand without squeezing it. Grip the forward support knob with the other hand, and gently squeeze the trigger. Do not apply downward pressure on the saw. Allow the saw blade to perform the cutting process, using the saw's own weight as downward pressure. Cut all the way through the conduit, and do not allow the cut portion to break off from the conduit in the vise.

Step 9 Check the finished cut for straightness and smoothness. Metal conduits must be deburred or reamed after cutting. That procedure is covered in a later section.

Step 10 If using a pipe or tubing cutter, follow the instructions in the previous section.

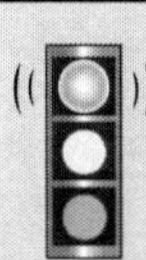

WARNING!
Cut conduit edges can be razor sharp. Wear proper hand protection and avoid injuries.

2.6.3 Cutting FMC and LFMC

Cutting flexible conduit requires practice in order to achieve a clean cut on the end of the flex. The following steps may be practiced in order to achieve a quality finished cut on flexible metal conduit or liquidtight flexible metal conduit.

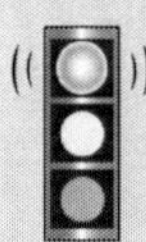

WARNING!
Always wear safety goggles and hand protection as needed when cutting any conduit. Be aware of sharp, jagged edges on cut flexible metal conduit.

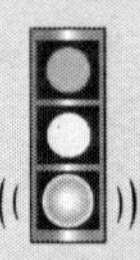

NOTE
Only a hand hacksaw should be used when cutting flexible metal conduit or liquidtight flexible metal conduit for better control because a complete (all-the-way-through) cut is not necessary. This will be explained in the following steps

Use the following procedure when cutting standard flexible metal conduit (FMC):

Step 1 A blade with 24 or 32 teeth per inch is usually recommended for cutting flexible metal conduit. The teeth must point to the front of the saw.

Step 2 Properly measure and mark the flex, leaving enough measured stub for the fitting.

Step 3 Tightly support the flex on either a flat surface with one hand or in a pipe vise, with the marked area facing up and extending 2" or 3" from the supported edge.

Step 4 Place the hacksaw blade at a 45-degree angle across the mark on the flex.

Step 5 Pull the hacksaw back to create a starting groove in the flex. Push the hacksaw forward in a cutting motion.

Step 6 Repeat this backward and forward cutting motion until a cut is made that is approximately one quarter of the way through the diameter of the flex. Do not cut completely through the diameter of the flex. Cut only one quarter of the way through the diameter.

Step 7 Stop cutting and remove the hacksaw.

Step 8 Support the cut portion of the flex between both hands with the cut in the flex facing upward, and gently break or snap the flex in a downward movement with both hands. The flex should separate into two parts, without major jagged edges.

Step 9 If any jagged edges do protrude from the finished end, remove these edges using the appropriate size cutting pliers (dikes) before installing fittings.

Use the following procedure when cutting liquidtight flexible metal conduit (LFMC):

Step 1 A blade with 24 or 32 teeth per inch is usually recommended for cutting liquidtight flexible metal conduit. The teeth must point to the front of the saw.

Step 2 Properly measure and mark the flex, leaving enough measured stub for the fitting.

Step 3 Use a utility knife to make a circled cut all the way through the plastic coating of the LFMC on the measured mark, completely around the flex.

Step 4 Place the hacksaw blade at a 45-degree angle across the mark on the flex.

Step 5 Pull the hacksaw back to create a starting groove in the flex. Push the hacksaw forward in a cutting motion.

Step 6 Repeat this backward and forward cutting motion until a cut is made approximately one quarter of the way through the diameter of the flex. Do not cut completely through the diameter of the flex. Cut only one quarter of the way through the diameter.

Step 7 Stop cutting and remove the hacksaw.

Step 8 Support the cut portion of the flex between both hands with the cut in the flex facing upward. Gently break or snap the flex in a downward movement with both hands. The flex should separate into two parts, without major jagged edges.

Step 9 If any jagged edges do protrude from the finished end, remove these edges using the appropriate size cutting pliers (dikes) before installing fittings.

Step 10 Trim away any jagged edges of the plastic coating with a utility knife or cutting pliers.

2.6.4 Cutting RNC (PVC) and ENT

Rigid nonmetallic conduit (PVC or RNC) and electrical nonmetallic tubing (ENT) can be cut using the same tools for each. A hand hacksaw is often used to cut PVC conduit, mainly because of its availability and versatility.

Hand-operated PVC conduit cutters, shown in *Figure 35*, are also used to cut PVC conduit and ENT up to 1¼". Except for the smallest tool shown, these operate in a ratchet-type motion. With each squeeze of the handles, the cutter blade cuts deeper into the conduit and remains in that position until the handles are released and squeezed again, or until a cut is made completely through the conduit or tubing.

2.7.0 Deburring Conduit

Most conduit, regardless of the type, tends to leave a burr on the inside diameter whenever it is cut. The burr is more distinct and damaging to conductors in rigid metal conduit cut with a pipe cutter than it is when cut with a hacksaw or bandsaw because the cutter wheel pushes in on the conduit wall while cutting. However, all burrs should be removed before the conduit is installed in the raceway system regardless of the type of material or the method of cutting. Even a plastic burr left on the inside of PVC conduit can cause damage to conductor insulation.

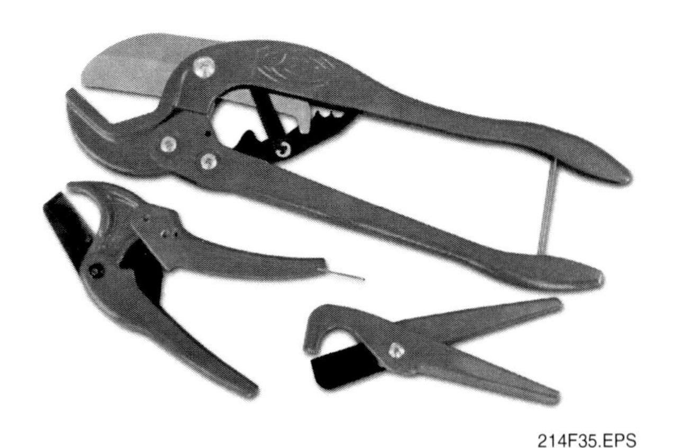
214F35.EPS

Figure 35 ◆ Hand-operated PVC conduit cutters.

2.7.1 Deburring RMC and IMC

Rigid metal conduit (RMC) and intermediate metal conduit (IMC) are typically cut with either a hacksaw or a pipe cutter. Both methods of cutting cause inside diameter burring that must be removed before the conduit is installed in the raceway system.

There are two methods normally used in deburring rigid metal conduit or IMC: using a half-round file and using a pipe reamer.

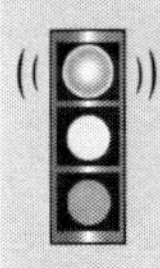

WARNING!
Deburring creates metal shavings and sharp edges. Both dangers must be guarded against by using the appropriate personal protective equipment, like safety glasses or goggles, and hand protection, such as gloves.

A half-round file is used because it does not have the tendency to cut a groove into the wall of the conduit. Round files and flat files do not fit the curvature of the conduit and are thus more likely to damage the conduit. Half-round files are available in a variety of radii. If a selection is available, choose a file with the radius that best fits the curvature of the inside of the conduit.

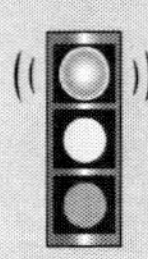

WARNING!
If using a file to deburr, make sure a safety handle has been installed on the tail of the file to prevent puncture injuries to the hand or arm from the file's tail.

Following these steps when deburring using a half-round file:

Step 1 Verify that the end of the conduit has a clean, square cut. If the cut of the conduit end is not square, re-cut the conduit before reaming it.

Step 2 Secure the conduit in a vise, making sure the vise is also secure to prevent it from tipping over during deburring.

Step 3 Use the rounded side of the file (*Figure 36*) and apply even forward strokes to the burred edge of the conduit. Do not file in a backward and forward motion. Files only cut on the forward stroke.

Step 4 Continue all the way around the inside diameter of the conduit, carefully checking with a finger the amount of burr remaining.

Step 5 Do not overfile. When the burr is evenly removed from the inside of the conduit, and the inside wall of the conduit is flush with the cut edge, stop filing.

One of the most common methods used to deburr RMC and IMC is using a handheld ratchet pipe reamer. There are generally two types of handheld ratchet pipe reamer heads available: the straight-fluted type and the spiral-fluted type (*Figure 37*). Either type works well in hand reaming rigid and intermediate metal conduit, as long as the conduit is secured in a stationary vise and not turned in a power vise.

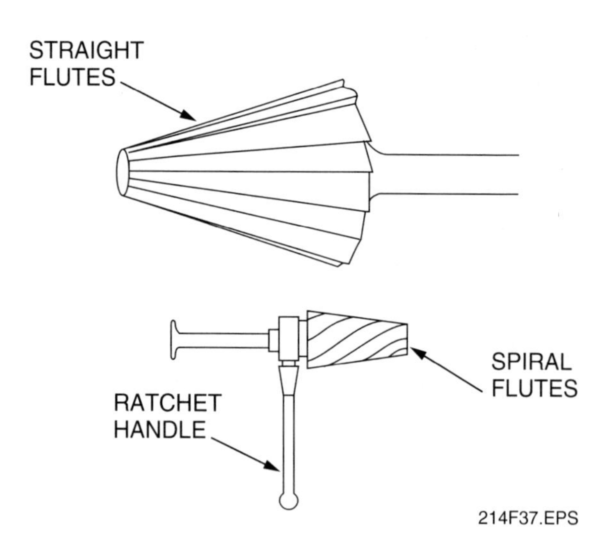

Figure 37 ◆ Reamers.

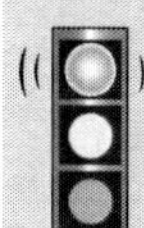

WARNING!

If the conduit that is to be reamed is chucked in a power threader, which is highly discouraged, and turned while a spiral-fluted hand-held reamer is held stationary without a tool rest, the reamer will have a very dangerous tendency to act as a large, self-tapping screw. It will rapidly thread itself into the inside of the conduit while the conduit is turning. This can easily cause injury to the operator should he or she try to stop the threading action. If power reaming is used, only a straight-flute reamer should be employed. It must be restrained with a tool rest on the threader.

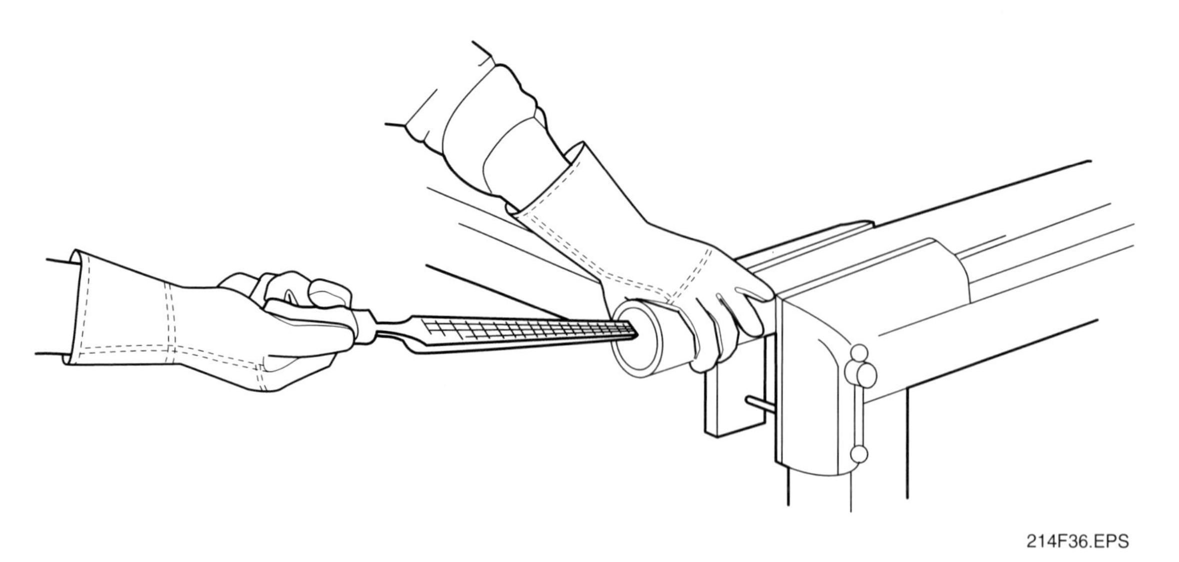

Figure 36 ◆ Using a half-round file.

The following steps should be followed when reaming using a handheld pipe reamer:

Step 1 Verify that the end of the conduit has a clean, square cut. If the cut of the conduit end is not square, recut the conduit before reaming it.

Step 2 Secure the conduit in a stationary vise, making sure the vise is also secure to avoid tipping over during reaming.

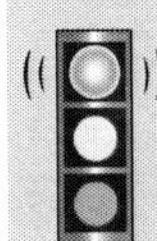

WARNING!

Avoid using a power pipe threader when deburring with a handheld pipe reamer.

Step 3 Refer to *Figure 38*. Insert the reamer head into the conduit with the lever arm positioned at 3 o'clock. Grasp the lever arm of the ratchet reamer with one hand while applying slight inward pressure to the shaft with the other hand. Gently push the lever arm downward (clockwise).

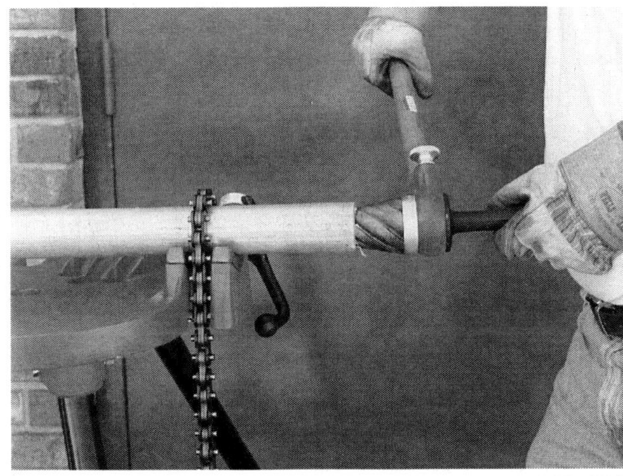

214F38.EPS

Figure 38 ◆ Reamer use.

Step 4 Slightly release the inward pressure, while raising (ratcheting) the lever arm back up to the 3 o'clock position.

Step 5 Repeat this procedure, stopping after two or three ratchet movements to check the amount of the burr remaining. Continue until the burr is removed and the inside wall of the conduit is flush with the cut edge.

2.7.2 Deburring EMT

Electrical metal tubing (EMT) does not have sufficient wall thickness to allow using a handheld pipe reamer. However, it does require deburring after cutting. This deburring can be accomplished through a variety of methods. A favorite method of electricians and instrumentation personnel responsible for the installation of EMT is to insert the jaw end of lineman's pliers into the conduit and rotate them back and forth. The semi-sharp edge of the pliers acts as a reamer and removes most of the burr. Other methods include using a half-round file, as discussed in the section addressing deburring of RMC and IMC. The same steps should be followed for EMT if the half-round file method is used for deburring. It may not be necessary to secure smaller sizes of EMT conduit in a vise, because it weighs much less than RMC or IMC.

There are also some specialty tools available for reaming EMT. One example is shown in *Figure 39*. This particular tool can be used on three different sizes of EMT. It is designed to slip over the end of a square-shank screwdriver and is secured in place with setscrews. The tool is inserted into the end of the EMT (*Figure 40*), and the handle of the screwdriver is rotated back and forth until the desired amount of deburring is achieved.

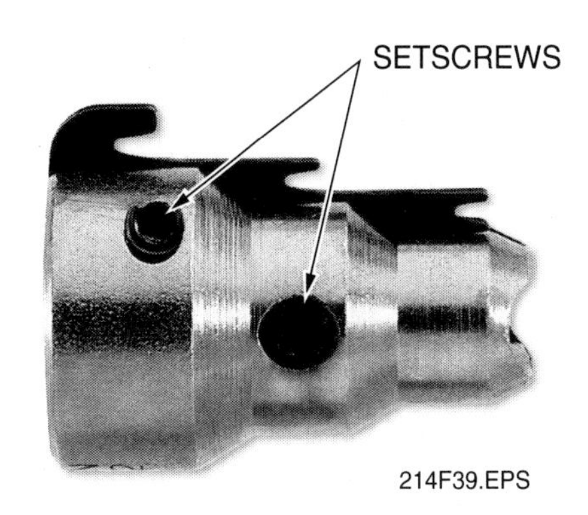

214F39.EPS

Figure 39 ◆ EMT deburring tool.

214F40.EPS

Figure 40 ◆ EMT deburring tool ready for use.

2.7.3 Deburring FMC and LFMC

Like EMT, flexible metal conduit (FMC) and liquidtight flexible metal conduit (LFMC) do not have sufficient wall thickness to deburr with a handheld pipe reamer. In fact, if these two flexible metal conduits are cut properly according to the procedures previously listed, it is not necessary to deburr them. It is merely necessary to remove any remaining sharp edges.

The FMC and LFMC fittings are typically designed with either an insulating insert or a spiral sleeve. The insert is threaded into the end of the flex before installing the fitting itself. The spiral sleeve screws directly into the end of the flex so the conductors do not contact the cut edge of the flex. They are installed through the sleeve.

2.7.4 Deburring RNC (PVC) and ENT

Rigid nonmetallic conduit (PVC) should also be deburred. Some of the same methods used to deburr EMT can also be used to deburr PVC conduit, except for the file method. The PVC material is generally too soft to be deburred with a file. A utility knife or pocket knife can often be used to clean out any burrs remaining on the inside edge of PVC conduit, in addition to using the lineman pliers method previously mentioned. It is important that any sharp edges in any type of conduit be removed before installing the conduit to prevent damage to the conductors during installation. *Figure 41* shows the proper way to position the knife when deburring PVC.

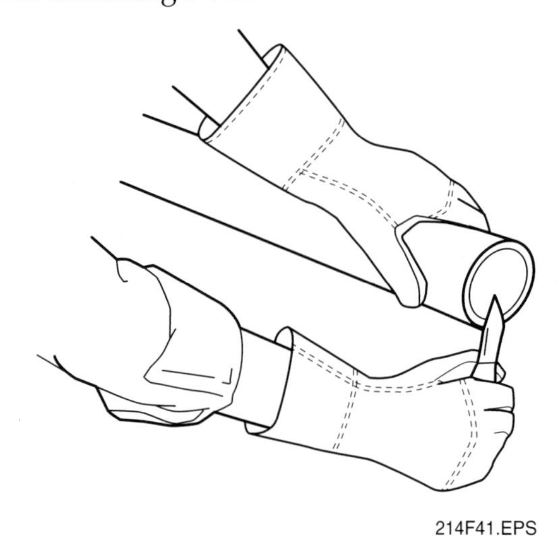

214F41.EPS

Figure 41 ◆ Deburring PVC with a knife.

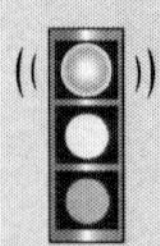

WARNING!

Always push the knife away from your body. Never pull the knife toward you. Wear eye protection to avoid injury from plastic shavings.

Electrical nonmetallic tubing (Smurf) does not generally require any deburring. However, should obvious burrs be present after cutting, use the knife method illustrated in *Figure 41* to remove them.

2.8.0 Threading Conduit

After conduit is cut and reamed, it will normally be threaded so it can be properly joined. Only rigid metal conduit and IMC have walls thick enough for threading. PVC is rarely threaded.

The tool used to cut threads in conduit is called a die. Conduit dies are made to cut a taper of ¾" per foot. The number of threads per inch varies from 8 to 18, depending on the diameter of the conduit.

The threading dies are housed in a die head. The die head can be used with a hand-operated ratchet threader or with a portable power-driven threader. A threader and die head are shown in *Figure 42*.

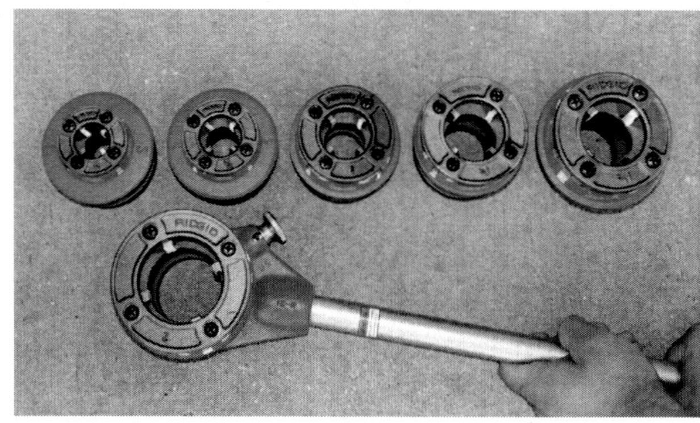

214F42.EPS

Figure 42 ◆ Hand-operated ratchet threader.

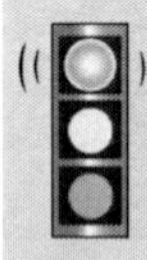

WARNING!

Threading metal conduit results in metal shavings and sharp edges. Wear appropriate personal protective equipment, such as safety glasses or goggles, and hand protection, such as gloves, to guard against these dangers.

To thread conduit with a manual hand-operated threader, perform the following steps:

Step 1 Insert the conduit in a properly stabilized vise. Tighten the vise enough to securely hold the conduit so it does not turn.

Step 2 Place supports under the conduit, if necessary.

Step 3 Select and obtain the correct die and die head.

Step 4 Insert the die securely in the die head. Make sure the proper die is in the proper numbered slot on the head.

Step 5 Determine the correct thread length to cut for the size of conduit being threaded.

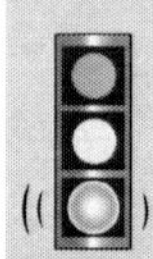

NOTE

Be sure to use proper lubrication when threading.

Step 6 Apply threading lubricant to the end of the conduit.

Step 7 Place the threader with die over the end of the conduit and apply slight pressure inward, while turning the threader clockwise with the ratchet handle.

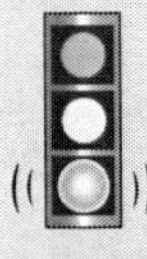

NOTE

When the teeth of the die take hold of the conduit, the die can be turned without pressing the die against the conduit. Back off the head slightly after each turn to clear chips.

Step 8 After a few turns, add more lubricant to the conduit through the holes in the die (*Figure 43*).

214F43.EPS

Figure 43 ◆ Lubricating the threading die.

Step 9 Reverse the knob on the ratchet to back the die head off the threads.

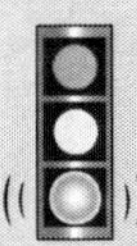

NOTE

Threads should be cut only to the required length because overcutting leaves threads exposed to corrosion.

Step 10 Remove the cut shavings and chips from the conduit with a clean cloth and wire brush.

Step 11 Inspect the threads to make sure they are clean, sharp, and properly made.

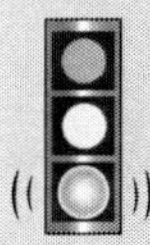

NOTE

Conduit should be re-reamed after threading to remove any burrs and edges. Follow reaming instructions.

Step 12 Clean all lubricant from the inside and outside of the conduit with a clean rag.

NOTE

It is important that all lubricant and other materials be removed from the conduit to prevent damage to the conductors. Don't use your hands or fingers to clean out the conduit.

Conduit can also be threaded with an electric-driven portable threader known as a power pony. An electric portable threader is shown in *Figure 44*. The steps for threading conduit with an electric portable threader are the same as with a hand threader.

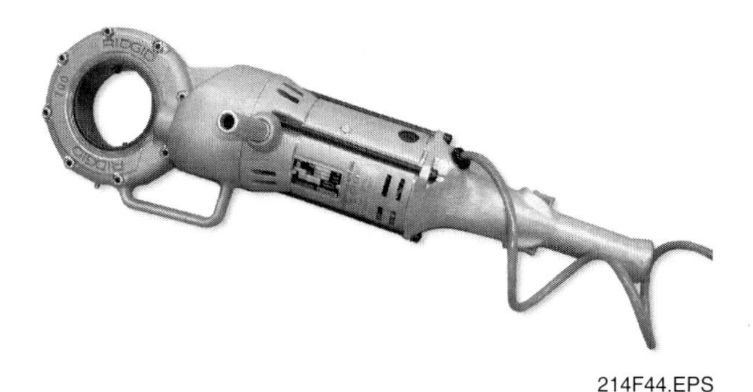

214F44.EPS

Figure 44 ◆ Handheld threader power head.

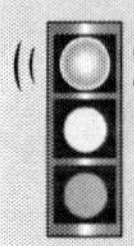

WARNING!

A handheld threader has very high torque and a high gear ratio. It can be used on pipe that is installed or clamped in a portable pipe vise. If used with a portable vise, a support arm furnished with the tool must be clamped to the pipe to counter the torque developed by the tool. Once the threader engages the conduit and begins cutting threads, releasing the trigger will not immediately stop the rotating action. It is important to have training in operating a power pony before using it. Unless a support arm is used and tightly clamped to the pipe, severe injuries (such as broken bones), as well as very costly equipment damage, can result from improperly using a power pony to cut threads.

Portable power threading machines are often used on larger conduit and where a large amount of threading is required. Most threading machines hold and rotate the conduit while the die is fed onto the conduit for threading. They may be used with hand cutting, reaming, and threading equipment, or they may be equipped with optional fixed cutting, reaming, and threading attachments along with lubricating recovery systems. One type of portable power threader is shown in *Figure 32*.

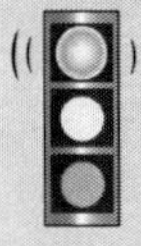

WARNING!

Always secure any loose clothing, long hair, cloths, cords, or any other item that can become entangled in rotating equipment. Due to the high torque of the threading machine, the rotating action cannot be immediately stopped, and serious personal injury and equipment damage can result.

2.9.0 Joining Conduit

Once the conduit is selected, measured, cut, deburred, and threaded, the conduit is joined to other conduit or to boxes and enclosures. The steps used to join conduit are unique in many ways for each type of conduit, yet common in just as many ways. They are covered in the following sections. Quality in joining conduit is as important in the overall conductor protection and code compliance plan as are all the other tasks already learned.

2.9.1 Joining Metal Conduit

Rigid metal conduit and intermediate metal conduit can be joined using the same type of conduit coupling. It is important that no type of thread sealant or any other compound that may interfere with the conducting properties of the threaded joint be applied to the threads. The metal conduit system, in addition to providing a protective path for the conductors, also serves as a grounding path for any fault current. The electrical integrity of the connections must be maintained for this purpose. Couplings should be made wrench tight and not merely hand tight. Most conduit installers use groove joint (ChannelLock®) pliers instead of a wrench to install conduit systems. Care should be taken not to overtighten the couplings and connectors. Overtightening could deform the threaded edge of the conduit if it flares out, creating a conductor-damaging sharp edge or burr inside the fitting. Couplings and connectors should be tightened to the conduit approximately one-quarter turn and no more past the point where the conduit bottoms out in the fitting.

Electrical metal conduit (EMT) can be joined by using either setscrew couplings or raintight couplings. When using setscrew couplings to connect two sections of EMT, the screwdriver should fit the screw slot properly and snugly. Do not attempt to tighten the screws beyond the point of snug, as these screws are generally made of a relatively soft metal that will break if excessive torque is applied. Raintight couplings should be installed using slip joint pliers. Make sure the compression rings are installed on the conduit before installing the fitting and tightening the compression nut. An installed raintight EMT coupling that has a compression ring missing will not hold the conduit sections together, nor will it provide raintight protection.

CAUTION

EMT conduit installers will often slip EMT conduit into a raintight coupling or connector without making sure the compression rings are located behind the compression nuts. It is common to find these fittings on the shelf or in truck boxes with compression rings missing. Always check these fittings for correctly positioned compression rings before slipping the EMT conduit into a raintight fitting.

Boxes and enclosures are installed in conduit systems to hold devices, **splices**, terminations, and **taps**, and to protect any nearby combustible material from potential fire caused by heat or sparks. Some of these boxes and enclosures are equipped with threaded **bosses** to accept the male conduit threads, especially in hazardous locations, as well as some outdoor locations. Other boxes and enclosures are equipped with knockout holes that require the conduit to have locknuts and bushings installed in order to properly make the connection. The following steps may be followed when connecting RMC and IMC to a box or enclosure through a knockout hole:

Step 1 Make sure the threads on the conduit are clean and not flattened or distorted.

Step 2 Determine any offset bend (kick) that may be required to line up the center of the conduit with the center of the knockout hole and make the necessary bends. *Figure 45* demonstrates the correct entrance angle for the conduit into the box.

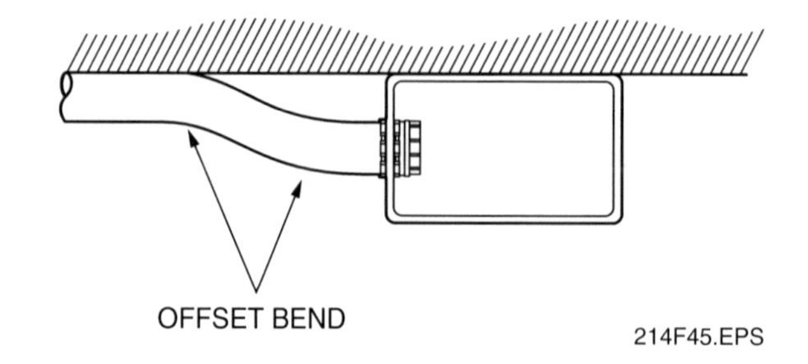

Figure 45 ◆ Correct entrance angle for conduit.

Step 3 Box connectors require locknuts inside and outside the box. Install the external locknut approximately three quarters of the way onto the conduit threads, making sure the concave (dished-in side) of the locknut is facing the box. Place the conduit with the external locknut into the knockout hole. Install the interior locknut on the conduit threads inside the box, again making sure that the concave (dished-in side) of the locknut is facing the box. Leave enough of the conduit threads exposed (about ¼") beyond the interior locknut to allow for a conduit bushing to be installed. Tighten the external locknut until a snug connection is made.

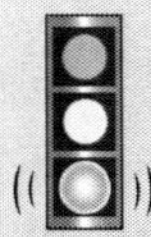

NOTE

If a grounding-type locknut is required, install it as the interior locknut.

Step 4 Install a conduit bushing on the interior threads. Tighten it snugly. *Figure 46* shows the conduit inserted into the box with both locknuts and the bushing.

2.9.2 Joining PVC Conduit

PVC conduit is usually installed in areas that are subjected to unusual atmospheric or organic contact, such as underground and corrosive environments. It is mandatory that PVC joints be made up correctly, using methods designed to provide a permanent and leakproof seal.

Three components that should always be used in joining PVC conduit are a compatible PVC cleaner such as acetone, a compatible primer, and a compatible solvent cement. To determine the compatibility of the chemical agents with the PVC material, identify the type of PVC material by the markings on it and read the labels on the cleaner, primer, and cement, or on the data sheets that are shipped with these items. Follow these general steps to join PVC conduit:

WARNING!

PVC primer, solvent cement, and acetone cleaners are hazardous. Follow the manufacturer's directions carefully. A materials safety data sheet (MSDS) is available from the manufacturer of these products and must be referenced before using any of these chemical agents. An MSDS lists the product's content, personnel dangers, and the safety equipment or personal protective equipment (PPE) that is necessary in using the product. Always wear safety glasses when joining PVC conduit using these chemical agents.

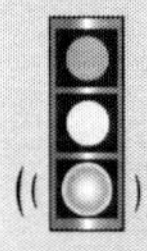

NOTE

Specifications for joining PVC conduit may be found in *ASTM D-2846*, which covers joining of pipe and fittings.

Step 1 Make sure the primer and solvent cement are compatible with the type of PVC being joined.

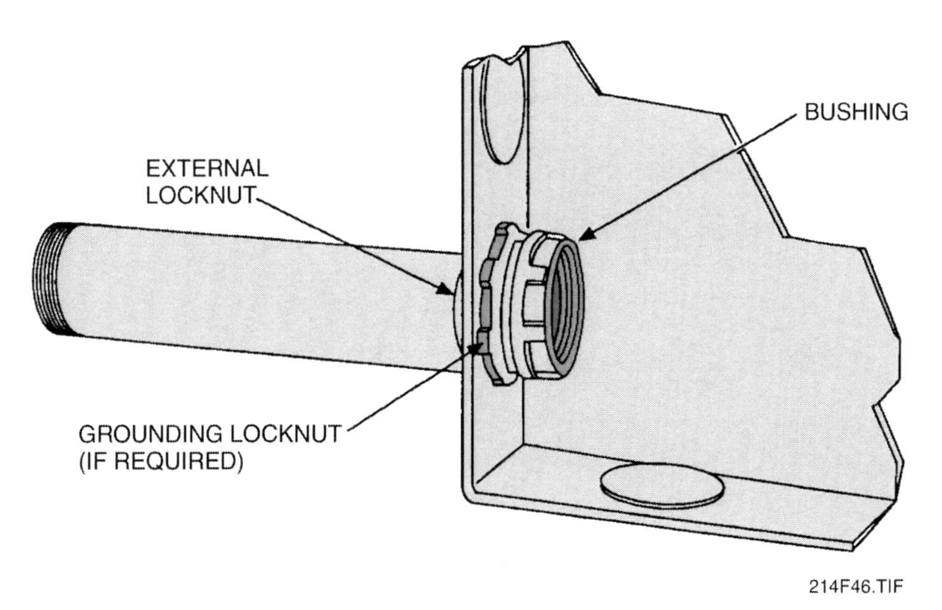

Figure 46 ◆ Assembly of conduit to a box.

Step 2 Clean the exterior of the conduit and the interior of the coupling with an approved cleaner, such as acetone, using a clean, lint-free rag.

Step 3 Apply primer to the conduit surface and to the inside of the coupling to be joined. Allow it to dry before applying cement.

Step 4 Apply a thin, even coat of PVC cement all the way around the outside of the conduit (*Figure 47*) and on the inside of the coupling.

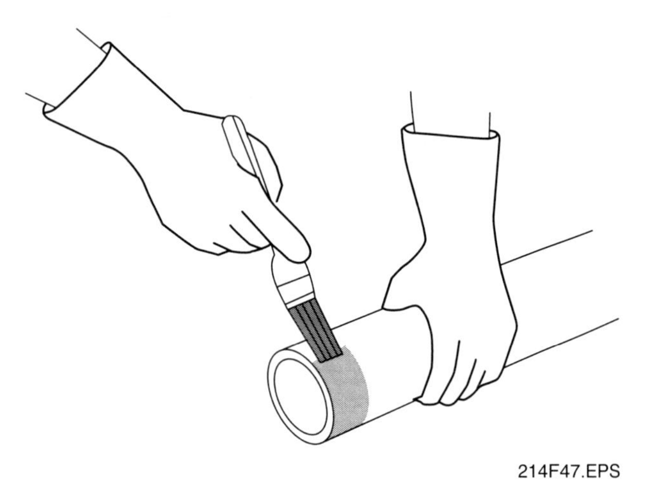

Figure 47 ◆ Applying solvent cement.

Step 5 Immediately push the conduit into the coupling as far as it will go. Twist the conduit approximately one-half turn in either direction to distribute the cement evenly. *Figure 48* shows how to hold the conduit when joining two sections together.

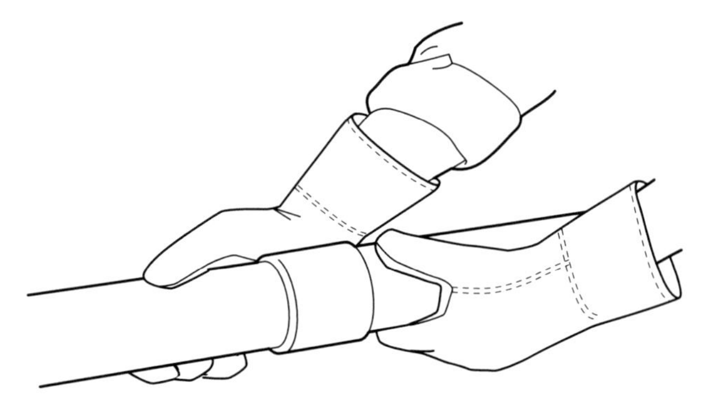

Figure 48 ◆ Inserting conduit into the fitting.

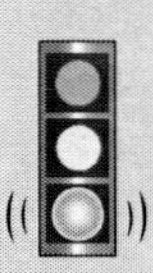

NOTE

Solvents evaporate faster in warm weather. Work quickly to keep the cement from setting up before the joint is assembled. Keep the cement as cool as possible. Try to stay out of direct sunlight.

Step 6 Hold the conduit and fitting in an immobile position for approximately 15 or 20 seconds to allow it to set. It is best to allow the freshly connected joint to set for another 10 minutes or so before installing it or installing conductors in it.

3.0.0 ◆ METAL WIREWAYS

Wireways are sheet metal troughs with hinged or screw-on removable covers. Like other types of raceways, wireways are used to house electric wires and cables. Wireways are available in standard 1', 2', 3', 4', 5', and 10' lengths. The availability of these various lengths of wireway allows runs of any number of feet to be made up without cutting the wireway ducts. Metal wireways are dealt with specifically in ***NEC Article 376***, while nonmetallic wireways are addressed in ***NEC Article 378***.

Wireways are sometimes referred to as auxiliary gutters and are addressed specifically in ***NEC Article 366***. Even though the component parts of wireways and auxiliary gutters are identical, instrumentation personnel should be familiar with the differences in their use. Auxiliary gutters are used as parts of complete assemblies, such as switchboards, distribution centers, and control equipment. However, an auxiliary gutter may only contain conductors or busbars, even though it looks like a surface metal raceway that may contain devices and equipment.

Unlike auxiliary gutters, wireways are used to carry conductors between points located considerable distances apart.

As listed in ***NEC Section 376.22***, the sum of the cross-sectional area of all contained conductors at any cross section of a wireway shall not exceed 20% of the interior cross-sectional area of the wireway. The derating factors of ***NEC Section 310.15(B)(2)(a)*** shall be applied only when the number of current-carrying conductors, including any neutral conductors that carry current, exceeds 30. Signaling circuit conductors or controller conductors between a motor and its starter, and used for starting duty only, are not considered current-carrying conductors.

It is also noted in ***NEC Article 376.56*** that conductors, together with splices and taps, must not fill the wireway to more than 75% of its cross-sectional area. Each wireway is also rated for a maximum permitted conductor size for any single conductor. No conductor larger than the rated size is to be installed in any wireway.

In many situations, it is necessary to make extensions from the wireways to wall receptacles and control devices. In these cases, ***NEC Sections 376.70*** and ***378.70*** require that these extensions be made using any wiring method presented in ***NEC Chapter 3*** that includes a means for equipment grounding. Finally, as required in ***NEC Sections 376.120*** and ***378.120***, wireways must be marked in such a way that their manufacturer's name or trademark is visible.

As you can see in *Figure 49*, a wide range of fittings is required to connect wireways to one another and to fixtures such as switchboards, power panels, and conduits.

Wireway components such as trough crosses, 90-degree internal elbows, and tee connectors serve the same function as fittings on other types of raceways. The fittings are attached to the duct using slip-on connectors. All attachments are made with nuts and bolts or screws. When assembling wireways, always place the head of the bolt on the inside and the nut on the outside so the conductors will not touch a sharp edge. It is usually best to assemble sections of the wireway system on the floor, then raise the sections into position. An exploded view of a section of wireway is shown in *Figure 50*. Both the wireway fittings and the duct come with screw, hinged, or snap-on covers to permit conductors to be laid in or pulled through.

The NEC specifies that wireways shall be used only for exposed work. Therefore, they cannot be used in underfloor installations. Generally, they are used for indoor work. If they are used for outdoor work, they must be made of approved raintight construction. Wireways must never be installed where they are subject to physical damage.

Wireway troughs must be installed so they are supported at distances not exceeding 5'. When specially approved supports are used, the distance between supports must not exceed 10', according to ***NEC Article 376.30***.

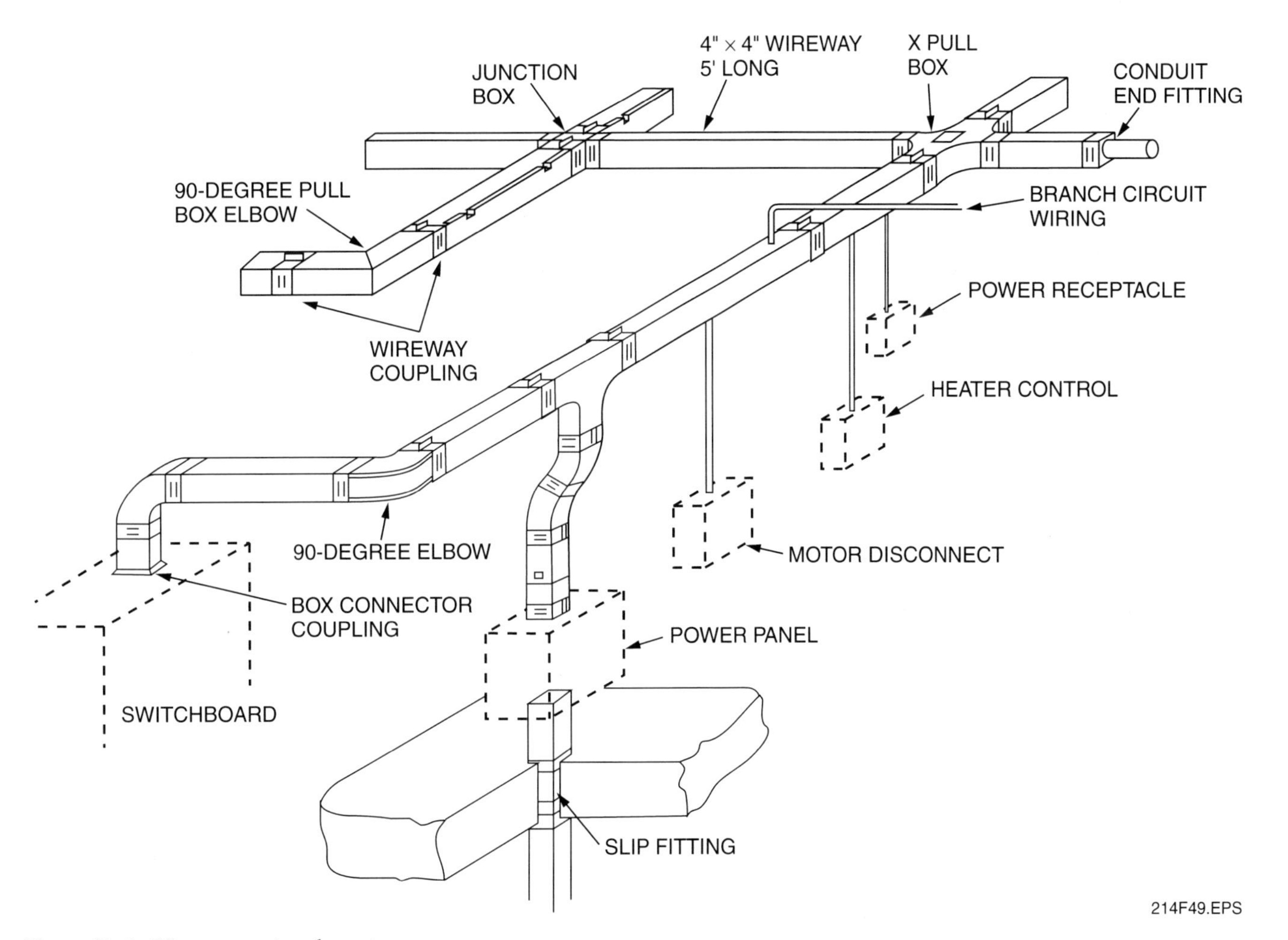

Figure 49 ◆ Wireway system layout.

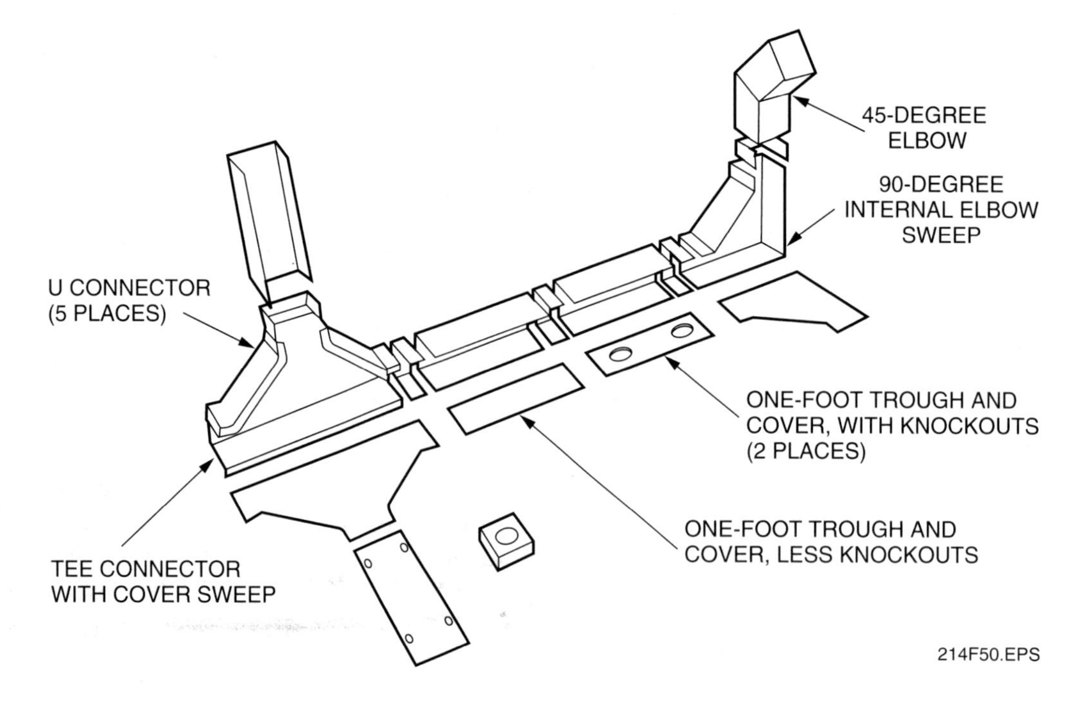

Figure 50 ◆ Assembly sections of wireway.

3.1.0 Wireway Fittings

Many different types of fittings are available for wireways, especially for use in exposed, dry locations. The following sections explain fittings commonly used in the instrumentation craft.

3.1.1 Connectors

Connectors are used to join wireway sections and fittings. Connectors are slipped inside the end of a wireway section and are held in place by small bolts and nuts. Alignment slots allow the connector to be moved until it is flush with the inside surface of the wireway. After the connector is in position, it can be bolted to the wireway. This helps to ensure a strong rigid connection. Connectors have a friction hinge that helps hold the wireway cover open when needed. A connector is shown in *Figure 51*.

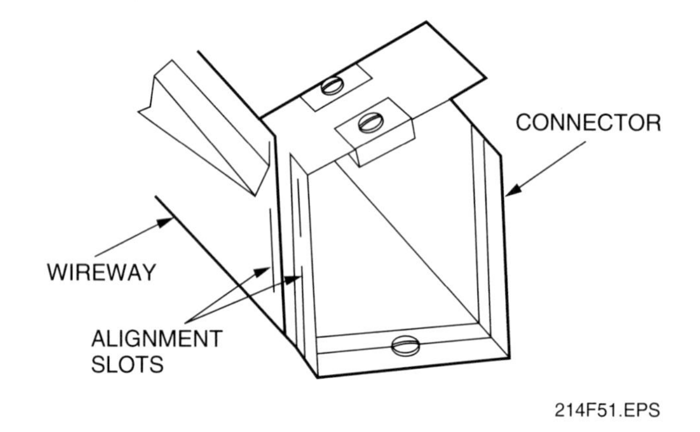

Figure 51 ◆ Connector.

3.1.2 End Plates

End plates, or closing plates, are used to seal the ends of wireways. They are inserted into the end of the wireway and fastened by screws and bolts. End plates contain knockouts, so conduit or cable may be extended from the wireway. An end plate is shown in *Figure 52*.

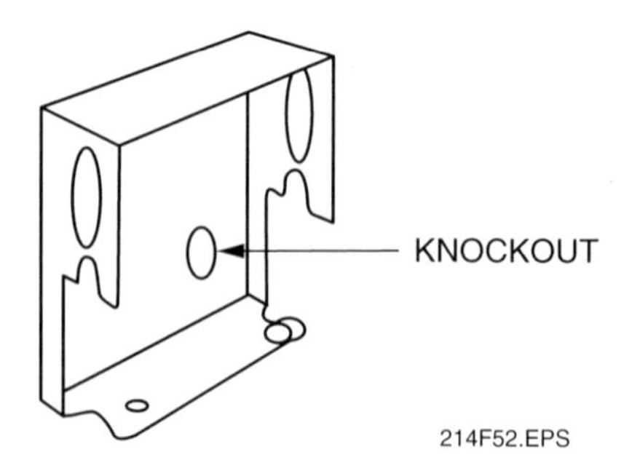

Figure 52 ◆ End plate.

3.1.3 Tees

Tee fittings are used when a tee connection is needed in a wireway system. A tee connection is one in which one conductor is taped or spliced into another to form a tee. The tee fitting's covers and sides can be removed for access to splices and taps. Tee fittings are attached to other wireway sections by standard connectors. A tee is shown in *Figure 53*.

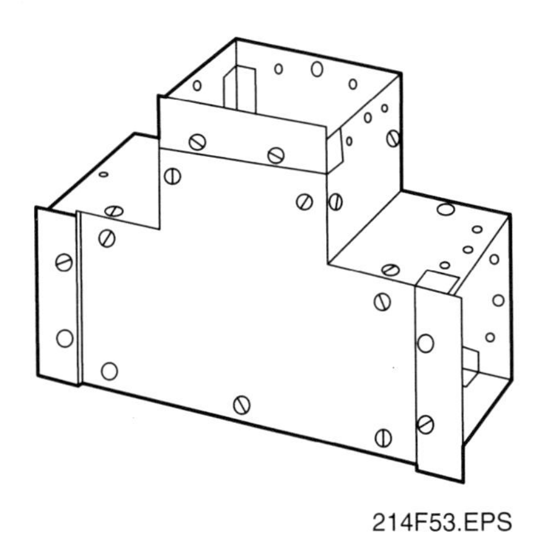

Figure 53 ◆ Tee.

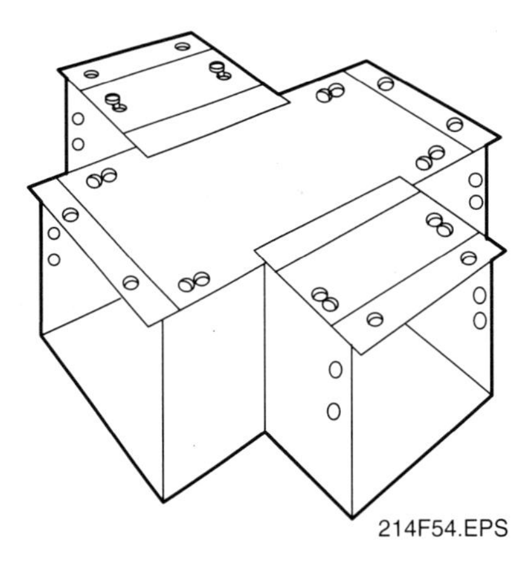

Figure 54 ◆ Cross.

3.1.4 Crosses

Crosses have four openings and are attached to other wireway sections with standard connectors. The cover of a cross is held in place by screws and can be easily removed for laying in wires or for making connections. A cross is shown in *Figure 54*.

3.1.5 Elbows

Elbows are used to make a bend in the wireway. Elbows are available in angles of 22½, 45, or 90 degrees and are either internal or external. They are attached to wireway sections with standard connectors. Covers and sides can be removed for wire installation. The inside corners of elbows are rounded to prevent damage to conductor insulation. An inside elbow is shown in *Figure 55*.

3.1.6 Nipples

Nipples are short lengths of wireway that are inserted between standard lengths when needed. They are commonly available in lengths of 3", 6", and 9". Standard connections are used to connect nipples to other wireway sections. A nipple is shown in *Figure 56*.

3.1.7 Telescopic Fittings

Telescopic, or slip fittings, may be used between lengths of wireway. Slip fittings are attached to standard lengths by setscrews and usually adjust from ½" to 11½". A slip fitting has a removable cover for installing wires and is similar in appearance to a nipple.

3.2.0 Wireway Supports

Wireways should be securely supported every 5' or 10', depending on code requirements presented in ***NEC Sections 376.30*** and ***378.30***.

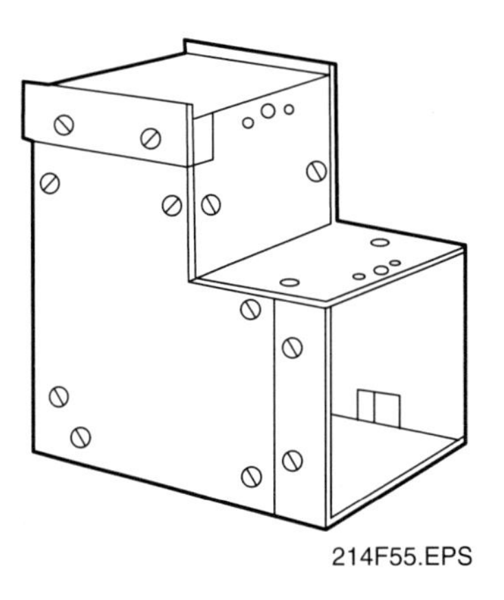

Figure 55 ◆ 90-degree inside elbow.

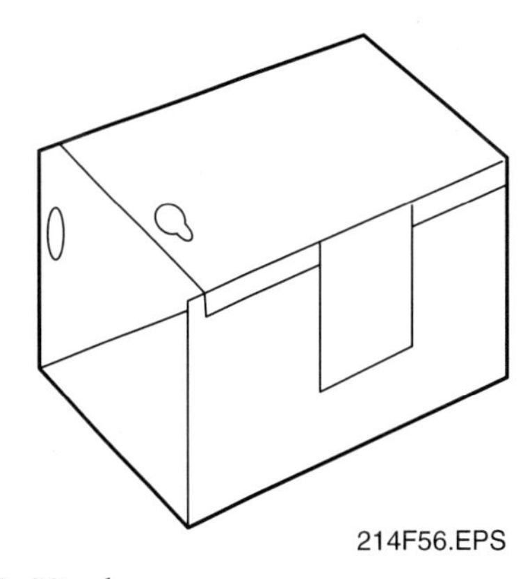

Figure 56 ◆ Nipple.

If possible, wireways can be mounted directly to a surface. Otherwise, wireways are supported by hangers or brackets.

3.2.1 Suspended Hangers

In many cases, the wireway is supported from a ceiling, beam, or other structural member. In such installations, a suspended hanger such as that shown in *Figure 57* may be used to support the wireway.

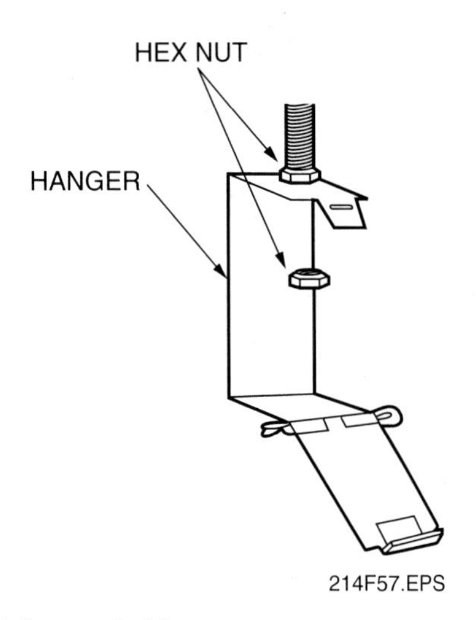

Figure 57 ◆ Suspended hanger.

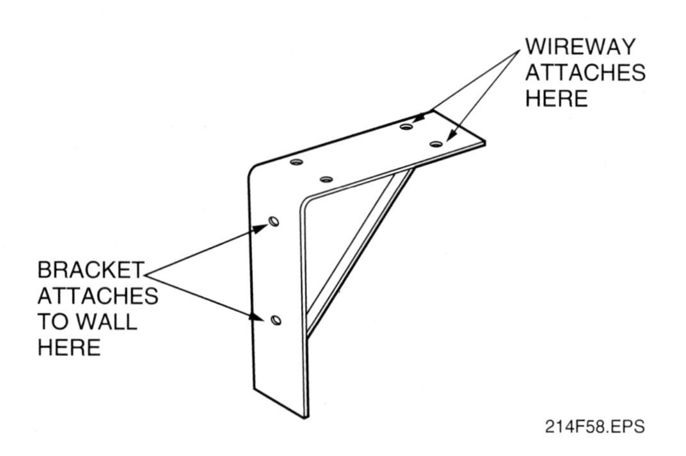

Figure 58 ◆ Gusset bracket.

The wireway is attached to or laid in the hanger. The hanger is suspended by a threaded rod. One end of the rod is attached to the hanger with hex nuts. The other end of the rod is attached to a beam clamp or anchor.

3.2.2 Gusset Brackets

Another type of support used to mount wireways is a gusset bracket. This is an L-type bracket that is mounted to a wall. The wireway rests on the bracket and is attached by screws or bolts. A gusset bracket is shown in *Figure 58*.

3.2.3 Standard Hangers

Standard hangers are made in two pieces. The two pieces are combined in different ways for different installation requirements. The wireway is attached to the hanger by bolts and nuts. A standard hanger is shown in *Figure 59*.

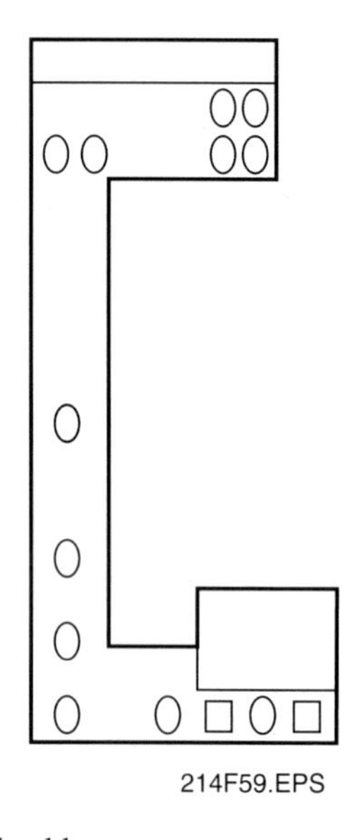

Figure 59 ◆ Standard hanger.

3.2.4 Trapeze Hangers

When a larger wireway must be suspended, a trapeze hanger may be used. A trapeze hanger is made by suspending a piece of strut from a ceiling, beam, or other structural member. The strut is suspended by threaded rods attached to beam clamps or other ceiling anchors, as shown in *Figure 60*.

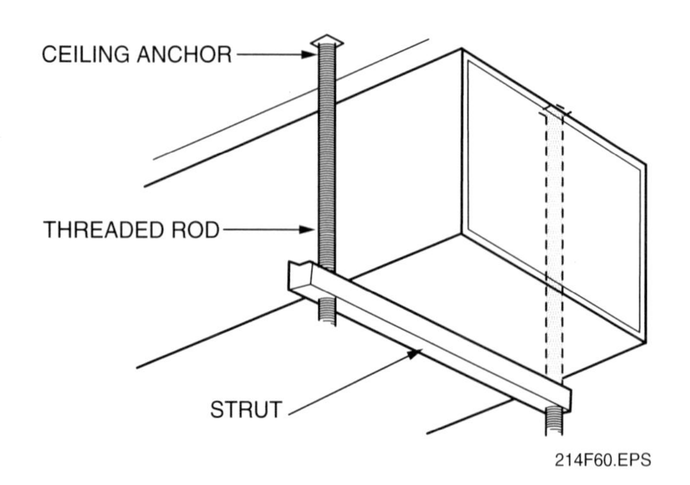

Figure 60 ◆ Trapeze hanger.

3.2.5 Nonmetallic Wireways

Nonmetallic wireways are similar to metal wireways and are flame retardant. Because they are nonconductive, a separate equipment grounding conductor is required in ***NEC Section 378.60*** if equipment grounding is required.

The derating requirements also differ from metallic wireway requirements. As specified in *NEC Section 378.22*, the derating listed in *NEC Section 310.15(B)(2)a* is applied to all current-carrying conductors up to and including the 20% fill limit.

4.0.0 ◆ SURFACE METAL AND NONMETALLIC RACEWAYS

Surface metal raceways consist of a wide variety of special raceways designed primarily to carry power and communication wiring to locations away from the ceilings or walls of building interiors.

Installation specifications of both surface metal raceways and surface nonmetallic raceways are listed in detail in *NEC Articles 386* and *388*. These fixtures must be installed in dry interior locations. The number of conductors and the amperage and cross-sectional area of conductors, as well as regulations for combination raceways, are specified in *NEC Tables 310.16* through *310.19* and *NEC Articles 386* and *388*.

Surface metal raceways are designed to protect conductors that run to outlets or devices not located near the walls and outlets in a room.

Surface metal and nonmetallic raceways have been divided into subgroups based on the specific purpose for which they are intended. There are three small surface raceways that are primarily used to carry power cables from one point to another. In addition, there are six larger surface raceways that have a much wider range of applications. Typical cross sections of the first three smaller raceways are shown in *Figure 61*.

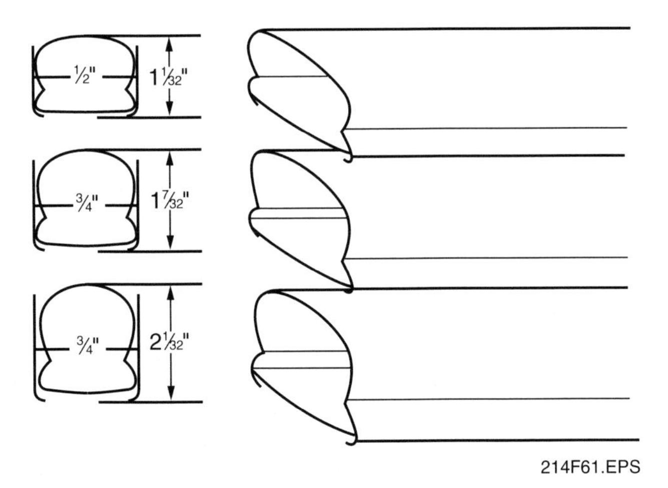

Figure 61 ◆ Smaller surface raceways.

Additional surface metal raceway designs are referred to as pancake raceways, because their flat cross sections resemble pancakes. Their primary use is to extend power and lighting, telephone, or signal wires along the floor to locations away from the walls of a room without embedding them under the floor. A pancake raceway is shown in *Figure 62*.

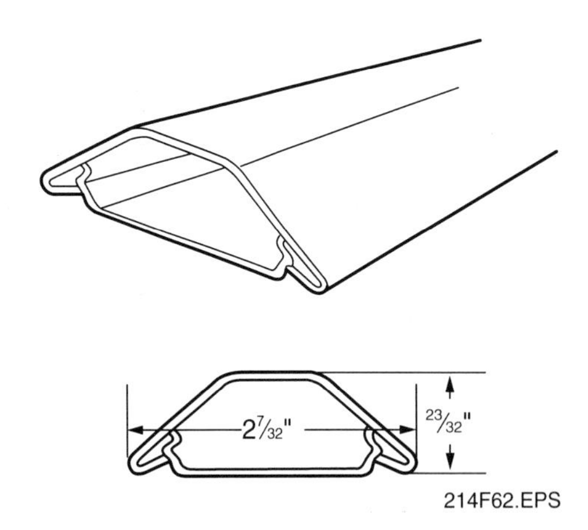

Figure 62 ◆ Pancake raceway.

There are also surface metal raceways available that house two or three different conductor raceways. These are referred to as twinduct or tripleduct. These raceways permit different circuits, such as power and signal, to be placed within the same strip. *Figure 63* shows an example of these raceways.

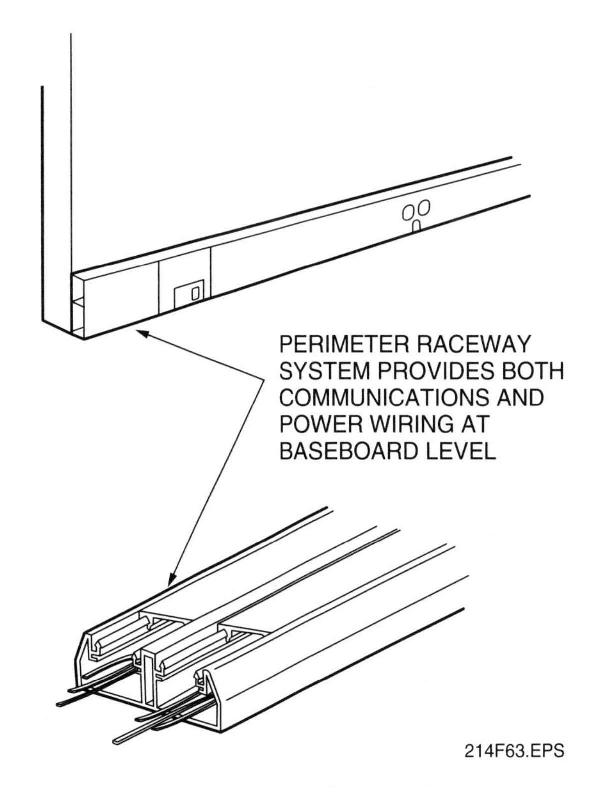

Figure 63 ◆ Dual-purpose surface metal raceway.

The number and types of conductors permitted to be installed, as well as the capacity of a particular surface raceway must be calculated and matched with NEC requirements as discussed previously. ***NEC Tables 310.16*** through ***310.19*** are used in the same manner in which they are used for wireways. For surface raceway installations with more than three conductors in each raceway, particular reference must be made to ***NEC Section 310.15(B)(2)***.

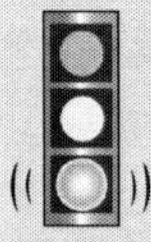

NOTE

Inserts must be installed so they are flush with the finished grade of the floor.

Underfloor raceway was developed to provide a practical means of bringing conductors for lighting, power, and signaling to cabinets and consoles. Underfloor raceways are available in 10′ lengths and widths of 4" and 8". The sections are made with inserts spaced every 24". The inserts can be removed for outlet installation. These are explained in ***NEC Article 390***.

Junction boxes are used to join sections of underfloor raceways. Conduit is also used with underfloor raceways by using a raceway-to-conduit connector (conduit adapter). A typical underfloor raceway with fittings is shown in *Figure 64*.

5.0.0 ◆ CABLE TRAYS

Cable trays serve as raceways for conductors and tubing. A cable tray has the advantage of easy access to conductors, and thus lends itself to installations where the addition or removal of conductors is a common practice. Cable trays are fabricated from aluminum, steel, and fiberglass. They are available in two basic forms: ladder and trough. Ladder tray, as the name implies, consists of two parallel channels connected by rungs. Trough consists of two parallel channels (side rails) having a corrugated, ventilated bottom or a corrugated, solid bottom.

Cable tray is commonly available in 12' and 24' lengths. It is usually available in 6", 9", 12", 18", 24", 30", and 36" widths and is provided with 4", 6", and 8" load depths.

Cable tray is permitted in those locations listed in ***NEC Section 392.3***. Some manufacturers provide an aluminum cable tray that is coated with PVC for installations in caustic environments. A typical cable tray system with fittings is shown in *Figure 65*.

Wire and cable installation in cable tray is defined by the NEC. Read ***NEC Article 392*** to become familiar with the requirements and restrictions made by the NEC for safe installation of wire and cable in a cable tray.

Metallic cable trays that support electrical conductors must be grounded as required by ***NEC Article 250***. Where steel and aluminum cable tray systems are used as an equipment grounding conductor, installers shall comply with all of the provisions of ***NEC Article 392***.

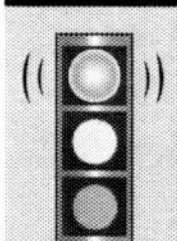

WARNING!

Do not stand on, climb in, or walk on a cable tray. It is not designed to support people.

5.1.0 Cable Tray Fittings

Cable tray fittings are part of the cable tray system. They provide a means of changing the direction or dimension of the different trays. Some of the uses of horizontal and vertical tees, horizontal and vertical bends, horizontal crosses, reducers, barrier strips, covers, and box connectors are shown in *Figure 65*.

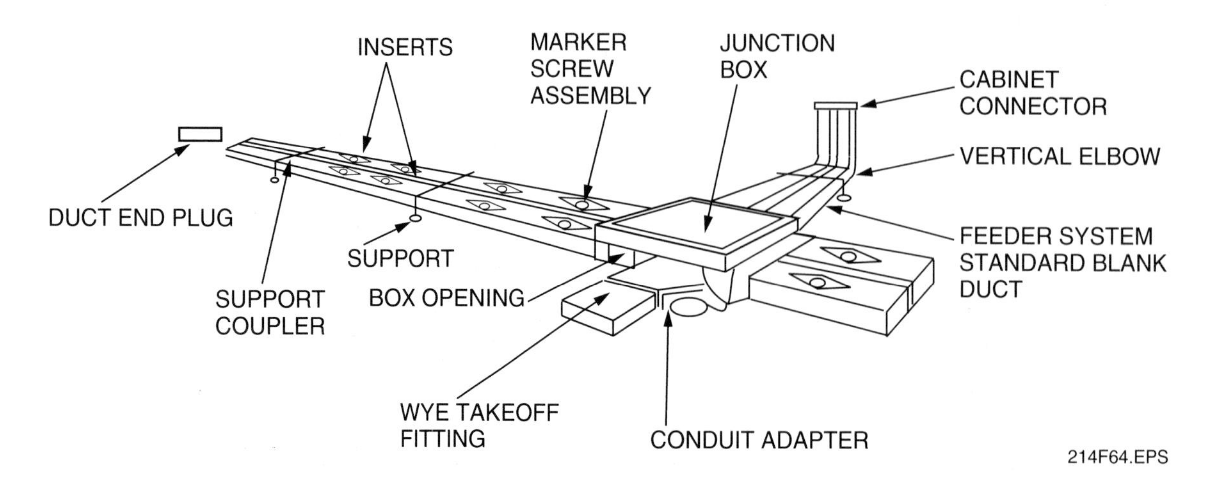

Figure 64 ◆ Underfloor raceway duct.

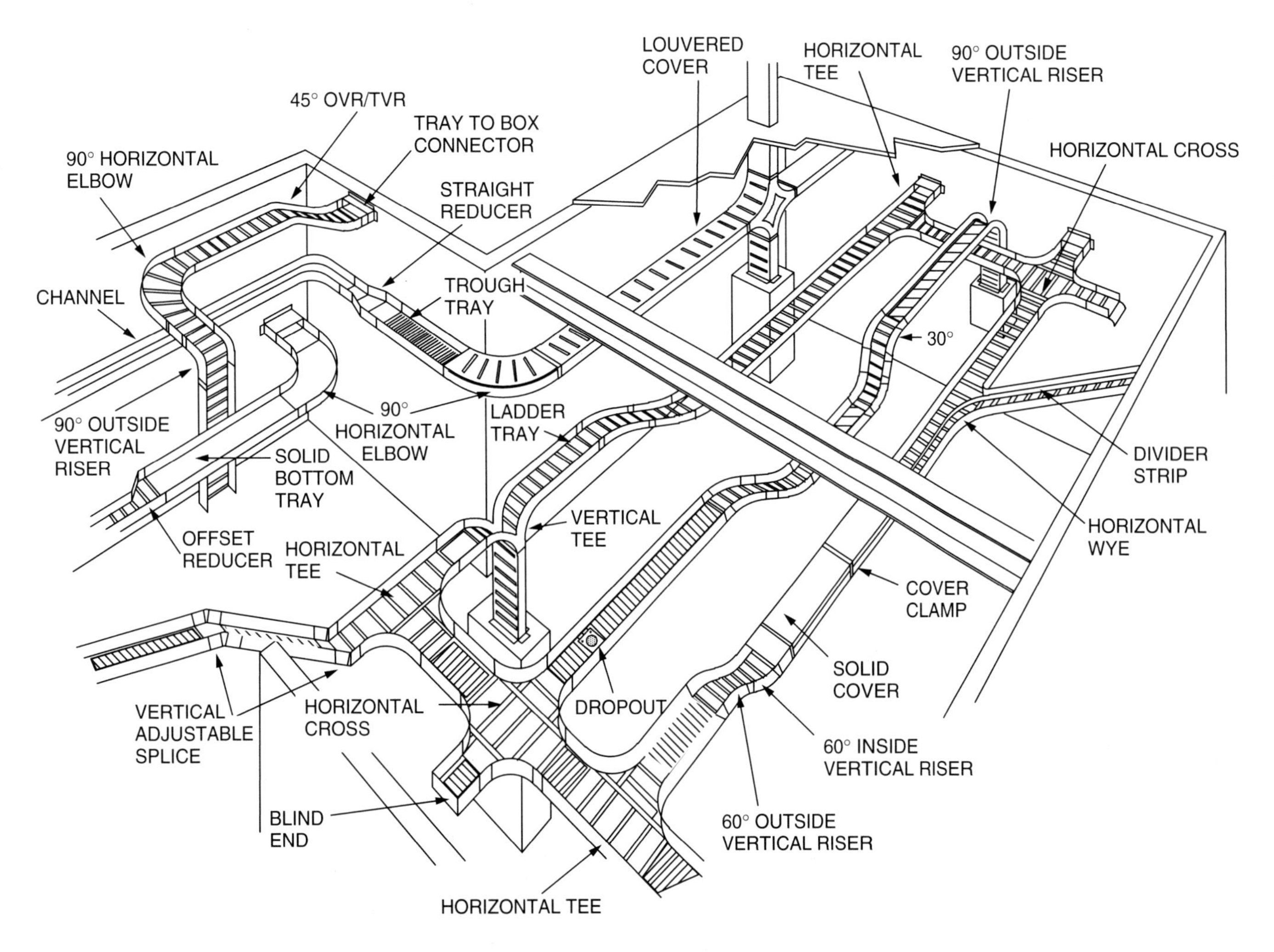

Figure 65 ◆ Cable tray system.

5.2.0 Cable Tray Supports

Cable tray is usually supported in one of four ways: direct rod suspension, trapeze mounting, wall mounting, or pipe rack mounting.

5.2.1 Direct Rod Suspension

The direct rod suspension method of supporting cable tray uses threaded rods and hanger clamps. One end of the threaded rod is connected to the ceiling or other overhead structure, usually using beam clamps. The other end is connected to hanger clamps that are attached to the cable tray side rails. A direct rod suspension assembly is shown in *Figure 66*.

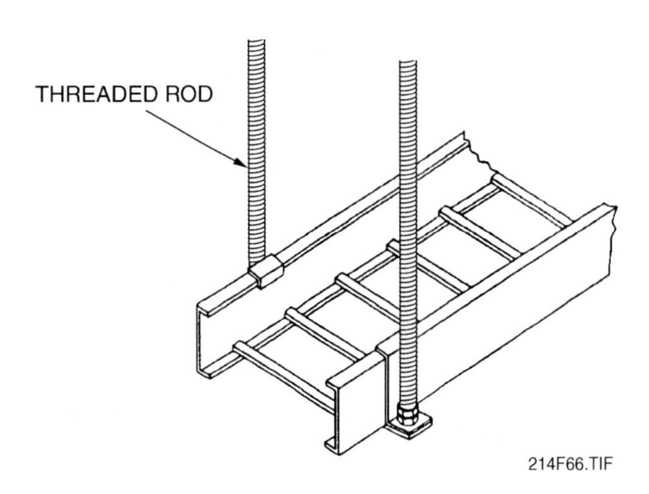

Figure 66 ◆ Direct rod suspension.

5.2.2 Trapeze Mounting

Trapeze mounting of cable tray is similar to direct rod suspension mounting. The difference is in the method of attaching the cable tray to the threaded rods. A structural member, usually a steel channel or strut, is connected to the vertical supports to provide a crossbar similar to a swing or trapeze. The cable tray is mounted to the structural member. Often, the underside of the channel or strut is used to support conduit. A trapeze mounting assembly is shown in *Figure 67*.

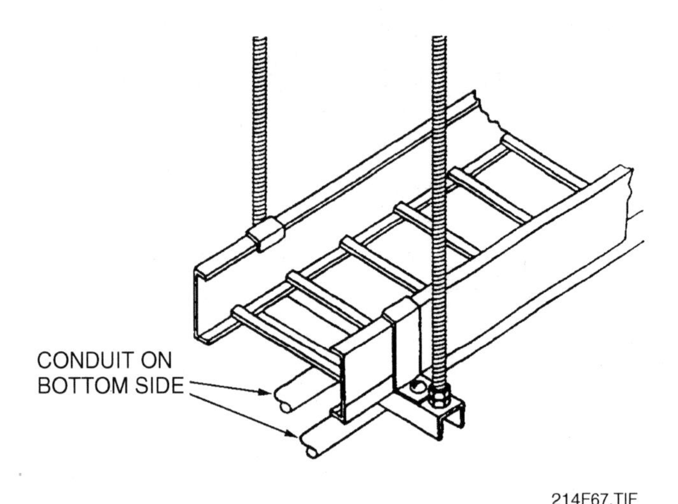

Figure 67 ◆ Trapeze mounting.

5.2.3 Wall Mounting

Wall mounting is accomplished by supporting the cable tray with structural members attached to the wall. This method of support is often used in tunnels and other underground or sheltered installations where large numbers of conductors interconnect equipment separated by long distances. A wall mounting assembly is shown in *Figure 68*.

5.2.4 Pipe Rack or Bridge Mounting

Pipe racks are structural frames used to support piping that interconnects equipment in outdoor industrial facilities. If the rack connects areas and buildings, such as control rooms, it is often referred to as a bridge. Usually, some space on the rack is reserved for conduit and cable tray. Pipe rack or bridge mounting of cable tray is often used when power distribution and instrumentation wiring is routed over a large area.

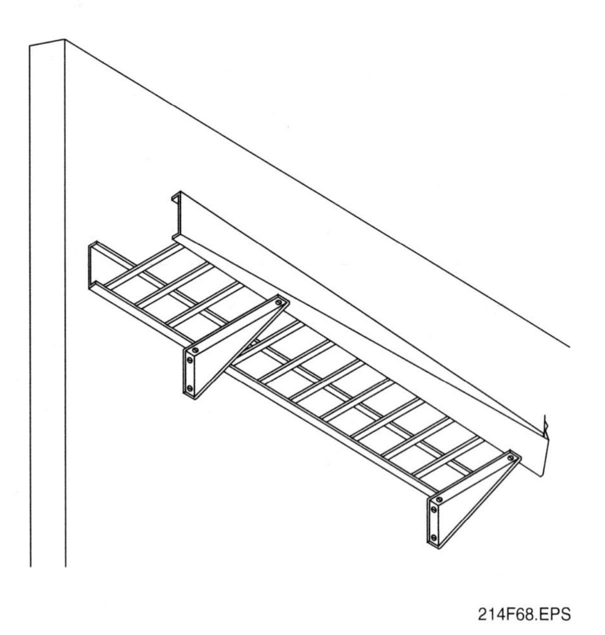

Figure 68 ◆ Wall mounting.

6.0.0 ◆ STORING RACEWAYS

Proper and safe methods of storing raceway (conduit, wireways, raceways, and cable tray) may sound like a simple task, but improper storage techniques can result in wasted time and damage to the raceways, as well as personal injury. There are correct ways to store raceways that will help avoid costly damage, save time in identifying stored raceways, and reduce the chance of personal injury.

Storage racks are commonly used for storing conduit. The racks provide support to prevent bending, sagging, distorting, scratching, or marring of conduit surfaces. Most racks have compartments in which different types and sizes of conduit can be separated for ease of identification and selection. The storage compartments in racks are usually elevated to prevent damage that might occur at floor level. Conduit that is stored at floor level is easily damaged or contaminated.

The ends of stored conduit should be sealed to prevent interior contamination. Conduit ends can be capped, taped, or plugged to seal them.

Always inspect raceway before storing it to make sure it is clean and not damaged. Make sure that the raceway is stored securely such that if one section is removed, others do not fall.

Raceways should be covered with a tarpaulin or other suitable covering to prevent exposure to the elements. It should also be separated from noncompatible materials, such as hazardous chemicals and corrosive atmosphere.

Raceway should always be stored off the ground. Stepping on or running over raceway ruins the raceway.

7.0.0 ◆ HANDLING RACEWAYS

Raceway is made to specifications and must meet NEC requirements. It can easily be damaged by careless handling. From the time raceway is delivered to a job site until installation is complete, proper and safe handling techniques will help greatly in ensuring quality raceway installations. There are a few basic guidelines for handling raceway that will help avoid damaging or contaminating it.

- Never drag raceway off a delivery truck or off other raceway.
- Never drag raceway on the ground or floor. Dragging raceway can damage the ends.
- Keep the thread protection caps on when handling or transporting conduit raceway.
- Keep raceway away from any material that might contaminate it during handling.
- Flag ends of long lengths of raceway during transportation.
- Never drop or throw raceway during handling.
- Never hit raceway against other objects during transportation.
- Always use two people when carrying long and heavy pieces of raceway.

Summary

This module introduces various types and uses of conduit, wireways, cable tray, and surface metal raceways. Mechanical skills, knowledge of NEC requirements, and a thorough understanding of raceways and the appropriate fittings and support devices combine to provide the skills needed to work with raceway systems. This module provides the foundation for training needed to perform quality raceway installation work.

Review Questions

1. A method used to hold and protect electrical conductors is known as a(n) _____.
   a. installation
   b. application
   c. raceway
   d. coupling

2. Two basic types of conduit are _____.
   a. plastic and steel
   b. aluminum and steel
   c. metal and nonmetallic
   d. metal and fiberglass

3. The majority of outdoor industrial installations such as those located in petrochemical facilities and power generation stations will require _____.
   a. rigid metal conduit
   b. electrical metallic tubing
   c. underground PVC
   d. electrical nonmetallic tubing

4. Because the wall thickness of EMT is less than that of rigid metal conduit, it is often referred to as _____.
   a. ENT
   b. thinwall
   c. Smurf
   d. intermediate

5. Aluminum conduit must not be buried in soil or concrete that contains high levels of _____.
   a. calcium chloride
   b. oxidation
   c. hydrogen sulfide
   d. aggregate

6. Rigid metal conduit can use the same couplings as _____.
   a. raintight EMT
   b. LFMC
   c. IMC
   d. FMC

7. The NEC considers any section of conduit not greater than 24" in length a(n) _____.
   a. nipple
   b. offset
   c. section
   d. raceway

8. Because of the nonconductive properties of PVC, a(n) _____ must be used.
   a. oxidizing compound at joints
   b. insulated grounded wire
   c. shield wire
   d. bare grounding wire

9. Smurf® is a common name for _____.

a. EMT
b. ENT
c. LFMC
d. RNC

10. Of the following conduit systems, the most costly to install is _____.

a. EMT
b. ENT
c. RMC
d. FMC

11. When cutting RNC or IMC, the tool that leaves the greatest burr inside the conduit is a _____.

a. half-round file
b. hacksaw
c. bandsaw
d. pipe cutter

12. A favorite method of electricians and instrumentation personnel for deburring EMT is to use _____.

a. lineman's pliers
b. a square-shank screw driver
c. a pipe die
d. a tee handle

13. According to ***NEC Section 376.22***, the maximum cross-sectional area of all conductors at any cross section of a metal wireway is _____.

a. 20%
b. 30%
c. 45%
d. unlimited

14. The number of small surface raceway types used for extending power cables from one point to another is _____.

a. one
b. three
c. five
d. six

15. One of the common lengths for cable trays is _____.

a. 6'
b. 8'
c. 9'
d. 12'

Refer to *Figure 1* for Question 16.

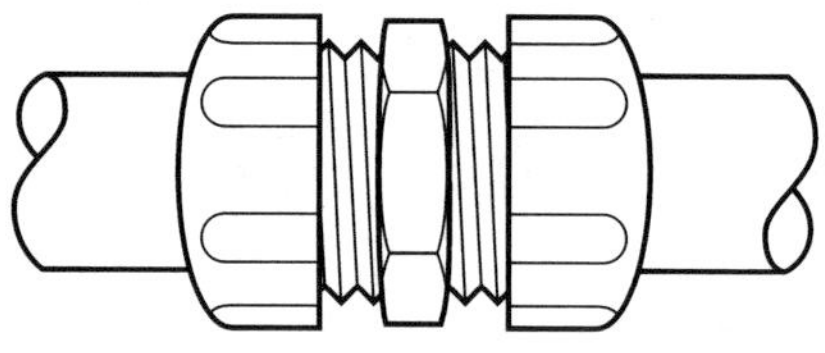

214E01.EPS

FIGURE 1

16. The fitting shown in *Figure 1* is a(n) _____.

a. flare fitting
b. swage fitting
c. compression coupling
d. EMT coupling

Refer to *Figure 2* for Question 17.

214E02.EPS

FIGURE 2

17. The couplings shown in *Figure 2* are commonly used with _____ conduit.

a. EMT
b. RMC
c. FMP
d. PVC

Refer to *Figure 3* for Question 18.

214E03.EPS

FIGURE 3

18. The conduit body shown in *Figure 3* would be designated _____.

a. LL
b. LR
c. LB
d. LC

Refer to *Figure 4* for Question 19.

214E04.EPS

FIGURE 4

19. The conduit body shown in *Figure 4* is designated Type _____.

a. L
b. T
c. Double T
d. X

20. True or False? A knockout punch is required to remove the knockout on all metal boxes.

a. True
b. False

Refer to the following for Questions 21–25.

Match the NEC article with its content:

a. ***NEC Article 362***
b. ***NEC Article 342***
c. ***NEC Article 344***
d. ***NEC Article 352***
e. ***NEC Article 348***

21. IMC _____

22. ENT _____

23. RMC _____

24. FMC _____

25. RNC _____

GLOSSARY

Trade Terms Introduced in This Module

***Approved*:** Meeting the requirements of the authority having jurisdiction.

***Bond*:** To tie all metal parts of a system together electrically so they can be grounded.

***Boss*:** A reinforced female-threaded hub that is part of a box or enclosure that receives threaded conduit or nipples.

***Cable tray*:** A rigid structure used to support electrical conductors.

***Conductors*:** Wire or cable used to carry electrical current.

***Conduit*:** A round raceway, similar to pipe, that houses conductors.

***Condulet*:** A conduit fitting equipped with a gasketed cover that permits the installation of conductors.

***National Electrical Code® (NEC)*:** A book that describes minimum electrical safety requirements.

***Raceway*:** Any channel designed and used for holding wires, cables, or other conductors.

***Raintight*:** Protected against exposure to beating rain.

***Running threads*:** Threads cut on a piece of conduit or nipple that extend from one end to the other with no interruption, cut from only one direction.

***Splice*:** The connection of two or more conductors.

***Tap*:** An intermediate point on a main circuit at which another wire is connected to supply electrical current to another circuit.

***Underwriters Laboratory (UL)*:** An agency that approves electrical components and equipment.

***Wireway*:** A steel trough designed to carry electrical wire and cable.

Additional Resources

This module is intended to present thorough resources for task training. The following reference works are suggested for further study. These are optional materials for continued education rather than for task training.

National Electrical Code, 2002. Quincy, MA: National Fire Protection Association, Inc.

National Electrical Code Handbook, 1990. New York, NY: McGraw-Hill Publishing Company.

Ugly's Electrical References, 1990. Houston, TX: United Printing Arts.

ANSI/ASME Standards.

Figure Credits

Veronica Westfall	214F03, 214F05 (top), 210F06, 214F07, 214F08, 214F09, 214F10, 214F12, 214F13, 214F15, 214F18, 214F21, 214F24, 214F26, 214F34, 214F35, 214F38, 214F42, 214F43
Chuck Rogers	214F04
Walter Johnson	214F05 (bottom)
Killark, a Division of Hubbell Inc. (Delaware)	214F11
Ridge Tool Co.	214F32 (bottom)
Gerald Shannon	214F29, 214F32 (top and middle), 214F44
Klein Tools, Inc.	214F39, 214F40

NCCER CURRICULA — USER UPDATE

NCCER makes every effort to keep its textbooks up-to-date and free of technical errors. We appreciate your help in this process. If you find an error, a typographical mistake, or an inaccuracy in NCCER's curricula, please fill out this form (or a photocopy), or complete the online form at **www.nccer.org/olf**. Be sure to include the exact module ID number, page number, a detailed description, and your recommended correction. Your input will be brought to the attention of the Authoring Team. Thank you for your assistance.

Instructors – If you have an idea for improving this textbook, or have found that additional materials were necessary to teach this module effectively, please let us know so that we may present your suggestions to the Authoring Team.

NCCER Product Development and Revision
13614 Progress Blvd., Alachua, FL 32615

Email: curriculum@nccer.org
Online: www.nccer.org/olf

❑ Trainee Guide ❑ AIG ❑ Exam ❑ PowerPoints Other ______________________________

Craft / Level: Copyright Date:

Module ID Number / Title:

Section Number(s):

Description:

Recommended Correction:

Your Name:

Address:

Email: Phone:

Index

Index

Abbreviations, 2.20, 2.21, 2.22–2.24, 2.30
Absolute Celcius scale. *See* Kelvin scale
Absolute Fahrenheit scale. *See* Rankine scale
Absolute standards, 5.9
Absolute value, 1.21–1.22
Absolute zero, 9.13, 9.17
Absorption, 10.10, 10.19, 11.22
AC. *See* Alternating current
Acceleration, 1.5, 1.6, 1.33
Accelerometers, 5.41
Accessibility of instrumentation, 12.3, 13.10, 13.13, 13.33
Accuracy, 1.21, 5.4–5.6, 13.41
Acetone, 3.2, 3.24, 3.25, 11.13, 14.27
Acetylene, 3.2, 3.19–3.20, 3.23, 3.24, 3.25
Acids, 11.6. *See also* pH
Acoustic signals, 13.17, 13.55
Activity, of compounds, 11.7
Actuators
  air- or fluid-to-lower, 7.17
  air- or fluid-to-raise, 7.17
  cylinder, 7.16
  definition, 7.1, 7.31
  diaphragm, 4.9–4.10
  electric, 7.16
  electric proportional, 4.11–4.12
  in limit switch, 9.8
  manual control, 4.10
  nameplate, 7.23
  piston, 7.16
  pneumatic control, 4.9–4.10
  principles of operation and construction, 7.16–7.19
  solenoid, 4.12–4.13
  spring and diaphragm, 7.16, 7.17, 7.31
  symbols, 2.39–2.40, 2.54–2.55
Adapter, socket-weld, 13.15, 13.17
Adsorption, 10.11, 10.19, 11.19
Aftercooler, 10.14
Air, dielectric constant, 5.22
Air bubbler method, 11.15
Air compressor, 9.6
Air pollution control devices, 10.1–10.3
Air quality issues, 10.14, 11.21
Air-to-close, 7.18, 7.19
Air-to-open, 7.18, 7.19
Alarms, 8.2, 11.21
Alkalinity, 11.7, 11.25. *See also* pH
Alternating current (AC), 3.8, 3.14
Alumel®, 1.29, 5.28
Alumina, 11.19, 11.25
Aluminum, 3.8, 5.33, 14.4, 14.5, 14.34
American National Standard Hose Thread, 7.25
American National Standard Pipe Thread, 7.25
American National Standards Institute (ANSI), 2.9, 2.22, 2.35, 3.26, 7.23
American Society for Testing and Materials (ASTM), 7.23, 7.31, 11.11, 14.27
American Society of Mechanical Engineers (ASME), 2.9
American Welding Society (AWS), 3.9, 3.15, 3.19
Ammeter, 6.25, 11.11
Amplifier, 4.7, 5.2, 5.4
Amplitude, in closed loop system, 4.16–4.17, 4.49
AMU. *See* Atomic mass unit
Analyzer, pH, 11.10
Anchoring systems, 13.10–13.12
Annubar tubes, 4.6, 5.17
Annular configuration, 8.12, 8.23
Annunciate, 6.16, 6.31
Anode, 9.9, 9.17, 11.6, 11.11
ANSI. *See* American National Standards Institute
Antifreeze, 11.15
Application errors, 5.7
Approved, 14.8, 14.41
Arc, welding, 3.7, 3.11, 3.12, 3.13
Arc blow, 3.8, 3.35
Arc flash, 3.5
Arc gap, 3.12
Archimedes' Principle of Flotation, 11.14, 13.32
Arc length, 3.12, 3.35
Arc voltage, 3.12, 3.35
Area, 1.2–1.3, 1.33
Argon, 3.14, 3.16, 3.18, 10.11, 11.20
Argon-arc welding. *See* Gas tungsten arc welding
Armature, relay, 8.3, 8.4, 8.5, 8.23
ASME. *See* American Society of Mechanical Engineers
ASTM. *See* American Society for Testing and Materials
Atmospheric pressure, 1.10–1.11, 1.12, 1.33
Atom, 11.1, 11.25
Atomic mass, 11.1, 11.4–11.5
Atomic mass unit (AMU), 11.5, 11.25
Atomic weight, 11.5, 11.25
Avogadro's number, 11.5, 11.25
AWS. *See* American Welding Society

Backer, 12.12
Backlash, 5.7
Backpressure
  and nozzle in controller, 6.10, 6.12, 6.13, 6.14, 6.15
  and relay function, 8.10
  symbol, 7.26
Baffles off the nozzle, 8.10, 8.23
Bakelite, 14.13
Balancing section, of DP cell, 5.42, 5.44
Barium titanate, 5.40
Baseline wiring diagram method, 2.5, 2.6
Bases, 11.6. *See also* pH
Batching, 13.22, 13.30, 13.31
Bellows
  derivative follow-up, 6.12
  in motion balance transmitter, 5.47–5.48
  pressure devices, 4.5, 5.33–5.34
  proportioning, 6.3, 6.4, 6.6
  in recorder, 6.24
  reset, 6.4, 6.6, 6.10, 6.11, 6.14
  in Taylor® controller, 6.10
Beta, 5.12
Bias, 5.7
Bill of materials, 2.15, 2.17
Blending process, 4.21, 13.22
Blowdown, 13.49, 13.50, 13.55
Blowout, 3.25, 13.51
Blueing, 12.7, 13.4
Blueprints, 2.1
Bluff body, 13.17, 13.18, 13.55
Boilers, 3.11
Boiling point, 11.6
Bonding, for bypass electrical path, 13.21, 14.3
Bonds, chemical, 11.4
Bonnet, 7.3, 7.7, 7.31
Boosters, 4.7, 4.8, 8.10, 8.11, 8.14–8.15
Boss, 14.26, 14.41
Bourdon measuring element, 6.21
Bourdon tube, 4.5, 5.3, 5.31–5.32, 5.35, 5.48, 6.24
Boxes
  conduit, 14.12–14.13, 14.26–14.27
  junction, 13.35, 13.37, 14.29, 14.34
  pull, 14.29
Brackets, 13.2, 14.32
Brass, 5.26
Brazing, 3.28, 3.35
Bridge
  for cable trays, 14.36
  circuitry, 5.39, 11.20
  conductivity, 11.11–11.12
Bridgewall markings, 7.23
Bronze, 5.33
Bubble, use as symbol, 12.1, 12.2–12.3
Bubbler, 4.5
Buffer solution, 11.10, 11.25
Buna-N, 7.9, 7.10, 7.11
Buoyancy, 13.32
Bushings, 7.16, 14.7, 14.8, 14.26, 14.27

Cable bundle, 2.5, 2.35
Cable-mounted instrument supports, 13.33–13.35, 13.36
Cables. *See also* Raceways
  fiber optic, 9.13, 9.14
  high tensile strength, 13.34
  to pH probe, 13.32
  welding, 3.8–3.9, 3.16
Cable tray, 14.1, 14.34–14.36, 14.41
Cable wiring diagram method, 2.5
Cadmium selenide, 9.10
Calcining, 10.3, 10.19
Calcium chloride, 10.10, 14.4
Calculator, 1.18–1.20
Calibration
  check, 5.7
  data listed on loop sheet, 2.13
  definition, 5.5, 5.10, 5.53
  drift and re-calibration, 5.8
  overview, 5.10
  of pH analyzer/controller, 11.10
  split-range, 4.42, 4.43
  use of significant digits, 5.10
Callouts, 2.17
Capacitance, 5.21, 5.22, 5.34, 8.18, 13.29
Capacitor, 3.12, 5.21–5.25, 13.29
Capillary column, 11.19–11.20
Carbon atom, as reference for atomic mass, 11.5
Carbon dioxide, 3.11, 3.16, 3.18, 10.11, 11.22. *See also* Gas metal arc welding
Carbon monoxide, 10.11
Carbon steel, 3.20, 13.2
Cascade control system, 4.17, 4.18–4.20, 6.1
Catalytic combustion detector, 11.20
Cathode, 9.9, 9.17, 11.6, 11.11
Celcius scale, 1.13, 1.14, 1.28–1.29
Cell, types, 11.12–11.13
Cell constant, 11.11
Cellulose, phenolic-impregnated, 10.4, 10.19
Centimeter-gram-second system (CGS), 1.1
Centrifugal force, 10.3
Centrifuge, 10.3, 10.19
Ceramic tube, 5.30
Cermet tube, 5.30
Certification, 3.11, 3.28
CGS. *See* Centimeter-gram-second system
Chamber-installed displacers, 13.32, 13.33
Chamfered, 5.13, 5.53
Channels. *See* Instrument control channels
Chart paper, circular, 6.22
Checklist. *See* Field checklist
Chemical composition analysis, 11.18–11.22
Chemical process control, 4.17
Chemical reactivity, 11.2, 11.4
Chemistry, basic principles, 11.1–11.7, 11.8
Chromatography, 11.19, 11.25
Chrome, 13.17, 13.41
Chromel®, 1.29, 5.28
Chromium-molybdenum electrode, 3.20
Chuck, 14.17, 14.22
Circle, area, 1.4
Circuit, 9.1, 9.17. *See also* Diagrams, single- and three-line
Circumference, 13.5, 13.55
Clamps, 3.12, 14.14, 14.15
Closed loop system, 4.13–4.18, 4.46, 4.49
Cohesiveness, 10.4, 10.19
Cold working pressure (CWP), 7.24, 7.31
Color codes, 2.20, 5.29
Combination set, 12.6
Combustion control, 4.20
Communications, 8.7, 14.33
Compass, 12.7
Compound, 11.2
Compressed air, 9.6, 10.3–10.4, 10.5, 10.9, 10.13. *See also* Dryers
Compressibility, 1.26, 13.26
Computer-controlled welding, 3.12, 3.13

Computers, video display from recorder, 6.22, 6.23
Concentration, 11.5–11.6
Concentricity, measuring, 13.15, 13.16
Concrete, 13.10–13.11, 14.3
Condensate, 5.13, 7.7, 10.9–10.10, 10.13, 13.31. *See also* Dryers
Conditioning piping, 13.18, 13.55
Conductivity
  calculation, 11.11
  definition, 11.10
  increased with electrolytes, 11.6
  measurement, 11.11–11.13
  thermal, 11.20–11.21
Conductor, 14.1, 14.41. *See also* Raceways
Conduit
  adapter, for raceways, 14.34
  bend requirements, 14.15–14.16
  boxes, 14.12–14.13
  cutting, 14.16–14.21
  deburring, 14.21–14.24
  definition, 14.41
  electrical. *See* Tubing, electrical metallic; Tubing, electrical nonmetallic
  fittings, 14.5–14.9
  flexible, 14.9–14.10, 14.12, 14.20–14.21, 14.23–14.24
  installation, 14.14–14.16
  intermediate metal, 14.2, 14.9, 14.16, 14.21–14.23, 14.26–14.27
  joining, 14.26–14.28
  layout. *See* Raceways, drawings
  liquidtight flexible metal, 14.9–14.10, 14.20–14.21
  metal, 14.2–14.11, 14.16, 14.20–14.24, 14.26–14.27
  nipples, 14.11
  nonmetallic, 14.11–14.12, 14.21, 14.24, 14.27–14.28
  overview, 14.1, 14.2
  rigid, 14.4–14.9, 14.11–14.12, 14.16, 14.21–14.23, 14.24, 14.26–14.28
  size requirements, 14.15
  supports, 14.13–14.14, 14.15
  threading, 14.22, 14.24–14.26
Conduit bodies, 14.5–14.6
Condulets, 14.5, 14.41
Connector heads, 13.41–13.43
Connectors
  raintight, 14.2, 14.3, 14.10, 14.26
  wireway, 14.30
Constant, 8.12, 8.23
Constantan, 5.28
Contactor, 8.2, 8.23
Contacts
  abbreviations and symbols, 2.20–2.21, 2.26
  butt, 9.2
  knife blade, 9.2
  overview, 9.2
Contaminants, 5.13, 10.1, 10.19, 13.20, 13.25. *See also* Filters
Continuous cycling. *See* Proportional plus integral plus derivative control
Control channels. *See* Instrument control channels
Controlled variable, 4.3–4.4, 4.13, 4.15, 4.16–4.17
Controllers
  ABB Automation MICRO-DCI™, 6.19, 6.20
  in cascade control, 4.18, 4.19, 6.1
  in closed loop system, 4.13, 4.14, 4.15, 4.16
  definition, 4.3, 6.31
  electronic, 6.16–6.20
  flow, 4.20, 4.21, 4.44
  four modes of control, 6.1
  Foxboro-Eckardt, 6.16
  function of the, 4.3–4.4, 6.1, 6.28
  level, 4.45
  Masoneilan® 12000 Level, 6.7, 6.8
  motor, wiring diagram, 2.4
  in on-off control, 4.22, 4.23, 6.2–6.3, 6.17
  overview, 4.8
  pH, 11.10
  pneumatic, 6.2–6.16
  pressure, 4.42
  proportional-plug-rate, 6.9
  ratio flow, 4.21
  relationship to positioner output, 7.20
  Taylor® Model 1414R, 6.9–6.16
  temperature, 4.41
  types and function. *See* Modulating control
  and valve positioner, 4.10
Control loops
  cascade control system, 4.17, 4.18–4.20
  closed loop system, 4.13–4.18
  demand on the system, 4.4, 4.26, 4.33, 4.39
  flow, 4.43–4.45
  level, 4.45
  open loop system, 4.13
  overview, 4.13, 4.46
  pressure, 4.41–4.43
  ratio control system, 4.20–4.22
  system delay, 4.25, 4.36, 6.18, 8.16
  temperature, 4.39–4.41
  use of photocell switches, 9.11
Control modes
  applications. *See* Control loops
  modulating control. *See* Modulating control
  on-off (two-position), 4.22–4.26, 6.2–6.3, 6.17
  overview, 4.22, 4.46
Control room, 8.10
Control valve (final control element)
  in closed loop system, 4.14, 4.15
  in on-off system, 4.24, 6.2
  overview, 4.3, 4.8–4.13
Converter, 5.4, 7.17, 10.5
Conveyer lines, 8.16
Copper
  in bimetallic thermometer, 5.27
  in resistance temperature detector, 1.28, 13.44
  solder joint fittings, 3.26, 3.29, 3.32
  straps, to mount sensing bulb, 13.37
  in thermocouple, 5.28, 13.43
  welding cable, 3.8
Copper-constantan, 1.29
Coriolis force, 13.24
Corner taps, 5.13, 5.14
Coulomb, 5.21, 5.53
Covalent bond, 11.4, 11.25
Crosses, 14.31
C-type Bourdon tube, 5.31–5.32, 6.24
Custody transfer, 13.22
Cutting outfit, 3.22
CWP. *See* Cold working pressure
Cyclonic filtration, 10.3, 10.4
Cylinders
  definition, 3.35
  in oxyacetylene welding, 3.21, 3.23–3.25
  porous filler materials in, 3.25
  safety issues, 3.22
  of shielding gas, 3.16
  volume, 1.4
Dampeners, pulsation, 5.35
Damper, symbol, 2.54

Data collection, 5.14, 5.53, 8.7
Data sheets, instrumentation, 2.17–2.18, 2.19, 2.35
DC. *See* Direct current
D control. *See* Derivative control
Deadband, 9.6, 9.17
Deadtime, 13.31, 13.55
Deburring process, 14.21–14.24
Deformation, 7.9, 7.31
Dehydrators. *See* Dryers
Deliquescent, 10.19
Demand, 4.4, 4.26, 4.33, 4.39
Density
  calculation, 11.14
  definition, 1.8, 1.33
  of gas, 1.18
  and mass flowmeters, 13.24
  measurement, 11.14–11.16, 11.26–13.28, 13.38–13.39, 13.40
Departments, abbreviations, 2.30
Derivative action, 6.12–6.13, 6.31
Derivative control (D), 4.31–4.32
Desiccant, 10.10, 10.19
Detectors, 5.12–5.30
  angular position density, 13.27
  capacitance-type
    level, 5.21–5.25, 13.29–13.30
    pressure, 5.34
  for chemical analysis, 11.20–11.22
  definition, 5.1–5.2
  direct *vs.* inferred measurements, 4.4
  function, 5.3
  hydrogen flame ionization, 11.20, 11.21
  infrared, 9.13
  ion chamber, 5.26
  ionization, 11.20
  orifice plates, 4.6, 4.21, 5.12–5.15, 5.44, 5.45, 13.14–13.17
  overview, 4.4, 4.5–4.7
  parameters monitored by, 5.3
  photovoltaic, 9.12
  proximity, 8.2, 9.14–9.15
  radiation, 5.26, 13.28–13.39, 13.40
  reflective, 9.11–9.12, 9.14
  resistance temperature. *See* Resistance temperature detectors
  vibrating plate, 13.28
  vibrating U-tube, 13.28
Dew point, 10.11, 10.14
Diagrams
  access by personnel, 2.17
  applications, 2.18–2.19
  block, 2.2
  flow, 2.17, 2.18, 2.35
  instrumentation data sheets, 2.17–2.18, 2.19, 2.35
  instrumentation panel layout, 12.3–12.10
  logic/ladder, 2.13–2.14
  loop sheets, 2.11–2.13, 2.35
  pH measuring instrument, 11.9
  schematic, 2.7
    reed relay, 8.2
    Taylor® controller, 6.9, 6.10, 6.11, 6.12, 6.13
  single- and three-line, 2.2–2.4
  standardized design methods and symbols, 2.20–2.32
  wiring, 2.4–2.7, 2.26, 2.36
Diaphragms
  annular configuration, 8.12
  in automatic-drain air filter, 10.4, 10.5
  in booster, 8.10, 8.11
  capacitance, 4.5, 5.34
  in DP cell, 5.44
  in explosion-proof switches, 9.4
  to measure level, 4.5
  in metallic pressure gauge, 5.32–5.33
  piezoelectric, 4.5
  potentiometric, 4.5
  in regulators, 10.7, 10.8
  in strain gauge, 4.5, 5.40
  in Taylor® controller, 6.14, 6.15
  in transmitters, 5.49, 5.50
  variable sensing, 4.5
Diaphragm seals, 5.34–5.35
Diatomaceous earth, 11.19, 11.25
Dielectric constant, 5.22
Differential gap, 6.2, 6.31
Differential pressure, 1.17, 5.15–5.16
Differential-pressure cell, 1.25–1.28, 5.3
Differentiator, 4.31
Diode, 9.10, 9.17
Dipstick, 1.24, 4.4
Dip transfer welding. *See* Gas metal arc welding
Direct-acting positioners, 7.20
Direct current (DC), 3.8, 3.14, 4.10
Direct rod suspension, 14.35
Disc
  of a butterfly valve, 7.14, 7.15, 7.16
  definition, 7.31
  of a globe valve, 7.1, 7.2, 7.3, 7.4, 7.6, 7.7
  identification, 7.24
  of a plug valve, 7.11
Dispersion, 13.27
Displacement, 4.6, 5.3, 5.10, 5.53, 13.32
Disturbance
  definition, 4.2, 4.49
  process demand, 4.4, 4.26, 4.33, 4.39
Dividers, 12.6–12.7, 13.4
Documentation, of calibration, 5.9, 5.10
Double-pole, double-throw switch (DPDT), 8.19
Downstream, 7.18
DP. *See* Relays, double-pole
DP cell, 5.41–5.46
DPDT. *See* Double-pole, double-throw switch
Drain hole, 5.13
Drawings
  applications, 2.18–2.19
  construction, 2.1
  electrical, 2.2–2.7
  graphic styles, 2.30–2.32
  instrumentation, 2.7–2.18, 2.19
  location, 2.14, 2.15, 2.35
  numbering systems, 2.28, 2.30
  orthographic *vs.* isometric, 2.30–2.32, 2.35
  piping and instrumentation, 2.8–2.10, 2.35
  raceway, 2.7, 2.8, 2.9, 2.35
  standardized design methods, 2.20–2.32
Drift, 5.7–5.8, 5.53
Driver, 13.25
Droop, 10.8, 10.19
Dryers
  absorbent (deliquescent), 10.10, 10.19
  adsorptive desiccant
    heatless, 10.12, 10.13
    heat-reactivated, 10.11, 10.12
  principles of operation, 10.9–10.12, 10.16
  refrigerated, 10.10–10.11
  selection, 10.13–10.14
DT. *See* Relays, double-throw
Duct, 14.29
Dyne, 1.6, 1.7, 1.33

Ear protection, during welding, 3.6
Earth, 1.5, 1.8, 1.10
ECKO. *See* Eddy current killed oscillator principle
Eddy current, 9.14, 9.17, 13.18
Eddy current killed oscillator principle (ECKO), 9.14–9.15
Edge filtration, 10.4, 10.19
Effluent, 13.30, 13.55
Elastomeric materials, 7.8, 7.9, 7.13, 7.14, 7.31
Elbows, 13.34, 13.35, 14.29, 14.31
Electrical drawings, 2.2–2.7, 2.20–2.27, 2.28, 2.29
Electrical elements, 5.11
Electrical metal tubing (EMT), 14.2–14.4, 14.19–14.20, 14.23
Electrical nonmetallic tubing (ENT), 14.11, 14.12, 14.21
Electrical shock, 12.12
Electrode holder, 3.8, 3.9
Electrodes
  classification, 3.15, 3.19, 3.20
  GMAW, 3.18–3.19
  grounding, 13.21
  GTAW, 3.14–3.15
  in magnetic flowmeter, 13.20
  to measure conductivity, 11.11, 11.12
  to measure pH, 11.8, 11.9, 11.10
  orbital welding, 3.10–3.11
  pH, 13.31
  polarity, 3.8
  reference, 11.9
  SMAW, 3.9
  tungsten, 3.10, 3.14–3.15
Electrolytes, 11.6–11.7, 11.12
Electromagnetic coil, 6.25
Electromagnetic field (EMF), 5.17–5.18, 5.19, 5.53, 9.15
Electromechanical relays (EMR), 8.2–8.9
Electron, 11.1, 11.2, 11.25
Electronic control loops
  flow, 4.44–4.45
  level, 4.45
  pressure, 4.42
  temperature, 4.41
Elements (chemical), 11.1–11.2, 11.25
Elements (system)
  final control. *See* Control valve
  primary, 2.41–2.50, 5.2, 5.10, 5.11, 5.53, 13.15
  secondary, 5.10, 5.30–5.35, 5.54
EMF. *See* Electromagnetic field
Emissions, 4.20, 10.1–10.3
Empirical data, 5.14, 5.53
Empty-tube feature, 13.21
EMR. *See* Electromechanical relays
EMT. *See* Electrical metal tubing
Emulsion, 13.27
End plates, 14.30
Energy
  in closed loop system, 4.15
  to constrict flow, 5.12
  in heat transfer process, 4.2–4.3
  kinetic, 7.2, 7.31
  loss through volumetric flowmeters, 13.22
Energy balance, 4.2, 4.3
English system, 1.1
ENT. *See* Electrical nonmetallic tubing
Equilibrium, 6.8, 8.11, 8.23
Equipment
  gas metal arc welding, 3.15–3.16
  gas tungsten arc welding, 3.13–3.14
  location drawings, 2.14, 2.35
  orbital welding, 3.10–3.11
  oxyacetylene welding, 3.21–3.25
  power, in one-line diagrams, 2.3
  safety, 12.3
  shielded metal arc welding, 3.7–3.9
Erosion, 7.3, 7.5
Error, measurement, 5.6, 5.53
Error signal
  amplification of, 4.12
  in D control, 4.31
  definition, 4.4, 4.12
  in I control, 4.31
  mechanism of production, 4.21–4.22
  in on-off control, 4.22
  in PD control, 4.36
  in PI control, 4.32, 4.34, 4.35
  in PID control, 4.38
Error signal burst, 6.19, 6.31
Ethane, 10.11
Ethylene, 10.11
Explosion, 3.21, 3.23, 3.25
Explosion-proof apparatus, 9.4, 9.6, 13.43, 14.8
Exponent, 11.7, 11.25
Extensions
  pipe, 13.2
  right-angle, for vertical leg, 13.34
  right-angle elbow, 13.34, 13.35
  square-leg, 13.34
  thermowell, 13.43
Eye protection, during welding, 3.5–3.6

Fabrication
  definition, 12.5, 12.15
  pipe stand, 13.2–13.9
  templates, 12.8, 12.10
Facing, flange, 13.45–13.46
Fahrenheit scale, 1.13, 1.14, 1.28–1.29
Fail-as-positioned, 7.18
Fail closed, 7.17
Fail open, 7.17
Farad, 5.21, 5.53, 13.29
Faraday's Law of Magnetic Induction, 5.17, 13.19
FC. *See* Controllers, flow
Feedback
  angle motion, 6.8
  definition, 4.8, 4.49
  moment-balance, 6.9
  positive and negative, 4.13
Feedback control, 4.13–4.18, 6.8–6.9. *See also* Cascade control
Feedforward, 4.13, 6.8–6.9, 6.31
Ferro-cyanide, 12.8
Ferrous, 3.35
FFC. *See* Controllers, ratio flow
FFY. *See* Flow ratio relay
Fiberglass, 5.30, 14.11, 14.34
Fiber optics, 9.13–9.14
Field checklist, on loop sheet, 2.13
Field-mounted instruments
  flanges, 13.45–13.47
  in-line-mounted, 13.13–13.28, 13.29, 13.51
  insertion-mounted, 13.39–13.45, 13.51
  manifolds, 13.47–13.51
  overview, 13.1
  stand-mounted, 13.2–13.13, 13.51
  strap-mounted, 13.32–13.39, 13.40, 13.51
  vessel-mounted, 13.28–13.32, 13.33, 13.51
File, 14.21, 14.22
Filler metal or rod, 3.6–3.7, 3.13, 3.15, 3.19, 3.21, 3.35

Filters
in combination with dryers, 10.11
compressed air, 10.3–10.4, 10.5
electronic/electrostatic, 10.1–10.3
principles of operation, 10.1–10.4, 10.5, 10.16
symbols, 10.15
upstream, with flowmeters, 13.25
on welding shield, 3.5–3.6
Filtration element, 10.3, 10.5, 10.19
Final control element. *See* Control valve
Fittings
butt welded, 3.26, 3.28, 3.29, 3.30–3.31, 13.47
cable tray, 14.34, 14.35
conduit, 14.5–14.9
copper solder joint, 3.26, 3.29, 3.32
elbow, 13.34, 13.35, 14.29, 14.31
for electrical metal tubing, 14.3–14.4
explosion-proof, 14.8
liquidtight, 14.10
plastic-coated, 14.5
seal-off, 14.8–14.9
slip, 14.29, 14.31
socket-welded, 3.26, 3.27–3.28, 13.15, 13.17, 13.47
tee, 14.30–14.31
telescopic, 14.31
wireway, 14.29–14.31
Flame, welding, 3.20, 3.21
Flame cutting, 3.20
Flame ionization detector, 11.20, 11.21
Flanges
butt-welded, 13.47
facings, 13.45–13.46
flowmeter, 13.19, 13.23
gaskets, 13.46, 13.47
methods of joining, 13.46–13.47
and orifice plate installation, 13.14
pressure ratings, 13.45
screwed-joint, 13.47
sizes, 13.45
socket-welded, 13.47
and templates, 12.9
Flange taps, 5.13, 5.14
Flapper
effects of diaphragm movement, 5.42, 5.43, 5.44
in pneumatic controller, 6.2, 6.4, 6.16
relationship with coil movement, 5.36, 5.37
Flashback arrestors, 3.22, 3.23
Flexible Control Strategy, 6.19
Flexible gates, 7.6
Flexible metal conduit (FMC), 14.9–14.10, 14.20–14.21, 14.23–14.24
Flexible wedge, 7.5–7.6
Float
in automatic-drain filter, 10.4, 10.5
basic principles, 5.3
in hydrometer, 11.14
in magnetic indicator, 6.27, 6.28
in variable-area flowmeter, 13.25
Float chamber, 4.28
Floodlights, 9.13
Flooring, anchoring systems for, 13.10–13.11
Flow
of air across a filter, 10.5
backflow and eddy currents, 9.14, 9.17, 13.18
common detectors, 4.6–4.7, 5.17–5.21, 5.44–5.45, 9.8. *See also* Flowmeters
control loops, 4.43–4.45
detection by differential pressure switch, 9.8
diagrams, 2.17, 2.18, 2.35
laminar *vs.* turbulent, 1.16, 5.12, 5.53
linear, 7.4, 7.31
ratio control of, 4.20–4.22
total, 1.14, 1.33
units, 1.14–1.15
and valve position, 4.9
zero, avoidance of, 13.20–13.21
Flowmeters
Coriolis mass, 13.24–13.25
differential pressure, 5.16, 13.14–13.17
installation, 13.14–13.26
magnetic, 5.17–5.18, 13.17, 13.19–13.21
mass, 4.7, 13.23–13.25, 13.26
measurement of velocity by, 1.17, 13.17–13.22
nutating-disc, 13.23
oval-gear, 13.23
pitot tube, 4.6, 5.16–5.17
positive displacement. *See* Flowmeters, volumetric
rotameter. *See* Flowmeters, variable-area
rotary-vane, 13.22–13.23
for shielding gas, 3.16
turbine, 4.6, 13.17–13.18
ultrasonic, 4.6, 5.18–5.21, 13.17, 13.21–13.22
variable-area, 13.25–13.26
volumetric, 1.14, 13.22–13.23
vortex shedding, 13.17, 13.18–13.19
Flow nozzle, 4.6, 4.43, 4.45, 13.14–13.17
Flow rate, 1.15–1.18, 1.33
Flow ratio relay (FFY), 4.20, 4.21
Flow restrictor, 5.3
Flow straightener, 13.18
Flow-to-close, 7.2, 7.5
Flow-to-open, 7.2
Flow transmitter (FT), 4.20, 4.21, 4.43, 4.44
Fluid, 1.4–1.5, 11.13
Fluorinated polymers, 7.9
Flux, 3.9, 3.35
FMC. *See* Flexible metal conduit
Food, dairy and beverage industry, 3.11, 11.15
Force
conversion, 1.7, 5.10
the physics of, 1.5–1.7, 1.8–1.10
from pressure applied to bellows, 6.10, 6.16
Force balance, 5.42, 8.12
Force bar, 5.42, 5.43, 5.49
Force-bridge, 8.11, 8.12
FR. *See* Recorders, flow
Fractionation, 11.19
Framing square, 12.6, 13.4
Freezing conditions, 10.10
Freezing point, 11.6
Frequency shift method, 5.21
Friction, 1.15–1.16, 4.23, 5.7, 5.10, 7.19
FT. *See* Flow transmitter
Fulcrum, 4.28–4.30, 4.49
Full flow taps, 5.14
Futbol, 13.51
FY. *See* Square-root extractor

Gain, 8.12, 8.23
Gain control. *See* Proportional control
Galling, 7.3, 7.31
Galvanized, 14.4
Galvanometer, 6.25, 6.31
Gamma ray, 13.38
Gas
flow, 5.12, 10.10, 10.11

inert, 3.10, 3.11, 3.35, 8.2. *See also* Shielding gas
measurement of velocity, 1.18
noble, 11.2, 11.26
purge, 3.11, 3.35
seal-off fittings for, 14.8–14.9
thermal conductivity analysis, 11.20–11.21
zinc oxide fumes, 10.3
Gas chromatography, 11.19–11.20
Gas cylinders. *See* Cylinders
Gaskets, flange, 13.46, 13.47
Gas metal arc welding (GMAW), 3.15–3.19, 3.20
Gasoline, 11.13, 11.16
Gas tungsten arc welding (GTAW), 3.10, 3.13–3.15
Gauges
overview, 6.24–6.25
pressure, 1.11, 1.33, 5.22–5.23, 5.32–5.33, 5.39–5.40
strain, 4.5, 5.38–5.40
Gauntlet, 3.3, 3.35
GAW. *See* Gram atomic weight
Gear train, 4.12, 8.17
Geiger-Muller tube, 5.26
GFCI. *See* Ground Fault Circuit Interrupter
Glass, 5.18, 5.26, 9.13, 11.9
Glass box principle, 2.31
Globular transfer, 3.18, 3.19
Gloves. *See* Personal protective equipment
GMAW. *See* Gas metal arc welding
GMW. *See* Gram molecular weight
Goggles. *See* Personal protective equipment
Gram atomic weight (GAW), 11.5, 11.25
Gram molecular weight (GMW), 11.5, 11.25
Graphic styles, 2.30–2.32
Graphite, 7.10
Grease, 7.12–7.13
Grinder, 13.4
Ground Fault Circuit Interrupter (GFCI), 13.6, 13.10, 13.11, 14.16
Grounding, 13.21, 14.3, 14.27
Grout, 13.11, 13.12
GTAW. *See* Gas tungsten arc welding
Gutter, auxiliary, 14.28

Half cells, 11.11
Hammers, 3.6, 12.7–12.8
Hangers, 14.14, 14.15, 14.31–14.32
Hastelloy®, 5.12, 5.53, 13.18, 13.41
Hazardous locations, 8.10, 14.26
Head loss. *See* Pressure drop
Head pressure. *See* Pressure, hydrostatic
Heads, connector, 13.41–13.43
Heat exchanger
aftercooler in dryers, 10.14
with open-loop control, 4.14
overview of process, 4.2, 4.3, 4.4
with PD controller, 4.36
with PI controller, 4.33
tube-type, 4.41
Heat tracing, 13.35, 13.37–13.38
Heat transfer process, 4.2–4.4
Heliarc welding. *See* Gas tungsten arc welding
Helical-type Bourdon tube, 5.32
Helium, 3.11, 3.14, 3.18
Heli-welding. *See* Gas tungsten arc welding
Hexane, 11.22
Hubs, 14.6–14.7
Hydrocarbons, analysis, 11.21
Hydrogen ion, and pH, 11.7, 11.8
Hydrometer, 11.14–11.15, 11.25, 13.27
Hydroxyl ion, and pOH, 11.7, 11.8
Hysteresis, 4.23, 4.49, 5.7, 5.8

I control. *See* Integral control
ID. *See* Pipe, outer and inner diameters; Tubing, outer and inner diameters
IMC. *See* Intermediate metal conduit
Impervious, 10.4, 10.19
Inconel®, 13.41
Indicators. *See also* Gauges
electronic, 6.26, 6.27
in-line hydrometer, 13.27
magnetic, 6.27–6.28
overview of types, 6.24–6.28
pH, 11.8
pneumatic, 6.26–6.27
seven-segment display on, 6.26, 6.27
thermal, 6.27, 6.28
Inductance, 4.5
Inductor/capacitor (LC), 9.15
Industrial process IR sensors, 9.13
Inert gas, 3.10, 3.11, 3.35, 8.2. *See also* Shielding gas
Inertia, 5.7, 5.10
Influent, 13.30, 13.55
Infrared devices, 9.13
Infrared methods of chemical analysis, 11.22
Infrared radiation (IR), 9.10, 9.11, 9.17
In-line-mounted instruments, 13.13–13.28, 13.29
Insertion-mounted instruments, 13.39–13.45
Installation
capacitance probe, 13.30
conduit, 14.14–14.16
drawings, 2.14, 2.16–2.17, 2.35
field-mounted instruments. *See* Field-mounted instruments
flange, 13.45–13.47
flowmeter, 13.14–13.26
instrument data sheets, 2.17
manifold, 13.49–13.51
panel-mounted instruments. *See* Panel-mounted instruments
rotameter, 13.26
sensing bulb, 13.37
thermowell, 13.43
Instrumentation. *See also* Calibration
accessibility, 12.3, 13.10, 13.13, 13.33
data sheets, 2.17–2.18, 2.19, 2.35
drift out of tolerance, 5.7
field-mounted. *See* Field-mounted instruments
location drawings, 2.14, 2.15, 2.35
panel-mounted. *See* Panel-mounted instruments
safety issues, 12.3
symbols, 2.27–2.28
tag numbers, 2.10, 2.12
welding, 3.28
Instrumentation, Systems, and Automation Society (ISA), 2.8, 2.28, 2.35
Instrumentation drawings, 2.7–2.18, 2.19
Instrument control channels, 4.4–4.13, 4.49, 5.1–5.4
Instrument Society of America (ISA), 5.28
Insulation, thermocouple wire, 5.30, 13.43–13.44
Integral control (I), 4.30–4.31, 6.4, 6.6
Integral windup. *See* Reset windup
Integrator, 4.31
Intermediate metal conduit (IMC), 14.2, 14.9, 14.16, 14.21–14.23, 14.26–14.27
Invar®, 5.26
Ion, 11.4, 11.6

Ionic bond, 11.4, 11.25
Ionization theory, 11.6
IR. *See* Infrared radiation
Iron, 3.9, 3.35, 5.28, 7.12, 8.15, 12.8
Iron-constantan, 1.29
Iron core, in electrical indicator, 6.25
ISA. *See* Instrumentation, Systems, and Automation Society; Instrument Society of America
Isometric drawing, 2.30–2.32, 2.35

Junction box, 13.35, 13.37, 14.29, 14.34

Kelvin scale, 1.13–1.14
Kerosene, 11.13
Kick, 14.14, 14.26
Kinetic energy, 7.2, 7.31
Knockouts, 14.12–14.13, 14.26
Kynar®, 5.24

Laboratory standards, 5.9
Ladder diagrams, 2.13–2.14
Ladder tray, 14.34
Lag time. *See* Control loops, system delay
Laminar flow, 1.16, 5.12, 5.53, 13.17
Lamps, color abbreviations, 2.20
Lapping, 7.31
Lay out, 12.4, 12.15
Layout
  cable tray system, 14.35
  for panel-mounted instruments, 12.3–12.10
  for wireway system, 14.29
Layout dye, 12.7, 13.4
LC. *See* Inductor/capacitor
LCD. *See* Liquid crystal display
LDR. *See* Light-dependent resistor
Leakage, 7.12, 8.7, 13.22
Least significant digit (LSD), 1.22, 1.23–1.24
LED. *See* Light-emitting diode
Left-hand thread, 3.23, 3.35
Legend, 2.35
Length
  conversion to and from metric, 1.2–1.3, 1.21, 1.24, 1.25, 10.3
  definition, 1.1
Lens, on welding shield, 3.5
Level (parameter)
  cascade control of, 4.19–4.20
  common detectors, 4.5, 5.21–5.25, 13.28–13.32
  control loops, 4.45
  of float, in a tank, 4.28, 4.30
  Masoneilan® 12000 Level Controller, 6.7, 6.8
  measurement, 5.26, 5.45, 13.28–13.32, 13.38–13.39, 13.40
  nuclear detection, 5.26
  on-off control of, 4.22, 4.24
  ultrasonic measurement, 5.25–5.26
Level (tool), 13.11
Lever, 4.28–4.30, 4.49, 6.10, 6.12
LFMC. *See* Liquidtight flexible metal conduit
Lift, 7.4, 7.31
Light absorption, 11.22
Light-dependent resistor (LDR), 9.10
Light-emitting diode (LED), 8.9, 8.23
Light-sensitive devices, 9.10–9.14
Limiter, high- and low-pressure, 8.12–8.14
Linear flow, 7.4, 7.31
Linear-variable differential transformer (LVDT), 5.40–5.41
Lineless (wireless) wiring diagram method, 2.5, 2.7
Link, positioning, 6.3
Liquid crystal display (LCD), 11.10
Liquidtight flexible metal conduit (LFMC), 14.9–14.10, 14.20–14.21, 14.23–14.24
Liter, 1.4, 1.33
Lithium chloride, 10.10
Litmus, 11.6, 11.25, 13.31
Load capacity, 2.2, 2.35
Local control, 6.1
Locknuts, conduit, 14.7–14.8, 14.26, 14.27
Logarithm and logarithmic scale, 11.7, 11.25
Logic diagrams, 2.13–2.14
Loop sheets, 2.11–2.13, 2.35
Lower range value, 1.26
LSD. *See* Least significant digit
Lubricants, 7.12–7.13, 13.51, 14.19, 14.24, 14.25
LVDT. *See* Linear-variable differential transformer

Magmeter. *See* Flowmeters, magnetic
Magnet
  in bimetallic thermometer, 5.27
  in drivers, 13.25
  in electrical indicators, 6.25
  in electromechanical relays, 8.2, 8.3, 8.4, 8.8–8.9
  in flowmeters, 4.6, 13.18, 13.23
Magnetic lines of flux, 5.17, 5.53
Magnitude, 5.2, 5.53
Maintenance
  layout tools, 12.6, 12.7
  pH probe, 13.31
  relay replacement, 8.2
  solid-state timer replacement, 8.19
Manifolds
  five-valve blowdown, 13.49, 13.50
  five-valve equalizing, 13.48–13.49
  installation, 13.49–13.51
  single-valve, 13.47–13.48
  three-valve equalizing, 13.48, 13.49
  two-valve, 13.48
Manometer, 11.16
Manual control
  actuator, 4.10
  electronic controller, 6.16
  overview, 6.1
  Taylor® controller, 6.14–6.16
Manufacturers Standardization Society (MSS), 7.23
Markings, valve, 7.23–7.26
Mass, 1.1
  advantages over volume in flow measurement, 13.24
  atomic, 11.1, 11.4–11.5
  conversion to and from metric, 1.5, 11.13
  definition, 1.1, 1.33
  in flow rate, 1.15
  vs. weight, 1.5–1.6
Mass density, 11.13
Mass flow rate, 1.14
Mass-inertia effect, 5.10, 5.53
Master (primary) control loop, 4.18, 4.19, 4.21
Materials safety data sheet (MSDS), 14.27
Mathematics
  calculators and instrumentation applications, 1.18–1.20
  errors in, 5.7
  metric system. *See* Metric system
  overview, 1.1–1.2
  significant digits, 1.22, 5.10
  technical applications, 1.20–1.29

Measurement
  chemical composition, 11.18–11.22
  conductivity, 11.10–11.13
  definition, 1.1, 1.33
  density and specific gravity, 11.14–11.16
  direct *vs.* inferred, 4.4, 5.8–5.9
  pH, 11.7–11.10
  pipe, 13.4–13.9
  precision *vs.* accuracy, 1.21
  pressure, 1.5, 1.6–1.8, 12.5
  smoke density, 11.21
  standards and elements of, 5.9–5.11
  temperature, 5.46, 5.48
  terminology, 5.4–5.8
  viscosity, 11.16–11.18
  wet *vs.* dry, 1.4–1.5
Measurement error, 5.6, 5.53
Measuring element. *See* Primary element
Measuring section, in motion balance transmitter, 5.46
Mechanical elements, 5.11
Mechanical recorders, 6.20–6.21
Memory, in a calculator, 1.20
Mercury, 1.10, 5.27, 11.13, 11.16
Metal fatigue, 5.8
Metal grating floor, anchoring systems, 13.11–13.12
Metals
  in bimetallic thermometer, 5.26, 5.27
  cutting, 3.20, 12.11
  in thermocouples, 5.28
Metal transfer, types, 3.16, 3.18, 3.19
Meter (measuring devices)
  abbreviations, 2.20
  conductivity, 11.11
  density, 11.14–11.16, 11.26–13.28
  electrical, 6.25
  hydrometer, 11.14–11.15, 11.25, 13.27
  pH, 11.9–11.10
  radiation, 5.26, 13.28–13.39, 13.40
  sound velocity, 13.27–13.28
  turbine, 4.6, 13.17–13.18
  viscosity, 11.16–11.18
Meter (unit), 1.1, 1.2–1.3, 1.9
Meter-kilogram-second system (MKS), 1.1
Methane, 10.11
Metric system
  conversion factors, 1.2–1.6, 1.7, 1.14, 1.24, 10.3, 11.13
  conversion methods, 1.20–1.24
  definition, 1.33
  measurements, 1.2–1.18
  prefixes, 1.2
  transition to, 1.1–1.2
  as world standard, 1.29
Microcircuit industry, 3.12
Micron, 10.3
Microprocessor-based instruments, 11.10
Microswitches, 9.9
Micro-wire welding. *See* Gas metal arc welding
MIG welding. *See* Gas metal arc welding
Minimum amplitude criterion, 4.17, 4.18
Minimum area criterion, 4.17
Minimum disturbance criterion, 4.17
MKS system. *See* Meter-kilogram-second system
Modular assembly positioners, 7.21
Modulating control, 4.26–4.39, 4.40, 4.46
  D (rate) control. *See* Derivative control
  I (reset) control. *See* Integral control
  P (gain) control. *See* Proportional control
  PD control. *See* Proportional plus derivative control
  PI control. *See* Proportional plus integral control
  PID control. *See* Proportional plus integral plus derivative control
Moisture. *See* Condensate
Molarity, 11.5–11.6, 11.25
Mole, 11.5, 11.6, 11.25
Molecule, 11.2
Molybdenum, 13.17, 13.41
Moment, 6.9, 6.31
Monel®, 5.12, 5.53, 13.41, 13.51
Monitoring, 5.6, 8.7, 11.21
Most significant digit (MSD), 1.22
Motion balance pneumatic transmitters, 5.46–5.48
Motion detectors, 9.13
Motor coil, 4.11
Motors
  controller wiring diagram, 2.4
  on electric proportional valve actuator, 4.12
  on recorder, 6.21, 6.22
  in rotating spindle viscometer, 11.17
  three-phase, power circuit, 2.4, 2.36
  on time clock, 8.17
MSD. *See* Most significant digit
MSDS. *See* Materials safety data sheet
MSS. *See* Manufacturers Standardization Society
Mule, 14.17, 14.18
Multi-point function, 2.12, 2.35
Mushrooming, 12.7, 12.15

Nameplate, 7.23–7.26
National Association of Relay Manufacturers (NARM), 8.7, 8.8
National Electrical Code (NEC)
  cable trays, 14.34
  conduit specifications
    boxes, 14.13
    installation, 14.14–14.15, 14.16
    metal, 4, 14, 14.2, 14.3, 14.8, 14.9, 14.10, 14.11
    nonmetal, 14.11, 14.12
    overview, 14.1–14.2
  definition, 14.41
  explosion-proof heads, 13.43
  hazardous location classification, 2.17
  metal wireways, 14.28–14.29, 14.32, 14.33, 14.34
  nonmetallic wireways, 14.32–14.33
  raceways, 14.33, 14.34
National Electrical Manufacturers Association (NEMA), 2.2, 2.35
NBS. *See* U.S. National Bureau of Standards
NC. *See* Relays, normally closed
NDIR. *See* Non-dispersive infrared analyzer
NEC. *See* National Electrical Code
Needle plug, 7.4
NEMA. *See* National Electrical Manufacturers Association
Neoprene, 7.9, 7.10, 7.11
Neutral zone, 4.23, 4.24, 4.26
Neutron, 11.1, 11.2, 11.5, 11.26
Newton (unit), 1.5, 1.6, 1.7, 1.33
Newton's Law of Motion, 1.5, 5.41
Nibbler, power, 12.11
Nickel
  in bimetallic thermometer, 5.26, 5.27
  in electrodes, 3.20
  in pressure capsules, 5.33
  in resistance temperature detectors, 1.28, 13.44
  in thermowell, 13.41
Nipples, 14.11, 14.31
Ni-Span C® pressure capsules, 5.33

Nitrogen, 3.16, 3.18, 4.42, 4.43, 10.11
NO. *See* Relays, normally open
Noble gas, 11.2, 11.26
Noble metal, 5.28, 5.53
Noise, 3.6, 3.7, 4.7, 4.49, 7.2, 13.19
Nominal, 7.25
Nonconsumable, 3.13, 3.35
Non-dispersive infrared analyzer (NDIR), 11.22
Nonferrous, 3.35
Notes on P&IDs, 2.30, 2.31
Nozzles
  baffles off the, 8.10, 8.23
  flapper
    effects of coil movement, 5.36, 5.37
    effects of diaphragm movement, 5.42, 5.43, 5.44, 5.45, 5.47
    in pneumatic controllers, 6.2, 6.4, 6.12, 6.13, 6.14
  flow, 4.6, 4.43, 4.45, 13.14–13.17
  and relay function, 8.10
NRC. *See* Nuclear Regulatory Commission
Nuclear level detection, 5.26, 13.38–13.39, 13.40
Nuclear Regulatory Commission (NRC), 5.26
Nucleus, 11.4, 11.26
Numbering systems for drawings, 2.28, 2.30
Nylon, 7.9, 7.10

OD. *See* Pipe, outer and inner diameters; Tubing, outer and inner diameters
Offset bend, 14.14, 14.26
Offset error, 4.32, 4.33, 4.36, 4.37, 4.38
Off shore drilling rigs, 14.4
OFW. *See* Oxyfuel welding
Ohmmeter (measuring device), 6.25–6.26
Ohm-meter (unit), 11.10
Ohm's law, 11.11
Oil, 11.16
1-inch water column, 1.10
One-line diagrams. *See* Diagrams, single- and three-line
On-off control, 4.22–4.26, 6.2, 6.17
Open loop system, 4.13, 4.46, 4.49
Operand, 8.11, 8.23
Operator system, 4.3
Orbital welding, 3.9–3.13, 3.28, 3.33
Orifice plates, 4.6, 4.21, 5.12–5.15, 5.44, 5.45, 13.14–13.17
Orifice plate taps, 5.14, 5.15
Orthographic drawing, 2.30–2.31
Oscillations
  in closed-control loop, 4.15, 4.16
  dampened, 4.15, 4.49
  in on-off control, 4.22, 4.23, 4.25, 4.26
  in PD control, 4.36–4.37
  in PI control, 4.33, 4.34
  and sound propagation, 5.19, 5.21
  in vibrating plate detector, 13.28
Oscillator
  definition, 9.17
  in a proximity sensor, 9.14, 9.15
  in a transducer, 5.36, 5.37, 5.38, 5.39
Overlapping, 3.1, 3.35
Overload, 9.1, 9.17
Overshoot, 4.16, 4.25, 4.26, 4.35, 4.37, 7.19
Oxidation, 9.2
Oxyfuel welding (OFW), 3.19–3.25
Oxygen
  characteristics and properties, 3.23
  desiccant dryers for, 10.11
  regulator, 3.24
  in the welding process, 3.16, 3.18, 3.20, 3.21

Packing, in valve stem, 7.2, 7.31
Packing or stuffing box, 7.8, 7.16
PAM. *See* Plant asset management
Panel-mounted instruments
  installation, 2.14, 12.10–12.13
  layout, 12.3–12.10
  P&ID drawings for, 12.1–12.3
Paper
  circular chart, 6.22
  dielectric constant, 5.22
  use in pipe measurement, 13.5–13.6
Parallax effect, 5.6, 5.53, 6.24, 11.9
Parallel disc, 7.6, 7.7, 7.8
Parts per million (ppm), 11.5, 11.6, 11.26
Pascal, 1.5–1.6, 1.7–1.8, 1.33
Pass, 3.13, 3.35
Pattern. *See* Template
P control. *See* Proportional control
PD control. *See* Proportional plus derivative control
P/E converter, 7.17
Periodic table of the elements, 11.2, 11.3
Personal protective equipment (PPE)
  cutting conduit, 14.16, 14.19, 14.20
  deburring conduit, 14.22, 14.23, 14.24
  power tools, 13.6, 13.10
  PVC solvents and acetone, 14.27
  threading conduit, 14.24
  welding, 3.2–3.5, 3.6
Petrochemical industry, 3.11–3.12
PFD. *See* Process flow drawings
pH
  basic principles, 11.7, 11.8
  definition, 6.31, 11.7
  measurement, 6.26, 11.7–11.10, 13.30–13.32
Pharmaceutical industry, 3.11
Phase shift, 4.14, 4.15
Phenolic-impregnated cellulose, 10.4, 10.19
Phonographic, 7.31
Photocell, 11.21
Photoelectric devices, 9.10–9.14, 9.15
Photoresistor, 9.10
Physics
  laminar *vs.* turbulent flow, 1.16, 5.12, 5.53
  Newtonian, 1.5, 1.6, 1.8
Picofarad, 5.22, 5.53
PI control. *See* Proportional plus integral control
P&ID. *See* Piping and instrumentation drawings
PID control. *See* Proportional plus integral plus derivative control
Piezoelectric crystal, 4.5–4.6, 5.19, 5.25, 5.40, 5.53, 13.21
Pilot hole, 12.12, 12.15
Pipe
  conditioning, 13.18, 13.55
  diameter, relationship with velocity, 1.16–1.17
  drain and vent holes, 5.13
  mating, 13.19
  measuring and cutting, 13.4–13.9
  outer and inner diameters, 3.12
Pipe caps, 13.8
Pipe cutter, 14.17–14.19
Pipe extension, 13.2
Pipe rack, 14.36
Pipe reamer, 14.21
Pipe stand. *See* Field-mounted instruments, stand-mounted
Pipe taps, 5.14, 5.15
Piping and instrumentation drawings (P&ID)
  notes on, 2.30, 2.31

numbering system for, 2.28, 2.30
overview, 2.8–2.10, 2.35
for panel-mounted instruments, 12.1–12.3
relationship to flow diagrams, 2.17
Pitot tube, 4.6, 5.16–5.17
Pitot-venturi tube, 5.16
Plant asset management (PAM), 7.21, 7.31
Plasma welding, 3.10
Plastic process control, 4.17
Platinum, 1.29, 5.28, 13.43, 13.44
PLC. *See* Programmable logic controllers
Plug. *See* Valve plug
Plug-type valve disc, 7.3, 7.4–7.5
Plumb, 13.11, 13.12
Pneumatic control loops
cascade, 4.18, 4.19
flow, 4.43–4.44
level, 4.45
pressure, 4.42
temperature, 4.39–4.41
Pneumatic input/output section, of DP cell, 5.44
pOH, 11.7, 11.8, 11.26
Point-to-point wiring diagram method, 2.5
Polarity, 1.20, 3.8, 4.14, 5.29
Polyethylene, 5.24
Polymerize, 10.4, 10.19
Polyurethane, 5.18
Polyvinyl chloride
boxes, 14.13
conduit, 14.11–14.12, 14.21, 14.24, 14.27–14.28
in thermocouple wire insulation, 5.30
wall brackets, 13.2
Port, of a plug valve, 7.1–7.12
Positioners
analog, 7.20–7.21
current-to-air valve, 4.10–4.11
definition, 7.31
digital, 7.21
direct- or reverse-acting, 7.20
force balance electropneumatic, 10.6
function, 7.1, 7.19
I/P, 7.20–7.21
Masoneilan® Model 4700, 7.21
modular assembly, 7.21
overview, 4.10–4.11
pneumatic, 7.20
principles of operation and construction, 7.19–7.22
process precautions, 7.21–7.22
reverse-acting, 7.20
Positive displacement. *See* Displacement
Power failure or interruption, 2.54–2.55, 13.32
Power monitoring, 8.7
Power pony. *See* Threaders
Power source
electronic controller, 6.16
electronic switches, 9.9
electrostatic precipitator, 10.1
gas metal arc welding, 3.15, 3.16
gas tungsten arc welding, 3.13–3.14
orbital welding, 3.10
welding machine, 3.7–3.8, 3.9
PPE. *See* Personal protective equipment
ppm. *See* Parts per million
PR. *See* Recorders, pressure
Precipitator, electrostatic, 10.1–10.3
Precision, 1.21, 1.22, 5.5–5.6, 5.53
Pressure. *See also* Backpressure
absolute, 1.11, 1.33
atmospheric, 1.10–1.11, 1.12, 1.33
common detectors, 4.5, 5.30–5.35
comparison of units, 1.13
control loops, 4.41–4.43
conversion factors, 1.7–1.8, 1.13
conversion to and from metric, 1.5–1.6, 1.26, 1.28
definition, 1.1, 1.33
in density meters, 11.15–11.16
differential, 1.17, 1.25–1.26, 4.5, 4.20
in direct-operated regulators, 10.7–10.8
flange ratings, 13.45
in gas cylinders, 3.2, 3.23, 3.25
gauges, 1.11, 1.33, 5.22–5.23, 5.39–5.40, 12.5
hydrostatic, 1.8–1.13, 1.25–1.26, 1.29, 1.33
lockup, 10.8, 10.19
measurement
with DP cell, 5.46
and mass flowmeter values, 13.24
with motion balance transmitter, 5.48
physics, 1.6–1.8
recommended, for solder and brazed joints, 3.32
units, 1.5–1.6, 1.7
vapor, 1.12
working, and regulators, 3.23
Pressure capsules, 5.33
Pressure drop, 1.15–1.16, 10.5, 13.14
Primary element, 2.41–2.50, 5.2, 5.10, 5.11, 5.53, 13.15
Primary standards, 5.9
Probes
capacitance, 5.22, 5.23, 5.24, 13.29–13.30, 13.40
conductance, 1.25
in level instruments, 13.28–13.32
pH, 11.9–11.10, 13.30–13.32
Process control theory
control applications, 4.39–4.45
control loops, 4.13–4.22
control modes, 4.22–4.39
instrument channel components, 4.4–4.13
overview, 4.1–4.2, 4.46
process characteristics, 4.2–4.3
the process control system, 4.3–4.4
Process flow drawings (PFD), 2.17, 2.18
Process measuring section, of DP cell, 5.42
Process mounted installation, 2.14, 2.35
Programmable logic controllers (PLC), 2.13–2.14, 8.7, 10.11, 12.2
Proportional band, 6.4, 6.31
Proportional control (P)
with electronic controllers, 6.17–6.18
with pneumatic controllers, 6.3–6.4, 6.5–6.6
principles, 4.27–4.30, 4.33, 4.37, 4.40
Proportional plus derivative control (PD), 4.35–4.38, 4.40
Proportional plus integral control (PI), 4.32–4.35, 4.40, 6.4, 6.6, 6.18
Proportional plus integral plus derivative control (PID), 4.38–4.39, 4.40, 6.1, 6.7–6.9, 6.18–6.20
Proportioning. *See* Scaling
Proton, 11.1, 11.2, 11.5, 11.26
Proximity sensors, 8.2, 9.14–9.15
Pryout, 14.12, 14.13
PT. *See* Transmitters, pressure
Pumps, 4.43, 4.44, 7.14
Punches
center, 12.7, 13.4
knockout, 12.10, 14.13
prick, 12.7

Purge assembly, 13.32
Purge gas, 3.11, 3.35
Pyrometer, 6.27, 6.28

Quality criteria, closed loop system, 4.17–4.18
Quantitative, 5.8, 5.53
Quartz, 5.22, 5.40

Raceways
  cable trays, 14.34–14.36
  conduit. *See* Conduit
  definition, 14.41
  drawings, 2.7, 2.8, 2.9, 2.35
  dual-purpose, 14.33
  handling, 14.37
  metal wireways, 14.28–14.33
  overview, 14.1–14.2
  pancake, 14.33
  storage, 14.36–14.37
  surface metallic and nonmetallic, 14.33–14.34
  underfloor, 14.34
Racks, 14.36
Radiation detectors, 5.26, 13.28–13.39, 13.40
Radioactive material, safety issues, 13.39
Radio frequency applications (RF), 13.29
Radius taps, 4.20, 4.49, 5.13, 5.14
Raintight, 14.2, 14.3, 14.26, 14.41
Random errors, 5.7
Range rod, 5.49, 5.50
Rankine scale, 1.13, 1.14
Rate control. *See* Derivative control
Ratio control system, 4.20–4.22
Ratio flow controller. *See* Controllers, ratio flow
RC. *See* Resistance/capacitance network
Reamer, 14.21, 14.22–14.23
Recorders
  ABB Commander™, 6.22, 6.23
  definition, 6.31
  electronic, 6.21–6.23
  flow, 4.20, 4.21, 4.44
  level, 4.45
  paperless, 6.22
  pneumatic, 6.23–6.24
  pressure, 4.42
  types, 6.20–6.24
Recording methods, 5.6
Rectifiers
  silicon-controlled, 8.8, 8.18, 8.23, 9.9–9.10
  transformer, 10.2
Reference accuracy, 5.4–5.5
Regulators
  on a cylinder, 3.16, 3.22, 3.23–3.25
  direct-operated, 10.7–10.8, 10.9
  electrical/electropneumatic, 10.5–10.6
  pilot-operated, 10.8–10.9, 10.10
  pneumatic, 10.6–10.9, 10.10
  principles of operation, 10.4–10.9, 10.10, 10.16
  selection guidelines, 10.13
  self-actuated, 2.51–2.53
  symbols, 10.15
Relative error, 1.21–1.22, 5.6, 5.53
Relays
  abbreviations, 2.20, 2.21
  booster, 8.10, 8.11, 8.14–8.15
  codes and symbols, 8.6, 8.7, 8.8
  computing, 8.11–8.15
  control, 8.2, 8.7
  dashpot timer, 8.15–8.17
  definition, 8.1
  double-break, 8.6
  double-pole, 8.3, 8.5, 8.6
  double-pole, double-throw, 8.19
  double-throw, 8.3, 8.6
  electrical, 8.2–8.10
  electromechanical, 8.2–8.9
  four-pole, 8.3, 8.6
  general purpose, 8.2, 8.3, 8.4–8.7, 8.8
  high- and low-pressure limiter, 8.12–8.14
  high- and low-pressure selector, 8.12–8.14
  latching, 8.4, 8.23
  liquid level, 8.7
  normally closed, 8.2, 8.6, 8.19, 9.6, 9.8
  normally open, 8.2, 8.6, 8.19, 9.6, 9.8
  phase-failure, 8.7
  pneumatic
    adding, subtracting, and inverting, 8.11–8.14, 8.15
    basic principles, 5.44, 5.47, 5.48
    multiplying and dividing, 8.11, 8.12
    scaling and proportioning, 8.12, 8.13
  power, 8.2
  Pre-Act™, 6.12
  ratio, 8.15
  reed, 8.2, 8.3–8.4
  replacement, 8.2
  reset, 6.11
  reversing, 8.15
  single-break, 8.6, 8.7
  single-pole, 8.5, 8.6
  single-pole, single-throw, 8.19, 9.10, 9.13
  single-throw, 8.6, 8.7
  solid-state, 8.2, 8.7–8.10, 8.23
Remote control, 6.1
Repeatability, 5.5, 5.7, 8.20
Repeater, 8.10, 8.23. *See also* Transmitters
Reproducibility, 5.5, 5.7
Reset control. *See* Integral control
Reset point, 9.6, 9.17
Reset value, 4.31
Reset windup, 4.31, 4.34, 4.35, 6.4
Resistance/capacitance network (RC), 8.18
Resistance temperature detectors (RTD)
  with capacitance detector, 5.24
  function, 4.5, 5.3
  overview, 13.44–13.45
  in temperature control loop, 4.41
  temperature conversions with, 1.28–1.29
  with thermowell, 13.40, 13.45
Resistivity, 11.10, 11.11
Resistors, 9.10
Response line, 6.3
Responsiveness, 5.8, 5.54
Reynolds number, 5.12, 5.54
RF. *See* Radio frequency applications
Rheostat, 6.25–6.26
Rhodium, 5.28, 13.43
Right-hand rule for induced EMF, 5.17
Right-hand thread, 3.23, 3.35
Rigid metal conduit (RMC), 14.4–14.9, 14.16, 14.21–14.23, 14.26–14.27
Rigid nonmetallic conduit (RNC), 14.11–14.12, 14.21, 14.24, 14.27–14.28
Ringelmann scale, 11.21
RMC. *See* Rigid metal conduit
RMC unions, 14.6, 14.7

RNC. *See* Rigid nonmetallic conduit
Rochelle salt, 5.40
Rod-out tool, 13.51
Rotameters, 4.7, 13.25–13.26
Rotors, in compressed air filter, 10.3, 10.4
Rounding off, 1.23
RTD. *See* Resistance temperature detectors
Rules, steel, 12.6
Running threads, 14.11, 14.41
Rust, 10.3, 10.10

Safety cap, 3.25
Safety issues
  acetylene pressure limits, 3.23
  conduit installation, 14.21, 14.22, 14.23, 14.24, 14.25, 14.26
  cylinder, 3.22
  file, 14.21
  grounding, 3.7
  instrumentation, 12.3
  mercury, 5.26
  metal grating flooring, 13.12
  ohmmeter, 6.26
  positioners, 7.21–7.22
  power tools, 12.12, 13.6, 13.10, 13.11, 14.16
  radioactive material, 13.39
  scribers, 12.5
  use of oxygen as fuel, 3.21
  welding, 3.2–3.6, 3.7
Safety plugs, 3.25
Salts, 11.6
Saws
  hand hacksaw, 14.16, 14.17, 14.19, 14.20–14.21
  power, 12.11, 13.4, 14.16, 14.17, 14.20
Scaling, 8.12, 8.23
Scanning devices, 9.11–9.12
Schematic diagrams. *See* Diagrams, schematic
SCR. *See* Rectifiers, silicon-controlled
Scribers, 12.5, 13.4
Sealing compound, 14.8–14.9
Seat
  ball valves, 7.9, 7.10
  butterfly valves, 7.14, 7.15
  definition, 7.31
  globe valves, 7.2, 7.3, 7.5, 7.8
  identification, 7.24
Secondary element, 5.10, 5.30–5.35, 5.54
Secondary standards, 5.9
Security systems, 8.2, 8.7, 9.13, 9.14–9.15
Selector, high- and low-pressure, 8.12–8.14
Self-actuated regulators, valves, and other devices, 2.51–2.53
Semiconductor strain gauge, 5.39
Semicone plug, 7.4
Sensing bulb, 13.35, 13.37–13.38, 13.55
Sensitivity, 5.8, 13.19, 13.55
Sensors. *See* Detectors
Sequence operations, 8.16
Servomotor, 6.22
Setpoint
  in cascade control, 4.18, 4.19, 6.1
  in D control, 4.31
  definition, 4.49
  effects of change, 4.26, 4.40
  in feedback control, 4.13–4.14
  in I control, 4.31
  in on-off control, 4.22, 4.23, 4.24–4.25, 6.2
  overview, 4.3, 4.4
  in PD control, 4.37
  in PI control, 4.32, 4.33, 4.34, 4.35
  in PID control, 4.38, 4.39
  range in pressure switch, 9.6
  in ratio control, 4.21
Setscrew, 13.13, 14.3, 14.4, 14.10
Seven-segment display, 6.26, 6.27
Shading coil, 8.5
Shears, power, 12.11
Shielded metal arc welding (SMAW), 3.6–3.9, 3.14, 3.28
Shielding gas
  function of, 3.16, 3.35
  gas metal arc welding, 3.15, 3.16, 3.18
  gas tungsten arc welding, 3.10, 3.13, 3.14
  orbital welding, 3.11
Ships, 14.4
Short arc welding. *See* Gas metal arc welding
Short circuit, 9.1, 9.17
Short-circuiting transfer, 3.16, 3.19
Shutdown, 13.19
Shutoff, 7.11, 7.17
Sight glass, 1.24–1.25
Signal comparison, 6.16
Signal conditioner, 4.7–4.8, 5.4
Signal conversion, 5.35, 6.16
Signal isolation, 6.16
Signal line, 12.3, 12.4
Significant digits, 1.22, 5.10
Silica gel, 11.19
Silicon, 5.39. *See also* Rectifiers, silicon-controlled
Single-pole, single-throw switch (SPST), 8.19, 9.10, 9.13
Sintered, 5.35, 5.54
Size designation, of valves, 7.25
Slag, 3.5, 3.9
Slave (secondary) control loop, 4.18, 4.19, 4.21
Sludge, 5.14
Slugs, 11.13
Slurry, 4.45, 5.16, 7.8, 13.24, 13.27
SMAW. *See* Shielded metal arc welding
Smoke density measurement, 11.21
Smurf, 14.12
Soapstone, 3.21, 3.22, 3.35
Sockets, 8.4, 8.19
Solar cells, 9.12
Solar module/panel/array, 9.12
Solenoids, 4.12–4.13, 8.3, 8.5, 8.15
Solute, 11.5, 11.26
Solvent cement, for PVC conduit, 14.27–14.28
Sound propagation, 5.19–5.20, 5.25
Sound velocity, 13.27–13.28
Span of measurement, 1.26
Spatter, 3.8, 3.15, 3.18, 3.35
Specification sheet, 2.14, 2.36
Specific gravity, 6.7, 11.13, 11.14–11.16
Spiral-type Bourdon tube, 5.32
Splice, 14.26, 14.41
Split-range configuration, 4.10, 4.42, 4.43, 8.15
Split wedge, 7.6, 7.7
Spray transfer, 3.18, 3.19
Spring, opposing, 6.3
Spring-to-close, 7.18
Spring-to-open, 7.18
SPST. *See* Single-pole, single-throw switch

Square-root extractor (FY), 4.20, 4.21, 4.44
Squares (carpentry), 12.6, 13.4
Squares (mathematical), 1.18–1.20
SSR. *See* Relays, solid-state
Stability in control systems, 4.15–4.16, 4.36, 4.37, 6.8
Stainless steel
  diaphragm seals, 5.34
  flowmeter, 13.18
  pressure capsules, 5.33
  pulsation dampeners, 5.35
  punching out, dies for, 12.10
  strapping, 13.38
  thermowell, 13.41
  wall brackets, 13.2
Stand alone, 12.2
Stand-mounted instruments
  in drawings, 2.14
  fabricating the stand, 13.2–13.9
  floor-mounted, 13.2
  mounting instruments on stands, 13.13
  securing the stand, 13.9–13.12
  wall-mounted, 13.2
Standoff support, 14.14
Starters, 8.2
Startup, 4.34
Steam systems, 7.6
Steam working pressure (SWP), 7.24
Steel. *See also* Stainless steel
  cable trays, 14.34
  electrodes, 3.20
  galvanized, 14.4
  rules, 12.6
  socket weld fittings, 3.27–3.28
  specific gravity, 11.16
  squares, 12.6, 13.4
  valves, 7.12
Sticker, calibration, 5.10
Stick welding. *See* Shielded metal arc welding
Storage, raceways, 14.36–14.37
Straightedge, 12.8
Strain gauges, 4.5, 5.38–5.40
Strap-mounted instruments, 13.32–13.39, 13.40
Straps, 13.38, 14.13–14.14
Striker, 3.21, 3.22, 3.35
Strut support, 14.14
Stuffing box, 7.8, 7.16
Summing junction, 4.14
Supply parameter, 4.4
Supports
  cable-mounted instrument, 13.33–13.35, 13.36
  cable tray, 14.35–14.36
  conduit, 14.13–14.14, 14.15
  standoff, 14.14
  strut, 14.14
  wireway, 14.21–14.33
Suspension, direct rod, 14.35
Switches. *See also* On-off control; Relays
  abbreviations and symbols, 2.20–2.21, 2.26
  classifications, 9.2–9.3, 9.15
  definition, 9.1
  differential-pressure, 9.8, 9.9
  electronic, 9.9–9.10
  level, 9.4, 9.5
  limit, 9.8–9.9
  microswitches, 9.9
  panel-mounted, 9.4, 9.5
  photocell, 9.10–9.12, 9.13
  pressure, 9.6–9.8
  reed, 8.2
  selection, 9.1
  synchronous time, 8.17–8.18
  three-way, 9.3, 9.4
  types, 9.3–9.10
SWP. *See* Steam working pressure
Symbology
  definition, 2.36
  P&ID, 12.2–12.3
Symbols
  actuator, 2.39–2.40, 2.54–2.55
  damper, 2.54
  electronic, 2.2–2.7, 2.20–2.27, 2.28, 2.29
  elements (chemical), 11.2
  in event of power failure, 2.54–2.55
  filters, 10.15
  general instrument or function, 2.37–2.38
  instrumentation, 2.27–2.28, 12.4
  line, 12.3, 12.4
  on location or connection section, 2.12
  primary elements, 2.41–2.50
  programmable logic controllers, 12.2
  regulators, 10.15
  relays, 8.6, 8.7, 8.8
  self-actuated regulators, valves, and other devices, 2.51–2.53
  standardized design methods for diagrams, 2.20–2.32
  use of, 12.1
  use of the bubble, 12.1, 12.2–12.3
  valves, 2.54, 2.55, 7.26
  valve threads, 7.25
  wiring diagram, 2.26
Systematic errors, 5.7
System failure, 2.54–2.55, 4.41, 5.10, 13.32

Tachometer, 10.5, 10.19
Tag, calibration, 5.10
Tag numbers, 2.10, 2.12, 2.36
Tanks, 1.25–1.28, 6.4, 6.7, 10.9
Tap, 14.26, 14.41
Tap locations, 5.13–5.14
TC. *See* Thermocouples (TC)
Tee fittings, 14.30–14.31
Teflon®, 5.18, 5.22, 5.24, 5.30
Temperature
  common detectors, 4.5, 5.26–5.30
  control loops, 4.39–4.41. *See also* Heat exchanger
  conversions, 1.14, 1.28–1.29
  definition, 1.1, 1.33
  and desiccant efficiency, 10.10
  effects on dielectric constant, 5.22
  effects on viscosity, 11.18
  flame from combustion of acetylene and oxygen, 3.20
  and limit switches, 9.8
  measurement, 5.46, 5.48
  physics of, 1.13
  recommended, for solder and brazed joints, 3.32
  scales, 1.13–1.14
  thermocouple wire limitations, 5.30, 13.44
Template
  creation, 12.8, 12.10
  definition, 12.8

development, 12.3–12.4
manufacturers', 12.9
Terminal board, 8.18
Theory of Ionization, 11.6
Thermal, 13.37, 13.55
Thermal conductivity gas analysis, 11.20–11.21
Thermal transients, 7.6, 7.31
Thermistor, 4.5, 5.24, 11.12, 11.20, 11.26
Thermocouples (TC)
in closed loop systems, 4.14, 4.15
construction, 5.29–5.30
definition, 5.28
in electropneumatic regulators, 10.5
in heat exchange control system, 4.3
metal, 5.28
overview, 5.28–5.30
temperature conversions with, 1.28–1.29
in thermowell, 13.40
voltage as indirect measurement, 4.4, 4.5, 5.3
wire, 5.28–5.29, 13.43–13.44
Thermometers
bimetallic strip, 4.5, 5.26–5.28
IR handheld, 9.13, 9.14
Thermostat, 4.23, 13.35, 13.37–13.38, 13.55
Thermowell, 4.41, 13.40–13.41, 13.42
Thoria, 3.15
Threaders, 14.18, 14.22, 14.24, 14.25, 14.26
Threads
on conduit, 14.9, 14.18, 14.22, 14.24–14.26
running, 14.11, 14.41
of a valve, 7.25
Three-phase systems, 2.4, 2.36
Throttling, 7.2, 7.31
TIG. *See* Gas tungsten arc welding; Tungsten inert gas process
Time, definition, 1.1
Timers and time clocks
dashpot, 8.15–8.17, 8.23
operating graph, 8.19
pneumatic, 8.15–8.17
preset cam, on dryer, 10.11
solid-state, 8.18–8.20
synchronous time switches, 8.17–8.18
Tools
for cutting conduit, 14.16, 14.17–14.19, 14.21
for deburring conduit, 14.21
for instrument panel installation, 12.10–12.11
for instrument panel layout, 12.4–12.8
for instrument stand fabrication, 13.3–13.5
power, safety issues, 12.12, 13.6, 13.10, 13.11, 14.16
Torches, 3.22, 3.28
Torch tips, 3.23
Torque, 7.9, 7.16, 14.25
Torricelli tube, 1.12
Total flow, 1.14, 1.33
Tourmaline, 5.40
Traceability, 5.9
Training, 3.11, 3.28
Transducers
common, 4.7
components, 10.6
current-to-pneumatic. *See* Transducers, I/P
definition, 5.54
function, 5.2, 5.4, 5.35
I/P, 5.35–5.35, 5.36, 5.37, 6.2, 6.31
operation, 5.36–5.38
P/I, 5.36–5.38
piezoelectric, 5.40
pneumatic-to-current. *See* Transducers, P/I
types, 5.35
on ultrasonic flowmeter, 13.21–13.22
voltage-divider pressure, 5.40
Transformer, linear-variable differential, 5.40–5.41
Transmitters
definition, 5.54
D/P, 13.13
electronic, 5.48–5.50
flow, 4.20, 4.21, 4.43, 4.44
force balance differential pressure electronic, 5.48–5.49
force balance differential pressure pneumatic, 5.41–5.46, 8.10–8.11
function, 5.2, 5.4
level, 4.45
motion balance pneumatic, 5.46–5.48
overview, 4.7–4.8
pneumatic, 5.41–5.48
pressure, 4.42, 5.35
Rosemount® Model 1151 Alphaline® pressure, 4.7, 4.8
temperature, 4.3, 4.18, 4.19, 4.40, 4.41
on ultrasonic flowmeter, 13.21
variable capacitance cell differential pressure electronic, 5.49–5.50
Trapeze hanger, 14.14, 14.32
Trapeze mounting, 14.36
Travel angle, 3.7
Triac power control circuit, 4.41, 4.44, 4.49, 8.8
Trim collar, 12.9, 12.11
Trim identification, for valves, 7.24
Trough, 14.34
TT. *See* Transmitters, temperature
Tubing
electrical metallic, 14.2–14.4, 14.19–14.20, 14.26–14.27
electrical nonmetallic, 14.11, 14.12, 14.21
outer and inner diameters, 3.12
Tungsten arc welding. *See* Gas tungsten arc welding
Tungsten inert gas process (TIG), 3.10. *See also* Gas tungsten arc welding
Turbulent flow, 1.16
Two-position control, 4.22–4.26
Type L copper tube, 3.26, 3.35

U-bolt, 13.13, 13.35
UL. *See* Underwriters Laboratory
Ultrasonic level measurement, 5.25–5.26
Undershoot, 4.25, 4.26
Underwriters Laboratory (UL), 14.11, 14.41
Unit effective area, 8.11, 8.23
Units, mathematical, 1.1, 1.21, 1.33
Upper range value, 1.26
Upstream, 7.18
U.S. National Bureau of Standards (NBS), 5.9

Vacuum, 1.11–1.12, 1.13, 1.33, 5.22, 8.2
Valence shell, 11.2, 11.4, 11.26
Valve actuators. *See* Actuators
Valve body, 7.3
Valve plug, 4.8, 4.10, 4.49, 7.4–7.5, 7.17, 7.31
Valve plug stem, 4.8, 4.9

Valve positioners. *See* Positioners
Valves
  angle, 7.2–7.3, 7.26, 7.31
  ball, 7.8–7.10, 7.31
  bar-stock instrument, 7.16
  bleed, 13.48
  butterfly, 7.13–7.16, 7.31
  bypass, 4.44, 7.6, 13.20
  check, 7.12, 7.26, 7.31, 10.12, 13.18
  control. *See* Control valve
  equalizer, 13.47, 13.48–13.49, 13.55
  functions, 7.1
  gain, 6.11
  gate, 4.8, 7.2, 7.5–7.8, 7.26, 7.31
  globe, 4.8, 7.1–7.5, 7.26, 7.31
  knife gate, 7.8, 7.9
  manifold valve assemblies, 13.47–13.51
  markings and nameplate information, 7.23–7.26
  multiport, 7.11
  needle, 7.16
  pilot, 4.11
  pinch, 4.45
  plug, 7.10–7.13, 7.31
  position of
    and float level, 4.28, 4.30
    and flow rate, 4.9
  purge, 10.12, 10.13
  rating designations, 7.24
  selection, 7.22
  shutoff, 7.11
  solenoid, 4.24, 4.25
  symbols, 2.54, 2.55, 7.26
  three-way, 7.7–7.8
  throttle. *See* Valves, globe
  types and applications, 7.23
  vent, 10.7, 13.49
  venturi type ball, 7.10
Valve stem, 7.3, 7.5, 7.14, 7.16, 7.24
Valve trim, 7.2
Variables
  controlled, 4.3–4.4, 4.13, 4.15, 4.16–4.17
  uncontrolled, 4.20
VDC. *See* Transmitters, force balance differential pressure electronic
Velocity
  critical, 1.16
  definition, 1.33
  factors which influence, 1.16–1.17
  flowmeters, 1.17, 13.17–13.22
  and Newton's First Law of Motion, 1.5
  relationship with pipe diameter, 1.17
Vena contracta taps, 4.20, 5.13, 5.14
Vent, three-position, 7.6
Vent hole, 5.13
Venturi tube, 4.6, 5.15–5.16, 13.14–13.17
Vessel-mounted instruments, 13.28–13.32, 13.33
Vibration, 13.19, 13.25, 13.28, 13.41, 14.2
Viscometers
  basic principles, 11.16–11.17
  rotating spindle, 11.17
  vibrating reed, 11.18
Viscosity, 1.15, 11.16–11.18, 13.22
Viscous, 5.15, 5.54
Vise, pipe, 14.18, 14.19, 14.22, 14.23
Voltmeter, 6.25
Volume
  conversion to and from metric, 1.3–1.4, 1.5
  definition, 1.33
  flow rate measurement by, 1.14, 1.15, 13.22–13.23
  of liquids, and hydrostatic pressure, 1.9–1.10
  wet volume measurements, 1.4–1.5
Volumetric, 11.13, 11.26
Vortex, 13.55
Vortex shedding, 4.6, 13.17, 13.18–13.19
V-port plug, 7.4

Wall-mounted installation, 2.14, 13.2, 14.36
Wall thickness, 3.12, 3.13
Washers
  fender, 13.11, 13.12, 13.55
  flat, 13.12
  lock, 13.13
Waste treatment and management, 5.26, 13.30
Water
  condensate, 5.13, 7.7, 10.9–10.10. *See also* Dryers
  dielectric constant, 5.22
  effects of temperature on viscosity, 11.18
  as incompressible fluid, 1.26
  physical and chemical characteristics, 11.7, 11.10, 11.13, 11.14, 11.16
  prevention from conduit system, 14.3
Water column, 10.1, 10.19
Waveforms
  in closed-loop systems, 4.16, 4.17
  in on-off control, 4.23, 4.24, 4.25
  in PD control, 4.37, 4.40
  in PI control, 4.33, 4.34, 4.35, 4.40
  in PID control, 4.38, 4.39, 4.40
Wedge flow element, 4.6
Wedge types, in gate valves, 7.5–7.7
Weight, 1.5, 1.14, 1.15, 11.13. *See also* Mass
Weld bead, 3.7
Weld head, orbital, 3.10
Welding
  applications for instrumentation, 3.28
  butt weld fittings, 3.26, 3.28, 3.29, 3.30–3.31, 13.47
  clothes, 3.2–3.4, 3.5
  current for, 3.8, 3.13, 3.16, 3.17
  fittings, 3.25–3.28, 3.29–3.32
  gas metal arc, 3.15–3.19, 3.20
  gas tungsten arc, 3.10, 3.13–3.15
  historical background, 3.1
  orbital, 3.9–3.13, 3.28, 3.33
  oxyfuel or oxyacetylene, 3.19–3.25
  plasma, 3.10
  principles of, 1.1–1.2
  safety, 3.2–3.6, 3.7
  shielded metal arc, 3.6–3.9
  socket-weld fittings, 3.26, 3.27–3.28
  speed, 3.12–3.13
  stick or stick electrode, 3.7. *See also* Welding, shielded metal arc
  types of metal transfer, 3.16, 3.18, 3.19
Welding blankets, 3.8
Welding gun, 3.16
Welding machines
  GTAW, 3.14
  orbital welding, 3.9, 3.10
  power source, 3.7–3.8
Welding shield or helmet, 3.5, 3.6, 3.8

Wheatstone bridge, 11.20
White space, 2.5, 2.36
Wire, thermocouple, 5.28–5.29, 5.30, 13.43–13.44
Wire drawing, 7.3
Wire feed, 3.16, 3.17
Wireways, metal, 14.1, 14.28–14.33, 14.41
Wire welding. *See* Gas metal arc welding
Wiring diagrams
  general purpose relay, 8.4
  motion detector, 9.13
  overview, 2.4–2.7, 2.36
  solid-state timer, 8.19
  switches, 9.3, 9.4, 9.9, 9.11
Working standards, 5.9
Work lead, 3.9
Wrenches
  regulator, 3.22
  torch, 3.22

Zero, absolute, 9.13, 9.17
Zero feature, 13.21
Zeroing of the instrument, 13.47
Zero screw, 6.12
Zinc oxide fumes, 10.3
Zirconia, 3.15